Institute of Physics Publishing

AA003266

Joint 21st AIRAPT and 45th EHPRG International Conference on High Pressure Science and Technology 2007

Journal of Physics: Conference Series Vol. 121

September 17-21, 2007
Catania, Italy

Printed from e-media with permission by:

Curran Associates, Inc.
57 Morehouse Lane
Red Hook, NY 12571
www.proceedings.com

ISBN: 978-1-60560-417-6

Some format issues inherent in the e-media version may also appear in this print version.

Copyright (2007) by the Institute of Physics Publishing.

All rights reserved.

For permission requests, please contact the Institute of Physics Publishing at the address below.

Institute of Physics Publishing
Dirac House, Temple Back
Bristol BS1 6BE UK

Tel 44 (0)117 929 7481
Fax 44 (0)117 929 4318

Institute of Physics Publishing

Joint 21st AIRAPT and 45th EHPRG International Conference
on High Pressure Science and Technology
2007

TABLE OF CONTENTS

THEORY, MODELLING, SIMULATION, AND MOLECULAR DYNAMICS

GEOLOGICAL, MARINE AND PLANETARY STUDIES

BIOLOGY AND BIOTECHNOLOGY AT HIGH PRESSURE

HIGH PRESSURE ENGINEERING AND EQUIPMENT

INDUSTRIAL APPLICATIONS

FOOD SCIENCE

METROLOGY

NANOMATERIALS

PHASE EQUILIBRIA AND TRANSITIONS, KINETICS AND TRANSPORT PHENOMENA

ELECTRICAL AND MAGNETIC PROPERTIES

SPECTROSCOPY AT HIGH PRESSURE

SUPERCONDUCTIVITY, MAGNETISM, AND STRONGLY CORRELATED SYSTEMS AT HIGH PRESSURE

HIGH PRESSURE SYNTHESIS OF POLYMERS, METALS AND COMPOSITES

Journal of Physics: Conference Series **121** (2008) 001001 doi:10.1088/1742-6596/121/0/001001

Foreword

R Pucci[1,2], **G G N Angilella**[1,2,3], **F Siringo**[1,2,3]

[1] Dipartimento di Fisica e Astronomia, Università di Catania, 64, Via S. Sofia, I-95123 Catania, Italy

[2] CNISM, UdR Catania, 64, Via S. Sofia, I-95123 Catania, Italy

[3] INFN, Sez. Catania, 64, Via S. Sofia, I-95123 Catania, Italy

E-mail: giuseppe.angilella@ct.infn.it

This volume collects the Proceedings of the Joint 21st AIRAPT and 45th EHPRG International Conference on *High Pressure Science and Technology,* that was held at the Department of Physics and Astronomy, University of Catania, Italy, in September 17–21, 2007.

This international conference qualifies as one of the largest in its series. The conference was attended by about 450 participants from some 30 countries, for about 430 submitted abstracts divided in 18 subtopics, *viz.*

- Theory, modelling, simulation, and molecular dynamics,
- Phase equilibria and transitions, kinetics and transport phenomena,
- Thermophysical properties,
- Electrical and magnetic properties,
- Spectroscopy at high pressure,
- Superconductivity, magnetism, and strongly correlated systems at high pressure,
- Superhard materials,
- Nanomaterials,
- Dynamic pressure, pressure jumps and shock waves,
- Solvothermal reactions of organic and inorganic substances,
- High pressure synthesis of polymers, metals and composites,
- Geological, marine and planetary studies,
- Biology and biotechnology at high pressure,
- High pressure engineering and equipment,
- Industrial applications,
- Pressure effects on chemical reaction kinetics,
- Food science,
- Metrology.

From the above list of topics, ranging from physics to chemistry, from biosciences to geology, from food science to metrology, industrial applications, and virtually all sectors of natural sciences and technology, it should be clear the paramount role nowadays played by high pressure as a tool to investigate the properties of matter under extreme conditions. It is therefore hoped

© 2008 IOP Publishing Ltd

Journal of Physics: Conference Series **121** (2008) 001001 doi:10.1088/1742-6596/121/0/001001

that these proceedings will be of relevance in setting the state of the art in the field of high pressure at the time in which the conference was held.

The University of Catania had already hosted the 36th EHPRG Meeting on *Molecular and low dimensional systems under pressure* (1998), and two Archimedes Workshops on *Solids under pressure* (1990), and on *Density functional theory* (1985), respectively. Actually, density functional theory emphasizes the fundamental role of the electron *density* and solids under pressure are systems at very high *density* indeed.

Now, it is believed that it was Archimedes who first introduced the concept of *density.*

Archimedes of Syracuse (287–212 BC) laid the foundations of mechanics as a science. In his *Equiponderance of the planes,* he expounds the theory of the centre of gravity and of the lever. He was so impressed by the efficiency of the lever, that Plutarch reports him to have remarked

Give me a place to stand and with a lever I will move the whole world.

In his book on *Floating bodies,* he treats hydrostatics, and there he describes the principle known by his name. The legend goes that King Hiero, having entrusted some gold to his goldsmiths to forge a crown, suspected them of alloying it with silver. He asked then Archimedes to test his suspicions. The concept of density was still unknown at that time. The story continues that, while bathing, Archimedes noticed that he displaced water equal in volume to his own body, and he realized that for equal weights, the lighter alloy would displace more water than the denser gold. He is said to have jumped out of the bath and run home naked shouting "eureka" ("I have found"). It was so that his principle, later deduced mathematically, was born.

Archimedes regarded this discovery, and the inventions for which he is noted, *viz.* the compound pulleys, the Archimedean screw, and the burning mirrors, as mere diversions. He considered himself as a pure geometrician and regarded his determination of π as his greatest achievement. By inscribing and circumscribing a circle with n-gons and, in modern language, letting n tending to infinity, he showed π to lie between $3 + \frac{10}{70}$ and $3 + \frac{10}{71}$.

He was killed by a Roman soldier after the storming of Syracuse in 212 BC. The legend narrates that his tomb was discovered and restored in 75 BC by Cicero, who was then quæstor in Sicily.

Archimedes should be regarded as the first modern physicist. His works show a combination of mathematics and observation. Restricted but definite problems are considered, hypotheses made and their consequences first deduced and then experimentally verified. We believe that Archimedes's scientific methodology can be of great inspiration also to contemporary researchers even though, of course, the scientific activity cannot be any more based only on the efforts of a single individual, but must be eager for international exchange and cooperation.

During the conference, the Bridgman Award was awarded to Takehiko Yagi, the Jamieson Award to Ignacio Hernández, and the EHPRG Award to Agnès Dewaele.

During the conference, the late George Samara (1936-2006) was commemorated. During his long career, George Samara was well recognized for his pioneering use of high pressure to resolve fundamental questions about the structure and properties of materials. While a major portion of his scientific work was devoted to phase transitions in ferroelectric materials, he also worked extensively on research on deep-level defects and their role in lattice relaxations in semiconductors, ionic conductivity, dielectric properties, ferromagnetic properties, and insulator-metal transitions. In recent years he was very active in nanoscience.

Soon after the conference ended, Igor Goncharenko (1965-2007) passed away. Igor Goncharenko was an outstanding experimental physicist, certainly recognized as a world leader for neutron scattering studies at high pressure. He demonstrated the use of neutrons to study small samples in diamond and sapphire anvil cells up to pressure in the half-megabar range. He was currently trying to measure neutron diffraction on an hydrogen sample in the 100 GPa range. He achieved some seminal contributions on the high pressure properties of hydrides, simple molecular systems and spin systems. Igor Goncharenko was an active member of the high-pressure crystallography community and a Committee member of the the EHPRG.

These proceedings are dedicated to George Samara and Igor Goncharenko.

Acknowledgments

We would like to thank the AIRAPT Excecutive Committee, the EHPRG Committee, the conference international Advisory Committee, all the Chairpersons, all the participants and the authors of all the proceedings papers, reporting the most recent results of their research activity, as well as all the anonymous referees, who have contributed to the scientific success of this conference.

We also acknowledge partial financial support from the Faculty of Sciences and the University of Catania, as well as from the Ente Regionale per il diritto allo Studio Universitario, the Azienda Provinciale per il Turismo di Catania, the Assessorato al Turismo del Comune di Catania, the Archivio di Stato di Catania, and all the exhibitors.

Special thanks are due to Mario Presti, who kindly designed the conference logo (Fig. 1).

The tremendous task of organizing such a large and complex event would not have been possible without the help of all the colleagues and staff members of the Department of Physics and Astronomy and INFN, Sez. Catania, especially including Filippo Arriva, Piero Ferlito, Alfredo La Rocca, Natale Marino, Gino Platania, and Carlo Rocca, the organizational effort of the travel agency Palenque Viaggi and its staff, including Damiano Arena, Giusi Conti, and Davide D'Ascari.

Last, but not least, we should like to warmly thank the gratuitous help of our young collaborators, students, and *friends,* who shared our troubles, doubled our passion, and added their invaluable cheerfulness to the common effort of the scientific and organizational success of the event, and in welcoming all the participants. These include, especially, Alessio Cardillo and Luca Marotta, as well as Pietro Artoni, Gabriele Bellocchi, Massimo Camarda, Andrea Canino, Gigi Cappello, Giacomo Cremona, Nadia Formica, Salvo Galatà, Ivan Herrera, Danilo Jaccarino, Davide Lodato, Ivano Lodato, Giuseppe Marino, Francesco Noto, Francesco M. D. Pellegrino, Alessandro Parisi, Giorgio Piccitto, and Piergiorgio Stella.

To everyone, our most sincere and warmest *thanks!*

Figure 1. 21st AIRAPT & 45th EHPRG International Conference on *High Pressure Science and Technology:* The conference logo.

Journal of Physics: Conference Series **121** (2008) 012001 doi:10.1088/1742-6596/121/1/012001

The role of anharmonicity in a variety of phase transitions controlled by pressure, including melting, brittle-to-ductile transition, and the liquid-vapour critical point

G G N Angilella[1], N H March[2,3], C C Matthai[4] and R Pucci[1]

[1] Dipartimento di Fisica e Astronomia, Università di Catania, and CNISM, UdR Catania, and INFN, Sez. Catania, 64, Via S. Sofia, I-95123 Catania, Italy

[2] Oxford University, Oxford, UK

[3] Department of Physics, University of Antwerp, Groenenborgerlaan 171, B-2020 Antwerp, Belgium

[4] Department of Physics and Astronomy, University of Cardiff, Cardiff, UK

E-mail: giuseppe.angilella@ct.infn.it

Abstract. A variety of phase transitions controlled by pressure are addressed, in which anharmonicity plays a central role. After a brief discussion of the melting temperature of some transition metals as a function of pressure, the brittle-to-ductile transition is treated. Finally, quantum critical points are briefly referred to, in relation to singular behaviour of the Grüneisen ratio.

Key words: melting and brittle-to-ductile transition; liquid-vapour phase transition; Grüneisen ratio.

1. Background and outline

In earlier work [1, 2], some attention has been focussed on both point and extended defect properties, particularly of the transition metals [3]. Here we extend the above work, which concerned ambient pressure properties, to treat variation of the melting temperature, T_m say, with pressure P. Some attention has also been devoted to a precursor phase change, namely the brittle-to-ductile transition (BDT) [4, 5, 6], again in relation to T_m at ambient pressure. Proposals will be made here concerning the potential interest in studying the BDT in transition metals, and also in chemically bonded Si and in graphite, as a function of P.

These considerations are developed somewhat in sections 2 and 3 below, where an analysis on $T_m(P)$ for transition metals by Burakovsky and Preston [7] has already focussed on the relation of a key fingerprint of anharmonicity, namely Grüneisen's constant, in gaining insight into the variation of $T_m(P)$ with pressure in metals like Pd and Pt. Section 3 proposes explanations of possible correlation between $T_m(P)$ and the BDT transition temperature $T_{\mathrm{BDT}}(P)$, both in transition metals and in Si and graphite. Then section 4 considers briefly the critical point in relation to melting temperature, especially in the alkali metals. In section 5, future directions are pointed out in relation to experiments on the variation of Grüneisen's constant with pressure.

© 2008 IOP Publishing Ltd

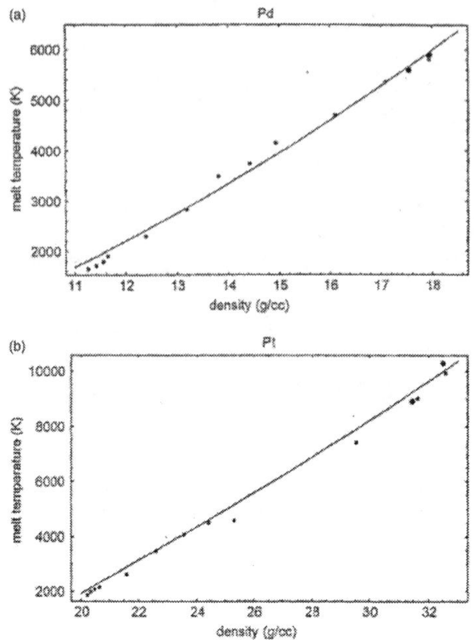

Figure 1. Shows melting temperature of Pd (upper panel) and Pt (lower panel) as a function of density. Redrawn after Burakovsky and Preston [7].

2. Melting temperature of transition metals and variation with pressure

Matthai and Bacon [8] have made computer simulation calculations on metals for a variety of point defect properties. In the interpretation of their simulation results for various point defects, they establish an intimate relation with Grüneisen's constant. This is highly relevant to the present theme, since an intimate connection between the monovacancy formation energy, E_{1f} say, and the thermal energy $k_B T_m$, has been known to materials scientists for a long time. The approximate result

$$\frac{k_B T_m}{E_{1f}} \sim 10 \tag{1}$$

works surprisingly well at ambient pressure, though lack of experimental data on E_{1f} under hydrostatic pressure P leaves open whether the correlation remains simple as a function of P.

Taking up the emphasis of Matthai and Bacon [8] on Grüneisen's constant, we note next the study of Burakovsky and Preston [7] on the melting temperature of metals like Pd and Pt as a function of pressure. Their work involves, in fact, also the variation of Grüneisen's constant itself with pressure. We have reproduced in Fig. 1 $T_m(P)$ *vs* P for Pd(Pt): Burakovsky and Preston [7] get fairly good agreement between their phenomenological theory and the experimental results. Since some emphasis is placed on the role of Grüneisen's constant here, we note briefly that Burakovsky and Preston [7] model the density dependence of the Grüneisen parameter Γ as

$$\Gamma(\rho) = \frac{1}{2} + \frac{\gamma_1}{\rho^{1/3}} + \frac{\gamma_2}{\rho^q}, \tag{2}$$

where γ_1, γ_2 and q (> 1) are constants.

Journal of Physics: Conference Series **121** (2008) 012001 doi:10.1088/1742-6596/121/1/012001

3. Is the brittle-to-ductile (BDT) transition temperature $T_{\mathrm{BDT}}(P)$ connected with melting temperature $T_m(P)$: with reference to transition metals and to Si.

The ductile-brittle transition is exhibited in bcc metals, such as low carbon steel, which become brittle at low temperature or at very high strain rates. In metals, it is generally agreed that the mechanism for plastic deformation at room temperature is by dislocation motion. The stress required to move a dislocation depends on the atomic bonding, crystal structure, and obstacles such as solute atoms, grain boundaries, precipitate particles and other dislocations. If the stress required to move the dislocation is too high, the metal will fail instead by the propagation of cracks and the failure will be brittle. Thus, either plastic flow (ductile failure) or crack propagation (brittle failure) will occur, depending on which process requires the smaller applied stress.

In fcc metals, the flow stress, *i.e.* the force required to move dislocations, is not strongly temperature dependent. Therefore, dislocation movement remains high even at low temperatures and the material remains relatively ductile. In contrast to fcc metal crystals, the yield stress or critical resolved shear stress of bcc single crystals is markedly temperature dependent, in particular, at low temperatures. The temperature sensitivity of the yield stress of bcc crystals has been attributed to the presence of interstitial impurities on the one hand, and to a temperature dependent Peierls-Nabarro force on the other. However, the crack propagation stress is relatively independent of temperature. Thus the mode of failure changes from plastic flow at high temperature to brittle fracture at low temperature.

A soft brittle-to-ductile transition (BDT) is one in which there is a gradual increase in dislocation activity at the crack tip as the temperature is increased until fracture. The transition region is fairly wide and the transition temperature is consequently hard to define precisely. Most single crystals display this form of transition. In a sharp BDT, the stress intensity rises sharply as the transition temperature is reached and there is an abrupt, rapid increase in dislocation activity. This form of transition is observed in dislocation-free Si crystals. As most BDT are soft, it means that it is the mobility of dislocations that plays a central role in determining the transition temperature.

In fracture measurements, it has been found that the loading rate, $\dot{K}$, is related to transition temperature, T_{BDT}, through some activation energy for dislocation generation or dislocation mobility, Q_{BDT}, by an exponential relation

$$\dot{K} = A \exp\left(-Q_{\mathrm{BDT}}/k_{\mathrm{B}}T_{\mathrm{BDT}}\right). \qquad (3)$$

Thus, comparing the measured activation energy with the energies associated with the various defect energies, it is possible to get an understanding about the mechanism for the transition from brittle to ductile behaviour. However, there has been some uncertainty associated with this. Thus, for example, Gumbsch *et al.* found the activation energy in tungsten to be about 0.2 eV [9] suggesting that plasticity in tungsten is controlled by the glide of non-screw (edge) dislocations. By contrast, the more recent studies by Giannattasio and Roberts [10] on tungsten point to a much higher activation energy $\sim$ 1 eV which is equal to the energy of double-kink formation on screw dislocations.

On a macroscopic level, fracture is related to elastic properties of a material. Based on Griffith's theory of micro-cracks and Cottrell's model of fracture [11], Petch [12] developed an equation for the transition temperature

$$T_{\mathrm{BDT}} \sim \log C - \log\left[\frac{4\beta G\gamma}{D} - D\right] - \log d^{-1/2}, \qquad (4)$$

where d is the grain diameter, β is a constant related to the resistance of the lattice to deformation, C and D are constants; G is the shear modulus and γ the effective surface energy of

Table 1. BDT and melting temperatures for 8 elements at ambient pressure.

Element	T_{BDT} [K]	T_m [K]
Si	1040	1687
Ge	700	1211
Fe	150	1808
V	150	2163
Nb	75	2741
Ta	120	3269
Mo	225	2890
W	400	3680

the crack surfaces. The important feature here is the dependence of T_{BDT} on the shear modulus. Burakovsky *et al.* [13] constructed a model of melting as a dislocation-mediated phase transition and showed that the melting temperature can be expressed in the form

$$k_{\mathrm{B}}T_m = \Omega G \times S_B, \tag{5}$$

where G is the shear modulus, and S_B denotes a structure dependent factor, while Ω is the atomic volume.

Thus, the role of dislocations and its connection with both the melting temperature T_m and the BDT transition temperature T_{BDT} suggest a connection between the two quantities. In this vein, Klein and March [4] have discussed the BDT transition as a precursor phase transition to melting, at ambient pressure. Indeed, it has been accepted by materials scientists that for dislocation free single crystals

$$T_{\mathrm{BDT}} \sim \frac{2}{3}T_m. \tag{6}$$

However, the bcc transition metals show much lower BDT transition temperatures (see Table 1) due to presence of dislocations and in these materials, the activation energy for dislocation glide or disclinations might be the determining factor. The brittleness of metals such as Fe at high pressures is of interest because of their use in pressurized containers as well in understanding materials in the earth's core. So, it is of interest to gain an understanding of the pressure dependence of the BDT temperature, $T_{\mathrm{BDT}}(P)$. Following the work of Matthai and March [14] relating the pressure dependence of the point defect energy to that of the melting temperature, it is worthwhile asking if the pressure dependence of the precursor BDT temperature $T_{\mathrm{BDT}}(P)$ also correlates with $T_m(P)$ as a function of pressure P.

Khantha *et al.* [5] proposed a model for the BDT based on the Kosterlitz-Thouless theory of the melting transition. In their model, they found that the rapid rise in free dislocation density occurs at the transition temperature

$$T_{\mathrm{BDT}} = \frac{1}{2k_{\mathrm{B}}}\left[\frac{q^2}{\epsilon(r_c)} - \frac{\sigma b r_c}{2}\right], \tag{7}$$

where σ is the stress, b the Burger's vector and r_c the average size of the dislocation loops at the critical temperature. ϵ is the polarizability and the quantity q is related to the shear (G) and bulk moduli (B), $q^2 = GB/(G+B)$. Now, in fracture theory, it is known (see Ref. [15], p. 291) that when hydrogen collects in microcracks it exerts an internal pressure, P, which may be added directly to the external stress σ to produce a total stress $(P+\sigma)$ for propagation. In

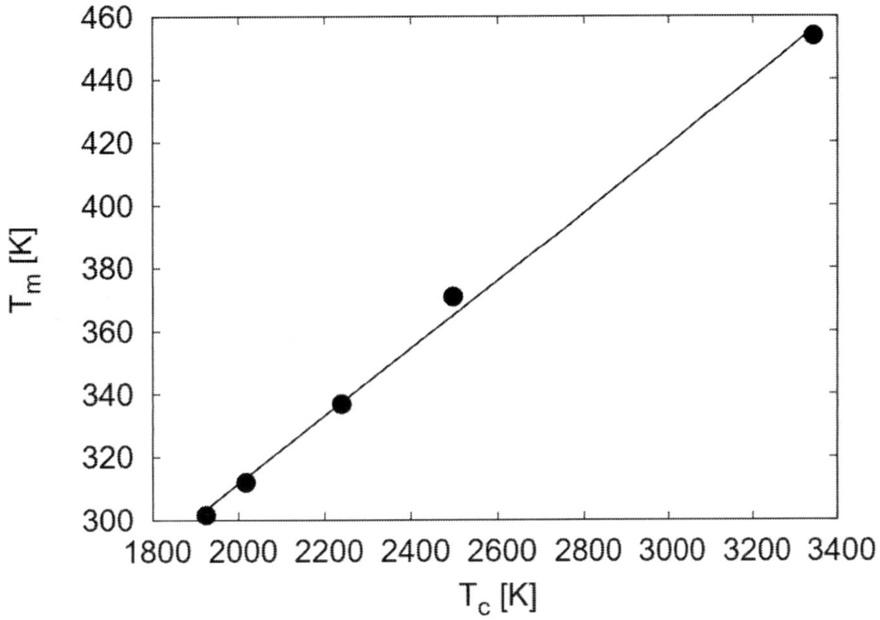

Figure 2. Melting temperatures T_m *vs* critical temperatures T_c of the first five alkali metals Li, Na, K, Rb, and Cs (bottom to top) at ambient pressure. Solid line is a best fist to the experimental data, $T_m = AT_c + B$ [17], with $A \approx 1/8$.

determining the effect of hydrostatic pressure on the transition temperature, we can apply this result by adding it to the external stress leading to

$$T_{\rm BDT}(P) = \frac{1}{2k_{\rm B}} \left[\frac{q^2(P)}{\epsilon(r_c)} - \frac{(\sigma + P)br_c}{2} \right]. \tag{8}$$

The pressure dependence of the BDT transition temperature appears to be more complicated than that of the melting temperature. However, it is clear from the above equation that a major contributory term in the slope of $T_{\rm BDT}(P)$ at ambient pressure is

$$\frac{\partial T_{\rm BDT}}{\partial P} \sim -\frac{br_c}{2}. \tag{9}$$

Thus, in contrast with the slope of the melting curve, the slope of the BDT temperature is predicted to be negative, *i.e.*, the BDT temperature decreases with increasing pressure. This is consistent with the findings of Churbaev *et al.* [16] for polycrystalline Mo and W. Of course, further experimental studies of $T_{\rm BDT}$ at high pressure will be essential in showing future directions for theory, and how the above very basic considerations are to be refined.

4. Critical point temperature T_c in relation to melting temperature T_m, especially for the alkali metals

Leys *et al.* [18] and March *et al.* [19], for the alkali metals, have stressed the correlation which exists between the critical temperature T_c and the melting temperature T_m. This can be written in the form $T_m = AT_c + B$ [17], with $A = 1/8$ (see Fig. 2), which is the inverse of the coordination number for the body-centred cubic lattice, as is appropriate for most solid alkali metals, including Rb and Cs. This is relevant to the present study since Rosenfeld [20] has argued for a correlation between critical point behaviour and anharmonicity. Unfortunately, the critical points of other metals are experimentally rather inaccessible, with of course the notable exception of Hg.

5. Summary and future directions

The main results of the present study set out above concern classical phase transitions involving high pressure. We have emphasized the importance of further experimental work, especially on the BDT temperature $T_{\rm BDT}(P)$ under pressure, and its possible relation to $T_m(P)$. Below we stress the interest for the future of quantal phase transitions, for reasons we elaborate on briefly below.

5.1. Quantum critical points and the behaviour of Grüneisen's constant

Quite recently, there has been considerable interest in the way that the Grüneisen's parameter behaves near quantum critical points. The Grüneisen's ratio Γ [21, 22, 23] involves both the molar specific heat $c_P = (T/N)(\partial S/\partial T)_P$ and the thermal expansion α by

$$\alpha = \frac{1}{V}\left(\frac{\partial V}{\partial T}\right)_{P,N} \equiv -\frac{1}{V}\left(\frac{\partial S}{\partial P}\right)_{T,N}, \tag{10}$$

where a Maxwell thermodynamic relation has been invoked to reveal that α relates to the pressure derivative of the entropy S, whereas c_P measures its response to temperature. Following Zhu *et al.* [23], we define Γ as simply

$$\Gamma = \frac{\alpha}{c_P} = -\frac{1}{V_m T}\frac{\partial S/\partial P}{\partial S/\partial T}, \tag{11}$$

V_m denoting the molar volume V/N. Older definitions of the so-called effective or thermodynamic Grüneisen parameter are equivalent to $\Gamma c_P/\kappa_T c_V$, where κ_T denotes the isothermal compressibility. However, neither κ_T nor the specific heat ratio is expected to diverge at a quantum critical point (QCP), whereas (at least for magnetic QCPs) Γ in Eq. (11) is indeed singular. This is to be contrasted with classical phase transitions. Close to such transitions, Γ will be constant.

5.1.1. Some experimental results on heavy Fermion compounds In heavy Fermion compounds, we briefly note that there is a competition between magnetic interactions and the Kondo effect, which can be controlled by pressure. There is a considerable sensitivity to pressure P, due to the exponential dependence of the Kondo temperature on the parameters of the specific heavy Fermion material.

Such work on quantal phase transitions, and its relation to anharmonicity, seems to afford a promising parallel to the classical phase transitions discussed in sections 2–4 above.

Acknowledgments

GGNA and NHM acknowledge that this study was initiated during their stay at the Department of Physics, University of Antwerp. Therefore they thank Professors D. Lamoen, C. Van Alsenoy, and D. Van Dyck for warm hospitality and for the stimulating environment. NHM also wishes to acknowledge that his contribution to this study was brought to fruition during a visit to DIPC, San Sebastian. NHM thanks Professors P. M. Echenique and A. Rubio for generous hospitality. Finally, we thank Professors D. J. Bacon and S. G. Roberts for valuable comments relating to the BDT transition.

Journal of Physics: Conference Series **121** (2008) 012001 doi:10.1088/1742-6596/121/1/012001

References

[1] March N H 2006 *Phys. Chem. Liq.* **44** 477
[2] Matthai C C and March N H 2006 *Phys. Chem. Liq.* **44** 329
[3] March N H 1989 *Phys. Rev. B* **40** 3356
[4] Klein D J and March N H 2004 *Phys. Chem. Liq.* **42** 545
[5] Khantha M, Pope D P and Vitek V 1994 *Phys. Rev. Lett.* **73** 684
[6] Klein D J and March N H 2007 *Phys. Chem. Liq.* **45** 609
[7] Burakovsky L and Preston D L 2004 *J. Phys. Chem. Solids* **65** 1581
[8] Matthai C C and Bacon D J 1985 *J. Nucl. Mater.* **135** 173
[9] Gumbsch P, Riedle J, Hartmaier A and Fishmeister H F 1998 *Science* **282** 1293
[10] Giannattanasio A and Roberts S G 2007 *Phil. Mag.* **87** 2589
[11] Cottrell A H 1958 *J. Trans. Am. Inst. Min. Metall. Petrol Eng.* **212** 192
[12] Petch N J 1958 *J. Philos. Mag.* **3** 1089
[13] Burakovsky L, Preston D L and Silbar R R 2000 *Phys. Rev. B* **61** 15011
[14] Matthai C C and March N H 2007 *Philos. Mag.* **114** 22
[15] Smallman R E and Bishop R J 2002 *Modern Physical Metallurgy and Materials Engineering* (Amsterdam: Elsevier (Butterworth-Heinemann))
[16] Churbaev R V, Dobromyslov A V and Taluts G G 1992 *J. Phys. Met. Metallogr.* **73** 212
[17] Leys F E, March N H and M-L Zhang 2004 *Phys. Chem. Liquids* **42** 217
[18] Leys F E, March N H, Angilella G G N and Zhang M L 2002 *Phys. Rev. B* **66** 073314
[19] Balaban A T, Klein D J, March N H, Tosi M P and Ausloos M 2005 *Chem. Phys. Chem.* **6** 1741
[20] Rosenfeld Y 2000 *Phys. Rev. Lett.* **84** 4272
[21] Grüneisen E 1912 *Ann. Phys. (Leipzig)* **39** 257
[22] Landau L D and Lifshitz E M 1959 *Statistical Physics* (Oxford: Butterwith-Heinemann)
[23] Zhu L, Garst M, Rosch A and Si Q 2003 *Phys. Rev. Lett.* **91** 066404

High pressure electrical resistivity in La doped $CeIn_3$

T Ebihara, Y Iwamoto, T Terasima, A Iyo[1], T Kagayama[2] and K Shimizu[2]

Department of Physics, Shizuoka University, Shizuoka 422-8529, Japan
[1] National Institute of Advanced Industrial Science and Technology, Tsukuba, Ibaraki 305-8568, Japan
[2] KYOKUGEN, Osaka University, Toyonaka, Osaka 560-8531, Japan

E-mail: sptebih@ipc.shizuoka.ac.jp

Abstract. Antiferromagnetic ordering temperature of heavy fermion material $CeIn_3$ is at 10K. In $CeIn_3$ when the magnetic ordering is suppressed by high pressure above 2.5GPa, magnetically mediated superconductivity emerges. Substituting La for Ce reduces the Néel temperature to facilitate reaching critical pressure at which the Néel order disappears. Pressure-Néel temperature phase diagram was mapped to determine critical pressure where the Néel temperature is suppressed to absolute zero.

1. Introduction

Superconductivity in magnetic material emerges near the quantum critical point (QCP) where magnetic ordering is well suppressed to arise quantum fluctuation. Near the QCP, quantum fluctuation survives long in time and distance to show superconductivity in some materials. Examples of external parameter to suppress magnetic ordering are alloying (x), magnetic field (H) and pressure (p).

High pressure is a powerful tool to tune electronic state of condensed matter. The high pressure technique is applied to induce superconductivity in magnetic material like heavy fermion system. Heavy fermion antiferromagnets $CeIn_3$ and $CePd_2Si_2$ are typical examples to show magnetically mediated superconductivity under pressure in heavy fermion system[1]. In such materials, the $4f$ electron of Ce is expected to owe quantum criticality or superconductivity. Because number of $4f$ electrons is controllable with controlling La concentration in a Ce compound, strength of magnetic interactions is weaken and the Néel temperature becomes lower with increasing La concentration [2, 3]. The lower Néel temperature indicates lower critical pressure near which superconductivity might emerge. In this paper we report high pressure electrical resistivity measurements to determine critical pressure in $Ce_{0.85}La_{0.15}In_3$.

2. Experiment

Single crystals of $Ce_{0.85}La_{0.15}In_3$ were grown with the self-flux method. X-ray diffraction measurement was performed to confirm the sample single-phase. Magnetization and magnetic susceptibility measurements at ambient pressure were done with MPMS-SQUID magnetometer. Electrical resistivity at ambient and high pressures with piston cylinder cell below 3GPa were performed with the standard four probe method in ^{4}He cryostat or dilution refrigerator.

© 2008 IOP Publishing Ltd

Journal of Physics: Conference Series **121** (2008) 012002 doi:10.1088/1742-6596/121/1/012002

3. Results and Discussion

Figure 1 shows temperature dependences of electrical resistivities at 0, 0.5, 1.0 and 1.5GPa in $Ce_{0.85}La_{0.15}In_3$. The anomaly from the Néel order is observed at 8.4K at ambient pressure. La doping to $CeIn_3$ reduces number of $4f$ electrons and widen distance between Ce ions. Because those two effects cause weakening of RKKY interaction, the Néel temperature could be reduced in $Ce_{0.85}La_{0.15}In_3$.

To $Ce_{0.85}La_{0.15}In_3$, we applied pressure for observing pressure dependence of Néel temperature shift. The Néel temperature is decreased to 7.4K at 0.5GPa. Temperature dependences of electrical resistivity at 1.0 and 1.5GPa in Fig.1 show the anomaly was broadened to be hump and the anomaly decreases with applying pressure. Those anomalies, corresponding to Néel ordering, are indicated with black arrows in Fig.1.

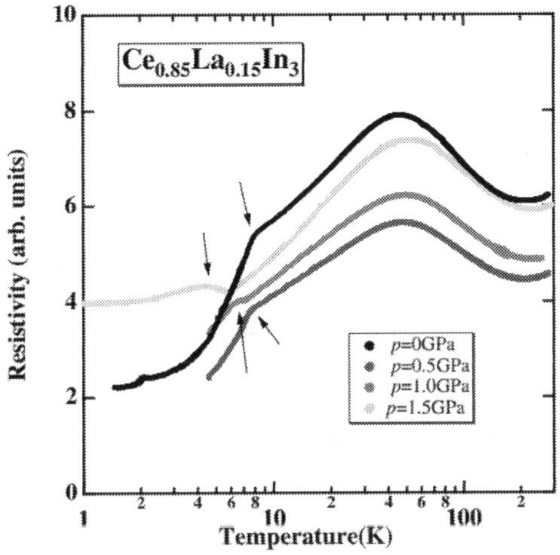

Figure 1. Temperature dependence of electrical resistivity in $Ce_{0.85}La_{0.15}In_3$ at ambient and high pressures. Anomalies that are relevant to Néel ordering are indicated black arrows.

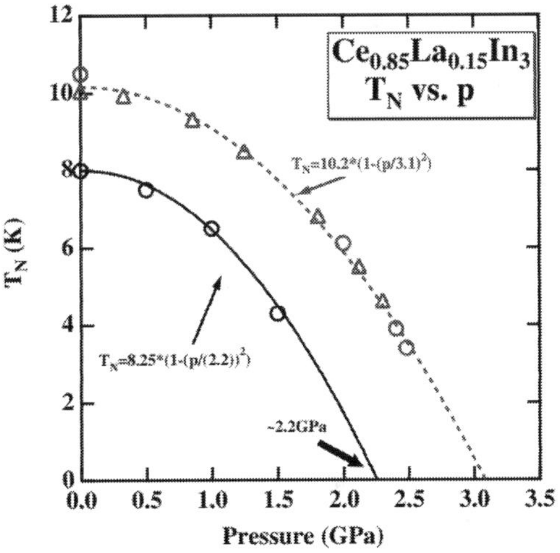

Figure 2. Pressure dependence of Néel ordering temperature in $Ce_{0.85}La_{0.15}In_3$. Black open circles represent determined T_N by us in $Ce_{0.85}La_{0.15}In_3$. Blue symbols show read data from Ref.[1]. Black solid and blue dashed lines show a quadratic fitting curves.

Figure 2 shows p-T phase diagram. Black open circles represent measured Néel temperatures in $Ce_{0.85}La_{0.15}In_3$. Blue symbols show read data from Ref. [1]. Black solid and blue dashed lines represent fitting curve as the function of $T_N = T_N^* \times [1-(p/p_c)^2]$ where T_N^* is Néel temperature and p_c means critical pressure. The T_N^* is 8.3K for $Ce_{0.85}La_{0.15}In_3$ and T_N^* is 10K for $CeIn_3$.

The critical pressures for $CeIn_3$ and for $Ce_{0.85}La_{0.15}In_3$ are 3.1 and 2.2GPa, respectively. With applying pressure to $Ce_{0.85}La_{0.15}In_3$, the Néel temperature reduces to lower one following to the quadratic function.

Although we are not able to directly compare the behavior of $Ce_{0.85}La_{0.15}In_3$ with that of Kondo lattice compound of $CeIn_3$ because alloying introduces disorder or randomness, pressure suppresses Néel order both in $Ce_{0.85}La_{0.15}In_3$ and $CeIn_3$. Because reduction of number of $4f$ electrons and widening distance between spins weaken the RKKY interaction, 15 percent alloying reduces the Néel temperature from 10K for $CeIn_3$ to 8K for $Ce_{0.85}La_{0.15}In_3$. It is shown that the critical pressure also reduces from 3.1GPa for $CeIn_3$ to 2.2GPa for $Ce_{0.85}La_{0.15}In_3$ as seen in the p-T phase diagram of $Ce_{0.85}La_{0.15}In_3$. Lower critical pressure could be caused the Néel order is suppressed easier in $Ce_{0.85}La_{0.15}In_3$ than in $CeIn_3$.

4. Summary

The Néel temperature reduces by both alloying and applying pressure. Fifteen percent alloying lets T_N reduced from 10K to 8K because of reduction of RKKY interaction. At $Ce_{0.85}La_{0.15}In_3$, pressure dependence of T_N shows quadratic reduction as seen in $CeIn_3$. The p-T phase diagram of $Ce_{0.85}La_{0.15}In_3$ was drawn to determined the critical pressure to be at 2.2GPa.

5. Acknowledgements

T. Ebihara expresses many thanks to supports from Grant-in-Aid for Scientific Research on Priority Areas "High Field Spin Science in 100T" (No.451) and Grant-in-Aid for Scientific Research (C) provided by Ministry of Education, Culture, Sports, Science and Technology (MEXT). The Asahi Glass Foundation also supported a part of this work done by T. E. Work by A. I. was performed under support by Scientific Research on Priority Areas "Invention of Anomalous Quantum Materials" (No.436) from MEXT, Japan. Work at KYOKUGEN was performed under auspices from Grant-in-Aid for Scientific Research 21COE, MEXT.

6. References

[1] Mathur N D,Groshe F M, Julian S R, Walker I R, Freye D M, Haselwimmer R K W and Lonzarich G G 1998 *Nature* **394** 39

[2] Iwamoto Y, Ebihara T, Harrison N, Jaime M, Silhanek A, Tezuka K, Morishita K, Terashima T and Iyo A 2007 *J.Mag.Mag.Matt.* **310** 300

[3] Pagliuso P G, Moreno N O, Curro N J, Thompson J D, Hundley M F, Sarrao J L, Fisk Z, Christianson A D, Lacerda A H, Light B E and Cornelius A L 2002 *Phys.Rev.B* **66** 54433

Effects of nonlinear elasticity and electromechanical coupling on pressure coefficient of the light emission in group-III nitride quantum structures

Łepkowski S P[1,2]

[1] Unipress - Institute of High Pressure Physics, Polish Academy of Sciences, ul. Sokołowska 29, 01-142 Warszawa, Poland

[2] Faculty of Mathematics and Natural Sciences, Cardinal S. Wyszyński University, ul. Dewajtis 5, 01-815 Warszawa, Poland

slawek@unipress.waw.pl

Abstract. We present theoretical studies of the effects of nonlinear elasticity and electromechanical coupling on the pressure coefficients of the light emission, dE_E / dP_{ext}, in GaN/AlN and $In_{0.2}Ga_{0.8}N$/GaN superlattices with compressively strained quantum wells. We show that effect of the nonlinear elasticity, manifesting itself in pressure dependence of elastic constants, leads to significant reduction of the values of dE_E / dP_{ext} in the considered structures. On the other hand, the effect of the electromechanical coupling, i.e., co-existence of the direct and converse piezoelectric effect, causes increase of the dE_E / dP_{ext} in these structures. However, the contribution to dE_E / dP_{ext}, originating from the nonlinear elasticity dominates over the contribution coming from the effect of electromechanical coupling. Interestingly, the influence of the effect of electromechanical coupling on dE_E / dP_{ext} diminishes when the nonlinear elasticity is taken into account.

1. Introduction

In the recent years, optical properties of wurtzite InGaN/GaN and GaN/AlGaN quantum wells (QWs) under hydrostatic pressure have been extensively studied [1-3]. These studies have revealed intriguingly small values of the pressure coefficients of the light emission, dE_E / dP_{ext}, in the structures grown along the *c*-axis. It has been shown that small values dE_E / dP_{ext} originate from the pressure-induced increase of piezoelectric field in these structures [1-3]. However, calculations of dE_E / dP_{ext}, based on the linear elastic and piezoelectric theory, have agreed with the experimental results only for GaN/AlGaN QWs with small aluminum concentration [3]. For InGaN/GaN QWs and for GaN/AlGaN QWs with aluminum concentrations larger than 20%, second-order elastic and piezoelectric effects have to be taken into account. First of all, it has been found that the nonlinear piezoelectricity, manifesting itself in dependence of piezoelectric constants on the volume-conserving strain [4], is essential for proper description of the pressure tuning of the built-in electric fields in wurtzite nitride QWs grown along the *c*-axis [1,2]. This effect is of primary importance for

© 2008 IOP Publishing Ltd

determination of the dE_E / dP_{ext} in these structures. Recently, the influence of two other second-order electromechanical effects on dE_E / dP_{ext} in nitride QWs has been studied. Namely, it has been shown that the effect of the nonlinear elasticity, originating from the pressure dependence of elastic constants, causes significant reduction of the values of dE_E / dP_{ext} in nitride QWs [5]. On the other hand, the effect of the electromechanical coupling, i.e., co-existence of the direct and converse piezoelectric effect, causes increase of the dE_E / dP_{ext} in these structures [6]. It should be noted, that in [5] and [6], these two effects have been studied assuming the strain dependence of piezoelectric constants of nitrides, i.e., taking into account also the effect of nonlinear piezoelectricity.

In this work, the effects of nonlinear piezoelectricity, nonlinear elasticity and electromechanical coupling on the dE_E / dP_{ext} in nitride quantum structures are studied altogether. We focus on superlattices (SLs) with compressively strained wells and unstrained barriers. The choice of SLs is motivated by the well-defined periodic boundary conditions for the electric fields [6]. We show that the contribution to the dE_E / dP_{ext}, originating from the nonlinear elasticity dominates over the contribution coming from the electromechanical coupling effect.

2. Pressure-tuning of strain and electric field in nitride superlattices

Let us indicate all material tensors, fields and parameters by indices 'a' for the wells and 'b' for the barriers. Then, one can describe the pressure tuning of the nonzero elements strain tensor and piezoelectric field in a SL with unstrained barriers using the following equations [5,6]

$$\varepsilon_{xx,a}(P_{ext}) = \frac{a_b}{a_a}\left\{1+\frac{C_{11,b}(P_b)-C_{13,b}(P_b)}{[C_{11,b}(P_b)+C_{12,b}(P_b)]C_{33,b}(P_b)-2[C_{13,b}(P_b)]^2}P_{ext}\right\}-1 \tag{1}$$

$$\varepsilon_{xx,b}(P_{ext}) = \frac{C_{11,b}(P_b)-C_{13,b}(P_b)}{[C_{11,b}(P_b)+C_{12,b}(P_b)]C_{33,b}(P_b)-2[C_{13,b}(P_b)]^2}P_{ext} \tag{2}$$

$$\varepsilon_{zz,a} = -\frac{2C_{13,a}(P_a)}{C_{33,a}(P_a)}\varepsilon_{xx,a} + \frac{e_{33,a}}{C_{33,a}(P_a)}E'_{z,a} + \frac{P_{ext}}{C_{33,a}(P_a)}\left(e_{33,a}E''_{z,a}-1\right) \tag{3}$$

$$E_{z,a} = E'_{z,a} + P_{ext}E''_{z,a} \tag{4a}$$

where

$$E'_{z,a} = \frac{L_b\left\{P_{sp,z,b} + 2\left[e_{31,b} - \frac{C_{13,b}(P_b)}{C_{33,b}(P_b)}e_{33,b}\right]\varepsilon_{xx,b} - P_{sp,z,a} - 2\left[e_{31,a} - \frac{C_{13,a}(P_a)}{C_{33,a}(P_a)}e_{33,a}\right]\varepsilon_{xx,a}\right\}}{L_b\kappa_a + L_a\kappa_b} \tag{4b}$$

$$E''_{z,a} = \frac{L_b}{L_b\kappa_a + L_a\kappa_b}\left[\frac{e_{33,a}}{C_{33,a}(P_a)} - \frac{e_{33,b}}{C_{33,b}(P_b)}\right] \tag{4c}$$

$$\kappa_a = \lambda_a + \frac{e_{33,a}^2}{C_{33,a}(P_a)}. \tag{4d}$$

Formulas for $\varepsilon_{zz,b}$, $E_{z,b}$, and κ_b can be obtained from equations (3) and (4a)-(4d) by changing indices 'a' and 'b'. In the above equations, e_{ij} are the piezoelectric constants, $C_{ij}(P)$ are the pressure-dependent elastic constants, P_{ext} denotes the external hydrostatic pressure, L is the thickness of the layer, P is the total hydrostatic pressure originating from the external and internal stresses [6].

3. Results and discussion

Using the above model of pressure tuning of the strain and electric field, we compute the electronic states in the SLs as a function of P_{ext} and then calculate dE_E/dP_{ext} [8,9]. We consider GaN/AlN SLs grown on AlN buffers and $In_{0.2}Ga_{0.8}N$/GaN SLs grown on GaN buffers. We focus on the SLs with equal widths of wells L_a and barriers L_b since in these structures, the electric fields do not depend on the thickness of the layers (see equations (4a)-(4b)).

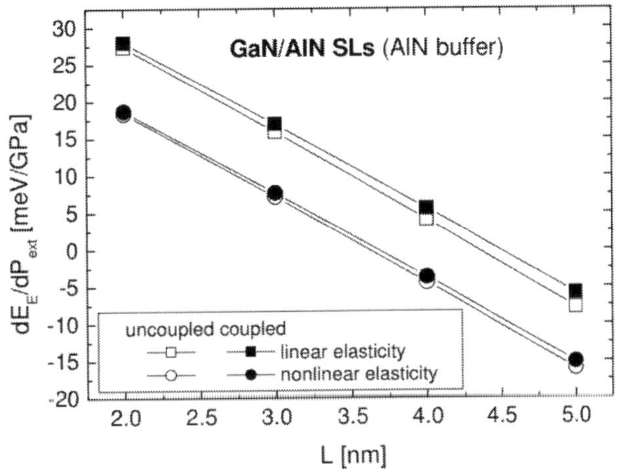

Figure 1. The values of dE_E/dP_{ext} for GaN/AlN SLs with $L = L_a = L_b$. Solid lines are added to guide the eyes.

In figures (1) and (2), we show the values of dE_E/dP_{ext} obtained for the considered structures as a function of the common length $L = L_a = L_b$. For each structure, four values of dE_E/dP_{ext} have been calculated. Namely, we have considered the following cases: (i) the effects of nonlinear elasticity and electromechanical coupling are taken into account (full circles), (ii) the effect of nonlinear elasticity is excluded (full squares), (iii) the effect of electromechanical coupling is excluded (empty circles), (iv) both effects of nonlinear elasticity and electromechanical coupling are not taken into account (empty squares). In all the cases, we have taken into account the strain dependence of piezoelectric constants of nitrides, i.e., the nonlinear piezoelectricity.

In both figures (1) and (2), one can see that the values of dE_E/dP_{ext} decrease linearly with L. This effect originates from the increase of electric field in the QWs with P_{ext} [5,6]. Let us focus on the contributions to dE_E/dP_{ext} coming from the effects of nonlinear elasticity and electromechanical coupling. The effect of the nonlinear elasticity results in significant reduction of the values of dE_E/dP_{ext} in GaN/AlN and $In_{0.2}Ga_{0.8}N$/GaN SLs. The effect of the electromechanical coupling causes slight increase of dE_E / dP_{ext} in GaN/AlN SLs. For $In_{0.2}Ga_{0.8}N$/GaN SLs, this effect is very small (see figure 2). Comparing the differences between full and empty symbols in figure 1, one can also observe that these differences are larger for squares than for circles. This means that the influence of the electromechanical coupling effect on dE_E / dP_{ext} diminishes when the nonlinear elasticity is taken into account.

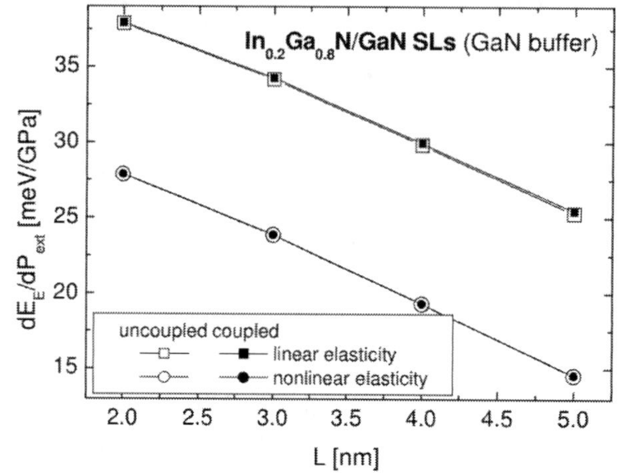

Figure 2. The values of dE_E / dP_{ext} for $In_{0.2}Ga_{0.8}N$/GaN SLs with $L = L_a = L_b$. Solid lines are added to guide the eyes.

4. Conclusions

We have studied the influence of the effects of nonlinear elasticity and electromechanical coupling on the dE_E / dP_{ext} in nitride SLs. Our calculations have revealed that (i) the effect of the nonlinear elasticity leads to significant reduction of the values of dE_E / dP_{ext} and (ii) the effect of the electromechanical coupling causes increase of dE_E / dP_{ext}. The contribution to dE_E / dP_{ext}, originating from the nonlinear elasticity dominates over the contribution coming from the electromechanical coupling effect. Interestingly, the influence of the electromechanical coupling effect on dE_E / dP_{ext} diminishes when the nonlinear elasticity is taken into account.

Acknowledgments

This work has been supported by the Polish State Committee for Scientific Research, Project No. 1P03B03729. Additionally, the author acknowledge the use of computer facilities of the ICM UW.

References

[1] Vaschenko G, Patel D, Menoni C S, Gardner N F, Sun J, Gotz W, Tome C N and Clausen B 2001 *Phys. Rev.* B **64** 241308(R)
[2] Vaschenko G, Patel D, Menoni C S, Ng H M and Cho A Y 2002 *Appl. Phys. Lett.* **80** 4211
[3] Lepkowski S P, Teisseyre H, Suski T, Perlin P, Grandjean N and Massies J 2001 *Appl. Phys. Lett.* **79** 1483
[4] Shimada K, Sota T, Suzuki K and Okumura H 1998 *Japan. J. Appl. Phys.* **37** L1421
[5] Lepkowski S P, Majewski J A and Jurczak G 2005 *Phys. Rev.* B **72** 245201
[6] Lepkowski S P and Majewski J A 2006 *Phys. Rev.* B **74** 035336

Determining the structure of compressed calcium in phase V by the *ab-initio* molecular dynamics simulation

T. Ishikawa[1], A. Ichikawa[2], H. Nagara[3], K. Kusakabe[4] and N. Suzuki[5]

[1 3 4 5] Graduate School of Engineering Science, Osaka University, 1-3 Machikaneyama, Toyonaka, Osaka 560-8531, Japan
[2] Department of Physics, Graduate School of Science, University of Tokyo, 7-3-1 Hongo, Bunkyo-ku, Tokyo 113-0033, Japan

E-mail: [1] `ishikawa@aquarius.mp.es.osaka-u.ac.jp`

E-mail: [2] `ichikawa@bio.phys.s.u-tokyo.ac.jp`

E-mail: [3] `nagara@mp.es.osaka-u.ac.jp`

E-mail: [4] `kabe@mp.es.osaka-u.ac.jp`

E-mail: [5] `suzuki@mp.es.osaka-u.ac.jp`

Abstract. We searched for the structure of calcium in phase V by a metadynamics based on the first-principles calculation, and found a new structure, where the positions of the calcium atoms in the unit cell have a zigzag pattern. Calculating the x-ray diffraction pattern of the new structure and comparing it with the experimental patterns of Ca-IV and Ca-V, we conclude that the new zigzag structure is a candidate of the structure of Ca-V.

1. Introduction

High-pressure studies have revealed that many elements have low symmetrical and complex structures under pressure. Amazingly, some of the complex structures appear above 100GPa. The complex structures are reviewed by McMahon and Nelmes and classified into four categories: incommensurate composite structures, incommensurately modulated structures, modulated layer structures, and other complex structures [1]. Group 2 element Ba is the first element found to have the incommensurate composite structure in high pressure. The structure of Ba in phase III is hexagonal close-packed structure and it transforms to a host-guest structure Ba-IVa at 12.6 GPa which has the incommensurate composite structure [2] [3]. Ba-IVa transforms to hcp (Ba-V) again via other three Ba-IV-type structures (Ba-IVb, Ba-IVc and Ba-IVd) at 45 GPa. Another group 2 element Sr also has complex structures and the β-tin structure taking the helical distortion appears at 35 GPa (Sr-IV) [4]. On further compression, Sr-IV transforms to the incommensurate composite structure (Sr-V) [5].

Calcium, which is the group 2 element similar to Ba and Sr, also has complex structures in high pressures . Calcium has the fcc structure at ambient pressure. The fcc structure transforms to bcc at 20 GPa and then it has the transition to a simple cubic (sc) structure at 32 GPa [6]. The sc structure is rare in compressed elements and it is also observed in P above 10 GPa (P-III) [7]

© 2008 IOP Publishing Ltd

and As above 25 GPa (As-II) [8]. In P the sc structure transforms to a complex structure at 107 GPa (P-IV). This structure has not been identified but it is theoretically predicted to be an incommensurately modulated structure [9]. In As sc transforms to a monoclinic incommensurate composite structure at 46 GPa (As-III) [10].

In 2005 Yabuuchi *et al.* observed that the sc structure of Ca-III transforms to Ca-IV at 113 GPa and further to Ca-V at 139 GPa [11]. The structures of Ca-IV and Ca-V are complex structures . In these phases calcium shows the high superconducting transition temperature. In 1996 Okada *et al.* measured an electrical resistance up to 150 GPa and found that T_c linearly increased and reached about 15 K at 150 GPa. The transition temperature T_c shows a tendency of further increase above the pressure [12]. In 2006 Yabuuchi *et al.* observed that the transition temperature T_c dramatically increased in the Ca-III phase and it reached to 20 K above 100 GPa. On further compression, T_c was increasing in Ca-IV and Ca-V and exceeded 25 K at 161 GPa [13]. In elements the highest T_c had been found in lithium and it had been 20 K and calcium updates the highest record. Thus, in order to clarify the origin of the interesting behavior on the superconductivity of calcium, it is crucial to determine the structures of Ca-IV and Ca-V. Very recently we theoretically determined these structures by an *ab-initio* metadynamics simulation: Ca-IV has a tetragonal lattice with a helical modulation and Ca-V a orthorhombic one with a zigzag modulation [14].

In this work, we focused on the zigzag structure of Ca-V. We compared the x-ray diffraction pattern of the structure obtained by the metadynamics simulation with that obtained by a conventional molecular dynamics (MD) simulation. We compared the enthalpies among sc, the hexagonal close-packed (hcp) structure and the zigzag structure. We also investigated the atomic volume and the electronic states.

2. Computational details

To explore the structures of calcium in Ca-IV and Ca-V, we performed the *ab-initio* MD and metadynamics simulations. For the simulation, we used density functional theory in a generalized gradient approximation (GGA) and an ultrasoft pseudopotential, where we employed the expression by Perdew and Wang for the exchange-correlation energy functional [15]. Using the full-potential augmented plane-wave + local orbitals method as embodied in the WIEN2k code [16], we checked a localization of electrons of the calcium atom. 2p level is 330.97 eV below the Fermi level and it is deep in energy. But 3s level lies 43.33 eV and the charge is not completely confined inside an atomic sphere. Thus, for the pseudopotential used in this study, we regard 10 electrons, 3s, 3p and 4s, as the valence electrons. We performed the k-space integration using $8 \times 8 \times 8$ mesh points in the first Brillouin zone and set the energy cut-off of the plane wave basis at 24 Ry.

First we performed a conventional constant-pressure MD at 120 GPa to find an initial local minimum on the Gibbs free energy surface, and then explored other local minima around the initial one by the metadynamics [17] which is a method to fill the potential wells with a Gaussian potential. In the method, we treated 6 independent variables of a cell matrix $\mathbf{h} = (\vec{a}, \vec{b}, \vec{c})$ as reaction coordinates. We updated the simulation cell by the steepest-descent method with a stepping parameter δh: $\mathbf{h}_{ij}^{t+1} = \mathbf{h}_{ij}^t + \delta h \mathbf{F}_{ij}^t / |\mathbf{F}_{ij}^t|$ where t is the number of the update of $\mathbf{h}$ and $\mathbf{F}$ is a driving force for the update. If we defined the Gibbs potential and the Gaussian potential as $G_o(\mathbf{h}^t)$ and $G_g(\mathbf{h}^t) = \sum_{t'<t} \prod_{i,j} W \exp(-[\mathbf{h}^t - \mathbf{h}^{t'}]_{ij}^2 / 2\delta h^2)$, respectively, then the driving force $\mathbf{F}^t$ is the sum of $\mathbf{F}_o^t = -\partial G_o(\mathbf{h}^t)/\partial h_{ij}$ and $\mathbf{F}_g^t = -\partial G_g(\mathbf{h}^t)/\partial h_{ij}$. We defined an update of the cell as a meta-step and ran 100 MD steps for the atoms between each meta-step. For the MD with fixed cell we used the PWscf code [18].

3. Simulated results

We started the conventional constant-pressure MD from the sc structure. At 120 GPa the sc structure was unstable and it easily transformed to other structure. The volume of the obtained structure decreased about 4.5 %. The unit cell is a orthorhombic, a = 3.05 Å, b/a = 1.08 Å, c/a = 1.44 Å, and it consists of 4 atoms which locate at $(-0.08, 0, -0.08)$, $(0.08, 0, 0.42)$, $(0.42, 0.5, 0.08)$ and $(0.58, 0.5, 0.58)$. Comparing the x-ray diffraction pattern of the structure with the experimental ones, we found that the orthorhombic structure was neither the structure of Ca-IV nor that of Ca-V (Fig.2).

Then, we explored other local minima by the metadynamics. We started the simulation from the orthorhombic unit cell obtained by the above conventional constant-pressure MD and set the Gaussian parameters W and δh to 10 mRy and 20 mÅ, respectively. Cell parameters γ, a and b began to change around 270 meta-steps and then we switched the Gaussian potential off by setting $W = 0$ mRy to confirm whether the system had escaped from the initial potential well. As a result the orthorhombic structure transformed to a new structure having a monoclinic distortion. The angle γ increased from 90° to approximately 92° and a and b were equal to each other. When we switched the Gaussian potential off before 270 meta-steps, the structure returned to the initial orthorhombic structure.

We found that the primitive unit cell of the new monoclinic lattice has the dimensions of a = 3.16 Å, b/a = 1.00, c/a = 1.39, γ = 92.66° at 120 GPa and consists of four atoms. The atomic positions are Ca(1)=(−0.10, 0.10, −0.06), Ca(2)=(0.10, −0.10, 0.44), Ca(1')=(0.40, 0.60, 0.06) and Ca(2')=(0.60, 0.40, 0.56). It can be viewed as the lattice consisting of two zigzag chains. Calcium atoms on each chain are shown by red or blue colors in Fig.1. This structure can be redefined as an orthorhombic lattice by the choice of the double sized unit cell. The space group of the zigzag structure is *Cmca* (D_{2h}^{18}), which is orthorhombic, with atoms occupying 8f sites. The 8f sites parameters can be defined as $(y, z) = (1/4 + d_1, -1/4 + e_1)$ with $d_1 = -0.10$ and $e_1 = 0.06$, after shifting origin.

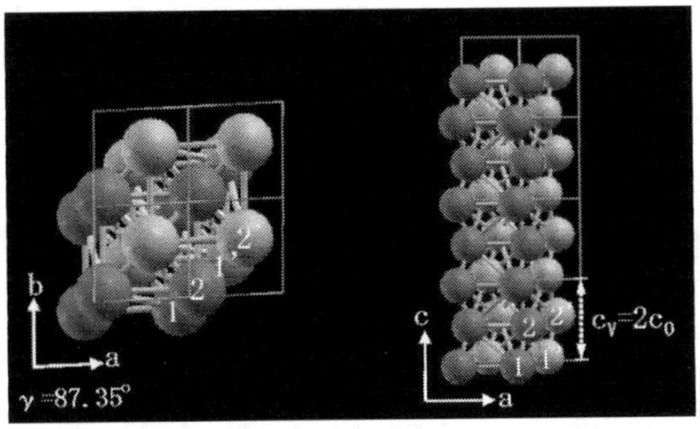

Figure 1. The zigzag structure obtained by the *ab-initio* metadynamics simulation. The left figure is a view along c axis and the right one a view along b axis. This structure has two zigzag chains in the unit cell and each chain is indicated by red or blue color. The number on the calcium atoms in the unit cell show their positions corresponding to those given in the text. In the right figure, c_0 is a lattice constant along c axis for an unmodulated structure.

4. Comparison of the x-ray diffraction patterns

Figure 2 shows the comparison of the x-ray diffraction patterns of the crystal structures theoretically obtained with the experimental diffraction patterns. The first and second figures from the top are the diffraction patterns of Ca-V and Ca-IV, respectively [11]. In these patterns the diffraction peaks from the sample are indicated by arrows and g represents reflections from a metal gasket. The third is the diffraction pattern of the structure obtained by the conventional constant-pressure MD, which is used as the initial structure for the metadynamics simulation. The bottom is the diffraction pattern of the structure obtained by the metadynamics simulation. At the transition from Ca-IV to Ca-V the most distinct change of the diffraction pattern is an appearance of the peak around $2\theta = 17°$. Both of the structures obtained by the theoretical simulations have the peak and the Miller index is $(hkl) = (111)$. But for the initial structure the diffraction patterns around $2\theta = 20°$ and 25° agree with neither those of Ca-IV nor those of Ca-V. On the other hand the diffraction pattern of the final structure obtained by the metadynamics simulation shows good agreement with that of Ca-V.

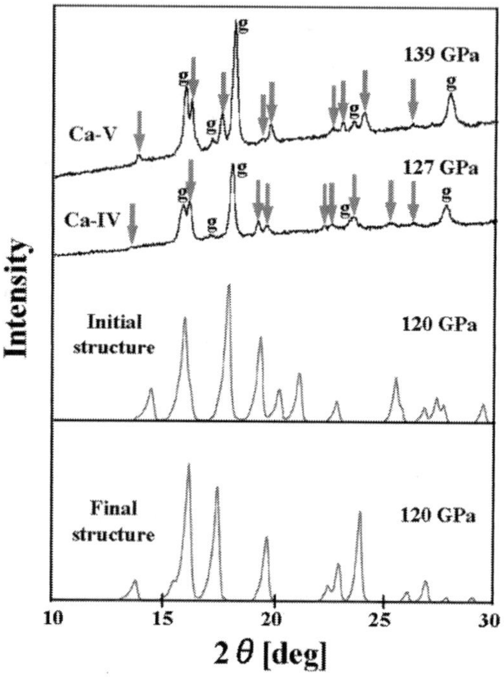

Figure 2. Comparison of the x-ray diffraction patterns among the structures theoretically obtained, Ca-IV and Ca-V. The first and second diffraction patterns from the top are the diffraction patterns of Ca-V and Ca-IV, respectively [11]. In these diffraction patterns, reflections from Ca sample are indicated by arrows and those from a metal gasket are by g. The third figure is the diffraction pattern of the structure obtained by the *ab-initio* conventional MD at 120 GPa. This structure is also the initial one for the *ab-initio* metadynamics simulation. The bottom is the diffraction pattern of the final structure for the metadynamics simulation.

5. Energy calculations of the zigzag structure of Ca-V

The structure of Ca-V has two types of distortion: (a) the displacement in the opposite direction along c axis between two zigzag chains which are identified by different colors and (b) the zigzag distortion along c axis (Fig.1). From the structure of Ca-V, we made three model structures, (i) a structure without the distortions (a) and (b), (ii) a structure with the distortion (a) only and (iii) a structure with the distortion (b) only, and compared the total energy of the structure of Ca-V with those of the model structures at the same volume determined at 100 GPa (Fig.3). If we treat the structure (i) as the reference for the comparison of the total energy, the total energies of the structure (ii) and the structure (iii) are +1.55 mRy/atom and +1.78 mRy/atom, respectively. However, the total energy of the structure of Ca-V, which has both the distortions (a) and (b), is lower than that of the structure (i) and the energy difference is −8.97 mRy/atom. Thus it is crucial for the stability of the candidate structure to take two distortions (a) and (b) simultaneously.

We compared the enthalpies among sc, the zigzag structure and hcp (Fig.4). The enthalpy of sc is the lowest of the three at 80 GPa. The enthalpy of the zigzag structure becomes lower than that of sc in the higher pressure region. The enthalpy difference between sc and the zigzag structure is 0.1 mRy/atom at 80 GPa. The hcp structure is a candidate structure of higher pressure phases of calcium [19], but the comparison of the enthalpies shows that hcp does not appear in the pressure range from 80 to 120 GPa. Atomic volume of the zigzag structure is smaller than that of sc and the volume decrease is 4.10 % at 80 GPa. The volume decrease is attributed to the decrease of c/a due to the zigzag pattern along c axis.

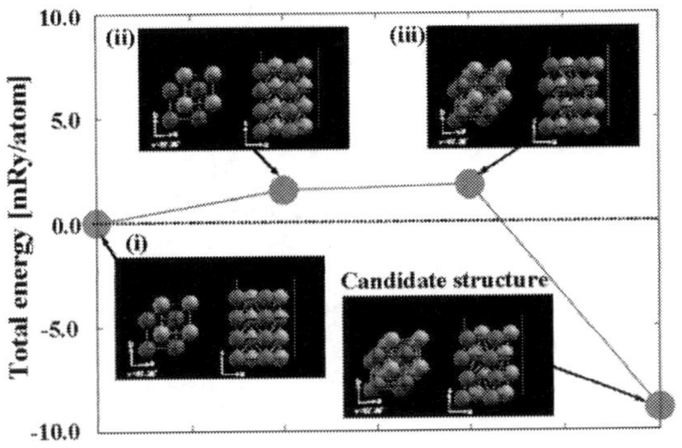

Figure 3. Comparison of the total energies between the structure of Ca-V obtained by the *ab-initio* metadynamics simulation and three model structures ((i), (ii) and (iii)) at same volume determined at 100 GPa. The structure of Ca-V has two types of distortion, (a) the displacement in the opposite direction along c axis between two zigzag chains which are identified by different colors and (b) the zigzag distortion along c axis. The structure (i) is the structure without the distortions (a) and (b). The structure (ii) and the structure (iii) have the distortion (a) only and the distortion (b) only, respectively.

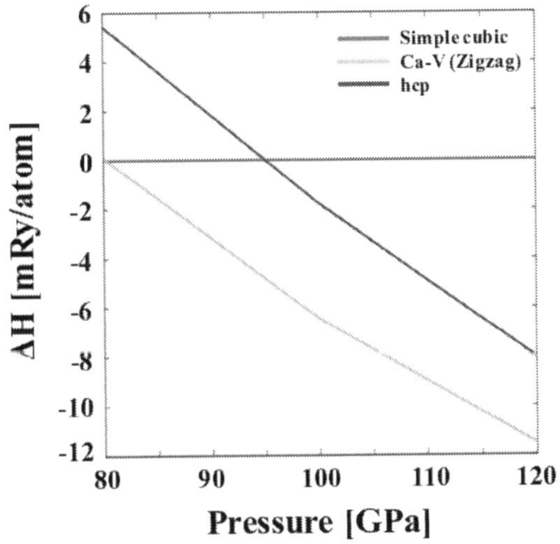

Figure 4. Comparison of the enthalpies among sc, hcp and the zigzag structure. The hcp structure has been one of the candidate structures of the higher pressure phases of calium [19].

In the sc phase electrons of the s-character decreases and that of the d-character increases with compression. This s-d transfer advances more by the transformation from sc to Ca-V. Around the intersection of the enthalpy curve of sc with that of the Ca-V (80 GPa), the s-electrons and p-electrons decrease from 22.25 to 21.78 % and from 58.38 to 58.05 %, respectively, while the d-electrons increases from 19.38 to 20.17 %.

6. Summary

In this study, we showed the structures of calcium in phase V, which is obtained by the *ab-initio* metadynamics simulation. The structure of Ca-V is the complex structure with the zigzag pattern. This structure can not be obtained by the conventional MD only in our research. The x-ray diffraction pattern of the zigzag structure shows very good agreement with the experimental pattern of Ca-V. The validity of this prediction is also supported by the comparison of the enthalpies among sc, hcp and the zigzag structure. Since we have to impose a periodic boundary condition on a system for theoretical calculations, there remains a possibility that Ca-V has an incommensurate structure. But the good agreement of the diffraction pattern shows that the incommensurate structure should be very close to the predicted commensurate structure.

Journal of Physics: Conference Series **121** (2008) 012004 doi:10.1088/1742-6596/121/1/012004

Acknowledgments

The authors thank Y. Nakamoto and K. Shimizu for valuable discussion on the x-ray powder patterns. This work was partially supported by a Grant-in-Aid for Scientific Research in Priority Areas "Development of New Quantum Simulators and Quantum Design" (No.17064013) of The Ministry of Education, Culture, Sports, Science, and Technology, Japan, and by JSPS Research Fellowships for Young Scientists "Determination of pressure-induced structures and Research on the pressure-induced superconductivity by the first-principles calculation" (No.19·1723).

[1] McMahon M I and Nelmes R J 2006 *Chem. Soc. Rev.* **35** 943
[2] Nelmes R J, Allan D R, McMahon M I and Belmonte S A 1999 *Phys. Rev. Lett.* **83** 4081
[3] Takemura K 1994 *Phys. Rev.* B **50** 16238
[4] Bovornratanaraks T, Allan D R, Belmonte S A, McMahon M I and Nelmes R J 2006 *Phys. Rev.* B **73** 144112
[5] McMahon M I, Bovornratanaraks T, Allan D R, Belmonte S A and Nelmes R J 2000 *Phys. Rev.* B **61** 3135
[6] Olijnyk H, Holzapfel W B 1984 *Phys. Lett.* **100A** 191
[7] Kikegawa T and Iwasaki H 1983 *Acta Crystallogr.* Sect. B:Struct. Sci. **39** 158
[8] Kikegawa T and Iwasaki H 1987 *J. Phys. Soc. Jpn.* **56** 3417
[9] Ishikawa T, Nagara H, Kusakabe K and Suzuki N 2006 *Phys. Rev. Lett.* **96** 095502
[10] Greene R G, Luo H and Ruoff A L 1995 *Phys. Rev.* B **51** 597
[11] Yabuuchi T, Nakamoto Y, Shimizu K and Kikegawa T 2005 *J. Phys. Soc. Jpn.* **74** 2391
[12] Okada S, Shimizu K Kobayashi T C, Amaya K and Endo S 1996 *J. Phys. Soc. Jpn.* **65** 1924
[13] Yabuuchi T, Matsuoka T, Nakamoto Y and Shimizu K 2006 *Special Issue Rev. High Press. Sci. Technol.* **16** 331
[14] Ishikawa T, Ichikawa A, Nagara H, Geshi M, Kusakabe K and Suzuki N 2008 *Phys. Rev.* B **77** 020101(R)
[15] Perdew J P 1991 *Electronic Structure of Solids '91*, ed P Ziesche and H Eschrig (Berlin: Akademie Verlag), p 11
[16] Blaha P, Schwarz K, Madsen G K H, Kvasnicka D and Luitz J, WIEN2k, An Augmented Plane Wave + Local Orbitals Program for Calculating Crystal Properties, Karlheinz Schwarz, Tech. Wien, Austria, 2001, ISBN 3-9501031-1-2
[17] Martoňák R, Laio A and Parrinello M 2003 *Phys. Rev. Lett.* **90** 75503
[18] Baroni S, Dal Corso A, de Gironcoli S and Giannozzi P, The PWscf code is available through: ⟨ http://www.pwscf.org/ ⟩
[19] Ahuja R, Eriksson O, Wills J M and Johansson B 1995 *Phys. Rev. Lett.* **75** 3473

Journal of Physics: Conference Series **121** (2008) 012005 doi:10.1088/1742-6596/121/1/012005

Ab initio and classical Molecular Dynamics calculations of the high-pressure melting of Ne

L. Koči[1], R. Ahuja[1,2], A. B. Belonoshko[2,3]

[1]Condensed Matter Theory Group, Physics Department, Uppsala University, Box 530, S-751 21, Uppsala, Sweden

[2]Applied Materials Physics, Department of Materials Science and Engineering, The Royal Institute of Technology, S-100 44 Stockholm, Sweden

[3]Condensed Matter Theory, Alba Nova University Center, Physics Department, The Royal Institute of Technology, S-100 44 Stockholm, Sweden

E-mail: love.koci@fysik.uu.se

Abstract. Classical molecular dynamics (CMD) calculations are fast but are heavily dependent on the potential feasibility. On the other hand, first-principles (*ab initio*) molecular dynamics (AIMD) does not use any empirical knowledge, but can be extremely time consuming. As both techniques have been applied to study melting at extreme conditions, a comparison of the methods is motivated. Furthermore, when melting is studied with MD, the use of coexistent solid and liquid structures (two-phase) in the initial simulation configuration, instead of a only a solid structure (one-phase), can have a significant impact. In this work, comparisons have been made between CMD and AIMD methods applied to one- and two-phase systems for the melting of Ne at high pressure.

1. Introduction

First-principles (*ab initio*) MD (AIMD) is based on the density functional theory (DFT) and the foundation of this theory was developed during the 1960's. [1] Today, the application of *ab initio* methods are common in material studies. [2, 3, 4, 5, 6] Although numerical algorithms and the increasing performance of computer systems continuously are improving, *ab initio* calculations are still computationally expensive. Classical MD (CMD), on the other hand, is much faster and can treat millions of atoms. This is due to the use of empirical interatomic potentials where electronic effects are incorporated implicitly. However, it can be problematic to develop potential parameters that accurately describe the physical properties of the material studied. As no *a priori* potential is used in AIMD, possible sources to uncertainties by the potential choice and the parameter fitting can be avoided.

In this work, the high-pressure melting of Ne has been studied by classical and *ab initio* methods. Furthermore, comparisons have been made between homogeneous and heterogeneous melting: when the melting initiates with a liquid nucleation in the bulk, spreading throughout the solid structure, the transition is said to be homogeneous. Heterogeneous melting, on the other hand, begins at the surfaces, where atoms start to diffuse. Homogeneous melting can show big hysteresis effects, as a crystal can be superheated considerably above (and undercooled below) the equilibrium melt temperature (T_m). In MD, homogeneous melting can be resembled by the

© 2008 IOP Publishing Ltd

heating of a crystal whereas heterogeneous melting can be modeled by the use of coexisting solid and liquid structures in the simulation box.

2. Models and technical details

The Lennard-Jones (L-J) potential,

$$\Phi(r) = 4\epsilon[(\frac{\sigma}{r})^{12} - (\frac{\sigma}{r})^{6}] \quad , \tag{1}$$

has shown to reproduce thermodynamic properties of gases such as argon, [7, 8] xenon, [9] and krypton. [10] Here, we have examined the potential parameters for Ne with $\epsilon/k = 35.1$ K (k is the Boltzmann constant) and $\sigma = 2.72$ Å from Bellisent-Funel *et al.* [11] The L-J potential is implemented in the *Moldy* package [12] which was used for the simulations. Periodic boundary conditions (PBC) were applied to all six surfaces of the simulation box. One advantage when using PBC is the absence of surface effects, with the consequence that the melting of a one-phase system is homogeneous. In the calculations, the size of the time steps dt was set to 0.5 fs and the radius of the cutoff was 6 Å. A strict cutoff was applied, meaning that all interactions between pairs of sites within the radius were included.

The *ab initio* DFT calculations in this work were performed using the Vienna *Ab Initio* Simulation Package (VASP) [13, 14] in the NVT ensemble. Ultrasoft pseudopotentials were used to represent the ionic cores. The local-density approximation (LDA) was chosen as exchange correlation function, as it has been reported to perform well when calculating the ground-state properties of other noble gases as e.g. Xe. [15] The cutoff energy was 450 eV and the size of the time step was 0.5 fs. In conjunction with other recent studies, only the Γ point was used for the Brillouin zone integrations to avoid too long simulation times. [16, 17]

3. Results

By the use of a link cell method, [18] the CMD calculations in this work scale $O(N)$ whereas the *ab initio* calculations performed with the VASP package scale $O(N^3)$. Shown in table 1 are the CPU times for the first 10 MD steps for Ne when using AIMD or CMD. As there is a significant system size difference when comparing similar calculation times for the two methods, the purpose of the table is to emphasize the AIMD cost and the CMD efficiency.

Table 1. CPU times for the 10 first steps in a MD calculation as a function of method, processors and atoms.

Method	♯ processors	♯ atoms	CPU time (s)
AIMD	12	2	1.5
		4	3.6
		16	37.7
		32	159.1
		54	507.2
		108	746.8
CMD	1	6912	1.1
		16384	2.9
		32000	5.6
		108000	20.9
		186624	36.6
		500000	101.8

To initiate the setup for the two-phase simulations, two perfect fcc lattices containing 864 atoms each were constructed. One of these lattices, shown as 1) in Fig. 1, was simulated at a high temperature to transform into a molten configuration. The solid and molten structures were put together with a small spacing in a simulation box, letting the lower part of the box be solid and the upper part be liquid, as shown in 2). Starting from this configuration, the MD simulations result in a monophase. If the temperature is above the melting temperature T_m, the phase will become liquid (3a) whereas the phase will solidify if the temperature is below T_m (3b). By narrowing the interval, the melting temperature can be estimated at the specific pressure in the simulation.

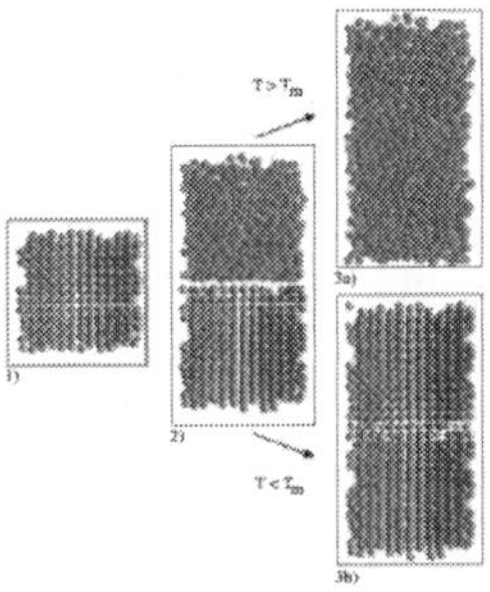

Figure 1. 864 atoms in a fcc structure are shown in 1). By melting an identical lattice, a two-phase setup can be constructed, as shown in 2). If $T > T_m$, the phase will become liquid (3a) whereas the phase will solidify if $T < T_m$ (3 b).

Belonoshko [19] and Tepper and Briels [20] have explained the two-phase simulation method in detail, and its successful application for a number of systems has been reported.[5, 21, 22, 23] For the rare gases in particular, the technique has recently been employed to study the melting of Xe. [24]
Shown in Fig. 2 is the melting of Ne at high *PT* conditions comparing different methods. The melting for the 1728 atom system (864+864, two-phase, NPT) and the 108 atom configuration (one-phase, NVT) with the L-J interaction from Bellisent-Funel *et al.* [11] were found by means of the radial distribution function. The simulations were performed for 40 000 and 3000 time steps, respectively. There is a discrepancy between the small and short one-phase runs and the big and long two-phase runs, as the former tend to overshoot the melting temperature. According to previous studies, [25, 26] this is expected. There is, however, another reason why to perform classical calculations of a small, one-phase system for a limited number of steps: a direct comparison can be made to the AIMD system, where these limitations often are necessary in order to avoid long computation times.
Compared to the results from the big two-phase system, there is also a melting temperature overshoot from the AIMD calculations. The CMD and AIMD melting results from this work are compared with one of the few predicted melting curves of neon at extreme conditions from Vos *et al.* [27] Clearly, there is a big discrepancy between this prediction and the AIMD one-phase

calculations, as the latter need significantly higher temperatures to melt the crystal. However, this melting follows the one-phase L-J curve quite accurately. Although the two systems have been calculated with the same limitations, this results is highly interesting. This could also indicate the conformity between the AIMD and CMD two-phase calculations, where the melting accuracy of the previous method has been reported. [5, 21, 22, 23]

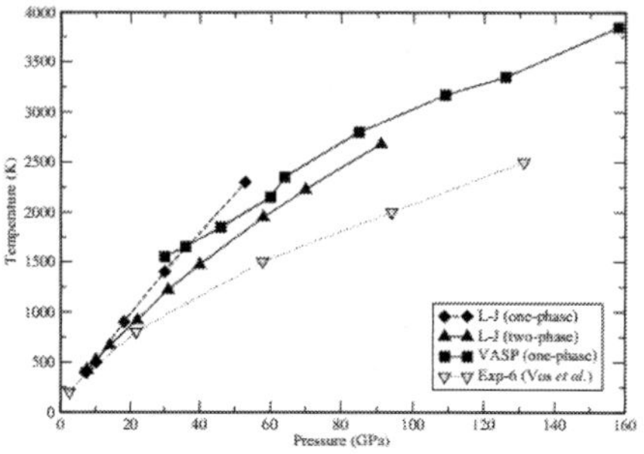

Figure 2. The melt line for the 1728 atom system (864+864, two-phase) with the L-J interaction from Bellisent-Funel *et al.* [11] was found by means of the radial distribution function. The simulations were performed for 40 000 time steps. Furthermore, the melt line from one-phase 108 atom simulations over 3000 time steps with the same potential is shown. The AIMD melting points were found from visual inspection of the atomic diffusion and the mean square displacement. The melting curve from Vos *et al.* [27] is also inserted.

4. Discussion and conclusion

Ahuja *et al.* [25] have reported the risk of overheating one-phase simulations in MD to find the melting temperature. However, bigger two-phase AIMD calculations, where liquid and solid configurations are simulated together, are extremely time consuming. In this paper, we have investigated the melting at high pressure of Ne by the use of classical and *ab initio* methods. Also, by comparing one- and two-phase CMD calculations, we have shown that the melting temperatures found by the one-phase configurations are slightly overestimated due to overheating. Although the high-pressure Ne melting curve determined by AIMD one-phase simulations is high compared to previous classical predictions, [27] it is found close to the CMD one-phase results in this work. Therefore, considering the documented accuracy of the two-phase method, the true melting curve could be found closer to the CMD two-phase results.

5. Acknowledgments

L. K. wishes to thank K. Refson for the *Moldy* simulation package and the Swedish National Infrastructure for Computing (SNIC) for the computer time. This work was supported by the Swedish Research Council (VR) and the European Science Foundation (ESF) in the framework of the *EuroMinSci* project.

[1] P. Hohenberg and W. Kohn. *Phys. Rev.*, 136:864, 1964.
[2] III H. F. Schaefer. *Properties of Complex Inorganic Solids.* Dover Publications, New York, 2004.
[3] L. Stixrude, R. E. Cohen, and D. J. Singh. *Phys. Rev. B*, 50:6442, 1994.
[4] P. Söderlind, J. A. Moriarty, and J. M. Wills. *Phys. Rev. B*, 53:14063, 1996.
[5] A. B. Belonoshko, R. Ahuja, and B. Johansson. *Phys. Rev. Lett.*, 84:3638, 2000.
[6] A. P. Jardine, S. Dworski, P. Fouquet, G. Alexandrowicz, D. J. Riley, G. Y. H. Lee, J. Ellis, and W. Allison. *Science*, 304:1790, 2004.
[7] J. A. Barker and D. Henderson. *Rev. Mod. Phys.*, 48:587, 1976.
[8] B. E. Clements, C. E. Campbell, P. J. Samsel, and F. J. Pinski. *Phys. Rev. A*, 44:1139, 1991.
[9] E. E. Polymeropoulos, P. Bopp, J. Brickmann, L. Jansen, and R. Block. *Phys. Rev. A*, 31:3565, 1985.
[10] J. M. Houlrik, D. P. Landau, and S. J. Knak Jensen. *Phys. Rev. E*, 50:2007, 1994.
[11] M. C. Bellissent-Funel, U. Buontempo, A. Filabozzi, C. Petrillo, and F. P. Ricci. *Phys. Rev. B*, 46:4605, 2002.
[12] K. Refson. *Moldy User's Manual.* 2003.
[13] G. Kresse and J. Hafner. *Phys. Rev. B*, 47:558, 1993.
[14] G. Kresse and J. Furthmueller. *Phys. Rev. B*, 54:11169, 1996.
[15] A. B Belonoshko, O. LeBacq, R. Ahuja, and B. Johansson. *J. Chem. Phys.*, 15:7233, 2002.
[16] P. A. Fedders. *Phys. Rev. B*, 64:165206, 2001.
[17] H. W. Sheng, W. K. Luo, F. M. Alamgir, J. M. Bai, and E. Ma. *Nature*, 439:419, 2006.
[18] B. Quentrec and C. Brot. *J. Comp. Phys.*, 13:430, 1975.
[19] A. B. Belonoshko. *Geochim. Cosmochim. Acta*, 58:4039, 1994.
[20] H. L. Tepper and W. J. Briels. *J. Cryst. Growth*, 230:270, 2001.
[21] A. B. Belonoshko, R. Ahuja, O. Eriksson, and B. Johansson. *Phys. Rev. B*, 61:3838, 2000.
[22] D. Alfè, M. J. Gillan, and G. D. Price. *J. Chem. Phys.*, 116:6170, 2002.
[23] A. Laio, S. Bernard, G. L. Chiarotti, S. Scandolo, and E. Tosatti. *Science*, 287:1027, 2000.
[24] A. B. Belonoshko, S. Davis, A. Rosengren, R. Ahuja, B. Johansson, S. I. Simak, L. Burakovsky, and D. L. Preston. *Phys. Rev. B*, 74:054114, 2006.
[25] R. Ahuja, A. B. Belonoshko, and B. Johansson. *Phys. Rev. E*, 57:1673, 1998.
[26] A. Belonoshko, N. V. Skorodumova, A. Rosengren, and B. Johansson. *Phys. Rev. B*, 73:012201, 2006.
[27] W. L. Vos, J. A. Schouten, D. A. Young, and M. Ross. *J. Chem. Phys.*, 94:3835, 1990.

Journal of Physics: Conference Series **121** (2008) 012006 doi:10.1088/1742-6596/121/1/012006

Bonding in the ε-Phase of High Pressure Oxygen

John S. Tse[1*], Yansun Yao[1], Dennis .D. Klug[2], Serge Desgreniers[3]

[1] Department of Physics and Engineering Physics, University of Saskatchewan, Saskatoon S7N 5E2, Saskatchewan, Canada

[2] Steacie Institute for Molecular Sciences, National Research Council of Canada, Ottawa K1A 0R6, Ontario, Canada

[3] Department of Physics, University of Ottawa, Ottawa K1N 6N5, Ontario, Canada

Email: John.Tse@usask.ca

Abstract. The electronic basis of the formation of $(O_2)_4$ clusters in the newly discovered structure in the ε phase of solid O_2 is elucidated through the analysis of the electron topology from first-principles calculations using the Bader atoms-in-molecules method. It is shown that the intra-cluster O_2 interaction is consistent with a donation-acceptor bond rather than the spin-paring of two open-shell molecules. Distortion of the $(O_2)_4$ cluster with increasing pressure is also investigated. It is shown that these structural changes eventually lead to a structural transition at 96 GPa, as observed experimentally.

1. Introduction

Solid oxygen has a complex phase diagram consisting of high-pressure phases with unusual physical properties [1]. When cooled at ambient pressure, three phases (α, β, and γ) of solid oxygen have been identified. Under compression at room temperature a sequence of structural transformations leads to four distinct phases, namely, β-, δ-, ε-, and ζ–O_2. At 300K, the rhombohedral β phase transforms first to δ–O_2 at 5.4 GPa [1]. The δ phase has an orthorhombic structure with a long-range anti-ferromagnetic order [2]. At 9.6 GPa, δ–O_2 transforms to the ε phase [3]. The monoclinic ε phase [4] is stable up to 96 GPa where it transforms into metallic ζ phase [5, 6]. The ζ phase shows superconductivity with a critical temperature of 0.5K at 105 GPa [7] but its structure still remains elusive [6]. Despite intense efforts over decades [1], the crystal structure of the ε phase was only recently resolved by x-ray diffraction [8, 9]. The ε phase reveals a remarkable monoclinic structure consisting of $(O_2)_4$ clusters. At 17.5 GPa the intermolecular distance within the cluster of 2.20 Å is significantly longer than the intramolecular distance of 1.20 Å which is almost identical to that of the free molecule. The very long intermolecular distance, however, raises important questions on the nature of the bonding between O_2 in the cluster [10]. A recent molecular orbital study investigated the stability of an isolated $(O_2)_4$ cluster as compared to S_8, but the nature of intermolecular and inter-cluster bonding was not discussed [11].

This paper presents results on the nature of chemical bonding in ε-oxygen through a topological analysis of charge density using the Bader atoms-in-molecules (AIM) method [12, 13]. The basic idea

© 2008 IOP Publishing Ltd

of AIM analysis is to extract bonding information from the electron density $\rho(\boldsymbol{r})$. The analysis of its electron density gradient $\nabla\rho(\boldsymbol{r})$ helps to define an atom within a molecule or solid through the "zero-flux surface" condition. The analysis of the electron density extrema, i.e., at critical points, located at $\boldsymbol{r}_{cp}$ for which $\nabla\rho(\boldsymbol{r}_{cp})=0$, allows the characterization of the nature of bonding. Once critical points are located and their curvatures are analyzed, (*i.e.*, the second derivative- Laplacian, $\nabla^2\rho(\boldsymbol{r}_{cp})$), they can be further classified according to the eigenvalues (λ_i) of the Hessian matrix. Other properties derived from the topological analysis of the charge density, such as the local energy, $H(\boldsymbol{r}_{cp})$ [14, 15], have been used to characterize the interactions. It has been shown that the charge density topology is not sensitive to small changes in the electron density [16]. AIM has been applied with great success for the characterization of a variety of chemical bonds [17]. A recent review on the application to solids has been given by Gatti [13].

2. Computational Methods

The results reported here were obtained from the pseudopotential plane wave electronic codes VASP [18]. The projected augmented wave (PAW) potential [19] of oxygen atoms from the pseudopotential library was used. The Bader atoms-in-molecules (AIM) analysis was performed with the program *InteGrity* [20]. The total charge density was calculated using the latest version of VASP code where both the atomic core and pseudo-valence charge densities were explicitly included. The convergence of the calculated results has been carefully checked against the mesh size used to generate the charge densities. A 120×120×120 mesh was found to be adequate. In addition to VASP, calculations on the isolated $(O_2)_4$ were also performed with the localized basis set code *Gaussian03* [21] employing a 6-311G(d,p) basis set for oxygen.

Even though ε-oxygen was found to be non-magnetic from neutron scattering experiments, weak itinerant magnetism cannot be ruled out [4]. To investigate this assumption spin-unrestricted calculations were performed on the known *C2/m* structure at 17 GPa [8 ,9] with an initial density matrix constructed with all spin-up *2p* electrons. The wave function and geometry were found to converge to the spin-restricted solution. Thus the spin restricted method was used in ensuing calculations. Structural parameters of the *C2/m* phase at selected pressures from 9 to 110 GPa were fully optimized. The results obtained are similar to those reported recently using the same theoretical approach [22]. The calculated equation of state over the entire pressure range apparently is in good agreement with experiment [9]. However, at pressures over 60 GPa, the calculated unit cell parameters, although still reproduce the experimental trend, a sudden reduction in the monoclinic angle (β) than observed (80 GPa) was found [22]. The discrepancy can be traced to a deficiency of the GGA approach which underestimates the band gap energy and predicted, prematurely, metallization in the *C2/m* structure at 50 GPa. As reported elsewhere [23] a quasiparticle theory is needed to provide a correct description of the electronic structure. In this study, discussions on the chemical bonding in ε-oxygen will focus on the low pressure regime.

3. Results and Discussions

A comparison of the structure of the $(O_2)_4$ cluster in ε-oxygen with an *isolated* $(O_2)_4$ reveals some unique features of the bonding in the solid state. Geometry optimization of an isolated cluster led to a square prism structure (D_{4h} point group symmetry) with calculated intramolecular O-O bond lengths of 1.201 Å from *Gaussian03* (1.216 Å, from VASP) and intermolecular $O_2 \ldots O_2$ distances of 2.120 (2.141) Å. Although, the intramolecular O_2 bond length is apparently in good agreement with experiment, the intermolecular $O_2 \ldots O_2$ distance of the isolated cluster, however, is significantly *shorter* than the observed separation of 2.34 Å at 10 GPa in the solid state. A much shorter $O_2 \ldots O_2$ distance predicted for the isolated cluster in the absence of external compression is surprising. Furthermore, at 10 GPa, the minimum contact distance between $(O_2)_4$ clusters in the solid state structure is 2.71 Å. Therefore, any significant inter-cluster interactions can be ruled out. In the free cluster, as has been characterized previously [1], the $O_2 \ldots O_2$ bond is derived from the paring of two electrons each from the π^* orbital of the participating O_2 molecule [11]. Thus, this can be described as a true spin-coupled two-electron covalent bond. The fact that a much longer intermolecular distance was observed in ε-oxygen, leads to the inevitable conclusion that the nature of chemical bonding must be difference from the isolated cluster.

We have attempted to analyze the charge density differences and electron localization function of the $(O_2)_4$ clusters in the *C2/m* structure but failed to uncover obvious hints of covalent interactions between the O_2 forming the cluster. A detailed analysis of the charge density topology is necessary to characterize the intermolecular and intra-cluster interactions. For this purpose, accurate 3-D electron densities for ε-oxygen were generated from first principles calculations. The (3,-1) bond critical points (BCP) derived from AIM analysis at 24 GPa are summarized in Table 1. As expected, the BCP's of the O_2 molecules have the largest value of local density $\rho(\mathbf{r}_{cp})$ (~0.55) and local energy $H(\mathbf{r}_{cp})$ [12, 14, 15] (-1.10 a.u.) and are indicative of genuine O-O covalent bonds. Significantly, intra-cluster BCP's are found between O_2 molecules forming the $(O_2)_4$ cluster. The charge density at the critical point $\rho(r_{cp})$ ~0.06 with a *positive* Laplacian and a *negative* $H(r_{cp})$ (-0.009 a.u.) are *unambiguous signs of covalent interactions between O_2 molecules,* being however much weaker than that expected for a normal covalent bond [17].

So what is the nature of the chemical bond in ε-oxygen? A positive Laplacian is perhaps surprising as it indicates "closed shell" interactions [12, 24] between the two O atoms of the adjacent O_2 molecules in the $(O_2)_4$ unit instead of the pairing of two "open-shell" spin-uncoupled orbitals forming a pair of two electron bonds. If the latter description were correct, the intermolecular distance would be close to that of a single σ-bond as predicted for the isolated cluster. The analysis, however, is consistent with the observed very long $O_2 \cdots O_2$ separation of about 2.34 Å at 10 GPa and 2.17 Å at 17 GPa. Results indicate that the O_2 molecules forming the rhomb-shaped clusters in the ε phase are stabilized *via* donor-acceptor resonance structures, *i.e.* the donation of a *paired* π^* electron from one O_2 into the empty π^* orbital of the other. In this interpretation, the O_2 converted from a spin-uncoupled state of a free molecule into a spin-paired state (two electrons occupying one of the π^* orbital) to facilitate interactions with the neighbouring O_2 forming the cluster. This description is consistent with the fact that the magnetic moment of solid O_2 has already decreased from 2 μ_B at the zero pressure structure to less than 0.5 μ_B to the δ phase *preceding* the δ→ε transition [4]. Therefore, interactions between O_2 in the cluster cannot be due to simple spin pairing of electrons from the isolated molecules [4]. A crystallographic evidence supports this interpretation: the O_2 bond distance of 1.20 Å determined at 17.6 GPa [8] is *longer* than 1.180 Å found in β O_2 at 5.5 GPa [25] and 1.175 Å in δ O_2 at 9.6 GPa and 297 K [1]. A longer O-O bond length observed at higher pressure suggests that two valence electrons occupy a single π^* orbital thus reducing the bond order and increasing the length of the molecular bond. BCP's are also identified between oxygen atoms of different clusters. As shown in Table 1, the electron density at these inter-cluster BCPs (0.028) are much smaller. Moreover, both the Laplacian and local energy density $H(r_{cp})$ are positive indicating closed shell van der Waals interaction between the clusters [15, 24]. Ring and cage CP's were also identified but they are of minor significance and will not be discussed here.

When ε-oxygen is compressed, the intra-cluster O_2 distance (d_1) and inter-cluster distance (d_3) decrease rapidly [8, 9]. Although the theoretical calculations reproduce this trend, as commented above (*vide supra*), the agreement between theoretically predicted and experimentally observed lattice parameters is not as good at pressure higher than 60 GPa [22]. Nevertheless, the shape of the $(O_2)_4$ cluster is found to distort from a square prism to a diamond shape prism with increasing pressure. The O···O···O bond angle decreases gradually from 88° at 20 GPa to 80° near the $\varepsilon \rightarrow \zeta$ transition at 97 GPa [8, 9]. The distortion leads to changes in the structure and the electron density topology. Table 2 gives results of the AIM analysis on an over-pressurized *C2/m* structure at 105 GPa. As indicated by $\rho(r_{cp})$ and $\nabla^2\rho(r_{cp})$ the intra-cluster and inter-cluster interactions of 0.065 and 0.062, respectively, are now almost equivalent. The $(O_2)_4$ clusters in the ***ab*** plane start to connect to each other. At this pressure, the calculated electronic band structure shows that the *C2/m* is metallic. The linking of $(O_2)_4$ clusters may be the precursor for the $\varepsilon \rightarrow \zeta$ phase transition.

4. Conclusion

The nature of the interactions between O_2 molecules and the preference of the formation of $(O_2)_4$ clusters in the *C2/m* structure of the ε phase have been elucidated from the analysis of the electron charge density using Bader atoms-in-molecules theory. It has been shown that overall interactions, particularly, weak non-bonding interactions between $(O_2)_4$ clusters help to stabilize the structure. Changes in the geometry of the $(O_2)_4$ cluster with pressure have been examined. It is shown that the lower pressure square prism cluster of ε-oxygen deforms into a diamond-shape prism at pressure close to the observed insulator-metal transition. It is shown that the enhanced interactions between the clusters lead to a calculated structural phase transition at 96 GPa.

Acknowledgements

One of the authors (JST) wishes to thank Dr. I. Goncharenko for a discussion on the collapse of magnetic moment in high-pressure solid oxygen.

References

[1] Freiman Y A and Jodl H J 2004 *Phys. Rep.* **401** 1
[2] Schiferl D et a.l 1983 *Acta Cryst.B* **39** 153
[3] Nicol M, Hirsch K R and Holzapfel H B 1979 *Chem. Phys. Lett.* **68** 49
[4] Goncharenko I N 2005 *Phys. Rev. Lett.* **94** 205701
[5] Desgreniers S, Vohra Y K and Ruoff A L 1990 *J. Phys. Chem.* **94** 1118
[6] Akahama Y, Kawamura H, Häusermann D and Hanfland M 1995 *Phys. Rev. Lett.* **74** 4690
[7] Shimizu K, Suhara K, Ikumo M, Eremets M I and Amaya K 1998 *Nature* **393** 767
[8] Lundegaard L F, Weck G, McMahon M I, Desgreniers S and Loubeyre P 2006 *Nature* **443** 201
[9] Fujihisa H et al 2006 *Phys. Rev. Lett.* **97** 085503
[10] Militzer B and Hemley R J 2006 *Nature* **443** 150
[11] Steudel R and Wong M W 2007 *Angew. Chem. Int. Ed.* **46** 1768
[12] Bader R F W 1990 *Atoms in Molecules – A Quantum Theory* (Oxford : Oxford University Press)
[13] Gatti C 2005 *Z. Kristallogr.* **220** 399
[14] Cremer D and Kraka E 1984 *Croat. Chem. Acta.* **57** 1259
[15] Cremer D and Kraka E 1984 *Agnew. Chem.* **23** 627
[16] Brown I D 2002 *Struct. Chem.* **13** 339
[17] Bader R F W 1985 *Acc. Chem. Res.* **18** 9
[18] Kresse G and Furthmüller J 1996 *Comp. Mater, Sci.* **6** 15
1996 *Phys. Rev. B* **54** 11169
Version 4.6.31 released on 08 Feb 07 was used in the charge density calculations.
[19] Kresse G and Joubert D 1999 *Phys. Rev. B* **59** 1758
[20] Katan C, Rabiller P, Lecomte C, Guezo M, Oison V and Souhassou M 2003 *J. Appl. Cryst.* **36** 65
[21] Frisch M et al 2003 *Gaussian 03* (Pittsburgh, PA: Gaussian, Inc.)
[22] Ma Y, Oganov A R and Glass C W 2007 *Phys. Rev. B* **76** 64101
[23] Tse J S, Klug D D, Yao Y and Desgreniers S 2007 *submitted for publication*
[24] Bone R G A and Bader R F W 1996 *J. Phys. Chem.* **100** 10892
[25] Schiferl D, Cromer D T and Mills R L 1981 *Acta. Cryst. B* **37** 1329

Table Captions

1. Bond critical point data (a.u.) located in the *C*/*2m* structure of $(O_2)_4$ at 25 GPa.
2. Bond critical point data (a.u.) located in the *C*/*2m* structure of $(O_2)_4$ at 105 GPa.

Table 1. Bond critical points for *C*2/*m* ε-oxygen at 24 GPa

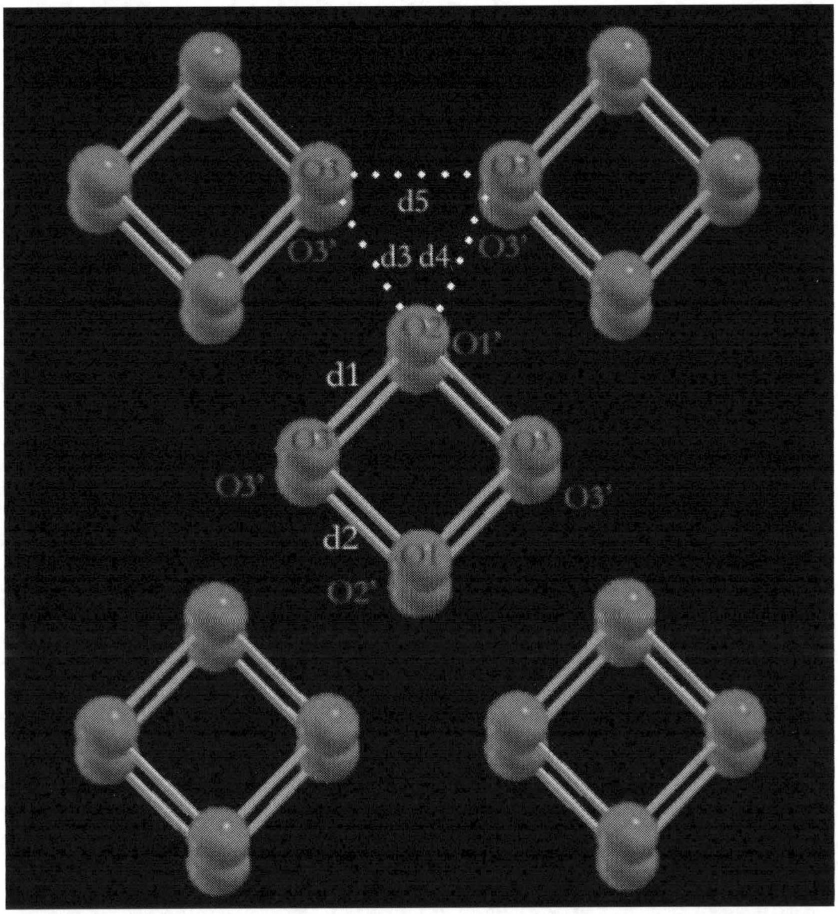

Bond	Type	$\rho(r_{CP})$	$\nabla^2\rho(r_{CP})$	λ_1	λ_2	λ_3	$H(r_{CP})$
$O_3 - O_3$	Molecular	0.5501	-0.536	-1.555	-1.527	2.546	-1.109
$O_1 - O_2$	Molecular	0.5468	-0.459	-1.518	-1.510	2.569	-1.087
$O_3 \ldots O_2$ (d1)	Intra-cluster	0.0622	0.231	-0.098	-0.091	0.421	-0.009
$O_3 \ldots O_1$ (d2)	Intra-cluster	0.0619	0.232	-0.098	-0.091	0.420	-0.009
$O_3 \ldots O_2$ (d3)	Inter-cluster	0.0276	0.099	-0.025	-0.024	0.148	0.003
$O_3 \ldots O_2$ (d4)	Inter-cluster	0.0194	0.089	-0.020	-0.019	0.128	0.003
$O_3 \ldots O_2$ (d5)	Inter-cluster	0.0163	0.080	-0.017	-0.017	0.122	0.004

Table 2. Bond critical points for an over-pressurized *C2/m* ε- oxygen at 105 GPa

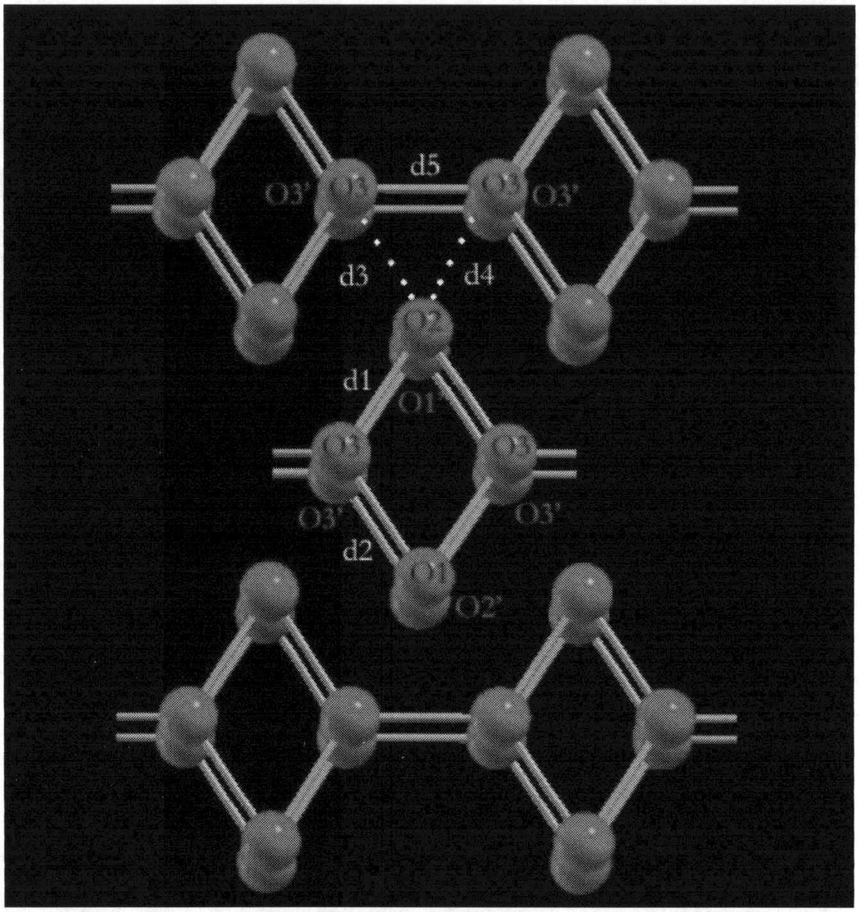

Bond	Type	$\rho(r_{CP})$	$\nabla^2\rho(r_{CP})$	λ_1	λ_2	λ_3	$H(r_{CP})$
$O_3 - O_3$	Molecular	0.5882	-0.737	-1.702	-1.657	2.622	-1.253
$O_1 - O_2$	Molecular	0.5866	-0.663	-1.659	-1.642	2.638	-1.238
$O_3 \ldots O_3$ (d5)	Inter-cluster	0.0652	0.291	-0.100	-0.093	0.484	-0.006
$O_3 \ldots O_2$ (d1)	Intra-cluster	0.0619	0.280	-0.094	-0.086	0.456	-0.005
$O_3 \ldots O_1$ (d2)	Intra-cluster	0.0619	0.282	-0.093	-0.084	0.459	-0.004
$O_3 \ldots O_2$ (d3)	Inter-cluster	0.0484	0.219	-0.066	-0.057	0.341	-0.001
$O_3 \ldots O_2$ (d4)	Inter-cluster	0.0449	0.204	-0.058	-0.051	0.312	0.001

Anomalous Static Electronic Screening in Compressed Lithium

A. Rodriguez-Prieto[1,2], **M. Martinez-Canales**[1,2], **V. M. Silkin**[2], **and A. Bergara**[1,2,3]

[1]Materia Kondentsatuaren Fisika Saila, Zientzi Fakultatea, Euskal Herriko Unibertsitatea, 644 Posta kutxatila, E-48080 Bilbo, Basque Country, Spain
[2]Donostia International Physics Center (DIPC), Manuel de Lardizabal Pasealekua, E-20018 Donostia, Basque Country, Spain
[3]Centro Mixto CSIC-UPV/EHU, 1072 Posta kutxatila, E-20080 Donostia, Basque Country, Spain

E-mail: a.bergara@ehu.es

Abstract. Lithium has long been considered a simple metal. However, under pressure lithium departs radically from this simple behavior, as it strongly enhances its superconducting transition temperature before phase transitions to complex structures emerge. Interestingly, these features are not unique of lithium, as have been also observed in other simple elements, but lithium becomes a good system to characterize their physical origin. In this article we will analyze anomalies arising in the static electronic susceptibility of lithium under pressure, where clear connections between anisotropies in the Fermi surface, phonon instabilities, and enhanced electron-phonon coupling will be shown.

1. Introduction

At normal conditions of pressure and temperature lithium adopts a compact and highly symmetric *bcc* structure, and main physical properties can be characterized within the nearly free electron-like model. However, under pressure the contribution associated to the non-local character of the pseudopotential rises[1, 2], the nearly free electron-like approximation breaks and lithium undergoes several phase transitions to more complex structures[3, 4]: *bcc* to fcc at 8 GPa, fcc to $hR1$ at 39 and $hR1$ to $cI16$ (with 16 atoms per unit cell) at 41 GPa.

In spite of lithium at normal conditions superconducts at the very low $T_c = 0.4$ mK [5, 6], the transition temperature to the superconducting state increases up to around 17 K when the applied pressure rises to 30 GPa [7, 8, 9], becoming the element with one of the highest transition temperature[10], which even rises the interest in characterizing physical properties of compressed lithium. Interestingly, this complex behavior is not only characteristic of lithium, but has been also observed in other simple elements[11]. However, considering the more simple electronic configuration of lithium, it becomes a good system to understand in detail the physical origin of these quite general features arising under pressure.

In this article we will analyze anomalies arising in the static electronic susceptibility of lithium under pressure, where clear connections between anisotropies in the Fermi surface, phonon instabilities, and enhanced electron-phonon coupling will be shown. Section 2 describes the

© 2008 IOP Publishing Ltd

computational methods that we have used in this study. In Section 3 we present our results and, finally, conclusions are presented in Section 4.

2. Computational Methods

We have used a plane-wave implementation of Density Functional Theory (DFT) within the Local Density Approximation (LDA), as implemented in the Vienna Ab Initio Simulation Package (VASP) [12, 13]. Although at high pressure an important core overlap is observed, we have checked that the Projector Augmented Wave (PAW) approximation[14], which fully treats all the electrons, and ultrasoft pseudopotentials [15] give similar results. Figures of three dimensional Fermi surfaces have been generated by XCrySDen [16] and phonon frequencies have been computed with Density Functional Perturbation Theory (DFPT), as implemented in the plane-wave self-consistent field (PWscf) code [17]. Generally, a sample of the Brillouin zone (BZ) with a Monkhorst-Pack mesh of $20 \times 20 \times 20$ has been demonstrated to be enough to reach electronic convergence. However, a $60 \times 60 \times 60$ mesh has been required to calculate electron-phonon linewidths and a much denser grid ($96 \times 96 \times 96$) is considered for the non interacting electronic susceptibility. At the same time, energetic convergence has been achieved by implementation of a 430 eV energy cutoff.

3. Results

An essential feature determining electronic properties of any metal is the topology of the Fermi surface (FS). Although at ambient pressure lithium is well described by a nearly free electron model, which is reflected by its almost perfect spherical FS, as a result of the increasing non-local character of the pseudopotential, it deforms significantly under compression[18, 19, 20]. Fig. 1 shows the Fermi surface of *fcc* lithium at 30 GPa and the analysis of its cross sections along several planes. Although the cross section of the FS along the ΓKX plane shows an almost perfect circle, it becomes highly distorted in the ΓKL and ΓWX planes, with cooper-like necks along ΓL and well defined nestings in the ΓX, ΓK and ΓW directions. Hence, a high number of valence electrons will be coupled with a momentum transfer of $2k_\mathrm{F}$ along these directions and Peierls distortions might be favored.

The interaction between the BZ boundary and the FS develops extended necks along the *fcc* ΓL direction where a pseudogap opens and, therefore, reduces the overall electronic energy. In fact, the electronic band structure strongly deviates from the nearly free electron-like behavior, and lithium shows a remarkable covalent character associated to the increasing electronic localization at atomic interstitials (Fig. 2). For example, the occupied bandwidth along ΓX is almost constant under pressure (3.93 eV at 8 GPa and 4.1 eV at 30 GPa) and even decreases along ΓL (3.67 eV at 8 GPa and 3.5 eV at 30 GPa), whereas the bandgap at L considerably increases (4.8 eV at 8 GPa and 7 eV at 30 GPa).

In order to deeper analyze the influence of the Fermi surface deformation on the physical properties of compressed lithium, we have calculated diagonal contributions to the *ab initio* static electronic susceptibility matrix of the non interacting electron system:

$$\begin{aligned}\chi^0_{\mathbf{G},\mathbf{G}'}(\mathbf{q}) &= \frac{1}{\Omega}\sum_{\mathbf{k}}^{BZ}\sum_{n,n'}\frac{f_{\mathbf{k}n}-f_{\mathbf{k}n'+\mathbf{q}}}{\varepsilon_{\mathbf{k},n}-\varepsilon_{\mathbf{k+q},n'}} \\ &\times\langle\psi_{\mathbf{k},n}|e^{-\mathrm{i}(\mathbf{q}+\mathbf{G})\cdot\mathbf{r}}|\psi_{n',\mathbf{k+q}}\rangle \\ &\times\langle\psi_{\mathbf{k+q},n'}|e^{\mathrm{i}(\mathbf{q}+\mathbf{G}')\cdot\mathbf{r}}|\psi_{\mathbf{k},n}\rangle. \qquad (1)\end{aligned}$$

Here $\mathbf{k}$ is a wave vector belonging to the first BZ, $f_{n\mathbf{k}}$ represents the Fermi-Dirac distribution function, Ω is the normalization volume, $\mathbf{G}$ and $\mathbf{G}'$ are reciprocal-lattice vectors, and if $\mathbf{k}+\mathbf{q}$ lies outside the first BZ, it gets folded back into the first BZ by addition of the appropriate reciprocal-lattice vector. The first summation in Eq. (1) was performed with the use of $96\times96\times96$ **k**-mesh, which correspond to 442368 points in the BZ. The second sum runs over the occupied and all unoccupied valence bands up to an energy of 50 eV above Fermi level.

Fig. 3 displays the static electronic susceptibility along ΓX, ΓK and ΓW directions in comparison with the electron gas approximation (Lindhard function). Screening corresponding to the *ab initio* calculation becomes less effective than in the electron gas approximation. The main reason of this difference is associated to the lowering under pressure of the occupied bandwidth [2, 3] and the opening of a pseudogap at N (necks in the FS). Additionally, although we have not seen any special feature on the static susceptibility along the ΓK direction, interesting kinks emerge at $\mathbf{q}_{\rm nest}^{\Gamma W} = 2\pi/a(4/3, 2/3, 0)$ and $\mathbf{q}_{\rm nest}^{\Gamma X} = 2\pi/a(19/14, 0, 0)$ corresponding to the above mentioned nestings along ΓW and ΓX, respectively.

In order to analyze the effect of these kinks in the phonon spectra, Fig. 4 shows the phonon frequencies of *fcc* lithium at 30 GPa along high symmetric directions. While longitudinal phonon frequencies increase with pressure, a remarkable softening is observed in the lowest transverse mode (T_1), specially along ΓK, where it becomes unstable at $\mathbf{q}_{\rm anom}^{\Gamma K} = 2\pi/a(-2/3, 2/3, 0)$. Interestingly, this instability is connected to the kink in the electronic susceptibility at $\mathbf{q}_{\rm nest}^{\Gamma W}$. In order to compare both momenta we have to take into account that $\mathbf{q}_{\rm nest}^{\Gamma W}$ extends out from the first BZ, which after folding it back with the reciprocal lattice vector $\mathbf{G} = 2\pi/a(-2, 0, 0)$ results to be $\mathbf{q}_{\rm anom}^{\Gamma K}$: $\mathbf{q}_{\rm anom}^{\Gamma K} = \mathbf{q}_{\rm nest}^{\Gamma W} + \mathbf{G}$. On the other hand, it is also interesting to analyze the remarkable Kohn anomaly in the longitudinal mode along the ΓX direction at $\mathbf{q}_{\rm anom}^{\Gamma X} = 2\pi/a(-9/14, 0, 0)$, which can be also associated to the kink in the electronic susceptibility at $\mathbf{q}_{\rm nest}^{\Gamma X} = 2\pi/a(19/14, 0, 0)$. As $\mathbf{q}_{\rm nest}^{\Gamma X}$ also extends out from the first BZ, in order to compare both momenta we need to fold it back by adding an adequate reciprocal lattice vector. Considering $\mathbf{G} = 2\pi/a(-2, 0, 0)$, $\mathbf{q}_{\rm nest}^{\Gamma X} + \mathbf{G} = \mathbf{q}_{\rm anom}^{\Gamma X}$.

As shown in Fig. 5, these two anomalies in the phonon spectra are also manifested in the electron-phonon linewidth, where the observed maxima in each direction correspond to these anomalous features. The maximum of the electron-phonon linewidth corresponding to the transverse mode at $\mathbf{q}_{\rm anom}^{\Gamma K}$ already has been well characterized and consider to be the main responsible of the observed enhanced superconducting transition in lithium under pressure[21, 22, 23, 24, 25]. However, the observed maximum along ΓX associated to the longitudinal mode, exactly located at $\mathbf{q}_{\rm anom}^{\Gamma X}$, has not been considered before, because, among other things, in this case a very dense mesh ($60 \times 60 \times 60$) has been required to reach the convergence and properly include nesting features in the Fermi surface.

On the other hand, although the kink at $\mathbf{q}_{\rm nest}^{\Gamma X}$ in the electronic susceptibility is enough to explain the strong Kohn anomaly in the longitudinal mode at $\mathbf{q}_{\rm anom}^{\Gamma X}$, besides the correlation between the kink at $\mathbf{q}_{\rm nest}^{\Gamma W}$ and the observed phonon instability in the transverse branch at $\mathbf{q}_{\rm anom}^{\Gamma K}$, this phonon instability cannot be completely explained by this kink, but without considering crystal local-field effects a divergent behavior of the electronic susceptibility would be required. Phonon frequencies can be related to the dielectric matrix, $\epsilon_{\mathbf{G},\mathbf{G}'}(\mathbf{q})$, via the Pick-Cohen-Martin formula[26]:

$$\begin{aligned} D_{\alpha\beta}(\mathbf{q}) &= \tilde{D}_{\alpha\beta}(\mathbf{q}) - \tilde{D}_{\alpha\beta}(0), \\ \tilde{D}_{\alpha\beta}(\mathbf{q}) &= \sum_{\mathbf{G},\mathbf{G}'} |\mathbf{q}+\mathbf{G}|^2\, V_c(\mathbf{q}+\mathbf{G})\, (\mathbf{q}+\mathbf{G})_\alpha \\ &\quad \times \epsilon^{-1}_{\mathbf{G},\mathbf{G}'}(\mathbf{q})\, V_c(\mathbf{q}+\mathbf{G}')\, (\mathbf{q}+\mathbf{G}')_\beta, \end{aligned} \tag{2}$$

$V_c(\mathbf{q}) = 4\pi/|\mathbf{q}|^2$ being the Coulomb potential, $D_{\alpha,\beta}(\mathbf{q})$ is the dynamical matrix and α, β are

Cartesian indices. The dielectric matrix in Eq. (2) includes umklapp processes via the so-called crystal local-field effects. In terms of the full electronic susceptibility, $\chi_{\mathbf{G},\mathbf{G}'}(\mathbf{q})$, the dielectric matrix is given by the following expression:

$$\epsilon^{-1}_{\mathbf{G},\mathbf{G}'}(\mathbf{q}) = \delta_{\mathbf{G},\mathbf{G}'} + V_c(\mathbf{q}+\mathbf{G})\chi_{\mathbf{G},\mathbf{G}'}(\mathbf{q}), \tag{3}$$

and within the random phase approximation (RPA), where exchange and correlation effects between the electrons are neglected, χ can be obtained from the non-interacting electronic susceptibility, χ^0:

$$\chi_{\mathbf{G},\mathbf{G}'}(\mathbf{q}) = \chi^0_{\mathbf{G},\mathbf{G}'}(\mathbf{q}) + \sum_{\mathbf{G}''} \chi^0_{\mathbf{G},\mathbf{G}''}(\mathbf{q}) V_c(\mathbf{q}+\mathbf{G}'')\chi_{\mathbf{G}'',\mathbf{G}'}(\mathbf{q}). \tag{4}$$

It is clear that an eigenvalue of the dynamical matrix in Eq. (2) can soften, and eventually go to zero, even if the macroscopic dielectric function,

$$\epsilon(\mathbf{q}) = \frac{1}{\epsilon^{-1}_{\mathbf{0},\mathbf{0}}(\mathbf{q})}, \tag{5}$$

or the diagonal contribution to the non-interacting electronic susceptibility, does not diverge[27]. Although a detail understanding of the contribution associated to crystal local-field effects will require a deeper analysis, Fig. 6 shows the effect of including umklapp processes in the diagonal contribution of the full electronic susceptibility matrix of *fcc* lithium at 30 GPa along ΓW. Interestingly, even though these diagonal elements just include crystal local-field effects through the inverse of the determinant associated to $\chi^0_{\mathbf{G},\mathbf{G}'}(\mathbf{q})$ required to obtain the full χ matrix in Eq. (4), their contribution becomes specially relevant at $\mathbf{q}^{\Gamma W}_{\rm nest}$. Actually, this important contribution of crystal local-field efects associated to the non diagonal elements in the electronic susceptibility can be also explained in terms of the increasing covalent character of lithium under pressure.

Another important difference between both phonon anomalies is their polarization. While at $\mathbf{q}^{\Gamma X}_{\rm anom}$ just the longitudinal mode is involved, the instability at $\mathbf{q}^{\Gamma K}_{\rm anom}$ corresponds to one of the transverse modes. This difference can be associated to the required reciprocal lattice vector, $\mathbf{G}$, to fold back $\mathbf{q}^{\Gamma X}_{\rm nest}$ and $\mathbf{q}^{\Gamma W}_{\rm nest}$ into the first BZ. While $\mathbf{q}^{\Gamma X}_{\rm nest}$ is paralell to $\mathbf{G}$ this is not the case of $\mathbf{q}^{\Gamma W}_{\rm nest}$, which also contributes to increase the influence associated to non-diagonal elements in Eq. (2), mainly characterizing features associated to the transverse modes.

4. Conclusions

Fermi surface of lithium becomes highly distorted under pressure, which is characterized by extended nestings. These nestings originate peaks in the static electronic susceptibility. As a result of the enhancement of crystal local-field effects associated to the increasing electronic inhomogeneity under pressure, phonon spectra of compressed *fcc* lithium is characterized by a well defined transverse phonon softening along ΓK correlated to the nesting associated kink in the electronic susceptibility at $\mathbf{q}^{\Gamma W}_{\rm nest}$. This softening, besides preluding the observed transition to complex structures, increases the electron-phonon coupling under pressure. Interestingly, a similar transverse soft-mode has been also observed in other simple elements correlated to their superconducting transition under pressure (e.g., P, Ca, Ga, S, and Y). On the other hand, the observed nesting along ΓX induces a well defined Kohn anomaly in the longitudinal mode at $\mathbf{q}^{\Gamma X}_{\rm anom}$, which also enhances the electron-phonon coupling. Additionally, band structure deformation of lithium under pressure not only modifies the static but also the dynamic response function is expected to be strongly pressure dependent (for example, a new low energy long-lived interband collective mode has been predicted to arise[28]).

We acknowledge partial support by the University of the Basque Country, the Basque Hezkuntza, Unibersitate eta Ikerketa saila and Spanish MCyT (Grant No.FIS 2004-06490-C03-01), and the European Community 6th Network of Excellence NANOQUANTA (NMP4-CT-2004-500198). Computational resources for this work were provided by the SGI/IZO-SGIker at the UPV/EHU (supported by the Spanish Ministry of Education and Science and the European Social Fund).

[1] Boettger J C and Trickey S B 1985 *Phys. Rev.* B **32** 3391
[2] Bergara A, Neaton J B, and Ashcroft N W 2000 *Phys. Rev.* B **62** 8494
[3] Neaton J B and Ashcroft N W 1999 *Nature* **400** 141
[4] Hanfland M, Syassen K, Christensen N E, and Novikov D L 2000 *Nature* **408** 174
[5] Juntunen K I and Tuoriniemi J T 2004 *Phys. Rev. Lett.* **93** 157201
[6] Tuoriniemi J T, Juntunen K I, Uusvuori J, Pentti E, Salmela A and Sebedash A 2007 *Nature* **447** 187
[7] Shimizu K, Ishikawa H, Takao D, Yagi T, and Amaya K 2002 *Nature* **419** 597
[8] Struzhkin V, Eremets M I, Gan W, Mao H K, and Hemley R J 2002 *Science* **298** 1213
[9] Deemyad S and Schilling J S 2003 *Phys. Rev. Lett.* **91** 167001.
[10] Ashcroft N W 2002 *Nature* **419** 569
[11] Schilling J S 2006 *Treatise on High Temperature Superconductivity* (Hamburg:Springer Verlag)
[12] Kresse G and Hafner J 1993 *Phys. Rev.* B **47** 558
[13] Kresse G and Furthmüller 1996 *Phys. Rev.* B **54** 11169
[14] Blöchl P E 1994 *Phys. Rev.* B **50** 17953
[15] Vanderbilt D 1990 *Phys. Rev.* B **41** 7892
[16] Kokalj A 1999 *J. Mol. Graphics* **17** 176
[17] Baroni S, Corso A del, Gironcoli S de, and Gianozzi P *http://www.pwscf. org*
[18] Rodriguez-Prieto A and Bergara A 2005 *Phys. Rev.* B **72** 125406
[19] Rodriguez-Prieto A, Bergara A, Silkin V M, and Echenique P M 2006 *Phys. Rev.* B **74** 172104
[20] Rodriguez-Prieto A and Bergara A 2007 *High Press. Res.* **26** 461
[21] Tse J S, Ma Y, and Tutuncu H M 2005 *J. Phys.: Condens. Matter* **17** S911
[22] Maheswari S U, Nagara H, Kusakabe K, and Suzuki N 2005 *J. Phys. Soc. Jpn.* **74** 3227
[23] Kasinathan D, Kunes J, Lazicki A, Rosner H, Yoo C S, Scalettar R T, and Pickett W E 2006 *Phys. Rev. Lett.* **96** 047004
[24] Profeta G, Franchini C, Lathiotakis N N, Floris A, Sanna A, Marques M A L, Luders M, Massidda S, Gross E K U, and Continenza A 2006 *Phys. Rev. Lett.* **96** 47007
[25] Shi L and Papaconstantopoulos D A 2006 *Phys. Rev.* B **73** 184516
[26] Pick R M, Cohen M H, and Martin R M 1970 *Phys. Rev.* B **1** 910
[27] Johannes M D, Mazin I I, and Howells C A 2006 *Phys. Rev.* B **73** 205102
[28] Silkin V M, Rodriguez-Prieto A, Bergara A, Chulkov A V, and Echenique P M 2007 *Phys. Rev.* B **75** 172102

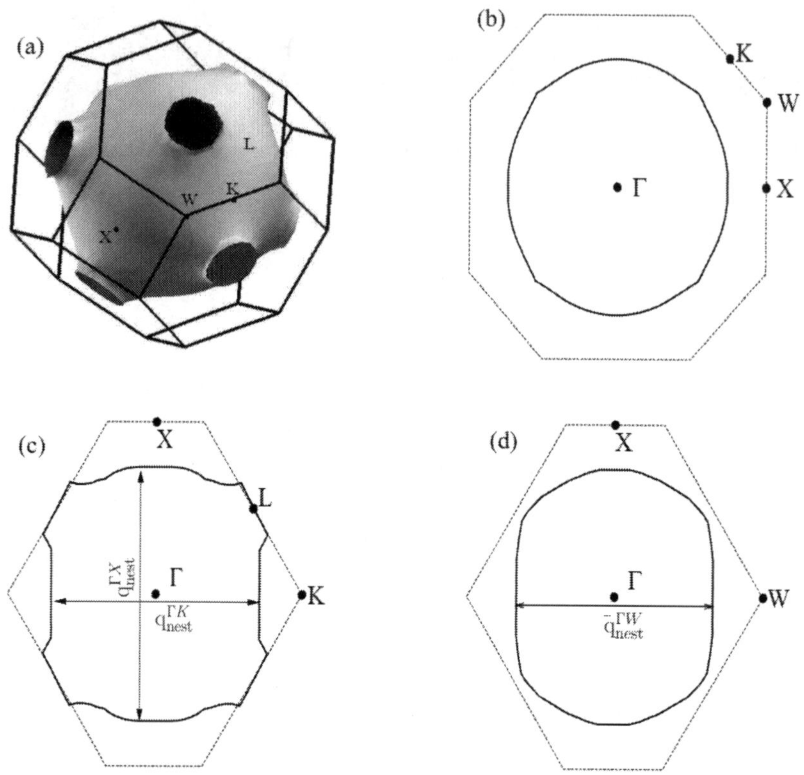

Figure 1. (a) Fermi surface of *fcc* lithium at 30 GPa and its cross sections (solid lines) along (b) ΓKX (100), (c) ΓKL (110) and (d) ΓWX (210) planes. Brillouin zone boundary is represented by dotted lines. The cross section along the ΓKX plane shows an almost perfect circle, whereas both in the ΓKL and ΓWX planes present clear nestings in the ΓX, ΓK and ΓW directions.

Figure 2. Electronic charge density of *fcc* lithium at P=30 GPa. Yellow areas represent minimum electronic charge density around the atoms, whereas the maximum electronic charge density (purple) locate at the interstitials, as a consequence of the increasing covalent character of lithium under pressure.

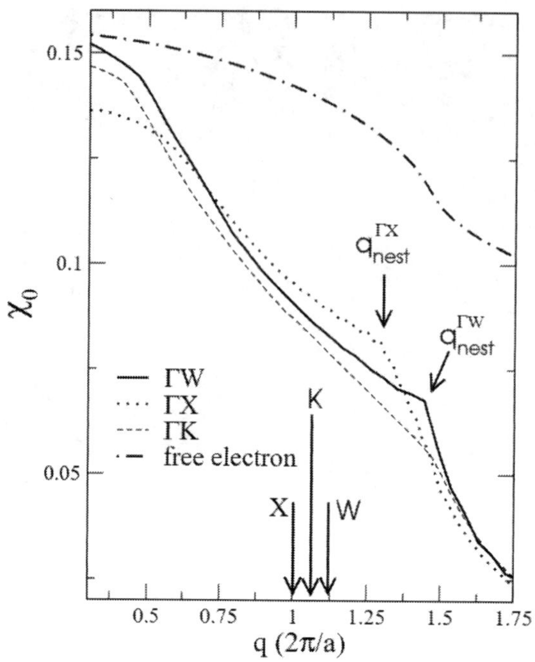

Figure 3. Diagonal contribution of the static non-interacting susceptibility, $\chi^0_{\mathbf{0},\mathbf{0}}(\mathbf{q})$, along ΓX (dotted line), ΓK (dashed line) and ΓW (solid line) directions at P=30 GPa, and (dashed-dotted line) the free electron gas approximation (Lindhard function) calculated with the *ab initio* effective mass, $m^* = 2m_e$, so that both different approximations present the same value at $\mathbf{q} = 0$. For clarity, the location of X, W and K points in the BZ are shown in the figure and, although no special feature is seen along ΓK, both kinks emerge along ΓX and ΓW with the nesting momenta, $\mathbf{q}^{\Gamma X}_{\mathrm{nest}}$ and $\mathbf{q}^{\Gamma W}_{\mathrm{nest}}$, extending out from the first BZ. On the other hand, band effects are shown to decrease the electronic screening compared to the electron gas approximation.

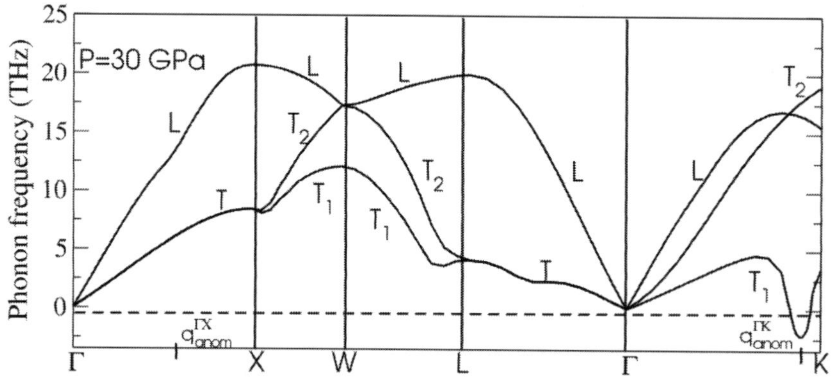

Figure 4. Phonon frequencies of *fcc* lithium at P=30 GPa. Besides the general softening of the transverse T_1 branch, it is remarkable the well defined instability of the T_1 branch along ΓK $\mathbf{q}^{\Gamma K}_{\mathrm{anom}}$. The longitudinal branch along ΓX also shows a well defined Kohn anomaly at $\mathbf{q}^{\Gamma X}_{\mathrm{anom}}$. Both anomalous features are connected to the above mentioned nesting induced kinks in the electronic susceptibility.

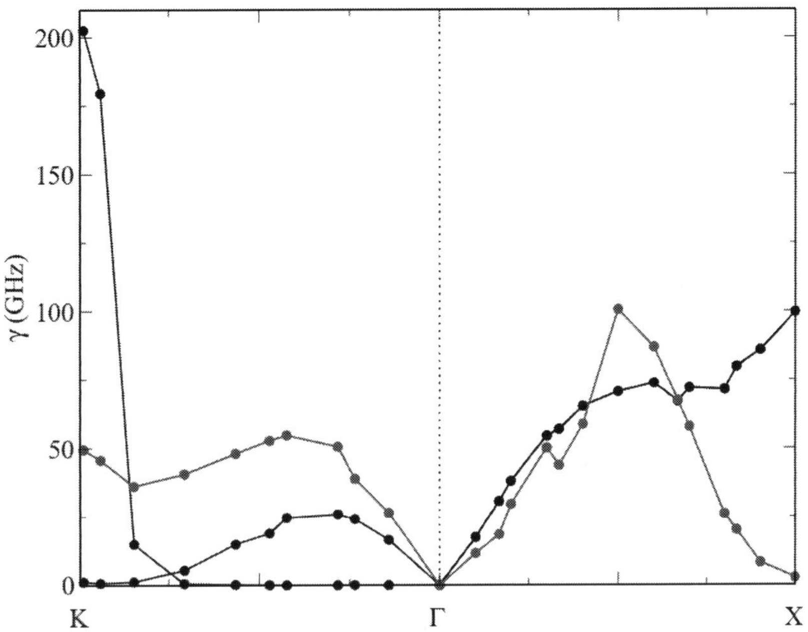

Figure 5. Electron-phonon linewidth of *fcc* lithium at 30 GPa along ΓK and ΓX directions. Blue (black) color corresponds to the longitudinal (transverse) mode. In order to correctly capture features in the Fermi surface and get the results converged, a $60 \times 60 \times 60$ Γ-centered grid has been required.

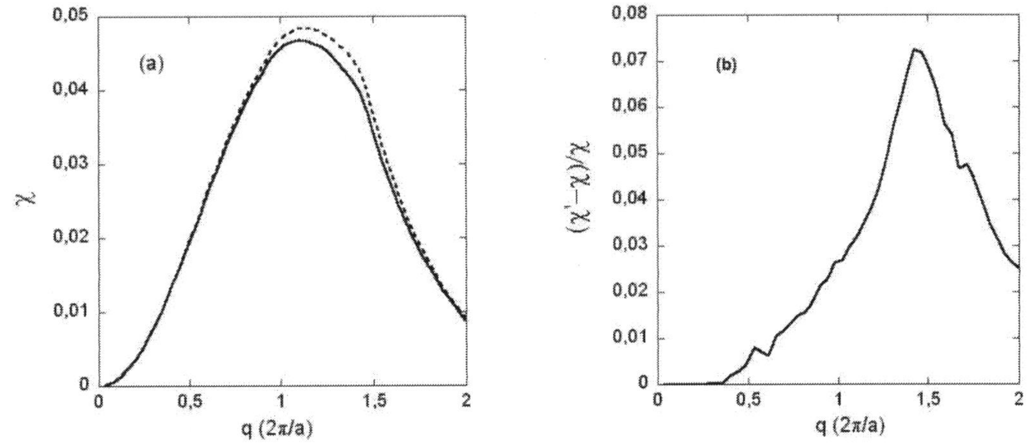

Figure 6. (a) Diagonal contribution of the static *ab initio* full electronic susceptibility, $\chi_{\mathbf{0},\mathbf{0}}(\mathbf{q})$, of *fcc* lithium at 30 GPa along ΓW with (solid line) and without (dashed line) including crystal local field effects. (b) Relative difference between both approximations, the maximum relative contribution associated to crystal local-field effects being located at $\mathbf{q}_{\text{nest}}^{\Gamma W}$.

Journal of Physics: Conference Series **121** (2008) 012008 doi:10.1088/1742-6596/121/1/012008

Electronic structure of sulfur under high pressure

Atsushi Nishikawa

Chiba-Keizai College, 4-3-30 Todoroki-cho, Inage-ku, Chiba 263-0021, Japan

E-mail: `nisikawa@chiba-kc.ac.jp`

Abstract. We have calculated the bandstructures and the total energies of the high-pressure phases of sulfur by means of the full potential linearized APW (FP-LAPW) method to investigate the structural stability. The tetragonal sulfur-II phase with spiral chain structure is a semiconductor with narrow energy gap, whereas the monoclinic sulfur-III with incommensurate modulation is metallic and the modulation yields a large dip near the Fermi-level of the density of states and contributes to the lowering of the bandstructure energy.

1. Introduction

Sulfur, selenium, and tellurium of the group-VIb elements undergo pressure-induced succesive structural phase transitions and also show insulator-metal transition under pressure. Numerous high-pressure forms are reported for these elements, and they have the similar structure sequence under high pressure, and it is found very recently that they have the same incommensurately modulated monoclinic structure under pressure.

Sulfur forms a stable orthorhombic structure, sulfur-I (S-I), consiting of S_8 ring molecules at ambient conditions[1], and this orthorhombic form undergoes a gradual transition to an amorphous form above 20 GPa on compressing at room temperature[2]. Recrystallization is occured on further compression around 35 GPa, and the sulfur-II (S-II) is a body-centered tetragonal structure with unique spiral chains of 4_1 and 4_3 screw, and is made up of 16 atoms in the unit cell[3].

A further phase transition to sulfur-III (S-III) is observed at 83-86 GPa, and though S-III was reported to have a base-centered orthorhombic structure[4],[5], S-III is very recently found to be isostructural with Te-III and Se-IV, and has a body-centered monoclinic structure with incommensurate modulation[6],[7]. Hejney et al. also found a previously unreported phase between S-II and S-III on pressure decrease, and this phase is the same triclinic structure as Te-II and Se-III[6]. On further pressure increase S-III transforms to the rhombohedral β-Po structure at 162 GPa, and this phase is stable to at least 212 GPa[4],[5]. The electrical property of S-II is a semiconductor, and that of S-III is a metal[8].

The purpose of this report is to investigate the electronic struture and the structural stability of S-II and S-III, and this is the first calculation on the complex structure of S-II from first principles.

2. Calculation

We have investigated the band structures and the total energies of high pressure phases of sulfur by means of the full-potential linearized augumented plane wave (FP-LAPW) method

© 2008 IOP Publishing Ltd

implemented in WIEN2k code [9] within the local spin density approximation (LSDA) [10]. The plane-wave cut-off energy is set for 14 Ry, and we have used 99 – 385 k-sampling points in the irreducible Brillouin zone for the self-consistent cycle and the total energy is converged within 0.1 mRy.

We have first of all calculated the electronic structure of the body-centered tetragonal structure of S-II. The lattice constants and the positions of atoms are taken from the experiments[3].

We then calculated the electronic structure of the unmodulated monoclinic structure of S-III[6],[7]. The observed incommensurate monoclinic structure of S-III is characterized by the incommmnsurate wave vector $\vec{q} = (0, q, 0), q = 0.28 \sim 0.30$. Thus we adopt the $1 \times 3 \times 1$ super cells for a model of the modulated structure of S-III, the b axis of which is tripled in length of the original monoclinic cell with some displacements of atoms in the ac plane to investigate the effect of the modulation on the band structure.

3. Results and Discussion

We have shown the electronic band structure and the density of states of the body-centered tetragonal structure of S-II in Fig. 1. The space group is D_{2h}^{20} ($I4_1/acd$), and the lattice constants are set to a = 7.841 Å and c = 3.100 Å. Sulfur atoms are in the 16f postions of $(x, x, 1/4)$ with $x = 0.136$ in the unit cell [3]. The calculated bands have a narrow indirect gap $E_g \sim 0.025$ Ry (0.34 eV). The lower eight bands are mainly originating from 3s-orbitals of S, and above bands are mainly originating from 3p-orbitals and they also contain 3d-orbitals slightly. This valence bands have splitted into two parts and the Fermi-level is just located in the energy gap, and thus sulfur-II phase is a semiconductor.

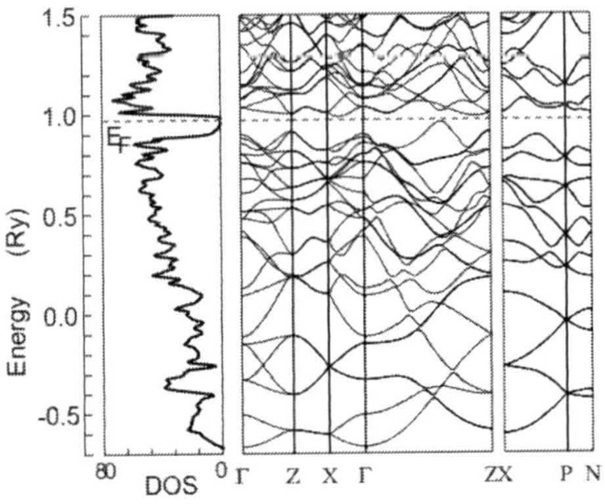

Figure 1. Band structure and density of states of the body-centerd tetragonal S-II. The lattice parameters are taken from the experiments; a = 7.841 Åand c = 3.100 Å. The broken lines in the band gap denote the Fermi-level.

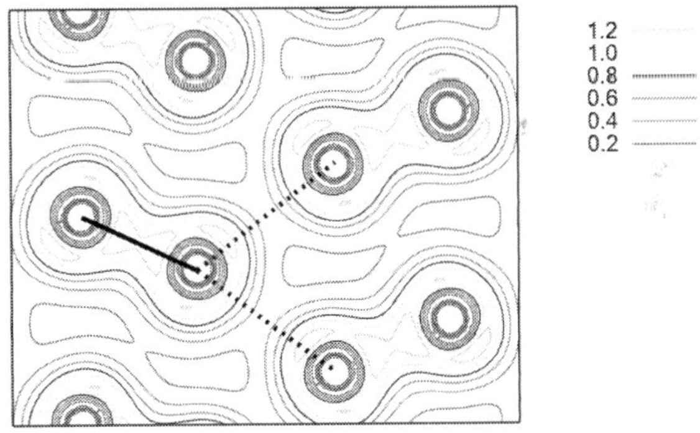

Figure 2. Electron density in the (100) plane of S-II. The numbers give the values of the density along lines of constant density in units of electrons per cubic angstrom. The first nearest neighbor S-S distance (1.95 Å) is depicted with sticks and the second nearest neighbor distance (2.64 Å) is drawn with dashed lines

The valence electron density plotted in the (100) plane is shown in Fig. 2. The first nearest neighbor distance is 1.95 Å along the chains, and the remarkable bonding charge are appeared between the nearest neighbor intrachain atoms. On the other hand there is no bonding between the inter-chain atoms, corresponding to the second nearest neighbor distance of 2.64 Å. Thus covalent bonds remain strong in the intrachain bonding of S-II. Note that the S-S separation in

Journal of Physics: Conference Series **121** (2008) 012008 doi:10.1088/1742-6596/121/1/012008

the ring molecule of S_8 in S-I is 2.05 Å, and the first nearest neighbor distance is almost same between S-I and S-II.

We have next shown the electronic band structure and the density of states of the body-centered monoclinic sulfur in Fig. 3. The lattice constants are set to $a = 2.802$ Å, $b = 3.455$ Å, $c = 2.208$ Å, and $\alpha = 90°, \beta = 113.15°, \gamma = 90°$ with two atoms in the unit cell. The lower two bands are mainly originating from $3s$-orbitals of S and the above bands are mainly originating from $3p$-orbitals. The latter bands have some peaks in the density of states and the Fermi-level is located in the middle of the broad peak. Though there are two first nearest neighbor atoms in the distance of 1.95 Å, and four second nearest neighbor atoms in the distance of 2.64 Å in different chains in S-II, in monoclinic S-III without modulation two atoms are in the distance of 2.208 Å and four atoms are in the distance of 2.225 Å. Therefore the coordination number is nearly six.

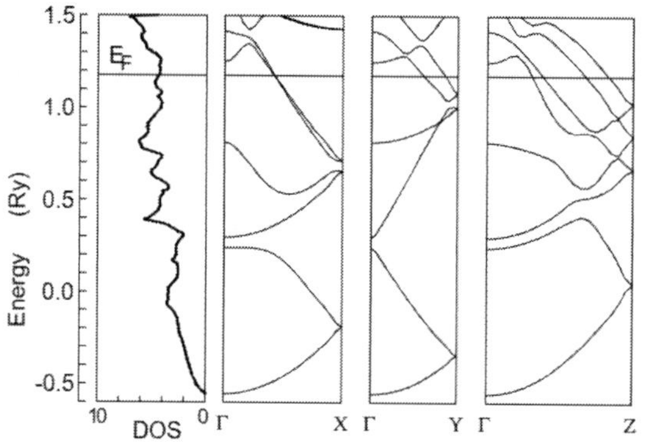

Figure 3. Band structure and density of states of the monoclinic S-III. The lattice parameters are taken from the experiments. The horizontal lines at about 1.2 Ry denote the Fermi-level.

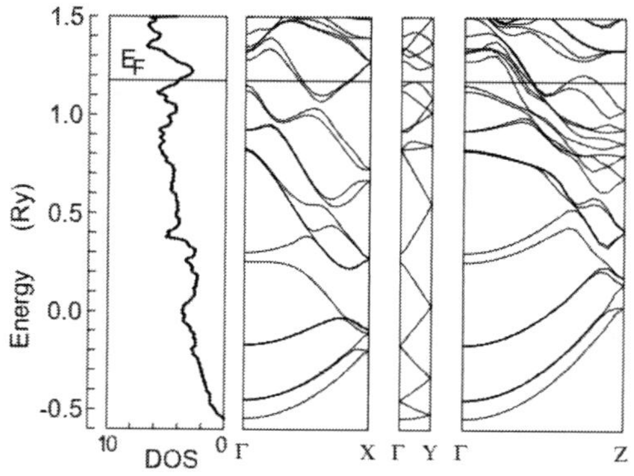

Figure 4. Band structure and density of states of the modulated structure of S-III. The modulation period along b axis is three times as long as the b lattice parameter of the monoclinic S-III.

We have then calculated the electronic structure and the total energy of the modulated structure, the lattice constants of which are set to $a = 2.802$ Å, $b = 10.365$ Å, $c = 2.208$ Å, and $\alpha = 90°, \beta = 113.15°, \gamma = 90°$ with six atoms in the unit cell, and the band structure and the density of states are shown in Fig. 4.

Since the b axis is tripled in length, the bands on the k_y axis are folded, and the modulation also affects the distances to the nearest-neighbor atoms. While the distance to the two nearest neighbor atoms at 2.208 Å corresponds to the lattice constant of c-axis and is therefore unaffected by the modulation, the four atoms at 2.225 Å assume different distances in the modulated structure in the range 2.089 Å to 2.425 Å. Thus the band gap yields on the k_y axis (Γ - Y) near the Fermi-level. This yields a large dip in the density of states at 1.25 Ry just above the Fermi level. The Fermi-level is just located in the dip, and it suggests that the band structure energy of the modulated structure should be lower than that of the original monoclinic structure. The calculated total energy of the modulated structure is slightly lower than that of the monoclinic S-III.

In summary, S-II is a semiconductor with strong covalent bonds in the intrachain atoms, and S-III is metallic and the modulation contributes to the lowering of the band structure energy and yields the structural stability.

Acknowledgments
Part of the calculation in this research were performed using supercomputing resources at Information Synergy Center, Tohoku University.

References
[1] Donohue J 1974 *The Structure of the Elements* (New York, Wiley)
[2] Luo H and Ruoff A L 1993 *Phys. Rev.* **48** 569
[3] Fujihisa H, Akahama Y, Kawamura H, Yamawaki H, Sakashita M, Yamada T, Honda K, and Bihan T L 2004 *Phys. Rev.* B **70** 134106
[4] Luo H, Greene R G, and Ruoff A L 1993 *Phys. Rev. Lett.* **71** 2943
[5] Akahama Y, Kobayashi M, and Kawamura H 1993 *Phys. Rev.* B **48** 6862
[6] Hejny C, Lundegaard L F, Falconi S, McMahon M I , and Hanfland M 2005 *Phys. Rev.* B **71** 020101
[7] Degtyareva O, Gregoryanz E, Somayazulu M, Mao H K, and Hemley R J 2005 *Phys. Rev.* B **71** 214104
[8] Luo H, Desgreniers S, Vohra Y K, and Ruoff A L 1991 *Phys. Rev. Lett.* **67** 2998
[9] Blaha P, Schwarz K, Madsen G K H, Kvasnicka D and Luitz J 2001 *WIEN2k An Augumented Plane Wave + Local Orbitals Program for Calculating Crystal Properties* (Wien: Techn. Universitat Wien)
[10] Perdew J P and Wang Y 1992 *Phys. Rev.* B **45** 13244

Journal of Physics: Conference Series **121** (2008) 012009 doi:10.1088/1742-6596/121/1/012009

Melting curve and Hugoniot of molybdenum up to 400 GPa by *ab initio* simulations

C. Cazorla[1,2], M. J. Gillan[1,2], S. Taioli[3] and D. Alfè[1,2,3]
[1]London Centre for Nanotechnology, UCL, London WC1H OAH, U.K.
[2]Department of Physics and Astronomy, UCL, London WC1E 6BT, U.K.
[3]Department of Earth Sciences, UCL, London WC1E 6BT, U.K.

Abstract. We report *ab initio* calculations of the melting curve and Hugoniot of molybdenum for the pressure range 0 – 400 GPa, using density functional theory (DFT) in the projector augmented wave (PAW) implementation. We use the "reference coexistence" technique to overcome uncertainties inherent in earlier DFT calculations of the melting curve of Mo. Our calculated melting curve agrees well with experiment at ambient pressure and is consistent with shock data at high pressure, but does not agree with the high pressure melting curve from static compression experiments. Our calculated $P(V)$ and $T(P)$ Hugoniot relations agree well with shock measurements. We use calculations of phonon dispersion relations as a function of pressure to eliminate some possible interpretations of the solid-solid phase transition observed in shock experiments on Mo.

1. Introduction

The melting curves of transition metals at pressures up to the megabar region are highly controversial, particularly for b.c.c. metals. Diamond anvil cell (DAC) measurements find that the melting temperature $T_{\rm m}$ increases by only a few hundred K over the range 1 - 100 GPa [1, 2], while shock experiments indicate an increase of several thousand K over this range [3, 4, 5]. There have been several *ab initio* calculations on transition-metal $T_{\rm m}$(P) curves, and the predictions agree more closely with the shock data than with the DAC data [6, 7, 8, 9]. A challenging case is molybdenum, where there are very large differences between DAC and shock data [12], and where the shock data reveal *two* transitions, the one at high pressure ($\sim$ 380 GPa) being attributed to melting, and the one at low pressure ($\sim$ 210 GPa) to a transition from b.c.c. to an unidentified structure [4]. We report here on new *ab initio* calculations of $T_{\rm m}(P)$ for Mo, and on the $P(V)$ and $T(P)$ relations on the Hugoniot. We also report preliminary information that may help in searching for the unidentified high-P solid phase of Mo.

We use density functional theory (DFT), which gives very accurate predictions for many properties of transition metals, including Hugoniot curves [10]. DFT molecular dynamics (m.d.) was first used to study solid-liquid equilibrium 12 years ago [11], and several different techniques are now available for using it to calculate melting curves. In such calculations, no empirical model is used to describe the interactions between the atoms, but instead the full electronic structure, and hence the total energy and the forces on the atoms, is recalculated at each time step. There have been earlier DFT calculations on the melting of Mo, but the techniques used were prone to superheating effects [6, 9]. In the present work, we use the so-called "reference coexistence " techniques [19, 17, 14], which does not suffer from this problem.

© 2008 IOP Publishing Ltd

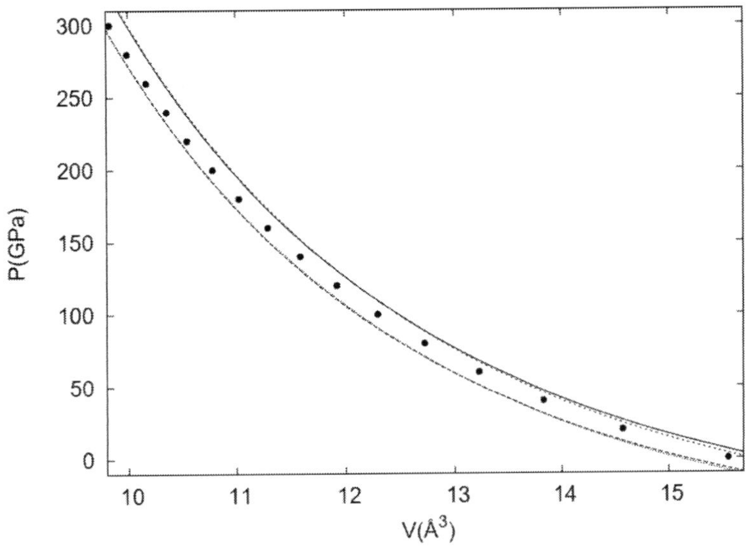

Figure 1. Comparison between PAW and FP-LAPW results for the GGA(PBE) and LDA(CA) approximations for E_{xc}. Solid and dashed curves show GGA(PBE) and LDA(CA) FP-LAPW results, respectively; short-dashed and dotted curves show GGA(PBE) and LDA(CA) PAW calculations, respectively. Solid dots show experimental data [5].

Our work has several aims. First, we want to improve on the reliability and accuracy of the predicted melting curve of Mo obtained from DFT; second, we use DFT to predict the $P(V)$ and $T(P)$ relations on the shock Hugoniot; third, we want to identify the unknown solid phase of Mo observed in shock experiments. Our tests on the accuracy of DFT for Mo, and our extensive calculations of the Mo melting curve will be reported in detail elsewhere [18], so we present only a summary here. However, our very recent calculations on the shock Hugoniot will be presented in more detail. These are important, because temperature is very difficult to measure in shock experiments [35] and DFT gives a way of supplying what is missing in the shock data. Our search for the unidentified solid structure of Mo is at the exploratory stage, but we present results for phonon frequencies as a function of pressure, which allow us to rule out some possibilities.

In the following, we summarise our tests on the accuracy of the DFT techniques (Sec. 2), and outline the reference coexistence technique. In Sec. 3 we present our results for the DFT melting curve and Hugoniot of Mo up to 400 GPa, and the study of the phonon frequencies. Discussion and conclusions are in Sec. 4.

2. Techniques and tests

Our DFT calculations are performed mainly with the projector augmented wave (PAW) implementation [20], using the VASP code [21, 25], since PAW is known to be very accurate. The main uncontrollable approximation in DFT is the form adopted for the exchange-correlation functional E_{xc}. To test the accuracy of PAW, and the effect of E_{xc}, we have compared our predictions for the $P(V)$ relation of the b.c.c. Mo crystal against low-T experimental results (Fig. 1). The pressure predictions from PAW using the Perdew-Burke-Ernzerhof (PBE) [28] and local-density approximation (LDA) [27] forms of E_{xc} deviate by ~ 1.5 % in opposite directions from the experimental data, but we adopt the PBE form, because the deviations in this case are rather constant. The PBE results of Fig. 1 were obtained with 4s states and below in the

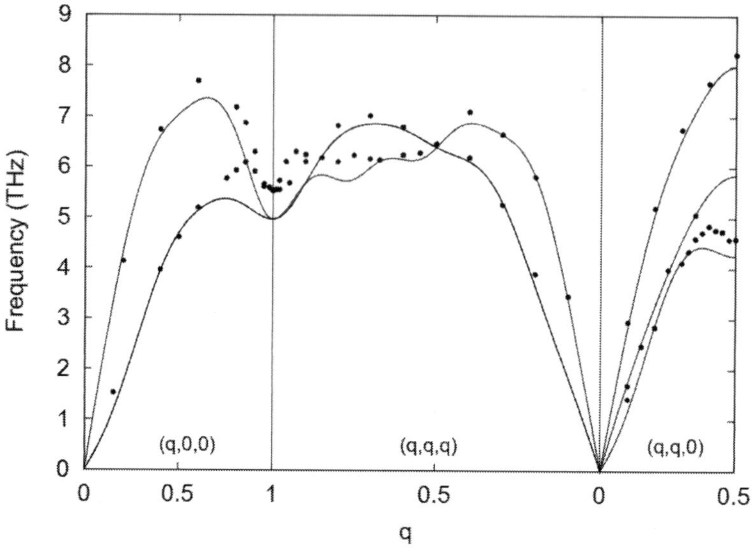

Figure 2. Comparison of calculated (curves) and experimental (dots) phonon dispersion relations of Mo at zero pressure. Experimental data are from Ref. [26].

core and all other states in the valence set. Inclusion of 4s states in the valence set makes no appreciable difference to the PBE results. We also tested the PAW implementation itself by repeating the $P(V)$ calculations with the even more accurate full-potential linearized augmented plane-wave (FP-LAPW) technique [22, 23, 24], using the WIEN2k code [33]. As shown in Fig. 1, PAW and FP-LAPW results are almost identical. Further confirmation for the accuracy of the PAW implementation and the PBE functional comes from our comparisons of the calculated phonon dispersion relations for the ambient-pressure b.c.c. crystal with experiment (Fig. 2).

The "reference coexistence" technique for calculating *ab initio* melting curves has been described elsewhere [18], but we recall the main steps. First, an empirical reference model is fitted to DFT m.d. simulations of the solid and liquid at thermodynamic states close to the expected melting curve. Then, the reference model is used to perform simulations on large systems in which solid and liquid coexist, to obtain points on the melting curve of the model. In crucial third stage, differences between the reference and DFT total energy functions are used to correct the melting properties of the reference model, to obtain the *ab initio* melting curve. In the present work, the total energy function of the reference model is represented by the embedded-atom model (EAM) [30, 31], consisting of a repulsive inverse-power pair potential, and an embedding term describing the d-band bonding. The detailed procedure for fitting $U_{\rm ref}$ to DFT m.d. data will be reported elsewhere [18], but we note that we needed to re-fit the model in different pressure ranges.

The simulations on solid and liquid Mo in stable thermodynamic coexistence using the fitted reference model employed systems of 6750 atoms, and we checked that the results do not change if even larger systems are used. The protocols for preparing these simulated systems, and for achieving and monitoring stable coexistence were similar to those used in earlier work on the melting curve of Cu [17]. The procedures for correcting the melting curve of the reference model, which depend on calculations of the the free energy differences between the reference and DFT systems, have been described and validated in earlier work [18].

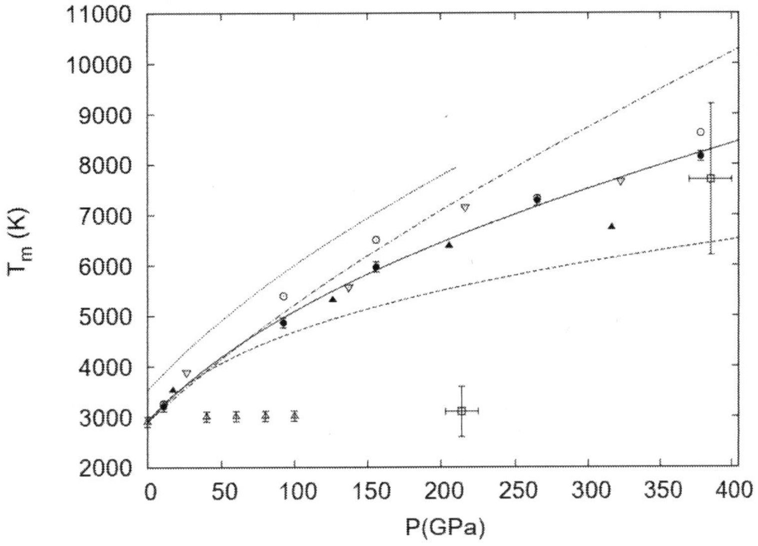

Figure 3. Calculated *ab initio* melting curve (filled circles and solid line) of this work compared with previous results: generalized pseudopotential calculations of Moriarty [6](dotted line), dislocation-mediated models of Belonoshko *et al.* [9](long-dashed line) and Verma *et al.* [29](dashed-dotted line); experimental shock-wave [4] and DAC [1] measurements are shown with empty squares and triangles, respectively. Filled and inverted-empty triangles show solid and liquid *ab initio* molecular dynamics calculations of Belonoshko *et al.* [9], respectively. Empty circles show results of this work obtained with the EAM model without free-energy corrections.

3. Results

3.1. Melting properties

Fig. 3 shows the melting curve of our reference EAM model, and the melting curve obtained from this by correcting for the differences between the DFT and reference total-energy functions; earlier *ab initio*-based calculations of the Mo melting curve due to Moriarty and to Belonoshko *et al.* are also indicated [6, 9]. We also show points on the melting curve from DAC and shock measurements [1, 4]. The differences between reference and corrected melting curves are only a few hundred K, so that the corrected curve should be very close to the melting curve that would be obtained from the (PBE) exchange-correlation functional if no statistical-mechanical approximations were made. Because we avoid approximations of earlier *ab initio*-based work, our present results should be a more accurate representation of the DFT melting curve. Up to 100 GPa, the differences between our results and earlier DFT work are rather small, and we confirm that DFT gives a much higher melting slope than that given by DAC experiments. We obtain an accurate value of $dT_{\rm m}/dP$ at $P = 0$ by fitting our melting curve to the Simon equation $T_m = a(1 + P/b)^c$, with $a = 2894$ K, $b = 37.2$ GPa, $c = 0.433$. The resulting $P = 0$ value of $T_{\rm m} = 2894$ K is close to the accepted experimental value $T_{\rm m} = 2883$ K. Our $dT_{\rm m}/dP$ value of 33.7 K GPa^{-1} at $P = 0$ agrees with an older experimental value of 33.3 K GPa^{-1} [32]. Our DFT melting curve is consistent with the point obtained at $P \simeq 370$ GPa from shock measurements. However, we stress that the temperature of the "experimental" point was not measured, but estimated by considering superheating corrections to the shock-wave data [12, 13]. Because of this, it is important to try and corroborate the estimated experimental temperature, and this can be done by *ab initio* calculations, as we explain next.

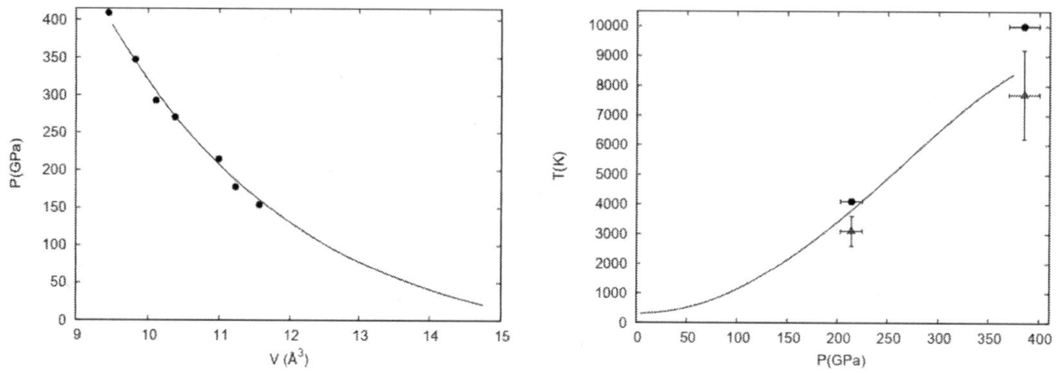

Figure 4. *Left*: $P_{\rm H}(V_{\rm H})$ *ab initio* relation on the Hugoniot of Mo up to pressure of 400 GPa; dots reproduce experimental data of Refs. [4, 5]. *Right*: $T(P_{\rm H})$ *ab initio* relation on the Hugoniot of Mo up to pressure of 400 GPa; dots reproduce experimental data of Refs. [4, 5] and triangles the same results but corrected for superheating effects as given by Ref. [12]. Error bars show the uncertainties in pressure and temperature.

3.2. Hugoniot curves

The pressure $P_{\rm H}$, volume $V_{\rm H}$ and internal energy $E_{\rm H}$ in the shocked state are related to the initial volume V_0 and internal energy E_0 by the well-known Rankine-Hugoniot formula [34]:

$$\frac{1}{2}P_{\rm H}\left(V_0 - V_{\rm H}\right) = E_{\rm H} - E_0 \; . \tag{1}$$

Since the internal energy and pressure are given in terms of the Helmholtz free energy F by $E = F - T(\partial F/\partial T)_V$ and $P = -(\partial F/\partial V)_T$, we can calculate the Hugoniot from our DFT simulations, provided we can calculate F as a function of V and T. So far, we have done this only for the b.c.c. crystal in the harmonic approximation, in which $F(V,T) = F_{\rm perf}(V,T) + F_{\rm harm}(V,T)$. Here, $F_{\rm perf}$ is the free energy of the rigid perfect crystal, including thermal electronic excitations, and $F_{\rm harm}$ is calculated from the phonon frequencies $\omega_{\mathbf{q}s}$ ($\mathbf{q}$ the wavevector, s the branch). We calculate $F_{\rm harm}$ in the classical limit, in which $F_{\rm harm} = 3k_{\rm B}T\ln(\beta\hbar\bar{\omega})$ per atom, with $\beta = 1/k_{\rm B}T$ and $\bar{\omega}$ is the geometric average of phonon frequencies over the Brillouin zone. The methods used for to calculate $F_{\rm perf}(V,T)$ and the frequencies $\omega_{\mathbf{q}s}$ were similar to those used in our earlier work on Fe (Ref. [16]). For a set of temperatures, we calculated $F_{\rm perf}$ at a set of volumes, and fitted the volume dependence with a third-order Birch-Murnaghan equation [36]. The temperature dependence of the coefficients in this equation were then fitted with a third-order polynomial. The phonon frequencies were calculated at 12 volumes in the range $15.6 - 9.2$ Å^3/atom, as explained elsewhere [18]. The volume dependence of the average $\bar{\omega}$ was then fitted with a third-order polynomial.

To obtain $P_{\rm H}(V_{\rm H})$ and $T(P_{\rm H})$ from our fitted free energy, for each value of $V_{\rm H}$ we seek the T at which the Rankine-Hugoniot equation is satisfied, and from this we obtain $P_{\rm H}$. For V_0 and E_0, we used values from our GGA calculations; we checked that use of the experimental V_0 made no significant difference. Comparison of our calculated $P_{\rm H}(V_{\rm H})$ with measurements of Hixson *et al.* [4, 5] (left panel of Fig. 4) shows excellent agreement. In the right panel, we compare our $T(P_{\rm H})$ with both uncorrected results of Hixson *et al.* and also with results corrected for superheating, and our results confirm their temperature estimates.

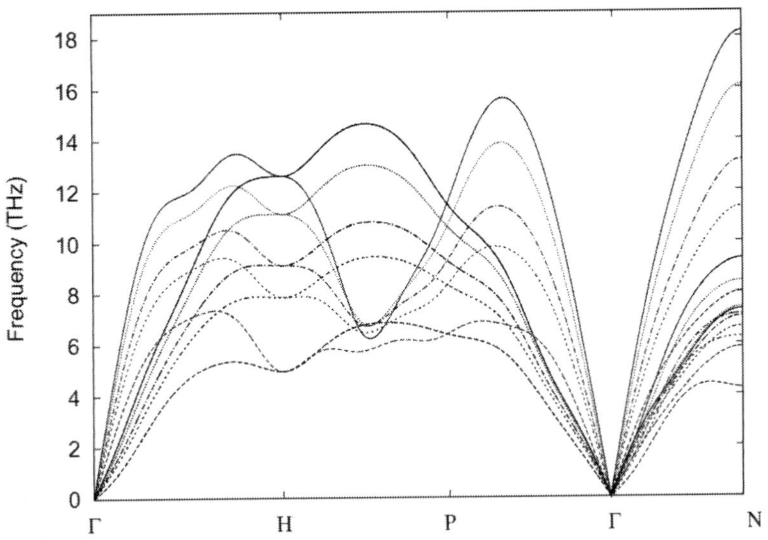

Figure 5. Comparison of the calculated phonon dispersion relations of Mo at different pressures: 0 GPa (– – –), 77 GPa (- - - -), 136 GPa (— · —), 274 GPa (· · · · · ·) and 400 GPa (——).

3.3. Solid-solid phase transition

Efforts have been made to identify the high-P/high-T structure of Mo indicated by shock experiments [4]. Hixson *et al.* used their theoretical prediction of b.c.c. → h.c.p. phase transition at $P \sim 320$ GPa and $T = 0$ K to suggest the h.c.p structure. However, later calculations locate this transition at higher pressures ($420 \leq P \leq 620$ GPa), and temperature stabilisation of h.c.p. over b.c.c. below melting seems improbable. Furthermore, there are recent claims that under pressure b.c.c Mo transforms first to f.c.c. rather than h.c.p. [37]. Very recently, Belonoshko *et al.* [9] reported *ab initio* simulations on the f.c.c and A15 structures at temperatures and pressures near those of interest (namely, $P \sim 210$ GPa and $T \sim 4000$ K), and concluded that both structures are unlikely high-T phases of Mo. We have performed *ab initio* calculations similar to those of Belonoshko *et al.* on the h.c.p. and ω structures ($P \sim 250$ GPa and $T = 5000$ K). We find that temperature neither favours any of them respect to b.c.c., thus they must be excluded as well. Recently, a 2nd-order phase transition from cubic to rhombohedral has been observed in vanadium at $P = 69$ GPa [38]. This appears to be related to an earlier *ab initio* prediction of a phonon softening in V [39]. This suggests that a similar structural transition might occur in Mo. We have used our calculated phonon dispersion relations of Mo over the range 0 – 400 GPa to test this. Fig. 5 depicts our results at 0, 77, 136, 274 and 400 GPa, but we see that no phonon anomaly like that reported for V is observed at any wavevector. This indicates that we should rule out structures based on small distortion of b.c.c.

4. Discussion

An important outcome of the present work is improved DFT calculations of the melting curve of Mo over the pressure range 0 – 400 GPa. In particular, our techniques allow us to avoid the superheating errors which appear to affect an independent recent DFT study of Mo melting [9]. The accuracy of our calculations is confirmed by the very close agreement with experiment for the melting temperature T_{m} and the melting slope dT_{m}/dP at ambient pressure. The results fully confirm that the increase of T_{m} by ~ 2000 K over the range 0 – 100 GPa predicted by DFT is about 10 greater than that deduced from DAC measurements. A second important outcome is that our calculations of the temperature along the shock Hugoniot support earlier temperature estimates based on experimental data but corrected for possible superheating effects [12]. This allows us to compare more confidently the point on the melting curve at $P \simeq 380$ GPa derived from shock data with our predicted melting curve, and we confirm that T_m at this pressure is *ca.* 8650 K. This is far above any reasonable extrapolation of the DAC data. Concerning the search for the unknown crystal structure of Mo indicated by shock experiments to exist above *ca.* 220 GPa, we have been able so far only to rule out some possibilities. Our calculations of the phonon dispersion relations in the b.c.c. structure over the range 0 – 400 GPa reveal no softening of any phonons, and no indication of any elastic instability. This means that the new crystal structure does not arise from small distortion of b.c.c. This is interesting in the light of the recent discovery of the elastic instability of b.c.c. V above $P \simeq 70$ GPa, predicted initially by DFT, and observed very recently in x-ray diffraction experiments [38]. It seems that the structural transition in Mo is of a different kind.

The large conflict between the melting curve of Mo derived from DAC measurements on one side and from shock experiments and DFT calculations on the other side must be due either to a misinterpretation of the DAC data or to a combination of serious DFT errors and misinterpretation of shock data. We note here that similar disagreements between DFT calculations and DAC measurements are also found for Ta [40] and Fe [41, 35, 42, 43]. Given the accuracy of DFT that we have been able to demonstrate (low-temperature $P(V)$ curve, Hugoniot $P(V)$ curve, ambient-P phonons), we think there is little evidence for significant errors in DFT, which is also in good accord with shock data. A possible explanation might be that the large temperature gradients and non-hydrostatic stress in DAC experiments might give rise to flow of material giving the appearance of melting, even well below the thermodynamic melting temperature. We also note recent evidence that temperature measurement in DAC experiments may be subject to previously unsuspected errors [44], though probably not of the size needed to resolve the conflict by themselves.

Acknowledgments

The work was supported by EPSRC grant EP/C534360, which is 50% funded by DSTL(MOD), and by NERC grant NE/C51889X/1. The work was conducted as part of a EURYI scheme award to DA as provided by EPSRC (see www.esf.org/euryi).

References

[1] Errandonea D., Schwager B., Ditz R., Gessmann C., Boehler R. and Ross M. 2001 *Phys. Rev.* B **63** 132104
[2] Errandonea D., Somayazulu M., Häusermann D. and Mao D. 2003 *J. Phys: Condens. Matter* **15** 7635
[3] Brown J. M. and Shaner J. W. *Shock Waves in Condensed Matter 1983*, ed. J. R. Assay, R. A. Graham and G. K. Straub (Amsterdam, Elsevier, 1983) p. 91
[4] Hixson R. S., Boness D. A., Shaner J. W. and Moriarty J. A. 1989 *Phys. Rev. Lett.* **62** 637
[5] Hixson R. S. and Fritz F. N. 1992 *J. Appl. Phys.* **71** 1721
[6] Moriarty J. A. 1994 *Phys. Rev.* B **49** 12431
[7] Moriarty J. A., Belak J. F., Rudd R. E., Söderlind P., Streitz F. H. and Yang L. H. 2002 *J. Phys: Condens. Matter* **14** 2825
[8] Wang Y., Ahuja R. and Johansson B. 2002 *Phys. Rev.* B **65** 014104
[9] Belonoshko A. B., Simak S. I., Kochetov A. E., Johansson B., Burakovsky L. and Preston D. L. 2004 *Phys. Rev. Lett.* **92** 195701
[10] Gillan M. J., Alfè D., Brodholt J., Vočadlo and G. D. Price 2006 *Rep. Prog. Phys.* **69** 2365
[11] Morris J. R., Wang C. Z., Ho K. M. and Chan C. T. 1994 *Phys. Rev. B* **49** 3109
[12] Errandonea D. 2005 *Physica B* **357** 356
[13] Luo S. -N. and T. J. Ahrens 2003 *Appl. Phys. Lett.* **82** 1836
[14] Alfè D., Gillan M. J. and Price G. D. 1999 *Nature* **401** 462
[15] Vočadlo L. and Alfè D. 2002 *Phys. Rev. B* **65** 214105
[16] Alfè D., Price G. D. and Gillan M. J. 2001 *Phys. Rev. B* **64** 045123

[17] Vočadlo L., Alfè D., Price G. D. and Gillan M. J. 2004 *J. Chem. Phys.* **120** 2072
[18] Cazorla C., Gillan M. J., Taioli S. and Alfè D. 2007 *J. Chem. Phys.* **126** 194502
[19] Alfè D., Gillan M. J. and Price G. D. 2002 *J. Chem. Phys.* **116** 6170
[20] Blöchl P. E. 1994 *Phys. Rev. B* **50** 17953
[21] Kresse G. and Joubert D. 1999 *Phys. Rev. B* **59** 1758
[22] Andersen O. K. 1975 *Phys. Rev. B* **12** 3060
[23] Koeling D. D. and Arbman G. O. 1975 *Journal of Phys. F* **5** 2041
[24] Singh D. 1994 Kluwer Academic Publishing ISBN 0-7923-9421-7
[25] Kresse G. and Furthmüller J. 1996 *Phys. Rev. B* **54** 11169
[26] Zarestky J., Stassis C., Harmon B. N., Ho K. M. and Fu C. L. 1983 *Phys. Rev. B* **28** 697
[27] Ceperley D. M and Alder B. I. 1980 *Phys. Rev. Lett.* **45** 566
[28] Perdew J. P., Burke K. and Ernzerhof M. 1996 *Phys. Rev. Lett.* **77** 3865
[29] Verma A. K., Rao R. S. and Godwal B. K. 2004 *J. Phys.:Condens. Matter* **16** 4799
[30] Daw M. S. and Baskes M. I. 1984 *Phys. Rev. B* **29** 6443
[31] Finnis M. W. and Sinclair J. E. 1984 *Phil. Mag. A* **50** 45
[32] Shaner J. W., Gathers G. R. and Minichino C. 1977 *High Temp. High Pressures* **9** 331
[33] Blaha P., Schwarz K., Madsen G. K., Kvasnicka D. and Luitz J. 2001 *WIEN2k: An Augmented Plane Wave plus Local Orbital Program for Calculating Crystal Properties* Technical University of Vienna
[34] Poirier J. P. in *Introduction to the Physics of the Earth's Interior* (Cambridge University Press, Cambridge, England, 1991), see e.g., Chap. 4.
[35] Yoo C. S., Holmes N. C., Ross M., Webb D. J. and Pike C. 1993 *Phys. Rev. Lett.* **70** 3931
[36] Birch F. 1978 *J. Geophys. Res.* **83** 1257
[37] Jona F. and Marcus P. M. 2005 *J. Phys.: Condens. Matter* **17** 1049
[38] Ding Y., Ahuja R., Shu J., Chow P., Luo W. and Mao H. 2007 *Phys. Rev. Lett.* **98** 085502
[39] Suzuki N. and Otani M. 2002 *J. Phys.: Condens. Matter* **14** 10869
[40] Taioli S., Cazorla C., Gillan M. J. and Alfè D. 2007 *Phys. Rev. B* **75** 214103
[41] Alfè D., Price G. D. and Gillan M. J. 2004 *Journal of Phys. and Chem. of Solids* **65** 1573
[42] Brown J. M. and McQueen R. G. 1986 *Journal of Geophys. Res.* **91** 7485
[43] R. Boehler 1993 *Nature* **363** 534
[44] Benedetti L. R., Guignot N. and Farber D. L. 2007 *J. of Appl. Phys.* **101** 013109

Ab-initio melting curve and principal Hugoniot of tantalum

S. Taioli[1], C. Cazorla[2,3], M. J. Gillan[2,3], and D. Alfè [1,2,3]

[1] Department of Earth Sciences, University College London, Gower Street, London WC1E 6BT, UK
[2]Department of Physics and Astronomy, University College London, Gower Street, London WC1E 6BT, UK
[3]London Centre for Nanotechnology, University College London, Gordon Street, London WC1H 0AH, UK

Abstract.
We report first principles calculations of the melting curve and principal Hugoniot ($P-V$ curve) of body centered cubic (bcc) tantalum in the pressure range 0-300 GPa. A description of lattice dynamics and thermal properties of bcc Ta using finite temperature density functional theory (DFT) is presented. The approach works within the projector augmented wave (PAW) implementation of DFT and explicitly treats in valence the 5p, 6s and 5d electrons. The principal Hugoniot ($P-V$ curve), obtained using the Rankine-Hugoniot equation, is investigated using the generalized gradient approximations (GGA). Very good agreement with the shock experiments is obtained with GGA in all the range of pressure. We also report the temperature-pressure relation on the shock Hugoniot and the full *ab-initio* melting curve of Ta.

1. Introduction

The past 10 years have seen remarkable developments in two branches of condensed matter physics. One is the improvement in the experimental set-up for studying phase transitions at high pressures by the use of shock-waves (Brown and Shaner 1984, Nellis et al. 2003) or laser-heated diamond-anvil cells (DAC) (Errandonea et al. 2001, Errandonea et al. 2003); the other is the improvement of the computational tools and computer power for calculating a number of materials properties from first principles, notably in transition metals and transition metals compounds. As well as contributing to our understanding of the high pressure phenomena and the control of the materials under extreme conditions, such advances also allow unprecedented access to the electronic and optical properties of matter.

Nevertheless, at high pressures and temperatures, when the behavior of materials is often anything but simple, the large discrepancies in the experimental results using different approaches cast a shadow on the reliability of the measurements, notably in the melting properties. The need of coherent experimental and theoretical results has been fuelled by both the fundamental desire to investigate the thermophysical properties of interacting many-fermion systems under extreme conditions and by the possibility to predict novel configurations that give materials entirely new properties.

Shock compression of materials is a well established (Zeldovich and Raizer 2002) technique, successfully applied to a broad range of materials such as aluminum, iron,

© 2008 IOP Publishing Ltd

copper and tantalum, to reach hundreds of gigapascals of pressure and temperatures of thousands of degrees.

Transition metals, due to their stability under extreme conditions, have been widely used as both anvils and impactors in shock wave experiments. Tantalum is a transition metal with a body-centered cubic (bcc) phase, stable even at high pressure of hundreds of GPa in the bulk material (Moriarty et al. 2002). It presents a high chemical and thermodynamical stability, having a melting point at ambient pressure of about 3270 K and it is widely used in the microelectronics industry for producing integrated circuits.

As well as improving experimental lay-out, a large effort has been devoted to theoretical investigations, which, particularly in the case of transition metals, are still the subject of discussion. While the complexity of treating correctly the exchange correlation energy makes the high pressure phenomena very difficult to solve from first principles, theoretical investigations can shed some light to argue about the reliability of the experiments.

In a previous paper (Taioli et al. 2007) a general description of lattice dynamics, thermal properties and melting of tantalum using DFT was presented. The equation of state at zero temperature and the phonon dispersion curves were investigated using both local density approximation (LDA) or generalized gradient approximations (GGA). The main goal of this work is to calculate the principal Hugoniot (P-V curve) of tantalum by solving the Rankine-Hugoniot equation, though we also report the melting curve of Ta obtained from first principles calculations.

The paper is organized as follows. In the next section we report the technical details of the calculations. In section 3 the strategy to calculate the thermal equation of state and tests on the reliability of the techniques are investigated. In section 4 the pressure-volume and pressure-temperature relations on the shock Hugoniot are discussed, and the melting curve of tantalum is also presented. Finally, conclusions are given in section 5.

2. Technical details

The present calculations have been performed with VASP (Kresse and Furthmuller 1996). In the finite-temperature formulation of DFT, in which the electronic excitations due to the temperature are taken into account, the relevant quantity is the electronic free energy, which has the following form

$$F_{static}(V, T_{el}) = E(V, T_{el}) - TS(V, T_{el}) \tag{1}$$

where E is the sum of kinetic, electron-nucleus, Hartree and exchange-correlation terms, and $S = -k_B T_{el} \sum_i [f_i \ln f_i + (1 - f_i) \ln(1 - f_i)]$ is the electronic entropy, with k_B being the Boltzmann constant, f_i the partial Fermi occupation of orbital i and T_{el} the electronic temperature. The ion-electron interaction has been treated using the projector augmented wave (PAW) technique (Blöchl 1994, Kresse and Joubert 1999), with single particle orbitals expanded in plane-waves with a cutoff of 224 eV, which ensures convergence of the structural parameters of Ta, like the equilibrium volume and the bulk modulus, to better than 0.05% . In a previous paper (Taioli et al. 2007) we tested different exchange-correlation functionals and various PAW potentials for bcc Ta. A very good agreement with the experimental zero temperature equation of state was found when using PBE and a PAW potential with inclusion of the $5p$ electrons in valence, and therefore here we used the same exchange-correlation and

Journal of Physics: Conference Series **121** (2008) 012010 doi:10.1088/1742-6596/121/1/012010

PAW potentials. Brillouin zone (BZ) sampling was performed using $16 \times 16 \times 16$ Monkhorst-Pack (MP) grid of **k**-points (Monkhorst and Pack 1976) on a primitive cell, which gives a converged value of the electronic free energy to within less than one meV. In agreement with previous studies (Boettger 2001, Bercegeay and Bernard 2005), we found that spin-orbit effects have a very small effect on the structural properties of Ta and have been neglected here altogether.

3. Components of the free energy

In a shock wave experiment the pressure P_H, internal energy E_H and volume V_H are related by the Rankine-Hugoniot relation (Poirer 1991) as follows:

$$\frac{1}{2}(P_H + P_0)(V_0 - V_H) = E_H - E_0 \tag{2}$$

where V_0, E_0 and P_0 are the equilibrium volume, the internal energy and the pressure at the starting conditions. The measurement of temperature in shock experiments is usually difficult, however it can be indirectly obtained by integrating the following thermodynamic relation (Brown and McQueen 1986)

$$dT_H = -T_H(\gamma/V_H)\,dV + [(V_0 - V_H)\,dP_H + (P_H - P_0)\,dV]/(2\,c_V) \tag{3}$$

where γ and c_V are the Grüneisen parameter and the constant-volume specific heat, respectively. Eqs. 2 and 3 can be solved if the Helmholtz free energy of the system is known as function of volume and temperature. We write this as a sum of two contributions:

$$F(V,T) = F_{static}(V,T) + F_{vib}(V,T) \tag{4}$$

where F_{static} has been defined in Eq. 1 and $F_{vib}(V,T) = F_{harm}(V,T) + F_{anharm}(V,T)$ is the contribution due to the vibrations of the ions, written as a sum of harmonic and anharmonic parts. Here we will neglect the anharmonic component of the free energy altogether. In the following two subsections we describe how we calculated these two contributions.

3.1. The perfect crystal

For any fixed T we calculated $F_{static}(V,T)$ at a number of different volumes, and fitted each isotherm to a third order Birch-Murnaghan equation of state (BMEOS). The parameters of the BMEOS $E_0(T)$, $V_0(T)$, $K_0(T)$ and $K_0'(T)$, have then been fitted to third order polynomials in T, and we report them in Table 1 for some values of T.

3.2. The harmonic crystal

In the high temperature limit, the vibrational free energy can be written as:

$$F_{harm}(V,T) = \frac{1}{N_{\mathbf{q}}}\,k_B T \sum_{\mathbf{q},s} \ln\left[\frac{\hbar\omega_{\mathbf{q}s}(V)}{k_B T}\right] \tag{5}$$

where k_B is the Boltzmann constant, the $\omega_{\mathbf{q}s}$ are the vibrational frequencies, $N_{\mathbf{q}}$ is the number of **q** points in the sum taken over phonon wave vectors **q** in the BZ and branches s. In principle the phonon frequencies $\omega_{\mathbf{q}s}$ depend on temperature because of electronic excitations, however, here we neglected this dependence on T. The

Table 1. Birch-Murnaghan fitted parameters for bulk bcc Ta as a function of temperature.

T (K)	$E_0(T)$ (eV)	$V_0(T)$ (Å^3)	$K_0(T)$ (GPa)	$K_0'(T)$
300	-11.73208	18.3419	190.88	3.8521
1408	-11.76216	18.3737	189.159	3.8672
3522	-11.91492	18.5003	184.749	3.8873
5400	-12.13376	18.6860	178.91	3.9003
7279	-12.42425	18.9318	171.857	3.9076
9157	-12.79214	19.2652	162.19	3.9255

phonon frequencies $\omega_{\mathbf{q}s}$ have been calculated using the PHON code (Alfè 1998), which implements the small displacement method (Kresse et al. 1995).

We find it useful to express the harmonic part of the Helmholtz free energy in terms of the geometric average $\bar{\omega}$:

$$\ln \bar{\omega} = \frac{1}{N_{\mathbf{q}}} \sum_{\mathbf{q},s} \ln(\omega_{\mathbf{q}s}) \tag{6}$$

which allows to write

$$F_{harm}(V,T) = k_B T \ln\left(\frac{\hbar\bar{\omega}}{k_B T}\right). \tag{7}$$

We calculated $\bar{\omega}$ at a number of different volumes, and then fitted $\ln\bar{\omega}$ to a third order polynomial in V. In Fig. 1 we report the calculated $\ln\bar{\omega}$ as a function of volume.

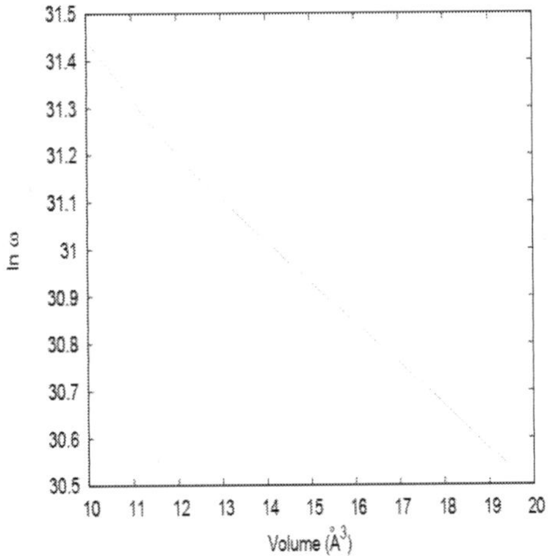

Figure 1. Geometric-mean phonon frequency $\ln\bar{\omega}$ of Ta as function of atomic volume V. $\bar{\omega}$ is in units of rad s^{-1}.

The technical parameters for the calculations of the phonon frequencies (size of the displacement, size of the supercell, number of electronic **k**-points, number of **q**-points for evaluating the sum in Eq. 6) were chosen in such a way to obtain free energies converged to within 1 meV/atom at $T = 300$ K.

Journal of Physics: Conference Series **121** (2008) 012010 doi:10.1088/1742-6596/121/1/012010

4. Thermal equation of state on the Hugoniot and melting curve of Ta

Once the Helmholtz free energy of the system is known as function of V and T it is straightforward to solve the Rankine-Hugoniot equation 2, which provides $P_H(V)$ and $T_H(P)$ curves.

We compare our results with the available shock wave measurements of Nellis et al. (Mitchell and Nellis 1980, Nellis et al. 2003). As shown in the left panel of Fig. 2, the agreement is very good, with discrepancies of less than 5 GPa in the worse case. These discrepancies are similar to those found for the room temperature static P(V) curve previously published (Taioli et al. 2007), and can be regarded as giving an indication of the accuracy of the GGA approximation. The temperature-pressure relation on the shock Hugoniot is compared with the experiments (Mitchell and Nellis 1980, Nellis et al. 2003) in the right panel of Fig. 2. The close agreement of our curve with experimental points supports the estimates of the temperature on the Hugoniot in these experiments. We also note that first principles calculations of the Hugoniot of Ta have been already reported by Cohen and Gülseren (Cohen and Gülseren 2001), and our results agree closely also with those calculation. As an independent test on

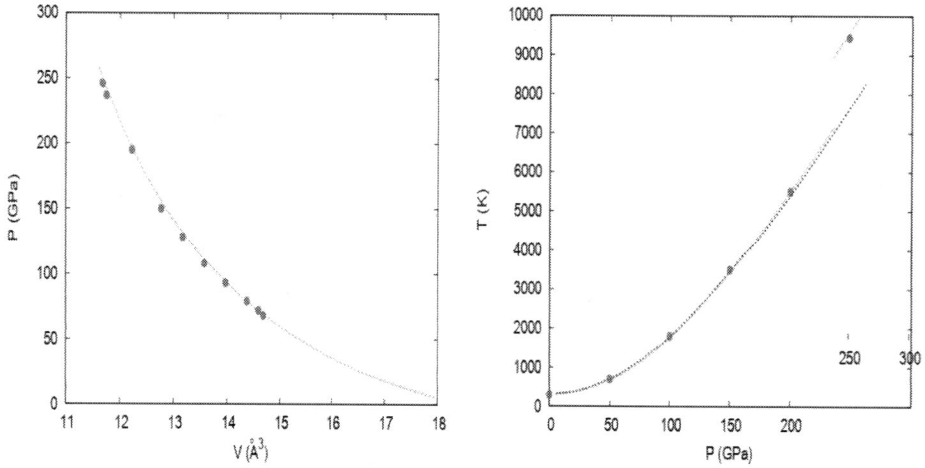

Figure 2. Principal Hugoniot (dashed line in the left panel) and temperature on the shock Hugoniot calculated using Eq. 2 (dashed line in the right panel) and Eq. 3 (point line in the left panel), respectively. Full circles show experimental data (Nellis et al. 2003).

the results we also calculated the $T_H(P)$ curve by integrating Eq. 3, which can easily be performed after γ and c_V have been calculated as function of V and T. The values of these thermodynamical quantities on the Hugoniot are reported on the left (γ) and right (c_V) panels of Fig. 3, and the $T_H(P)$ curve is shown in the right panel of Fig. 2. The $T_H(P)$ curves obtained with the two different numerical methods are very close to each other. The calculated heat capacity along the principal Hugoniot for Ta agrees well with first-principles results obtained by Wang who used the full-potential linearized augmented plane wave method (Wang et al. 2002).

To calculate the melting curve of Ta we applied the method of the coexistence of phases, applied with the help of an auxiliary reference system. The full *ab-initio* melting curve has then been obtained by calculating free energy differences between the reference system and the *ab-initio* system.

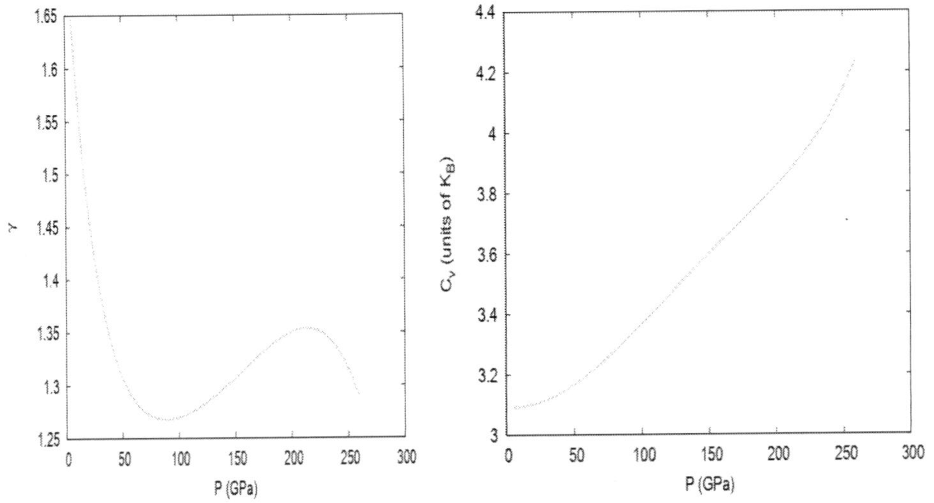

Figure 3. Variation with pressure of the Grüneisen parameter (left) and of the total constant-volume specific heat per atom (right, units of k_B) on the Hugoniot.

The technical details of the method and the melting curve of Ta have been reported in a previous paper (Taioli et al. 2007), here we only show the melting curve in Fig. 4, where we compare our results with the available DAC and shock wave experiments and previous theoretical results by Moriarty et al. (Moriarty et al. 2002) and Strachan et al. (Strachan et al. 2004). The interpolated value of the melting temperature at zero pressure (T_m = 3270 K) is in very good agreement with DAC experiments (Errandonea et al. 2001) (T_m = 3270 K), but as pressure is increased the agreement with DAC experiments quickly deteriorates, and at ~ 100 GPa the difference between our DFT calculations and these DAC experiments is already about 2000 K. At 307 GPa our calculated value $T_m = 9783 \pm 85$ K, is in good agreement within the statistical error bars with shock waves experiments (Brown and Shaner 1984), which estimate $T_m = 8500 \pm 1500$ K.

5. Discussion and conclusions

Our work reports a detailed *ab-initio* study of the principal Hugoniot ($P-V$ curve) and of the temperatures on the shock Hugoniot ($P-T$ curve). The principal Hugoniot has been obtained by solving the Rankine-Hugoniot equation using free energies obtained by first principles calculations; the temperature on the shock Hugoniot has been calculated by both solving the Rankine-Hugoniot equation and integrating the differential equation 3, the latter being the standard tool employed in shock experiments to deduce temperatures on the Hugoniot. A very good agreement has been found with the most recent shock data on tantalum (Nellis et al. 2003) in the range of pressures (0-300 GPa) and temperature (300-10000 K) investigated. We have also reported the values of the specific heat c_V and the Grüneisen parameter γ on the Hugoniot.

The agreement between the calculated Hugoniot and the experimental results also supports the reliability of the *ab-initio* melting curve of Ta, reported already in a previous paper (Taioli et al. 2007). Our results support other theoretical data

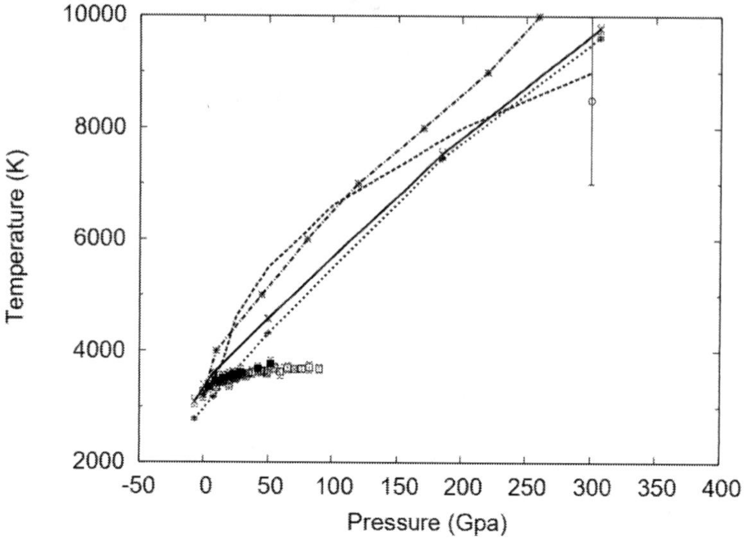

Figure 4. Phase diagram of Ta obtained using an auxiliary reference system with the coexistence MD approach (point line), and the resulting *ab-initio* curve obtained after adding free energy corrections (solid line). Also reported are DAC melting (Errandonea et al. 2001) (filled and open squares), shock melting (Brown and Shaner 1984) (open circle) and calculations from Ref. (Moriarty et al. 2002) (point-dashed line) and Ref. (Strachan et al. 2004) (dashed-line).

previously obtained on melting properties of tantalum, though our melting curve is somewhat lower. Furthermore, our melting curve is in good agreement with DAC experiments in the very low pressure regime, but starts to differ with increasing pressure. At high pressure our calculated melting curve disagrees substantially from DAC experiments, but is in good agreement with the shock datum at 300 GPa. This disagreement with DAC measurements has been already found previously (Moriarty et al. 2002) and it is a common feature in transition metals which melt from a bcc crystal structure, such as Ta and Mo for example. These experiments would call into question the reliability of DFT-based calculations at high pressure and temperature. However, although one can argue about the correctness of the different approximations for the exchange-correlation functional (LDA, GGA), which in some cases such as MgO (Alfè 2005), Al (Vočadlo and Alfè 2002, Alfè 2003) and silicon (Sugino and Car 1995, D. Alfè and M. J. Gillan 2003) can make noticeable differences in the calculated melting temperatures, we believe that it is very unlikely for DFT to be so much in error in the present case, given the large quantity of experimental data accurately predicted by this level of theory. We suggest that at least some of the DAC experiments may suffer from a imperfect diagnostic tool to detect melting, like the use of the intensity of X-ray diffraction peaks (Errandonea et al. 2003) which might lead to underestimate the melting temperature. Another possible problem in DAC experiments is the temperature measurement. Recently, an analysis of the spectroradiometrics effects of the dispersion and adsorbance properties of diamond has found that chromatic effects induced by the diamond windows can be substantial, and may lead to an underestimate of melting temperatures of several hundred degrees (Benedetti et al. 2007). Applying this analysis to the melting curve of Fe Benedetti et al. (Benedetti et al. 2007) concluded that, for example, the melting curve of Boehler (Boehler 1993) and that of Saxena and Dubrovinski (Saxena and Dubrovinski 2000) are at least a few hundred degrees too low at megabar pressures.

Acknowledgements

This work was supported by NERC grant NE/C51889X/1 and by EPSRC grant EP/C534360, which is 50% funded by DSTL(MOD). This work was conducted as part of a EURYI award to DA as provided by EPSRC (see www.esf.org/euryi). Allocation of computer time on the HPCx service was provided by NERC, and on the UCL facilities by Research Computing.

References

D. Alfè, program available at http://chianti.geol.ucl.ac.uk/~dario.
D. Alfè and M. J. Gillan and G. D. Price 2002 *J. Chem. Phys.* **116** 6170
D. Alfè 2003 *Phys. Rev. B* **68** 064423
D. Alfè and M. J. Gillan 2003 *Phys. Rev. B* **68** 205212
D. Alfè 2005 *Phys. Rev. Lett.* **94** 235701
L. R. Benedetti and N. Guignot and D. L. Farber 2007 *J. App. Phys.* **101** 013109
C. Bercegeay and S. Bernard 2005 *Phys. Rev. B* **72** 214101
F. Birch 1947 *Phys. Rev.* **71** 809
F. Birch 1978 *J. Geophys. Res.* **83** 1257
P. E. Blöchl 1994 *Phys. Rev. B* **50** 17953
R. Boehler 1993 *Temperatures in the Earth's core from melting point measurements of iron at high static pressures, Nature, Sec. 7* **363** 534
J. C. Boettger 2001 *Phys. Rev. B* **64** 035103
J. M. Brown and J. W. Shaner 1984 *Shock waves in Condensed Matter* **J. R. Asay et al., New York: Elsevier**
J. M. Brown and R. G. McQueen 1986 *J.Geophys. Res.* **91** 7485
R. E. Cohen and O. Gülseren 2001 *Phys. Rev. B* **63** 224101
D. Errandonea and B. Schwager and R. Ditz and C. Guessmann and R. Boehler and M. Ross 2001 *Phys. Rev. B* **63** 132104
D. Errandonea and M. Somayazulu and D. Häusermann and D. Mao 2003 *J. Phys. : Condens. Matter* **15** 7635
G. Kresse, J. Furthmüller and J. Hafner, *Europhys. Lett.* **32**, 729.
G. Kresse and J. Furthmuller 1996 *Phys. Rev. B* **54** 11169
G. Kresse and D. Joubert 1999 *Phys. Rev. B* **59** 1758
A.C. Mitchell and W. J. Nellis 1980 *J. Appl. Phys.* **52** 3363
H. J. Monkhorst and J. D. Pack 1976 *Phys. Rev. B* **13** 5188
J. A. Moriarty and J. F. Belak and R. E. Rudd and P. Søderlind and F. H. Streitz and L. H. Yang 2002 *J. Phys. : Condens. Matter* **14** 2825
F.D. Murnaghan 1944 *Proc. Natl. Acad. Sci. USA* **30** 244
W. J. Nellis and A.C. Mitchell and D. A. Young 2003 *J. Appl. Phys.* **93** 304
J. P. Poirer 1991 *Introduction to physics of the Earth's Interior* **Cambridge University Press, Cambridge, England** Chap. 4
S. K. Saxena and L. S. Dubrovinski 2000 *Amer. Mineral.* **85** 372
A. Strachan and T. Çağin and O. Gülseren and S. Mukherjee and R. E. Cohen and W. A. oddard III 2004 *Model. Simul. Mater. Sci. Eng* **12** 445
O. Sugino and R. Car 1995 *Phys. Rev. Lett.* **74** 1823
S. Taioli and C. Cazorla and M. J. Gillan and D. Alfè 2007 *Phys. Rev. B* **75** 2141031
L. Vočadlo and D. Alfè 2002 *Phys. Rev. B* **65** 214105
I. B. Zeldovich and Y. P. Raizer 2002 *Physics of Shock Waves and High-temperature Hydrodynamic Phenomena* **Dover Publications, Mineola, New York**
Y. Wang and R. Ahuja and B. Johansson 2002 *High Pressure Research* **22** 485

Journal of Physics: Conference Series **121** (2008) 012011 doi:10.1088/1742-6596/121/1/012011

The role of the intermolecular potential in determination of equilibrium and dynamic properties of molecular nitrogen (N_2) properties: MD simulations

Paweł Strąk[1], **Stanisław Krukowski**[2,3]

[1] Faculty of Physics, Warsaw University of Technology, 00-672 Warsaw, Koszykowa 75, Poland

[2] Institute of High Pressure Physics, Polish Academy of Sciences, 01-142 Warsaw, Sokołowska 29/37, Poland

[3] Interdisciplinary Centre for Materials Modelling, Warsaw University, 02-106 Warsaw, Pawińskiego 5a, Poland

E-mail: stach@unipress.waw.pl

Abstract. High precision approximation of the intermolecular potential is necessary for molecular dynamics (MD) simulations of the physical properties of molecular nitrogen (N_2). Recently, Quantum Mechanical (QM) high precision calculations (coupled-cluster singles-and-doubles with a perturbative triples corrections [CCSD(T)]) were used to determine $N_2 - N_2$ intermolecular interaction. Using these data, the potential, covering all orientations of the pair of the nitrogen molecules, was obtained in the form invariant with respect of rotation of the reference system. The new N_2-N_2 potential remains in reasonably good agreement with the isotropically scaled potential function of van der Avoird et al, based on the results of Hartree–Fock calculations. Nitrogen equation of state, at pressures up to 30 GPa (300 kbar) and at temperatures from room temperature to 2000K, was obtained by MD simulations, employing the following potentials: the new CCSD(T) function, the nonscaled and scaled expression by van der Avoird et al., and four-atom interaction by Johnson et al. It was shown that the new potential gives considerably better agreement with the experimental data than in all other cases. MD simulations were also used to determine shear viscosity of nitrogen. It was shown that the influence on the potential variation on the MD obtained values of the viscosity is much stronger than on the equation of state, especially at higher densities. The results were compared with the experimentally measured values of the viscosity showing considerable difference between the date obtained from experiment and derived form MD simulations.

1. Introduction

Since nitrogen is ubiquitous in normal life and also, it is used in many important technologies, the investigation of its properties is a task of great importance. Both thermodynamic and dynamic properties of nitrogen were therefore studied in the past and they are intensively investigated at present. In particular, nitrogen-nitrogen ($N_2 - N_2$) intermolecular interaction potential was studied quite intensively. This was partially related to the fact that the intermolecular nitrogen-nitrogen

© 2008 IOP Publishing Ltd

interaction is relatively weak and, in majority of instances, its energy is a tiny fraction of the total system energy, expressed as all electron energy, which is typical for quantum mechanical calculations. Therefore, extremely precise QM calculations are necessary for determination of the interaction potential at the precision level required for reliable simulation of its physical properties. For long time this task was simply too difficult and therefore, the semi-empirical methods, employing model potentials, were used [1]. These model potential functions include: the 6-12 Lennard Jones interaction with the added electrostatic quadrupole moments [2], de Boer atom-atom diatomic model [3], the Kihara core model [4] and Berne-Pechukas overlap potential [5]. Using these results and the analogy with hydrogen, Raich and Gillis constructed an anisotropic interaction of the pair of nitrogen molecules as sum of three terms: the electrostatic quadrupole-quadrupole interaction, long range dispersion terms and shape dependent repulsive overlap or valence contribution [1]. This approach was very fruitful in the following years as generalizations of Raich – Gillis formulation was later generally used to obtain fully anisotropic nitrogen molecular pair-potential [1].

Using the route devised by Raich and Gillis, Berns and van der Avoird employed Hartree-Fock calculations to obtain the short range part of the interaction potential [6]. The electrostatic long range part, expressed as series of multipole electrostatic interactions, was obtained from Hartree-Fock calculations by researchers belonging to the same group [7]. The dispersion terms were also put in the form of multipole series [7]. At approximately the same time, Bohm and Ahlrichs obtained very important result, showing that the pair-wise-interaction is a good approximation of the full nitrogen intermolecular potential [8].

A different route was used by Johnson et al. [9]. They treated the nitrogen molecule as a sum of the two atoms linked by the bond. In their formulation, the N_2-N_2 potential is a sum of four N-N potentials for the force acting between the atoms belonging to different molecules and the two additional terms describing the N-N intramolecular interactions. The intramolecular contribution term was parameterized using the N_2 molecule dissociation energy, the bond length and the vibration frequency. The interaction energy with the other molecule atoms was approximated using exponential-six function. The three different parameterizations of the exponential-six function were used: (i) by fitting to Hay's configuration-interaction potential [10], (ii) by adjusting to Hugoniot data for N_2 [11], and from nonlinear fit to Pack's electron gas calculations [10]. These three selections differ by about 20% in energy [9]. These formulations can account nitrogen interaction anisotropy, but their precision and validity, especially at high pressures, are hard to asses.

Employing earlier results, van der Avoird, Wormer and Jensen (AWJ) formulated full potential, including electrostatic, dispersion and overlap terms in the form invariant with respect to rotation of the coordinate system [12]. The dispersion interaction was expressed using Tang and Toennis damping function [13]. By comparing their own results and those of Bohm and Ahlrichs, AWJ found that the Hartree-Fock approximation significantly underestimates the overlap energy. Therefore they introduced isotropic scaling factors for the slope of the interaction potential and the exponents of damping function of Tang and Toennis. They have also found that this adjusted potential better recovers several nitrogen properties, such as: lattice parameters for α - nitrogen at very low pressure ($p = 0$), the sublimation (or cohesion) energy, isothermal compressibility of α - nitrogen at pressure and temperature close to zero, phonon frequencies obtained from the infrared, Raman and neutron scattering experiments [14]. Despite the above successes, the magnitude of the two arbitrarily introduced scaling factors, close to 1.5, and the fact that these factors were rotationally invariant, poses serious question whether the AWJ potential is close to the exact one, both in the radial and the angular dependence. The obtained results suggest also that the high pressure phase γ is not well described if the angular dependence is not correct. Similarly, the results more sensitive to the anisotropy of the potential, like scattering measurements or Raman line-widths, indicate that the AWJ potential seriously underestimates the anisotropy of nitrogen potential [15].

Recently, Stallcop and Partridge have performed ab initio calculations for the pair of two rigid N_2 molecules in coupled-cluster singles and doubles approach with perturbation account of the triple excitations, usually denoted as CCSD(T) approach [16]. They obtained several versions of nitrogen intermolecular potentials, depending on the basis sets used. The potential was again adjusted to recover the measured value of the second virial coefficient B(T). Thus the basic flaw of AWJ approach was not removed. The results published by Stallcop and Partridge are not complete and cannot be used independently for the simulations of the properties of nitrogen. Therefore, a high precision CCSD(T) calculations were necessary to derive the interaction energy for large number of different configurations of the pair of N_2 molecules. These results were obtained and recently used to determine the intermolecular potential in the system rotation invariant form [17].

For the pressures below 30 GPa, at room or higher temperatures, nitrogen is supercritical fluid, inert chemically. At extremely high pressures nitrogen looses its very stable molecular structure. In fact, the results of shock wave compression experiments show that, at about 33 GPa and 6000K, nitrogen undergoes continuous molecular to non-molecular transition [18]. The problems related to the transition to atomic form are outside the scope of this work, and will not be discussed here.

The results presented below were obtained using molecular dynamic (MD) modeling of the nitrogen in the molecular form. Several potential were used: the one proposed by Johnson et al. [9], the scaled and non-scaled potential of van der Avoird et al. [12], and the recently derived CCSD(T) based function [18]. The two basic physical properties: the equation of state and the shear viscosity will be considered. The results will be critically compared with the existing experimental data. The conclusions regarding the form of the nitrogen potential will be drawn.

2. Different approximations of the N_2-N_2 intermolecular potential

Molecular nitrogen (N_2) bonding energy, equal to 937.6 kJ/mole (9.72 eV/molecule), is the highest of all diatomic molecules [17]. This is due to the existence of triple chemical bond between N atoms which creates the region of high electronic density between the two nitrogen atoms. Therefore, either special model potential could be used, or very precise QM calculations are necessary to recover the nitrogen intermolecular interaction with high precision.

An example of fairly successful implementation of model function is the potential proposed by Johnson et al. which combines the separate atom-atom interaction of the atoms belonging to the different molecules with the bond energy describing the intramolecular interactions [9]. The three different versions of the potential were proposed by these authors. The bond energy was the same in all three versions. It was selected to be of the Morse form:

$$\phi_{bond}(r) = D_e\left\{1-\exp\left[-\xi\left(\frac{r}{r_e}-1\right)\right]\right\} \tag{1}$$

where D_e is the classical dissociation energy, r_e – is the equilibrium interatomic distance, and ξ, related to the cubic anharmonicity of the potential α by the following relation: $\xi = (2/3)\alpha$, is given by:

$$\xi = \left(\frac{m\omega_e^2 r_e^2}{4D_e}\right)^{1/2} \tag{2}$$

The following values were used: D_e = 114 950 K = 158.7057 $*10^{-20}$ J, $\hbar\omega_e$ = 3393.5 K = 4.6852 $*10^{-20}$ J, r_e = 1.097 685 Å, m= 14.0067 amu and ξ = 951 86. The first quoted numbers are those originally used by Johnson et al. [9] (the energy was expressed in temperature units), the second ones are their equivalents in standard units, used in the simulations presented here.

The interaction of the atoms belonging to different molecules was cast in the exp-6 form:

$$\phi_{atom}(r) = \frac{\varepsilon}{\alpha-6}\left\{6\exp\left[-\alpha\left(\frac{r}{r_0}-1\right)\right]-\alpha\left(\frac{r}{r_0}\right)^6\right\} \tag{3}$$

The three versions of the potential differ by the selection of the values of the parameters: ε, r_0 and α. In our simulations, we used the third version presented by Johnson et al., i.e. the one obtained from Pack electron-gas calculations [10], as it recovers the experimental data better than the two others. In this version, the following parameter values were adopted: $\varepsilon = 26.5$ K $= 36.587 * 10^{-23}$ J, $r_0 = 3.877$ Å and $\alpha = 12.896$. The relaxed bond length, i.e. the distance between the N atoms, used to fit the above parameters, was selected to be: $r'_e = 1.17397$ Å [9].

The other choice that avoids the use of the model potential containing the parameters, adjusted to the experimental results, is to resort to ab intio total energy calculations in order to derive the interaction potential. This entails application of sophisticated computational scheme, capable to account the electron-electron correlations with sufficiently high precision. The N_2-N_2 system is, in fact, the closed shell system, composed of the two nitrogen molecules, weakly interacting at majority of typical distances. The total system energy of the system is close to 7 keV, which is six orders of magnitude larger than typical N_2-N_2 interaction energy, which may be estimated using the interaction energy at the minimum of the potential, about $1.6* 10^{-22}$ J (10 meV or equivalently 80 cm^{-1}). The comparison of these data indicates that extremely high precision calculations are necessary for correct description of the potential at various distances.

The above requirements essentially exclude Hartree-Fock self consistent field (SCF) approximation, characterized by large numerical errors [19,20]. As it turned out, the SCF results could be used providing that appropriate scaling is used. This route was used by van der Avoird et al. (AWJ) [12]. They used the SCF potential and the molecule orientation independent scaling factors to readjust the potential to the experimental data. In order to verify the validity of the procedure, the two versions of AWJ potential, i.e. scaled and nonscaled, were used in our simulations.

By its nature, the SCF single determinant wavefunctions do not account Coulomb interactions completely. These interactions determine the correlations of the opposite spin electrons [20]. A natural extension of SCF approach is to use either configuration interaction (CI) [21,22] or coupled cluster (CC) [23-25] schemes which account electron-electron correlations by the use of series excitation operators, modifying single determinant SCF wavefunctions. In the case when sufficiently high excitation are included, these two methods are characterized by high numerical precision, necessary for proper determination of nitrogen intermolecular potential.

An important advantage of CC over CI is that the CC method is size consistent, i.e. in the case of non-interacting molecules, the total CC wavefunction may be written as a product of wavefunctions of separate molecules at any truncation of the series. This is not valid for CI approximation; therefore we are convinced that, for our purposes, the CC method is better suited, especially at larger distances between nitrogen molecules.

We used Gaussian commercial package in Rev 03 [26], in coupled-cluster singles-and-doubles with a perturbative triples corrections [CCSD(T)] approximation, which, at present, is a standard choice for high precision QM calculations. According to the investigations of a single N_2 molecule by Kowalski and Piecuch [27], and by Ruden et al. [28], the contribution of higher order terms is about 18.8 cm^{-1} and 0.03 pm for the harmonic oscillation frequency and the equilibrium distance, respectively [28]. While the energy error seems to be large, as compared to the intermolecular interaction energy, it was obtained on top of the triple bond N_2 binding energy, equal to 9.72 eV/molecule. The numerical errors scale with the binding energy; therefore we expect that the error for the intermolecular interaction energy is at least one order of magnitude smaller, which is sufficient for reliable determination of the N_2 interaction potential.

The QM results were obtained using aug-cc-pVQZ basis set, for the total number of 315 different configurations of the pair of N_2 molecules. These configurations, marked as points in the four dimensional space in R, θ_A, φ_A, and θ_B coordinates, create 3 x 6 x 5 x 6 mesh, which is sufficient for determination of the full anisotropic potential. The total number of points was 540 but, due to the symmetry, their number could be reduced to 315, which was used in the approximation procedure by

Berns and van der Avoird [6]. We used identical scheme, so that the number of points is smaller than that in the procedure used by AWJ [12] for determination of the short range overlap interaction.

The N_2-N_2 potential is expressed in the standard form, i.e. in the form of interaction of the two nitrogen molecules, denoted as (A) and (B):

$$V\left(\vec{R},\hat{r}_A,\hat{r}_B\right)=\sum_{L_A,L_B,L}\upsilon^{L_A,L_B,L}(R)A_{L_A,L_B,L}\left(\hat{R},\hat{r}_A,\hat{r}_B\right) \tag{4}$$

where $\vec{R}=\left(R,\hat{R}\right)=(R,\Theta,\Phi)$ is the vector pointing from the center of molecule A to the center of molecule B, $\hat{r}_A=(\theta_A,\phi_A)$ and $\hat{r}_B=(\theta_B,\phi_B)$ are the unit vectors, describing the orientation of the molecules A and B respectively, and $\hat{R}=(\Theta,\Phi)$ is the unit vector, describing the orientation of the vector $\vec{R}$. The angular dependence of the potential was described using the set of the spherical harmonics:

$$A_{L_A,L_B,L}\left(\hat{R},\hat{r}_A,\hat{r}_B\right)=\sum_{M_A,M_B,M}\begin{pmatrix}L_A & L_B & L\\ M_A & M_B & M\end{pmatrix}Y_{L_AM_A}(\hat{r}_A)Y_{L_BM_B}(\hat{r}_B)Y_{LM}\left(\hat{R}\right) \tag{5}$$

where $Y_{lm}(\hat{r})$ denote the spherical harmonics and the symbol in bracket is Wigner 3-j coefficient. This form of the potential function assures invariance of the energy with respect to the rotation of the system as a whole. The dependence on the intermolecular distance R was expressed using v coefficients, i.e. the sum of the electrostatic, overlap and dispersion contributions:

$$\upsilon^{L_AL_BL}=\upsilon^{L_AL_BL}_{elec}(R)+\upsilon^{L_AL_BL}_{overlap}(R)+\upsilon^{L_AL_BL}_{disp}(R) \tag{6}$$

The electrostatic part is expressed in multipole series, as:

$$\upsilon^{L_AL_BL}_{elec}(R)=\delta_{L_A+L_B,L}C^{L_AL_B}R^{-L_A-L_B-1} \tag{7}$$

where $\delta_{L_A+L_B,L}$ is Kronecker delta and C^{L_A,L_B} are the coefficients, listed in the Table 1. Since we limited the number of the points as in the procedure used in Ref 6 [6], the coefficients for the (L_A, L_A, L) equal to (6,4,10) and (6,6,12) were set equal to zero.

Table 1. Coefficients - electrostatic and dispersion interaction

	ELECTROSTATIC		**DISPERSION**	
L_A,L_B,L	**$C^{LA,LB}$ (kJ mol^{-1} nm$^{LA+LB+1}$)**	**$C_6^{LA,LB,L}$ (kJ mol^{-1} nm^6)**	**$C_8^{LA,LB,L}$ (kJ mol^{-1} nm^8)**	**$C_{10}^{LA,LB,L}$ (kJ mol^{-1} nm^{10})**
0, 0, 0	0	$4{,}119 \cdot 10^{-3}$	$3{,}795 \cdot 10^{-4}$	$3{,}491 \cdot 10^{-5}$
2, 0, 2	0	$2{,}165 \cdot 10^{-4}$	$9{,}780 \cdot 10^{-5}$	$1{,}380 \cdot 10^{-5}$
2, 2, 0	0	$5{,}454 \cdot 10^{-6}$	$1{,}458 \cdot 10^{-6}$	$3{,}150 \cdot 10^{-7}$
2, 2, 2	0	$6{,}518 \cdot 10^{-6}$	$-2{,}179 \cdot 10^{-6}$	$-8{,}662 \cdot 10^{-7}$
2, 2, 4	$1{,}624 \cdot 10^{-3}$	$5{,}247 \cdot 10^{-5}$	$1{,}133 \cdot 10^{-5}$	$5{,}778 \cdot 10^{-6}$
4, 2, 6	$6{,}771 \cdot 10^{-5}$	0	0	0
4, 4, 8	$5{,}411 \cdot 10^{-6}$	0	0	0
6, 2, 8	$8{,}024 \cdot 10^{-7}$	0	0	0

Similarly, these terms vanish for London dispersion interaction, defined as:

$$\upsilon_{disp}^{L_A L_B L}(R) = -\sum_{n=6,8,10} f_n^{L_A L_B L}(R) C_n^{L_A L_B L} R^{-n} \tag{8}$$

where Tang and Toennis damping function is:

$$f_n^{L_A L_B L}(R) = 1 - \left[\sum_{k=0}^{n} a_k^{L_A L_B L} \cdot R^k\right] \cdot \exp\left(-\alpha^{L_A L_B L} \cdot R - \beta^{L_A L_B L} \cdot R^2\right) \tag{9}$$

$$a_k = \sum_{i=k/2}^{k} \frac{\alpha^{2i-k} \beta^{k-i}}{(2i-k)!(k-i)!} \tag{10}$$

The coefficients are listed in Table 2. The sum runs over all integer i values (in the case when k is odd, the initial i value is the integer resulting from division of k by 2). The short range or overlap interaction is given by:

$$\upsilon_{overlap}^{L_A L_B L}(R) = F^{L_A L_B L} \exp\left(-\alpha^{L_A L_B L} R - \beta^{L_A L_B L} R^2\right) \tag{11}$$

where the coefficients were obtained by the use of CCSD(T) results.

Table 2. Coefficients - overlap interaction – CCSD(T)

L_A, L_B, L	$F^{LA,LB,L}$ (kJ mol^{-1})	$\alpha^{LA,LB,L}$ (nm^{-2})	$\beta^{LA,LB,L}$ (nm^{-2})
0, 0, 0	$2.8478\cdot10^4$	$1.1064\cdot10^1$	$3.6899\cdot10^1$
2, 0, 2	$9.0024\cdot10^3$	8.8428	$3.8176\cdot10^1$
2, 2, 0	$1.0841\cdot10^2$	-2.9439	$5.1960\cdot10^1$
2, 2, 2	$-4.0285\cdot10^2$	1.0930	$4.7670\cdot10^1$
2, 2, 4	$6.5419\cdot10^3$	9.4469	$3.9388\cdot10^1$
4, 0, 4	$1.6028\cdot10^3$	8.9254	$3.8277\cdot10^1$
4, 2, 2	$9.3398\cdot10^{-1}$	$-1.5597\cdot10^1$	$6.5891\cdot10^1$
4, 2, 4	$-4.1600\cdot10^1$	$-6.7843\cdot10^{-1}$	$4.9233\cdot10^1$
4, 2, 6	$2.7633\cdot10^3$	$1.2160\cdot10^1$	$3.7111\cdot10^1$
4, 4, 0	$-5.6779\cdot10^{-8}$	$-8.2570\cdot10^1$	$1.4890\cdot10^2$
4, 4, 2	$-1.5435\cdot10^{-8}$	$-9.1007\cdot10^1$	$1.5929\cdot10^2$
4, 4, 4	$5.5700\cdot10^{-3}$	$-3.0125\cdot10^1$	$8.3531\cdot10^1$
4, 4, 6	-6.1088	-2.0943	$5.1267\cdot10^1$
4, 4, 8	$2.1840\cdot10^3$	$1.5694\cdot10^1$	$3.5594\cdot10^1$
6, 0, 6	$7.4753\cdot10^2$	$1.7922\cdot10^1$	$2.6677\cdot10^1$
6, 2, 4	$2.5770\cdot10^{-2}$	$-1.7763\cdot10^1$	$6.6740\cdot10^1$
6, 2, 6	$-1.8191\cdot10^1$	8.6153	$3.6955\cdot10^1$
6, 2, 8	$3.0364\cdot10^3$	$2.3517\cdot10^1$	$2.3421\cdot10^1$

* For $R > 14a_o$ β for negative values are replaced by zero.
($a_o = 5{,}291\cdot10^{-11}$ m i.e. $R > 0{,}7406$ nm)

The comparison of the Johnson, the non-scaled and the scaled AWJ potentials, and also the present fully ab initio derived potential, is presented in Fig.1. Both radial and angular dependence of these potentials is shown. Generally, the new potential is much closer to the scaled AWJ function. This confirms the general validity of AWJ scaling. It has to be pointed out that the uniform angle independent scaling is oversimplified, as shown by the various relation of the scaled and new potential for different orientation of the molecules. The angular dependence is more important for the determination of the dynamic properties of nitrogen, like viscosity, thermal conductivity, etc. The angularly averaged values are more relevant for the determination of pressure, i.e. for the equation of state.

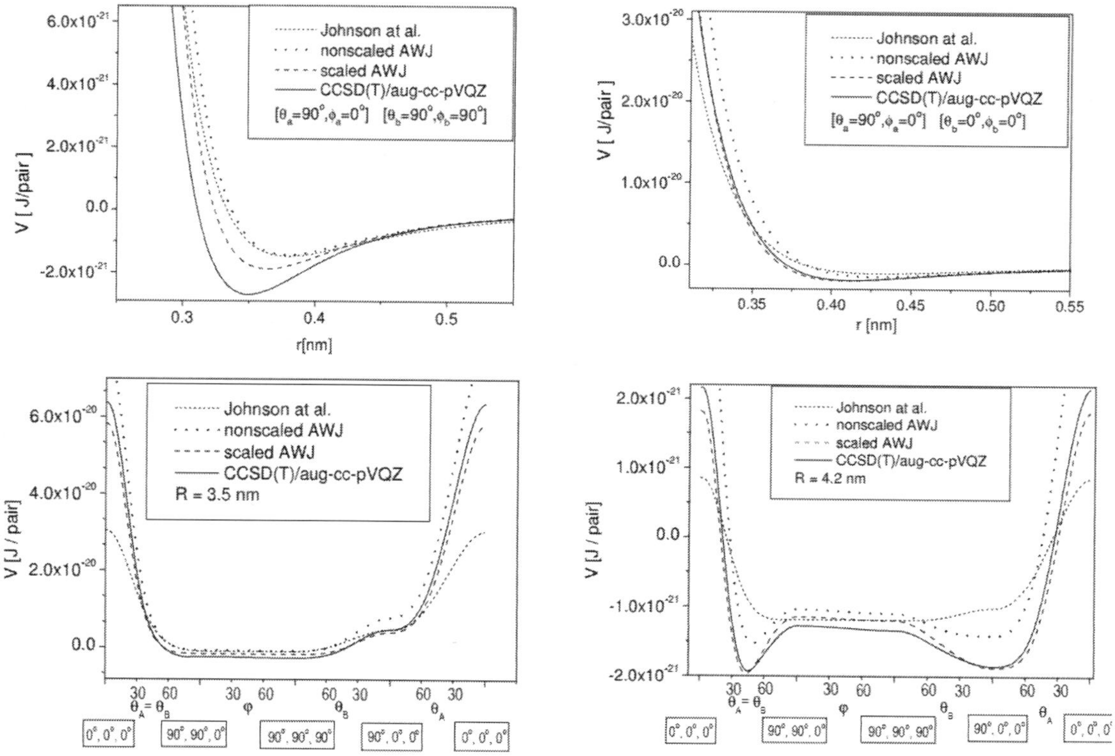

Figure 1. Comparison of the Johnson et al.[9], nonscaled AWJ, scaled AWJ [12], and fully ab initio derived potential: upper row – radial dependence, lower row – angular dependence

3. Equation of state

Nitrogen equation of state was simulated by MD procedure, using a box, which has periodic boundary conditions (PBC) imposed at all walls. The pressure – temperature data, for 300K and 1200K, are shown in Fig.2. Presently developed, scaled and nonscaled AWJ and Johnson potentials were used. For comparison, we have plotted the experimental data collected and critically assessed by Jacobsen et al. [29]. These results, presented in Fig. 2 prove that the best agreement with the experimental data is for the case of the new potential. For lower temperatures, i.e. for 300K - 1200K range, the best MD results recover the experimental data of Jacobsen et al. with the precision better than 1%. For the highest temperature, used in our simulations, i.e. for T = 2000K, the error is about 2.5%. The increase of the error is due to the fact that rigid rotor approximation is less precise at very high temperatures when the nitrogen molecules are excited to higher vibrational states.

We can also compare the present data with the previously reported results [30] where, using AWJ potential, we obtained the agreement of 4% and 7% at high and low temperatures, respectively. In the present work we obtained definitely better agreement than the previously reported, which we attribute to the new interaction potential. This confirms that the new nitrogen-nitrogen intermolecular interaction potential definitely better reflects the magnitude of the real interaction. The increase of the error at high temperatures is related to deviation form rigid rotor approximation. Unfortunately, the determination of the interaction potential without rigid rotor approximation is definitely beyond our present numerical capabilities.

For higher pressures, i.e. above 1GPa, the only data regarding nitrogen equation of state were those obtained by Johnson et al. [9]. We have used their potential in our MD simulations to compare with the results obtained using other potentials. The data obtained using Johnson potential are in better agreement that nonscaled AWJ potential, but they are definitely worse that those obtained using both scaled AWJ and newly developed potentials.

Table 3. Coefficients c_i for the new equation of state of nitrogen (Eq. 12a-c).

i	c_i	p_{xi}	p_{yi}	i	c_i	p_{xi}	p_{yi}
1	$1{,}3923\cdot10^4$	0	0	36	$1{,}2249\cdot10^4$	4	4
2	$4{,}1643\cdot10^4$	1	0	37	$3{,}4797\cdot10^3$	6	4
3	$1{,}0143\cdot10^4$	0	1	38	$-1{,}7816\cdot10^4$	7	4
4	$5{,}6175\cdot10^4$	2	0	39	$5{,}9612\cdot10^3$	5	6
5	$2{,}1570\cdot10^4$	1	1	40	$-1{,}3060\cdot10^3$	7	6
6	$-5{,}9437\cdot10^2$	0	2	41	$5{,}0350\cdot10^3$	3	4
7	$4{,}3124\cdot10^4$	3	0	42	$1{,}3381\cdot10^4$	2	5
8	$1{,}3886\cdot10^4$	2	1	43	$9{,}5754\cdot10^2$	5	3
9	$-2{,}6985\cdot10^3$	1	2	44	$-2{,}6779\cdot10^4$	6	3
10	$1{,}6797\cdot10^4$	4	0	45	$-3{,}9784\cdot10^4$	4	5
11	$-3{,}7499\cdot10^3$	3	1	46	$-3{,}1512\cdot10^3$	2	7
12	$-5{,}6862\cdot10^3$	2	2	47	$2{,}8167\cdot10^3$	8	3
13	$1{,}6845\cdot10^2$	5	0	48	$3{,}3869\cdot10^4$	6	5
14	$3{,}4035\cdot10^3$	1	3	49	$-1{,}1001\cdot10^4$	6	6
15	$-8{,}7323\cdot10^3$	4	1	50	$-4{,}9037\cdot10^3$	2	3
16	$-5{,}1870\cdot10^3$	0	4	51	$2{,}6733\cdot10^4$	4	3
17	$-1{,}0624\cdot10^3$	3	2	52	$1{,}7940\cdot10^3$	9	1
18	$-3{,}9074\cdot10^3$	6	0	53	$-1{,}3852\cdot10^4$	5	5
19	$1{,}7163\cdot10^3$	5	1	54	$1{,}7599\cdot10^4$	3	7
20	$-1{,}2353\cdot10^4$	1	4	55	$-5{,}0971\cdot10^3$	9	3
21	$4{,}9783\cdot10^3$	6	1	56	$6{,}8713\cdot10^3$	9	4
22	$-7{,}4409\cdot10^2$	5	2	57	$7{,}3557\cdot10^2$	10	4
23	$4{,}0763\cdot10^3$	0	6	58	$-1{,}4906\cdot10^2$	1	5
24	$-3{,}0792\cdot10^4$	3	5	59	$9{,}2570\cdot10^3$	7	2
25	$-4{,}2914\cdot10^3$	7	3	60	$-6{,}7223\cdot10^2$	10	5
26	$2{,}1791\cdot10^3$	0	3	61	$-1{,}0674\cdot10^2$	8	1
27	$-1{,}6429\cdot10^3$	0	5	62	$-3{,}8851\cdot10^3$	9	2
28	$-7{,}5549\cdot10^2$	2	6	63	$-1{,}2313\cdot10^4$	8	4
29	$2{,}0842\cdot10^4$	7	5	64	$8{,}4419\cdot10^3$	8	6
30	$-1{,}6747\cdot10^3$	4	2	65	$9{,}9384\cdot10^2$	8	0
31	$-1{,}6268\cdot10^3$	7	0	66	$3{,}4963\cdot10^2$	9	0
32	$1{,}2804\cdot10^4$	3	3	67	$1{,}3813\cdot10^3$	10	2
33	$-1{,}4132\cdot10^3$	7	1	68	$-8{,}7471\cdot10^2$	10	3
34	$2{,}4826\cdot10^3$	6	2	69	$-4{,}9918\cdot10^2$	10	0
35	$8{,}5344\cdot10^3$	1	6				

The equation of state was constructed in the form of scaled function of the density and the temperature [30], as:

$$P(x, y) = \sum_{i=1}^{69} c_i x^{p_{xi}} y^{p_{yi}} \tag{12a}$$

where:

$$x = 1 + \frac{2(\rho - 1.899)}{1.899 - 0.0003528} \tag{12b}$$

$$y = 1 + \frac{(T - 2000)}{1000} \tag{12c}$$

In this equation pressure is expressed in bars, density in g/cm^3 and temperature in Kelvins. The coefficients c_i, obtained by adjusting these functions to the CCSD(T) numerical data, are listed in Table 3.

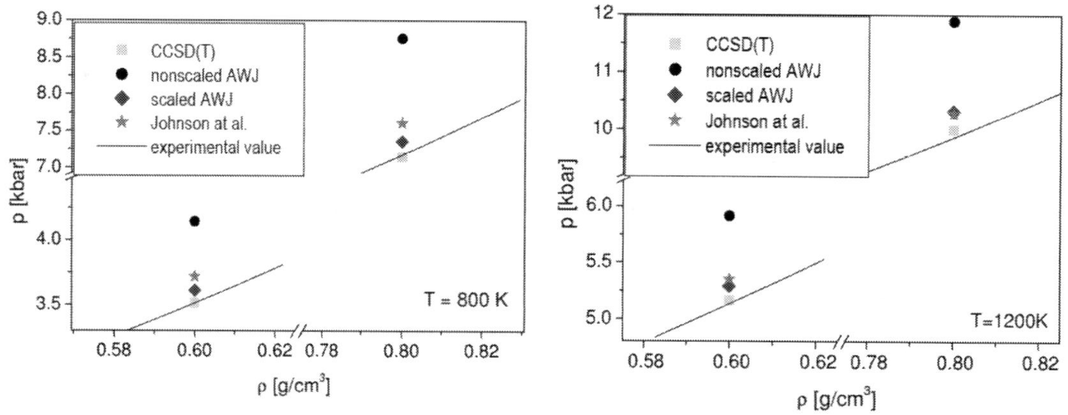

Figure 2. Comparison of the equation of state obtained from MD simulations for different nitrogen interaction potentials and the experimental data by Jacobsen et al. [29].

Comparison of the MD results and the experimental data confirms that the radial dependence of the interaction potential is well recovered by our CCSD(T) calculations, and also that the previous versions of the potential describe the radial dependence with reasonably good precision. An important results, with respect to the angular dependence of the potential, could be derived from the comparison of the experimental and MD data on the nitrogen viscosity, especially at higher densities. This will be discussed in the next Section.

4. Viscosity

MD simulations of the nitrogen viscosity were carried out using two different approaches. In the first case, the equilibrium molecular dynamic (EMD) procedure is based on Green-Kubo relation for the velocity-velocity correlation functions. In this approach the equilibrium time-lag correlation functions should be obtained. The box shape of the simulated volume with the periodic boundary conditions imposed at all walls, was used in MD procedure. These simulations provide precise and reliable data. Unfortunately, they require massive computations, and were not widely used in our simulations.

The EMD approach is based on Green – Kubo formula for the shear viscosity [31]:

$$\eta = \lim_{t\to\infty}\left(\frac{V}{kT}\right)\int_0^t \left\langle \sigma_{xy}(s)\sigma_{xy}(0)\right\rangle ds \qquad (13)$$

where the symbol <…> denotes average in the phase space and the stress shear tensor σ_{xy} is expressed by standard definition, as:

$$\sigma_{xy} = -\left(\frac{1}{V}\right)\sum_i \left(m_i v_{ix} v_{iy} + f_{ix} y_i\right) \quad (14)$$

In the above formula f_{ix}- denotes x component of the force, acting on i-th molecule where the index (i) runs over all molecules in the system.

In the second approach nonequilibrium molecular dynamics (NEMD) is used. This scheme is based on the application of thermodynamic force which drives the system out of equilibrium to the stationary regime. In this approach a linear relation between the thermodynamic force and the response of the system is assumed. As a rule, the coefficient in this relation plays a role of generalized susceptibility. Thus a precise determination of the force and the estimate of the response of the system is crucial. Generally, this is a difficult task, and therefore NEMD derived values are burdened by larger errors [31].

A typical implementation of the NEMD method enforces the simple planar Couette flow, in which the tangential shear stress and the velocity gradient are evaluated. The ratio of these two quantities gives the shear viscosity. In our simulations we impose the velocity gradient by adding and subtracting the specified velocity (or momentum) of the molecules crossing the upper and lower surface, respectively. In both, parallel to the flow directions, periodic boundary conditions are imposed. The stress component is evaluated from the momentum transfer in the unit time, for the unit area. The velocity gradient is obtained by division of the velocity difference at the opposite surfaces by the distance between them.

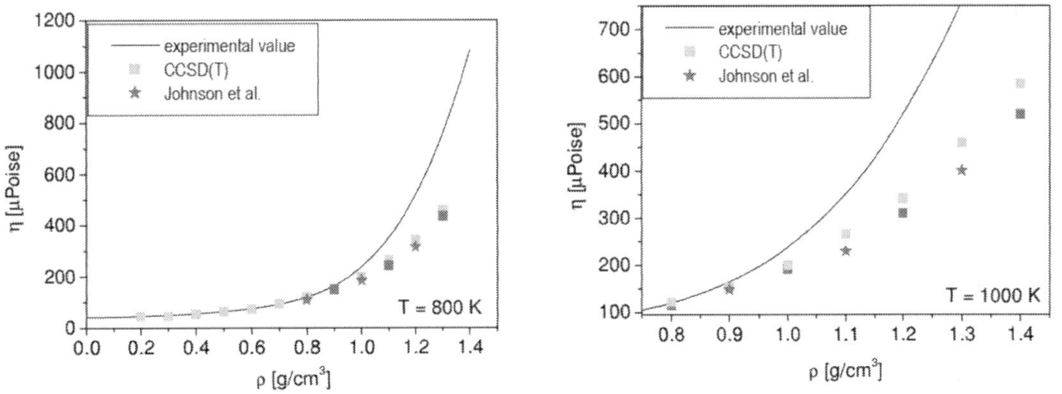

Figure 3. Comparison of the shear velocity obtained from MD simulations using two different potentials and the experimental data by Lemmon and Jacobsen [32].

The MD data on shear viscosity were obtained for the two cases: Johnson et al. [9] and new CCSD(T) derived potentials [18]. These MD data and also the nitrogen viscosity derived from the critical review of experimental data by Lemmon and Jacobsen [33] are presented in Fig. 3. The comparison of these data indicates that the viscosity obtained in the kinetic regime, i.e. for low densities, where the interaction could be well approximated by instantaneous collision events, recovers the experimental data for all investigated potentials. This indicates that the MD simulation procedure is basically valid. However, for higher densities, the MD values are about 30% lower than the experimental data. There exists some differences between two used potential, but in general the change related to the selected potential is relatively small. Slightly better agreement is obtained for our ab initio derived potential. Thus these data are not satisfactory and they indicate that the treatment of the nitrogen intermolecular interaction is not adequate and needs further investigations.

5. Summary

The results that were reported in this paper can be summarized as follows:

- nitrogen intermolecular potential was obtained using results of high precision CCSD(T) calculations
- equation of state of nitrogen at pressures up to 30 GPa was obtained by MD simulations
- the results of MD simulations of the nitrogen equation of state are generally in good agreement with the experimental data for all used potentials: Johnson et al. [9], nonscaled and scaled AWJ [12] and our newly developed ab inito derived potential [18]. This confirms that the radial dependence of the nitrogen potential is properly described.
- The best agreement with the experimental data for the equation of state is obtained for new CCSD(T) potential, while the scaled AWJ and Johnson potential give slightly inferior results and the nonscaled AWJ potential is characterized by the largest deviation from the experimental data.
- nitrogen viscosity experimental data are well recovered in kinetic regime for all considered potentials
- for higher densities the MD results are in considerable disagreement with the experimental data for all potentials
- further studies of the nitrogen intermolecular potential anisotropy and its role in the transport properties of nitrogen is necessary.

Acknowledgement

The results presented in this paper has been supported by Poland's Ministry of Science and Higher Education grants no R003601 and N202 042 32/1171. The calculations were made using computing facilities of Interdisciplinary Centre for Mathematical and Computational Modelling of Warsaw University (ICM UW).

References

[1] Raich J C and Gillis N S 1977 *J. Chem. Phys.* **66**, 846
[2] Kuan T S, Warshel A and Schnepp O 1970 J. Chem. Phys. **52**, 3012
[3] Raich J C, Gillis N S, and Anderson A B 1974 *J. Chem. Phys.* **61**, 1399
[4] Kihara T 1953 *Rev. Mod. Phys.* **25**, 831
[5] Berne B J and Pechukas P 1972 *J. Chem. Phys.* **56**, 4213
[6] Berns R M. van der Avoird A 1980 *J. Chem. Phys.* **72**, 6107
[7] Mulder F, van Dijk G. and van der Avoird A, 1980 *Mol. Phys.* **39**, 407 ()
Mulder F, van der Avoird A and Wormer P E S, 1979 *Mol. Phys.* **39**, 159
[8] Bohm H J and Ahlrichs R 1985 *Mol. Phys.* **55**, 445
[9] Johnson J D, Shaw M S and Holian B L 1984 *J. Chem. Phys.* **80**, 1279
[10] Hay J P, Pack R T and Martin R L. 1984 *J. Chem. Phys.* **81**, 1360 ().
[11] LASL Data ed. by .P. Marsh (University of California, Berkeley 1980) pp. 113-114
[12] van der Avoird A, Wormer P E S, Jansen A P J, 1986 *J. Chem. Phys.* **84**, 1629
[13] Tang K T and Toennies J P 1984 *J. Chem. Phys.* **80**, 3726
[14] Kochin B C 1960 *J. Chem. Phys.* **33**, 882
Heberlein D C, Adams E D and Scott T A 1970 *J. Low Temp. Phys.* **2**, 449

Fondere F, Obriot J, Marteau Ph, Allavena M and Chakroun H 1981 *J. Chem. Phys.* **74**, 2675
Thiery M M and Fabre D 1976 *Mol. Phys.* **32**, 257
Kjems J K and Dolling G 1975. *Phys. Rev. B* **11**, 1639

[15] Cappelletti D, Vecchiocattivi F, Pirani F, McCourt F RW, 1996 *Chem. Phys. Lett.* **248**, 237
Huo W M, Green S 1996 *J. Chem. Phys.* **104**, 7572

[16] Stallcop J R and Partridge H 1997 *Chem. Phys. Lett.* **281**, 212

[17] Barin I 1994 *Thermochemical Data of Pure Substances 3rd edition*, (Weinheim: VCH)

[18] Strak P, Krukowski S 2007 *J. Chem. Phys.* **126**, 194501

[19] Szabo A and Ostlund N S 1989 *Modern Quantum Chemistry: Introduction to Advanced Electronic Structure Theory*, 1st ed., (New York: McGraw-Hill)

[20] R. J. Bartlett and J. F. Stanton, in *Reviews in Computational Chemistry*, K. B. Lipkowitz and D. B. Boyd, Eds., VCH Publishers, New York, 1994, Vol. 5, Chap. 2, pp. 65-169. Applications of post-Hartree-Fock methods: A tutorial.

[21] Shavitt I 1977 in *Methods of Electronic Structure Theory*, ed H. F. Schaefer (New York: Plenum Press) pp. 189-275. The method of configuration interaction

[22] Sherrill C D and Schaefer H F *Adv. Quantum Chem.*, in press. The configuration interaction method: Advances in highly correlated approaches.

[23] Bartlett R J 1995 *Modern Electronic Structure Theory*, Vol. 2 of *Advanced Series in Physical Chemistry*, ed D R Yarkony (Singapore: World Scientific), Chap. 16, pp. 1047-1131

[24] Lee T J and Scuseria G E 1995 *Quantum Mechanical Electronic Structure Calculations with Chemical Accuracy*, e S R Langhoff (Dordrecht: Kluwer Academic Publishers) pp. 47-108

[25] Harris F E, Monkhorst H J and Freeman D L 1992 *Algebraic and Diagrammatic Methods in Many-Fermion Theory*, (New York: Oxford Press)

[26] Gaussian 03, Revision C.02, Gaussian, Inc., Wallingford CT, 2004

[27] Kowalski K and Piecuch P 2000 *J. Chem. Phys.* **113,** 5644

[28] Ruden T A, Helgaker T, Jorgensen P and Olsen J 2004 *J. Chem. Phys.* **121** 5874

[29] Jacobsen R T, Stewart R B, Jahangiri M 1986 *J. Phys. Chem. Ref. Data* **15**, 735

[30] Krukowski S, Strak P 2006 *J. Chem. Phys.* **124**, 134501

[31] Ryckaert J-P, Bellemans A, Ciccotti G and Paolini G V 1989 *Phys. Rev.* **A 39**, 259

[32] Lemmon E W and Jacobsen R T 2004 *Int. J. Thermophys.* **25**, 21

Joint 21st AIRAPT and 45th EHPRG Int. Conf. on High Pressure Science and Technology IOP Publishing
Journal of Physics: Conference Series **121** (2008) 012012 doi:10.1088/1742-6596/121/1/012012

Monte Carlo calculations of thermodynamic properties of deuterium under high pressures

P.R. Levashov[1], V.S. Filinov[1], A. Boţan[1], M. Bonitz[2], V.E. Fortov[1]

[1] Joint Institute for High Temperatures, Izhorskaya 13-2, Moscow 125412, Russia
[2] Cristian-Albrechts-Universität zu Kiel, ITPA, Leibnizstr. 15, 24098 Kiel, Germany

E-mail: pasha@ihed.ras.ru

Abstract. Two different numerical approaches have been applied for calculations of shock Hugoniots and compression isentrope of deuterium: direct path integral Monte Carlo and reactive Monte Carlo. The results show good agreement between two methods at intermediate pressure which is an indication of correct accounting of dissociation effects in the direct path integral Monte Carlo method. Experimental data on both shock and quasi-isentropic compression of deuterium are well described by calculations. Thus dissociation of deuterium molecules in these experiments together with interparticle interaction play significant role.

1. Introduction

During last decades significant efforts have been made for the investigation of thermophysical properties of dense plasma. The importance of this activity is mainly connected with the creation of new experimental facilities. Powerful current generators and ultrashort lasers are used for production of very high pressures which can not be reached in traditional explosive devices and light-gas guns. Such experiments give information about different properties of strongly coupled plasma which can be exploited for the creation of wide-range equations of state and allows one to verify various theoretical approaches and numerical methods.

In this work we calculate shock Hugoniots of gaseous deuterium of different initial density corresponding to experiments [1, 2] as well as compression isentrope. We use two different methods: direct path integral Monte Carlo (DPIMC) [3] and reactive Monte Carlo (REMC) [4]. Brief description of these approaches together with calculation details are given below.

2. Method and simulation details

2.1. Path integral Monte Carlo method

The main idea of the path integral Monte Carlo method is in representaion of a density matrix of a quantum system of particles and, consequently, thermodynamic properties, as path integrals. This first-principle method which can be applied at any coupling parameters and significant degeneracy parameters, takes into account Fermi statistics and spin effects [3]. It has been used for calculation of thermodynamic properties of hydrogen and hydrogen-helium plasmas including hypothetical phase transition region. We have also applied this method to fully quantum electron-hole plasma in semiconductors.

Dissociation of deuterium molecules occurs under compression and essentially influences thermodynamic properties; therefore one must take into account this effect. As the path integral

© 2008 IOP Publishing Ltd

Monte Carlo method operates with deuterons and electrons one should apply a criterion at which conditions a group of chosen deuterons and electrons can be considered as, for instance, an atom or molecule. A strict criterion of this kind is hardly possible to derive. However one can demonstrate the existence of bound states in calculations by analyzing the pair distribution functions. A comparison with chemical picture model [5] also gives good agreement which is an evidence of correctness of dissociation and ionization effects in DPIMC.

2.2. Reactive Monte Carlo method

Complete description of the REMC method can be found elsewhere [4, 6]. We consider only molecular hydrogen dissociation reaction and its inverse recombination reaction: $H_2 \Leftrightarrow 2H$. Ionization can be neglected at temperatures lower than the hydrogen ground state energy ($\propto$ 13.6 eV) and at moderate densities. Effective pair potentials between different sorts of particles have been considered by the Buckingham-EXP6 potential, corrected at small distances [7].

REMC simulations have been performed in the canonical ensemble for hydrogen and deuterium with a proper account of molecular dissociation. Accordingly, there are 3 types of Monte Carlo steps in a simulation box: particle movement, molecular dissociation into atoms and atoms recombination to a molecule. The expressions for probabilities of two last steps contain internal partition functions of atoms and molecules. All electrons in all particles are considered to be in the ground state. Further, the internal partition function of atom consists of the translational partition function only, for molecule besides it — of the rotational and vibrational partition functions. To calculate rotational and vibrational energy levels E_{nl} of a hydrogen molecule we solve the Schrödinger equation numerically in the central-symmetric field, described by Murrell-Sobie potential [8].

We calculate thermodynamic functions such as pressure P and configurational part of internal energy U^{int} by averaging over all Monte Carlo configurations [9].

To calculate an isentrope one should be able to determine entropy which is defined by the thermodynamic relation: $S = -(\partial F(T,V,N)/\partial T)_V$, where $F = \sum_{i=1}^{2} N_i\mu_i - PV$ — the free energy for a two-component system. We can obtain chemical potentials μ_i of each component in the mixture with the help of the test particle method [10]: $\mu = \mu^{id} + \mu^r$, where μ^{id} is the ideal gas chemical potential: $\mu^{id} = k_BT\log(\Lambda^3N/V)$. Here Λ is the de Broglie wavelength of a particle.

The residual chemical potential μ^r can be evaluated as [11]:

$$\mu^r = -k_BT\log[L_a\exp(-\Delta U/k_BT)].$$

Here ΔU — the change in configurational energy produced by the insertion of one additional particle, L_a — the ratio of allowed (nonoverlapped) insertion intervals along trajectories which traverse (parallel to any of the axes) the simulation box from side to side, to the length of the box [11]. The test particle is then inserted randomly into some point in the allowed intervals and the change in configurational energy ΔU is evaluated. The main advantage of this approach is the possibility of calculation of chemical potentials at high densities, where the usual test particle method tends to fail. Chemical potentials are calculated separately for atoms and molecules; the free energy is then determined as indicated above.

3. Results

Calculations were performed in a cubic simulation box with periodic boundary conditions, and with a cutoff radius equal to half of the box length. The initial configuration of particles was taken to be the fcc lattice for every input density $\rho = N_H m_H/V + N_{H_2}m_{H_2}/V$ with $N_H = N_{H_2} = 250$. The system was equilibrated by $2 \cdot 10^7$ steps and additional 10^7 steps were used for the calculation of thermodynamic values. Averaging of 20 blocks was used to calculate statistical error, which did not exceeded 2% for pressure and energy.

In Fig. 1 shown are three shock Hugoniots of gaseous deuterium with experimental conditions corresponding to those in experimental data: $\rho_0 = 0.1335$ g/cm^3, $P_0 = 1.57$ GPa (I), $\rho_0 = 0.153$ g/cm^3, $P_0 = 2.03$ GPa (II) [1]; $\rho_0 = 0.171$ g/cm^3, $P_0 = 2.72$ GPa (III) [2]. The initial energy in calculations by both methods was $E_0 = -15.875$ eV/atom and consisted of ionization energy of deuterium atom $I = -13.6$ eV/atom and half of dissociation energy of deuterium molecule $D_0/2 = -2.275$ eV/atom. Simulation parameters (density and temperature) in both approaches were chosen in such a way that the calculated values of pressure P and energy E satisfied the Hugoniot equation $E = E_0 + 0.5(P + P_0)(1/\rho_0 - 1/\rho)$; here index 0 denotes the values corresponding to the initial state.

It is worth to mention that at low pressures (up to 15 GPa) the calculated REMC shock adiabat is in very good agreement with light gas gun experiments [12]. All three shock Hugoniots agree with data [1,2] within experimental accuracy; adiabat III occupies intermediate position between points [13, 14] and [15]. As in the REMC method ionization was not taken into account calculations were performed up to temperatures 60 kK. On the other hand, computational difficulties prevented DPIMC from use at low temperatures. Thus, these two methods supplement each other; at intermediate pressures these two completely different approaches are in very good agreement (see Fig. 1).

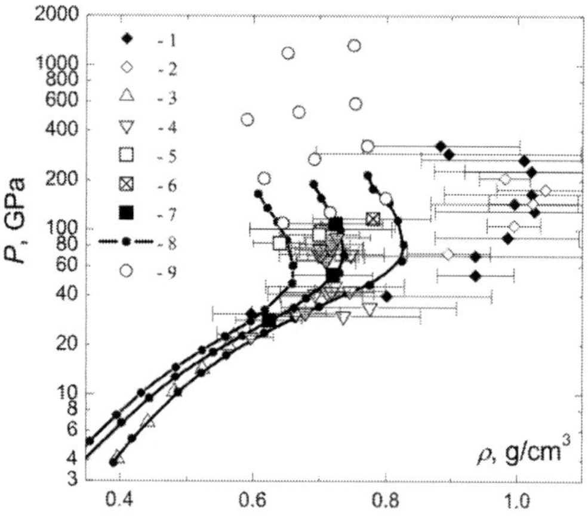

Figure 1. Shock Hugoniots of gaseous deuterium. Experiment: 1, 2 — [15] (NOVA); 3 — [12] (light-gas gun); 4 — [13] (Z-pinch); 5, 6 — [1] (hemisphere, gas); 7 — [14] (hemisphere, liquid). This work: 8 — REMC, 9 — DPIMC, from left to right $\rho_0 = 0.1335$ g/cm^3, 0.153 g/cm^3, 0.171 g/cm^3.

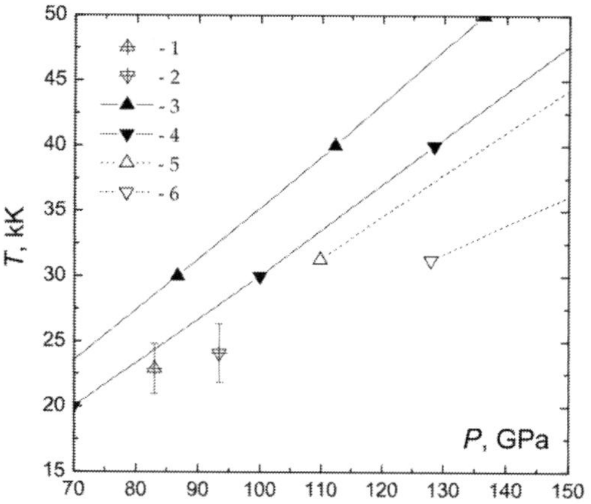

Figure 2. Temperature on shock Hugoniots of gaseous deuterium with initial densities 0.1335 g/cm^3 (triangles) and 0.153 g/cm^3 (down triangles). 1, 2 — experiment [1]; 3, 4 — REMC, this work; 5, 6 — DPIMC, this work.

In Fig. 2 we show the result of calculation of temperatures along 2 Hugoniot curves of deuterium in comparison with experimental data [1]. One can see that the REMC method shows a little overrated temperatures, whereas extrapolation of calculation by PIMC reveals good agreement with experiment [1] at low temperatures.

Dramatic rise of electrical conductivity of deuterium by 4–5 orders magnitude [16] at pressures ~1 Mbar and densities about 1 g/cm^3 indicates that one might also expect peculiarities in thermodynamic properties. Indeed, there are experimental results on quasi-isentropic compression of deuterium in cylindrical explosive chamber which show 30% density jump at pressure about 1.4 Mbar [2]. Using free energy of deuterium from simulations by REMC and Widom's test particle methods we calculated the compression isentrope of deuterium. The results are shown in Fig. 3 in comparison with the experimental data [2]. Excellent agreement proves that no special assumptions about phase transition are needed to explain such a remarkable behavior of experimental points and dissociation effects mostly contribute to this phenomenon. In the nearest future we plan to calculate the same isentrope by alternative computationally cheap Zel'dovich's method.

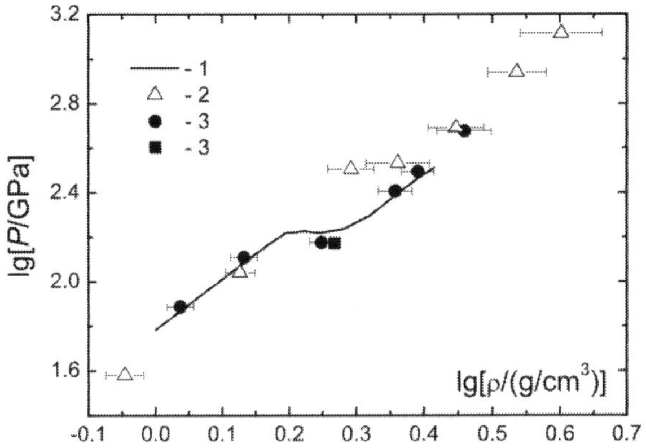

Figure 3. Quasi-isentropic compression of deuterium. This work: 1 — REMC. Experiment: 2 — [17], 3 — [2].

Conclusions and Acknowledgments

We want to conclude that new experimental data on shock and quasi-isentropic compression of deuterium are in good agreement with calculations by two different Monte Carlo methods. The coincidence between two methods at intermediate pressure indicate that the direct path integral Monte Carlo method describes dissociation effects sufficiently well. Moreover, good agreement between REMC calculations of deuterium quasi-isentrope shows that 30% density jump is caused by dissociation of deuterium molecules.

We thank Dr. Igor Iosilevskiy for valuable discussions. This work was supported by the Council on Grants from the President of the Russian Federation, Project No. NSh-3683.2006.2.

References

[1] Grishecheckin S K, Gruzdev S K, Gryaznov V K *et al.* 2002 *JETP Letters* **76** 433–5
[2] Fortov V E, Ilkaev R I, Arinin V A *et al.* 2007 *Phys. Rev. Lett* **99** 185001
[3] Zamalin V M, Norman G E and Filinov V S 1977 *The Monte-Carlo Method in Statistical Thermodynamics* (Moscow: Nauka)
[4] Johnson K J, Panagiotopoulos A Z and Gubbins K E 1994 *Mol. Phys.* **81** 717
[5] Filinov V S, Levashov P R, Bonitz M and Fortov V E 2005 *Contrib. Plasma Phys.* **45** 258–265
[6] Bezkrovniy V, Schlanges M, Kremp D and Kraeft W D 2004 *Phys. Rev. E* **69** 061204
[7] Juranek H and Redmer R 2000 *J. Chem. Phys.* **112** 3780–3786
[8] Liu L X, Su W L *et al.* 2002 *Int. J. Quantum Chem* **87** 1–11
[9] Allen M P and Tildesley D J 1987 *Computer Simulation of Liquids* (Oxford: Clarendon)
[10] Widom B 1963 *J. Chem. Phys.* **39** 2808
[11] Arrieta E, Jedrzejek C and Marsh K N 1991 *J. Chem. Phys.* **95** 6838
[12] Nellis W J, Mitchell A C, van Theil M, Devine G J, Trainor and R J Brown N 1983 *J. Chem. Phys.* **79** 1480
[13] Knudson M D, Hanson D L *et al.* 2004 *Phys. Rev. A* **69** 144209
[14] Boriskov G V, Bykov A I, Il'kaev R I *et al.* 2005 *Phys. Rev. B* **71** 092104
[15] Collins G W Da Silva L B, Celliers P *et al.* 1998 *Science* **281** 1178–1181
[16] Weir S T, Mitchell A C and Nellis W J 1996 *Phys. Rev. Lett.* **76** 1860–1863
[17] Grigor'ev S B, Kormer S B, Mikhailova O L, Tolochko A P and Urlin V D 1972 *JETP Lett.* **16** 286–290

Solid methane behaviours under high pressure at room temperature

H Hirai[1], K Konagai[1], T Kawamura[2], Y Yamamoto[2], T Yagi[3]
[1] University of Tsukuba, Tsukuba, Ibaraki, 305-8572, Japan.
[2] National Institute of Advanced Industrial Science and Technology, Tsukuba, Ibaraki, 305-8569, Japan.
[3] Tokyo University, Kashiwanoha, Kashiwa, Chiba, 277-8581, Japan.

Abstract. High pressure behaviours of solid methane were examined in a pressure range of 0.5 GPa to 86 GPa at room temperature using diamond anvil cell. In-situ X-ray diffractometry and Raman spectroscopy revealed existence of two high pressure phases above 35 GPa. The Raman spectra showed clear peak splitting and appearance of new vibration modes at 35 GPa and 62 GPa. The XRD patterns of the high pressure phases were basically similar to those of the known phases B and HP. These results indicate that the observed high pressure phases have a similar fundamental structure but have different molecular orientations. In addition, an intermediate metastable phase was found between phases A and B. This phase can be thought as that: the configuration of the methane molecules almost attains as that in phase B, and the rotation of molecules still remains as that in phase A.

1. Introduction

Solid methane exhibits a wide variety of phases due to molecular orientational ordering. Many phases with different molecular orientations were reported in low temperature region [e.g. 1-5]. After that, also in high pressure region at room temperature, X-ray diffractometry using synchrotron radiation and Raman spectroscopy revealed phase changes as follows. Fluid methane crystallizes to phase I (cubic) at 1.7 GPa [6,7], phase I transforms to phase A (rhombohedral or tetragonal) at 5.2 GPa [3,8,9], phase A to phase to B (cubic or hcp) at 10-18 GPa [3,10], phase B to a high-pressure phase (HP) (cubic or hcp) at 25 GPa [4,10], HP retains at least up to 37 GPa 10]. Recently, an optical study made up to 288 GPa reported change of refractive index suggesting insulator-semiconductor transformation [11]. Under high temperatures and high pressure, a few shock compression studies reported dissociation of the methane molecule to form diamond at 20 GPa and 2000 K [12], while a theoretical study predicted polymerization of solid methane to produce ethane and propane at 100 to 300 GPa and 4000 K rather than dissociation of the methane molecule [13]. An experiment using laser-heating diamond anvil cells observed the formation of both polymerized hydrocarbons and diamond at 19 GPa and 2000 to 3000 K [14]. Thus, solid methane behaviours under high pressure have not yet been clarified both at room temperature and high temperatures.

In this study high pressure experiments at room temperature of solid methane were carried out using combined Raman spectroscopy with X-ray diffractometry (XRD), and existence of another two high-pressure phases were shown above 35 GPa. And, an intermediate metastable phase was found

© 2008 IOP Publishing Ltd

between the known phase A and phase B. The nature of the intermediate phase was discussed in related to the sluggish transition between phases A and B.

2. Experimental

A lever-and-spring type diamond anvil cell was used in the high-pressure experiments. The ruby fluorescence method was used for the pressure measurements. The accuracy of the present measurement system is 0.1 GPa, taking the resolution of the spectrometer and the analytical procedure into account. The samples of methane were filled into the gasket hole by a gas loading vessel in which methane gas (99.99%) was compressed to supercritical fluid at approximately 0.15 GPa at room temperature (Critical point of methane is 190.6K, 4.60MPa, so the filling condition is above the point). The XRD study was performed using synchrotron radiation on BL-18C and BL-13A at the Photon Factory, High Energy Accelerator Research Organization (KEK). A monochromatized beam with a wavelength of 0.6198 Å was used. XRD and Raman studies and optical observations were conducted for twelve samples at room temperature in a pressure range from 0.5 to 86 GPa.

3. Results

Figure 1 shows a series of optical micrographs with pressure change. The microscopic observation detected crystallization from fluid at 1.7 GPa and the transition at 5.2 GPa. Above the pressure range up to 86 GPa any transition or change in refractive index was not observed with increasing and decreasing pressure by a conventional optical observation (Figure1). The XRD study

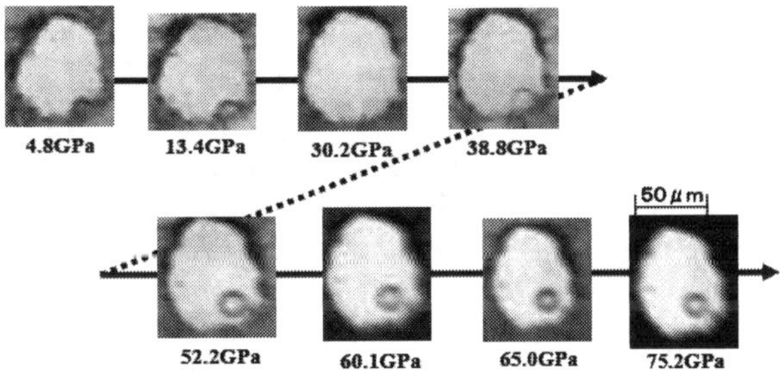

Figure 1. Optical micrographs with pressure change.

confirmed crystallization to phase I (fcc) at 1.7 GPa and the transition from phase I to phase A (rhombohedral) at 5.2 GPa. The transition from phase A to B was very sluggish as reported before [2,3]. The process of the transition was traced by peak splitting of a C-H symmetric stretching mode of the methane molecule, ν_1, in the Raman spectra as reported previously by IR spectroscopy [3]. After confirming completion of the A-B transition by Raman and XRD, higher pressure was loaded to the samples. Figure 2(a) shows changes in XRD patterns with pressure after the A-B transition. In the XRD patterns of phase B, five distinct diffraction lines were observed, which agreed with the previous XRD study [10]. The XRD lines were indexed as a simple cubic system with a lattice parameter of a= 8.047 Å at 17.4 GPa using a software indexing XRD pattern, LTC. Deviations between the observed d-values and the calculated ones were very small within 0.1 %. At higher pressures above the A-B transition, the diffraction patterns were basically unchanged up to 86 GPa, so the XRD patterns were also indexed as the same simple cubic lattice. But, above 63.5 GPa, several diffraction lines shift slightly to lower angle, comparing with the continuous shift of those from Re gasket to higher angle (Figure 1(a)). The result suggested some structural change at approximately 62 GPa. Fig. 2(b) shows variation of volume as the simple cubic lattice with pressure change. In the figure 2(b), changes in slope of volume with pressure were not clearly distinguished at approximately 35 GPa and 62 GPa. That is similar to that of the B-HP transition at 25 GPa reported by Umemoto et al. [10]. On the other hand, distinct changes were observed by Raman spectroscopy at these pressures, as described below.

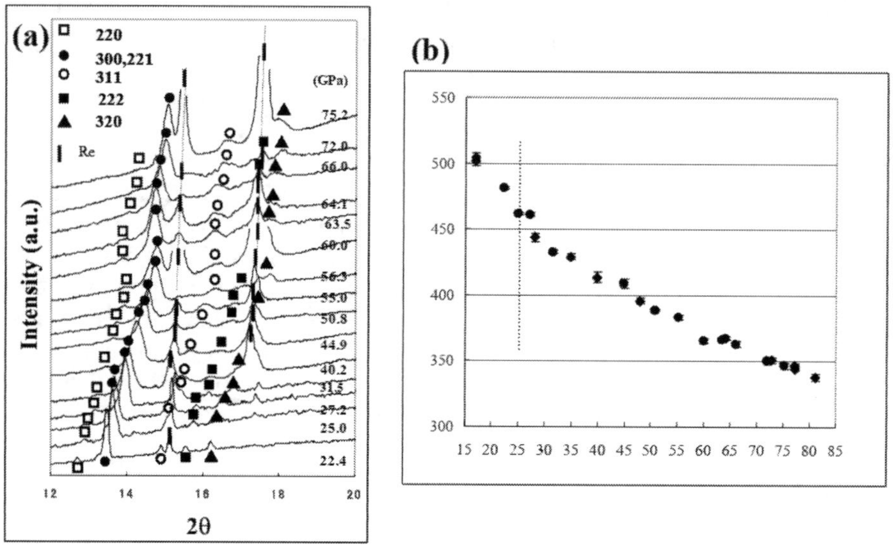

Figure 2. (a) XRD patterns with pressure change. (b) Variations of volume with pressure.

Raman spectra of C-H symmetric stretching modes of methane molecules, ν_1 and ν_3, with pressure change are given in Figure 3(a, b). The variations in the Raman frequency shift of these vibration modes with pressure change are shown in Figure 3(c). Peak splitting and changes in peak intensities were clearly observed at approximately 35 GPa and 62 GPa in Figure 3(a, b). The changes of each peak were easily traced throughout the pressure range in Figure 3(c). In addition, changes in slope of the Raman frequency shift with pressure were distinguished at 25 GPa, 35 GPa and 62 GPa. All changes observed in increasing pressure reverted in decreasing pressure. Above phase B, the transitions proceed in a relatively shorter time than the A-B transition.

The transition from Phase A to B was sluggish. At 17.4 GPa the transition started 4 hours after the pressure had reached this value and the transition was completed after 10 hours. In the case of 12.7 GPa, more than 10 hours and heating at about 330 K were necessary to accomplish this transition. The process of change had been detected by IR spectroscopy [3]. In this study, the process of change was tracked by XRD combined with Raman spectroscopy. The XRD study revealed the existence of an intermediate phase, or rather, a "state", having another diffraction pattern different both from phase A and phase B. This phase is tentatively referred to in this paper as phase pre-B. Figure 4 (a) shows the XRD patterns of phases A, pre-B and B. The diffraction pattern of phase pre-B was totally different from A, but rather related to that of phase B. Four diffraction lines were observed, which were indexed as a simple cubic system. The lattice parameter for phase pre-B was almost similar but a little smaller than that of phase B. The diffraction lines that appeared and their intensities were different from those of phase B. This indicated that the fundamental structure of phase pre-B is similar or closely related to that of phase B. The transition from phase A to pre-B occurred within a short time, approximately one hour after increasing pressure; on the other hand, the transition from phase pre-B to B was sluggish, as described above. During peak splitting observed in Raman spectra (Figure 4 (b)), both diffraction patterns of phase pre-B and B were superimposed. At the completion of the transition, only the pattern of phase B was observed. When phase pre-B was compressed without transition to phase B, the diffraction patterns were basically unchanged up to 81 GPa, and the transition to phase B did not occur (Figure 5(a)). The Raman spectrum of phase pre-B showed a single peak each for ν_1 and ν_3, which was similar to that of phase A (Figures 4(b), 5(b)). This spectrum was retained up to 78 GPa without any peak splitting (Figure 5(b)).

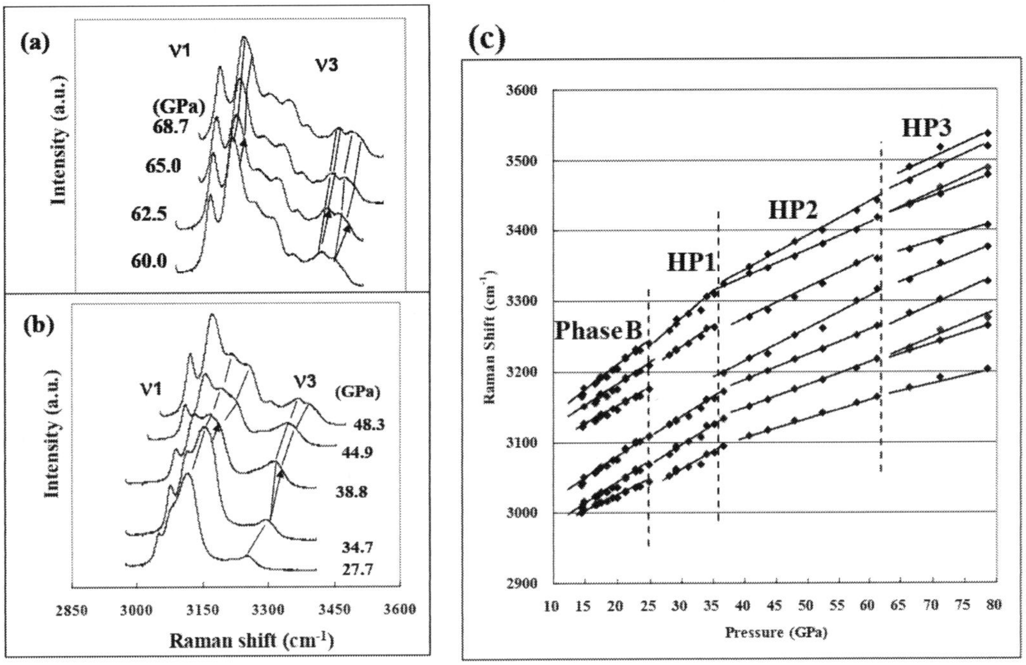

Figure 3. (a, b) Raman spectra with pressure change. (c) Variation of Raman frequency shift.

4. Discussion

The phase changes of solid methane were observed at approximately at 5.2, 12-18, 25, 35, 62GPa in the present XRD and Raman studies. The former three changes correspond to the known ones as one from phase I to A, from phase A to B and from phase B to HP, respectively. Thus, existence of two additional high pressure phases above 35 GPa up to 86 GPa were revealed in this study. The phase above 25 GPa was previously named phase HP [3, 10], thus the high-pressure phases above 25 GPa were tentatively called HP1, HP2 and HP3 in the order of pressure in this paper (Figure 3(c)). The phase changes at 25 GPa, 35 GPa and 62 GPa were clearly recognized by the changes in the intramolecular vibration modes by Raman spectroscopy. On the other hand, by the X-ray diffractometry they were detected by the relatively slight changes such as the peak shifts at about 62 GPa. Therefore, the observed phase changes above phase B are thought to be caused by the changes in the state of the methane molecules accompanied by no fundamental structural change.

The determination of high-pressure structures has long been an issue [3-5, 10, 15]. The structure for phase I has been established [5], while different reports for structures above phase A exist. The detailed IR and Raman studies by Bini et al. demonstrated that the structure evolved from fcc (phase I) to hcp (phase B and HP) through a tetragonal system (phase A) [3, 4]. On the other hand, the XRD studies showed a rhombohedral structure for phase A [9] and a cubic structure for phases B and HP [10]. The XRD patterns we obtained for phases, B, HP1, HP2 and HP3 were similar to each other and they were indexed as the same simple cubic lattice. Thus, the fundamental structures, i.e. the positions of the molecules, of these phases are thought to resemble to each other. On the other hand, the intramolecular vibration modes were considerably different among these phases. The changes observed in Raman spectra from phase B to HP1, HP2 and HP3 can be interpreted as methane molecules showing different manners of orientational ordering in one phase to another in a similar fundamental structure.

As for the sluggish A-B transition, Bini et al. described, on the basis of their detailed IR studies, that complex reconstruction of molecules may be performed at the A-B transition, thus a prolonged

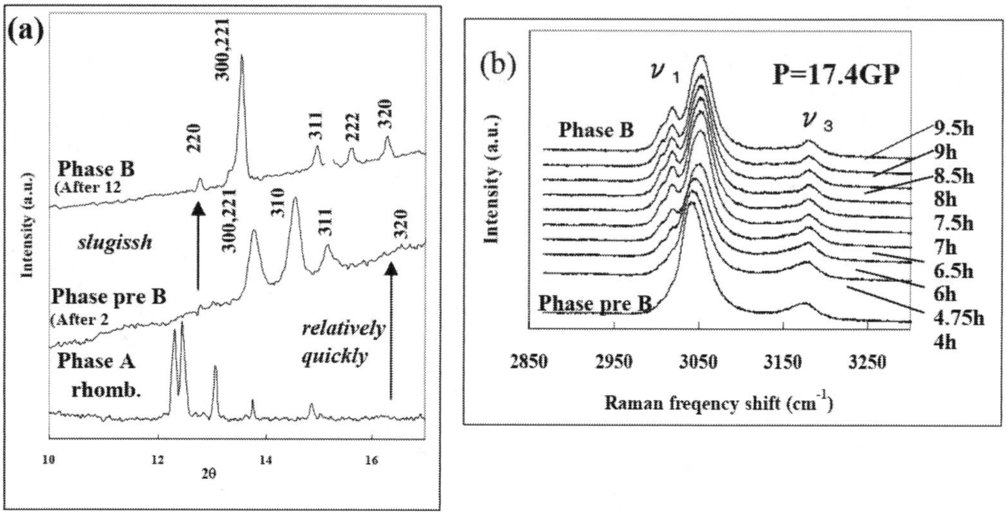

Figure 4. (a) XRD patterns of phases of A, pre-B and B. (b) Raman spectra from phase pre-B to B

period of time is required for it [4]. In our study we found an intermediate metastable phase, pre-B, between phases A and B. The transition from A to pre-B was accomplished relatively quickly, unlike the transition from phase pre-B to B, which was very sluggish. The XRD pattern of phase pre-B was similar to that of phase B, indicating that the fundamental structure might be somewhat related to that of phase B. This means that broad configurations of methane molecules as phase B might already be built up at the state of phase pre-B, i.e., immigration of the molecules might be almost completed in a relatively short time span. A longer period might be required for the molecules to arrange themselves into suitable and stable orientations at each molecular position. The phase pre-B can be explained as follows: the configuration of the methane molecules almost attains as that in phase B, and the rotation of molecules still remains as that in phase A without establishing an orientation as observed in phase B.

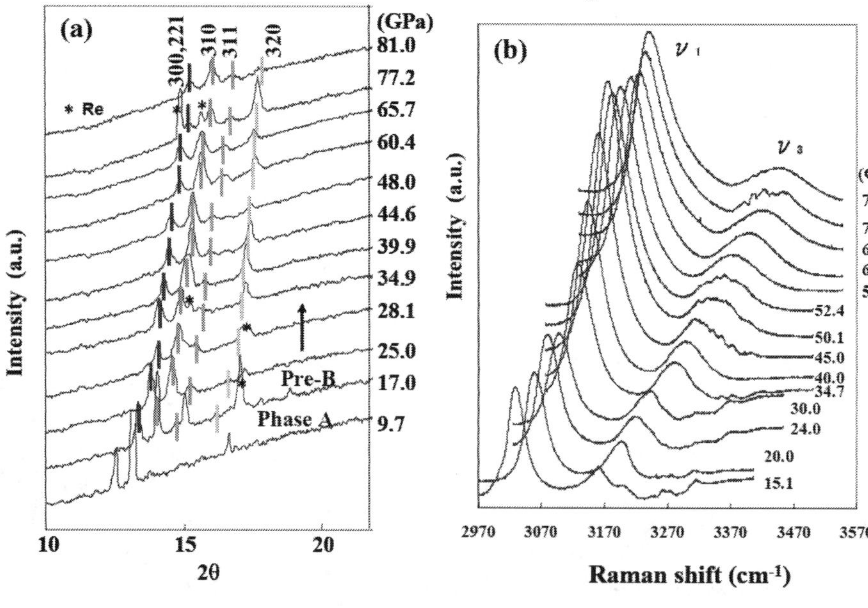

Figure 5. (a) XRD patterns and (b) Raman spectra of phase Pre-B with pressure change.

The phase changes of solid methane are rather different from analogous materials such as solids CF_4 and CCl_4. Solid CF_4 transforms at 1.86 GPa from a rhombohedral structure to C2/n and at 2.86 GPa it goes to $P2_1/c$ at room temperature [16, 17]. Solid CCl_4 transforms at 0.4 GPa from a rhombohedral or fcc structure to C2/c and at 0.7 GPa to $P2_1/c$ [18, 19]. These phase changes of the analogous materials proceed at lower pressures. The high-pressure structures, monoclinic structures, differ from those of solid methane, although the lower-pressure structures, fcc and rhombohedral, are the same. Methane molecules are small and can be regarded as almost spherical in shape rather than the CCl_4 and CF_4 molecules. Electron negativities differ among these constituent atoms, which results in different multipole interactions among these tetrahedral-shaped molecules. Such differences in the chemical and physical properties of the methane molecule might lead to the different phase change behaviour from the analogous materials.

References

[1] Press W J 1972 *Chem. Phys.* **56** 2597
[2] Herbert P, Polian A, Loubeyre P and Toullec R Le, 1987 *Phys. Rev. B* **36** 9196
[3] Bini R, Ulivi L, Jodl H J and Salvi P R 1995 *J. Chem. Phys.* **103** 1353
[4] Bini R and Pratesi G 1997 *Phys. Rev. B* **55** 14800
[5] Thiery M –M, Fabre D and Kobashi K 1985 *J. Chem. Phys.* **83** 6165
[6] Hazen R M, Mao H K, Finger L W and Bell P M 1980 *Appl. Phys. Lett.* **37** 288
[7] Yagi T and Suzuki H 1990 *Proc. Jpn. Acad. B* **66**167
[8] Wu Y H, Sasaki S and Shimizu H 1995 *J. Raman spectroscopy* **26** 963
[9] Nakahata I, Matsui N, Akahama Y and Kawamura H, 1999 *Chem. Phys. Lett.* **302** 359
[10] Umemoto S, Yoshii T, Akahama Y and Kawamura H 2002 *J. Phys. Cond. Mattr.* **14** 10675
[11] Sun L, Ruoff A L, Zha C-S and Stupian G 2006 *J. Phys. Chem. Solid* **67** 2603
[12] Nellis W J, Ree F H, van.Thiel M and Mitchell A C 1981 *J. Chem. Phys.* **75** 3055
[13] Ancilotto F, Chiarotti G L, Scandolo S and Tosatti E 1997 *Science* **275**128.
[14] Benedetti L R, Nguyen J H, Caldwell W A, Liu H, Kruger M and Jeanloz R 1999 *Science* **286** 100
[15] El Kader M S A 2002 *Chem. Phys.* **277** 77
[16] Fitch A N and Cockroft K 1993 *Z. Kristallogr.* **203** 29
[17] Sasaki S, Ikeda Y and Shimizu H 1991 *J. Phys. Soc. Jpn.* **60** 1560
[18] Cohen S, Powers R and Post R 1979 *Acta Cryst. B* **35** 1670
[19] Piermarini G J and Braun A B 1973 *J. Chem. Phys.*, **58** 1974

Journal of Physics: Conference Series **121** (2008) 102002 doi:10.1088/1742-6596/121/0/102002

Temperature dependence of EXAFS Debye-Waller factor in the high pressure perovskite $SrGeO_3$

Takafumi Hashimoto[1]*, Akira Yoshiasa[1], Masahiko Sugahara[1], Hiroshi Arima[2], Hiroshi Fukui[3], Kei-ichiro Murai[4] and Maki Okube[5]

[1]Graduate School of Science and Technology, Kumamoto University, Kumamoto 860-8555, Japan

[2]Graduate School of Science, Osaka University, Osaka 560-0043, Japan

[3]Institute for Study of the Earth's Interior, Okayama University, Misasa, Tottori, 682-0193, Japan

[4]Department of Advanced Materials, Institute of Technology and Science, The University of Tokushima, 21 Minami-josanjima, Tokushima, 770-8506, Japan

[5]Materials and Structures laboratory, Tokyo Institute of Technology, 4259, Nagatsuta, Yokohama, 226-8503, Japan

E-mail: takafumi@es.sci.kumamoto-u.ac.jp

Abstract. The temperature dependence of EXAFS Debye-Waller factors in $SrGeO_3$ perovskite, which synthesize in a cubic anvil type apparatus under 9GPa and 1000K, was investigated with the cumulant expansion method. The Sr and Ge K-edge EXAFS spectra were measured in transmission mode at various temperatures from 30 to 300K. Effective pair potentials, $V(u) = \alpha u^2/2$, were evaluated by harmonic model, the potential coefficient α for Sr-O, Ge-O, Ge-Sr, Sr-Sr and Ge-Ge distances in $SrGeO_3$ perovskite were 2.6 e/VÅ^2, 10.3 e/VÅ^2, 6.5 e/VÅ^2, 4.3 e/VÅ^2 and 8.6 e/VÅ^2, respectively. Ge-O bond has the largest value. The coefficients for Sr-O bond and Sr-Sr distances are small values in composition with other cubic perovskite-type compounds. It is suggested that the amorphouszation occurs at relatively lower temperature of 500K in $SrGeO_3$ perovskite because these values are very small. In $A^{II}+B^{IV+}O_3$-type perovskite, the potential coefficient for the bond distance which associated B site atom is larger than that for the others distances and this tendency is also seen in high pressure perovskite such as $CaGeO_3$.

1. Introduction

Germanate compounds as analogue of silicate minerals are very useful, because germanate has the same structure of silicate high pressure phase at much lower pressures. $SrGeO_3$ perovskite is analogue material of $CaSiO_3$ perovskite and $MgSiO_3$ perovskite, which are thought as major constituent minerals in lower mantle. Thus, the detail study for structural and thermal properties of perovskite

© 2008 IOP Publishing Ltd

would provide useful information of the lower mantle. The previous crystal structure analyses of $SrGeO_3$ perovskite was carried out by means of powder X-ray diffraction [1], and then X-ray diffraction and X-ray absorption fine structure (XAFS) observations were performed to investigate the back-transformation at high temperature through amorphous phase of $SrGeO_3$[2]. In this study, we have studied thermal property of $SrGeO_3$ perovskite such as thermal vibration of local structure by XAFS.

2. Experimental and analysis

We have carried out the synthesis of $SrGeO_3$ perovskite at high pressure and high temperature by a cubic anvil type apparatus at 8GPa-1000K with wollastonite type $SrGeO_3$ as starting material. The perovskite-type $SrGeO_3$ crystal was identified by X-ray diffraction. The appropriate amount of fine powder sample and boron nitride powder was mixed and pressed into pellet of <0.2 mm in thickness and 10.0 mm in diameter. The sample had edge-jump with 0.7 ($\Delta\mu d$), where μ is the linear absorption coefficient and d is the thickness. The measurements of Sr and Ge K-edge EXAFS spectra were carried out in the transmission mode at beam line BL-9C of the Photon Factory in KEK, Tsukuba (Proposal No. 2006G082). All X-ray absorption measurements in the temperature range from 30K to 300 K were made under a helium atmosphere. The synchrotron radiation was monochromatized by a Si(111) double crystal monochromator. Details of the measurement and analysis were given in reference [3, 4].

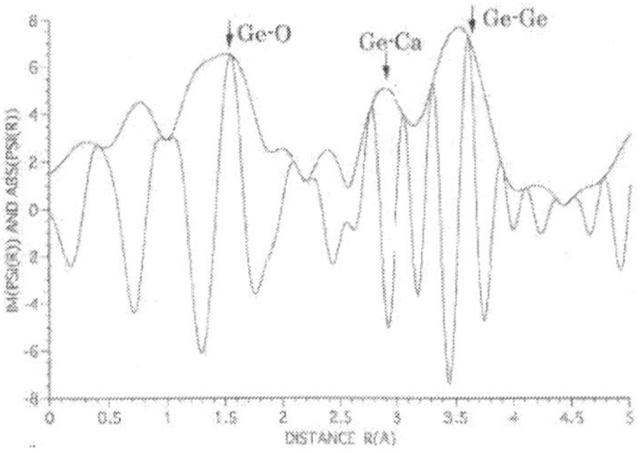

Figure 1. Fourier transforms of the Ge K-edge EXAFS for $SrGeO_3$ perovskite at 270K. No phase shift corrections are made.

The EXAFS interference function, $\chi(k)$, was extracted from the measured absorption spectra using the standard procedure [5]. Figure 1 shows examples of Fourier transforms of the Ge and Sr K-edge EXAFS spectra in range of $3.5< k <14$. In quantitative analysis, we carried out the Fourier-filtering technique and a nonlinear least-squares fitting method by comparing the observed $\chi(k)_{exp}$ and calculated $\chi(k)_{calc}$. We used the EXAFS formula in the single scattering theory with the cumulant expansion up to the third order term. Because the third-order term was negligible, the refinement was performed with harmonic model. The harmonic approximation requires a cumulant up to only second term. The reliability of fit parameters,

$$R = \sum_{S} \left| k_s^3 \chi(k_s)_{exp} - k_s^3 \chi(k_s)_{calc} \right| / \left| k_s^3 \chi(k_s)_{exp} \right|$$

between the experimental and calculated EXAFS function was less than 0.03. The quality of fit between $k^3\ \chi_{exp}$ and $k^3\ \chi_{calc}$ is shown in Fig. 2.

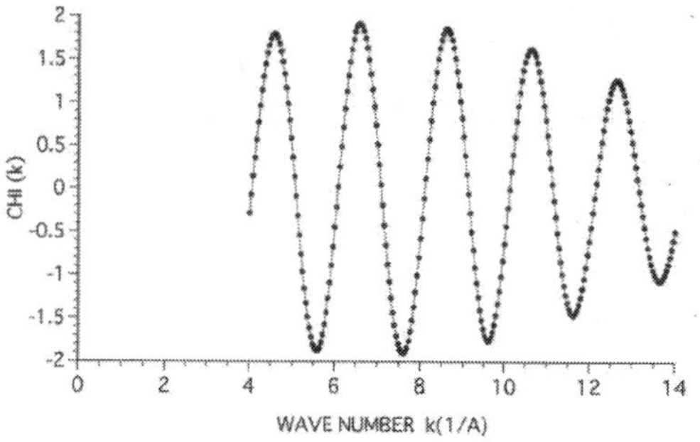

Figure 2. Fourier filtered EXAFS spectra (dotted curve) and least-squares fit (solid curve) for the first nearest Ge-O distances in $SrGeO_3$ perovskite at 270K.

3. Result and discussion

Figure 3 shows the temperature dependence of the EXAFS Debye-Waller factor for (a) Sr-O and Ge-O (b) Sr-Sr and Ge-Ge in $SrGeO_3$ perovskite. The Debye-Waller factor σ2 includes the effects of the static disorder and dynamic disorder. The static disorder is the configuration disorder, while the dynamic disorder is caused by thermal vibration of atoms [6]. In addition, there is a correlation in calculation between σ2 and the mean free path of the photoelectron. If we assume that straight line with the same gradient passes through the origin, the deviation of the absolute value is attributed to static disorder and to correlation between σ2 and mean free path The contribution of the thermal vibration, $\sigma_{thermal}$, can be estimated under the assumption of the classical statistical dynamics by the temperature dependence of σ2. A steep slope of the temperature dependence of this parameter represents a weak bonding. And the gradient for the obtained experimental σ2 is equal to k_B/α, where k_B is the Boltzmann constant and α is the potential coefficient. We can evaluate the harmonic effective pair potential $V(u) = \alpha u^2/2$, where u is the deviation of the bond distance from the location of the potential minimum.

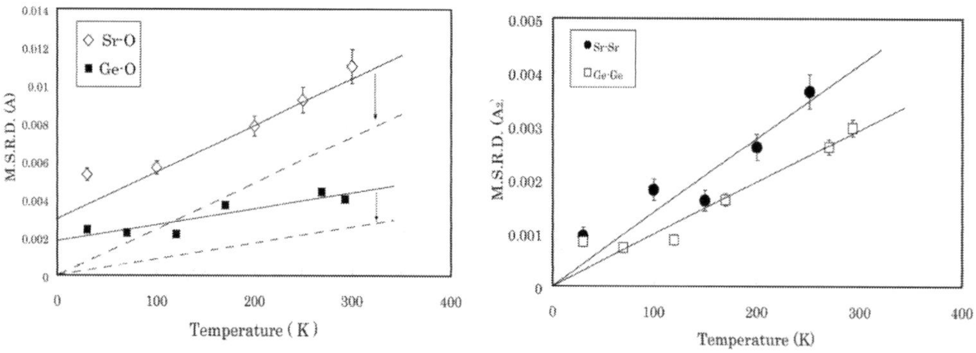

Figure 3. Temperature dependence of the EXAFS Debye-Waller factor for (a) Sr-O and Ge-O (b) Sr-Sr and Ge-Ge in $SrGeO_3$ perovskite. The deviation (arrows) is attributed to the static disorder.

Calculated potential parameters for $A^{II+}B^{IV+}O_3$-type perovskite $SrGeO_3$, $CaTiO_3$ and $CaGeO_3$ [8], are listed in Table 1. The potential constants α for Sr-O, Ge-O, Ge-Sr, Sr-Sr and Ge-Ge distances in $SrGeO_3$ perovskite are 2.6 eV/Å^2, 10.3 eV/Å^2, 6.5 eV/Å^2, 4.3 eV/Å^2 and 8.6 eV/Å^2, respectively. Ge-O bond has the largest value and then Ge-Ge distance. These coefficients for Sr-O bond and Sr-Sr distance are very small values in composition with other cubic perovskite-type compounds. If a certain bond or cation-cation distance has small value, the bond is soft or large correlation displacement [7].

The amorphouszation occurs at relatively lower temperature of 500K in $SrGeO_3$ perovskite might be caused by such a small values associated with Sr. It is interesting to show that the gradients of σ2 for the Sr-O distances are larger than those for the Ge-Ge distances in $SrGeO_3$ (Fig. 3): the potential coefficients for the Sr-O distances are smaller than those for the longer Ge-Ge distances though the potential coefficient decreases usually as a result of the larger bond distance. The A cation-Oxygen bonds in perovskite type compounds have relatively larger mean square amplitudes of vibration than those for B cation-B cation distances as judged by each value of potential coefficient. In $A^{II+}B^{IV+}O_3$-type perovskite, the potential coefficient for the bond distance which associated with B site atom is larger than that for the others distances and this tendency is also seen in high pressure perovskite such as $CaGeO_3$.

Table 1. Calculated potential parameters, α (eV/ $Å^2$), for the A-O, B-O and B-B distances in ABO_3 perovskite compounds

	$SrGeO_3$*	$CaGeO_3$**	$CaTiO_3$**
A-O	2.6	4.9	4.3
B-O	10.3	9.8	6.9
B-B	8.6	7.4	8.0

*this study, **Yoshiasa et al.[8]

References

[1] Shimizu, Y., Syono, Y. and Akimoto, S. (1970) *High Temperature-High Pressure,* **2**, 113-120.

[2] D. Andrault, J.-P. Itie, and F. Farges (1996) *American Mineralogist*, **81**, 822-832.

[3] Yoshiasa, A., Koto, K., Maeda, H. and Ishii, T. (1997) *Jpn. J. Appl. Phys.*, **36**, 781-784

[4] Yoshiasa, A., Tamura, T., Kamishima, O., Murai, K., Ogata, K. and Mori, H. (1999) *J. synchrotron rad.*, **6**, 1051-1058

[5] Maeda, H. (1987) *J. Phys. Soc. Jpn.*, **56**, 2777-2787

[6] Willis, B.T.M. and Pryor, A.W. (1975) Thermal vibrations in crystallography, Cambridge University Press.

[7] Yoshiasa, A., Nagai, T., Ohtaka, O., Kamishima, O. and Shimomura, O. (1999) *J. synchrotron rad.*, **6**, 43-49

[8] Yoshiasa, A., Nakajima, K., Murai K. and Okube, M. (2001) *J. synchrotron rad.*, **8**, 940-942

Journal of Physics: Conference Series **121** (2008) 112001 doi:10.1088/1742-6596/121/1/112001

Pressure effects on catalytic properties and structural stability of human paraoxonase

C Cléry-Barraud, J Leva, N El Bakdouri, F Renault, P Masson and D Rochu

Département de Toxicologie, Centre de Recherches du Service de Santé des Armées, BP 87, 38702 La Tronche cedex, France

E-mail : cclerybarraud@crssa.net

Abstract. Human paraoxonase (HuPON1) is a candidate as catalytic bioscavenger for pre-treatment and therapy of poisoning by organophosphate compounds. HuPON1 is a hydrophobic protein associated with a partner, the human phosphate binding protein (HPBP) in plasma high density lipoproteins. The relationship between the composition and the size of multimeric states of HuPON1 is not well understood. Moreover the effect of HPBP's presence on enzyme catalytic mechanisms and stability is unclear. We investigated the effect of hydrostatic pressure and temperature on structural stability and activity of different PON1 preparations (hybrid recombinant PON1, natural HuPON1 free of its partner or in the presence of 50% w/w HPBP). We showed that PON1 exists under several multimeric forms and that the binding of HPBP amends the size of the hetero-oligomeric states and exerts a stabilizing effect on the activity of PON1.

1. Introduction

Human paraoxonase (HuPON1; EC 3.1.8.1) is a 354-amino acid calcium-dependent protein associated with high-density lipoprotein (HDL) in plasma. HuPON1 is also tightly associated with Human Phosphate Binding Protein (HPBP). Little progress have been made in identifying natural substrate(s) of HuPON1 and defining its physiological function(s). HuPON1 hydrolyses a broad range of esters such as arylesters, lactones and organophosphates (OP). This promiscuous enzyme with antiatherogenic and detoxification properties is considered as a promising pharmaceutical catalytic scavenger [1], and is the subject of intensive research. The three dimensional structure of HuPON1 is still unknown, but a monomeric crystal of a bacterially-expressed mouse-rat-rabbit-human PON1 chimera obtained through gene shuffling experiments recently provided the structure of a recombinant PON1 (rePON1) [2]. Besides, the X-ray structure of HPBP was also solved [3]. Recent experiments suggested that HuPON1 and HPBP exist in different oligomeric states, but the relevance of this association on mechanisms and stability is unclear. High pressure (HP) is a suitable and efficient tool for dissociating complex biological systems and perturbing macromolecule conformation [4]. In the present work, we have undertaken the study of HuPON1-HPBP molecular interactions using high pressure biochemical methods to determine the content of multimeric complexes of HuPON1, and to clarify the contribution of HPBP in activity/stability of HuPON1.

© 2008 IOP Publishing Ltd

2. Materials and methods

2.1. Enzymes and chemicals

HuPON1 from pooled plasma was purified to obtain free HuPON1 or HuPON1 in the presence of 50% HPBP (HuPON1/HPBP) [5]. The enzyme was stored at –20 °C in 25 mM Tris/HCl, 1 mM $CaCl_2$, 0.1 % Triton-X100, pH 8. The recombinant PON1 (rePON1) dubbed G3C9-8His was kept at 4°C in Tris 50 mM, NaCl 50 mM, $CaCl_2$ 1 mM, tergitol 0.1%. The protein differs by ≈ 16% of amino acids from the human enzyme.

2.2. Structural stability studies

The equipment of electrophoresis under HP was described elsewhere [6]. The empiric relationship of Ferguson describes the protein relative mobility (m) *vs* acrylamide concentration (T%) in non-denaturing polyacrylamide gel electrophoresis (PAGE):log m = log m_0 $-K_R$ x T% where K_R, the retardation coefficient, depends of the protein size and shape. For a spherical protein, K_R is related to the equivalent molecular radius [7].

2.3. Functional stability studies

Enzymatic parameters (k_{cat} and K_m) with phenylacetate (arylesterase activity) and paraoxon (OP-hydrolase activity) were determined by Michaelis-Menten analysis. Kinetic data with coumaranone (lactonase activity) were fitted to the Michaelis-Menten equation for $[S]<<K_m$. We determined the apparent activation volume ($\Delta V^{\#}_{k_{cat}/K_m}$) with $\Delta V^{\#}_{k_{cat}/K_m}$ = -RT x (δln (k_{cat}/K_m)/ $\delta P)_T$, where R is the gas constant (82 ml. atm. K^{-1}.mol^{-1}).

3. Results and discussion

3.1. Oligomeric states in PON1 preparations at atmospheric pressure

The estimation of the molecular weight (MW) of unknown samples referring to a calibration curve generated with standard proteins was obtained by plotting Ln (MW) versus $K_R^{1/2}$. HuPON1 exists in multiple oligomeric forms and HPBP associates with HuPON1 to form large hetero-oligomers retaining arylesterase activity. At atmospheric pressure, dimeric, trimeric and tetrameric HuPON1-containing forms were present in free HuPON1 preparation, whereas dimeric, tetrameric and hexameric forms coexisted in HuPON1-HPBP sample.

3.2. Pressure and temperature effects on structural stability of PON1

As shown in figure 1, PON1 structure was insensitive to pressure (up to 180-200 MPa) and temperature (up to 65°C) in all preparations (except free HuPON1 above 40°C). Stabilizing role of HPBP towards heat, already shown by capillary electrophoresis at varying temperature and differential scanning calorimetry [8], was confirmed by studying the pressure/temperature effects. Pressure (from 50 MPa) (2C) and temperature (10-65°C) (2F) promoted dissociation of homo-oligomers into the dimeric PON1 form, but not into monomer. There was no pressure-induced dissociation of HuPON1/HPBP complex up to 200 MPa under these experimental conditions. K_R values showed large fluctuations in the presence of HPBP. This suggests that HPBP is either more sensitive to pressure and/or that interface interactions are sensitive to pressure.

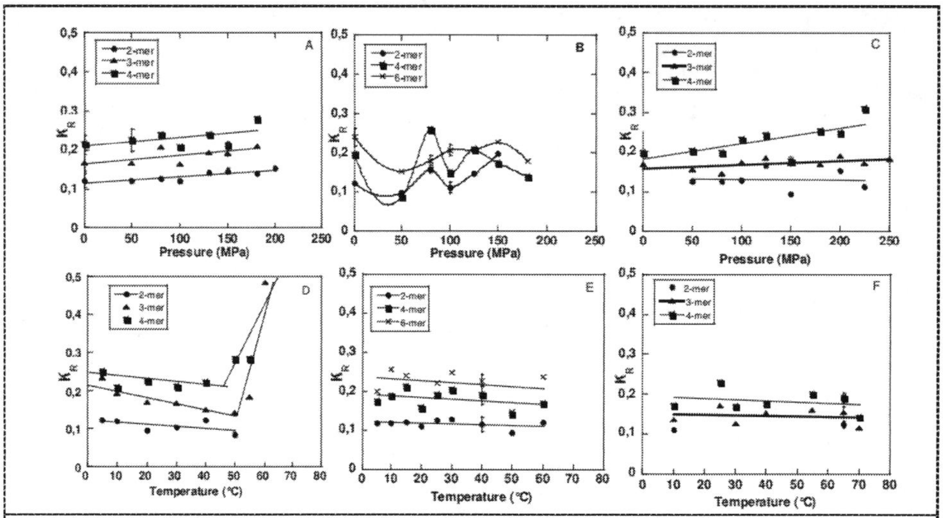

Figure 1. Changes in retardation coefficient, K_R with pressure or temperature for HuPON1 (A,D), HuPON1/HPBP (B,E), and rePON1 (C,F).

3.3. Determination of kinetic parameters at atmospheric pressure

The catalytic efficiency parameters (k_{cat}/K_m) obtained at atmospheric pressure are presented in table 1.

Table 1. Catalytic efficiency (k_{cat}/K_m) for the three PON1 preparations

k_{cat}/K_m ($min^{-1}.M^{-1}$)	HuPON1	HuPON1/HPBP	RePON1 (G3C9)
Arylesterase activity	$(3 \pm 0.4) \times 10^6$	$(6 \pm 0.3) \times 10^6$	$(8 \pm 0.5) \times 10^6$
Lactonase activity	ND*	$(0.8 \pm 0.1) \times 10^6$	$(7.9 \pm 0.4) \times 10^6$
OP-hydrolase activity	ND*	$(2 \pm 0.2) \times 10^3$	$(9 \pm 0.7) \times 10^4$

ND*: not determined. Activity buffer was 50 mM tris, 1 mM NaCl, pH 8 at 25°C and atmospheric pressure. (mean values ± S. E., results in triplicate).

The catalytic efficiency for hydrolysis of arylester was of the same order for the three PON1 preparations (~10^6 $min^{-1}.M^{-1}$). The fact that the catalytic efficiency for both hydrolysis of lactone and OP was higher for rePON1 than for HuPON1/HPBP suggests that the numerous mutations at the surface and near the active site of rePON1 favor the enzyme activity/stability. The HuPON1 kinetic parameters for these activities could not be determined due to a too low enzyme response and a loss in stability. In this respect, rePON1 proved to be a satisfactory model to mimic the "physiological" HuPON1, i.e., in the presence of HPBP.

3.4. Pressure effect on functional stability of PON1

Figure 2 shows the pressure dependence of Ln (k_{cat}/K_m) of arylesterase, lactonase and OP-hydrolase activities for the three PON1 preparations. For both arylesterase and lactonase activities, plots were biphasic, quite similar and allowed the determination of two activation volumes ($\Delta V^{\#}$), depending of the pressure range. The three PON1 preparations were slightly affected by pressure, up to the pressure break (P_t), corresponding to an associated $\Delta V^{\#}$ value close to zero $ml.mol^{-1}$. Beyond this pressure break, the $\Delta V^{\#}$ value was positive for both activities, suggesting a sudden pressure-induced inhibition of the reaction.

The pressure dependence of Ln (k_{cat}/K_m) for OP-hydrolase activity was quite different for both rePON1 and HuPON1/HPBP. The $\Delta V^{\#}$ value determined for rePON1 was close to zero up to 250 MPa, suggesting no pressure-induced unfolding of the enzyme. Moreover, the pressure dependence of k_{cat}/K_m for HuPON1/HPBP OP-hydrolase activity was linear up to 200 MPa. The observed negative $\Delta V^{\#}$ value indicated a reaction widely favored by pressure (5-fold more active at 200 MPa than at atmospheric pressure).

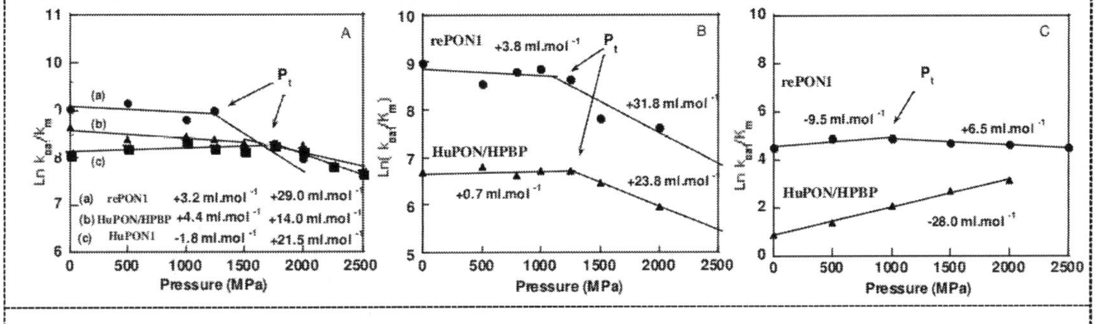

Figure 2. Activation volume ($\Delta V^{\#}_{kcat/Km}$) associated with arylesterase (A), lactonase (B) and OP-hydrolase (C) activities by HuPON1 (■), HuPON/HPBP (▲) and rePON1 (●).

Conclusion

All these results provide evidence that there was no pressure-induced unfolding of PON1. The non-linear pressure dependence of k_{cat}/K_m could be caused by a change in the rate determining step and/or local modification of the enzyme active site. This assumption is strengthened by the fact that the pressure effect on PON1 was similar for phenylacetate and coumaranone hydrolysis, but was different for paraoxon hydrolysis. This suggests that the arylester and the lactone are similarly positioned in the active site, whereas the OP binds on another subsit. This corroborates Tawfik's statement [9]. Further studies are in progress to determine catalytic mechanisms of PON1.

Acknowledgments

This work was supported by DGA/DSP/STTC PEA n°010807 to P.M. and German Bundesministerium der Verteidigung (M/SAB/1/7A004) to D.R.

References

[1] Rochu D, Chabrière E and Masson P 2007 *Toxicology* **233** 47
[2] Harel M *et al.* 2004 *Nat. Struct. Mol. Biol.* **11** 412
[3] Morales R *et al.* 2006 *Structure* **14** 601
[4] Balny C , Masson P and Heremans H 2002 *Frontiers in high pressure biochemistry and biophysic*, eds C Balny, P Masson and K Heremans (Amsterdam, Elsevier) 3
[5] Renault F, Chabrière E, Andrieu J-P, Dublet B, Masson P and Rochu D 2006 *J. Chromatogr. B* **836** 15
[6] Masson P 1997 in *High Pressure Techniques in Chemistry and Physics : a practical approach.* Holzapfel W B & Isaacs N S IRL Press, Oxford pp353.Ferguson K A 1964 *Metabolism* **13** 985
[8] Rochu D, Renault F, Cléry-Barraud C, Chabrière E, Masson P 2007 *Biochim. Biophys. Acta* **1774** 874
[9] Kersonsky O and Tawfik D S 2005 *Biochemistry* **44** 6371

Islet amyloid polypeptide and high hydrostatic pressure: Towards an understanding of the fibrillization process

D H J Lopes, V Smirnovas, and R Winter *

University of Dortmund, Department of Chemistry, Physical Chemistry I - Biophysical Chemistry, D-44227 Dortmund, Germany

* E-mail: roland.winter@uni-dortmund.de

Abstract. Type II Diabetes Mellitus is a disease which is characterized by peripheral insulin resistance coupled with a progressive loss of insulin secretion that is associated with a decrease in pancreatic islet β-cell mass and the deposition of amyloid in the extracellular matrix of β-cells, which lead to islet cell death. The principal component of the islet amyloid is a pancreatic hormone called islet amyloid polypeptide (IAPP). High-pressure coupled with FT-IR, CD, ThT fluorescence spectroscopic and AFM studies were carried out to reveal information on the aggregation pathway as well as the aggregate structure of IAPP. Our data indicate that IAPP pre-formed fibrils exhibit a strong polymorphism with heterogeneous structures very sensitive to high hydrostatic pressure, indicating a high percentage of ionic and hydrophobic interactions being responsible for the stability the IAPP fibrils.

1. Introduction

Amyloid formation and accumulation are implicated in a number of diseases, such Alzheimer's and Parkinson disease, Type II Diabetes Mellitus (TIIDM) and prion disorders [1, 2], and seem to be the central factors in the development of the symptoms of these diseases. Proteins can form different types of insoluble aggregate. One the one hand, amorphous aggregates may be formed, which have multiple protein conformations with rather ill-defined intermolecular interactions, in contrast to protein crystals that can generally be characterized by a single protein conformation with a well-defined intermolecular packing arrangement. On the other hand, also some ordered non-crystalline protein structures such as amyloid fibrils share these properties. In fact, amyloid can be thought of as a one-dimensional crystal in which packing in the plane perpendicular to the direction of fibrillar growth is not homogeneous [3]. Amorphous aggregates generally form rapidly when the protein concentration exceeds the solubility limit. Conversely, protein crystal formation requires time, owing to the kinetic barrier imposed by nucleus formation, which is the rate determinant step. Ordered non-crystalline protein polymer structures such as amyloid share this requirement for nucleation. Nucleus formation requires a series of association steps that are thermodynamically unfavourable (equilibrium constant for nucleation $K_n \ll 1$) because the resultant intermolecular interactions barely compensate the entropic costs of association. Once the nucleus has formed, further addition of monomers becomes thermodynamically favourable (equilibrium constant for growth process $K_g \gg 1$), however, because the monomers contact the growing polymer at multiple sites, resulting in rapid polymerization-like growth.

A key step in the fibrillation process is the transition of a protein from its native structure to a β-sheet arrangement [1,3,4]. Upon formation of amyloid fibrils, the protein molecules that compose the fibrils (partially) lose their native conformation and generally adopt ordered, stacked cross-β-sheet structures. These observations suggest that prevention of the ability of amyloidogenic proteins to nucleate or later, adopt a β-sheet conformation would be useful as a way to interfere with the amyloid self-assembly process. Despite the recent advances in the theoretical and experimental techniques to understand these aggregation processes, the underlying mechanisms have proved challenging to study. This is essentially due to the irreversibility of protein aggregation under ambient temperature and

© 2008 IOP Publishing Ltd

pressure conditions, which complicates or even prohibits the analysis of the underlying kinetic, thermodynamic, and structural parameters.

IAPP is a 37 amino acid residue peptide hormone [5] that is co-synthesized and co-secreted with insulin by pancreatic β-cells [6]. Several functions have been associated with the soluble form of this hormone [6-10], including the control of hyperglycemia by restraining the rate at which dietary glucose enters the bloodstream. For reasons that are still not fully understood, IAPP aggregates in the extra cellular matrix of the β-cells forming fibrillar amyloid deposits. These deposits are present in approximately 95 % of TIIDM patients and are strongly associated with islet β-cell degeneration and loss [11,12].

It has been proposed that the IAPP aggregation process has two distinct phases: a lateral growth of oligomers followed by longitudinal growth into mature fibrils [13-15], and it has been demonstrated that the initial stages of IAPP fibril formation are driven by the increase of solvent-exposure of hydrophobicity patches [13]. The structural changes behind the fibrillization process are still poorly understood, however.

High hydrostatic pressure (HHP) has been widely used as a tool to understanding protein folding [16-22]. High pressure tends to destabilize proteins due the fact that the protein-solvent system in the unfolded state occupies a small volume than the system in the native state. In a similar way, pressure leads to the dissociation of oligomeric proteins. These effects seem to be caused by a combination of factors. The presence of cavities within the folded proteins or in the interface of oligomers favour unfolding or dissociation of these structures [16,17]. The dissociation of electrostatic interactions also leads to a marked reduction in the overall volume caused by electrostrictive effects of the water molecules around the unpaired charged residues [23]. In a similar way, solvation of polar groups results in a decrease in volume. These effects compensate for the increase in volume as the crystalline-like state of the protein interior is disrupted, exposed to solvent and hydrated upon unfolding.

The use of the pressure in studies of protein folding and stability can offer a great number of advantages [16,17]. Pressure affects only the volume of the system in focus, while the temperature denaturation, for example, involves changes in both volume and thermal energy of the system. Also, temperature-induced aggregation generally results in an irreversible aggregation process, which may be explained by less disfavoured hydrophobic interactions at higher temperatures. On the contrary, denaturation by high pressure generally is a reversible process. Furthermore, high pressure treatment can lead to dissociation of aggregated structures and may result in formation of monomers and natively folded structures. Also, measurement of the activation volume for pressure-induced kinetic folding or unfolding reactions is one of the few methods yielding structural information regarding the transition states [24]. Hence, in addition to co-solvent and temperature perturbation, pressure dependent studies shed new light on alternative folding/aggregation pathways and their intermediate states.

So far, there are no reports regarding the effects of HHP on IAPP. Owing to the fact that high hydrostatic pressure acts to disfavour hydrophobic and electrostatic interactions that cause protein aggregation, this parameter can be used as efficient tool to reveal new important information on the nucleation and growth processes of the protein. High-pressure coupled with Fourier-transform Infrared (FT-IR) spectroscopic studies and atomic force microscopy (AFM) measurements complemented by Thioflavin T (ThT) fluorescence and circular dichroism (CD) experiments were carried out to reveal the changes in IAPP aggregate and fibril formation under pressure-perturbation. These results lead to a better understanding of the aggregation pathways and the possible amyloidogenic states of IAPP and may hence contribute to a better understanding of the pathogenesis of type II diabetes mellitus. Moreover, the results obtained may prove useful for the identification and molecular characterization of toxic intermediate states that may be used as targets for parallel searching of compounds that can interfere with their formation.

Journal of Physics: Conference Series **121** (2008) 112002 doi:10.1088/1742-6596/121/1/112002

2. Materials and Methods

Synthetic human IAPP was obtained from Calbiochem-Novabiochem (Bad Soden, Germany). NaH_2PO_4 buffer was purchased from Gibco BRL, 2,2,2-trifluoroethanol (TFE) was obtained from Aldrich. ThT, D_2O and all other reagents were from Sigma. The reagents were of the highest analytical grade available. Pre-distilled water was filtered and deionised through a Millipore purification system.

Incubation under pressure was carried out in a pressure cell equipped with sapphire optical windows, similar to that originally described by Paladini and Weber [25,26]. The temperature of the pressure cell was controlled by means of a jacket connected to a circulating bath. Pressure was increased stepwise in increments of 150 bar until 3.5 kbar. The windows of the pressure cell were flushed with nitrogen at low temperatures to prevent water condensation.

IAPP aggregation and fibril formation was determined by CD measurements using a Jasco J-720 spectropolarimeter. CD spectra were collected over the wavelength range from 190 to 260 nm at 0.2 nm intervals with a spectral bandwidth of 0.8 nm and 1 s integration time. The cell holder was maintained at 25 ± 0.1 °C. For the CD measurements, IAPP stock solutions were kept in 100 % TFE at 4 °C and diluted prior to the experiments to give the chosen final concentration in 10 mM NaH_2PO_4 (pH 7.4), with residual 1 % TFE. CD spectra, taken at least four times, are presented after subtracting the spectra of buffer and smoothing, using the Origin 7 scientific graphic and analysis software. The data are expressed as molar ellipticity $[\theta]$ in units of deg cm^2 $dmol^{-1}$.

ThT spectroscopic measurements were carried out using an ISS K2 multifrequency phase fluorometer (ISS Inc., Champaign, IL) coupled with a water circulating bath to keep the chosen temperature constant. The samples were prepared as for the CD measurements and ThT was added to a final concentration 8 times higher than that of the protein. The fluorescence signal (excitation at 450 nm) was recorded between 450 and 600 nm using 1 nm slits both for the emission and excitation measurements.

For the FT-IR measurements, CaF_2 transmission windows with 0.05 mm Teflon spacers were used. The temperature in the cell was controlled through an external water-circuit. FT-IR spectroscopy has proven to be a powerful technique to determine the secondary structure elements [17,27-29]. The amide I' band (between 1600 and 1700 cm^{-1}) was recorded, which is mainly associated with the carbonyl stretching vibration of the amide groups and which is directly related to the backbone conformation and hydrogen bonding pattern of the protein [29]. For scanning the samples before and after pressure treatment, we used 10 mM NaH_2PO_4 with 1 % residual TFE in D_2O at pD 7.4 and a protein concentration of 250 μM (~0.1 % w/w). The FT-IR spectra were collected on a Nicolet 5700 FT-IR spectrometer equipped with a liquid nitrogen–cooled MCT detector. For each spectrum, 256 interferograms of 2 cm^{-1} resolution were co added. The sample chamber was continuously purged with dry air. From the spectrum of each sample, a corresponding buffer spectrum was subtracted. All the spectra were baseline-corrected and normalized for the amide I' band area. Data processing was performed with GRAMS software.

For the atomic force microscopy (AFM) measurements, samples were diluted with deionised water to yield a final concentration of 1 μM. 30 μL were applied onto freshly cleaved muscovite mica and allowed to dry. Data were acquired in the tapping mode on a Multi Mode TM SPM AFM microscope equipped with a Nanoscope IIIa Controller from Digital Instruments. As AFM probes, Silicon SPM Sensors "NCHR" (force constant, 42 N/m; length, 125 mm; resonance frequency, 300 kHz) from Nanosensors were used [30,31].

3. Results

In order to investigate the stability of IAPP towards high pressure, either fresh peptide or pre-formed fibrils were subjected to pressures up to 3.5 kbar and the changes were monitored by different techniques to yield information about the transformation process and the structures evolving at various levels of complexity. The protein solutions were prepared from stock solutions in a water buffer with 1 % residual TFE. Such condition has been shown to give raise to or to synchronize the formation of pre-assembled β-sheet structures which are rich in fibrils and protofibrils [32].

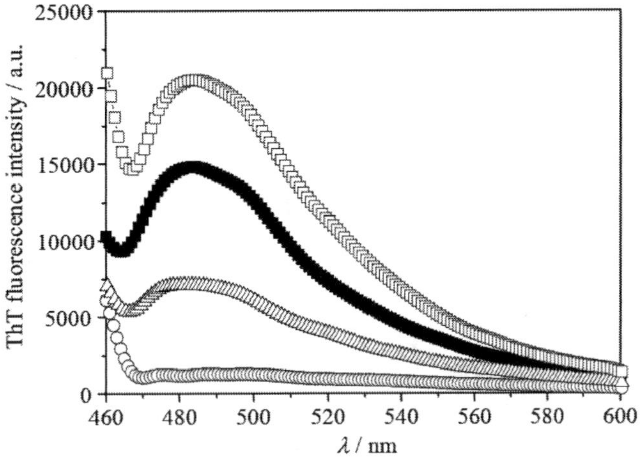

Figure 1. ThT fluorescence intensity values (excitation at 450 nm) of IAPP (10 μM in 10 mM NaH_2PO_4 buffer (pH 7.4) containing 1 % TFE at T = 25 °C) subjected to different treatments: (■) sample at ambient pressure after 1 d, (○) sample at ambient pressure after 3 d (precipitation occurs), (Δ) sample pressurized at 3.5 kbar for after 3 d, (□) sample after subsequent pressure release. Non-specific background fluorescence was subtracted from the samples by using appropriate blanks not containing the protein.

The first step was to investigate the ability of IAPP to bind ThT – a fluorophore, widely used to detect amyloid structure in proteins [33] – under different pressure conditions. First, ThT analysis (ratio IAPP/ThT 1:8) of the amyloid content was performed 5 min after hydration of the protein (figure 1). The sample showed a substantial emission at 482 nm, characteristic of ThT bound to IAPP amyloid [34], already after such a short time, in agreement with previous studies on IAPP [35]. Then, the sample was split into two aliquots, one kept at ambient pressure and the other was subjected to 3.5 kbar for three subsequent days. After this incubation period, both samples were checked again for ThT fluorescence emission. The IAPP sample aggregated for three days at ambient pressure was not able to bind further ThT, and almost no fluorescence was detected which is due to the fact that the aggregate precipitated. However, when the sample was subjected to pressures of 3.5 kbar for 3 days, further binding of ThT was observed (figure 1). Under these conditions, the amyloid is probably partially disrupted and smaller species, such as IAPP oligomers and protofibrils, which can bind ThT [36], are formed being able now to bind more ThT. As mature fibrils are normally very resistant even to extremes of high pressure [17,33], we might conclude that these IAPP aggregates formed at ambient pressure conditions do not consist (only) of mature, densely packed "cores", but contain a significant amount of species which are sensitive to HHP.

In a second step, we were looking for corresponding changes in the secondary structure of the samples. For this purpose, far-UV CD spectroscopy has been used, which is a sensitive technique that allows assessing the secondary and tertiary structure of proteins in solution [37-39] and hence permits evaluating changes in the conformation of the IAPP before and after pressure treatment. A line shape analysis of the CD spectra can be used for distinguishing early conformational events that precede fibril formation. As for the ThT measurements, we analyzed the secondary structure content of the sample 5 min after sample preparation. As shown in figure 2, formation of a typical β-sheet conformational transition state is clearly visible. Calculation of the secondary structure content [38,39] from the CD spectra of this solution using the protein reference spectra by Sreerama et al. [40] suggests that IAPP at early stages of aggregation at 25 °C still contains substantial amounts of random coil (64.1 %), 20.2 % of β-sheets and turns and some α-helical structures (~15.7 %). These data are in reasonable agreement with previous reports for soluble IAPP by Kayed et al. [13] (random coil: 71 %, β-sheets and turns: 21 %, α-helices: 8 %). CD spectra recorded after 3 days at room temperature and ambient pressure show a decrease in the α-helical content and a concomitant increase in β-sheet structures indicated by an increase in of modulus of the ellipticity at ~220 nm, which is due to increasing aggregate formation. On the other hand, the CD spectrum of the pressurized IAPP sample looks different, with a similar minimum but shifted to about 222 nm, however the overall CD ellipticity has decreased drastically, most likely due to precipitation of the IAPP aggregate (as can be seen by light scattering).

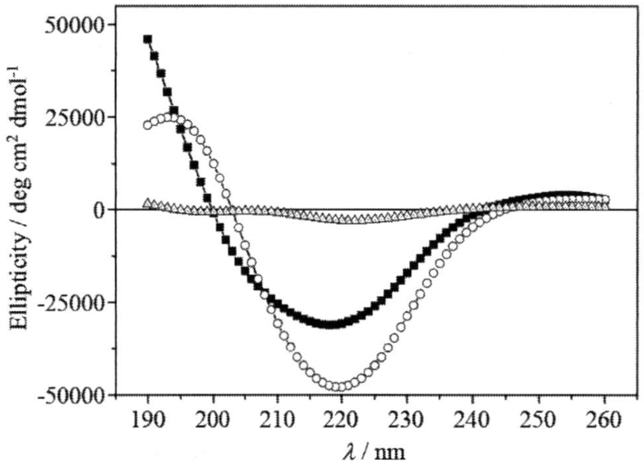

Figure 2. IAPP amyloid formation followed by CD spectroscopy. Far-UV CD spectra of IAPP solutions (10 μM in 10 mM NaH_2PO_4 buffer (pH 7.4) containing 1 % TFE, T = 25 °C) subjected to different treatments: (■) sample at ambient pressure after 1 d, (○) sample at ambient pressure after 3 d (precipitation occurs), (Δ) sample pressurized at 3.5 kbar for after 3 d. Non-specific background was subtracted from the samples by using appropriate blanks not containing the protein.

In a third step we were looking deeper into the secondary conformational changes using FT-IR spectroscopy, a common tool used for monitoring α-helix to β-sheet transitions and to disentangle between different β-sheet structures that accompany protein aggregation processes [27,33,41], and which can also be used for pressure-dependent spectroscopic studies. Figure 4 shows FT-IR spectra of various IAPP samples: freshly dissolved IAPP and IAPP aggregated for 3 days at ambient pressure and at 3.5 kbar, respectively (all spectra were taken at room temperature, 25 °C). The FT-IR spectrum of the freshly prepared IAPP sample at a concentration as high as 250 μM exhibits a IR band appearing at ~1622 cm^{-1}, likely of intermolecular parallel β-sheet structures, and a pronounced peak at 1673 cm^{-1}, which is due to turns and residual trifluoracetic acid (TFA) in the sample. Such IR pattern is indicative of the aggregated state of IAPP according to data of other amyloidogenic proteins, like insulin [41]. After the incubation period of 3 days at ambient pressure and temperature, the infrared spectrum shows similar features, with a broadening and a small shift of the amide I' band to slightly smaller wavenumbers (~1618 cm^{-1}), however. Very likely, a distribution of different sizes and shapes of IAPP aggregate is found initially, and the overall size of the structures evolving increases with increasing IAPP concentration. After the incubation period it seems that the β-strands are realigning from a non-perfectly packed initial state to more ordered structures with stronger H-bonding, leading to the observed shift of the amide I' band to 1618 cm^{-1} accompanied by an increase of the β-turn content.

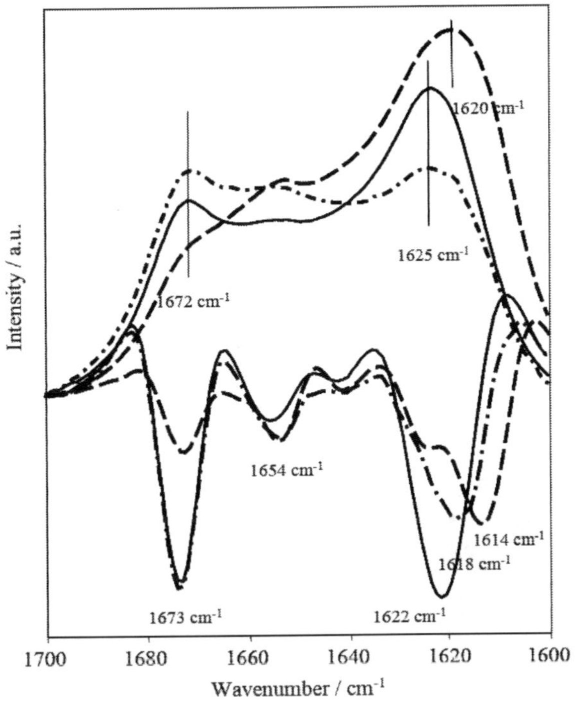

Figure 3. FT-IR spectra (at the top of the figure) of IAPP: (--) 5 min after preparation, (-.-.-) after 3 d of aggregation in 10 mM NaH_2PO_4 (pD 7.4) at ambient pressure, (- - -) after 3 d at 3.5 bar. All data were taken at T = 25 °C. At the bottom of the figure, also the second derivative spectra are shown, which allow a more accurate determination of peak positions.

When IAPP was aggregated under pressure, the main amide I' IR band shifts to significantly smaller wavenumbers (~1614 cm^{-1}), indicating that the pressure treatment leads to the formation of a different aggregate structure, probably with stronger intermolecular H-bonding between β-sheet strands, or to a larger population of these species.

In order to reveal the different morphological structures formed, additional AFM measurements were carried out for this IAPP concentration (figure 4). Most of the aggregate structures seen in the sample that was not subjected to HHP are short (< 1 μm) fibrils with an average of 5 - 15 nm diameters as determined from the AFM height profile. Such rather short fibrillar structures probably appear due to a comparably fast nucleation process at this high protein concentration, whereas the elongation process seems to be comparably slow. The pressure treated sample still contains fibrils, but also a significant amount of smaller oligomeric particles (of 0.5 - 1.5 nm size).

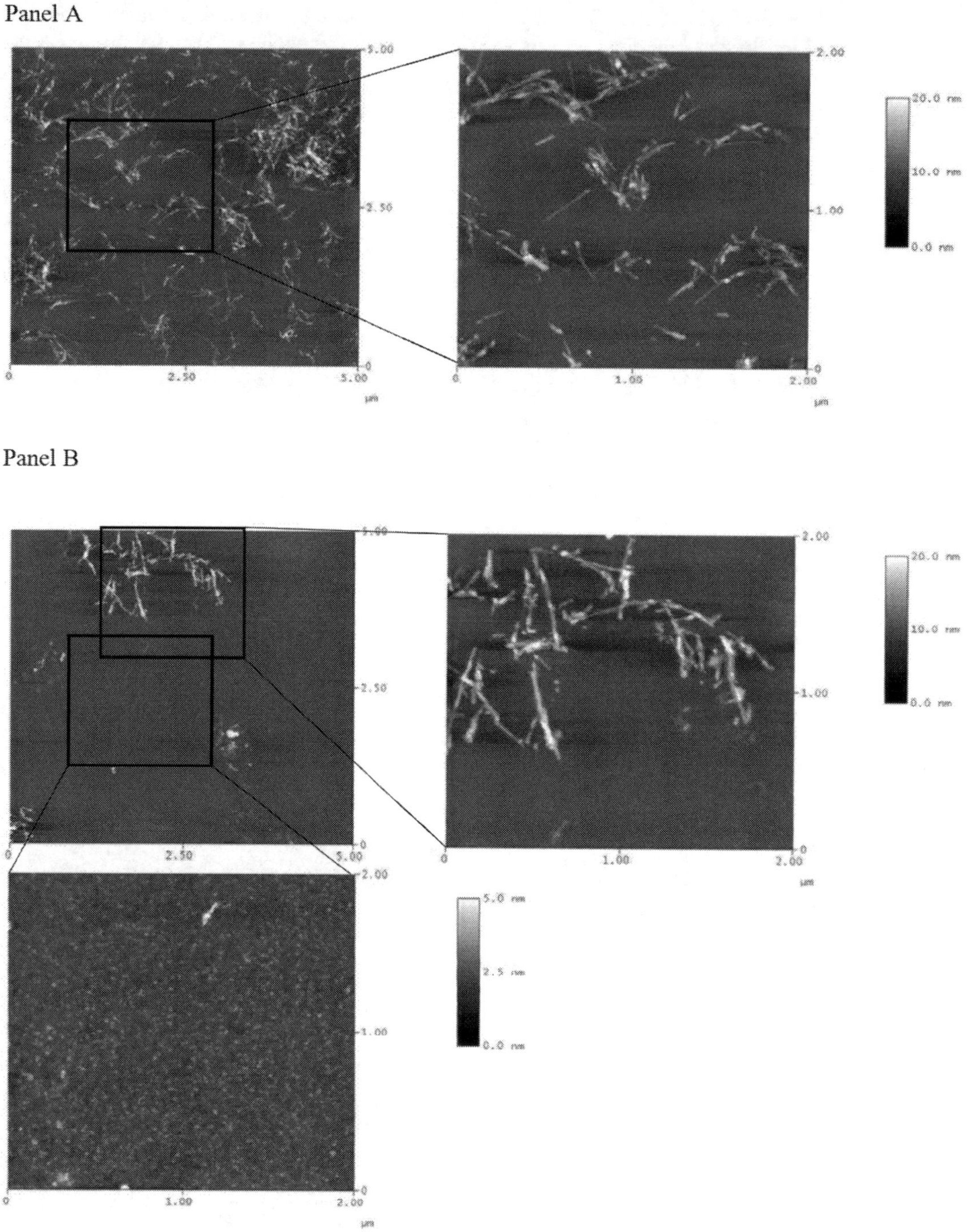

Figure 4. AFM height images of 250 µM IAPP aggregates after 3 d without (panel A) and with pressure treatment at 3.5 kbar (panel B). The same samples analyzed by FT-IR and CD spectroscopy were diluted with deionized water to yield a final concentration of 1 µM. 30 µL were applied onto freshly cleaved muscovite mica and allowed to dry before the AFM analysis.

Taking the FT-IR spectroscopic and AFM results together, we may conclude that the sample subjected to high pressure treatment displays less fibrillar β-sheet structures and a larger population of smaller amorphous aggregates with a different H-bonding pattern. To support these conclusions, the samples without and with 3 d pressure treatment were centrifuged at 16.000 rpm at 4 °C for 20 min in order to remove any insoluble material and the protein concentration of the supernatant was determined from the UV absorbance at 274.5 nm, using a molar extinction coefficient of 1440 $M^{-1}cm^{-1}$ [13]. Protein was detected only for the pressurized IAPP (~17 % of the overall protein concentration). Immediately after the absorbance measurements, also far-UV CD spectra were taken. The HHP treated sample showed a strong negative band at about 200 nm and a positive band at 230 nm, characteristic for the presence of unordered structures (data not show).

The different fibrillar and non-fibrillar amorphous/oligomeric morphologies found for the high pressure treated sample indicate that not all IAPP aggregate structures are equally sensitive to pressure, hence suggesting the existence of pressure resistant fibrils with densely packed cores and a population that can be dissociated by HHP.

4. Conclusions

Here we report on the first high pressure work on IAPP aiming to reveal new information on its aggregation pathways and aggregate/fibril structures. The conceptual framework for using such pressure axis experiments is as follows: The interior of proteins is largely composed of rather efficiently packed residues (with void volume on the order of 0.5 %), more likely hydrophobic than those at the surface. High hydrostatic pressure induces conformational fluctuations due to a decrease in the strength of hydrophobic interactions, finally leading to partially pressure-induced unfolding through transfer of water molecules into the protein interior, gradually filling cavities and leading to the dissociation of close hydrophobic contacts and subsequent swelling of the hydrophobic protein interior [28,42]. According to the literature, aggregation of the C-terminal domain of IAPP (amino acid sequences 20-29 and 30-37) is thought to be most likely driven by hydrophobic interactions. The distinctly amyloidogenic region 20-29 has been pointed out to be a key fibril-forming region [43]. Hence it is conceivable that the C-terminus of IAPP, spanning residues 20 to 37, forms a contiguous β-sheet. The secondary structure predictions of human IAPP indicate that there is one potential α-helical region between amino acid residues 8-14, and three potential β-strand regions. A β-turn has been predicted at Asn31, which would result in two adjacent β-strands (32-37 and 24-29); a third β-strand is proposed to exist in region 18-23 [44]. According to these data, it may be concluded that IAPP has not only one but several amyloidogenic cores that are interacting to form an organized aggregate structure, and that hydrophobic interactions may drive the initial stage of the aggregation process. As HHP is acting to weaken or even prevent hydrophobic self-organization, we may expect that the amyloidogenic "cores" cannot be arranged and packed the same way upon aggregation as they are able to do under ambient pressure conditions, thus leading to formation of a more heterogeneous population of fibrils and smaller amorphous/oligomeric species.

Taken all data together, a hypothetical model for IAPP fibril formation may be suggested: i) IAPP undergoes fast nucleation (due to several amyloidogenic "cores"), largely driven by hydrophobic interactions. Hence, formation and packing of fibrils is not perfect and mixed-registry β-sheet structures might exist, in particular at high protein concentrations, which can partially be dissociated by pressure leading to smaller aggregate structures and oligomers. HHP already of pressures as low as 3.5 kbar is sufficient to weaken and (at least partially) disrupt the hydrophobic cores thus leading to formation of a heterogeneous population of fibrillar aggregates with IR amide I' bands in the low wave number region (which is typical of a more strongly H-bonding pattern of intermolecular β-sheets) and a large amount of non-fibrillar smaller aggregates and oligomers, as detected by AFM (with IR amide I' bands in the larger wave number region around 1620 cm^{-1}). ii) Our data also indicate that the pre-formed IAPP fibrils are sensitive to high hydrostatic pressure, similar to amorphous aggregates and inclusion bodies [45-47]. Considering the fact that high hydrostatic pressure is an effective means in disturbing ionic and hydrophobic interactions but not hydrogen bonds, we can conclude that these former two types of interaction are important for the stability of IAPP fibrillar aggregates, as also suggested in work using denaturing agents [13]. The results may be physiologically relevant and may have important implications regarding the stability of the IAPP fibrils against degradation, a fact that can contribute to the development of antagonists of the pathogenesis of type II diabetes mellitus.

Acknowledgment
Financial support from the Deutsche Forschungsgemeinschaft (DFG), the Fonds der Chemischen Industrie, and NRW/EU (Europäischer Fonds für regionale Entwicklung) is gratefully acknowledged.

References

[1] Kelly J W 1998 *Curr. Opin. Struct. Biol.* **8**, 101
[2] Dobson C M 2001 *Trans. R. Soc. Lond. B. Biol. Sci.* **356**, 133.
[3] Lansbury P T Jr. 1992 *Biochemistry* **31,** 6865
[4] Serpell L C, Sunde M, Benson M D, Tennent G A, Pepys M B and Fraser P. E. 2000 *J. Mol. Biol.* **300**, 1033
[5] Lorenzo A and Yankner B A 1994 *Proc. Natl. Acad. Sci. USA* **91**, 12243
[6] Cooper G J S, Day A J, Willis A C, Roberts A N, Reid K B M and Leighton B 1989 *Biochim. Biophys. Acta* **1014**, 247; Cooper G J S, Willis A C and Leighton B 1989 *Nature* **340**,272
[7] Ohsawa H, Kanatsuka A, Yamaguchi T, Makino H and Yoshida S 1989 *Biochem. Biophys. Res. Commun.* **160**, 961
[8] Fehmann H C, Weber V, Goke R, Goke B, Eissele R and Arnold R 1990 *Biochem. Biophys. Res. Commun.* **167**, 1102
[9] Silvestre R A, Peiro E, Degano P, Miralles P and Marco J 1990 *Regul. Pept.* **31,** 23.
[10] Johnson K H, O'Brien T D, Betsholtz C and Westermark P 1992 *Diabetes Mellitus. Lab. Invest.* **66**, 522
[11] Westermark P and Wilander E 1978 *Diabetologia* **15**, 417
[12] Hayden M R 2002 *Mol. Med.* **99**, 495
[13] Kayed R, Bernhagen J, Greenfield N, Sweimeh K, Brunner H, Voelter W and Kapurniotu A 1999 *J. Mol. Biol.* **287**, 781
[14] Padrick S B and Miranker A D 2001 *J. Mol. Biol.* **308**, 783; Padrick S B and Miranker A D 2002 *Biochemistry* **41**, 4694
[15] Green J D, Goldsbury C, Kistler J, Cooper G J S and Aebi U 2004 *J. Biol. Chem.* **279**, 12206
[16] Weber G and Drickamer H G 1983 *Q. Rev. Biophys.* **16**, 89
[17] Winter R, Lopes D, Grudzielanek S and Vogtt K 2007 *J. Non-Equilib. Thermodyn.* **32**, 41; Ravindra R and Winter R 2003 *Chem. Phys. Chem.* **4**, 359
[18] Torrent J, Alvarez-Martines M T, Harricane M-C, Heitz F, Liautard J-P, Balny C and Lange R 2004 *Biochemistry* **43**, 7162
[19] Silva J L, Cordeiro Y and Foguel D 2006 *Biochim. Biophys. Acta* **1764**, 443
[20] Panick G, Vidugiris G J A, Malessa R, Rapp G, Winter R and Royer C A 1999 *Biochemistry* **38**, 4157; Desai G, Panick G, Zein M, Winter R and Royer C A 1999 *J. Mol. Biol.* **288**, 461; Panick G and Winter R 2000 *Biochemistry* **39**, 1862; Zein M and Winter R 2000 *Phys. Chem. Chem. Phys.* **2**, 4545
[21] Ulmer H M, Herberhold H, Fahsel S, Gänzle M G, Winter R and Vogel R F 2002 *Appl. Environm. Microbiology* **68**, 1088
[22] Cordeiro Y, Kraineva J, Ravindra R, Lima L M T R, Gomes M P B, Foguel D,. Winter R and Silva J L 2004 *J. Biol. Chem.* **279**, 32354; Dzwolak W, Jansen R, Smirnovas V, Loksztejn A, Porowski S and Winter R 2005 *Phys. Chem. Chem. Phys.* **7**, 1349
[23] Kauzman W 1959 *Adv. Protein Chem.* **14**, 1
[24] Perrett S and Zhou J M 2002 *Biochim. Biophys. Acta* **1595**, 210

[25] Paladini A A Jr and Weber G 1981 *Biochemistry* **20**, 2587
[26] Herberhold H, Marchal S, Lange R, Scheyhing C H, Vogel R F and Winter R 2003 *J. Mol. Biol.* **330**, 1153
[27] Panick G, Malessa R and Winter R 1999 *Biochemistry* **38**, 6512
[28] Herberhold H, Royer C A and Winter R 2004 *Biochemistry* **43**, 3336
[29] Heremans K and Smeller S 1998 *Biochim. Biophys. Acta* **1386**, 353
[30] Jansen R, Dzwolak W and Winter R 2005 *Biophys. J.* **88**, 1344
[31] Jansen R, Grudzielanek S, Dzwolak W and Winter R 2004 *J. Mol. Biol.* **338**, 203
[32] Tatarek-Nossol M, Yan L M, Schmauder A, Tenidis K, Westermark G and Kapurniotu A 2005 *Chem. Biol.* **12**, 797
[33] Grudzielanek S, Jansen R and Winter R 2005 J. Mol. Biol. **351**, 879; Grudzielanek S, Smirnovas V and Winter R 2006 *J. Mol. Biol.* **356**, 497
[34] LeVine H III 1999 *Meth. Enzymol.* **309**, 274
[35] Lopes D. H. J, Colin C, Degaki T, Sousa A C V, Vieira M N N, Sebollela A S, Bloch C Jr, Martinez A M, Ferreira S T and Sogayar M C 2004 *J. Biol. Chem.* **279**, 42803
[36] Jiang Y, Li H, Zhu L, Zhou J M and Perrett S 2004 *J. Biol. Chem.* **279**, 3361
[37] Greenfield N J 1996 *Anal. Biochem.* **235**, 1
[38] Lobley A, Whitmore L and Wallace B A 2002 *Bioinformatics* **18**, 211
[39] Whitemore L and Wallace B A 2004 *Nucleic Acids Res.* **32**, 668
[40] Sreerama N and Woody R W 2000 *Anal. Biochem.* **287** 252; Sreerama N, Venyaminov S Y and Woody R W 2000 *Anal. Biochem.* **287**, 243
[41] Dzwolak W, Ravindra R, Lendermann J and Winter R 2003 *Biochemistry* **38**, 11347; Dzwolak W, Smirnovas V, Jansen R and Winter R 2004, *Protein Sci.* **13**, 1927
[42] Hummer G, Garde S, Garcia A E, Paulaitis M E and Pratt L R 1998 *Proc. Natl. Acad. Sci. USA* **95**,1552
[43] Ma Z, Westermark G T, Sakagashira S, Sanke T, Gustavsson A, Sakamoto H, Engstrom U, Nanjo K and Westermark P 2001 *Amyloid* **8**, 242
[44] Jaikaran E T, Higham C E, Serpell L C, Zurdo J, Gross M, Clark A and Fraser P E 2001 *J. Mol. Biol.* **308**, 515
[45] Gorovits B M and Horowitz P M 1998 *Biochemistry* **37**, 6132
[46] Foguel D and Silva J L 2004 *Biochemistry* **43**, 11361
[47] St John R J, Carpenter J F and Randolph T W 1999 *Proc. Natl. Acad. Sci. USA* **96**, 13029

Microdomain Ca^{2+} dynamics in mammalian muscle following prolonged high pressure treatments

S Schnee[1,+], F vWegner[1,+,§], S Schürmann[1], H Ludwig[2], RHA Fink[1] and O Friedrich[1,§]

[1] Medical Biophysics, Institute of Physiology & Pathophysiology, University of Heidelberg, INF326, 69120 Heidelberg, Germany, [2] Physical Chemistry, Institute of Molecular Biotechnology & Biopharmacy, University of Heidelberg, INF 366

E-mail: oliver.friedrich@physiologie.uni-heidelberg.de
[+,§]: These authors contributed equally to this study

Abstract. High pressure (HP) applications are an important thermodynamic tool to influence cellular processes. Especially processes that undergo large volume changes, e.g. opening or closing of ion channels, are in particular susceptible to HP treatments. Such volume changes are extremely difficult to assess for intracellular ion channels, like ryanodine receptors (RyR) residing in the membrane of organelles. In skeletal muscle, RyR act as Ca^{2+} release channels. We previously showed that plasmalemmal Na^+ and Ca^{2+} ion channels were irreversibly altered after prolonged 20 MPa treatments. Here, changes in microdomain Ca^{2+} levels due to elementary Ca^{2+} release events (ECRE) were monitored using confocal fluorescence microscopy. We studied ECRE in mammalian skeletal muscle following 3 h HP treatments up to 30 MPa to clarify whether RyR induced intracellular microdomain Ca^{2+} dynamics was more susceptible to HP treatment compared to surface membrane ion currents. ECRE frequencies exponentially declined with pressure. ECRE amplitudes and rise times (RT) were quite robust towards HP treatments. In contrast, spatial and temporal ECRE extension showed a tendency towards larger values up to 20 MPa but declined for higher pressures. Activation volumes for pressure-induced persistent ECRE alterations were zero for RT but showed a bimodal behavior for event duration. It seems that although ECRE frequencies are markedly reduced, ECRE morphology is less affected by HP. In particular, RyR opening time is practically unaltered and the observed morphological ECRE changes might reflect alterations in local Ca^{2+} buffers and Ca^{2+} concentration profiles rather than involvement of RyR in mammalian skeletal muscle.

1. Introduction

The vast majority of the earth's biosphere is governed by high hydrostatic pressure environments that can exceed pressures as high as 100 MPa for the deepest ocean trenches [1]. It is still a mystery what precise mechanisms have evolved to enable life exposed to such high pressures and moreover, which compensatory strategies are present in organisms migrating various depth levels, e.g. deep sea fish. High pressure is known to have major effects on biological membranes, e.g. to induce fluid- to gel-like phase transitions [1]. The fluidity of membranes is a prerequisite for their physiological function and depends on factors like structural polymorphism, water content, pH, ionic strength, content of saturated/unsaturated phospholipids, cholesterol, membrane proteins, temperature and pressure itself. Deep-sea animals have been implicated in a regulation of membrane composition in response to environmental challenges termed

© 2008 IOP Publishing Ltd

'*homeoviscous adaptation*' [2, 3, 4]. Such regulation involves changes in phospholipid or cholesterol fractions, chain length and unsaturation as well as protein content. As a result of this plasticity, simple phase transitions are not common in natural membranes. Another mechanism that has been encountered in deep sea animals is the active production of osmolytes to increased environmental pressures [5, 6]. However, the aforementioned mechanisms are not applicable to diving mammals, e.g. whales or dolphins, or humans in a diving scenario. While diving mammals use sophisticated management of body airspaces or increased oxygen stores in muscle [7], it is predominantly impaired central nervous system function that limits pressure exposure in humans (high pressure nervous syndrome, HPNS) [8].
High pressure (HP) applications to cells and tissues have gained increasing interest in high pressure bioscience in the last years as they allow to elucidate specific interactions that directly or indirectly affect integrative cell functions, e.g. membrane excitability of nerve cells [9] or contractile properties of muscle [10]. Additionally, prolonged HP treatments are suitable to define principle pressure limits for the reversibility of pressure induced specific impairments of cellular and subcellular processes [11, 12].
The activation of skeletal muscle is a complex cascade involving membrane excitation, excitation-contraction (ec) coupling and motor protein activation. Ca^{2+} ions play a crucial role in this finely tuned machinery and perturbations of Ca^{2+} homeostasis are associated with various disease states [13]. Especially the ec-coupling process, involving direct protein-protein interactions of ion channels within the outer cell membrane and within the intracellular membranes of the sarcoplasmic reticulum (SR) are associated with fluctuations in microdomain Ca^{2+} concentrations that take place on a millisecond time scale. Ryanodine receptor Ca^{2+} channels (RyR) within the SR membrane are the site of Ca^{2+} release in muscle and elementary Ca^{2+} release events (ECRE) can be visualized using fluorescent Ca^{2+} dyes in conjunction with confocal microscopy [14, 15, 16].
We have previously shown that there is a rather sharp pressure limit for the reversibility of high pressure induced impairment of contractility [17], membrane function [18] and intracellular global Ca^{2+} homeostasis [19] around 20 MPa to 25 MPa in 3 h long-term pressure exposures. Organelle membranes, such as the SR, have been suggested to be much more susceptible to HP compared to plasma membranes due to their lower cholesterol content [20].
In the present experimental study, we were in particular interested whether prolonged high pressure treatments up to 30 MPa affect elementary Ca^{2+} release events in mammalian skeletal muscle recorded in the post-decompression phase. For this, we evaluated ECRE morphology and frequencies using confocal laser scanning microscopy. Our results show quite unexpected results of relatively stable spatio-temporal morphological parameters of isolated ECRE but lower overall ECRE frequencies following the HP treatment. The results suggest that albeit a larger pressure sensitivity of the SR membrane, channel clusters herein seem to be more pressure resistant compared to ion channels of the plasma membrane.

2. Methods

2.1. Prolonged high pressure treatment of mammalian muscle

For this study, *extensor digitorum muscle* (edl) from adult C57 mice were used. All experiments complied with the guidelines laid down by the Local Animal Care Committee. The preparation of muscles is described in detail in [17]. Excised edl muscles were transferred into small Eppendorff caps completely filled with Ringer solution (see 2.3) and sealed for high pressure treatment. The caps were then inserted into the high pressure vessel that had been pre-cooled to 4 °C. The pressure vessel consisted of an autoclave (PN2500, Wepuko Hydraulics, Metzingen, Germany; effective chamber volume: 4 ml) that was connected to a manual hand spindle (SITEC-Sieber Engineering, Ebmatingen, CH) for pressurization. Water was used as pressurization medium (10 MPa per 0.053% volume change, [17]). A thermoelement to monitor temperature was coupled

into the vessel and pressure was digitally monitored from a probe within the pressure tubes. To avoid adiabatic heating during pressurization (that additionally was minimized by a very slow pressurization protocol, see below), the autoclave was immersed in a basin filled with a water-glycerol-ethanol mixture kept at a constant temperature of 4 °C [17].
Pressurization was carried out using a standardized slow pressurization protocol [17, 18, 19]. Briefly, pressure was increased at a rate of 0.2 MPa per min up to 6 MPa and then at a rate of 0.6 MPa per min until the end pressure of either 10 MPa, 15 MPa, 20 MPa or 30 MPa was reached. This was shown to maximize survival of individual cells [17]. Pressure was kept constant for 3 h and decompression was performed using the reverse procedure.

2.2. Single fibre preparation and post-decompression confocal Ca^{2+} fluorescence microscopy
Single muscle fibres from the the high pressure treated muscle were mechanically dissected while bathing the muscle in relaxing solution (see 2.3). This solution prevents contraction of muscle fibres due to the very low concentration of Ca^{2+} ions and the presence of adenosine-triphosphate (ATP) that relaxes muscle fibres [15]. Control muscles were kept at atmospheric pressure and at 4 °C for the same time period as needed for the complete high pressure protocol and then further processed the same way as high pressure treated muscle. Single fibres were chemically permeabilized using the membrane detergent saponin (Sigma Aldrich, Taufkirchen, Germany, 0.01 % w/v dissolved in relaxing solution) for 2 min. After washout, the solution was exchanged to an internal solution containing high K-glutamate (125 mM, see below). The Ca^{2+} dye Fluo-4 was added to the internal solution to a final concentration of 90 μM and incubated for $\sim$ 15 min before confocal recording of ECRE. The dye was excited with the 488 nm laser line of an Ar^+ ion laser and the fluorescence signal was collected between 515 nm and 560 nm using an inverted laser scanning microscope (Leica SP2, Leica Microsystems, Mannheim, Germany). ECRE recordings in single fibres usually started 30 min after the decompression so that the post-decompression phase was considered a new steady-state [18].

2.3. Physiological solutions
Isotonic Ringer solution contained (mM): NaCl 145, KCl 5, $CaCl_2$ 2.5, $MgCl_2$ 1, glucose 10, Hepes 10, pH 7.4. Relaxing solution contained: K-glutamate 125, Na_2ATP 5, Na_2CP (CP: creatine phosphate) 10, $CaCl_2$ 0.13, $MgCl_2$ 6, EGTA 1, glucose 10, Hepes 10, pH 7.0. Internal K-glutamate solution contained: K-glutamate 124, Na_2ATP 5.3, Na_2CP 10.6, $CaCl_2$ 0.19, $MgCl_2$ 7.6, EGTA 0.96, glucose 10, sucrose 10, Hepes 10, pH 7.0. In the latter solutions, small amounts of sucrose were added to adjust osmolarity to 310-320 mOsm in order to avoid shrinking or swelling of muscle fibres between solution exchanges that would otherwise affect triad microstructure and ECRE [21]. The composition of the solution was calculated using the program React II (provided by Dr. G.L. Smith, University of Glasgow). For both relaxing and internal solution, nominal free $[Ca^{2+}]$ was 100 nM and $[Mg^{2+}]$ was 1 mM.

2.4. ECRE recording protocols and data analysis
ECRE frequency and morphology was determined from image time series (XYT) taken from the fibre middle and repetitive recordings of one spatial dimension along a scan line within the fibre area (XT, 'line-scans', see Fig.1). XYT series consisted of 50 to 200 consecutive images (512x512, 0.8 fps). Although present within the fibre and the external space at equal concentrations, Fluo-4 is known to unspecifically bind to sarcomeric proteins, thus increasing the signal contrast within the fibre [22]. From such images (Fig.1), the cell borders can clearly be visualized. The fibre area scanned in the XYT series was calculated from the voxel size at a given zoom factor. XT images were acquired as 512x512 images with a speed of 2.5 ms per line. ECRE (i.e. Ca^{2+} sparks and embers [15]) were automatically detected in the line-scans using an automated denoising, event detection and analysis algorithm based on the '*à trous*'

implementation of the discrete wavelet transform. The IDL source code (IDL Res. Systems, Boulder, CO, USA) and its application both on synthetic and experimental data in different experimental environments has been published elsewhere [23, 24]. Images were background normalized to obtain a F/F_0 representation. Following wavelet-based denoising, automated event detection was performed using a hard-thresholding technique applied to the wavelet scales, from which the original event was reconstructed [23]. The ECRE spatial morphology and kinetics were quantified by fitting a gaussian to the F/F_0-X plot from which the ECRE amplitude A and the full-width at half-maximum (FWHM) as a measure for the spatial extent of the ECRE were extracted (Fig.1B). The rise-time (RT) was defined as the time from 10 % over baseline F/F_0 to reach the maximum amplitude and the full-duration at half-maximum (FDHM) was a measure for the ECRE duration [25]. In XYT series, ECRE were detected using a modified version of the denoising and detection algorithm that included background estimation over all images [26]. Due to the fast characteristics of ECRE in mammalian muscle with durations between ten and several tens of ms, isolated ECRE did not span multiple images (0.8 Hz frame rate). ECRE were counted and the frequency was expressed as events per s per mm^2 fibre area. All XYT series were included in the analysis. Note that due to the low overall ECRE frequencies, some XYT series contained no events. This, and the fact that during the scanning of the image ECRE may occur randomly in other parts of the image will be missed (ECRE duration $\ll$ image scan duration), will render our mean values for ECRE frequency statistics a lower limit for true ECRE frequencies. Differences in parameters following each pressure treatment compared to controls were assessed using Student's t-test whereas differences among pressure groups were tested for using either one-way ANOVA or pairwise multiple comparison procedures (Holm-Sidak method). $P < 0.05$ was considered significant. Outliers were detected in the final bin of data of each pressure using a Grubbs-test at the $P = 0.05$ level.

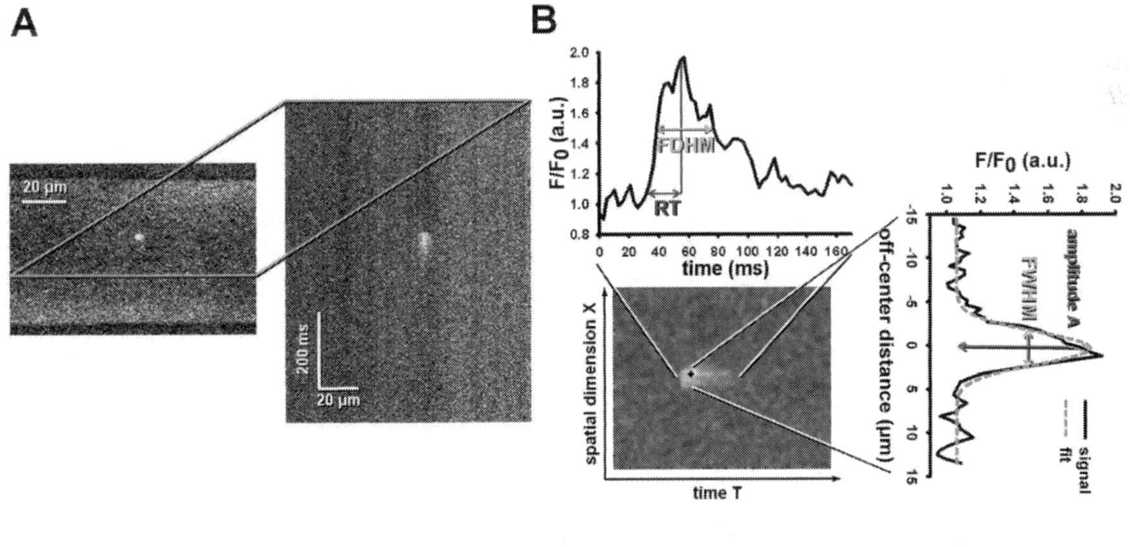

Figure 1. Example of ECRE morphology analysis from line-scans. **A,** example image shows Fluo-4 fluorescence of a single fibre showing an ECRE and the localization of the scan line used for XT imaging. Note that the XYT- and line-scan shown was cut from the 512x512 image for better visualization of the Ca^{2+} spark. **B,** Fluorescence analysis from a line-scan using the event detection algorithm after wavelet based denoising of the image. Cross within the spark indicates region of peak fluorescence.

2.5. Activation volumes for irreversible, pressure-induced alteration of ECRE dynamics

The ECRE dynamics of activation are described by the rise time RT, the ECRE duration (as an apparent measure for activation and inactivation processes) and is represented by the FDHM. From the pressure dependence of these parameters, activation volumes for the pressure-induced alteration of channel kinetics that persist in the post-decompression phase [12] can be formally calculated according to the equation [27]:

$$\left(\frac{\partial ln\kappa}{\partial p}\right)_T = -\frac{\Delta V^{\ddagger}}{RT} \tag{1}$$

where κ denotes the appropriate rate constants κ_{RT} and κ_{FDHM} related to RT^{-1} and $FDHM^{-1}$, respectively. R and T have their usual thermodynamic meanings. Null hypothesis was tested in case of very small activation volumes.

3. Results

3.1. ECRE frequency following prolonged high pressure treatments

Fig.2A shows a selection of images from a representative XYT series recorded in a single fibre following a 3 h 10 MPa high pressure treatment. At two time points, ECRE within the fibre can be detected (arrows). Fig.2B shows the frequency distribution in several image series from different single fibres for the pressure treatments indicated. ECRE frequency was (mean $\pm$SEM) $36.0 \pm 15.2 \frac{ECRE}{mm^2 \cdot s}$ under control conditions (atmospheric pressure and 4 °C for 3 h, n= 38), $13.7 \pm 2.3 \frac{ECRE}{mm^2 \cdot s}$ following 10 MPa (n= 55 XYT scans), $5.4 \pm 1.6 \frac{ECRE}{mm^2 \cdot s}$ following 15 MPa n= 39), $19.0 \pm 4.4 \frac{ECRE}{mm^2 \cdot s}$ following 20 MPa (n= 26) and $2.5 \pm 2.7 \frac{ECRE}{mm^2 \cdot s}$ following 30 MPa (n= 21) treatments. ECRE frequencies following high pressure treatments showed a tendency for smaller values compared to control conditions that became, however, only significant for 15 MPa treatments ($P < 0.03$). Among pressure treatments, 10 MPa and 20 MPa treatments showed pairwise significantly larger ECRE frequencies compared to 15 MPa and 30 MPa treatments. The decline in ECRE frequency could be described by an exponential with a pressure constant of 13.6 MPa (dashed line in Fig.2B, $r^2 = 0.76$).

3.2. ECRE morphology following prolonged high pressure treatments

Fig.3 shows example distribution histograms for FWHM (A, left panel) and FDHM (B, left panel) in control fibres and following prolonged high pressure treatments (A: 15 MPa, B: 20 MPa). In both cases, a right-shift of the histograms towards larger spatial and temporal morphology values could be observed . This trend was also seen when evaluating all pressures up to 20 MPa (right panels in Fig.3 A & B). Interestingly, for pressures larger than 20 MPa (i.e. 30 MPa) this behavior was reversed and values sharply declined. In contrast, amplitudes and rise-times (RT) were less affected by pressure and were not significantly different from controls with the exception of amplitudes following 15 MPa treatments that were significantly increased. Values among the pressure treated groups were never significantly different as indicated by the bars in Fig.3. For the ECRE morphologies shown in the figure, n= 139 ECRE under control conditions, n= 16 ECRE following 10 MPa, n= 12 following 15 MPa, n= 17 following 20 MPa and n= 3 following 30 MPa treatments were evaluated. Note that for pressures above 20 MPa survival of muscle cells sharply declined during the prolonged 3 h HP treatment in agreement with our previous observations in mammalian muscle [17, 18].

3.3. Activation volumes for pressure induced alteration of ECRE kinetics

Fig.4 shows the rate constant vs. pressure relationships according to eq.1 for the rise-time (RT) and FDHM of ECRE. From this, the activation volumes $\Delta V^{\ddagger}$ for the pressure-driven irreversible alteration of release channel function that persisted within the post-decompression phase were

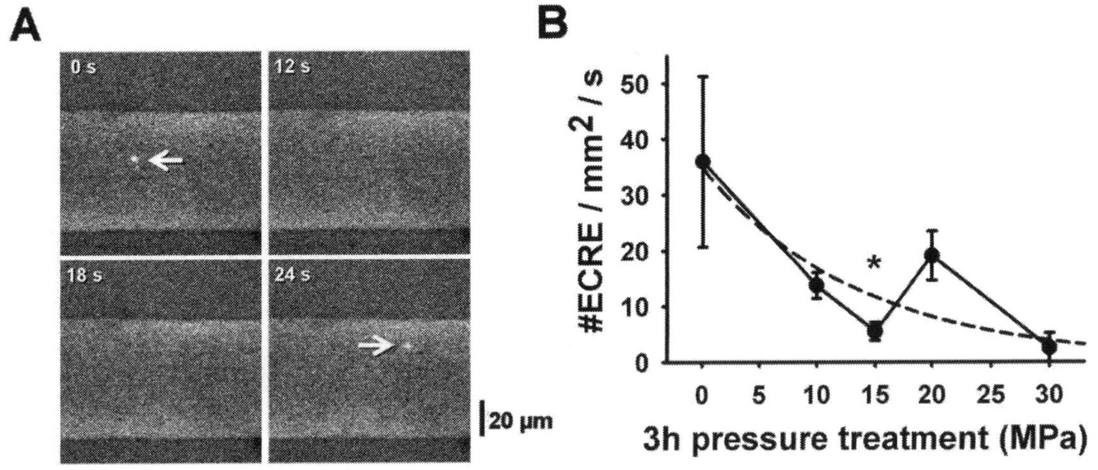

Figure 2. Pressure dependence of ECRE frequency recorded in the post-decompression phase following 3 h high pressure treatments. **A,** example XY Fluo-4 fluorescence images taken at the indicated time points in a single fibre following a 3 h 10 MPa treatment. Arrows: ECRE. **B,** ECRE frequency-pressure relation. *: $P < 0.05$ vs. control.

calculated. $\Delta V^{\ddagger}$ was virtually zero for RT indicating no alteration of ECRE activation. On the contrary, FDHM showed $\Delta V^{\ddagger}$ values of $-18.3\AA^3$ when considering the whole pressure range. The negative value of $\Delta V^{\ddagger}$ would suggest a negative volume change of the underlying process that is accelerated by pressure. However, FDHM tended to increase with pressure up to 20 MPa resulting in temporally more broadened ECRE. Therefore, we assumed a two-step process for pressure-induced alteration in ECRE mechanisms underlying FDHM similar as previously described by us for the activation of tubular Ca^{2+} currents [12]. When applying eq.1 to the pressure range up to 20 MPa, $\Delta V^{\ddagger}$ was $\sim +60\AA^3$ and reversed for larger pressures ($\Delta V^{\ddagger} \sim -225\AA^3$, Fig.4B).

4. Discussion

In the present study we present, to our knowledge for the first time, a detailed analysis of microdomain Ca^{2+} dynamics in mammalian skeletal muscle following prolonged 3 h HP treatments. Spontaneous elementary Ca^{2+} release events (ECRE) in permeabilized mammalian skeletal muscle fibres have been extensively investigated and characterized in the last couple of years [15, 24, 28, 29].

4.1. Pressure effects on ECRE frequencies

Under control atmospheric conditions, ECRE frequencies were somewhat lower than previously described ($\sim 40s^{-1}mm^{-2}$ in the present study compared to $\sim 150s^{-1}mm^{-2}$ [15]). Due to the stochastic character of ECRE appearance in XYT images, this frequency is affected by various factors like the scanning speed, proportion of fibre area within the image, cytosolic Ca^{2+} and Mg^{2+} concentration [25, 28] or filling of the SR. Following HP treatments, ECRE frequency declined exponentially. The pressure constant of ~ 14 MPa detected in the post-decompression phase is larger than the one found for global decline in resting cytosolic Ca^{2+} fluorescence (~ 5 MPa) but smaller than the decline in mitochondrial Ca^{2+} fluorescence (~ 20 MPa) detected during pressurization using high pressure microscopy [19]. However, these conditions cannot be directly compared. From our previous interpretations, we suggested a pressure-induced Ca^{2+}

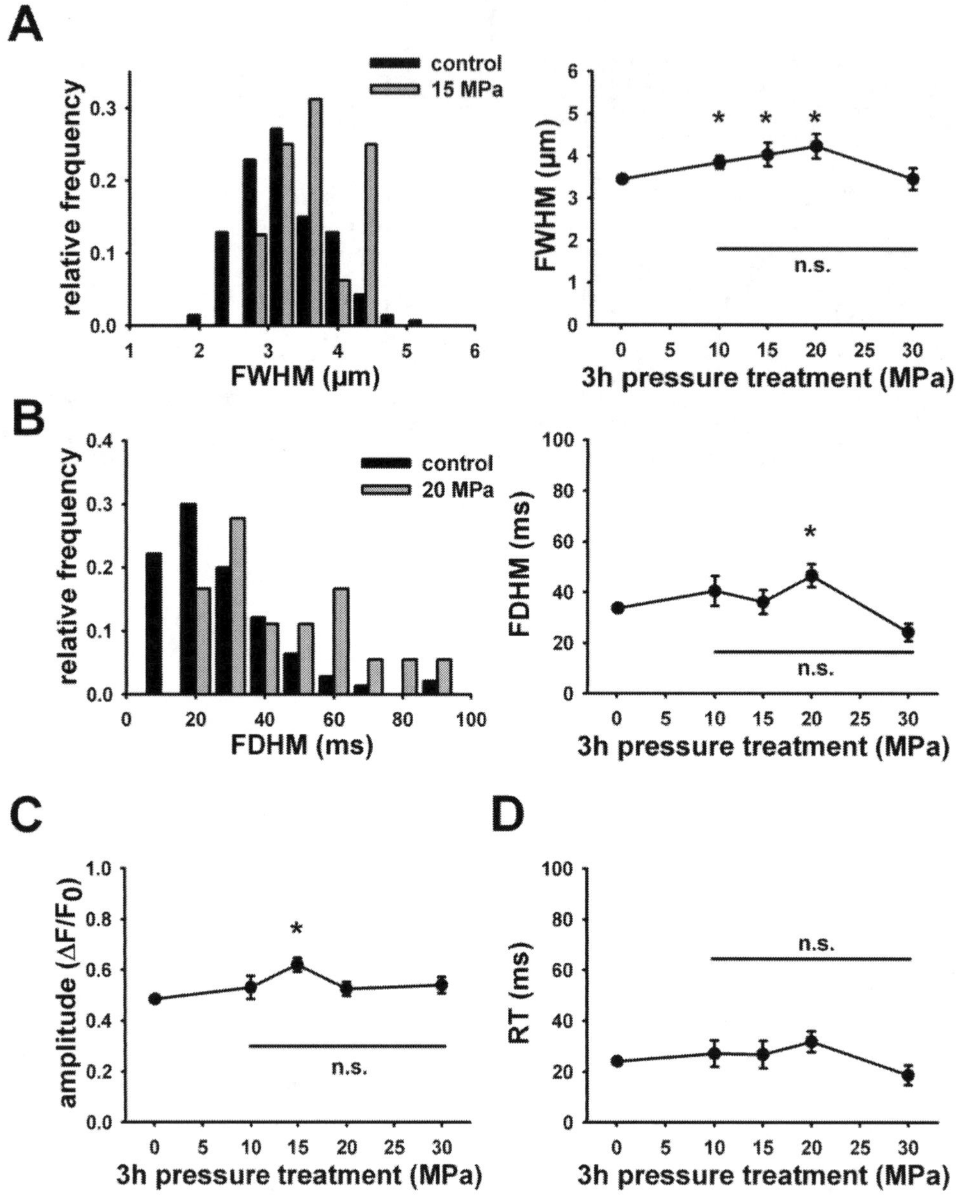

Figure 3. Pressure dependence of ECRE morphologies recorded in the post-decompression phase following 3 h high pressure treatments. Histograms of data distribution (left) and pressure dependence of mean values (right) for FWHM (**A**) and FDHM (**B**). **C, D,** pressure dependence of mean values for ECRE amplitudes and RT, respectively. *: $P < 0.05$ vs. control.

leakage from the SR during the HP treatment, either via increased ECRE frequency, reduced SR pump activity or a combination of both [19]. Similar results were also obtained in pressurized SR vesicles from rabbit muscle [30]. It should be noted that in the current study, ECRE were detected in the post-decompression phase. Therefore, a lower overall ECRE frequency would be in agreement with a lower filling of the SR under these conditions.

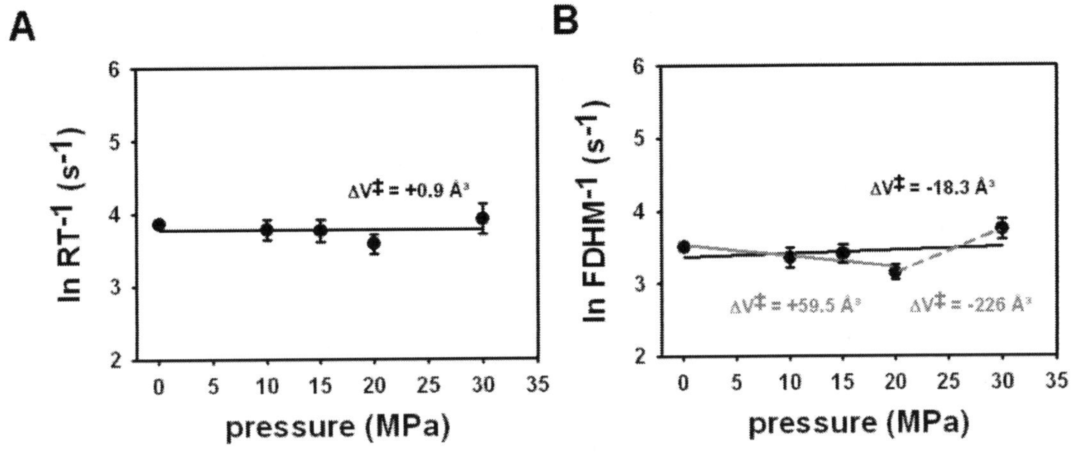

Figure 4. Apparent activation volumes for pressure driven irreversible alteration of ECRE RT (**A**) and FDHM (**B**) detected in the post-decompression phase after 3 h prolonged HP treatment. See text for details.

4.2. Pressure effects on ECRE morphologies

ECRE parameters of chemically permeabilized fibres under control conditions had amplitudes similar to those given by Zhou et al. [25] but smaller than originally described by us in a previous study [15]. This can be partly explained by a smaller dynamic range for Ca^{2+} fluorescence in the present study due to larger photomultiplier voltages used. This would also result in amplified background fluorescence and reduced signal range. Subsequently, as a portion of low-amplitude ECRE will be left undetected, ECRE frequencies might actually be higher than in the current setting. However, this does not affect our interpretations regarding pressure as settings were kept constant.

Rise times (RT) that reflect the RyR channel open time [28] were similar to previous values [15, 25]. The relative pressure resistance of ECRE amplitudes and RT that is also reflected by the latter's minute activation volume suggest no persistent irreversible alteration of RyR opening and maximum Ca^{2+}-release into the narrow, diffusion restricted triad space [31]. This parallels recent findings from plasma membrane Na^{+} channels that also showed no pressure-induced alteration in activation and even inactivation kinetics despite a reduction in number of functional channels [12]. However, tubular Ca^{2+} channels were much more susceptible to the pressure treatment with markedly slowed activation and inactivation kinetics. The increase in FDHM for ECRE with pressure may also reflect a tendency towards increased inactivation kinetics for RyR, however, this cannot unambiguously be judged at the present state as FDHM is also determined by RT and cytosolic Ca^{2+} buffer properties. Similarly, the spatial and temporal broadening of ECRE found after HP treatments might also reflect altered buffer properties rather than inactivation kinetics of the RyR itself, although further research is needed to address this point.

4.3. Pressure sensitivity of RyR

From the decrease in ECRE frequency a reduction in number of functional RyR by pressure similar to that found for Na^{+} and Ca^{2+} channels can be suggested. However, the pressure sensitivity of RyR seems to be more similar to surface Na^{+} channels but much less pronounced compared to L-type Ca^{2+} channels (DHPR) in our preparation. This is particularly interesting as DHPR and RyR are mechanically linked via a direct protein-protein interaction to confer membrane excitation to SR Ca^{2+} release. Another point to be addressed is the relatively lower

cholesterol content of endoplasmic reticulum membrane [20]. This would suggest a much higher pressure sensitivity of this organelle and its embedded ion channels. The contrary seems to be true at least for the RyR itself which did not show a major breakdown of function following prolonged high pressure treatments.
Our results are the first to show that ion channels in an intracellular environment may be relatively more protected from high hydrostatic pressure effects compared to surface ion channels in mammalian skeletal muscle.

References

[1] R Winter and W Dzwolak. Exploring the temperature-pressure configurational landscape of biomolecules: from lipid membranes to proteins. *Philos Transact A Math Phys Eng Sci*, 363(1827):537–562, 2005.
[2] HMW Okland, IS Stoknes, JF Remme, M Kjerstad, and M Synnes. Proximate composition, fatty acid and lipid class composition of the muscle from deep-sea teleosts and elasmobranchs. *Comp Biochem Physiol B Biochem Mol Biol*, 140(3):437–443, Mar 2005.
[3] JR Hazel and EE Williams. The role of alterations in membrane lipid composition in enabling physiological adaptation of organisms to their physical environment. *Prog Lipid Res*, 29(3):167–227, 1990.
[4] AG Macdonald and AR Cossins. The theory of homeoviscous adaptation of membranes applied to deep-sea animals. *Symp Soc Exp Biol*, 39:301–322, 1985.
[5] AL Samerotte, JC Drazen, GL Brand, BA Seibel, and PH Yancey. Correlation of trimethylamine oxide and habitat depth within and among species of teleost fish: an analysis of causation. *Physiol Biochem Zool*, 80(2):197–208, 2007.
[6] PH Yancey. Organic osmolytes as compatible, metabolic and counteracting cytoprotectants in high osmolarity and other stresses. *J Exp Biol*, 208(Pt 15):2819–2830, 2005.
[7] PJ Butler and DR Jones. Physiology of diving of birds and mammals. *Physiol Rev*, 77(3):837–899, 1997.
[8] AE Talpalar and Y Grossman. CNS manifestations of HPNS: revisited. *Undersea Hyperb Med*, 33(3):205–210, 2006.
[9] AA Harper, AG Macdonald, and KT Wann. The action of high hydrostatic pressure on the membrane currents of Helix neurones. *J Physiol*, 311:325–339, 1981.
[10] NS Fortune, MA Geeves, and KW Ranatunga. Contractile activation and force generation in skinned rabbit muscle fibres: effects of hydrostatic pressure. *J Physiol*, 474(2):283–290, 1994.
[11] L Smeller. Pressure-temperature phase diagrams of biomolecules. *Biochim Biophys Acta*, 1595(1-2):11–29, 2002.
[12] O Friedrich, KR Kress, M Hartmann, B Frey, K Sommer, H Ludwig, and RHA Fink. Prolonged high-pressure treatments in mammalian skeletal muscle result in loss of functional sodium channels and altered calcium channel kinetics. *Cell Biochem Biophys*, 45(1):71–83, 2006.
[13] MW Berchtold, H Brinkmeier, and M Muentener. Calcium ion in skeletal muscle: its crucial role for muscle function, plasticity, and disease. *Physiol Rev*, 80(3):1215–1265, 2000.
[14] SQ Wang, C Wei, G Zhao, DXP Brochet, J Shen, LS Song, W Wang, Di Yang, and H Cheng. Imaging microdomain Ca2+ in muscle cells. *Circ Res*, 94(8):1011–1022, 2004.
[15] WG Kirsch, D Uttenweiler, and RHA Fink. Spark- and ember-like elementary Ca2+ release events in skinned fibres of adult mammalian skeletal muscle. *J Physiol*, 537(Pt 2):379–389, 2001.
[16] H. Cheng, W. J. Lederer, and M. B. Cannell. Calcium sparks: elementary events underlying excitation-contraction coupling in heart muscle. *Science*, 262(5134):740–744, Oct 1993.
[17] KR Kress, O Friedrich, H Ludwig, and RHA Fink. Reversibility of high pressure effects on the contractility of skeletal muscle. *J Muscle Res Cell Motil*, 22(4):379–389, 2001.
[18] O Friedrich, KR Kress, H Ludwig, and RHA Fink. Membrane ion conductances of mammalian skeletal muscle in the post-decompression state after high-pressure treatment. *J Membr Biol*, 188(1):11–22, 2002.
[19] O Friedrich, F vWegner, M Hartmann, B Frey, K Sommer, H Ludwig, and RHA Fink. 'In situ' high pressure confocal Ca(2+)-fluorescence microscopy in skeletal muscle: a new method to study pressure limits in mammalian cells. *Undersea Hyperb Med*, 33(3):181–195, 2006.
[20] P Mentre, L Hamraoui, G Hui Bon Hoa, and P Debey. Pressure-sensitivity of endoplasmic reticulum membrane and nucleolus as revealed by electron microscopy. *Cell Mol Biol (Noisy-le-grand)*, 45(3):353–362, 1999.
[21] CA Martin, N Petousi, S Chawla, AR Hockaday, AJ Burgess, JA Fraser, CLH Huang, and JN Skepper. The effect of extracellular tonicity on the anatomy of triad complexes in amphibian skeletal muscle. *J Muscle Res Cell Motil*, 24(7):407–415, 2003.

[22] S Hollingworth, C Soeller, SM Baylor, and MB Cannell. Sarcomeric Ca2+ gradients during activation of frog skeletal muscle fibres imaged with confocal and two-photon microscopy. *J Physiol*, 526 Pt 3:551–560, 2000.

[23] F v Wegner, M Both, and RHA Fink. Automated detection of elementary calcium release events using the a trous wavelet transform. *Biophys J*, 90(6):2151–2163, 2006.

[24] F v Wegner, M Both, RHA Fink, and O Friedrich. Elementary Ca2+ release events in mammalian skeletal muscle: effects of the anaesthetic drug thiopental. *J Muscle Res Cell Motil*, 27(5-7):315–326, 2006.

[25] J Zhou, G Brum, A Gonzalez, BS Launikonis, MD Stern, and E Rios. Ca2+ sparks and embers of mammalian muscle. Properties of the sources. *J Gen Physiol*, 122(1):95–114, 2003.

[26] F v Wegner, M Both, RHA Fink, and O Friedrich. Fast xyt imaging of elementary calcium release events in muscle with multifocal multiphoton microscopy and wavelet denoising and detection. *IEEE Trans Med Imaging*, in press, 2007.

[27] R Meyer and SH Heinemann. Temperature and pressure dependence of Shaker K+ channel N- and C-type inactivation. *Eur Biophys J*, 26(6):433–445, 1997.

[28] J Zhou, BS Launikonis, E Rios, and G Brum. Regulation of Ca2+ sparks by Ca2+ and Mg2+ in mammalian and amphibian muscle. An RyR isoform-specific role in excitation-contraction coupling? *J Gen Physiol*, 124(4):409–428, 2004.

[29] L Csernoch, J Zhou, MD Stern, G Brum, and E Rios. The elementary events of Ca2+ release elicited by membrane depolarization in mammalian muscle. *J Physiol*, 557(Pt 1):43–58, 2004.

[30] A Okamoto, A Suzuki, Y Ikeuchi, and M Saito. Effects of high pressure treatment on Ca2+ release and Ca2+ uptake of sarcoplasmic reticulum. *Biosci Biotechnol Biochem*, 59(2):266–270, 1995.

[31] YH Jiang, MG Klein, and MF Schneider. Numerical simulation of Ca(2+) "Sparks" in skeletal muscle. *Biophys J*, 77(5):2333–2357, 1999.

Evidence for high-pressure-induced rupture of hydrogen bonds in LH2 photosynthetic antenna pigment-protein complexes

Liina Kangur[1], **Kristjan Leiger**[1], **and Arvi Freiberg**[1, 2]
[1]Institute of Physics, University of Tartu, Riia 142, Tartu 51014, Estonia
[2]Institute of Molecular and Cell Biology, University of Tartu, Riia 23, Tartu 51010, Estonia

Abstract. The bacteriochlorophyll *a*-containing LH2 light harvesting complex is an integral membrane protein that catalyzes the photosynthetic process in purple photosynthetic bacteria. The LH2 complexes from *Rhodobacter sphaeroides* show characteristic strong absorbance at 800 and 850 nm due to the bacteriochlorophyll *a* molecules confined in two separate areas of the protein. Using these cofactors as intrinsic probes to monitor changes in membrane protein structure, we investigate the response to high hydrostatic pressure up to 2.1 GPa of LH2 complexes embedded into natural membrane environment or extracted with detergent. We demonstrate that high pressure does induce significant alterations to the tertiary structure of the protein in proximity of the protein-bound bacteriochlorophyll *a* molecules, including breakage of the hydrogen bond they are involved in. The membrane-embedded complexes appear more resilient to damaging effects of the compression than the complexes extracted into detergent environment. This difference has tentatively been explained by more compact structure of the membrane-embedded complexes.

1. Introduction

Membrane proteins are ubiquitous in life-supporting processes, such as signal transmission, immune response, photosynthesis, and respiration, to name but a few (see [1] for a review). The peripheral antenna of purple photosynthetic bacteria, or light-harvesting complex 2 (LH2), is one of the best-characterized membrane proteins. The crystal structure of isolated LH2 (from *Rhodopseudomonas (Rps.) acidophila* strain 10050) [2] reveals a highly symmetric ring of nine pigment-protein subunits, each containing two (α and β) helical membrane-spanning polypeptides, three non-covalently bound bacteriochlorophyll *a* (Bchl) molecules, and a carotenoid pigment. The α- and β-polypeptides form two concentric cylinders providing, respectively, inside and outside support to the cofactor rings between them (see figure 1). A ring of 18 tightly coupled Bchl cofactors in a waterwheel-like arrangement are seen in the lumenal part of the photosynthetic membrane (bottom side of figure 1). It is responsible for the intense near infrared absorption of the LH2 complex at about 850 nm (thus the name, B850). The position of the B850 cofactors relative to each other is determined by hydrogen bonds to the surrounding protein as well as by coordinative bonds with the central magnesium ion of the Bchls. Another ring of 9 Bchl molecules is located towards the polar cytoplasmic part of the membrane (top side of figure 1), being in charge with the absorption band at 800 nm (B800). Both the B850 and B800 bands are related to the Q_y electronic transition in the component Bchl molecules.

High-pressure modulation of structure and function of proteins is an established technique (see [3] for a review). Providing fundamental insights about phase properties, high-pressure measurements can contribute into understanding of various elements that help to stabilize, purify, and storage the membrane proteins. Only few high-pressure studies have so far concerned the LH2 antenna complexes [4-7]. These early works, which generally were limited to 0.6 GPa, have shown that increasing the pressure at physiological temperatures causes substantial red shift and broadening of the Bchl absorption bands. A loss of the 800 nm-absorbing Bchls in the purified LH2 complexes from *Rhodobacter (Rb.) sphaeroides* 2.4.1 and *Rps. acidophila* 10050 suggested pressure-induced alterations to the tertiary structure of the protein in proximity to the membrane/cytosol interface [6].

© 2008 IOP Publishing Ltd

However, no significant perturbation of the oligomerization state of the polypeptides, distortion of the protein-bound B850 Bchl molecules, or breakage of the hydrogen bond they are involved in was observed.

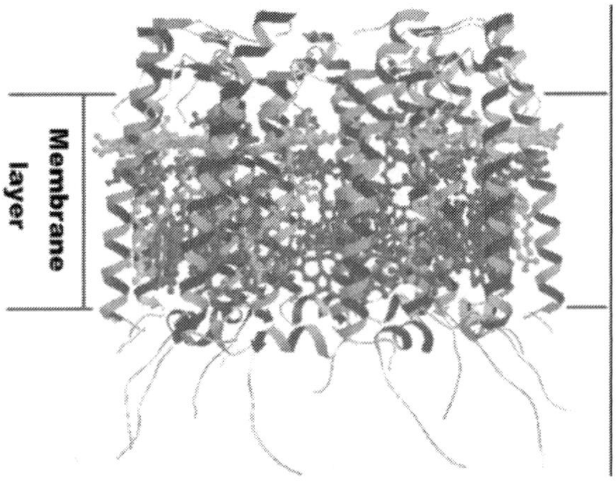

Figure 1. Structural view of the LH2 pigment-protein complex. For clarity, only the B850 (red) and B800 (green) Bchl pigment cofactors and transmembrane α- helices are shown. The lipid bilayer covering hydrophobic parts of the complex is also schematically presented.

That the oligomeric protein does not dissociate at pressures as high as 0.6 GPa is surprising. From literature, it is known that most of the multi-chain soluble proteins unfold at room temperature already below 0.2 GPa [8]. This process is driven by a decrease in volume, which results from both the release of intra-molecular voids and the expose of the interior of the protein to solvent. Thus motivated, we make a step forward and investigate a spectral response of the LH2 complex to unusually high for the proteins hydrostatic pressures up to 2.1 GPa. A comparison is made between the electronic absorption spectra of Bchl pigments in LH2 complexes from *Rb. sphaeroides* either isolated or embedded in the native membranes.

2. Experimental

Membrane vesicles (chromatophores) of genetically engineered *Rb. sphaeroides* bacteria in which only the LH2 antenna protein is synthesized were prepared as described earlier [9]. The protein stocks were stored at liquid nitrogen temperature and thawed before using. The samples were diluted with a buffer containing 20 mM Tris, 1 mM EDTA, 0.1 M NaCl (pH 8.0) to obtain an optical density of about 0.4 at the B850 absorption band maximum in the assembled high-pressure cell. As shown in [10], the pH of the Tris buffer is almost pressure insensitive. About 1% (w/w) lauryldimethylaminoxide (LDAO) in the solution with the purified LH2 complexes was used to ensure well-isolated complexes in detergent micelles [9]. No degradation of the samples was observed within 23 hours at ambient pressure and temperature.

A micro-spectroscopy setup was arranged around a Zeiss Axioskop 2 microscope, using a 30 W power-stabilized halogen incandescent lamp as a light source. The light coming through the sample in a diamond anvil cell (DAC, D-02 from Diacell Products Ltd.) was collected by the microscope's objective and focused into a 200 μm diameter glass fiber. The absorption spectra were recorded either with a BWTek BTC111E series CCD spectrometer or a 0.3 m spectrograph (Shamrock SR-303i, Andor), combined with an electrically cooled CCD camera (EEV 30-11, Andor).

The sample solution was injected into a 0.3 mm-diameter hole in about 0.35-mm thick stainless steel gasket of the DAC. After the DAC closure the pressure was increased with an average rate of 15-150 MPa per minute in different experiments. The pressure in DAC was determined following a shift of a 690.3 nm emission line of a Sm^{2+}-doped SrFCl micro-crystalline pressure sensor mounted into the sample volume. The pressure sensitivity of 1.1 nm/GPa of this sensor is linear over a broad pressure range [11]. The luminescence of the sensor crystal was recorded by a 0.85 m SPEX 1402 spectrometer. The absorption spectra measured at different pressures were corrected for the background absorbance of the DAC, filled with only the buffer solution and the detergent.

3. Results and discussion

3.1. Absorption spectra of LH2 complexes at ambient pressure

Figure 2A compares the absorption spectra of isolated and membrane-soaked LH2 complexes of *Rb. sphaeroides* measured at ambient pressure and temperature over the spectral range of 550 – 1000 nm. As seen, the spectra of both samples, revealing two intense, well-separated B800 and B850 bands of expected stoichiometric integral intensity ratio of ~1/2, match perfectly well.

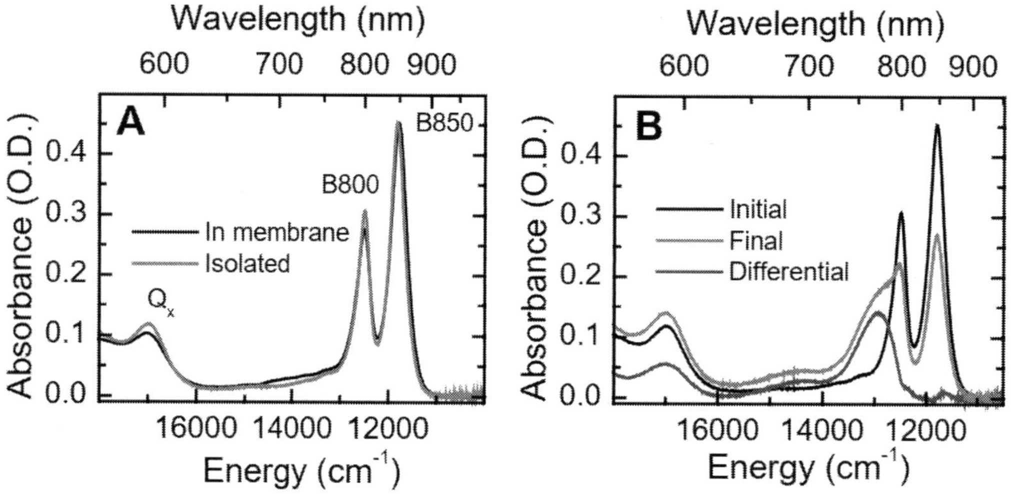

Figure 2. (A) Absorption spectra of membrane-associated and LDAO-isolated LH2 complexes from *Rb. sphaeroides* at ambient pressure. (B) The spectra of LDAO-isolated LH2 complexes recorded at ambient pressure before and after the measurement cycle of figure 3. The difference spectrum (blue curve) was acquired by subtracting the initial spectrum (black curve) from the normalized (by the B850 peak intensity) final spectrum (red curve) under a simplifying assumption that the B800 and B850 bands decay proportionally under pressure.

Figure 2B demonstrates an incomplete recovery of the spectrum of the detergent-isolated complexes after pressure release, subsequent to about 2.5 hours trials of figure 3A where maximum pressure of 2.1 GPa was attained. The difference spectrum that was acquired by subtracting the initial ambient-pressure spectrum from the final spectrum after pressure release shows a maximum at 773.4 nm. The 750-780 nm spectral region is commonly related to non-functional (solubilized) pigments [12], suggesting that the difference spectrum in figure 2B is due to individual Bchl molecules in contact with the buffer-detergent solvent phase. Assuming that the absorbance of the pigments either solubilized or organized into the B850 aggregate is roughly the same, we estimate that in total about 40% of the LH2 complexes in figure 2B have been broken down by the pressure treatment. This is in stark contrast with the membrane-protected particles that largely maintain their intactness throughout the pressure handling (data not shown).

3.2. Pressure-induced variations of the absorption spectra

Pressurizing of the membrane-associated LH2 complexes causes a continuous, quasi-linear shift of the B800 and B850 absorption bands to longer wavelengths (red shift), accompanied with some band broadening. In the low-pressure region where comparison is feasible, the results of the present work are rather similar to the previous data [5,6]. The red shift rate of the B850 band (−626± 40 cm^{-1} GPa^{-1}) more than three times surpasses that of the B800 band (-204 ± 23 cm^{-1} GPa^{-1}). According to figure 3B there is considerable slow-down of the shift rate of both the B800 and B850 bands at ~0.6 GPa.

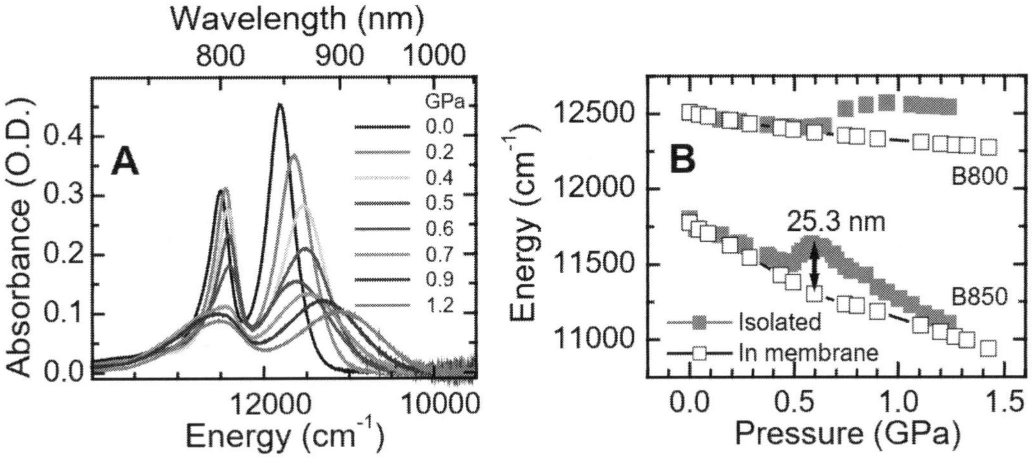

Figure 3. (A) Pressure dependence of the absorption spectra of LDAO-isolated LH2 complexes from *Rb. sphaeroides*. (B) Absorption band positions for the membrane-associated (open rings) and the purified isolated (filled squares) LH2 complexes as a function of pressure.

For the detergent-isolated LH2 complexes, the pressure dependence of the absorption spectra is shown on figure 3A. At low pressures up to ~0.5 GPa the spectra behave similar to the ones of the membrane-associated complexes. However, toward higher pressures striking, previously unobserved, differences appear. Firstly, the usual red shift of the B850 band is abruptly, between 0.5 and 0.6 GPa, replaced by a considerable blue shift that in its maximum deviation from the absorption peak position of the membrane samples reaches 332 cm^{-1} or 25.3 nm. Past ~0.6 GPa the blue shift of the B850 band is yet again overtaken by a red shift. The spectra of the isolated and membrane-associated complexes approach each other once more at pressures above ~1.3 GPa. Secondly, as already mentioned, a partial breakdown of purified complexes takes place during the pressure treatment. In figure 3A, this is demonstrated by simultaneous decrease of both the B800 and B850 bands and concomitant increase of the short-wavelength shoulder of the B800 band. The pressure damage intensifies past ~0.7 GPa, as evidenced by the strongly deviated courses of the B800 band in purified and membrane complexes (figure 3B). The damage is permanent, so that the final spectrum measured after passing the whole pressure cycle is rather unlike the spectrum before squeezing (see figure 2B).

A blue shift at ambient pressure of the B850 absorption band was previously observed for modified LH2 complexes of *Rb. sphaeroides* [13]. By site-directed mutagenesis analysis, this shift was correlated with a breakage of both 2-acetyl carbonyl hydrogen bonds that coordinate the two Bchls in the basic unit of the B850 annular to the surrounding protein (see Introduction). It was also demonstrated that the changed hydrogen bonding pattern leads to ~26 nm blue shift in the B850 absorbance measured at 77 K. A comparison with the present 25.3 nm pressure-induced shift in figure 3B strongly implies that essentially all hydrogen bonds to the B850 molecules are broken by external high hydrostatic pressure.

4. Concluding remarks

The lipid bilayer membrane is a natural background for the LH2 protein and it is quite expected that the membrane-embedded complexes appear more resilient to the damaging effects of compression by external hydrostatic pressure than the complexes extracted into the fatty detergent environment. For the well-isolated antenna complexes we have clearly demonstrated that high pressure does induce significant distortion not only of the binding site of the protein-bound 800 nm-absorbing Bchl molecules as found earlier [6], but also of the 850 nm-absorbing Bchls, including rapture of the hydrogen bonds they are involved in. While in the membranes the pressure-induced alterations to the tertiary structure of the LH2 protein seem to concentrate in proximity of the membrane/cytosol interface at the N-terminal side, in isolated species the whole bulk of the protein is involved.

A change of the pressure-induced absorption band shift rate for the membrane-associated LH2 complexes around 0.6 GPa (figure 3B) implies significant decrease of the compressibility of the samples at higher pressures. The dependences for isolated and membrane complexes, being close below 0.5 GPa, approach each other again at pressures around 1.3 GPa. It is thus quite probable that also in the membranes the pressure-induced distortions of the hydrogen bond pattern take place that, however, are more gradual and less shattering than observed for the isolated complexes.

Acknowledgements

The Estonian Science Foundation grant No. 7002 supported this work. The authors are grateful to R. Jaaniso for graciously making available the pressure sensor crystals and the supporting infrastructure, to K. Reivelt for a loan of the BWTek spectrometer, to A. Ellervee for assistance with DAC instrumentation, and to K. Timpmann for providing the LH2 samples and for useful discussions.

References

[1] Palazzo G 2006 *Curr. Opin. Colloid Interface Sci.* **11** 65

[2] Papiz M Z, Prince S M, Howard T, Cogdell R J and Isaacs N W 2003 *J. Mol. Biol.* **326** 1523

[3] Scharnagl C, Reif M and Friedrich J 2005 *Biochim. Biophys. Acta* **1749** 187

[4] Tars M, Ellervee A, Kukk P, Laisaar A, Saarnak A and Freiberg A 1994 *Lithuanian J. Phys.* **34** 320

[5] Timpmann K, Ellervee A, Pullerits T, Ruus R, Sundström V and Freiberg A 2001 *J. Phys. Chem. B* **105** 8436

[6] Gall A, Ellervee A, Sturgis J N, Fraser N J, Cogdell R J, Freiberg A and Robert B 2003 *Biochemistry* **42** 13019

[7] Braun P, Gebhardt R, Kwa L and Doster W 2005 *Brazilian J. Med. Biol. Research* **38** 1273

[8] Boonyaratanakornkit B B, Park C B and Clark D S 2002 *Biochim. Biophys. Acta* **1595** 235

[9] Timpmann K, Woodbury N W and Freiberg A 2000 *J. Phys. Chem. B* **104** 9769

[10] Neuman R C, Kauzmann W and Zipp A 1973 *J. Phys. Chem.* **77** 2687

[11] Lorenz B, Shen Y R and Holzapfel W B 1994 *High Pressure Research* **12** 91

[12] Miller J F, Hinchigeri S B, Parkes-Loach P S, Callahan P M, Sprinkle J R, Riccobono J R and Loach P A 1987 *Biochemistry* **26** 5055

[13] Fowler G J S, Visschers, R W, Grief G G, Van Grondelle R and Hunter C N 1992 *Nature* **355** 848

Joint 21st AIRAPT and 45th EHPRG Int. Conf. on High Pressure Science and Technology IOP Publishing
Journal of Physics: Conference Series **121** (2008) 112005 doi:10.1088/1742-6596/121/1/112005

High pressure modulated transport and signalling functions of membrane proteins in models and *in vivo*

Rudi F. Vogel, Kai Linke, Holger Teichert, Matthias A. Ehrmann

Technische Universität München, Technische Mikrobiologie, Weihenstephaner Steig 16, 85350 Freising, Germany

rudi.vogel@wzw.tum.de

Abstract. Cellular membranes serve in the separation of compartments, recognition of the environment, selective transport and signal transduction. Membrane lipids and membrane proteins play distinct roles in these processes, which are affected by environmental chemical (e. g. pH) or physical (e. g. pressure and temperature) changes. High hydrostatic pressure (HHP) affects fluidity and integrity of bacterial membranes instantly during the ramp, resulting in a loss of membrane potential and vital membrane protein functions. We have used the multiple drug transporter LmrA from *Lactococcus lactis* and ToxR, a membrane protein sensor from *Photobacterium profundum*, a deep-sea bacterium, and *Vibrio cholerae* to study membrane protein interaction and functionality in proteolioposomes and by the use of *in vivo* reporter systems, respectively. Both proteins require dimerization in the phospholipid bilayer for their functionality, which was favoured in the liquid crystalline lipid phase with ToxR and LmrA. Whereas LmrA, which resides in liposomes consisting of DMPC, DMPC/cholesterol or natural lipids, lost its ATPase activity above 20 or 40 MPa, it maintained its active dimeric structure in DOPC/DPPC/cholesterol liposomes up to 120 MPa. By using a specific indicator strain in which the dimerisation of ToxR initiates the transcription of lacZ it was demonstrated, that the amino acid sequence of the transmembrane domain influences HHP stability of ToxR dimerization *in vivo*. Thus, both the lipid structure and the nature of the protein affect membrane protein interaction. It is suggested that the protein structure determines basic functionality, e.g. principle ability or kinetics to dimerize to a functional complex, while the lipid environment modulates this property.

1. Introduction

Cellular membranes are complex associations of macromolecules consisting of lipids and (glyco)proteins. The lipids in these systems are mostly phospholipids with hydrophilic (ionic) head groups and lipophilic fatty acid chains. This amphiphilic nature forces them to spontaneously associate to form bilayers or vesicles. These phospholipid bilayers comprise compartments of biological systems, which allow cellular functions in a controlled cytoplasmic space. The biochemical nature of the phospholipid side chains determines the basic characteristics of biological membranes with respect to their phase behavior, which can adopt several physical states between liquid crystalline and gel. These are influenced by physical (temperature, pressure) and biochemical (lipid composition, pH, ions) parameters. While pure lipids and artificial mixtures of few different lipids show distinct parameter sets for phase transitions and areas of co-existence, complex natural lipid mixtures exhibit smooth phase transitions. In the phospholipid bilayers of bacterial membranes a large variety of

© 2008 IOP Publishing Ltd

proteins are inserted, serving vital cellular functions of recognition, signal transfer or transport of metabolites and ions.

The effect of high hydrostatic pressure (HHP) on cellular systems has been extensively studied with foodborne bacteria, directed at the inactivation of pathogens and food spoiling bacteria. The vast majority of publications in this area describe bacterial inactivation in model and food systems along sigmoid inactivation curves revealing a threshold pressure before any effect can be seen and a very small number of survivors within the population, which were somehow protected or intrinsically resistant to HHP. Below the threshold pressure sublethal damage of bacteria has been described, which involves a cease of metabolic functions [17], ribosomal biosynthesis [14], impaired membrane integrity and fluidity [17] or lateral tension [16]. This initially reversible loss of HHP induced membrane functionality has been described to be causative for the killing of cells as a function of the holding time upon which environmental factors exert detrimental effects. While low pH acts synergistically with pressure to induce bacterial inactivation, molar concentrations of ions or neutral solutes stabilize bacterial membranes to near complete pressure resistance [12].

HHP has also widely been applied to investigate protein structure. The pressures used are very high (up to 1 GPa and beyond) and models used mostly employ high protein concentrations in buffer solutions. Therefore, the information obtained cannot easily be transferred to the conditions of a biological system. Furthermore, most proteins used were soluble and investigated in liquid systems, while information on membrane proteins in their natural lipid environment is scarce. With whole cells it has been shown that membrane fluidity affects the activity of membrane transport proteins [18]. It was therefore the aim of our study to elucidate the role of the phospholipid membrane´s ability to influence membrane protein function in response to environmental changes, which affect membrane biophysical properties. Two systems were developed and employed. Proteoliposomes with the multiple drug transporter LmrA from *Lactococcus lactis* were used to investigate the influence of membrane lipid composition and fluidity on membrane protein functionality. On the other hand, a reporter system was developed which allowed investigation of membrane protein interaction as a function of protein structure *in vivo*. In both systems, HHP was used to adjust and vary membrane physical parameters, namely transitions from liquid crystalline to gel phases and induction of raft formation.

2. Materials and Methods

2.1. HHP dependence of LmrA activity in proteoliposomes

2.1.1. Production of LmrA. The plasmid pNHlmrA containing the His-tagged *lmr*A gene under the control of a nisin-inducible promoter was transformed into *Lactococcus lactis* [9]. Expression of LmrA was induced by nisin addition, and LmrA protein was purified by affinity chromatography using Ni^{2+} sepharose beads following a modified protocol from van den Berg van Saporea *et al.* [20].

2.1.2. Production of membrane vesicles. Vesicles were constructed from natural glycolipids of *Lactobacillus plantarum*, dimyristoylphosphatidylcholine (DMPC), DMPC/10% (mol/mol) cholesterol and dioleoylphosphocholine (DOPC)/dipalmitoylphosphocholine (DPPC)/cholesterol (1:2:1). Lipid mixtures were swelled in 50 mM HEPES pH=7.0 and agitated to form multi lamellar vesicles (MLV) and extruded through a carbon membrane filter with 200nm pores to obtain large unilamellar vesicles (LUV). Liposomes were destabilized with DDM as detergent, LmrA was added to a lipid/protein ratio of 20:1 and DDM was adsorbed in two successive steps.

2.1.3. Activity of LmrA. Vesicles were subjected to pressures of 20 to 120 MPa in 50 mM HEPES-KOH buffer pH=7,0 immediately after addition of ethidium bromide as transport substrate and ATP as energizing substrate. After 100 min of pressurization in the presence of the substrates functionality and activity of LmrA were determined along the release of P_i using a colorimetric method based on the protocol from Lanzetta et al. [8] and compared with a phosphorous standard. All experiments were compared with parallel experiments in the same vessel in the presence of ortho-vanadate. Readings were subtracted to exclude artifacts and selectively get the vanadate sensitive ethidium assisted ATPase activity of LmrA. Ortho-vanadate inhibits the activity of LmrA [19] [21].

2.2. HHP dependence of ToxR dimersation *in vivo*

2.2.1. Construction of an Escherichia coli ToxR reporter strain. The *omp*L promoter of *Photobacterium profundum* SS9 was was fused to an *E. coli lac*Z encoding sequence. This fragment was inserted in vector pLDR10, which allowed, in combination with helper vector pLDR8, the integration into the λ attachment site *att*B of *E. coli* Top10. This strain was designated POLA. The ToxR gene was cloned into the arabinose-inducible plasmid vector pBADK-myc-hisB. The resulting vector was termed pBADK-toxR-his. The reporter strain POLA was then transformed with pBADK-toxR-his.

2.2.2. *Determination of ToxR activity.* Cells were grown at 37°C for 22 h at pressures from 0.1 to 35 MPa. Upon binding of the *P. profundum* ToxR dimer to the *omp*L promoter, *lac*Z is expressed, whose activity can be determined in an enzymatic assay using the chromogenous substrate ONPG. Pressure induced dissociation of the ToxR dimer results in decrease of the *lac*Z transcription and finally lower ß-galactosidase activity.

2.2.3. Chimeric ToxR proteins. To deterimine the effect of the transmembrane domain structure on ToxR dimerization, ToxR chimeric proteins obtained from *Vibrio cholerae* were used. The proteins were encoded on the plasmid ToxRIV+TMS [5] in which the cytoplasmic domain of ToxR is linked to the periplasmic maltose binding protein (MalE) moiety via different trans membrane segments (TMS). The following TMS were used: AZ2: Simplified version of a membrane-spanning leucine zipper interaction domain (LLAALLALLAALLALL), EG4: Mutation of the simplified version of the leucin zipper (LLAALAAALAALAAAL), GpA13: Wild type glycophorin A transmembrane segment (LIIFGVMAGVIGT), G83A: Mutant of wild type glycophorin A transmembrane segment (LIIFGVMAAVIGT) [6][7]. As these ToxR constructs originated from *V. cholerae*, the *V. cholerae* *ctx* promoter was used to initiate *lac*Z transcription in the indicator cells.

3. Results and Discussion

3.1. HHP dependence of LmrA activity in proteoliposomes

LmrA is an ATP-dependent multiple drug transporter from *Lactococcus lactis*. It requires dimerization in the phospholipid bilayer for its functionality [4]. Its substrate spectrum reaches from hydrophobic drugs to ionic molecules, e.g. the ethidium cation, which are transported outwards from the interior of the membrane [2]. The hydrophilic part of this transporter comprising an ATPase activity reaches into the cytoplasm and its transport energy results from the cleavage of ATP to ADP and inorganic phosphate (Pi). In the proteoliposomes used in this work, LmrA was inserted after the formation of the vesicles pushing in an unidirectional mode, with the ATPase function located outside of the vesicle, which renders it experimentally accessible. This enables easy substrate addition and observation of the released phosphate. For the activity of LmrA, dimerization of two LmrA molecules is required. Recent studies established that the minimal unit in the functional LmrA transporter is the homodimer [21]. To exclude any side effects, we determined the vanadate-sensitive ATPase activity of LmrA. As it has been demonstrated for an ATP dependent calcium-transporter that HHP did not affect the inhibitory capability of vanadate [15], vanadate sensitive ATPase activity of LmrA under HHP likely reflects changes in enzyme activity. Several types of lipids were used to investigate their influence on LmrA dimerization/activity. In vesicles from natural phospholipids from *Lactobacillus plantarum* LmrA activity was slightly reduced at 20 MPa and ceased above 40 MPa. In vesicles consisting solely of

DMPC employing saturated fatty acid chains LmrA activity ceased above 20 MPa. In DMPC/cholesterol vesicles, ATPase activity was strongly reduced above 20 MPa but still present at a low level up to 40 MPa. As an example, figure 1 shows the LmrA activity in DOPC/DPPC/cholesterol vesicles. LmrA activity was slightly reduced at 20 and 40 MPa, as observed for natural phospholipid mixtures. At 80 MPa the activity significantly increased again, and it was reduced above 120 MPa. Under these conditions the coexistence of lipid phases of such raft mixtures has been described [22]. LmrA activity may therefore be maintained at elevated pressures in transporter dimers residing in the DOPC islands, which are still in the liquid crystalline phase at these pressures because of the double bonds in the fatty acid chains.

Figure 1: Activity of LmrA in raft mixture (DOPC/DPPC/Cholesterol 1:2:1). Large unilamellar vesicles (LUV) with LmrA 20:1 (w/w) were used to measure the vanadate sensitive ethidium assisted ATPase activity under HHP.

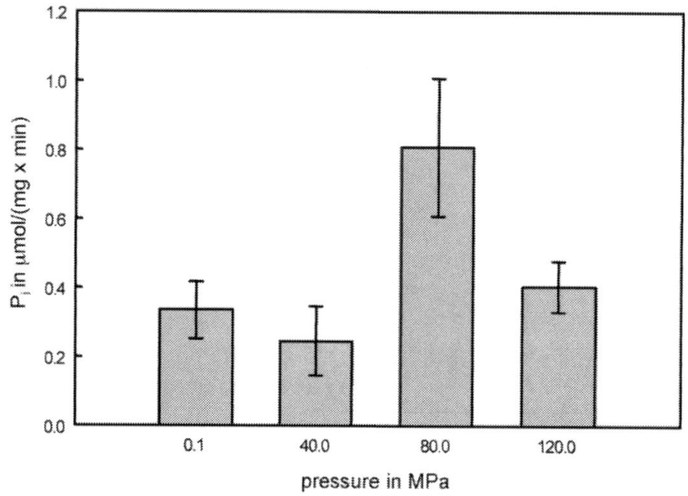

3.2. HHP dependence of ToxR dimersation *in vivo*

It is essential for bacteria to recognize changes in environmental conditions to adapt their metabolism, stress tolerance or virulence. The ToxRS system is one of many systems residing in the bacterial membrane to receive environmental signals and initiate an intracellular signal cascade. The principle of sensing is the formation of a heterotrimer consisting of two ToxR and one ToxS molecules. This complex specifically binds to promoter regions and thus initiates transcription of the respective genes [11][13]. In *P. profundum*, the ToxRS system is involved in HHP adaptation as it regulates the formation of outer membrane pore proteins (Omp) by switching between induction of OmpH and OmpL formations. Under HHP conditions of the deep sea, the ToxRS complex is dissociated and OmpH is formed, while OmpL is induced at ambient pressure when the ToxRS complex is formed and binds to the OmpL promoter [1] [23] [24]. In *Vibrio cholerae* the ToxRS system is involved in the formation of the cholera toxin by binding to the *ctx* promoter [10].

3.2.1. *Photobacterium profundum ToxR.* The reporter system used in this work only employs ToxR, which can dimerize in the reporter strain under ambient pressure without the aid of ToxS, because it is overexpressed as compared to the natural situation [13]. In this reporter strain, LacZ (ß-galactosidase) activity indicating ToxR dimerization is increased by 20% at 10 MPa and remains unchanged at 20 MPa and 0.1 MPa. At pressures above 20 MPa it is successively reduced with increasing pressure. Above 40 MPa LacZ activity is no longer observed. Above this pressure, ribosomal synthesis starts to be affected, and therefore any measurements loose their significance.

3.2.2. *Chimeric Vibrio cholerae ToxR*. The interaction of proteins, namely their binding capability to each other depends on the biochemical nature of the contacting peptide chains. In membrane proteins the determinative domains for protein/protein interaction reside in the hydrophobic part of the membrane. The chimeric proteins used in this work consisted of the intracellular promoter-binding domain of ToxR and the periplasmic MalE protein, which helped correct orientation in the membrane. These domains were linked by 4 different TMS, which have previously been characterized with respect to their ability to dimerize in *E. coli* reporter strains [6][7]. Figure 2 shows the expression of *lacZ* as a result of HHP dependent ToxR hybrid dimerization. The basic levels of dimerization at 0.1 MPa, as determined by lacZ expression, varied for the different constructs and confirmed previously published data at ambient pressure. The behaviour of the construct with the leucin zipper (AZ2) mostly resembled that one obtained with ToxR from *P. profundum* characterized by a slight increase and subsequent decrease of dimerization with increasing pressure. The modified leucin zipper (EG4) and glycophorin segment (G83A) showed constantly decreasing dimerization with increasing pressure. Most interestingly, the dimerization of the construct containing the unmodified glycophorin TMS (A13) was little affected by the pressures applied. These results demonstrate, that the nature of the protein domain involved in dimerization of ToxR influences its dimerization independently of the lipid phase in which the protein resides. The different behaviour of the ToxR hybrids, namely the relative pressure stability of the A13 construct, also indicates that HHP induced inhibition of other cellular functions, including ribosomal synthesis, does not interfere with the *in vivo* reporter system at the pressures used here.

Figure 2: ß-Gal-activity indicating ToxR dimerization in *E. coli*-FHK12+pToxRIV+AZ2 [white bar], *E. coli*-FHK12+pToxRIV+EG4 [hatched bar], *E. coli*-FHK12+pToxRIV+GpA13 [cross-hatched bar] and *E. coli*-FHK12+pToxRIV+GpAG83A [back-hatched bar] after 24 h of growth at HHP.

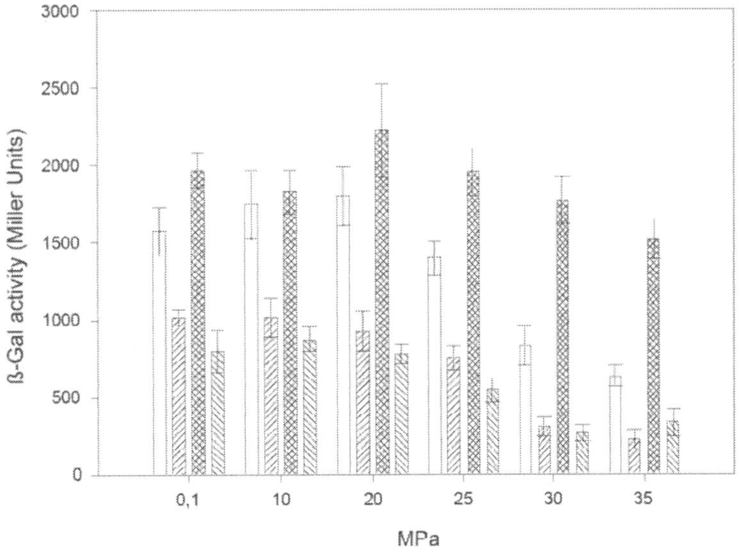

4. Conclusion

In this work, HHP and *in vivo* reporter systems were combined for the first time to investigate the structure function relationship of integral membrane proteins involved in transport and cellular signaling. We could show that the interaction of membrane proteins is pressure dependent *in vivo* and is maximal in the systems used at pressures of 10 – 20 MPa. *In vitro* experiments with raft mixtures of selected lipids suggest lipid sorting to coexisting islands allowing the proteins to find an area with the optimal fluidity required for their specific function. Also, changes in the protein structure affect dimerization of membrane proteins. Not only are basal levels of functionality altered, but pressure dependent activity is observed again indicating the influence of membrane physical structure on the protein functionality. Certainly, the hen/egg question is obsolete when mirrored in the protein/lipid field in search of causative responsibility for membrane protein activity changes. Both the lipid structure and the nature of the protein affect membrane protein interaction. It is suggested that the protein structure determines basic functionality, e. g. principle ability or kinetics to dimerize to a functional complex, and may even affect principle structure of its lipid environment. The lipid environment modulates this property. Therefore, the lipid structure state needs to be considered as a major effector on protein functionality in living systems. Based on these findings it is tempting to speculate that environmentally caused changes in the membrane´s physical structure selectively modulates membrane protein activity and, possibly, the sorting of proteins to membrane areas which enable their maximal activity, thus helping cells to adapt to the environmental change.

References

[1] Bidle KA and Bartlett DH, 2001. RNA arbitrarily primed PCR survey of genes regulated by ToxR in the deep-sea bacterium *Photobacterium profundum* strain SS9. J Bacteriol 183: 1688-93.
[2] Borges-Walmsley MI, McKeegan KS and Walmsley AR, 2003. Structure and function of efflux pumps that confer resistance to drugs, Biochem J 376:313–338.
[3] Casadei MA, Manas P, Niven G, Needs E and Mackay BM. 2002. Role of membrane fluidity in pressure resistance of *Escherichia coli* NCTC 8164. Appl Environ Microbiol 68:5965–5972.
[4] Federici L, Woebking B, Velamakanni S, Shilling RA, Luisi B, and van Veen HW, 2007. New structure model for the ATP-binding cassette multidrug transporter LmrA. Biochem Pharmacol 74:672-678.
[5] Gurezka R and Langosch D. 2001. *In vitro* selection of membrane-spanning leucine zipper protein-protein interaction motifs using POSSYCCAT. J Biol Chem276:45580-7.
[6] Gurezka R, Laage R, Brosig B, Langosch D 1999. A heptad motif of leucine residues found in membrane proteins can drive self-assembly of artificial transmembrane segments. J Biol Chem 274:9265-70.
[7] Langosch D, Brosig B, Kolmar H, Fritz HJ 1996. Dimerisation of the glycophorin A transmembrane segment in membranes probed with the ToxR transcription activator. J Mol Biol 263:525-30.
[8] Lanzetta PA, Alvarez LJ, Reinach PS, and Candia OA, 1979. An improved assay for nanomole amounts of inorganic phosphate, Anal Biochem 100:95-97.
[9] Margolles A, Putman M, van Veen HW, and Konings WN, 1999. The purified and functionally reconstituted multidrug transporter LmrA of *Lactococcus lactis* mediates the transbilayer movement of specific fluorescent phospholipids, Biochemistry 38:16298-16306.
[10] Miller VL and Mekalanos JJ, 1984. Synthesis of cholera toxin is positively regulated at the transcriptional level by toxR. Proc Natl Acad Sci U S A, 81:3471-5.
[11] Miller VL, DiRita, VJ, and Mekalanos JJ, 1989. Identification of toxS, a regulatory gene whose product enhances toxR-mediated activation of the cholera toxin promoter. J Bacteriol 171:1288-93.
[12] Molina-Guiterrez A, Rademacher B, Gänzle MG and Vogel RF. 2002. Effect of sucrose and sodium chloride on the survival and metabolic activity of *Lactococcus lactis* under high-pressure conditions. Trends High Pressure Biosci Biotechnol 19:295-302
[13] Ottemann KM and Mekalanos JJ, 1996. The ToxR protein of Vibrio cholerae forms homodimers and heterodimers. J Bacteriol 178:156-62.
[14] Pavlovic M, Hörmann S, Vogel, RF and Ehrmann MA. 2005. Transcriptional response reveals translation machinery as target for high pressure in *Lactobacillus sanfranciscensis*. Arch Microbiol 184:11-17.
[15] Ronzani N, Stephan L and Hasselbach W, 1991. Pressure effects on the binding of vanadate to

the sarcoplasmic reticulum calcium-transport enzyme. Eur J Biochem 201:265-271.

[16] Schwab C and Gänzle MG. 2006. Effect of membrane lateral tension on the expression of fructosyltransferases in *Lactobacillus reuteri*. System Appl Microbiol 29:89-99.

[17] Ulmer H, Gänzle MG and Vogel RF. 2000. High pressure effects on survival and metabolic activity of *Lactobacillus plantarum*. Appl Environ Microbiol 66:3966–3973.

[18] Ulmer HM, Herberhold H, Fahsel S, Gänzle MG, Winter R and Vogel RF. 2002. Effects of pressure induced membrane phase transitions on HorA inactivation in *Lactobacillus plantarum*. Appl Environ Microbiol 86:1088–1095.

[19] Urbatsch IL, Sankaran B, Weber J, Senior AE, 1995. P-glycoprotein is stably inhibited by vanadate-induced trapping of nucleotide at a single catalytic site. J Biol Chem, 270:19383-19390.

[20] van den Berg van Saparoea HB, Lubelski J, van Merkerk R, Mazurkiewicz PS, and Driessen AJM, 2005. Proton motive force-dependent Hoechst 33342 transport by the ABC transporter LmrA of *Lactococcus lactis*, Biochemistry 44:16931-16938.

[21] van Veen H, Margolles A, Müller M, Higgins CF and Konings WN, 2000. The homodimeric ATP-binding cassette transporter LmrA mediates multidrug transport by an alternating two.site (two-cylinder engine) mechanism. EMBO J 19:2503 – 2514.

[22] Veatch SL and Keller SL, 2003. Separation of liquid phases in giant vesicles of ternary mixtures of phospholipids and cholesterol. Biophys J 85:3074–3083.

[23] Welch TJ and Bartlett DH, 1996. Isolation and characterization of the structural gene for OmpL, a pressure-regulated porin-like protein from the deep-sea bacterium *Photobacterium* species strain SS9. J Bacteriol 178:5027-31.

[24] Welch TJ and Bartlett DH, 1998. Identification of a regulatory protein required for pressure-responsive gene expression in the deep-sea bacterium *Photobacterium* species strain SS9. Mol Microbiol 27:977-85.

Acknowledgements
The authors are indebted to Dieter Langosch Technische Universität München, Germany, to provide ToxR/MalE hybrids with different TMS and the German Research Foundation (DFG) for financial support during projects VO582/3-1 and 3-2.

High-pressure cell for a SQUID magnetometer with a plug for *in situ* pressure measurements

J Sanchez-Benitez, S Tancharakorn, MK Hutchison, KV Kamenev[1]
Centre for Science at Extreme Conditions and School of Engineering & Electronics, University of Edinburgh, Mayfield Road, Edinburgh EH9 3JZ, UK

E-mail: K.Kamenev@ed.ac.uk

Abstract. We describe the design of a miniature high-pressure cell built for a Magnetic Property Measurement System based on the Superconducting Quantum Interference Device technology magnetometer. The key feature of the cell is the plug for *in situ* pressure measurements with feed-through wires connected to a manganin pressure sensor, which has a known pressure dependence of electrical resistivity. By monitoring pressure continuously during magnetisation measurements in the range of operational temperatures the true pressure can always be established.

1. Introduction

Magnetic Property Measurement System (MPMS) from Quantum Design (USA) is one of the most popular commercially available magnetometers utilising Superconducting Quantum Interference Device (SQUID) technology [1]. The MPMS can resolve magnetic moment changes as small as 10^{-8} emu over a wide range of temperatures and magnetic fields. Despite the major constraint of the instrument - the diameter of the inner bore of the MPMS is only 9 mm - there has been several pressure cells of the piston-cylinder type built for MPMS and pressure can now also be added as a variable for sensitive magnetisation measurements [2-8].

One of the problems however that needs to be resolved when the pressure cell is designed is that of the pressure measurement. There are several methods of doing this in the existing designs. However, all of them rely on a measurement of pressure at one particular temperature and then the pressure is *estimated* at any other temperature based on a certain assumption. This temperature correction is needed because of the pressure change on cooling or heating caused by a difference in thermal expansion of the pressure cell and the pressure-transmitting medium (normally, oil). In this work we take on the challenge to design the pressure cell in which pressure can be *measured* at any given temperature.

In the existing pressure cells for SQUID magnetometer the most frequently used method is by using a superconducting (SC) "manometer" like lead, tin or indium [9]. In these materials Tc is calibrated as a function of pressure and so by measuring the SC transition from a tiny speck of Pb, Sn or In one can accurately measure the pressure inside the pressure cell. The limitations of these method are that (i) it provides the pressure value for the temperature of the SC transition (typically at around 3-7 K); (ii) it produces a strong magnetic signal, which screens the sample and often makes it impossible to do an accurate measurements of the sample's magnetic susceptibility at temperatures

[1] To whom any correspondence should be addressed.

© 2008 IOP Publishing Ltd

below the SC transition. The alternative method for measuring pressures is by either measuring elongation of the cell [9, 10] or by deriving the pressure from the load on the piston if a hydraulic press is used for applying pressure. This provides the pressure value at room temperature and eliminates the need for an SC "manometer" if measurements are to be done at very low temperatures. Combining the two methods described above it is possible to establish the absolute pressure change between the ambient and low (around SC transition) temperatures [10] which results from the difference in the thermal expansion coefficients of the pressure transmitting medium and the material of the pressure cell. However, determining the pressure in the vast temperature region between 300 K and about 3-7 K can be difficult. For simplicity it is often assumed that pressure changes linearly with pressure and the correction for the pressure is made based on this assignment. We show below that there is a deviation from linear dependence and using this approach can introduce a large error into the data analysis.

In large piston-cylinder cells there are two common ways for measuring pressure continuously as temperature is varied. One of them is with calibrated strain-gauges [11]. However implementing this method in the SQUID pressure cell design will be difficult as the strain-gauge will need to be positioned at the point of the maximum deformation which is in the middle of the cylindrical body of the cell, i.e. close to the sample position inside the cell, and thus will interfere with the magnetic signal from the sample. The other method is to measure pressure by measuring the resistance of a manganin coil located inside the pressure cell. Manganin is an alloy of 86% copper, 12% manganese, and 2% nickel. It has been developed by Edward Weston as the material with a very low temperature coefficient of resistance. Manganin also has its electrical resistivity calibrated as a function of pressure at any given temperature. There are several ways to calculate pressure from its resistivity [12, 13] of which the simplest one is the following:

$$R(P) = R(P_{amb})\,(1 + \alpha P)$$

where $R(P)$ and $R(P_{amb})$ are resistivities of the manganin at pressure P and ambient pressures respectively, $\alpha = 0.002465$ kbar^{-1} and P is pressure in kbar.

Manganin pressure sensors are normally produced as small coils and introduced into pressure cells by means of an electrical plug. The design of the plug is typically based on a reversed Amagat cone [14]. The conical window itself can be made of steels, alloys or other metals in which case the wires are electrically insulated from the plug by use of pyrophilite or resin epoxy [15, 16]. In some cases the Amagat window is made exclusively of pyrophilite or epoxy [17 - 20]. However, all of the above plugs have been produced for rather large pressure cells. The bore of a SQUID pressure cell is typically 2.5 - 3.0 mm in diameter and the outer diameter is restricted by the 9 mm bore of the magnetometer. Introducing a plug with electrical feed-throughs into such a small bore was the first major challenge we needed to overcome in the design process.

2. Design

The conventional design of a feed-through plug for large-volume pressure cells typically consists of two parts - the plug itself and a backing screw used to push the plug into the bore and engage the seals. The advantage of using this 2-piece method is that as the screw turns the plug is engaged into the translational movement only, thus avoiding the twisting action at the neck of the plug once the metal seal is engaged. However, our estimates have shown that the space restrictions would make it impossible to implement the 2-piece design for a plug with feed-throughs in the case of a miniature pressure cell for the SQUID magnetometer. The reason for this is that it would be difficult to pull out the plug out of the pressure cell during the disassembly because it is virtually impossible to reach to it and to grip it firmly inside the hole in the cylinder of the cell which is less than 6 mm in diameter.

We therefore decided to use a 1-piece design in which the plug and the screw are made as a single part. The design of the plug required careful calculations based on the strength of materials involved in order to avoid the above mentioned problem of the neck of the plug being effectively "welded" to the bore of the cell once the metal seals are engaged. The drawing of the plug is shown in Fig. 1.

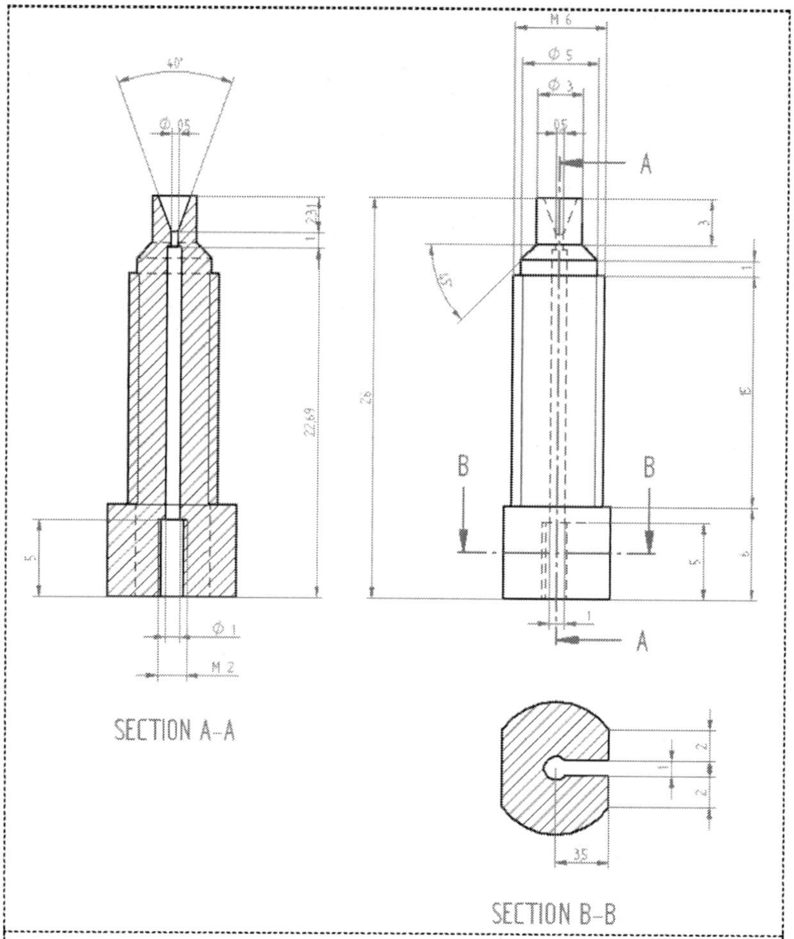

Figure 1. The drawing of the plug for electrical feed-throughs.

The plug is made of BERYLCO-25 alloy (NGK Berylco U.K. Ltd). The feed-throughs are made of up to 20 enamelled copper wires of 0.1 mm diameter. Although only 4 wires are required for the resistivity measurement we have found having spare wires useful in case of any possible breakages. The wires are introduced into the pressure cell through a narrow neck of 0.5 mm in diameter, which at the top end opens up into a conical hole for a reversed Amagat cone and at the bottom end becomes a 1.0 mm diameter channel for taking the wires out. Once the wires are introduced into the plug, the conical neck is sealed with Stycast 2850 FT epoxy (Emerson & Cuming Inc.) with catalyst 24 LV, as this has proved to be the most reliable at low temperatures. The epoxy is allowed to go down the channel under the force of gravity without any extra suction force at the opposite end of the plug. This has been found to give the most uniform distribution of the epoxy without air bubbles trapped in it. The external end of the plug has a 5 mm long and 1 mm wide slot cut to take the wires out through the side of the plug and an M2 thread is made inside the channel to hold the pressure cell (once assembled) on the MPMS sample rod. A copper seal is put on the 3 mm neck to provide the seal between the cylinder of the pressure cell and the plug.

The cross-sectional view of the assembled pressure cell is shown in Fig. 2. The pressure is applied in a hydraulic press and is generated in the oil pressure-transmitting medium by a piston at the end opposite to the plug. The pusher shown in Fig. 2 is removed once the pressure is applied. The oil used in our experiments is Daphne 7373 (IDEMITSU-ILS) and the maximum pressure that can be achieved in this pressure cell is 10 kbar. The copper wires terminate on a 4-pin plug that connects to a 4-pin socket assembled on an adapter connected to the standard MPMS sample rod as shown in Fig. 3. The rod is hollow and sealed at the opposite end to eliminate any leak of air into the sample chamber.

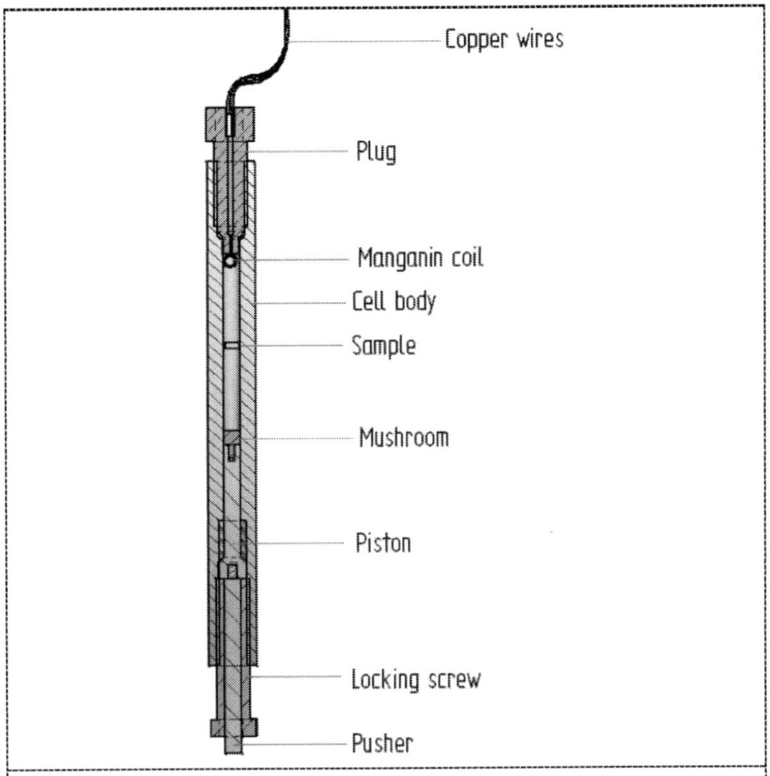

Figure 2. The cross-sectional view of the assembled pressure cell.

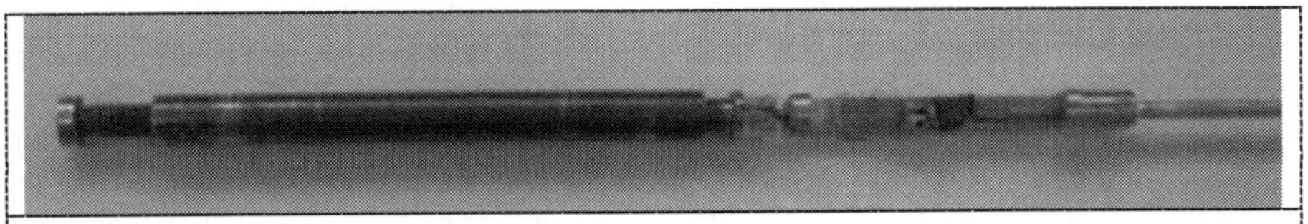

Figure 3. Assembled pressure cell attached to the MPMS sample rod through an adapter with a 4-pin socket mounted on it.

3. Tests & Measurements

There are several ways in which the data acquisition can be arranged and the resistivity data collected from the manganin pressure sensor. One of them is through the use of RP013 resistivity option for MPMS provided by Quantum Design. It allows to incorporate the resistivity measurements through the MultiVu software that runs the magnetisation measurements in the MPMS and write the resistivity data into the same data file as the rest of the acquired data such as magnetisation, temperature, magnetic field, etc. The alternative way is to use any data acquisition software for collecting the manganin resistivity data and to run it in parallel with MultiVu. This would produce two data files which can then be combined in one set of data based on the time column which can be included into both files.

We have measured the resistivity of the manganin sensor in a pressure cell assembled with a sample mounting table and a sample inside as a function of temperature in the MPMS at a number of pressures. Because of the limited space available for this paper we will report the results of the pressure dependence of the magnetic properties of the studied material elsewhere. The purpose of the test was to establish how the pressure actually changes inside the pressure cell at temperatures between 300 K and 2 K. The manganin coil resistance measured as a function of temperature at starting pressures (as applied at room temperature) of 0, 5 and 10 kbar is shown in Figure 4.

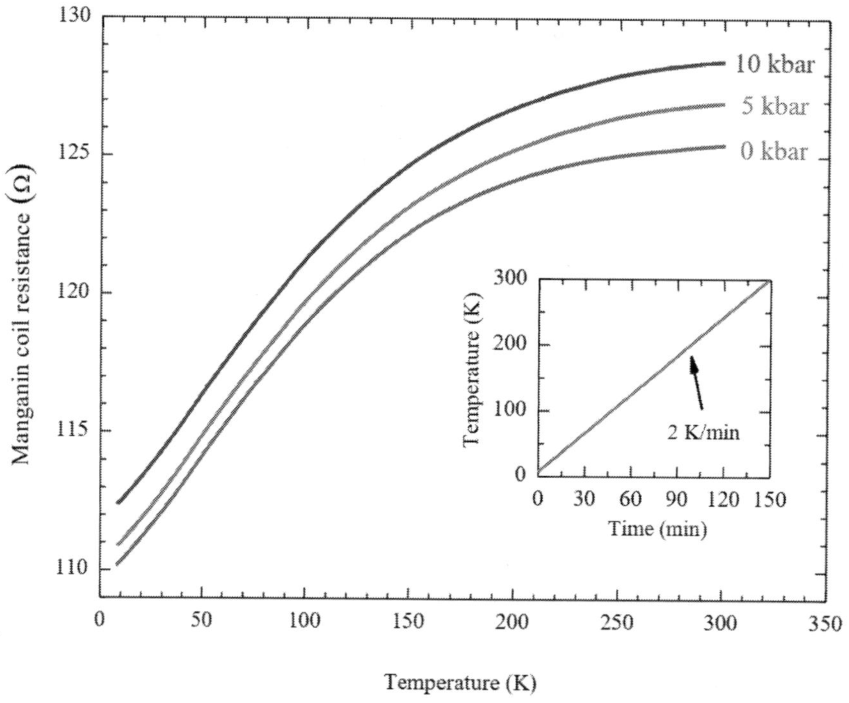

Figure 4. Temperature dependence of the electrical resistivity of the manganin coil measured at three different pressures. The inset shows the temperature change as a function of time during these measurements confirming the uniform temperature change rate of 2 K/min.

The measurements were performed at a number of temperature change rates and the optimum value was found to be 2 K/min, which is sufficient for the pressure cell to remain in thermal equilibrium with the environment during heating/cooling runs. The data were then treated using the equation quoted above and the results of the analysis for the initial pressure of 5 kbar is presented in Fig. 5. The solid line in Fig. 5(a) shows the actual pressure inside the pressure cell as measured by manganin gauge. The overall drop in pressure between 300 K and 2 K is just under 2.5 kbar. This has been observed for the Daphne 7373 oil before [6,10], however from the shape of the line it is clear that the linear approximation for the pressure change (dashed line in Fig. 5(a)) is inadequate and can result in an error of up to 0.6 kbar in pressure determination (Fig. 5(b)).

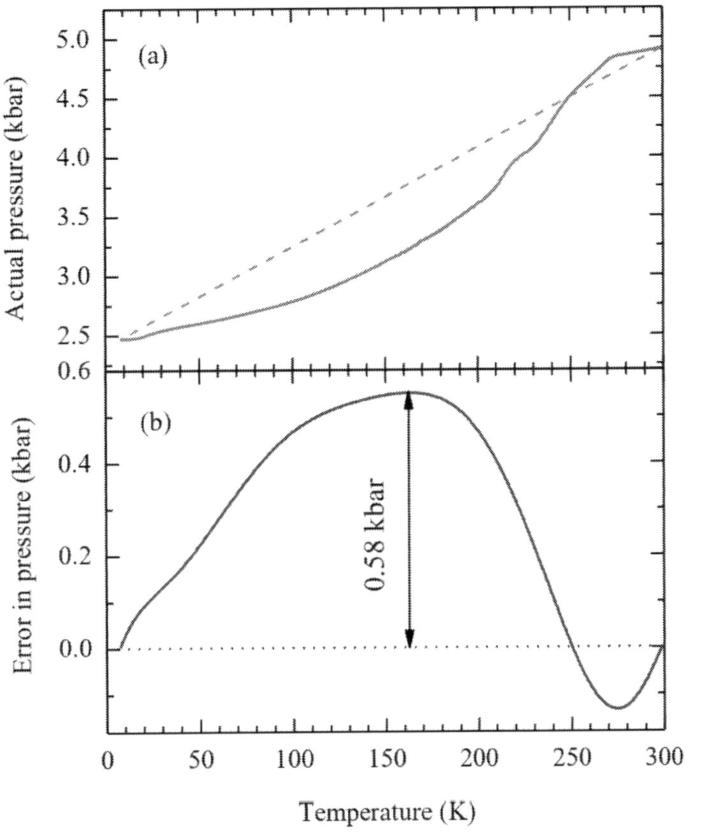

Figure 5. (a) the actual temperature dependence of the pressure inside the pressure cell (solid line) and the linear approximation (dashed line); (b) the difference between the two lines in the top graph.

It is hard to interpret the shape of the P(T) curve but it is clear that some features in it are due to the pressure transmitting fluid changing its state from liquid to solid. Other features are due to changes in the rest of the sample environment such as for example the thermal expansion of various parts of the pressure cell assembly, e.g. sample mounting table, piston, seals or the sample itself in the end. These changes can be specific to each individual measurement and thus the use of the *in situ* measurement of pressure can greatly improve the accuracy of the data especially in cases when the material studied is sensitive to pressure or when a fluid with unknown properties is used as pressure-transmitting medium.

We should also comment on the change of the resistivity of the manganin wire in magnetic field due to the magnetoresistance effect. It has been found recently that the magnetoresistance of the manganin is very low (+0.69% in the field of 10 T at temperature of 4 K) [21] and can in most cases be neglected. Alternatively, a correction can be made to take this change into account in calculating the actual pressure inside the pressure cell.

In conclusion, we have designed a pressure cell with a plug for *in situ* pressure measurement which has been tested to pressures of 10 kbar. It has been shown that the pressure inside the pressure cell has a complicated temperature dependence due to the thermal expansion coefficients of the materials used in the construction of the cell and that of the sample itself. This paper aims to provide sufficient information for the researchers interested in building and using the plug and the pressure cell.

4. References

[1] Quantum Design, web-site: http://www.qdusa.com
[2] Reich S and Godin T 1996 *Meas. Sci. Technol.* **7** 1079
[3] Diederichs J, Gangopadhyay A K and Schilling J S 1996 *Phys. Rev. B* **54** R9662
[4] Uwatoko Y, Hotta T, Matsuoka E, Mori H, Oki T, Sarrao J L, Thompson J D and Mori N 1998 *Rev. High Pressure. Sci. Technol.* **7** 1508
[5] Kamishima K, Hagiwara M and Yoshida H 2001 *Rev. Sci. Instrum.* **72** 1472
[6] Kamarád J, Machátová Z and Arnold Z 2004 *Rev. Sci. Instrum.* **75** 5022
[7] Mcell 10 pressure cell manufactured by easyLab Technologies; web-site: http://www.easylab.co.uk/mcell.html
[8] Uwatoko Y, Fujiwara T, Hedo M, Tomioka F and Umehara I 2005 *J. Phys.: Condens. Matter* **17** S1011
[9] Eiling A and Schilling J S 1981 *J. Phys. F: Metal Phys.* **11** 623-39
[10] Kamenev K V, Tancharakorn S, Robertson N and Harrison A 2006 *Rev. Sci. Instrum.* **77** 073905
[11] Kamarád J, Kamenev K V and Arnold Z 1996 *High Pressure in Science and Technology,* ed W A Trzeciakowski (Singapore: World Scientific) pp. 51–53
[12] Andersson O E and Sundqvist B 1997 *Rev. Sci. Instrum.* **68** 1344
[13] Atanov Yu A and Ivanova E M, Translated from 1971 *Izmeritel'naya Tekhnika* **2** pp. 46–48
[14] Bridgman P W 1931 *The Physics of High Pressure* (London: G. Bell and Sons, Ltd.)
[15] Downs J L and Payne R T 1969 *Rev. Sci. Instrum.* **40** 1278
[16] Schirber J E and Shanfeldt D W 1968 *Rev. Sci. Instrum.* **39** 270
[17] Hammons B E 1971 *Rev. Sci. Instrum.* **42** 1889
[18] Lazarre F 1953 *J. Phys. Radium* **14** 213
[19] Dugdale J S and Hulbert J A 1958 *Can. J. Phys.* **35** 472
[20] Petrova A E and Stishov S M 2006 *Instr. and Exp. Techniques* **49** 592
[21] Abrecht M, Adare A and Ekinb J W 2007 *Rev. Sci. Instrum.* **78** 046104

Acknowledgments

This work was supported by EPSRC Grants EP/E06471X/1 & EP/D503744/1. We also thank the Royal Thai Government for a studentship (ST).

Journal of Physics: Conference Series **121** (2008) 122002 doi:10.1088/1742-6596/121/2/122002

A technique for precise electrical-transport measurements under pressure above 10 GPa

Masashi Ohashi

Division of Civil and Enviromental Engineering, Kanazawa University, Kakuma-machi, Kanazawa 920-1192, Japan

E-mail: ohashi@t.kanazawa-u.ac.jp

Abstract. We report a technique for the precise measurement of electrical resistivity at high pressures up to 15 GPa by using Bridgman anvils. The relatively large size of the pressure chamber (1.0 mm in diameter) allows the use of large specimens and simple experimental procedures rather than using a standard diamond anvil cell. A SUS310 gasket is pressed by two tungsten carbide anvils. A sample with typical dimensions of approximately $0.5\times0.2\times0.1\text{mm}^3$ is placed in a small hole of the gasket. In order to obtain a quasi-hydrostatic pressure, the pressure chamber is filled with a 1:1 mixture of Fluorinert FC70 and FC77 as the pressure transmitting medium. Electrical leads are introduced through shallow grooves milled into the anvil. The grooves are filled with a mixture of alumina powder for insulation. Accurate data of the resistance values of Bi and Fe at room temperature are available. We observe sharp transitions for Bi at 2.55, 2.7 and 7.7 GPa. The electrical resistance of Fe shows a sudden increase due to a structural transition near 14 GPa.

1. Introduction

It is well known that the application of high pressure induces various types of electronic transitions in materials. One of the typical examples is the pressure-induced quantum phase transition[1]. Many interesting electronic properties are often found near the quantum critical point under high pressures at low temperatures. In view of these pressure-induced phenomena, the electrical resistance measurement under high pressures has been recognized as one of the powerful tools to investigate the electronic properties.

However, all pressure transmitting medium freeze upon cooling and subject the immersed sample to shear stress of varying strengths. In order to maintain the sample under hydrostatic pressures at low temperatures, a wide variety of techniques have been used[2]. In general, a large working space and a fluid pressure transmitting medium are needed to generate hydrostatic pressures. Piston-cylinder devices have been used so far for precise measurements because of their large working space in which quasi-hydrostatic pressure is generated; it enables the measurements of not only the electrical resistance and the ac magnetic susceptibility but also the thermal expansion by using a strain gage method[3]. However, the generated pressure is limited at the most up to 3 GPa.

A diamond anvil cell is widely used for measurements that require higher pressures. By using helium as the pressure transmitting medium, purely hydrostatic pressures can be generated[4]. However, it is unconvenient to study the electronic properties since the sample space is extremely small; therefore, it is difficult to attach electrical leads.

© 2008 IOP Publishing Ltd

Recently, Môri[5] and Nakanishi *et al.*[6] developed a Bridgman anvil apparatus thet yielded nearly hydrostatic pressures in which a large working volume of $\phi 1.0 \times 0.9$ mm^3 and a fluid pressure transmitting medium were realized in a Teflon capsule between anvils made of tungsten carbide (WC). However, the generated pressure is limited at the most up to 8 GPa.

In the present study, we describe a new technique for the precise measurement of the electrical resistivity under high pressures up to 15 GPa by using Bridgman anvils. The relatively large size of the pressure chamber (1.0 mm in diameter) with a fluid pressure transmitting medium allows to not only maintain nearly hydrostatic pressures but also simplifies the experimental procedures as compared to those of the standard diamond anvil cell.

2. Experimental setup

Figure 1 shows the schematic cross section of the present high-pressure cell, which consists of two anvils made of (A) WC, (B) a metal gasket, and (C) a ring to tighten the gasket.

The anvils were grounded with a taper angle of 10° and terminated at a flat tip with a diameter of 3.0 mm diameter. Four shallow grooves are milled into the lower anvil in order to introduce electrical leads into the pressure chamber. The grooves are filled with alumina powder, which ensures electrical insulation as well as sealing off the pressure chamber.

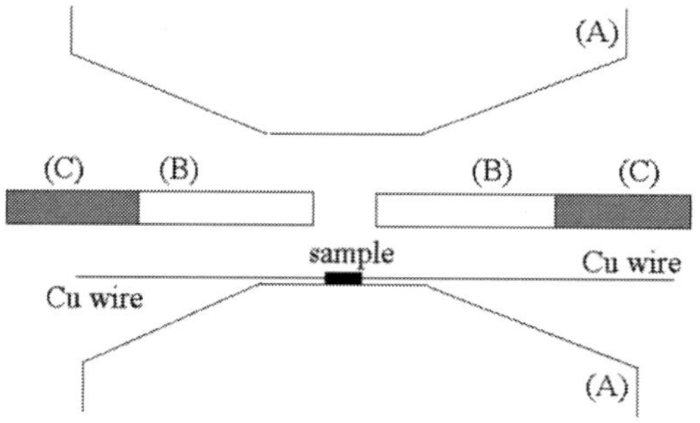

Figure 1. Schematic cross section of the high-pressure cell. See the text for details.

The metallic gasket is made of stainless steel (SUS304 or SUS 310) with a thickness of 0.5 mm and diameter of 8.0 mm. A sample hole with a diameter of 1.0 mm is prepared as the pressure chamber. A sample with typical dimensions of approximately $0.5 \times 0.2 \times 0.1$mm^3 is placed in the hole. The gasket is tightened by using a metal ring – made of alloys such as SNCM8 alloy, NiCrAl alloy and so on – with a Rockwell hardness on the C-scale (HRC) of $\sim$ 50; the ring prevents the movement of the gaskets when it is compressed by the anvils. After precompression, a 1:1 mix of Fluorinert FC70 and FC 77 is filled in the pressure chamber as the pressure transmitting medium. To compress the metal gasket, a force is generated and applied to the WC anvils. Resistivity measurements are carried out using the standard four-probe technique.

3. Pressure calibration

Pressure in the cell was calibrated according to the phase transitions of Bi and Fe at room temperature. Figure 2 shows the results of the resistance of Bi at room temperature by using our pressure apparatus. It is wellknown that Bi has three phase transitions: Bi I-II at 2.55 GPa, Bi II-III at 2.7 GPa, and Bi III-IV at 7.7 GPa[7]. As shown in Fig. 2, all the three transitions are observed in our result for the electrical resistance. The sharpness of the transitions shows that a hydrostatic pressure up to 8 GPa is obtained in the pressure chamber.

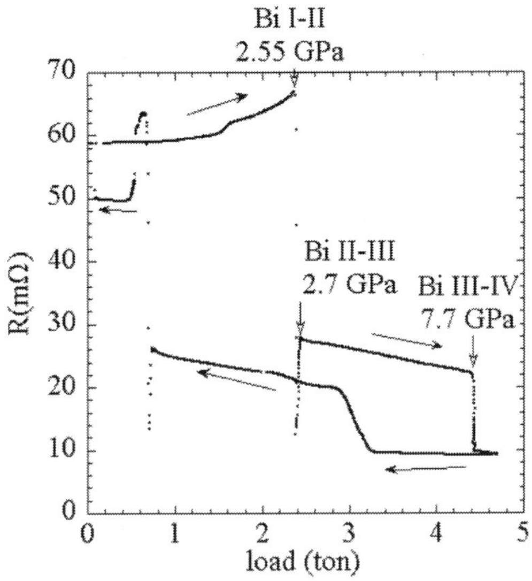

Figure 2. Load dependence of the resistance of Bi at room temperature.

Next, we show the results for the resistance of Fe at room temperature in Fig. 3. At first, the electrical resistance R decreases as the pressure load is applied. After showing a minimum at 5.9 ton, R shows a sudden increase near 7 ton. Although we stopped applying the pressure load at 8 ton and waited for several hours, R continued to increase. Since the increase in R reflects the change in the electron-phonon scattering, it arises from a first-order transition from the bcc α-phase to the fcc ϵ-phase near 13$\sim$14 GPa[8, 9]. As shown in the results, the transition is sluggish and observed over a much broader pressure gradient. In order to exactly calibrate the pressures, the measurements according to other standard materials are needed.

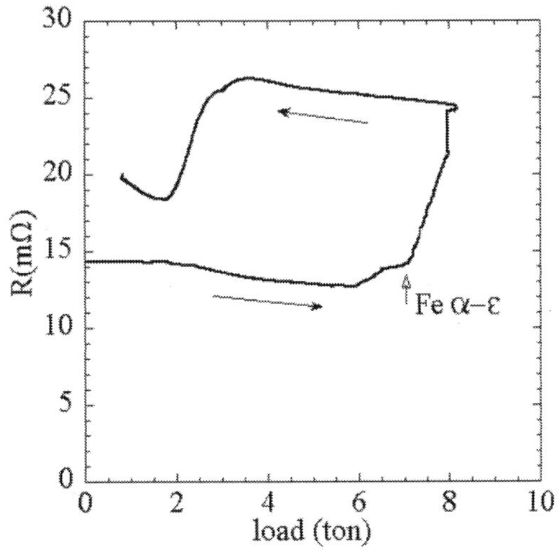

Figure 3. Load dependence of the resistance of Fe at room temperature.

Journal of Physics: Conference Series **121** (2008) 122002 doi:10.1088/1742-6596/121/2/122002

4. Summary

We have succeeded in developing a new high-pressure apparatus for measuring the electrical resistivity under hydrostatic pressures up to ~15 GPa. Because of its compact size – nearly as large as a conventional piston-cylinder cell, it can be easily combined with a cryostat and a superconducting magnet. To verify the performance of our pressure apparatus, the electrical resistances of Bi and Fe have been measured at room temperature. The measurements according to other standard materials will be needed to calibrate the pressure at low temperatures.

Acknowledgment

This work was supported by a Grant in Aid for Scientific Research from the Japanese Ministry of Education, Science and Culture.

References

[1] Ohashi M, Oomi G, Koiwai S, Hedo M and Uwatoko Y 2003 *Phys. Rev. B* **68** 144428.
[2] Honda F, Kaji S, Minamitake I, Ohashi M, Oomi G, Eto T and Kagayama T 2002 *J. Phys.: Condens. Matter* **14** 11501.
[3] Ohashi M, Tashiro A, Oomi G, Maeda S and Zheng X G 2006 *Phys. Rev. B* **73** 134421.
[4] Thomasson J, Okayama Y, Sheikin I, Brison J P and Braithwaite D 1998 *Solid State Commun.* **106** 637.
[5] Môri N, Nakanishi T, Takeshita N, Goto H, Yomo S and Okayama Y *Physica B* **265** 263.
[6] Nakanishi T, Takeshita N and Môri N *Rev. Sci. Instrum.* **73** 1828.
[7] Andersson G, Sundqvist B and Bäckström G *Jpn. J. Appl. Phys.* **65** 3943.
[8] Wang F M and Ingalls R 1998 *Phys. Rev. B* **57** 5647.
[9] Jaccard D, Holmes A T, Behr G, Inada Y and Onuki Y: *Phys. Lett. A* **299** 282.

Journal of Physics: Conference Series **121** (2008) 122003 doi:10.1088/1742-6596/121/2/122003

A new pressure regulation device using high pressure isolation valves

R Bruyère, A Prat, C Goujon, and J-L Hazemann

Institut Néel / CNRS et Université Joseph Fourier, Dépt MCMF, BP 166, F-38042 GRENOBLE Cedex 9, FRANCE

E-mail: remy.bruyere@grenoble.cnrs.fr

Résumé. A new high pressure equipment has been developed in the context of the French High Pressure Network. It includes a new concept of high pressure isolation valves and a new device of pressure regulation. The main feature of the valves is the absence of the stuffing box and a working mode based on the elastic deformation of materials. The regulation device uses these high-performance valves and proposes a compact system allowing the programming of pressure cycles up to 2000 bar with a high reliability and flexibility. The setup is regularly used for *in situ* experiments in high pressure and high temperature cells at ESRF.

1. Scientific interest : structural changes of supercritical fluids

The understanding of the processes of molecular interactions in fluids at elevated pressures and temperatures is of crucial importance, in particular in the case of aqueous solutions. The growing interest in these processes stems from both the fundamental and applied insights that their understanding brings to the geochemical, biochemical, and industrial communities. It must be emphasized in the case of aqueous solutions that the interparticle interactions under elevated temperature and pressure conditions are very sensitive to the structure of the solvent and, in particular, to the hydrogen bonding strength. As a consequence, studying the structural and electronic evolution of solute particles in aqueous solutions is an efficient indirect method to get information on the solvent itself.

In this framework, a new line of research has been initiated in the Institut Néel (CNRS, Grenoble). It consists of understanding the relation between physical properties and structural organization of liquids under elevated conditions of pressure and temperature, up to and beyond their critical point. Such a challenging program is now made possible by the availability of the extremely bright and well focused X-ray beams at ESRF, combined with our expertise in the design of high temperature and high pressure cells. New optical high pressure/high temperature (HP/HT) cells for independent 1-2000 bar and 30-1700°C variations have been developed [1]. From a geochemical and geophysical point of view, the pressure range of our instruments is very well adapted for *in situ* studies of hydrothermal fluids, sub- and supercritical natural solutions. This range is complementary to the great pressures obtained with Paris-Edimburgh (PE) presses or Diamond Anvil Cells (DAC), necessary to investigate the pressure issues in the mantle crust science. These cells are dedicated to X-Ray Absorption Spectroscopy on FAME beamline at ESRF, Small Angle X-Ray Scattering on D2AM at ESRF, Inelastic X-Ray Scattering (biocat at APS) and Visible Raman Scattering techniques. It allows the study of aqueous solutions (T <

© 2008 IOP Publishing Ltd

600°C and P < 2000 bar) and liquid metals and glasses (T < 1700°C and P < 2000 bar). The technology of our set-up is based on a helium pressurization of the vessel in which the sample and its heating system are placed. Different sample container configurations exist, depending on the used techniques, on the investigated samples and on the conditions of temperature and pressure.

The main effects in the structural changes of supercritical fluids correspond to the thermal density fluctuations undergone by all fluids when approaching their critical point. These fluctuations are at the origin of the divergence of the isothermal compressibility and are characterized by the diverging correlation length. This length corresponds to a distance over which the density fluctuations are correlated, which can be interpreted as an average size of the dynamical clusters of solvent molecules. Thus, in the case of HP/HT aqueous solutions it appears essential to understand how this inhomogeneous distribution of solvent molecules influences solute solvation and reactivity. To follow such phenomenon, the required precision in the measurement of pressure is less than 1 bar and less than 1°C for the temperature. To allow an easy and precise prospection of the supercritical phase diagram with such a precision range, a new pressure regulation device was needed to work with our dedicated cell for aqueous samples.

This paper presents the technological products stemmed from the collaboration between the Institut Néel and the French High Pressure Network (http ://reseauhp.org). The high pressure valves, on which the regulation system is based, are firstly described. Then the features and the advantages of the regulation device are given. This experimental set-up is especially suitable and regularly used for *in situ* experiments or kinetic studies.

2. High pressure isolation valves

A very high pressure gas valve (up to 15000 bar) was firstly developed and patented [2]. Its working, based on an innovating concept, is presented on figures 1 and 2. The main part of the valve (the body) is a steel cavity of 10 cm long, crossed by a tungsten carbide needle (the stem) of 1 mm diameter. On a side of the valve, the stem is fixed to the body. On the other side, the stem tip is in contact with the seat through a plane surface (fig. 1). To open the valve, oil is introduced around the body at a certain pressure (maximum of 300 bar). This hydraulic pressure involves an elastic deformation of the body, which undergoes an elongation up to 0.25 mm. The contact between the stem tip and the seat is then suspended, and the gas under high pressure (up to 15000 bar) can escape through an exhaust tubing of 0.4 mm inside diameter (fig. 2). It is noteworthy that, since it is an isolation valve, it can operate in a closed assembly. In that case, the gas transfer can also occur in the opposite direction (from the "exhaust" towards the seat). To close the valve, the hydraulic pressure is released and the contact between the stem tip and the seat is recovered. A photo of the 15000 bar valve is presented on figure 3. The main advantage of the system is the absence of a stuffing box (the seal generally used in classic valves) which wears out very fast under high pressure and involves many problems such as leaks, pollution and friction. The concept based on the elastic deformation of materials to open and close the valve also avoids the hysteresis effects and implies a perfect proportionality between the hydraulic pressure and the opening of the valve. It ensues a very fine precision of regulation, of 1 bar at 15000 bar.

From this realization, a second valve was designed to propose a competitive product for industrial applications. This second model works up to 2000 bar and is about 30 cm long (fig. 4). It presents the same features except that the operations of opening and closure are controlled by compressed air instead of oil (0 to 5 bar). The transfer of this technology towards the industry is in progress. Presently, a third model is also being studied : with a working mode similar to the one of the 2000 bar valve (compressed air), it will allow to work up to 5000 bar for industrial uses.

Journal of Physics: Conference Series **121** (2008) 122003 doi:10.1088/1742-6596/121/2/122003

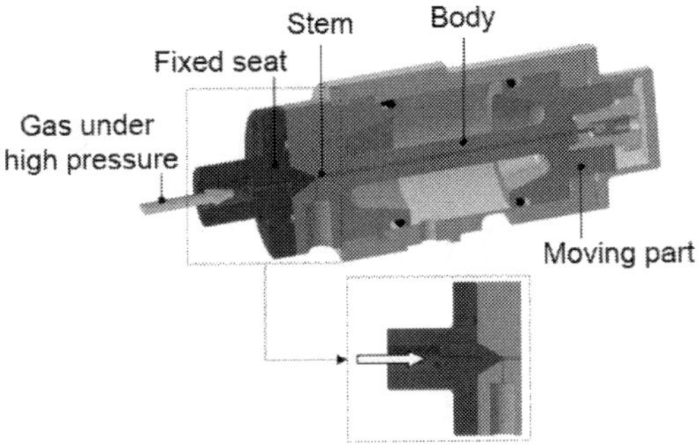

Fig. 1. Closed position of 15000 bar valve.

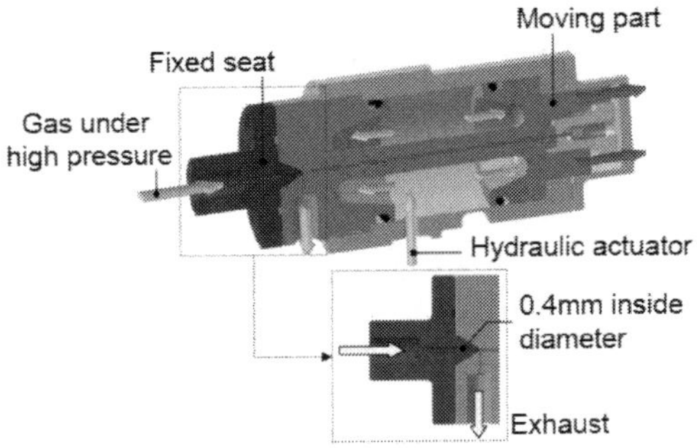

Fig. 2. Opened position of the 15000 bar valve.

Fig. 3. Photo of the 15000 bar valve.

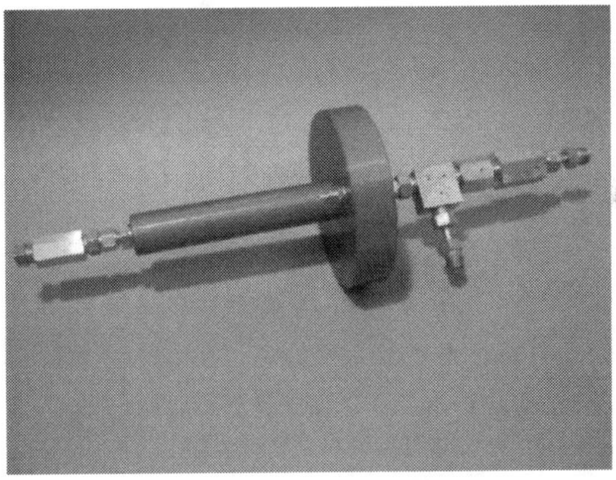

Fig. 4. Photo of the 2000 bar valve.

3. Regulation device

The innovating features of the 2000 bar valve (fig. 4) were used to develop a new device of gas regulation. The system is able to provide a stable pressure of Helium from 0 to 2000 bar in small volume (5 cm^3). It is contained in a rack of 19" wide, weighs 30 kg and can therefore be moved easily (fig. 5). A PID controller combining two feedback loops controls the pressure in the experimental cell via a RS232 connection. The assembly diagram is presented on figure 6. The pressure is generated by a pneumatic or electric compressor with metal membranes. The first loop is a proportional action with wide band and controls the pressure in a buffer volume of 50 cm^3, which absorbs the fluctuations produced at the starting of the compressor. The user selects the pressure of the buffer volume (setpoint 1) at a value 15% higher than the pressure desired in the experimental cell (setpoint 2) and this pressure is maintained by the controller via a static relay (0/10V output). To reduce the compressor startings, the pressure difference between the lowest limit (to restart the compressor) and the setpoint 1 is optimized. The pressure in the buffer volume is measured by a sensor with metal gauges (4/20mA) then is returned to the controller. The exhaust is provided by a high pressure valve connected to the buffer volume.
The innovating concept of the device concerns the second loop. Two 2000 bar valves are activated by compressed air coming from proportional valves (5 bar), themselves controlled by the proportional 0/10 V output signal of the controller. One valve is placed between the buffer volume and the experimental cell for the inlet and the second one is after the experimental cell for the exhaust. The use of valves without hysteresis effects, the presence of a proportional signal

on the whole control system and the optimization of the PID parameters of the second loop lead to an accuracy of 0.05‰ (0.1 bars at 2000 bar). As for the first loop, a sensor measures the pressure in the experimental cell, and returns the signal to the controller. The security of the system is ensured by the controller alarms (maximum value, rate variation) on one hand, and by the presence of a rupture disc in the circuit on the other hand. The system also offers flexibility in process programming, allowing a wide range of pressure cycles, such as ramps (typical rate of 30 bar/min), dwells and sudden rises.
This regulation device presents many advantages for its reliability and its quality of regulation as well as for its ease of installation and use. Since it works with Helium, it is also compatible with all non corrosive gas. Future developments can be suggested such as the adaptation to other volumes, or the addition of a third loop to control, for example, the temperature in the experimental cell. The transfer of this technology towards the industry is also in progress.

Fig. 5. Regulation device.

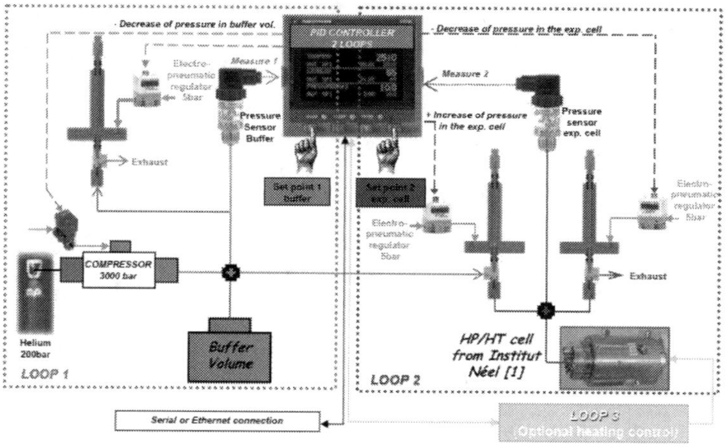

Fig. 6. Assembly diagram of the regulation device.

4. Conclusion

A new concept of high pressure isolation valve has recently been developed and patented. Two models of valves have been conceived providing high reliability up to 15000 bar and 2000 bar, respectively. In these high-performance valves, the stuffing box has been removed and the opening and closure operations are controlled by the elastic deformation of the valve body, avoiding hysteresis effects. Based on the use of the 2000 bar valve, a new regulation device has been conceived to control high pressure cycles (up to 2000 bar) in small volumes. It is presently used on FAME beamline at ESRF for the study of supercritical fluids in HP/HT cells. This light and compact system presents advantages for its ease of installation and use as well as for its high reliability. A dual output controller maintains simultaneously the pressure in a buffer volume and in the experimental cell through coupled proportional actions. The absence of hysteresis of the high pressure valves and the presence of a 100% proportional signal lead to an accuracy of 0.05‰.

5. References

[1] Testemale D, Argoud R, Geaymond O and Hazemann J-L 2005 *Rev. Sci. Instrum.* **76** 043905
[2] Argoud R, Michel JP, Prat A and Roux J 2005 *CNRS Patent* **WO2005080839**

Fabrication of a NiCrAl Pressure Cell and the Application to a Two-Leg Ladder Compound $Sr_2Ca_{12}Cu_{24}O_{41}$

Naoki Fujiwara, Takehiko Matsumoto[a], Kazuko Koyama-Nakazawa[a], Yoshiya Uwatoko[a], Akihiko Hisada, Yousuke Fujimaki[b], and Shinichi Uchida[b]

Graduate School of Human and Environmental Studies, Kyoto University, Kyoto, Kyoto 606-8501, Japan,

[a]Institute for Solid State Physics, University of Tokyo, Kashiwa, Chiba 277-8581, Japan
[b]Department of Physics, University of Tokyo, Hongo, Tokyo 113-0033, Japan

The corresponding author's e-mail address : naokif@mbox.kudpc.kyoto-u.ac.jp

Abstract. A hybrid NiCrAl pressure cell was fabricated to measure magnetic quantities above 3 GPa. A pressure of 4 GPa was achieved and the pressure cell was found to be reusable even after pressurizing trial up to 4GPa. The inner diameter of the cylinder expanded 2.5% at 4 GPa. The pressure cell was applied to the ^{63}Cu-NMR measurement at an optimum pressure of 3.8 GPa in a pressure-induced superconductor $Sr_2Ca_{12}Cu_{24}O_{41}$.

1. Introduction

Pressure has played an important role as a clean parameter for strongly correlated electron systems such as 3d-transition-metal oxides, heavy fermion systems, and organic conducting systems. These materials exhibit a variety of phases with applied pressure. The phases include charge-ordering (CO), magnetic ordering, paramagnetic metallic, and superconducting (SC) states. In the case of pressure-induced superconductivity, a spin- or charge-ordering phase often appears at the lower pressure side on a pressure-temperature (P-T) phase diagram. Among these compounds, some have been precisely investigated using various techniques under pressure, whereas the others have not. It is an important factor whether a piston-cylinder-type pressure cell is applicable or not for researches with various experimental techniques. A large sample space is available only for a piston-cylinder-type pressure cell. The simplest pressure cell is made of CuBe, whose upper limit is at most 1.5 GPa. The upper limit of a Ti-alloy pressure cell is at most 2 GPa [1-2]. The upper limit of a hybrid pressure cell with an outer CuBe sleeve and inner WC is at most 3GPa [3]. Recently, NiCrAl or MP35N cylinders have attracted much attention as post-WC cylinders, because they are nonmagnetic and possess high tensile and yield strengths [4-6]. NiCrAl contains fewer impurities than MP35N, and is suitable for magnetic measurements [7, 8]. Microscopic investigation of superconductivity in $Sr_{14-x}Ca_xCu_{24}O_{41}$ became possible owning to the development of a NiCrAl pressure cell [9], because superconductivity emerges at a pressure of over 3 GPa [10]. The cuprate spin-ladder system is a representative that invokes the significance of the NiCrAl pressure cell. To promote researches on a wide variety of materials, the precise characterization of the NiCrAl pressure cell is of great importance. In the present work, we

© 2008 IOP Publishing Ltd

fabricated a hybrid NiCrAl pressure cell and investigated the pressure limit and the efficiency. We also applied the pressure cell to the NMR measurement of $Sr_{12}Ca_2Cu_{24}O_{41}$ at an optimum pressure of 3.8 GPa .

2. Experimental Apparatus

2.1. Pressure Cell

Fig. 1 shows a scheme of a hybrid pressure cell [11]. The length of the NiCrAl cylinder is 30 mm, and the inner and outer diameters are 6 and 16 mm, respectively. The cylinder is covered by a CuBe sleeve with an outer diameter of 40 mm. The outside of the cylinder and the inside of the sleeve are tapered off, and the cylinder is inserted into the sleeve by a force from a press. A mixture of fluorinated liquids, Fluorinert FC-70 and FC-77, was prepared as a pressure-mediation liquid. This mixture remains hydrostatic up to 10GPa. A Teflon tube with an inner volume of 4.4-mm diameter $\times$15-mm length was filled with the Fluorinert mixture. Both ends were closed by a WC piston and a CuBe plug. The inner liquid was sealed with CuBe rings. An optical fiber and an NQR coil were equipped together in the sample space. We used a fiber of 0.25-mm outer diameter and a coil of 2-mm length with a winding of 25 turns. Ruby powder was fixed using a quick-drying glue on the top of the fiber. The position of the ruby powder was 2 mm above the top of the CuBe plug.

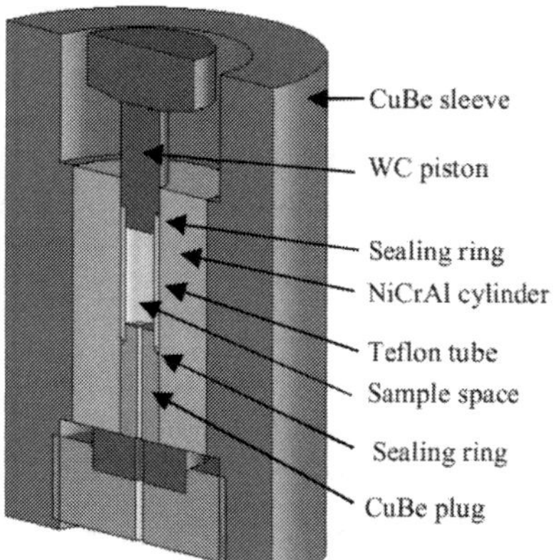

Fig. 1. Illustration of a NiCrAl pressure cell

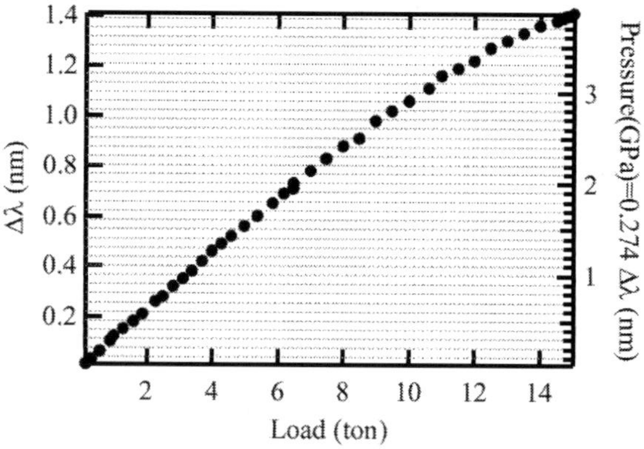

Fig. 2. Load dependence of the R1 shift.

2.2. Pressure Calibration

Ruby fluorescence is often adopted for diamond anvil cells. In a conventional method using a diamond anvil cell, only a piece of ruby is inserted in the cell. In our measurements, the ruby fixed on the top of the optical fiber was inserted into the sample space. The shift of the R1 transition shows a linear relation with pressure. Ruby fluorescence has been confirmed to be a reliable calibration method, and pressure-calibration curves have been presented in several studies [12-14]. The pressure range we measured is rather close to that performed by G. P. Piermarini, They measured pressure up to 19.5 GPa and obtained the fitting curve as

$$P(GPa) = 0.274\Delta\lambda(nm)\,. \tag{1}$$

We measured the R1 shift up to 15 tons. The load dependence of the R1 shift is shown in Fig. 2.

Pressure estimated from Eq. (1) was plotted on the vertical axis on the right side. A pressure of 4 GPa was achieved without damages and the pressure cell was found to be reusable after releasing the load.

2.3. Strain of the Cylinder

As for the deformation of the cylinder during pressurizing process, no appreciable expansion was observed for the first trial up to 10.7 tons. The load can be safely applied to the pressure cell at least up to 3.2 GPa. On the other hand, the inner diameter partly expanded over 2% when a load of 15 tons was applied. The maximum expansion is roughly located at the center of the sample space. The expansion has a relation with the applied pressure rather than the applied load.

3. Application of NiCrAl Cell

We used the pressure cell for the NMR measurements of $Sr_{12}Ca_2Cu_{24}O_{41}$. Fig. 3 shows the results of relaxation rate $1/T_1$ measured at 3.8 GPa with the data measured at 3.5 GPa and ambient pressure. The measurements were performed for the central transition of ^{63}Cu ($I = -1/2 \Leftrightarrow 1/2$) on the ladder sites. As shown in Fig. 3, $1/T_1$ was decomposed into an activated term and a Korringa-type T-linear term;

$$\frac{1}{T_1} = A\exp\left(-\frac{\Delta}{T}\right) + BT \qquad (2)$$

where the spin gap is expressed as Δ in Eq. (2). The values were estimated to be 120, 170 and 190 K for 3.8, 3.5 GPa and ambient pressure, respectively. The value of Δ decreased with applying pressure. However, a spin gap was maintained as a large value even at the optimum pressure. The value of B remarkably increased at 3.8 GPa, whereas those at 3.5 GPa and ambient pressure were almost the same. The value of B at 3.8 GPa was 45% larger than that at 3.5 GPa. The spin gap and B shows remarkable change when pressure is increased toward an optimum pressure of 3.8 GPa.

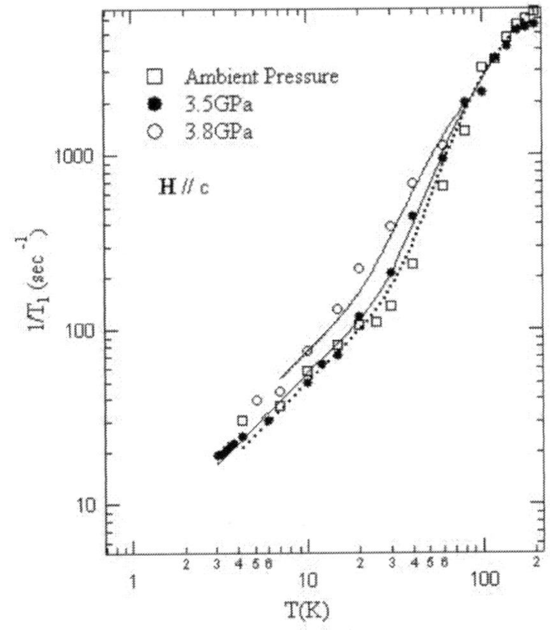

Fig. 3. Relaxation rate $1/T_1$ of ^{63}Cu on the ladder site.

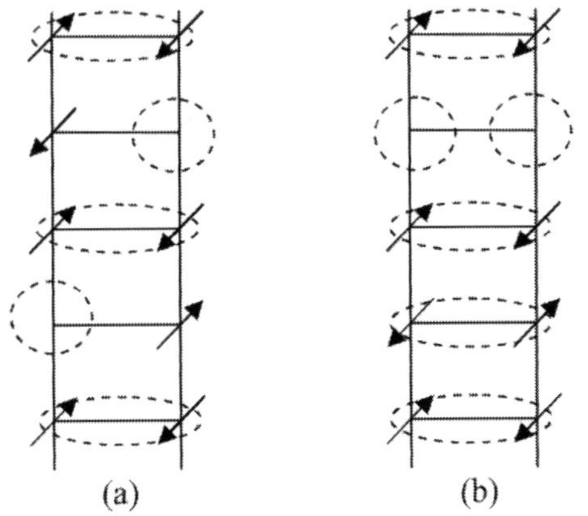

Fig. 4. Illustration of spins and charges at the normal state just above T_c. Arrow and circles represent spins with S=1/2 and a hole. (a) Free motion of a hole or a spin. (b) Hole pair formation.

Journal of Physics: Conference Series **121** (2008) 122004 doi:10.1088/1742-6596/121/2/122004

The T-linear contribution is important factor to determine the mechanism of superconductivity as well as conducting properties at the normal state. In the case of conventional metals, a Korringa-type T-linear term is proportional to square of the density of states. The present system is a strongly correlated electron system, and therefore direct comparison with the conventional metals may be difficult. However, the T-linear term would be caused by a mechanism similar to the conventional metals because superconductivity is followed by the T-linear term below T_c [9]. The two contributions in Eq. (2) are independent of each other although both show remarkable pressure dependence around the optimum pressure. The value of Δ is maintained as a hundred Kelvin even at 3.8 GPa, where T_c becomes a maximum. The spin state is hardly destroyed because this arises from the spin excitation due to the exchange coupling of about 2000 K. The T-linear term is involved with free motion of unpaired spins or holes as shown in Fig. 3 (a). Formation of hole pairs on the rungs would merely separate a spin-ladder system into two parts as shown in Fig. 3 (b), and therefore any relaxation mechanisms other than the activated behavior are hardly expected. In the normal metallic state, single-hole motion contributes to the appearance of the T-linear term and such single holes would make pairs in the superconducting state. The model shown in Fig. 3 (a) was already proposed in the previous paper [9]. We extended the experiments at an optimum pressure, and confirmed the T-linear term originates from spin and/or charge fluctuations.

Acknowledgements

The present work was partially supported by Grant-in-Aid for the Ministry of Education, Science and Culture, Japan,.

References

[1] I. R. Walker, Cryogenics, **45** (2005) 87.
[2] K. Kamishima,M. Hagiwara, H. Yoshida, Rev. Sci. Instrum., **72** (2001) 1472.More references
[3] H. Fujiwara, H. Kadomatsu, K. Tohma, Rev. Sci. Instrum., **51** (1980) 1345.
[4] M. I. Eremets, High pressure Experimental methods (Oxford, Science Pub. (1996))
[5] I. R. Walker, Rev. Sci. Instrum., **70** (1999) 3402.
[6] A. Uchida, M. Kosaka, N. Mori, T. Matsumoto, Y. Uwatoko. J.L. Sarrao, J.D. Thompson Physica B, 312-313 (2002) 339-40.
[7] Y. Uwatoko, T. Hotta, E. Matsuoka, H, Mori, T. Oki. J.L. Sarrao, J.D. Thompson, N. Mōri and G. Oomi, Rev. High Pressure Sci. Technol. **7** (1998) 1508
[8] Y. Uwatoko, S. Todo, K. Ueda, A. Uchida, M. Kosaka, N. Mori, and T. Matsumoto, J. Phys. Condens. Matter, **14** (2002) 11291
[9] N. Fujiwara, N. M$\hat{o}$ri, Y. Uwatoko, T. Matsumoto, N. Motoyama, and S. Uchida, Phys. Rev. Lett. **90** (2003) 137001.
[10] M. Uehara, T. Nagata, J. Akimitsu, H. Takahashi, and N. Mori, and K. kinoshita, J. Phys. Soc. Jpn. **65** 2764 (1996)
[11] N. Fujiwara, T. Matsumoto, K. Koyama-Nakazawa, A. Hisada, and Y. Uwatoko, Rev. Sci. Instrum., **78** (2007) 073905.
[12] G. J. Piermarni, S. Block, J. D. Barnett and R. A. Forman, J. App. Phys. **46** (1975) 2774.
[13] H. K. Mao, P. M. Bell, J. W. Shaner, and D. J. Steinberg, J. App. Phys. **49** (1978) 3276.
[14] R. J. Hemley, C. S. Zha, A. P. Jephcoat, H. K. Mao, and L. W. Finger, and D. E. Cox, Phys. Rev. B **39** (1989) 11820

Journal of Physics: Conference Series **121** (2008) 132001 doi:10.1088/1742-6596/121/3/132001

Increase of linear size uniformity of synthetic diamond grinding powder

G Ilnitskaya[1], **G Bogatyreva, N Novikov, A Shepelev, G Nevstryev, O Leshchenko**
ISM – Institute of Superhard Materials, National Academy of Sciences of Ukraine,
2, Avtozavodskaya St., Kiev, 04074, Ukraine

E-mail: olesh@ism.kiev.ua

Abstract. It has been shown in article that increase of powder uniformity is achieved by carrying out additional sorting of powders in the sizes and form of grains. The sorting process includes the following stages: chemical rounding of diamond grains by strong oxidants; special grain size classification on sieves with mesh sizes corresponded to geometrical progression of twentieth (R-20) and fortieth (R-40) number series; additional separation of diamond grains in narrow graininess by grain form. Diamond powder of AC6 100/80 series has been produced by this technology, as a result of that the content of basic fraction runs up 85 %.

1. Introduction

Increase of efficiency of diamond tools applied in material engineering is important problem today. One of the problem solutions is an increase of quality of diamond powders applied in tooling due to more accurate control of synthesis mechanism as well as technological regimes of powder production.

Diamond is most progressive tool material, especially in the fields where exacting requirements to workpiece quality as well as working accuracy and roughness of workpiece surface of details are demanded. One of the most effective instruments to achieve higher class of workpiece surface and to increase diamond tooling wear-resistance is an increase of size uniformity of granulometric composition and form of diamond grains of powder applied in tooling.

Using diamond powders with higher size uniformity is especially effective in those industry branches where precise form and high grade of workpiece finish are required [1-3].

On the basis of studies carried out in the ISM[2] it have been ascertained that for increase of efficiency of abrasive wheels in polishing of cut glass the especial diamond powders are required. These powders have the following properties: uniformity by linear sizes (length (a) and width (b) of grain projection), content of main fraction by grain size is upwards of 80% and form factor (a/b) is at the most 1.3.

The production engineering of such diamond grinding powders of various grain sizes has been developed in the ISM. The process includes the following stages: grain size separation on sieves with mesh sizes corresponded to geometrical progression of twentieth (R-20) and fortieth (R-40) number series with $\sqrt[20]{10} = 1.122$ and $\sqrt[40]{10} = 1.059$ ratio correspondingly; special chemical processing and

---

[1] To whom any correspondence should be addressed.
[2] ISM – Institute of Superhard Materials, National Academy of Sciences of Ukraine, Kiev, Ukraine

© 2008 IOP Publishing Ltd

separation by grain form on vibrating table [4-6]. The powders produced by this technology are uniform by linear sizes, by strength as well as content of volume and surface defects.

2. Experimental methods

Experiments have been carried out with various diamond grinding powders synthesized in the Ni–Mn–C system. The average grain sizes are 400/315, 125/100 and 80/63 μm. The diamond powders have been studied with the object to the uniformity of some basic characteristics (strength, grain composition etc.).

It is suggested to assess the powder uniformity by the concentration of grits whose characteristic in the given property corresponds to the mean (nominal) characteristic of the powder under study. Here we discuss the uniformity of diamond powders with respect to the basic technical characteristics: strength ($C_{un.str.}$) and grain composition ($C_{un.\ lin.size}$).

We have studied the diamond powders in the following sequence. At first, grain size separation on sieves with mesh sizes corresponded to geometrical progression of twentieth (R-20) number series. Produced narrow fractions of 125/115 and 72/63 have been processed by special chemical treatment of strong oxidants, after that the powders have been separated by grain form on vibrating table.

In all produced fractions we have determined strength (P, N), main fraction content (%), coefficient of strength uniformity ($C_{un.str.}$, %) [6], form factor ($C_f = a/b$), coefficient of linear size uniformity ($C_{un.lin.size}$, %) – a ratio of number of average size grains ($(a+b)/2$) to number of all grains.

3. Results and discussion

3.1. Results of additional sieve classifying

The results of studies of properties of diamond grinding powders of AC125 400/314, AC6 125/100 and AC15 80/63 series after additional sieve classifying on R-20 sieves are presented in Table 1.

Table 1. Properties of diamond grinding powders of AC125 400/314, AC6 125/100 and AC15 80/63 series after additional sieve classifying on R-20 sieves.

Grain size	P, N	Series	Main fraction, %	$C_{un.str.}$, %	C_f, convent. units	$C_{un.\ lin.size}$, %
315/355	218.4	AC125	95.2	46.5	1.11	44.6
355/400	216.7	AC125	93.4	47.8	1.09	47.2
400/315 init.	211.5	AC125	82.5	32.0	1.18	33.6
125/115	7.2	AC6	90.0	25.1	1.32	35.4
115/100	6.4	AC6	92.0	24.4	1.31	32.6
125/100 init.	6.8	AC6	72.0	14.5	1.44	24.4
80/72	11.2	AC20	92.0	22.1	1.40	39.6
72/63	10.9	AC20	92.0	18.5	1.39	37.5
80/63 init.	10.8	AC20	72.0	12.6	1.42	14.7

As appears from the above the strength uniformity of diamond grinding powders of 125/115 narrow class (from 125/100 standard grain size) increases from 14.5% to 25.1%, whereas linear size uniformity increases from 24.4% to 35.4%. The similar results have been obtained for powders of 400/315 and 80/63 grain sizes.

3.2. Results of chemical processing and separation by grain form

The results of studies of properties of diamond grinding powders of AC6 125/115 and AC15 72/63 series after chemical processing and separation by grain form on vibrating table are presented in Table 2.

Table 2. Properties of diamond grinding powders of AC6 125/115 and AC15 72/63 series after chemical processing and separation by grain form on vibrating table.

Grain size	№ of separat. product	P, N	Series	Main fraction, %	$C_{un.str.}$, %	C_f, convent. units	$C_{un.\ lin.size}$, %
125/115	1	10.5	AC6	93	33.1	1.27	39.6
	2	8.4	AC6	92	28.6	1.44	33.1
	3	4.5	AC6	89	21.4	1.75	28.6
	Initial	7.2	AC6	90	24.4	1.32	35.5
72/63	1	11.8	AC20	93	34.5	1.39	45.6
	2	11.0	AC20	94	31.2	1.46	34.6
	3	9.6	AC20	92	17.8	1.71	22.1
	Initial	10.9	AC20	92	18.6	1.39	37.5

As appears from the above the strength of diamond powder of AC 72/63 series increases from 10.5N to 4.5N that is correlated with an increase of form factor and at the same time the strength uniformity increases from 14.5% (initial powder) to 31.2% (finished product). The linear size uniformity increases from 24.4% (initial powder) to 30.6% (product of number 1).

3.3. Results of diamond abrasive wheel test in-polishing of cut glassware surface

Diamond powders of AC15 80/63 series produced by form factor sorting have been used for production of abrasive wheels with B2-01-1 metal-polymeric binder. The wheels have been tested in-polishing of cut glassware surface.

The following operating indices of efficiency have been studied at work of these abrasive wheels: relative (q_p) and specific (q_v) removal of diamond (wear-resistance of the wheels) and roughness of finished surface (R_a). The indices have been determined by the procedures [7].

For verification of results reproducibility the sorting of AC15 80/63 diamond powders, production of abrasive wheels with these powders and polishing tests of these abrasive wheels have been carried out repeatedly.

The results of these tests are presented in Table 3.

Table 3. Correlation between characteristics of diamond grinding powders of AC15 80/63 series and grinding wheel efficiency (results reproducibility).

Sample	Technological characteristics			Operating characteristics		
	Main fraction, %	C_f, convent. units	$C_{un.\ lin.size}$, %	q_p, mg/g	q_v, mg/g^3	R_a, µm
1	92	1.1	45.6	11.52	308.93	0.17 / 0.18
2	94	1.2	38.7	18.79	274.33	0.19 / 0.20
3	92	1.3	35.3	21.16	168.19	0.25 / 0.20
1	93	1.1	46.2	11.23	303.97	0.28 / 0.32
2	92	1.2	37.3	18.76	273.90	0.20 / 0.22
3	92	1.3	36.1	20.82	163.96	0.15 / 0.18

As appears from the above when coefficient of linear size uniformity increases by 10%, the removal of diamond decreases by 1.5-1.8 times and roughness of finished surface diminishes by 1.4-1.6 times.

The results of tests of abrasive wheels with grinding powders of various series are presented in Table 4.

Table 4. Correlation between characteristics of diamond grinding powders of various series and grinding wheel efficiency.

Series, Grain size	Sample	Technological characteristics			Operating characteristics		
		Main fraction, %	C_f, convent. units	$C_{un. lin. size}$, %	q_p, mg/g	q_v, mg/g^3	R_a, μm
AC20 80/63	**1**	93	1.1	56.7	8.88	129.65	0.18
	2	94	1.2	46.2	22.50	328.50	0.24
	3	92	1.3	55.4	25.40	370.84	0.29
AC6 100/80	**1**	88.6	1.25	43.2	3.58	60.01	0.12
	2	80.5	1.29	39.6	4.11	73.29	0.15
	3	73.1	1.36	32.1	5.02	129.65	0.16
AC6 80/63	**1**	88.6	1.25	43.2	4.12	70.54	0.14
	2	80.5	1.29	39.6	5.34	84.12	0.15
	3	73.1	1.36	32.1	7.09	145.34	0.25

As appears from the above when main fraction content increases, wear-resistance of the wheels increases too and relative removal of diamond q_p decreases.

4. Conclusion

1. Production of diamond powders uniformed by strength characteristics and linear sizes is provided by classifying processes. It is the result of using increasingly advanced methods of diamond separation by corresponding physical-mechanical characteristics of regular synthetic diamond crystals.
2. As a result of carried out studies the following correlation dependences have been determined: as main fraction content and diamond grain form factor are increased the wear-resistance (q_p, q_v) of abrasive wheels is increased too, while roughness of finished surface (R_a) is decreased.
Maximal reduction of specific diamond consumption in grinding is amount to 2.5 times for diamond of AC20 series and 1.4 times for diamond of AC6 series, while roughness of finished surface is decreased by 1.6 and 1.3 times correspondingly.

References

[1] Bokiy G B, Bezrukov G N, Klyuev Yu A, Naletov A M and Nepsha V I 1986 *Natural and synthetic diamonds* (Moscow: Nauka)
[2] DSTU 3292-95 1996 *Synthetic diamond powders General technical terms* (Kiev: Gosstandart Ukrainy)
[3] Stihov L V 1978 *Determination of sizes and grain size of diamond in different sieve classes* (Moscow: VNIIAlmaz works) pp 50–56
[4] Nevstruev G F and Ilnitskaya G D 2004 Method of production of narrow grain size grinding powders Patent 69947 A *Promyslova vlasnist* **9**
[5] Nevstruev G F and Ilnitskaya G D 2004 Method of production of narrow grain size abrasive grinding powders Patent 69949 A *Promyslova vlasnist* **9**
[6] Novikov N V, Nevstruev G F, Ilnitskaya G D, Kushch V I and Kolodnitskiy V N 2006 Assessment of superhard material powders quality Part 2 Practical application of quality characteristic assessment method *Superhard Materials* **6** 58–67
[7] Shepelev A A and Dubrova A E 2006 Assessment of operating quality indices of superhard material abrasive wheels *Modern processes of machining by SHM tools and surface condition of machine elements* (Kiev:) pp 89–96

Joint 21st AIRAPT and 45th EHPRG Int. Conf. on High Pressure Science and Technology IOP Publishing
Journal of Physics: Conference Series **121** (2008) 142005 doi:10.1088/1742-6596/121/4/142005

Relative permittivity behavior and temperature changes in linoleic acid during the phase transition.

R Kościesza, R M Siegoczyński, A J Rostocki, D B Tefelski, A Kos, and W Ejchart

Faculty of Physics, Warsaw University of Technology, Koszykowa 75, 00-662 Warszawa, POLAND.

E-mail: raphael@if.pw.edu.pl

Abstract. In our earlier works several fatty liquids (edible oils and unsaturated fatty acids) which exhibit existence of a new phase induced by high pressure were presented. Conclusion of those experiments is that C=C bonds existence in these liquids plays a dominant role in a new phase occurrence. Relative permittivity in pure acids investigated till now seems to behave in specific way. That is why we decided to investigate linoleic acid ($C_{18}H_{32}O_2$) under high pressure. In our experiment such quantities as: electric capacity, pressure and temperature were recorded. The experimental setup gives us also a possibility to conduct optical investigations. We observed a transmitted and scattered beams of close infrared light (λ = 800nm) in directions 0° and 90° towards the incident beam. Due to the rapid grow of temperature and the rapid change of transmitted and scattered beams we may say that observed phenomenon is a first order phase transition and a proof for the significant change of liquid structure. This paper contains time dependencies of permittivity, temperature, transmitted and scattered light intensity and also permittivity vs. pressure changes during the phase transition in linoleic acid and first of all measured data analysis which lets us explain the transition reasons.

1. Introduction

Conclusions that we came to in experiments with oleic acid and other fatty liquids showed us that investigations of liquids which contain C=C bond have a reasonable basis. Their increasing role in food and petrol industry and medical application (its isomers prevent breast and colon cancer [1]) is just another reason to keep investigations on. In this paper we are especially interested in the linoleic acid. It is one of unsaturated fatty acids. As well as in liquids investigated by us till now we can induce a high-pressure phase transition in linoleic acid too. We can observe occurrence of the new phase by measuring the scattering light and the temperature inside the chamber. There are no significant changes of the relative permittivity which lets us to conclude that we deal with a transition of investigated medium to the new phase. We have investigated a few fatty acids and the insignificant changes of electric permittivity[2] seems to be characteristic in general.

2. Experiment

The pressure in our high pressure apparatus is generated by piston-cylinder system with multiplication. This system is supplied by a computer-controlled pump with regulated velocity of pressure application. The chamber has three sapphire windows for optical investigations. Optical signal from the modulated, infrared pin diode is registered by two detectors. Then it is conditioned in

© 2008 IOP Publishing Ltd

lock-in and amplifier. The source wavelength is well-chosen to avoid molecular absorption and to be well transmitted by the sapphire windows. Electric quantities such as capacity, voltage (from copper-constantan thermocouple and light detectors) and resistance (from manganin coil) are measured by multimeter and RLC bridge. All data obtained during the experiment is collected and converted by acquisition system designed in our lab.

3. Results

As it was shown in an earlier paper [3], a new high pressure phase occurrs in linoleic acid. Main pointer of phase transition is a rapid increase of a scattered light and rapid decrease of transmitted light simoultaneously. Such case is shown in figure 1. This paper considers relative permittivity and temperature behavior during the phase transition. Changes of these parameters are presented in figure 2.

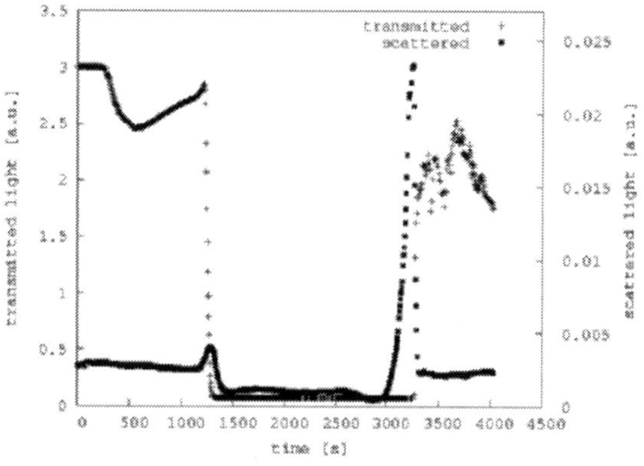

Figure 1. Scattered and transmitted light intensity.

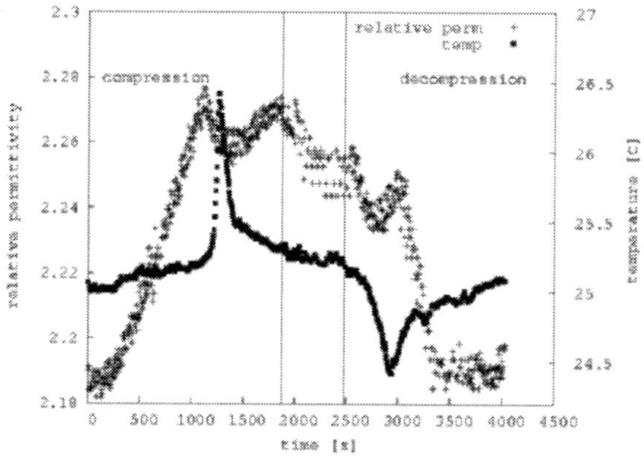

Figure 2. Time dependencies of relative permittivity and temperature.

Comparing the light and permittivity-temperature graphs we may notice that sudden changes of beams intensity are correlated with rapid temperature changes. Small changes of relative permittivity during the transition were observed. Linoleic acid is a dipole liquid that was placed in electric field during the experiment. Figure 2 exemplifies thermal effects on electric dipoles mutual orientation. At the begginig of the phase transition increase of temperature destroys parallel orientation of single, free dipoles which are turned according to the electric field. Random orientation of dipoles leads to a decrease of total dipole moment and decrease of the relative permittivity. During the return to the initial phase temperature decreases and free dipoles align according to the electric field again. Analysis of the relative permittivity changes was our main tool which let us to investigate phase transitions in fatty liquids [4-8]. To find out more about the phenomenon that we deal with, the authors decided to consider investigated liquid using the Fröhlich theory of dielectrics [9]. Such analysis seems to be the most reasonable one, for it gives the best view on squared average macroscopic dipole moment in our liquid, which depends on the relative permittivity changes that is correlated with pressure [10]. This relation is described by equation 1.

$$\langle mm^* \rangle = \frac{3\varepsilon_0 kT}{N(p)} \frac{(\varepsilon(p)-\varepsilon_\infty)(2\varepsilon(p)+\varepsilon_\infty)}{3\varepsilon(p)} \tag{1}$$

<mm*> is an average product of the dipole moment **m*** of single cell of Fröhlich cavity and the whole cavity dipole moment **m,** ε is a relative permittivity, ε_∞ can be approximated as a squared refractive index of the linoleic acid, N is molecules concentration per volume unit. The figure 3 presents structural factor **<mm*>** and molecules concentration changes during phase transition.

Journal of Physics: Conference Series **121** (2008) 142005 doi:10.1088/1742-6596/121/4/142005

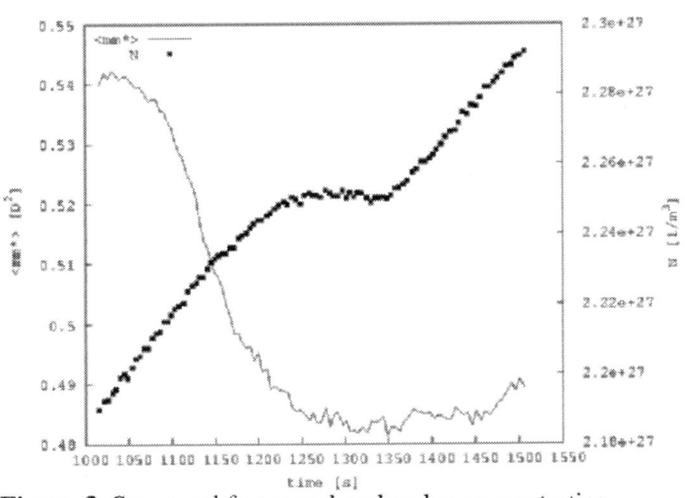

Figure 3. Structural factor and molecules concentration during phase transition.

Total dipole moment is decreasing while concentration of molecules in capacitor is still increasing. It leads to insignificant changes of relative permittivity in time of the transition. Small decrease of ε during transition is caused by earlier mentioned thermal influences and dipoles interactions. In spite of the electric dipoles mutual interactions play insignificant role in mechanism causing the phase transition we may observe the loop of hysteresis (figure 4), which proves that some of them turned anti-parallel one to each other during the whole compression-decompression cycle.

Width of loop of hysteresis is proportional to the number of free electric dipoles that are not creating dimers with other molecules of acid. In our case, the structure of liquid is changing but its electric properties are kept. Strong scattering state and visible temperature changes are our main pointers of a new phase occurring. All our measurements were done in the room temperature. From thermodynamic point of view we had foster conditions for rapid grow of a scattering centers composed of interacting acid molecules [11]. As in the case of oleic acid these centers may be structures of short range order similar to the smectic phase in liquid crystals [12]. Such crystal-like structures have other refractive index and density than surrounding liquid. Number, diameter and packing of the cristalites grow due to pressure increase, and this situation forces the change of acid compressibility (figure 5).

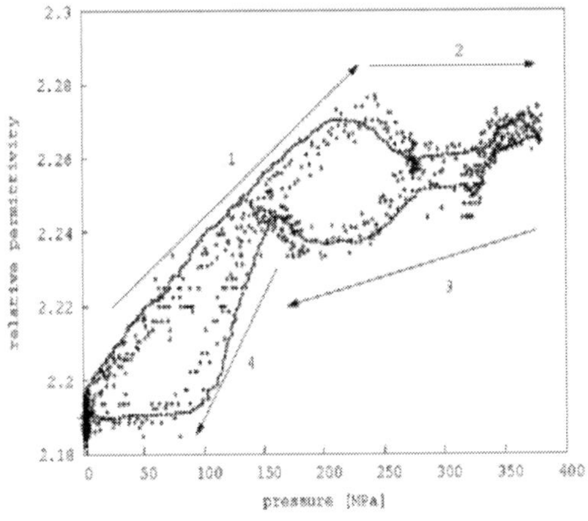

Figure 4. Change of relative permittivity in compression – decompression cycle.

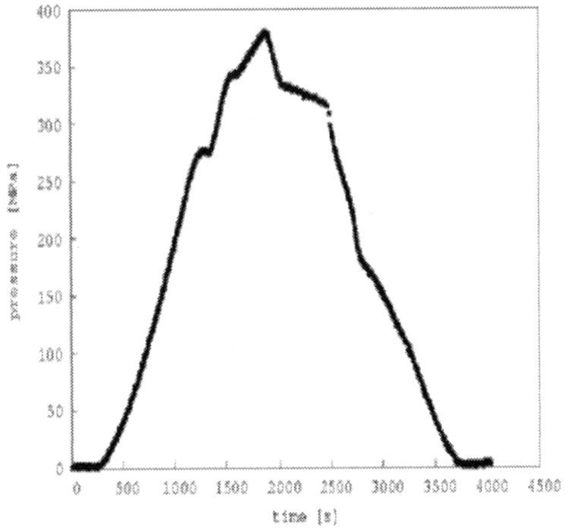

Figure 5. Time dependency of the pressure.

In this figure we can clearly see the changes of the compressibility by observing the pressure curve slope. Every change of slope in time of compression as well as decompression means change of compressibility and packing of cristalites. What is more interesting is that there are no visible changes of scattered and transmitted light intensity according to the changes of optical properties of investigated medium. Due to compression, liquid is getting more dense so the cristalites do not grow any more, optical signal falls and finally it is getting stable. Multiple light reflections and multiple scattering inside the chamber attenuate the optical signal.

4. Conclusions

First order phase transition in linolec acid was observed. New high-pressure phase occurrence is strongly correlated with scattered light and temperature increase. The most interesting thing is that no significant changes of electric permittivity were noticed, what is well explained using Fröhlich theory of dielectrics. High pressure treatment of linoleic acid is causing significant structure changes at molecular level. It leads to creating the scattering centers structure of which is similarly ordered as in liquid crystals smectic phase. As it is shown in figure 5 there are few intermediate states which have different physical properties such as compressibility and pacing of molecules.

5. Acknowlegdements

The work was supported by the Polish Ministry of Science and Higher Education under Grant No. 134/SPB/COST/KN/DWM 105/ 2005-2007

References

[1] Whigham L D, Cook M E, Atkinson R L 2000 Pharmacological Research 42 No 6, pp. 503-510
[2] Kos A, Tefelski D B, Siegoczyński R M, Rutkowski R, Ejchart W and Wiśniewski R 2005 High Pressure Res 75 p 56
[3] Siegoczyński R M and Wiśniewski R 1994 Proc. Int. Conf. on XXXII. Annual Meeting EHPRG
[4] Kościesza R, Rostocki A J, Kos A, Tefelski D B, Siegoczyński R M and Zych L2007 High Pressure Res 27 p 55
[5] Rostocki A J, Kościesza R, Tefelski D B, Kos A, Siegoczyński R M and Chruściński L2007 High Pressure Res 27 p 47
[6] Rostocki A J, Wiśniewski R and Wilczyńska T 2007 J Mol Liq 135 p 122
[7] Vereshchagin S F, Kuznethov L F and Alameda T I 1956 J Experimental and Theoretical Physics 3 p 661
[8] Siegoczyński R M, Wiśniewski R and Ejchart W 1999 Physica B 265 p 274
[9] Fröhlich H 1958 Theory of Dielectrics (Oxford:Oxford University Press)
[10] Siegoczyński R M 1998 Reports of Institute of Physics (Warsaw: Publishing House of Warsaw University of Technology)
[11] Siegoczyński R M, Kos A, Tefelski D B, Kościesza R and Ejchart W 2007 High Pressure Res 27 p 51
[12] Przedmojski J and Siegoczyński 2002 Phase Transitions 75 p 581

Pulsed pressure treatment for inactivation of *Escherichia coli* and *Listeria innocua* in whole milk

S Buzrul[1,2,3], A Largeteau[1], H Alpas[2], G Demazeau[1]

[1] ICMCB, CNRS, Université Bordeaux 1, site de l'ENSCPB, 87 avenue du Dr. A. Schweitzer, 33608 PESSAC cedex, France,
[2] Food Engineering Department, Middle East Technical University, 06531, Ankara, Turkey

E-mail: sbuzrul@metu.edu.tr

Abstract. *E. coli* and *L. innocua* in whole milk were subjected to continuous pressure treatments (300, 350, 400, 450, 500, 550 and 600 MPa) at ambient temperature for 5, 10, 15 and 20 min. These treatments underlined that at moderate pressure values (300, 350 and 400 MPa), increasing the pressurization time from 5 to 20 min did not improve cell death to a great extent. Therefore, pulsed pressure treatments (at 300, 350 and 400 MPa) for 5 min (2.5 min × 2 pulses, 1 min × 5 pulses and 0.5 min × 10 pulses), 10 min (5 min × 2 pulses, 2 min × 5 pulses and 1 min × 10 pulses), 15 min (5 min × 3 pulses, 3 min × 5 pulses and 1.5 min × 10 pulses) and 20 min (10 min × 2 pulses, 5 min × 4 pulses, 4 min × 5 pulses and 2 min × 10 pulses) were applied. As already observed in continuous pressure experiments, in pulsed pressure treatments the inactivation level is improved with increasing pressure level and in addition with the number of applied pulses; however, the effect of pulse number is not additive. Results obtained in this study indicated that pulsed pressure treatments could be used to pasteurize the whole milk at lower pressure values than the continuous pressure treatments. Nevertheless, an optimization appears definetely necessary between the number of pulses and pressure levels to reach the desirable number of log-reduction of microorganisms.

1. Introduction

The application of high hydrostatic pressure (HHP) in food preservation generally ranges from 100 to 1000 MPa; however, when high pressure equipment are operated over a long time at more than 600 MPa pressurization, creep phenomenon can be seen and HHP equipment may be damaged [1]. Moreover, the cost of pressure equipment rises more than linearly with the maximum operating pressure, therefore it is wise to study the interaction of the variables (pressure, temperature, time etc.) in HHP processing and determine the minimum conditions to obtain desirable levels of microbial inactivation while maintaining a maximum degree of sensory and nutritional quality [2, 3].

HHP treatment can consist of a single exposure period (one pressure cycle; i.e., longer holding time) or the application of multi-pulsed pressure in shorter periods (equal total duration but shorter holding times for each pulse). The question that arises is whether one form is more effective than the other or whether they are equivalent. If differences between the outcomes of the two treatments are

[3] To whom any correspondence should be addressed.

© 2008 IOP Publishing Ltd

observed, are they results of differences in the inactivation mechanism, or just manifestation of small differences in the exposure time, caused by inevitable slight differences in the compression (come-up) and decompression (pressure release) times [4].

Many researchers tried to give some answers to these questions. López-Caballero *et al.* [5] demonstrated that step-pulsed pressurization (400 MPa at 7 °C in two 5-min pulses) produced no apparent advantages over continuous pressurization (400 MPa at 7 °C for 10 min) on the inactivation of microbial flora of the oysters. Alemán *et al.* [6] observed that step-pulsed pressurization is more effective than continuous pressurization in reducing the microbial load in pineapple. Studies on spores were also tried; Hayakawa *et al.* [1] achieved complete inactivation of 6 log initial load of *Bacillus stearothermophilus* spores by use of oscillatory pressure treatments (6 pulses of 5-min pressurization with 600 MPa at 70 °C). Rigaldie *et al.* [7, 8] have successfully applied pulsed HHP treatment for the inactivation of pharmacopeia strains including the spores of *Bacillus subtilis*.

The objective of this study was to observe the differences between continuous and pulsed pressure treatments for the inactivation of *Escherichia coli* and *Listeria innocua* in whole milk and further to explore the potential usability of pulsed pressure treatment.

2. Materials and methods

2.1. Preparation of bacterial species

The microorganisms used were *E. coli* ATCC 11775 and *L. innocua* ATCC 33090 (both from Périgueux, France - E.R.A.P laboratory). The strains were maintained on tryptic soy agar plus 0.6 % yeast extract (TSAYE) (Merck, Darmstadt, Germany) slants. For growth, a loopful of each organism was transferred to tubes of tryptic soy broth supplemented with 0.6 % yeast extract (TSBYE) (Merck, Darmstadt, Germany) at 37 °C for 15–21 h and transferred to fresh broth every 48 h for use in this study.

Ultrahigh-temperature (UHT) whole milk (pH 6.64) was obtained from a local market (Bordeaux, France). It was inoculated with pure cultures from the stationary phase to obtain about 10^7–10^8 colony forming units (CFU) mL^{-1} milk. All samples were inoculated at least 2 h before treatment to allow cells to acclimatize to the new environment. One mL of un-inoculated milk was transferred onto TSAYE plates (Sterilin, Staffordshire, UK) and incubated at 37 °C for 48 h to make sure that the milk samples were sterile.

2.2. HHP treatment

The cell suspensions were dispensed in 6 mL-portions in sterile plastic vials (Nunc, Roskilde, Denmark) in duplicate. Air bubbles were avoided. The vials were vacuum sealed in sterile plastic bags (Fischer Scientific, PA) and kept at 4 °C prior to pressurization that did not exceed 1 h. Pressurization of samples was carried out using a computer controlled high pressure unit with 3 L sample compartment, capable of operating at up to 800 MPa and designed by NFM – Technologies (Le Creusot, France) and FRAMATOME (Paris, France), marketed by CLEXTRAL (Firminy, France). The pressure transmitting fluid was ethylene glycol. Although the maximum pressure increase and release rate of this equipment was 375 MPa/min, the pressure increase and pressure release time were both set to 300 MPa/min for safety considerations. Pressure and temperature were measured using sensors inside and outside the high pressure vessel and all data were stored in the computer system. *E. coli* and *L. innocua* in whole milk were subjected to continuous pressure treatments (300, 350, 400, 450, 500, 550 and 600 MPa) at ambient temperature (about 22 °C) for 5, 10, 15 and 20 min.. Pulsed pressure treatments were applied at 300, 350 and 400 MPa for a total duration of 5 min (2.5 min × 2 pulses, 1 min × 5 pulses, and 0.5 min × 10 pulses), 10 min (5 min × 2 pulses, 2 min × 5 pulses and 1 min × 10 pulses), 15 min (5 min × 3 pulses, 3 min × 5 pulses and 1.5 min × 10 pulses) and 20 min (10 min × 2 pulses, 5 min × 4 pulses, 4 min × 5 pulses and 2 min × 10 pulses). The pressure holding time did not include the process come-up or depressurization times. Temperature increases during pressurization due to adiabatic heat were predetermined using a K-type thermocouple inside the pressure vessel. Compression heating during pressure was taken into consideration in the experiment so that temperature of the pressure transmitting fluid during pressure was controlled at near ambient temperature (~ 22 °C) [the temperature increase of ethylene glycol (pressure transmitting fluid) and whole milk were about 4.0 °C/100 MPa and 3.4 °C/100 MPa, respectively].

Immediately after pressure treatment, the vials were transferred to ice-water mixture. Cells suspensions from each vial was serially diluted with 0.1 % peptone water (Biokar, Beauvais, France) and each dilution (from each sample) were surface plated in duplicate giving four plates per dilution on prepoured TSAYE. With samples containing less than 25 CFU mL^{-1} in a 1:10 dilution, 1 ml of undiluted cell suspension was plated in three plates (0.3, 0.3 and 0.4 ml). The plates were incubated at 37 °C for 72 to 96 h prior to counting colonies to allow injured cells to form visible colonies. The entire experiment was repeated once more and the averages were determined.

3. Results and discussion

3.1. Continuous HHP treatment

Figure 1 shows the continuous HHP treatments between 300 and 600 MPa for *E. coli* and *L. innocua* in whole milk. Results indicated that, at a constant moderate pressure (300, 350 and 400 MPa), increasing pressurization time from 5 to 20 min did not increase cell death to a great extent. Inactivation of both microorganisms were about one $\log_{10}$ at 400 MPa for 20 min. Another observation was: as the pressure increases the inactivation increases. In fact, 7 $\log_{10}$ reduction (complete inactivation) for both microorganisms is only possible at 600 MPa for 15 min and at 550 MPa at 20 min.

(a) (b)

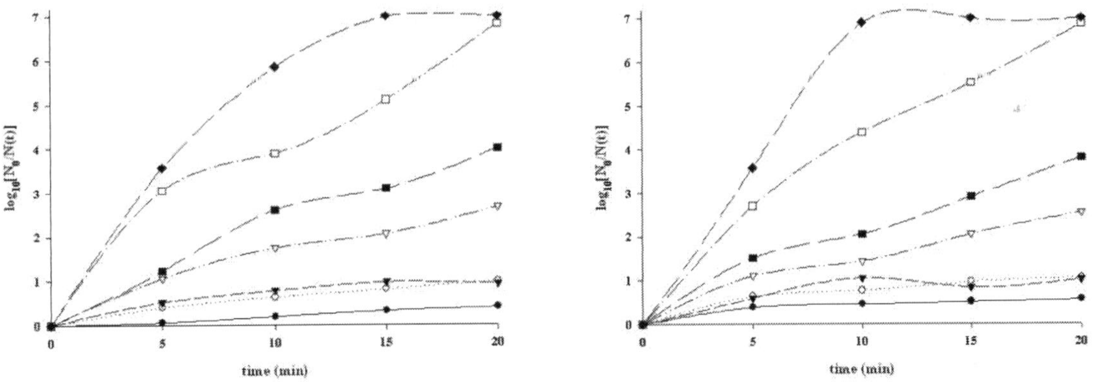

Figure 1. Inactivation of *E. coli* (a) and *L. innocua* (b) in whole milk at 300 (●), 350 (○), 400 (▼), 450 (▽), 500 (■), 550 (□) and 600(♦) MPa for 5, 10, 15 and 20 min.

3.2. Effect of pressure level in pulsed HHP treatment for a constant duration and pulse number

Figure 2 shows 20 min pulsed HHP treatment (4 min × 5 pulses) at 300, 350 and 400 MPa. As already observed in continuous pressure treatment, in pulsed HHP treatment inactivation increases with increasing pressure level; however inactivation differences between pressure levels were more pronounced in pulsed HHP treatment. Moreover, 4 $\log_{10}$ reduction is possible at 400 MPa for 4 min × 5 pulses while in continuous HHP treatment only one $\log_{10}$ reduction was observed at 400 MPa for 20 min as mentioned before. Note that even at 300 MPa inactivation is more than one $\log_{10}$ for both microorganisms.

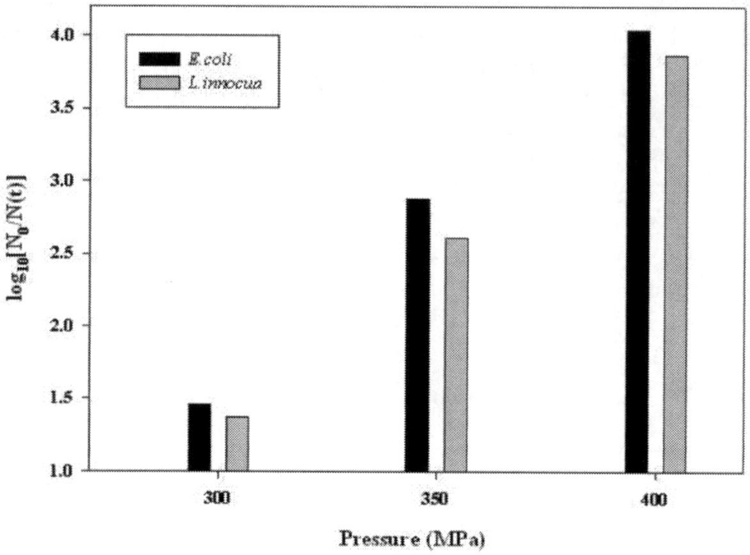

Figure 2. Inactivation of *E. coli* and *L. innocua* in whole milk at 300, 350 and 400 MPa for 4 min × 5 pulses

3.3. Effect of pulse number for a duration constant of 5 min for each pulse

Figure 3 shows the pulsed HHP treatment at 400 MPa for 5 min × 1 pulse (continuous 5 min), 5 min × 2 pulses, 5 min × 3 pulses and 5 min × 4 pulses. Figure indicates as the pulse number increases inactivation is improved. Again a comparison can be made with continuous HHP treatment and can be observed that increasing time 5 min (5 min × 1 pulse) to 20 min (5 min × 4 pulses) increases cell death significantly ($P < 0.05$) in pulsed HHP treatment. It was also observed that the effect of pulse number is not additive.

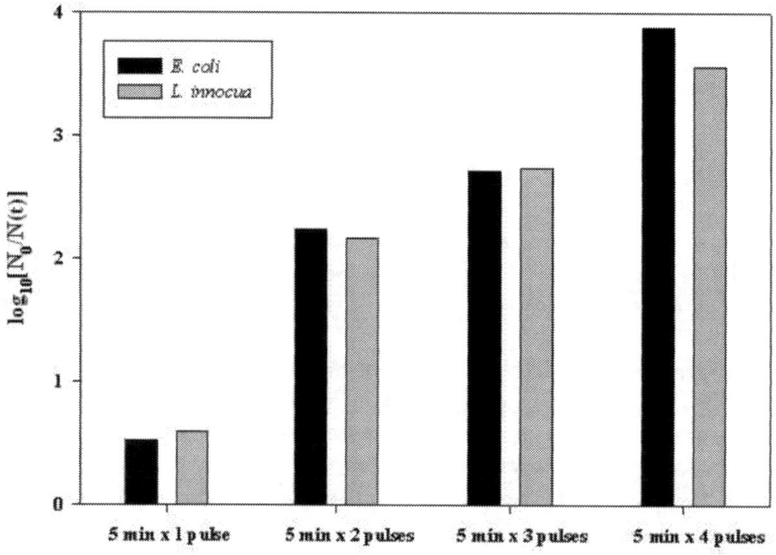

Figure 3. Inactivation of *E. coli* and *L. innocua* in whole milk at 400 MPa for 5 min × 1 pulse (continuous 5 min), 5 min × 2 pulses, 5 min × 3 pulses and 5 min × 4 pulses.

3.4. Effect of pulse holding time

Choosing a total holding time of 20 min at 400 MPa, different combinations of pulse holding time and number of pulses were applied (20 min × 1 pulse, 10 min × 2 pulses, 5 min × 4 pulses, 4 min × 5 pulses and 2 min × 10 pulses). Figure 4 shows the results of these treatments. As the number of pulses increases (up to 10 pulses) and the pulse holding time decreases inactivation increases. More than 4 $\log_{10}$ reduction is possible at 400 MPa for 2 min × 10 pulses for both *E. coli* and *L. innocua*. In the literature, Donsì *et al.* [9] observed that, for the inactivation of *Saccharomyces cerevisiae* in orange juice, pulsed HHP treatment is more effective than isostatic (continuous) HHP treatment if the pulse holding time is longer than 60 sec.

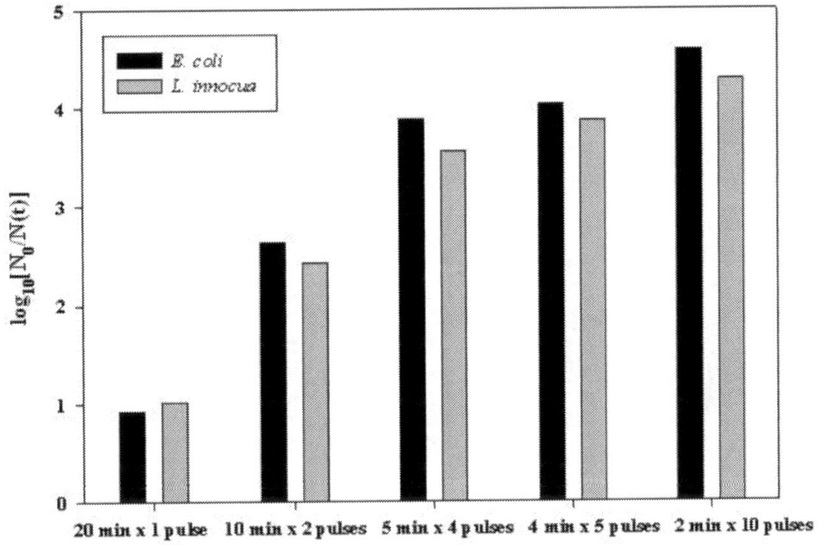

Figure 4. Inactivation of *E. coli* and *L. innocua* in whole milk at 400 MPa for 20 min × 1 pulse (continuous 20 min), 10 min × 2 pulses, 5 min × 4 pulses, 4 min × 5 pulses and 2 min × 10 pulses.

4. Conclusions

Results obtained in this study indicated that pulsed HHP treatments could be used to pasteurize the whole milk at lower pressure values than the continuous pressure treatments. Note that, the total duration of a pulsed HHP treatment at 300 MPa for 2 min × 10 pulses is 40 min (20 min of holding time + 10 min compression time + 10 min decompression time) which is too long for an industrial application. Therefore an optimization appears necessary between the number of pulses and pressure levels to reach the desirable number of log-reduction of microorganisms compatible with an industrial application.

5. References

[1] Hayakawa I, Kanno T, Yoshiyama K, Fujio Y 1994 *J Food Sci* **59** 164.
[2] Buzrul S, Alpas H 2004 *FEMS Microbiol Lett* **238** 29.
[3] Buzrul S, Alpas H, Bozoglu F 2005 *Food Res Int* **38** 151.
[4] Peleg M 2002 *J Food Sci* **67** 896.
[5] López-Caballero M E, Pérez-Mateos M, Montero P, Borderías A J 2000 *J Food Prot* **63** 196.
[6] Alemán G D, Ting E Y, Farkas D F, Mordre S C, Hawes A C O, Torres J A 1998 *J Sci Food Agric* **76** 383.

[7] Rigaldie Y, Largeteau A, Lemagnen G, Grislain L, Demazeau G 2002 Adv *High Press Biosci Biotech* **2** 283. R Winter Ed. Proceedings of the 2nd International Conference on High Pressure Bioscience and Biotechnology, Dortmund.
[8] Rigaldie Y, Largeteau A, Lemagnen G, Grislain L, Demazeau G 2002 Adv *High Press Biosci Biotech* **2** 316. R Winter Ed. Proceedings of the 2nd International Conference on High Pressure Bioscience and Biotechnology, Dortmund.
[9] Donsì G, Ferrari G, Maresca P 2007 *J Food Eng* **78** 984.

Acknowledgments
This research work was developed in the scope of a thesis in co-tutelle (S B) between the University Bordeaux 1 "Sciences and Technologies" and Middle East Technical University (METU). The French Embassy in Ankara, Turkey is deeply acknowledged for this support. Authors would like to thank Mr. Pierre Tyndiuk for his help during pressurization experiments.

Journal of Physics: Conference Series **121** (2008) 142002 doi:10.1088/1742-6596/121/4/142002

New pressure and temperature effects on bacterial spores

Alexander Mathys[1], Volker Heinz[2], Dietrich Knorr[1]

[1]Berlin University of Technology, Department of Food Biotechnology and Food Process Engineering, Koenigin-Luise-Str. 22, D-14195 Berlin, Germany, Tel: +49 30 314-71248, Fax: +49 30 832 76 63

[2]German Institute of Food Technology, p.o.box: 1165, D-49601, Quackenbrueck, Germany, Tel: +49 5431 183 232, Fax: +49 5431-183114

E-mail for correspondence: alexander.mathys@tu-berlin.de

Abstract. The mechanism of inactivation of bacterial spores by heat and pressure is still a matter of discussion. Obviously, the change of the dissociation equilibrium under pressure and temperature plays a dominant role in inactivation of microorganisms. Heat and pressure inactivation of *Geobacillus. stearothermophilus* spores at different initial pH-values in ACES and phosphate buffer confirmed this view. Thermal inactivation in ACES buffer at 122°C resulted in higher logarithmic reductions. Contrary, after pressure treatment at 900 MPa with 80°C phosphate buffer showed higher inactivation. These results indicated the different dissociation equilibrium shifts in buffer systems by heat and pressure. Due to preparation, storage and handling of highly concentrated spore suspensions the clumping and the formation of aggregates can hardly be avoided. Consequently, the impact of the agglomeration size distribution on the quantitative assessment of *G. stearothermophilus* spore inactivation was determined by using a three-fold dynamic optical backreflexion measurement. Two limiting cases have been discriminated in mathematical modelling: three dimensional, spherical packing for maximum spore count and two dimensional, circular packing for minimum spore count of a particular agglomerate. Thermal inactivation studies have been carried out in thin glass capillaries, where by using numerical simulations the non isothermal conditions were modelled and taken into account. It is shown that the shoulder formation often found in thermal spore inactivation can sufficiently be described by first-order inactivation kinetics when the agglomeration size is considered. In case of high pressure inactivation agglomerations could be strongly changed by high forces at compression and especially decompression phase. The physiological response of *Bacillus licheniformis* spores to high pressure was investigated using multiparameter flow cytometry. Spores were treated by high pressure at 150 MPa with 37°C, and then dual stained with the fluorescent dyes SYTO 16 and propidium iodide. For pressure treated spores four distinct populations were detected by flow cytometry, and for these we suggest a three step model of inactivation involving a germination step following hydrolysis of the spore cortex, an unknown step, and finally an inactivation step with physical compromise of the spore inner membrane. An understanding of these effects and mechanisms will aid the safety assessment of pressure assisted thermal sterilisation, in turn facilitating the adoption by industry and commercialisation of such processes.

© 2008 IOP Publishing Ltd

1. Impact of dissociation equilibrium shift on spore inactivation

Spore inactivation studies were performed in two buffer systems (phosphate [PBS] and N-(2-Acetamido)-2-aminoethanesulfonic acid [ACES]). For a more obvious presentation of the inactivation difference the $\log_{10}$-reductions from each individual buffer solution were subtracted (Fig. 1). By comparing the difference in $\log_{10}$-reduction ([$\log_{10}$ (N/No)](PBS-ACES)) in phosphate and ACES maximum values up to 2 $\log_{10}$ after thermal (122°C, Fig. 1a) and 1.5 $\log_{10}$ after pressure (900 MPa at 80°C, Fig. 1b) treatment could be observed. These data indicated the different dissociation equilibrium shift (pK_a-shift) in buffer systems by heat and pressure [1]. At lower initial pH-values a higher spore inactivation was observed, because of increased differences between pH and pK_a-value. Direct comparison of different treatments and detailed amounts of pK_a-shifts is difficult because of the unknown temperature dependence of the pK_a-value at higher temperatures and especially various inactivation mechanisms, which were detected by different non-linear $\log_{10}$-reduction (shoulder and tailing).

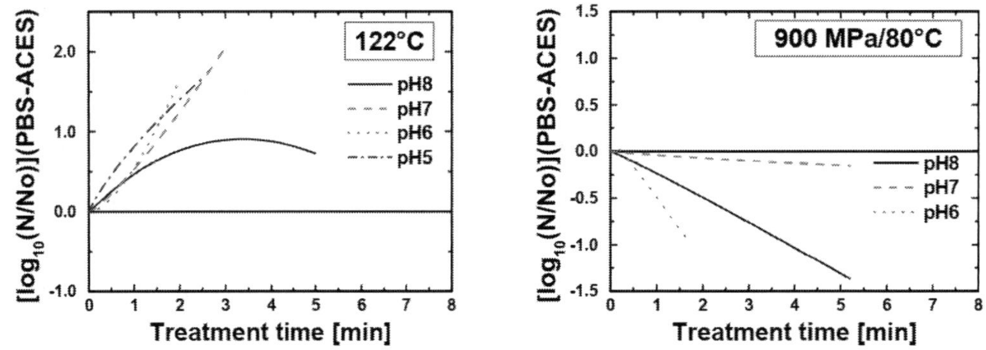

Fig. 1): Difference in $\log_{10}$-reduction of *G.stearothermophilus* spores in phosphate (PBS) and ACES ([$\log_{10}$ (N/No)](PBS-ACES)) after thermal (122°C, Fig. 1a) and pressure (900 MPa at 80°C, Fig. 1b) treatment at different initial pH-values.

2. Physiological response of spores to high pressure

The physiological response of *Bacillus licheniformis* spores to high pressure and thermal inactivation was investigated using multiparameter flow cytometry. Spores were treated by high pressure at 150 MPa (37°C), and then dual stained with the fluorescent dyes SYTO 16 (permeable DNA dye, indicator of spore germination) and propidium iodide (impermeable nucleic acid dye, indicator of membrane permeabilisation).On the basis of the comparisons among different samples, a population assignment was made (Fig. 2a). Sub-population 1 was assigned as the dormant (culturable) spore sub-population, and sub-population 4 as the inactivated (non-culturable) spore sub-population. Results indicated that sub-population 2 represented the germinated (culturable, but heat sensitive) sub-population, and sub-population 3 a second, unknown heat-sensitive sub-population, of unknown culturability. Sub-population 2 is presumed to be cortex hydrolysed spores with an intact inner membrane. The nature of sub-population 3 is difficult to presume. One possibility is that sub-population 3 represents a population with a hydrolysed cortex and some damage to the inner membrane, thus allowing PI to enter the spores and partially displace SYTO 16 and reduce the observed green staining. SYTO 16 may also be quenched in sub-population 3 by fluorescence resonance energy transfer to PI, as has been reported for other SYTO dyes [2].

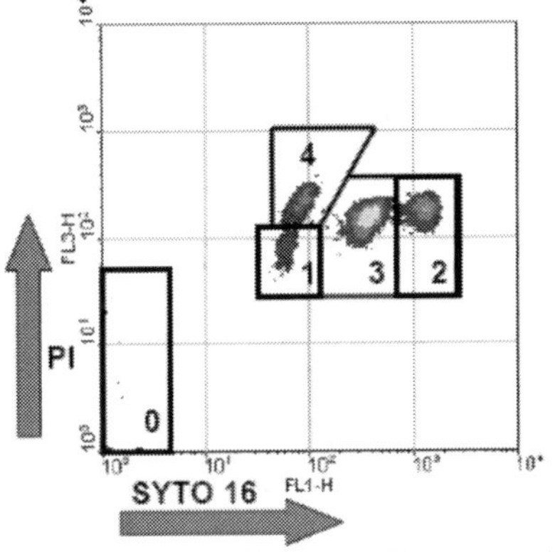

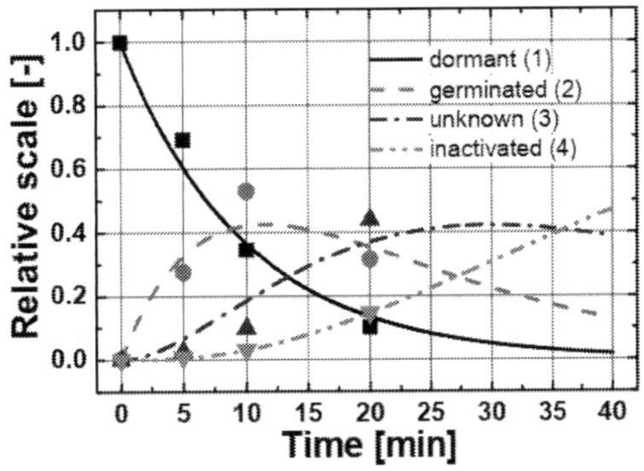

Fig. 2a): Sub-population assignment for density plot diagrams of *B.licheniformis* spores in sodium citrate buffer after treatment at 150 MPa with 37°C for 20 min: (0 = noise), 1 = dormant, 2 = germinated, 3 = unknown, 4 = inactivated [3].
2b) Heterogeneous population distribution in a predicted three-step-model (N1→N2→N3→N4, lines) for *B.licheniformis* spores after pressure treatment at 150 MPa with 37°C in sodium citrate buffer fitted with experimentally determined flow cytometric measurements; dormant (■, black), germinated (●, green), unknown (▲, blue) and inactivated (▼, red) [3].

Using the sub-population assignment described (Fig. 2 a), a three step model of inactivation based on a series of chemical reactions with associated rate constants k_i (i =1, 2, 3) is suggested. The whole model includes: a germination step ($N_1 \rightarrow N_2$), an unknown step ($N_2 \rightarrow N_3$) and finally the inactivation step ($N_3 \rightarrow N_4$). The velocity v_i of each step is related to the rate constants k_i of the reaction and the concentration of the participating entities N_j (j =1, 2 ,3 ,4), where Eq. 1-4

$$dN_1/dt = -k_1N_1 = -v_1 \quad (1)$$
$$dN_2/dt = k_1N_1 - k_2N_2 = v_1 - v_2 \quad (2)$$
$$dN_3/dt = k_2N_2 - k_3N_3 = v_2 - v_3 \quad (3)$$
$$dN_4/dt = k_3N_3 = v_3 \quad (4)$$

represent a differential equation system of this assumptions in which changes in the species concentrations N_j with time t proceed according to the velocities v_i of the reactions that form or remove them. After a multiparameter fit of the whole differential equation system with the sub-population values from each individual population, all three rate constants were obtained. The predictive three step model for the physiological mechanism of inactivation under pressure at 150 MPa showed good regression with the experimental results (Fig. 2b). The modelling enabled the assessment of the continuous population distribution and an extrapolation of the experimental data. Variations between the model and experimental data occurred because of partial overlap of some sub-populations.

3. Impact of spore agglomeration on heat inactivation

The agglomeration size distribution in suspensions of *Geobacillus stearothermophilus* spores was determined by using a three-fold dynamic optical back-reflexion measurement (3D ORM) (Fig. 3). Since 3D ORM accurately yields the maximum length extension of an agglomerate, but provides no information on the packing density. After idealization to the spherical form, the assumed radius of the spores is 1.125 μm based on the X-ray microscopy determinations [4]. The assumption for the agglomerate geometry is divided into three approaches (one, two or three dimensional). The regular three-dimensional arrangements of sphere (3D) with the highest density are cubic close packing or hexagonal close packing. After Kepler [5] both arrangements have an average density of $\pi(\sqrt{18})^{-1}$. In the case of two dimensions (2D) the cross-sectional area can be derived with the measured equivalence diameter. The honeycomb circle packing with a density of $\pi(\sqrt{12})^{-1}$ is the unique densest lattice sphere packing in two dimensions [6]. For only one dimension (1D), a chain formation with the product of spore diameter and spore number was supposed. The different geometrical approaches show large differences in number of spores per agglomerate. Thermal inactivation data have been modelled (Fig. 3b) using first-order inactivation kinetics [7], superimposed by the agglomeration size (Fig. 3a). Thermal inactivation studies have been carried out in thin glass capillaries, where by using numerical simulations the non isothermal conditions were modeled and taken into account. It is shown that the shoulder formation often found in thermal spore inactivation can sufficiently be described with the suggested approach (Fig. 3b).

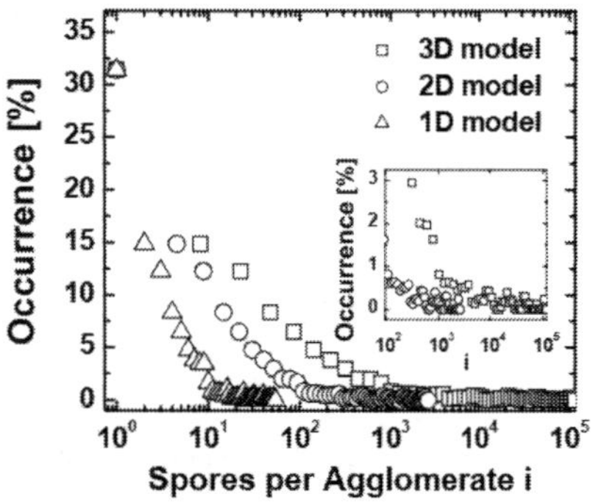

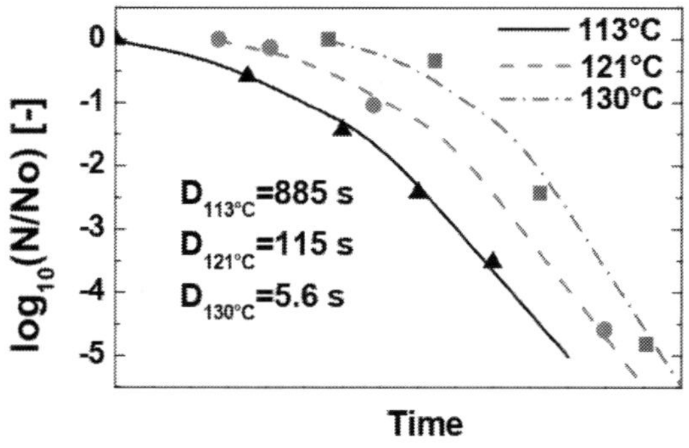

Fig. 3a): Agglomeration size distribution of a *G.stearothermophilus* spore suspension with different geometrical assumptions for the agglomerates and an inset, which shows agglomerations sizes between 10^2-10^5 spores per agglomerate [7].
3b): Experimental plots for ▲113°C, ●121°C, ■130°C and predicted data (lines) after thermal inactivation of *G.stearothermophilus* ATCC 7953 spores. Inactivation curves were modelled with spherical (3D) assumption of the agglomerate geometry [7].

4. Conclusion

The mechanistic background of the inactivation of bacterial spores is still matter of discussion [8]. Different influences and effects were investigated. The change of the dissociation equilibrium under high pressure plays a major role in sensitive reactions e.g. inactivation of microorganisms and/or denaturation of proteins. Heat and pressure inactivation of *G. stearothermophilus* spores at different initial pH-values in ACES and phosphate buffer confirmed this view. The experimental results indicated the different pK_a-shifts in buffer systems by heat and pressure. By using flow cytometry a simple and very fast (within 20 min of processing) method for the rapid assessment of spore physiological state could be developed. This high throughput method offers substantial benefits with regards acquisition of large data sets, which are required to predict and model bacterial spore inactivation by high pressure and heat, and to determine the stochastic nature of such inactivation. After particle measurement and thermal treatment large influence of the agglomeration in spore suspensions on the inactivation could be detected. The high occurrences of middle size agglomerates had the main influence on the curvilinear form in our study. Practical experiments are difficult in this context, because of the necessary defined separation of the agglomerates. Thus it appears that agglomerations in spore suspensions need to be considered by modelling of the thermal inactivation. However, at this time no information about agglomeration of spores under pressure is given. Detail studies to find possible explanations are under way at present.

Applying the above models, spore inactivation can be better anticipated in planning experimental designs. Such extensive empirical data, coupled with an improved understanding of the mechanism(s) of inactivation, will ultimately be required to demonstrate the benefits and risks of high pressure thermal processing to industry.

Acknowledgments
This research was supported by German Ministry of Economics and Technology (BMBF, Grant No. 0330089A), the European Project NovelQ (015710-2 NOVELQ), Food Science Australia and the Commonwealth Scientific and Industrial Research Organisation Food Futures National Research Flagship.

References

[1] **Distèche, A.** (1959). pH Measurements with a Glass Electrode Withstanding 1500 kg/cm^2 Hydrostatic Pressure. Review of Scientific Instruments, **30(6)**, 474-478.

[2] **Stocks, S. M.** 2004. Mechanism and use of the commercially available viability stain, *Bac*Light. Cytometry Part A **61A:**189-195.

[3] **Mathys, A., B. Chapman, M. Bull, V. Heinz, and D. Knorr.** 2007. Flow cytometric assessment of *Bacillus* spore response to high pressure and heat. Innovative Food Science & Emerging Technologies. accepted.

[4] **Mönch, S., V. Heinz, P. Guttmann, and D. Knorr.** 1999. X-ray microscopy in food science.Presented at the International Conference on X-Ray Microscopy-XRM 99, University of California, Berkeley, USA, 1.- 6.08.1999.

[5] **Kepler, J.** 1611. Ioannis Kepleri S. C. Maiest. Mathematici Strena Seu De Niue Sexangula. Gottfried Tampach, Francofurti ad Moenum.

[6] **Lagrange, J. L.** 1773. Recherches d´arithmetique. Nouv. Mem. Acad. Roy. Sc. Belle Letteres. Oeuvres III:265-312.

[7] **Mathys, A., V. Heinz, F. H. Schwartz, and D. Knorr.** 2007. Impact of agglomeration on the quantitative assessment of *Bacillus stearothermophilus* heat inactivation. Journal of Food Engineering **81:**380-387.

[8] **Heinz, V. and D. Knorr.** 2002. Effects of high pressure on spores, p. 77-114. *In* M. E. G. Hendrickx and D. Knorr (ed.), Ultra high pressure treatments of foods. Kluwer Academic/ Plenum Publishers, New York.

Effect of heat/pressure on cyanidin-3-glucoside ethanol model solutions

Margarita Corrales[1], Ralf Lindauer, Peter Butz and Bernhard Tauscher.

Institute of Chemistry and Biology, Federal Research Centre for Nutrition and Food, 76131 Karlsruhe, Germany

E-mail: margarita.corrales@bfel.de

Abstract. The stability of cyanidin-3-glucoside (Cy3gl) in 50% ethanol model solutions under heat/pressure treatments was investigated. Cy3gl was rapidly degraded when solutions were subjected to a heat/pressure treatment. The higher the pressure and the temperature used, the higher the degradation. Moreover, the degradation was increased according to increasing holding times. Parallel to the degradation of Cy3gl several hydrolytic products were formed and identified by LC-DAD/ESI-MS. The degradation of Cy3gl was well fitted to a first order reaction (R=0.99). This study pointed out the rate of susceptibility of Cy3gl in model solutions to degrade when exposed to a heat/pressure treatment and the trigger effect of high hydrostatic pressure to hydrolyse Cy3gl. By contrast, the degradation of anthocyanins in a food matrix (red grape extract solutions) was negligible after a heat/pressure process at 600MPa, 70°C during 1h ($P>0.05$).

1. Introduction

Anthocyanins (anthos: sprossom, flower, Kyáneos: darkblue) belong to the group of water soluble pigments and are responsible for the red, blue and purple hues of flowers and fruits in nature. Anthocyanins belong to the group of flavanoids whose chemical structure is constituted by 3 aromatic rings with a positive charge on the C-ring and a glucoside moiety at the C_3 which confers them certain stability compared to their aglycon forms (anthocyanidins) (Figure 1). Different pH values modify their chemical structure manifesting different colours from yellow to black [1]. Other factors such as oxygen, enzymes, copigments, metal ions, ascorbic acid, sulfur dioxide, sugars and sugar degradation products may influence their stability and colouring properties [2]. Thus, the thermal instability of anthocyanins has been widely reported. Anthocyanin degradation at high temperatures follows a first-order reaction. Hrazdina, [3] pointed out that the decomposition of anthocyanins upon heating leaded to a chalcone structure which further decomposed to coumarin glucoside derivatives with a loss of the B-ring. Moreover, Adams et al., [4] 1973 reported that the aglycon sugar bonds were highly susceptible to hydrolysis even at acid pH. Later on Simpson, [5] suggested that the thermal degradation of anthocyanins could occur via two mechanisms: (1) hydrolysis of the 3-glucoside linkage to form a labile aglycon; and (2) hydrolytic opening of the pyrilium ring to form a substituted chalcone, which further degrade into brown insoluble compounds. Anthocyanin degradation is generally correlated to a loss or change in color.

While the influence of temperature on anthocyanins has been extensively investigated, effects of pressure, also in combination with temperature on anthocyanin model solutions are still un-

© 2008 IOP Publishing Ltd

known. High hydrostatic pressure has been used in the food industry as pasteurisation and sterilisation method primarily in Japan, United States and in less term in Europe. Pressure pasteurisation is feasible at room temperature since it inactivates micro-organisms avoiding undesirable changes such as vitamin loss and taste or colour modifications. Moreover, the effect of pressure and temperature for sterilisation at pressures up to even more than 1000MPa and temperatures around 80-120°C during very short times has been established yielding high quality foods [6]. The influences of moderate heat/pressure treatments are still under research and suppose a promising field since heat/pressure combined treatments may deliver food products of greater quality than thermal treated ones [7]. Thus, several studies have demonstrated high hydrostatic pressure as an effective pasteurization method for different anthocyanin enriched products which have retain their colour and flavour characteristics during longer time [8-10]. However complex matrices complicate the accurate estimation of the effect of high hydrostatic pressure on individual compounds. For a closer knowledge of the influence of high hydrostatic pressure on anthocyanins, cyanidin-3-glucoside (Cy3gl) in ethanol model solutions were subjected to different heat/pressure treatments and their stability in this work was reported. Cy3gl is the predominant anthocyanin found in red raspberries, blackberries and blood orange [11, 12]. Furthermore, the influence of a heat/pressure treatment was also studied in an anthocyanin enriched matrix.

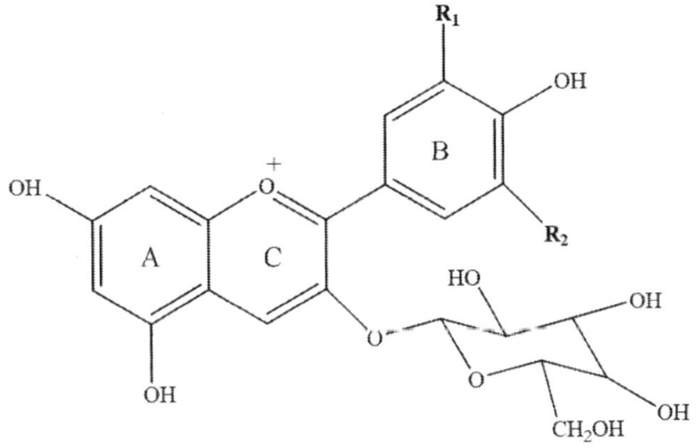

Anthocyanins	R_1	R_2
Cyanidin (Cy)	OH	H
Delphinidin(Dl)	OH	OH
Malvidin (Mv)	OCH_3	OCH_3
Pelargonidin (Pl)	H	H
Peonidin (Pd)	OCH3	OCH_3
Petunidin (Pt)	OH	OCH_3

Figure 1. Chemical structure of anthocyanins and substitution pattern.

2. Material and Methods

2.1. Materials

Cyanidin-3-O-glucoside (Cy3gl) was provided by Extrasynthese, (Genay, France). Dornfelder *vitis vinifera* red skins were supplied by wineries from the region of Palatine, Germany. Solvents for analysis were purchased by Merck (Darmstadt, Germany).

2.2. Methods

2.2.1. Preparation of Cy3gl ethanol model solutions. The model solutions were prepared with 1.3mM Cy3gl in 50% ethanol solution. Anthocyanins from Dornfelder *vitis vinifiera* red grape skins were extracted weighing 10g of dark grape skins in 100mL 50% ethanol solution assisted by ultrasonics (Bandelin, Sonorex RK 100H, Walldorf, Germany).

2.2.2. Analysis of anthocyanins and degradation products by LC-DAD/ESI-MS. Samples were analysed using an Agilent Technologies LC/MSD Series 1100 (binary solvent delivery, autosampler, UV-Vis Diode Array Detector (DAD), electrospray ionization (ESI); (Agilent Technologies, Palo Alto, CA).The separation was performed with an Aqua column (80A 250 x 4.6mm i.d.; 5µm), operated at 20°C. Solvents were (A): 0.5% acetic acid/water (v/v) and (B): 0.5% acetic acid/50% water/50% acetonitrile (v/v/v). Conditions were adapted from the method described by Kammerer, Schieberle and Carle, [13]. The elution gradient was: from 10-15% B (10min), 15% (3min), 15-25% (10min), 25-55% (30min) and 55-100% (6min) at a flow rate of 1.0 mL min^{-1}. The injection volume was 20μLl and detection was monitored at 280, 320, 370 and 520nm.

For anthocyanin analysis, solution (A) was water/formic acid/acetonitrile (87:10:3; v/v/v) and solution (B): water/formic acid/acetonitrile (40:10:50; v/v/v). The gradient used was: from 10-25% B (10min), 15% B isocratic (3min), from 15-25% B (7min), from 25-55% B (30min), from 55-100% B (1min), 100% B isocratic (5min), from 100-10% B (0.1min) at a flow rate of 1.0mL min^{-1}. The injection volume was 20μL and detection was monitored at 370, 280 and 520nm.

Spectra were recorded in negative ion mode for reaction hydrolytic products and in positive ion mode for anthocyanin analysis in the range *m/z* 50-1000. The mass spectrometer was programmed to do an MS^2 scan of the most abundant ion in the full mass. Nitrogen was used both as drying gas at a flow rate of 11.0 l min^{-1}, and as nebulising gas at a pressure of 60psi. The nebuliser temperature was 350°C.

2.2.3. Heat/Pressure treatments. A device constituted by a three series of thermostated microautoclaves connected by valves (i.d.=16mm, ~10mL, 700MPa) were used for high hydrostatic pressure treatments (aad GmBH, Frankfurt, Germany). Pressure was generated by an air-driven pump in combination with a pressure intensifier and the pressure-transmitting medium was a mixture of water and glycol (80:20; v/v).
Experiments at 800MPa were carried out in a hydraulic press U 101 Unipress (Polish Academy of Sciences, Warsaw, Poland) which was manually operated by a twin hydraulic piston.
The temperature in the vessels was controlled by a thermostat Polystat from Huber (Offenburg, Germany). Samples were pressurized in polyethylene ampoules (250μL) and heat sealed.
Experiments were carried out in triplicate and significant differences among the means were estimated by the analysis of the variance, one way ANOVA ($P<0.05$).

3. Results and Discussion

Experiments carried out at different high hydrostatic pressure intensities at 20°C and 30°C did not affect the stability of Cy3gl after 0.5h treatment. Although when temperature was risen up to 50°C a higher degradation of Cy3gl could be observed up to 400MPa (Figure 2). The higher the pressure and the temperature, the higher was the degradation of Cy3gl. In addition, longer holding times also increased Cy3gl degradation rate. The degradation of Cy3gl fitted well a first-order reaction of the form:

$$\ln(At / Ao) = -kt \quad (1)$$

Where *At* was the concentration at *t* time, *Ao* initial concentration, *k* reaction constant and *t* time. The half life could be determined from the slope by the expression:

$$T_{1/2} = -\ln 0.5 / k \quad (2)$$

Cy3gl degradation rate constant in heat-pressurized samples (70°C/600MPa) was $0.33h^{-1}$ whereas in heated samples (70°C) was $0.02h^{-1}$. The half life ($T_{1/2}$) in heat-pressurized samples was 0.38h and 5.77h for heated samples. The latter values are within the range determined by Kirca and Cemeroglu, [11] for blood orange juice at 70°C; 2.0h for 69°Brix concentrate and 6.3h for 11.2°Brix concentrate. By contrast the half life of the reaction in heat-pressurized samples was remarkably shorter indicating a faster degradation.

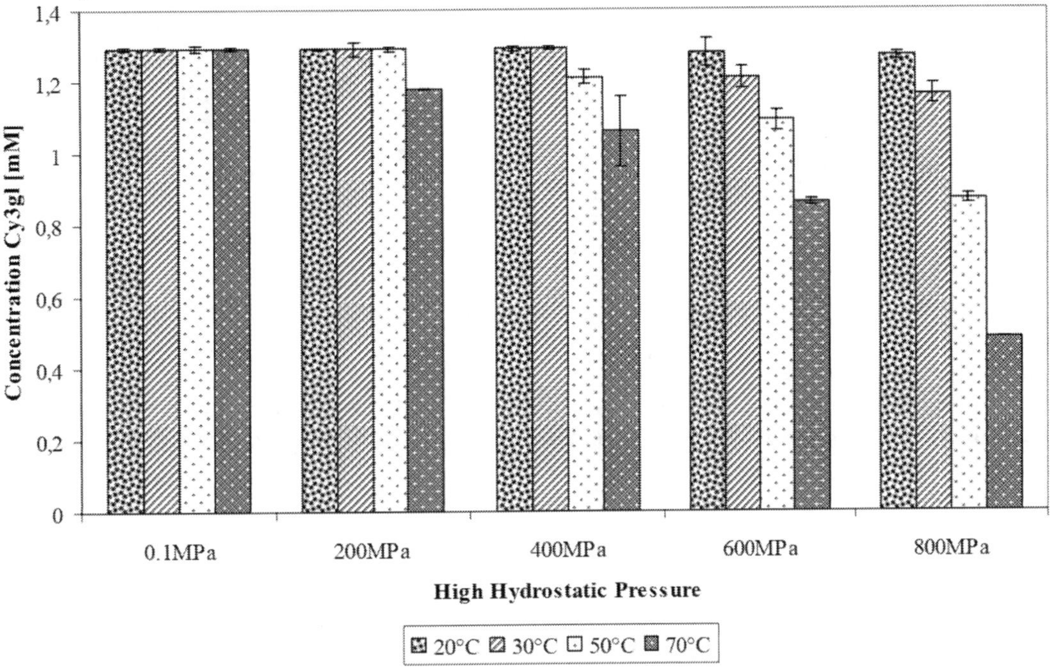

Figure 2. Effect of high hydrostatic pressure intensity (MPa) and temperature (°C) on Cy3gl degradation. Treatment holding time: 0.5h.

When ln(*k*/*k*o) was plotted against pressure (Figure 3) the activation volume (dV) could be determined from the slope by the following expression:

$$dV = -RT\frac{d\ln(k/ko)}{dP} \qquad (k/ko \sim V/Vo) \qquad (3)$$

The activation volume at 70°C was $-2.8cm^3mol^{-1}$ relatively low compared to chemical reactions mainly influenced by high hydrostatic pressure where activation volumes are often in the range of $-20cm^3$ [7]. Hence, the decisive role of the temperature in the results herein achieved.

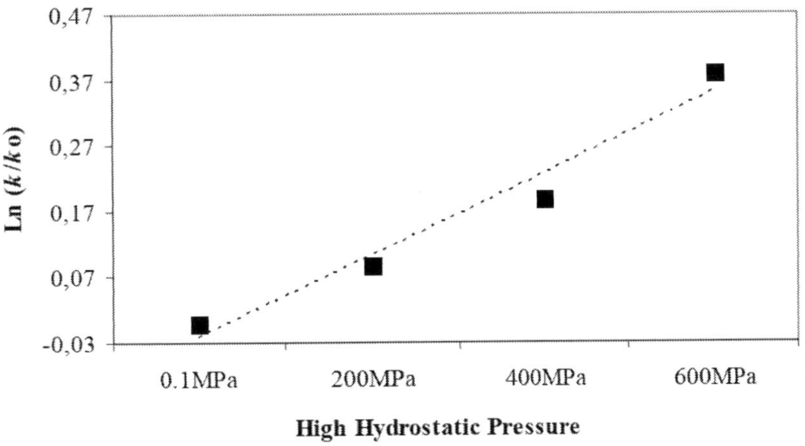

Figure 3. Effect of high hydrostatic pressure on the degradation rate of Cy3gl in 50% ethanol model solutions. Treatments carried out at 70°C during 0.5h.

According to Cy3gl degradation, some hydrolytic products could be identified by LC-DAD at 280nm. The concentration of hydrolytic products was remarkably higher when samples were treated at 600MPa/70°C/0.5h and higher at 800MPa/70°C/0.5h when the highest disappearance of Cy3gl was also determined. As a result of the degradation, 9 hydrolytic products could be found and identified by LC-DAD/ESI-MS (Figure 4). One of the main components from the degradation had a mass of $[195 + H]^+$ which corresponded to a first deglycosidation of the flavilium moiety yielding a chalcone which further transformed into a coumarin derivative with a loss of the B-ring. Quercetin $[301 + H]^+$ was also identified which is according to studies reported by Tanchev and Ioncheva, [14]. Other compounds such as phloroglucinaldehyde and protocatechuic acid also indicated in the literature for anthocyanin thermal degradation were not found. Due to the low concentration of formed peaks an accurate identification of further products could not be discerned. Results here achieved suggest the formation of other degradation products different from those obtained in heat treatments. Only NMR experiments might indicate the chemical structure and possible degradation pathways of Cy3gl under a heat/pressure treatment.

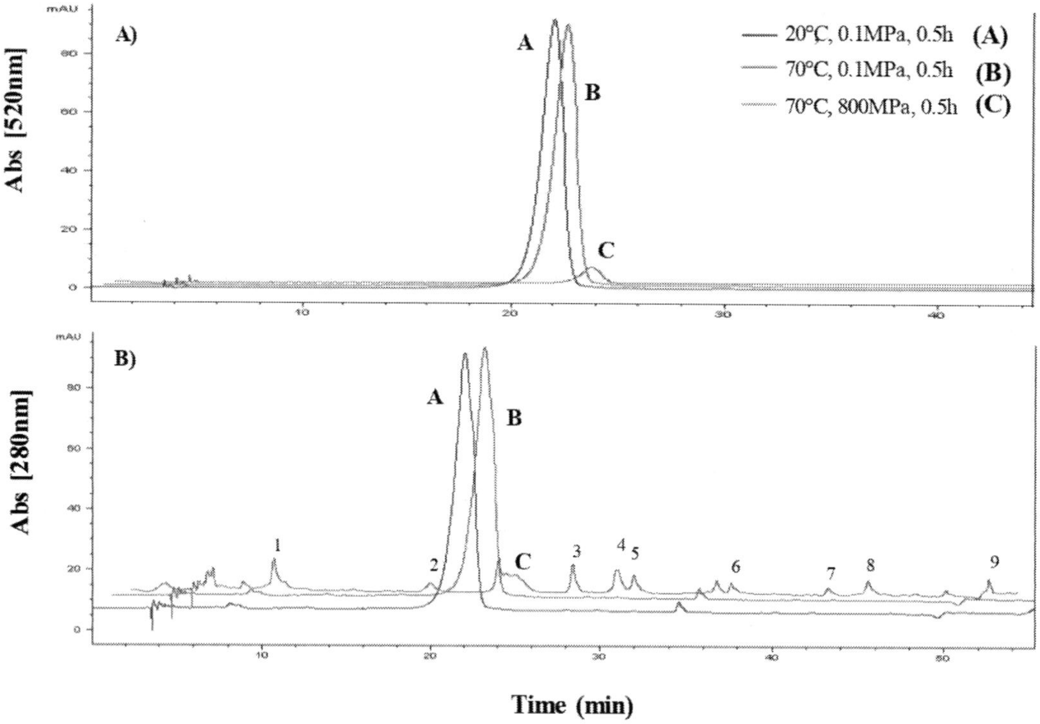

Figure 4. A) LC-DAD Chromatogram of Cy3gl at different heat/pressure conditions (λ=520nm) and B) LC-DAD Chromatogram of Cy3gl derived products at different heat/pressure conditions (λ=280nm). 1. M^+=195; 2. M^+=217; 3. M^+=319; 4. M^+=unknown; 5. M^+=545; 6. 464; 7. M^+=301; 8. M^+=348; 9. M^+=unknown.

Moreover the effect of a high hydrostatic pressure treatment was tested in a solution containing red grape extracts from Dornfelder *vitis vinifera*. After a high hydrostatic pressure treatment at 600MPa/70°C/1h only a small loss in the predominant anthocyanin compounds was appreciable (P<0.05) (Figure 5). As a result when samples were subjected during a shorter treatment (10min) any anthocyanin concentration loss was detected (data not shown). These results demonstrated that in a matrix anthocyanin degradation did not occur as in model solutions. The presence of other substances with antioxidant properties (e.g: phenolic acids) and the synergistic effects among them may exert a protective effect which avoids anthocyanin rapid degradation and enhances thus product integrity.

4. Conclusions

Although Cy3gl model solutions were sensitive to different heat/pressure treatments, the treatment of grape extract solutions presented similar characteristics in terms of anthocyanin content as their control patterns. Even heated pressurised extracts at 600MPa, 70°C during 1h did not exhibit drastic losses of anthocyanins probably due to the action of intrinsic protective substances. These results are highly valuable since a moderate pasteurization method may be applied for anthocyanin enriched foods without altering their chemical characteristics and improving product shelf life. Nonetheless, further experiments in reference to the effect of heat and pressure on other compounds present in targeted matrices must be further evaluated. Only the establishment of an appropriate pasteurization method might guarantee the quality and safety of treated products.

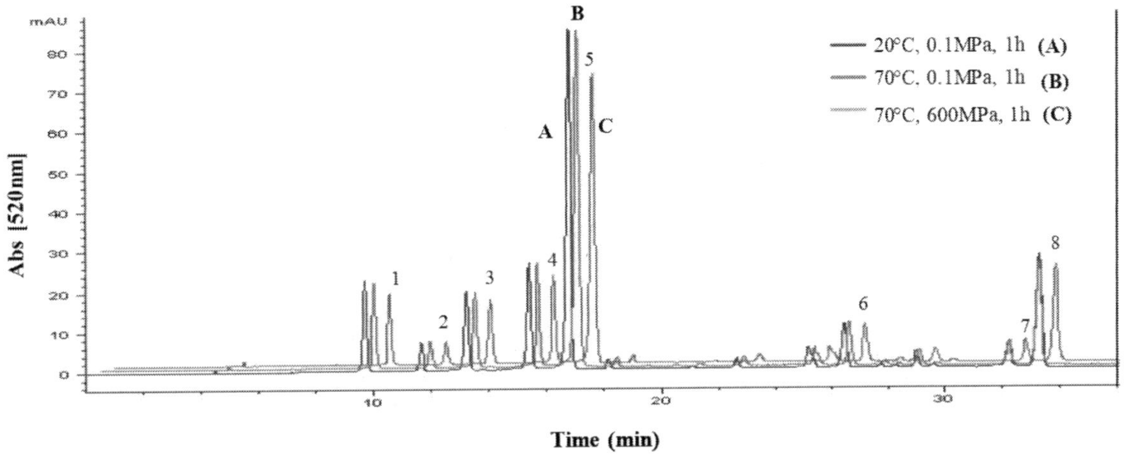

Figure 5. LC-DAD Chromatograms of Dornfelder *vitis vinifera* extracts at different heat/pressure treatments. 1. Dl3gl, M^+=465; 2. Cygl, M^+=449; 3. Pt3gl, M^+=479; 4. Pn3gl, M^+=463; 5. Mv3gl, M^+=493; 6. Dl3acgl, M^+=507; 7. Pt3acgl, M^+=521; 8. Pn3acgl, M^+=505; 9. Mv3acgl, M^+=535; 10. Cy3pcmgl, M^+=595; 11. Pt3pcmgl, M^+=625; 12. Pn3pcmgl, M^+=609; 13. Mv3pcmgl, M^+=639. (gl=glucoside; acgl=acetylglucoside; pcmgl=p-coumarylglucoside)

References

[1] Bayer E, Egeter H, Fink A, Nether K, and Wegmann K 1966 Angew. *Chem. Int.* Ed. Engl. **5** 791.
[2] Jackman RL, Yada RY, Tung MA and Speers RA 1987 *J. Food Biochem.* **11** 201.
[3] Hrazdina G 1971. *Phytochem.* **10** 1125.
[4] Adams JB 1973 *J Sci. Food Agric.* **24** 747.
[5] Simpson KL 1985 *Chemical Changes in Natural Food Pigments* (New York: Van Nostrand Reinhold)
[6] Meyer R, Cooper KLL, Knorr D and Lelieveld HLM 2000 *Food Technol* **54** 67
[7] Tauscher B 1995 *Z. Lebensm Unters Forsch.* **200** 3
[8] Suthanthangjai W, Kajda P and Zabetakis I. 2004 *Food Chem.* **90** 193
[9] Kouniaki S, Kajda P and Zabetakis I. 2004. *Flav. Frag. J.* **19** 281
[10] Zabetakis I, Leclerc N and Kajda P 2000 *J. Agric. Food Chem.* **48** 2749
[11] Kirca A and Cemeroglu B 2003 *Food Chem.* **81** 583
[12] Rommel A, Wrolstad RE and Heatherbell DA 1992 *J Food Sci* **57** 385.
[13] Kammerer D, Claus A, Carle R and Schieberle A 2004. *J. Agric Food Chem.* **52** 4360
[14] Tanchev SS and Ioncheva N 1976. *Nahrung.* **18** 747

Joint 21st AIRAPT and 45th EHPRG Int. Conf. on High Pressure Science and Technology IOP Publishing
Journal of Physics: Conference Series **121** (2008) 142004 doi:10.1088/1742-6596/121/4/142004

The investigation of the dynamics of the phase transformation in triolein and oleic acid under pressure

D B Tefelski, R M Siegoczyński, A J Rostocki, A Kos, R Kościesza and K Wieja

Faculty of Physics, Warsaw University of Technology, Koszykowa 75, 00-662 Warszawa, PL

E-mail: tefelski@if.pw.edu.pl

Abstract. An aim of our work is the understanding of processes happening during phase transformations under the pressure in triglycerides and unsaturated fatty acids. Particles of investigated liquids possess the double bond between carbon atoms, which causes the bended shape of the particle and makes its free rotation impossible. This property causes low temperatures of melting point and high temperatures of boiling and also investigated by us phase transformations. For study of the dynamics of phase transformation in these liquids we measured light transmission and light scattering at 90 degrees angle, temperature, permittivity and internal pressure versus time. We applied pressure using computer controlled pump with a stepping motor, which makes increase of the pressure steady. The phase transformation in oleic acid lasts several seconds, in triolein it lasts several minutes. We think that the elongated time of phase transformation is caused by a hooked shape of particles of triolein and the dynamics of that process is determined by the tangling of particles. We checked the influence of smaller particles of oleic acid on the phase transformation by investigating the mixture of these liquids.

1. Introduction

Interesting phase transformations firstly were found in castor oil [1, 2]. Formally optical observation were the first where such phenomena was observed, then they were followed by the other observations like temperature changes, permittivity, volume changes, x-ray diffraction [3]. Understanding processes which occur during high pressure treatment of compounds of oils is very important because of its meaning in food and fuel industry. Observations which were performed and described in [4] have been repeated with much higher precision. Data acquisition and control of experiments are now fully automated. Pump with stepping motor allowed us to apply pressure in very accurate way. Light transmission and light scattering [5] studies lead us to pay attention to rod-like compound models such as, for example presented in this work [6].

2. Measurement configuration

Our experiments were performed with use of high pressure optical chamber. This steel chamber has three optical entrances which are made of sapphire cylinders in configuration, allowing light observation in transmission geometry and at 90 degrees (scattered light). The windows are sealed using Bridgeman sealing system with use of brass gaskets. There is also an electrical interface which also uses Bridgeman sealing. The high pressure electrical interface has several thin wires which are wound on thread cone-shaped element and filled with epoxy resin. Inside the chamber there is a cylindrical space which is filled with examined liquid. Sensors like thermocouple,

© 2008 IOP Publishing Ltd

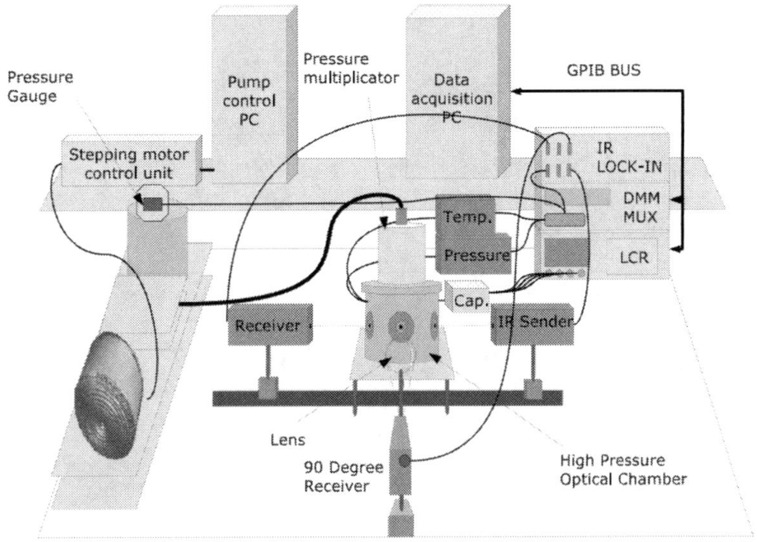

Figure 1. Measurement system

manganine wire pressure sensor and capacitor for dielectric permittivity investigation are placed there. A simple piston moves in this space causing pressure increase in the examined liquid. A force is transmitted by pressure multiplication device mounted on top of the pressure chamber which is based on a different diameter of pistons. The pressure multiplication index is 25. The pressure at low values (up to 30MPa) is generated by pressure pump driven by stepping motor which is controlled by computer. This provides the steady increase of the pressure. Sensors from high pressure chamber are connected to corresponding measuring devices. Capacitor is connected through the high pressure interface by four wire method to an precision LCR meter (HP4284A). Thermocouple is connected to digital multi-meter (DMM) with multiplexed input (HP34970A) through a device providing constant temperature for a second thermocouple terminal. This provides a precision of temperature measurement, regardless of ambient temperature variations. A manganine wire in form of coil is also connected to input of HP34970A. Our experiment set has optical devices: Infrared Sender device with IR LED ($\lambda = 800nm$), two Infrared Receivers with p-i-n photo-diodes, lens for focusing scattered light and a modular phase–sensitive detection system. This phase–sensitive detector (Lock-in) built on Faculty of Physics Warsaw University of Technology makes an SNR (Signal to Noise Ratio) high increase and high stability in long time observations in presence of external disruptions possible. For Lock-in adaptation, emitted light is modulated with frequency of 2931Hz. The wavelength ($800nm$) of emitted light was chosen specially to avoid molecular absorption. Because the intensity of scattered light was very small the positive lens was used to focus light on the detector photo-diode. The Lock-in system is connected with HP34970A. Both measuring devices: HP4284A and HP34970A are controlled via GPIB Bus by data acquisition system on personal computer with National Instruments LabVIEW software, which is prepared for long term acquisition, necessary in some classes of examined liquids. Permittivity was measured in similar way to the one described in [7]. Complete acquisition system is presented in figure 1.

3. Measurement conditions

Experiments are performed in standard conditions with ambient temperature $t_c = 25°C$, ambient pressure of $P_a = 101.325kPa$ and relative humidity of $RH = 40\%$. Data from all sensors is collected in a period of 2 seconds. Capacitance was measured using Auto Balancing Bridge Method. A frequency used in this method was 1kHz and the amplitude of measurement signal was 1V. The examined liquids were chemically pure. For a mixture, liquids were mixed in magnetic stirrer. Before the experiments, the samples were stored in cool and dark place for preventing them from degradation.

4. Examined liquids

Triolein $(C_{17}H_{33}COO)_3C_3H_5$ is a triglyceride and unsaturated fat. It consists of flat type molecule of glycerine and three rod like molecules of oleic acid. It has a hooked shape. Oleic acid $CH_3\,(CH_2)_7\,CH = CH\,(CH_2)_7\,COOH$ is a mono-unsaturated omega-9 fatty acid with rod-like shape bended on double bound. Triolein and its mixtures with methyl alcohol were examined in works [8, 9, 10], but not in optical chamber. Oleic acid was examined in works [11, 12].

5. Experiment progress

During experiment we increased the pressure with constant speed of piston displacement. We used stepping motor with 60Hz which is about 0.25MPa/s, 120Hz which is about 0.5MPa/s and 240Hz which is about 1MPa/s. The pressure was increased to 450MPa in examined liquid, and then we waited till all measurable parameters changed. After that, the pressure was decreased with the same speed. Data with light transmission, scattered light and pressure is shown in figure 2 for pure triolein and figure 3 for pure oleic acid. All experimental data from sensors were collected simultaneously by automated acquisition system.

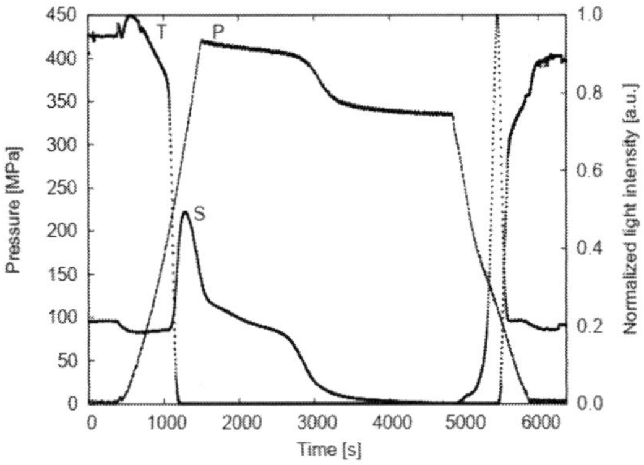

Figure 2. (P) Pressure change, (T) light transmission, (S) scattered light at 90° versus time in triolein v=120Hz (0.5MPa/s)

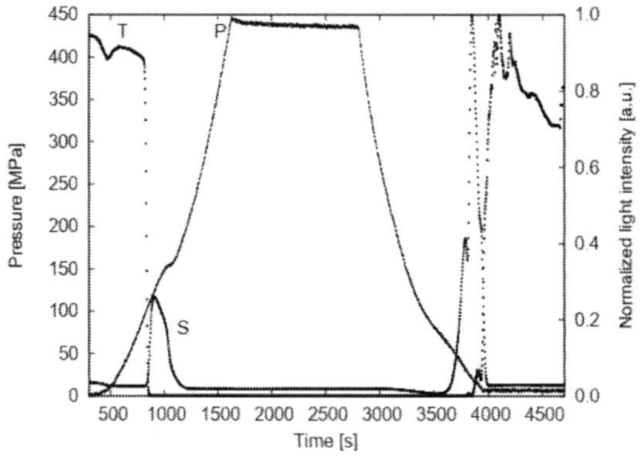

Figure 3. (P) Pressure change, (T) light transmission, (S) scattered light at 90° versus time in oleic acid v=120Hz (0.5MPa/s)

6. Observation

In figure 4 we see large optical hysteresis. A noticeable fact is that when we add more then 25% oleic acid to triolein, when phase transformation begins the pressure tends to decrease. We can also see different character of light transmission changes in triolein and in oleic acid. Phase transformation in triolein took much more time as in oleic acid and it begins slowly with distinct slope. Phase transformation in oleic acid begins abruptly and the beginning slope is very slight. The dependence of the transmitted light on gradually applied pressure was discussed in paper [13]. According to the Mie's theory, strong decrease of light transmission is connected with an increase of the sizes of the light scattering objects. It is because particles accumulate and build up inceptions of new structure. Such new clustered objects cause increased scattering, which is visible on scattered light at 90°. Later because of solidification, both transmitted and scattered light diminishes. In pure triolein, there is a characteristic slow scattering light decay in similar shape like in the pressure curve. In all mixtures of triolein and oleic acid and in all examined speeds we did not observe such elongated scattered light behavior. It can be explained that phase transformation (creation of inceptions of new structure) of oleic acid masks the phase transformation behavior of triolein. Simply it begins earlier. Leaving light changes out of account, the pressure changes show the moment of original structure breakdown. In triolein this happens when compression stage is completed and we just wait (time near 3000s in figure 2. In oleic acid it is visible during compression stage at about 150MPa pressure (time near 1000s in figure 3).

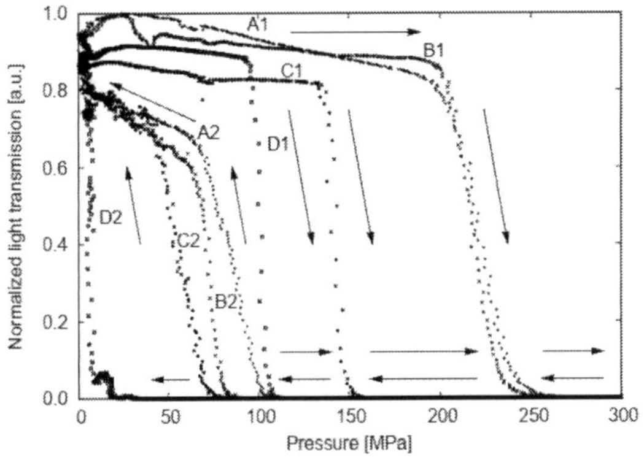

Figure 4. Light transmission versus pressure. Labels with 1 - pressure increase, with 2 - pressure decrease. A - triolein, B - triolein + oleic acid 75:25, C - triolein + oleic acid 50:50, D - oleic acid.

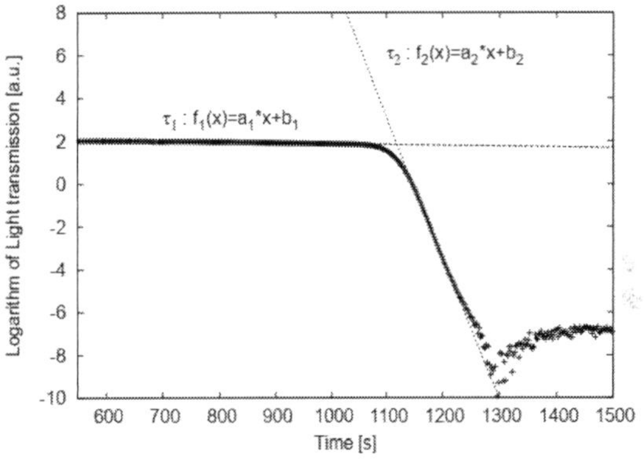

Figure 5. Logarithm of light transmission in pure triolein and fitted lines to fragments representing two relaxation times in Edwards model of motion in potential and "hopping". Slopes are inverse negative τ_1 and τ_2.

7. Applying the model

We have considered using the Edwards [6] model of rod-like compounds to this group of liquids. So after light transmission observation we choose two characteristic parameters from light transmission. They appear to behave as exponential decay. We have fitted line functions to logarithm of light transmission and obtained slopes corresponding to harmonic move in the potential τ_1 and "hopping" τ_2 which is time characteristic for opening and closing gates as obstacles in compound movement. Then having this parameters, we calculated the α parameter which is related with Kramer's probability of escape and further is a part of diffusion equation. According to the Edwards calculations, diffusion is connected with α parameter by $D = D_0\,(1-\alpha)$ and $\frac{\tau_2}{\tau_1} = \frac{\alpha}{1-\alpha}$ which transformed is $\alpha = \frac{\tau_2}{\tau_1+\tau_2}$ Therefore slopes from fitted lines $(y = ax + b)$ to logarithm of light transmission (which is shown in figure 5 are inverse negative τ_1 and τ_2 parameters. $\tau_1 = -\frac{1}{a_1}$, $\tau_2 = -\frac{1}{a_2}$. After calculations, obtained parameters are shown on table 1. The α parameter versus speed of applied pressure for triolein, oleic acid and their mixtures is presented in the figure 6.

8. Conclusion

It is interesting that the α parameter decreases proportionally to speed of applied pressure. It has the highest value for pure triolein and tends to have the smallest value for pure oleic acid, so diffusion in triolein is the smallest and diffusion during phase transformation tends

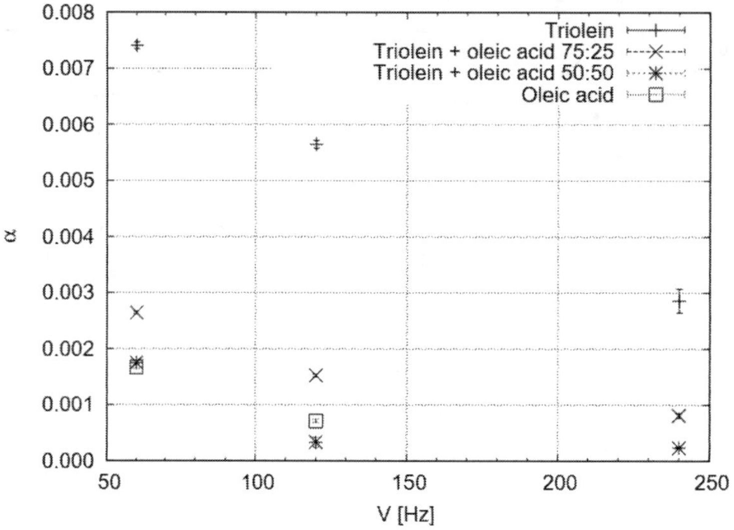

Figure 6. α parameter versus speed of pressure change.

Table 1. τ_1, τ_2, α parameters and their uncertainties obtained from fitting linear functions to fragments of logarithm of light transmission in examined liquids at particular speed of applied pressure.

Examined liquid	τ_1 [s]	$\Delta\tau_1$ [s]	τ_2 [s]	$\Delta\tau_2$ [s]	α	$\Delta\alpha$
Triolein v=60Hz	4710	20	35.2	0.2	0.00741	0.00006
Triolein v=120Hz	2680	20	15.22	0.06	0.00565	0.00007
Triolein v=240Hz	3100	200	8.79	0.05	0.0028	0.0002
Triolein + oleic acid 75:25 v=60Hz	10270	60	27.21	0.07	0.00264	0.00002
Triolein + oleic acid 75:25 v=120Hz	8670	60	13.3	0.1	0.00153	0.00002
Triolein + oleic acid 75:25 v=240Hz	8200	300	6.7	0.1	0.00081	0.00004
Triolein + oleic acid 50:50 v=60Hz	5940	40	10.4	0.2	0.00175	0.00004
Triolein + oleic acid 50:50 v=120Hz	20000	5000	6.63	0.01	0.00033	0.00008
Triolein + oleic acid 50:50 v=240Hz	15600	600	3.73	0.04	0.00024	0.00001
Oleic acid v=60Hz	6130	20	10.2	0.1	0.00167	0.00003
Oleic acid v=120Hz	6900	200	4.90	0.07	0.00071	0.00003

to increase within faster pressure applying. So this shows that movement of more complex hook-shaped compound like triolein is harder than simpler rod-like compound like oleic acid. Addition of oleic acid increases mobility in such mixture, because there is more space between complex compounds which tend to entangle. Also we see that addition of oleic acid changes behavior of observed scattered light dramatically, so the dynamics of phase transformation of such configuration changes and oleic acid phase transformation tend to mask triolein behavior. Generally as a basis of approach towards diffusion, we used speed of appearing and growing of the scattering objects as a dependency of pressure. Further research in this direction is legitimate. Exploiting the Edwards model should be verified in other conditions like temperature dependence.

9. References

[1] Siegoczyński R M, Wiśniewski R and Ejchart W 2003 Changes of structure in castor oil under pressure *High Pressure Res.* **23** 105–9

[2] Siegoczyński R M, Wiśniewski R and Ejchart W 2003 On the structure of a pressure-induced new phase in castor oil *J. Mol. Liq.* **107** 257–61

[3] Przedmojski J and Siegoczyński R M 2002 X-ray diffraction investigation of oleic acid under high pressure *Phase Transit.* **75** 581–5

[4] Siegoczyński R M 1998 *Reports of the Institute of Physics* (Warsaw: Publishing House of the Warsaw Technical University of Technology) **46**

[5] Siegoczyński R M 1995 Light scattering studies of oleic acid under high pressure *High Pressure Res.* **13** 89

[6] Edwards S F and Vilgis Th 1986 The dynamics of the glass transition *Phys. Scripta* **T13** 7–16

[7] Manohar R, Gupta M and Shukla J P 2000 Phase transition studies of some cholesteric liquid crystals and their mixtures using dielectric, optical transmittance and density measurement techniques *J. Phys. Chem. Solids* **61** 1465–73

[8] Wiśniewski R, Siegoczyński R M, Długosz A, Przewłocki M, Szymański M and Trzeciecki M 2001 Investigations of triolein under high pressure *High Temp–High Press.* **33** 231–6

[9] Kos A, Tefelski D B, Kościesza R, Rostocki A J, Roszkiewicz A, Ejchart W, Jastrzębski C and Siegoczyński R M 2007 Certain physico-chemical properties of triolein and methyl alcohol-triolein mixture under pressure *High Pressure Res.* **27** 39–42

[10] Tefelski D B, Rostocki A J, Kos A, Kościesza R and Siegoczyński R M 2007 Pressure induced volume changes in triolein with addition of methyl alcohol *High Pressure Res.* **27** 57–61

[11] Siegoczyński R M, Kos A, Tefelski D B, Kościesza R and Ejchart W 2007 Temperature effect upon the pressure-induced phase transformation in oleic acid *High Pressure Res.* **27** 47–50

[12] Kos A, Tefelski D B, Siegoczyński R M, Rutkowski R, Ejchart W, Wiśniewski R 2005 Changes of liquid structure under pressure in oleic acid *High Pressure Res.* **25** 51–6

[13] Siegoczyński R M, Wiśniewski R and Ejchart W 1999 Optical effects observed in pressure-induced phase transition for some liquid molecular system *Physica B* **265** 272–6

Effect of high pressure on mesophilic lactic fermentation streptococci

A Reps[1], M Kuźmicka and K Wiśniewska
Chair of Food Biotechnology, University of Warmia and Mazury, ul. Heweliusza 1, 10-724 Olsztyn, Poland

[1] e-mail: arnold.reps@uwm.edu.pl

Abstract. The research concerned the effect of high pressure on mesophilic lactic fermentation streptococci, present in two cheese-making commercial inocula produced by Christian-Hansen. Water solutions of inocula were pressurized at 50-800 MPa, at room temperature, for 30-120 min. Pressurization at 50-100 MPa slightly increased or reduced the number of lactic streptococci, depending on the inoculum and pressurization time. Pressurization at 200 MPa caused a reduction in the number of streptococci by over 99.9%, whereas the pressure of 400 MPa and above almost completely inactivated streptococci. Pressurization also reduced the dynamics of microorganism growth and acidification, to the degree depending on the pressure.

1. Introduction.

High pressure technology (100-1000 MPa) is a modern method of food processing, used for non-thermal preservation of raw materials and food products as well as for modification of their organoleptic and technological properties [1].

Microorganisms demonstrate diverse resistance to high pressure [2] which is a strain characteristic [3]. Resting spores are particularly resistant to high pressure, as they are not completely inactivated even at 1000 MPa [4]. The pressure-caused inactivation of microflora was determined by: the pressure and time of its application [5], temperature [6,7] environmental pH [8, 9] and environmental composition [10].

Potential possibilities of using high pressure technology in the dairy industry seem interesting.

Pressurization of milk can replace the process of pasteurization [2], and it can additionally result in the reduction of rennet coagulation time, an increase in the efficiency of a cheese-making process [11] and improvement of organoleptic qualities of produced cheese [2].

Pressurization of ripenning cheese can accelerate its ripening [12] and shape its sensory and physicochemical characteristics [13].

It is also possible to apply prior pressurization to inocula in order to obtain attenuated cultures of lower acidifying activity but of preserved lipo- and proteolytic activity [14].

2. The aim of the experiment.

The aim of this experiment was to characterize the effect of high pressures on mesophilic lactic streptococci, present in two commercial cheese-making inocula.

© 2008 IOP Publishing Ltd

3. Materials and methods.

3.1. The preparation of research material for pressurization.
The subject of the research were lyophilized commercial inocula produced by Christian-Hansen: FD-DVS Flora Danica and FD-DVS CHN-19, used for direct inoculation of milk in cheese production.
Lyophilized inocula were rehydrated for two hours by shaking water suspensions of cells at room temperature. Water solutions were prepared in such a way so that they should contain lactic streptococci in the amount of about 10^8 cfu/cm^3.

3.2. Pressurization of water solutions of inocula.
Inocula solutions were pressurized at 50, 100, 200, 400, 600 and 800 MPa for 30, 60 and 120 min at room temperature, using U 101 apparatus produced by UNIPRESS EQUIPMENT Warszawa. A part of inocula was left unpressurized as a control inocula.

3.3. Determination of the number of lactic acid streptococci in the inoculum. The number of streptococci was determined on M17 agar according to Terzaghi (Merck). Cultures were incubated at 30°C/48 h.

3.4. Comparison of methods used for testing the number of streptococci population: by the colony-count method and the method of measuring optical density (OD). The usefulness of the OD method (according to 3.5) for determining the dynamics of microorganism growth was assessed by determining the optical density of solutions of inocula prepared according to 3.1 and their subsequent dilutions by sterile distilled water, and determining at the same time the number of streptococci in the above mentioned diluted solutions according to 3.3.

3.5. Determination of the dynamics of the growth of lactic streptococci and acidification of liquid nutrient medium. A liquid nutrient medium – M17 broth according to Terzaghi (Merck) – was inoculated with the same number of living cells. At four-hour intervals, the optical density of the sample was determined at 600 nm in Beckman DU 650 spectrometer. The cultures were continued at 30°C until the streptococci reached a stationary phase. Parallel to this determinations the active acidity of the culture was determined by means of pH measurement.

4. Results and discussion

4.1. The effect of high pressures on the number of lactic streptococci.
Pressurization of FD-DVS Flora Danica inoculum at 50 MPa/30 and 60 min increased the number of lactic streptococci in comparison with the control sample, by 36.8% and 13.2%, respectively. Extending the time of pressurization to 120 min decreased the number of streptococci by 34.2 %. Pressurization of the inoculum solution at 100 MPa/30, 60 and 120 min caused a reduction in the number of streptococci by 15.0%, 32.5% and 37.5%, respectively.

Pressurization of FD-DVS CHN-19 inoculum at 50 MPa/30, 60 and 120 min increased the number of streptococci by 37.8%, 22.4 and 20.4%, respectively. Pressurization at 100 MPa/30 min also increased the streptococci population by 27.3% and at 100 MPa/60 and 120 min reduced the number of streptococci by 37.5% and 43.2%, respectively.

The fact, that lower pressures can cause an increase in the number of lactic acid streptococci, can be related to the disruption of "chains" of streptococci into smaller fragments, which would cause the increase in the number of units capable of forming colonies.

Pressurization of FD-DVS Flora Danica inoculum at 200 MPa/30, 60 and 120 min resulted in a reduction of bacteria population by 94.8%, 98.8% and 99.995%, respectively. Pressurization of FD-DVS CHN-19 inoculum at 200 MPa/30, 60 and 120 min resulted in a reduction in bacteria population by 94.8%, 98.5% and 99.965%, respectively. Pressurization of inocula at 200 MPa/120 min had similar effects as pasteurization.

Pressurization of FD-DVS Flora Danica inoculum at 400 and 600 MPa/30-60 min resulted in a reduction of the number of streptococci to below 100 cfu/cm^3, and after 120 min of pressurization, no presence of living bacteria was found. After pressurization of the inoculum at 800 MPa, no streptococci was observed, regardless of the pressurization time. Pressurization of FD-DVS CHN-19 inoculum at 400, 600 and 800 MPa/30 min reducted the number of streptococci to below 100 cfu/cm^3, and in solutions pressurized for 60 and 120 min, no presence of lactic streptococci was found.

A similar sensitivity of bacteria from dairy inocula to high pressures was observed by Krasowska et al. [14] who found that pressurization of lyophilized inoculum, containing lactic streptococci and bacilli at 200-800 MPa/15 min decreased the number of streptococci by 0.67-6.00 logarithmic orders.

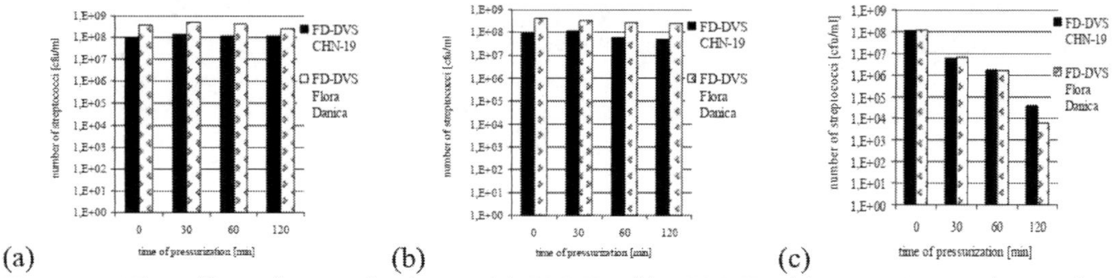

Figure 1. The effect of pressurization at (a) 50 MPa (b) 100 MPa (c) 200 MPa on the number of streptococci

4.2. Comparison of the methods of testing the increase in the number of streptococci population by a colony-count method and the method of measuring optical density.

The examined dilutions of FD-DVS Flora Danica inoculum were characterized by OD from 0.0006 to 0.4626, which corresponded to the number of bacteria between 2.10×10^7 cfu/cm^3 and 9.35×10^9 cfu/cm^3; while for FD-DVS CHN-19 inoculum it was between 0.0010 to 0.5041, which corresponded to the number of bacteria between 5.75×10^7 cfu/cm^3 and 9.9×10^9 cfu/cm^3. The relationship between OD and the number of streptococci determined using a colony-count method was directly proportional, linear and the correlation coefficient was very high, (0.9992 and 0.9988, respectively). Therefore, the usage of a faster OD method seems justified.

4.3. Determination of the dynamics of the growth of lactic streptococci in a liquid nutrient medium.

Both control inocula reached a stationary phase after 12 h of incubation and pH decreased with the increase in the streptococci population. For FD-DVS Flora Danica pH in a stationary phase it was 6.09 (in relation to 7.2 at the beginning of culture), while the extension of the culturing time to 20 h lowered pH to 5.76. For FD-DVS CHN-19 pH in a stationary phase it was 6.16 (in relation to 7.2 at the beginning of culture) – while the extension of culturing time to 20 h lowered pH to 5.70.

Pressurization of the inoculum at 50 and 100 MPa, regardless of the time of pressurization, resulted, however, in the extension of time necessary for reaching a stationary phase by 4 h. For FD-DVS Flora Danica pH was lowered during the culturing time from 7.12 to 6.09 and from 7.13 to 6.17, respectively. For FD-DVS CHN-19 pH was lowered during the time of culturing from 7.07 to 5.97 and from 7.12 to 6.17, respectively. It appears to support the thesis that pressurization at low pressures rather causes the disruption of streptococci chains than stimulates the development of lactic streptococci. This increases the number of colonies obtained on M17 agar, but at the same time, it reduces the ability of cells to multiply.

Pressurization of the inoculum solution at 200 MPa, regardless of the time of pressurization, caused an extension of time necessary for reaching a stationary phase to 20 hours of culturing. For FD-DVS Flora Danica pH was lowered during the culturing time from 7.13 to 6.15. For FD-DVS CHN-19 pH was lowered during the time of culturing from 7.14 to 5.87.

Pressurization of the inoculum solution at 400, 600 and 800 MPa, regardless of the time of pressurization, resulted in an almost complete inhibition of streptococci growth and nutrient medium acidification. Pressurization at 400 MPa and higher inactivates or damages bacteria cells to such an extent that within 24 h they are not able to continue growth in a liquid nutrient medium.

Impeded functioning and multiplication of pressurized bacteria cells can be to a high extent related to the damage to the cell walls and cell membrane during pressurization [15].

4.4. Comparison of the resistance to high pressures in the case of FD-DVS Flora Danica and FD-DVS CHN-19 inocula.

The inocula under examination were slightly different as regards their resistance to high pressures. FD-DVS CHN-19 inoculum was slightly more resistant, which is proved by the fact that pressurization at 50 MPa (regardless of the time) and at 100 MPa/30 min did not result in a reduction of streptococci in their composition, while 120-minute pressurization at 50 MPa and 30-minute at 100 MPa were sufficient to lower the number of streptococci in FD-DVS Flora Danica inoculum.

There were no significant differences observed in the rate of development of streptococci from various inocula in a liquid nutrient medium, but in most cases the lowering of pH value during the incubation was more obvious in the case of FD-DVS CHN-19.

5. Conclusions

Pressurization of dairy inocula, to the degree depending on the pressure applied, has a significant effect on the number, growth dynamics and acidification of a nutrient medium by lactic acid streptococci present in the composition of both cultures.

A significant reduction in the number of streptococci (over 99.9%) can be obtained by applying a pressure of at least 200 MPa.

Pressurization of inocula at 400 MPa and higher pressures significantly reduces or completely inactivates streptococci present in the inocula.

During the study, financial support was granted from the European Grant Fund and the state budget under the Integrated Regional Operational Programme 2004-2006.

6. References

[1] Trujillo A J, Capellas M, Saldo J, Gervilla R and Guamis B 2002 *Innov. Food Sci. Emerg. Technol.* **3** 295–307
[2] Buffa M, Guamis B, Royo C and Trujillo A J 2001 *Food Microbiol.* **18** 45-51
[3] Alpas H, Kalchayanand N, Bozoglu F, Sikes A, Dunne C P and Ray B 1999 *Appl. Environ. Microbiol.* **65** 4248-51
[4] Moerman F, Mertens B, Demey L and Huyghebaert A 2001 *Meat Sci.* **59** 115-25
[5] Kołakowski P, Reps A and Babuchowski A 1998 *Pol. J. Food Nutr. Sci.* **48** 473–83
[6] Mills G, Earnshaw R and Patterson M F 1998 *Lett. Appl. Microbiol.* **26** 227-30
[7] Reyns K M F A, Sootjens C C F, Cornalis K, Weemaes C A, Hendrickx E and Michiels C W 2000 *Int. J. Food Microbiol.* **56** 199-210
[8] Jordan S L, Pascual C, Bracey E and Mackey B M 2001 *J. Appl. Microbiol.* **91** 463-9
[9] O'Reilly C E, O'Connor P M, Murphy P M, Kelly A L and Beresford T P 2002 *Int. Dairy J.* **12** 915–22
[10] Gervilla R, Capellas M, Ferragut V and Guamis B 1997 *J. Food Prod.* **60** 33–7
[11] Huppertz T, Fox P F and Kelly A L 2004 *Innov. Food Sci. Emerg. Technol.* **5** 1-8
[12] O'Reilly C E, Kelly A L, Oliveira J C, Murphy P M, Auty M A E and Beresford T P 2003 *Innov. Food Sci. Emerg. Technol.* **4** 277-84
[13] Messens W, Foubert I, Dewettinck K and Huyghebaert A 2000 *Milchwiss.* **55** 328-32
[14] Krasowska M, Reps A, Jankowska A, Arabas J, Porowski S and Proszek *A* 2003 *High Press. Res.* **23** 93-6
[15] Malone A S, Shellhammer T H and Courtney P D 2002 *Appl. Environ. Microbiol.* **68** 4357-63

The effect of high pressures on the yoghurt from milk with the stabilizer

A Reps, [1] A Jankowska, K Wiśniewska
Chair of Food Biotechnology, University of Warmia and Mazury, Heweliusza str 1., 10-724 Olsztyn, Poland
E-mail: arnold.reps@uwm.edu.pl

Abstract. The effect of high pressures on the microbiological and physio-chemical properties of yoghurt was investigated. The best results were obtained when the yoghurt was manufactured from milk with the addition of MYO 752 stabilizer (starch, gelatin, pectin) selected from 10 stabilizers. Yoghurt manufactured with the addition of 0, 6% MYO 752 stabilizer was processed at the pressure of 400-600 MPa/15 min. in the range of 50 MPa. Pressurization caused a total reduction of number of *Lactobacillus delbrueckii ssp. bulgaricus* and reduced the number of Streptococcus *thermophilus* by 1-2 orders of magnitude. Pressurized and non-pressurized yoghurts characterized of a homogenous consistency and typical plain yoghurt taste. The decrease of the number of living bacteria was observed in yoghurts during the storage. The acidity of pressurized yoghurts remained on the some level at the temperature of 4°C and 20°C. The more intensive antibacterial activity of microflora was observed in yoghurts storaged at 20°C in comparison with yoghurts storaged at 4°C. Disadvantageous changes of the pressurized yoghurts consistency were not found. The taste and aroma of yoghurts remained without any changes.

1. Introduction

High pressure can be applied in yoghurt technology to preserve milk intended for the production of yoghurt and to preserve a final product. Yoghurt preserved by means of high pressure can be also a good carrier of probiotic bacteria [1].One of problems that the producers of yoghurt must face, particularly in the case of low-fat yoghurt, is to obtain its proper texture. It has been found that the application of high pressure in the pre-treatment of cow milk intended for yoghurt improves the firmness of yoghurt curd and reduces its syneresis. The best results have been obtained under the pressure of 350 or 500 MPa applied for 15 minutes, at 25 or 55°C [2, 3].The pressure above 200 MPa significantly inhibits the souring activity of yoghurt bacteria. The application of pressures of 300-400 MPa significantly reduced the number of living cells of *Lactobacillus delbrueckii ssp. bulgaricus*, whereas the number of *Streptococcus thermophilus* was only slightly lowered [1].Pressurization of yoghurt in the pressure range of 400-600 MPa has a negative effect on its consistency; increased whey leak was observed. The objective of the study was to determine the impact of the addition of stabilizing agents used in the industrial production of yoghurts on the quality of pressurized yoghurt. Experiments were additionally aimed at determining whether the pressurization process allows decreasing the dose of stabilizing agent added to yoghurt. The most advantageous of the examined stabilizers was selected for the further part of the research. Yoghurts produced from milk with the addition of the experimentally determined amount of a stabilizer, one from ten, were subjected to the pressure of 400-600 MPa, and afterwards they were stored for four weeks at 4 and 20°C.

[1] corresponding author

© 2008 IOP Publishing Ltd

2. Research materials and methods

The research material was yoghurt obtained from commercial yoghurt inoculum: YC-X16 (Chr. Hansen). Yoghurt was prepared with the addition of the following stabilizers, in the following amounts:YO 752 1.2%, YO 747 1.0%, YO749 0.2% (Danisco Biolacta sp. z o. o., Poland);IMG Super 1.8%, MPG Super 1.7%, KBM 100-Super 1.45%, MGP-2 1.80% (Superior, Poland);JMB1 1%, Novation Starch 1.5%. (Dairy Science Institute, Poland).

2.1 Pressurization of yoghurt with the addition of selected stabilizers

Yoghurt (a 100-ml portion) with the addition of selected stabilizers was pressurized in hydraulic pressure generator UNIPRESSEQUIPMENT (High Pressure Research Centre Polish Academy of Science, Warsaw) at 550 MPa/15 min/18°C; the temperature of pressurization was constant. Active acidity (pH) was determined and an organoleptic assessment was performed before the pressurization and directly after the treatment, as well as after four weeks of storage at 4°C. Additionally, the viscosity of yoghurts was tested before and directly after the pressurization, by means of rotating rheometer Rheostress 1 (Haake Gm BH), with the use of Rheo Win Pro Job Manager Archema 2000 software, Edition Version 2.67. The consistency factor was determined, as well as flow behaviour index. Based on the results obtained, YO 752 stabilizer was selected for the subsequent part of the research.

2.2. Pressurization of yoghurt produced from milk with the addition of YO 752 stabilizer.

Yoghurt (a 100-ml portion) prepared from milk with the addition of 0.6% YO 752 was pressurized in the pressure range of 400-600 MPa/15 min. every 50 MPa/18°C; the temperature of pressurization was constant. The following values were determined in yoghurt before pressurization, directly after the treatment and after each one of the four weeks of storage at 4 and 20°C: the number of bacteria, using a colony-count method *Lactobacillus delbrueckii ssp. bulgaricus* in MRS broth (Merck) and *Streptococcus thermophilus* in M17 broth (Merck); active acidity (pH) and titratable acidity (°SH); antibacterial properties, using a modified well-diffusion method, the size of the inhibition growth zones of eleven test strains was measured (mm); an organoleptic assessment was performed, according to the Polish Standard PN- A – 86086:2002.

3. Results and discussion

3.1. The effect of pressurization on the quality of yoghurt with the addition of selected stabilizers.

The experiment involved selected stabilizers used in the production of yoghurt in Poland. It is important in yoghurt that the stabilizer should prevent whey syneresis. Unpressurized yoghurt with the addition of stabilizers had a homogenous, firm curd, stringy consistency, white colour and smooth cross section of "porcelain sheen". Differences were found in pressure-treated yoghurts and they mostly concerned the appearance of curd and the consistency of the product. A slight whey separation was observed with a grainy consistency, and in some cases, a slightly ragged curd. In yoghurts with the addition of YO 752, IMG-Super and MGP-2 Super stabilizers, the curd after the pressurization was homogenous, with slight whey separation, and its taste was refreshing. These yoghurts preserved their organoleptic qualities during the four week storage at 4°C.

As the storage period continued, the deterioration of product quality was observed in the case of some yoghurts. In yoghurt with the addition of YO 747 IMB stabilizer and Novation Starch, after four weeks of storage, a slight wheying off was found. Similar changes were also observed in pressurized yoghurts with the addition of IMB 1 and YO 747.

While assessing taste and smell, it has been established that most unfavourable qualities during four weeks of storage were found for yoghurt with the addition of YO 752 stabilizer.

The acidity of yoghurts with the addition of the analysed stabilizers before pressurization ranged from pH 4.30 – in the case of yoghurt with the addition of YO 749 stabilizer – to pH 4.47 in case of the addition of Novation Starch. After four weeks of storage at 4°C, there was also a slight increase in acidity observed, both in unpressurized yoghurts and in those subjected to the pressure treatment.

The consistency factor and flow behaviour index are the characteristics of the rheological properties of yoghurt. Higher values of consistency factor results in higher viscosity and firmer consistency of yoghurt. The lowest value of the coefficient k=0.29 was demonstrated by pressurized yoghurt, without the addition of stabilizer, which indicates a low capacity to water binding. Therefore, the highest degree of wheying off was established in this yoghurt. The value of flow behaviour index n for all yoghurts was within the range of 0.55-0.92. In all yoghurts after pressurization, there was a decline observed in coefficient K and flow behaviour index n, which corresponded with the results of the organoleptic assessment, based on which it was found that yoghurts after pressurization were characterized by higher whey syneresis.

3.2. *The effect of pressurization on yoghurt produced from milk with the addition of YO 752 stabilizer.*

3.2.1. Viability of yoghurt bacteria.

In unpressurized yoghurt, directly after production, the presence of 2.57×10^9 cfu/g *Streptococcus thermophilus* and 1.90×10^6 cfu/g *Lactobacillus delbrueckii* ssp. *bulgaricus* was established. During four weeks of storing unpressurized yoghurt at 4°C, there was a reduction observed in the number of *Lactobacillus delbrueckii* ssp. *bulgaricus* by one order of magnitude after the first week and of *Streptococcus thermophilus*. While storing yoghurts at 20°C, there was a change observed in the number of *Streptococcus thermophilus* by two orders of magnitude, and the number of *Lactobacillus delbrueckii* ssp. *bulgaricus* was reduced by three orders of magnitude to 8.00×10^3 cfu/g. After pressurization and during the storage of yoghurts pressurized in the pressure range of 400-600 MPa/15 min., no presence of *Lactobacillus delbrueckii* ssp. *Bulgaricus* was found.

The pressure of 400 MPa had no significant effect on the change in the number of *Streptococcus thermophilus*. There were no significant changes established as regards changes in the number of these bacteria in the first week of storage at 4°C. In the subsequent week of storage, there was a decrease observed in their number by about two orders of magnitude. In the third and fourth week of storing yoghurts pressurized at 400-500 MPa at 4°C, no presence of *Streptococcus thermophilus* were found, except from the yoghurt that was subjected to the pressure of 500 MPa. At 20°C, the presence of streptococcus was observed only in the first week of storage. The pressure of 550 and 600 MPa resulted in a decrease in the number of yoghurt streptococci by one order of magnitude. While storing at 20°C, yoghurts subjected to the pressure of 550 and 600 MPa, the number of streptococci was reduced to the level below 10^3 cfu/g within a week. During the subsequent storage, no presence of *Streptococcus thermophilus* was established in the analysed yoghurt. The established high sensitivity of *Lactobacillus delbrueckii* ssp. *bulgaricus* to pressures over 400 MPa corresponds with the results of research conducted by de Ancos et al. [1].It was found that the addition of stabilizer, as opposed to fruit additives [4], does not have a protective effect on the bacilli of *Lactobacillus delbrueckii* ssp. *bulgaricus* during the pressurization and the storage. The higher the storage temperature, the more intense was the reduction of bacteria. Similarly, Canganella et al. [5] observed a significant drop in the number of *Streptococcus thermophilus* after 25 days of storage at 12°C, both in natural yoghurt and yoghurt with the addition of fruit.

3.2.2. The effect of pressurization and storage conditions on the acidity of yoghurt.

The acidity of yoghurt directly after preparation was 4.40 pH and 44°SH, respectively. As the result of pressurization, depending on the operating pressure, there was a drop observed in acidity by 0.01-0.06 pH. Changes in yoghurt acidity could be caused by the effect of high pressure on the structure of albumen and mineral components of milk. [6].The acidity of unpressurized yoghurt during the four-week storage at 4°C gradually increased from 4.40 to 4.21 pH and from 44°SH to 50°SH, and at 20°C – to 3.91 pH and 58.5°SH.However, in pressurized yoghurts, during storage at 4 °C and 20°C, the pH value was maintained at a similar level, with a slight growing tendency.

3.2.3. The effect of pressurization and storage conditions on the antibacterial activity

Milk fermentation bacteria synthesize many substances of antibacterial properties. This bacillus is also capable of producing large amounts of H_2O_2, to which bacteria of the *Clostridium* species are particularly sensitive. *Streptococcus thermophilus* demonstrates wide spectrum of activity by producing bacteriocins of antibacterial properties in relation to Gram-positive and Gram negative bacteria [7]. Unpressurized yoghurt demonstrated the best antibacterial properties, both directly after production and during the storage. As regards 11 test strains, unpressurized yoghurt revealed the best antibacterial proprieties in relation to four strains. In the yoghurt stored at 20°C, unlike yoghurt stored at 4°C, a stronger antibacterial activity was observed in relation to all test strains. Under high pressure, a weakening of the antibacterial activity in pressurized yoghurt was observed. Based on the results obtained, it can be stated that the amount of pressure applied influences the antibacterial properties of yoghurt. The amount of pressure was inversely proportional to antibacterial properties demonstrated by yoghurt directly after pressurization and during the storage, which is also confirmed by [8].

3.2.4. The effect of pressurization and storage conditions on the organoleptic qualities of yoghurt. Yoghurts subjected to the pressure of 400 and 450 MPa were characterized by white colour, homogenous, firm curd with slight whey separation. The consistency of yoghurts was thick and stringy. While stored at 4°C, yoghurt was characterized by a typical aldehyde taste, was slightly sour, without any strange aftertaste, and at 20°C no clear yoghurt taste could be sensed, but it was slightly sour with noticeable sweetish aftertaste. Yoghurt was subjected to pressures of 500, 550 and 600 MPa/15 min. At both storage temperatures, it preserved white colour and homogenous curd with light leak of whey. At 4°C, yoghurts characterized by homogenous consistence, slight yoghurt taste, were sweetish and slightly sour. Yoghurts stored at 20°C were characterized by runny, but homogenous consistency. However, no typical yoghurt taste could be sensed. The addition of a stabilizer had a favourable effect on the improvement of the consistency of the examined yoghurts, by significantly reducing syneresis. The most favourable organoleptic qualities were found in yoghurt subjected to the pressures of 400 and 450 MPa.

4. Conclusions

Streptococcus thermophilus present in the composition of yoghurt inocula are characterized by greater sensibility to high pressures, as compared with the bacillus of *Lactobacillus delbrueckii ssp. bulgaricus*. The amount of pressure applied has an effect on the viability of yoghurt bacteria. The addition of a stabilizer has a favourable effect on the consistency of pressurized yoghurt by reducing syneresis. The stabilizer applied does not have a preserving effect on the viability of yoghurt bacteria during pressurization. High pressure has a negative effect on antibacterial properties of yoghurt. Pressurization impedes the process of yoghurt souring during storage. The addition of stabilizer helps to obtain yoghurts of good organoleptic characteristics and significantly extended durability. Improvement of antibacterial properties of yoghurt after pressurization is feasible upon the addition of the probiotic bacteria *Lb. acidophilus* and *Bifidobacterium sp.*

5. References

[1] Jankowska A., Wiśniewska K, Reps A *2005 High Pressure Research*. 25, **1** 57-62
[2] Needs E C, Capellas M, Bland A, Macdougal D, Paul G, 2000 *J. Dairy Res.* **67** 329-348
[3] Ancos de B Cano M P Gomez R 2000 *Int. Dairy J.* **10** 105-11
[4] Walker MK, Farkas D F Loveridge V, Meunier-Goddik L, 2006 *Int. J of Food Sci. and Tech.* **41** 464-7
[5] Canganella F, Giontella D, Nespica M L, Massa S, Trovatelli L D 2000 *Annals of Microbiology*. 50 **43-53**
[6] Huppertz T, Kelly A L, Fox P F 2002 *Int. Dairy J.* **12** 561-572.
[7] Ivanova I, Miteva V, Stefanova T, Pantev A, Budakov I, Danova S, Moncheva P, Nikolova I, Dousset X, Boyaval P 1998 *Int. J of Food Microbiol.* **42** 147-158
[8] Warmińska-Radyko I, Łaniewska-Moroz Ł, Reps A, Krzyżewska A 2002 *Trends in High Pressure Bioscience and Biotechnology*. **19** 343-8

Journal of Physics: Conference Series **121** (2008) 142008 doi:10.1088/1742-6596/121/4/142008

Multiple phase transitions in palmitoleic acid under the pressure

A Kos, R M Siegoczyński, A J Rostocki, D B Tefelski, R Kościesza, K Wieja

Faculty of Physics, Warsaw University of Technology, Koszykowa 75, 00-662 Warszawa, PL

E-mail: aldona@if.pw.edu.pl

Abstract. Long unsaturated carbon chain liquids exhibit a phase transition under a high pressure. We have investigated the phase transition in palmitoleic acid as a representative of this group not researched yet. A temperature, a relative permittivity and a dissipation factor have been measured, while the pressure was applied. We have also observed the transmitted and the scattered IR light. These parameters allowed us to find phase transitions that occurred due to the pressure. Especially the behavior of the temperature indicates several phase transitions during compression. Above certain pressure the liquid becomes opaque, it is connected with most significant phase transition that occurs around 177 MPa. We have observed asymmetry of intensity of the scattered light, which is much stronger during decompression. Observed multiple phase transitions are not present in other acids investigated by us.

1. Introduction

Our research group has been investigating pressure-induced phase transitions in fatty acids and triacylglycerols of those acids. This paper concerns phase transition in palmitoleic acid (Z)-9-hexadecenoic acid, which is monounsaturated fatty acid. So far we have researched few fatty acids [1, 2, 3]. We have observed phase transition in each of them. In particular the oleic acid (Z)-9-octadecenoic acid is interesting because of structural similarity to palmitoleic acid and its significant role as a potential carcinogen (breast cancer [4]). We have expected similar phase transition in both. However it turned out that it is somewhat different.

2. Experimental setup

The block diagram of experimental system is given in figure 1. We have performed measurements of several parameters of palmitoleic acid during the compression and decompression. The following parameters were observed: the temperature, the pressure, the relative permittivity and the dissipation factor, forwarded and scattered IR laser light.

The phase transition occurs in high pressure optical chamber with thermocouple sensor, manganine wire pressure sensor and capacitor for dielectric permittivity test. Capacitor is connected by four wire method to LCR meter, according to paper [5]. Investigated liquid is compressed by pump with computer controlled stepping motor. We have applied pressure with constant speed 0.33 MPa/s. In the experiment we have used palmitoleic acid 90-95% (GC) supplied by Fluka.

© 2008 IOP Publishing Ltd

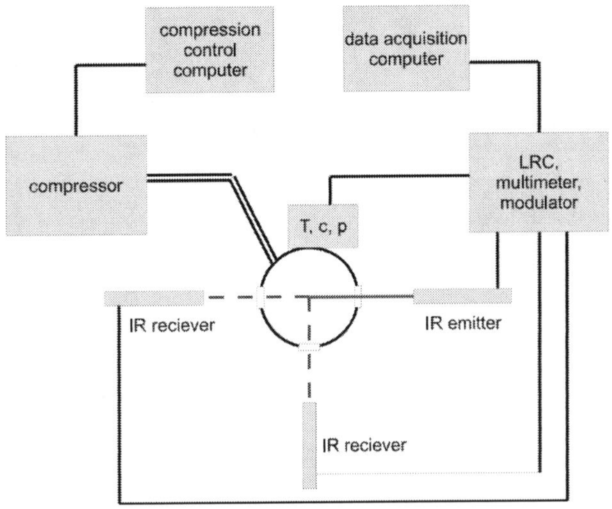

Figure 1. Measurement system.

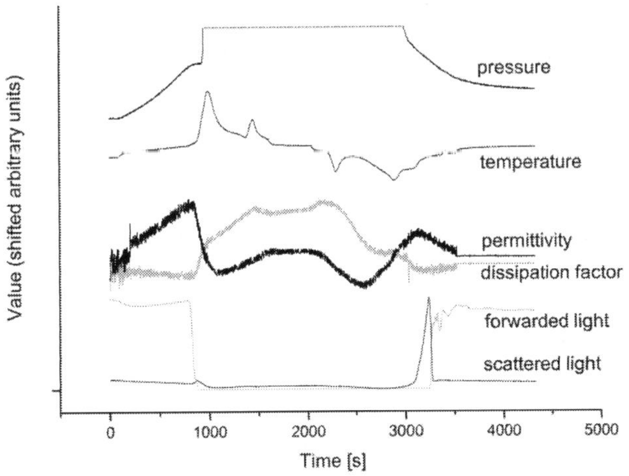

Figure 2. Results of compression and decompression of palmitoleic acid. The graphs are labeled on the right side. The values are in arbitrary units and are normalized to have fixed range. The graphs were shifted to separate them and have a clear picture. The values of pressure in the dashed line area were not measured correctly. It is worth to note that the intensity of scattered light is four orders of magnitude smaller than the intensity of forwarded light.

3. General results

The results of our experiments are given in figure 2.

The sudden drop of forwarded light above a threshold pressure indicates the phase transition [6, 7]. Due to the pressure-induced aggregation of the liquid molecules, the light scattering mechanism changes. Rayleigh scattering is dominant in normal liquids, where the molecules are

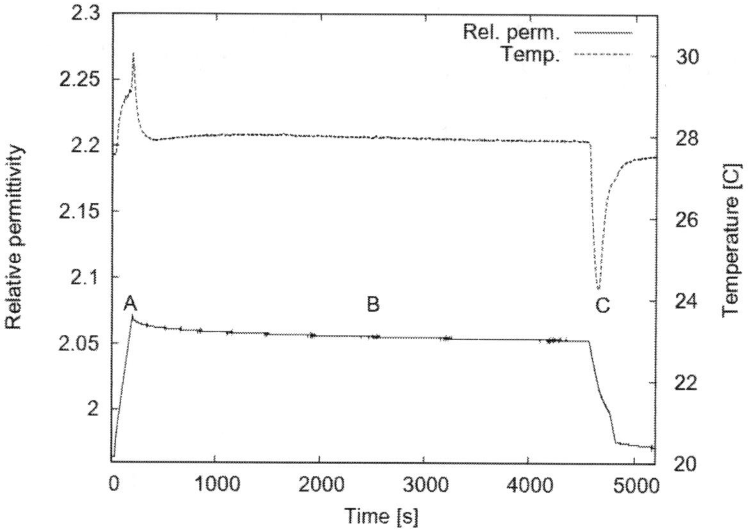

Figure 3. Results of compression and decompression of oleic acid. Changes of relative permittivity and temperature versus time. A - compression, B - stagnation stage after pressure applying, C - decompression

much smaller than the wavelengths. The large aggregates are comparable in size with wavelength, so the Mie effect takes place of Rayleigh scattering [1]. As a result the liquid becomes opaque. The high peak of scattered light during decompression in figure 2 can be explained as follows. During compression, the liquid quickly changes from transparent to opaque. Fully opaque liquid simply blocks all light, both transmitted and scattered. During decompression, the transition is smoother, going through half-opaque stage. This stage both scatters and transmits light, so that scattered light is not blocked.

The temperature peak that closely follows the drop of forwarded light confirms the phase transition and shows that it is exothermic [8]. Peaks of temperature indicate that the phase transition is multiple. The oleic acid does not display multiple temperature peaks (figure 3). This shows basic difference between phase transitions in both.

Permittivity rises along with the pressure until the phase transition due to increasing the concentration of dipoles. During phase transition it suddenly drops. Oleic acid does not display such behavior [5] (figure 3). This further confirms the differences between their behavior under the pressure.

The pressure versus time curve flattens around the phase transition point despite the compressor still applying pressure. The cause of this phenomenon, called by us collapse [9], is change of volume of the liquid undergoing phase transition.

4. Conclusions

We have observed a phase transition in the palmitoleic acid under the pressure. Unlike in the very similar oleic acid the transitions turns out to be multiple and the permittivity drops during phase transition. That may be explained as restriction of movement of dipoles, although the mechanism is unknown yet.

5. Acknowledgments

The work was partially supported by the Polish Ministry of Science and Higher Education under Grant No. 134/SPB/COST/KN/DWM 105/ 2005-2007.

6. References

[1] Siegoczyński R M 1998 *Reports of the Institute of Physics* (Warsaw: Publishing House of the Warsaw Technical University of Technology) **46**

[2] Siegoczyński R M 1995 Light scattering studies of oleic acid under high pressure *High Pressure Res.* **13** 89

[3] Kościesza R, Siegoczyński R M, Rostocki A J, Tefelski D B, Kos A and Ejchart W 2007 Relative permittivity behavior and temperature changes in linoleic acid during phase transition *in preparation EHPRG Catania 2007*

[4] Yu F-L, Greenlaw R, Fang Q, Bender W, Yamaguchi K, Xue B H, Yu C-C 2004 Studies on the chemopreventive potentials of vegetable oils and unsaturated fatty acids against breast cancer carcinogenesis at initiation. *Eur J Cancer Prev.* **13** 239

[5] Manohar R, Gupta M and Shukla J P 2000 Phase transitions studies of some cholesteric liquid crystals and their mixtures using dielectric, optical transmittance and density measurement techniques *J. Phys. Chem. Solids* **61** 1465-73

[6] Kos A, Tefelski D B, Siegoczyński R M, Rutkowski R, Ejchart W and Wiśniewski R 2005 Changes of liquid structure under pressure in oleic acid *High Pressure Res.* **25** 51-56

[7] Siegoczyński R M, Wiśniewski R and Ejchart W 2002 Non-conventional behaviour of castor oil under high pressure *High Pressure Res.* **22** 715-19

[8] Siegoczyński R M, Kos A, Tefelski D B, Kościesza R and Ejchart W 2007 Temperature effect upon the the pressure-induced phase transition in oleic acid *High Pressure Res.* **27** 47-50

[9] Siegoczyński R M, Kos A, Tefelski D B and Kościesza R 2007 Molecular Collapse - Modification of the Liquid structure induced by the pressure in oleic acid *in preparation*

Journal of Physics: Conference Series **121** (2008) 142009 doi:10.1088/1742-6596/121/4/142009

High Pressure Changes of the Castor Oil Viscosity By Ultrasonic Method

A J Rostocki[1], R M Siegoczyński[1], P Kiełczyński[2] and M Szalewski[2]

[1] Faculty of Physics, Warsaw University of Technology, Koszykowa 75, 00-662 Warszawa, PL
[2] Institute of Fundamental Technological Research, Polish Academy of Science, Świętokrzyska 21, 00-049 Warszawa, PL

E-mail: arostock@if.pw.edu.pl

Abstract. The pressure change of viscosity of castor oil have been measured by ultrasonic method within the range of pressure up to 0.9 GPa. For the measurement, the authors have applied a new ultrasonic method based on Bleustein-Gulyaev (B-G) waves. For the lower pressures (up to 0.3 GPa) the results have been compared with earlier results obtained by falling body method, whereas for the higher pressure range results were compared with those obtained by the flow type viscometer. The measurements have shown: 1. Exponential rise of viscosity with pressure up to 0.4 GPa according to the Barus formula. 2. Extraordinary increment of viscosity at constant pressure during phase transition. 3. The decomposition of the high pressure phase during the decompression process have shown very large hysteresis of viscosity on pressure. 4. After the decompression process the viscosity lasts higher then a initial value for several hours.

1. Introduction

The aim of this paper was: 1. To test the new ultrasonic method of viscosity measurement based on Bleustein-Gulyaev waves at the high-pressure conditions. 2. To compare the results with the results of viscosity on pressure measurements of castor oil obtained by other methods. 3. To observe the isobaric changes of castor oil viscosity during the phase transition process. 4. To measure the viscosity of high pressure phase of castor oil. Castor oil was one of the most extensively examined under high pressure natural oils, for its wide use in some laboratories as a pressure transmitting medium in dead-weight piston gauges for the range up to 1 GPa [1]. For this reason its rheologic properties were investigated in limited range several years ago [2, 3]. It was also a subject of our investigation since the beginning of our works. The strong, exponential dependence of viscosity on pressure known in the literature as "Barus formula" suggested that the small difference in the composition of liquid might cause the big difference of the viscosity at high pressure. For investigation of those effects the high pressure viscometer of the falling ball type with rotating pressure chamber up to 1.5 GPa and electronic time measurement system has been built [2]. Although the pressure range of the chamber was up to 1.5 GPa, the measurement of viscosity of castor oil was limited to about 300 MPa (above that pressure the ball was not falling dawn at all). In 1986-87 the high pressure phase transition in castor oil was observed [4, 5]. The results of measurement of refractive index [6], compressibility [7], and dielectric permittivity [8] suggested that new high pressure phase of castor oil should also have significantly higher viscosity. Since the falling body method did not give the result for the pressure range above 300 MPa the

© 2008 IOP Publishing Ltd

flow type method based on the Hagen-Poiseuille's rule has been applied [9, 10, 11]. The new high-pressure phase of castor oil was found to be strongly non-Newtonian liquid. Thus the viscosity obtained from Hagen-Poiseuille's relation was giving only effective (apparent) viscosity. In such case the complete rheological characteristics should be found what makes all measurement very complicated. Moreover all those complications made the measurement during phase transition process almost impossible. In this situation various ultrasonic methods were considered and finally the one based on Bleustein-Gulyaev waves was applied for the new measurement of castor oil viscosity.

2. Viscosity measurement using B-G waves

The Bleustein - Gulyaev (B-G) wave is a shear horizontal acoustic surface wave having only one component of mechanical displacement, parallel to the propagation surface and perpendicular to the direction of propagation. The energy of the B-G wave is concentrated in the vicinity of the waveguide surface. The viscoelastic liquid covering the waveguide surface loads it mechanically. The mechanical impedance of a layer of liquid loading the surface of the B-G wave waveguide is equal to the characteristic shear impedance of the liquid Z_L for plane waves:

$$Z_L = (\rho_L \cdot G_L)^{1/2} \tag{1}$$

where $G_L = G' + jG''$ is the complex shear modulus of the liquid defined as the ratio (T/S) of the shear stress T to the shear strain S, ρ_L- is the liquid density and $j = (-1)^{1/2}$. In general, liquid loading of the sensor surface changes the phase velocity v and the attenuation α of the B-G wave. The complex propagation constant γ of the B-G wave changes:

$$\frac{\Delta\gamma}{\beta} = \frac{\Delta\alpha}{\beta} - j\frac{\Delta v}{v_0} \tag{2}$$

where $\gamma = \alpha + j\beta$, $\beta = \omega/v$, v_0 is the phase velocity of the nonperturbed B-G wave on the free surface, and ω is the angular frequency of the B-G wave. One can prove that the change in the complex propagation constant γ of the B-G wave produced by viscoelastic liquid loading is as follows [14, 15]:

$$\Delta\gamma = -j\left(\frac{|v_3|^2_{x_2=0}}{4P}\right) Z_L = -jKZ_L \tag{3}$$

where v_3 is the B-G wave amplitude on the waveguide surface ($x_2 = 0$), P is the mean power on the unit width of the B-G wave, K is the characteristic quantity for each B-G wave waveguide and depends solely on the material parameters of the waveguide and frequency. Knowing the change in the complex propagation constant $\Delta\gamma$ from the experiment, we can calculate the complex shear impedance of the liquid $Z_L = R_L + jX_L$. Subsequently by separating the real and imaginary parts of Eq.(1), we obtain the following components of the complex shear modulus of the liquid:

$$G' = \frac{R_L^2 - X_L^2}{\rho_L} \tag{4}$$

$$G'' = \frac{2R_L X_L}{\rho_L} \tag{5}$$

In the next step, the rheological model describing the investigated liquid should be determined. Using the chosen model and Eqs.(4) and (5), we came to the formulas for shear elasticity μ and viscosity η of a viscoelastic liquid.

$$\mu = \frac{(R_L^2 + X_L^2)^2}{\rho_L\left(R_L^2 - x_L^2\right)} \tag{6}$$

$$\eta = \frac{(R_L^2 + X_L^2)^2}{2\rho_L R_L X_L \omega} \tag{7}$$

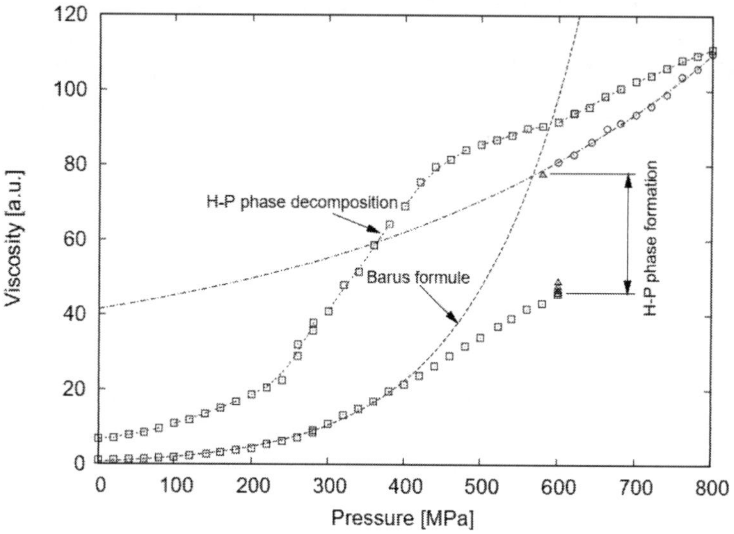

Figure 1. Pressure changes of castor oil viscosity measured by B-G ultrasonic waves method.

3. Experimental set up

The real advantage of the method presented here is that contrary to the other methods of the viscosity measurement it does not require any special high-pressure equipment. Any high pressure equipment which can contain the ultrasonic B-G transducer can be used. In our case it was a rectangular bar of PZT type zirconium ceramics covered at one side with $200\mu m$ layer of silver at one side with a piezoceramic transducer bonded at the end of ceramics. The size of it was $6\times10\times40mm$. The pressure was generated in a thick-walled cylinder of 17 mm internal diameter with a simple piston detaily presented in [16]. This piston-cylinder system was operated by a 20-tonne laboratory hydraulic press, what limited the maximum pressure to about 1 GPa. For pressure measurement, a typical 500Ω manganin transducer was used. Its resistance was measured with a precision HP 34970 multimeter. Pressure measurement accuracy was better than 0.1%. The experiments were done at room temperature. Temperature was measured by the copper - constantan thermocouple placed inside the chamber. For measuring viscosity using B-G wave [15], the sending - receiving piezoelectric transducer was driven by the TB-1000 pulser - receiver computer card. The B-G wave impulse generated by this transducer was reflected in multiple ways between two opposite edges of the piezoceramic waveguide. The signals received by the transducer were amplified by the pulser -receiver card and sent into PDA-500 digitizer card. This card sampled and digitized the input analog signals. The stored signals were then analyzed by computer software. For each measurement, the ultrasonic signal was averaged 1024 times in order to improve the signal-to-noise ratio. A computer program which controls the set-up operation was written in C++ language. Time of flight of the ultrasonic pulses was evaluated by applying the cross - correlation method.

4. Results of the measurement

The pressure was generated by the hand operated pump in 20 MPa steps then kept constant for about 2-5 minutes. During that time the pressure was carefully observed. That allowed to identify pressure drop due to the first order phase transition to observe whether the system is reaching thermodynamic equilibrium. After approaching 0.6 GPa the pressure was kept constant for about 20 hours to enable the phase transformation. During the phase transition the small drop of pressure and big increment of viscosity was observed. The results of the viscosity measurement are shown in figure 1. As one can notice the results up to about 400 MPa are almost tangential to the exponential curve which represents Barus formula $\eta = \eta_0 \exp(\alpha p)$. Above 400 MPa the experimental points are rising slower than the theoretical prediction. Finally at 600 MPa when the pressure rise was stopped for about 20 hours the viscosity has risen to the new value characteristic for the high pressure phase of castor oil . The further increment of viscosity was rather linear function of pressure.

5. Conclusions

1. Exponential rise of viscosity with pressure up to the phase transition at 0.6 GPa according to the Barus formula. 2. Extraordinary increment of viscosity at constant pressure during phase transition. 3. The decomposition of the high pressure phase during the decompression process have shown very large hysteresis of viscosity on pressure. 4. After the decompression process the viscosity lasts higher than the initial value for several hours.

6. Acknowledgements

This work was supported by the Ministry of Science and Information Society Technologies under Grant No. 3 T10C 006 28.

7. References

[1] Wiśniewski R, Rostocki A J, Rajski K and Bock W 1980 *High Pressure - Generation, Measurement and Application* A monograph (Warsaw: State Publishers for Science and Technology (WNT)) ISBN 83 204 0184 4 (in Polish)
[2] Rostocki A J 1975 High pressure viscometer up to 15 kbar. 1975 *Proc. of III Czechoslovak Conference on High Pressure Technique in Science and Technology* (Nove Mesto na Morave), 189-193
[3] Rostocki A J, Wiśniewski R, Rajski K and Birks A W 1976 *Materials of 14th Annual Meeting of EHPRG* (Dublin: 6 8 August 1976) 17
[4] Siegoczyński R M 1998 *Reports of the Institute of Physics* (Warsaw: Publishing House of the Warsaw Technical University of Technology) **46**
[5] Siegoczyński R M, Jędrzejewski J and Wiśniewski R, 1989 *High Pressure Res.* **1** 225
[6] Siegoczyński R M, Jędrzejewski J and Wiśniewski R, 1991 *High Pressure Res.* **8** 439
[7] Wiśniewski R, Jędrzejewski J, Siegoczyński R M and Tkacz A 1994 *High Pressure Res.* **11** 385
[8] Wiśniewski R, Jędrzejewski J and Siegoczyński R M 1994 *High Pressure Res.* **13** 41
[9] Wiśniewski R and Gałąź R 1993 *Patent PL No. 171272*
[10] Wiśniewski R and Siegoczyński R M 2003 *High Pressure Res.* **23** 211
[11] Wiśniewski R, Siegoczyński R M and Rostocki A J 2005 *High Pressure Res.* **25** 63
[12] Collings A F and McLaughlin E 1971 *Trans. Faraday Soc.* **67** 340
[13] Ruttle S G R and Stephenson M I 1975 *A high-pressure ultrasonic viscometer* Proc. Ultrasonic International Conference 224-227
[14] Kiełczyński P, Pajewski W and Szalewski M 2003 *IEEE Transactions on Ultrasonics Ferroelectrics and frequency control* **50** no 3 230-6
[15] Kiełczyński P, Pajewski W, Szalewski M and Balcerzak A 2004 *Rev.Sci. Instrum.* **75** 2363
[16] Rostocki A J, Wiśniewski R and Wilczyńska T 2007 *J.Mol.Liq.* **135** 120

A viscosity measurement during the high pressure phase transition in triolein

R M Siegoczyński[1] , A J Rostocki[1], P Kiełczyński[2] and M Szalewski[2]

1 Faculty of Physics, Warsaw University of Technology, Koszykowa 75, 00-662 Warszawa, PL

2 Institute of Fundamental Technological Research, Polish Academy of Science, Świetokrzyska 21, 00-049, PL

E-mail: siego@if.pw.edu.pl

Abstract. The high-pressure properties of triolein, a subject of extensive research at the Faculty of Physics of Warsaw University of Technology (WUT) have been enhanced by the results of viscosity measurement within the pressure range up to 0.8 GPa. For the measurement the authors have adopted a new ultrasonic method based on Bleustein-Gulyaev waves, successfully developed earlier for the low pressures in the Section of Acoustoelectronics of the Institute of Fundamental Technological Research. The measurements have shown: 1. Exponential rise of viscosity with pressure up to 0.5 GPa. 2. Extraordinary increment of viscosity at constant pressure during phase transition. 3. Further exponential rise of viscosity with pressure of the high-pressure phase of triolein. 4. The pressure exponents of the viscosity of both phases were different (the high-pressure phase had much smaller exponent). 5. The decomposition of the high pressure phase due to the slow decompression have shown very large hysteresis of viscosity on pressure dependence.

1. Introduction

High-pressure research of the physical properties of natural oils has been stimulated by the fast development of such technologies as biodiesel production and high-pressure food processing and food conservation. Most of the natural oils like rapeseed oil, soybean oil, sunflower oil etc. consist of triglycerides of various fatty acids. The high-pressure properties of such oils were a subject of several experiments performed in high-pressure laboratory of WUT [1]. Starting from the investigation of castor oil where the first order phase transition was discovered [2-5] the experiments were extended on the other natural oils leading to the discoveries of similar phase transitions in rapeseed oil [6] soybean oil [7] and sunflower oil [8] - all of them known as a raw materials for biodiesel production [12,13]. The statistical character of the molecular composition of those oils have made several difficulties for the interpretation of the phenomenon observed at high pressure. Further - more systematical experiments went into two directions: a. Investigation of well-defined triacylglycerol (triglyceride) structures with R_1=R_2=R_3 (see fig 1 and 2) at high pressure. b. Investigation of same high pressure properties of fatty acids which residuals appear in form of long chains as R_1, R_2,R_3 in natural oils. Table 1 shows approximate content of those fatty acids in vegetable oils. From the above table one can see that most of the chains R_1 R_2 and R_3 consist of Oleic acid (18:1), linoleic acid (18:2) and linolenic acid (18:3). As a first liquid of group "a" a triolein (all 3 chains made of oleic acid) was selected as a model structure.

© 2008 IOP Publishing Ltd

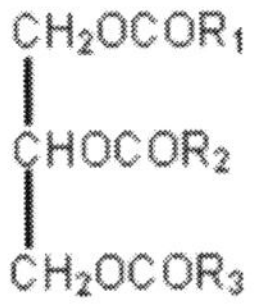

Figure 1. Triacylglycerol.

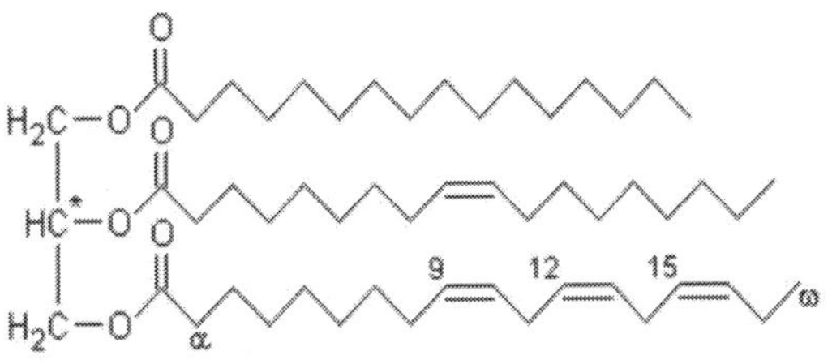

Figure 2. 1–linolenyl–2–palmitoyl–3–oleylglycerol.

Table 1. Composition of vegetable oils.

Fatty acids→ / *Vegetable oils↓*	14:0	16:0	16:1	18:0	18:1	18:2	18:3	20:0	others
castor oil	0	1.1	0	3.1	4.9	1.3	0	0	89.6
rapeseed oil	0	3.5	0	0.9	54.1	22.3	8.2	0	10.0
soybean oil	0	10.9	0	0.2	24.3	52.6	6.0	0	6.0
sunflower oil	0.08	6.8	0.1	4.0	30.7	57.0	0.1	0.3	0.1
linseed oil	0	5.0	0.3	2.5	18.9	18.1	55.1	0	0.1
triolein	0	0	0	0	100	0	0	0	0

2. High pressure properties of triolein

The first high-pressure studies of triolein has been focused on the optical properties of it showing similar phase transition [9] as in castor oil. The further experiments were done on the measurement of dielectric permittivity $\varepsilon(p)$ dependence [10] and also $\Delta V/V$=f(p) dependence [11]. All of them have shown discontinuous change of investigated parameters as a function of pressure, typical for the first order phase transition. Moreover the triolein became very good model-liquid for the whole group of triglycerides since it was found that the phase transition takes places within only 10-20 minutes after the application of sufficient pressure (about 500MPa). That allows for more detailed and accurate studies than similar phase transition in vegetable oils taking even up to 10 days period from the application of sufficient pressure to the start of the phase transition. Since for the industrial application, the reological properties of the discussed oils are one of the most important [12,13] (the whole transestrification process is performed for only one reason - to reduce the viscosity of fuel about 6-10 times), the measurements of triglycerides viscosity at high pressure condition became one of the most important targets of our research at Faculty of Physics WUT.

3. Viscosity measurements at high pressure

There are several methods of the viscosity measurement at high pressure conditions [14-33]. The most popular is one based on the Stokes method with falling ball or the other falling sinker[14-19]. This method was also used in our lab [15-16] for the viscosity measurement but its application is very limited to a few hundred MPa. Another group of viscometers based on the Hagen-Poiseuille's formula for capillary flow [20-25] was also used at WUT for the measurement of the castor oil viscosity [24, 25]. There are some constructions of Couette type high-pressure viscometers [26,

27], but the problem of the rotation transmission into high pressure chamber is limiting the pressure range of this type devices. The most promising for high-pressure application is a group of viscometers based on ultrasonic transducers of various types.

4. Application of Bleustein-Gulyaev method

Looking for the best method for observation of viscosity change during phase transition several methods, also ultrasonic have been considered. Among them the shear horizontal acoustic waves seemed to be optimal. Finally Bleustein-Gulyaev (B-G) method was selected. The Bleustein-Gulyaev (B-G) wave is a shear horizontal acoustic surface wave having only one component of mechanical displacement, parallel to the propagation surface and perpendicular to the direction of propagation. The viscoelastic liquid covering the waveguide surface loads it mechanically. The mechanical impedance of a layer of liquid loading the surface of the B-G wave waveguide is equal to the characteristic shear impedance of the liquid ZL for plane waves:

$$Z_L = \sqrt{(\rho_L \cdot G_L)} \tag{1}$$

where $G_L = G' + jG''$ is the complex shear modulus of the liquid. In general, liquid loading of the sensor surface changes the phase velocity v and the attenuation α of the B-G wave. Knowing the change in the complex propagation constant $\Delta\gamma$ from the experiment, we can calculate the complex shear impedance of the liquid $Z_L = R_L + jX_L$. Subsequently by separating the real and imaginary parts of Eq.(1), we obtain the formulas for shear elasticity μ and viscosity η of a viscoelastic liquid.

$$\mu = \frac{(R_L^2 + jX_L^2)^2}{\rho(R_L^2 - jX_L^2)^2} \quad (2) \qquad \eta = \frac{(R_L^2 + jX_L^2)^2}{2\rho_L R_L X_L \omega} \quad (3)$$

5. Experimental

For the pressure generation the standard thick walled cylinder of 17mm internal diameter with simple piston, described in [6] has been used. The piston cylinder assembly was working with 20 tonne hydraulic press, driven by hand operated pump. Inside the chamber the ultrasonic arrangement was mounted. The maximum pressure of this arrangement is limited to about 1.2 GPa due to the hydraulic press working range. For the pressure measurement, the typical 500Ω manganin transducer has been used. The resistance of it was measured by precise HP 34970 multimeter. An accuracy of the pressure measurement was better than ± 0.5 MPa. Inside the high pressure chamber there is the viscosity sensor made of lead zirconnate oxide (PZT) piezoceramics. The dimensions of the sensor were 6x10x40 mm. One long side of it was covered by 200μm layer of silver and there was a piezoceramic sending-receiving transducer bonded at the end of silver layer at one side. The transducer was driven by the TB-1000 pulser-receiver computer card. The B-G wave impulse generated by this transducer was reflected in multiple ways between two opposite edges of the piezoceramic waveguide. The signals received by the transducer were amplified by the pulser-receiver card and sent into the PDA-500 digitizer card. All experiments were carried out at the temperature 293 K. The temperature was controlled by copper-constantan thermocouple placed inside the chamber.

6. Results of the experiment

The pressure was generated by the hand operated pump in 20 MPa steps then kept constant for about 2 minutes. During that time the pressure was carefully observed, whether the system is reaching thermodynamic equilibrium. Up to about 500MPa the viscosity was increasing exponentialy according to known Barus formula $\eta(p) = \eta_0 exp(\alpha p)$. After approaching 0.7 GPa the compression was stopped to enable the phase transformation to go undisturbed. During the phase transition pressure drop of 100MPa was observed. Despite that pressure drop the viscosity have shown the further rise. After the termination of the phase transformation process,

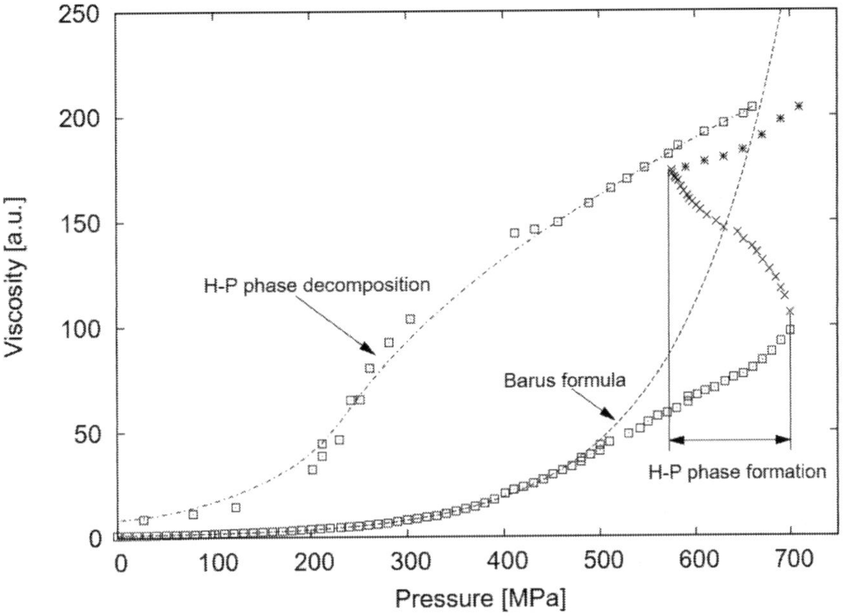

Figure 3. Pressure changes of triolein viscosity measured by B-G ultrasonic S-H waves method.

the further much slower increament of viscosity than during phase transition was observed. The changes of viscosity due to the decompression process (upper curve) have shown large hysteresis.

7. Conclusions

The measurements have shown: 1. Exponential rise of viscosity with pressure up to 0.5 GPa. 2. Extraordinary increament of viscosity at constant pressure during phase transition. 3. Further exponential rise of viscosity with pressure of the high pressure phase of triolein. 4. The pressure exponents of the viscosity of both phases were different (the high pressure phase had much smaller exponent). 5. The decomposition of the high pressure phase due to the slow decompresion have shown very large hysteresis of viscosity on a pressure dependence.

8. Acknowledgements

This work was supported by the Ministry of Science and Information Society Technologies under Grant No. 3 T10C 006 28.

9. References

[1] Siegoczyński R M. 1998, Reports of the Institute of Physics, Publishing House of the Warsaw University of Technology **46**
[2] Siegoczyński R M, Jędrzejewski J and Wiśniewski R 1989 *High Pressure Res* **1** pp. 225-301
[3] Siegoczyński R M, Jędrzejewski J and Wiśniewski R 1991 *High Pressure Res* **8** pp. 439-441
[4] Wiśniewski R, Jędrzejewski J, Siegoczyński R M, and. Tkacz A 1994 *High Pressure Res* **11** pp. 385-391
[5] Wiśniewski R, Jędrzejewski J and Siegoczyński R M, 1994 *High Pressure Res* **13** pp. 41-45
[6] Rostocki A J, Wiśniewski R, and Wilczyńska T 2007 *J.Mol.Liq* **135** pp. 120-122

[7] Rostocki A J, Kościesza R, Tefelski D B, Kos A, Siegoczyński R M and Chruściński Ł2007 *High Pressure Res* **27** pp. 43-46
[8] Rostocki A J, Siegoczyński R M, Galas J, Kościesza R and Tefelski D B, paper presented on 5th ISCEPP Conference, Segovia 24-27 June 2007
[9] Siegoczyński R M, Wiśniewski R, and Ejchart W ed. W. Trzeciakowski 1996 *Proc. Int. Conf. on High Pressure Science & Technology, World Scientific* pp. 834-836
[10] Kos A, Tefelski D B, Kościesza R, Rostocki A J, Roszkiewicz A, Ejchart W, Jastrzębski C and Siegoczyński R M, 2007 *High Pressure Res* **27** pp. 39-42
[11] Tefelski D B, Rostocki A J, Kos A, Kościesza R and Siegoczyński R M 2007 *High Pressure Res* **27** pp. 57-61
[12] Srivastava A and Prasad R 2000 *Renewable & Sustainable Energy Reviews* **4** pp. 111-133
[13] Demirbas A 2005 *Progress in Energy and Combustion Science* **31** pp. 466- 487
[14] Irwing J B and Barlow A J 1971 *An automatic high pressure viscometer J. Phys. E* **4** pp. 232-236
[15] Rostocki A J 1975 High pressure viscometer up to 15 kbar *Proc. of III Czechoslovak Conference on High Pressure Technique in Science and Technology Nove Mesto na Morave* pp. 189-193
[16] Rostocki A J, Wiśniewski R, Rajski K and Birks A W 1976 *Proc. Int. Conf. on XIV Annual Meeting of EHPRG* p 17
[17] Vergne Ph 1990 *High Temperatures - High Pressures* **22** pp.613-621
[18] Bair S A 2004 Tribology Transactions **47** pp. 356-360
[19] Schaschke C J, Abid S, Heslop M J 2007 *High Pressure Res* **27** pp. 33-37
[20] Novak J O and Winer W O 1968 *Rheology Trans. ASME* **90** p 580
[21] Wiśniewski R and Rostocki A J 1978 Industrial Laboratory of Primary Pressure Standards, *Proc.International Symposium on Design and Technology of Extreme Parameters* Bratislava pp. 211-217
[22] Wiśniewski R and Gałąź R Patent PL No. 171272 on 1993.07.08
[23] Couch M A, Binding D M 2000 *Polymer* **41** pp. 6323-6334
[24] Wiśniewski R and Siegoczyński R M 2003 *High Pressure Res.* **23** pp. 211-215
[25] Wiśniewski R, Siegoczyński R M and Rostocki A J 2005 *High Pressure Res* **25** pp. 63-70
[26] Bair S and Winer W O 1993 *Tribology Transactions* **36** pp. 721-725
[27] Jimenez J A and Kostic M 1994 *Rev.Sci. Instrum.* **65** pp. 229-241
[28] Mason W P, Baker W O, McSkinin H J and Heiss J H 1949 *Phys.Rev.* **75** 936
[29] Collings A F and McLaughlin E 1971 *Trans.Faraday Soc* **67** pp. 340-352
[30] Ruttle S G R, Stephenson M I, A high-pressure ultrasonic viscometer 1975 *Proc.Ultrasonic International Conference* pp. 224-227
[31] Alig I, Lellinger D, Sulimma J, Tadjbakhsch 1997 *Rev.Sci. Instrum.* **68** pp. 1536-1542
[32] Kiełczyński P, Pajewski W and Szalewski M 2003 *IEEE Transactions on Ultrasonics Ferroelectrics and frequency control* **50** pp. 230-236
[33] Kiełczyński P, Pajewski W, Szalewski M and Balcerzak A 2004 *Rev.Sci. Instrum* **75** pp. 2363-2367

The effect of high hydrostatic pressure on black truffle (*Tuber melanosporum*) flavour compounds

C Verret[1*], P Ballestra[1], C Cruz[1], P Pardon[2], A Largeteau and A H El Moueffak[1]
Joint 21st AIRAPT and 45th EHPRG Int. Conference - Abstract N_0. 0358
[1] Equipe de Recherche Agroalimentaire Périgourdine (ERAP) IUT de Périgueux Université Bordeaux IV site universitaire F24019 Périgueux Cedex France
[2] Laboratoire d'analyses de l'Institut du pin Université Bordeaux I France
[3] Groupe Haute Pression ICMCB – ENSCPB Université Bordeaux I France

* E-mail : moueffak@u-bordeaux4.fr

Abstract. The effects of high hydrostatic pressure (HHP), at 4°C or -18°C, on black truffle flavour compounds, alteration enzymes (lipoxigenase (LOX), peroxidase (POD) and polyphenoloxidase (PPO)) and microbiological qualities were evaluated. The choosen analytes for this study are six alcohols, three aldehydes, one ketone and on sulfur component. The highest flavour stability was observed when samples were pressurized at 300 MPa / 4°C / 10 min. All the treatments induced a drastic decrease of LOX activity and a slight decrease of POD activity. On the other hand, the PPO was not inactivated by pressurization at sub-zero (200 MPa / -18°C / 10 min) and was strongly increased after the 300 MPa / 4°C / 10 min treatment. Pressurization at 300 and 550 MPa lead to an almost complete *Pseudomonas fluorescens* reduction (6 and 6.5 log destruction, respectively) whereas pressurization at -18°C (200MPa) allowed to obtain only 3 log reduction.

Keywords : black truffle, volatile aromatic compounds, microbiological and enzymatic properties, high pressure

1. Introduction

Fresh black truffles have well recognized organoleptic qualities and can be processed through two ways : fresh or sterilized. Truffles are highly fragile and thermal processing leads to a certain exudation detrimental to the yield and the organoleptic properties. Food high pressure treatment can inactivate vegetative micro-organisms and enzymes. Contrary to thermal treatment, high pressure processing maintains the organoleptic quality of foods [1] [2].

The aim of this study was to evaluate the changes in the organoleptic and microbiological qualities in black truffles after pressure processing. The studied parameters are the volatile compounds, the alteration enzymes and bacterial charge.

2. Materials and methods

The black fresh truffles were harvested in Dordogne (France).

The samples, placed in a plastic pouch, were hydrostatically pressurized in a 5 L sample compartment by means of a hyperbar experimental apparatus Framatome (France). HHP treatments (200 MPa / - 18°C / 10 min, 300 and 550 MPa / 4°C / 10 min) were carried out on vacuum packaged black truffles (20 g). The untreated and treated samples were stored at 4°C until further experiments.

The volatile component extraction was obtained by headspace solid-phase microextraction (HS-SPME). The analysis carried out by gas chromatography-mass spectrometry (GC-MS) allowed the identification of 20 volatile aromatic compounds in black truffle [3] [4]. The LOX, POD and PPO were done following the methods of Ben-Aziz [5], Crelier [6] and Gomes and Ledward [7].

© 2008 IOP Publishing Ltd

Microbiological analysis focused on one of the main bacterias of black truffles : *Pseudomonas spp.* [8].

3. Results and discussion

3.1. Effects of high pressure on volatile compounds

The choosen analytes for our studies were six alcohols (ethanol, 2-methyl propanol, 2-methyl butanol, 3-methyl butanol, 1-hexanol and 1-octen-3-ol), three aldehydes (hexanal, 2-methyl butanal and benzaldehyde), one ketone (2-butanone) and one sulfur component (dimethylsulfure) [3]. The concentration of most volatile components decreased after HHP treaments (loss of ~ 50%), in particular, for 550 MPa / 4°C / 10 min and 200 MPa / - 18°C / 10 min (figure1). On the other hand, the concentration of 1-octen-3-ol, hexanal, 2-methyl butanal and benzaldehyde considerable increased with all HHP treatments (table 1). These components can not be synthetized during the short time of treatment. The increase of their volatile compounds could be explained by an improved detection after the HHP treatment. The highest flavour stability was observed when samples were treated with the 300 MPa / 4°C / 10 min combination [9].

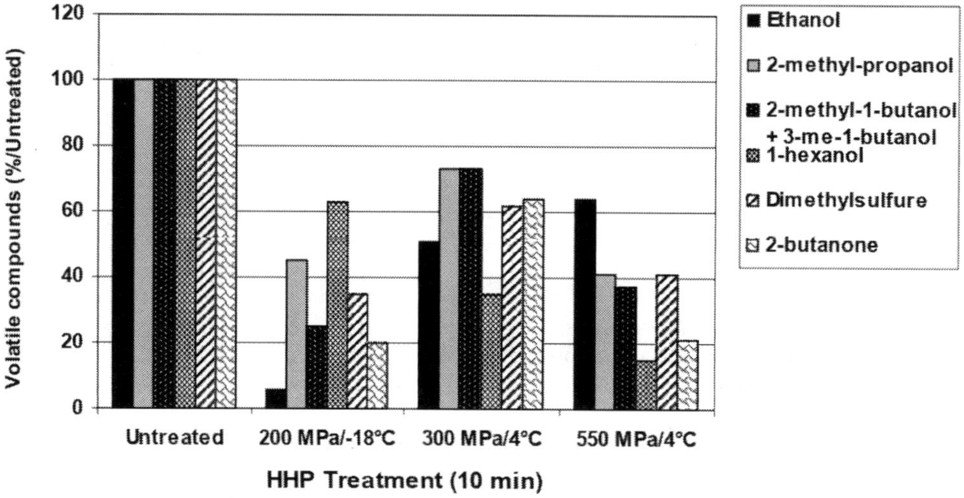

Figure 1. The effect of HHP treatments on the majority volatile components of black truffle.

Table 1. The effect of HHP treatments on the other volatile components of black truffle.

	Area of peaks			
	Untreated	200 MPa/-18°C/10 min	300 MPa/4°C/10 min	550 MPa/4°C/10 min
1-octen-3-ol	$3.84\ 10^6$	$9.27\ 10^7$	$3.89\ 10^8$	$4.86\ 10^8$
hexanal	$4.95\ 10^6$	$5.05\ 10^6$	$1.50\ 10^7$	$1.16\ 10^7$
Benzaldéhyde	$3.68\ 10^7$	$1.23\ 10^8$	$9.87\ 10^7$	$5.02\ 10^7$
2-methyl-butanal	$1.35\ 10^7$	$3.32\ 10^8$	$7.82\ 10^7$	$1.11\ 10^8$

3.2. Effects of high pressure on enzymatic activities

All the treaments induced a drastic decrease of LOX activity (figure 2), a reduction of 60% was observed for 200 MPa / - 18°C / 10 min treatment and this activity was not detected after 300 and 550 MPa /4°C / 10 min treatments. A slight decrease of POD activity was observed for all HHP treaments (10 at 20% of reduction) whereas the PPO activity was stable after 200 MPa / - 18°C / 10 min treatment and this activity was increased by the other HHP treatments. These results are in accordance with previous works [8].

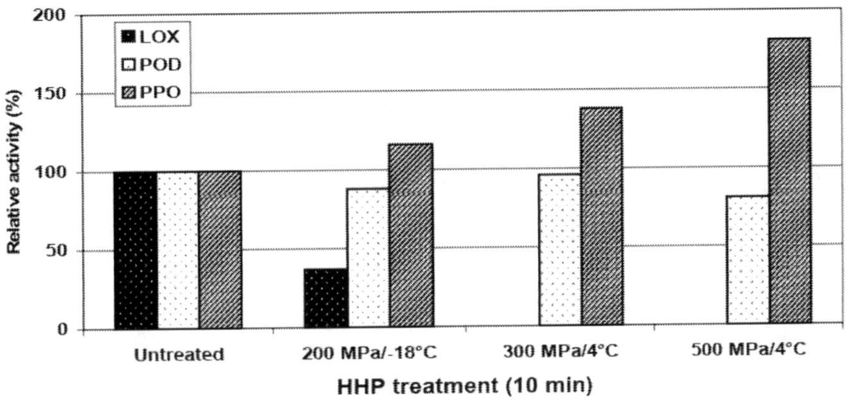

Figure 2. The effect of HHP treatments on the enzymatic activities of black truffle.

3.3. Effects of high pressure on Pseudomonas

An almost complete reduction of *Pseudomonas spp.* was produced by 300 and 550 MPa / 4°C / 10 min treaments (6 – 6.5 log of destruction). Pressurization at sub-zero temperature was induced only 3 log reduction of *Pseudomonas spp.* (table 2).

The HHP treatment at 300 MPa / 4°C / 10 min) was shown to be the most interessing process to stabilize black truffle. These results should be confirmed by a preservation study.

Table 2. effect of HHP treatments on the *Pseudomonas* charge of black truffle.

	Untreated	200 MPa/-18°C/10min	300 MPa/4°C/10min	550 MPa/4°C/10min
log CFU / g	7.58	4.26	1.60	1.10

Achnowledgements
This research was supported by the FEDER program, the "Conseil Général de la Dordogne" and the "Communauté d'Agglomération Périgourdine".

References

[1] Knorr D 1993 *Food Technol.*, **47**, 156.
[2] El Moueffak A H *et al* 1996 In Hayashi, R. And Balny, C. (Ed.), *HP Biosci. Biotechn.*, 401.
[3] Pelusio F *et al* 1995 *J. Agric. Food Chem.*, **43**, 2138.
[4] Diaz P *et al* 2003 *J. Chromatography A*, **1017**, 207.
[5] Ben-Aziz A *et al* 1970 *Anal. Biochem.*, **34**, 88.
[6] Crelier S *et al* 1999 In Ludwig, H. (Ed.), *Advances HP Biosci. Biotechn.*, **413**, Springer-Verlag, Berlin.
[7] Gomes M R A and Ledward D A 1996 *Food Chem.*, **56**, 1.
[8] Ballestra P *et al* 2003 *HP Sci. Techn. AIRAPT 19th – 41th EHPRG*, Bordeaux.

Temperature control for high pressure processes up to 1400 MPa

Kai Reineke[1], Alexander Mathys[1], Volker Heinz[2], Dietrich Knorr[1]

[1] Berlin University of Technology, Department of Food Biotechnology and Food Process Engineering, Koenigin-Luise-Str. 22, D-14195 Berlin, Germany, Tel: +49 30 314-71248, Fax: +49 30 832 76 63
[2] German Institute of Food Technology, p.o.box: 1165, D-49601, Quackenbrueck, Germany, Tel: +49 5431 183 232, Fax: +49 5431-183114

E-mail for correspondence: alexander.mathys@tu-berlin.de

Abstract.

Pressure- assisted sterilisation is an emerging technology. Hydrostatic high pressure can reduce the thermal load of the product and this allows quality retention in food products. To guarantee the safety of the sterilisation process it is necessary to investigate inactivation kinetics especially of bacterial spores. A significant roll during the inactivation of microorganisms under high pressure has the thermodynamic effect of the adiabatic heating. To analyse the individual effect of pressure and temperature on microorganism inactivation an exact temperature control of the sample to reach ideal adiabatic conditions and isothermal dwell times is necessary. Hence a heating/cooling block for a high pressure unit (Stansted Mini-Food-lab; high pressure capillary with 300 µL sample volume) was constructed. Without temperature control the sample would be cooled down during pressure built up, because of the non-adiabatic heating of the steel made vessel. The heating/cooling block allows an ideal adiabatic heat up and cooling of the pressure vessel during compression and decompression. The high pressure unit has a pressure build-up rate up to 250 MPa s^{-1} and a maximum pressure of 1400 MPa. Sebacate acid was chosen as pressure transmitting medium because it had no phase shift over the investigate pressure and temperature range. To eliminate the temperature difference between sample and vessel during compression and decompression phase, the mathematical model of the adiabatic heating/cooling of water and sebacate acid was implemented into a computational routine, written in TestPoint. The calculated temperature is the setpoint of the PID controller for the heating/cooling block. This software allows an online measurement of the pressure and temperature in the vessel and the temperature at the outer wall of the vessel. The accurate temperature control, including the model of the adiabatic heating opens up the possibility to realise an ideal adiabatic heating and cooling as well as isotherm dwell time of the sample under high pressure. This research offers a useful tool to investigate the additive effect of high pressure and thermal treatment on the inactivation of microorganisms.

Keywords: temperature control; sterilisation; adiabatic heating; inactivation

1. Introduction

The food industry is interested in improving the quality of products and protecting their ingredients. To reach this aim and to guarantee the food safety isostatic high pressure is an up coming technology. At pressures higher than 400 MPa a significant inactivation of vegetative bacteria, yeast or virus is observed even at ambient temperatures within treatment times of several minutes [1]. This effect is

© 2008 IOP Publishing Ltd

already used for the industrial pasteurisation of food. A combination of isostatic high pressure and thermal treatment can reduce the thermal load of the product and this allows quality retention in food products. To guarantee the safety of the sterilisation process it is necessary to investigate inactivation kinetics especially of bacterial spores. A significant roll during the inactivation of microorganisms under high pressure has the thermodynamic effect of the adiabatic heating.

The adiabatic heat of compression which occurs during the pressure build-up can maximize the benefits of high pressure technology.

All compressible materials change temperature during physical compression, depending on their compressibility and specific heat [2]. Aqueous systems show a volume contraction of approx. 20% when pressurized to 800 MPa [3]. The specific work of compression can be expressed as

$$W_{compr} = -\int V \cdot dp \qquad \text{Eqn.1}$$

In water, the specific work of compression W_{compr} is not higher than 55 kJ in the case when 1 kg of water is compressed up to 1400 MPa.

This value has been obtained upon integration of equation (1) for isothermal situations using the functional relation V=f(p) for pure water given by NIST formulation [4].

Following the first and the second laws of thermodynamic and by re-arrangement of the Maxwell equations [5], the heating during the compression and the cooling during the decompression can be described as a function of thermo-physical properties of the compressible product. The temperature increase upon compression in adiabatic- isentropic- situations is obtained:

$$\frac{\partial T}{\partial p} = \frac{\beta}{\rho \cdot c_p} \cdot T \qquad \text{Eqn.2}$$

where compressibility β, density ρ and the specific heat c_p are pressure-temperature dependent [4]. By using equation 2 the calculation of the thermal profile during the compression phase is possible. The differences in the adiabatic heating between some pressure transmitting media are shown in figure 1.

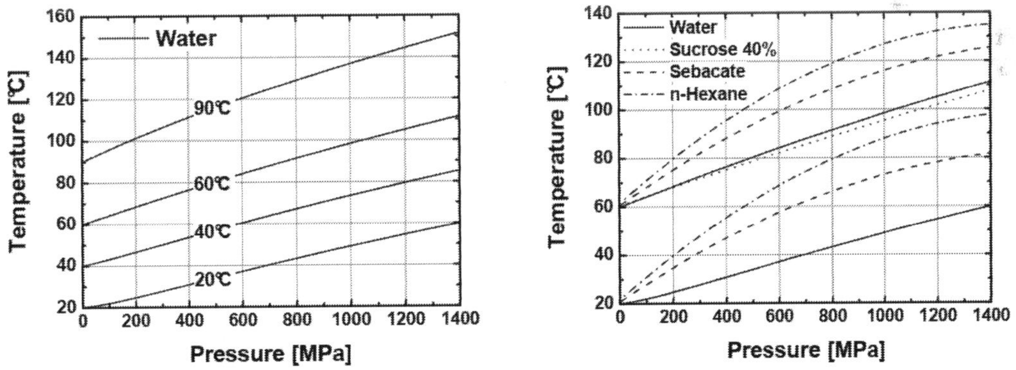

Figure 1. Adiabatic heat of compression in water, sucrose solution with 40% solid content, sebacate and n-hexane [3].

The adiabatic heating of the shown fluid food system sucrose are obtained with the help of equations for estimating physical properties of mixture of pure substances. The main ingredient in most food is water and thus the thermodynamic properties of water can be utilized to estimate the temperature increase upon compression of high moisture food [4].

2. Material and Methods

The adiabatic heating of water based on equation 2 was calculated with the help of a computational routine (MathCAD 2001i Professional, MathSoft Engineering & Education, Inc., USA). The equation of adiabatic heating of Di-2-ethyl-hxyl sebacate (Nr. 84822; Fluka; Steinheim) is based on experimental data [4]. Di-2-ethyl-hexyl sebacate was chosen as the pressure transmitting media because it has no phase shift from liquid to solid in the researched pressure and temperature range of 0.1 to 1400 MPa and 20°C to 140°C. In Figure 2 the used Mini Foodlab FPG15620 (Stansted Fluid Power Ltd) high pressure unit is shown.

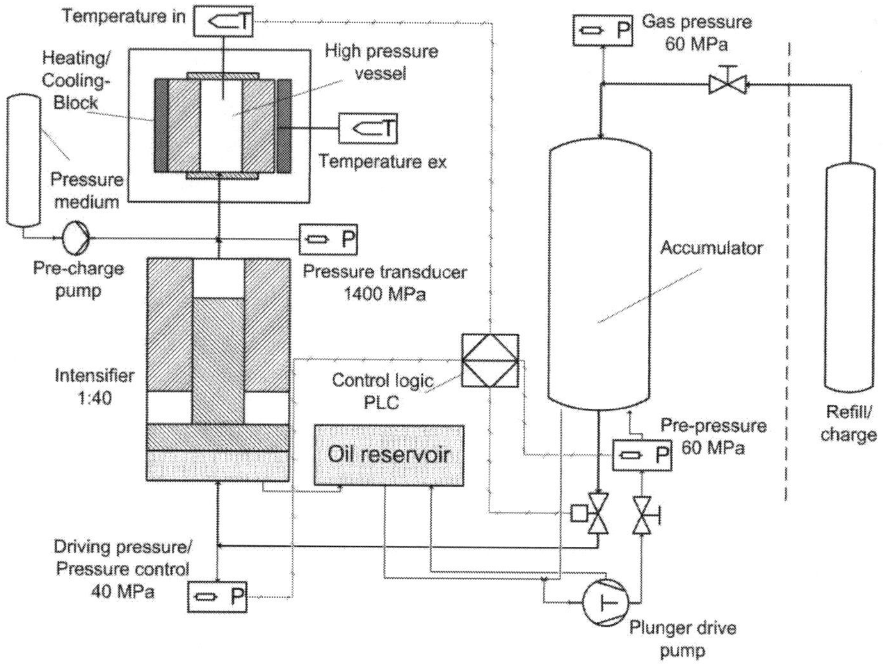

Figure 2. Mini Foodlab FPG15620 with heating-cooling block.

The HP unit has a pressure build-up rate up to 250 MPa s^{-1} and two pressure cells (300 μL, 10 mL). The heating-cooling block for the 300 μL cell, manufactured in copper, allows quick temperature changes of the high pressure vessel. A computational routine written in TestPoint (Capital Equipment Corporation, Massachusetts, USA) for the temperature control, realized adiabatic pressure build-up phases and isothermal dwell times.

3. Results and Discussion

To eliminate the temperature difference between sample and vessel during compression and decompression phase, the mathematical model of the adiabatic heating/cooling of water and sebacate acid was implemented into a computational routine, written in TestPoint. The calculated temperature of water is the setpoint of the PID controller for the heating/cooling block. Furthermore it is possible to heat up or cool down the vessel to an arbitrary temperature.
This software allows an online measurement of the pressure and temperature in the vessel and the temperature at the outer wall of the vessel. The acquired temperature and pressure data are shown in user interface of the program furthermore it is possible to save and export this data.

4. Conclusions

The accurate temperature control, including the model of the adiabatic heating opens up the possibility to realise an ideal adiabatic heating and cooling as well as an isotherm dwell time of the sample under hydrostatic high pressure. This research offers a useful tool to investigate the additive effect of hydrostatic high pressure and thermal treatment on the inactivation of microorganisms, denaturation of proteins or enzymes and other reaction studies.

References

[1] Hendrickx, M. and D. Knorr, *Ultra High Pressure Treatment of Foods*. Food Engineering Series, ed. G.V. Barbosa-Canovas. 2002, New York: Kluwer Academic/ Plenum Publisher.
[2] Ting, E., et al., *Determining thermal effects in high pressure processing*. Journal of Food Technology, 2002. **56**(2): p. 31-35.
[3] (NIST)., N.I.o.S.a.T., *NIST Standard Reference Database 10; Steam Properties Database Version 2.21.*
[4] Ardia, A., *Process Considerations on the Application of High Pressure Treatment at Elevated Temperature Levels for Food Preservation*, in *Department of Food Biotechnology and Food Process Engineering*. 2004c, Berlin University of Technology: Berlin. p. 94.
[5] Perry, R.H., *Perry's Chemical Engineers'Handbook*. Vol. 6. 1984: McGraw- Hill Book Co.

Journal of Physics: Conference Series **121** (2008) 152001 doi:10.1088/1742-6596/121/5/152001

The Influence of Hydraulic Fluids on Pressure Balance Calibration Results

Claudiu L Giusca, Ian Severn, John Greenwood

National Physical Laboratory, Hampton Road, Teddington, UK, TW11 0LW

claudiu.giusca@npl.co.uk

Abstract. The effective area of a hydraulic pressure balance is typically characterised using the cross-floating technique, where the pressure balance is compared against a standard instrument. This method usually involves operating both instruments in the pressure transmitting fluid of the standard, even if the tested device is normally operated with a different fluid. We describe novel experiments to determine the influence of various commonly used oils on the measured effective area of the balances. Measurements were made using five different oils in the pressure range 28 MPa to 200 MPa. Our results show that in some cases the contribution to the effective area when using different oils was nearly 60 ppm. Although a difference in the effective area with respect to the liquid viscosity was intuitively expected, the results appear to show a greater dependency on the liquid density, which contradicts the current assumptions used to formulate the theory describing the pressure balance.

1. Introduction

One of the methods by which pressure can be measured is in terms of the force that acts on a surface of known area. The instrument using this approach is known as the pressure balance [1], in which the downward force, mainly due to known masses supported on the cylindrical piston, balances the pressure acting on the other end of the piston [2]. Pressure balances are widely used in the range 3.5 kPa to 500 MPa.

A formal theory of the pressure balance, based on the laws of fluid dynamics, was developed in the mid 1960s [3]. This theory has proven to be extremely successful, however, in recent years, with the introduction of modern manufacturing techniques and improved tolerances, a certain amount of anomalous behaviour has been observed [4],[5],[6],[7]. The most likely explanation is that the assumptions used to formulate the theoretical model for the pressure balance are no longer adequate.

Although the basic theory does not predict a change of the piston-cylinder's effective area when using different liquids as pressure transmitting media, a difference related to the liquid viscosity, due to its contribution to the pressure distribution in the piston-cylinder gap, would be expected. As indicated by earlier work [1], if any such changes do occur, they will influence the effective area of the piston-cylinder at high pressures, thus having an effect on the 'distortion coefficient'.

In general, working calibration laboratories use Sebacate oil (di-2-ethylhexylsebacate) in hydraulically operated devices, but laboratories also operate and calibrate instruments using other oils. The common approach however, is to calibrate the hydraulic pressure balance (HPB) piston-cylinder assembly (PCA) in the same fluid as is used for the standard instrument.

© 2008 IOP Publishing Ltd

Journal of Physics: Conference Series **121** (2008) 152001 doi:10.1088/1742-6596/121/5/152001

This paper describes some of the work performed within the scope of National Measurement System Directorate Engineering Measurement Programme, Project 2.19: *Improved Realisation of Conventional High Pressure Scale* which was initiated as a response to growing concerns about the use of different oils in pressure balances [8].

2. Experimental details

Following a survey to discover what types of lubricating liquids are used as pressure transmitting fluids by different laboratories, four oils were selected for evaluation.

The oils used in this work, together with some of the relevant characteristics (density and dynamic viscosity), are listed in Table 1.

Table 1 Oils used in this work

Fluid	Fluid density (kg m^{-3}) at 20 °C and 1 013.25 Pa	Viscosity (mPa s)
DH-Budenberg - HP Fluid	989	131
Hydraulic Mineral Oil (Shell Tellus T15) - ISO VG15	862	26
Hydraulic Mineral Oil (Shell Tellus T37) - ISO VG37	870	76
Univis P12 (early NPL oil) (di-2-ethylhexylsebacate supplied by Esso)	921	22.5
CP Hall Sebacate (current NPL oil) (di-2-ethylhexylsebacate supplied by CP Hall)	915	21.5

The experiment involved calibrating several test instruments using the oils detailed in Table 1 as the pressure-transmitting medium, against a standard instrument, which used CP Hall Sebacate.

The setup consists of two independent pressure systems and a High Pressure Sensor (HPS) with parts-per-million (ppm) resolution, linked together via two constant-volume valves (V1 and V2) as shown in Figure 1, in which the two valves kept the oils separated during tests.

Valves V3 and V4, the ram screw and oil reservoir at the left of valve V1 were used to create pressure in the *'Test'* system isolated by valve V1. The valve V7 was used to drain the oil in the *'Test'* system.

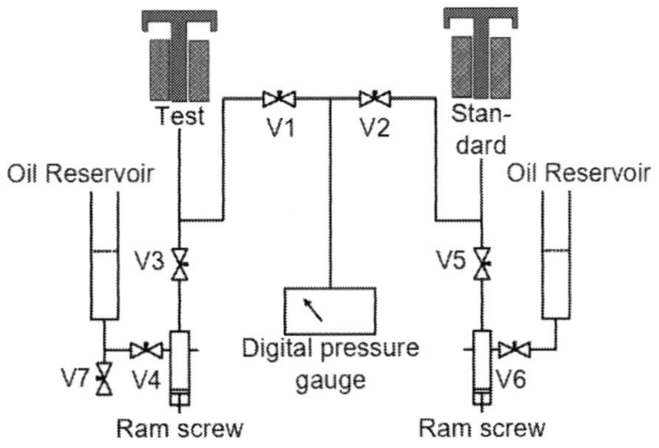

Figure 1 Experimental setup (V1 to V7 are isolating valves)

Valves V5 and V6, the ram screw and oil reservoir at the right of valve V2 were used to create pressure in the *'Standard'* system. The HPS and the space between V1 and V2 were filled with CP Hall Sebacate and were initially pressurised with V2 open and V1 closed.

The experiment used nine PCAs (Ruska 2481) in three nominal sizes of effective area (84 mm^2, 16.8 mm^2, 8.4 mm^2), covering the hydraulic pressure range 2.8 MPa to 280 MPa.

For each nominal size (family), one PCA was selected as *'Standard'* instrument and two assemblies as *'Test'* instruments.

For each family of PCAs, one of the *'Test'* assemblies was calibrated in re-entrant mode, and the other *'Test'* assembly was calibrated in free-deformation mode. The *Standard* assembly was operated in re-entrant mode. A detailed account of PCA selection is given in the NPL report [8].

All piston cylinder assemblies under test were calibrated initially in CP Hall Sebacate and subsequently in the oils introduced in Table 1. For reconfirmation, a final calibration was performed using CP Hall Sebacate. The initial calibration results using CP Hall Sebacate in both PCAs were used as reference values. Six calibrations were therefore performed for each PCA at the same four nominal pressures.

3. Results

The effective area of a '*Test*' pressure balance is usually determined by cross-floating the balance against a similar '*Standard*' pressure balance.

To avoid the cancelling-out effects of the traditional cross-floating technique, an alternative technique was used in which the measurements were made using two separate oils simultaneously [8]..

Initially, the effective area of the *'Test'* pressure balance was determined by comparing against a *'Standard'* pressure balance using the same oil (CP Hall Sebacate). After the initial calibration, the oil used by the *'Test'* pressure balance was changed and the new effective area was determined.

Thus the differences in effective area when using different oils, ΔA_T, will be given by:

$$\Delta A_T = p_{St}^{-1} \times \Delta F_T \qquad (1)$$

where p_{St} is the pressure generated by *'Standard'* and ΔF_T is the difference in force applied on tested piston due to the difference in effective area of the *'Test'* pressure balance when changing the oil.

Some results of the calibrations performed in different oils are plotted in the graphs presented in Figure 2 a) and Figure 2 b). The graphs show the change in effective area with respect to the initial CP Hall Sebacate calibration, together with the typical uncertainties [8] of the 8.4 mm^2 family, for the free-deformation and re-entrant assembly respectively.

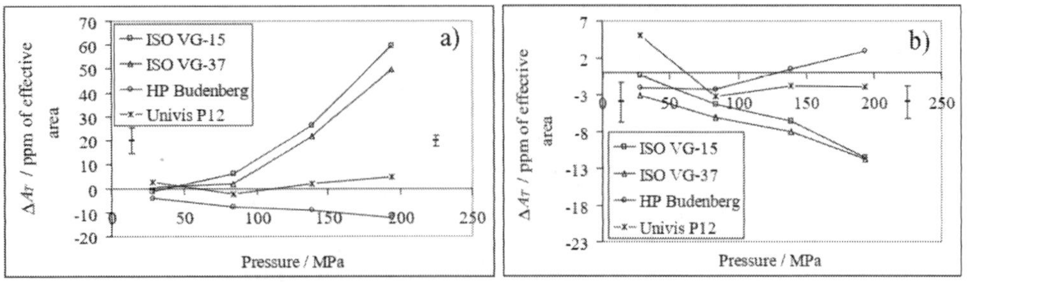

Figure 2 a) Difference in 8.4 mm^2 free-deformation assembly effective area w.r.t. CP Hall Sebacate; b) Difference in 8.4 mm^2 re-entrant assembly effective area w.r.t. CP Hall Sebacate. Typical error bars shown at $k = 1$.

These results show the largest change in effective area for both ISO VG15 and ISO VG37 oil. For the free-deformation assembly the change seems to be positive for oils with lower densities than CP Hall Sebacate and negative for oils with bigger densities than CP Hall Sebacate, therefore change ΔA_T is anti-correlated with oil density. However, for the re-entrant assembly the effect is positively correlated. The effect is consistent over most of pressure range, and at highest pressures (138.6 MPa and 193.8 MPa) the differences are significant when compared with the uncertainties.

It is predicted, on the basis of the current results, that the calibration of an instrument similar to the 8.4 mm^2 re-entrant assembly, using CP Hall Sebacate when it is designed to use ISO VG37, will introduce a possible relative systematic error of 60 ppm in effective area at 200 MPa. In the work reported here the relative uncertainty would be ± 68 ppm ($k = 2$) of effective area. It follows that, as calibration providers, we should either calibrate instruments using the prescribed operating oil or take extra account of the different oil when evaluating the uncertainties.

4. Conclusions

Significant systematic errors are found in the effective area when using different fluids as pressure transmitting media in a pressure balance. The change in effective area relative to CP Hall Sebacate is dependent on the pressure, and therefore on gap size, although at this time we cannot make any quantitative predictions relating the size of the effect to the initial gap. Although a difference related to the liquid viscosity (due to its contribution to the pressure distribution in the piston-cylinder inner space) was intuitively expected [1], the experiment instead appeared to show an effective area dependency on the liquid density.

Existing theories do not provide a mechanism for a density dependent effect. Efforts to validate any new theoretical model (that attempts to describe the behaviour seen in this work) are likely to require a selection of oils having a wider range of density and viscosity. Additionally, the experimental work will need to be extended to assemblies having better metrological characteristics than those used here (e.g. better repeatability and higher pressures ranges) and a wider range of PCA gap sizes.

Using piston cylinder assemblies with smaller gaps reduces the magnitude of the effect. When cross-floating two pressure balances with the *same* fluid many of the systematic effects, such as a change in effective area, are largely cancelled out - it is possible that larger effects would be seen if different oils were used to calibrate high precision electronic instruments.

References

[1] Dadson R S, et al. 1982 *The pressure balance-Theory and practice* HMSO London ISBN 0-11-480048-0
[2] Peggs G N, et al. 1983 *High pressure measurement techniques* Applied Science Publishers Ltd London & New York ISBN 0-85334-189-3
[3] Dadson R S, et al. 1965 *Metrologia* **1**, 55-67
[4] Sokhan V P, et al. 2005 *Molecular Simulation* **31** 535 – 542
[5] Welch B E. et al. 1989 *BIPM Monographie* **1** 81-94
[6] Legras J C, et al. 1993/94 *Metrologia* **30** 721-725
[7] Fitzgerald M P and Sutton C M, 2005 *Metrologia* **42** S239-S241
[8] Giusca C L, Severn I, Greenwood J 2007 NPL Report DEPC-EM 009 - *The effect of hydraulic fluids on pressure balances* HMSO ISSN 1744-0254

Joint 21st AIRAPT and 45th EHPRG Int. Conf. on High Pressure Science and Technology IOP Publishing
Journal of Physics: Conference Series **121** (2008) 162001 doi:10.1088/1742-6596/121/6/162001

Hydrostatic pressure effect on the optical spectra of glass-embedded $CdS_{1-x}Se_x$ nanocrystals

V S Shusta[1], A G Slivka[1], O O Gomonnai[1], Yu M Azhniuk[2], V V Lopushansky[2]

[1] Uzhhorod National University, Pidhirna Str. 46, Uzhhorod 88000, Ukraine

[2] Institute of Electron Physics, Ukr. Nat. Acad. Sci., Universytetska Str. 21, Uzhhorod 88017, Ukraine

e-mail: yu.azhniuk@gmail.com

Abstract. Effect of hydrostatic pressure on the optical absorption spectrra of $CdS_{1-x}Se_x$ nanocrystals embedded in borosilicate glass matrix, is studied. A pressure-induced blue shift of the optical absorption edge and confinement-related maxima is observed. Energy band gap pressure variation coefficients are determined, their compositional and size behaviour being discussed in view of possible structural difference within the nanocrystal ensemble.

1. Introduction

Size-dependent features in electronic structure of cadmium chalcogenide nanocrystals (quantum dots), revealed in their optical spectra, have made them an object of extensive scientific interest due to specific physical effects as well as to broad applications [1–3]. Size variation of CdSe nanocrystals (NCs) enables, for example, their luminescence to be tuned within practically the whole visible spectral range [2, 3]. Mixed $CdS_{1-x}Se_x$ NCs, grown in most cases by solid-state precipitation in glass matrices [4–6], offer additional oportunities for tuning due to variation of both the NC size and chemical composition.

Thorough studies of pressure effect on the optical properties were carried out for CdSe [7–12] and CdS [10, 13, 14] NCs, obtained by colloidal synthesis in organic polymers [7–13] as well as embedded in glass matrices [10, 14]. At higher pressures CdSe and CdS NCs undergo a phase transition from hexagonal to cubic phase [7–11, 13, 14]. Pressure effects for mixed $CdS_{1-x}Se_x$ NCs are much less investigated [10, 13, 15, 16], the studies were carried out only for a very limited set of compositions.

Here we report on the studies of $CdS_{1-x}Se_x$ NCs ($0.4 < x \le 1$) embedded in a borosilicate glass matrix under hydrostatic pressure by optical absorption spectroscopy.

2. Experimental

$CdS_{1-x}Se_x$ quantum dots were grown in a borosilicate glass by solid-state precipitation technique generally similar to that of [4–6]. The chemical composition of the NCs was determined within $\Delta x = \pm 0.03$ from Raman spectroscopy [17]. The average NC size depended on the heat treatment temperature and duration and was determined from the absorption spectra. Plates with area of near 5×5 mm^2 and thickness down to 0.15 mm were prepared for optical measurements.

The spectra were measured at room temperature in a three-window optical pressure cell, benzene being used as a pressure medium. The pressure was varied from 0 to 0.4 GPa in increasing and decreasing mode. A LOMO MDR-2 monochromator with a FEU-106 phototube was used.

© 2008 IOP Publishing Ltd

3. Results and discussion

Figure 1 presents the variation of optical absorption spectra of $CdS_{1-x}Se_x$ NCs of the same average radius (2.2–2.3 nm), but different chemical composition, under hydrostatic pressure. The pressure effect on the optical absorption spectra of CdSe NCs of different average size is shown in Figure 2. In the obtained absorption spectra once can clearly observe maxima, related to the charge-carrier confinement, whose spectral position depends on the NC size and composition.

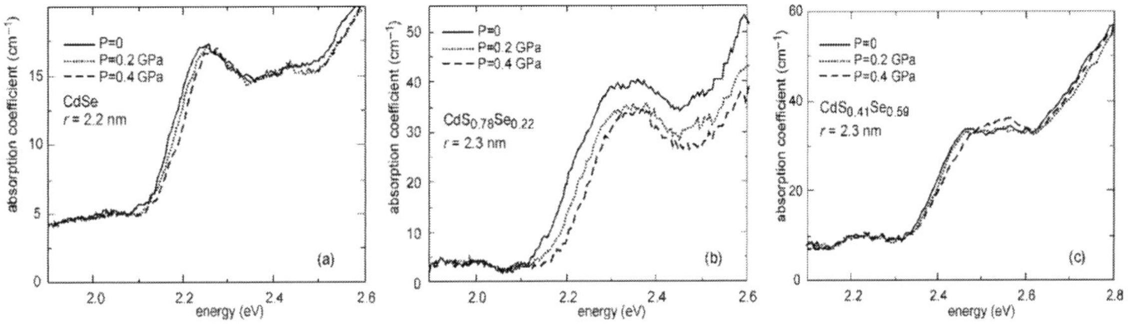

Figure 1. Effect of hydrostatic pressure on the optical absorption spectra of $CdS_{1-x}Se_x$ NCs embedded in a borosilicate glass.

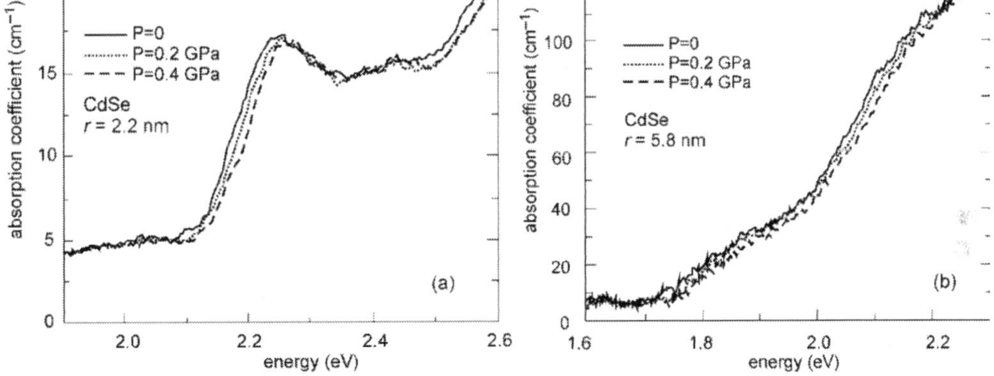

Figure 2. Optical absorption spectra of the glass-embedded CdSe NCs of different average size under hydrostatic pressure.

While scaling the absolute values of absorption coefficient α in the spectra, the value for the samples (borosilicate glass with the NCs embedded), not of the NCs themselves is meant. The actual α value for the NCs is by about two orders of magnitude higher, since the effective thickness of the NCs themselves is considerably smaller than the sample thickness, taking into account the concentration of the semiconductor phase in the samples which follows from the content of the relevant components in the initial mixture (below 1 %). An important point concerns the determination of the band gap value E_g in the NCs. Generally it is calculated from the experimental spectra by extrapolation of the $\alpha^2(h\nu)$ plot for allowed direct optical transitions. However, especially in disordered materials, when the density-of-states tails smear the absorption edge shape, often a substitutive parameter E_g^{α} is introduced, determined as the energy position of the fixed absorbance value α_f [18]. Usually for bulk materials α_f is taken at 10^2–10^3 cm^{-1}. In composites (semiconductor NCs in a dielectric medium), due to the above discussed difference in the absorption of the samples and the actual absorption of the NCs, the value $\alpha_f = 20$ cm^{-1} , corresponding to the steepest part of the edge, is often chosen for defining E_g^{α} [16].

For all the samples under study the hydrostatic pressure increase results in a blue shift of the absorption edge and the confinement-related maxima (Figures 1, 2), the corresponding data being summarized in Table 1.

Table 1. Parameters obtained from the optical studies of the glass-embedded $CdS_{1-x}Se_x$ NCs.

Composition	Average radius (nm)	Band gap pressure coefficient dE_g^{α}/dp (eV/GPa)
$CdS_{0.59}Se_{0.41}$	2.3	0.034
$CdS_{0.4}Se_{0.6}$	3.0	0.041
$CdS_{0.22}Se_{0.78}$	2.3	0.070
$CdS_{0.18}Se_{0.82}$	4.7	0.037
CdSe	2.2	0.045
CdSe	5.8	0.064

The measurements were performed at increasing and decreasing pressure and no hysteresis was observed in the investigated pressure range. This indicates that $CdS_{1-x}Se_x$ NCs are in direct contact with the glass matrix. It is consistent with the known fact that at normal conditions $CdS_{1-x}Se_x$ NCs in borosilicate glass already sustain hydrostatic pressure from the host matrix $\Delta p \approx 0.5$ GPa caused by the difference in thermal expansion coefficients of the NCs and the matrix [19]. Hence, for the correct determination of the average NC size from the absorption spectra one should introduce the glass matrix pressure-induced band gap increment $\Delta E_g^p = dE_g^{\alpha}/dp \cdot \Delta p$ into the expression for the energies of the confinement-related maxima known from the effective mass approximation [1]:

$$E_{(n_e,l_e),(n_h,l_h)} = E_b + \Delta E_g^p + \frac{\hbar^2}{2r^2}\left[\frac{\varphi_{n_e,l_e}^2}{m_e^*} + \frac{\varphi_{n_h,l_h}^2}{m_h^*}\right]. \tag{1}$$

where E_b is the bulk energy gap for the same crystal composition, r is the NC radius, $\varphi_{n,l}$ are spherical Bessel function roots for the corresponding quantum numbers, m_e^* and m_h^* are the effective masses of electrons and holes, respectively. Since no data on the effective masses are available for $CdS_{1-x}Se_x$, the corresponding parameters can be estimated by interpolation using the known values of $m_e^*=0.18m_0$ and $m_h^*=0.53m_0$ for x=0 [20] and $m_e^*=(0.11–0.13)m_0$ and $m_h^*=(0.44–0.63)m_0$ for x=1 [21, 22] as well as the chemical composition values obtained from the Raman spectra [17]. The average NC radii, estimated from Eq. (1), are given in Table 1.

The obtained band gap pressure coefficient values are generally consistent with those reported earlier for $CdS_{1-x}Se_x$ NCs, ranging from 0.027 to 0.061 eV/GPa [7, 10, 12, 16]. No monotonous compositional trend could be observed for the NCs of the same average size (Figure 1). Neither can one claim unambiguously for the size dependence of the band gap pressure coefficient. While for CdSe the measured band gap pressure coefficient increases with the NC size, the situation is quite opposite for the mixed NCs with $x \approx 0.8$ (See Table 1). The observed spread of the measured band gap pressure coefficient values and the absence of its compositional and/or size dependence for $CdS_{1-x}Se_x$ NCs under study can be related with the possibility of coexistence of hexagonal and cubic nanoparticles in the ensemble of the glass-embedded NCs, which is quite probable for the case of the NC size below 1.5–2.5 nm even without additional external pressure [9, 10]. An efficient tool to distinguish between the pure hexagonal (wurtzite) and cubic (rocksalt) NC structure is Raman spectroscopy due to the absence of the first-order Raman spectrum for the rocksalt structure [13, 14]. However, in the case of coexistence of the two phases it can hardly be applied since the measured spectrum always contains the wurtzite-phase phonons. The Raman line intensities in NCs are subject to resonance effects and, hence, cannot be used for quantitative estimates. If the presence of NCs with zinc-blende cubic structure is assumed, it will be also difficult to confirm its coexistence with wurtzite since the LO phonon frequencies for the two phases are very close to each other [25], and in NCs for both of them the effects of phonon confinement and surface phonons on the Raman band frequencies are larger than the difference between the phonon frequencies for the two phases [26]. Therefore, Raman spectroscopy cannot be applied to confirm the phase coexistence in this particular case.

4. Conclusions

The performed studies of influence of hydrostatic pressure on the optical absorption edge of $CdS_{1-x}Se_x$ ($0.4 < x \leq 1$) NCs embedded in a borosilicate glass matrix have shown a pressure-induced blue shift of the absorption edge and the confinement-related features. The identical character of the spectra obtained in the pressure increase and decrease mode as well as the absence of hysteresis confirm that in borosilicate glass $CdS_{1-x}Se_x$ NCs are in direct contact with the host matrix. The obtained band gap pressure coefficients dE_g/dP lie within 0.034 to 0.070 eV/GPa and do not show a distinct dependence on the NC composition or size. The obtained values agree well with those obtained earlier for colloidal and glass-embedded cadmium chalcogenide NCs [7, 10, 12, 16] as well as with the corresponding values for bulk CdS and CdSe [23, 24]. The absence of a pronounced compositional and size dependence of the band gap pressure coefficients for the glass-embedded $CdS_{1-x}Se_x$ NCs under investigation can be explained by possible coexistence of hexagonal and cubic NCs within the ensemble for the samples with the NC average size below 2.5 nm, inducing additional uncertainties into the average NC parameters.

References

[1] Woggon U 1996 *Optical Properties of Semiconductor Quantum Dots*, (Berlin: Springer).
[2] Murray C B, Kagan C R and Bawendi M G 2000 *Annu. Rev. Mat. Sci.* **30** 545.
[3] Nalwa H S (Ed.) 2002 *Nanostructured Materials and Nanotechnology* (San Diego: Academic Press).
[4] Borrelli N F, Hall D, Holland H and Smith D 1987 *J. Appl. Phys.* **61** 5399.
[5] Ekimov A 1996 *J. Luminesc.* **70** 1.
[6] Yükselici M H 2001 *J. Phys.: Condens. Matter* **13** 6123.
[7] Tolbert S H and Alivisatos A P 1993 *Z. Phys. D* **26** 56.
[8] Tolbert S H, Herhold A B, Johnson C S and Alivisatos A P 1994 *Phys. Rev. Lett.* **73** 3266.
[9] Tolbert S H and Alivisatos A P 1995 *J. Chem. Phys.* **102** 4642.
[10] Schroeder J and Persans P D 1996 *J. Luminesc.* **70** 69.
[11] Wickham J N, Herhold A B and Alivisatos A P 2000 *Phys. Rev. Lett.* **84** 923.
[12] Li J, Li G-H, Xia J B, Zhang J, Lin Y and Xiao X 2001 *J. Phys.: Condens. Matter.* **13** 2033.
[13] Schroeder J, Hwang L-W, Silvestri M R, Lee M and Persans P D 1996 *J. Non-Cryst. Sol.* **203** 217.
[14] Makino T, Arai M, Matsuishi K, Onari S and Arai T 1998 *J. Phys.: Condens. Matter* **10** 10919.
[15] Silvestri M and Schroeder J 1995 *J. Phys.: Condens. Matter* **7** 8519.
[16] Shusta V S, Gomonnai A V, Slivka A G, Gerzanich O I, Azhniuk Yu M and Lopushansky V V 2001 *J. Optoelectron. Adv. Mater.* **2** 515.
[17] Azhniuk Yu M, Milekhin A G, Gomonnai A V, Lopushansky V V, Yukhymchuk V O, Schulze S, Zenkevich E I and Zahn D R T 2004 *J. Phys.: Condens. Matter* **16** 9069.
[18] Mott N F and Davis E A 1979 *Electronic Processes in Non-Crystalline Materials* (Oxford: Clarendon Press).
[19] Kunets V P 1998 Ukr. Fiz. Zh. **43** 64.
[20] Lippens P E and Lannoo M 1989 *Phys. Rev. B* **39** 10935.
[21] Cohen R and Sturge M D 1982 *Phys. Rev. B* **25** 3828.
[22] Trallero-Giner C, Debernardi A, Cardona M, Menendez-Proupin E and Ekimov A I 1998 *Phys. Rev. B* **57** 4664.
[23] Edwards A L and Drickamer H G 1961 *Phys. Rev.* **122** 1149.
[24] Cervantes P, Williams Q, Côté M, Zakharov O and Cohen M 1996 *Phys. Rev. B* **54** 17585.
[25] Zelaya-Angel O, Castillo-Alvarado F L, Avendaño-Lopéz J, Escamilla-Esqivel A, Contreras-Puente G, Lozada-Morales R and Torres-Delgado G 1997 *Solid State Commun.* **104** 161.
[26] Azhniuk Yu M, Hutych Yu I, Lopushansky V V, Raevskaya A E, Stroyuk A L, Kuchmiy S Ya, Gomonnai A V and Zahn D R T 2007 *J Phys.: Conf. Ser.* **79** 012017.

Joint 21st AIRAPT and 45th EHPRG Int. Conf. on High Pressure Science and Technology IOP Publishing
Journal of Physics: Conference Series **121** (2008) 162002 doi:10.1088/1742-6596/121/6/162002

High-pressure Raman study of nano-channel materials: $BaAl_2Si_2$ and $BaSi_6$

Yusuke Narita[1], Tetsuji Kume[1], Shigeo Sasaki[1], Hiroyasu Shimizu[1], Shoichi Maekawa[2], Takashi Kikudome[2], and Shoji Yamanaka[2]

[1] Department of Materials Sci. and Tech., Gifu University, 1-1 Yanagido, Gifu, 501-1193, Japan

[2] Department of Applied Chemistry, Hiroshima University, Higashi-Hiroshima 739-8527, Japan

E-mail: kume@gifu-u.ac.jp.

Abstract. Structural phase transitions were explored for new nano-channel materials α-$BaAl_2Si_2$ and $BaSi_6$ by means of high pressure Raman experiments, on the basis of the change in the Raman spectra which cover the frequencies from 40-600 cm^{-1}. In spite of the structural similarity between $BaSi_6$ and α-$BaAl_2Si_2$, the different behaviours were found under high pressure; for α-$BaAl_2Si_2$, there is an irreversible phase transition around 16-17 GPa, and for $BaSi_6$ the reversible spectral change without distinct phase transition was observed up to 23 GPa.

1. Introduction

Recently, many authors have extensively investigated guest-host materials such as clathrates[1] and skutterudites[2], in which the guest atoms are interstitially trapped in the host frame. These materials have increasingly received a great deal of attention from aspects of their characteristic properties such as superconductivity, wide band-gap, high thermoelectric power (due to the behavior of phonon-glass and electron-crystal (PGEC)), and pressure stability. In recent years, new materials that consist of Ba as the guest atom and Si and/or Al as the host frame have been synthesized under the high temperature and pressure[3,4]. The new guest-host materials $BaAl_2Si_2$[3] and $BaSi_6$[4] are shown in figure. 1. For $BaAl_2Si_2$, the two phases (α and β phase) can be synthesized under the different conditions. The α phase takes an orthorhombic structure with space group *Cmcm* and has a channel structure accommodating the guest Ba atoms. On the other hand, the β phase shows a layered structure with a tetragonal symmetry (*I4/mmm*). The β phase can be made under high pressure (5 GPa) and high temperature (1200°C). The $BaSi_6$, which is the binary compound of Ba and Si with the channel structure (orthorhombic, *Cmcm*)[figure 1(c)], was synthesized from the Ba_8Si_{46} clathrate by treatments of high temperature and high pressure. Thus, the $BaSi_6$ can be considered as a high-pressure phase in the Ba-Si binary system. The pressurization of Ba_8Si_{46} at room temperature is known to cause the amorphization above ~40 GPa, which means that the thermal energy is required for forming $BaSi_6$.

Up to now, high pressure studies for semiconductor clathrates have been performed by several authors[5-15]. The structural stabilities and phase transitions depending on the guest-host interaction were investigated with XRD[5-8,13-15] and Raman[9-13] experiments as well as theoretical

© 2008 IOP Publishing Ltd

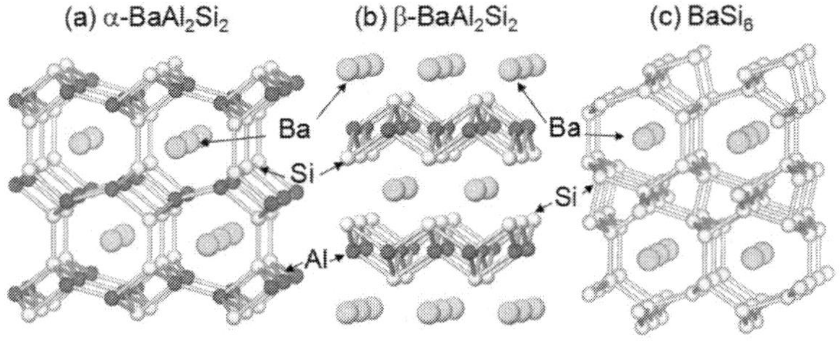

Figure 1. Crystal structures of (a) α-$BaAl_2Si_2$, (b) β-$BaAl_2Si_2$, and (c) $BaSi_6$.

calculation[14]. For the new channel materials, however, there are few high pressure experiments for understanding the high pressure stabilities. As seen in figure 1, the crystal structure of $BaSi_6$ is similar to that of α-$BaAl_2Si_2$. It is interesting to compare the structural stability under high pressure between these channel materials.

In this work, we carried out high pressure Raman study of α-$BaAl_2Si_2$ and $BaSi_6$ in order to investigate the phase transition and structural stability. The Raman spectra at 1 atm are compared with that of Ba_8Si_{46}, and the vibrational properties are discussed. The high pressure experiments demonstrate an irreversible phase transition at 16 GPa for α-$BaAl_2Si_2$ and no sign of the phase transition for $BaSi_6$ up to 23 GPa.

2. Experimental

The preparation for the α– and β–$BaAl_2Si_2$ was carried out in the same manner described elsewhere[3]. By arc-melting a ternary mixture of Ba, Al, and Si elements in a molar ratio of 1:2:2, a sample of high-pressure phase β-$BaAl_2Si_2$ was prepared under HTHP conditions. The typical pressing and heating modes were as follows; the sample was first compressed to 5 GPa, and heated to 1200°C in 1 h, and then kept for 1 h at its temperature, followed by cooling to 700°C in 2 h. The sample was further cooled down to room temperature in a few minutes. For preparation of α-$BaAl_2Si_2$, the separate arc-melted samples were annealed in h-BN cell vacuum-sealed in a quartz glass tube at ambient pressure. The tube was kept in a furnace at 1200°C for 1 day and cooled down to 800°C for 2 days, followed by cooling to room temperature for 2 days. This annealed sample takes α-$BaAl_2Si_2$[3]. The binary compound $BaSi_6$ [4] was prepared by the treatment of the Ba_8Si_{46} clathrate compound under a pressure of 15 GPa at 1000°C, or from a stoichiometric mixture of $BaSi_2$ and Si by treatment under similar high-pressure and high-temperature conditions.

Raman spectra were measured in a back-scattering geometry with a micro-Raman spectrometer (JASCO NR 1800) equipped with a triple polychromator and a charge-coupled device (CCD) detector. The 532 nm line of a solid laser (Coherent Verdi2W) was used for the excitation. The spectral resolution was about 1 cm^{-1}. High-pressure experiments were carried out by using a diamond anvil cell (DAC) with a metal gasket. The hole of the gasket serving as the sample chamber was set to about 100 μm in diameter. A single-phase sample of 20-30 μm in size was placed into the chamber of DAC, together with a ruby chip for pressure

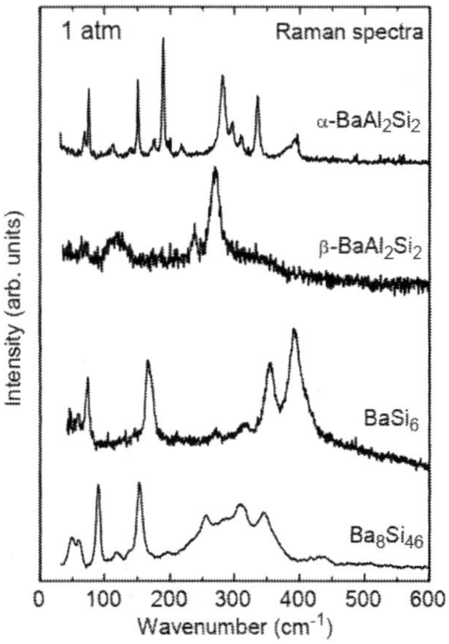

Figure 2. Raman spectra of α-$BaAl_2Si_2$, β-$BaAl_2Si_2$, $BaSi_6$, and Ba_8Si_{46} at 1 atm.

measurements. For fine Raman measurements, we used the dense argon as the pressure-transmitting medium[16] that is free from Raman signals.

3. Result and Discussion

Figure 2 shows the Raman spectra of α- and β-$BaAl_2Si_2$ and $BaSi_6$ under ambient pressure. For comparison, the Raman spectrum of Ba_8Si_{46} clathrate[9] is also shown in this figure. According to group theory, it is expected that α-$BaAl_2Si_2$ have 15 Raman active fundamental modes ($5A_g+3B_{1g}+2B_{2g}+5B_{3g}$) including three Ba modes($A_g+B_{1g}+B_{3g}$). The β-$BaAl_2Si_2$ is expected to have only four Raman active modes and there is no Raman active Ba vibration in contrast to the α phase. As shown in figure 2, the Raman peaks are observed in the frequency region below 100 cm^{-1} for α-$BaAl_2Si_2$. These low frequency modes can be considered as the Ba vibrations, on the basis of the similarity with Ba_8Si_{46} for which the Ba vibrations are observed below 100 cm^{-1}. On the other hand, the β-$BaAl_2Si_2$ does not show distinct peak in the low frequencies. These findings are consistent with the prediction by the group theory. For $BaSi_6$ with similar nano-channel structure to that of α-$BaAl_2Si_2$, it is expected that 21 modes are observable in the Raman scattering. Among the 21 modes, 5 modes are predicted to be Ba related vibrations. The number of peaks actually observed in Raman spectrum are found to be much lower than the prediction (figure 2). The peaks located below 100 cm^{-1} are considered to be guest Ba vibrational mode.

Figure 3 shows Raman spectra of $BaAl_2Si_2$ measured when the α phase was pressurized up to 17 GPa. On compression process, the spectrum became weak in intensity at 16.2 GPa. At the same time, a new peak appeared at 71 cm^{-1}. The sharp peaks at 90, 100 cm^{-1} disappeared above 17.2 GPa. These spectral changes suggest strongly that a structural change occurs at 16 GPa. On the other hand, upon the decompression process after the phase transition (figure 3(b)), the spectra of α-$BaAl_2Si_2$ did not appear. Consequently, we conclude that the phase transition at 16 GPa is not reversible. The Raman spectrum of the recovered sample at 1 atm is similar to that of β-$BaAl_2Si_2$; the mode of 350 cm^{-1} in the recovered sample corresponds to weak shoulder observed in the *β* phase (figure 2), and 110 cm^{-1} band

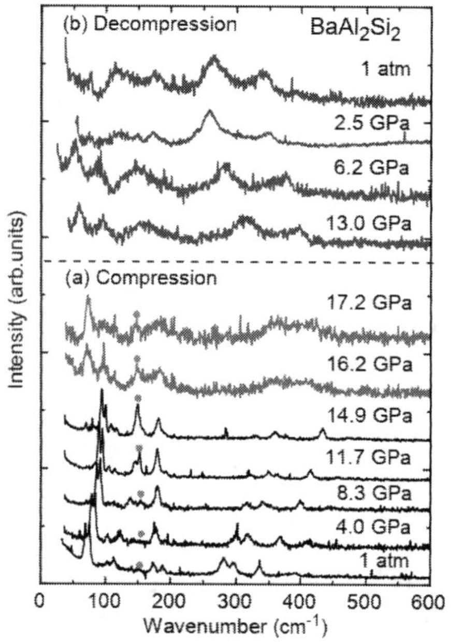

Figure 3. Raman spectra obtained when α-$BaAl_2Si_2$ is pressurized. (a) and (b) correspond to compression and decompression process, respectively.

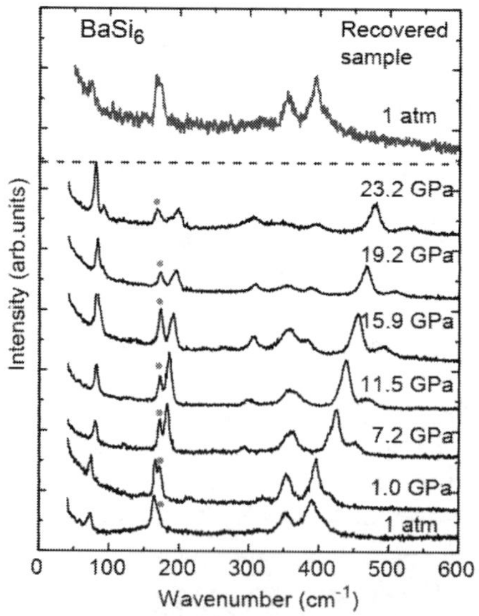

Figure 4. Pressure dependence of Raman spectra of $BaSi_6$ up to 23 GPa.

is observed also in the *β* phase. However, we observed additional peaks around 70 cm^{-1} and 180 cm^{-1} for the recovered sample. The lower frequency mode is probably due to a Ba vibration. Since the Ba vibrations are Raman inactive in the β-$BaAl_2Si_2$, the structure of the recovered sample is not exactly the same as the *β* structure, but closely related with that of the β phase. It can be considered as the structure of the recovered sample that the framework is almost the same as the *β* phase, but the Ba position is different. In order to clarify the above, it is highly required to perform an XRD experiments.

Figure 4 shows the Raman spectra obtained for $BaSi_6$ under high pressures up to 23 GPa. The Raman spectra indicate a gradual change and no obvious sign of the phase transition. The recovered sample after the pressurization shows exactly the same spectrum as before pressurization. Although the most obvious spectral change upon compression is the peak splitting around 160 cm^{-1}, this is not related with the structural change because this band initially has a shoulder, i.e., consists of two components at 1 atm. The splitting observed for the 100 and 350 cm^{-1} peaks around 7-15 GPa can also be explained by the same argument. Therefore, in contrast to the case of $BaAl_2Si_2$, obvious change related to the phase transition cannot be detected for $BaSi_6$ up to 23 GPa. Even if there exist some structural changes, those are thought to be due to small distortions of the framework or displacements of the guest Ba position because Raman spectral feature remains almost the same.

In figure 3, we find that, for the α-$BaAl_2Si_2$ the frame vibration around 154 cm^{-1} (solid circles) shows a negative frequency shift against pressure. Also for $BaSi_6$ (figure 4), the peak located at about 170 cm^{-1} (solid circles) indicates the negative shift. The softening behavior of the frame vibration was observed also for various clathrate compounds[9-13]; in the case of Ba_8Si_{46}, for example, the peak of 153 cm^{-1} indicates the softening[9]. The origin of the softening for Si clathrates has been discussed as follows. Since, in the corresponding frequency region, the diamond Si also has softening modes assigned to zone-boundary TA phonons[17], the softening modes for the Si clathrates were believed to arise from the sp^3 tetrahedral bonding nature. As seen in figure 1, α-$BaAl_2Si_2$ and $BaSi_6$ basically consist of tetrahedral bonding networks although they are distorted. Therefore, the present softening behavior of the host framework modes can originate from the tetrahedral bonding network.

4. Summary

We present Raman spectra of nano-channel materials α-$BaAl_2Si_2$ and $BaSi_6$ at high pressures and room temperature, which show that the structural transformation of α-$BaAl_2Si_2$ at 16-17 GPa was irreversible and, the α to β phase transition was strongly suggested. On the other hands, for $BaSi_6$, no obvious sign for the structural change was observed up to 23 GPa.

References

[1] Bobev S and Sevov S C 2000 *J. Solid State Chem.* **153** 92 and references are therein.
[2] Keppens V *et al.* 1998 *Nature* **395** 876
[3] Yamanaka S, Kijiyama M, Sivakumar S N, and Fukuoka H 2004 *High Press. Res.* **24** 481
[4] Yamanaka S and Maekawa S 2006 *Z. Naturforsch.* **61**b 1493
[5] San-Miguel A *et al.* 1999 *Phys. Rev. Lett.* **83** 5290
[6] San-Miguel A *et al.* 2002 *Phys. Rev. B* **65** 054109
[7] San-Miguel A and Toulemonde P 2005 *High Press. Res.* **25** 159
[8] San-Miguel A *et al.* 2005 *Europhys. Lett.* **69** 556
[9] Kume T *et al.* 2003 *Phys. Rev. Lett.* **90** 155503
[10] Kume T, Koda T, Sasaki S, Shimizu H, and Tse J S 2004 *Phys. Rev. B* **70** 052101
[11] Shimizu H *et al.* 2003 *Phys. Rev. B* **68** 212102
[12] Shimizu H *et al.* 2005 *Phys. Rev. B* **71** 094108
[13] Shimizu H *et al.* 2007 *J. Appl. Phys.* **101** 063549
[14] Tse J S *et al.* 2002 *Phys. Rev. Lett.* **89** 195507
[15] Yang L *et al.* 2006 *Phys. Rev. B* **74** 245209
[16] Shimizu H, Tashiro H, Kume T, and Sasaki S 2001 *Phys. Rev. Lett.* **86** 4568
[17] Weinstein B A and Piermarini G J 1975 *Phys. Rev. B* **12** 1172

High pressure Raman studies of carbon nanotube materials

D. Christofilos[1*], J. Arvanitidis[1,2], E. Efthimiopoulos[1,3], X. Zhao[4], Y. Ando[4], T. Takenobu[5], Y. Iwasa[5], H. Kataura[6], S. Ves[3] and G. A. Kourouklis[1]

[1]Physics Division, School of Technology, Aristotle University of Thessaloniki, 54124 Thessaloniki, Greece

[2]Department of Applied Sciences, Technological Educational Institute of Thessaloniki, 57400 Sindos, Greece

[3]Physics Department, Aristotle University of Thessaloniki, 54124 Thessaloniki, Greece

[4]Department of Materials Science and Engineering, Meijo University, Nagoya 468-8502, Japan

[5]Institute for Materials Research, Tohoku University, 2-1-1 Katahira, Aoba-ku, Sendai 980-8577 and CREST, Japan Science and Technology Corporation, Kawaguchi 332-0012, Japan

[6]National Institute of Advanced Industrial Science and Technology (AIST), 1-1-1 Higashi, Tsukuba, Ibaraki 305-8562, Japan

[*]E-mail: christof@vergina.eng.auth.gr

Abstract. Raman spectroscopy is employed to study the pressure response of single-(SWCNTs), double- (DWCNTs) and multi-wall (MWCNTs) carbon nanotube materials. In DWCNTs, the outer tubes exhibit a similar pressure response to SWCNTs, where the tube sensitivity to pressure application is diameter dependent (higher for large diameter tubes). On the other hand, the inner tubes in DWCNTs and the innermost tubes in MWCNTs appear to be much less sensitive to pressure application, indicating the pressure screening inside the outer tubes. The pressure screening effect is strongly enhanced in MWCNTs.

1. Introduction

The mechanical and transport properties of carbon nanotube materials, crucial for technological applications, are determined by the structural characteristics of the individual tubes and their mutual interaction, either in a nanotube bundle (intertube interaction) and/or within a double- or multi-shell system (intratube interaction). High pressure application on these systems is a valuable tool for the investigation of their mechanical and structural stability, while resonance Raman spectroscopy allows the selective probing of different tubes [1,2]. This is possible due to the size-dependent peculiar electronic structure of carbon nanotubes, originating from the quasi one-dimensional electron confinement [3,4], in combination with the inversely proportional relation between the radial breathing

© 2008 IOP Publishing Ltd

mode (RBM) frequency and the tube diameter [5]. In this work, we study the effect of high-pressure on SWCNT, DWCNT and MWCNT materials by means of Raman spectroscopy, in order to compare the structural stability of the coaxial tubes in the various carbon nanotube systems.

2. Experimental

The bundled SWCNT sample (Carbon Nanotechnologies Incorporated, BuckyPearls™), with tube diameters 0.8-1.2 nm, has been prepared by the high pressure catalytic decomposition of carbon monoxide (HiPCO). The bundled DWCNT material has been synthesized by means of the C_{60}-peapod conversion process, following Bandow's procedure [6]. The inner tubes diameters are 0.6-0.9 nm while those of the outer tubes 1.3-1.6 nm. The infrared irradiation purified MWCNTs were prepared by DC arc discharge evaporation of graphite rods in pure hydrogen gas. They consist of 10-15 concentric tubes with outermost and innermost tube diameters of ~10 and ~1 nm, respectively [7].

Raman spectra were recorded in the back-scattering geometry using a micro-Raman, triple grating system (DILOR XY) equipped with a cryogenic CCD detector. High pressure Raman measurements were carried out using a Mao-Bell type diamond anvil cell (DAC) with the 4:1 methanol-ethanol mixture as the pressure transmitting medium, while the ruby fluorescence technique was used for pressure calibration. The 676.4 nm line of a Kr^+ laser was focused on the sample by means of a 20x objective, while the laser power was kept below 4 mW, measured directly before the DAC.

3. Results and discussion

The Raman spectra in the RBM frequency region of the studied carbon nanotube materials at lower (0.1-0.2 GPa, bottom panels) and higher pressures (2.4-2.7 GPa, top panels) are illustrated in figure 1.

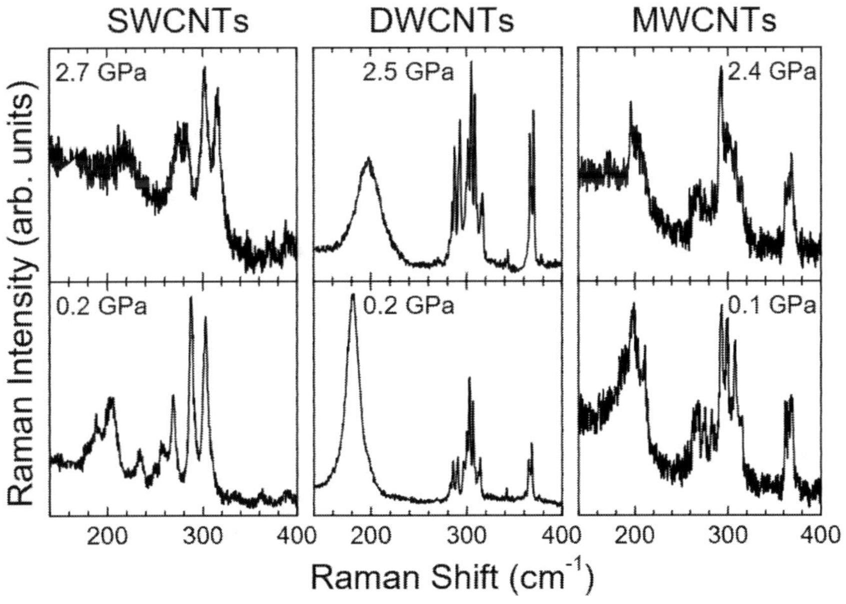

Figure 1. Raman spectra of SWCNTs, DWCNTs and MWCNTs at different pressures.

The various RBM peaks appearing in the spectra correspond to different tubes that are in resonance with the excitation wavelength. In the DWCNT spectra, the low frequency, relatively broad, RBM band corresponds to the outer tubes, while the numerous narrow peaks at higher frequencies to the smaller diameter inner tubes. The narrow lineshape of the latter reflects the structural perfection of the inner tubes, well protected by the outer ones. The large number of the inner tube RBMs originates from their splitting due to their intratube interaction-dependent frequency upshift [8]. A similar situation is also anticipated for the MWCNT innermost tubes. Usually, the observation of their RBMs

Journal of Physics: Conference Series **121** (2008) 162003 doi:10.1088/1742-6596/121/6/162003

is not possible due to the large number of encapsulating tubes. In our case, the smaller number of coaxial tubes allows their appearance in the Raman spectrum, though with low signal to noise ratio. From the RBM frequencies (ω) and the relation $\omega = 223/d_t + 10$ [9], we can make a rough estimation for the diameters d_t of the resonantly probed carbon tubes in the investigated samples. The calculated diameters are 0.8-1.2 nm for SWCNTs, 0.6-0.9 nm (~1.3 nm) for the inner (outer) tubes of DWCNTs and 0.6-1.3 nm for the innermost tubes of MWCNTs.

The pressure dependence of the frequencies of some characteristic RBMs in the Raman spectrum of the studied nanotube materials is plotted in figure 2. With increasing pressure, all the presented RBMs shift towards higher frequencies at different rates. Their pressure slopes are 5.2-7.2 cm^{-1}/GPa for SWCNTs, 6.7 cm^{-1}/GPa (9.3 cm^{-1}/GPa at low pressures) for the DWCNT outer tubes, 1.0-1.1 cm^{-1}/GPa for the DWCNT inner tubes and 0.1-0.3 cm^{-1}/GPa for the MWCNT innermost tubes. The pressure-induced frequency shift of the SWCNT RBMs is larger for the lower frequency Raman peaks (larger tube diameters) and is accompanied by their gradual broadening and intensity attenuation (figure 1). These observations are in line with earlier experimental studies [10] and theoretical predictions for bundled SWCNTs [11], which suggest that the tubes suffer significant pressure induced cross-section deformations with carbon nanotubes of smaller diameter exhibiting higher structural stability. We emphasize that if a shift of the resonance conditions was the only pressure effect, it would have led to the decrease of the RBM signal of some tubes but it should have enhanced the signal originating from other tubes, initially out of resonance, which is not the present case. Therefore the observed behavior should be attributed to the pressure-induced increase of the intertube interactions and the theoretically expected cross-section deformation of the tubes.

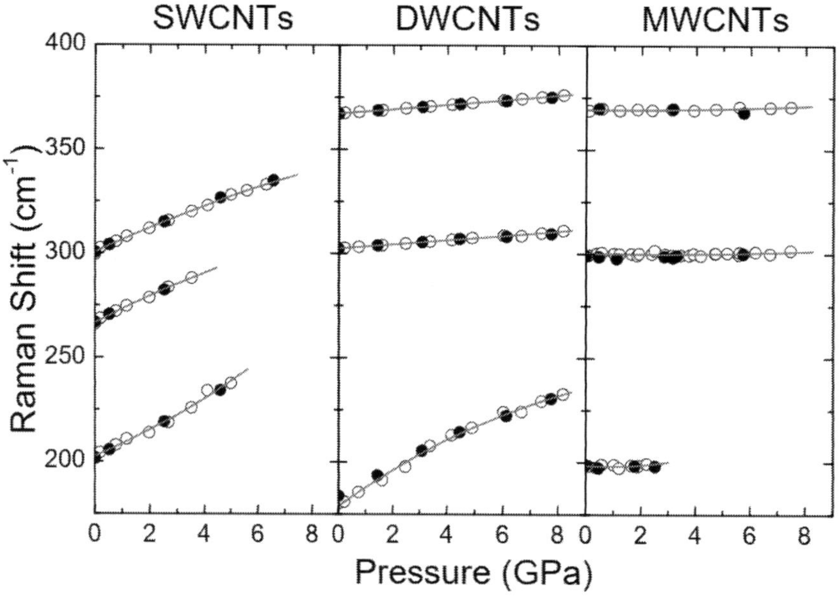

Figure 2. Pressure dependence of the RBM frequencies in SWCNTs, DWCNTs and MWCNTs. Open (solid) symbols denote data acquired for increasing (decreasing) pressure.

The RBM band attributed to the outer tubes of DWCNTs exhibits also an intensity decrease and broadening at elevated pressures, while its pressure slope is comparable to that of SWCNTs with similar size. Nevertheless, it persists to higher pressures than the corresponding in SWCNTs, possibly due to the presence of the inner tube in the former material that structurally supports the outer tube against pressure-induced cross-section deformations. The presence of the inner tubes in DWCNTs could also explain the sublinear pressure response of the outer tube RBM frequency by assuming that with increasing pressure the intratube interaction becomes progressively stronger further supporting the outer tubes (reduced slope at elevated pressures). On the other hand, the inner tube RBM peaks remain narrow and intense up to the highest pressure attained in our experiments (~9 GPa). This behavior reflects the protection of the encapsulated inner tubes from external perturbations, in contrast to the pressure-induced structural distortions encountered for SWCNTs and the outer tubes in DWCNTs. As the inner tubes suffer an effective pressure smaller than that exerted on the outer ones, the observed changes in the profile of the relative intensities can be mainly attributed to pressure-induced changes of their resonance conditions as any other cross-section-related changes should be minimal inside the outer tubes. The substantially smaller pressure slopes of the inner tube RBM frequencies as compared to those of the outer tubes in DWCNTs and the SWCNTs of similar diameters, further supports the scheme of pressure screening inside carbon nanotubes. Our detailed high pressure Raman studies, by means of several excitation wavelengths, combined with theoretical calculations suggest that the pressure screening in DWCNTs depends strongly on the intratube interaction and consequently on the inner-outer tube separation [12]. Finally, the experimental data presented in figure 2 clearly reveal that the addition of more graphene shells in the case of MWCNTs results in an even stronger reduction of the pressure slopes of the RBM frequencies. Thus, the pressure screening becomes much more efficient in this case, where the innermost tubes are well protected against external perturbations in the interior of several coaxial tubes. Then, differences in the pressure slopes of the RBM frequencies should also be related to the different number of shells comprising the MWCNTs. Note that, in accordance to the pressure screening scheme, the relative intensities among the different RBM peaks in the spectrum do not change significantly as compared with the changes observed for the DWCNTs and SWCNTs.

Upon pressure release, the frequency shifts of all the observed Raman peaks are fully reversible and the original frequency positions are restored after total pressure release (figure 2). However, the RBM intensities corresponding to the larger SWCNTs and the DWCNT outer tubes are attenuated and the peaks remain broader after release of the pressure (not shown). These divergences suggest the existence of residual pressure-induced deformations in these tubes.

4. Conclusions

The pressure response of carbon nanotube materials consisting of one or more coaxial rolled graphene sheets is deduced from the evolution of the RBM frequencies and their lineshape modifications upon pressure application. The larger tubes in SWCNTs are more vulnerable to pressure application than the smaller ones, while the outer tubes in DWCNTs exhibit similar pressure response to the SWCNTs of the same size. On the contrary, the inner tubes in DWCNTs appear to be less sensitive to pressure application, reflecting the pressure screening in the interior of carbon nanotubes. This interpretation is further supported by our results on MWCNTs for which no Raman data on the high pressure RBM response of the innermost tubes existed until now. The significant reduction of the RBM pressure slopes (almost two orders of magnitude with respect to those of SWCNTs of similar size) and the milder changes of the RBM intensity profile constitute clear evidences of the protection-enhancement provided by increasing the number of coaxial nanotubes.

Acknowledgements

This study was partly supported by the Industrial Technology Research Grant Program in 2006 from New Energy and Industrial Technology Development Organization (NEDO) of Japan. The financial support from the Hellenic Ministry of National Education and Religious Affairs and the European Union through the "Pythagoras II" program is also acknowledged.

References

[1] Loa I 2003 *J. Raman Spectrosc.* **34** 611
[2] Dresselhaus MS and Eklund PC 2000 *Adv. Phys.* **49** 705
[3] Kataura H, Kumazawa Y, Maniwa Y, Umezu I, Suzuki S, Ohtsuka Y and Achiba Y 1999 *Synth. Metals* **103** 2555

[4] Wang F, Dukovic G, Brus LE and Heinz TF 2005 *Science* **308** 838
[5] Rao AM, Richter E, Bandow S, Chase B, Eklund PC, Williams KA, Fang S, Subbaswamy KR, Menon M, Thess A, Smalley RE, Dresselhaus G and Dresselhaus MS 1997 *Science* **275** 187
[6] Bandow S, Takizawa M, Hirahara K, Yudasaka M and Iijima S 2001 *Chem. Phys. Lett.* **337** 48
[7] Zhao X, Ando Y, Qin LC, Kataura H, Maniwa Y and Saito R 2002 *Chem. Phys. Lett.* **361** 169
[8] Pfeiffer R, Simon F, Kuzmany H and Popov VN 2005 *Phys. Rev. B* **72** 161404
[9] Fantini C, Jorio A, Souza M, Strano MS, Dresselhaus MS and Pimenta MA 2004 *Phys. Rev. Lett.* **93** 147406
[10] Venkateswaran UD, Masica DL, Sumanasekera GU, Furtado CA, Kim UJ and Eklund PC 2003 *Phys. Rev. B* **68** 241406
[11] Elliott JA, Sandler JKW, Windle AH, Young RJ and Shaffer MSP 2004 *Phys. Rev. Lett.* **92** 095501
[12] Christofilos D, Arvanitidis J, Kourouklis GA, Ves S, Takenobu T, Iwasa Y and Kataura H 2007 *Phys. Rev. B* **76** 113402

A simple model for the interpretation of the pressure response of the radial breathing modes in double-wall carbon nanotubes

D. Christofilos and J. Arvanitidis*

Physics Division, School of Technology, Aristotle University of Thessaloniki, 54124 Thessaloniki, Greece

E-mail for correspondence: jarvan@vergina.eng.auth.gr

Abstract. We present a simple phenomenological model that allows the qualitative description of the mechanical properties of single- (SWCNTs) and double-wall carbon nanotubes (DWCNTs) under the application of pressure. The pressure screening effect for the inner tube of a DWCNT and the mechanical support provided by the inner to the outer tube are reflected in the pressure induced strain of the individual tubes constituting the system. These effects are experimentally identifiable in the pressure response of the radial breathing modes (RBMs) of carbon nanotubes, probed by Raman spectroscopy. The pressure response depends both on the diameters of the individual tubes constituting the DWCNT system as well as their diameter difference.

1. Introduction

The fascinating mechanical and transport properties of carbon nanotubes are determined by the structural characteristics of the individual tubes and their tube-tube van der Waals interaction. Raman spectroscopy has been proven very useful in the study of carbon nanotubes [1], while their structural stability has been successfully addressed by high pressure Raman spectroscopy [2,3]. Double-wall carbon nanotubes constitute a model system for studying nanotube encapsulation effects. A pressure screening effect for the inner tube in a DWCNT is suggested by the experimental high pressure behaviour of the Raman active modes in DWCNTs [4,5]. This is further supported by calculations predicting an enhancement of the structural rigidity of the tubes in the DWCNT system as compared to their rigidity as individual tubes [6,7]. It has been also observed that the normalized pressure derivatives Γ ($=\omega^{-1}\partial\omega/\partial P$) for the outer tube RBMs are significantly larger than those for the inner ones. In addition, for both inner and outer tubes, the overall tendency is size dependent with smaller nanotubes (higher RBM frequencies) attaining smaller Γ-parameter values. However, the data for the inner tubes are grouped in quasi-linear distributions where the previous tendency is inversed, namely the Γ-parameter values within a group increase for increasing RBM frequency [8].

A simple model, based on two harmonically coupled harmonic oscillators, has been recently reported for the description of the size dispersive properties of the RBM frequencies in DWCNTs [9]. This model can be solved analytically providing a better physical insight of the system, while at the same time, it gives very similar results with those calculated by more elaborate models [10]. However, it is not very appropriate for the interpretation of the high pressure Raman results, since it does not

© 2008 IOP Publishing Ltd

include non-linearity, which is inherent in the high pressure response of the real system. In this work, in order to interpret the experimental findings concerning the pressure response of the RBMs in DWCNTs, we have modeled the system by two anharmonically coupled anharmonic oscillators.

2. Phenomenological model

The cross-section of a single-wall carbon nanotube (SWCNT) is modeled as a regular polygonal arrangement of carbon atoms (ring), at an equilibrium distance of l= 1.42 Å (figure 1). The restoring forces for the fully symmetric radial and the high frequency tangential vibrations of the atoms can be easily identified and they are obviously diameter dependent. These vibrational modes correspond to the RBM and G^- mode of a carbon nanotube and we refer to them as radial breathing like mode (RBLM) and G^--like mode (GLM). When the diameter tends to infinity, the restoring force for the RBLM tends to zero. At the same time, the GLM should attain the frequency of the tangential mode of graphite (~ 1580 cm^{-1}), which can be used for the calibration of the model. The model becomes more realistic by assuming a Lennard-Jones potential instead of a harmonic interaction between neighbouring carbon atoms (represented by the force constant k of the spring in figure 1). Pressure application in the real tube can be approximated by assuming that the pressure acts in the radial direction on an effective area of 1.42x1.42 $Å^2$ for each atom. Assuming that the radial symmetry is preserved, the RBLM can be described by a simple anharmonic oscillator.

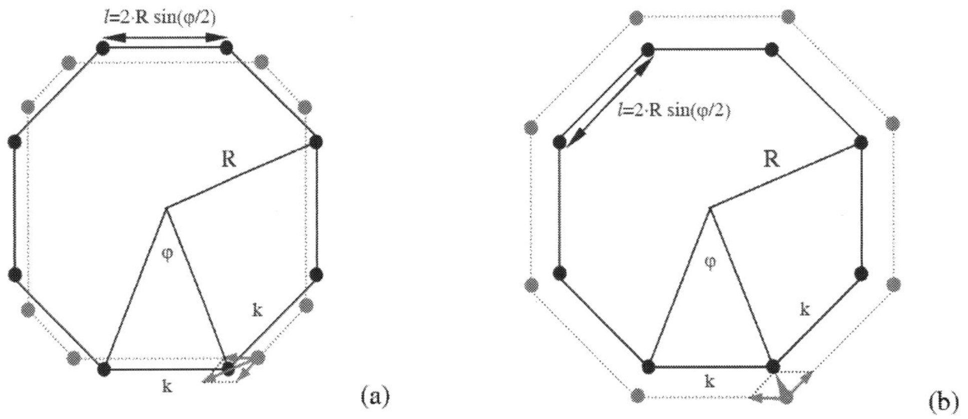

Figure 1: (a) The GLM and (b) the RBLM of a single-ring. In each case, red arrows represent the components and the resultant restoring force exerted on a carbon atom.

The case of a DWCNT would then be represented by two concentric rings mutually interacting by a Lennard-Jones potential that is assumed to be only a function of the radius difference of the two rings. The equilibrium radius difference is set to the interlayer spacing of turbostratic graphite (3.44 Å), while for very large radius we demand that the out-of-phase vibration of the oscillators should attain the value of 127 cm^{-1}, which is the layer-layer vibration of graphite (B_{2g} mode) [11]. When the diameter difference of the two rings constituting the double-ring system is other than 6.88 (2x3.44) Å, then the system is allowed to relax before any pressure application is assumed. As a consequence, the inner ring expands (contracts) and the outer ring contracts (expands) when the diameter difference is larger (smaller) than 6.88 Å. Within the context of our phenomenological model, the DWCNT system is actually modeled by two anharmonically coupled anharmonic oscillators.

3. Results and discussion

In figure 2, we present the pressure induced strain (P= 1 GPa) of various diameter inner and outer rings constituting the relaxed double-ring system. It is clear that the strain experienced by an inner or outer ring in the double-ring system is always smaller than the corresponding value for a single-ring of the same diameter. This is the manifestation of the pressure screening effect for the inner tubes and the mechanical support that they provide to the outer tubes. The strain for larger double-rings is larger, reflecting the diameter-dependent rigidity of the individual rings constituting the system (smaller for larger diameter tubes). Moreover, when the diameter difference is sufficiently large, the response of the outer ring tends to that of a single-ring with the same diameter (no support effect), while the pressure screening effect for the inner ring is maximized. On the other end, for small diameter differences, the pressure application is more efficiently transmitted to the inner tube (reduction of the screening effect), while the outer ring is better supported by the inner. Finally, it is interesting to note that the strain of the inner tube for large diameter differences tends to zero from positive values. As described above, when the diameter difference is large, the inner ring slightly expands due to the small ring-ring interaction. Upon pressure application, the diameter of the outer ring is reduced. Consequently, the diameter difference between the inner and the outer ring is also reduced, leading to an increase of the inner-outer ring interaction that causes further expansion of the inner ring (positive strain).

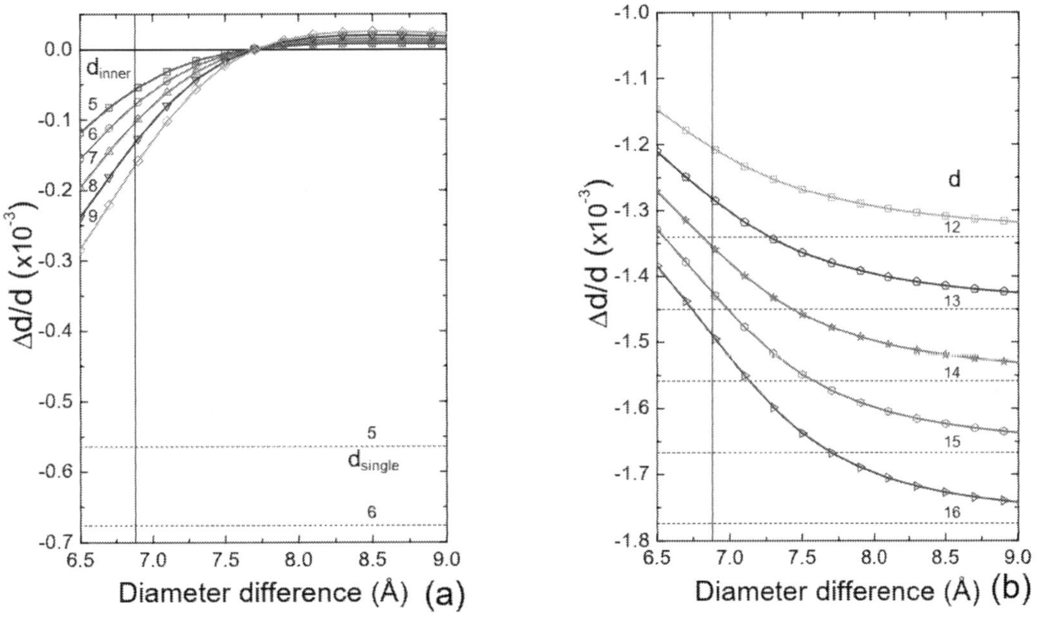

Figure 2 Strain of (a) the inner and (b) the outer rings constituting the relaxed double-ring system as a function of their initial diameter difference. Vertical lines denote the turbostratic constraint of graphite, while horizontal dotted lines represent the strain value for single rings with corresponding diameter values.

In figure 3a we present our recent experimental results concerning the Γ-parameter values for the RBMs of the inner tubes in a DWCNT material [8]. The pressure response of SWCNTs, represented by the appropriately labeled curve, is also included for comparison [12]. We can use our simple model described above to calculate the Γ-parameter values of the RBLMs as a function of the diameters of the involved rings and their diameter difference (figure 3b). Taking into account the simplicity of our model, we do not aim at the quantitative agreement between the model calculations and the experimental results. Although the order of magnitude is quite similar, it is apparent that the single ring is stiffer and its pressure response is less strongly diameter-dependent than the real system. However, the comparison to be made is between the tendency exhibited by the model calculations for the pressure response of the single- and the double-rings and the corresponding experimental trend.

The resemblance between the two graphs is striking and allows the interpretation of the experimental results on the basis of the model calculations. The pressure screening effect for the inner tube is reflected in the smaller Γ-parameter value compared to that for a SWCNT of the same diameter. In addition, the overall tendency of reducing Γ-parameter values for higher RBM frequencies (tubes of smaller diameter) is a manifestation of the higher rigidity of the smaller tubes as indicated by their strains (figure 2).

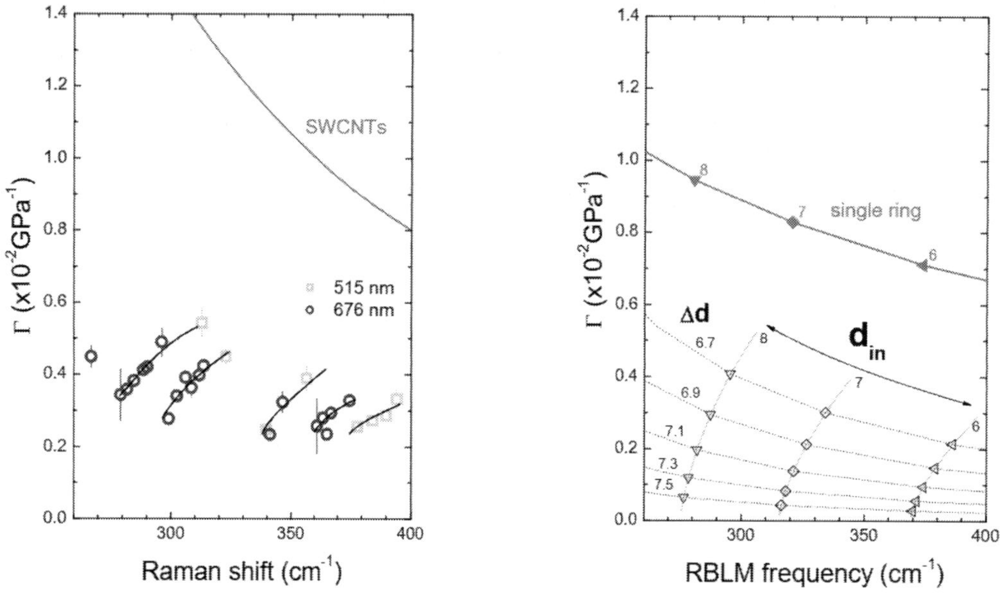

Figure 3 (a) Experimental results for the Γ-parameter values of the inner tube RBMs in DWCNT as a function of their RBM frequency [8]. The fitting curve for the experimentally obtained Γ values for SWCNTs is also included [12]. (b) The corresponding diagram based on the model described in the text.

More importantly, the grouping of the Γ-parameter values in discrete distributions, wherein the Γ-parameter values increase with increasing RBM frequency, clearly reflects the dependence of the pressure screening effect on the inner-outer tube spacing in a DWCNT. Consequently, each distribution can be considered as originating from one inner tube (or more with very similar diameters) encapsulated in outer tubes of various diameters.

4. Conclusions

In conclusion, we have used a simple phenomenological model in order to demonstrate the consequences of tube encapsulation on the pressure response of the tubes constituting a DWCNT system. The dependence of the pressure induced strain on the size of the involved tubes and their separation is reflected in the behavior of their RBM frequencies upon pressure application. Therefore, high pressure Raman spectroscopy can be a powerful tool in the characterization of DWCNT materials and the estimation of the inner-outer tube spacing.

Acknowledgements

The financial support by the European Social Fund and National Resources (YPEPTH-EPEAEK, Greece) through Pythagoras II is greatly acknowledged.

References

Journal of Physics: Conference Series **121** (2008) 162004 doi:10.1088/1742-6596/121/6/162004

[1] Dresselhaus MS and Eklund PC 2000 *Adv. Phys.* **49** 705
[2] Loa I 2003 *J. Raman Spectrosc.* **34** 611
[3] Venkateswaran UD, Rao AM, Richter E, Menon M, Rinzler A, Smalley RE and Eklund PC 1999 *Phys. Rev. B* **59** 10928
[4] Puech P, Hubel H, Dunstan DJ, Bacsa RR, Laurent C and Bacsa WS 2004 *Phys. Rev. Lett.* **93** 095506
[5] Arvanitidis J, Christofilos D, Papagelis K, Andrikopoulos KS, Takenobu T, Iwasa Y, Kataura H, Ves S and Kourouklis GA 2005 *Phys. Rev. B* **71** 125404
[6] Gadagkar V, Maiti PK, Lansac Y, Jagota A and Sood AK 2006 *Phys. Rev. B* **73** 085402
[7] Yang X, Wu G and Dong J 2006 *Appl. Phys. Lett.* **89** 113101
[8] Arvanitidis J, Christofilos D, Papagelis K, Takenobu T, Iwasa Y, Kataura H, Ves S and Kourouklis GA 2005 *Phys. Rev. B* **72** 193411
[9] Dobardzic E, Maultzsch J, Milosevic I, Thomsen C and Damnjanovic M 2003 *Phys. Stat. Sol. (b)* **237** R7
[10] Popov VN, and Henrard L 2002 *Phys. Rev. B* **65**, 235415
[11] Reich S and Thomsen C 2004 *Phil. Trans. R. Soc. Lond. A* **362** 2271
[12] Venkateswaran UD 2004 *Phys. Stat. Sol. (b)* **241** 3345

A paper submitted to Airapt 21 (366 338upBHwu revised)

The effects of high pressure on the ferroelectric properties of nano-BaTiO3 ceramics

J L Zhu[1], S Lin[1], S M Feng[1], F Y Li[1], L J Wang[1] C Q Jin[1,*], X H Wang[2], and L T Li[2]
[1]The Institute of Physics, Chinese Academy of Sciences, P.O.Box 603, Beijing 100080, China
[2]Department of Materials Sciences, Tsinghua University , Beijing, China

*E-mail for correspondence: Jin@aphy.iphy.ac.cn

Abstract. Temperature dependence of the dielectric constant of nano-grain BaTiO3 (BTO) ceramic has been investigated under hydrostatic pressure up to 5 GPa. We show that the paraelectric-ferroelectric phase transition temperature, which is represented by T_C (Curie temperature point), decreases with pressure at a rate of $dT_C/d\sigma$ = -40.0 K/GPa for coarse grain ceramic and -34.3 K/GPa for ceramic with 60nm size grains. For the paraelectric-ferroelectric phase transition temperature, the variation of inter-granular stress with reducing grain size is considered the main factor causing the decrease of T_C and the change of $dT_C/d\sigma$. Adding the hydrostatic-pressure and inter-granular stress contributions to a Ginsburg-Landau-Devonshire (GLD) type thermodynamic theory provides a satisfactory explanation to our experimental results. Based on the GLD theory, a phase diagram, reflecting the relationship among the transition temperature, grain size and hydrostatic-pressure, has been established, which matches well with our experimental results.

1.Introduction

As a classic ferroelectric material, BTO has been studied for more than half a century and has found many applications, especially for making multilayer ceramic capacitors that made the miniaturization of electronic devices possible [1]. As the layer thickness of the multilayer structure becomes thinner and thinner, the grain size required for the BTO ceramic is down to the order of submicrons, sometimes even nano-meters. A strong size effect begins to appear at such small scale that affects the physical properties of the material and the performance of devices. In recent years, size effect has become one of the most important topics in physics and materials science communities.

When the grain size is in the nano-meter scale, BTO ceramic exhibits some distinct properties. Many experimental investigations have been carried out on this topic, including electrical impedance spectroscopy, Raman scattering, X-ray diffraction, Scanning electron microscope (SEM),

© 2008 IOP Publishing Ltd

Transmission electron microscopy, etc. These studies mainly focused on the dielectric properties and microstructures of BTO ceramic [2-4]. There are also many theoretical investigations on size effects based on different theoretical models [5-13]. It was found that the permittivity of BTO ceramic changes nonlinearly with the reduction of grain size. At room temperature, the dielectric constant increases with the reduction of the grain size until a maximum value is reached, then, it begins to decrease on further decreasing of the grain size [14,15]. At the same time, 90° ferroelectric domains, in which the directions of polarizations are perpendicular to each other, and which are formed to relieve the transformation strain in the tetragonal phase of BTO, begin to disappear as the grain size of BTO ceramic reaches nano-meter scale [15,16]. Eventually, each nano-gain becomes a single domain crystal. Consequently, the internal compressive stresses produced during sintering process and by the paraelectric to ferroelectric phase transition will not be fully released due to the lack of 90° ferroelectric domains in nano-grain BTO ceramic. Such internal compressive stresses will lower the cubic to tetragonal phase transition temperature T_C (~ 130 °C in coarse grain BTO ceramic with grain size >1μm).

It is well known that high pressure suppresses spontaneous polarization in the ferroelectric phase of BTO and can induce transition to the cubic phase at room temperature because the cubic phase has smaller volume. Therefore, increasing pressure will result in a shift of T_C to lower temperature and may significantly influence the peak dielectric constant. Such effect has been observed in nano-grain BTO ceramic. Therefore, high pressure can be an effective tool to study the size effect, especially for nano-grain BTO ceramic.

The internal compressive stresses and internal shear stresses caused by the reduction of 90° ferroelectric domains and the incomplete sintering process can be sensitive to the applied hydrostatic-pressure. However, up to date, there are no documented studies on the progressive decrease of T_C in nano-BTO ceramics under high pressure. Samara *et al* [17]. and W. J. Merz *et al* [18-21] have systematically reported the change of the paraelectric–ferroelectric phase transition of coarse grain BTO in the 1950s and 1960s, respectively. They reported that the T_C of BTO decreased linearly with pressure at a rate from –35 to –76 K/GPa. Samara investigated the dielectric constant, polarization and Curie point of single crystal and ceramic BTO under hydrostatic pressure up to 2.5GPa. He suggested that the T_C shift increases with the c/a ratio of the tetragonal lattice. Recently, several authors also reported studies on the ferroelectric phase transition of perovskite-like materials under high pressure [22-23]. T. Ishidate [22] measured the temperature-pressure phase diagram of BTO by means of dielectric measurements under an applied pressure up to 10 GPa and temperature down to 4 K. He showed that all the three ferroelectric phase can be kept when temperature down to 4 K. The existence of the ferroelectric phase under high-pressure [23] showed some fundamental difference with that of the ferroelectric phase under low and ambient pressure. In this work, we report the measurement of the temperature dependence of the dielectric constant of coarse-grain as well as nano-grain BTO ceramic under different hydrostatic pressure from ambient up to 5 GPa from room temperature to 220 °C. Our aim here is to find out the pressure effect on the dielectric properties and the phase transition temperature of nano-grain and coarse-grain BTO ceramic to shed some lights on the fundamental principle of size effects.

2. Experimental setup

All samples were prepared by the two-stage [24] solid-state reaction method under high pressure. The pellets with a diameter of 5 mm were pressed uniaxially under 4 MPa pressure using 10 nm BTO starting powders, followed by an initial calcination at 600 °C for 12 h in ambient pressure; then, the sintered samples were heated up to 1000 °C for a few seconds (to increase density) under 5 GPa pressure using a cubic anvil type apparatus, followed by a second treatment at 1000 °C, 900 °C, and 830 °C for 30 min at the same pressure to prepare 5000 nm, 100 nm and 60 nm grain size BTO ceramic samples, respectively. The purpose of the initial sintering and the first step sintering under pressure is for better densification. The grain size is mainly controlled by the second step sintering under pressure at different temperatures.

The temperature dependence of the dielectric constant of BTO ceramic was measured under pressure at 10^5 Hz using an electrical impedance analyzer (HP 4192A). The temperature and dielectric constant values were recorded by a personal computer in an automated measurement set-up.

The electrodes on the samples were prepared by applying a thin coating of Ag paint on the two parallel planes of each sample. The dielectric constant between room temperature and 220 °C under pressure up to 5 GPa was measured on cubic anvil apparatus. During the experiment, the sample was placed in the center of a pyrophyllite cube (25×25×25mm3 in size) surrounded by a pressure transmitting medium (boron nitride). To maintain the hydrostatic pressure environment, the sample was shaped into a smaller but more regular geometry with 3mm in diameter and 1mm in thickness. A Cu wire of 0.1 mm in diameter encapsulated by a Teflon tube was used to make a flexible contact with the sample at the diagonal corners of the pyrophyllite cube and external leads were screened by a metal net. The sample was in a graphite tube and two metal pellets attached to the top and bottom anvils served as the heater and also as the contacts for the electric circuit. The pyrophyllite cube was placed in the center cavity of the cubic anvil apparatus, which could generate pressure up to 5 GPa.

3. Experimental results and discussions

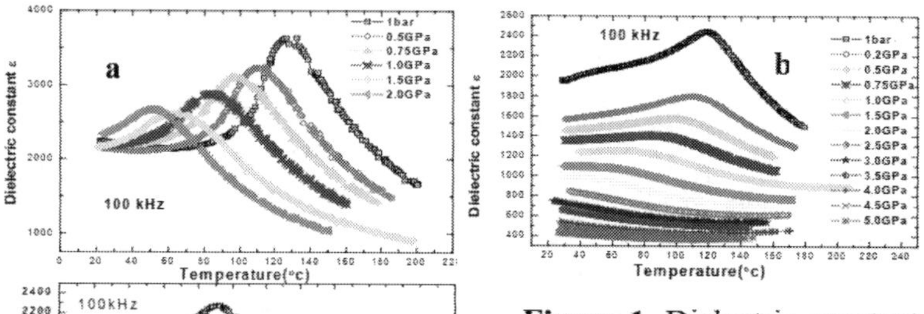

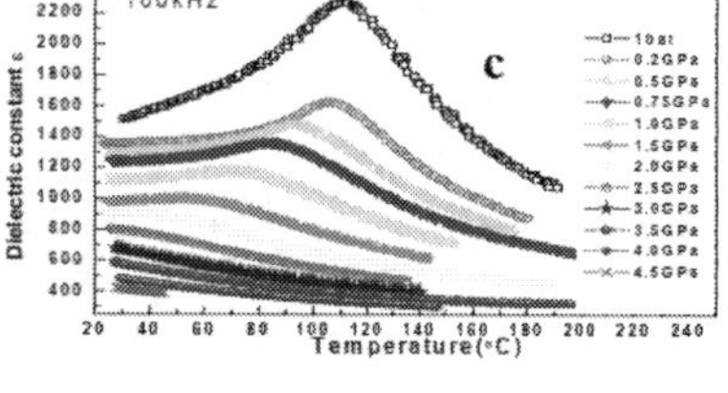

Figure 1. Dielectric constant vs. temperature under different pressures at 100 kHz for BTO with different grain size: (a) 5000 nm, (b)

Figure 1 shows the dielectric constant of BTO ceramic as a function of temperature measured at 100 kHz under different hydrostatic pressure. The grain size of the samples was determined by the mean intercept length measured by SEM. The grain sizes are: (a) 5000 nm, (b) 100 nm, (c) 60 nm, respectively. For the paraelectric-ferroelectric phase transition, it was found that the peak value of dielectric constant is strongly suppressed by decreasing grain size, which may be explained by the "dead layer" model for the grain boundaries that have very low dielectric constant [25].

The measured T_C is 128 °C, 120 °C, and 111 °C for the 5000 nm, 100 nm, and 60 nm grain-size BTO ceramic samples, respectively. As it has been reported, when decreasing the grain-size, T_C shifts to low temperature. Although the dielectric peak becomes more diffuse for the nano-grain BTO ceramic, a dielectric anomaly indicating the phase transitions are still clearly visible for all samples as shown in Figure 1. The diffuse transition peak in nano-grain BTO ceramic is attributed to both the grain size distribution and the presence of inhomogeneous internal stresses [26-30]. In nano-grain ceramic, nonuniform grain size distribution results in a distribution of the Curie temperature T_C because each grain will have different phase transition temperature. This will cause a diffuse peak in the total effective dielectric constant vs. temperature curve. Internal stresses at grain boundaries can produce spatial variations of the local polarization that can also lead to a distribution of T_C [31].

The grain size distribution is much wider in nano-grain BTO ceramics. In addition, smaller grain size results in higher internal stresses in the ceramic so that the distributions of the polarization and the structure phase-transition temperatures T_C will be even less uniform in smaller grain BTO ceramic. The progressive decrease of T_C is complicated by the corresponding electric and mechanic boundary conditions in nano-grain BTO ceramic. We suggest that the internal compressive stresses and the electrostatic screen effect are major factors in reducing T_C. Since the internal compressive stresses in the ceramics are believed to be compressive type because of the volume increase at the cubic to tetragonal phase transition, while compressive stress favors the cubic structure [32], therefore, the behavior of nano-grain BTO ceramic is very much like a system under a hydrostatic pressure. A GLD type theoretical model developed based on this idea has successfully explained the size effect in nano-grain BTO ceramic [33].

Compressive pressure favors smaller volume lattice structure, therefore, a trend of reducing c/a and reducing T_C is expected [18]. As shown in Figure 2, the transition temperature T_C decreases with increasing pressure up to 2 GPa. The dielectric peaks of the smaller nano-grain BTO ceramic samples at the phase-transition temperatures become more diffuse than the coarse grain-size samples. The grain size dependence of the transition temperature T_C under different pressure is illustrated in Figure 2. In the pressure range from 0 to 2GPa, T_C decreases linearly with pressure [22]. The rate of change $dT_C/d\sigma$= -40.0 K/GPa, -37.6 K/GPa, -34.3 K/GPa, respectively, for 5000 nm, 100 nm, 60 nm grain size BTO ceramic samples. This reduction of T_C is consistent with the increase of bulk moduli with decreasing grain size in nano-grain BTO ceramics.

Figure 3 give the pressure dependence of the dielectric constant at 30 °C measured at 100 kHz. One can see that the pressure effect on the dielectric constant is quite different in the coarse-grain and nano-grain BTO ceramic. For comparison, the dielectric constant data of coarse-grain BTO ceramic from Ref.[17] are also shown in Figure 3. From our results and the data of Ref.[17], the dielectric

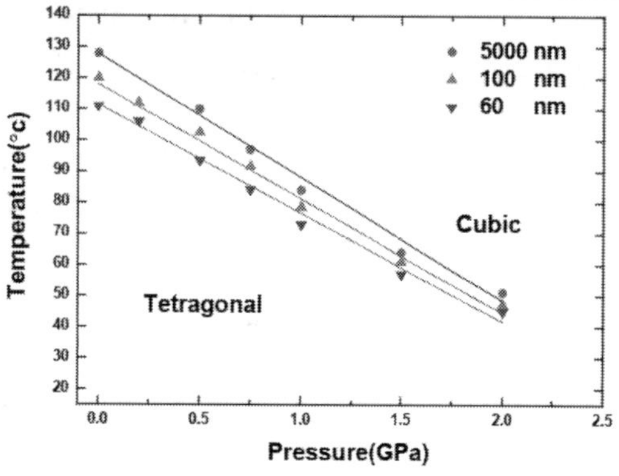

Figure 2. Dependence of the transition temperatures T_C on pressure and grain size. Solid lines are theoretical results and the symbol points are experimental data.

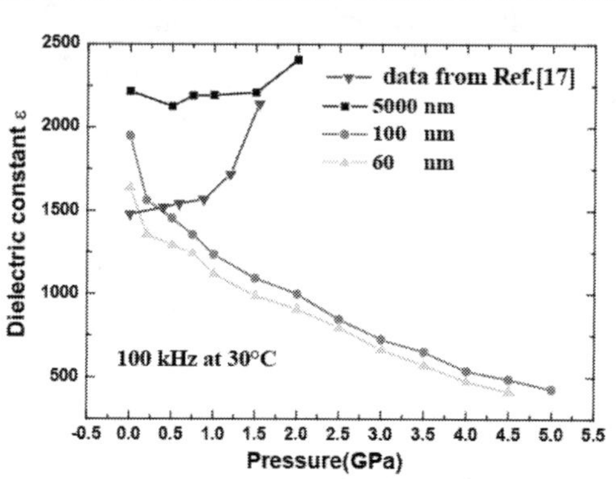

Figure 3. Dependence of the dielectric constant on pressure at 100 kHz at 30° C. The experimental data from ref. [17] are listed for comparison.

constant of coarse-grain BTO ceramic has the tendency to increase with increasing pressure at 30 °C despite the difference of the absolute value of the two sets of data. The change of the dielectric constant at 30 °C with pressure is a result of the competition between the dielectric constant increase caused by the downward T_C shift and amplitude depression caused by external pressure. In coarse-grain BTO ceramic, the T_C shift effect due to pressure is bigger than pressure suppression of dielectric constant; hence, the total dielectric constant increases. On the contrary, for the two nano-grain BTO ceramic samples, the dielectric peak is diffuse and becomes more diffuse under high pressure, therefore, the pressure suppression of dielectric constant is bigger than the dielectric constant increase due to pressure induced T_C shift. Therefore the total effective dielectric constant at 30 °C decreases with applied pressure. In addition, the coexistence of tetragonal and orthorhombic phases in nano-grain BTO ceramic may also contribute to the diffuse dielectric peak. [16]

4. Modified GLD theory

The GLD theory has been successfully used to study BTO single crystals [34-35]. In order to study the size effect in nano-grain systems, Lin *et al* [33] recently proposed a modified GLD free energy by considering the combined effects of applied stresses and grain size. They proposed that some of the GLD free energy coefficients depend on grain size and internal stresses so that the critical temperature T_C for the ferroelectric phase transition in BTO ceramic will be size dependent as well as dependent on the distribution of internal stresses accumulated inside the ceramic resulting from the lacking 90° domains to release the transformation strain.

In this paper, we just discuss the application of GLD theory for the paraelectric to ferroelectric phase transition. For nano-grain systems, each grain is almost single domain so that we can use a quasi-1-D model to describe the cubic paraelectric to the tetragonal ferroelectric phase transition. The generalized GLD free energy is given by [33]:

$$G_1 = \alpha_1(T)P^2 + \alpha_{11}P^4 + \alpha_{111}P^6 + \alpha_{1111}P^8 \quad (1)$$

Here P is the polarization, α_1 $(T,)$, α_{11}, α_{111} and α_{1111} are the dielectric stiffness and higher-order stiffness coefficients at constant stress. Based on Eq. (1), considering the interaction between external hydrostatic pressure σ and the internal compressive stresses in the nano-grain BTO ceramic, a modified GLD model is given below.

$$G_1 = \alpha_1 {}^*(T,\sigma,d)P^2 + \alpha_{11}P^4 + \alpha_{111}P^6 + \alpha_{1111}P^8 - 3(\frac{1}{2}s_{11} + s_{12})\sigma^2 - (Q_{11} + Q_{12})\sigma P^2 \quad (2)$$

where s_{ij} is the elastic compliance coefficients at constant polarization; and Q_{ij} is the cubic electrostrictive constant.

$$\alpha_1 {}^*(T,\sigma,d) = \alpha_1(T) + \frac{K + S\sigma}{10d} \quad (3)$$

is assumed to depend on temperature T, hydrostatic pressure and grain size d. To unify the units, the constant S has been set to be 2.4 m^5/C^2. We should notice that the internal compressive stresses are included in d, which presents the grain size effect [33].

By fitting the experimental results, K=8.186×109J·m^2/C^2, and d takes the value of 5000 nm, 100 nm, 60 nm. Other parameters [36] of the elastic Gibbs free energy were taken from Refs.[35,37].

Figure 2 gives the dependence of the transition temperature T_C on pressure and grain size. The solid lines are theoretical results and the symbol points represent experimental data. The calculated results using the modified GLD theory fit well with the experimentally measured data. From results in Figure 3, we may draw the following conclusions:

(1) For a given pressure, T_C decreases with the reduction of the grain size d. For example, T_C =111.46 °C and 111 °C for calculated and experimentally measured transition temperatures of BTO ceramic with 60 nm grains, which is about 17 °C lower than the transition temperature of coarse-grain BTO ceramic.

(2) External pressure has influence on internal compressive stress distribution as described by Eq.(2). From the physical viewpoint, mutual interaction of external pressure and internal compressive stresses is a competition between two mechanical forces. Some authors have reported the influence of stresses on nano-powder and nano-ceramic BTO ceramic [5,15]. In this work, we have quantified the influence by systematically measured the dielectric property change under high-pressure. A GLD theoretical calculation based on Eq. (2) was carried out to provide a phenomenological interpretation to the experimental results.

(3) The smaller is the grain size, the stronger is the modification of external pressure on internal compressive stresses. If removing the external pressure in Eq. (3), the calculated four lines will have the same slope $dT_C/d\sigma$ so that these lines will be parallel to each other. The interaction between external pressure and internal compressive stresses is the main factor that changes the slope $dT_C/d\sigma$ in nano-grain BTO ceramic. We must mention here that the internal shear stresses, which are considered to be the reason for the coexistence of tetragonal and orthorhombic phases in nano BTO ceramic

[15,16,38-39], were not included in our GLD equation since here we only treat the tetragonal to the cubic phase transition under a hydrostatic pressure, which does not involve shear deformation. Therefore, this GLD model should be further improved; i. e., the depolarization field and internal shear stresses have to be considered in this model.

5. Summary and Conclusions

We have shown experimentally and theoretically that the reduction of grain size in BTO ceramic produced internal compressive stresses that affect the nature of the paraelectric-ferroelectric phase transition. By applying an external hydrostatic pressure from ambient to 5 GPa, we have quantified the changes of T_C as well as the amplitude of the dielectric constant of nano-grain BTO ceramic with pressure. An interesting finding is the opposite trend of the pressure dependence of the dielectric constant in nano-grain and course-grain BTO ceramic. Our results showed that the paraelectric-ferroelectric phase transition temperature T_C becomes lower with reducing grain size.

In the pressure range 0~2 GPa, TC decreases linearly with pressure. The experimentally measured rates of change for 5000 nm, 100 nm, 60 nm BTO ceramics are: $dT_C/d\sigma$= -40.0 K/GPa, -37.6 K/GPa, -34.3 K/GPa, respectively. The variation of $dT_C/d\sigma$ in different grain size ceramic is mainly caused by the modification of the internal compressive stresses by the external high pressure. We also demonstrated that the grain size effect as well as the applied pressure effect could be well described by a modified GLD phenomenological theory.

Acknowledgments:

This research was supported by NSF & MOST of China through research projects.

References

[1] Haertling G H 1999 *J. Am. Ceram. Soc.* **82**, 797

[2] Venkateswaran U D, Naik V M and Naik R 1998 *Phys. Rev. B.* **58**, 14256

[3] Boulos M, Guillemet-Fritsch S, Mathieu F, Durand B, Lebey T and Bley V 2005 *Solid State Ionics* **176** 1301

[4] Wu Y C, Lee J S and Lu H Y 2007 *Journal of Electroceramics,* **18** 13

[5] Uchino K, Sadanaga E and Hirose T 1989 *J. Am. Ceram. Soc.*, **72**, 1555

[6] Zhong W L 1994 *Phys. Lett. A* **189**, 121

[7] Aoyagi S, Kuroiwa Y, Sawada A, Kawaji H and Atake T 2005 *J. Therm. Anal. Cal.* **81**, 627

[8] Yen F S, Hsing H I and Chang Y H 1995 *Jpn. J. Appl. Phys.* **34**, 6149

[9] Junquera J and Ghosez P 2003 *Nature (London)* **422**, 506

[10] Kim Y S, Kim D H, Kim J D, Chang Y J, Noh T W, Kong J H, Char K, Park Y D, Bu S D, Yoon J G and Chung J S 2005*Appl. Phys. Lett.* **86**, 102907

[11] Anliker M, Brugger H R and Kanzig W 1954 *Helv. Phys. Acta* **27**, 99

[12] Böttcher R, Klimm C, Michel D, Semmelhack H C, Völkel G, Cläsel H J and Hartmann E 2000 *Phys. Rev. B* **62**, 2085;Erdem E, Boettcher R, Clasel H-J, Hartmann E, Klotzsche G and Michel D 2005 *Ferroelectrics* **316**, 43

[13] Glinchuk M D and Morozovskaya A N 2003 Phys. Stat. Sol (b) **238**, 81; Glinchuk M D and Bykov P I 2004 *J. Phys: Condens. Matter* **16**, 6779

[14] Waser R 1997 *Integr. Ferroelectr.* **15** 39

[15] Arlt G, Henning D and With G de 1985 *J. Appl. Phys.* **58** 1619

[16] Li B R, Wang X H, Li L T, Zhou H, Liu X T, Han X Q, Zhang Y C, Qi X W and Deng X Y 2004 *Mater. Chem. Phys.* **83**, 23

[17] Samara G A 1966 *Phys. Rev.* **151** 378

[18] Merz W J 1950 *Phys. Rev.* **77**, 52

[19] Klimowski J 1962 *Phys. Status Solidi* **2**, 456

[20] Leonidova G G and Polandov I N 1962 *Fiz. Tverd. Tela* **4**, 2613

[21] Minomura S, Kawakubo T, Nakagawa T and Sawada S 1964 *J. Appl. Phys.* **3**, 562

[22] Ishidate T, Takahashi S Abe H and Môri N 1997 *Phys. Rev. Lett.* **78** 2397

[23] Íñiguez J and Vanderbilt D 2002 *Phys. Rev. Lett.* **89** 115503

[24] Chen I W and Wang X H 2000 *Nature* **404** 168

[25] Basceri C, Streiffer S K, Kingon A I and Waser R 1997 *J. Appl. Phys.* **82**, 2497

[26] Kinoshita New K and Yamaji A 1976 *J. Appl. Phys.* **47**, 371

[27] Kanata T, Yoshikawa T and Kubota K 1987 *Solid State Commun.* **62**, 765

[28] Zhong W, Zhang P, Wang Y and Ren T L 1994 *Ferroelectrics* **160**, 55

[29] Deng X Y, Wang X H, Wen H, Kang A G, Gui Z L and Li L T 2006 *J. Am. Ceram. Soc.* **89**, 1059

[30] Buscaglia M T, Viviani M, Buscaglia V, Mitoseriu L, Testino A, Nanni P, Zhao Z, Nygren M, Harnagea C, Piazza D and Galassi C 2006 *Phys. Rev. B* **73** 064114

[31] Buscaglia M T, Buscaglia V, Viviani M, Petzelt J, Savinov M, Mitoseriu L, Testino A, Nanni P,

Harnagea C and Zhao Z 2004 *Mats Nygren Nanotechnology* **15** 1113

[32] Zhao Z, Buscaglia V, Viviani M, Buscaglia M T, Mitoseriu L, Testino A, Nygren M, Johnsson M and Nanni P 2004 *Phys. Rev. B* **70,** 024107

[33] Lin S, Lü T Q, Jin C Q and X H Wang 2006 *Phys. Rev. B.* **74**,134115

[34] Bell A J and Cross L E 1984 *Ferroelectrics*, **59**, 197

[35] Li Y L, Cross L E, and Chen L Q 2005 *J. Appl. Phys.* **98**, 064101

[36] The list of the parameters of *G* (in SI units, the temperature *T* in 0C) used in the calculations. For BaTiO3: α_1(T)=4.124 × 105(T-118)Jm/C^2 ; α_{11}=-2.097 × 10^8Jm^5/C^4; α_{111}=1.294 × 10^9Jm^9/C^6; α_{1111}=3.863 × 10^{10}Jm^{13}/C^8; s_{11}=8.3 × $10^{-12}$$m^2$/N; s_{12}=-2.7 × $10^{-12}$$m^2$/N;$Q_{12}$=-0.043$m^4$/$C^2$; Q_{11}=0.1024m^4/C^2

[37] Pertsev N A, Zembilgotov A G and Tagantsev A K 1998 *Phys. Rev. Lett.* **80**, 1988

[38] Frey M H and Payne D A 1996 *Phys. Rev. B* **54** 3158

[39] Zembilgotov A G, Pertsev N A and Waser R 2005 *J. Appl. Phys.* **97**, 114315

Comparative studies of structural transition between AlN nanocrystals and nanowires

W W Lei, D Liu, J Zhang, Q L Cui[1], G T Zou

National Laboratory of Superhard Materials, Jilin University, Changchun 130012, P. R. China

E-mail: cql@jlu.edu.cn

Abstract. The structural transition of AlN nanocrystals and nanowires were investigated simultaneously under pressures up to 37.2 GPa by *in situ* angle dispersive high-pressure x-ray diffraction using synchrotron radiation source and a single diamond anvil cell. The size of hexagonal AlN nanocrystals and the diameter of nanowires are 45 nm on average. A pressure-induced wurtzite to rocksalt phase transition starts at 21.5 GPa and completes at 27.8 GPa for the nanocrystals and nanowires, respectively. The high-pressure behaviors of AlN nanocrystals the same as the AlN nanowires might arise from the similar size and diameter in nanocrystals and nanowires. Hexagonal AlN nanocrystals (45 nm) display an apparent volumetric contraction as compared to the AlN nanocrystals (10 nm) which might induce the difference of transition pressure.

1. Introduction

III-nitride compounds have attracted increasing attention in recent years due to their significant applications, such as optoelectronic and field-emission devices. As an important candidate of these compounds, Aluminum nitride (AlN) has various unique properties such as excellent thermal conductivity, high chemical resistance, high melting point, wide band gap and small electron affinity, which have gained considerable interests [1-5]. As for bulk AlN, high-pressure studies have proved that a pressure-induced wurtzite to rocksalt phase transition occurs at the pressure of about 20.0 GPa [6, 7]. In recent years, the high-pressure studies on nanomaterials have stimulated great enthusiasm in order to improve our understanding of the structural stability and the mechanical properties of nanomaterials [8-13]. The high-pressure measurements on AlN nanocrystals with an average crystal size of 10 nm show that the wurtzite to rocksalt phase transition starts at 14.5 GPa [14], the transition pressure is significantly lower than the bulk material. They indicated that the reduction of phase transition pressure observed results from the particle size below the critical size of 15 nm and the volumetric expansion. Since then, a similar behavior has been observed form numerous nanocrystals (e.g. TiO_2, γ-Fe_2O_3 and CeO_2) [11-13]. However, the high-pressure behavior, in particular the particle size larger than the critical size of 15 nm and the volumetric contraction, of nanocrystals is less known. Apparently more detailed experimental and theoretical works are needed to fully understand this high-pressure behavior of nanocrystals. As for AlN nanowires, our previous study [15] showed that the transition pressure at 24.9 GPa for the wurtzite to rocksalt phase transition is larger than the bulk material, we attributed distinct high pressure behaviors in AlN nanowires to unique geometry. However, there are no direct comparative high-pressure studies between AlN nanocrystals and AlN nanowires simultaneously in a single diamond anvil cell.

In this paper, we report a high-pressure study between AlN nanocrystals and nanowires simultaneously using an *in situ* angle dispersive synchrotron radiation technique in a single diamond anvil cell. The size in our AlN nanocrystals is markedly larger than the critical size of 15 nm and display an apparent volumetric contraction. This comparative subtle approach would eliminate

[1] Author to whom any correspondence should be addressed.

© 2008 IOP Publishing Ltd

systematic errors arising from instrument response using different techniques [8] and thus would allow effective detection of the indeed different high-pressure behaviors between AlN nanocrystals and AlN nanowires. The size of hexagonal AlN nanocrystals and the diameter of nanowires are 45 nm on average larger than the critical size of 15 nm. An onset pressure of 21.5 GPa from the wurtzite to rocksalt phase transition is observed in AlN nanocrystals and nanowires, respectively. Furthermore, the transition to the rocksalt phase is completed up to 27.8 GPa, which is very swift compared with the pressure-induced transition in bulk and Nanocrystals (10 nm) AlN. Such same high pressure behaviors in AlN nanocrystals and nanowires might be attributed to them with the similar size and diameter. The transition pressure of AlN nanocrystals (45 nm) is significantly larger than that in AlN nanocrystals (10 nm) and is also little larger than that in bulk AlN because of volumetric contraction.

2. Experimental details

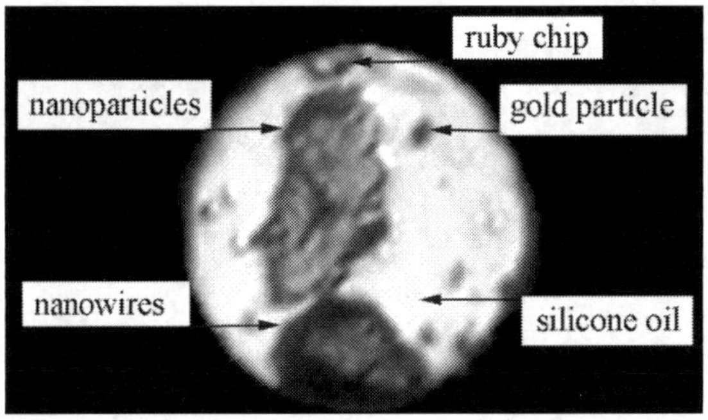

Figure 1. In-situ high-pressure optical micrograph of the AlN nanocrystals and nanowires in a single diamond anvil cell.in a single diamond anvil cell.

The AlN nanocrystals studied here were previously obtained by the dc arc plasma method through the reaction of metal aluminum (Al) with the mixture gas of nitrogen (N_2) and ammonia (NH_3) [16]. The high-density AlN nanowires were synthesized by a direct reaction of Al with 20 kPa N_2 in a closed chamber using arc discharge method [17]. The AlN nanocrystals and nanowires used here have an average diameter of 45 nm. In each case, the sample composition and phase purity were confirmed by the corresponding electron diffraction pattern analysis and powder x-ray diffraction. The pressure was generated in a diamond anvil cell with the diamonds having a culet size of 400μm. The AlN nanocrystals and nanowires with the liquid quasihydrostatic pressure-transmitting medium (Silicone oil) [18, 19] were loaded into a 100μm hole drilled in a T301 stainless steel gasket, as shown in Fig. 1. A ruby chip and some gold particles were put in the sample chamber as pressure marker.
In situ high pressure x-ray diffraction measurements were performed with angle dispersive synchrotron x-ray source (0.4074 Å) of the Advanced Photon Source at the Argonne National Laboratory, Argonne, IL. The diffraction data were collected using MAR165 CCD detector. The two-dimensional (2D) data sets were integrated using the software package Fit2D to yield conventional one-dimensional (1D) powder diffraction patterns [20].

3. Results and discussion

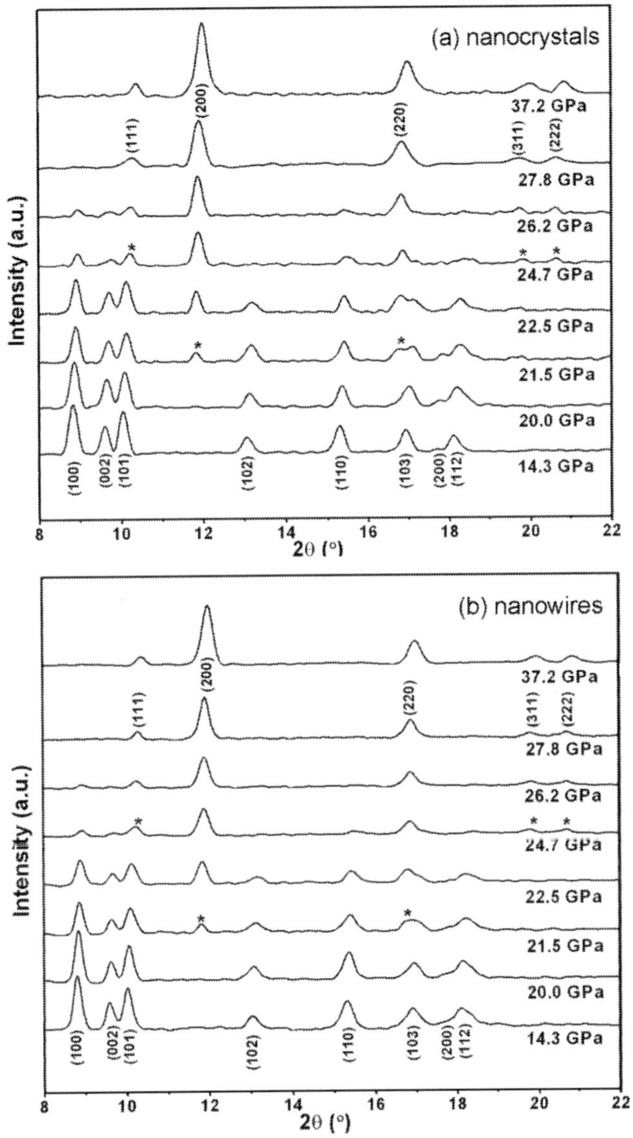

Figure 2. X-ray diffraction patterns of AlN nanocrystals and nanowires, respectively. The asterisk (*) denotes the occurrence of the new peaks of the rocksalt phase AlN.

Figure 2 (a) shows the XRD patterns of AlN nanocrystals up to 37.2 GPa. At ambient, all the diffraction peaks can be indexed to a pure wurtzite AlN crystal with lattice constants of a=3.110 Å and c=4.963 Å. The lattice parameters with corresponding unit cell volume V_0=41.56 Å^3 are obviously smaller than those in bulk AlN macrocrystals [a=3.111 Å, c=4.978 Å, V_0=41.72 Å^3 (Ref. 21)]. Figure 2 (b) shows the XRD patterns of AlN nanowires up to 37.2 GPa, all the diffraction peaks can be indexed to a pure wurtzite AlN crystal with lattice constants of a=3.125 Å and c=4.998 Å. The lattice parameters with corresponding unit cell volume V_0=42.26 Å^3 are obviously larger than those in bulk AlN macrocrystals [21]. As for AlN nanocrystals and nanowires, the wurtzite phase keeps stable up to 20.0 GPa, as shown in Figure 2 (a) and (b). When the pressure reaches 21.5 GPa, a new peak of the high-pressure phase starts to appear, indicating an occurrence of the phase transition. With increasing pressure, additional diffraction peaks assigned to the high-pressure phase arise remarkably, and meantime the diffraction peaks representing the wurtzite phase become weak. When the pressure elevates to 27.8 GPa, Figure 2 (a) and (b) show all diffraction peaks from the wurtzite phase disappear and the five obvious diffraction peaks from high pressure phase can be indexed to a cubic rocksalt phase with the unit cell parameters of a=3.95Å and V_0=61.54 Å^3 and a=3.91 Å and V_0=59.78 Å^3 (four units of AlN per unit cell), respectively. As compared with the high pressure parameters in bulk AlN, AlN nanocrystals and AlN nanowires (Table 1), the transition pressure in our samples is elevated and complete phase transition is quite swift. The effect of hydrostaticity might be responsible for the small pressure range (~6GPa) of the two coexistent phases.

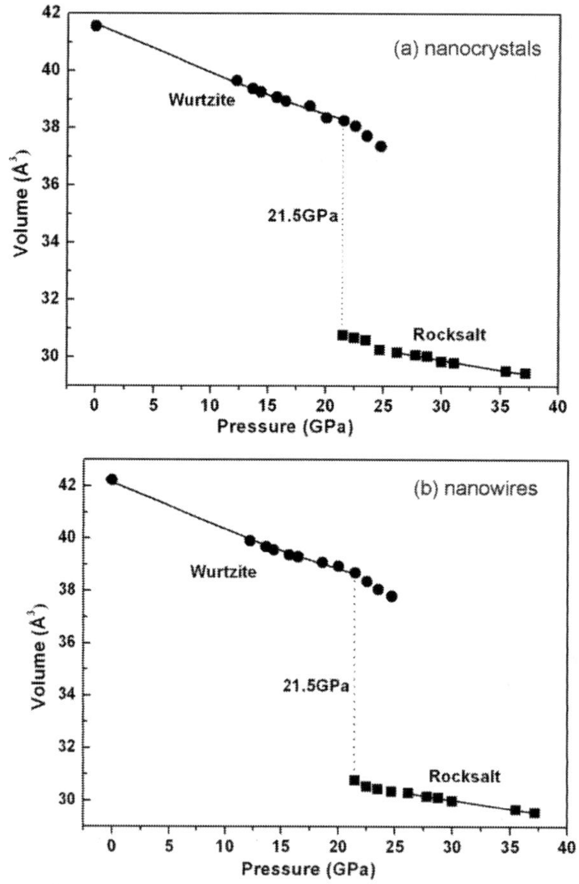

Figure 3. Pressure-volume relation of AlN nanocrystals and nanowires, respectively. The square and dot represent the wurtzite and the rocksalt phase, respectively. The line represents the fit to the Birch-Murnaghan equation.

Figure 3 (a) and (b) show the pressure-volume data of AlN nanocrystals and nanowires for both the wurtzite and rocksalt phases, respectively. It is seen that the wurtzite phase in each case still exists after the phase-transition pressure at 21.5 GPa, but the volumetric change of the wurtzite phase with the pressure became discontinuous after the phase-transition pressure. Similarly, the volume of rocksalt phase also has a big collapse at the initial transformation to rocksalt phase due to the existence of a preponderant wurtzite phase. When the pressure elevates to 26.2 GPa, the rocksalt phase is absolutely preponderant. Therefore, in the following fitting, we confine the wurtzite phase in the pressure range of 0–21.5 GPa and the rocksalt phase in the range of 26.2–37.2 GPa. The data were fitted with a third-order Birch-Murnaghan equation of state:

$$P = 3/2B_0\,[(V_0/V)^{7/3} - (V_0/V)^{5/3}]\,\{1 + 3/4(B_0' - 4)] \times [(V_0/V)^{2/3} - 1]\}, \qquad (1)$$

where B_0 and B_0' are isothermal bulk modulus and its pressure derivative at ambient temperature, respectively. In order to facilitate the comparison of the B_0 values, we have followed the standard procedure of setting $B_0' = 4.0$ for AlN nanocrystals and nanowires. The fitting results for wurtzite phase of AlN nanocrystals and nanowires give the bulk modulus B_0=215.8±8.5 GPa and B_0=208.8±8.8 GPa, respectively, which are close to that of bulk AlN crystals (B_0=207.9±6.3 GPa). The bulk modulus of rocksalt phase of AlN nanocrystals and nanowires are B_0=312.6±22.7 GPa, and B_0=324.9±15.8 GPa, respectively, which are a little larger than that in bulk AlN (295±17 GPa) [7]. Therefore, this indicates that the reduction of particle size can significantly lead to an enhancement of the bulk modulus.

Table 1. Transition pressure (P_{TR}), completion of phase transition (P_{CO}), volume collapse at transition (P_{CO}), unit cell volume, and equation of state parameters (B_0 and B_0') of wurtzite phase of AlN nanocrystals and nanowires, respectively.

AlN morphology	P_{TR} (GPa)	Pco (GPa)	ΔV (%)	$V0$ (Å3)	Bo (GPa)	B_0'	Ref.
Bulk	20.0-22.9	31.4	17.9	41.72	207.9±6.3	6.3±0.9	6, 7
Nanocrystals(10 nm)	14.5	36.9	20.5	42.09	321±19	4	13
Nanowires (45 nm)	24.9	45.4	20.0	42.26	303.0±3.9	4	14
Nanocrystals(45 nm)	21.5	27.8	19.5	41.56	215.8±8.5	4	a
Nanowires (45 nm)	21.5	27.8	20.0	42.26	208.8±8.8	4	a

[a]This study.

We summarize our results for AlN nanocrystals and nanowires and compare them with reference data in table 1. As for AlN nanowires, our previous study reveal that P_{TR} in AlN nanowires increases greatly with the diameter of 45 nm might be due to their unique geometry. The high-pressure behaviors of AlN nanocrystals (45 nm) the same as the AlN nanowires might arise from the similar size and diameter. It was reported that a significant reduction of the phase transition pressure (P_{TR}) in AlN nanocrystals with the size of 10 nm was mainly attributed to the size-induced volume expansion at ambient conditions and softening of both the Poisson ratio and shear modulus [14]. Similarly, the phase-transition pressure has been found to decrease with the particle sizes smaller than the critical size of 15 nm for the TiO_2, γ-Fe_2O_3 and CeO_2 [11-13]. Previous studies indicate that, below the critical size, the nanocrystals typically display an apparent volumetric expansion and softening of the Poisson ratio and shear modulus, but above the critical size, they exhibit a reverse volumetric contraction [22-24]. The size in our AlN nanocrystals is markedly larger than the critical size of 15 nm and display an apparent volumetric contraction, rather than an expansion. Therefore, it can be estimated that the Poisson ratio and shear modulus might not be softened occurred in AlN nanocrystals (45 nm). It was reported that the CdSe nanocrystals with an apparent volumetric contraction exhibit an increased transition pressure [25, 26]. Thus, it is logically conclude that the P_{TR} in AlN nanocrystals (45 nm) is significantly larger than that in AlN nanocrystals (10 nm) and is also little larger than that in bulk AlN because of volumetric contraction. The higher surface energy contribution gives a considerable impact on the enhanced bulk modulus. In addition, the large bulk modulus in AlN nanocrystals and nanowires also indicate that the compressibility is less than that of bulk AlN.

4. Conclusions

In summary, we have investigated the AlN nanocrystals and nanowires to 37.2 GPa simultaneously at room temperature and at a quasihydrostatic condition by using high-energy synchrotron X-ray diffraction. The size of AlN nanocrystals and the diameter of nanowires are 45 nm on average and exhibit a noticeable volumetric contraction and volumetric expansion, respectively. A pressure-induced wurtzite to rocksalt phase transition starts at 21.5 GPa and completes at 27.8 GPa for the nanocrystals and nanowires phases, respectively, which is very swift compared with the pressure-induced transition in bulk AlN. The effect of quasihydrostaticity might be responsible for the small pressure range ($\sim$6 GPa) of the two coexistent phases. The high-pressure behaviors of AlN nanocrystals the same as the AlN nanowires might arise from the similar size and diameter. The bulk modulus for the wurtzite phases were determined to be 215.8±8.5 GPa and 208.8±8.8 GPa for the nanocrystals and nanowires, respectively, larger than that of the bulk AlN crystals with a B_0 of 207.9±6.3 GPa. In contrast to the case in AlN nanocrystals (10 nm), the phase transition in AlN nanocrystals (45 nm) is significantly great. It is suggested that the volumetric contraction results in the increasing of the phase transition pressure.

Acknowledgments
This work was financially supported by National Basic Research Program of China (Nos. 2005CB724400 and 2001CB711201) and Natural Science Foundation of China (No. 50072005).

References

[1] U Mazur 1990 *Langmuir* **6** 1331Another reference
[2] C I Wu, E Kahn, E S Hellman and D N Buchanan 1998 *Appl. Phys. Lett.***73** 1346
[3]S P Grabowski, M Schneider, H Nienhaus, W Monch, R Dimitrov, O Ambacher and M Stutzmann 2001 *Appl. Phys. Lett.* **78** 2503
[4] M Akiyama, N Ueno, K Nonaka and H. Tateyama 2003 *Appl. Phys. Lett.* **82** 1977
[5] Y Taniyasu, M Kasu, and T Makimoto 2004 *Appl. Phys. Lett.* **84** 2115
[6] M Ueno, A Onodera, O Shimomura and K. Takemura 1992 *Phys. Rev. B* **45** 10123
[7] S Uehara, T Masamoto, A Onodera, M Ueno, O Shimomura and K. Takemura 1997 *J. Phys. Chem. Solids* **58** 2093
[8] J Z Zhang, Y S Zhao and B Palosz 2007 *Appl. Phys. Lett.* **90** 043112
[9] H Wang, J F Liu, Y He, Y Wang, W Chen, J Z Jiang, J S Olsen and L Gerward 2007 *J. Phys.: Condens. Matter* **19** 156217
[10] Rekhi S, Saxena S K and Lazor P J 2001 *Appl. Phys. Lett.* **89** 2968
[11] Jiang J Z, Olsen J S, Gerward L and Morup S 1998 *Europhys. Lett.* **44** 620
[12] Wang Z, Saxena S K, Pischedda V, Liermann H P and Zha C S 2001 *Phys. Rev.* B **64** 012102
[13] Z W Wang, S K Saxenal, V Pischeddal, H P Liermann and C S Zha 2001 *J. Phys.: Condens. Matte*r **13**, 8317
[14] Wang Z W, Tait K, Zhao Y S, Schiferl D and Zha C S 2004 *J. Phys. Chem.* B **108** 11506
[15] L H Shen, X F Li, Y M Ma, K F Yang, W W Lei, Q L Cui and G T Zou 2006 *Appl. Phys. Lett.* **89** 141903
[16] H D Li, G T Zou, H Wang, H B Yang, D M Li, M H Li, S Yu Y Wu and Z F Meng 1998 *J. Phys. Chem. B* **102** 8692
[17] L H Shen, X F Li, J Zhang, Y M Ma, F Wang, G Peng, Q L. Cui and G T Zou 2006 *Appl. Phys. A: Mater. Sci. Process.* **84** 73
[18] H Z Liu, J S Tse and H K Mao 2006 *J. Appl. Phys.***100** 093509
[19] H Z Liu, J Z Hu, J F Shu, D Häusermann and H K Mao 2004 *Appl. Phys. Lett.***85** 1973
[20] A P Hammersley, S O Svensson, M Hanfland, A N Fitch and D Häusermann 1996 *High Press. Res.* **14** 235
[21] Powder Diffraction File No. 76-0566, G. A. Jeffrey, G. S. Parry, and R. L. Mozzi 1956 *J. Chem. Phys.* **25** 1024
[22] T D Shen, C C Koch, T Y Tsui, G M Pharr 1995 *J. Mater. Res.* **10** 2892
[23] P Villain, P Goudeau, P O Renault, K F Badawi 2002 *Appl. Phys. Lett.* **81** 4365

[24] J E Spanier, R D Robinson, F Zhang, S W Chan, I P Herman 2001 *Phys. Rev. B* **64** 245407
[25] S T Tolbert, A P Alivisatos 1994 *Science* **265** 373
[26] J Y Zhang, X Y Wang, M Xiao, L Qu, X Peng 2002 *Appl. Phys. Lett.* **81** 2076

Studies of Novel Deuterides RMn_2D_6 (R – Rare Earth) compressed in DAC up to 30 GPa

[1*]S.M. Filipek, [2]H. Sugiura, [3]V. Paul-Boncour, [1]R. Wierzbicki, [4]R.S. Liu, [4]N. Bagkar

[1]Institute of Physical Chemistry PAS., ul. Kasprzaka 44, 01-224 Warsaw, Poland

[2] Graduate School of Integrated Sci., Yokohama City University, 22-2 Seta, Kanazawa-ku, Yokohama 236-0027, Japan

[3] Chimie Metallurgique des Terres Rares, ICMPE, GLVT, 2-8 rue Henri Dunant, 94320 Thiais Cedex, France

[4] Department of Chemistry, National Taiwan University, Taipei 106, Taiwan, Republic of China

*Corresponding author: smf@ichf.edu.pl

Abstract. The exposure of RMn_2 (C15 or C14 cubic Laves phase, where R = Y; Dy; Ho or Er) to high deuterium pressure leads to formation of novel, unique YMn_2D_6, $DyMn_2D_6$, $HoMn_2D_6$ and $ErMn_2D_6$ deuterides with cubic *Fm-3m* symmetry. In spite of different structures and molecular volumes of parent RMn_2 compounds, the molar volumes of RMn_2D_6 deuterides are almost identical. In this paper, we present results of studies on RMn_2Dx (where R = Y, Dy, Ho and Er) submitted to compression up to 30 GPa in diamond anvil cell (DAC) combined with energy dispersive X-ray diffraction. The EOS (equation of state) parameters of the above four RMn_2D_6 samples and YMn_2D_x, with $x \leq 4$ are compared.. The EOS parameters of YMn_2D_6 are very similar to those of other RMn_2D_6 but very different than those of interstitial deuterides YMn_2D_x ($x \leq 4$). The phase transition or segregation was not detected in RMn_2D_6 up to 30 GPa.

1. Introduction

The earlier investigations related to the influence of hydrostatic pressures on the properties of various groups of metal hydrides [1 - 4] were performed either in the piston-cylinder devices or in the Bridgman anvils high pressure apparatus. The introduction of more advanced techniques, such as the diamond anvil cell (DAC) made possible the *in situ* measurements of structural and optical properties of the matter in the range of pressures extended even above 100 GPa. However, hydrides derived from metals belonging to different groups of the periodic table were studied mainly at pressures not higher than 30 GPa. Besides the derivation of EOS (equation of state) parameters, a number of pressure induced structural changes were observed as, for example, in the alkaline metal hydrides [4-6], transition metals hydrides [7-9] and rare earth hydrides [10-13].

Despite large number of reports about properties of Laves hydrides, only few papers are related to their structural changes under high pressure [14-16]. Actually, Laves phases and particularly their hydrides belong to the most intensively investigated systems mainly due to their interesting structural

© 2008 IOP Publishing Ltd

and magnetic properties, which can be modified by dissolution of hydrogen and changing its concentration. In several Laves phase intermetallics, the hydrogen absorption capacity can reach or exceed 5 hydrogen atoms per formula unit, for example, YFe_2H_5 [17], YMn_2H_6 [18,19] and $ErMn_2D_6$ [20] which are synthesized recently by our group. In contrast to YFe_2H_5 which has orthorhombic structure derived from its parent C15 lattice, the YMn_2D_6 or $ErMn_2D_6$ undergo a complete structural reorganization. The neutron diffraction experiments [19] have shown that YMn_2D_6 has a fluorite K_2PtCl_6 type structure, which is cubic *Fm-3m* formed in a dramatic reconstruction of the initial *Fd-3m* space group. The refinement of neutron diffraction pattern indicated that manganese atoms are located in two different positions labeled as Mn1 and Mn2 in which Mn2 atoms are surrounded by six hydrogen/deuterium atoms. To the best of our knowledge, this type of hydride has never been obtained before by direct reaction of an intermetallic compound with gaseous hydrogen (deuterium).

It is noteworthy to mention that the distances between Mn2 atom and surrounding deuterium atoms in YMn_2D_6 are 1.65 Å, which are close to those observed in other M_2TH_6 hydrides (where M and T represent two different metals) [21]. It suggests the covalent character of the Mn2-D bonding in YMn_2D_6 [22]. The additional evidence for covalent bonding comes from the unusual thermal stability of YMn_2H_6 which is much higher than that of hydrides with lower hydrogen (deuterium) concentration. However, the change of electronic structure from metallic to covalent bonding was not supported by the X ray absorption (XAS) measurements in which the Mn $L_{2\text{-}3}$ near edge and Mn K-edge near edge structure of YMn_2D_6 [18] was close to that observed for $YMn_2H_{4.5}$. On the other hand, the high pressure studies of YMn_2D_x hydrides (with x = 1.15; 2; 3.4; 4 and 6) performed in DAC [15, 23] have shown that the compression of YMn_2D_6 has very unique character completely different from that of YMn_2D_x hydrides with smaller hydrogen concentration. This result gives another support for different type of Mn-D bonding in YMn_2D_6.

In the present investigation, we report the synthesis of novel RMn_2D_6 deuterides isostructural with YMn_2D_6 and $ErMn_2D_6$ starting from $DyMn_2$ and $HoMn_2$ intermetallics. The structural and magnetic properties of $DyMn_2D_6$ and $HoMn_2D_6$ will be published in detail elsewhere. The objective of this paper is to focus on the determination of EOS parameters and search for possible pressure induced phase transitions in these novel deuterides.

2. Experiments

The RMn_2 samples were prepared by induction melting of an appropriate amount of metals under vacuum or in argon atmosphere. The quality of samples was checked by XRD and microprobe analysis. Samples were located in the high pressure apparatus and pretreated at 100^0 C in vacuum before the deuterium charging. Then the deuteriation was carried out at pressure and temperature conditions as described elsewhere [19].

Powder samples were pressurized in DAC filled with alcohol as a pressure transmitting medium. Pressure was estimated according to the ruby scale. X-ray diffraction pattern were measured by energy dispersive type X-ray diffraction system with intrinsic Ge detector. Tungsten tube operated at 47.5 kV and 27.5 mA was used as a X-ray source. Energy axis of energy spectrum and slit angle in diffraction were calibrated by using fluorescence lines and diffraction lines in cubic InAs patterns.

3. Results and discussion.

The isostructural group of RMn_2D_6 deuterides crystallizing in *Fm-3m* space group has been formed under high deuterium pressure and high temperature conditions starting from RMn_2 (R = Y, Dy, Ho, Er) intermetallics exhibitting various crystal structures (C15 for YMn_2, C14 + C15 mixture for $DyMn_2$ and $HoMn_2$ and C14 for $ErMn_2$). The structutal parameters of the parent Laves phases and their corresponding RMn_2D_6 deuterides are summarized in the table 1. The unusual, very stable *Fm-3m* structure can be formed from compounds having different crystal structures (cubic C15 or hexagonal C14) and different molecular volumes (in descending order; YMn_2, $DyMn_2$, $HoMn_2$ and $ErMn_2$ due to the lanthanide contraction). It is surprising that in contrast to the parent compounds the volumes of RMn_2D_6 deuterides have nearly the same values of 75.2 ± 0.7 Å^3, which suggest that the volume expansion during the formation of RMn_2D_6 is inversely proportional to the initial molar volume of the corresponding RMn_2 compound. The case where the absorption of the same amount of hydrogen in isostructural compounds induces so different volume changes (12.1 Å^3 for $ErMn_2D_6$ and only 8,87 Å^3 for YMn_2D_6) is very unusual and can suggest that the stable $(Mn^ID_6)^{5-}$ complex anions play a dominant role in the structural properties of these compounds. It indicates that the "negative pressure" effects related to hydrogen absorption are different for different compounds. Thus it was reasonable to expect differences in compression behaviour of these deuterides.

Especially, one could expect higher compressibility of the deuterides formed with the greatest increase in the molecular volume.

Table 1. Structural properties of RMn_2 and RMn_2D_6 compounds (R= Er, Y, Ho and Dy)

Compound	Structure; percentage	A (Å)	c(Å)	V(Å^3)	Z	V/Z (Å^3)	$\Delta V/V_0$ (%)^a
YMn_2	C15; 100%	7.6791		452.82	8	56.60	
YMn_2D_6	Fm-3m, 100%	6.7083		301.89	4	75.47	33.33^c
$DyMn_2$	C14; 4%	5,3606	8,7343	217,36	4	54,34	
	C15; 95%	7.5872		436,76	8	54,59	
$DyMn_2D_6$	Fm-3m; 100%	6.720		303.52	4	75.88	39,63^b 38.98^c
$HoMn_2$	C14; 19%	5,3269	8,6885	213,51	4	53,38	
	C15; 80 %	7,5430		429,17	8	53,65	
$HoMn_2D_6$	Fm-3m; 100%	6.677		297.78	4	74.44	39,46^b 38.77^c
$ErMn_2$	C14, 100%	5.2949	8.6446	209.90	4	52.47	
$ErMn_2D_6$	Fm-3m, 100%	6.6796		298.04	4	74.51	42.00^b

^{a}Variation of the molecular volume: $\Delta V/V_0 = (V_{deuteride} - V_{parent})/V_{parent}$
^{b}Value calculated with respect to the parent C14 structure.
^{c}Value calculated with respect to the parent C15 structure.

The energy dispersive X-ray patterns of RMn_2D_6 deuterides taken at different pressures are shown in figure 1. For all investigated samples, a continuous decrease of the lattice parameter was observed with increasing pressure excluding possibility of phase transition. The dependence of pressure on the molar volumes of YMn_2D_6, $DyMn_2D_6$, $HoMn_2D_6$ and $ErMn_2D_6$ are plotted in the figure 2 together with EOS curves derived by using the second order Birch-Murnagham approach. The EOS parameters presented in table 2 together with our previous data for YMn_2D_x ($0 \leq x \leq 4.3$) [15, 23] give evidence that the compression of YMn_2D_6 differs from that of pure YMn_2 and YMn_2D_x ($0 \leq x \leq 4$) for which the bulk modulus changes discontinuously around 8 GPa. This was observed not only for $YMn_2D_{1.1}$ and YMn_2D_2 (which display reversible spinodal decomposition around 5 GPa) but also for both $YMn_2D_{3.4}$ (C15 cubic) and YMn_2D_4 (rhombohedral) deuterides.

A similar change of bulk modulus in $GdMn_2$, $GdMnH_1$ and $GdMn_2H_2$ was later reported by Palasyuk *et al.*[16] who also observed changes around 5 GPa suggesting a phase transition in both hydrides but could not give consistent explanation. The speculation on C15 – C14 phase transition seems to be unreasonable because of the reversible character of these changes. By taking into account of these reversible changes and almost identical volume-pressure relationship, it could be interpreted rather as spinodal decomposition found by us for $YMn_2D_{1.1}$, and YMn_2D_2 deuterides. On the other hand the magnetic properties of the RMn_2 compound are very sensitive to the change in distances and an increase in pressure is known to reduce the magnetic ordering temperature and also the spin fluctuation [26]. In these compounds the spin fluctuations lead to a large thermal expansion and a strong pressure dependence of the cell volume [27]. A suppression of spin fluctuation above 8 GPa could therefore explain the modification of the bulk modulus in RMn_2 compounds.

Anyhow, it would be interesting to confirm whether the discontinuous change of bulk modulus around 8 GPa is limited only to YMn_2, $GdMn_2$ and their interstitial hydrides/deuterides or is more common for RMn_2 based compounds.

The dependence of pressure on the molar volume of YMn_2D_6 has different character. For whole investigated range of pressure (up to 35 GPa) only one set of EOS parameters is sufficient to fit the experimental data. The same was confirmed also for $DyMn_2D_6$, $HoMn_2D_6$ and $ErMn_2D_6$.

Table 2. EOS parameters of RMn_2 and RMn_2D_x deuterides (R = Y, Dy, Ho and Er)

K_0 and *K_0'* are bulk modulus and its first derivative

Sample	*Vo* cm3/mol	*Ko* [GPa]	*Ko'*	*n*	Remarks	Refer.
YMn_2	34.02	21.6	13.2	21	*Cubic* C15 (low pressure segment *P*<8GPa)	[15,23]
	31.24	83.1	7.13	29	*Cubic* C15 (high pressure segment *P*>8Gpa)	
$YMn_2D_{3.4}$	40.11	60.28	4.0 fixed	11	*Cubic C15* (low pressure segment *P*<8GPa)	
	37.80	133.9	4.0 fixed	20	*Cubic* C15 (high pressure segment *P*>8Gpa)	
YMn_2D_4	41.65	46.07	4.0 fixed	27	*Rhombohedral* (low pressure segment)	
	38.22	153.0	4.0 fixed	17	*Rhombohedral*(high pressure segment)	
YMn_2D_6	45.35	68.05	6.28		*cubic* Fm-3m (whole range)	
$DyMn_2$	33,01	63.2	6.1		Cubic (C15)	[24]
$DyMn_2D_6$	45.69	83.8	4.59	60	*cubic* Fm-3m (whole range)	
$HoMn_2$	32,31				*C15 + C14*	
$HoMn_2D_6$	44.83	73.2	5.85	23	*cubic* Fm-3m (whole range)	
$ErMn_2$	31,6				*C14*	
$ErMn_2D_{4.6}$	40.65	79.2 (2)	4 fixed		*hexagonal*	[28]
$ErMn_2D_6$	44.97	76.1	5,20	50	*cubic* Fm-3m (whole range)	

n *–number of points*

The RMn_2D_6 family of deuterides is unique in this aspect that both interstitial (RMn_2D_x; $x \leq 5$) or complex hydrides (RMn_2D_6) can be formed from the same intermetallic compound. In order to get more insight, the EOS parameters (within the high pressure segment (HP)) of the YMn_2 are compared (table 2) with those of three single phase deuterides with different deuterium content and different crystal structure ($YMn_2D_{3.4}$ – Cubic C15; YMn_2D_4 – rhombohedral; YMn_2D_6 Cubic Fm-3m). It was observed that for both interstitial hydrides the bulk modulus increases with deuterium concentration. For YMn_2D_4 its value is more than 80 % higher than for parent YMn_2. As we discussed elsewhere [25] a large increase of bulk modulus follows the formation of ionic hydrides (alkali metals and rare earths). A large decrease (by 50 %) in bulk modulus was observed during the formation of covalent hydrides, such as AlH_3 and CuH. On the other hand, the bulk modules of transition metals and their hydrides do not differ much. It is interesting that during formation of interstitial deuterides $YMn_2D_{3.4}$ and YMn_2D_4

a large increase in bulk modulus occurs, whereas further increase in deuterium concentration up to 6 D/f.u. (f.u.= formula unit) causes the reduction in bulk modulus slightly below the value of the parent YMn_2 in the high pressure range. This can be used as an additional support for the different character of YMn_2D_6 compound (complex deuteride) compared to other YMn_2 based deuterides (interstitial deuterides). The pressure dependence of volume for $ErMn_2H_{4.6}$ which crystallizes in a hexagonal structure has been studied [28]. They also found a change from a LP to HP phase at 3GPa. The bulk modulus of their HP phase was 79.2 GPa, which is slightly higher from that observed for our $ErMn_2D_6$ (76.1 GPa).

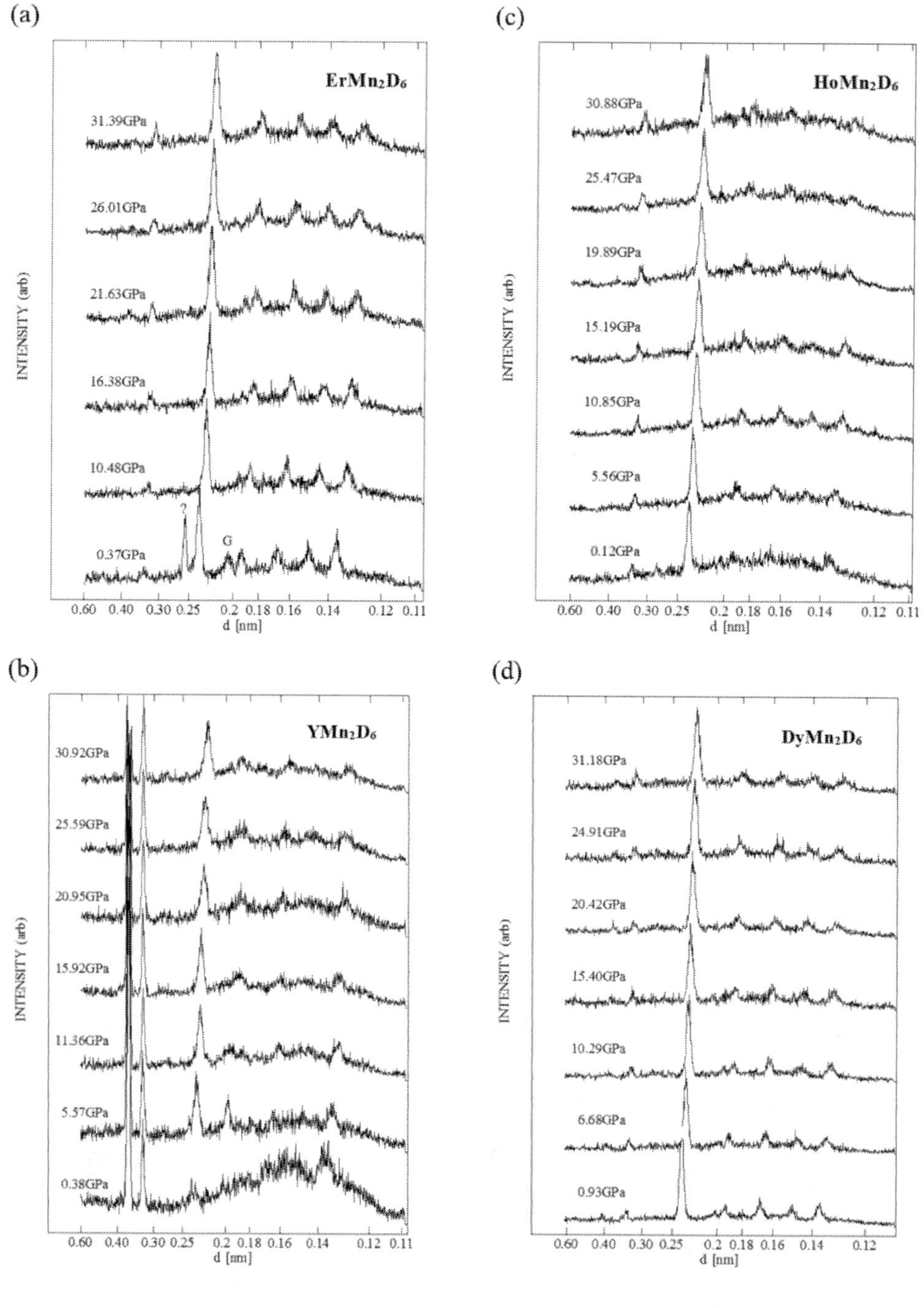

Fig. 1: Energy dispersive X-ray diffraction patterns of $ErMn_2D_6$ (a); YMn_2D_6 (b); $HoMn_2D_6$ (c) and $DyMn_2D_6$ taken at different pressures

As can be seen from table 2 and figure 2, the volume – pressure relationships for all RMn_2D_6 are located very close each other. The values of their bulk modulus are of the similar order and, in the case of YMn_2D_6 and $DyMn_2D_6$, it does not differ much from those of the HP values for parent compounds. This is one more unique property of this group of deuterides.

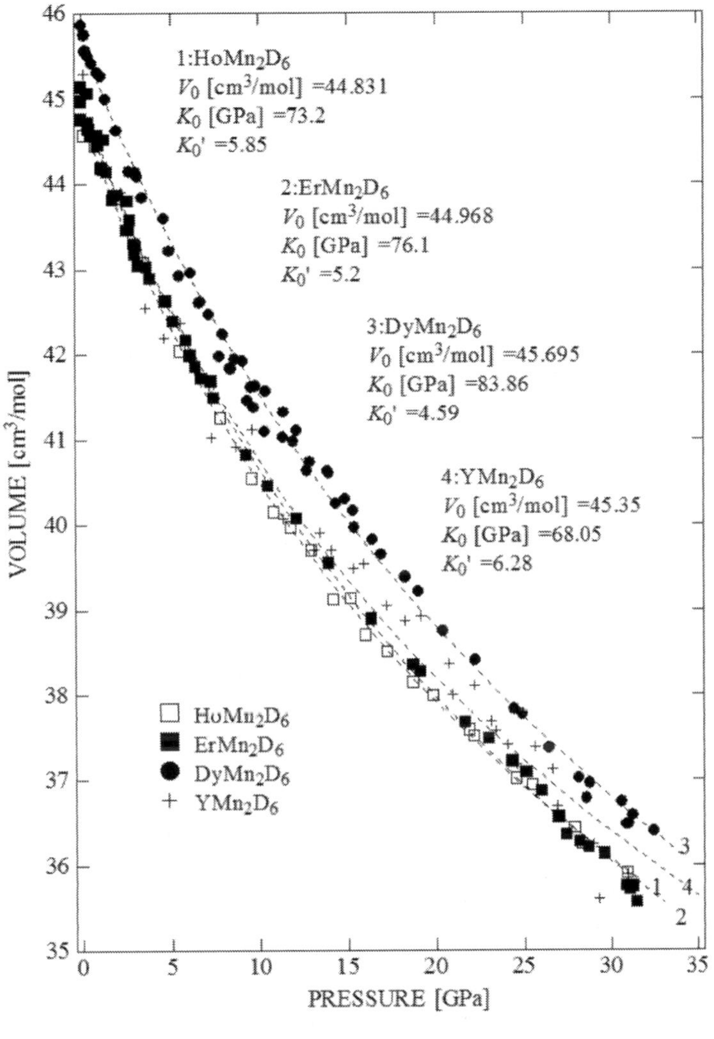

Fig.2. Pressure dependence of the molar volume of $ErMn_2D_6$, YMn_2D_6, $HoMn_2D_6$ and $DyMn_2D_6$.

4. Conclusions

A series of novel RMn_2D_6 deuterides were studied by high pressure energy dispersive X-ray diffraction. The initial volumes of four RMn_2D_6 investigated compounds and their EOS parameters, obtained by fitting the experimental data derived from a second order Birch-Murnagham equation, have similar numerical values. In the case of YMn_2D_6 and $DyMn_2D_6$ the EOS parameters are of the same order as their parent compounds. For interstitial YMn_2D_x ($x \leq 4$) bulk modulus increases with the deuterium concentration but falls down for YMn_2D_6 to a value of the parent compound. The syntheses oriented to find more Laves derived complex hydrides as well as DAC experiments on other series of RMn_2 based hydrides and deuterides are currently underway.

Acknowledgements.
This work was supported by State Committee for Scientific Research under the grant number KBN-

3T/08C03128 and 08C03128 and National Science Council of Taiwan under the grant number 96-2113-M-002-014-MY2.

References

[1] Filipek S M, Wakamori K, Sawaoka A, Saito S 1976 Proc. *Dai-17-kai Koatsu Toronkai (17th Conf. on High Pressure, Sapporo, Japan, Octob.1976)*, 1B14.
[2] Sawaoka A, Wakamori K, Filipek S M 1977 Proc. *Spring Meeting of the Phys. Soc. of Japan, (April 1977, Yamaguchi, Japan)*
[3] Wakamori K, Sawaoka A, Filipek S M, Baranowski B 1982 *J.Less-Common Metals*, **88** 217.
[4] Ponyatovskii E G and Bashkin I O 1985 *Z.Phys Chem (NF)* **146** 137.
[5] Hochheimer H D, Strossner K, Honle W, Baranowski B, Filipek S M 1985 *Z.Phys.Chem.(N.F.)*, **143** 139
[6] Duclos S J, Vohra Y K, Ruoff A L, Filipek S M and Baranowski B 1987 *Phys.Rev.* **36** (14) 7664
[7] Tkacz M, Baranowski B 1992 *Polish J. Chem.* **66** 1301
[8] Tkacz M, Baranowski B 1994 *Fresenius J.Anal.Chem.* **349** 97 - 101
[9] Filipek S M, Sugiura H, Skoskiewicz T 2000 *AIRAPT-17 Proc. Science and Technology of High Pressure eds* Murli H. Manghnani, William J. Nellis and Malcolm F. Nicol (Universities Press, Hyderabad, India 2000) 550-554
[10] Sawaoka A, Wakamori K, Filipek S M and Baranowski B 1980 *Trans. Jpn. Inst. Met.* **21** 41.
[11] Palasyuk T, Tkacz M, Vajda P 2005 *Solid State Commun* **133** 226
[12] Palasyuk T, Tkacz M 2005 *Solid State Commun* **133** 481
Palasyuk T, Tkacz M 2005 *Solid State Commun* **141** 302
Palasyuk T, Tkacz M 2005 *Solid State Commun* **141** 354
[13] Machida A, Ohmura A, Watanuki T, Ikeda T, Aoki K, Nakano S and Takemura K 2006 *Solid State Commun.* **138** 436
[14] Sugiura H, Marchuk I, Paul-Boncour V, Percheron-Guegan A, Kitazawa T, Filipek S M 2003 *J. Alloys Compd.* **356-357** 32
[15] Sugiura H, Paul-Boncour V, Percheron-Guégan A, Marchuk I, Hirata T, Filipek S M, Dorogova M 2004 *J. Alloys and Compd*, **367** 230.
[16] Palasyuk T, Figiel H and Tkacz M 2004 *J. Alloys Compd.* **375 62 - 66**
[17] Paul-Boncour V, Filipek S M, Percheron-Guegan A, Marchuk I, Pielaszek J 2001 *J. Alloys Compd.* **317-318** 83-87
[18] Wang C-Y, Paul-Boncour V, Kang C-C, Liu R-S, Filipek S M, Dorogova M, Marchuk I, Hirata T, Percheron-Guegan A, Sheu H-S, Jang L-Y, Chen J M and Yang H D, 2004 *Solid State Commun.* **130** 815-820
[19] Paul-Boncour V, Filipek S M, Dorogova M, Bourée F, André G, Marchuk I, Percheron-Guégan A and Liu R-S 2005 *J. Sol .State Chemistry* **178** 356 – 362
[20] Paul-Boncour V, Filipek S M, André G, Bourée F, Guillot M, Wierzbicki R, Marchuk I, Liu R-S, Villeroy B, Percheron-Guégan A, Yang H D and Pin S C 2006 *J Phys.;Condens. Matter* **18** 6409
[21] Orgaz E, Gupta M 1993 *J.Phys.* **5** 6697-6718
[22] Didisheim J J, Zolliker P, Yvon K, Fischer T, Schefer J, Gubelmann M and Gupta M 1984 *Inorg. Chem.* **23** 1953 - 57
[23] Sugiura H, Filipek S M, Paul-Boncour V, Marchuk I, Liu R-S, Pyun S I 2006 *Nukleonika* **51** (Supplement 1) S73-S77.
[24] Reiss G, Energiedispersive Roentgenbeugungsuntersuchungen an einigen Laves-Phasen unter hohem Druck, February 2000 Dissertation, University of Paderborn
[25] Filipek S M, Journal of Adv. Science *accepted for publication*
[26] Hauser R, Bauer E, Gratz E, Haufler T, Hilscher G and Wiesinger G 1994 *Phys. Rev. B* **50** 13493
[27] Shiga M, Wada H, Nakamura Y, Deportes J and Ziebeck K R A 1988 *Journal de Physique* **C8** 241.
[28] Makarova O L, Goncharenko I N and Le Bihan T 2004 *Sol. State Comm.* **132** 329,

Garnet growth in natural high pressure regime: a key tool in unravelling the pressure-temperature path of rocks involved in mountain belt orogeny

G Angì[1], R Cirrincione[2], E Fazio[1], G Ortolano[1], A Pezzino[1], R Punturo[1]

[1] Dipartimento di Scienze Geologiche, University of Catania, Corso Italia, 57, 95129 Catania, Italy

[2] Dipartimento di Scienze della Terra, University of Calabria, Via P. Bucci, 87036 Arcavacata di Rende (CS), Italy

Corresponding author's e-mail address: punturo@unict.it

Abstract. A study of a natural system concerning the growth of minerals (garnet) into metapelite host rocks from Calabria (southern Italy), which underwent pressure up to 1.2 GPa, is presented. Through the application of a thermodynamic model it was possible to determine P and T conditions achieved during metamorphic evolution of these rocks useful to reconstruct the related tectonic settings of poly-orogenic metamorphic complexes.

1. Introduction

Subduction of oceanic lithosphere and continental collision, occurring during the plate convergence, represent the most significant examples of natural system in which crustal rocks can achieve high pressure (HP) metamorphic conditions. In this scenario, entire crustal sectors are shifted to deeper levels up to 60-70 km, where they constitute the roots of orogenic mountain belts. In such environments, new crystalline phases as well as indicative polymorphs occur. At the same time, other minerals change their composition during growth. This is strictly dependent on environmental conditions, and as a consequence compositional trends within mineral phases reveal information about experienced P and T metamorphic conditions.

Garnet represents one of the most common minerals used as tracer of P-T changes. It is normally found both in gneissic- and schistose-type metamorphic rocks derived from argillaceous protolith, as well as in magmatic-derived rock, always due to the effects of high-pressure and high-temperature conditions.

Garnet is a marker of changes occurring during metamorphic evolution, because of its very sensitive composition to pressure and temperature variations. In addition, cation diffusion within its crystalline framework is sufficiently slow to preserve compositional differences within a grain. For the above reasons, it represents a key mineral in interpreting the genesis of many metamorphic rocks via geothermobarometry applications.

The structural formula of garnet is characterised by three different sites ($X_3Y_2Z_3O_{12}$) occupied by cations in the following ways: a) X is 8-coordinated (distorted cubic) site that incorporates Ca, Mn,

© 2008 IOP Publishing Ltd

Fe^{2+} and Mg; b) Y is a small 6-coordinated (octahedral) site that incorporates Al, Fe^{3+} and Cr^{3+}; c) Z is a tetrahedral site where generally Si allocates (Figure 1a).

Most of common garnets are characterised by an ideal solid solution among the following endmembers: almandine ($Fe_3Al_2Si_3O_{12}$), pyrope ($Mg_3Al_2Si_3O_{12}$), spessartine ($Mn_3Al_2Si_3O_{12}$) plus grossular ($Ca_3Al_2Si_3O_{12}$), even if the relatively high fO_2 values can lead to the formation of Fe^{3+} endmember (andradite - $Ca_3Fe_2Si_3O_{12}$) as schematically reported in Figure 1b.

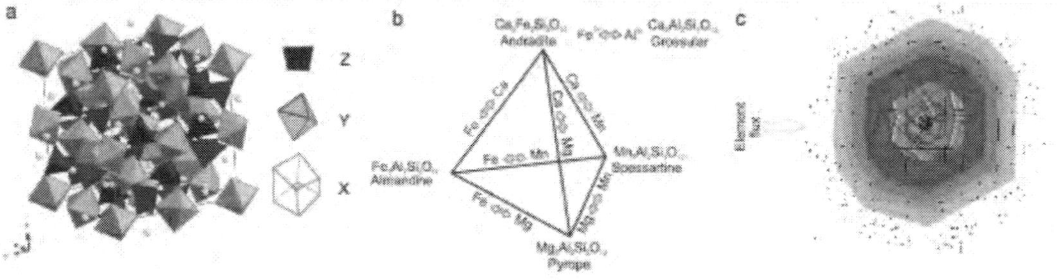

Figure 1. a) 3-D crystal framework of garnet. The octahedral and tetrahedral sites are occupied by Al, Fe^{3+}, Cr^{3+} and Si, respectively. Circles represent the bivalent cations (Ca, Mn, Fe^{2+}, Mg) placed in the distorted cubic site; **b)** Composition space of garnet showing the compositional exchange vectors of the garnet solid solution series; **c)** Schematic representation of a possible relative concentration gradient around a garnet porphyroblast.

2. Garnet as a natural geobarometer

Natural garnet growth depends on several parameters, which operated during the tectono-metamorphic evolution of the Earth crust. Compositional changes during growth (i.e. zoning) preserved in garnet crystals are interpreted as an evidence for fractional crystallisation, which changed the chemical composition of the matrix around the crystals during its growth [1] (Figure 1c). The chemical composition of the matrix around garnet crystals depends on the amounts of elements fractionated into the mineral and on the element flux into the matrix volume [2, 3]. If fractional crystallisation of garnet in a given matrix volume is faster than the transport of garnet nutrients (Figure 1c) into this volume, the composition of the matrix reservoir is progressively depleted in elements fractionated into the garnet crystal [4]. These phenomena are significantly controlled by the variation of $\Delta P/\Delta T$ ratio, strictly connected to different geothermal gradient occurring during the evolution of plate geodynamics. Garnet zoning, developed during crystal growth, represents the response of a natural system (here identified by the bulk rock chemistry considered to be in equilibrium – i.e. the effective bulk composition described in ref [5]) to P-T variations.

Since garnet constitutes a very widespread mineral in rocks metamorphosed at HP conditions, its chemical variations become a useful tool for geobarometric estimates, able to unravel the tectono-metamorphic history of the mountain belts.

It is worth noting that the above considerations are based on the formal thermodynamic relationship among pressure, temperature and mineral composition that forms the basis of all calculations of mineral equilibria as a function of these variables (Equation 1).

$$\Delta G_R = 0 = \Delta H_{298,1} + \int_{298}^{T} \Delta C_P \cdot dT + \int_{1}^{P} \Delta V \cdot dP - T\left(\Delta S_{298,1} + \int_{298}^{T} \frac{\Delta C_P}{T} \cdot dT\right) + RT\ln K \quad (1)$$

The terms ΔG, ΔH, ΔS, ΔV, ΔCP are the free energy, enthalpy, entropy, volume and heat capacity of reaction, respectively. K represents the constant of equilibrium, which is function of the activity coefficient of the reactants/products ratio of a reaction. At equilibrium, the free energy of reaction must be zero.

There are two general methods applied for solving the metamorphic petrology problems:

1. Calculations by using a single equilibrium relation of the form of Equation 1. In some applications two or more are combined and solved simultaneously for parameters such as pressure and temperature. Conventional geothermometry and geobarometry are examples of this type of approach.
2. Calculations that utilize a set of equations of the form of Equation 1 that completely define the phase equilibria of the system. These types of calculations permit simultaneous calculations of the compositions of all minerals in the same assemblage at equilibrium at any values of P and T.

In the former case, the goal of geothermobarometry is to infer the conditions at which one rock equilibrated, by determining the equilibrium constant to be used in the equation for calculation of T (at fixed P) and P (at fixed T) or both. The basic method of geothermobarometry is very straightforward. Values of ΔH, ΔS, ΔCp, and ΔV are known from experimental calibrations or published thermodynamic data. Within a rock, the value of the equilibrium constant is obtained by determining the compositions of coexisting minerals (i. e. mineral paragenesis) by means of an electron microprobe (EPMA). Analytical data are then used to draw a line of constant equilibrium on a P-T diagram and it is inferred that the sample must have equilibrated somewhere along this line. If two different equilibria can be evaluated then two lines of constant equilibrium can be drawn on the P-T diagram and the intersection of the two lines defines a single pressure and temperature value of equilibrium.

In the latter case it is possible to infer calculations that incorporate only the equations of chemical equilibrium (Equation 1), and calculations that incorporate both the equations of chemical equilibrium and the equations of mass balance. In this last case, it can be possible to ascertain the intensive variables T, P and mineral composition, as well as the mineral abundances within the system (modes). Moreover, the equilibrium assemblage as a function of rock bulk composition can be examined.

Furthermore, theoretical predictions of representative equilibrium assemblages, supported by petrological and chemical observations, can provide a series of P-T paths by using equilibrium-phase relations calculated by free energy minimisation for a system of specific bulk composition (i.e., pseudosection calculation method - [6, 7, 8]). For each sample, the P-T history is detailed by combining petrological and rock-textural information of the studied garnets.

Even if the above method seems to entirely overcome the intrinsic contradictions related to conventional geothermobarometry, mainly based on exchange reactions among minerals in different domains of a single rock [9] calculated at specific experimental conditions, several sets of limitations must be considered in order to achieve reliable results.

In particular it is important to take into account: a) the reliability of the thermodynamic data from available datasets; b) errors in the effective bulk composition used in the calculation [5]; c) extreme simplification of solid solution models of some minerals like garnet.

In the present study we considered three garnet bearing rocks with comparable bulk chemical composition (Table 1a), collected into three distinct areas of the southern sector of Calabria-Peloritani Orogen (Figure 2). The P-T evolution is achieved by intersection of garnet isopleths (i.e. iso-compositional curves) of garnet bearing metapelites using the thermodynamic modelling via P-T pseudosections [8].

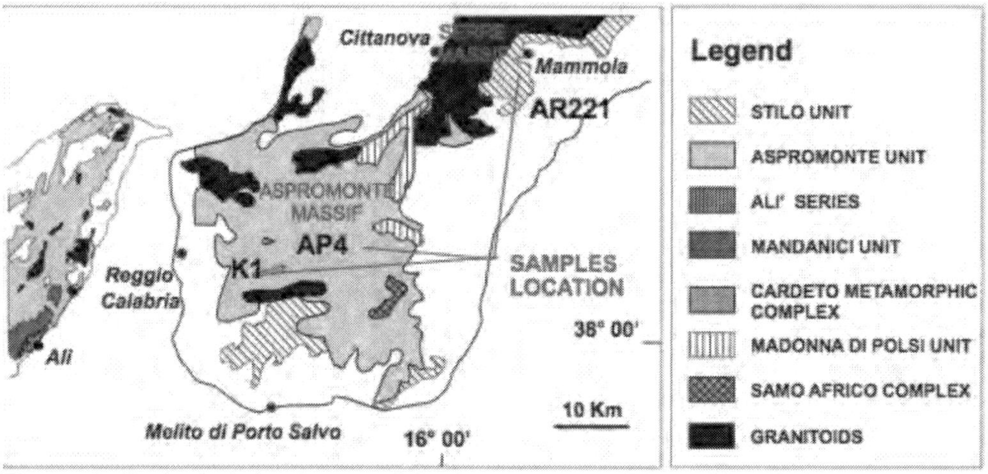

Figure 2. Geological sketch map of the Aspromonte and Serre Massifs in southern Italy (modified after [12]) and location of collected garnet bearing samples.

Table 1a. XRF-bulk rock analyses of metapelite samples in the simplified chemical system MnNCKFMASH.

Location	Serre Massif	Aspromonte Massif	
Sample	AR221	K1	AP4
Lithotypes	Paragneiss	Micaschist	Paragneiss
MnO	0.19	0.06	0.11
Na_2O	3.43	0.73	1.00
CaO	3.91	0.45	0.16
K_2O	3.38	4.66	4.00
FeO^{tot}	7.11	6.94	7.85
MgO	2.81	2.33	1.76
Al_2O_3	22.18	22.75	20.23
SiO_2	56.99	62.08	64.89

Table 1b. End-member values from chemical composition of garnet point analyses.

Sample	Garnet zone	Point analyses	Alm%	Grs%	Sps%	Pyr%
AR221	Grt1	A1	56	27	12	3
	Grt2	A2	64	23	8	5
	Grt3	A3	61	9	23	7
K1	Grt1	B1	69	12	15	3
	Outer Grt1	B2	77	14	2	7
	Grt2	B3	85	6	7	2
AP4	Grt1	C1	84	3	13	10
	Inner Grt2	C2	55	22	20	3
	Outer Grt2	C3	55	24	18	3

3. The solution of geological problems via thermodynamic modelling approach: a case study

The Calabria Metamorphic basement represents a natural laboratory to investigate and describe several types of garnet bearing rocks which often reflect a multi-stage metamorphic evolution developed during different orogenic cycles.
As already evidenced, in our study we considered three garnet typologies, which belong to metapelite samples from the Aspromonte and Serre Massifs located in the southern sector of the Calabria Peloritani Orogen (Figure 2).

Table 2. Summary tables of the solid solution models with brief commentary

Name and phase of solution model	End members	
Biotite - Bio (HP)	Mn-biotite - Mnbi	$KMn_3AlSi_3O_{10}(OH)_2$
	Annite - Ann	$KFe_3AlSi_3O_{10}(OH)_2$
	Phlogopite - Phl	$KMg_3AlSi_3O_{10}(OH)_2$
	Eastonite - East	$KMg_2Al_3Si_2O_{10}(OH)_2$
	mnts_i	1east+2/3mnbi-2/3phl
	sdph_i	1east+2/3ann-2/3phl
Garnet - Gt (HP)	Almandine - Alm	$Fe_3Al_2Si_3O_{12}$
	Grossular - Grs	$Ca_3Al_2Si_3O_{12}$
	Spessartine - Sps	$Mn_3Al_2Si_3O_{12}$
	Pyrope - Pyr	$Mg_3Al_2Si_3O_{12}$
Chlorite - Chl (HP)	Mn-Chlorite - Mnchl	$Mn_{10}Al_4Si_6O_{20}(OH)_{16}$
	Daphnite - Daph	$Fe_{10}Al_4Si_6O_{20}(OH)_{16}$
	Amesite Ames	$Mg_8Al_8Si_4O_{20}(OH)_{16}$
	Clinochlore - Clin	$Mg_{10}Al_4Si_6O_{20}(OH)_{16}$
Epidote - Ep (HP)	Clinozoisite - Czo	$Ca_2Al_3Si_3O_{12}(OH)$
	Fe-epidoto - Fep	$Ca_2Fe_2AlSi_3O_{12}(OH)$
White Mica - Pheng (HP)	Celadonite - Cel	$KMgAlSi_4O_{10}(OH)_2$
	Fe-Celadonite - Fcel	$KFeAlSi_4O_{10}(OH)_2$
	Muscovite - Mu	$KAl_3Si_3O_{10}(OH)_2$
Plagioclase - AbPl	Anortite - An	$CaAl_2Si_2O_8$
	High albite - Abh	$NaAlSi_3O_8$
Feldspar - feldspar	K-feldspar - Kfs	$KAlSi_3O_8$
	Anortite - An	$CaAl_2Si_2O_8$
	High albite - Abh	$NaAlSi_3O_8$
Glaucophane - Gl	Glaucophane - Gl	$Na_2Mg_3Al_2Si_8O_{22}(OH)_2$
	Fe-Glaucophane - Fgl	$Na_2Fe_3Al_2Si_8O_{22}(OH)_2$
Chloritoid - Ctd (HP)	Mn-Chloritoid - Mnctd	$MnAl_2SiO_5(OH)_2$
	Fe-Chloritoid - Fctd	$FeAl_2SiO_5(OH)_2$
	Mg-Chloritoid - Mctd	$MgAl_2SiO_5(OH)_2$
Staurolite - St (HP)	Mn-Staurolite - Mnst	$Mn_4Al_{18}Si_{7.5}O_{48}H_4$
	Fe-Staurolite - Fst	$Fe_4Al_{18}Si_{7.5}O_{48}H_4$
	Mg-Staurolite - Mst	$Mg_4Al_{18}Si_{7.5}O_{48}H_4$

The Aspromonte and Serre Massifs represent two fragments of the upper Paleozoic suture produced by the plate convergence of North America, Europe, Asia and Gondwana. This collision, which have caused the so called Hercynian orogenesis (from 390 to 310 Ma), brought to the formation of a super-terrane known as Pangea and was responsible for the production of the main metamorphic complexes outcropping in the two Massifs.

The successive geodynamic evolution of this chain sector was characterised by thinning of continental crust and consequent opening of Tethys Ocean around 250 Ma. The subsequent closure of the Tethys gave rise to the formation of the Alpine orogenic belt (from 100 to 25 Ma).

The dominant features characterising these orogenic events are: High T/Low P and Low T/High P ratio for the Hercynian and the Alpine orogeny, respectively.

In this scenario, the Hercynian geological relics are still well preserved in the metamorphic basement of the Serre Massif. Differently, the present-day geological framework of the Aspromonte Massif is the result of the subsequent Alpine orogenesis, which in part reworked the relic Hercynian terrains, affecting at the same time part of the overlaying Mesozoic sedimentary cover.

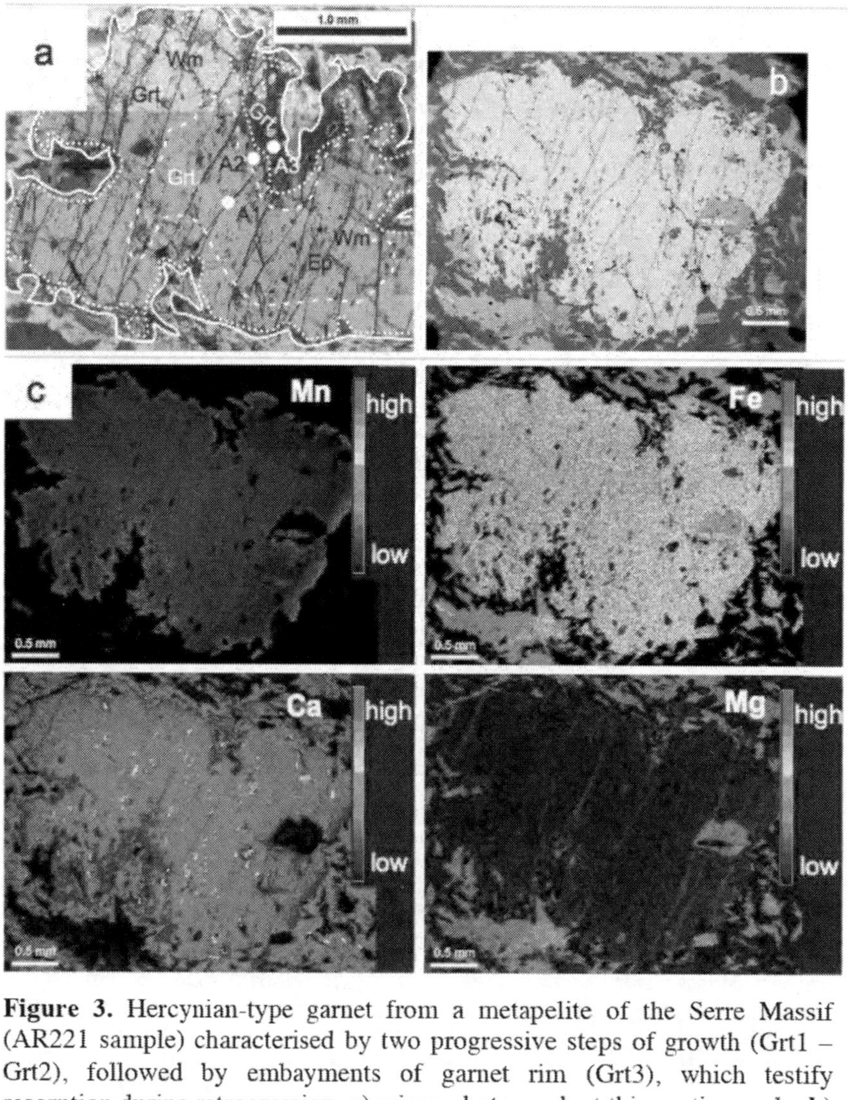

Figure 3. Hercynian-type garnet from a metapelite of the Serre Massif (AR221 sample) characterised by two progressive steps of growth (Grt1 – Grt2), followed by embayments of garnet rim (Grt3), which testify resorption during retrogression. **a)** micro-photograph at thin section scale, **b)** BSE image of investigated garnet, **c)** EDS X-ray maps (Ca, Mn, Mg and Fe) of selected garnet.

In this composite tectonic setting, the investigation of compositional variation in garnet, represents one of the most reliable key tool to unravel the different effects of the Alpine- (high $\Delta P/\Delta T$) and Hercynian-orogenesis (low $\Delta P/\Delta T$).

The P-T evolution of the collected garnet bearing metapelites was yielded by means of P-T pseudosections (phase diagram calculated at constant bulk composition). In particular, for each specimen we started with a multiple approach, which lead to computation of microscopic observations at the thin section by means of optical petrographic microscopy, and continued with the use of Scanning Electron Microscope equipped with EDS spectrometer for the acquisition of BSE images and elaboration of X-ray maps of garnets (Figures 3, 5, 7). Indeed, X-ray fluorescence (XRF) spectrometry was employed to define the bulk-rock chemistry useful to calculate the specified pseudosection for each sample, using free energy minimisation approach by ref [8]. This approach, based upon the approximation that the variable solid solution compositions of a single mineral phase can be simplified by a series of arbitrarily defined components (i.e. pseudocompounds), treats with solutions up to 3 independent mixing sites and up to 3 species mixing on each site [10].

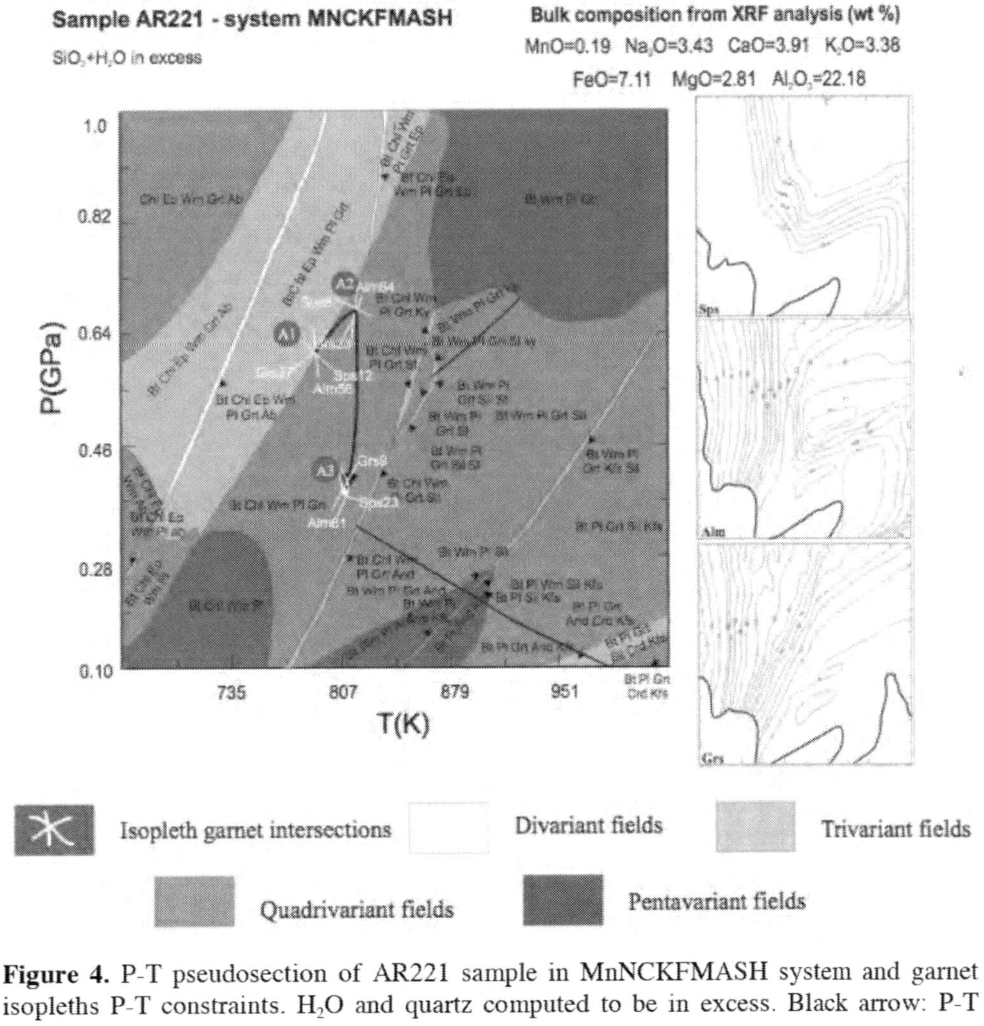

Figure 4. P-T pseudosection of AR221 sample in MnNCKFMASH system and garnet isopleths P-T constraints. H_2O and quartz computed to be in excess. Black arrow: P-T path.

Journal of Physics: Conference Series **121** (2008) 022002 doi:10.1088/1742-6596/121/2/022002

The accuracy of this approximation is related to the choice of the appropriate solid solution model (Table 2) and the relative spacing of the pseudocompounds in the compositional range (i.e. solid solution model dataset), which can be controlled by the user.

The preservation of concentric compositional shells developed during the multi-stage garnet growth of the collected samples permitted us to infer evolution of the host rocks, constraining the P-T values by means of the intersection of garnet isopleths.

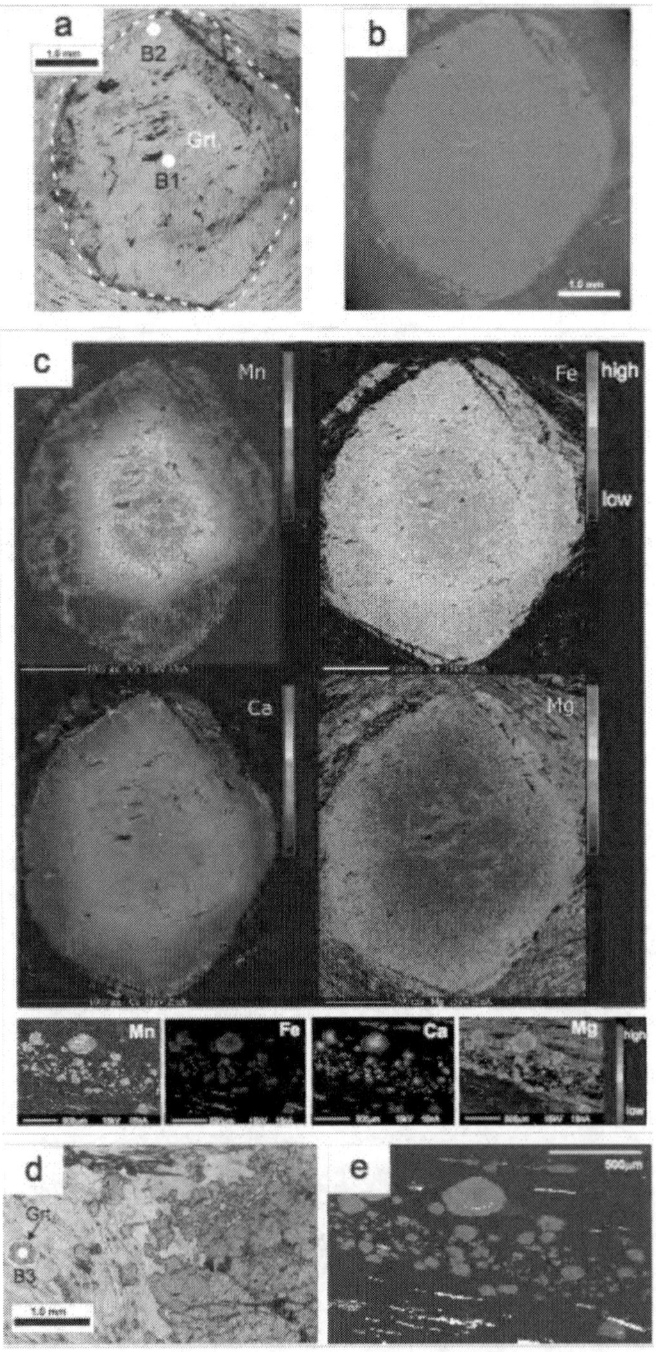

Figure 5. Alpine-type garnet from a metapelite of the Aspromonte Massif (K1 sample), characterised by a first garnet generation represented by large crystals (Grt1), followed both by rim crystal growth and small garnets with different composition (Grt2). **a)** photograph at thin section scale, **b)** BSE image of investigated garnet, **c)** EDS X-ray maps (Ca, Mn, Mg and Fe), which show a compositional zoning patterns for Grt1 garnet, and a concentric chemically zoned composition of Grt2 garnet.

All of the examined rocks exhibit very similar chemical composition; this fact minimises the effects of bulk composition and allows one to make a comparison between garnets belonging to rocks that, even though similar in composition (Table 2), experienced different metamorphic histories.

The first example is a Hercynian-type garnet from a metapelite specimen collected in the Serre Massif (sample AR221 – Figure 2). In this sample, garnet is characterised by a first growth step represented by inner core grown at expenses of epidote, chlorite and white mica, followed by a subsequent growth stage represented by outer core grown at expenses of chlorite and white mica. Petrographic evidence of embayments in garnet rim suggests resorption during the P-T decreasing probably linked to the exhumation of the rock (Figure 3).

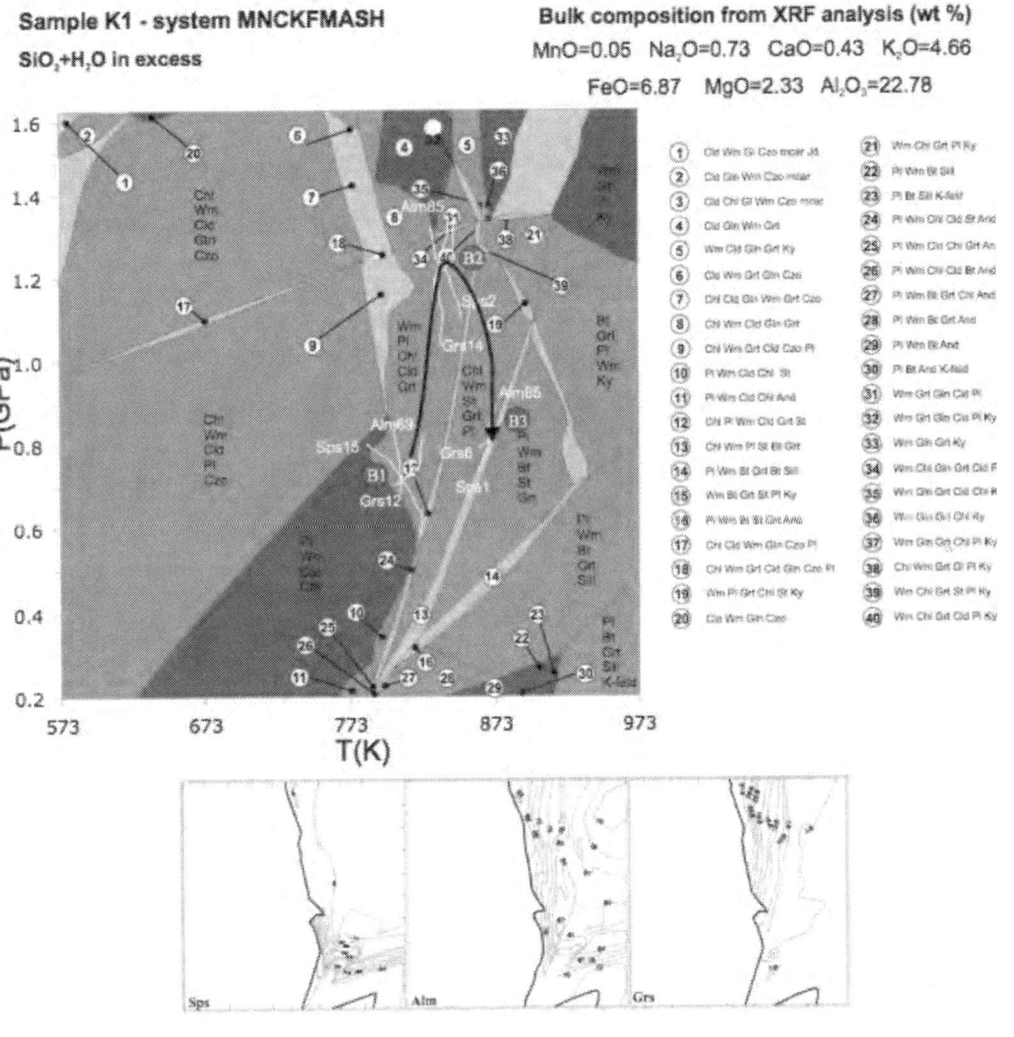

Figure 6. P-T pseudosection of K1 sample in MnNCKFMASH system and garnet isopleths P-T constraints. H_2O and quartz computed to be in excess. Black arrow: PT path. Legend on figure 4.

Intersection among isopleths of garnet core composition, evolving from inner core composition of $Alm_{56}Grs_{27}Sps_{12}Pyr_3$, to an outer core composition of $Alm_{64}Grs_{23}Sps_8Pyr_5$, allowed to constrain P-T values ranging from 0.61 to 0.68 GPa and from 789K to 825K pertaining to a crustal thickening event probably related to the early Hercynian compressional event. The rim compositions ranging from

$Alm_{64}Grs_{13}Sps_{17}Pyr_7$ to $Alm_{61}Grs_9Sps_{23}Pyr_7$, provided pressure values of 0.56 GPa and 0.40 GPa at peak or slightly decreasing temperature suggesting a *quasi*-isothermal decompression event at about 0.28 GPa as the result of uplift and exhumation (Figure 4).

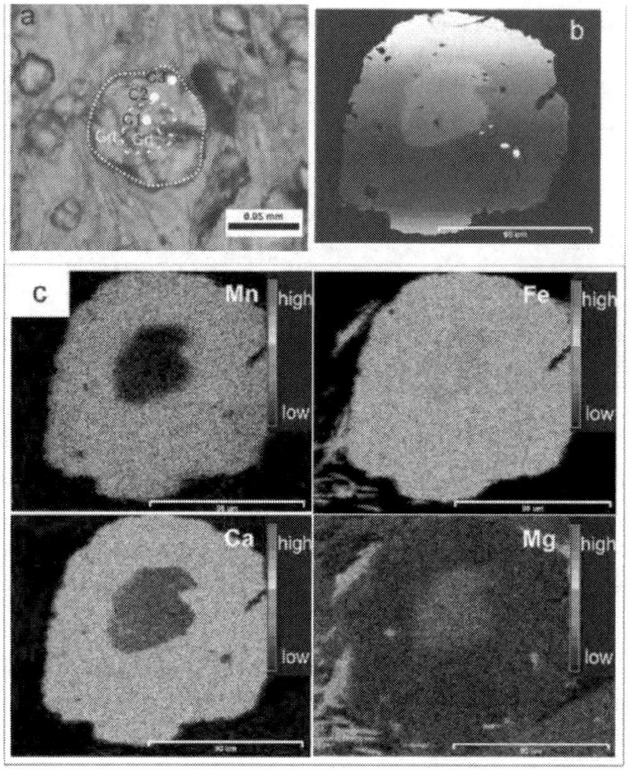

Figure 7. Example of poly-orogenic garnet nucleation from a metapelite of the Aspromonte Massif (AP4 sample), displaying a relic metamorphic evolution preserved in a small Hercynian garnet cores (Grt1), overprinted by a subsequent Alpine type garnet overgrowth (Grt2). **a)** photograph at thin section scale, **b)** BSE image of investigated garnet, **c)** EDS X-ray maps (Ca, Mn, Mg and Fe) of selected garnet, clearly showing the evident double discontinuous stages of crystallisation.

The second example (sample K1 – Figure 2) corresponds to an Alpine-type garnet because of the high ΔP/ΔT ratio revealed by its thermobarometric evolution. In particular, at least two stages of growth are well documented in this rock: the first garnet generation Grt1 ($Alm_{65\text{-}74}Sps_{8\text{-}15}Grs_{7\text{-}15}Pyr_{2\text{-}12}$) is generally represented by large cores, actually preserved in the inner portion of crystals. X-ray map images show clearly changes in composition between adjacent concentric shells (Figure 5), allowing to constrain the P-T conditions of the prograde trajectory to 0.7GPa at 823K (Figure 6).

The outer shell of Grt1-type crystal associated with the last growth has been interpreted as the nucleation region of the new garnet crystals Grt2 ($Alm_{65\text{-}88}Sps_{3\text{-}6}Grs_{4\text{-}28}Pyr_{6\text{-}13}$), which testified the peak equilibrium reaction at P=1.0GPa and T=843K (Figure 6).

The third and last sample (sample AP4 – Figure 2), consisting of a Hercynian-type garnet that was subsequently overgrowth by an Alpine-type one, is characterised by a relic mineral association preserved as rare small cores of garnet in equilibrium with ilmenite inclusions and epidote. Intersection among isopleths of relic garnet cores ($Alm_{84}Grs_3Sps_{13}Pyr_{10}$), constrains the earliest P-T values (0.60±0.02 GPa – 920±20 K) interpreted as remnants of Hercynian metamorphic evolution (low ΔP/ΔT).

A subsequent pervasive Alpine type high-pressure overprint characterised by new growth of idioblastic syn- to post- kinematic garnet (up to 100 μm) and/or overgrowth on the earliest Hercynian garnet cores, evolving toward the rim, to lower spessartine content compositions (Figure 7).

Isopleth intersections of new garnet, which evolve from outer core composition ($Alm_{55}Sps_{20}Grs_{22}Pyr_3$) to rim composition ($Alm_{55}Sps_{18}Grs_{24}Pyr_3$), infer P-T values ranging from 0.78 to 0.87 GPa and from 750 to 770K, highlighting the relatively high pressure overprint interpreted as the prograde path of the Alpine metamorphism (Figure 8).

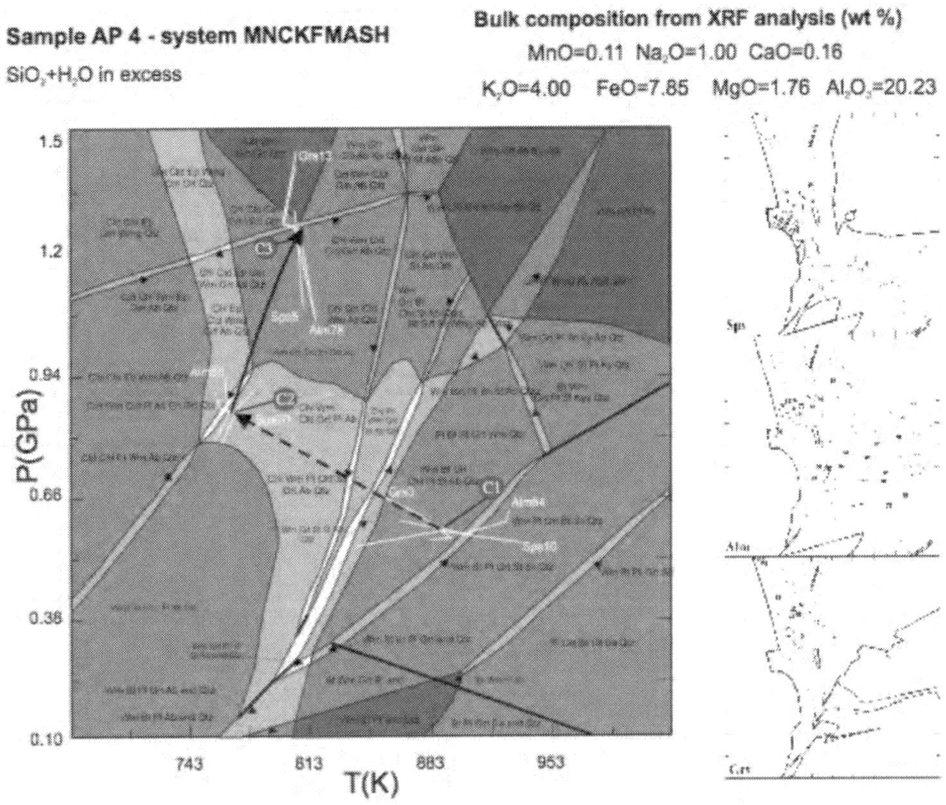

Figure 8. P-T pseudosection of AP4 sample in MnNCKFMASH system and garnet isopleths P-T constraints. H_2O and quartz computed to be in excess. Black arrow: PT path. Black dashed arrow: PT gap between relic Hercynian and Alpine garnet growth. Legend on Figure 4.

4. Final remarks and perspectives

This work is an example of thermodynamic modelling of several types of natural garnet growths useful to delineate the P-T evolution of some representative garnet bearing metapelites via P-T pseudosection method. The present day knowledge of the thermodynamic coefficients of mineral end member constituents, implemented within several internally consistent thermodynamic datasets [11], allows us to obtain a series of P-T equilibrium estimates, which highlighted how natural garnet can be considered a reliable tracer of fossil geothermal gradient.

The above evidences highlight, at the same time, how this approach is useful to solve several types of geological problems, which anyway have to be previously related to geological and petrological constraints.

As a final remark it is worth noting that thermodynamic modelling of comparable metapelite systems permitted to discriminate between various typologies of tectono-metamorphic conditions, characterised by peculiar $\Delta P/\Delta T$ ratios. The first two examples, AR221 and K1, are characterised by relatively LP/HT (Hercynian) and HP/LT (Alpine) metamorphic history, respectively. Differently, the third metapelite specimen shows evidence of both Hercynian and Alpine metamorphism, testified by concentric poly-stage growth evolving from Low Pressure core to High Pressure garnet overgrowth.

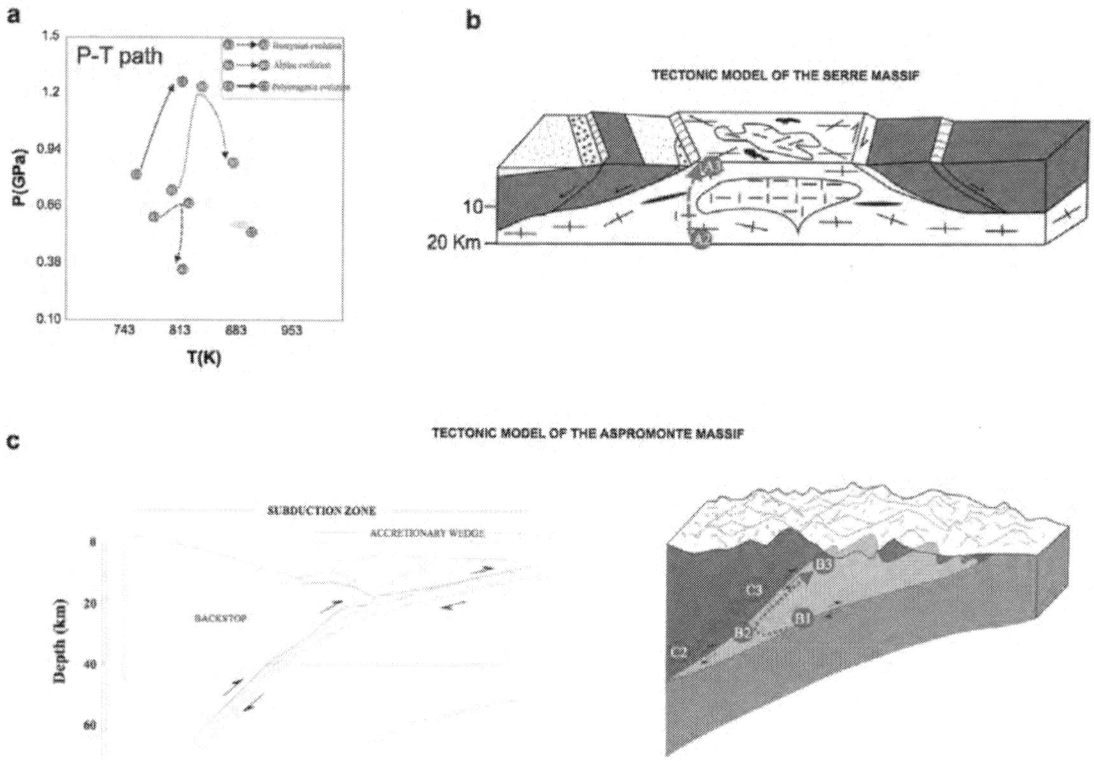

Figure 9. **a)** P-T trajectories reconstructed from garnet composition of three different stages of growth for samples AR221 (circles A1, A2, A3), K1 (circles B1, B2, B3) and AP4 (circles C1, C2, C3). **b)** the tectonic model of the Serre Massif is shown with supposed trajectory followed by the sample AR221. **c)** tectonic model of the Aspromonte Massif. Different positions probably occupied by samples K1 and AP4 during their metamorphic evolution are shown.

Nevertheless, although yielded results are reliable, they have to be considered susceptible of several fonts of errors, mainly derived from: a) incomplete realisation of thermodynamic internally consistent dataset; and b) over-simplification of the solid solution models of natural minerals.

As an example, in our case, garnet is simplified as characterised only by the solid solution between four end members (Table 1b), although in nature this mineral is characterised by six different endmembers miscibility solid solution.

Indeed, since natural environments are typically open systems, it is difficult to provide a reliable correspondence between synthetic and natural systems due to the complexity of natural one, because of the huge number of parameters, which concomitantly operate during the real metamorphic evolution.

A synthetic system doesn't take into account the competitiveness between two or more different phases, which simultaneously grow in the same area of interest, as usually occurs in rocks. In this context, we think that a model garnet growth which takes into account, the free energy minimisation effects at the equilibrium conditions as well as the geometry of garnet nucleation and the effects of the surrounded matrix chemical composition, could provide reliable P-T constraints within a more detailed thermodynamic modelling of natural systems.

References

[1] B. A. Marmo, G. L. Clarke, R. Powell, J. Metamorphic Geol. **20** 151(2002).
[2] W. D. Carlson, Contrib. Mineral. Petrol. **103** 1 (1989).
[3] C. B. Chernoff and W. D. Carlson, J. Metamorphic Geol. **15** 421(1997).
[4] M. K. Schmolke, M. R. Handy, J. Babist, P. O'Brien, Contrib. Mineral. Petrol. **16** 181 (2005).
[5] K. Stüwe, Contrib. Mineral. Petrol. **129** 43 (1997).
[6] R. Powell and T. J. B. Holland, American Mineralogist **78** 1174 (1993).
[7] C. de Capitani, Mineral. Gesellschaft **6** 1 (1994).
[8] J. A. D. Connolly and K. Petrini, J. Metamorphic Geol. **20** 697 (2002).
[9] M. R. St-Onge, Journal of Petrology **28** 1 (1987).
[10] J. A. D. Connolly, Am. J. Sci. **290** 666 (1990).
[11] T. J. B. Holland and R. Powell, J. Metamorphic Geol. **16** 309 (1998).
[12] T. Graessner and V. Schenk, J. Metamorphic Geol. **17** 157 (1999).

Joint 21st AIRAPT and 45th EHPRG Int. Conf. on High Pressure Science and Technology IOP Publishing
Journal of Physics: Conference Series **121** (2008) 022003 doi:10.1088/1742-6596/121/2/022003

High pressure Raman spectroscopic study of low-cristobalite $GaPO_4$

C. H. Chio[1], S. K. Sharma[1], L. C. Ming[1a] , Y Nakamoto[2], and S. Endo[3]
[1]Hawaii Institute if Geophysics and Planetology, University of Hawaii, Honolulu, HI 96822, U.S.A.
[2]Center for Quantum Science and Technology under Extrem Conditions, Osaka University, Toyonaka, Osaka, 560-8531, Japan
[3]Department of Environmental Security System, Chiba Institute of Science, Choshi, Chiba 288-0025, Japan

[a] ming@soest.hawaii.edu

Abstract. Gallium orthophosphate of low-cristobalite structure has been studied by Raman spectroscopy up to 28.5 GPa. Three phase transitions have been identified: 1.6, 3.1 and ~7 GPa. Raman spectra (200 to 700 cm^{-1}) acquired above 3 GPa show signs of six-coordinated gallium ions, in agreement with earlier studies that the orthorhombic $C222_1$ structure transforms into the orthorhombic *Cmcm* structure. As pressure increases above 8 GPa, the Raman peaks associated with the low-cristobalite framework structure diminish gradually and replaced by peaks that are possibly related to isolated phosphate units. The spectra also show that the structure of the quenched sample retains that of the high-pressure phase prior to decompression.

1. Introduction

Among different materials that are isostructural to the SiO_2 polymorphs, the low-quartz type $AlPO_4$, $GaPO_4$ and their solid solutions have been extensively studied for the past two decades. The piezoelectric properties of these materials have found important industrial applications [1-3]. Numerous studies regarding to the temperature and pressure stabilities of these low-quartz homeotypes have been reported using a variety of spectroscopic [4-7] and diffraction methods [8-12]. In comparison, studies of the low-cristobalite polymorphs are not as extensive. Under ambient conditions, low-cristobalite $AlPO_4$ and $GaPO_4$ have orthorhombic lattices of space group $C222_1$ [13]. Temperature-induced phase transition into the cubic high-cristobalite phase [14-16], and pressure-induced phase transition into a second orthorhombic *Cmcm* phase have been reported [17, 18]. The latter increases the coordination number of the trivalent cation from four to six. In tetragonal low-cristobalite SiO_2, Palmer and Finger [19] and Palmer et al. [20] reported phase transition into a monoclinic phase at ~1.5 GPa, a modest pressure relative to other phase transitions reported earlier [21]. A similar orthorhombic→monoclinic phase transition has also been proposed in a recent x-ray diffraction study [18]. In the present study, low-cristobalite $GaPO_4$ (abbreviated as *lc*-$GaPO_4$ hereafter) is studied by Raman spectroscopy up to 28.5 GPa for a better understanding of the details related to the reported phase transition(s).

© 2008 IOP Publishing Ltd

2. Experimental Methods

High pressure Raman spectroscopic study was carried out using a Mao-Bell type diamond anvil cell, which contained a pair of type-II anvils with a culet of 500 μm. Without the use of pressure-transmitting medium, powdered *lc*-$GaPO_4$ obtained from Bell Laboratories was loaded together with several ruby chips into a hole of 200 μm diameter, drilled on a stainless steel gasket (250 μm thick). Pressure was calibrated using the ruby fluorescence method. The sample was excited with the 514.5-nm radiation from an argon laser. Sample excitation and collection of Raman photons were carried out in a back-scattering geometry with a 10x long working-distance objective. Raman photons were dispersed by a Spex Triplemate spectrometer equipped with a CCD detector. To avoid sample heating, the power of the incident laser beam was kept below 15 mW. Raman spectra were calibrated using the atomic emission lines from a neon discharge lamp, and spectral resolution was 5 cm^{-1}.

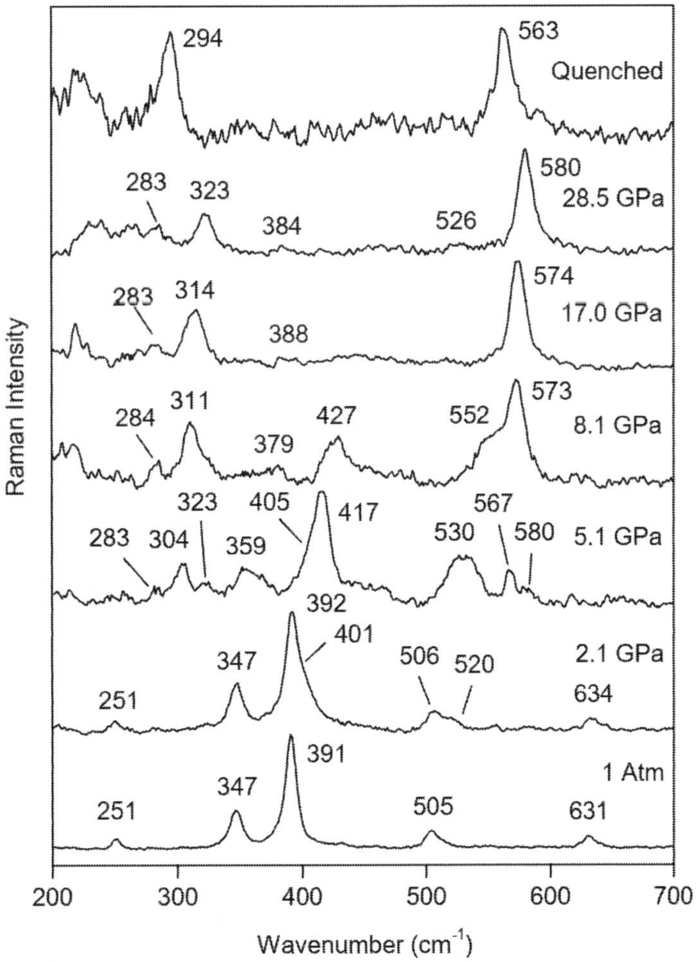

Figure 1. Raman spectra (200 to 700 cm^{-1}) of *lc*-$GaPO_4$ at pressures up to 28.5 GPa.

3. Results and discussion

Between 200 and 700 cm^{-1}, the Raman spectra of *lc*-$GaPO_4$ at various pressures are shown in Fig. 1, and the pressure dependence of selected vibrational modes are displayed in Fig. 2. The Raman spectra of *lc*-$GaPO_4$ at ambient condition, 5.1 and 14.1 GPa agree with those reported by Robeson et al. [17] at similar pressure levels. To the best of the authors' knowledge, the symmetries and assignments of the *lc*-$GaPO_4$ Raman peaks have yet to be studied in detail. In this report, the Raman peaks are interpreted based on a framework of corner-sharing PO_4^{3-} and GaO_4^{5+} tetrahedral units. The strongest Raman peak of *lc*-$GaPO_4$ is located at 391 cm^{-1}, which is likely attributed to stretching vibrations of the bridging O atoms (ν_s(T-O-T), T = Ga and P) along the axis bisecting the Ga-O-P angle. In $GaPO_4$ of low-quartz structure, this peak is located at 458 cm^{-1} [22,23]. In comparison, the ν_s(Si-O-Si) Raman peak in low-cristobalite and low-quartz SiO_2 is located ~420 [24,25] and ~465 cm^{-1} [26,27], respectively. The wavenumber-pressure plots (Fig. 2) indicate phase transitions at 1.6, 3.1 and ~7 GPa, and possibly an additional transition between 14 and 17 GPa. The high-pressure phases associated with these changes are labeled in this report as I', II, III and III', respectively.

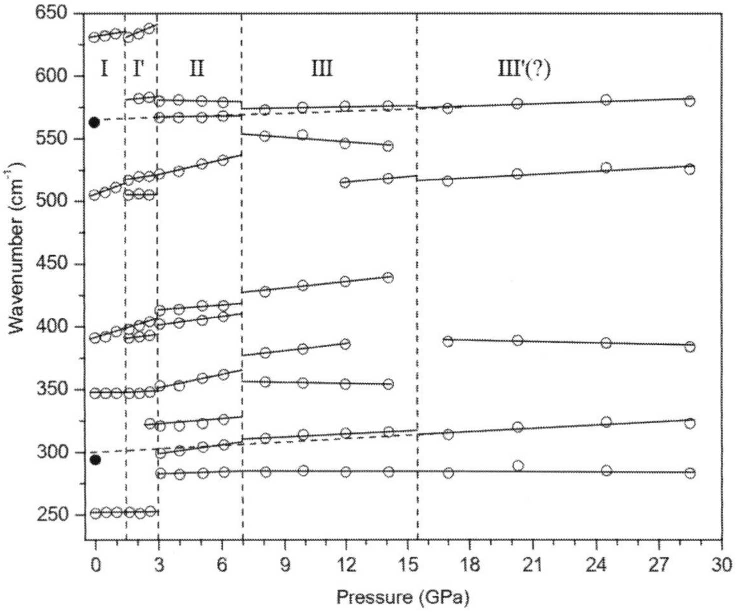

Figure 2. Plots showing Raman wavenumbers of various vibrational modes as a function of pressure in *lc*-$GaPO_4$. Open and closed circles indicate increasing and decreasing pressures, respectively. Up to three phase transitions have been identified: 1.6, 3.1 and between 6.1 and 8.1 GPa. A possible fourth transition between 14.1 and 17.0 GPa is also indicated.

At ~1.6 GPa, the Raman peaks at 391 and 505 cm^{-1} each becomes asymmetric with high-wavenumber shoulder peaks. The gradual increase in intensity of these shoulder peaks with pressure is possibly resulted from pressure-induced orientation disorder. In addition, a weak Raman peak appears at ~580 cm^{-1} at 2.1 GPa. The phase transition at ~3 GPa is highlighted by the appearance of several weak Raman peaks between 275 and 325 cm^{-1} and the disappearance of the 631 cm^{-1} peak. A new peak at ~570 cm^{-1} also appears close to the 580 cm^{-1} peak, and their intensities increase with the applied pressure. In SiO_2 polymorphs with six-membered rings of corner-sharing SiO_4 units (cristobalite, quartz and tridymite), the Raman peaks between 750 and 800 cm^{-1} are attributed largely to the motions of the Si atom [28]. Similarly, it is believed that the 631 cm^{-1} peak in *lc*-$GaPO_4$ is related to motions of Ga and P atoms in the framework structure. Between 6.1 and 8.1 GPa, the Raman spectra clearly indicate structural changes which have not been observed in X-ray diffraction study [18]. In the spectrum acquired at 8.1 GPa, the intensity of the 427 cm^{-1} peak weakens substantially, and two peaks at 311 and 573 cm^{-1} peak become the strongest in the spectrum. In addition, the 533 cm^{-1} peak, observed at 6.1 GPa, moves up ~20 cm^{-1} and partially overlaps with the 573 cm^{-1} peak. Above 14.1 GPa, there are fewer number of Raman peaks visible in the spectra. This might be the result of deteriorating spectral intensity, or indicate a pressure-induced phase transition (into phase III'). Among them, two peaks at 314 and 574 cm^{-1} are the most intense. No further changes in the spectral pattern are visible up to the highest pressure reached in the current study (28.5 GPa). After the sample has been quenched from 28.5 GPa to ambient condition, the Raman spectrum shows two strong peaks located at 294 and 563 cm^{-1}. Extrapolation of the best-fit regression lines from 28.5 GPa to ambient condition closely matches with the wavenumbers of those two peaks, which indicates that the high-pressure phase is retained (Fig. 2).

The nature of these pressure-induced changes in the Raman spectra is yet to be fully understood. Ming et al. reported no substantial changes in the X-ray diffraction pattern and the molar volume of *lc*-$GaPO_4$ between 6.1 and 8.0 GPa. The Raman spectrum of the quenched phase is also different from that of Robeson et al. [17], which indicates partial reversion of the high-pressure phase back to *lc*-$GaPO_4$ after decompression from 22.0 GPa. The difference is likely resulted from different experimental conditions: hydrostatic (Ref. [17]) versus non-hydrostatic (present study) compression process. Such effects have been discussed in the high-pressure Raman study of low-quartz $AlPO_4$ [4]. Moreover, it is interesting to point out that the Raman spectrum of *lc*-$GaPO_4$ at 17.0 GPa (and above) reported in this study resembles that of low-quartz $GaPO_4$ at 16.2 GPa [6], but with different peak positions. The authors in Ref. [6] proposed the structure of the high-pressure phase is orthorhombic with *Cmcm* space group, and the phase transition likely occurred at ~11 GPa.

In Fig. 2, the Raman peaks at 323 and 580 cm^{-1} observed at 28.5 GPa can be traced back to their first appearances at ~3 GPa (299 and 567 cm^{-1}, respectively). Moreover, because the fading of the 631 cm^{-1} peak is believed to reflect changes in the vibrational motions of Ga and P atoms, the spectral variations at ~3 GPa likely indicate a phase transition into orthorhombic *Cmcm* structure (II) [18,29]. As mentioned earlier, the strongest Raman peak in *lc*-$GaPO_4$ is attributed to the symmetric stretching vibrations of the bridging O atoms. The fading of this Raman peak and the simultaneous strengthening of two peaks at 311 and 573 cm^{-1} above 8.1 GPa (III and III') might have been caused by structural rearrangement of the corner-sharing polyhedra framework. It is suspected that these two peaks are attributed to the internal vibrational modes of the PO_4^{3-} tetrahedral, suggesting that interactions between Ga and O atoms have been significantly reduced.

Acknowledgments
This work was partially supported by an NSF grant DMR-0102215.

References

[1] Reiter C, Krempl P W, Thanner H, Wallnofer W and Worsch P M 2001 *Annal. Chim. Sci. Mat.* **26** 91
[2] Thanner H, Krempl P W, Wallnöfer W and Worsch P M 2002 *Vacuum* **67** 687
[3] Elam J W and Pellin M J 2005 *Anal. Chem.* **77** 3531
[4] Gillet P, Badro J, Varrel B and McMillan P 1995 *Phys. Rev.* B **51** 11262
[5] Pinan-Lucarre J-P, Ouillon R and Ranson P 1998 *J. Raman Spectrosc.* **29** 1019
[6] Peters M J, Grimsditch M and Polian A 2000 *Solid State Commun.* **114** 335
[7] Angot E, Le Parc R, Levelut C, Beaurain M, Armand P, Cambon O and Haines J 2006 *J. Phys.: Condens. Matter* **18** 4315
[8] Muraoka Y and Kihara K 1997 *Phys. Chem. Minerals* **24** 243

[9] Badro J, Itie J P and Polian A 1998 *Eur. Phys. J.* B **1** 265
[10] Sharma S M, Garg N and Sikka S K 2000 *J. Phys.: Condens. Matter* **12** 6683
[11] Jacobs K, Hofmann P, Klimm D, Reichow J and Schneider M 2000 *J. Solid State Chem.* **149** 180
[12] Haines J, Cambon O, Fraysse G and van der Lee A 2005 *J. Phys.: Condens. Matter* **17** 4463
[13] Mooney R C L 1956 *Acta Cryst.* **9** 728
[14] Ng H N and Calvo C 1977 *Can. J. Phys.* **55** 677
[15] Nicola J H, Scott J F and Ng H N 1978 *Phys. Rev.* B **18** 1972
[16] Achary S N, Mishra R, Jayakumar O D, Kulshreshtha S K and Tyagi A K 2007 *J. Solid State Chem.* **180** 84
[17] Robeson J L, Winters R R and Hammack W S 1994 *Phys. Rev. Lett.* **73** 1644
[18] Ming L C, Nakamoto Y, Endo S, Chio C H and Sharma S K 2007 *J. Phys.: Condens. Matter* in press
[19] Palmer D C and Finger L W 1994 *Amer. Mineral.* **79** 1
[20] Palmer D C, Hemley R J and Prewitt C T 1994 *Phys. Chem. Minerals* **21** 481
[21] Tsuchida Y and Yagi T 1990 *Nature* **347** 267
[22] Nakamura M, Orihara H, Ishibashi Y, Kim P and Hirano S 1990 *J. Phys. Soc. Jpn.* **59** 1831
[23] Peters M J, Grimsditch M and Polian A 2000 *Solid State Commun.* **114** 335
[24] Cherukuri S C, Pye L D, Chakraborty I N, Condrate R A, Ferraro J R, Cornilsen B C and Martin K 1985 *Spectrosc. Lett.* **18** 123
[25] Palmer D C, Hemley R J and Prewitt C T 1994 *Phys. Chem. Minerals* **21** 481
[26] Shapiro S M, O'Shea D and Cummins H Z 1967 *Phys. Rev. Lett.* **19** 361
[27] Sharma S K, Philpotts J A and Matson D W 1985 *J. Non-Cryst. Solids* **71** 403
[28] Kihara K, Hirose T and Shinoda K 2005 *J. Miner. Petrol. Sci.* **100** 91
[29] Murashov V V, Dubrovinsky L S, Tse J S and LePage Y 1995 *J. Phys.: Condens. Matter* **7** 8279

Journal of Physics: Conference Series **121** (2008) 022004 doi:10.1088/1742-6596/121/2/022004

Thermopower and phase transition in $YbPd_2Si_2$ under ultra high pressure

Sergey V. Ovsyannikov [1,2,*], **Vladimir V Shchennikov** [2], **Tetsuya Fujiwara** [3] and **Yoshiya Uwatoko** [1]

[1] The Institute for Solid State Physics, The University of Tokyo, 5-1-5 Kashiwanoha, Kashiwa, Chiba 277-8581 Japan

[2] High Pressure Group, Institute of Metal Physics of Russian Academy of Sciences, Urals Division, GSP-170, 18 S. Kovalevskaya Str., Yekaterinburg 620041, Russia

[3] Faculty of Sciences, Yamaguchi University, Yamaguchi 753-8512, Japan

*highpressgroup@mail.ru

Abstract. We report the results of investigation of the thermopower (Seebeck effect) of a $ThCr_2Si_2$-structured heavy fermion single-crystalline $YbPd_2Si_2$ compound (itterbium-palladium-silicon, 1-2-2) under ultra high pressure P up to 22 GPa. At ambient conditions the thermopower S was found to be negative. Under pressurization S inverted its sign and exhibited a bend near 6 GPa. The compressibility of $YbPd_2Si_2$ also showed a bend near 6 GPa corroborating a supposition about phase transformation.

1. Introduction

Heavy fermion systems – intermetallic compounds on a basis of rare earth and actinide elements attract a steady attention due to unusual properties and rich phase diagrams. They are model systems for investigation of novel physical phenomena, such as unconventional superconductivity, magnetically mediated [1] and of other types [2].

One of the oldest and promising series of heavy fermion systems is "1-2-2", RM_2X_2 (R = Ce, Yb, U; M – transition metal, X = Si, Ge). A discovery of a heavy fermion superconductivity in a one of the RM_2X_2 materials, $CeCu_2Si_2$ below $T_c \sim 0.5$ K [1] motivated further investigations of heavy fermion materials, including also synthesis and characterization of new "1-2-2" compounds. Application of high-pressure allows varying the density of materials and thus, changing the intensity configuration of different electronic and magnetic interactions [3]. Owing to very high sensitivity of properties (superconductivity in particular) to quality of samples, the existing experimental data are to some extent contradictory that leads to distinguishing interpretations. Therefore, for understanding of physics of these compounds and also for attribution of found states to certain interactions, a behaviour of each of them seems to be important.

At ambient conditions, a moderate heavy fermion system – $YbPd_2Si_2$ (γ = 203 mJ/(mole·K^2) [4]) adopts a tetragonal $ThCr_2Si_2$-type structure with a space group *I4/mmm* [5] (Fig. 1), that is typical for the RM_2X_2 family. The lattice parameters of $YbPd_2Si_2$ are as following: a = 4.099 Å and b = 9.870 Å [6]. For a number of compounds this lattice was found to be stable against pressure application, for

© 2008 IOP Publishing Ltd

Journal of Physics: Conference Series **121** (2008) 022004 doi:10.1088/1742-6596/121/2/022004

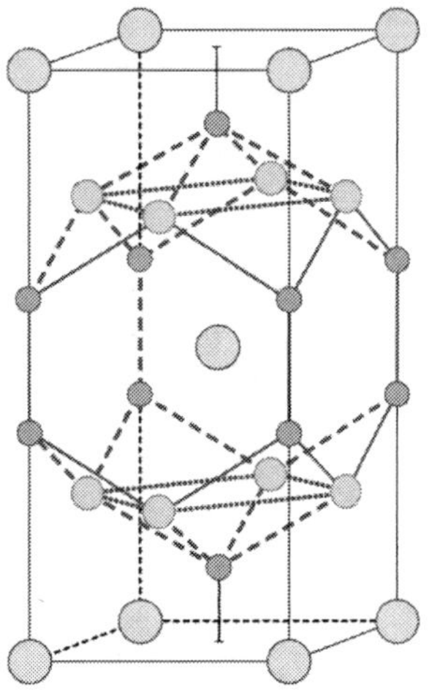

Fig. 1. The body-centred tetragonal crystal structure of *I4/mmm* space group ($ThCr_2Si_2$ structural type) [5] of the majority of RM_2X_2 compounds (R = Ce, Yb; M – transition metal, X = Si, Ge).

example, to $P \sim 60$ GPa in $CeCu_2(Si,Ge)_2$ [7]. Recently, a new isostructural transformation concurrent with a volume collapse was discovered in this family on example of $CeCu_2Ge_2$, of which lattice parameters, a and c dropped by ~ 0.7 % at 10 K near 15 GPa [8].

In the present work we undertook an ultra-high pressure study of $YbPd_2Si_2$ single crystal to 22 GPa in order to probe both a stability of the ambient state and P dependencies of the thermopower. We employed the original automated high-pressure setup designed for registration of phase transformation [9]. The setup permitted simultaneous registration of several parameters of a sample and an environment under almost continuous alteration of pressure during pressurization and decompression [9]. A high efficiency of this setup was proved by a discovery of the new "hidden" semimetal high-pressure phases in ZnTe [10] and GaAs [11] which are "invisible" for the conventional structural techniques.

2. Experiment

The measurements of the thermopower (Seebeck effect) S under ultra-high quasi-hydrostatic pressure P were performed in a synthetic diamond anvil cell of the Bridgman type [12]. A sample of sizes ~ 200-250 × 200-250 × 20–30 μm^3 was placed in a hole in a gasket from the lithographic stone (soft $CaCO_3$–based material) (Fig. 2). The ratio of the gasket's thickness to the working diameter of the anvils was lower than 0.05, providing a quasi-hydrostatic character of compression. The synthetic diamond anvils were characterised by a high electrical conductivity [12] and were used both as electrical outputs to a sample and a heater-cooler pair (upper anvil was heated) [13-15]. To account for a possible contribution into S from the anvils themselves, Pb samples ($S \approx -1.27$ μV/K) were measured in the same conditions. A sample's contraction was registered by a digital dilatometer attached to the plungers [16,17]. The single-crystalline samples of $YbPd_2Si_2$ were grown by a flux method and were preliminary characterized by a conventional x-ray diffraction technique.

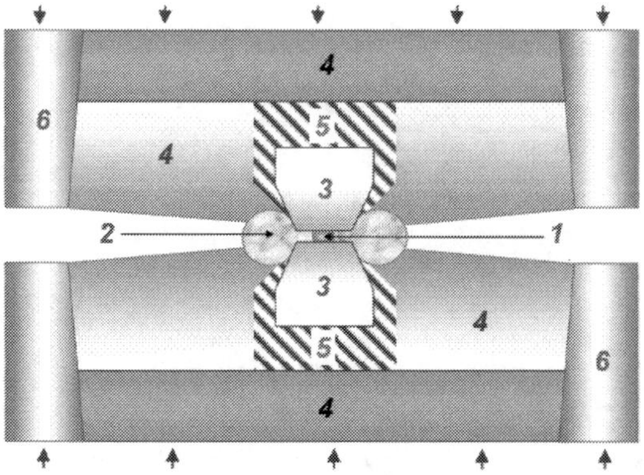

Fig. 2. A high-pressure cell with the sintered diamond anvils, developed in the High Pressure Group (IMP) [9], ***1*** – a sample; ***2*** – a gasket; ***3*** – anvil insets in high-pressure plungers; ***4*** – supporting hard-alloy matrixes (plungers), usually made of tungsten carbide; ***5*** – a special hard-durable part of a plunger (synthesized together with ***3***); ***6*** – supporting rings. A ring-like bulge of the gasket ***2*** provides a supporting pressure P_s (up to 10 GPa) around the tips of the anvils. High quasi-hydrostatic pressure P (from to 30 GPa) is produced in the central part of the gasket around a sample. The arrows show external applied P.

3. Results and discussion

The pressure dependencies of the thermopower *S(P)* of $YbPd_2Si_2$ are presented in Fig. 3. The negative sign of the thermopower at ambient conditions agreed with the data of Ref. [18] reporting the Seebeck coefficients for a number of Yb-based heavy fermion systems, $S \sim -(7 \div 35)$ μV/K; for $YbPd_2Si_2$ Ref. [6] established $S \approx -18$ μV/K. A pressure-driven decrease in $|S|$ (Fig. 3) was observed in other Yb-based compounds too [19]. This decrease in $|S|$ was concurrent with a lowering in the electrical resistance (not shown), and thus could be induced by an increase in the electron concentration under pressure (Fig. 3). Previous studies on $YbPd_2Si_2$ [20] reported on a decrease in the electrical resistivity with *P*. The *S(P)* curves (Fig. 3) exhibited a bend near 6 and 3 GPa on pressurization and decompression cycles, respectively. Below 1.5 GPa some peculiarity in the *S(P)* curves was noticed. The curves of a sample's contraction exhibited the kinks (Fig. 3, inset) indicating a change in the compressibility. So, one can presumably address the feature near 6 GPa to a reversible phase transition concurrent with structural changes. Above this crossover point *S* became positive (Fig. 3).

To the present, no structural transitions in the RM_2X_2 compounds from the $ThCr_2Si_2$ lattice to a different one were reported on a *P* range to 20 GPa and above, for instance, in $CeCu_2Si_2$ [21,7], in $CeCu_2Ge_2$ [22,7,8], in $CeNi_2Ge_2$ [7], and in $YbCu_2Si_2$ [23]. Then, a probable explanation of the feature found (Fig. 3) might be a pressure-driven valence transition which may be concurrent with a volume collapse ΔV owing to an abrupt change of ionic radius of Yb [24] (Fig. 1). In the majority of heavy fermion compounds, a valence of Yb changes between 2+ and 3+ limits. Yb^{2+} has a filled 4*f* shell and thus has no magnetic moment contrary to magnetic Yb^{3+} [25]. This circumstance explains both the differences in properties of compounds with Yb^{2+} and Yb^{3+} ions, and anomalies of properties when the valence is intermediate [26]. A rising of Yb valence indeed had to lead to an increase in the electron concentration resulting in a decrease in $|S|$ (Fig. 3).

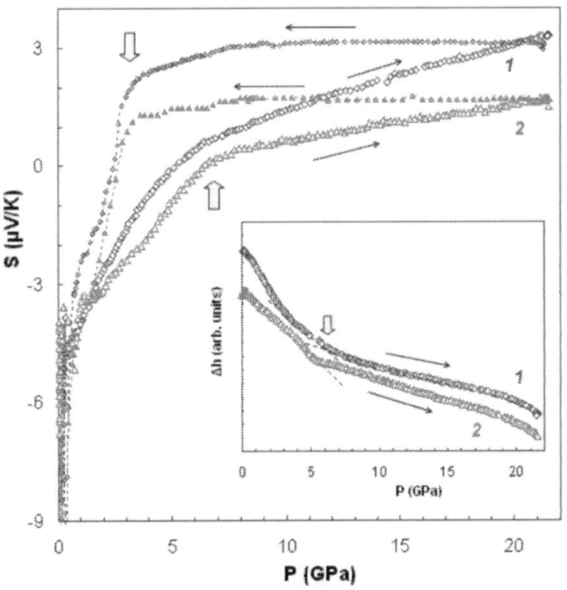

Fig. 3. The pressure *P* dependencies of the thermoelectric power (Seebeck effect) *S* of $YbPd_2Si_2$ single crystal at T=293 K for two cycles (*1*, *2*) of pressurization (big open symbols) and decompression (small closed ones). In the inset, the curves of a relative sample's contraction Δh (reflect a change in the compressibility) corresponding to these *S(P)* curves are shown.

At ambient conditions an effective valence of Yb in $YbPd_2Si_2$ was found to be 2.8 [26], while application of external pressure can lead it to a trivalent limit. Ref. [25] qualitatively demonstrated an increase in Yb valence under $P \sim 4$–5 GPa, meanwhile, a trivalent state was not achieved. In other Yb-based compounds, such as $YbInCu_4$ and $YbMn_2Ge_2$ a valence transition under *P* was also established [27-29]. A reason why such a potentially small change in Yb valence results in a significant modification of properties might consist, for example, in appearance of magnetic properties in initially non-magnetic material [26].

4. Conclusions

By the high-pressure thermopower technique [30,31] a phase transition in $YbPd_2Si_2$ was established near 6 GPa at room temperature. Presumably, this crossover may be addressed to a valence transition which can lead both to a volume collapse effect (ΔV) owing to a difference in the ionic radii of Yb^{3+} and Yb^{2+} ions (Fig. 1) [24], and probably to an appearance of magnetic properties.

Acknowledgements
The work was supported by the RFBR (Gr. No. 07-08-00338) and the Presidium of RAS.

References

[1] Steglich F, Aarts J, Bredl C D, Lieke W, Meschede D, Franz W and Schafer H 1979 *Phys. Rev. Lett.* **43** 1892
[2] Yuan H Q, Grosche F M, Deppe M, Geibel C, Sparn G and Steglich F 2003 *Science* **302** **2**104
[3] Mathur N D, Grosche F M, Julian S R, Walker I R, Freye D M, Haselwimmer R K W and Lonzarich G G 1998 *Nature* **394** 39
[4] Dhar S K, Sampathkumaran E V and Vijiyaraghavan R 1987 *Solid State Commun.* **61** 479
[5] Ban Z and Sikirica M 1965 *Acta Crystallographica* **18** 594
[6] Casanova R, Jaccard D, Marcenat C, Hamdaoui N and Besnus M J 1990 *J. Magn. Magn. Mat.* **90-91** 587
[7] Tsuduki S, Onodera A, Ishida K, Kitaoka Y, Onuki A, Ishimatsu N and Shimomura O 2005 *Solid State Commun.* **134** 747
[8] Onodera A, Tsuduki S, Ohishi Y, Watanuki T, Ishida K, Kitaoka Y and Onuki Y 2002 *Solid State Commun.* **123** 113
[9] Shchennikov V V, Ovsyannikov S V, Derevskov A Y and Shchennikov Jr V V 2006 *J. Phys. Chem. Sol.* **67** 2203 and references within
[10] Ovsyannikov S V and Shchennikov V V 2004 *Solid State Commun.* **132** 333
[11] Ovsyannikov S V and Shchennikov V V 2006 *J. Phys.: Condens. Matter* **18** L551
[12] Ovsyannikov S V and Shchennikov V V, Misiuk A and Shchennikov Jr V V 2004 *Solid State Commun.* **132** 545
[13] Ovsyannikov S V, Shchennikov V V, Kar'kin A E and Goshchitskii B N 2005 *J. Phys.: Condens. Matter* **17** S3179
[14] Shchennikov V V and Ovsyannikov S V 2007 *physica status solidi (b)* **244** 437
[15] Ovsyannikov S V and Shchennikov V V 2007 *Appl. Phys. Lett.* **90** 122103
[16] Ovsyannikov S V, Shchennikov V V, Titov A N and Uwatoko Y 2007 *physica status solidi (b)* **244** 174.
[17] Ovsyannikov S V, Shchennikov V V and Goshchitskii B N 2007 *Mater. Sci. Eng. A* **462** 427.
[18] Alami-Yadri K, Jaccard D and Andreica D 1999 *J. Low Temp. Phys.* **114** 135
[19] Link P, Jaccard D and Lejay P 1996 *Physica B* **225** 207
[20] Nakano T, Hedo M, Uwatoko Y and Sampathkumaran E V 2004 *Solid State Commun.* **132** 325
[21] Spain I L, Steglich F, Rauchschwalbe U and Hochheimer H D 1986 *Physica BC* **139-140** 449
[22] Wassilew-Reul C, Kunz M, Hanfland M, Hausermann D, Geibel C and Steglich F 1997 *Physica B* **230-232** 310.
[23] Winkelmann H, Abd-Elmeguid M M, Micklitz H, Sanchez J P, Vulliet P, Alami-Yadri K and Jaccard D 1999 *Phys. Rev. B* **60** 3324
[24] Broto J M, Gonzalezjimenez F, Fert A, Sanchez J, Besnus M J and Kappler J P 1988 *J. Magn. Magn. Mat.* **76-77** 289
[25] Moser J, Munch K H and Kalvius G M 1988 *Hyperfine Interactions* **40** 405
[26] Sampathkumaran E V, Frank K H, Kalkowski G, Kaindl G, Domke M and Wortmann G 1984 *Phys. Rev. B* **29** 5702
[27] Hedo M, Uwatoko Y, Matsumoto T, Sarrao J L and Thompson J D 2003 *Acta Phys. Pol. B* **34** 1193
[28] Fujiwara T, Uwatoko Y, Fujii H, Koyama K, Motokawa M and Shigeoka T 2004 *J. Magn. Magn. Mat.* **272** 599
[29] Kurita N, Kano M, Hedo M, Uwatoko Y, Sarrao J L, Thompson J D and Tozer S W 2006 *Physica B* **378-380** 104
[30] Ovsyannikov S V, Shchennikov V V and Goshchitskii B N 2005 *JETP Letters* **81** 167.
[31] Shchennikov V V and Ovsyannikov S V 2006 *JETP Letters* **84** 21.

On the conception of minimal metal conductivity in metal-insulator phase transitions

S.F.Gabibov, M.I.Daunov, I.K.Kamilov, A.B.Magomedov

Institute of Physics of Daghestan Science Center, Russian Academy of Sciences, 367003, Makhachkala, M.Yaragskogo str.,94, Russia

E-mail: i.daunov@mail.ru

Abstract. The agreement between the conception of minimal metal conductivity and experiment has been shown. It is found, that the smoothness of metal-insulator transition observed in crystal semiconductors is caused either by disorder and then we need the percolation theory to be used for data interpreting at T=0 K, or by small value of critical concentration of carriers. The critical concentration depends on the compensation level, classical and "natural" widening of defect level or correlated distribution of impurities. According to experimental data, the characteristic parameters of transition in n-$CdSnAs_2$, p-$CdSnAs_2$<Cu> and p-Ge has been obtained by the extrapolation to T=0 K.

1. Introduction

While considering a metal-insulator phase transitions in Mott and Anderson crystal semiconductors, an interesting theoretical problem concerning to the type of phase transitions and to the electronic conductivity behaviour in the vicinity of transition point is appearing. Still in his early works, analyzing a nature of the metal conductivity, Mott had suggested that the metal-insulator transition is not continuous [1, 2]. His conclusion was based on Ioffe's and Regel's statement that a free pass length can not be less than electron wave length [3]. However, until today many aspects remain vaguely enough about the nature of metal - insulator phase transitions.

2. Results and discussion

2.1. In [4], the minimal metal conductivities in the conduction band (where the Anderson transition takes a place) and in the deep acceptor band (the Mott transition) have been obtained from pressure dependences of electric conductivity $\sigma(P)$ and Hall coefficient $R_H(P)$ in heavily compensated quasi-gapless semiconductor p-$CdSnAs_2$<Cu>, extrapolating the temperature interval 2÷5 K to T=0 K [5]. However in some cases the idea of minimal metal conductivity has not been confirmed [6, 7].

In [8], the smoothness of the metal-insulator phase transition has been thought as a consequence of large scale fluctuations of the impurity distribution. A transition to the conducting state in bulk samples is defined as a coherence of metalized regions, i.e. when atoms of the metalized regions of a sample, form an infinite cluster penetrating the whole sample. It can be said, that the metal-insulator transition is determined by a solution of a continual problem of the percolation theory, according to which the conductivity as a function of governing parameter grows from zero value continuously. The transition occurs in weakly doped semiconductors with chaotically distributed impurities and in

© 2008 IOP Publishing Ltd

heavily doped compensated semiconductors, while we have studied a transition peculiarities analysing data for the temperature dependence of electric conductivity $\sigma(T)$ at $T\leq 4$ K and $T\to 0$K. It should be noted, that the role of disorder in the whole physics of the phenomenon rises with the temperature decrease [9] and near T=0 K may become determinative.

Sometimes the smoothness of the transition may be caused by a smallness of electron critical concentration (as an instance the n-type of semiconductor is considered): $n_c=N_i\cdot(1-K_c)$, where N_i is a concentration of primary impurity, K_c is a critical coefficient of compensation. As it has been shown in [10], the coefficient K_c is close to 1 in heavily doped semiconductors as well as in weakly doped semiconductors containing hydrogen-like impurities with N_i near to N_{hd}. (N_{hd} is a critical impurity concentration when impurity and intrinsic bands unite).

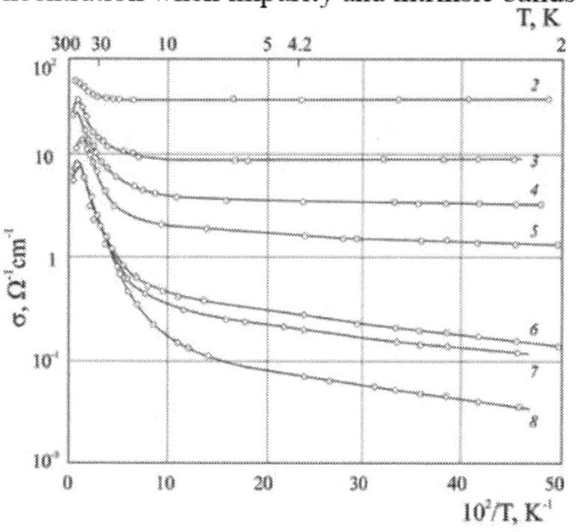

Figure 1. Dependence of conductivity from temperature in weakly doped and compensated GaAs crystals of *n*-type with shallow hydrogen-like donors. Hopes on the nearest centers [14-15]. Sample numbers correspond to concentrations: *1*- $4.5\cdot 10^{17}$; *2*- $1.7\cdot 10^{17}$; *3*- $3\cdot 10^{16}$; *4*- $1.3\cdot 10^{16}$; *5*- $1.3\cdot 10^{16}$; *6*- $7.1\cdot 10^{15}$; *7*- $6.4\cdot 10^{15}$; *8*- $3.5\cdot 10^{15}$ см$^{-3}$ [14-15].

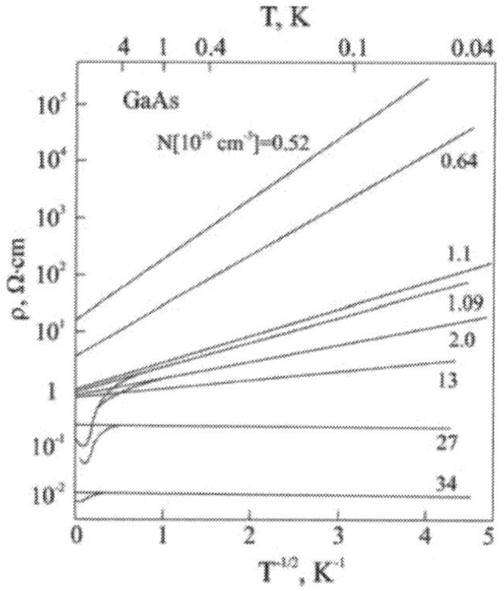

Figure 2. Dependence of resistivity from temperature in weakly doped and compensated GaAs crystals of *n*-type with shallow hydrogen-like donors [16]. Hopes with variable length, Efros-Shklovsky law.

Besides, as a result of smallness of the ionization energy for shallow donors in narrowband weakly doped *n*-InSb and *n*-InAs, the impurity band originated from fluctuating classical field is overlapped with the intrinsic band at $N_d<N_M$ ($N_M\cdot a^3_B\approx 0.02$, a_B is an effective Bohr radius). The impurity states closest to the edge of intrinsic band, merge with the states of intrinsic band continuum and become hybridized (so called "natural" broadening of the level) [10-13]. Appeared free extra carriers screen the potential of the impure center. The ionization energy of localized levels decreases. With increase in N_d , a process assumes an avalanche-like character, giving the fast transition from weakly doped state to the heavily doped state when the impurity band merge with the conduction band completely. It should be noted that "metal" conductivity in n-type narrowband semiconductors is revealed when impurity concentrations $N_M<N_M$. For instance, in *n*-InSb N_M exceed N_M by about two orders of magnitude. Thus, in narrowband *n*-type semiconductors the critical concentration of electrons is essentially underestimated at the absence of compensation.

At last, typical profile of the random potential is forming in heavily doped compensated semiconductors under the correlated distribution of the charged impurities, which decreases the carries critical concentration for several orders. Obviously, then we always have the inequity $N_c^{eff}<N_c$ to be

held. For example, in $CdSnAs_2$<Cu> N_c^{eff}, and n_c is by two orders lesser than N_c during Anderson localization [4, 12] (Table).

2.2. According to experimental data on hopping conduction in *n*-GaAs, *p*-Ge, *p*-$CdSnAs_2$<Cu> (Fig.1-4), the parameter T_0 in expression $\sigma(T)=exp[-(T_0/T)^p]$ tends to zero (see Fig.3, 4) at fixed temperatures when approaching to the transition point at temperatures close to T=0.04 K (Fig.2). Thus, the conductivity σ tends to 0 with $T\to 0$ K and if $T_0>0$, insensibly to the type of the hopping conductivity [7, 8]. If $T_0=0$, the minimal conductivity can be found from experimental data, extrapolating to $T=0$ K. The characteristic parameters of the transition are presented in the Table. Critical concentrations of electrons (holes) are obtained from the Hall coefficient $n_c(p_c)=(|R_H|e)^{-1}$ data.

Table

Minimal metal conductivities (σ_{min}) and effective mobilities (μ_{Hmin}), critical concentrations of electrons (n_c), p_c holes of Mott [1, 2] and Anderson metal-dielectric transition [3].

Semiconductors	σ_{min}, $\Omega^{-1}\cdot cm^{-1}$	p_C, n_C, cm^{-3}	μ_{Hmin}, $cm^2\cdot V^{-1}\cdot s^{-1}$
n-GaAs [14,15] [1)] $K_C\approx 0$	9.5	$3.0\cdot 10^{16}$	2000
p-Ge [16] [1)] $K_C=0.4$	4.2	$1.5\cdot 10^{17}$	175
p-$CdSnAs_2$<Cu> [4] [2)]	$6.1\cdot 10^{-2}$	$2.5\cdot 10^{15}$	155
p-$CdSnAs_2$<Cu> [4] [3)]	$4.3\cdot 10^{-3}$	10^{13}	2750

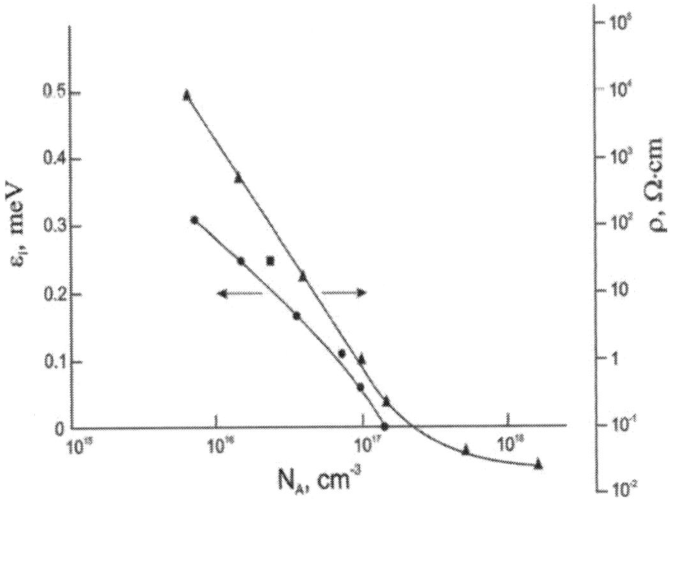

Figure 3a. Concentration N_A dependences of activation energy and hopping conductivity through the nearest shallow hydrogen-like acceptors of *p*-Ge crystals (T=1.25 K, see Table in [17]).

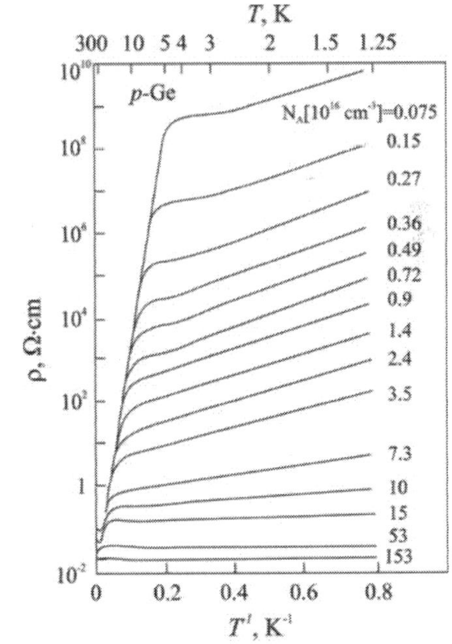

Figure 3b. Conductivity versus temperature for *p*-Ge crystals with shallow hydrogen-like acceptors at T=1.25 K (see Table in [17]).

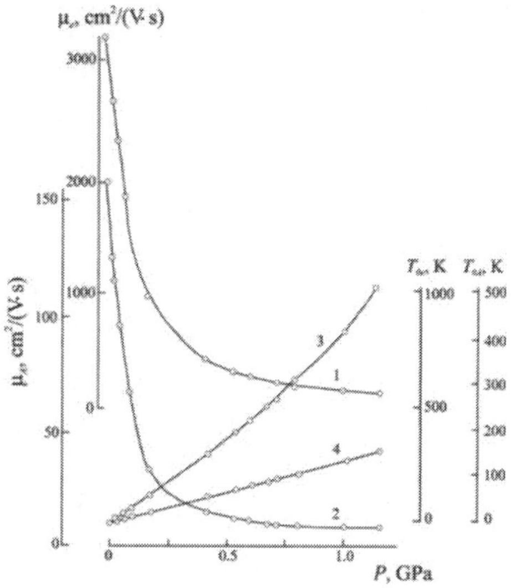

Figure 4. Pressure dependences for: 1) conduction band electron mobility μ_e, 2) deep acceptor band hole mobility μ_A, 3) parameter of hopping conductivity with altering step (Mott's law) of electrons T_{0e}, and (4) parameter of hopping conductivity of acceptor band holes T_{0A} (T=4.2 K, sample #2 p-$CdSnAs_2$<Cu> [4]).

3. Conclusion

The problem of minimal metal conductivity is stipulated by the complicated situation. On the one hand we do not have a strict, theoretically founded solution to this problem [18], but on the other hand every particular case requires a thorough analysis. A smoothness ("continuous") of the electric conductivity change in the vicinity of the critical point may be due to: 1.) the smallness of carrier critical concentration at T=0 K, measured experimentally at nearly apparatus error limit, 2.) the large scale fluctuations of the impurity distribution, which can play a determinative role at $T\to 0$ K. In latter case an interpretation of the experiment should be carried out in the frame of percolation theory. According to the theory, the electric conductivity increases from zero value after formation of an infinite cluster, made of "metalized" atoms and penetrating the whole bulk sample. It is worth to be noted, that experiments carried out at hydrostatic pressures also indicate the "continuity" of structural phase transition in solid states. The smoothness of the phase transition observed in these experiments also is caused by a disorder of the medium [19].

Acknowledgements. The work is supported by the Russian Foundation for Basic Research (Project no. 07-02-00238).

References

[1] Mott N F *Adv. Phys.* 1967 **16** 49.
[2] Mott N F *Phil. Mag.* 1974 **29** 613.
[3] Ioffe A F, Regel A R *Prog. Semicond.* 1960 **4** 237.
[4] Kamilov I K, Daunov M I, Elizarov V A, and Magomedov A B *Zh. Eksp. Teor. Fiz.* 1993 **104** 1 2436 [*JETP* 1993 **77** 92].
[5] Daunov M I, Kamilov I K, Elizarov V A, Magomedov A B and Danilov V I *Dokl. Akad. Nauk* 1997 357 612 [*Phys. Dokl.* 1997 42 530].
[6] Paalanen M A, Rasenbaum T F, Thomas G A, Bhatt R N *Phys. Rev. Lett.* 1982 **48** 1284.
[7] Gantmaher V F *Electrons in disordered metters.* M.: «Fizmatlit» 2005 -232 p.
[8] Zabrodski A G, Nemov S A, and Ravich Yu I *Electronic Properties of Disordered Systems* (Nauka, St. Petersburg, 2000) [in Russian].
[9] Daunov M I, Kamilov I K, Magomedov A B *Zh. Eksp. Teor. Fiz.* 1997 **111** 2 562.
[10] Daunov M I, Kamilov I K, Gabibov S F *Fiz. Tekh. Poluprovodn.* 2006 **40** 5 536 [*Semiconductors* 2006 **40** 5 521].
[11] Kamilov I K, Daunov M I, Elizarov V A, and Magomedov A B *Zh. Eksp. Teor. Fiz.* 1991 **54** 10 589 [*JETP Lett.* 1991 **54** 594].
[12] Daunov M I, Kamilov I K, Magomedov A B *Phys. stat. sol.(b)* **211** 513.
[13] Daunov M I, Kamilov I K, Elizarov V A *Fiz. Tverd. Tela* 1995 **37** 2276 [*Phys. Solid State* 1995 **37** 1245].
[14] Emeliyanenko O V, Lagunova T S, Nasledov D N, and Talalakin G N *Fiz. Tverd. Tela* 1965 **7** 1315 [*Sov. Phys. Solid State* 1965 **7** 1063].
[15] Emeliyanenko O V, Lagunova T S, and Nasledov D N *Fiz. Tverd. Tela* 1961 **3** 198 [*Sov.Phys. Solid State* 1961 **3** 144].
[16] Reztsch R, Frieland K J, Ionov A N et al. *Phys. Rev.* 1989 **B40** 1216.
[17] Fritzsche H, Guevas M *Phys. Rev.* 1960 **119** 1238.
[18] Shklovski B I and Efros A L *Electronic Properties of Doped Semiconductors* (Nauka Moscow 1979; Springer New York 1984).
[19] Daunov M I, Kamilov I K, Arslanov R K, Daunova D M, Gabibov S F *J. Phys.: Condens. Matter.* 2003 **15** 2335.

Journal of Physics: Conference Series **121** (2008) 022006 doi:10.1088/1742-6596/121/2/022006

High pressure as an external factor, effectively governing a chaotic potential of crystalline semiconductors

I.K.Kamilov, M.I.Daunov, S.F.Gabibbov, A.B.Magomedov

Institute of Physics of Daghestan Science Center, Russian Academy of Sciences, 367003, Makhachkala, M.Yaragskogo str.,94, Russia

E-mail: i.daunov@mail.ru

Abstract

It has been revealed from pressure, temperature, magnetic and electric fields dependences of kinetic coefficients in doped compensated Ge<Au,Sb>, p-type InSb, InAs, $CdSnAs_2$<Cu> and HgTe, that the observed anomalies of electron transport in the listed above materials are linked to the presence of large-scale fluctuations of ionized impurity concentrations. It is shown, that the influence of chaotic potential begotten by these fluctuations becomes more noticeable when the temperature decreases or the pressure rises.

1. Introduction

It is known [1], that smooth large scale fluctuations of random spatial potential are being formed in doped compensated semiconductors, when concentration of free charge carriers is small in comparison with the concentration of ionized impurity centers.The influence of chaotic potential increases while temperature decreases. Obviously, it is of current importance to estimate: 1. to what extent a chaotic potential may affect the energy spectrum of charge carriers, and therefore the properties of crystal semiconductors. 2. The propriety of application of corresponding correlations in interpretation of experiments for every given case if the following inequalities take a place: $\varepsilon_F \gg \gamma$ in degenerated semiconductor and $k_BT \gg \gamma$ in nondegenerated semiconductor. Here ε_F – Fermi energy, γ - typical value of random potential amplitude, k_B – Bolzman's constant, T –temperature in K. The problem becomes more complicated if we deal with correlation of distribution of impurity in doped compensated semiconductors, which stipulates a nonelectronic nature of screening [1, 2].

2. Results and discussion

Typically, pressure coefficients of energy gaps $\Delta\varepsilon$ do not depend on pressure in semiconductors [3]. As an example, let's take so-called quasi-zero-gap semiconductors (p-InAs, p-$CdSnAs_2$<Cu> etc.) [4] (Figure1) which have deep impurity band near the edge of intrinsic band. It has been concluded that the use of dispersion law for ideal semiconductor in a semiconductor with random potential is correct until the dependence of energy gap on hydrostatic pressure $\Delta\varepsilon(P)$, calculated from electron transport experiments and corresponding correlation relations, is close to the linear. Increasing deviation from linearity in $\Delta\varepsilon(P)$ dependence with the temperature fall and a decrease of carrier concentration testifies a considerable influence of random potential on dispersion law.

© 2008 IOP Publishing Ltd

Let's consider the situation of quasi-gapless semiconductor p-$CdSnAs_2$<Cu> [2], where a deep acceptor level (in some cases shredded in an acceptor band) generated by cadmium vacancies, is located at the distance $\varepsilon_A = \varepsilon_A^0 + \alpha T - \beta P$ from the bottom of the conduction band (where $\varepsilon_A^0 = -30$ meV, $\alpha = 4.6 \cdot 10^{-2}$ meV/K, $\beta = \partial\varepsilon_g/\partial P = 120$ meV/GPa, and $\varepsilon_g = 0.28$ meV - width of forbidden band). Energy is counted from the bottom of the conduction band and direction inside the band is chosen to be positive. The density of states for the discussed model, in the case of the occupation parameter for acceptor band is $K_A = N_A^-/N_A \approx 0$, is depictured in Fig.1.

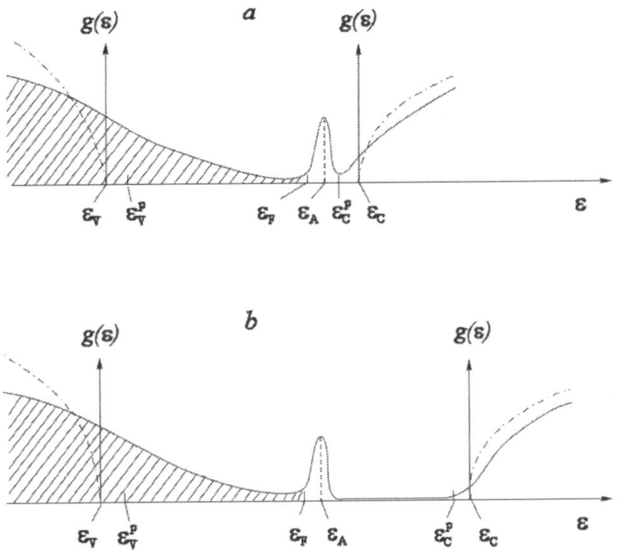

Figure 1. Density of states (schematically) of heavily doped compensated semiconductor with deep acceptor band p-$CdSnAs_2$<Cu> at atmospheric (a) and hydrostatic (b) pressures, where acceptor band occupation parameter $K_A \approx 0$. Hereε_v, ε_c – undisturbed energies of the top of valence and the bottom of the conduction bands; ε_a, ε_F – energies of deep acceptor level and Fermi level; ε_v^P, ε_c^P – energies of percolation levels for holes and electrons correspondingly. Stroke dotted line – the shape of density states for an ideal crystal. Filled states are shown shaded.

Characteristic parameters of sample #10 p-$CdSnAs_2$<Cu> [2] are given in the Table 1 and Fig. 2. Here n, μ_e and p_A, μ_A are concentration and mobility of electrons of the conduction band and holes of the accepter band respectively.

Table 1. Parameters of sample # 10

T, K	P, GPa	R_0, cm^3/C	ρ_0, $\Omega\cdot cm$	p_A, cm^{-3}	μ_A, $Cm^2/V\cdot s$	p_A/n,	μ_c/μ_A,	K_A
280	10^{-4}	-650	0.286	5.6	130	16.5	30	0.08
	1.5	-300	2.50	5.4	22	1450	180	–
	∞	530	3.80	5.4	10	∞	0	0.10

Hereinafter, the characteristic parameters of charge carriers and pressure coefficients of energy gap β are calculated from temperature and magnetic field dependences of Hall coefficients, resistivity and well known values of band parameters. It is seen from Fig. 2, that pressure coefficient β decreases at P>1 GPa, that is stipulated by the influence of chaotic potential on energy spectrum at T=295 K.

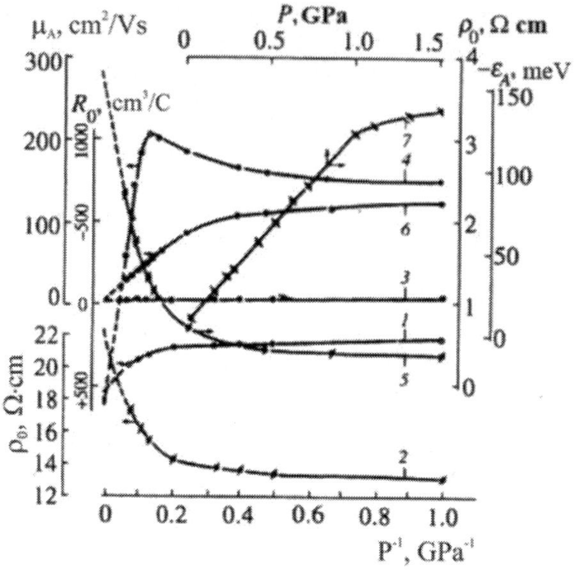

Figure 2. The pressure dependences of Hall coefficient R_0 (curves 1, 4), resistivity ρ_0 (2, 5), mobilities of acceptor band holes μ_A (3, 6) in vanishing small magnetic field at T=77.6 K (1–3) and T=280 K (4–6). (7)- the pressure dependences of the deep acceptor level ε_A at 280 K (sample #10 [2]).

Experimental data on pressure coefficients for the energy deep impurity centers of gold [5, 6] and copper [7] in germanium are presented in Table 2 and Fig. 3. A detailed analysis of these data has been performed earlier in [5].

Table 2. Pressure coefficients of the four levels due to gold in germanium

Level	Type	E_C–E_i	$\partial(E_{CL}-E_i)/\partial P$	T (K)
E_4	acceptor	0.043	21	45–49.3
E_3	acceptor	0.19	29 (44**)	112–194
E_2	acceptor	0.60	44	170–273
E_1	donor	0.71	49	49.5
E_v		0.75	50	

We will note briefly only two factors responsible for the decreases in the pressure coefficients $\gamma_{cL}=d(E_{cL}-E_i)/dP$ and $\gamma_v=d(E_i-E_v)/dP$ in n- and p- regions, respectively. First of all, it is crucial to use the Gibbs statistics consistently, which has not been correctly done in the analysis carried out by Pel' and others [7]. Holland and Paul [6] have taken into account the Gibbs statistics and, moreover, measured the pressure dependence of the electrical resistivity in Ge: Au at low temperatures with the aim of obtaining more reliable results. However, slightly soluble impurities (such as Au, Hg and Cu) in doped and compensated germanium can form cluster with a high probability. As the temperature decreases and the pressure increases, the concentration of free charge carriers decreases and the amplitude of a random potential and correspondingly, the degree of fluctuation bending of energy bands increases. In this situation, if the energy levels of deep impurity centers are located in the band gap, a transition to a state which is similar to a heavily doped, completely compensated semiconductor can occur [1]. This with necessity must lead to a decrease in the pressure coefficient for the energy gaps calculated from the pressure dependence of the carrier concentration, as is confirmed by the data presented in Fig. 3 and in the Table 2. It is also evident that the effect of the random potential increases with a decrease both in the temperature and in the energy separation between the deep impurity level and the edge of the intrinsic band.

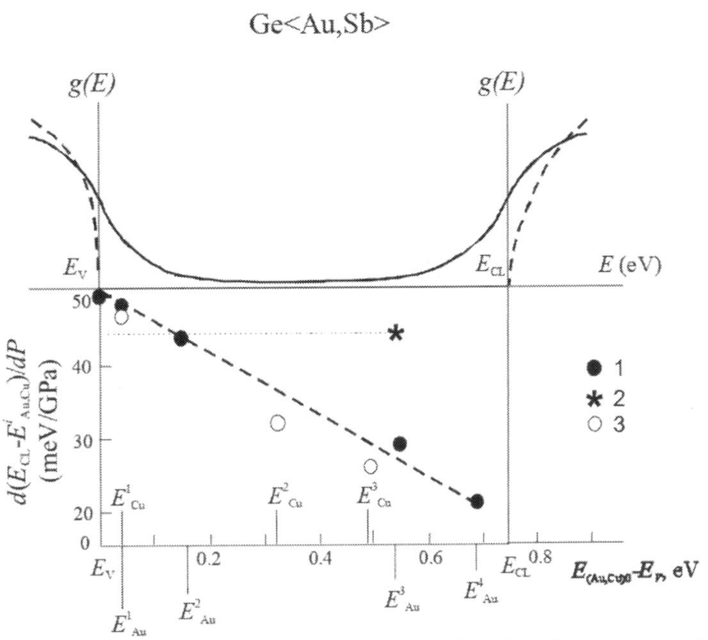

Figure 3. State density $g(E)$ in a heavily doped compensated semiconductor (germanium) and pressure coefficient of Au levels according to data [6] (1) and [5] (2), and Cu levels according to [7] (3).

In the above listed semiconductors the value of pressure coefficient of energy gap decreases, because dispersion law for ideal semiconductor was applied for the quantitative analysis. It testifies considerable influence of chaotic potential.

Unlike usual semiconductors, the influence of chaotic potential in gapless semiconductors (with inversed band structure: $\varepsilon_g=\varepsilon_{\Gamma6}-\varepsilon_{\Gamma8}<0$), for instance in p-HgTe, leads to an excess of pressure coefficient $\beta=\partial\varepsilon_g/\partial P$. Coefficient $\beta=\partial\varepsilon_g/\partial P$ for lightly doped sample p-HgTe-1 has been calculated according to the experimental data [8]. In slightly doped sample p-HgTe-1 at T=295 K and T=77.6 K β=121 meV/GPa, at T=4.2 K β=220 mev/GPa; in heavily doped sample p-HgTe-2 β=85 mev/GPa at T=295 K, and β=210 mev/GPa at T=77.6 K [9].

3. Conclusion

Hydrostatic pressure affects effectively on the electron spectrum of crystal semiconductors and allows one to determine the extent of the influence of a large scale fluctuation potential on the charge carrier energy spectrum. The latest allows one then to examine the propriety of corresponding equations for quantitative analysis of experimental results.

Acknowledgements. This work has been supported by the Russian Foundation for Basic Research (project no. 07-02-00238a).

References

[1] Shklovski B I and Efros A L *Electronic Properties of Doped Semiconductors* (1979 Nauka, Moskow,; 1984 Springer, New York).
[2] Daunov M I, Kamilov I K, Magomedov A B 1997 *Zh. Eksp. Teor. Fiz.* **111** 562 [1997 *JETP* **84** 309].
[3] Matulionis A, Pozela J, Simulyte E, Juciene V *Pressure transducers on the basis of graded-gap semiconductors* 1980 Mokslas Vilnus p 141
[4] Daunov M I, Kamilov I K, Elizarov VA and Magomedov A B 1997 *Dokl. Akad. Nauk* **357** 612 [1997 *Phys. Dokl.* **42** 530].
[5] Daunov M I, Kamilov I K, Gabibov S F 2001 *Semiconductors* **35** 59.
[6] Holland M G, Paul W 1962 *Phys. Rev.* **128** 43.
[7] Pel' E G, Fistul' V I, Jagshigel'diev A, Jakovenko A G 1980 *Semiconductors* **14** 1220.
[8] Stankiewicz J, Giriat W 1976 *Phys. Rev. B.* **13**(2) 665.
[9] Daunov M I, Broyda E L 1973 *Phys. Stat. Sol. (b)* **55**, K155.

Pressure-induced structural change in liquid GeTe and Liquid GeSe

Masatoshi Tomomasa, Takuya Higaki, Takazumi Hayakawa, Ayano Chiba, Kazuhiko Tsuji

Department of Physics, Keio University, 3-14-1, Hiyoshi, Yokohama, 223-8522, Japan

E.mail: tsuji@phys.keio.ac.jp

Abstract. X-ray diffraction has been measured up to 19 GPa for liquid GeTe and up to 15 GPa for liquid GeSe using synchrotron radiations. Static structure factor *S(Q)* and pair distribution function *g(r)* was obtained. Both *S(Q)* of liquid GeTe and liquid GeSe change their shape with increasing pressure, indicating anisotropic contraction of the local structure. A sharp change in *S(Q)* and *g(r)* was observed in liquid GeTe between 1.8 GPa and 4.0 GPa. Pressure dependences of *S(Q)* and *g(r)* of liquid GeSe change around 6 GPa. In liquid GeSe a small pre-peak was observed around 1.2 $Å^{-1}$ up to 1.7 GPa which suggests the presence of the medium range order of covalent bond network. With increasing pressure the Peierls-type distortion becomes small and the bond angle decreases; the local structure becomes a rhombohedral structure. These pressure dependences of the structure of liquid GeTe and liquid GeSe are different from those of the crystals. These results are compared with those of other liquids and discussed in relation to the bonds between atoms in the liquids.

1. Introduction

The pressure dependence of the local structure of liquid group 14 elements and liquid III-V compounds is different from that of the crystal. The structural change occurs continuously in a wide pressure region below the transition pressure of the crystal. The local structure at high pressures is different from that of the crystal [1-6]. On the other hand, in liquid CdTe a structural transformation occurs in a narrow pressure width around the transition pressure of the crystal [7, 8]. In these liquids atoms form three dimensional networks of the covalent bonds and the pressure-induced structural changes depend on the ionicity in the chemical bonds between atoms. On the other hand, in liquids with low-dimensional networks of covalent bonds, such as liquid iodine, liquid selenium and liquid tellurium, the length of covalent bonds increases with increasing pressure in spite of the volume contraction, which arises from the increase of the intermolecular or interchain interaction with decreasing distance between atoms with weak bonds [1, 9, 10].

In crystalline GeTe and GeSe, atoms form two dimensional networks of covalent bonds with coordination number of three at the normal pressure and the room temperature and at high temperature it transforms to a rocksalt (B1) structure with coordination number of six. In liquid GeTe and liquid GeSe, the local structure at the normal pressure is known to be different from the rocksalt structure but

© 2008 IOP Publishing Ltd

Journal of Physics: Conference Series **121** (2008) 022007 doi:10.1088/1742-6596/121/2/022007

similar to the ternary coordinated structure [11-13]. The presence of the Peierls distortion was reported and the strong covalent bonds between intralayer atoms and the relatively weak bonds between interlayer atoms coexist. Although the transition pressures in the crystalline phase are 18 GPa in GeTe and 25 GPa in GeSe [14], an anisotropic local contraction is expected in the liquids with increasing pressure below these transition pressures. X-ray diffraction was measured at high pressures up to 19 GPa for liquid GeTe and up to 15 GPa for liquid GeSe.

2. Experimental

Pressure was generated by the cubic-anvil apparatus MAX80 installed on the AR-NE5C beamline at the Photon Factory, KEK, and the double stage apparatus SPEED1500 installed on the BL04B1 beamline at SPring-8. By using synchrotron radiations with high brightness, high energy and small divergence, x-ray diffraction was measured with an energy-dispersive method. A powdered mixture of c-GeTe or c-GeSe and BN was placed in a BN capsule. MgO was also used for a sample capsule. Measurements were made at temperatures about 50 K above the melting point of each pressure. Energy-dispersive diffraction profiles were measured at 10 different scattering angles, 2θ, between 2.5° and 20° (MAX80) and between 3° and 15° (SPEED1500) with white x-rays of 30–140 keV. From the measured diffraction intensities, the static structure factors $S(Q)$ of liquid metals were deduced up to Q=17 Å^{-1} after some corrections. Details of the procedures are given elsewhere [15, 10].

3. Results and discussion

Figures 1-4 show the static structure factor $S(Q)$ and $g(r)$ of liquid GeTe and liquid GeSe at various pressures. In liquid GeTe, $S(Q)$ changes its shape with increasing pressure. If the local structure contracts uniformly, $S(Q)$ expands towards higher Q without changing its height. The change in the shape of $S(Q)$ indicates an anisotropic structural change with pressure. The height of the first peak increases with changing its position towards higher Q value. On the other hands, the height of the subpeak around 3.2 Å^{-1} decreases. Figure 5 shows the height of $S(Q)$ at Q=2.30 Å^{-1}. Its value increases markedly between 1.8 GPa and 4.0 GPa and its slope is different below 1.8 GPa and above 4 GPa. The second peak consists of at least two subpeaks. The shape of the second peak changes with increasing

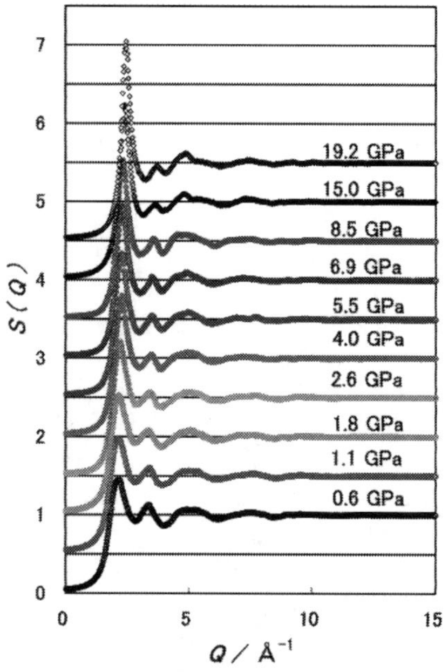

Figure 1. $S(Q)$ of liquid GeTe at various pressures

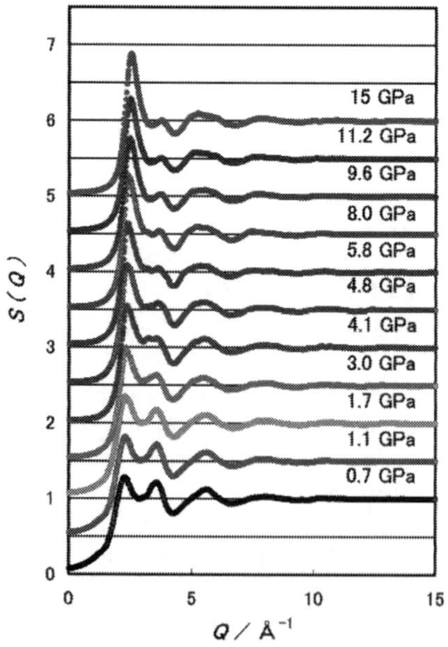

Figure 2. $S(Q)$ of liquid GeSe at various pressures

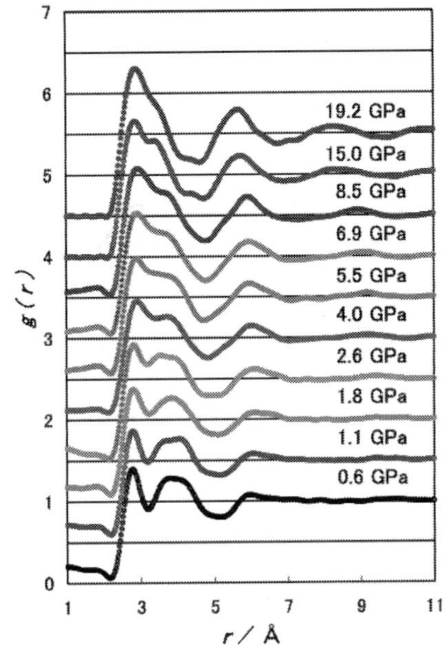

Figure 3. *g(r)* of liquid GeTe at various pressures

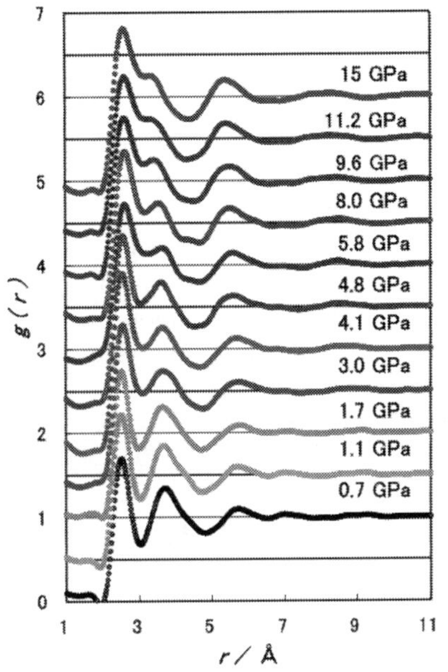

Figure 4. *g(r)* of liquid GeSe at various pressures

pressure. With increasing pressure, the subpeak at the low Q side becomes higher with shifting its position towards lower Q and the subpeak at high Q side becomes lower. Because it is difficult to determine the second peak position, Q_2, in Figure 6 the center of the positions where the second peak of $S(Q)$ intersects the line of $S(Q)=1$, is plotted against pressure. Q_2 decreases with increasing pressure up to 3 GPa in spite of volume contraction. Above 3 GPa, the slope becomes small but still remains negative. These suggest that the pressure-induced structural transformation occurs around 3 GPa. A similar change in $S(Q)$ was observed in liquid GeSe. The pressure dependence of $S(Q)$ changes around 6 GPa. As shown in Figure 6, the sign of the pressure derivative of Q_2 changes around 6 GPa in liquid GeSe. In the low Q region of $S(Q)$ of liquid GeSe, there is a small peak around 1.2 Å^{-1}. In several compound-forming liquids, the first sharp diffraction peak has been observed and they are considered the indication of the medium range order of the cluster. A small peak around 1.2 Å^{-1} was observed up to 1.7 GPa and its height decreases with increasing pressure. This peak suggests that the networks of the covalent bonds similar to the crystal with two dimensional zigzag planar structure remains in the

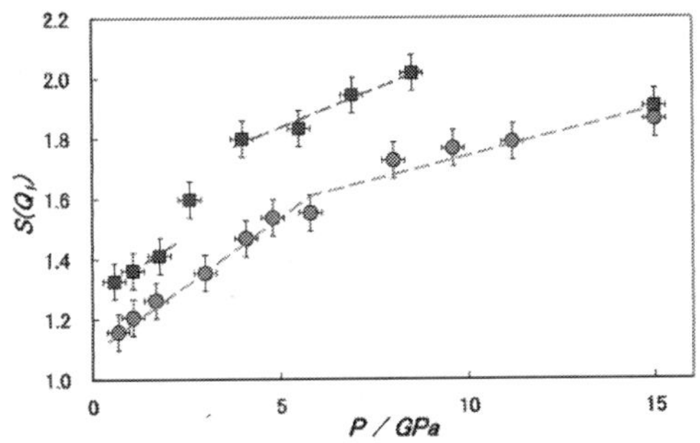

Figure 5. Pressure dependence of the height of $S(Q)$ of liquid GeTe (square) at Q=2.30 Å^{-1} and liquid GeSe (circle) at Q=2.48 Å^{-1}. The dashed lines are only to guide the eyes.

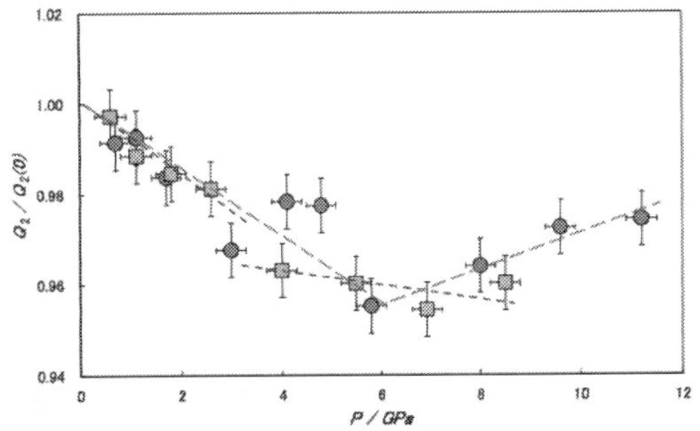

Figure 6. Pressure dependence of the second peak position Q_2 of $S(Q)$ for liquid GeTe (square) and liquid GeSe (circle). The dashed lines are only to guide the eyes.

low pressure region.

Figure 3 shows *g(r)* of liquid GeTe at several pressures. The main peak consists of two subpeaks. The sharp subpeak around 2.8 Å indicates the presence of the covalent bonds between atoms in the liquid. With increasing pressure, it shifts towards larger *r* in spite of the volume contraction, although its rising position does not shift. The subpeak at larger *r* shifts towards smaller *r*. The height of *g(r)* at the first minimum increases with increasing pressure (Figure 7). The pressure dependence of $g(r_m)$ changes at about 3 GPa. With increasing pressure, the second peak shifts towards smaller *r*. Similar change was observed in liquid GeSe. The pressure dependence of *g(r)* changes at about 6 GPa.

The coordination number, *CN*, was deduced from twice of the integrated value of radial distribution function to the first peak position of *g(r)*. With increasing pressure, *CN* of liquid GeTe increases from 2.5 at 0.6 GPa to 3.2 at 2.6 GPa and increases to 4.2 at 4.0 GPa, and then it increases gradually to 5.1 at 8.5 GPa. In liquid GeSe, *CN* increases from 2.5 at 0.7 GPa to 3.9 at 15 GPa. The crystal structure just below the melting curve is rocksalt structure for GeTe and GeSe in the pressure region of present measurements. In the low pressure region, *CN* of liquid GeTe and liquid GeSe are smaller than six for rocksalt structure and almost same as three for rhombohedral GeTe structure (same as A7 structure) or the orthorhombic GeS structure. It approaches to the value of the rocksalt structure at high pressures.

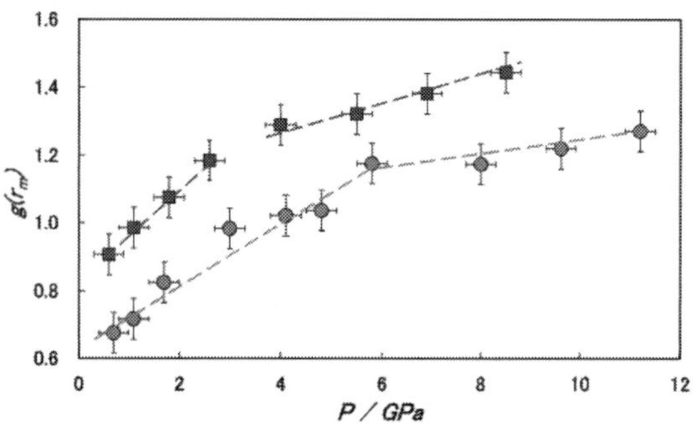

Figure 7. Pressure dependence of the height of *g(r)* of liquid GeTe (square) and liquid GeSe (circle) at the first minimum of *g(r)*. The dashed lines are only to guide the eyes.

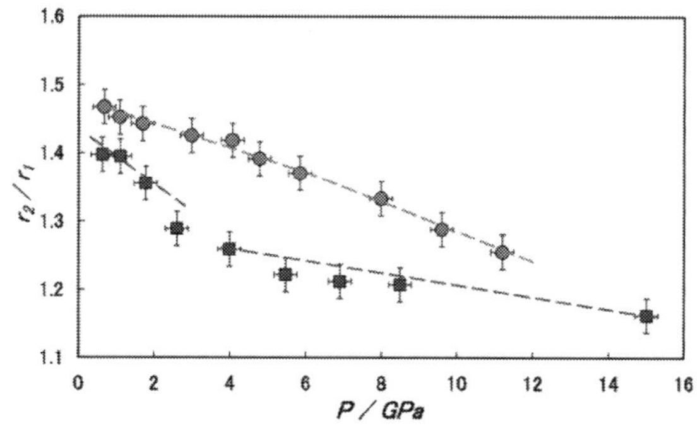

Figure 8. The ratio r_2/r_1 of liquid GeTe (square) and liquid GeSe (circle), where r_1 and r_2 are the first and the second peak positions of *g(r)*, respectively. The dashed lines are only to guide the eyes.

For liquid GeTe, the lower *r* side of the first peak of *g(r)* does not change with increasing pressure while the height of the first minimum increases with pressure. This change suggests the shortening of the long bonds between the atoms on the adjacent layer with increasing pressure. This change should arise from the weakening of the Peierls distortion. The position of the subpeak at larger *r* side shifts towards smaller *r* which suggests the shortening of the second neighbor distance of the atoms in the same layer. As the energy increase with bond contraction is large in the covalent bond, the structural change may occur by bond bending in order to decrease the free energy through pV term by volume contraction. The theoretical simulation suggests such bond bending in liquid P with increasing pressure [16]. On the structural change from the diamond structure to the beta-tin structure in group 14 elements and III-V compounds, the bond bending results in a large volume contraction. As shown in Figure 8 the r_2/r_1 ratio of liquid GeTe changes from 1.40 at 0.6 GPa to 1.16 at 15 GPa. If the peak positions indicate the intralayer atomic distances, the bond angle changes from 89° to 71°, although the first peak consists of three kinds of bond lengths; the first nearest neighbor and the second nearest neighbor distance in the same layer and the first neighbor distance between atoms on the adjacent layers. In liquid GeSe, the ratio changes from 1.46 at 0.7 GPa to 1.25 at 11.2 GPa. The corresponding bond angle changes from 94° to 77°.

4. Conclusion

X-ray diffraction measurements at high pressures using synchrotron radiations show that the pressure-induced structural transformation occurs at about 3 GPa in liquid GeTe and the pressure dependence of the local structure changes at about 6 GPa in liquid GeSe. The local structure of liquids at low pressures is ternary coordinated structure and is different from those of the crystals. With increasing pressure, the Peierls type distortion decreases and the bond angle becomes smaller than 90° without changing the covalent bond length. The pressure-induced structural changes in liquid GeTe and liquid GeSe are different from those of crystals, similar to liquid group 14 elements and liquid III-V compounds. These results suggest that the covalent bonds play an important role to the local structure of liquids in these IV-VI compounds.

Acknowledgements

This work is supported by the Grand in Aid from the Ministry of Education, Science and Culture of Japan under contract No.16340110. The synchrotron radiation experiments were performed at PF-AR, KEK (Proposal number 2005G138) and at the SPring-8 (Proposal number 2006A1083 and 2006B1208).

References

[1] Katayama Y and Tsuji K 2003 *J. Phys. Condens. Matter* **15** 6085
[2] Tsuji K, Hattori T, Mori T, Kinoshita T, Narushima T and Funamori N 2004 *J. Phys. Condens. Matter* **16** S989
[3] Hattori T, Kinoshita T, Taga N, Takasugi Y, Mori T and Tsuji K 2005 *Phys. Rev.* B **72** 064205
[4] Hattori T, Tsuji K, Kinoshita T and Narushima T 2004 *J. Phys. Condens. Matter* **16** S997
[5] Hattori T, Tsuji K, Taga N, Takasugi Y and Mori T 2003 *Phys. Rev.* B **68** 224106
[6] Funamori N Tsuji K 2002 *Phys. Rev. Lett.* **88** 255508
[7] Kinoshita T, Hattori T, Narushima T and Tsuji K 2005 *Phys. Rev.* B **72** 060102
[8] Hattori T, Kinoshita T, Narushima T, Tsuji K and Katayama Y 2006 *Phys. Rev.* B **73** 054203
[9] Tsuji K, Shimomura O, Tamura K and Endo H 1988 *Z. Phys. Chem. Neue Folge* **156** 495
[10] Funamori N and Tsuji K 2001 *Phys. Rev.* B **65** 014105
[11] Raty J Y, Gaspard J P, Bionducci M, Ceolin R and Bellissent R 1999 *J. Non-Cryst. Sol.* **250-252** 277
[12] Raty J Y, Godrevsky V V, Gaspard J P, Bichara C, Bionducci M, Bellissent R, Ceolin R, Chelikowsky R, and Ghosez P 2002 *Phys.Rev.* B **65** 115205
[13] Raty J Y, Godrevsky V V, Gaspard J P, Bichara R, Bionducci M, Bellissent R, Ceolin R , Chelikowsky J R and Ghosez P 2001 *Phys.Rev.* B **64** 235209
[14] Onodera A, Sakamoto I and Fujii Y 1997 *Phys. Rev.* B **56** 7935
[15] Tsuji K, Yaoita K, Imai M, Shimomura O and Kikegawa T 1989 Rev. Sci. Instrm. **60** 2425
[16] Morishita T 2002 *Phys. Rev.* B **66** 054204

Ultrasonic Study of Monomeric Fullerite C_{60} under Pressure

O F Yagafarov, E L Gromnitskaya, A G Lyapin, V V Brazhkin, and M V Kondrin

Institute for High Pressure Physics, Russian Academy of Sciences, Troitsk , Moscow region, 142190, Russia

oscar.yagafarov@gmail.com

Abstract. The ultrasonic study of the elastic moduli of polycrystalline fullerite C_{60} was carried out in the temperature range from liquid nitrogen to room temperature at pressures up to 1.4 GPa, including the regions of the orientational fcc-to-sc and glass transitions. The temperature dependencies of the elastic moduli at ambient pressure are in good agreement with previous studies. For the first time the pressure dependencies of the shear modulus and Poisson's ratio were determined. Particularly, the Poisson's ratio was found to increase from ≈0.3 at ambient pressures to ≈0.38 at $p = 1$ GPa. The observed high values of the Poisson's ratio, as well as the high values of pressure derivatives of the bulk modulus ≈23-25 are discussed in terms of the central *m-n* power potential and contribution of the non-central intermolecular forces due to intramolecular deformations.

1. Introduction

Fullerene C_{60} is the most stable molecule among other fullerenes, and fullerite C_{60} is an archetype among all-carbon molecular solids bonded by the van der Waals forces. For the past two decades, the physical and chemical properties of fullerenes have been a fruitful subject of study. The measurements of the elastic properties are an important tool to study any material. The knowledge of the elastic constants is crucial to verify and develop the force potentials for an intermolecular interaction in C_{60}. Many experimental techniques have been applied to study the bulk modulus B: the measurements of the lattice constant by the use of the X-ray and neutron diffraction methods and the direct measurement of compressibility. Some experiments have involved the use of an ultrasonic technique [1,2]. A comprehensive review on physical and elastic properties of molecular C_{60} is given in Ref. [3]. More recent results are presented in Refs. [2,4-6]. Although all these methods are comparably accurate, it is surprising that the results from different groups often vary by significant amounts [3]. Therefore, the issue of the C_{60} bulk modulus B still remains open. Also there is little information on the shear modulus G of C_{60} under pressure. Ref. [1] presented only the temperature dependence of G at atmospheric pressure.

At room temperature the C_{60} crystal structure is a face centered cubic one with space group *Fm3m* [7]. The studies on this structure [8-10] show that the molecular rotation is not completely free. On cooling, the rotation slows down and, below an orientational transition temperature T=260 K [11,12], the rotational motion is replaced by a combination of a rapid jumping motion between a number of well-defined orientational states and librational motion about the equilibrium orientation in each such state. The basic structure of the low-temperature phase is identical with that of the fcc phase except

© 2008 IOP Publishing Ltd

that the four molecules in the cubic unit cell have different orientations, changing the fcc symmetry to a simple cubic (sc) structure. In such ordered structures, electron-rich double bonds on one molecule face the electron-poor centres of hexagons and pentagons. Below the glass transition temperature T_g= 90 K [12] the remaining orientational disorder becomes frozen in creating orientational glass with the crystalline order of molecular centers.

In this work we present an ultrasonic and volumetric study of C_{60} on compression and heating, including fcc, sc and glass phases. The study provides accurate data on the elastic characteristics of these phases under pressure and their changes during transitions.

2. Experimental details

The measurements were carried out using an ultrasonic piezometer in the pressure and temperature ranges up to 1.4 GPa and 80 - 290 K, respectively, in accordance with the procedure described in Ref. [13]. The travel time of an ultrasonic wave was directly measured by the "Akustomer-1" device designed in the Institute for High Pressure Physics RAS. The measurements provided pressure and temperature dependencies of the longitudinal and transversal wave velocities in C_{60}. The hight of the sample and, subsequently, the density were determined independently. We estimated the accuracy in the determination of the bulk and shear moduli as 2 %.

The studied samples of C_{60} were cylinders of 7-10 mm in height and 16 mm in diameter. The initial compaction of the C_{60} powder into a polycrystalline sample was made under air-free conditions. The pressure cell used in the experiments was of a piston-cylinder type.

For the experiments to be accurate, it was very important that hydrostatic loading could be provided. We hope our experiments adequately met this condition, because C_{60} is a soft and plastic molecular material. Additionally, we tried to increase the hydrostaticity by placing a lead pad between the sample and cell walls. It was also important, that the ultrasonic wave length we used was smaller that the sample height by a factor of 5-10. When measuring the bulk modulus it is important that both the C_{60} phases have the cubic symmetry, which diminishes the role of a possible sample texture during compression

3. Results and discussion

First, we determined the elastic properties of C_{60} at room temperature (293 K) and atmospheric pressure. Precise direct measurements of ultrasound velocities allow the determination of the bulk modulus B_0, shear modulus G_0, Poisson's ratio σ_0, and Young's modulus E_0. The comparison of the B_0, G_0, and E_0 values is presented in Table 1. The moduli were obtained by the use of different techniques that gave adiabatic or isothermal values. However, it is known that the differences between adiabatic and isothermal elastic constants and moduli are less than 1% [14], thus, comparing the data of different experiments, including our adiabatic ultrasound ones, seems correct.

Table 1. Shear, bulk, and Young's moduli at room temperature and atmospheric pressure

	References				
	This work	Ref. [3]	Ref. [1]	Ref. [4]	Ref. [15]
G_0, GPa	4.14		4.85		3.75
B_0, GPa	10.02	from 6.4 to 14.8	10.8	15.6	8.4
E_0, GPa	10.93		12.6		9.8

The most important results are the dependencies of the shear and bulk moduli on pressure and temperature. The ultrasonic wave travel time and sample length in isothermal experiments were measured on compression and in isobaric experiments – on heating (the heating rate was ≈1 K/min). As a result, the pressure and temperature dependencies of the ultrasound velocity were found (see, e.g., the inset in Fig. 3).

The pressure dependencies $B(p)$ and $G(p)$ are presented in Fig. 1. The anomalies in changes in the above characteristics were identified as phase transition points in C_{60}. The high values of pressure derivatives of the bulk modulus (B'_p≈23–25) indicate that this molecular system may be rather

Journal of Physics: Conference Series **121** (2008) 022008 doi:10.1088/1742-6596/121/2/022008

sensitive to environmental conditions, and, hence, to non-hydrostaticity, presence of impurities, texture, and so on. This can explain the lack of coincidence of the data from different experiments on the bulk modulus. The values of both the bulk and shear modulus pressure derivatives are very slightly temperature-dependent. The pressure derivative of the bulk modulus for the fcc phase is very close to that for the sc phase. On the contrary, the pressure derivative of the shear modulus for the fcc structure is lower than that for the sc structure. From the data in Fig. 1 it is possible to calculate the Poisson's ratio σ, which at room temperature varies from ≈ 0.3 at atmospheric pressure to ≈ 0.38 at p=1 GPa. A very similar picture is observed for the change of σ at other temperatures for the sc phase.

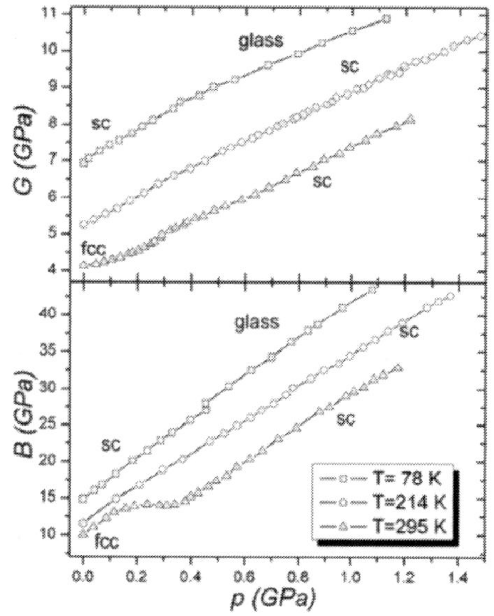

Figure 1. Pressure dependencies of shear and bulk moduli of C_{60} at different temperatures.

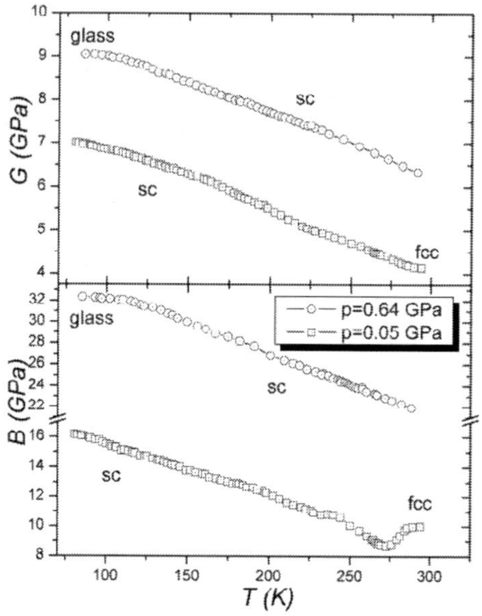

Figure 2. Temperature dependencies of shear and bulk moduli of C_{60} at different pressures.

The temperature dependencies $B(T)$ and $G(T)$ are presented in Fig. 2. It is interesting that the G modulus is almost insensitive to the sc-to-fcc structure transition, whereas the B modulus changes dramatically. For the glassy state the temperature derivatives of both G and B are lower than those for the sc structure. The picture of the change of B during the sc-to-fcc transition at p=0.05 GPa (close to atmospheric pressure) is qualitatively similar to the dependence from Ref. [2] for normal pressure.

Figure 3 shows our points of phase transitions established from the anomalies in the pressure and temperature dependencies of ultrasonic wave velocities. One can see a good agreement with the phase diagram previously proposed [3]. Typical anomalies of the ultrasound wave velocities at fcc-to-sc phase transition are shown in the inset in Fig. 3. In Figs. 1 and 2 one can see the examples of the corresponding anomalies for the elastic moduli, where for the bulk modulus the effects are much more pronounced. The hysteresis value of the pressure dependencies of ultrasound velocities (under compression and following decompression) was certainly lower than the transition width (see the inset in Fig. 3 and horisontal bars in the main part of Fig. 3). Also the anomalies at higher pressures were found. The points near the dashed line in Fig. 3 correspond to jumps in the pressure derivatives of sound velocities and volume. For example, the value of $(\partial v_t/\partial p)_{T=285K}$ derivative changes from ≈ 0.64 (before the transition) to ≈ 0.45 (after the transition). As regards the "isobaric points" at low temperatures (≈ 100 K), the temperatures correspond to the upper boundary of transition from the glassy state to sc-phase. These points were determined as points of finalization of derivative changes of sound velocities and/or volume (see examples in Figs. 1 and 2).

The C_{60} elastic moduli can be analyzed with respect to a simple central intermolecular potential (see, e.g., Ref. [14]). For the case of the *m-n* power potential $U(V)=A/V^m-B/V^n$ (here, A, B, m, and n are constants), it is well known that the pressure derivative of the bulk modulus equals $B'_p=m+n+2$ [16]. For the Lennard-Jones form of this potential, $U(r)=\alpha/r^a-\beta/r^b$ (r is the intermolecular distance), we have $B'_p=(a+b)/3+2$. For different sets of the *a-b* parameters approximated for molecular C_{60} [14,17], we can obtain $B'_p\approx19$–21, which is somewhat lower than the observed values in the current work. One can suppose that the difference between experimental and calculated values results from non-central contributions to the intermolecular potential and possibly from a stronger contribution of the repulsion part to the central potential under pressure. The first assumption can be easily verified. For the pure central interatomic (intermolecular in our case) potential, the Cauchy relation for the elastic constants under pressure [18] gives the following relation for the elastic moduli [19], $G=(3B-6p)/5$, where G is the shear modulus calculated by the Voigt's method. For cubically symmetric crystals the Voigt's and real shear moduli are usually quite close.

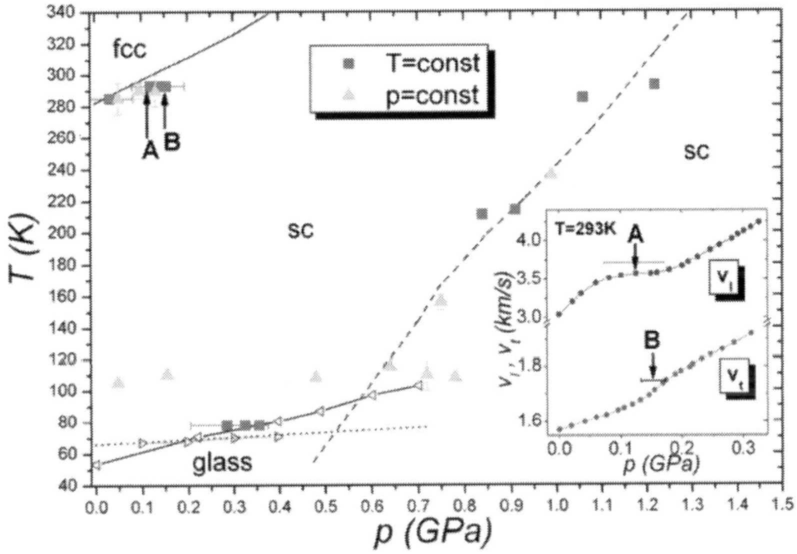

Figure 3. Phase diagram of C_{60}.Inset: typical dependencies of longitudinal (v_l) and transverse (v_t) wave velocities.

For our experimental data, the experimental values of the shear modulus near room pressure are by $\approx$1.5 times lower than those calculated from the bulk modulus for the central potential (for different temperatures). Moreover, the relation between the pressure derivatives of the elastic moduli for the central potential, $G'_p=(3B'_p-6)/5$, is not valid for our experimental data either. Indeed, $G'_p=3.2$–4.3 for the sc phase at different temperatures and pressures up to 1 GPa, while $(3B'_p-6)/5\approx15$ (or 9 if B'_p is calculated from the *m-n* potential). This means that the negative contribution of non-central forces to the shear modulus significantly increases with pressure. We suppose that this negative non-central contribution is associated with mutual intramolecular deformations due to their interaction. These deformations make easier any elastic shear deformation that provides additional ways for energy relaxations of the intramolecular deformations.

4. Conclusions

We present the study of the elastic moduli of polycrystalline fullerite C_{60} at pressures up to 1.4 GPa and at temperatures from $\approx$80 K to room temperature by an ultrasonic technique, which ensures quite accurate direct determination of the elastic properties. The obtained values of the bulk and shear moduli and their pressure derivatives allow us to determine a considerable contribution of non-central intermolecular forces increasing with pressure.

Acknowledgements
We thank S.V. Popova for helpful discussions. Financial support from The Russian Foundation for Basic Research (Project No. 05-02-16596; No. 07-02-01275), The Russian Science Support Foundation, and the Programs of the Presidium of Russian Academy of Sciences is gratefully acknowledged.

References

[1] Kobelev N, Nikolaev R, Soifer Ya, Khasanov S 1998 Phys. Solid State **40** 154
[2] Kobelev N P, Nikolaev R K, Sidorov N S and Soifer Ya M 2001 Phys. Solid State **43** 2344
[3] Sundqvist B 1999 Advances in Physics **48** 1
[4] Horikawa T, Kinoshita T, Suito K, Onodera A 2000 Solid State Commun. **114** 121
[5] Pintschovius L, Blaschko O, Krexner G and Pyka N 1999 Phys. Rev. B **59** 11020
[6] Kobelev N P 2002 Phys. Solid State **44** 195
[7] Fleming R M *et al.* 1991 Mater Res. Soc. Symp. Proc. **206** 691
[8] Dresselhaus M S, Dresselhaus G and Eklund P C 1996 *Science of Fullerenes and Carbon Nanotubes* (San Diego, California: Academic Press)
[9] Pintschovius L, Chaplot S L, Roth G and Heger G 1995 Phys. Rev. Lett. **75** 2843
[10] Chow P C *et al.* 1992 Phys. Rev. Lett. **69** 2943
[11] Heiney P A *et al.* 1991 Phys. Rev. Lett. **66** 2911
[12] David W I F *et al.* 1992 Europhys. Lett. **18** 219
[13] Gromnitskaya E L, Yagafarov O F, Stalgorova O V, Brazhkin V V and Lyapin A G 2007 Phys. Rev. Lett. **98** 165503
[14] Zubov V I, Tretiakov N P, Sanchez J F, Caparica A A 1996 Phys. Rev. B **53** 12080
[15] Kobelev N P, Moravsky A P, Soifer Ya, Bashkin I O, Rybchenko O G 1994 Phys. Solid State **36** 1491
[16] Aleksandrov I V, Goncharov A F, Zisman A N and Stishov S M 1987 Sov. Phys. JETP **66** 384
[17] Magomedov M N 2005 Phys. Sol. St. **47** 1090
[18] Brazhkin V V and Lyapin A G 1997 Phys. Rev. Lett. **78** 2493
[19] Brazhkin V V and Lyapin A G 2002 J. Phys.: Condens. Matter **14** 10861

High pressure synthesis of $BiCrO_3$, a candidate for multiferroism

C Goujon, C Darie, M Bacia, H Klein, L Ortega and P Bordet

Institut Néel / CNRS et Université Joseph Fourier, Dépt MCMF, BP 166, F-38042 GRENOBLE Cedex 9, FRANCE

E-mail: celine.goujon@grenoble.cnrs.fr

Résumé. The synthesis and the thermal stability of the high pressure and high temperature perovskite phase $BiCrO_3$ have been studied. $BiCrO_3$ can be synthesized in a wide range of pressure (1-6GPa) and temperature (690-865°C). The thermal behavior of $BiCrO_3$ depends on the atmosphere : it tends to be stable under inert gas whereas it decomposes at 650°C under O_2. The melting point of $BiCrO_3$ is determined by DTA under argon and a value of 1040°C is proposed. First SEM and TEM observations are also presented.

1. Introduction

Multiferroics are materials which combine two or all three properties of ferroelectricity, ferromagnetism and ferroelasticity in the same phase. For several years, this class of materials arouses increasing interest in solid-state physics as well as in different technological fields. However, although lots of materials are ferroelectric or ferromagnetic, only few of them present both magnetic and electric ordering simultaneously [1] : Ferroelectricity requires structural distorsion to allow electric polarization whereas magnetic ions tend to be stabilized in centrosymmetric structures. $BiFeO_3$ [2] and $BiMnO_3$ [3] have been reported to belong to this class of materials and have been widely investigated in the past few years. In these Bi-containing perovskite-type oxides, the origin of ferroelectricity is thought to be related to a complex distorsion of the Bi-O polyhedra resulting from the stereochemical activity of the Bi^{3+} cation lone pair. Recently, theoretical studies of $BiCrO_3$ have shown the interest of this material as a new multiferroic, which could be an ideal model for studying phase transitions in perovskite-structure ferroelectrics [4, 5]. However, $BiCrO_3$ has still not been extensively investigated. In the work of Sugawara et al. [6], the perovskite phase was synthesized between 3.5 and 5.5GPa and 700 and 800°C. The authors determined a structural phase transition at 137°C and suggested triclinic and monoclinic symmetries for the low and high temperature phases, respectively. As for magnetism, $BiCrO_3$ is inferred to be antiferromagnetic below -150°C. The recent study of Niitaka et al. [7], based on powder synchrotron diffraction patterns, confirms the structural transition temperature between 147 and 167°C but proposes C2 monoclinic and Pnma orthorhombic structures for the low and high temperature phases, respectively. It also shows the existence of a parasitic ferromagnetic ordering at -159°C and of an anomaly in the dielectric permittivity at about 167°C, temperature of the structural phase transition. A few papers also present the thermal behavior of $BiCrO_3$. Sugawara et al [6] reported a decomposition temperature of 400°C in air and 300°C in vacuum. Reznitskii [8] found that $BiCrO_3$ presents a polymorphic transition

© 2008 IOP Publishing Ltd

temperature at 882°C and melts at 928°C, in accordance with results of Masuno [9]. These latter values, however, have not been confirmed yet and recently, DTA experiments of Chen et al. [10] explain this melting event at 928°C as an eutectic-like behavior of the Bi_2O_3-Cr_2O_3 system rather than the melting point of $BiCrO_3$.
In this study, we firstly propose to determine more precisely the conditions to synthesize $BiCrO_3$. The effect of synthesis parameters, like temperature, pressure and time, are examined to determine optimized conditions. In a second part, we focus our interest on the thermal study of the phase up to the decomposition temperature, under different atmospheres. In a last part, first TEM observations of $BiCrO_3$ are presented.

2. Experimental

Polycrystalline samples of $BiCrO_3$ were synthesized by a solid state reaction under high pressure in a Conac anvil-type apparatus. Stoichiometric proportions of Bi_2O_3 (Aldrich, 99.9%) and Cr_2O_3 (Prolabo, 99.0%) were mixed, ground and packed into a gold capsule (6mm diameter x 6mm high). A wide range of synthesis conditions was studied : pressure, temperature and time were varied from 1 to 6GPa, 690 to 865°C and 1 to 7h, respectively. The desired pressure was first applied, then the samples were heated to the dwell temperature. After the reaction, they were quenched and the pressure was slowly released. Each obtained sample was characterized by x-ray diffraction (XRD) on a SIEMENS D5000 diffractometer operating in reflection mode and equipped with Co Kα radiation ($\lambda = 1.7890\mathring{A}$). More precise patterns were also collected on a BRUKER D8 diffractometer (Cu Kα1 radiation, $\lambda = 1.5406\mathring{A}$). Differential thermal analyses (DTA) were performed to study the thermal stability of the perovskite phase, either in oxygen or in argon atmosphere, by using a NETZSCH TASC 414/3 analyzer at a heating rate of 10°C/min. High temperature X-ray diffraction was also carried out by using an INEL diffractometer equipped with a home-made furnace and operating with Fe Kα1 radiation ($\lambda = 1.936\mathring{A}$). The experiment was performed in inert gas (He) and the heating rate was 1°C/min, from room temperature up to 850°C. The samples were either quenched or slowly cooled (1°C/min). Synthesized powders were observed by scanning electron microscopy (SEM JEOL 840) at 20kV and analyzed by energy-dispersive x-ray spectrometry (EDS). Transmission electron microscopy (TEM) observations were performed using a Philips CM300ST microscope working at 300kV and powder samples were dispersed on a carbon-coated copper grid.

3. Results and discussion

3.1. Optimization of the synthesis conditions

The different conditions tested for the synthesis of $BiCrO_3$ and the influence of these parameters on grain sizes and impurities are summarized in table 1. $BiCrO_3$ with perovskite-type structure was previously reported to be synthesized in the ranges 3.5-5.5GPa and 700-800°C [7][6]. The experiments of the present study show that the phase has a larger domain of stability and can be synthesized at a pressure "as low as" 1GPa. The diffraction patterns of C28-0, C28-4 and C28-2 samples are shown on Figure 1. At 6GPa (C28-0 sample), the high-pressure polymorph of bismuth oxide with hexagonal structure [11] is detected (peaks with open squares on figure 1). Our experiments therefore agree with an upper limit of ~5.5GPa. Concerning the influence of the temperature at 2GPa, 690°C appears to be the lowest limit of the domain of stability of $BiCrO_3$. Indeed, the monoclinic polymorph α-Bi_2O_3, which is one of the initial powders, is identified in the C28-4 sample (peaks with open triangles on figure 1).
The main impurities detected by XRD analyses (see table 1 and C28-2 diffraction pattern on figure 1) are the bismuth oxycarbonate $Bi_2O_2CO_3$ (or bismutite, open circles) and a phase which may be identified as a cubic bismuth oxide with body-centered symmetry γ-Bi_2O_3 [11], but which could also incorporate chromium cations (filled circles). SEM microanalysis shows an atomic bismuth/chromium ratio of 50/50 in most of the grains, in accordance with the

Tab. 1. Synthesis conditions of $BiCrO_3$.

Sample	P(GPa)	T(°C)	Dwell time (h)	Grain sizes (μm)	Impurities
C28-0	6	690	3	?	Hex. Bi_2O_3, Cr_2O_3, $(Bi_{1-x}Cr_x)_2O_3$ (cub.)
C28-1	4	750	3	0.5 to 2	$(Bi_{1-x}Cr_x)_2O_3$, $Bi_2O_2CO_3$
C28-2	2	805	3	2.5 to 5	$(Bi_{1-x}Cr_x)_2O_3$, $Bi_2O_2CO_3$
C28-3	1	825	3	4.5 to 7	$(Bi_{1-x}Cr_x)_2O_3$, $Bi_2O_2CO_3$
C28-4	2	690	3	1.5 to 3.5	α-Bi_2O_3 (mon.), $(Bi_{1-x}Cr_x)_2O_3$
C28-5	2	740	3	1.5 to 3.5	$(Bi_{1-x}Cr_x)_2O_3$, $Bi_2O_2CO_3$
C28-6	2	865	3	6 to 12	$(Bi_{1-x}Cr_x)_2O_3$, $Bi_2O_2CO_3$
C28-7	2	805	1	1.5 to 2.5	$(Bi_{1-x}Cr_x)_2O_3$, $Bi_2O_2CO_3$
C28-8	2	805	5	1.5 to 6.5	$(Bi_{1-x}Cr_x)_2O_3$, $Bi_2O_2CO_3$
C28-16	2	805	7	1.5 to 6.5	$(Bi_{1-x}Cr_x)_2O_3$, $Bi_2O_2CO_3$

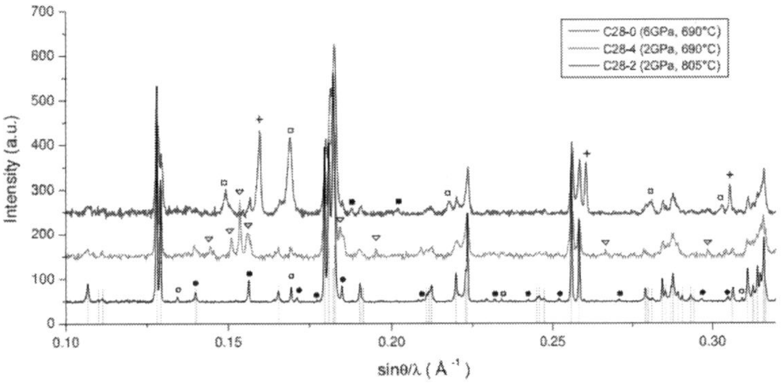

Fig. 1. XRD patterns of the C28-0, C28-2 and C28-4 samples. ▽ α-Bi_2O_3, □ hexagonal Bi_2O_3, ● $(Bi_{1-x}Cr_x)_2O_3$, ○ $Bi_2O_2CO_3$, ■ Cr_2O_3, + Si (standard reference peak). Diffraction peaks of $BiCrO_3$ are represented by green lines.

composition of $BiCrO_3$. A few analyses however show 80/20 or 90/10 ratios. Locally, by TEM, it was also possible to find Cr-rich zones and particles of chromium oxide. In the following, we will call this cubic phase $(Bi_{1-x}Cr_x)_2O_3$. The quantity of bismutite is similar in all samples (about 2 wt%). Its presence is not surprising, since it has already been observed during previous syntheses of $BiCrO_3$ [7] and $BiMnO_3$ [12] and may be related to the CO_2 contamination from the graphite heater [12]. On the contrary, the amount of $(Bi_{1-x}Cr_x)_2O_3$ may depend on the synthesis conditions. At 2GPa, the lowest quantity of this phase is obtained at 805°C (around 3 wt%). We therefore decide that, in our experimental setup, the optimized conditions for the synthesis of $BiCrO_3$ are 2GPa, 805°C and a mean dwell time of 3h (C28-2 sample).

In table 1 are also reported the grain sizes determined by SEM and illustrated by micrographs of the C28-1, C28-5 and C28-6 powders (figure 2). The grains display irregular and angular shapes. Some aggregates are observable although they are not predominant. The comparison of images (a) and (b) suggests a light influence of the pressure : a 2GPa decrease leads to slightly larger grain sizes. The influence of the temperature is stronger and clearly visible on images (b) and (c) : the higher the temperature, the larger the grain sizes. More generally, the evolution of grain sizes for all samples shows that at 2GPa, an increase of the grain size from 1.5μm up to 12μm is observed when the temperature is increased from 690°C to 865°C (table 1). It is also noticeable that longer dwell times tend to involve larger grain sizes, both observations being in accordance with diffusion laws.

(a) C28-1 : 4GPa, 750°C. (b) C28-5 : 2GPa, 740°C. (c) C28-6 : 2GPa, 865°C.

Fig. 2. SEM micrographs of (a) C28-1, (b) C28-5 and (c) C28-6 samples (3h of dwell time).

3.2. Structural transitions and thermal stability

The DTA of $BiCrO_3$ powders is illustrated on figure 3. The thermal behavior of the powders clearly depends on the atmosphere. Under O_2, three endothermic peaks and an exothermic one are observed, with onset temperatures of about 165°C, 900°C, 922°C and 650°C, respectively. XRD analyses, performed after heating of the powder at 850°C, showed that $BiCrO_3$ is decomposed. These measurements suggest that the exothermic peak at 650°C is related to the decomposition of $BiCrO_3$ under O_2. This temperature is slightly higher than the temperature proposed by Sugawara et al. [6]. The presence of peaks at 900°C and 922°C could be attributed to the melting temperatures of the decomposition products of $BiCrO_3$ in O_2. However, the behavior of $BiCrO_3$ under Ar is quite different. Although the drift of the baseline makes the graph less readable, three endothermic peaks are detected at about 160°C, 870°C and 1040°C. In this case, XRD analyses, performed after heating of the powder at 850°C, proved that $BiCrO_3$ is stable. Since 870°C is close to the melting temperature of Bi_2O_3 (825°C), the small endothermic peak detected at this temperature may be associated with the melting of the cubic $(Bi_{1-x}Cr_x)_2O_3$ phase, which is formed during the synthesis of $BiCrO_3$ in our experiments. Optical inspection of the sample after DTA indicated that it has molten during the experiment. The temperature of 1040°C can therefore be interpreted as the melting temperature of $BiCrO_3$.

High temperature XRD measurements, performed up to 850°C under helium atmosphere, also reveal four noticeable temperatures : 165°C and 560°C during the heating, 825°C and 610°C during the slow cooling. For each temperature segment, diffraction patterns were summed and are shown on figure 4. The change at 165°C, detected by DTA as well as by HT-XRD, is in agreement with the structural phase transition observed by Niitaka et al. [7] and therefore confirms the existence of a high temperature polymorph of $BiCrO_3$. We also verify the reversibility of this transition by both methods (DTA and HT-XRD), as can be seen by comparing diffraction patterns 1 and 3 on figure 4. At 560°C, the peak at 34.9° (identified as the major peak of the cubic $(Bi_{1-x}Cr_x)_2O_3$ phase) is shifted toward lower angles (34.6°), as indicated by the pink filled circle on figure 4 (diffraction pattern 4). This shift is combined with a slight decrease of the peak intensities of $BiCrO_3$ and could be associated with the incorporation of chromium cations in $(Bi_{1-x}Cr_x)_2O_3$. Three other new peaks also appear at 36.5°, 37.5° and 38.3° (see pink crosses on diffraction pattern 4, figure 4). Thus, it appears that, on heating under helium, a progressive decomposition of $BiCrO_3$ starts around 560°C. On cooling, two different behaviors are noteworthy :

- if a slow cooling is imposed, the decomposition of $BiCrO_3$ continues. At 825°C, which is the melting (and crystallization) temperature of Bi_2O_3, the intensity of the peak at 34.6° suddenly increases (orange arrow on figure 4, diffraction pattern 6), while the $BiCrO_3$ peaks are dramatically reduced. The intensities of the peaks at 36.5°, 37.5° and 38.3° progressively increase (purple arrows on diffraction pattern 7). At 610°C, two other new peaks are detected at 21.2° and 21.6°, as shown by the black arrows on figure 4 (diffraction pattern 7). $BiCrO_3$ is therefore almost entirely decomposed and the main product of decomposition is the $(Bi_{1-x}Cr_x)_2O_3$ phase. A small amount of α-Bi_2O_3 (monoclinic) and Cr_2O_3, the initial phases, are also identified. A few diffraction peaks remained not identified.
- if the sample is quenched, the phase decomposition is stopped and the $BiCrO_3$ phase is mainly recovered, with $(Bi_{1-x}Cr_x)_2O_3$ and a small amount of Cr_2O_3.

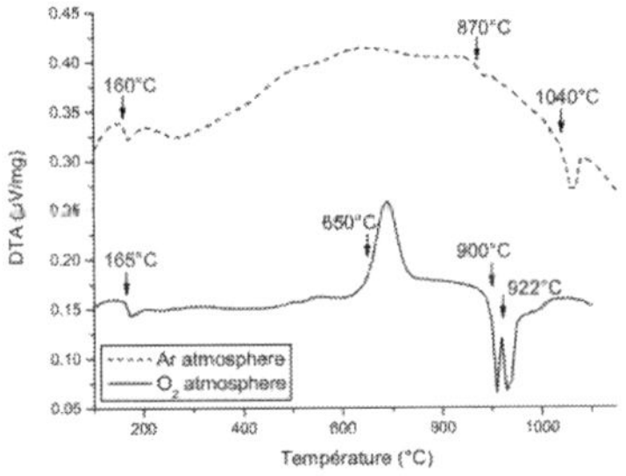

Fig. 3. DTA of $BiCrO_3$ powders under O_2 or Ar atmosphere.

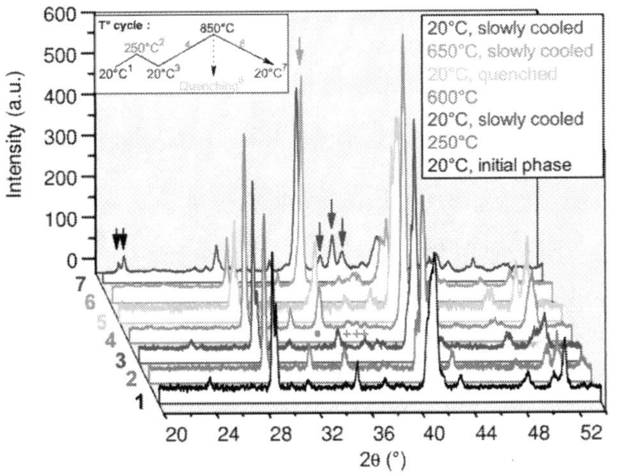

Fig. 4. High temperature XRD of $BiCrO_3$ powders in inert gas (He), presented in a 3D-plot. The inset shows the temperature profile used. * major peak of the cubic $(Bi_{1-x}Cr_x)_2O_3$ phase; + new peaks at 36.5°, 37.5° and 38.3° ; ↓ sudden increase of the major peak of the cubic $(Bi_{1-x}Cr_x)_2O_3$ phase; ↓ increase of the peaks at 36.5°, 37.5° and 38.3° ; ↓ new peaks at 21.2° and 21.6°

3.3. TEM characterization

TEM observations were performed on samples synthesized in optimal conditions (2GPa, 805°C, 3h (C28-2 sample)). Electron diffraction patterns of the $BiCrO_3$ phase indicate a twinning of the monoclinic unit cell. Fig. 5(a) shows a diffraction pattern obtained in the [1 3 1] zone axis. The vertical rows of diffraction spots next to the central row show a systematic alternation of strong and weak spots (families A and B, respectively). These two families of spots belong to two orientations of a twinned crystal. In the second row from the center, the diffraction spots of the two orientations coincide. The common plane in real space of the twin domains is (1 0

$\bar{1}$). Dark field images taken close to the [1 3 1] zone axis and selecting a spot from one or the other family show complementary contrast corroborating the twinning of the crystal (fig. 5(b) and (c)). The mean size of the twin domains can be estimated from these images to be less than 10 nm. In the diffraction pattern, streaky features of the spots in the [2 0 $\bar{2}$] direction can be interpreted as a consequence of the small domain size of the twins.

(a) Electron diff. pattern of the [1 3 1] zone axis of monoclinic $BiCrO_3$.

(b) Dark field image of $BiCrO_3$ selecting a spot of family A.

(c) Dark field image of $BiCrO_3$ selecting a spot of family B.

Fig. 5. TEM observations of $BiCrO_3$ (C28-2 sample).

4. Conclusion

Powder samples of $BiCrO_3$ have been obtained by a solid state reaction under high pressure and high temperature of the stoichiometric mixture of Bi_2O_3 and Cr_2O_3. The synthesis domain is large, from 1 to 6GPa and 690 to 865°C, at least. It is interesting to note that, very recently, Belik et al. [13] reported a single-phased $BiCrO_3$ synthesis at 6 GPa and higher temperature (1380°C). In our optimized conditions (2GPa, 805°C, 3h), the main impurities are the bismuth oxycarbonate $Bi_2O_2CO_3$ and a cubic bismuth oxide, of type $(Bi_{1-x}Cr_x)_2O_3$. The thermal stability of $BiCrO_3$ has been studied under O_2 and inert gas. In the first case (O_2), the phase is entirely decomposed at 650°C. In the second case (inert gas), $BiCrO_3$ is still present at 850°C and is preserved after a quenching. The melting of $BiCrO_3$ at ambient pressure has been observed by DTA and occurs at 1040°C. First TEM observations of $BiCrO_3$ indicate a twinning of the monoclinic unit cell, characterized by twin domains of about 10 nm and a common plane in real space of (1 0 $\bar{1}$).

5. References

[1] Hill N A 2000 *J. Phys. Chem.* B **104** 6694
[2] Teague J R, Gerson R and James W J 1970 *Solid State Com.* **8** 1073
[3] Moreira Dos Santos A, Parashar S, Raju A R, Zhao Y S, Cheetham A K, and Rao C N R 2002 *Solid State Com.* **122** 49
[4] Hill N A, Bättig P and Daul C 2002 *J. Phys. Chem.* B **106** 3383
[5] Baettig P, Ederer C and Spaldin N A 2005 *Phys. Rev. B* **72** 214105
[6] Sugawara F, Iiida S, Syono Y and Akimoto S I 1968 *J. Phys. Soc. Japan* **25** 1553
[7] Niitaka S, Azuma M, Takano M, Nishibori E, Takata M and Sakata M 2004 *Solid State Ionics* **172** 557
[8] Reznitskii L A 1974 *Inorg. Mater. (USSR)* 993
[9] Masuno K 1969 *J. Chem. Soc. Japan* **90** 1122
[10] Chen J I L, Mahesh Kumar M and Ye Z G 2004 *J. Solid State Chem.* **177** 1501
[11] Atou T, Faqir H, Kikuchi M, Chiba H and Syono Y 1998 *Mater. Res. Bull.* **33** 289
[12] Montanari E, Righi L, Calestani G, Migliori A, Gilioli E and Bolzoni F 2005 *Chem. Mater.* **17** 1765
[13] Belik A A, Tsujii N, Suzuki H and Takayama-Muromachi E 2007 *Inorg. Chem.* **46** 8746

Transport peculiarities and phase transition in diluted magnetic semiconductors $CdGeAs_2$: Mn at high hydrostatic pressure

Mollaev A. Yu.[1], Kamilov I. K.[1], Arslanov R. K.[1], Zalibekov U. Z.[1], Bashirov.R.R.[1], Novotorzev V. M.[2], Marenkin S. F.[2], Varnavskiy S.A.[2]

[1] Institute of Physics, Daghestan Scientific Center, Russian Academy of Sciences, 367003, M.Yaragskogo str., 94, Makhachkala, Russia

[2] Institute of common and inorganic chemistry of the Russian Academy of Sciences, Moscow, Russia

E-mail: a.mollaev@mail.ru

Abstract. Pressure dependences of resistivity ρ(P) and Hall coefficient R_H have been measured for the novel high-temperature ferromagnetic semiconductor p-$Cd_{1-x}Mn_xGeAs_2$ (x=0÷0.36) at the room temperature. Structural phase transitions with positions shifting towards low pressures, when percentage of manganese increases from 5.9 GPa for $CdGeAs_2$ to 4.8 GPa for p-$Cd_{0.64}Mn_{0.36}GeAs_2$, are found out. Some anomalies of dependence R_H(P), which we attribute to magnetic properties and the presence of impurities, are found out for the crystals with the greater percentage of manganese (x≥0.18).

1. Introduction

Today, the research of new functional magnetic materials for spintronics applications is included into a number of the major scientific fields worldwide[1]. It is possible to formulate some basic requests to which such materials should satisfy: they should be ferromagnetics with high enough Curie temperatures (well above the room range) and possess both magnetic and semiconducting properties. These requirements in the great extent are fulfilled for diamond-like semiconductors $A^{II}B^{IV}C_2^{V}$ where high mobility of charge carriers, their small effective masses and the greater ratio of electron mobility to the hole one are inherent. Controlled bringing in of transition elements like Mn, Fe, Cr, *etc.* to anionic and cationic sublattices of these compounds may switch semiconducting material into the ferromagnetic state with a high Curie temperature.

Recently, Marenkin *et al.* [2] had synthesized bulk $CdGeAs_2$-based compounds, including their single crystalline forms, where the content of manganese has been changed from x=0 to x=0.36. According to X-ray analysis all synthesized samples were homogeneous chalcopyrites with Curie temperature T_C reached to 355 K - a largest value for the $A^{II}B^{IV}C_2^{V}$:Mn family known up today. The present work is stimulated by our interest in high pressures research of CdGeAs2:Mn compounds and their electrophysical properties in the vicinity of polymorphic transformation.

© 2008 IOP Publishing Ltd

2. Experimental technique and crystals

Measurements were carried out for single crystalline and polycrystalline p-$Cd_{1-x}Mn_xGeAs_2$ samples at hydrostatic pressures up to 9GPa at the room temperature. The fluoroplastic 80 мм^3 capsule with 8 lead-in contacts was used as a sample containing cell. The cell was set in the solenoid with magnetic field intensity of H~5 kOe. Pressure was controlled with the manganin manometer, calibrated against to several reference points in the whole pressure range. Synthesis of samples has been carried out from high pure powders of monocrystals CdAs2 and Ge. Samples had the form of a bar with dimension of 3×1×1 mm, and the uniformity of samples was checked through the resistivity and Hall coefficient measurements by means of the four-probe method. In more details, the experimental technique is described in [3, 4].

3. Results of measurements and discussions

Pressure dependences of resistivity ρ and Hall coefficient R_H for all investigated samples p-$Cd_{1-x}Mn_xGeAs_2$ are presented in figure 1 and figure2. Structural phase transitions are found out for all samples around pressures P=5.9; 5.7; 5.5; 5.4; 5.2; 4.9; and 4.8 GPa, showing up on dependences of resistivity ρ(P) (figure1) and Hall coefficient R_H(P) (figure2) as abrupt declines. Also, the structural phase transitions are found out in the vicinity of pressures P=2.9; 2.8; 2.7; 2.7; 2.6; 2.5; 2.4; and 2.3 GPa when pressure falls. The position of phase transition shifts towards lower pressures with an increase in percentage of manganes. The similar picture is observed for position of phase transition at the falling pressure. The ratio of the conditional phase transition position at increasing pressure to the same value at falling pressure makes P_{PT}=$P`_{PT}$=2÷2.1, i.e. the difference is within an experiment inaccuracy.

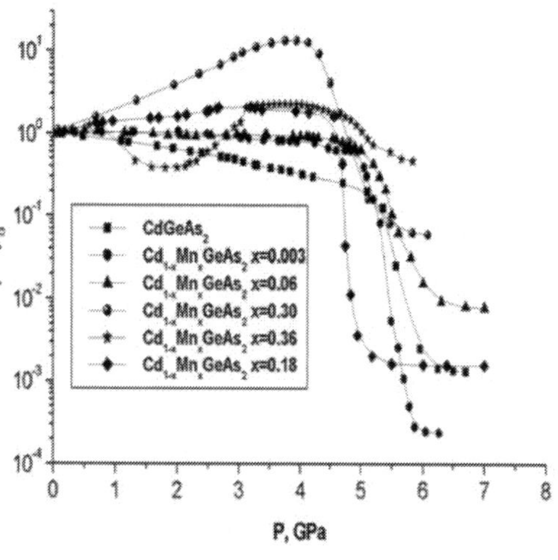

Figure 1. Pressure dependences of resistivity of the investigated samples at rise of pressure.

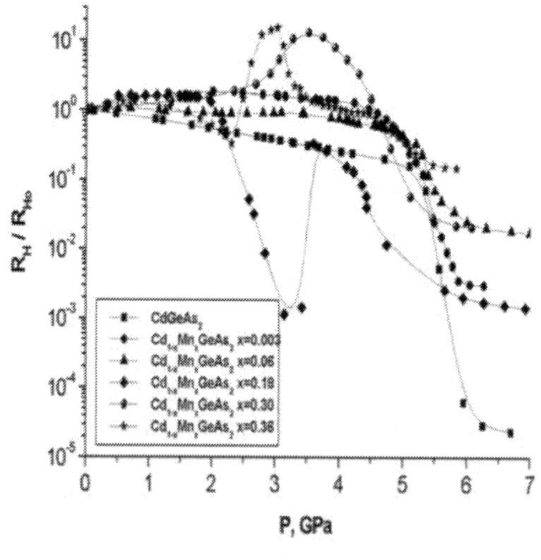

Figure 2. Pressure dependences of a Hall coefficient of the investigated samples at rise of pressure.

In figure3, pressure dependences of resistivity and a Hall coefficient for the base sample $CdGeAs_2$ are presented. It is seen, that the resistivity slowly decreases while pressure increases up to point P=5.3 GPa and then sharply falls almost for three orders of magnitude at P≥5.3 GPa, indicating the beginning of the phase transition, and after P> 6.5 GPa tends to flat asymptotics (saturation region). The resistivity of saturation region, normalized to the atmospheric pressure value, makes ρ/ρ_0=763. In the region of saturation, conductivity is σ =353 Ω^{-1} sm^{-1} that exceeds theoretically calculated values of the minimal metallic conductivity which should be in our case σ =200 Ω^{-1} sm^{-1} [5]. This allows us to assume the metallic type of conductivity has taken a place. The hysteresis is well seen when pressure falls and the phase transition is observed around P=2.9 GPa. Values of resistivity at atmospheric pressure and at falling up to P=0 pressure are identical, i.e. $\rho_0=\rho_0$` (where $\rho_0$`=ρ ($P\to0$)) , implying the reversibility of the transition.

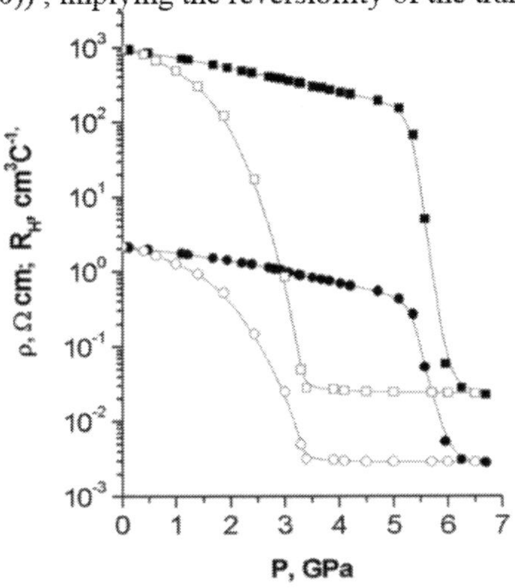

Figure 3. Pressure dependence of resistivity (circles) and a Hall coefficient (squares) for the base sample p-$CdGeAs_2$. Full symbol – pressure rise (compression), open symbol – pressure fall (decompression)

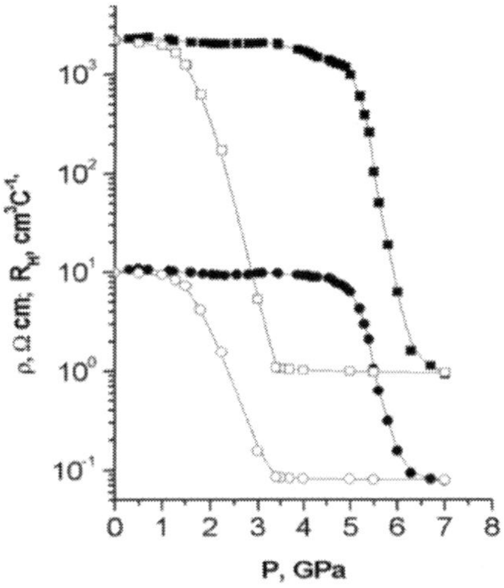

Figure 4. Pressure dependence of resistivity (circles) and a Hall coefficient (squares) for the sample p-$Cd_{0.94}Mn_{0.06}GeAs_2$. Full symbol – pressure rise (compression), open symbol – pressure fall (decompression).

The Hall coefficient also slowly decreases until P<5.1GPa, then after P=5.1GPa falls abruptly almost for 5 orders of magnitude and tends to flat asymptotics when P> 6.5 GPa. In the region of saturation, concentration of charge carriers is $\approx2.8\cdot10^{20}$ sm^{-3} that does not conflict with our assumption about metal type of conductivity in this region. At falling pressure, the wide hysteresis is observed and the phase transition occurs around P=2.9GPa, so that $R_{H0}=R_{H0}$`. From above mentioned, we suppose the reversible structural phase transition is observed for the base sample p-$CdGeAs_2$, accompanied with metal-insulator transition in the conductivity type. Considering pressure dependence of the resistivity for $Cd_{0.94}Mn_{0.06}GeAs_2$ (figure 4), note that resistivity varies very weakly until P<4.7 GPa, then sharply falls for 2 orders between P=4.7 GPa and P=6.1GPa, leaving the transition region with values ρ_0/ρ_H=12, σ_H=12.3 Ω^{-1} sm^{-1}, $\rho_0=\rho_0$`. Phase transition is well pronounced in the $\rho$(P) curve at $P_{PT}$=2.7 GPa (falling pressure) as well. The dependence of the Hall coefficient on pressure is similar. Concentration of carriers in the saturation region is $\approx10^{18}$ $sm^{-3}$, and $R_{H0}=R_{H0}$`. Thus, relying on values of conductivity and Hall coefficient values prior to and after phase transition, we conclude that a reversible structural transition takes a place in the sample p-$Cd_{0.94}Mn_{0.06}GeAs_2$.For the sample p-$Cd_{0.64}Mn_{0.36}GeAs_2$ (figure 5) at high hydrostatic pressure, the resistivity increases weakly up to P≈0.8 GPa, then falls and reaches a minimum at P≈2 GPa, and then tends to flat asymptotics after P≈3GPa and falls again at P≈4.1GPa, indicating thereby the presence of phase transition. The resistivity also tends to the flat asymptotics, and in the saturation region of conductivity we have $\sigma\approx17.8$ Ω^{-1} sm^{-1}, and $\rho_0=\rho_0$`.The dependence of Hall coefficient on pressure also has complex character. Hall coefficient does not vary almost up to P≈0.8 GPa, then increases until P<1.6GPa

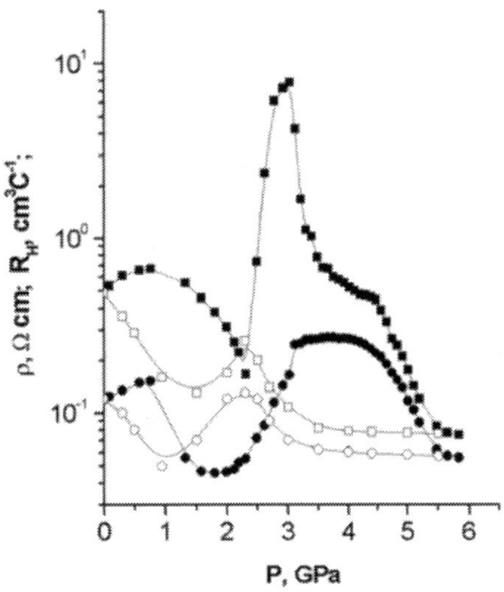

Figure 5. Pressure dependence of resistivity and a Hall coefficient for the sample p-$Cd_{0.64}Mn_{0.36}GeAs_2$. Full symbol – pressure rise (compression), open symbol – pressure fall (decompression)

reaching a maximum, then falls at P≈2.3GPa, then again grows and at P≈3 GPa reaches a maximum again and after P=5.6 GPa tends to flat asymptotics. In the field of saturation concentration of charge carriers $\approx 8\cdot 10^{19}$ sm-3, and σ reaches values typical for the metal conductivity. Thus, it is possible to assume, that in p-$Cd_{0.64}Mn_{0.36}GeAs_2$ the transition semiconductor-metal is observed. Found anomalies on curves ρ(P) and R_H(P) are most likely due to magnetic properties of samples or the presence of impurity (vacancy) sites.

4. The conclusion

Conditionally, all investigated samples fall into 3 groups: 1) the base sample $CdGeAs_2$ and weakly doped sample p-$Cd_{0.897}Mn_{0.003}$ $GeAs_2$; 2) samples $Cd_{0.947}Mn_{0.053}GeAs_2$ and $Cd_{0.94}Mn_{0.06}$ $GeAs_2$, for which anomalies of Hall coefficient pressure dependences are not observed, and 3) the group of samples, where anomalies of R_H(P) dependence are observed - $Cd_{0.82}Mn_{0.18}GeAs_2$ and $Cd_{0.64}Mn_{0.36}GeAs_2$. Apart there is the sample $Cd_{0.7}Mn_{0.3}GeAs_2$: no anomalies for pressure dependence of a Hall coefficient have been observed despite the high enough doping level. The discussed features are intricate interplay between native acceptors in as-grown basic $CdGeAs_2$ sample [6] and Mn^{2+} substitutions for Cd^{2+} in Mn-doped samples [7]. We are planning to conduct an additional study to clarify the nature of these anomalies.

In summary, for all samples the structural reversible phase transitions, which position shifts towards low pressures with the increase in percentage of manganese, are observed. To what structural modification the chalcopyrite phase comes transformed as a result of a phase transition we cannot conclude without *in situ* X-ray diffraction analysis.

The works is supported by Russian foundation for basic research (projects 05-02-16608) and project of Presidium of the Russian Academy of Science "Physics and mechanics of highly condensed matter and of the problem of inner structure of the Earth and planets".

References

[1] Yu-Jun Zhao, et al., Phys.Rev. **B65**, 094415 (2002).
[2] S.F.Marenkin, et al., Inorganic materials, **40**, 2, 135-137 (2004).
[3] L.G.Khvostantsev, et al.,, High Temp.-High Pressure, **9**, 6, 637-639 (1977).
[4] A.J.Mollaev, et al., Inorganic materials, **37**, 4, 405-408 (2001)
[5] N. Mott and E. Davis, *Electronic Processes in Non-Crystalline Materials* (Clarendon Press, Oxford, 1979) 2nd ed.
[6] Ravindra Pandey, *et al,*. J. Phys.: Condens. Matter **10,** 5525 (1998). Lihua Bai1, *et al,* J. Phys.: Condens. Matter **16**, 1279–1286(2004)
[7] S. M. Evans , *et al,* phys. stat. sol. (b), **243**, No. 15, 4070–4079 (2006)

Local structure and effective pair potential of rock-salt type and disordered rock-salt type AgI under pressure

Akira Yoshiasa[1], Hiroshi Arima[2], Hiroshi Fukui[2], Osamu Ohtaka[2], Osamu Kamishima[3], Maki Okube[4], and Yoshinori Katayama[5]

[1]Graduate School of Science, Kumamoto University, Kumamoto 860-8555, Japan

[2]Graduate School of Science, Osaka University, Toyonaka 560-0043, Japan

[3]Institute Multidisciplinary Research Adv Mat, Tohoku University, Miyagi 980-8577, Japan

[4]Materials and Structures Laboratory, Tokyo Institute of Technology, Yokohama 226-850 Japan.

[5]Japan Atomic Energy Agency , Hyogo 679-5148, Japan

e-mail: yoshiasa@sci.kumamoto-u.ac.jp

Abstract. X-ray absorption fine structure (XAFS) observations on iodine K edge of AgI have been performed under high pressure and high temperature up to 6.0 GPa and 1000K to investigate the local structure and effective potentials in high pressure phases. In EXAFS analysis, we have directly carried out the numerical integration of EXAFS function. The anharmonic effective pair potentials $V(u)=au^2/2+bu^3/3!$ for I-Ag bond have been determined under pressures. α-AgI and disordered rock-salt type phase have super-ionic conduction behaviors. Ag ions occupy both octahedral and tetrahedral sites in the disordered phase and twenty percent of Ag ions occupy the tetrahedral site as a maximum value at 2GPa. The transition between the rock-salt type and disordered rock-salt type phases is a broad disorder type within the same structure. From the viewpoint of the local structure analyses on XANES and EXAFS spectra, some sudden changes are recognized near the phase transition point. Pressure influences greatly the effective pair potential of anti-bonding side and anharmonicity decreases with increasing pressure. Phonon dispersion relation of the rock-salt type AgI under pressure has been derived from EXAFS experiments. Analysis of EXAFS Debye-Waller factor is useful because the force constant can be decided directly even at high pressure and high temperature.

1. Introduction

AgI is a much studied material because of the super ionic behavior of Ag ions in α-AgI phase with bcc sub-lattice structure and in the disordered rock-salt type phase with fcc sub-lattice structure. We have determined the precise P-T phase diagram of AgI (Fig. 1) by *in-situ* high-pressure high-temperature synchrotron experiments [1]. The wurtzite type structure of AgI undergoes a first-order phase transition to α-AgI phase. The rock-salt type phase transforms to the disordered rock-salt type phase by a higher-order phase transition under pressure. An *in-situ* neutron diffraction study under high pressure and high temperature showed that the super ionic behavior in the disordered rock-salt type phase is related to the diffuse transition with a small anomaly in lattice constant and a continuous increase in occupation of interstitial tetrahedral sites by Ag ion with increasing temperature [2].

© 2008 IOP Publishing Ltd

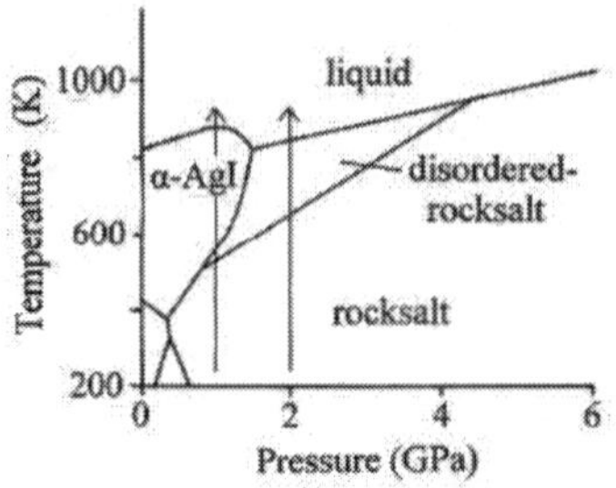

Figure 1. P-T phase diagram of AgI [1].The rock-salt type phase transforms to disordered rock-salt type phase with increasing temperature at 2.0 GPa and 620 K.

X-ray absorption fine structure (XAFS) spectra yields information on the local environments and bonding distances around a particular kind of atom under high pressure and high temperature. The usefulness of extended X-ray absorption fine structure (EXAFS) as a probe of vibration dynamics of local structure was discussed by many authors [3-5]. The analysis of EXAFS Debye-Waller factor provides an effective pair potential with temperature-independent and pressure-dependent shape [5-7]. We investigated the anharmonic effective pair potentials in AgI, AgBr and CuBr under high pressure using the EXAFS technique and revealed the pressure dependent effective pair potential and super ionic conduction mechanism in α-AgI type phases [7-10].

In this study, XAFS measurements of AgI under high pressure and high temperature up to 6.0 GPa and 1000 K have been performed to investigate the local structure in α-AgI, the rock-salt type and disordered rock-salt type phases. The transition between the rock salt-type and disordered rock-salt type phases has been discussed based on the local structure analysis. We have demonstrated the pressure dependence of anharmonic effective pair potentials for I-Ag bond in the rock-salt type phase. Phonon dispersion relation of the rock-salt type AgI under pressure has been derived from the EXAFS experiments.

2. Experimental and analysis

XAFS measurements were performed on a bending magnet beamline, BL14B1, at the SPring-8, Hyogo, Japan. High-pressure and high-temperature conditions were generated using a cubic-type 180 ton apparatus, SMAP2 [11], installed on the beamline. AgI powder with a nominal purity of 99.99 % was provided by Furuuchi Chemical Co. (Tokyo, Japan). The powder was finely ground and mixed with BN powder to dilute the sample. The powder mixture occupied half the volume of the sample chamber and the rest was filled with the pressure calibrant [10,12]. The chamber was put between a pair of graphite-disc heaters and encased in a cubic-shaped pressure-transmitting medium made of boron and epoxy resin. Details of the high-pressure cell assembly have been described elsewhere [11,12]. Generated pressure was determined with an accuracy of about 0.2 GPa [12]. Temperature was monitored with a thermocouple inserted between the two disc heaters.

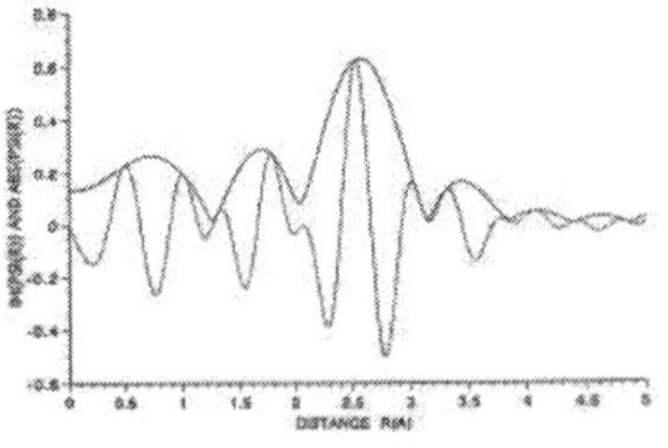

Figure 2. Fourier transforms of the I K-edge EXAFS for AgI at 2.0 GPa 673 K. No phase shift correction is made.

XAFS spectra near the I K-edge were measured in the transmission mode under high pressure up to 6.0 GPa and 1000 K The synchrotron radiation was monochromatized by a Si(311) double-crystal monochromator and focused vertically by two millers. The size of the X-ray beam was reduced to 0.3 x 0.3 mm^2. The intensities of the incident and transmitted beam were measured by ionization chambers.

The EXAFS interference function, $\chi(k)$, was extracted from the measured absorption spectrum using the standard procedure, where k denotes the wave number of photoelectrons: $k = [2m(E-E_0)/\hbar^2]^{1/2}$. The $\chi(k)$ was normalized using MacMaster coefficients according to the EXAFS workshop report [13]. Figure 2 shows examples of Fourier transforms of χk^3 in the range 3.0 $Å^{-1} < k <$ 13.0$Å^{-1}$ at 2.0 GPa 673 K.

We consider the vibration of cation and anion, interacting via an anharmonic pair potential V(u): $V(u)=\mathbf{a}u^2/2+\mathbf{b}u^3/3!$, where **a** and **b** are the potential coefficients and u is the deviation of the bond distance from the location of the potential minimum. In the parameter fitting, we have directly carried out the numerical integration of EXAFS function and evaluated the precise anharmonic effective pair potentials by comparing Fourier-filtered observed and calculated EXAFS interference functions, $\chi(k)$ [5,8]. On the nonlinear least-squares parameter fitting, according to the EXAFS workshop report [13], the equal deformation of observed and calculated EXAFS functions improves the accuracy of the parameter fitting. We could not determine the individual I-Ag distance for the octahedral and tetrahedral sites in the disordered rock-salt type phase separately because of large correlation among the parameters. Single shell model, therefore, was adopted. Because no evident pressure dependence for the experimental threshold energies and mean free path at this experimental resolving power was revealed, we assumed that the threshold energies and mean free path have negligible pressure dependence in this pressure range. The reliability of fit parameter, R, between the experimental and calculated EXAFS functions was less than 0.041. Details on the measurement and analysis have been shown elsewhere [7-10].

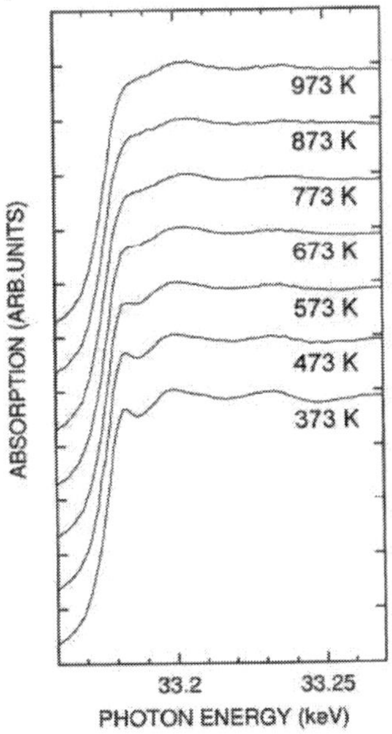

Figure 3. Experimental I K-edge XANES spectra of the rock-salt type (373, 473, 573 K), disordered rock-salt type (673 and 773 K) and liquid phases (873 and 973 K) at 2.0 GPa. The shape of the XANES changed suddenly near 573 K.

3. Results and discussion

3.1. Local structure changes near the phase transition points to super ionic conductors.

X-ray absorption near-edge structure (XANES) spectra are quite sensitive to the electronic states and three-dimensional atomic configuration around X-ray absorbing atoms. XANES spectra are useful for determining the coordination environments of I ions under high temperature and high pressure [9,11]. XANES spectra of the rock-salt type (373, 473, 573 K), disordered rock-salt type (673 and 773 K) and liquid phases (873 and 973 K) at 2.0 GPa are shown in Fig. 3, where the XANES spectra seem to change suddenly near the phase transition point between the rock-salt type and disordered rock-salt type phases though the edge features change gradually with increasing temperature. The edge features of the melt AgI are comparable to that of the disordered phase, but the main peak positions shift to lower energies. It is known that the melt AgI has two distinct local structures of tetrahedral and octahedral coordination [14].

The I-Ag distances are plotted as a function of temperature at 1.0, 4.0 and 6.0 GPa in Fig 4a. The distances from iodine to silver ions in the octahedral sites of the rock-salt type phase decreases with pressure. At 1.0 GPa, the rock-salt type phase transforms to α-AgI and the coordination number change six-fold to four-fold at 550 K. The I-Ag distances become rapidly short at the first-order phase transition point with increasing temperature.

Figure 4b shows the temperature dependence of I-Ag distances at 2.0 GPa. The rock-salt type phase transforms to the disordered rock-salt type around 620 K at this pressure. Ag ions occupy both the octahedral and tetrahedral sites of the fcc sub-lattice formed by I ions in the disordered phase. Our EXAFS analyses provide the average distances from iodine to silver ions in the tetrahedral and octahedral sites. From the viewpoint of the local structure analyses on XANES and EXAFS spectra (Figs. 3 and 4b), some discontinuous or sudden changes are recognized near the phase transition point. The tetrahedral site in the disordered rock-salt type phase seems to be occupied at higher temperature than 573 K by silver ions. The Ag ions which occupy the tetrahedral site increase rapidly between 573K and 673K. Twenty percent of Ag ions occupy the tetrahedral site as a maximum value at 2GPa (Fig. 4b).

3.2. Pressure dependence of anharmonic effective pair potential in rock-salt type AgI.

We have determined the anharmonic effective pair potential $V(u)=\mathbf{a}u^2/2+\mathbf{b}u^3/3!$. A smaller value of potential coefficient **a** represent a weaker bonding and a broader potential. The potential coefficients **a** at 2.0, 4.0 and 6.0 GPa are 1.66(5), 1.70(6) and 1.88(6) eV/Å^2, respectively. The coefficient **a** for the I-Ag bond in the rock-salt type AgI is significantly smaller than that of the tetrahedral site (2.5(1) eV/Å^2) in α-AgI at 1.0 GPa. The octahedrally coordinated ion in the rock-salt type structure has a broader effective pair potential and a larger mean square relative displacement of thermal vibration than the tetrahedrally coordinated ion. The rock-salt type AgI and AgBr take the similar pressure dependence [8].

The parameter **b** refers to the skewness of effective pair potential, arising from anharmonic motion. The potential coefficients **b** at 2.0, 4.0 and 6.0 GPa are -2.6(6), -2.1(5), -1.9(7) eV/Å^3, respectively. Figure 5 shows the anharmonic effective pair potentials for I-Ag bonds under the pressure. The effective pair potentials are influenced by pressure and become steeper with increasing pressure. Pressure influences greatly the potential of anti-bonding side and anharmonicity decreases with increasing pressure.

EXAFS provides the information about thermal vibration of local structure. The derived values refer to short-range correlation of the atomic motion. We try to estimate the phonon dispersion relations using the potential coefficient **a** by calculating the dynamical matrix with rigid-ion model [15], where unknown parameters such as the effective charge were used temporarily from a model adopted for stable phase at ambient conditions. Details of the calculation were given in Ref. 16. The phonon dispersion relation along [100] is shown in Fig. 6 along with the experimental data obtained by neutron inelasitic scattering [17]. The obtained results for AgI derived from the EXAFS are compared with those for AgBr at ambient pressure from the Neutron inelastic scattering. Neutron inelastic scattering experiments have many difficulties under the high pressure and high temperature. On the other hand, analysis of EXAFS Debye-Waller factor is useful because the force constant can be decided directly even at high pressure and high temperature. We can also use the EXAFS Debye-Waller factor analysis for the assignments of vibration mode in complicated systems under pressure.

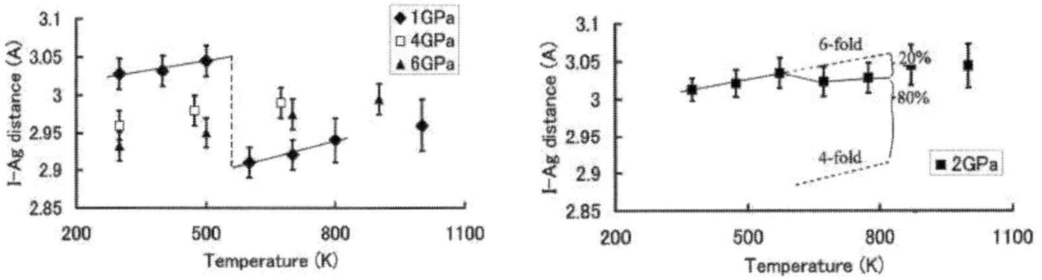

Figure 4. Temperature dependence of the I-Ag distances in AgI at 1.0, 4.0 and 6.0 GPa (a). Occupancies of Ag ions in the tetrahedral site were estimated by average I-Ag distances at 2.0 GPa (b).

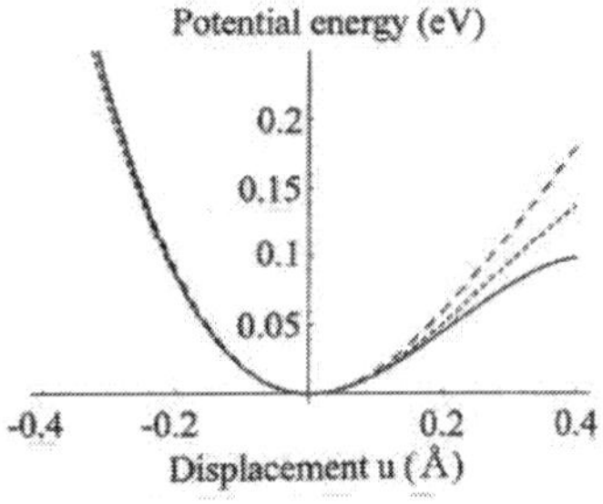

Figure 5. Anharmonic effective pair potentials for I-Ag bond in the rock-salt type phases at 2.0 GPa (straight line), 4.0 GPa (dotted line) and 6.0 GPa (dashed line).

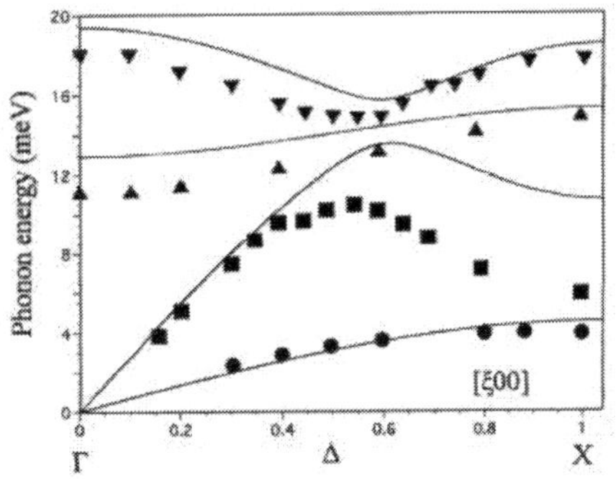

Figure 6. Comparison of the calculated phonon dispersions of the rock-salt type AgI at 6.0 GPa with the experimental data of rock-salt type AgBr at 0.1MPa [17]. The curves are derived from the EXAFS experiments.

Acknowledgments

We thank Y. Okajima for her technical support during the XAFS measurements. The results were achieved with BL14B1 of SPring-8 under the Facilities Utilization system of Japan Atomic Energy Agency

References

[1] Ohtaka O, Takebe H, Yoshiasa A, Fukui H and Katayama Y 2002 *Solid State Comm.* **123** 213
[2] Keen D A, Hull S, Hayes W and Gardner N J G 1996 *Phys. Rev. Lett.* **77** 4914
[3] Stern E A, Liviņš P and Zhang Z, 1991 *Phys. Rev.* **B43** 8850
[4] Dalba G, Fornasini P, Gotter R, Cozzii S, Ronchetti M and Rocca.F 1994 *Solid State Ionic,* **69** 13
[5] Yoshiasa A and Maeda H *Solid State Ionics* 1999 **121** 175
[6] Dalba G, Fornasini P and Rocca F 1993 *Phys. Rev.* **B47** 8502
[7] Yoshiasa A, Murai K, Nagai T and Katayama Y 2001 *Jpn. J. Appl. Phys.* **40** 2395
[8] Yoshiasa A, Fukui H, Arima H, Okube M, Katayama Y, Murai K, Sugahara M and Ohtaka O 2007 *The American Institute of Physics,Conf. Proc.* **882** 569
[9] Yoshiasa A, Okube M, Ohtaka O, Kamishima O and Katayama Y 2000 *Jpn. J. Appl. Phys.* **39** 6747
[10] Yoshiasa A, Nagai T, Murai K, Yamanaka T, Kamishima O and Shimomura O 1998 *Jpn. J. Appl. Phys.* **37** 728
[11] Utsumi W, Funamori K, Katayama Y, Okada T and Shimomura O 2002 *J. Phys.: Condens. Matter* **14** 10497.
[12] Yoshiasa A, Nagai T, Ohtaka O, Kamishima O and Shimomura O 1999 *J. Synchrotron Rad.* **6** 43
[13] Lytle F W, Sayers D E and Stern E A 1989 *Physica B,* **158** 701
[14] Arima H, Ohtaka O, Hattori T, Katayama Y, Utsumi W and Yoshiasa A 2007 *J. Phys.: Condens. Matter,* **19** 076104.
[15] Brüesch P 1982 *Phonons: Theory and Experiments I* Springer Sciences 34 in Solid-State Sciences.
[16] Kamishima O, Ishii T, Maeda H and Kashino S 1997 *Jpn. J. Appl Phys.* **36** 247
[17] Dorner B, von der Osten W and Bührer W 1976 *J. Phys.* **C9** 723

Thermodynamic Stability of Mixed Gas Hydrates Containing Hydrogen

Shunsuke Hashimoto, Takeshi Sugahara, Masato Moritoki, Hiroshi Sato, Kazunari Ohgaki

Division of Chemical Engineering, Graduate School of Engineering Science, Osaka University, 1-3, Machikaneyama, Toyonaka, Osaka 560-8531, Japan.

E-mail: ohgaki@cheng.es.osaka-u.ac.jp

Abstract. Phase equilibrium curves of H_2 + tetrahydrofuran and H_2 + tetra-*n*-butyl ammonium bromide mixed gas hydrates were measured. Each three-phase equilibrium curve converges at three-phase equilibrium point of pure tetrahydrofuran and tetra-*n*-butyl ammonium bromide hydrates, respectively. It is directly confirmed by use of Raman spectroscopy that H_2 is enclathrated in the hydrate cages by adding a small amount of tetrahydrofuran or tetra-*n*-butyl ammonium bromide. In both mixed gas hydrates, H_2 is selectively enclathrated in the small cage despite the concentrations of aqueous solution.

1. Introduction

H_2 has become the object of attention as a clean and promising energy resource. Recently, H_2 hydrate has been considered as a medium of H_2 storage and transportation. However, the pure H_2 hydrate is generated only in extremely high-pressure region of 100-360 MPa [1]. Mao *et al.* [2] and Mao and Mao [3] reveal that H_2 + H_2O mixtures generate the structure-II (s-II) hydrate at high pressure of 200 MPa and low temperature of ca. 80 K, where the hydrate cages are multiply occupied by two H_2 molecules in the S-cage and four in the L-cage. Much milder conditions are desired to utilize H_2 hydrate as a medium of H_2 storage and transportation.

Tetrahydrofuran (hereafter, THF), well known as a common solvent, generates s-II hydrate [4] below atmospheric pressures. THF can be enclathrated in the L-cage while it cannot occupy the S-cage [5]. THF has been widely used as an additive that would reduce the equilibrium pressure of other gas hydrates. Furthermore, the effect of THF was the highest among several additives [6, 7]. Florusse *et al.* [8] have reported the first results that the clusters of H_2 can be stabilized and stored at low pressures in the s-II THF hydrate. According to Florusse *et al.* [8], Lee *et al.* [9], and Strobel *et al.* [10], one or two H_2 molecules are enclathrated in the S-cage and one THF molecule in the L-cage. Lee *et al.* [9] have claimed that the H_2 molecule can occupy the L-cage as well as the S-cage at low THF mole fraction region. On the other hand, Strobel *et al.* [10] have reported that the cage occupancy of H_2 is independent of the THF concentration and H_2 never occupies the L-cage in low mole fraction region. Therefore, the cage occupancy of H_2 for H_2 + THF mixed gas hydrate is still unclear. Shimada *et al.* [12] have reported that isobaric phase equilibrium (*T*-*x*) relations for the tetra-*n*-butyl ammonium bromide (hereafter, TBAB) hydrate have the maximum temperature point (285.15 K) at atmospheric pressure with 40.5 wt% (3.7 mol%) water solution (the structure of TBAB hydrate is considered to be less certain, therefore, described as Type A). TBAB hydrate is a semi-clathrate hydrate where TBAB

© 2008 IOP Publishing Ltd

is incorporated with the water molecules to construct the hydrate cage. TBAB hydrate also has some empty S-cages [12], which is similar to THF hydrate. It has been reported that H_2 can be entrapped with hydrate cages of TBAB hydrate [13, 14]. In order to reveal the cage occupancy of H_2, it is necessary to investigate the composition effect on the thermodynamic stability and cage occupancy of H_2 for H_2-containing mixed gas hydrate.

In the present study, thermodynamic stabilities of H_2 + THF and H_2 + TBAB hydrates were measured. In addition, Raman spectra for each single crystal of H_2 + THF and H_2 + TBAB hydrates were measured under the three-phase coexisting conditions. Both measurements were performed for the several concentrations of THF or TBAB aqueous solutions.

2. Experimental Section

2.1. Materials

Research grade H_2 of mole fraction purity 99.9999 % was obtained from the Neriki Gas Co., Ltd. The maximum impurity was 0.2 ppm of nitrogen. Research grade THF of mole fraction purity 99.7 % was obtained from Merck, Ltd., Japan. Research grade TBAB of mole fraction purity 98.0 % (a major impurity is H_2O) was obtained from Wako Pure Chemical Industries, Ltd. The distilled water was obtained from the Wako Pure Chemical Industries, Ltd. All of them were used without further purifications.

2.2. Experimental apparatus and procedures

The experimental apparatus for the phase equilibrium measurements is the same as the previous one [15] except for the introducing line for the THF or TBAB aqueous solution under the conditions of three-phase (the aqueous solution, gas phase, and hydrate phase) coexisting. The inner volume and maximum working pressure of high-pressure cell were 150 cm^3 and 10 MPa, respectively. The cell had a set of windows for visually observing the phase behavior in the high-pressure cell. All of them were immersed in a temperature-controlled water bath. The contents were agitated using an up-and-down mixing bar driven by an exterior permanent magnetic ring. The high-pressure optical cell for the Raman spectroscopic analysis had a pair of quartz (highly pure) windows on both the upper and lower sides. This high-pressure optical cell is the same as previous one [16] except for the window material. The thermostated water was circulated constantly in the exterior jacket of high-pressure optical cell. A ruby ball was enclosed to agitate the contents by the vibration from outside. The system temperature was measured within a reproducibility of 0.02 K using a thermistor probe (Takara D-632), which was inserted into a hole in the cell wall. The probe was calibrated with a Pt resistance thermometer. The system pressure was measured by the pressure gauge (Valcom VPRT) calibrated by RUSKA quartz Bourdon tube gauge with an estimated maximum uncertainty of 0.01 MPa.

Either THF or TBAB aqueous solution prepared at a desired mole fraction (x) was introduced into the evacuated high-pressure cell. The contents were pressurized up to a desired pressure by supplying H_2 and then continuously agitated by a mixing bar. At first, the system temperature was decreased and kept constant at the point of pressure depression which is caused by mixed gas hydrate formation (the formation can be confirmed through the window of the cell). We cannot directly measure the equilibrium composition of aqueous phase under the coexistence of gas hydrate phase. Therefore, the content was heated very gradually and step by step (0.1 K each) until there was a negligibly small amount of gas hydrate left in order to minimize change of THF or TBAB concentration in the aqueous solution. When the last particle of hydrates disappeared on gradually heating, we adopted this point as the equilibrium point. The single crystal of gas hydrate prepared from the mixtures of H_2 + THF solution and H_2 + TBAB solution was analyzed through a quartz window by *in situ* Raman spectroscopy using a laser Raman microprobe spectrophotometer with multichannel CCD detector. The argon ion laser beam (wavelength: 514.5 nm, power: 100 mW) from the object lens was irradiated to the sample through the upper quartz window. The backscatter of the opposite direction was taken in with the same lens. The spectral resolution was about 1 cm^{-1}.

3. Results & Discussion

Phase equilibria of the H_2 + THF and H_2 + TBAB mixed gas hydrate systems for various concentrations of aqueous solutions are shown in Fig. 1. The stoichiometric concentration of the pure THF and TBAB hydrates is x = 0.056 and x = 0.037, respectively [11, 12]. The three-phase equilibrium curve of pure THF hydrate has a maximum temperature point at 277.45 K and 4.9 kPa [11]. The three-phase equilibrium curve of H_2 + THF hydrate, which is prepared from the THF stoichiometric aqueous solution, converges at the maximum temperature point. The equilibrium curve vertically rises up in the pressure range up to 3 MPa, which may be attributed to the hydrogen content in the hydrate. In the high-pressure region, the equilibrium curve agrees well with that of Florusse *et al.* [8]. The other three-phase equilibrium curves converge at each equilibrium point of pure THF hydrate prepared from the same mole fraction solution. They exhibit similar behavior to that of stoichiometric aqueous solution. Each three-phase equilibrium pressure increases continuously with the temperature increasing. The equilibrium temperature for the pure TBAB hydrate of stoichiometric mole fraction at atmospheric pressure is 285.15 K [12]. The three-phase equilibrium curves of H_2 + TBAB mixed gas hydrate are similar to those of H_2 + THF mixed gas hydrate, while the three-phase equilibrium curve of H_2 + TBAB mixed gas hydrate shifts to the higher temperature of about 8 K than that of H_2 + THF mixed gas hydrate.

In the present study, Raman spectroscopic measurements were carried out for each single crystal of H_2 + THF and H_2 + TBAB mixed hydrates at the various mole fractions of aqueous solution. The results are shown in Fig. 2. The experimental pressure range is from 11.0 to 14.0 MPa. The Raman spectra corresponding to the intramolecular vibration of H_2 and THF for various THF concentration in the aqueous solution are shown in Fig. 2(a) and (b). In the hydrate phase, the Raman peaks are detected at 919 and 1033 cm^{-1}, which correspond to the enclathrated THF molecules [17, 18]. On the other hand, the three peaks are detected at 883, 920, and 1036 cm^{-1} in the aqueous phase. The THF molecule was hydrogen-bonded with H_2O. The peak of 883 cm^{-1} corresponds to the C-O-C stretching vibration mode [8]. In the gas phase, four peaks corresponding to the rotation of H_2 molecules are observed at 354, 588, 816, and 1036 cm^{-1}, which are consistent with the reference data [19]. They are slightly shifted to 351, 584, and 814 cm^{-1} in the hydrate phase. The peak detected around 1036 cm^{-1}

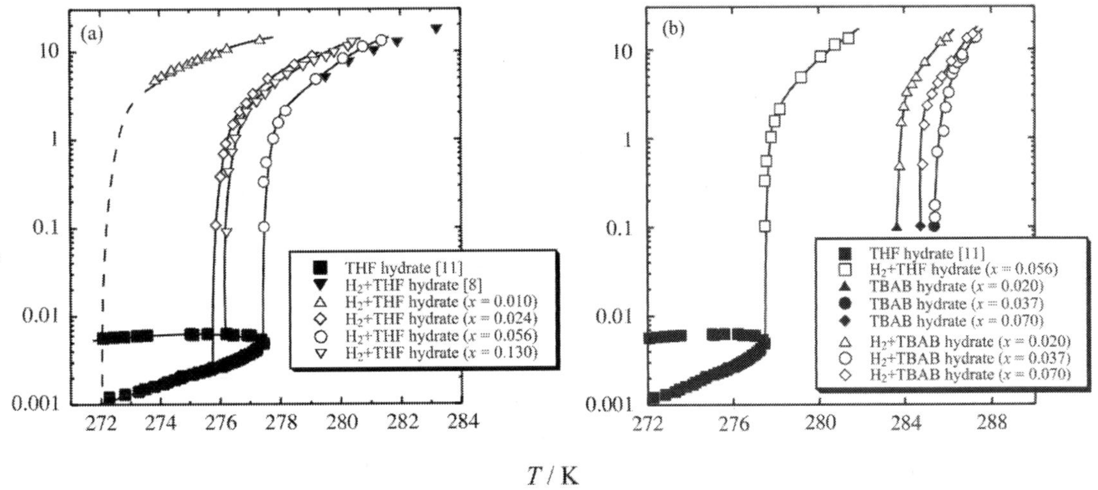

Figure 1. Three-phase equilibrium curves of H_2 + THF (a) and H_2 + TBAB (b) mixed gas hydrate systems. The solid lines are fitting lines for experimental data and the broken line is extrapolated line on the assumption that the three-phase equilibrium line for the THF mole fraction of 0.010 has a similar behavior with the other lines.

overlaps with that of THF. As shown in Fig. 2(b), four peaks corresponding to the H-H stretching vibration mode are detected at 4128, 4146, 4158, and 4165 cm^{-1} in the gas phase, while a broad single peak is detected at 4139 and 4131 cm^{-1} in the aqueous and hydrate phases, respectively. The Raman spectra in the present study are consistent with those of Florusse *et al.* [8]. On the other hand, our data are not consistent with the Raman spectra of H_2 in the S-cage measured by Mao *et al.* [2]. In the present experimental conditions, it is suggested that not a bimolecular cluster of H_2 but an H_2 molecule is entrapped by only the S-cage, and THF occupies the L-cage of s-II hydrate. In addition, the position and shape of all peaks derived from H_2 agree well with those of H_2 + THF mixed gas hydrate at the stoichiometric THF concentration. There is no remarkable change of Raman spectrum with the variations of THF concentration. Therefore, the cage occupancies of H_2 + THF mixed gas hydrate are independent of THF concentrations in the aqueous solutions under the present experimental condition. That is, the THF molecule occupies the L-cage completely while the H_2 occupies the S-cage of s-II hydrate. The results obtained in the present study are consistent with those of Strobel *et al.* [10] on the point that H_2 molecule cannot occupy the L-cage for the H_2 + THF mixed gas hydrate.Figure 2(c) shows the Raman spectra of H_2 for H_2 + TBAB mixed hydrate at the various mole fractions of TBAB. The experimental pressure range is from 12.0 to 14.0 MPa. In the H_2 + TBAB + water mixed system, four peaks corresponding to the H-H stretching vibration mode are observed at 4130, 4147, 4159, and 4166 cm^{-1} in the gas phase. In the aqueous and hydrate phases, a broad single peak is detected at 4140

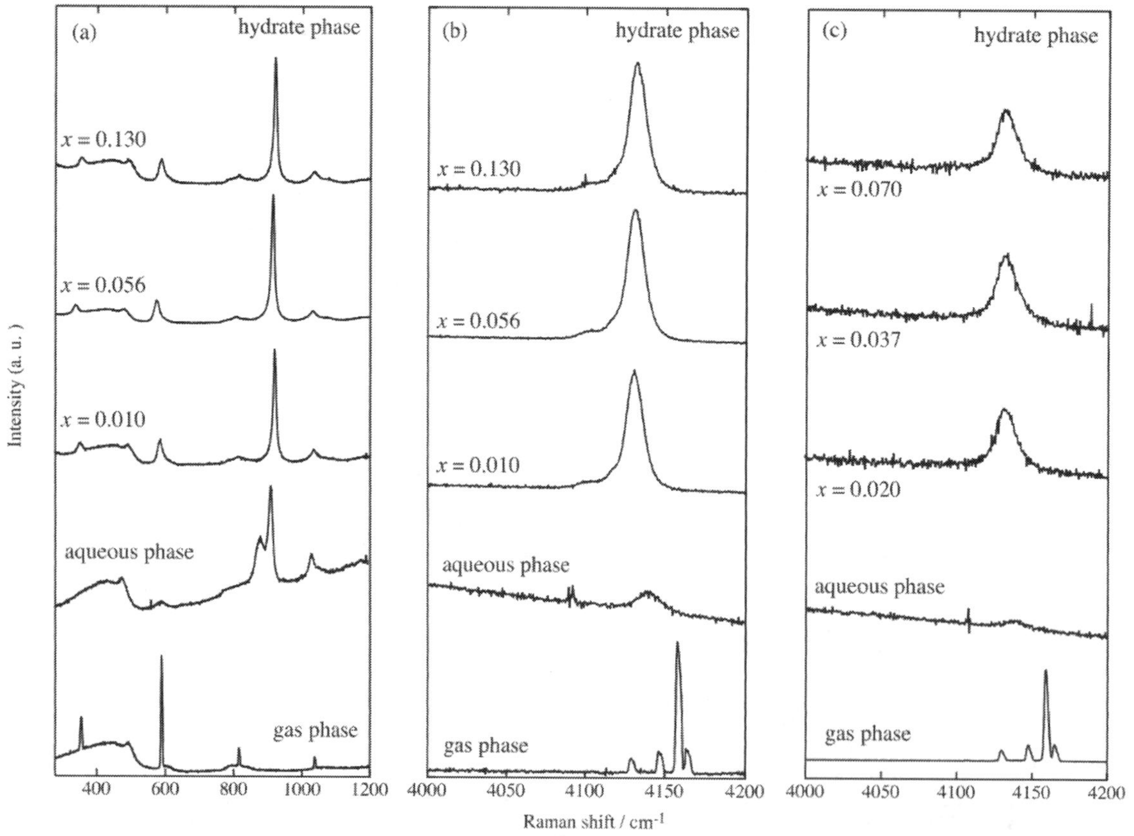

Figure 2. Raman spectra of the intramolecular vibration for THF (a) and H_2 (b) in the H_2 + THF mixed gas hydrates, and H_2 (c) in the H_2 + TBAB mixed gas hydrates at various mole fractions of aqueous solutions. Panel (a) contains the spectra corresponding to the rotation of H_2. The broad signals which appear in lower than 520 cm^{-1} and at (600, 810, and 1060) cm^{-1} correspond to the quartz windows of high-pressure optical cell.

and 4132 cm^{-1}, respectively. In addition, no peak is detected around 4100 cm^{-1} from the pure TBAB hydrate. It is suggested that the H_2 molecules are entrapped in the empty S-cages of TBAB hydrate. In addition, there is no remarkable change of Raman spectrum with the variations of TBAB concentration, which is similar to the results of H_2 + THF mixed gas hydrate. That is, H_2 would occupy the S-cage of TBAB semi-clathrate hydrate.

4. Conclusion

Phase equilibria for the H_2 + THF + water and H_2 + TBAB + water ternary systems were measured at the various mole fractions of aqueous solution. Each three-phase equilibrium curve of H_2 + THF hydrate converges at the three-phase equilibrium point of pure THF hydrate. The three-phase equilibrium curves of H_2 + TBAB mixed gas hydrate are similar to those of H_2 + THF hydrate, while they lie at about 8 K higher temperature than those of H_2 + THF hydrate. The Raman spectra in the single crystal for these mixed systems reveal that the H_2 molecule occupies the S-cage separately. The cage occupancy of H_2 molecule in these mixed gas hydrates does not depend on the variation of mole fraction in aqueous solutions.

Acknowledgements

This work was financially supported by New Energy Industrial Technology Development Organization (NEDO). Shunsuke Hashimoto shows his gratitude for the center of excellence (21th century COE) program "Creation of Integrated EcoChemistry of Osaka University". We gratefully acknowledge the Division of Chemical Engineering, Graduate School of Engineering Science, Osaka University for the scientific support by "Gas-Hydrate Analyzing System (GHAS)".

References

[1] Dyadin, Y. A.; Larionov, E. G.; Aladko, E. Y.; Manakov, A. Y.; Zhurko, F. V.; Mikina, T. V.; Komarov, V. Y.; Grachev, E. V. 1999 *J. Struct. Chem.* **40** 790-95.
[2] Mao, W. L.; Mao, H.; Goncharov, A. F.; Struzhkin, V. V.; Guo, Q.; Hu, J.; Shu, J.; Hemley, R. J.; Somayazulu, M.; Zhao, Y. 2002 *Science* **297** 2247-49.
[3] Mao, W. L.; Mao, H. 2004 *Proc. Nat. Acad. Sci. USA* **101** 708-10.
[4] Hawkins, R. E.; Davidson, D. W. 1966 *J. Phys. Chem.* **70** 1889-94.
[5] Gough, S. R.; Davidson, D. W. 1971 *Can. J. Chem.* **49** 2691-99.
[6] Kang, S. -P.; Lee, H.; Lee, C. -S.; Sung, W. -M. 2001 *Fluid Phase Equilib.* **185** 101-09.
[7] Seo, Y. -T.; Kang, S. -P.; Lee, H. 2001 *Fluid Phase Equilib.* **189** 99-110.
[8] Florusse, L. J.; Peters, C. J.; Schoonman, J.; Hester, K. C.; Koh, C. A.; Dec, S. F.; Marsh, K. N.; Sloan, E. D. 2004 *Science* **306** 469-71.
[9] Lee, H.; Lee, J. -W.; Kim, D. Y.; Park, J.; Seo, Y. -T.; Zeng, H.; Moudrakovski, I. L.; Ratcliffe, C. I.; Ripmeester, J. A. 2005 *Nature* **434** 743-46.
[10] Strobel, T. A.; Taylor, C. J.; Hester, K. C.; Dec, S. F.; Koh, C. A.; Miller, K. T.; Sloan, E. D., Jr. 2006 *J. Phys. Chem. B* **110** 17121-25.
[11] Makino, T.; Sugahara, T.; Ohgaki, K. 2005 *J. Chem. Eng. Data* **50** 2058-60.
[12] Shimada, W.; Ebinuma, T.; Oyama, H.; Kamata, Y.; Takeya, S.; Uchida, T.; Nagao, J.; Narita, H. 2003 *Japanese J. Appl. Phys.* **42** 129-31.
[13] Hashimoto, S.; Murayama, S.; Sugahara, T.; Sato, H.; Ohgaki, K. 2006 *Chem. Eng. Sci.* **61** 7884-88.
[14] Chapoy, A.; Anderson, R.; Tohidi, B. 2007 *J. Am. Chem. Soc.* **129** 746-47.
[15] Sugahara, T.; Murayama, S.; Hashimoto, S.; Ohgaki, K. 2005 *Fluid Phase Equilib.* **233** 190-93.
[16] Nakano, S.; Moritoki, M.; Ohgaki, K. 1998 *J. Chem. Eng. Data* **43** 807-10.
[17] Tulk, C. A.; Klug, D. D.; Ripmeester, J. A. 1998 *J. Phys. Chem. A* **102** 8734-39.
[18] Manakov, A. Y.; Goryainov, S. V.; Kurnosov, A. V.; Likhacheva, A. Y.; Dyadin, Y. A.; Larionov, E. G. 2003 *J. Phys. Chem. B* **107** 7861-66.
[19] Fink, U.; Wiggins, T. A.; Rank, D. H. 1965 *J. Mol. Spectrosc.* **18** 384-95.

Phase diagram studies on iron and nickel silicides: High-pressure experiments and *ab initio* calculations

D. Santamaría-Perez[1,2], D. Errandonea[3,*], A. Vegas[2], J. Nuss[4], M. Jansen[4], P. Rodríguez-Hernández[5], A. Muñoz[5], and R. Boehler[1]

[1]Max-Planck Institut für Chemie, Postfach 3060, D-55020 Mainz, Germany

[2]Instituto de Química-Física Rocasolano, CSIC, C/Serrano 119, 28006 Madrid, Spain

[3]MALTA Consolider Team, Departamento de Física Aplicada-ICMUV, Univiversitat de València, C/Dr.Moliner 50, 46100 Burjassot, Spain

[4]MPI für Festkörperforschung, Heisenbergstrasse 1, 70569 Stuttgart, Germany

[5]MALTA Consolider Team, Departamento de Física Fundamental II, Universidad de La Laguna, La Laguna, Tenerife, Spain

Email: daniel.errandonea@uv.es

Abstract. We performed high-pressure ADXRD studies on Fe_5Si_3 and Ni_2Si up to 75 GPa. No evidence of the occurrence of a phase transition was observed in them. Fe_5Si_3 was found to compress isotropically, but an anisotropic compression was observed in Ni_2Si. These results are supported by *ab initio* total-energy calculations, which for Fe_5Si_3 also predicted a transition at 283 GPa from the hexagonal *P6₃/mcm* phase to a cubic phase. High-pressure melting studies were conducted on FeSi up to 70 GPa. We found a change in the melting slope at 12 GPa, which is attributed to the intersection of the melting curve with the phase boundary between ε-FeSi and CsCl-type FeSi. Finally, an equation of state for Fe_5Si_3 and Ni_2Si is reported.

1. Introduction

Iron silicides are probable candidates of the origin of the ultra-low velocity zone at the base of the Earth's mantle. They may appear at the core-mantle boundary since liquid iron coexists with solid silicates. In addition, the high P-T phase diagram of FeSi is currently of interest since silicon is a possible major alloying element in the Earth's outer core. In order to address these issues, several studies on iron silicides under pressure have been done [1–3]. On the other hand, iron and nickel silicides have received attention since their magnetic and electronic properties make them prospective materials for optoelectronics [4]. Finally, these silicides also produce oxides in which the Fe:Si (Ni:Si) stoichiometry is maintained. For example, Fe_5Si_3 (the mineral xifengite) produces the garnet $Fe_5Si_3O_{12}$ whose Fe_5Si_3 subarray differs from the crystal structure of the alloy itself. As in some cases, the cation subarray of the oxide reproduces the structure of the high-pressure phase of the precursor alloy [5], it has been predicted that at high pressure the Fe_5Si_3 alloy would undergo a phase transition to a structure, which may reproduce that of the cation array in $Fe_5Si_3O_{12}$. Because of these reasons, iron and nickel silicides are interesting compounds from geophysical, technological, and crystallochemichal points of view.

[*] To whom any correspondence should be addressed.

© 2008 IOP Publishing Ltd

Here we present new studies of the high P-T phase-diagram of iron and nickel silicides. Angle dispersive x-ray diffraction (ADXRD) experiments were performed on Fe_5Si_3 and Ni_2Si up to 75 GPa. High-pressure melting studies were carried out in FeSi up to 70 GPa. We found that Fe_5Si_3 remains stable in the hexagonal *P6₃/mcm* phase (SG No. 193) up to the highest pressure reached in our experiments and Ni_2Si remains stable in the orthorhombic *Pbnm* structure (SG No. 62). We also found at 12 GPa a change in the slope of the melting curve of FeSi, which agrees with the occurrence of the phase transition between ε-FeSi and CsCl-type FeSi. The experimental results were supported by *ab initio* total-energy calculations. The calculations also predict the presence beyond 283 GPa of a new denser phase in Fe_5Si_3, which has the structure adopted by Fe and Si in $Fe_5Si_3O_{12}$. Finally, an equation of state (EOS) for Fe_5Si_3 and Ni_2Si was obtained. The reported results could have important implications for the differentiation processes of the planets and the composition of their cores.

2. Experimental details

Fe_5Si_3 and Ni_2Si were synthesized by solid state reaction from stoichiometric amounts of high-purity elements. Samples were prepared as finely ground powders from these materials. High-pressure ADXRD measurements were carried out in a 300-μm culet Mao-Bell DAC. The samples were loaded into a 100-μm-diameter hole drilled on a rhenium gasket. Silicone oil was used as pressure-transmitting medium. Pressure was determined using the ruby fluorescence method [6]. The ADXRD studies were performed at the 16-IDB beamline of the HPCAT at the Advanced Photon Source (APS). The x-ray beam (λ = 0.3931 Å) was focused down to 10 x 10 μm^2. Diffraction images were recorded with a Mar345 image-plate detector and were integrated and corrected for distortions using FIT2D. Indexing, structure solution, and refinements were performed using DICVOL and POWDERCELL.

In the melting experiments high-purity FeSi samples (Alpha Aesar) were compressed in a DAC equipped with 300-μm culet anvils with conical supports [7]. Samples were around 40 μm in diameter and 10 μm thick. The sample chamber was a 100-μm-diameter hole drilled in tungsten gasket. High-purity KBr, KCl, and CsCl were used as pressure media and to thermally isolate the samples from the diamonds [8, 9]. FeSi was heated at high pressure with a Nd:YVO_4 infrared laser (TEM_{00} mode, λ = 1.064 μm). Melting was detected by the laser-speckle method [10-12]. Each data point reported is the average of five experiments. Temperatures were determined by fitting a Planck function to the thermal emission spectra of the sample [9, 13] measured in the 500-900 nm spectral range. Pressure was measured before and after heating using the ruby fluorescence method [6]. During the experiments, the heated sample is subjected to a pressure relaxation, due to annealing of the pressure medium, and to a thermal pressure increase. Uncertainty in pressure is about 1 GPa [8, 14]. After each experiment, the samples were recovered and inspected, not showing signs of reaction or oxidation.

3. Overview of the calculations

The structural stability of the phases of Fe_5Si_3 and Ni_2Si was further investigated theoretically by means of total-energy calculations performed within the framework of DFT with the Vienna *ab initio* simulation package (VASP) [15]. A review of DFT-based total-energy methods as applied to the study of phase stability can be found in Ref. [16]. The exchange and correlation energy was described within the generalized gradient approximation [17]. We used ultrasoft pseudopotencials and we adopted the projector augmented wave scheme. We employed a basis set of plane waves up to a kinetic energy cutoff of 334.9 eV for Fe_5Si_3 and 336.9 eV for Ni_2Si, and Monkhorts-Pack grids for the Brillouin-zone integrations which ensure highly converged results (to about 1 meV per formula unit). At each selected volume for a given structure the external and internal parameters are relaxed through the calculations of the forces on the atoms and the components of the stress tensor. Valuable structural information for each phase was obtained from the calculated energy-volume curves after an EOS fitting.

4. Results and discussion

4.1 Structural studies of Fe_5Si_3

Fig. 1 shows our ADXRD data for Fe_5Si_3 at selected pressures and compares them with a diffraction pattern measured at atmospheric pressure (0.0001 GPa) outside the DAC. This pattern corresponded to the hexagonal *P6₃/mcm* structure, with no indication of any additional phase in it. The unit-cell parameters were a = 6.752(5) Å and c = 4.740(3) Å, in agreement with Ref. [3]. Under compression, the only changes we observed in the x-ray diffraction patterns are the typical peak broadening of DAC experiments [18, 19] and the appearance of a peak around 2θ = 11°. This peak has been assigned to a gasket line and can be easily identified since its pressure shift is smaller than that of the Fe_5Si_3 peaks. Regarding the Fe_5Si_3 peaks, we observed that they shift smoothly with compression and that all the Bragg reflections present in the diffraction patterns can be indexed with the $P6_3/mcm$ structure up to 75 GPa. From the diffraction data, we obtained the evolution with pressure of the volume and lattice parameters of Fe_5Si_3. These results are shown in Fig. 2, where we compare them with previous data obtained using NaCl as pressure medium [3]. Both experiments agree within their accuracy up to 20 GPa. Above this pressure, the former study slightly underestimates the decrease of the volume. This fact can be due to the larger non-hydrostatic stresses caused by the NaCl pressure medium [20, 21].

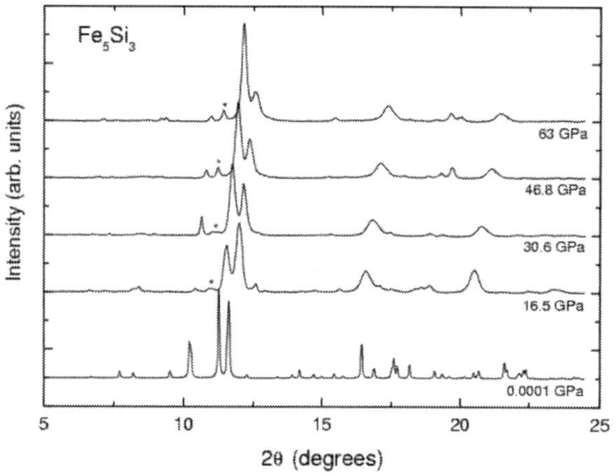

Figure 1: Room-temperature ADXRD data of Fe_5Si_3 at different pressures. In all diagrams the background was subtracted. The gasket peak is denoted by the symbol *. Pressures are indicated in the figure.

Our pressure-volume data shown in Fig. 2 have been fitted using a third-order Birch-Murnaghan EOS [22]. By fixing V_0 = 187.154 Å^3 we obtained B_0 = 215 ± 14 GPa and B_0' = 3.6 ± 0.6. The bulk modulus obtained from our data is slightly smaller than that reported in Ref. [3], but their difference is smaller than those found for other iron silicides [2, 23 – 25]. In our case, the differences in B_0 may be caused by two facts: 1) The data reported in Ref. [3] gives a smaller compressibility than the present data for P ≥ 20 GPa. 2) A Murnaghan EOS [26] was used in Ref. [3] to fit B_0 and B_0' overestimating the bulk modulus. Let us mention here that the B_0 of Fe_5Si_3 is larger than the B_0 of ε-FeSi (B_0 = 160 - 200 GPa) [2, 23, 24], CsCl-type FeSi (B_0 = 184 GPa) [27], α-$FeSi_2$ (B_0 = 182 GPa) [28], β-$FeSi_2$ (B_0 = 180 - 200 GPa) [25], and Fe_7Si_3 (B_0 = 199 -207 GPa) [2, 29]. The compressibility of iron silicides has been proposed to be correlated with the coordination number (CN) of the minority element [3]. In Fe_5Si_3 the silicon atoms have a CN = 9, whereas in the other alloys 6 ≤ CN ≤ 8. This fact makes Fe_5Si_3 the least compressible compound among the different iron silicides studied up to now.

Fig. 2 shows that the compression of Fe_5Si_3 is isotropic. The same fact was observed in Ref. [3] up to 30 GPa. Our experiments verify that the compression of Fe_5Si_3 remains isotropic up to 75 GPa. A quadratic fit to our data gives the following pressure dependence of the unit-cell parameters of Fe_5Si_3:

$$a = 6.76(1) - 9.5(6)\ 10^{-3}\ P + 3.5(7)\ 10^{-5}\ P^2 \quad \text{and}$$

$$c = 4.736(9) - 7.5(5)\ 10^{-3}\ P + 3.8(6)\ 10^{-5}\ P^2,$$

where a and c are given in Å and P is given in GPa. From these two relations it can be estimated that from 0.0001 GPa to 75 GPa a and c are reduced approximately a 7.5%.

We compare now our experimental and theoretical results. Fig. 3 shows the energy-volume curves for the different structures considered for Fe_5Si_3, from which the relative stability of the different phases can be extracted. This figure shows the hexagonal *P6₃/mcm* structure to be stable up to 283 GPa, which agrees with the absence of phase transitions observed in the experiments. At ambient pressure we obtained for the $P6_3/mcm$ structure V_0 = 175.7 Å^3, B_0 = 238.76 GPa and B_0' = 3.8. These values compares well with the experimental results, with differences within the typical systematic errors in DFT calculations. A similar degree of agreement exists for the calculated internal parameters x_1 = 0.2450 for the Fe 6g atoms, and x_2 = 0.6044 for the Si 6g atoms (experimental: 0.2307 and 0.597, respectively) and c/a ratio = 0.719. Our calculations also confirm that the compression of Fe_5Si_3 is isotropic up to 283 GPa; i.e. in the whole range of stability of the hexagonal phase of Fe_5Si_3. As pressure increases, the $P6_3/mcm$ structure becomes unstable and converts to a body center cubic phase (SG $Ia\bar{3}d$, No. 230) with Fe atoms in 24c and 16a positions, and the Si atoms in 24d positions. This phase is isomorphous to the structure adopted by the Fe_5Si_3 subarray in the $Fe_5Si_3O_{12}$ garnet. This fact is in good agreement with the hypothesis that proposes the existence of a correlation between oxidation and pressure [5]. The high-pressure phase only emerges as thermodynamically stable above a compression threshold of 283 GPa. From the common tangent construction or the enthalpy versus pressure plot, our calculations predict that Fe_5Si_3 becomes unstable in the $P6_3/mcm$ phase at 283 GPa against the cubic $Ia\bar{3}d$ phase. The transition is of first-order with a volume change of 1.1 %. The EOS fitting gives V_0 = 346.60, B_0 = 249.95 GPa and B_0' = 3.77 for the predicted high-pressure phase.

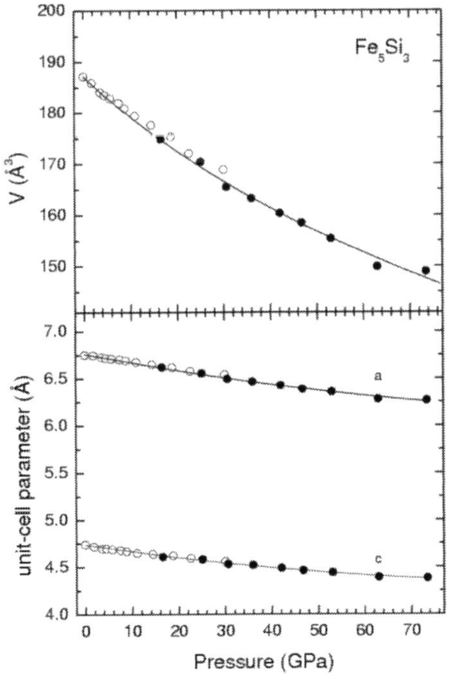

Figure 2: Volume and lattice parameters of Fe_5Si_3 under pressure. Solid circles: present data. Empty circles: data of Ref. [3]. Solid lines: fitted EOS and quadratic fits to a and c.

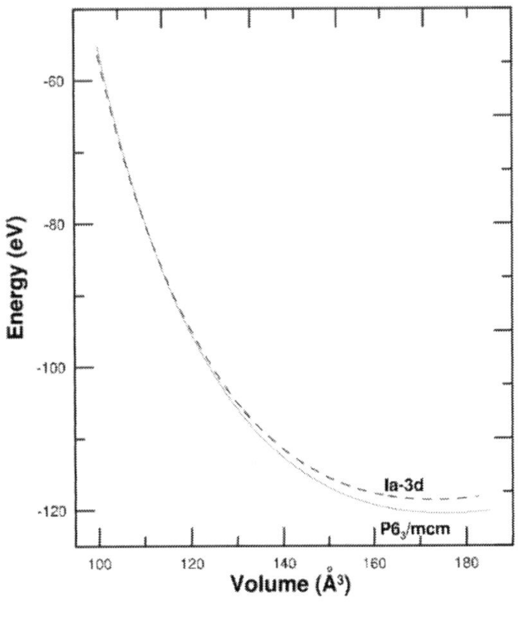

Figure 3: Total-energy versus volume from *ab initio* calculations for the analysed structures of Fe_5Si_3. (Volume and energy are per two formula units).

4.2 Structural studies of Ni_2Si

At atmospheric pressure, the diffraction pattern for Ni_2Si corresponded to the *Pbnm* structure, with no indication of any extra phase in it. The unit-cell parameters were a = 7.059(7) Å, b = 4.990(5) Å, and c = 3.719(4) Å, in agreement with Ref. [30]. Upon compression, we observed that the Ni_2Si peaks shift smoothly and that all of them can be assigned to the *Pbnm* structure up to 75 GPa. From our diffraction data, we obtained the pressure evolution of the volume and lattice parameters of Ni_2Si. These results are plotted in Fig. 4. They have been analysed using a third-order Birch-Murnaghan EOS [22]. By fixing V_0 = 131.049 Å^3 we obtained B_0 = 147 ± 5 GPa and B_0' = 4.5 ± 0.5.

Fig. 4 shows that the compression Ni_2Si of is highly anisotropic. In particular the c-axis is much less compressible that the other two crystalline axis. A quadratic fit to our data gives the following pressure dependence for the unit-cell parameters of Ni_2Si:

$$a = 7.00(5) - 2.6(6)\ 10^{-2}\ P + 2.0(3)\ 10^{-4}\ P^2\ ,$$
$$b = 5.085(9) - 1.0(3)\ 10^{-2}\ P + 4(1)\ 10^{-5}\ P^2\ , \qquad \text{and}$$
$$c = 3.278(6) - 2.4(4)\ 10^{-3}\ P + 3.6(9)\ 10^{-6}\ P^2,$$

where a, b, and c are given in Å and P is given in GPa. From these three relations it can be estimated that from 0.0001 GPa to 75 GPa a is reduced a 12.7%, b is reduced a 9.3%, and c is reduced a 3.9%. The last result implies that the linear incompressibility of Ni_2Si along the c-axis is similar to that of diamond. The structure of Ni_2Si can be described in terms of columns of face-sharing Ni_6Si trigonal prisms connected by edge-sharing to form layers perpendicular to the a-axis. As observed in other crystals [31], a consequence of this layered structure is the anisotropic compressibility of Ni_2Si.

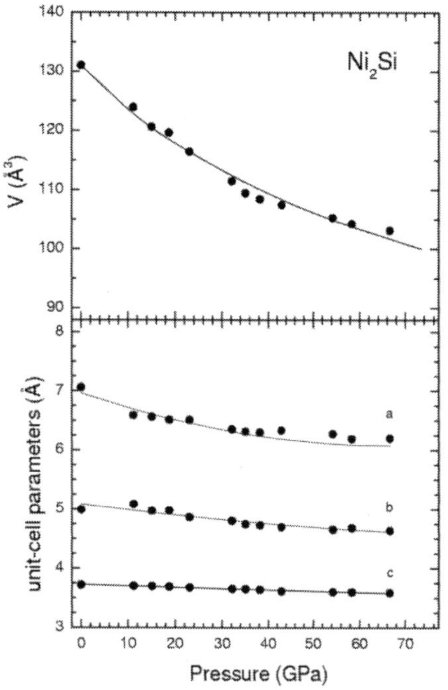

Figure 4: Volume and lattice parameters of Ni_2Si under pressure. Solid circles: present data. Solid lines: fitted EOS and quadratic fits to a, b, and c.

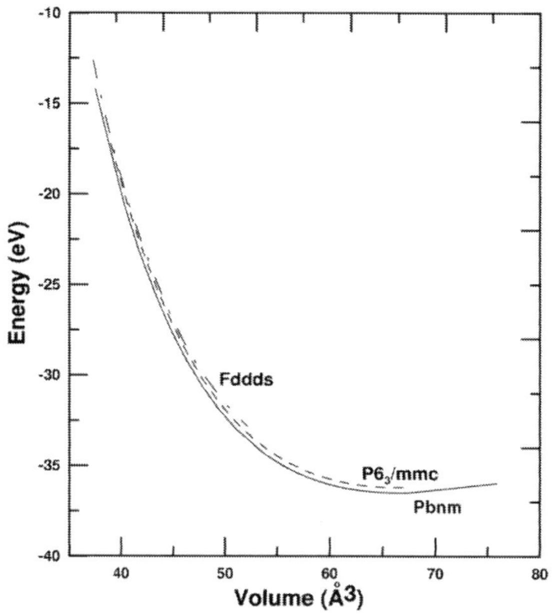

Figure 5: Total-energy versus volume from *ab initio* calculations for the analysed structures of Ni_2Si. (Volume and energy are per two formula units).

We compare now our experimental data with the results from our calculations. Fig. 5 shows the energy-volume curves for the different structures considered for Ni_2Si, from which the relative stability of the different phases can be extracted. We select some candidate structures, the *Fddds* structure (SG No. 70) and the *P6₃/mmc* structure (SG No. 194) adopted by other A_2X compounds like Si_2Ti and Ni_2In. Fig. 5 shows the orthorhombic *Pbnm* structure to be stable up to nearly 400 GPa, which agrees with the absence of phase transitions observed in the experiments. Our calculations also confirm that the compression of Ni_2Si is highly anisotropic. They also give for the *Pbnm* structure the following EOS parameters: V_0 = 133.44 Å^3, B_0 = 175.07 GPa and B_0' = 5. The calculated values of the internal parameters agree very well with the experimental values x_1 = 0.0606, y_1 =0.3304, x_2 =0.2039, y_2 =0.0404 for the Ni 4c atoms, and x_3= 0.3855, y_3= 0.2862 for the Si 4c atoms.

4.3 Melting studies of FeSi

Fig. 6 shows the melting curve of iron silicide. The data obtained using different pressure media are consistent among themselves, indicating that the observed melting temperatures are independent of the pressure medium employed. Initially, the melting curve is flat, with a melting temperature close to 1700 K. However, at 12 GPa, we found that a change in the slope of the melting curve of FeSi takes place. The melting slope beyond 12 GPa is 40 K/GPa, but the melting curve of FeSi considerable flattens at high pressures as occurs in many metals [11]. The melting slope at 70 GPa is 10 K/GPa. The observed change in the melting slope at 12 GPa indicates the presence of a triple point, which can be assigned to the intersection of the phase boundary between the ε-phase and CsCl-phase of FeSi [1] with the melting line (see Fig. 6). The melting slope can be calculated using the Clausius–Clapeyron equation $dT/dP = \Delta V/\Delta S$, where ΔV and ΔS are, respectively, the difference in molar volume and entropy of the solid and liquid coexisting phases. According to this equation, it is reasonable to expect that the volume difference between the two solid phases of FeSi [1] would produce a discontinuous change of melting slope at the solid-solid-liquid triple point [32, 33]. Finally, it is important to note that the melting curve for FeSi crosses the melting curve of iron at 32 GPa and 2400 K [14].

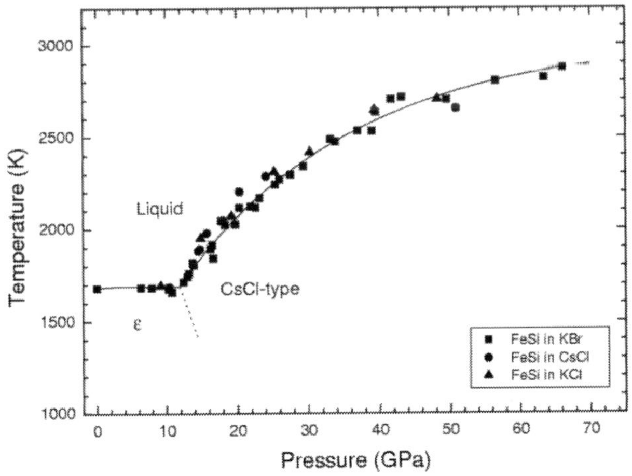

Figure 6: Melting curve of FeSi. Our melting data are represented by solid symbols. The different pressure media used in the experiments are indicated in the inset. The solid line is a fit to the experimental data. The dotted line schematically represents the phase boundary between ε-FeSi and CsCl-type FeSi.

5. Concluding Remarks

The high P-T phase diagrams of iron and nickel silicides have been studied by means of x-ray diffraction and melting experiments as well as by *ab initio* calculations. In the experiments we observed the absence of phase transitions in Fe_5Si_3 and Ni_2Si up to 75 GPa. We also found that the compression of Fe_5Si_3 is rather isotropic whereas the compression of Ni_2Si is highly anisotropic. The experimental results are supported by the total-energy calculations, which also predict the occurrence of a phase transition in Fe_5Si_3 at 283 GPa from the hexagonal *P6₃/mcm* phase to a cubic phase belonging to space group $Ia\bar{3}d$. On the other hand, the calculations predict that the *Pbnm* structure of Ni_2Si is the most stable structure at least up to 400 GPa. In the melting experiments, the melting curve of FeSi was determined up to 70 GPa showing a change in the melting slope at 12 GPa, likely due to the first-order phase transition between ε-FeSi and the CsCl-type FeSi structures. Finally an EOS was determined for Fe_5Si_3 (Ni_2Si) giving the following parameters: V_0 = 187.154 Å^3, B_0 = 215 ± 14 GPa, and B_0' = 3.6 ± 0.6 (V_0 = 131.049 Å^3, B_0 = 147 ± 5 GPa, and B_0' = 4.5 ± 0.5).

Acknowledgments. This work was made possible through financial support of the MCYT of Spain under Grants No. MAT2004-05867-C03-01/-02/-03 and CSD-2007-00045. D.E. acknowledges the support from the MCYT of Spain through the "Ramón y Cajal" program and the Alexander von Humboldt Foundation. He is also indebted to the Fundación de las Artes y las Ciencias de Valencia for granting him the IDEA prize. Use of the APS was supported by DOE-BES, under Contract No. DE-AC02-06CH11357. DOE-BES, DOE-NNSA, NSF, DOD-TACOM, and the W.M. Keck Foundation supported the use of the HPCAT. We thank S. Sinogeikin and Y. Meng, for their contribution to the success of the ADXRD experiments. We also thank O. Tschauner and R. Kumar from UNLV for sharing part of their beamtime with us.

References

[1] D.P. Dobson, L. Vocadlo, and I.G. Wood, 2002 *Am. Mineral.* **87** 784.
[2] J.F. Lin, A. J. Campbell, D.L. Heinz, and G.Y. Shen, 2003 *J. Geophysical Research* **108** 2045.
[3] D. Santamaria-Perez, J. Nuss, J. Haines, M. Jansen, and A. Vegas, 2004 *Solid State Sciences* **6** 673.
[4] K. Takarabe, T. Suemasu, Y. Ikura, and F. Hasegawa, 2002 *J. Appl. Phys. Jpn.* **39** L789.
[5] A. Vegas and M. Jansen, 2002 *Acta Cryst. B* **58** 38.
[6] H. K. Mao, J. Xu, and P. M. Bell, 1986 J. *Geophys. Res.* **91** 4673.
[7] R. Boehler and K. De Hantsetters, 2004 *High Pres. Res.* **24** 391.
[8] R. Boehler, 2000 *Rev. Geophys.* **38** 221.
[9] D. Errandonea, 2006 *J. Phys. Chem. Solids* **67** 2017.
[10] R. Boehler, 1993 *Nature* **363** 534.
[11] D. Errandonea, B. Schwager, R. Ditz, C. Gessmann, R. Boehler, and M. Ross, 2003 *Phys. Rev. B* **63** 132104.
[12] D. Errandonea, M. Somayazulu, D. Häusermann, and H.K. Mao, 2003 *J. Phys. Condens. Matter* **15** 7635.
[13] R. Boehler, 1986 *Geophys. Res. Lett.* **13** 1153.
[14] D. Errandonea, R. Boehler, B. Schwager, and M. Mezouar, 2007 *Phys. Rev. B* **75** 014103.
[15] G. Kresse and J. Furthmüller, 1996 *Comput. Mater. Sci.* **6** 15; 1996 *Phys. Rev. B* **54** 11169.
[16] A. Mujica, A. Rubio, A. Muñoz, and R. J. Needs, 2003 *Rev. Mod. Phys.* **75** 863.
[17] J. P. Perdew, K. Burke and M. Ernzerhof, 1996 *Phys. Rev. Lett.* **77** 3865.
[18] D. He and T. Duffy, 2006 *Phys. Rev. B* **73** 134106.
[19] D. Errandonea, R. Boehler, S. Japel, M. Mezouar, and L. R. Benedetti, 2006 *Phys. Rev. B* **73** 092106.
[20] D. Errandonea, Y. Meng, M. Somayazulu, and D. Häusermann, 2005 *Physica B* **355** 116.
[21] D. Errandonea, J. Pellicer-Porres, F. J. Manjón, A. Segura, Ch. Ferrer-Roca, R. S. Kumar, O. Tschauner, P. Rodriguez-Hernandez, S. Radescu, J. Lopez-Solano, A. Mujica, A. Muñoz, and G. Aquilanti, 2005 *Phys. Rev. B* **72** 174106.
[22] F. Birch, 1978 *J. Geophys. Res.* **83** 1257.
[23] I.G. Wood, W.I.F. David, S. Hull, and G.D. Price, 1996 *J. Appl. Crystallogr.* **29** 215.
[24] L. Vocadlo, G.D. Price, and I.G: Wood, 2000 *Acta Crystal. B* **56** 369.
[25] K. Takarabe, R. Teranishi, J. Oinuma, and Y. Mori, 2002 *J. Phys. Condens. Matter* **14** 11007.
[26] F.D. Murnaghan, 1949 *Proc. Nat. Acad. Sciences* **30** 244.
[27] D.P. Dobson, W.A. Crichton, P. Bouvier, L. Vocadlo, and I.G. Wood, 2003 *Geophys. Res. Letters* **30** 1014.
[28] K. Takarabe, T. Ikai, Y. Mori, H. Udono, and I. Kikuma, 2004 *J. Appl. Phys.* **96** 4903.
[29] N. Hirao, E. Ohtani, T. Kondo, and T. Kikegawa., 2004 *Phys Chem. Minerals* **31** 329.
[30] K. Toman, 1952 *Acta Crystal.* **5** 329.

[31] D. Errandonea, D. Martínez-García, A. Segura, A. Chevy, G. Tobias, E. Canadell, and P. Ordejón, 2006 *Phys. Rev. B* **73** 235202.
[32] R. Boehler, D. Errandonea, M. Ross, 2002 *High Pressure Res.* **22** 479.
[33] D. Errandonea, 2005 *Physica B* **357** 356.

Journal of Physics: Conference Series **121** (2008) 022014 doi:10.1088/1742-6596/121/2/022014

High pressure structural phase transition in zircon ($ZrSiO_4$)

W. Luo[1,2] **and R. Ahuja**[1,2]

[1]Condensed Matter Theory Group, Department of Physics, Uppsala University, BOX 530, S-751 21, Uppsala, Sweden

[2]Applied Materials Physics, Department of Materials Science and Engineering, The Royal Institute of Technology, S-100 44, Stockholm, Sweden

Abstract. We report on a full potential linear muffin-tin orbital calculation of the electronic structure of zircon at ambient conditions as well as under high pressure. The calculations reproduce the experimentally observed pressure induced phase transition from the zircon *to* scheelite structure. The calculated transition pressure of 8 GPa compares very well with the experimental value of about 10 GPa. Our calculated equation of state and the volume collapse associated with the crystallographic change are also in very good agreement with experiment.

PACS: 61.66.Bi, 64.30.+t, 71.25.Pi

The silicate mineral zircon $ZrSiO_4$ has attracted a considerable amount of interest from nuclear as well as earth scientists. In nuclear industry, it is being considered to be used as a nuclear waste storage material for disposal of weapons-grade plutonium.[1] ¿From earth science point of view, it is a host mineral for radioactive elements such as U and Th in the earth's crust. Understandig of the distribution of these sources of radiogenic heat in the deep interior of the earth is important for the understanding of the thermal evolution of the earth. High pressure behaviour of zircon could possibly provide relevant information regarding the distribution of U and Th amongst the high pressure phases in the earth's deep interior. Zircon is also used for various applications as a refractory bulk material. It is an excellent feedstock for the plasma spraying of protective coatings and free-standing bodies.[2] Zircon is a well characterized, naturally occurring phase that is extremely durable.[3] Ewing et al.[1] propose that zircon can be used as a waste form for the disposal of the more than 100 metric tons of plutonium that will result from the dismantling of nuclear weapons. This highly durable material, even under hydrothermal conditions, with its high waste loading and small volume allows deep, permanent disposal of the weapons plutonium in geologic environments in which the borosilicate waste-form glass would not be stable.

There is a good amount of experimental data available regarding the structure and high pressure behaviour of zircon but we are not aware of a single calculation of its electronic structure and high pressure behaviour. Knittle and Williams[4] have measured Raman spectra of $ZrSiO_4$ up to 37 GPa. They have shown that it undergoes a first order phase transition from the zircon to scheelite structure. They have also shown that this phase transformation is irreversible at 300 K and that the scheelite structure is quenchable at ambient pressure. Dawson et al.[5] have measured the vibrational spectrum of zircon. They have calculated the dielectric parameters $\epsilon^{'}$ and $\epsilon^{''}$ from the reflection spectrum as well as that they have

© 2008 IOP Publishing Ltd

Table 1. Calculated and experimental equilibrium volume (Å^3/f.u.), transition pressure (GPa), volume collapse (%) and bulk modulus (GPa) for zircon → scheelite.

Equi. vol.(zircon)		Equi. vol.(scheelite)		Trans. Press.		Volume Collapse		Bulk modulus	
Expt.	Calc.	Expt. [4]	Calc.	Expt. [4]	Calc.	Expt. [4]	Calc.	Expt. [4]	Calc.
65.9	66.6	58.8	61.8	10	8	9.9	7	227	234

also measured the Raman polarizability tensor and the angular dependence of the scattered intensity of severals bands. Kusaba et al.[6, 7] have studied the structural phase transition in $ZrSiO_4$ under shock compression. They have also analyzed the crystal structure of the high pressure scheelite structure using X-ray powder diffraction method. Liu[8] has used diamond-anvil pressure coupled with laser heating to investigate the high pressure phase transition in the pressure range 10-30 GPa. They have found that zircon has transformed to a scheelite type of structure around 12 GPa. It has also been shown that $ZrSiO_4$ decomposes into a mixture of ZrO_2 and SiO_2 in the pressure range 20-25 GPa. Hazen and Finger[9] have studied the crystal structure and compressibility of zircon up to 4.8 GPa. They have shown that zircon has a bulk modulus of 227 GPa and is less compressible than any other measured compound with tetrahedrally-coordinated silicon. Ozkan and Jamieson[10] have measured the pressure dependence of the elastic constants of zircon. They have associated the anomalous negative behaviour of the elastic constant, C_{66}, under pressure with a structural phase transition. Reid and Ringwood[11] have shown that $ZrSiO_4$ transforms completely to the scheelite structure at about 12 GPa with an increase in density of about 11% compared with the original zircon form.

In order to study the electronic structure and total energy of $ZrSiO_4$ we have used the full-potential linear muffin-tin-orbital (FPLMTO) method[12]. The calculations were based on the local-density approximation and we used the Hedin-Lundqvist[13] parametrization for the exchange and correlation potential. Basis functions, electron densities, and potentials were calculated without any geometrical approximation[12]. These quantities were expanded in combinations of spherical harmonic functions (with a cut-off $\ell_{max} = 8$) inside non-overlapping spheres surrounding the atomic sites (muffin-tin spheres) and in a Fourier series in the interstitial region. The muffin-tin sphere occupied approximately 55% of the unit cell. The radial basis functions within the muffin-tin spheres are linear combinations of radial wave functions and their energy derivatives, computed at energies appropriate to their site, principal as well as orbital atomic quantum numbers, whereas outside the muffin-tin spheres the basis functions are combinations of Neuman or Hankel functions.[14, 15] In the calculations reported here, we made use of pseudo-core $4s$, $4p$ and valence band $5s$, $5p$ and $4d$ basis functions for Zr, pseudo-core $2p$ and valence band $3s$, $3p$ and $3d$ basis functions for Si, and valence band $2s$, $2p$ and $3d$ basis functions for O, with corresponding two sets of energy parameters, one appropriate for the semi-core $4s$,$4p$ and $2p$ states, and the other appropriate for the valence states. The resulting basis formed a single, fully hybridizing basis set. This approach has previously proven to give a well converged basis[12]. For sampling the irreducible wedge of the Brillouin-zone we used the special k-point method[16]. In order to speed up the convergence we have associated each calculated eigenvalue with a Gaussian broadening of width 20 mRy.

In Fig. 1 we give the total energy as a function of volume for the zircon and the scheelite crystal structure plotted as a function of volume. From Fig.1 we can see that zircon is the stable phase at ambient pressure as it should be. It transforms to the scheelite phase at 8 GPa whereas the experimental pressure for the zircon → scheelite transformation is about 10 GPa.[4] The volume collapse at the transition was determined to be 10 % experimentally, which should be compared with our theoretical value of 7 %. In Table I, along with the transition pressure and volume change, we have also reported our calculated volume for the low (zircon) and high

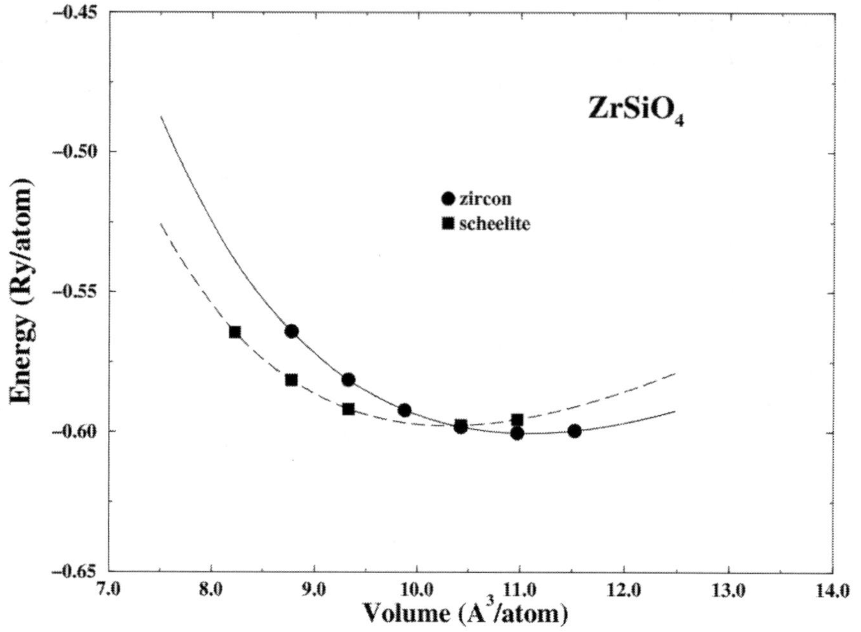

Figure 1. Total energy of the Zircon and Scheelite type crystal structures for $ZrSiO_4$ as a function of volume.

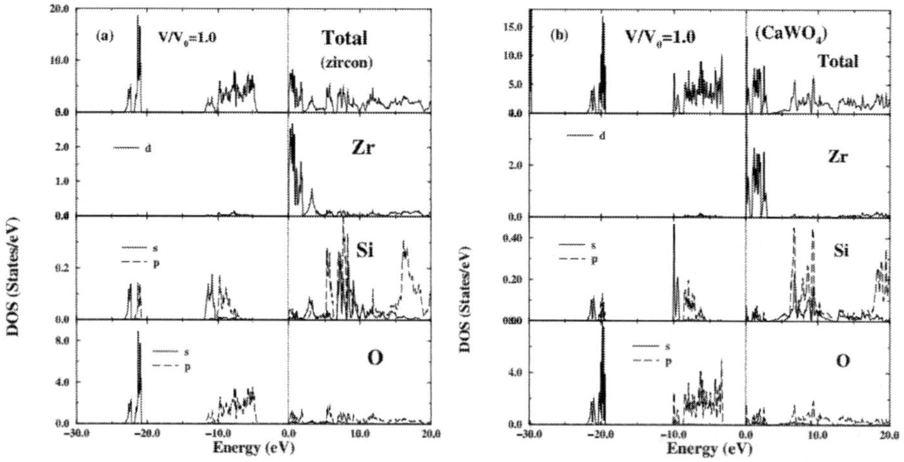

Figure 2. Calculated total density of states (DOS) for $ZrSiO_4$ in the zircon (a) and Scheelite (b) type crystal structures at experimental volume. The Fermi level is set at zero energy and marked by a vertical dashed line.

(scheelite) pressure phase and bulk modulus of zircon along with the experimental data. Our calculated volume for zircon differ by only 1% from the experimental volume whereas for the high pressure scheelite phase the difference between the calculcation and experiment is about 5%. The discrepency for the bulk modulus between experiment and calculations is around 3%. Thus the calculated properties for the zircon → scheelite transition are in quite good agreement with the experimental values.

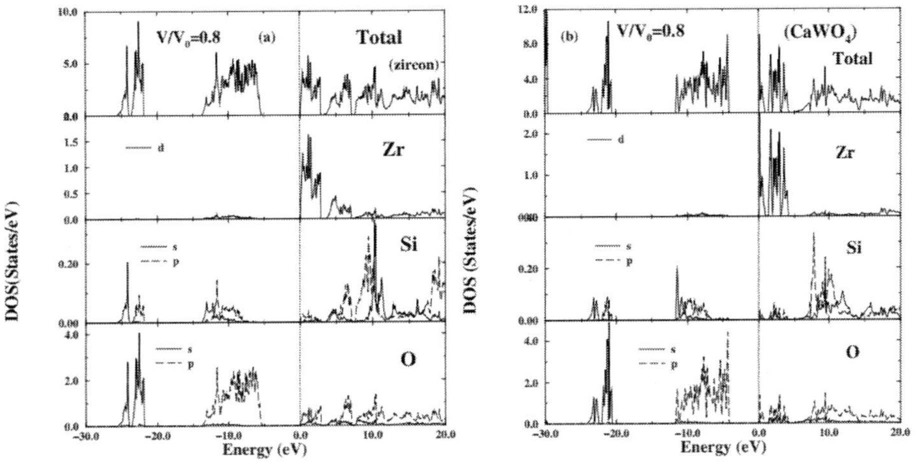

Figure 3. Calculated total density of states (DOS) for $ZrSiO_4$ in the zircon (a) and Scheelite (b) type crystal structures at compressed volume. The Fermi level is set at zero energy and marked by a vertical dashed line.

In Fig.2 (a) we present density of states (DOS) of zircon on a coarse energy scale to give an overall view of the electronic structure. In this figure we also show the orbital decomposition of the density of states. Starting about 22 eV below the minimum of the conduction bands, the DOS is mainly composed from $2s$ states and a small amount of Si $3s$ and $3p$ states. These states are associated mainly with bonding and antibonding pairs of orbitals on the oxygen. These states are split by about 0.8 eV and are seperated by 9.0 eV from a complex structure of O $2p$ states with a small admixture of Si $3s$ and $3p$ states and Zr $4d$ states. These states cover a range of about 8.0 eV and are seperated by a band gap of about 5 eV from the lowest conductions state. We are not aware of any experimental information about the bandgap of zircon. Usually LDA is known to underestimate the band gaps. The bottom of the conduction band consists mainly of Zr $4d$ states with a small hybridization with O $2p$. The higher lying conduction states *i.e.* states above 5 eV from the Fermi level (E_F) are dominated by O $2p$ states with a small mixture of Si s and p states and Zr d states. In Fig. 2(b) we show the DOS for the $CaWO_4$ type structure. This DOS is very similar to that for zircon except for small shifts of the peak positions. It can be noticed that the bandgap is reduced in the $CaWO_4$ structure. It is now around 4 eV in comparison to 5 eV in zircon. Thus there is a significant reduction in the the band gap. If one looks closely, the width of valence band in zircon is a little bit broader than for the $CaWO_4$ type structure. This fact also supports that zircon is more stabe than the $CaWO_4$ type structure at ambient conditions.

Fig. 3 (a) and (b) show the corresponding DOS for zircon and $CaWO_4$ structure at a compressed volume, V/V_o=0.8, where V_o is the experimental equilibrum volume. Under this compression the bands becomes broader for both structures. From these DOS one can see that in both structures, the bandgap increases with pressure. The lowest energy peak in the total DOS around 25 eV, which mainly originates from the O s states, is now broadened by about 1 eV in comparison to its value at V/V_o=1.0. Simlarily the bands below the bandgap which is mainly consituted by O p states are also broaden by one eV. Thus the gap between these states and the lowest lying states has now been reduced. The Zr d states are also broadened and a similar behaviour under pressure has been seen by us in the case of pure Zr.[17]

For the zircon $\rightarrow$ scheelite phase transition, the coordination of Zr, Si and O atoms remain unchanged. Increase in the density of the scheelite structure is the result of a more efficient packing of the coordination polyhedra. The tetrahedral sites are ouccpied by Si atoms where eight-coordinated sites formed by the interpenetration of two unequal size are occupied by Zr atoms in both structures. These tetrahedra are more uniform in the scheelite structure than for zircon. It is well known that when one compresses the solid it transforms to a denser structure. Thus under pressure zircon also shows the same behaviour since it transforms to the scheelite structure because scheelite is the denser phase than zircon.

In conclusion, by means of first principles theory we have reproduced most of the observed aspects of the zircon → scheelite phase transition. Our calculated volume, bulk modulus and transition pressures are in very good agreement with experiment. Thus LDA reproduces very well the observed features of this technologically and geophysically important material.

Acknowledgments
We are grateful to the Swedish Natural Science Research Council and TFR for financial support. We are also grateful to J.M.Wills for letting us use his FPLMTO code.

References
[1] Ewing R C, Lutze W, and Weber W J 1995 *J. Mater. Res.* **10** 243.
[2] Chraska P, Neufuss K, and Herman H 1997 *J. Thermal Spray Tech.* **6** 445.
[3] Hanchar J M and Miller C F 1993 *Chemical Geology* **110** 1.
[4] Knittle E and Williams Q 1993 *American Mineralogist* **78** 245.
[5] Dawson P, Hargreave M M and Wilkinson G R 1971 *J. Phys. C* **4** 240.
[6] Kusaba K, Syno Y, Kikuchi M, and Fukuoka K 1985 *Earth and Planetary Sci. Lett.* **72** 433.
[7] Kusaba K, Yagi T, Kikuchi M, and Syno Y 1986 *J. Phys. Chem. Solids* **47** 675.
[8] Liu L G 1979 *Earth and Planetary Sci. Lett.* **44** 390.
[9] Hazen R M and Finger L W 1979 *American Mineralogist* **64** 196.
[10] Ozkan H and Jamieson J C 1978 *Phys. Chem. Minerals* **2** 215.
[11] Reid A F and Ringwood A E 1969 *Earth and Planetary Sci. Lett.* **6** 205.
[12] Wills J M (unpublished); Wills J M and Cooper B R 1987 *Phys. Rev. B* **36** 3809; Price D L and Cooper B R 1989 *ibid.* **39** 4945.
[13] Hedin L and Lundqvist B I 1971 *J. Phys. C* **4** 2064.
[14] Andersen O K 1975 *Phys. Rev. B* **12** 3060.
[15] Skriver H L 1984 *The LMTO Method* (Springer, Berlin).
[16] Chadi D J and Cohen M L 1973 *Phys. Rev. B* **8** 5747; Froyen S 1989 *ibid.* **39** 3168.
[17] Ahuja R, Wills J M, Johansson B, and Eriksson O 1993 *Phys. Rev. B* **48** 16269.

Journal of Physics: Conference Series **121** (2008) 022015 doi:10.1088/1742-6596/121/2/022015

Optical and dielectric properties of $CuInP_2S_6$ layered crystals at high hydrostatic pressure

P.P.Guranich, A.G.Slivka, V.S.Shusta, , O.O.Gomonnai, I.P.Prits

Uzhhorod National University, Pidhirna Str. 46, Uzhhorod 88000, Ukraine

e-mail: optics@univ.uzhgorod.ua

Abstract. Effect of hydrostatic pressure on the birefringence and dielectric properties of $CuInP_2S_6$ crystals is studied. The pressure behaviour of the phase transition temperature confirms that the phase transition in these crystals belongs to the order/disorder type. An additional line of birefringence anomalies in the ferrielectric phase is revealed. Based on the studies of the optical and dielectric properties of $CuInP_2S_6$ crystals their (p,T) diagram was constructed.

1. Introduction

$CuInP_2S_6$ crystals belong to layered ferrielectrics, where ferrielectric polarization arises normally to the layers and results from anticollinear contributions due to copper ion ordering and indium ion displacement [1]. A first - order phase transition in these crystals at atmospheric pressure occurs at T=315 K. In the paraelectric phase the crystal structure belongs to monoclynic syngony: – C2/c , in the ferrielectric phase – Cc [1]. The spontaneous polarization value is P_s=2,55 μC/cm^2 [2]. Observation of a high-pressure induced first-order phase transition in $CuInP_2S_6$ crystals at p=4.0 GPa and room temperature from the monoclynic to a trigonal phase is reported, based on Raman scattering measurements [3].

Here we report on the studies of dielectric permeability and birefringence of $CuInP_2S_6$ crystals at hydrostatic pressure $p$atm$<p<$250 MPa in the temperature interval 77 K$<T<$400 K.

2. Experimental

$CuInP_2S_6$ crystals were grown by Bridgman technique. They were shaped in 0.2–2-mm thick plates, with silver paste or aquadag electrodes applied to the largest faces. Complex dielectric permeability was measured at the measuring field frequencies of 1 kHz and 1MHz.

Birefringence was studied by Senarmont technique at laser wavelength 0.63 μm. The light beam propagated along the normal to the layer. The measurements were performed in a three-window optical high-pressure chamber, benzene being used as a pressure medium.

3. Results and discussion

Figure 1,2 shows temperature dependences of $CuInP_2S_6$ dielectric permeability, obtained at measuring field frequencies of 1 kHz and 1 MHz. At atmospheric pressure the dielectric permeability maximum, corresponding to the phase transition temperature in the crystals under investigation, is achieved at the temperature $T_c\approx$315 K. The increase of the dielectric permeability value at 1 kHz (curve 1 in Fig. 1) in the range $T>$330 K results from the ionic conductivity of Cu ions [1]. It should be noted that for the samples under study a temperature hystheresis of the phase transition is observed, $\Delta T\approx$1.7 K. This value is essentially below that observed for the same crystals in [1] and coincides with the results of [4].

The performed studies of the dielectric properties of the $CuInP_2S_6$ crystals at atmospheric pressure have shown a considerable dependence of the dielectric permeability value on the sample thickness. This, in our opinion, can be responsible for a considerable data spread for the maximal dielectric permeability values (140$<\varepsilon_{max}<$900), obtained by different authors [1,4].

© 2008 IOP Publishing Ltd

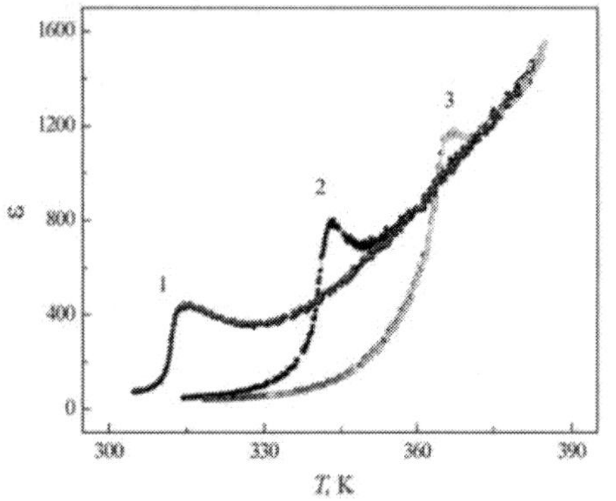

Figure 1. Temperature dependences of the dielectric permeability of $CuInP_2S_6$ crystals at the measuring field frequency 1 kHz at different values of hydrostatic pressure p: $p=p_{atm}$ (1), 128 MPa (2), 248 MPa (3).

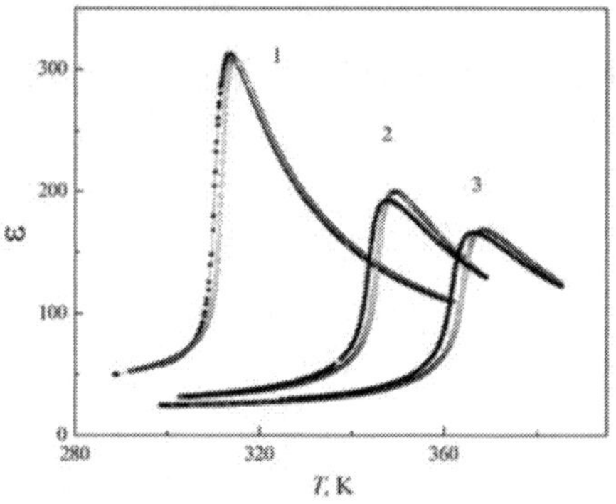

Figure 2. Temperature dependences of the dielectric permeability (open circles – heating, dark circles – cooling) at the measuring field frequency 1 MHz for $CuInP_2S_6$ crystals at different values of hydrostatic pressure p: $p=p_{atm}$ (1), 152 MPa (2), 249 MPa (3).

Temperature dependences of the dielectric permeability of $CuInP_2S_6$ crystals at the measuring field frequency f=1 kHz are shown in Fig. 1. With the hydrostatic pressure increase the curves shift towards higher temperatures. As seen from Fig. 1, the pressure increase practically does not affect the character of the ionic conductivity in the paraelectric phase. The pressure increase is accompanied with the increase of the step of ε at the phase transition. This results from the lack of contribution of ionic conductivity into the dielectric permeability in the ferrielectric phase. According to our calculations, in the pressure range near p=400 MPa the anomaly of the dielectric permeability will be completely masked by Cu ion conductivity.

The results of hydrostatic pressure effect on the temperature dependences of the dielectric permeability for $CuInP_2S_6$ crystals, obtained at the measuring field frequency f=1 MHz, are presented in Fig. 2.

The shift of the ε anomalies is accompanied by a decrease of its maximal values at the constant temperature hystheresis of the phase transition, what is the evidence for the character of the phase transition remaining unchanged. The Curie-Weiss constant value which in the paraelectric phase at atmospheric pressure is C_W=7.5·10^3 K, decreases with pressure. The coefficient $\frac{dC_W}{dp} = -2.8\text{K/MPa}$. The phase transition is accompanied by a maximum of the tanδ value, coinciding with the temperature of the dielectric permeability maximum. The increase of loss in the high-temperature range is due to the ionic conductivity of the Cu atoms.

Temperature dependences of the $CuInP_2S_6$ crystal birefringence at different values of hydrostatic pressure are shown in Figs. 3 and 4. The pressure increase results in a slight change of the birefringence in the high-temperature range and to its essential increase in the ferrielectric phase. At the temperature dependences of the anomalous part of the birefringence a step and a temperature hystheresis of ~2K are observed, what is the evidence for the phase transition in this material being of the first order. For proper ferroelectrics the anomalous part of birefringence is known to be proportional to the squared spontaneous polarization P_s^2: $\Delta n = M P_s^2$.

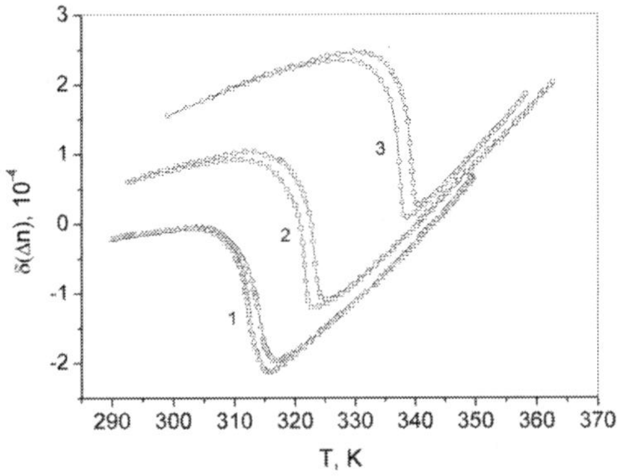

Figure 3. Temperature dependences of birefringence of $CuInP_2S_6$ crystals at different values of hydrostatic pressure p: $p=p_{atm}$ (1), 40 MPa (2), 120 MPa (3).

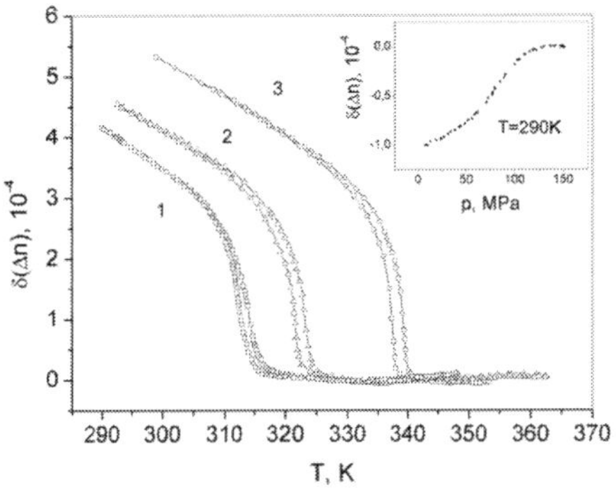

Figure 4. Temperature dependences of the anomalous part of the $CuInP_2S_6$ crystal birefringence at different values of hydrostatic pressure p: $p=p_{atm}$ (1), 40 MPa (2), 120 MPa (3). The insert shows the pressure dependence of the anomalous part of the $CuInP_2S_6$ crystal birefringence at constant temperature T=290 K.

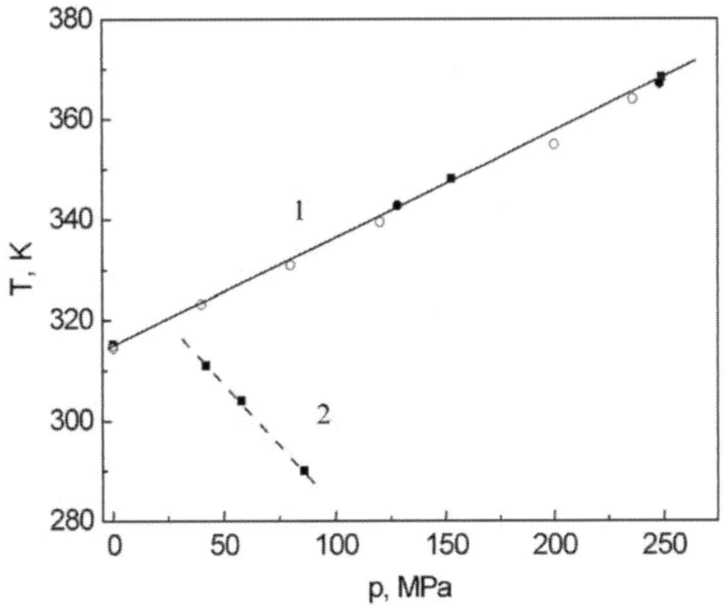

Figure 5. (p,T) phase diagram for $CuInP_2S_6$ crystals: *1* is the ferrielectric phase transition line, open circles being the birefringence data, dark symbols – dielectric data at the frequency f=1 kHz (circles) and f=1 MHz (squares), *2* is the line of birefringence anomalies in the ferrielectric phase.

Taking the spontaneous polarization value at normal conditions for $CuInP_2S_6$ crystals as P_s=2.55 μC/cm^2, one can estimate the proportionality coefficient M=0.587 m^4/C^2. In the range of the ferrielectric phase the studies of isothermal dependences of birefringence have revealed an additional anomaly in the form of a smeared jump (See the inset to Fig. 4). The position of this anomaly in the (p,T) space is extrapolated to the temperatures $T\approx 335$ K at atmospheric pressure. In this temperature range the authors of [1,2] observed the anomalies of the lattice parameters.

Based on the studies of the temperature and pressure dependences of the dielectric properties and birefringence of $CuInP_2S_6$ crystals, their (p,T) phase diagram was built, shown in Fig. 5. In the pressure range under investigation the increase of p results in a linear increase of the ferroelectric phase transition temperature at a rate $\frac{dT_c}{dp} = 210\text{K/GPa}$. This coefficient value is positive what is typical for order/disorder phase transitions and is high enough in comparison with other materials with such phase transition type [5]. Additional studies are required to elucidate the nature of the observed birefringence anomalies in the ferroelectric phase at high pressures.

4. Conclusions

Dielectric properties and birefringence of $CuInP_2S_6$ crystals at high hydrostatic pressures are studied. The pressure behaviour of the phase transition temperature confirms that the phase transition in these crystals belongs to the order/disorder type. An additional line of birefringence anomalies in the ferrielectric phase is revealed. The (p,T) phase diagram of $CuInP_2S_6$ crystals is built.

References

[1] A.Simon., J.Ravez, V.Maisonneive, C.Payen, V.Cajipe, "Paraelectric – ferroelectric transitsons in the lamellar thiophosphate $CuInP_2S_6$", Chem.Mater.**6**,1575(1994).

[2] V.Maisonneuve, V.Cajipe, A.Simon, V.Muhll, J.Ravez, "Ferrielectric ordering in lamellar $CuInP_2S_6$",Phys. Rev. B. **56**, 10860(1997).

[3] A.Grzechnik, V.Cajipe, C.Payen, P.Mcmillan, "Pressure-induced phase transition in ferrielectric $CuInP_2S_6$", Solid State Commun. **108**, 43(1998).

[4] Yu.M.Vysochanskii., A.A.Molnar., M.I Gurzan, V. Cajipe , "Phase transitions in $CuInP_2(Se_xS_{1-x})_6$ layered crystals", Ferroelectrics. **257**, 147 (2001)

[5] R.Blinc and B.Žekš, *Ferroelectrics and Antiferroelectrics*, North-Holland Press, 1974.

Decomposition of the ω-phase in the equiatomic TiZr alloy under high pressure

I O Bashkin[1], **V V Shestakov, M K Sakharov and E G Ponyatovsky**

Institute of Solid State Physics, Russian Academy of Sciences, Chernogolovka, Moscow district, 142432, Russia

E-mail: bashkin@issp.ac.ru

Abstract. The phase decomposition phenomenon is found in the hexagonal ω-phase of the Ti–Zr system under high pressure. The $\omega \rightarrow \omega_1 + \omega_2$ decomposition of the equiatomic TiZr alloy occurred due to long thermobaric treatment at $P = 5.5 \pm 0.6$ GPa and $T = 710 \pm 30$ K. The chemical compositions of the ω_1- and ω_2-phases recovered to ambient conditions were estimated from the X-ray data to be around $Ti_{20}Zr_{80}$ and $Ti_{83}Zr_{17}$. The experimental data were used to calculate the mixing energy and the top of the decomposition curve in the isobaric *T–C* diagram of this system. We find that the equilibrium *T–C* phase diagram of the Ti–Zr system at pressures above ~8 GPa is of the eutectoid type with the high-temperature β-phase and the low-temperature ω_1- and ω_2-phases.

1. Introduction

Two modifications are known for titanium and zirconium at atmospheric pressure, the low-temperature hexagonal close-packed (*hcp*) α-phase and the high-temperature body-centered cubic (*bcc*) β-phase. Both phases of the Ti–Zr system form continuous solid solutions, and characteristic of the *T–C* phase diagram is the monotectic point located at 52 at.% Zr [1] and 852 K [2]. Applied pressure induces a transition of the α-phase to the hexagonal ω-phase both in elemental Ti and Zr [3-5] and in the binary Ti–Zr alloys [2,6]. The ω-phase recovered from the high-pressure experiments at room and moderate temperatures has also been reported to form continuous solid solutions [2,6]. The ω-phase solid solutions showed the concentration dependence of the specific volume close to the Vegard's law, and the reported values of the specific volumes were by less than 1 % above the straight line drawn between the volumes of pure elements.

Earlier, the pressure effect on the phase diagram has been studied by means of the differential thermal analysis (DTA) carried out on the equiatomic TiZr alloy up to 7 GPa [2]. There were plotted the lines of the isoconcentrational $\alpha \leftrightarrow \beta$ and $\beta \leftrightarrow \omega$ transformations. The α–β transformation had a hysteresis of 70 K whereas the hysteresis of the β–ω transformation was close to zero, and the parameters of the α–β–ω triple point determined by DTA were as low as $P = 4.9$ GPa and $T = 733$ K. In contrast, the hysteresis of the

[1] To whom correspondence should be addressed.

© 2008 IOP Publishing Ltd

α–ω transformation was large, and the ω-phase recovered to ambient conditions remained intact at atmospheric pressure until heated to 700 K [2].

The thermodynamic calculation of the *T–P–C* phase diagram of the Ti–Zr system has been attempted under the assumption of the isoconcentrational α–β, α–ω and β–ω phase transformations [6]. The estimated value of the mixing energy of the ω-phase solid solutions was positive and rather high, W_m = 11.3 kJ/mole. This estimate as well as the positive deviation from the Vegard's law indicates that the ω-phase solid solutions tend to decompose under pressure. Therefore, one should assume that some $\omega_1+\omega_2$ two-phase region is the case in the equilibrium *T–P–C* phase diagram. This work was aimed at the experimental search of the $\omega \rightarrow \omega_1 + \omega_2$ decomposition reaction in the TiZr alloy at pressures above its triple point.

2. Experimental

The TiZr alloy was prepared from the metallic elements with the total impurity contents 0.04 at.% or less using multiple electron-arc melting in vacuum. The chemical composition of the final alloy, 49.6 at.% Ti and 50.4 at.% Zr, was measured with the electron microprobe, JXA-5. The composition was uniform over the ingot within ±0.4 at.%. This same alloy (the *hcp* α-phase with the lattice parameters a = 3.104 Å and c = 4.923 Å) was used previously for the study of the *T–P* diagram [2] and the structural high-pressure measurements [7]. To refine the grain structure and to minimize the sample texture in this work, a piece of the alloy was hydrogenated to the atomic ratio of H/Me ≈ 1.5, ground in a mortar to a characteristic powder size of 0.1 mm, and then hydrogen was evacuated by vacuum annealing at $T \leq$ 1020 K so that to avoid sintering of the powder. The experimental samples compacted of the TiZr powder at P = 0.7 GPa were shaped as pellets of 6 mm diameter and 1.5 mm high.

High pressure was generated using the quasi-hydrostatic toroid-type device with the operation range to 7 GPa. The device was calibrated against the high-pressure transitions in Bi and Tl and provided the accuracy of the pressure determination of ±5 %. To avoid the reaction of the TiZr alloy with the materials of the high-pressure cell, the sample was encapsulated into the NaCl medium, annealed in air to 500 K, and then the capsule with the sample was mounted into the operation zone of the cell.

The X-ray diffraction was measured at room temperature using the $CuK_{\alpha 1}$ irradiation at the *Siemens* D500 diffractometer provided with the SiO_2 monochromator of the incident beam and the Ni filter at the position-sensitive detector. The data were registered with a step of 0.02° typically at a rate of 0.15°/min.

3. Thermobaric treatment

The $\omega \rightarrow \omega_1 + \omega_2$ decomposition reaction should occur in the *T–P* region where the ω-TiZr phase becomes stable with respect to the α- and β-phases. From the kinetic reasons, the temperature of the thermobaric treatment had to be close, but below the ω–β transition temperature. There were carried out several experiments where the sample was mounted in the high-pressure cell together with the Bi reference (cf. the BiIV–VI transition line in Ref. [3]). This allowed estimation of the long-term behavior of the cell during the thermobaric treatment. The *T–P* parameters of the present thermobaric treatment are plotted in the earlier *T–P* diagram of the isoconcentrational equilibria in the TiZr alloy [2] (figure 1). In this diagram, the dashed line shows an assumptive position of the α–ω transition line whereas the actual transition line may be between the dashed line and the dotted line connecting the α–β–ω triple point and the onset of the ω–α transition observed by calorimetry [2]. With this choice of the *T–P* parameters, we could deliberately compensate their variation due to relaxation of the high-pressure cell during long high-pressure annealing.

We noted also in the preliminary experiments that the transformation to the $\omega_1+\omega_2$ two-phase state was more complete when the initial sample state was the ω-phase rather than the α-phase. For this reason, the present TiZr sample was brought first to the ω-phase by means of cycling the sample over the ω–β transition line, that is, between $T = 670$ and 820 K. In the following experiment, the ω-TiZr sample was maintained at $P = 5.5\pm0.6$ GPa and $T = 710\pm30$ K for 100 h, cooled and recovered to ambient conditions. The X-ray diffraction patterns of the sample states obtained in these experiments are compared in the next section.

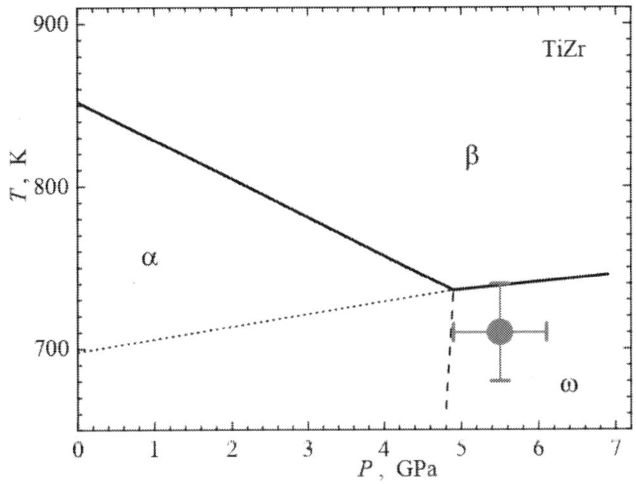

Figure 1. The *T–P* diagram of the isoconcentrational transitions in the TiZr alloy [2] and the parameters of the present thermobaric treatment (circle with the experimental bars). Solid transition lines in the diagram are determined from DTA [2], a tentative line of the α–ω transition is dashed, and the dotted line indicates the low-pressure limit for the position of the actual α–ω transition line.

4. X-ray characterization

Figure 2 represents the diffraction patterns of the TiZr alloy after the ω↔β thermal cycling at $P \approx 6$ GPa (at the top) and after the 100-hour thermobaric treatment (at the bottom). The first pattern is well described within the ω-phase structure (sp. gr. *P6/mmm*) with the lattice parameters $a = 4.844$ Å, $c = 2.996$ Å and $c/a = 0.618$, in good correspondence with the earlier data [2]. In addition to the ω-phase reflections, the pattern includes several very weak reflections that may be related to the residual α-phase with a rather distorted *hcp* lattice ($a = 3.088$ Å, $c = 5.006$ Å and $c/a = 1.621$).

The bottom pattern in figure 2 is more complex. A comparison of the low-angle peaks shows that the pattern represents a superposition of two sets of reflections with the intensities correlated within each set like those in the ω-phase. The peaks in the set with the lower 2ϑ values are not symmetric therefore the Rietveld refinement of the pattern is still in progress. A fit using the POWDERCELL program gives indexing of the main features within two ω-phases with the lattice parameters for the ω_1-phase equal to $a = 4.921$ Å, $c = 3.060$ Å and $c/a = 0.622$ and those for the ω_2-phase $a = 4.680$ Å, $c = 2.864$ Å and $c/a = 0.612$. In addition to the main reflections, a few broad low-intensity features are quite discernible in this pattern at $2\vartheta < 40°$. These features may be interpreted as an admixture of the α-phase with the lattice parameters of $a = 3.049$ Å, $c = 4.873$ Å and $c/a = 1.598$.

It seems that the asymmetry of the ω_1-phase reflections as well as the admixture of the α-phase are of a kinetic nature and result from the incomplete decomposition reaction. A longer thermobaric treatment is expected to give a larger degree of the transformation.

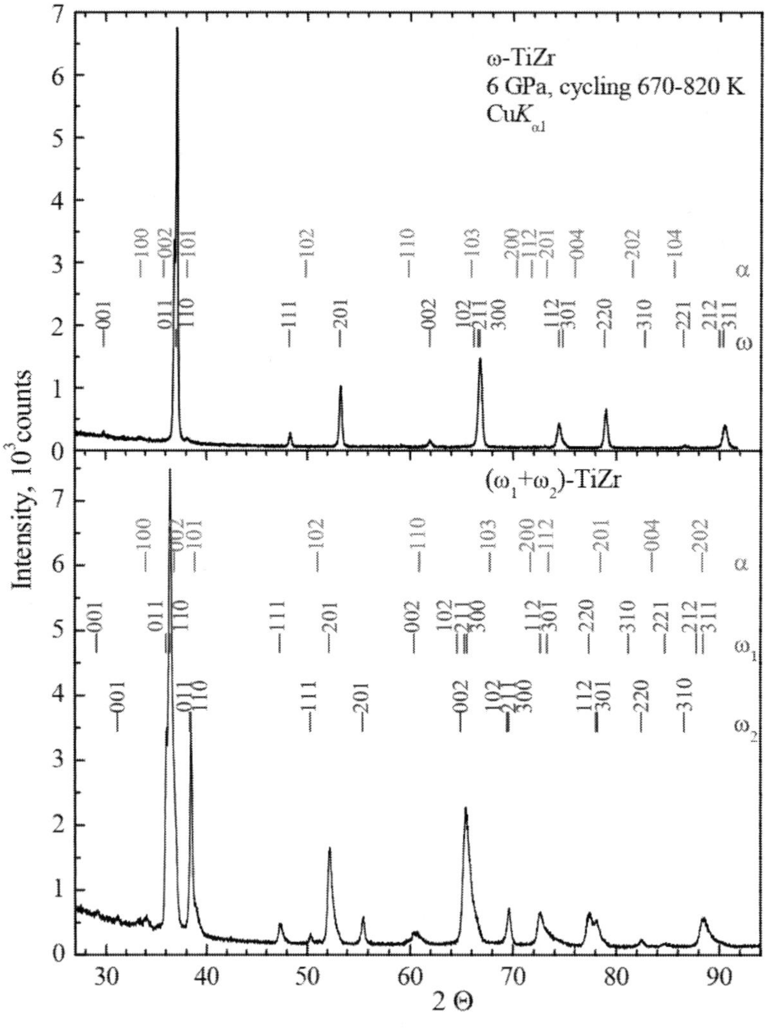

Figure 2. X-ray diffraction patterns measured on the ω-TiZr alloy (top) and on the (ω_1+ω_2)-TiZr alloy (bottom). Bar diagrams show the calculated peak positions for the phases involved.

5. Simulation of the high-pressure *T*–*C* phase diagram

The above lattice parameters of the ω_1- and ω_2-phases were used to calculate their specific volumes and to estimate their compositions using the Vegard's law. The obtained compositions were $Ti_{20.5}Zr_{79.5}$ for the ω_1-phase and $Ti_{83.1}Zr_{16.9}$ for the ω_2-phase. These values are rather symmetric with respect to the equiatomic composition therefore we applied the model of the regular solid solutions in order to calculate the curve of the $\omega \leftrightarrow (\omega_1 + \omega_2)$ equilibrium in the Ti–Zr system under pressure.

According to the model of the regular solid solutions,

$$W_m = T_0 \cdot R \cdot \frac{\ln \frac{1-C_1}{C_1}}{1-2C_1}$$

where W_m is the mixing energy, T_0 is the temperature of the phase equilibrium, $C_1 = 1 - C_2$ is the concentration of one of the phases in equilibrium, and R is the gas constant. This equation describes a symmetric curve with a maximum, and $C_1 = 0.5$ in the critical point at the top of the curve therefore the critical temperature is $T_{cr} = 0.5W_m/R$.

If we accept T_0 =710 K, $C_1 = 0.19$ for the present experiment, the equation returns $W_m = 13.8$ kJ/mole. The change of the W_m value is less than ±0.2 kJ/mole when the C_1 value is changed by ±0.01 (compare with the difference in the ω_1 and ω_2 compositions). The critical temperature is estimated then as $T_{cr} = 830$ K. The calculated $\omega \leftrightarrow (\omega_1 + \omega_2)$ equilibrium curve is represented in figure 3 in thick solid line with a dotted segment showing a part of the line that enters into the stability region of the β-phase. For simplicity, other elements of the isobaric *T–C* section of the phase diagram are related to 8 GPa, the pressure of the α–β–ω triple point in Ti [5]. Thus we eliminated the α-phase from consideration. The β–ω transition temperature in Zr at 8 GPa is estimated from the graphic data in Ref. [4]. The line of equal Gibbs potentials of the β- and ω-phases (dashed) is drawn through the end points for pure Ti and Zr and the temperature of the isoconcentrational β–ω transition in the TiZr alloy [2]. The choice of the equilibrium eutectoid temperature and the shape of the (β+ω_1) and (β+ω_2) two-phase boundaries is arbitrary; these curves are drawn in thin solid lines.

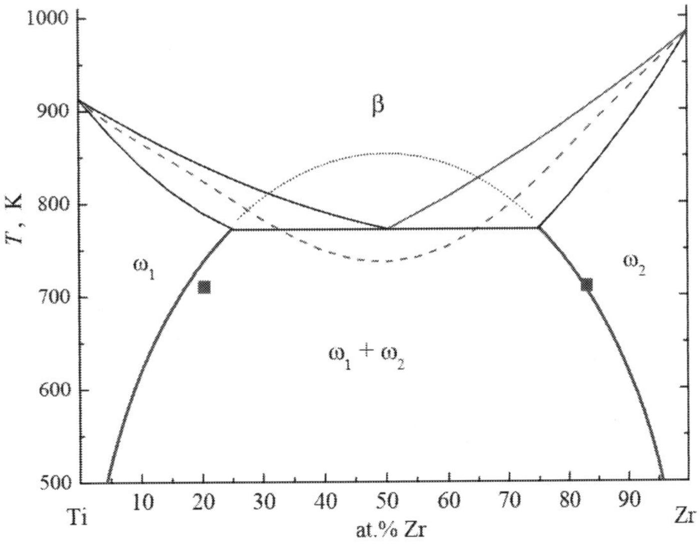

Figure 3. An outline of the equilibrium *T–C* phase diagram of the Ti–Zr system at $P \geq 8$ GPa. Thick solid lines show the calculated boundaries of the $\omega \rightarrow \omega_1 + \omega_2$ decomposition reaction; the end-capping high-temperature segment of the calculated curve is in the β-phase stability region (dotted line). The dashed curve represents the line of equal Gibbs potentials for the β- and ω-phases. Thin solid lines are drawn through the end points in compliance with the available experimental data. Squares show the experimental data.

So, the equilibrium *T–C* phase diagrams in some pressure range above the Ti and Zr triple points are of the eutectoid type, the α-phase stability regions should be taken into consideration at lower pressures.

6. Conclusion

The present experimental data clearly demonstrate that the equilibrium *T–P–C* phase diagram of the Ti–Zr system includes the $\omega_1+\omega_2$ two-phase region where ω-TiZr may persist only as a metastable phase due to slow diffusion kinetics of the components. The $\omega \rightarrow \omega_1+\omega_2$ decomposition reaction occurs even at pressures just above the triple point of the alloy.

Acknowledgements

This work was supported by the Russian Foundation for Basic Research, Grant No. 03-02-17005, and by the Russian Academy of Sciences, Program “Thermal Physics and Mechanics of the Extreme Action”.

References

[1] *Binary Alloy Phase Diagrams* 1986 vol 2 ed T B Massalski (Metal Park, Ohio: ASM) p 2142
[2] Bashkin I O, Pagnuev A Yu, Gurov A F, Fedotov V K, Abrosimova G E and Ponyatovskii E G 2000 *Phys. Solid State* **42** 170-6
[3] Tonkov E Yu and Ponyatovsky E G 2004 *Phase Transformation of Elements under High Pressure* (London-New York-Washington: CRC Press) p 223, 228
[4] Jayaraman A, Klement W and Kennedy G C 1963 *Phys. Rev.* **131** 644-9
[5] Bundy F P 1966 *New materials and Methods for Investigation of Metals and Alloys* (Moscow: Metallurgia) p 230-5
Bundy F P 1963 *G. E. Research Lab. Rept. No. 63-RL-348IC* (USA)
[6] Aksenenkov V V, Blank V D, Kulnitskii B A and Estrin E I 1990 *Fiz. Metallov Metalloved.* **69** 154-9
[7] Bashkin I O, Fedotov V K, Nefedova M V, Tissen V G, Ponyatovsky E G, Schiwek A and Holzapfel W B 2003 *Phys. Rev.* B **68** 054401

Pressure induced metallization in ACrO3 in perovskite compounds

L X Yang[1], Y W Long[1], C Q Jin[1,*], R C Yu[1], J S Zhou[2], J B Goodenough[2], H Z Liu[3], G Y Shen[3] and H K Mao[3]

[1] Institute of Physics, Chinese Academy of Science, P.O. Box 603, Beijing, China
[2] Texas Materials Institute, University of Texas, 1 University Station, C2201 Austin, TX 78712, USA
[3] Advanced Photon Source, Argonne National Lab, USA

*E-mail for correspondence: jin@aphy.iphy.ac.cn

Abstract. We studied the electrical conductivity of ACrO3 (A=Sr, Ca) perovskites performed at various pressures up to 40 GPa using diamond anvil cell techniques. The samples were synthesized under high pressure high temperature. Pressure induced metallizations were observed in both samples. However the xray diffraction experiments with synchrotron radiation source indicated no discernable crystal structural transition up to 60 GPa. Therefore the pressure induced metallizations were ascribed to electronic type phase transitions. It possibly came from the change of electronic structure due to an orbital ordering evolution induced by pressure.

1. Introduction

The transition-metal oxides show so many interesting physical properties such as the charge-ordering system $La_{1-x}Sr_xFeO_3$ [1]、 $La_{1-x}Sr_xNiO_4$[2] and the superconductive system $La_{2-x}Sr_xCuO_4$ [3]and the metal-insulator transition systems $R_{1-x}A_xTiO_3$[4]. The most acceptable explanations of the phenomenon just base the complexity electrical structures and the strong electron-electron interactions [5]. The most of transition-metal oxides with perovskite or a perovskite-related structure is usually very stable under high pressure and the effects of pressure of these materials just are the decrease of the volume and the transition of electrical structures. So the high pressure research of these materials can easily investigate the behavior of the electronic structures. $ACrO_3$ (A=Ca, Sr) perovskites are Mott solutions with tunable narrow energy gap and replacing Sr by Ca gives an increases the band gap. The x-ray diffraction shows that the crystal structures of Sr/$CaCrO_3$ keep stable in the pressure range 0-60 GPa at room temperature and the bulk modulus of $SrCrO_3$ (B_0 =178±5 GPa) is smaller than $CaCrO_3$'s (189±2GPa) [6] and it indicates that $CaCrO_3$ is more difficult to compress under pressure. The bonding instability has been confirmed by the pressure induced bond softening in $SrCrO_3$ [6]. So ACrO3 (A=Ca, Sr) is a perfect system to research the behavior of the electrical structures under pressure and we performed the electrical conductivity of $ACrO_3$ (A=Ca, Sr) performed at various pressures up to 40GPa using diamond anvil cell techniques.

2. Experiment

© 2008 IOP Publishing Ltd

A hardened Be-Cu DAC and a T301 stainless steel were used in our experiments. The pressure medium and the insulator layer is Al_2O_3 powder. Pressure was measured by the ruby fluorescence method. The electrical resistance was measured by a real four-probe method. All experiments were performed from 80K to 300K.

3. Results and Discussion

The pressure variation of the electrical resistance R (P) of Sr/$CaCrO_3$ at a special temperature (280K) is shown in figure. 1. As can be seen a precipitous decrease of Sr/$CaCrO_3$ all occur at low pressure but it mainly come from the decrease of the grain-boundary resistance and the contact resistance under pressure. So the change is not a reliable proof for the insulator to metal transition.

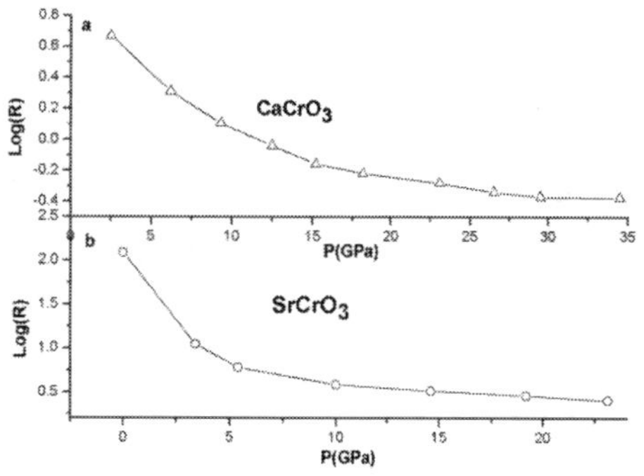

Figure 1. The pressure dependence of the electrical resistance of Sr/$CaCrO_3$ under different pressure at 280K.

In order to further explore the high pressure electrical state of Sr/$CaCrO_3$ we performed the R (T) measurement at various pressures. As shown in figure. 2, the R (T, 0GPa) curve of $SrCrO_3$ is typical of

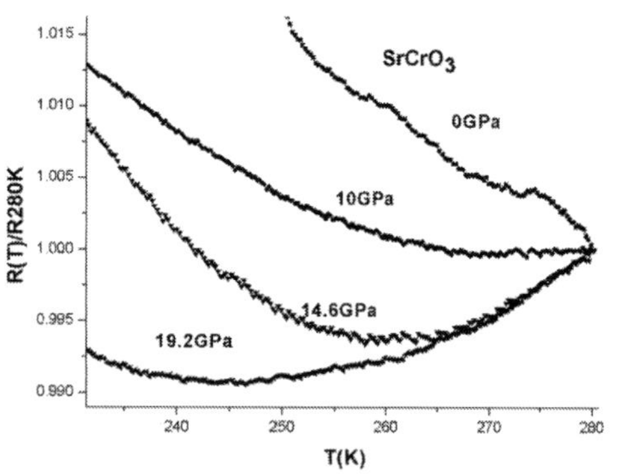

Figure 2. The temperature dependence of renormalized resistance of $SrCrO_3$.

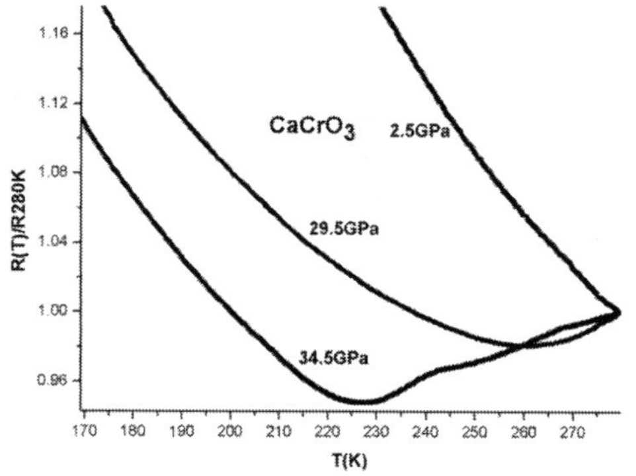

Figure 3. The temperature dependence of renormalized resistance of $CaCrO_3$.

an insulator and the R (T, 10GPa) curve with a minimum value at about 260K is a characteristic of the metal-insulator transition (MIT). The temperature of MIT (T_c) also decreases under higher pressure and it reaches 240K at the highest pressure 23.1GPa. $CaCrO_3$ has the resembled electronic properties under pressure like $SrCrO_3$'s and the start pressure and temperature of MIT is about 29.3GPa and 260.4K (figure. 3). The transition temperature T_c of $CaCrO_3$ also decreases following the increasing of pressure and at the highest pressure 40.4GPa it reaches about 235K. Our results of T_c of $SrCrO_3$ and $CaCrO_3$ vs pressure are displayed in figure. 4. We extrapolate the T_c (P) curves of

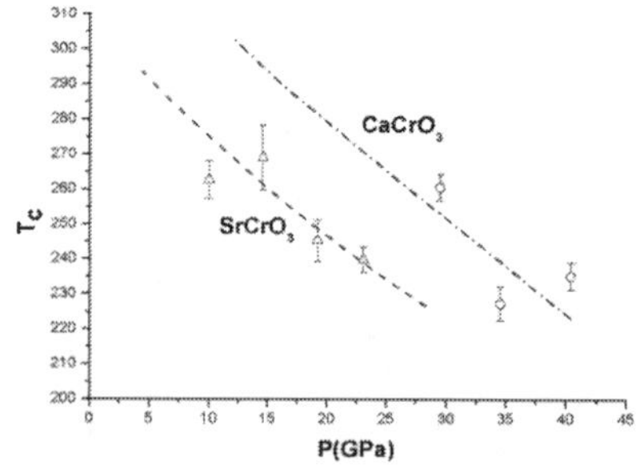

Figure 4. Pressure dependence of the metal-insulator transition temperature of Sr/$CaCrO_3$.

$SrCrO_3$ and $CaCrO_3$ to lower pressure and it would locate the T_c=290K for $SrCuO_3$ and $CaCrO_3$ near 5.5GPa and 16GPa. The transition pressure of $CaCrO_3$ is much larger than $SrCrO_3$ and it indicate that $CaCrO_3$ is more difficult to metallize under pressure.

4. Conclusions

We have performed the electrical conductivity measurements of Sr/$CaCrO_3$ in DAC. The MIT of Sr/$CaCrO_3$ under pressure all occur and at room temperature (290K) the transition pressure of $SrCrO_3$ and $CaCrO_3$ is 5.5GPa and 16GPa and it belongs to an electronic type phase transition. The MIT of CaCrO3 the same as $SrCrO_3$ comes from the pressure induced bond softening and the different band gaps of Sr/$CaCrO_3$ are the reason of the difference of the metallization pressure. So all the results of Sr/$CaCrO_3$ indicate that the pressure can change the band gap continuously and these compounds undergo an electronic type phase transition under pressure.

Acknowledgements

This work was supported by NSF & MOST through the research projects.

References

[1] Takano M and Takeda Y 1983 *Bull. Inst. Chem. Res.* **61** 406
[2] Tranquada J M, Buttrey D J, Sachan V and Lorenzo J E 1993 *Phys. Rev. Lett.* **73** 1003
[3] Radaelli P G, Jorgensen D, JSchultz A, Peng J L and Greene R L 1994 *Phys. Rev.* B **49** 15322
[4] Torrence J B, Lacorre P, Nazzal A I, Ansaido E J and Niedermayer C 1992 *Phys. Rev.* B **45** 8209
[5] Masatoshi I, Atsushi F, etal. 1998 *Rev. Mod. Phys.* **70** 1039
[6] Zhou J S, Jin C Q, Long Y W, Yang L X and Goodenough J B 2006 *Phys. Rev. Let.* **96** 046408

Joint 21st AIRAPT and 45th EHPRG Int. Conf. on High Pressure Science and Technology IOP Publishing
Journal of Physics: Conference Series **121** (2008) 022018 doi:10.1088/1742-6596/121/2/022018

Melting, density, and anisotropy of iron at core conditions: new X-ray measurements to 150 GPa

R. Boehler[1*], D. Santamaría-Pérez[1], D. Errandonea[2], and M. Mezouar[3]

[1] Max-Planck Institute für Chemie, Postfach 3060, D-55020 Mainz, Germany

[2] Dpto. Fís. Apl.-ICMUV, Univ. de València, C/Dr. Moliner 50, 46100 Burjassot, Spain

[3] ESRF, 6 Rue Jules Horowitz, BP 220, 38043, Grenoble Cedex, France.

E-mail: boe@mpch-mainz.mpg.de

Abstract. Melting temperatures of iron up to 130 GPa and compressibility to 150 GPa were measured by X-ray diffraction in the double-sided laser-heated diamond cell. New experimental features are the very tightly focused X-ray beam, in situ alignment of X-ray beam and hot spot, and rapid data acquisition. The melting temperatures are in good agreement with previous optical measurements supporting "low" melting temperatures in the core. Above 70 GPa only the ε (hcp)-Fe phase is found. The measured densities of the annealed iron samples yield an accurate equation of state. The c/a ratio of iron decreases with pressure but shows an increase with temperature and at core conditions is nearly the same as at low pressure.

1. Introduction

Understanding the Earth's thermal gradient and thermal history and the seismic density and structure of the Earth's core requires an accurate knowledge of the phase behavior, equation of state and melting temperature of Fe at the inner-outer core boundary (IOCB) at 330 GPa. The melting curve was determined using the laser-heated diamond-anvil cell (LHDAC) by a number of laboratories up to 200 GPa[1], and shock velocity measurements have found melting between 220 and 240 GPa[2]. Theoretical melting calculations have been made to 360 GPa[3-5]. However, the extrapolated measurements and the theoretical predictions of the melting temperatures at the IOCB (330 GPa) range from 4800 K to 7000 K (see figure 1).

© 2008 IOP Publishing Ltd

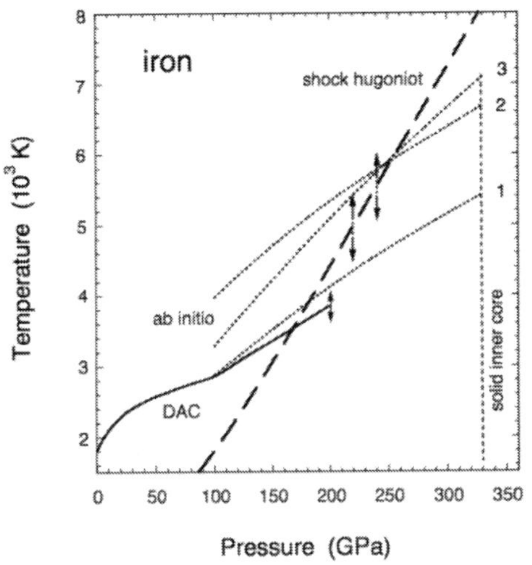

Figure 1- Iron melting curves. Results of LHDAC experiments [1] and shock melting experiments indicated by the arrows on the Hugoniot[2] and ab initio melting calculations[3-5] (1, 2 and 3, respectively) are shown. The melting temperatures at the inner-outer core boundary range from 4800 K to 7000 K.

At ambient temperature, iron has the bcc structure and transforms into hcp at 13 GPa[6]. The hcp phase was found to be stable up to about 300 GPa by X-ray diffraction[7]. Several years ago it was suggested, based on X-ray diffraction measurements in the laser-heated diamond cell, that the phase diagram of iron may be more complicated containing orthorhombic or dhcp phases. More recently, however, no such transitions were found[8,9]. It is highly likely that both the previously observed distortion from hcp to orthorhombic, and a change in stacking order hcp→dhcp were due to deviatoric stresses using hard pressure media in the diamond-cell. There is also a new report of the existence of a bcc-Fe phase above 230 GPa and 3400 K[10], which could be the stable structure in the Earth's inner core. In order to clarify these earlier discrepancies on the iron phase behavior and melting, we made new measurements at the laser-heating beamline ID 27 at the ESRF using a variety of pressure media.

Most of the iron melting measurements are at pressures below 100 GPa. The results from different laboratories using six different methods in the laser-heated diamond cell and shock experiments are now in very good agreement within the experimental uncertainty of about ±100 K[1,9,11-14]. Therefore the iron melting curve to 100 GPa is probably better constrained than that for any other material. For the detection of melting a number of physical property changes between solid and liquid have been used, such as discontinuous changes in the resistivity of thin wires[15] and changes in reflectivity and laser power[1,6,16] or motion on the sample surface that can be made visible with the speckle pattern of an argon-laser beam[12,17]. These methods were compared in some experiments with the first observation of microscopic melt features on the surface of the quenched sample for iron[1] and for a number of other metals[18]. The results for iron from these different measurements are in very good agreement. Additionally, the effect of different pressure media on the melting temperatures has been thoroughly checked[1,6]. These melting results have been recently confirmed using X-ray diffraction up to 58 GPa[9].

Additional evidence for "low" iron melting temperatures below 100 GPa comes from shock velocity measurements on pre-heated iron[11]. At 71±2 GPa and 2775±160 K, a very sharp drop in the shock velocity was observed which is due to melting. Temperatures in this work were calculated and possible overshoot of the equilibrium melt conditions was not taken into account.

Recently we made several improvements in order to increase the accuracy of melting measurements: Two new 40 Watt diode-pumped Nd:vanadate lasers (INAZUMA, Spectra-Physics) with perfect TEM00 mode and very high power stability, resulted in larger hot spots and reduced temperature fluctuations to less than 10 degrees. A more stable gasket, consisting of tungsten and diamond powder provided a larger cell volume resulting in a more efficient thermal insulation of the sample from the diamonds and thus, a reduction in the temperature gradients. This gasket also allowed the use of argon as a pressure medium to over 100 GPa. Moreover, we significantly improved the optical resolution by replacing the previous reflecting objective with achromatic lenses with variable numerical aperture to minimize chromatic aberration.

The only diamond cell measurements above 100 GPa are from Boehler[1] to 200 GPa and Saxena et al.[13] to 140 GPa . These two sets of measurements using similar methods to detect melting are in good agreement and well within the experimental uncertainty, suggesting a melting temperature at 200 GPa near 4000 ±200 K. However, both the theoretical estimates and shock data predict higher melting temperatures.

Above 100 GPa melting measurements in the diamond cell become exceedingly difficult and the detection of melting is limited to the observation of an increase in the reflectivity upon melting. We therefore started to systematically investigate the melting behavior of iron above 100 GPa by X-ray diffraction.

2. Experiment

The data reported here are from 20 different runs. Disks of high-purity iron with dimensions of 10-20 μm in diameter and 5-10 μm in thickness were loaded into a newly designed DAC with conical anvils[19] using tungsten gaskets. Culets varied from 200 to 300 μm. Ar, Xe, KCl and Al_2O_3 were used as pressure transmitting media. For the solid pressure media the high-pressure cells were dried under vacuum and subsequently flushed with Ar, prior to pressurizing. In the experiments with noble gases as pressure media, ruby plates with 10 μm thickness were used to thermally insulate the sample from the diamonds. Pressure was measured by the ruby fluorescence method[20]. These pressures were in good agreement with those obtained from the EOS of Ar, Xe, and Al_2O_3. The samples were heated from both sides with two 40 W Nd:YVO_4 lasers (λ= 1.06 μm wavelength). The laser beams were defocused to create the largest possible hotspot. At the highest pressures of this study temperature gradients in an area with 10 μm in diameter within this hotspot were within 100 K. Using the laser focusing, temperatures on both sides were adjusted to within 100 K. Emission spectra were measured from areas of 2 μm diameter in the spectral range 550 – 800 nm. The Planck's radiation function was fitted to the spectra assuming wavelength-independent emissivities. The measurements were performed at the ID-27 beamline of the European synchrotron radiation facility (ESRF) using X-ray wavelengths of 0.26472 and 0.3738 Å and beam diameters of 3 μm (FWHM) were used. Diffraction images were recorded every two seconds with a Mar345 CCD detector, while the power of the lasers was gradually increased.

Ideally, at melting, all diffraction lines of the sample should disappear and a diffuse scattering halo should appear. However, the determination of melting by X-ray diffraction is not that straightforward because at melting the absorption properties of the sample change, leading to instabilities. Additionally, the molten sample tends to disperse into the pressure medium. Therefore, sample confinement in a suitable pressure medium and rapid recording of the X-ray patterns are essential. We observed two different types of changes in the X-ray patterns when the melting temperature was reached: The first are fast changes in the position and intensity of the spots in the diffraction pattern (see figure 2). We interpret this as recrystallization in a partially molten sample caused by temperature instabilities, which are due to changes in the absorption. This phenomenon is observed every time melting is reached. The second is the expected halo, or diffuse scattering ring. This observation, however, is not always reproducible because it requires a larger of melt fraction the mobile, recrystallizing sample probed by the X-ray beam. The appearance of a diffuse scattering ring is accompanied by the disappearance of all iron peaks (see figure 3), indicating complete melting of the sample. After cooling the iron diffraction peaks reappear and the ring disappears.

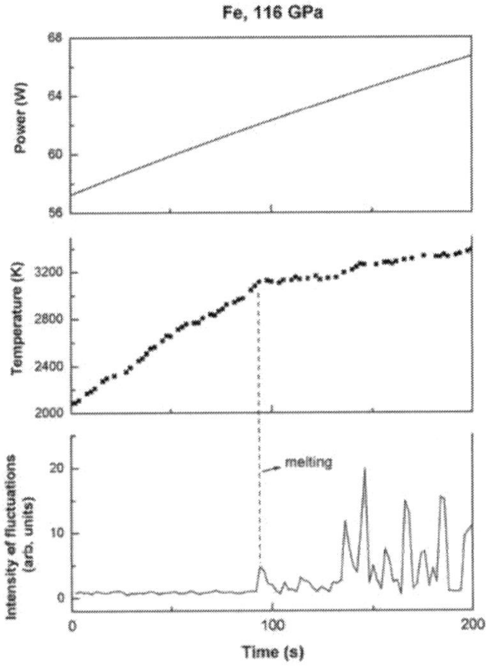

Figure 2 – Effect of recrystallization during melting of iron at 116 GPa. Top: Laser power, middle: maximum sample temperature, bottom: peak intensity fluctuation versus time. When the sample starts to melt (3160 K), large fluctuations in the peak intensities are observed and the temperatures remain nearly constant with increasing laser power.

At the onset of fast recrystallization, the temperature raises slower with increasing laser power. This phenomenon has been frequently observed in previous melting experiments. It is likely due to change in absorption in the liquid and a gradual increase in the melt fraction of the sample.

In order to study the first of these phenomena, the recrystallization process, the observed intensities on the CCD detector were integrated as a function of 2θ in order to give conventional, one-dimensional diffraction profiles. The area under every peak of iron was calculated for each diagram (i.e. Peaks of ε-hcp-Fe at t = 2 seconds: $I_{(100)}$, $I_{(002)}$, $I_{(101)}$, $I_{(102)}$ and $I_{(110)}$; at t = 4 s: $I'_{(100)}$, $I'_{(002)}$, $I'_{(101)}$, $I'_{(102)}$ and $I'_{(110)}$; etc…). If the total sum of all area changes of the iron peaks in consecutive diffractograms ($|I_{(100)} - I'_{(100)}| + |I_{(002)} - I'_{(002)}| + \ldots$) is plotted versus time, large fluctuations are observed when the melting point is reached (see Fig. 2). The melting temperature is then taken at the onset of these strong fluctuations assuming that the fast reorientation of crystallites is only possible in a partially molten sample. These melting temperatures are in agreement with those obtained from the disappearance of diffraction peaks and the occurrence of diffuse scattering described above (both phenomena take place on the temperature plateau).

Journal of Physics: Conference Series **121** (2008) 022018 doi:10.1088/1742-6596/121/2/022018

A B

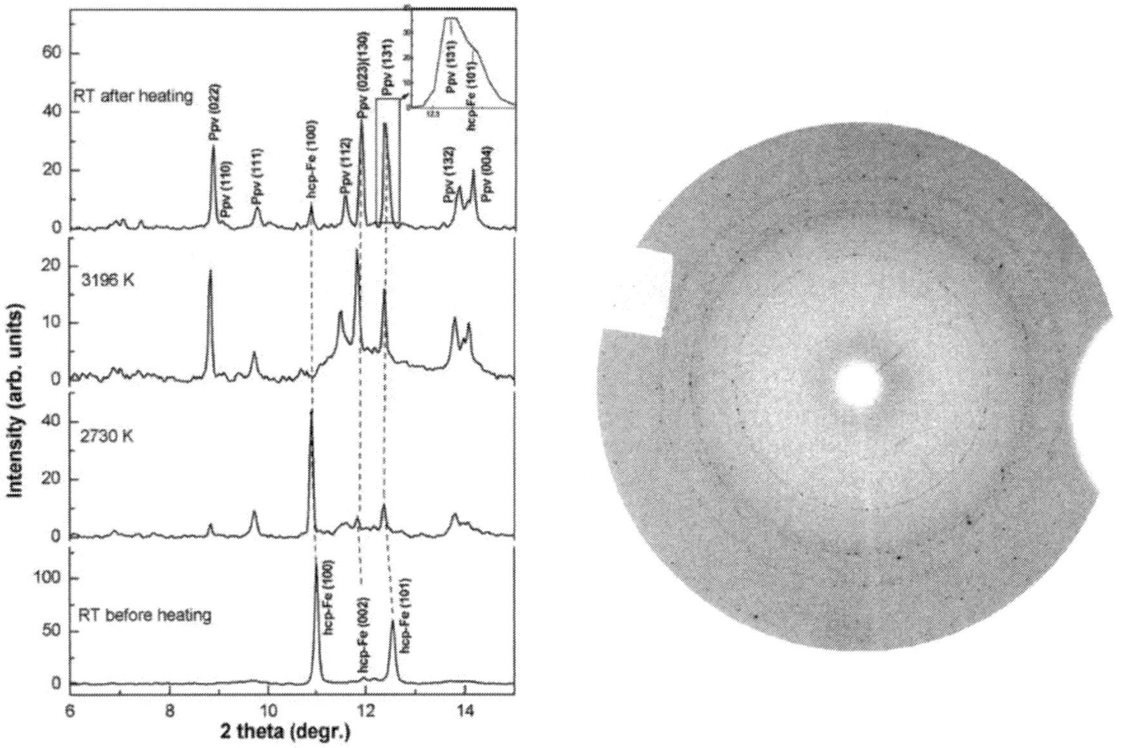

Figure 3 – A) Diffraction patterns of a 10 micron iron sample before (148 GPa), during and after (130 GPa) heating. The pressure medium was $MgSiO_3$ dry gel, which converted to the post-perovskite phase during heating. At 3136 K all hcp iron peaks disappear completely and a diffuse ring associated with liquid iron appears (clearly visible in the original spectrum). When the sample is cooled down, hcp-Fe peaks reappear and diffuse scattering disappears. In the present case, the intensities of the Fe peaks after heating are significantly smaller than before heating, most likely due to a thinning of the sample. B) Diffraction image at 3196 K. At this temperature only peaks belonging to the post-perovskite phase are present [21] and the diffuse ring is due to the fully molten iron sample . We did not detect any chemical reaction between iron and the pressure transmitting media from the diffraction patterns.

3. Results and discussion

3.1. Fcc – hcp boundary and melting

Figure 4 shows the new phase diagram of iron obtained from our experiments in comparison with previous X-ray and optical measurements[1,9,22]. The only phases observed for iron were (ε) hcp and (γ) fcc. Only in those experiments at conditions close to the ε - γ boundary peaks both phases coexist. The γ-ε- liquid triple point is at a pressure between that reported in reference [1] and [8] at 70 GPa and 2700 K.

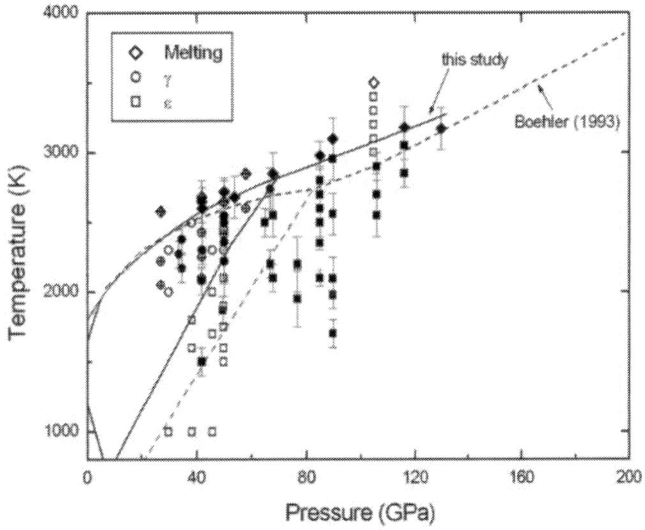

Figure 4 – Phase diagram of iron. The solid lines correspond to the present data (solid symbols) and the dashed lines are taken from Boehler[1]. X-ray data by Ma et al.[22] (empty symbols) and Shen et al.[9] (crossed symbols) are also shown.

Our data confirm that iron at high P-T conditions has the hcp structure and that the melting slope is small above 100 GPa. Compared to our previously published phase diagram [1] both the melting curve and the fcc-hcp boundary are shifted to slightly higher temperatures and lower pressures. This is most likely due to two factors: X-ray measurements tend to overestimate the temperature, because the heated surface of the sample is always slightly higher than the average sample probed by the X-ray beam. Second, the pressure measured optically in stiff pressure media was most likely overestimated due to annealing effects in close vicinity of the heated sample.

3.2. Equation of state of ε-Fe and its anisotropy

Volume measurements of the ε-Fe are presented in Fig. 5. Our pressure-volume data were fitted using a third-order Birch-Murnaghan EOS. The best-fit values for the characteristic parameters were V_0 = 11.23(2) Å^3/atom, B_0 = 160(6) GPa and B_0' = 5.6(2). The bulk modulus obtained from our data is only slightly smaller than that recently reported by Dewaele et al.[23] at room temperature with helium as pressure transmitting medium, but in general the compression measurements shown in figure 5 are in good agreement. The small discrepancy between the different EOS reported could be qualitatively explained by the effect of nonhydrostatic stresses in the sample (different pressure media) and the experimental uncertainty in the pressure measurements (different pressure gauges and pressure gradients).

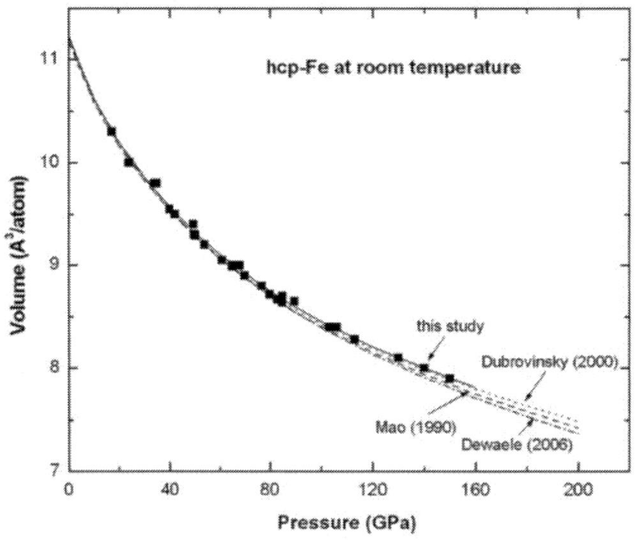

Figure 5 – Room temperature compression of iron in comparison with other recent measurements. The samples in this study are fully annealed due to several melting-freezing cycles but the compressibility is smaller than in the study by Dewaele et al.[23] in a helium pressure medium. All the data shown[7,23,24] are in agreement within the uncertainty of the pressure measurements.

Apart from the compressibility of ε-Fe the variation of the axial ratio c/a of hexagonal iron has gained renewed interest due to the observation that compressional seismic waves in the inner core travel faster along the Earth's rotation axis than in the equatorial plane[25], implying that the solid inner core is anisotropic. From the data plotted in Fig. 6, it can be seen that the compression of ε-Fe at room temperature is anisotropic, the relation c/a decreasing with pressure. The data were fitted using a Bolzmann sigmoidal function, giving the following pressure dependence of the axial ratio c/a:

$$c/a = 1.5963(1) + (0.0131/(1 + \exp((P-26(6))/47(2))))$$

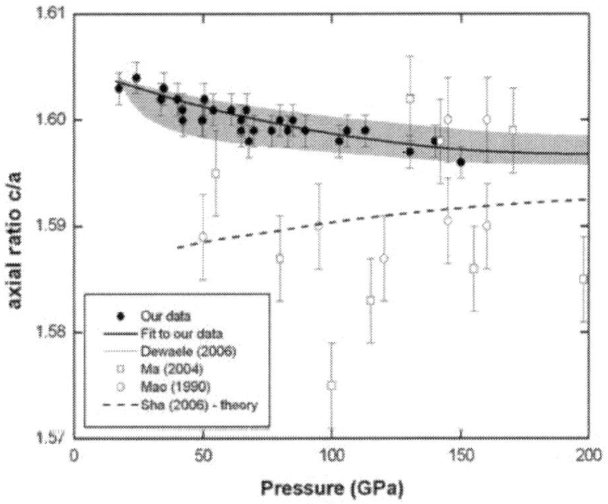

Figure 6 – Axial compression of ε-Fe at room temperature. Our results (solid symbols) are fully consistent with those of Dewaele et al.[23] using helium as a pressure medium.

In comparison with other experimental and theoretical studies, Ma et al.[22] and Mao et al.[7] found no systematic variation in c/a with pressure. Fig. 6 shows the large data dispersion in these studies. Lattice dynamics and thermodynamic calculations at high-pressure found that the equilibrium axial ratio of ε-hcp-Fe increases slightly as a function of pressure at ambient temperature[26]. However, our results are consistent with a recent study of Dewaele et al.[23], where the pressure effect led to a decrease in the c/a axial ratio.

The c/a variation in ε-Fe with temperature was also studied. Figure 7 allows us to compare our results with previous experimental data and with results from theoretical calculations. Both, experiments and theoretical calculations find that there is an increase of the c/a ratio with temperature at high-pressure. Previous experimental data initially shows an increasing trend with temperature, but at temperatures above 2300 K a slight decrease is observed. However, this decrease is not found in the theoretical prediction. Here, we report experimental data up to 3300 K at three different pressures: 84, 106 and 140 GPa. We find that the axis ratio increases continuously with temperature in all the cases, using linear fitting showing a linear decrease with pressure. Our results are in good qualitative agreement with Sha et al.[26] and Modak et al.[27].

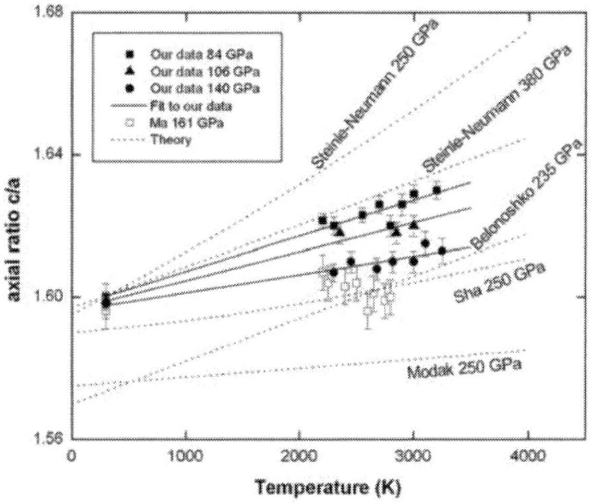

Figure 7 – Axial ratio of ε-Fe as a function of temperature for different pressures. The pressures of the present work are 84, 106 and 140 GPa and the data are represented by solid symbols. Empty symbols correspond to the experimental work of Ma et al.[22] Dashed lines are theoretical calculations: Sha et al.[26], Steinle-Neumann[28], Belonoshko et al.[5] and Modak et al.[27].

4. Conclusions.

The new measured melting curve of iron using X-ray diffraction agrees with previous optical measurements within the error bars but is systematically shifted to lower pressures. The fcc-hcp-liquid triple point is at 70 GPa and 2700 K. Our experiments confirm that the hcp (ε)-Fe is the only phase found at high pressures and high temperatures. The EOS of hcp-Fe yields the following parameters: V_0 = 11.23(2) Å^3/atom, B_0 = 160(6) GPa and B_0' = 5.6(2). Our results show that the compressibility of the lattice parameters is anisotropic at room temperature, the c/a ratio decreasing with pressure, in agreement with recent hydrostatic measurements[23]. Other conclusion from this work is that c/a increases with temperature but at the P-T conditions of the core the c/a ratio is close to that at low pressures.

5. Acknowledgments

This work was partially supported by ESRF. D.E. acknowledges the financial support from the Spanish MCYT (Ramón y Cajal program) and the A. von Humboldt Stiftung.

References

[1] R. Boehler, Nature **363**, 534 (1993).
[2] J.M. Brown and R.G McQueen, J. Geophys. Res. **91**, 7485 (1986); J. H. Nguyen and N. C. Holmes, Nature **427**, 339 (2004).
[3] D. Alfe, M. J. Gillan, and G. D. Price, Nature **401**, 462 (1999); A. Laio, G. L. Bernard, G. L. Chiarotti et al., Science **287**, 1027 (2000).
[4] D. Alfe, L. Vocadlo, G. D. Price et al., J.Phys: Condens. Matter **16**, S973 (2004).
[5] A. B. Belonoshko, R. Ahuja, and B. Johansson, Phys. Rev. Lett. **84**, 3638 (2000).
[6] R. Boehler, N. von Bargen, and A. Chopelas, J. Geophys. Res. **95** (B13), 21731 (1990).
[7] H. K. Mao, Y. Wu, L. C. Chen et al., J. Geophys. Res. **95**, 21737 (1990).
[8] G. Y. Shen, H. K. Mao, R. J. Hemley et al., Geophys. Res. Lett. **25** (3), 373 (1998).
[9] G. Y. Shen, V.B. Prakapenka, M.L. River et al., Phys. Rev. Lett. **92**, 185701 (2004).
[10] L. Dubrovinsky, N. Dubrovinskaia, O. Narygina et al., Science **316**, 373 (2007).
[11] T. Ahrens, K. G. Holland, and G. Chen, Geoph.Res.Lett. **29**, 54 (2002).
[12] A.P. Jephcoat and S.P. Besedin, in *Philosophical transactions: Development in high-pressure, high-temperature research and the study of the Earth's deep interior*, edited by A.P Jephcoat, R.J. Angel, and R.K. O'Nions (Royal society of London, London, 1996), Vol. 354, pp. 1333.
[13] S. K. Saxena, G. Shen, and P. Lazor, Science **264**, 405 (1994).
[14] C.S. Yoo, P. Söderlind, J.A. Moriarty et al., Phys.Lett.A **214**, 65 (1996).
[15] R. Boehler, Geophys. Res. Lett. **13**, 1153 (1986).
[16] S.K. Saxena, G. Shen, and P. Lazor, Science **260**, 1312 (1993).
[17] R. Boehler, M. Ross, and D. B. Boercker, Phys. Rev. B **53**, 556 (1996).
[18] D. Errandonea, B. Schwager, R. Ditz et al., Phys. Rev.B **63**, 132104 (2001).
[19] R. Boehler, Rev.Sci.Instr **77**, 115103 (2006).
[20] H.K. Mao, J. Xu, and P.M. Bell, J. Geophys. Res. **91**, 4673 (1986).
[21] M.K. Murakami, K. Hirose, and K. Kavamura, Science **304** (5672), 855 (2004).
[22] Y. Ma, M. Somayazulu, G. Shen et al., Phys. Earth Planet.Inter. **143**, 455 (2004).
[23] A. Dewaele, P. Loubeyre, F. Occelli et al., Phys.Rev.Lett. **97**, 215504 (2006).
[24] L. Dubrovinsky, S. K. Saxena, F. Tutti et al., Phys. Rev. Lett. **84**, 1720 (2000).
[25] A. Morelli, A.M. Dziewonski, and J.H. Woodhouse, Geoph.Res.Lett. **13**, 1545 (1986).
[26] X. Sha and R.E. Cohen, Phys.Rev.B **74**, 064103 (2006).
[27] P. Modak, A.K. Verma, R.S. Rao et al., J. Phys.: Condens. Matter **19**, 016208 (2007).
[28] G. Steinle-Neumann, L. Stixrude, R.E. Cohen et al., Nature **413**, 57 (2001).

The structure of liquid Sb under pressure

Ayano Chiba[1,2], Masatoshi Tomomasa[1], Takuya Higaki[1], Takazumi Hayakawa[1], Ryota Takahashi[1], Kazuhiko Tsuji[1]

[1]Department of Physics, Keio University, Yokohama 223-8522, Japan

[2]Rutherford Appleton Laboratory, Chilton, Didcot, Oxon OX11 0QX, United Kingdom

ayano@phys.keio.ac.jp

Abstract. Synchrotron x-ray diffraction measurements of liquid Sb under pressure have revealed that the structure in the low pressure range is not simple-cubic-like nor slightly Peierls-type distorted one, although the coordination number and the pair correlation function show some similarities to that of simple-cubic-like liquid. Structural changes were observed at around 3GPa and 6.5-9GPa. The changes of the peak positions and the coordination number are reported.

1. Introduction

The pressure dependence of the structure of liquids are currently attracting a lot of attention; liquid Phosphorus is a well-known example which shows a first-order liquid-liquid phase transition[1]. In this paper we study the structure of a heavier group-V element, liquid antimony (Sb), under pressure. For the crystalline phase, Sb is known to form A7 structure, which can be obtained by a Peierls-type distortion of a Simple Cubic (SC) structure[2]. In this A7 structure, the 6 nearest neighbors in the original SC structure splits in to 3 nearest neighbors and 3 next nearest neighbors. On melting, it is known to transform to a denser structure in which the coordination number is 6[3]. On the other hand, a lighter element Arsenic (As) is known to preserve the A7-like structure on melting, and it is suggested that this is a Peierls distortion in liquid phase[4]. In liquid Sb, the Peierls distortion is thus suggested to disappear[5]. Theoretical works supported this suggestion, although there is a small discrepancy in that the structure is not-distorted at all[6] or slightly Peierls-type distorted[7]. However, the observed pair correlation function with a double-peak-like hump and a largely asymmetric first peak[3] does not have a good similarity to that of the SC-like liquid, which should be the not-distorted structure for A7-like liquid. In addition, it is known that SC structure is not stable in the crystalline phase for Sb. Furthermore, it is difficult to guess the bonding character by an one-dimensional x-ray diffraction measurement. The measurements under pressure have an advantage in that we can extract the bonding character such as covalency by the pressure dependence.

In the p-T phase diagram, Sb is known to have a melting curve with slightly negative slope and almost flat curvature up to 5GPa[8]. On the other hand, a lighter element As is known to have a melting curve with large positive slope with a large negative curvature[8]. Since the crystalline phase of both these elements is known to be A7 structure, the difference of these melting curves should be due to the difference in the structure of the liquid phase. We investigate the pressure dependence of the structure of the liquid phase to clarify this point.

© 2008 IOP Publishing Ltd

There is one another interesting point to focus on. Recently the crystalline structure of Sb has been found to show a complex pressure sequence including incommensurate structures[9]. We measure the pressure range including the range of the triple point of A7, incommensurate Sb-II and liquid phase, and see the pressure sequence in the liquid phase.

2. Experimental

High-pressure high-temperature synchrotron x-ray diffraction measurements have been performed by use of the MAX80 system, installed at PF-AR, KEK, Japan. Measurements were performed at temperatures about 50K above the melting point of each pressure. The details of the measurement and analysis method is described in Ref.[10]. The momentum (Q) range was covered up to 18.3A^{-1}.

3. Results

Figure 1 shows the pressure dependence of the structure factor S(Q) for liquid Sb. The height of the first peak increases and that of the hump next to it decreases with pressure increase. At the pressure 3-4GPa, the hump at 3.2A^{-1} shows a relatively large decrease in its height, whereas the hump position shift toward larger Q continuously with pressure increase. The second peak around 5 A^{-1} changes its shape from a double-peak like character to a single peak with pressure increase; the shoulder at right side of the second peak largely decreases its intensity at around 3GPa.

Figure 2 shows g(r) at each pressure. The first peak, hump, and the second peak positions are shown in Fig. 3 (a), (b), (c), respectively. The first peak position, R_1, in (a) shows an elongation or preservation of the nearest neighbor distance in spite of the pressure increase in the pressure range roughly up to 5GPa. This fact shows that the bonding character is still strongly covalent in this pressure range. At 9GPa, the first peak shows a remarkable broadening in the low r side. This shows a change of the bonding character at 6.5-9GPa. In (b), the hump position, R_h, shifts toward lower r with a large slope. We found that this contraction of the hump position is almost proportional to the molar volume. We note that the hump positions at high pressure range are not shown because the hump merges in to the first peak in the pressure range. In (c), the second peak position, R_2, shows a kink in the slope as a function of pressure at 3GPa.

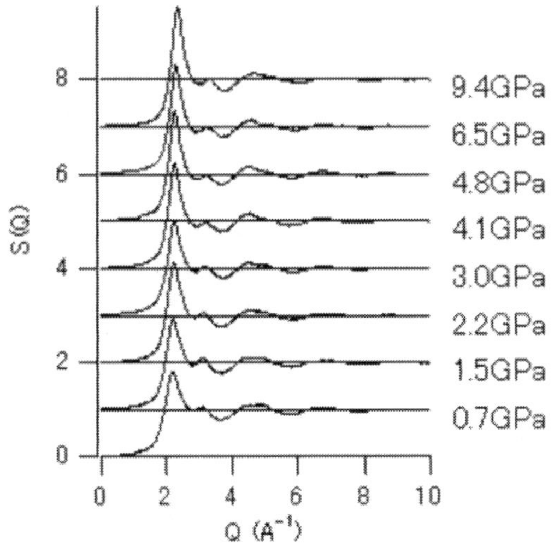

Figure 1: S(Q) of liquid Sb at high pressures

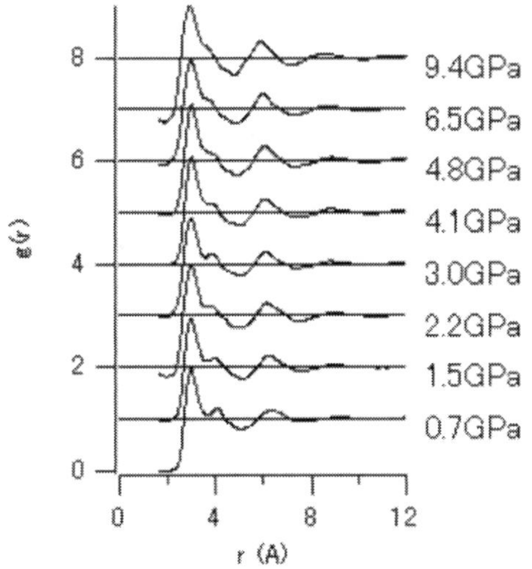

Figure 2: g(r) of liquid Sb at high pressures

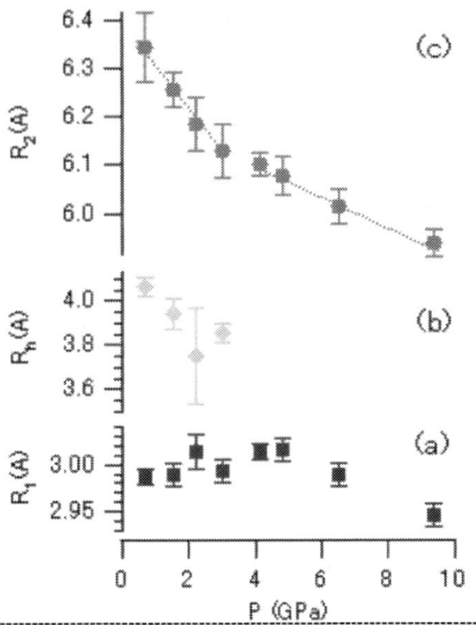

Figure 3. (a) The first peak position, R_1, (b) the hump position, R_h, and (c) the second peak position, R_2, of g(r) at each pressure. Doted lines in (a) are guide for the eyes.

4. Discussion

At the lowest pressure, the ratio of the second peak position to the first peak position, R_2/R_1, is 1.36, which is near the value for the SC-like structure, root 2 = 1.41. The coordination number at the lowest pressure was estimated to be around 5, which is acceptable as a SC-like liquid with vacancies. These facts show the structure in the low pressure range are somewhat SC-like for liquid Sb. However, the contraction behavior largely differ between the first peak position and the hump position. The preservation or elongation of the bond length clearly shows that there is still strong covalency in the system. The fact that the hump position largely contracts with pressure shows that the bond angle is changing largely with pressure. These contraction behaviors thus show that SC-like liquid structure is not stable at any pressure range, even though the structure may be somewhat SC-like. This will be the preferable contraction behavior in liquid state without periodicity, i.e., changing the bond angle is a more preferable response to the pressure than shortening the bond length. We note that SC structure is neither stable at any pressure range in the crystalline-phase pressure sequence for Sb. Even in the case of liquid As where the SC structure is stable in wide pressure range in crystal, we found that the SC-like liquid can not be observed in any pressure range[11].

We may thus suggest that the pressure sequence in the liquid phase do not have a stable SC-like structure, even though there may be a relationship to the SC-like structure. The points about this suggestion are the shape of g(r) and S(Q) with remarkable hump next to the first peaks, as well as their pressure dependences. For example, the first peak of g(r) does not sharpens but broadens up to 3GPa. When SC-like liquid is achieved, the first peak should sharpen, because the peak positions for 3 nearest neighbors and 3 next nearest neighbors are expected to merge to form a one peak. Although the shape of the peak due to the 3 nearest neighbors are expected to broaden with pressure because of the change of the bonding character, the fact that there does not appear a sharp minimum between the first peak and the hump positions shows that the structure is not SC-like. The broadening thus denies the suggestion that liquid Sb is nearly SC-like structure with a small Peierls-type distortion remaining in it. Also, we may say that the structure of liquid Sb is not SC-like at any pressure, although there are some similarity to it in low pressure range. The main difference is the asymmetric character of the first peak and an absence of the sharp minimum between the first peak and the hump positions.

The ratio R_2/R_1 at the lowest pressure is 1.36, which is smaller than 1.41, and decreases with pressure increase. This point also denies the expectation that the structure of liquid Sb at ambient pressure is weakly Peierls-distorted one.

At around 3GPa, an interesting changes were observed in S(Q) and g(r), as stated in the previous section. Since the triple point of the phase diagram for the two crystalline phases and the liquid phase locates at 5GPa, this change in the liquid phase was observed in rather lower pressure range than that for crystalline phase. We note that our result is consistent with the result in Ref.[12] by Umnov et al., although they observed other transitions not only the one at 3GPa.

Turning now to the coordination number(CN), it is estimated to be 4.7 at 0.7GPa. This estimation is by the definition of the integration up to the maximum position in the radial distribution function (RDF) and multiply 2. This value is correctly predicted by an ab initio molecular dynamics simulation[7]. Another definition, the integration up to the first minimum position of RDF, results in 5.3. This difference between two definitions reflect the asymmetric shape of the first peak. The coordination number increases with pressure, and becomes 7 at the highest pressure. CN thus changes from 5 to 7 by pressure increase up to 9GPa. This is in good contrast to the liquid As which shows CN nearly 3 at low pressure range[4,11]. Thus the difference of the melting curve is mainly due to the A7-like loosely packed structure in liquid As at low pressures. The A7-like liquid is less denser, and this results in the large slope of the melting curve, and it is thus compressible, which causes a large negative curvature in the curve. On the other hand, liquid Sb shows already a dense structure at low pressure range.

5. Conclusion

We measured the structure of liquid Sb under pressure up to 9GPa. The structure in the low pressure range was found to have some similarity to the simple-cubic-like liquid, but it was found to be different from it. The ratio of the second peak to the first peak of g(r), R_2/R_1, was found to be smaller than root 2 at the lowest pressure, 0.7GPa, and found to decrease with pressure. The width of the first peak of g(r) does not sharpen but broadens with pressure up to 3 GPa, and the shape of the first peak shows asymmetric character. The first minimum between the first peak and the hump positions of g(r) does not appear as a sharp minimum in any pressures. These facts show that the structure of liquid Sb does not take simple-cubic-like structure nor slightly Peierls-type distorted structure.

At around 3 and 6.5-9GPa, we observed structural changes. At 3-4GPa, we observed a change in the contraction behaviour and a change in the height of the hump of S(Q). This change is located in the lower pressure range than the triple point in the phase diagram. At 6.5-9GPa, the change was mainly observed in the low-r side of the first coordination shell, which indicates a change of the bonding character.

Acknowledgement

One of the authors (A.C.) is supported by JSPS Fellowships for Research Abroad.

References

[1] Katayama Y, Mizutani T, Utsumi W, Shimomura O, Yamakata M, and Funakoshi K 2000 Nature **403** 170.
[2] Littlewood P B CRC Critical Reviews in Solid State and Materials Sciences 1984 **11** 229.
[3] Gaspard J -P, Bellissent R, Menelle A, Bergman C, Ceolin R Il Nuovo Cimento D 1990 **12** 649,
[4] Bellissent R Bergman C Ceolin R Gaspard J P Phys. Rev. Lett. 1987 **59** 661. Gaspard J P, Marinelli F, Pellegatti A Europhys. Lett. 1987 **3** 1095.
[5] Raty J Y, Gaspard J P, Ceolin R, Bellissent R J. Non-cry. Solids 1998 **232-234** 59.
[6] Bichara C, Pellegatti A, Gaspard J.-P Phys. Rev. B 1993 **47** 5002.
[7] Seifert K, Hafner J, Kresse G J. Non-cry. Solids 1996 **205-207** 871.
[8] Klement W Jr, Jayaraman A, Kennedy G C Phys. Rev. B 1963 **131** 632.
[9] See, for e.g., Degtyareva O, McMahon M I, Nelmes R J Phys Rev B 2004 **70** 184119.
[10] Tsuji K, Yaoita K, Imai M Rev. Sci. Instrum. 1989 **60** 2425.
[11] Chiba A, Bennington S M, Hannon A, Tsuji K, submitted.
[12] Umnov A G, Brazhkin V V High Pressure Research 1995 **13** 233.

Sorption of high pressure fluid mixtures into semi-crystalline polymers: investigation of the system carbon dioxide/vinylidenefluoride/poly(vinylidenefluoride)

Andrea Cipollina, Alessandro Galia, Onofrio Scialdone, Giuseppe Filardo

Dipartimento di Ingegneria Chimica dei Processi e dei Materiali, Università di Palermo

galia@dicpm.unipa.it

Abstract. A novel experimental apparatus for measuring gas absorption has been adopted to study sorption of carbon dioxide CO_2 and vinylidenefluoride (VDF) in a semi crystalline polymer, namely poly(vinylidenefluoride) (PVDF). The experimental technique allows to calculate simultaneous absorption of different components in the polymer phase, through a gas chromatographic determination of the composition of the fluid phase. To take into account indirectly the polymer swelling, a suitable not-absorbable probe, namely argon, has been added to the fluid phase. Concentrations of molecules absorbed in the polymer phase were simply calculated through mass balance equations written for each component in the two phases. Several different operating conditions were investigated. Temperature was kept constant at 50°C, while pressure varied between 8 and 35 MPa. Sorption was measured for pure CO_2 and VDF and for mixtures containing around 0.20 w/w and 0.40 w/w of VDF in CO_2, to analyse how the presence of a different component could affect the sorption of the other. Results show that pure CO_2 presents a solubility more than three fold higher than pure VDF at the same pressure conditions. On the other side, when both gases were present in the mixtures, the sorption of CO_2 seems to be strongly inhibited by VDF, whereas this latter presents a solubility quite similar to the pure gas if operating at the same partial pressure.

1. Introduction

The interest in measuring sorption of dense gases in a polymer phase has been rising in the last years, due to the importance of characterizing the gas absorption during polymerization processes in supercritical fluids, in membrane separation technology, in the devolatilization of solvents in as-polymerized macromolecular matrixes, etc.. Indeed, several experimental methods have already been described in the literature to study sorption and swelling of polymers contacted with high pressure fluid systems. Some of them are based on the measurement of the pressure decay accompanying the sorption of the gaseous compound inside the matrix [1], others are substantially in-situ gravimetric techniques such as the quartz spring method originally adopted at pressures lower than 10 MPa by R. G. Wissinger and M. E. Paulatis [2] and later extended to more dense supercritical systems. [3] Similar approaches are based on the utilization of electro [4], micro [5] or mass suspension [6-7] balances to perform the weighting of the polymer sample directly in the dense fluid atmosphere. More sophisticated approaches based on the measurement of the resonant frequency of a suitable mechanical

© 2008 IOP Publishing Ltd

system made solidal to the polymer specimen and immersed in the supercritical phase have been proposed also. [8]

All these techniques share a common drawback even if at different level: all of them suffer of the uncertainty arising from the difficulty of evaluating the volume dilation of the polymer sample during the sorption measurements. This can be a relevant source of error in the evaluation of the solubility of the penetrant. Moreover they cannot be used to measure sorption of penetrants in multi-component systems as it is difficult to separate the effect of the absorption of different compounds.

On the basis of these considerations, the aims of the present work was to investigate the sorption of supercritical fluid mixtures containing carbon dioxide (CO_2) and vinylidenefluoride (VDF) in poly(vinylidenefluoride) (PVDF) using a gas chromatographic experimental approach. This experimental technique, already presented for the case of sorption of a pure component [9], allows for the sorption measurement taking into account intrinsically the effect of the swelling of the polymer and can be used in the presence of multi-component fluid systems. For this reason a small amount of Argon, a not absorbable gas to be used as GC internal standard, was added in the gaseous phase. The amount of CO_2 and VDF dissolved in the polymer is determined by solving the mass balance equations written for each component (CO_2, VDF and probe) using their relative concentration evaluated by gas chromatographic analyses of the gaseous mixture after attainment of equilibrium

2. Materials and method

PVDF Solef 1010, cristallinity 55%, under the form of cylindrical pellets (diameter d=3 mm, d/h=1), and VDF were kindly donated by Solvay Solexis. Prior to the insertion in the sorption vessel the polymer was pretreated in a vacuum oven by heating at 140 °C for 3 h and then slowly cooled down to room temperature (cooling time 3-5 hours). The adopted CO_2 was Air Liquide 99.998 pure, Ar was Air Liquide 5.0. Both gases were used without further purification. The gas chromatographic analyses were carried out by a Hewlett Packard Serie HP 6890 GC, equipped with a Carboxen TM-1000 separation column and a thermal conductivity detector (TCD), and Hydrogen 6.0 produced by a Parker electrolytic gas generator was used as carrier gas.

The experimental apparatus was assembled using commercial elements and is schematically depicted in Figure 1. It consists of a high pressure sorption cell (PARR vessel Model 2670 with nominal free volume 72 mL) fitted with a high pressure transducer (Sensotec Model: STJE/1890-20A) with accuracy ± 0,05 MPa. The sorption cell is inserted in a closed circuit prepared with 1/16 inch OD AISI 316 tubing and equipped with a six-ways high pressure sampling valve (Rheodyne valve Model 7000). The switching valve was added to allow the real time sampling of the fluid phase contacted with the polymer and to this purpose two ports of the valve are connected with a 6μl high pressure sampling loop while other 2 ports are connected to the carrier line (H_2) of the gas chromatograph. The GC carrier circuit was modified to use the hydrogen itself, available at 0.5 MPa from the gas generator, to entrain the fluid contained in the sample loop to the chromatographic column. Mixing in the sorption cell was obtained by a high pressure magnetic driven gear pump (5), Micropump® model GAH-V21-C, which allowed circulation flow rates up to 100 ml/min at operating conditions, thus guaranteeing very short mixing times.

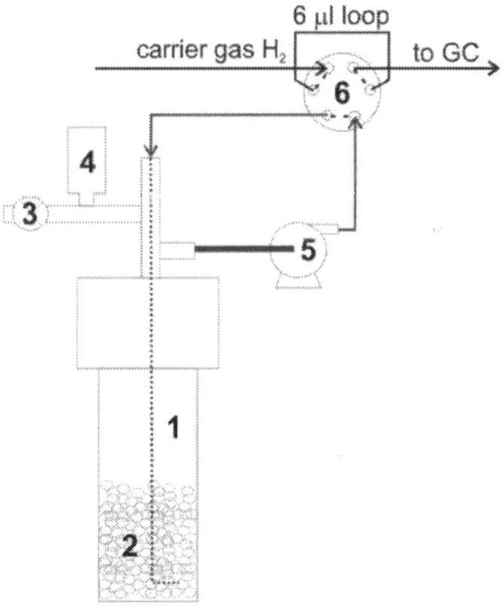

Figure 1. Sketch of the sorption cell and auxiliaries. 1.High pressure Parr cell; 2.polymer pellets; 3.loading and purging valve; 4.pressure transducer; 5.circulation pump; 6.Rheodyne sampling valve.

Journal of Physics: Conference Series **121** (2008) 022020 doi:10.1088/1742-6596/121/2/022020

3. Results and discussion

The amount of compounds dissolved in the polymer is determined from the evaluation of gaseous components concentrations (CO_2, VDF and probe) in the fluid phase, obtained considering a homogeneous system, by GC measurements. In particular, the amount of CO_2 and VDF absorbed into the polymer can be computed by the mass balance equations of the three components (eqs. 1-3) properly combined (eqs. 4, 5).

$$M^{\circ}_{CO2} = M^{P}_{CO2} + \omega^{g}_{CO2}[V\text{-}V^{\circ}_{P} - V^{sw}_{P}] \quad (1)$$

$$M^{\circ}_{VDF} = M^{P}_{VDF} + \omega^{g}_{VDF}[V\text{-}V^{\circ}_{P} - V^{sw}_{P}] \quad (2)$$

$$M^{\circ}_{Probe} = \omega^{g}_{Probe}[V\text{-}V^{\circ}_{P} - V^{sw}_{P}] \quad (3)$$

where V = reactor volume; V°_{P} = initial volume of the polymer; V^{sw}_{P} = swelling volume; M_i^0 = initial masses (weighted); M^P = mass absorbed in the polymer; ω^g = equilibrium concentrations in the fluid phase.

$$M^{P}_{CO2} = M^{\circ}_{CO2} - (\omega^{g}_{CO2} / \omega^{g}_{Probe})M^{\circ}_{Probe} \quad (5)$$

$$M^{P}_{VDF} = M^{\circ}_{VDF} - (\omega^{g}_{VDF} / \omega^{g}_{Probe})M^{\circ}_{Probe} \quad (6)$$

A crucial point for the accuracy and precision of our technique is the selection of the compound that must be used as a probe. An ideal probe should have the following features:

1) it must be absorbed in a negligible way into the polymer phase;

2) it must negligibly modify chemical potentials of the other species dissolved in the gas mixture;

3) it must be completely miscible with other components of the fluid phase under adopted experimental conditions and must be detectable by TCD using H_2 as a carrier.

Argon was selected as a probe since we have estimated, by Berens approach, that at 200 bar and 40°C its solubility in the polymer is in the range of 0.001÷0.002 gr/gr of PVDF. [9] Moreover, using the Peng & Robinson EOS, the effect of the noble gas on chemical potentials of the other components was found negligible provided that its mole fraction in the fluid phase is kept lower than 0.05. [9]

All the experiments were performed loading the polymer and the gaseous components and then heating the system at 50°C leaving it with a proper mixing for about 24 hrs in order to achieve equilibrium conditions. Eventually several GC analysis (at least 5 analysis) were performed to measure the fluid phase composition in order to verify the achievement of equilibrium.

Sorption values for the pure components were measured first and are reported in Figs. 2 and 3:

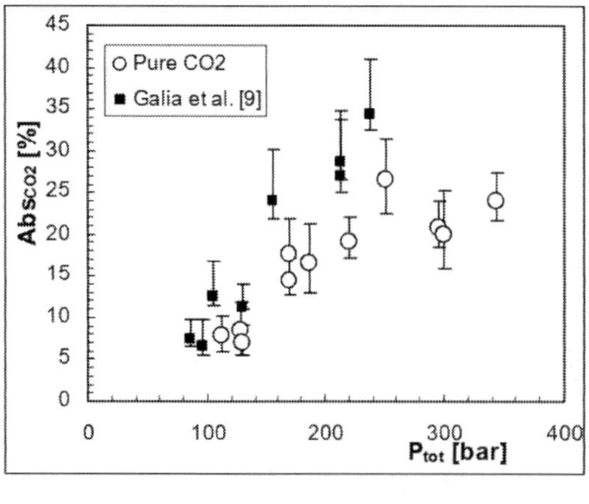

Fig.2 Sorption measurements for pure CO_2 at 50°C compared with literature data at 40°C.

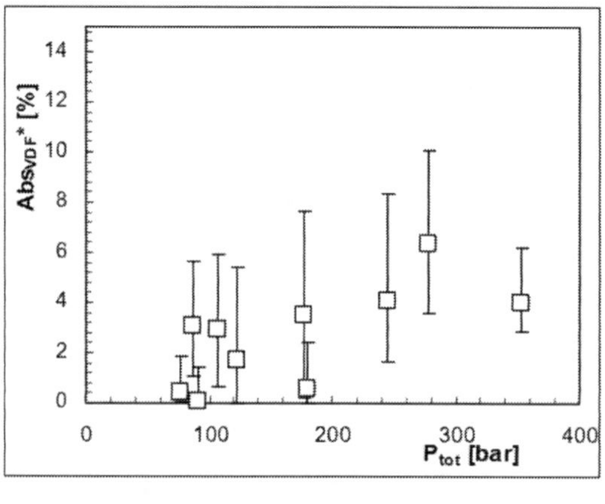

Fig.3 Sorption measurements for pure vinylidenefluoride at 50°C.

The two figures indicate that higher sorption values are achieved by pure CO_2, with a maximum of about 25 gr of CO_2 per 100gr of amorphous PVDF achieved at about 340 bar of total pressure. VDF presents a much smaller solubility in PVDF, achieving sorption values slightly higher than 5gr of VDF per 100gr of amorphous polymer at similar pressure conditions.

Sorption measurements for fluid mixtures were performed with mixtures containing CO_2 and VDF at mass ratios of 80/20 and 60/40, which are reported in Figs. 4 and 5 with reference to the sorption of CO_2 and VDF respectively. Interestingly, at the same values of partial pressure sorption of CO_2 is significantly reduced by the presence of VDF, which therefore acts as inhibitor. On the other side, VDF sorption values seem to be strongly dependent on the partial pressure, being not markedly influenced by the presence of CO_2.

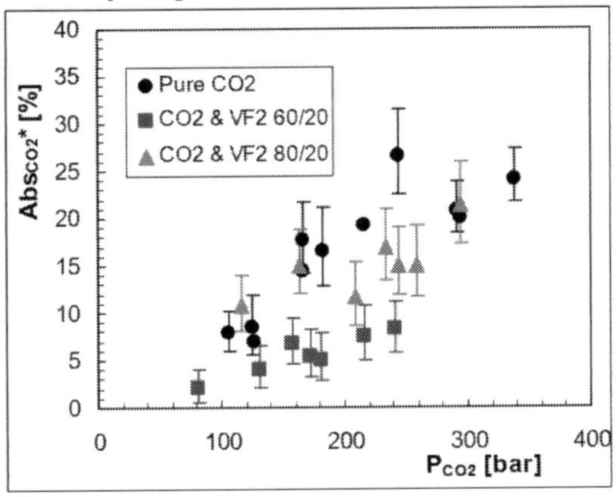

Fig.4 CO_2 sorption vs P_{CO2} for the pure component and the two mixtures (CO_2/VDF weight ratio of 60/40 and 80/20) at 50°C.

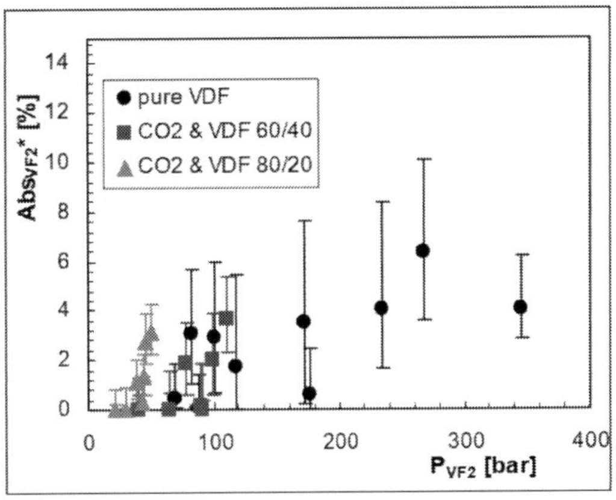

Fig. 5 VDF sorption vs P_{VDF} for the pure component and the two mixtures (CO_2/VDF weight ratio of 60/40 and 80/20) at 50°C

Conclusions

A novel experimental method to measure sorption of supercritical fluid mixtures inside polymers was used to study the solubility of supercritical CO_2 and VDF in PVDF. The technique is based on the use of a gas chromatograph to determine the equilibrium composition of the fluid phase contacted with the polymer. These data, obtained with the use of a purposely selected inert probe (Argon), are used to solve the algebraic system of mass balance equations of the components that partition between the two phases. Results have shown that both pure CO_2 and VDF adsorb in PVDF, although sorption values for CO_2 arrive up to about 25gr per 100gr of amorphous polymer, while solubility of VDF is only slightly higher than 5gr per 100gr of amorphous polymer at pressures around 35 MPa. When the two gases are present in mixtures lower sorption values were measured, indicating in particular an anti-solvent effect of VDF on CO_2 solubility, which is highlighted by the much lower sorption values achieved by CO_2 in mixtures when compared to the sorption of the pure gas at similar value of partial pressure.

References

[1] Stern S A and De Meringo AH 1978 *J. Polym. Sci., Polym. Phys. Ed.* **16** 735
[2] Wissinger R G and Paulatis M E 1987 *J. Polym. Sci. Part B: Polym. Phys.* **25** 2497
[3] Chang S H, Park S C and Shim J J 1998 *J. Supercrit. Fluids* **13** 113
[4] Kamiya Y, Hirose T, Mizoguchi K and Naito Y 1986 *J. Polym. Sci. Part B* **24** 1525
[5] Kikic I, Lora M, Cortesi A and Sist P 1999 *Fluid Phase Equilibria* **158-160** 913
[6] Von Schnitzler J and Eggers R 1999 *J. Supercrit. Fluids* **16** 81
[7] Rajendran A, Bonavoglia B, Forrer N, Storti G, Mazzotti M and Morbidelli M 2005 *Ind. Eng. Chem. Res.* **44** 2549
[8] Bonner D C and Cheng Y J 1975 *J. Polym. Sci. Polym. Lett. Ed.* **13** 259
[9] Galia A, Abduljwad M, Scialdone O and Filardo G 2006 *AIChE Journal* **52** 2243

High-pressure x-ray diffraction study on lithium borohydride using a synchrotron radiation

Satoshi Nakano[1], Atsuko Nakayama[2] and Takumi Kikegawa[3]

[1] National Institute for Materials Science, Tsukuba, Ibaraki 305-0044, Japan
[2] Dept. of Materials Science and Engineering, Meijo University, Nagoya 468-8502, Japan
[3] Photon Factory (PF), Institute of Materials Structure Science, High Energy Accelerator Research Organization (KEK), Ibaraki 305-0801, Japan

E-mail: NAKANO.Satoshi@nims.go.jp

Abstract. Lithium borohydride ($LiBH_4$) was compressed up to 10 GPa using a diamond-anvil-cell to investigate its high-pressure structure. *In-situ* x-ray diffraction profiles indicated a pressure-induced transformation at 1.1 GPa, which was consistent with the previous experimental observation such as Raman scattering spectroscopy. The high-pressure phase was indexed on a tetragonal symmetry of $P4_2/mmc$, which was not corresponding some structural models proposed by previous calculation studies. An unknown substance (presumably another Li-B-H compound), which was contained in the starting material, also transformed into its high-pressure phase at 0.6 GPa without any relation to the transformation of $LiBH_4$.

1. Introduction

Ternary light-element hydrides of alkali borohydrides are attractive as novel hydrogen-storage materials due to their gravimetric densities of hydrogen. Especially lithium borohydride, $LiBH_4$, has extremely high hydrogen content of 18 mass %.

Recently a densification of $LiBH_4$ has been simulated in some theoretical-calculation studies [1-3]. Also, very recently, Talyzin et al. [4] have reported that a high-pressure transformation of $LiBH_4$ was observed between 0.8 and 1.1 GPa using Raman scattering and x-ray diffraction measurement. However, the structure of high-pressure phase has not clarified in the experimental result yet.

In this study, we have examined *in-situ* high-pressure x-ray measurement using a synchrotron radiation to investigate the high-pressure structure of $LiBH_4$.

2. Experimental section

The samples (purity > 95%) were purchased from some chemical makers [5]. They were crushed with an agate mortar and loaded into a diamond-anvil-cell (DAC) in an argon atmosphere. The DAC consists of a pair of diamond anvils of 1/4 carat with a culet of 800 μm in diameter and a gasket (Inconel 700, 250 μm in thickness) with a hole of 400 μm in diameter. Some ruby balls were loaded with the sample into the gasket hole (a sample chamber) as a pressure marker. No pressure medium was used because any contamination should be avoided.

© 2008 IOP Publishing Ltd

Angle-dispersive x-ray diffraction (XRD) using synchrotron radiation was performed at a bending magnet beamline BL-18C of Photon Factory (PF) in High Energy Accelerator Research Organization (KEK). The x-ray beam was monochromatized to energy of 20 keV and introduced to the specimen in the DAC through a pinhole collimator with 60-100 μm in diameter. Each diffraction pattern was obtained by exposing an image-plate detector for 30-60 min. *In-situ* XRD measurement was performed up to the pressure of 10 GPa at room temperature.

The diffraction peaks in the x-ray profiles were individually analyzed by fitting of the peaks with Pearson VII function. The structure was determined from the consideration of site symmetries of the space groups and systematic absences of the diffraction peaks. Lattice parameters were refined with the peak positions using the least-squares procedure of the Appleman-Evans program [6].

3. Results and discussion

3.1. Starting material

The starting materials loaded into DAC were identified using XRD before compression. A XRD pattern of the starting material and its integrated profile are shown in figure 1 (a) and (b), respectively.

The diffraction rings shown with arrows in figure 1 (a) correspond to peaks with open circles in figure 1 (b), which were identified as a room-temperature (ambient-pressure) phase of $LiBH_4$ [7]. Besides the peaks of $LiBH_4$ many other peaks are also observed in the diffraction profile. In the previous publications, two kinds of phase for $LiBH_4$, room-temperature phase and high-temperature one at ambient pressure, have been reported [8]. However, the peaks do not correspond to the high-temperature phase of $LiBH_4$. Also, they could not be identified as any oxides/hydroxides of lithium and/or boron, which are included in ICDD (International Center for Diffraction Data). The unknown substance shows the diffraction rings with some strong spots as shown in figure 1 (a). It suggests that the unknown substance is crystalline and fine, and denies the assumption that it has been formed in the reaction of $LiBH_4$ with the atmospheric moisture. The same unknown peaks have been observed also in a previous report [9] but they have never been identified.

Namely, the starting material consists of the room-temperature phase of $LiBH_4$ and other unknown substance. All of the "$LiBH_4$" powders purchased from the chemical makers contained the same substance. The fact should be noted especially in experiments of spectroscopic works. In the present study, we carefully observed the influence of the substance in the high-pressure behavior of $LiBH_4$.

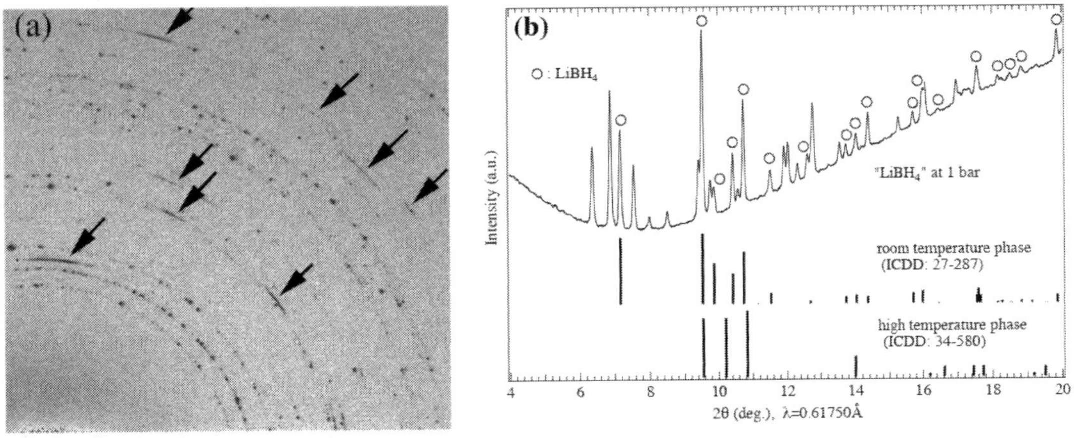

Figure 1. (a) XRD pattern of the starting material in DAC before compression and (b) its integrated profile. The background due to x-ray scattering from air and diamond windows is not subtracted. The diffraction rings emphasized with arrows in (a) and the peaks with open circles in (b) correspond to the ambient phase of $LiBH_4$.

3.2. Observation of high-pressure structural changes

In-situ high-pressure XRD patterns up to 2GPa are shown in figure 2. At 0.6 GPa, all of the unknown peaks disappeared at the same time and new peaks appeared, which means the unknown peaks should be assigned to a single substance and it was transformed into its high-pressure phase at the pressure. The substance might be another Li-B-H compound.

$LiBH_4$ was transformed to a high-pressure phase at about 1.1 GPa without any relation to the unknown substance. It is consistent to the previous report, in which a pressure-induced transformation has been observed using Raman scattering spectroscopy by Talyzin et al [4]. Additional transformation did not occur in the compression up to 10 GPa. In decreasing the pressure, the high-pressure phase was reversed into the ambient-pressure form of $LiBH_4$ at 0.5 GPa.

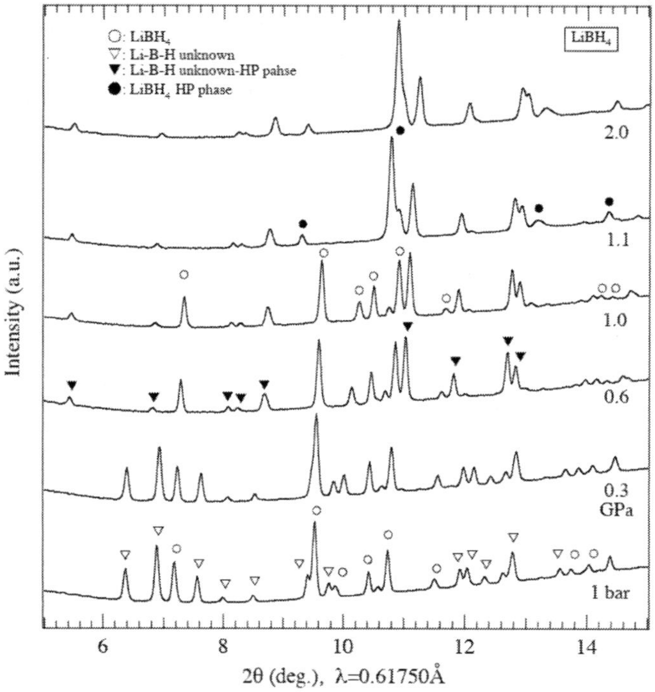

Figure 2. High-pressure XRD patterns of $LiBH_4$ pressurized up to 2GPa.

3.3. Structure of the high-pressure phase of $LiBH_4$

The structure of the high-pressure phase was analyzed with the XRD profiles. The previous calculation studies have proposed some space groups: monoclinic structure *Cc* [1], $P2_1/c$ [2] and tetragonal structure $P\bar{4}2_1c$ [3]. However, they could not explain the present diffraction profiles.

In our analysis of the XRD profiles obtained, the high-pressure structure was successfully indexed on a tetragonal symmetry with the dimensions of a = 3.801(1) Å, c = 6.509(4) Å at 1.1 GPa. The diffraction peaks at the condition and the indices are listed in Table 1. The systematic absences of the diffraction peaks suggested the space groups $P4_2/mmc$ (No. 131) or $P\bar{4}2c$ (No. 112). Previous Raman scattering results [4] has suggested that a tetrahedral molecular anion of $[BH_4]^-$ in $LiBH_4$ remains even in the high-pressure structure after the transformation. However, the coordination of the space group $P\bar{4}2c$ does not allow the tetrahedral molecule in the structure. From the consideration of site symmetries of the space groups and the existence of the tetrahedral anion, the crystal structure of high-pressure phase should be proposed as the space group $P4_2/mmc$.

Figure 3 shows a schematic drawing of the structure estimated. Positions of hydrogen atoms were presumed to be located between lithium and boron because the B-H length (1.04-1.25 Å at ambient pressure) is enough shorter than the Li-B distance in the high-pressure phase (2.50 Å at 1.1 GPa). In the future work, a high-pressure neutron diffraction measurement and a detailed analysis using Rietveld refinement should be performed in order to determine the atomic positions of hydrogen.

Table 1. Indices of the diffraction patterns of $LiBH_4$ high-pressure phase at 1.1 GPa.

Peak No.	d obs. (Å)	d calc. (Å) [a]	hkl
1	3.803	3.801	100
2	3.284	3.282	101
3	2.687	2.688	110
4	2.472	2.472	102
5	2.073	2.072	112

[a] These d-values were calculated on a tetragonal symmetry with the dimensions of a = 3.801(1) Å, c = 6.509(4) Å.

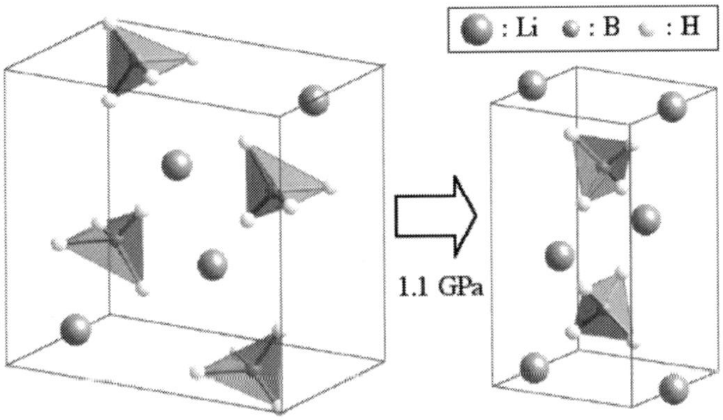

Figure 3. The structure of ambient-pressure phase of $LiBH_4$, orthorhombic *Pnma* Z=4, (left) and that of high-pressure phase proposed in the present study, tetragonal $P4_2/mmc$ Z=2, (right).

Acknowledgments
The authors are very grateful to Dr. Y. Kanke, National Institute for Materials Science, for his helpful support in the x-ray profile analysis.

References
[1] Łodziana Z and Vegge T 2004 *Phys. Rev. Lett.* **93** 145501
[2] Frankcombe T J, Kroes G-J and Züttel A 2005 *Chem. Phys Lett.* **405** 73
[3] Vajeeston P, Ravindran P, Kjekshus A and Fjellvåg H 2005 *J. Alloys Compd.* **387** 97
[4] Talyzin A V, Andersson O, Sundqvist B, Kurnosov A and Dubrovinsky L 2007 *J. Sold State Chem.* **180** 515
[5] Three companies: Alfa Aesar, Sigma-Aldrich Co., and Wako Pure Chemical Industries, Ltd.. Alfa Aesar had assaied the present sample lot as $LiBH_4$ 98% by means of a gas volumetric method.
[6] Appleman D E and Evans Jr. H T 1973 *U.S. Geological Survey, Computer Contribution* 20, U.S. National Technical Information Service, Document PB2-16188
[7] ICDD data 00-027-0287
[8] Soulié J-Ph, Renaudin G, Černý R and Yvon K 2002 *J. Alloys Compd.* **346** 200
[9] Orimo S, Nakamori Y, Kitahara G, Miwa K, Ohba N, Towata S and Zuttel A 2005 *J. Alloys Compd.* **404-406** 427.

Journal of Physics: Conference Series **121** (2008) 022022 doi:10.1088/1742-6596/121/2/022022

Inclusion complexes of triphenylphosphine derivatives and peracetylated-β-cyclodextrin in supercritical carbon dioxide

A Galia [1], E C Navarre [1], G Filardo [1], O Scialdone [1] and E Monflier [2]

[1] Dipartimento Ingegneria Chimica Processi e Materiali, Università di Palermo, Viale delle Scienze Ed. 6, 90128 Palermo, Italy

[2] Laboratoire de Physico-Chimie des Interfaces et Applications, Université d'Artois, FRE CNRS 2485, Faculté des Sciences Jean Perrin, Rue Jean Souvraz, SP 18-62307 Lens Cédex, France

galia@dicpm.unipa.it

Abstract. The supramolecular chemistry of peracetylated-β-CD (perAc-β-CD) as a host for triphenyphosphine derivatives has been studied in supercritical ($scCO_2$) using UV absorption spectroscopy. It was found that the association constant in $scCO_2$ at 40 °C and 300 bar is 10 to 1000 times smaller compared to analogous systems in aqueous solvent. Studies of the thermodynamics of the inclusion process found an enthalpy of association of –30 kJ/mole and an entropy of –55 J/moleK. This difference with respect to water is attributed to the absence of the hydrophobic effect in $scCO_2$ due to the much smaller polarity of $scCO_2$ versus water. To further explore the effect of the solvent on the association constant, values of solvent polarity intermediate between the limits of $scCO_2$ and water could be useful. To this end, methanol is added to the $scCO_2$ as a cosolvent to modify the overall polarity. By taking advantage of the tunability of the supercritical phase, a wide range of solvent polarity can be accessed by varying the temperature (35 to 50 °C), pressure (200 to 400 bar), and mole fraction of methanol. This study is pursued to obtain a better comprehension of the inclusion process in CO_2-based supercritical fluids to make the association constant of the phosphine–perAc-β-CD system comparable to those observed in aqueous systems.

1. Introduction

Viewed as an alternative solvent system for industrial chemical syntheses, supercritical carbon dioxide ($scCO_2$) presents many opportunities and challenges in its application. The insufficient solubility of common organometallic catalysts in $scCO_2$ impedes its use for many reactions employing homogeneous catalysis. [1] In practice, limited solubility is a common problem for a wide range of solutes in $scCO_2$ and is sometimes resolved by modifying the compound to make it more CO_2-philic. Predominantly, this entails the addition of fluorinated groups which renders many compounds quite expensive and difficult to apply on an industrial scale.

The use of native β-cyclodextrin as a molecular host in aqueous solutions is well established as a means for enhancing the solubility of otherwise poorly soluble organic compounds. By forming inclusion complexes, the cyclodextrin increases the concentration of the organic compound in the aqueous phase and consequently, can increase reaction rates. This effect has been studied for the

© 2008 IOP Publishing Ltd

inclusion interaction of water-soluble sulfonated triarylphosphines with β-cyclodextrin. It was found that values of the association constant (K_f) were often 10^3 to 10^4 with native β-cyclodextrin and up to 10^5 with modified analogues[2,3].

We have demonstrated an enhanced solubility of triarylphosphines in $scCO_2$ using a host-guest interaction between the phosphine and peracetylated-β-cyclodextrin (perAc-β-CD). [4]. Considering the high solubility of perAc-β-CD in $scCO_2$ [5], it could offer a means to increase the solubility of organometallic catalysts without changing their structure and potentially altering their activity. For this approach to be practical, the association constant should be large to encourage the solubility of the guest compound. Recently, studies of three candidate arylphosphines were conducted to determine which host-guest interactions would be most favorable in $scCO_2$. [6] The compounds diphenyl(4-phenylphenyl)phosphine, diphenyl[4-(4-tert-butylphenyl)-phenyl]phosphine, and diphenyl(4-adamantylphenyl)phosphine (AdTPP) were found to have association constants of near 10^2. The strongest host-guest interaction among these phosphines was the AdTPP for which the K_f was twice that of the other candidates.

The smaller K_f of these phosphines in $scCO_2$ relative to similar phosphines in water, was attributed to the absence of a solvent hydrophobic effect in $scCO_2$. Thus, a more polar supercritical phase may lead to a more favorable host-guest interaction. The tunable properties of the supercritical phase offer an opportunity to explore the influence of the solvent on the complexation process. A wide range of solvent polarity can be accessed by varying the temperature (35 to 50 °C), pressure (200 to 400 bar), and mole fraction of methanol when used as a modifier of the supercritical phase. As demonstrated by studies of the dielectric constant of methanolic $scCO_2$, large increases in the dielectric constant of the supercritical solution occur with the addition of a relatively small mole fraction of methanol. [7] If we take the dielectric constant of the solvent as an indicator of solvent polarity, significant enhancements of the polarity may be obtained with small amounts of methanol. For example, a 20% increase in dielectric constant, relative to pure $scCO_2$, is expected from the addition of 8 mole % of methanol. The present work investigates the complexation between AdTPP and perAc-β-CD in pure $scCO_2$ and in methanolic solutions of $scCO_2$ using the adjustable solvent polarity to probe the thermodynamics of the host-guest interaction.

2. Experimental

2.1. Materials
Peracetyl-β-cyclodextrin [$(C_{12}H_{16}O_8)_7$; M_w = 2017.8 g/mol] and HPLC-grade methanol purchased from Aldrich Chemicals, and carbon dioxide (99.998% purity) purchased from Air Liquide were used without further purification. Diphenyl(4-adamantylphenyl)phosphine (AdTPP) (M_w = 396.5 g/mol) was prepared as reported in the literature. [6]

2.2. Apparatus
The experiments were performed in a windowed stainless steel reaction vessel. The reactor was fitted with a pressure transducer monitored by a digital voltmeter. Temperature was measured with a K-type thermocouple with a resolution of 0.1 °C. A small magnetic stirring bar was placed into the reactor to promote mixing. The free volume of the vessel was determined to be 24.17 mL. The reaction vessel was placed in an insulated enclosure and heated by a forced-air heating element regulated by a temperature controller.

Samples of the phosphine and perAc-β-CD were weighed and deposited into the reaction vessel as solids. Methanol was added to the vessel using a syringe after introduction of the dry samples. The quantity of methanol was determined gravimetrically. The vessel was then charged to the required pressure with liquid CO_2 using a high-pressure hand pump after which the vessel was weighed again to determine the mass of CO_2. Typical operating conditions for the experiments were 30 ± 0.8 MPa at a temperature of 40 ± 0.2 °C. This gives a density of 0.9 g/mL.

Absorbance measurements were made using an Avantes SD2000 spectrometer equipped with a DH2000 deuterium light source. All connections between components and optical fibers were made with SMA 905 connectors. Light from the deuterium lamp was connected to the reaction vessel by an interface optic which collimates the input beam to an approximately 10 mm diameter spot inside the reactor and condenses the output beam to couple with the detector-side fiber optic. The sapphire optical windows on the reaction vessel are 20 mm diameter and approximately 20 mm thick such that the path length for absorbance measurements is 5 mm. The output beam was then sent the spectrometer. The spectrometer signal was recorded as transmitted intensity then converted to absorbance using a previously recorded blank spectra of carbon dioxide or methanol-CO_2 solution at equivalent operating conditions as the other experments. It was determined experimentally that the blank spectra were invariant with respect to temperature.

2.3. Mathematical models

The mathematical models used for the determination of the association constants are derivations of the Beer-Lambert Law for systems of two absorbers. Both have been described throughly in the literature. [6,8] Method 1 calculates K_f at each wavelength in the spectrum via a least-squares minimization. Method 2 is an algorithmic approach which uses the integrated area beneath the first derivative spectral curve. This approach thus calculates a single value of K_f over the selected spectral interval rather than individual values at each wavelength.

3. Results

Shown in figure 1 are absorbance spectra recorded at 45 °C for three molar ratios of AdTPP to perAc-β-CD in pure $scCO_2$. With increasing amounts of perAc-β-CD, a shoulder between 235 and 240 nm becomes more intense. This feature is attributed to absorbance arising from the complex. Relatively high ratios of perAc-β-CD were necessary to observe this change in the spectrum. The collection of data was limited on the one hand by too little difference between the phosphine and complex spectra

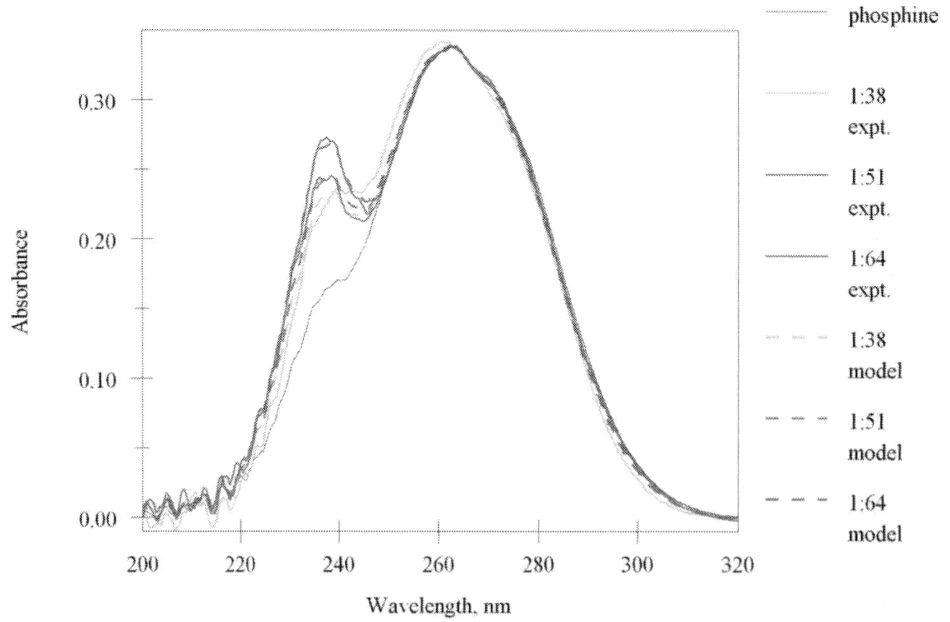

Figure 1. Experimentally measured absorbance spectra (solid lines) of 0.11 mM AdTPP with varying molar ratios of perAc-β-CD at 45 °C and 34 MPa, nominal density of 0.91 g/mL. Modelled spectra (dashed lines) are results from the calculation of K_f with Method 1.

below a ratio of 1:10 and the solubility limit of the perAc-β-CD above 1:70. Also shown in figure 1 are the calculated spectra resulting from Method 1 in which the three experimental spectra are fit to a single value of K_f. The match between the measured absorbance and the fitted result is quite good.

A series of spectra similar to those shown in figure 1 were recorded at 35, 40, 45, and 50 °C to determine the thermodynamic parameters of the inclusion process. The values of K_f determined at each temperature are presented in figure 2 as a Van't Hoff plot. From this relationship, the enthalpy and entropy of the process were determined to be −30 ± 2 kJ/mole and −55 ± 6 J/moleK, respectively. The thermodynamic values determined by the two methods of calculating K_f were equivalent within the experimental error. Compared to the thermodynamic properties found for *o*-, *m*-, and *p*-substituted monosulfonated triphenylphosphines in aqueous solutions (average ΔH° = −41 kJ/mole and ΔS° = −57 J/moleK), AdTPP has a similar behavior. [3] The entropy in both cases is effectively equivalent and more notably negative, signaling an entropy disfavored process. As in the case of the water soluble phosphines, AdTPP is only partially inserted into the perAc-β-CD cavity which leads to a reduction of its degrees of freedom that outweighs any positive entropic contribution from the displacement of solvent from host or guest. The AdTPP complex is enthalpy favored, however, its ΔH° is about 11 kJ/mole less negative than the water soluble analogs. This weaker interaction is reflected in the significantly small K_f values originally found for AdTPP in $scCO_2$.

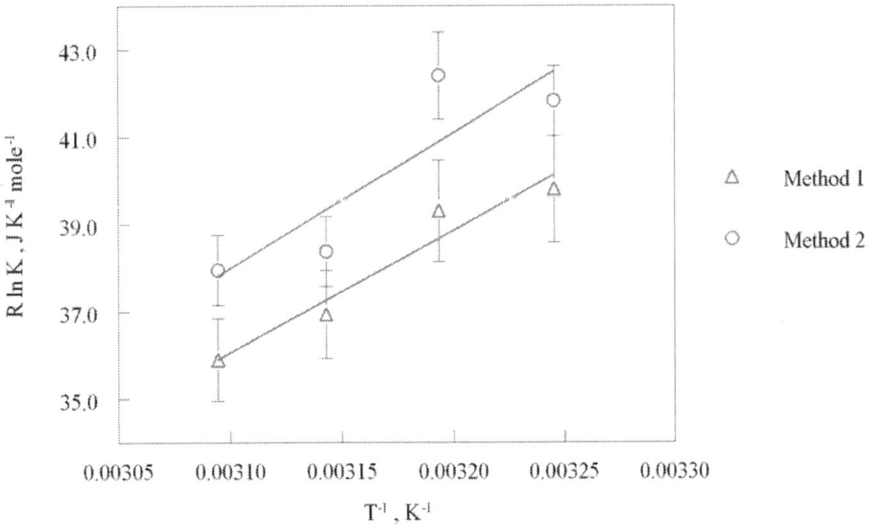

Figure 2. Van't Hoff plots for AdTPP.

Preliminary results of AdTPP inclusion reactions in methanolic $scCO_2$ show some important differences with respect to pure $scCO_2$. Three spectra from studies of AdTPP in 8 mole % methanolic $scCO_2$ are shown in figure 3. As before, a shoulder on the short-wavelength side of the spectrum appears in the presence of perAc-β-CD. In methanolic $scCO_2$ the absorbance maximum is about 30 % more intense with an equivalent amount of phosphine compared to figure 1. Additionally, the absorbance feature centered at 265 nm appears to decrease in intensity with high loadings of perAc-β-CD whereas in pure $scCO_2$ this feature was invariant. This part of the absorbance spectrum has been attributed to the two phenyl groups of the phosphine which are not inserted into the perAc-β-CD cavity. Thus, it is not yet clear what is the origin of this effect. From these sparse data, it is evident that the addition of methanol as a cosolvent has some effect on the host-guest interaction. It appears that there is a modest increase in the K_f and somewhat more negative values of both ΔH° and ΔS°.

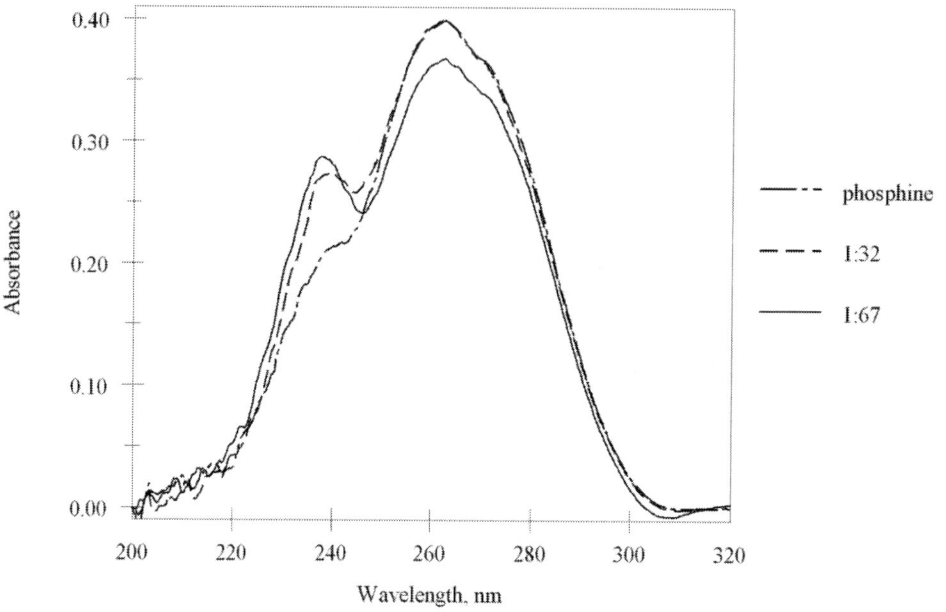

Figure 3. Absorbance spectra of 0.11 mM AdTPP with varying molar ratios of perAc-β-CD in 8 mole % methanolic $scCO_2$ recorded at 50 °C and 37 MPa, nominal density of 0.97 g/mL.

References

[1] Jessop P G, Hsiao Y, Ikariya T and Noyori R 1996 *J. Am. Chem. Soc.* **118** 344
[2] Canipelle M, Tilloy S, Ponchel A, Bricout H and Monflier E 2005 *J. Incl. Phenom* **51** 79
Caron L, Christine C, Tilloy S, Monflier E, Landy D, Fourmentin S and Surpateanou G 2002 *Supramol. Chem* **14** 11
Monflier E, Tilloy S, Mèliet C, Mortreux A, Landy D, Fourmentin S and Serpateanou G 1999 *New J. Chem* **23** 469
[3] Canipelle M, Caron L, Christine C, Tilloy S and Monflier E 2002 *Carbohydr. Res* **337** 281
[4] Filardo G, Di Blasi M, Galia A, Ponchel A, Bricout H, Sayede A D and Monflier E 2006 *J. Supercrit. Fluids* **36** 173
[5] Potluri V K, Xu J, Enick R, Beckman E and Hamilton A D 2002 *Org. Lett* **4** 2333
Potluri V K, Hamilton A D, Karanikas C F, Bane S E, Xu J, Beckman E J and Enick R M 2003 *Fluid Phase Equilib* **211** 211
[6] Galia A, Navarre E C, Scialdone O, Ferreira M, Filardo G, Tilloy S and Monflier E 2007 *J. Phys. Chem. B* **111** 2573
[7] Roskar V, Dombro R A, Prentice G A, Westgate C R and McHugh M A 1992 *Fluid Phase Equilibria* **77** 241
[8] Canipelle M, Caron L, Bricout H, Tilloy S and Monflier E 2003 *New J. Chem* **27** 1603

Journal of Physics: Conference Series **121** (2008) 022023 doi:10.1088/1742-6596/121/2/022023

Pressure-induced phase transitions in organic molecular crystals: a combination of X-ray single-crystal and powder diffraction, Raman and IR-spectroscopy

Elena V. Boldyreva[1,2*], Heidrun Sowa[3], Hans Ahsbahs[4], Sergei V. Goryainov[1,5], Vladimir V. Chernyshev[4], Vladimir P. Dmitriev[7], Yuri V. Seryotkin[1,3], Evgenia N. Kolesnik[1,2], Tatyana P. Shakhtshneider[1,2], Svetlana N. Ivashevskaya[1], Tatyana N. Drebushchak[1,2]

[1]*Research and Education Center "Molecular Design and Ecologically Safe Technologies", REC-008, Novosibirsk State University, Russia*
[2]*Institute of Solid State Chemistry SB RAS, Novosibirsk, Russia*
[3]*Goettingen University, Germany*
[4]*Marburg University, Germany*
[5]*Institute of Geology and Mineralogy SB RAS, Novosibirsk, Russia*
[6]*Chemistry Department Moscow State University, Russia*
[7]*Swiss-Norwegian Beamline ESRF, Grenoble,France*

[*]*E-mail for correspondence:* boldyrev@nsu.ru

Abstract. The contribution summarizes the results of recent studies of phase transitions induced by high pressure in a number of molecular organic crystals, such as polymorphs of paracetamol, chlorpropamide, polymorphs of glycine, L- and DL-serine, β-alanine. The main attention is paid to the following topics: (1) Reversible / irreversible transformations; (2) Different behavior of single crystals / powders; (3) The role of pressure-transmitting liquid; (4) The role of the kinetic factors: phase transitions on decompression, or after a long storage at a selected pressure; (5) Isosymmetric phase transitions; (6) The role of the changes in the hydrogen bond networks / intramolecular conformational changes in the phase transitions; (7) Superstructures / nanostructures formed as a result of pressure-induced phase transitions.

Keywords: phase transition; amino acid; hydrogen bond; molecular crystal; X-ray diffraction; spectroscopy

1. Introduction

Pressure-induced phase transitions in the crystals formed by small organic molecules are interesting in relation to the interplay of the thermodynamic versus kinetic factors affecting polymorphism of solid organic compounds. The studies of these transitions help in achieving a better understanding of the conformational flexibility, rotational dynamics, and intermolecular interactions in organic solids. They can find important applications in pharmaceutical industry and in molecular electronics [1-17]. The present contribution gives a concise overview of a few recent examples from our own experience.

© 2008 IOP Publishing Ltd

2. Experimental

The phase transitions were studied complementary by single-crystal X-ray diffraction (in laboratory), by X-ray powder diffraction (in laboratory and using a synchrotron source at SNBL, ESRF), by IR- and Raman-spectroscopy, and by optical microscopy. Details of the sample preparation, the types of the diamond anvil cells used, the instrumentation used for data collection and the software used for data processing was described in details in the original publications [18-34].

3. Examples of results.

3.1. Paracetamol

The high-pressure study of paracetamol was, to the best of our knowledge, the first study of a pharmaceutical under hydrostatic pressure by single-crystal X-ray diffraction [19]. The two polymorphs of this compound – the monoclinic (form I) and the orthorhombic (form II) ones – have been known since a very long time (Fig. 1) [35,36].

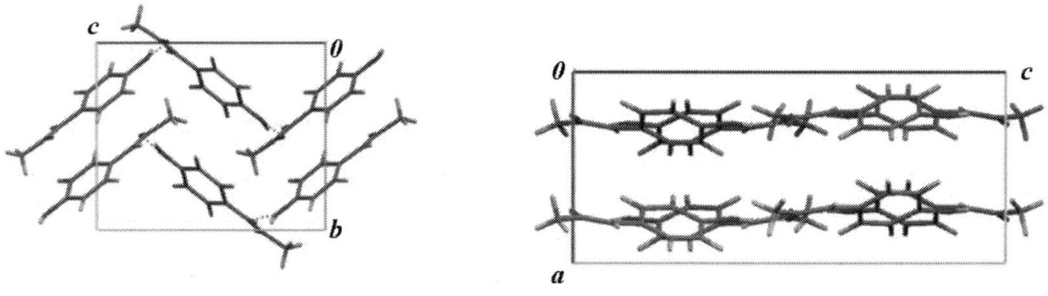

(a) (b)

Figure 1. Fragments of the crystal structures: (a) paracetamol I ($P2_1/n$); (b) paracetamol II (Pcab)

Polymorph I cannot be tabletted without excipients, polymorph II is easily tabletted, ans was supposed to be more compressible [37]. Our X-ray diffraction study has shown, however, that the bulk compressibility of the two forms is the same, but the anisotropy of lattice strain is radically different (Fig. 2) [19,20].

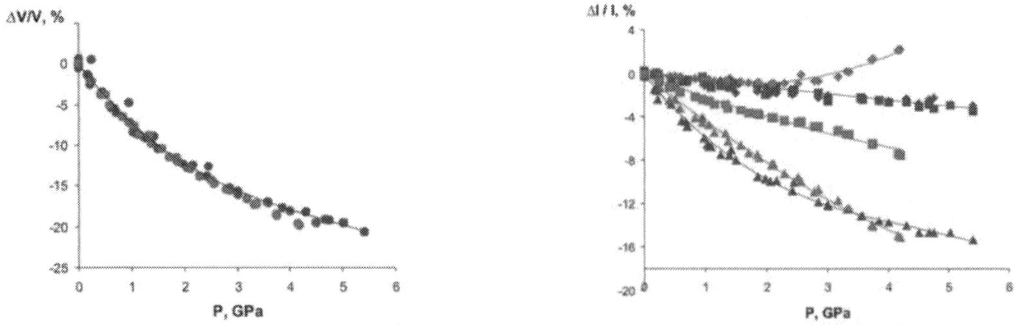

(a) (b)

Figure 2. Relative changes in cell volume (a) and relative linear strain along the principal axes of strain tensor (b) with increasing pressure; red: paracetamol I, blue: paracetamol II [19,20]

On grinding, polymorph II transforms into polymorph I [20]. Our X-ray diffraction experiments in situ have shown, that on hydrostatic compression, on the contrary, paracetamol I transforms into paracetamol II [20]. This transformation could be observed only if powder samples were taken, and not on direct compression, but on decompression down to about 1 GPa from a higher (above 4 GPa) pressure [20]. A few years later, paracetamol II was crystallized directly from solution at high hydrostatic pressure [12]. Theoretical calculations and experimental studies of paracetamol polymorphs at variable temperatures and pressures resulted in a conclusion, that form II becomes the stable polymorph at high pressures [38]. Thus, the transformation of paracetamol II to paracetamol I on grinding accounts for local heating at the contacts and / or shear strain, but not from the local pressure generation at the grain contacts. It is possible to avoid this transformation also during the pharmaceutical processing, if the presence of even traces of moisture in the system is excluded [39].

Since the single crystals of paracetamol I could be compressed without any phase transformations up to 4 GPa, the anisotropy of structural strain could be followed in relation to the pressure-induced changes in the molecular conformations and the compression of the intermolecular hydrogen bonds (Fig. 3) [19,40]. An interesting result is that although all the intermolecular hydrogen bonds are compressed with increasing pressure, the crystal structure *expands* in several crystallographic directions. This expansion is due to the flattening of the molecules and of the pleated molecular layers [19].

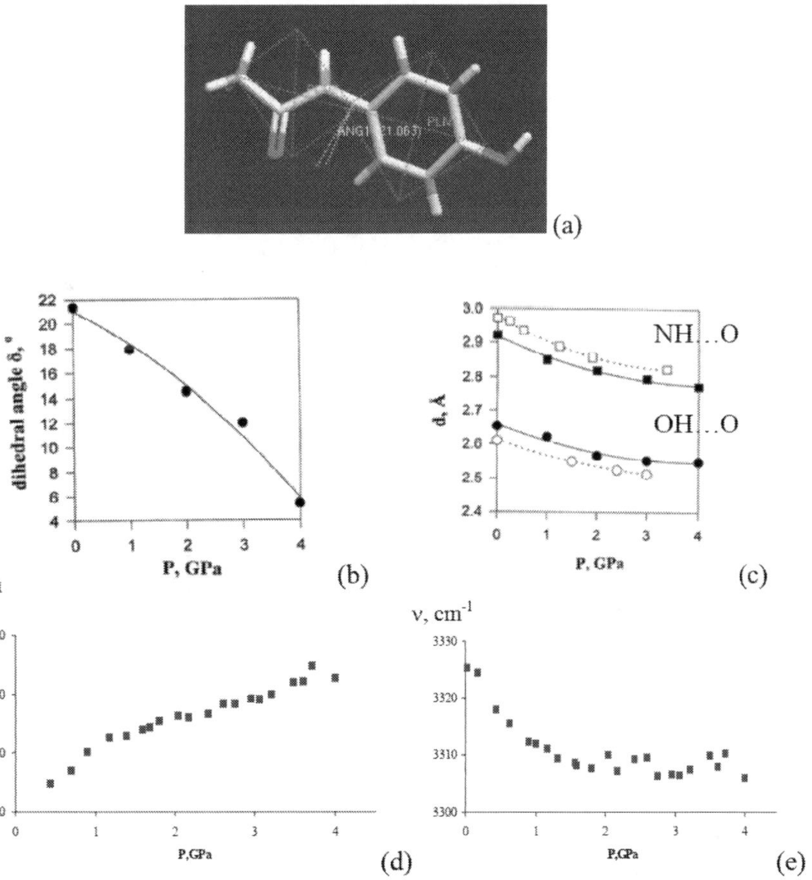

Figure 3. Dihedral angle between the phenyl-ring and the plane of the acetamido-group in paracetamol (a) and its change with pressure (b); shorteining of the distances between non-hydrogen atoms in the NH…OH and OH…O=C hydrogen bonds (c, full symbols) and corresponding shifts of the vibrational frequencies of the ν(OH) (d) and ν(NH) (e) versus pressure [19,20,40]. Open symbols in Fig. 3c correspond to NH…O bonds between NH_3- and NO_2-groups in $[Co(NH_3)_5NO_2]Cl_2$ [18].

The compression of the NH…O and the OH…O hydrogen bonds is comparable, however, it manifests itself differently in the IR-spectra: red shift with increasing pressure was measured for –NH stretching vibrations, whereas the frequency of the –OH stretching vibration shifted to the blue (Fig. 3, d,e). The reason is that the hydroxyl-group is simultaneously a proton donor in the -OH…O=C- hydrogen bond, and a proton acceptor in the –NH…OH hydrogen bond [40].

3.2. Chlorpropamide

The high-pressure study of another solid drug, chlorpropamide (4-chloro-N-((propylamino)-carbonyl)-benzenesulfonamide) (Fig. 5), provides an example of a liquid-mediated phase transition at high pressure.

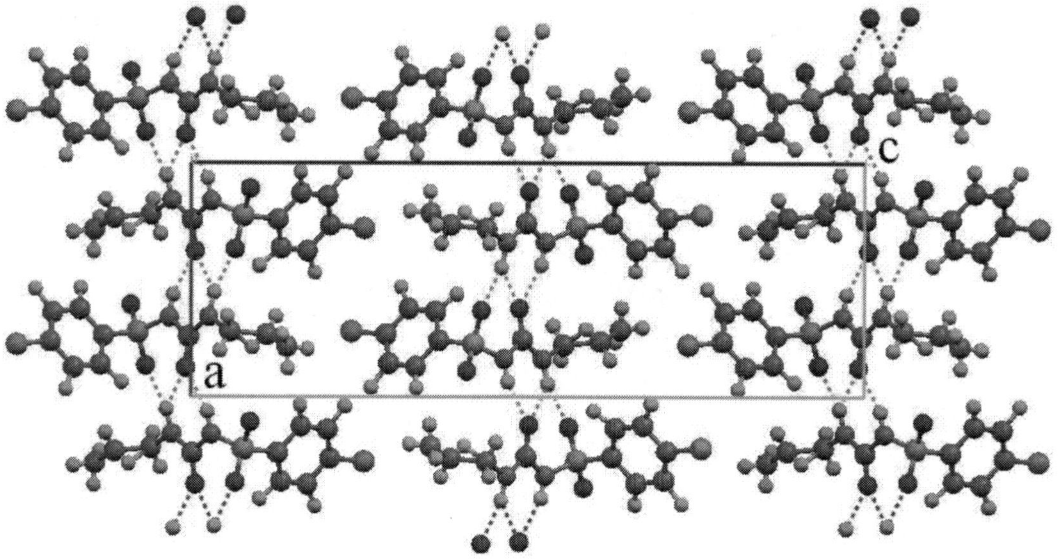

Figure 5. A fragment of the crystal structure of chlorpropamide A. Green – Cl, red – O, blue – N, yellow – S, grey – C, white - H

Chlorpropamide can be considered as a model compound to study polymorphic transformations on tabletting [41-43]. On tabletting, form-A has been reported to transform partially into another polymorph, form-C [41-44]. The density of the form-A (measured using an air comparison pycnometer) is larger, than the density of the form-C (measured in the same way), and this let Otsuka et al. conclude, that pressure increase on tabletting cannot account for the A-C transformation, since the molar volume of form-C is larger, than that of form-A [41,42]. Form-C is known to be formed from any other form of chlorpropamide on heating up to about 115 °C [45-47], and one might suppose, that the reported form-A to form-C transformation on tabletting could be a result of local heating, melting and/or recrystallization [41-43].

In a model study of the effect of hydrostatic pressure on form-A of chlorpropamide by Raman spectroscopy in a diamond anvil cell in situ, no “A-to-C” transformation with increasing pressure was observed, but two polymorphic transformations were claimed to occur at approximately 0.9 and 2.0 GPa [48]. Since the saturated solution of chlorpropamide in ethanol was used as a pressure-transmitting liquid in these experiments, one could not exclude the role of ethanol in these transformations.

In order to test, if any pressure-induced transformations can be induced by compressing a dry powder sample of chlorpropamide form-A, we have monitored the sample compressed in a diamond anvil cell in-situ by high resolution powder X-ray diffraction. In the absence of a pressure-transmitting liquid, obviously no A-C polymorphic transformation could be observed [33]. A polymorphic transformation of chlorpropamide form-A to form-C, that was reported during tabletting [41-43], must therefore be due to local heating effects. The pressure-induced phase transitions at 0.9 and 2.0 GPa studied by Raman spectroscopy on a sample of chlorpropamide form-A in its saturated ethanol solution [48] also appear to be solvent-mediated. This was confirmed by comparative X-ray diffraction studies of samples of chlorpropamide compressed in the saturated ethanol solution, or without any liquid at all [33,49].

An important conclusion from this study of chlorpropamide in relation to the tabletting problems, is the importance of taking into account, that liquids present in the system may have effect on the polymorphic transformations. Similar examples were described previously. No polymorphic transition could be induced by pressure in a Co(III)-coordination compound in a fluorinated oil, but it was observed easily at a relatively low pressure in alcohol [50]. Examples are described, when the effects of tabletting on a solid drug substance were also solvent-mediated. Thus, no phase transitions were observed on tabletting dry intact powder of indomethacin, but occurred if an ethanol slurry was compressed [51]. One should be especially careful, if the solid is soluble in the pressure-transmitting liquid at ambient conditions, as is the case of chlorpropamide and ethanol. The crystallization of new polymorphs and solvates of drug substances at high hydrostatic pressures has been described in several recent publications [12,14,15,52-54]. A similar process (a re-crystallization) can occur also when pressure induces polymorphic transformations in solids immersed into liquids.

3.3. Glycine

The polymorphs of glycine provide an example of a radical difference in the stability of several polymorphs of the same compound with respect to hydrostatic pressure, and of the pronounced kinetic effects in the polymorphic transformations (Fig. 6) [22-25,32,55-58]. The α-polymorph is stable at least up to 23 GPa (the highest pressure achieved in Raman spectroscopy experiments) [55], the β-polymorph transforms reversibly and without fragmentation of the single crystal into a high-pressure β′-polymorph at relatively low pressure, 0.76 GPa [24,56], the γ-polymorph transforms irreversibly into another high-pressure form, the δ-polymorph in a wide pressure range [23,25,32,56,58], and with a strong fragmentation of the starting single crystals [23,25,32,56]. On decompression, the δ-form is preserved down to 0.6 GPa, and then it transforms into another high-pressure form, the ζ-form, which is structurally related to other layered polytypes of glycine, such as the α-, the -β, the β′-, and the δ-forms [32].

The structural changes in the polymorphs of glycine with increasing pressure and on decompression were followed directly by X-ray diffraction [22,23,25,56,57], and manifest themselves also very well in Raman spectra [24,32] and IINS spectra [58]. For example, the frequency of the NH-stretching vibration in the γ-polymorph shifts to the red with increasing pressure as this phase is compressed, then increases jump-wise at about 100 cm^{-1}, when some of the hydrogen bonds formed by these groups in the crystal structure of the γ-glycine break during the phase transition into the δ-form, and the structure, in which the triple helices are linked with each other in a three-dimensional network, transforms into a layered structure (Fig. 7) [32]. The pressure-induced γ-to-δ- phase transition can be compared with the irreversible conformational transformations of collagen and other fibrillar proteins [59].

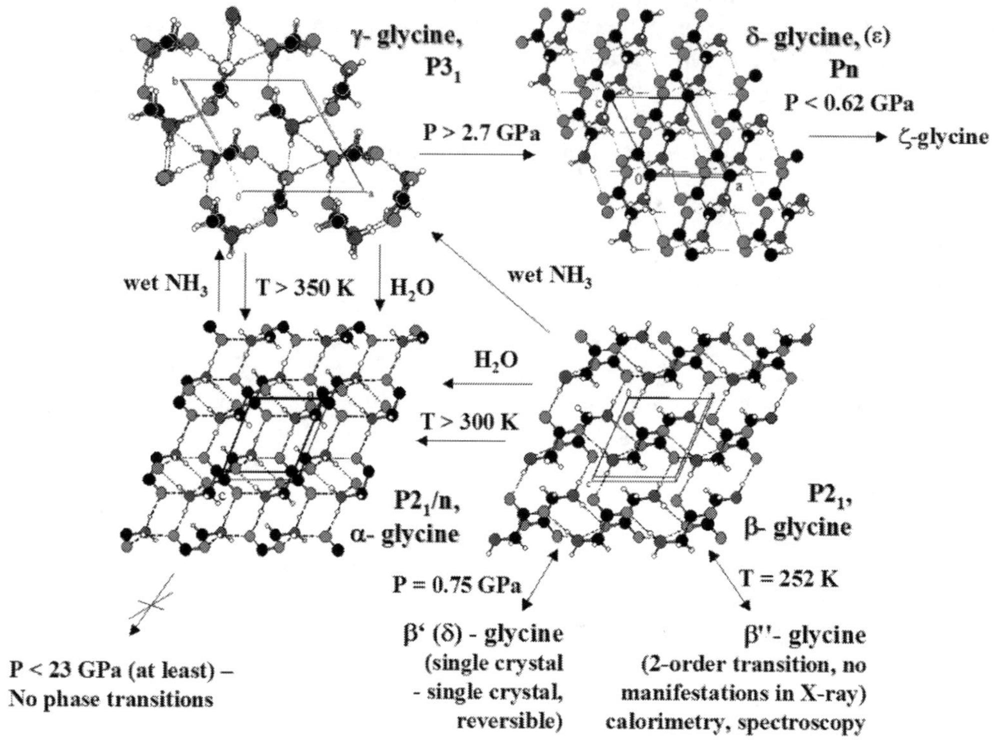

Figure 6. Schematic presentation of transitions between polymorphs of glycine [22-25,55,32]. Notation in brackets was suggested later in [56].

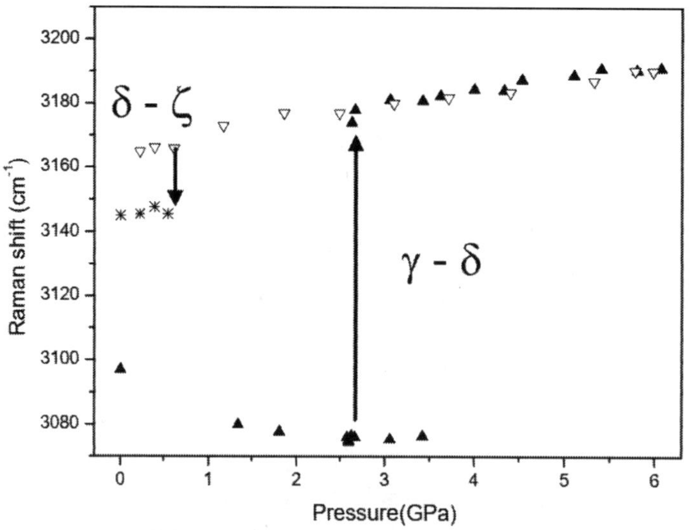

Figure 7. Frequency shifts of the ν(NH)-stretching vibration versus pressure, corresponding to the phase transitions of the γ-glycine into the δ-glycine on compression, and of the δ-glycine to ζ-glycine on decompression [32]

3.4. L- and DL-serine

The effect of pressure on the chiral and the racemic crystalline forms of another amino acid, serine, is very different. The bulk compressibility of the two crystals is the same up to about 5 GPa [27,30], but at higher pressures DL-serine is preserved as the same polymorph, whereas L-serine undergoes a cascade type reversible phase transition, during which an interface propagates rapidly through a crystal, which remains intact [26,60]. The space group symmetry of L-serine does not change on the transformation of L-serine-I into the L-serine-II, but the cell volume and the cell parameters change jumpwise, due to a jumpwise rotation of the side $-CH_2OH$ chains. At a higher pressure (8 GPa) one more cascade type phase transition from L-serine-II into the L-serine-III occurs, during which the side $-CH_2OH$ chains rotate cooperatively again (Fig. 8) [26,29,31,61]. The reversible polymorphic transitions in the crystals of L-serine at high pressures can be compared with the functioning of serine-serine zippers [62]. Pressure-induced reversible single-crystal to single-crystal isosymmetric phase transition resulting from a cooperative jumpwise rotation of oxalate-ions was observed recently in sodium oxalate [28].

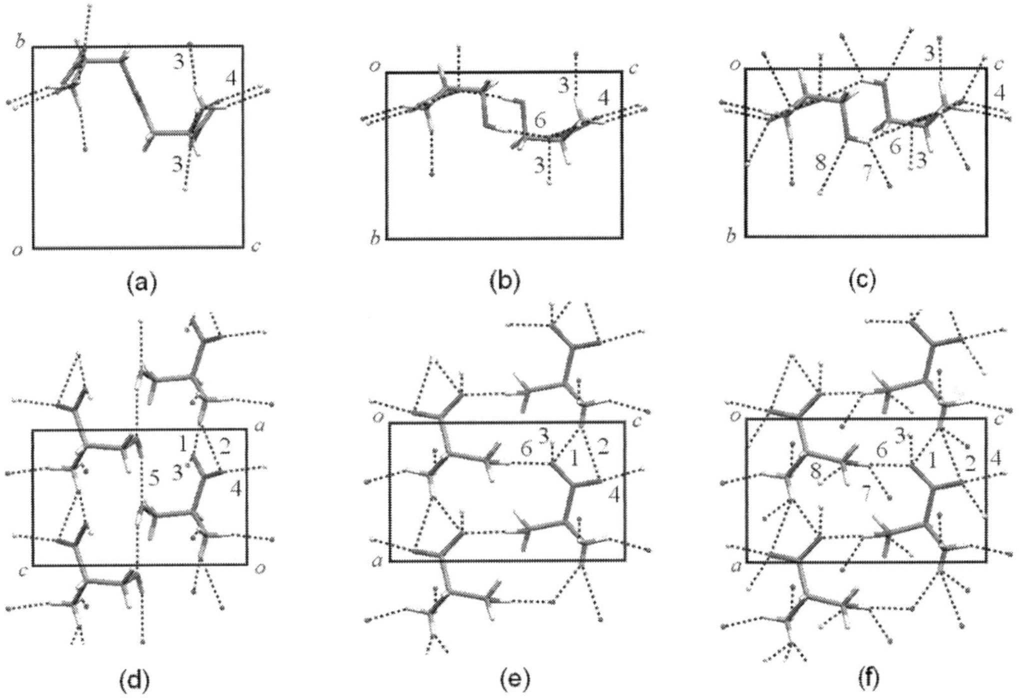

Figure 8. Fragments of the crystal structures of the polymorphs of L-serine: I at ambient pressure (a,d), II at 5.4 GPa (b, e), III at 8.0 GPa (c,f) [29]. Hydrogen bonds to neighboring molecules are shown by dashed lines, numeration of hydrogen bonds refers to table Tables 1.2 in reference 29.

The structure of L-serine-III obtained by compressing a powder sample could be well described, assuming the single-crystal structural model [31] with a superstructural tripling of ***a*** and ***c*** unit cell parameters [32]. Interestingly enough, a neutron powder diffraction study of the completely deuterated L-serine sample at pressures up to 8 GPa [61] gave the same "basic" structural model for the high-pressure phase III, as the single-crystal study [31], or the X-ray powder diffraction study [32], but did not reveal either the co-existence of several phases in the sample, or the formation of any super- or nanostructures [61]. It is difficult to judge, if the origin in this discrepancy is in the different choice of the techniques, or of the different samples – the neutron diffraction patterns [61] are rather noisy in the regions, where extra weak peaks were observed in the X-ray synchrotron diffraction spectra collected from a sample in a specially designed DAC without Be-background [63,64]. One can also suppose, that deuterated and non-deuterated samples may behave slightly differently, although the pressures reported for the two transitions in single crystals and powder samples are in a good agreement for deuterated and non-deuterated samples. One can also expect "simply" an irreproducible formation of the metastable non-equilibrium superstructures / nanostructures, sensitive to subtle changes in the sample characteristics and the compression conditions. This highlights once again the importance of kinetic factors and the role they play in obtaining reproducible results and complete phase transitions.

3.5. β-Alanine

An interesting behavior was observed recently for β-alanine (Fig. 9): the crystals of the ambient-pressure form transformed into a structurally related polymorph sharply at about 5.5 GPa for single crystals, and in an extended pressure range for powder samples. The single crystals were preserved, and the transformation was reversible, but only if the sample was first compressed in small (0.5 GPa) steps up to 8 GPa and then decompressed in similar steps down to ambient conditions within a day; if the sample was compressed up to 5.5 GPa and was kept at this pressure for about three days, another phase transition occured, giving a different high-pressure form, which was preserved on decompression down to 1.6 GPa, and then converted back to the ambient-pressure structure of β-alanine, and the latter transformation resulted in the fragmentation of single crystals giving a fine powder sample [34]. The processes could be followed both by Raman spectroscopy and by X-ray diffraction. As an example, see the changes in the Raman spectra shown in Fig. 10.

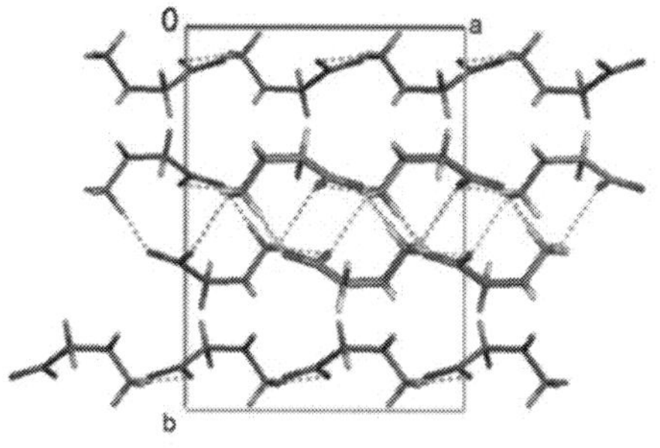

Figure 9. A fragment of the crystal structure of β-alanine

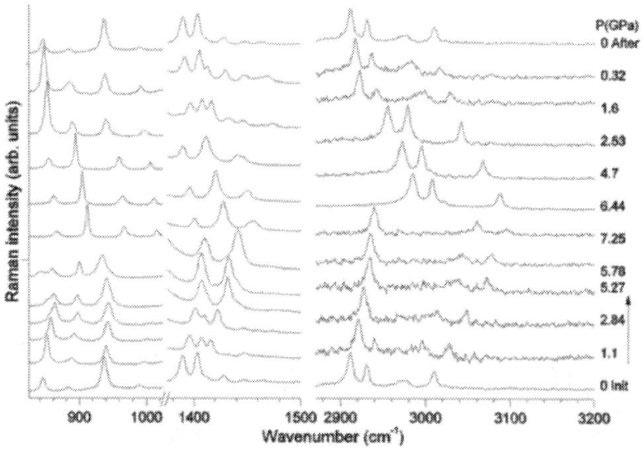

Figure 10. Selected fragments of the Raman spectra of β-alanine on compression and decompression. The sample was kept several days at 6.4 GPa [34]

4. Conclusions

The selected examples from our recent experience aim to show, how complex and fascinating the phase transitions induced by very moderate pressures (below 8-10 GPa) in the crystals of organic small molecules can be. Kinetic factors play an enormous role in these transformations, and therefore the polymorph formed at a particular P-T conditions depends strongly not on the equilibrium phase diagram only, but to no less extent from the pre-histrory of the sample (the choice of a starting polymorph), the hydrostatic liquid, the rate of compression and the details of the compression-decompression procedure [16]. Since the symmetry of the crystal structures is low, the space group is often preserved even when molecules change their conformation, shift, or rotate, and only a detailed study at multiple pressure points makes it possible to distinguish reliably between an isosymmetrical phase transition and an anisotropic compression of the same phase [8].

References

[1] Katrusiak, A. *Cryst. Res. Techn.* 1991, *26*, 523-531.
[2] Katrusiak, A. *Crystallogr. Rev.* 1996, *5*, 133–180.
[3] Hemley, R.J. & Dera, P. In: Reviews in Mineralogy and Geochemistry, 41, High-Temperature and High-Pressure Crystal Chemistry, Mineralogical Society of America, 2000, 335-419.
[4] Katrusiak, A. In: Frontiers of High Pressure Research II: Application of High Pressure to Low-Dimensional Novel Electronic Materials, Kluwer Academic Publishers, Dordrecht 2001 pp 73–85.
[5] Boldyreva, E.V. *J. Molec. Struct.* 2003, *647*, 159-179.
[6] Katrusiak, A. & McMillan, P.F., Eds.; *High-Pressure Crystallography*, Kluwer: Dordrecht, 2004.
[7] Boldyreva, E.V. *Russ. Chem. Bulletin* 2004, *7*, 1315-1324.
[8] Boldyreva, E.V. *Cryst. Engineering* 2004, *6*(4), 235-254.
[9] Boldyreva, E.V. *J. Mol. Struct.* 2004, *700*(1-3), 151-155.
[10] Boldyreva, E.V.; Drebushchak, T.N.; Shakhtshneider, T.P.; Sowa, H.; Ahsbahs, H.; Goryainov, S.V.; Ivashevskaya, S.N; Kolesnik, E.N.; Drebushchak, V.A.; Burgina, E.B. *Arkivoc* 2004, *XII*, 128-155.
[11] Katrusiak, A. In: International School of Crystallography. Diversity amidst Similarity: A Multidisciplinary Approach to Polymorphs, Solvates and Phase Relationships. 35th Course, a Nato ASI and EuroSummerSchool, 2004, Erice, Lecture Notes, 174–184.
[12] Fabbiani, F.P.A.; Allan, D.R.; David, W.I.F.; Moggach, S.A.; Parsons, S.; Pulham, C.R. *CrystEngComm*. 2004, *6*, 504-511.
[13] Boldyreva, E.V. In: Proceedings of the IV-th International Conference on High Pressures in Biosciences and Biotechnology, Tsukuba, Japan, September 25-26, 2006; Abe, F., Suzuki, A., Eds., Presented by J-STAGE: Saitama, Japan, 2006, Vol.1, No. 1, pp. 28-46.
[14] Fabbiani, F.P.A.; Pulham, C.R. *Chem. Soc. Rev.* 2006, *35* (10), 932-942.
[15] Fabbiani, F.P.A.; Allan, D.R.; David, W.I.F.; Davidson, A.J.; Lennie, A.R.; Parsons, S.; Pulham, C.R.; Warren, J.E. *Cryst. Growth Design* 2007, *7*(6), 1115-1124.
[16] Boldyreva, E.V. Cryst. Growth Design, 2007, *7(9)*, 1662-1668.
[17] Boldyreva, E.V. In: Boeyens, J.C.A. and Ogilvie, J.F. (Eds.), *Models, Mysteries and Magic of Molecules*, Springer, 2007, 169-194.
[18] Boldyreva, E.V.; Naumov, D.Yu.; Ahsbahs, H. *Acta Cryst.* 1998, *B54*, 798-808.
[19] Boldyreva, E.V.; Shakhtshneider, T.P.; Vasilchenko, M.A.; Ahsbahs, H.; Uchtmann, H. *Acta Cryst.* 2000, *B56* (2), 299-309.
[20] Boldyreva, E.V.; Shakhtshneider, T.P.; Ahsbahs, H. *J. Therm. Analys. Calorim.* 2002, *68*, 437-

452.
[21] Boldyreva, E.V.; Goryainov, S.V.; Smirnov, M.B.; Ahsbahs, H.; Chernyshev, V.V.; H.-P. Weber. *Doklady Phys. Chem.* 2003, *390*, N.6, 787-790 (Rus.), or 154-157 (Engl.).
[22] Boldyreva, E.V.; Ahsbahs, H.; Weber, H.-P., *Z. Kristallogr.* 2003, *218*, 231-236.
[23] Boldyreva, E.V.; Ivashevskaya, S.N.; Sowa, H.; Ahsbahs, H.; Weber, H.-P. *Dokl. Akad. Nauk* 2004, *396*, 358-361.
[24] Goryainov, S.V.; Kolesnik, E.N.; Boldyreva, E.V. *Physica B Condensed Matter* 2005, *357* (3-4), 340-347.
[25] Boldyreva, E.V.; Ivashevskaya, S.N.; Sowa, H.; Ahsbahs, H.; Weber, H.-P. *Z. Kristallogr.* 2005, *220*(1), 50-57.
[26] Kolesnik, E.N.; Goryainov, S.V.; Boldyreva, E.V. *Doklady Phys. Chem.* 2005, *404*, 61-64 (Rus.), or 169-172 (Engl.).
[27] Boldyreva, E.V., Kolesnik, E. N., Drebushchak, T.N., Ahsbahs, H., Beukes, J.A., Weber, H.-P. *Z. Kristallogr.* 2005, 220(1), 58-65.
[28] Boldyreva, E.V.; Ahsbahs, H.; Chernyshev, V.V.; Ivashevskaya, S.N.; Oganov, A.R. *Z. Kristallogr.* 2006, *221*, 186-197.
[29] Boldyreva E.V.; Sowa, H.; Seryotkin, Yu.V.; Drebushchak, T.N.; Ahsbhas, H.; Chernyshev, V.V.; Dmitriev, V.P. *Chem. Phys. Letters* 2006, *429*, 474-478.
[30] Boldyreva,E.V.; Kolesnik, E.N.; Drebushchak, T.N.; Sowa, H.; Ahsbhas, H.; Seryotkin. Yu.V., *Z. Kristallogr.* 2006, *221*, 150-161.
[31] Drebushchak, T.N.; Sowa, H.; Seryotkin, Yu.V.; Boldyreva, E.V., *Acta Cryst.* 2006, *E62*, o4052-o4054.
[32] Goryainov, S.V.; Boldyreva, E.V.; Kolesnik, E.N. *Chem. Phys. Letters* 2006, *419* (4-6), 496-500.
[33] Boldyreva, E.V.; Dmitriev, V.P.; Hancock, B.P. *Int. J. Pharm.* 2006, *327*, 51-57.
[34] Boldyreva, E.V.; Goryainov, S.V.; Seryotkin, Yu.V.; Ahsbahs, H.; Dmitriev, V.P. *Proceed. NSU, Ser. Physics* 2007, *2*(2), 30-35.
[35] Haisa, M.; Kashino, S.; Maeda, H. *Acta Crystallogr.* 1974, *B30*, 2510-2512.
[36] Haisa, M.; Kashino, S.; Kawai, R.; Maeda, H. *Acta Crystallogr.* 1976, *B32*, 1283-1285.
[37] Nichols, G.; Frampton, C. S., *J. Pharm. Sci.* 1998, *87*, 684-692.
[38] Espeau, P.; Ceolin, R.; Tamarit, J.-L.; Perrin, M.-A.; Gauchi, J.-P.; Leveiller, F. *J. Pharm. Sci.* 2005, *94* (3), 524-539.
[39] Boldyreva, E.V.; Drebushchak, V.A.; Paukov, I.E.; Kovalevskaya, Yu.A;, Drebushchak, T.N. *J. Therm. Analys. Calorim.* 2004, *77*, 607-623.
[40] Boldyreva, E.V.; Shakhtshneider, T.P.; Ahsbahs, H.; Uchtmann, H.; Burgina, E.B., Baltakhinov, V.P. *Polish J. Chem.* 2002, *76*, 1333-1346.
[41] Otsuka, M.; Matsumoto, T.; Kaneniwa, N. *J. Pharm. Pharmacol.* 1989, *41*, 665-669.
[42] Matsumoto, T.; Kaneniwa, N.; Higuchi, S.; Otsuka, M. *J. Pharm. Pharmacol.* 1991, *43*, 74-78.
[43] Otsuka, M.; Matsumoto, T.; Higuchi, S.; Otsuka, K.; Kaneniwa, N. *J. Pharm. Sci.* 1995, *84*, 614-618.
[44] Koivisto, M. & Lehto, V.-P. In: Polymorfi, Fysikaalisen farmasian XVII vuosittainen symposium: Vesi farmasiassa, 2006, Abstracts, 28.
[45] Simmons, D.L.; Ranz, R.J.; Gyanchandani. *Canad. J. Pharm. Sci.* 1973, *8*, 125-127.
[46] Burger, A.. *Sci. Pharm.*, 1975, *43*, 152-161.
[47] Al-Saieq, S.S. & Riley, G.S. *Pharm. Acta Helv.*, 1982, *57*, 8-11.
[48] Cao, W. Polymorphic Transformations under Compression. Thesis, 2002, Purdue University.
[49] Boldyreva, E.V.; Dmitriev V.P.; Ahsbahs, H., in preparation.
[50] Boldyreva, E.V.; Ahsbahs, H.; Uchtmann, H.; Kascheeva, N. E. *High Pressure Research* 2000, *17*, 79-99.
[51] Okumura, T.; Ishida, M.; Takayama, K.; Otsuka, M. *J. Pharm. Sci.* 2006, *95*, 689-700
[52] Fabbiani, F.P.A.; Allan, D.R.; Dawson, A.; David, W.I.F.; McGregor, P.A.; Oswald, I.D.H.;

Parsons, S.; Pulham, C.R. *Chem. Comm.* 2003, *9* (24), 3004-3005.

[53] Fabbiani, F.P.A.; Allan, D.R.; Marshall, W.G.; Parsons, S.; Pulham, C.R.; Smith, R.I. *J. Crystal Growth* 2005, *275*, 185-192.

[54] Fabbiani, F.P.A..; Allan, D.R.; Parsons, S.; Pulham, C.R. *Cryst. Eng. Comm.* 2005, *7*, 179-186.

[55] Murli, C.; Sharma, S.M.; Karmakar, S.; Sikka, S.K. *Physica B* 2003, *339*, 23-30.

[56] Dawson, A.; Allan, D.R.; Belmonte, S.A.; Clark, S.J.; David, W.I.F.; McGregor, P.A.; Parsons, S.; Pulham, C.R.; Sawyer, L. *Cryst. Growth Design* 2005, *5*(4), 1415-1427.

[57] Tumanov, N.A., Boldyreva, E.V., Ahsbahs, H., in preparation.

[58] Bordallo, H.; Boldyreva, E.; Koza, M., in preparation.

[59] Pain, R.H. (Ed.). Mechanisms of Protein Folding, Frontiers in Molecular Biology, Ser. Ed. B.D. Hames and D.M. Glover, Oxford University Press, Oxford, 2000.

[60] Moggach, S.A.; Allan, D.R; Morrison, C.A.; Parsons, S.; Sawyer, L. *Acta Cryst.* 2005, *B61*, 58-68.

[61] Moggach, S.A.; Marshall, W.G.; Parsons, S. *Acta Cryst.* 2006, *B62*, 815-825.

[62] Adamian, L, Liang, J. *Proteins* 2002, *47*, 209-218.

[63] Ahsbahs, H. *Z. Kristallogr.* 2004, *219*, 305-308.

[64] Sowa, H. & Ahsbahs, H. *J. Appl. Cryst.* 2006, *39*, 169-175.

Joint 21st AIRAPT and 45th EHPRG Int. Conf. on High Pressure Science and Technology IOP Publishing
Journal of Physics: Conference Series **121** (2008) 022024 doi:10.1088/1742-6596/121/2/022024

Pressure-Induced Phase Transition in PLZT: A Neutron Diffraction and Dielectric Study

George Samara, Eugene Venturini and Bruno Morosin
Sandia National Laboratories, Albuquerque, NM 87185, USA

E-mail: bmorosi@sandia.gov

Abstract. Zr-rich, La-doped lead zirconate titanate ceramic samples with composition near $Pb_{0.99}La_{0.01}(Zr_{0.91}Ti_{0.09})O_3$ (hereafter PLZT) were studied by time-of-flight neutron diffraction and dielectric measurements at 295 K and 1 bar and at 250 K versus increasing and decreasing pressure to 0.55 GPa. The diffraction data at 295 K show that the sample has the rhombohedral ferroelectric *R3c* structure ($F_{R(LT)}$) and remains in that phase upon cooling to 250 K. As pressure is increased at 250 K, a transition is observed above 0.3 GPa to essentially the antiferroelectric $CaTiO_3$ orthorhombic *Pnma* (A_O) phase. There appear two other peaks in the neutron spectra, suggesting a possible incommensurate cell for this A_O. After the initial drop of the $F_{R(LT)}$ content, the transformation remains incomplete to 0.55 GPa with ~10% of the sample retaining the $F_{R(LT)}$ low pressure phase. On pressure release, 91% of the $F_{R(LT)}$ phase is recovered with the remainder remaining in the A_O phase. Isothermal dielectric data at 250 K suggest a structural transition with an onset near 0.37 GPa for increasing pressure, consistent with the diffraction results.

1. Introduction

Lead lanthanum zirconate titanate ceramics have had a long history for their optical properties.[1] These ceramics are doped members of the perovskites $PbZrO_3$ and $PbTiO_3$ which form the solid solution series $PbZr_{1-x}Ti_xO_3$ over the entire composition range. The undoped solid solutions, commonly called PZTs, are rich in the variety of ferroelectric and nonferroelectric transitions that can be induced in them by variations in composition, temperature, electric field, or pressure (stress).[2] They also exhibit large spontaneous polarizations and piezoelectric coefficients. These factors along with their availability in high-quality ceramic form are responsible for their widespread technological usage in applications based on piezoelectric and ferroelectric properties.[2,3]

These perovskites are paraelectric (PE) and in the ideal cubic perovskite structure at high-temperature (T). On cooling, this phase transforms either to a rhombohedral ferroelectric (FE) phase or to an orthorhombic (*Pbam*) antiferroelectric (AFE), A_O. The rhombohedral ferroelectric region usually is divided into two phases: a high-T phase of symmetry *R3m* ($F_{R(HT)}$) and a low-T phase of symmetry *R3c* ($F_{R(LT)}$), which has a unit cell double that of the *R3m* phase. For certain PZT compositions this phase is induced by pressure (P).

Doping with other metals such as Nb enhances various processing parameters as well as electrical and dielectric properties, particularly the various phases the materials undergo on either increasing P or decreasing T. One such ceramic of composition $Pb_{0.99}(Zr_{0.95}Ti_{0.05})_{0.98}Nb_{0.02}O_3$, known as PZT95/5-2Nb was the subject of our recent study on the P and T induced changes of its crystal structure and dielectric properties.[4] The positions in the phase diagram of the $F_{R(HT)}$-$F_{R(LT)}$ and $F_{R(LT)}$-A_O transition and the triple point are shifted dramatically from that of pure PZT. Most, if not all, of these phase transitions are known to be associated with lattice dynamical instabilities, or soft phonon modes. Small ionic displacements accompany such transitions, and hence they are often called displacive transitions.

With significant La doping, ceramics of high optical clarity with electrical controlled transmission are possible.[1] PLZT ceramics with a composition near $Pb_{0.91}La_{0.09}(Zr_{0.65}Ti_{0.35})O_3$ are transparent with low optical loss for wavelengths from 400 to 6500 nm.[5] Many aspects of the various P and T-induced transitions in the PLZTs, including the details of the atomic displacement and changes in lattice parameters involved, remain unresolved. This paper reports the structure and dielectric

© 2008 IOP Publishing Ltd

behaviour of our PLZT, $Pb_{0.99}La_{0.01}(Zr_{0.91}Ti_{0.09})O_3$, primarily at 250 K, and compares it with results previously determined on PZT95/5-2Nb.[4]

2. Experimental Details

2.1. *Sample preparation.*

Oxide powders were precipitated from liquid solutions containing the appropriate quantities of Zr, Ti, Pb, and La ions, with 3.75% PbO excess to account for the partial evaporation of Pb during calcining and the subsequent sintering process. The powders were filtered and dried, and then calcined at about 1175 K for 16 hours. The powder was mixed with organic binders (2 wt% acrylic binder solution) to facilitate pressing, dried, pressed into billets and fired to remove the organic additives. The billets were sintered at about 1625 K for 6 hours in the presence of calcined powder of composition similar to that of the compact in a double-crucible configuration in order to minimize Pb loss to form the final ceramic.

2.2. *Neutron powder diffraction measurements*

Time-of-flight neutron powder diffraction data were collected on the Special Environment Powder Diffractometer (SEPD) [6] at the Intense Pulsed Neutron Source (IPNS) at Argonne National Laboratory. Only limited diffraction data were collected as a function of hydrostatic P (0 - 0.55 GPa) at 250 K after initial measurements at ambient. Hydrostatic P was generated with a helium gas P cell, and cooling was achieved using a closed-cycle helium refrigerator. The ±90° detector bank data were analyzed using the Rietveld refinement method with the GSAS (EXPGUI) suite [7,8]. The composition was fixed to the nominal one $Pb_{0.99}La_{0.01}(Zr_{0.91}Ti_{0.09})O_3$. Neutron scattering lengths used were Pb: 9.405, Zr: 7.16, Ti: -3.438, La: 0.83, and O: 5.803 fm. In the course of refining the structure parameters above the phase transition, it was apparent that the transition involved a different high P orthorhombic phase than that of *Pbam* observed in our PZT95/5-2Nb study.[4] Various known structure types were attempted before the more satisfactory fit was obtained employing a cell similar to that of antiferroelectric $CaTiO_3$, of orthorhombic *Pnma* symmetry for the A_O phase. In addition the possibility of an incommensurate cell was evident by two extra peaks, suggesting results similar to those of published HRTEM studies [9] on similar compositions. The latter were tentatively interpreted as a ferroelectric *R3c* to an incommensurate antiferroelectric phase.

2.3. *Dielectric Measurements*

The real (ε') and imaginary (ε'' or dielectric loss tanδ = ε''/ε') components of the dielectric response of our PLZT ceramic were measured versus T (250-400 K) and P (0-0.55 GPa) with an Agilent Model 4284A Precision LCR meter at frequencies from 10^2 to 10^6 Hz. The dielectric sample was a thin plate, 0.3x0.3x0.08 cm^3, with silver contacts fired on the broad faces. The P cell used helium gas as the P fluid and was placed inside a cryogenic dewar. The P was monitored by the change in resistance of a calibrated manganin coil, and the T was determined with a chromel/alumel thermocouple mounted adjacent to the sample holder. Typical AC voltages (<10 V/cm) generate a linear dielectric response and are insufficient to move any domain boundaries of the ferroelectric phases.

3. Results and Discussion

3.1. *Neutron Diffraction*

The diffraction study at 295 K and 1 bar shows that the sample is the rhombohedral ferroelectric *R3c* phase ($F_{R(LT)}$) and remains in that phase upon cooling to 250 K. As P is increased at 250 K, a transition is observed just above 0.3 GPa to essentially an antiferroelectric $CaTiO_3$ orthorhombic *Pnma* (A_O) phase, which because of the appearance of two other peaks in the spectra, may possibly be an incommensurate cell. Refined lattice and positional parameters and the weight fractions, f, for the

Table 1. Refined structural parameters for $F_{R(LT)}$ and A_O phases in ceramic sample of PLZT at 250 K. For $F_{R(LT)}$, space group *R3c* (No. 161), with Pb (& La) at 6a (0,0,z), Zr/Ti at 6a (0,0,z), and O at 18b (x,y,z = 1/12). For A_O, space group *Pnma* (No. 62), with Pb (& La) at 4c (x,1/4,z), Zr/Ti at 4b (0,0,1/2), O1 at 4c (x,1/4,z), and O2 at 8d (x,y,z). Numbers in parentheses are statistical standard deviations of the last significant digit; * denotes decreasing P from 0.552 GPa.

P, GPa	0	0.310	0.552	0.300*	0*
f(*R3c*)	1.0	1.0	0.13	0.13	0.92
R3c a, Å	5.83856(4)	5.8336(5)	5.8275(8)	5.84228(4)	5.8395(5)
R3c c, Å	14.4143(2)	14.375(4)	14.356(3)	14.416(2)	14.3970(2)
z(Pb)	0.2169(2)	0.2173(6)	0.2121(9)	0.2117(9)	0.2168(2)
z(Zr/Ti)	-.01352(6)	-.01493(9)	-.0149(9)	-.01550(9)	-.01331(3)
x(O)	0.1281(4)	0.1137(4)	0.107(3)	0.1131(3)	0.1275(3)
y(O)	0.3169(5)	0.3090(8)	0.293(3)	0.298(3)	0.3173(4)
f(*Pnma*)	0.0	0.0	0.87	0.87	0.08
Pnma a, Å			5.8451(6)	5.8412(6)	5.8565(8)
Pnma b, Å			8.1804(3)	8.1860(3)	8.1909(9)
Pnma c, Å			5.8540(7)	5.8565(7)	5.8649(7)
x(Pb)			0.0002(9)	0.0001(9)	0.010(7)
z(Pb)			-.0011(9)	-.0022(9)	0.012(9)
x(O1)			0.484(2)	0.483(2)	0.458(5)
z(O1)			0.0531(9)	0.053(4)	0.081(5)
x(O2)			0.253(9)	0.253(9)	0.254(7)
y(O2)			0.026(4)	0.026(4)	0.015(4)
z(O2)			-.263(3)	-.263(3)	-.243(6)

$F_{R(LT)}$ and A_O phases at 250 K before the transition and at maximum P as well as the corresponding results on P decrease are given in Table 1. Figure 1 shows the $F_{R(LT)}$ - A_O phase transition deduced from neutron diffraction data on a stack of ceramic disks cut from the same pellet as the thin disks used for the dielectric measurements. Plotted in Figure 1 is the weight fraction, f, of the $F_{R(LT)}$ phase as a function of P at 250 K as well as the initial ambient 298 K point. It is seen that the transition in the ceramic is not sharp and occurs over the range of ~3 to 0.55 GPa and that there is sufficient hysteresis as P is decreased that the majority of the A_O phase transforms below 0.3 GPa back to the $F_{R(LT)}$ phase. (On increasing P, a "back" smooth curve interpolation would yields ~0.35GPa for the phase transition onset.) This is in contrast to the much sharper transition observed in our previous PZT95/5-2Nb study.

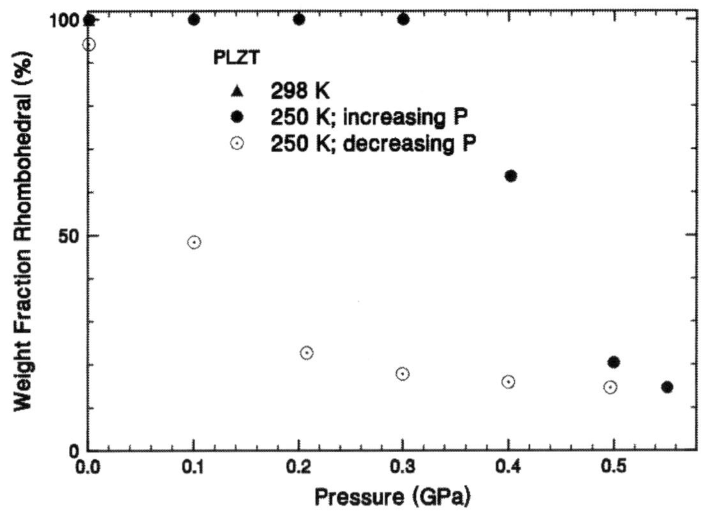

Figure 1. Weight fraction of $F_{R(LT)}$ phase as a function of increasing hydrostatic pressure at 250 K (filled circles). The open symbols with dot are for decreasing pressure. Ambient datum at 298 K (filled triangle). The remaining weight fraction is the A_O phase.

Furthermore, the present high P phase is a different high P orthorhombic phase than that of Pbam observed in PZT95/5-2Nb. Just as in the latter case after the initial drop of the $F_{R(LT)}$ content, the transformation remains incomplete to 0.55 GPa with ~10% of the sample remains in the $F_{R(LT)}$ low P phase. On P release, 91% of the $F_{R(LT)}$ phase is recovered with the remainder remaining in the A_O phase, an amount larger than in PZT95/5-2Nb.[4] (These data points also fit a smooth curve toward ambient P.) There is a similar large hysteresis in the phase recovery.

The value of the *a* and *c* cell parameters (hexagonal basis) of the $F_{R(LT)}$ phase are given in Figures 2. Dashed lines through filled circles are used for 250 K data with increasing P; solid lines through circles with dot for decreasing P. The corresponding 295 K, ambient P data are filled triangles. Note that the *a* cell parameter is particularly non-linear, as though the ceramic sample might be "clamped" more along that direction than perpendicular to it (along the *c* direction). Such clamping was also an issue in our previous PZT95/5-2Nb study.

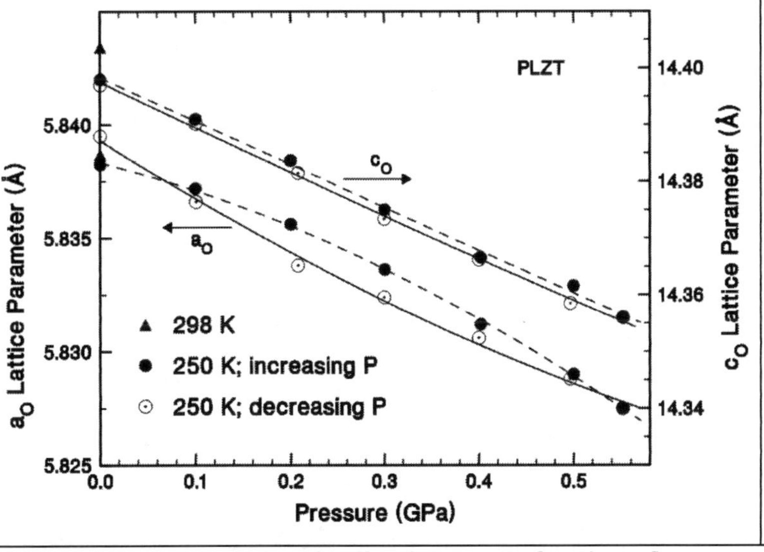

Figure 2. Relative change in the *a* and *c* cell parameters (hexagonal basis) of the $F_{R(LT)}$ phase. Dashed lines through filled circles are used for 250 K data with increasing pressure; solid lines through circles with dot for decreasing pressure. The corresponding 295 K, ambient P datum is a filled triangle.

Figure 3 shows the reduced cell volume as a function of pressure for the $F_{R(LT)}$ and A_O phases. Dashed/solid lines go through 250 K values (solid circles/circles with dots for increasing/decreasing P)

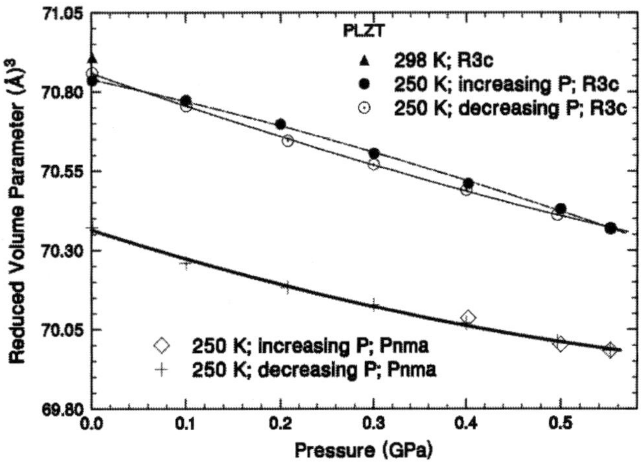

Figure 3. Reduced cell volume as a function of pressure. For the $F_{R(LT)}$ phase, dashed line (solid circles) through 250 K values) for increasing P; solid line for decreasing P values (circles with dots). For the A_O phase, dotted wide line is through 250 K values (open diamonds/plus signs used for increasing/decreasing P). Ambient shown as a filled triangle. Note the smaller cell for the A_O phase compared to the $F_{R(LT)}$ phase.

for the $F_{R(LT)}$ phase; for the A_O phase, heavy dashed line is through the corresponding T values (open diamonds/plus signs for increasing/decreasing P). Note the smaller cell for the A_O phase compared to the $F_{R(LT)}$ phase. The 0.7% volume contraction at the $F_{R(LT)}$ -to- A_O transition is of the same order as the 1% contraction observed in PZT95/5-2Nb.[4] The ambient value for the reduced cell volume is shown as a filled triangle.

3.2. *Dielectric Properties*

The isothermal data in Figure 4 show the in-phase dielectric constant ε' at 250 K and 10 kHz versus increasing and subsequent decreasing P. Increasing P produces a gradual rise in ε', reaching a maximum near 0.37(1) GPa; a similar increase in ε' with increasing P is observed for PZT95/5-2Nb ceramic samples in the $F_{R(LT)}$ phase. Dielectric data at other frequencies form parallel curves with no dispersion in the P for the ε' maximum. Based on the neutron diffraction results in Figure 1, 0.37 GPa is identified as the onset of the A_O phase with increasing P. The decrease in ε' above 0.37 GPa continues to the highest P of 0.57 GPa where the neutron data show that over 10% by weight of the sample remains in the $F_{R(LT)}$ phase.

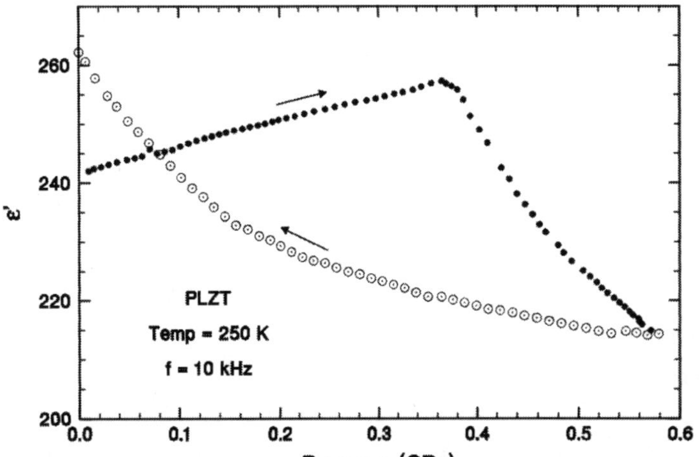

Figure 4. Dielectric constant ε' for a PLZT ceramic sample at 250 K and 10 kHz versus increasing (solid circles) and decreasing (open circles) pressure.

With decreasing P there is a gradual, continuous rise in ε' with no evidence for a maximum. Again, consistent with the neutron diffraction results in Figure1, these data reflect a continuous transformation of the A_O fraction of the sample back to the $F_{R(LT)}$ phase. However, part of the increase in ε' with decreasing P is attributed to the P dependence of ε' within the A_O phase. In contrast to an increase in ε' with increasing P in the ferroelectric $F_{R(LT)}$ phase, ε' is expected to show a decrease with increasing P in the antiferroelectric A_O phase based on dielectric data for PZT95/5-2Nb. Also note that the initial value of ε' at 10 kHz and 0.01 GPa in this isothermal experiment is ~242, while the final value at the same frequency and P is nearly 10% larger. This increase is attributed to a different ferroelectric domain structure in the recovered $F_{R(LT)}$ phase as well as residual internal strains (i.e., the clamping effect discussed above) following the P excursion. Hence, it is not possible to separate these effects and estimate the fraction of A_O phase from the dielectric data for either increasing or decreasing P.

Prior to the isothermal experiment at 250 K the dielectric sample was "annealed" by heating to 400 K (maximum T for the gas pressure cell) and then cooling to 250 K under ~0.01 GPa. During heating the dielectric data indicated a transition from the $F_{R(LT)}$ to the $F_{R(HT)}$ phase at 248(2) K; during cooling the reverse transition occurred at 242(2) K. Similar thermal hysteresis is observed for PZT95/5-2Nb ceramic samples, where this transition occurs at lower T. A second 400 K anneal at 0.01 GPa after the 250 K P experiment restored ε' to its original value of ~242 at 250 K.

4. Concluding Remarks

The P-induced $F_{R(LT)}$-A_O phase transition in PLZT has been shown to differ from that in PZT95/5-2Nb. Initial details of the ionic displacements accompanying the transformation have been determined. The present study shows the degree of incompleteness and irreversibility of the transformation, possibly resulting from strains experienced by the untransformed $F_{R(LT)}$ grains that exist as a minority phase in the high P A_O phase. However, these strains are not as large as that determined in our previous study of the PZT95/5-2Nb material due to the much smaller reduced volume difference between the $F_{R(LT)}$ and A_O phases.

Isothermal dielectric data to 1.5 GPar near room T (not shown) suggest two transitions with increasing P; based on published HRTEM studies[9] on similar compositions as a function of T (but P data are not available), these transitions are tentatively identified as ferroelectric *R3c* to an incommensurate antiferroelectric phase, then to an orthorhombic antiferroelectric phase at higher T. Though we have some dielectric P data at 298 K which would be consistent with such a latter phase transition near 1.0 GPa, this would exceed the current P maximum of our neutron diffraction cell. Even though in general decreasing T acts as increased P, at 250 K no evidence for this transition is seen to the current limits of either the neutron or dielectric data taken to date.

Acknowledgements: The following are acknowledged: Pin Yang and Chris DiAntonio for samples and discussions; Bruce Tuttle for discussion and support.

The work at Sandia National Laboratories was supported by the U.S. Department of Energy's National Nuclear Security Administration. Sandia is operated by Sandia Corporation, a Lockheed Martin Company, for the Department of Energy under Contract DE-AC04-94AL85000. Data were collected at IPNS, Argonne National Laboratory, which is supported by the U.S. Department of Energy, Office of Science, Contract No. W-31-109-ENG-38.

5. References

[1] Haertling G H and Land C E 1971 *J. Am. Ceram. Soc.* **54** 1
[2] Jaffe B, Cook, Jr. W R and Jaffe H 1971 *Piezoelectric Ceramics* (New York: Academic)
[3] Samara G A 1979 *High Pressure Science and Technology (Vol. 1)* ed K D Timmerhaus and. S. Barber, (New York: Plenum) p 177
[4] Avdeev M, Jorgensen J D, Short S, Samara G A, Venturini E L, Yang P and Morosin B 2006 *Phys Rev* **B73** 064105; references there in.
[5] Haertling, G H 1987 *Ferroelectrics* **75** 25
[6] Jorgensen J D, Faber Jr. J, Carpenter J M, Crawford R K, Haumann J R, Hittermann R L, Kleb R, Ostrowski G E, Rotella F J and Worlton T G 1989 *J. Appl. Crystallogr.* **22**, 321
[7] Larson A C and Von Dreele R B 1994 *Los Alamos Natl. Lab., LAUR* 86-748
[8] Toby B H 2001 *J. Appl. Cryst.* **34**, 210
[9] Dai X, Xu Z, Li J F and Viehland D 1996 *J. Mater. Res.* **11** 618

Journal of Physics: Conference Series **121** (2008) 022025 doi:10.1088/1742-6596/121/2/022025

Equation of state and phase diagram of tin at high pressures

K V Khishchenko

Joint Institute for High Temperatures, Russian Academy of Sciences, Izhorskaya 13/19, Moscow 125412, Russia

E-mail: konst@ihed.ras.ru

Abstract. A thermodynamic model of the free energy as a function of volume and temperature for tin with taking into account the polymorphic transformation and melting effects is proposed. Results of equation-of-state and phase-diagram parameters calculations for this metal are compared with available experimental data on static and shock compression at high pressures.

1. Introduction

Description of the thermodynamic properties and phase transitions of matter over a wide range of densities and temperatures is of essential interest for shock-wave physics [1–3].

In the present work a new multi-phase equation of state (EOS) for tin is presented. This EOS takes into account the polymorphic transformation and melting effects. Calculated results are compared with available experimental data at high pressures. The most essential static and shock compression experiments are described.

2. EOS model

The specific free energy of the substance is considered as a sum of three components [1],

$$F(V,T) = F_c(V) + F_a(V,T) + F_e(V,T), \tag{1}$$

describing the cold part of interparticle interaction at $T = 0$ K (F_c) and the thermal contributions by heavy particles (atoms, ions, nuclei) and electrons (F_a and F_e respectively). The first and second components of equation (1) have different forms for the solid and liquid phases, but the third is defined identically as in the EOS model [2, 4].

The cold interaction energy for each solid phase in the compression region ($\sigma_c \geqslant 1$, where $\sigma_c = V_{0c}/V$, V_{0c} is the specific volume at $P = 0$ and $T = 0$ K for the phase in question) is given by the relation [5],

$$F_c(V) = E_{0c} + a_0 V_{0c} \ln \sigma_c - 3V_{0c} \sum_{i=1}^{6} \frac{a_i}{i} \left({\sigma_c}^{-i/3} - 1 \right) + 3V_{0c} \sum_{i=1}^{2} \frac{b_i}{i} \left({\sigma_c}^{i/3} - 1 \right), \tag{2}$$

providing for the condition $F_c(V_{0c}) = E_{0c}$. As can be readily seen, differentiation of the energy (2) with respect to volume yields an equation for the pressure $P_c(V) = -\mathrm{d}F_c/\mathrm{d}V$, which is

© 2008 IOP Publishing Ltd

analogous to the relation proposed previously [6] as an expansion of the Thomas–Fermi model in powers of the atomic cell radius $r_c \sim {\sigma_c}^{-1/3}$.

The value of the coefficient b_2 in equation (2) is determined from the condition of coincidence with the model of degenerate ideal Fermi-gas of nonrelativistic electrons in the range of compressions above $\sigma_c \sim 10^3 \div 10^4$ [5]. In order to determine the coefficients b_1 and a_i in equation (2), one must solve the problem of minimization of the root-mean-square deviation of pressure at some points $V = V_n$, $n = 1, \ldots, N$, in the interval $\sigma_c = 50 \div 10^3$ from the results of calculation by the Thomas–Fermi model with corrections [7] subject to the conditions for the pressure, bulk modulus and its derivative with respect to pressure at $\sigma_c = 1$, $P_c(V_{0c}) = 0$, $B_c(V_{0c}) = -V\mathrm{d}P_c/\mathrm{d}V = B_{0c}$ and $B_c'(V_{0c}) = \mathrm{d}B_c/\mathrm{d}P_c = B_{0c}'$ respectively. The problem of conditional minimization is solved with the introduction of Lagrange factors [8]. The values of the parameters V_{0c}, B_{0c} and B_{0c}' are fitted by iterations so as to satisfy under normal conditions the tabular value of specific volume V_0 and the values of isentropic compression modulus $B_S = -V(\partial P/\partial V)_S = B_{S0}$ and its pressure derivative $B_S' = (\partial B_S/\partial P)_S = B_{S0}'$ determined by the data of static and shock compressibility measurements.

The energy on the cold curve of a solid phase in the rarefaction region ($\sigma_c < 1$) is given by the polynomial,

$$F_c(V) = V_{0c} \sum_{i=1}^{3} \frac{c_i}{k_i} {\sigma_c}^{k_i} + E_{sub}, \tag{3}$$

which provides for the value of the sublimation energy $F_c = E_{sub}$ at $V \to \infty$. The requirement of continuity of the function $F_c(V)$ and its three derivatives with respect to volume leaves only two free parameters, k_2 and k_3, in equation (3).

The thermal lattice component of the free energy for each solid phase is defined as follows,

$$F_a(V,T) = RT\left(3\ln\left(1 - e^{-\theta_a/T}\right) - D(\theta_a/T)\right), \tag{4}$$

where $R = k_B(Am_u)^{-1}$, k_B is the Boltzmann constant, m_u is the atomic mass unit (amu), A is the atomic mass (in amu), D is the Debye function [9], $\theta_a = \theta_s(V)$ is the characteristic temperature of phonon spectrum [2,4],

$$\theta_s(V) = \theta_{s0}\sigma^{2/3} \exp\left((\gamma_{s0} - 2/3)\,\frac{{B_s}^2 + {D_s}^2}{B_s} \arctan\frac{B_s \ln\sigma}{{B_s}^2 + D_s\left(\ln\sigma + D_s\right)}\right), \tag{5}$$

where $\sigma = V_0/V$, γ_{s0} is the value of Grüneisen coefficient of the solid phase under normal conditions.

The energy of the liquid phase at $T = 0$ K is determined by the relations

$$F_c^{(l)}(V) = F_c^{(s1)}(V) + V_{l0}\frac{2{\sigma_l}^2}{1 + {\sigma_l}^3}\left(d_0 + \frac{d_n}{n_l}{\sigma_l}^{n_l}\right) \text{ at } \sigma_l \geqslant 1, \tag{6}$$

$$F_c^{(l)}(V) = V_{l0}\sum_{i=1}^{3}\frac{d_i}{n_i}{\sigma_l}^{n_i} + E_{sub} \text{ at } \sigma_l < 1, \tag{7}$$

where superscripts $s1$ and l denote the high-pressure solid phase and the liquid, $\sigma_l = V_{l0}/V$, V_{l0} is the specific volume of the liquid phase at a chosen point on the melting curve of phase $s1$ ($P = P_{m0}$, $T = T_{m0}$), the coefficients d_0 and d_n define the volume change under phase transition of $s1$ to the liquid at this point. The parameters of (7) are determined from the requirement of continuity of the function (6)–(7) and its three derivatives with respect to volume at $\sigma_l = 1$.

The contribution of thermal movement of heavy particles to the free energy of the liquid phase is defined in the form of equation (4), where $\theta_a = \theta_l(V) + (T_a T {\sigma_l}^{2/3})^{1/2}$, T_a is a constant,

$$\theta_l(V) = \theta_{l0}{\sigma_l}^{2/3} \exp\left((\gamma_{l0} - 2/3)\,\frac{{B_l}^2 + {D_l}^2}{B_l}\arctan\frac{B_l \ln \sigma_l}{{B_l}^2 + D_l\,(\ln \sigma_l + D_l)}\right), \tag{8}$$

the parameter θ_{l0} defines the enthalpy change under melting of phase $s1$ at $P = P_{m0}$, γ_{l0} is the value of Grüneisen coefficient of the liquid phase at $\sigma_l = 1$.

3. Calculation results

Under the normal conditions ($P = 0.1$ MPa, $T = 298$ K), the white tin modification, β-Sn, is stable. The transformation of β-Sn to the body-centered tetragonal structure γ-Sn occurs under compression up to pressure of $P = 9.4 \pm 0.4$ GPa at $T = 298$ K [10, 11]. Further compression up to $P = 45\pm5$ GPa at room temperature leads to the transformation of γ-Sn to the body-centered cubic structure δ-Sn, which is stable up to maximal pressure of $P = 120 \pm 10$ GPa achieved in experiments with diamond anvil cells [12].

In the present work, EOS for β–γ–liquid tin system is constructed on the basis of the model developed. Calculated isotherm $T = 298$ K and principal Hugoniot of the metal in comparison with data on static [12, 13] and shock [14–18] compression are presented in figure 1. One can see that the multi-phase EOS proposed for tin adequately describes the measurement results for compressibility of the considered polymorphs and liquid phase of tin in a wide range of pressures.

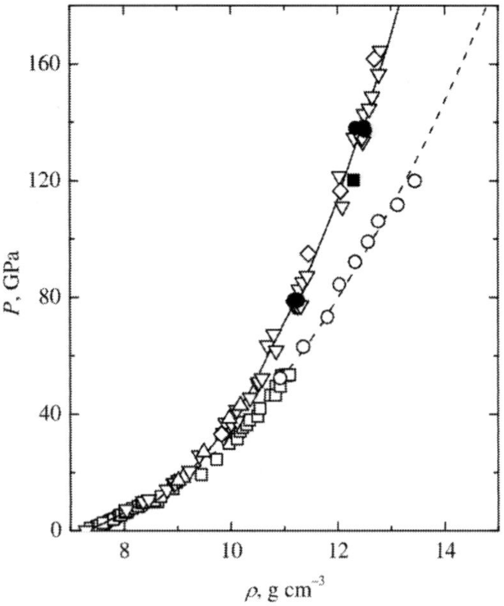

Figure 1. Static and shock compression of tin. Solid line corresponds to the calculated Hugoniot curve with accounting for the β–γ transition and melting, dashed line is the isotherm of γ-Sn at $T = 298$ K from present EOS. Experimental data: ○, [12]; □, [13]; ▽, [14]; ■, [15]; ◇, [16]; △, [17]; ●, [18].

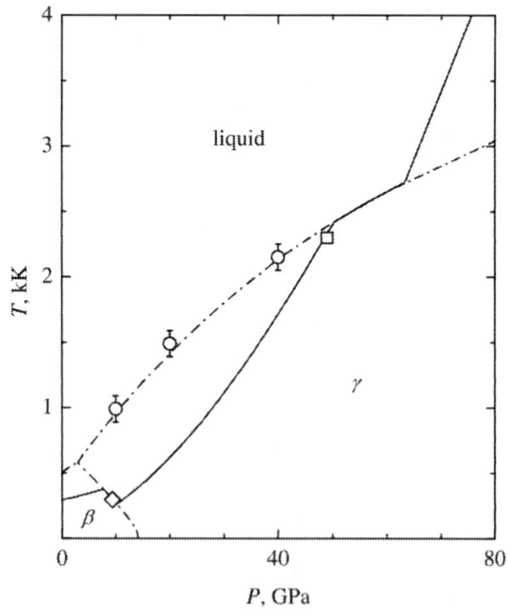

Figure 2. Phase diagram of β–γ–liquid tin system. Solid line is the calculated Hugoniot; dash-dot lines are the phase boundaries; ◇ is the measured point of β–γ transition at $T = 298$ K [10, 11]; □ is the melting point on the Hugoniot from EOS [19]; ○ marks points on the calculated melting curve [20].

Phase diagram of the β–γ–liquid tin system is shown in figure 2. The curves of the β–γ, β–liquid and γ–liquid phase equilibrium resulting from the present EOS are in a good agreement with available experimental data [10, 11] on the phase boundaries at pressures $P \lesssim 10$ GPa. The present melting curve of γ-Sn also agrees with results of calculations [19, 20]. The melting occurs on the obtained Hugoniot in the pressure range of $P = 50.3 \div 63.3$ GPa (see figure 2).

Thus the multi-phase EOS developed for tin provides a consistent description of the data of static and shock experiments, and it can be used in numerical simulation of processes at high pressures.

Acknowledgments
This work was done under financial support of the Russian Foundation for Basic Research, grant No. 06-02-17464.

References
[1] Zel'dovich Ya B and Raizer Yu P 1967 *Physics of Shock Waves and High-Temperature Hydrodynamic Phenomena* (New York: Academic Press)
[2] Bushman A V, Fortov V E, Kanel' G I and Ni A L 1993 *Intense Dynamic Loading of Condensed Matter* (Washington: Taylor & Francis)
[3] Kanel' G I, Razorenov S V and Fortov V E 2004 *Shock-Wave Phenomena and the Properties of Condensed Matter* (New York: Springer)
[4] Bushman A V and Fortov V E 1988 *Sov. Tech. Rev. B: Therm. Phys.* vol 1, ed A E Scheindlin and V E Fortov (New York: Harwood Academic) p 219
[5] Khishchenko K V 2004 *Tech. Phys. Lett.* **30** 829
[6] Kalitkin N N and Govorukhina I A 1965 *Sov. Phys. Solid State* **7** 287
[7] Kalitkin N N and Kuz'mina L V 1975 *Preprint* No 35 (Moscow: Institute of Applied Mathematics, USSR Academy of Sciences)
[8] Bushman A V, Fortov V E and Lomonosov I V 1991 *High Pressure Equations of State: Theory and Applications*, ed S Eliezer and R A Ricci (Amsterdam: North-Holland) p 249
[9] Landau L D and Lifshitz E M 1980 *Statistical Physics* (Oxford: Pergamon Press)
[10] Young D A 1991 *Phase Diagrams of the Elements* (Berkeley, CA: University of California Press)
[11] Tonkov E Yu and Ponyatovsky E G 2005 *Phase Transformations of Elements under High Pressure* (Boca Raton, FL: CRC Press)
[12] Desgreniers S, Vohra Y K and Ruoff A L 1989 *Phys. Rev.* B **39** 10359
[13] Liu M and Liu L 1986 *High Temp.-High Press.* **18** 79
[14] Marsh S P (ed) 1980 *LASL Shock Hugoniot Data* (Berkeley, CA: University of California Press)
[15] Al'tshuler L V, Krupnikov K K and Brazhnik M I 1958 *Sov. Phys. JETP* **7** 614
[16] Al'tshuler L V, Bakanova A A and Trunin R F 1962 *Sov. Phys. JETP* **15** 65
[17] Walsh J M, Rice M H, McQueen R G and Yarger F L 1957 *Phys. Rev.* **108** 196
[18] McQueen R G and Marsh S P 1960 *J. Appl. Phys.* **31** 1253
[19] Mabire C and Hereil P L 2000 *Shock Compression of Condensed Matter—1999*, ed M D Furnish, L C Chhabildas and R S Hixson (New York: AIP Press) p 93
[20] Bernard S and Maillet J B 2002 *Phys. Rev.* B **66** 012103

Journal of Physics: Conference Series **121** (2008) 032001 doi:10.1088/1742-6596/121/3/032001

High pressure materials for energy storage. The case of V_2O_5

M. E. Arroyo-de Dompablo[1*], U. Amador[2], J.M. Gallardo–Amores[1], C. Baehtz[3], N. Biskup[4] and E. Morán[1]

[1]Dpto de Química Inorgánica, Facultad de Ciencias Químicas, UCM, 28040-Madrid (Spain).

[2]Dpto de Química, Facultad de Farmacia, USP-CEU, 28668-Boadilla del Monte, (Spain).

[3]Forschungszentrum Dresden-Rossendorf, Institute of Ion Beam Physics and Material Research, PO 510119, 01314 Dresden, Germany

[4]Instituto de Ciencia de Materiales de Madrid, CSIC, 28049-Cantoblanco, (Spain)

e.arroyo@quim.ucm.es

Abstract. The high pressure polymorph of V_2O_5 is investigated as example of a high-pressure material (HP-V_2O_5) with potential applications in electrochemical devices, such as lithium batteries. HP-V_2O_5 reversibly intercalates up to 2 lithium ions down to 1 Volt vs Li^+/Li. The typical voltage-composition profile of HP-V_2O_5 shows distinct features than that of the ambient pressure polymorph, AP-V_2O_5. Remarkably, the specific capacity retention with cycling of HP-V_2O_5 // Li cells (250 Ah/Kg at a C/3.5 rate) is comparable to that of the ambient pressure polymorph. Furthermore, the measured resistivities at room temperature (10000 Ωcm in AP and 400 Ωcm in HP) reveal that HP-V_2O_5 is a better electronic conductor than AP-V_2O_5. The enhanced electronic conductivity of HP-V_2O_5 is an advantage for electrochemical applications.

1. Introduction

New synthesis strategies are needed to prepare novel electrode materials for lithium batteries. High pressure / high temperature routes are widely used within other fields of Solid State Chemistry to induce structural transformations of materials, conducting to novel polymorphs possessing structures not accessible at ambient pressure. However, compounds obtained under high pressure present structures denser than those corresponding to ambient pressure materials, therefore, they have been traditionally ruled out as host materials for insertion reactions. On the other hand, a high pressure polymorph generally presents a better electronic conductivity that the ambient pressure polymorph, which is an advantage in terms of electrochemical applications.

In the last years we have explored the electrochemical application of high-pressure materials [1-3]. V_2O_5, ranks among the most studied electrode for lithium batteries due to its high energy density and capacity retention upon cycling [4]. Under high pressure (40-80 kbar) V_2O_5 transforms into a layered polymorph, HP-V_2O_5, consisting on $V^{5+}O_6$ octahedra instead of $V^{5+}O_5$ square pyramids (figure 1) [5]. In this contribution we focus on the high-pressure polymorph of V_2O_5, as an example of a high-pressure material displaying a good performance as positive electrode for lithium batteries.

2. Results and discussion

Commercial (Aldrich) V_2O_5 was subjected at 4, 6 and 8 GPa and 800 ºC for 1 hour in a belt-type press. *In-situ* synchrotron X-ray diffraction (SXRD) experiments were performed at the B2 line at HASYLAB working in transmission mode, using a cell previously described [6], and radiation of wavelength λ=0.489646 Å. As detector a scintillation counter with an analysing crystal in front of it for high resolution is used. The treatment of the diffraction data was carried out using the FullProf program [7].

© 2008 IOP Publishing Ltd

The electrochemical experiments were performed in Swagelok cells using a disk of lithium metal as the negative and reference electrode. As electrolyte we used the commercial solution of $LiPF_6$ in EC (ethylene carbonate) and DMC (dimethyl carbonate) (1:1) Selectipur LP30® (Merck). The positive electrode consisted of a mixture of HP-V_2O_5, carbon black, and a binder (PVDF-HFP) in a 80:15:5 weight ratio conformed in 8 mm-diameter pellets of about 10 mg. Cell were assembled in a glove-box filled with argon and connected to a MacPile multichannel potentiostat- galvanostat. DC transport measurements in temperature region $77 < T < 300K$ have been done in both, 4-contact and 2-contact configurations.

3. Results and disccusion

V_2O_5 treated at 4-8 GPa and 800 °C turns into a black-reddish powder and, independently of the pressure used in the experiments, the patterns of all the samples can be indexed with the unit cell previously proposed for HP-V_2O_5 [4] (see Figure 1).

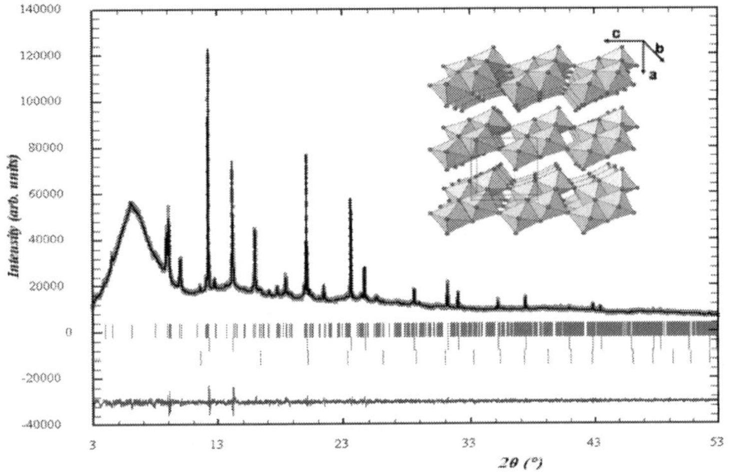

Figure 1: Graphic result (experimental, calculated and their difference) of the fitting of SXRD data of HP- V_2O_5; in the inset a schematic representation of its structure is depicted. Peaks from the sample holder (Al) and Li from the electrode are fitted as phase-II and phase-III. Final cell parameters for HP-V2O5 are a = 7.181(1) Å, b = 3.58747 (7) Å, c = 6.2998 (2)Å, β = 89.854(7),*S.G. P21/m.* R_p: 2.20, R_{wp}: 2.93, R_{exp}: 0.80, χ^2: 13.3.

Figure 2a compares the Open Circuit Voltage (OCV) measurements of two Li // HP-V_2O_5 cells using the AP-V_2O_5 (curve in red) and the HP-V_2O_5 (curve in green) as active materials. The single phases Li_x-V_2O_5 appear as abrupt voltage drops in the composition curve. Up to six Li_xHP-V_2O_5 single phases are observed, with approximate compositional stability domains of $0< x <0.02$, $0.23 < x <0.26$, $0.56 < x <0.59$, $0.76 < x < 0.9$, $1.57 < x < 1.65$ and $1.80 < x < 2$. These compositional limits are clearly different from those found on the $Li_xV_2O_5$ phases derived from the ambient pressure polymorph [4] . In particular, the single phase centred at x=1, which is very characteristics of AP-V_2O_5, is absent in the HP-V_2O_5. The polarization is of the same order in both cells, suggesting a similar kinetics of the lithium insertion reaction into the AP/HP host materials.

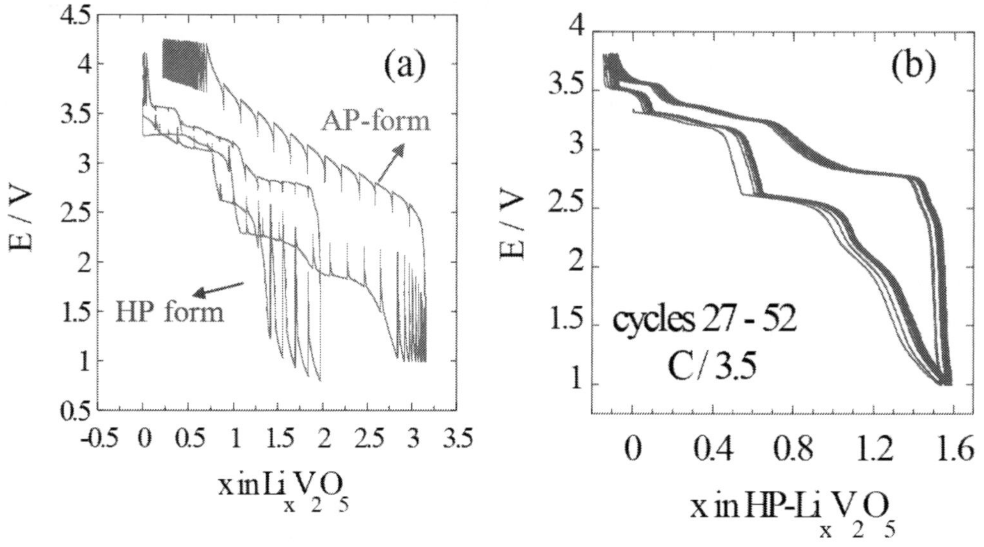

Figure 2.- (a) Open circuit voltage of Li // V_2O_5 cells discharged at j = 0.37 mA/cm^2 for 1 h (C/7 rate). Cells were kept at open circuit for 20 hours, or until the variation of voltage with time was lower than 0.01mV/h. (b) Cycling of Li // HP-V_2O_5 cell in a voltage window 3.8-1V.

Figure 2b shows a typical cycling experiment of a Li // HP-V_2O_5 cell performed between 1 and 3.8 V under constant current density equivalent to a C/3.5 rate. The specific capacity delivered by the cell is of about 250 Ah/Kg, and remains constant with cycling. Thus, lithium ions can be reversibly inserted / deinserted into / from *the high pressure polymorph of V_2O_5* with almost no capacity fade upon cycling. It should be noticed that HP-V_2O_5 displays the same reversible capacity than V_2O_5, though in a somewhat wider voltage range (see figure 2a). It is worth mentioning that none of the characteristic processes of V_2O_5 was observed in the V-x curves for any of the Li//HP-V_2O_5 cells studied. This discards an eventual reversion of the phase transformation induced by high pressure.

The electrical conductivity of an electrode material determines the polarization of the positive electrode, thus governing the power rate capability of the lithium cell. The improvement of electronic conductivity is particularly relevant in the case of electrode materials lacking a good electronic conductivity, such as $LiFePO_4$; by means of high pressure it might be possible to modify the electronic structure of a material, presumably improving the electronic conductivity. Hence, in the example under study, it is important to compare the conductivity of the AP/HP V_2O_5 polymorphs. The d.c. conductivity was measured on both polymorphs of V_2O_5 from room temperature down to T=77K [8]. As shown in figure 3, the room temperature resistivity decreases by more than one order of magnitude: from 10000 Ωcm in AP-V_2O_5 to 400 Ωcm in HP-V_2O_5. In the above temperature window, resistivities of both compounds show activated behavior with activation energies E_A of 0.232 and 0.068 eV for AP and HP polymorphs, respectively. The enhanced electronic conductivity of HP-V_2O_5 with respect to that of the ambient pressure form is a major advantage in terms of electrochemical applications.

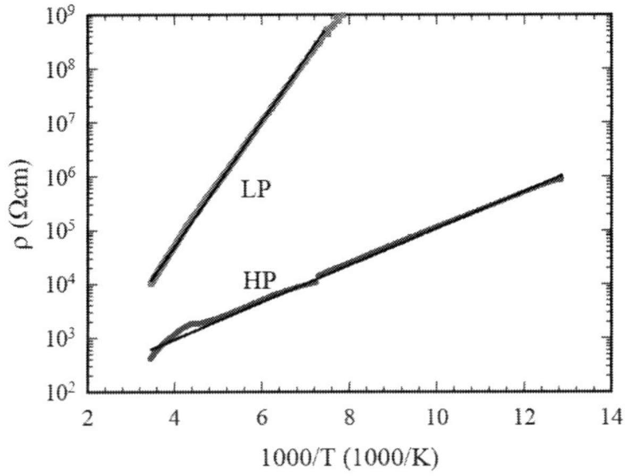

Figure 3.- Resistivity of AP-V_2O_5 and HP-V_2O_5 polymorphs versus inverse temperature. Black lines are fit to $\rho = \rho_0 \exp(E_A/kT)$.

4. Conclusions.

High pressure synthesis is a promising strategy to produce novel electrode materials, either polymorphs of well-known compounds or completely innovative materials. As an example, we have compared the electrode characteristics of the ambient pressure and the high-pressure polymorphs of V_2O_5. V_2O_5 (d = 3.32 g / cm^3) transforms under high pressure (40 kbar) into a polymorph, HP-V_2O_5, which possess a much denser structure (d = 3.80 g / cm^3). In spite of its higher density, HP-V_2O_5 also presents a layered structure, which facilitate the incorporation and mobility of cations in between the layers. Li // HP-V_2O_5 cells display a good cyclability and capacity retention with cycling of about 250 Ah/Kg. This behavior is comparable to that of the ambient pressure V_2O_5. Furthermore, the measured resistivites at room temperature (10000 Ωcm in AP and 400 Ωcm in HP) reveal that HP-V_2O_5 is indeed a better electronic conductor than AP-V_2O_5.These results demonstrate that materials obtained under high pressure can be suitable for electrochemical applications.

Acknowledgements
This work was supported by Universidad Complutense de Madrid (PR1/07-14911, PR34/07-1854), the Spanish MEC (MAT2007-62929, CSD2007-0045) and CAM (S-0505/PPQ-0093).

References

[1]. M.E. Arroyo y de Dompablo, J.M. Gallardo-Amores, U. Amador, and E. Moran, *Electrochem. Communications*, **9/6**, 1305, (2007)
[2]. M.E. Arroyo y de Dompablo, J.M. Gallardo-Amores, and U. Amador, *Electrochemical and Solid State Letters*, **8**(11), A564(2005)
[3]. O. Garcia Moreno, M. Alvarez Vega, F. Garcia Alvarado, J. Garcia Jaca , J.M. Gallardo-Amores, M.L. Sanjuan, and A. U., *Chem. Mater.*, **13**, 1570 (2001)
[4]. C. Delmas, H. Cognac-Auradou, J.M. Cocciantelli, M. Ménétrier, and J.P. Doumerc, *Solid State Ionics*, **69**, 257(1994)
[5]. V.P. Filonenko, M. Sundberg, P.E. Werner, and I.P. Zibrov, *Acta Cryst*, **B 60**, 375 (2004)
[6]. M. Knapp, C. Baehtz , H. Ehrenberg, and H. Fuess, *J. Synchrotron Rad*, **11**, 328 (2004)
[7]. J. Rodriguez-Carvajal. *Fullprof.* in *Satellite Meeting on Powder Diffraction of the XV Congress of the IUCr*. 1990. Toulouse, France.
[8] J.M. Gallardo-Amores, N. Biskup, U. Amador, K. Persson, G. Ceder, E. Moran and M.E. Arroyo y de Dompablo, *Chem. Mater.*, ***19***, 562, (2007)

Method for elastic uniaxial stress of single crystals: description and applications

N Ya Minina[1], N B Brandt[1], A M Savin[2] and E V Bogdanov[1]

[1]Physics Department, Moscow State University, 119992 Moscow, Russia
[2]Low Temperature Laboratory, Helsinki University of Technology, FIN-02015 TKK, Finland

E-mail: min@mig.phys.msu.ru

Abstract. The original method for a strong elastic uniaxial compression of single crystals up to $5 \div 6$ kbar, calibration procedure and its applications for semimetal and semiconductor samples are described in details for the first time. This method simplifies sample installation and adjustment; it prevents the frequent destruction of samples, which often takes place in conventional methods of uniaxial compression between two anvils.

1. Introduction

The well-known phrase "uniaxial stress breaks crystal symmetry and samples" combines both the advantage and disadvantage of this kind of strong external influence. The latter one causes the relatively rare use of the uniaxial stress technique in physical experiments. For the sake of avoiding a destructive shear stress, a sample between two anvils should have faces that are perfectly flat and parallel in addition to being placed strictly along the stress direction. In this paper the original method for a strong elastic uniaxial compression of single crystals up to 5-6 kbar and the calibration procedure are described in detail for the first time. This method permits the simplification of sample installation and adjustment. In addition, it prevents the frequent destruction of samples that takes place in conventional methods of uniaxial compression between two anvils. The method was developed in Moscow State University and has been successfully used in scientific research in Copenhagen University, Moscow State University, and Humboldt University in Berlin. Along with the method description, some results are discussed for the illustration of the method possibilities.

2. Method description

In the method under consideration (see figure 1a), a sample in the form of a parallelepiped is firmly fastened with epoxy in an elastic ring along the "Y" direction, and afterward a tensile stress is to be applied to the ring in the "X" direction. The ring transforms a tensile stress into a compressive one, applied to the both ends of the sample in the plane of the ring. Axial distribution of the stress in the ring and a rigid fastening in the ring prevent the sample of a premature destruction. Moreover, in case when plastic deformation of a single crystal develops by a sliding process, the fixation of the main cleavage planes of the crystal inside the ring moves the plastic flow limit to higher values, until it starts developing along secondary cleavage planes. This takes place, for example, in Bi, Sb, and Bi_xSb_{1-x} alloys in which the elastic limit at low temperatures is determined by the plastic flow along (111) planes. Securing the easiest (111) cleavage planes in the plane of the ring with Araldit epoxy makes it possible to increase the elastic loading by more than 10 times in comparison with the

© 2008 IOP Publishing Ltd

previous uniaxial stress experiments [1]. The strain tensor components in the elastic region, determined by X-Ray diffraction, reach ε_{xx} = 0.13%, ε_{yy} = -0.3%, ε_{zz} = 0.1% in Bi samples under tensile stress F = 300 N. The maximum pressure achieved by this technique is limited not by the sample destruction, but by the breaking of the ring in points of stress concentration.

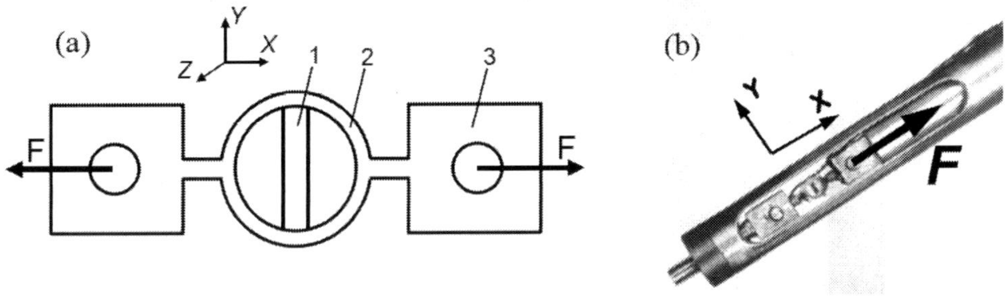

Figure 1. (a) Sample (1), ring (2), "holder" for the fixation in the stretching device (3). (b) Ring fixation in the stretching device.

With the help of a special stretching device, the lower part of which is depicted in figure 1b, the stress can be applied to the ring just at the liquid helium temperature, and the value of the stress can be determined by the elongation of the calibrated spring (this part is not depicted). This method is very convenient for magnetotransport and optical measurements and is applicable for single crystals with quite different elastic properties: (Bi [1], Sb [2], $Hg_{1-x}Cd_xTe$ [3], GaAs [4]). Concerning optical measurements, the device on figure 1b is open for illumination, and experiments under uniaxial stress are much easier than with devices for hydrostatic (or quasi hydrostatic) compression.

Calibration of the ring with Bi, Bi_xSb_{1-x} or GaAs samples inside was performed at liquid nitrogen temperature from the shift of X-Ray diffraction angles from different crystallographic planes in dependence on the stress F applied to the ring. In figures 2a and 2b, we represent the dependence of diffraction maxima and diffraction angles on the applied stress F. Reversible change of both characteristics under loading and unloading demonstrates the elastic nature of the strain. From calculations made on the base of X-Ray measurements, the useful empirical correlation (figure 3) was determined. It is used to find a compressive force F_s applied directly to the sample from the tensile force F = F_{ring}, which is applied to the ring (figure 1a). The calibration curve depicted in figure 3 permits evaluation of the direct compressive force for the samples of different cross sections A and Young Modulus E_s, if only the ring dimensions and elastic properties are preserved.

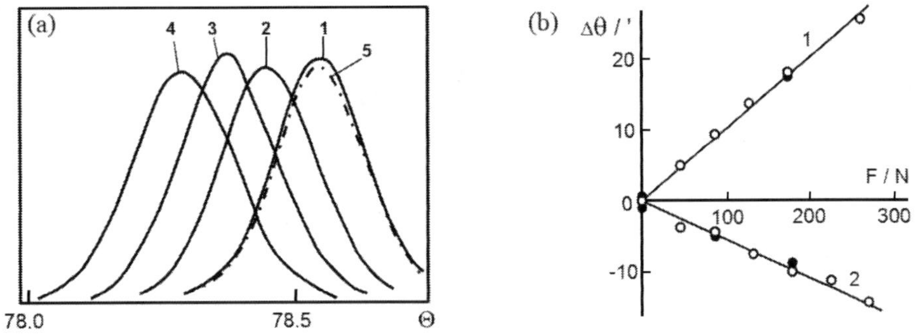

Figure 2. (a) X-ray diffraction maximum ($Bi_{0.95}Sb_{0.05}$ at T = 77K) corresponding to (555) plane at different force F applied to the ring: 1 – 0, 2 – 90 N, 3 – 180 N, 4 – 270 N, 5 – 0 after the pressure is removed . (b) Change of diffraction angle for planes 1 – (633) and 2 – (555) in $Bi_{0.95}Sb_{0.05}$ under loading (open circles) and unloading (filled circles) of the sample.

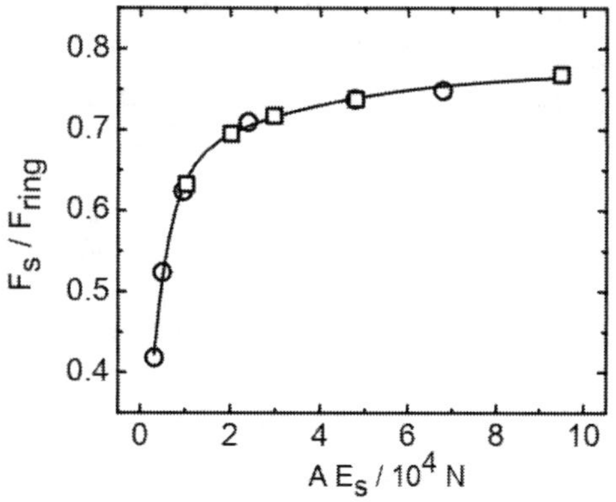

Figure 3. Ratio between compressive force F_S applied to the sample and tensile force $F = F_{ring}$ applied to the ring for samples with different cross sections A (squares) and samples with different elastic constants E_S (circles). Calculations are made on the base of X-ray measurements.

In our experiments the rings with inner diameter D_{in} = 3 mm, outer diameter D_{out} = 4.5 mm and thickness d = 0.75 mm were cut by the electro-erosion method from nonmagnetic alloy. The sample dimensions are determined by geometry of the ring. The data in figure 3 reflect X-Ray measurements on Bi, Bi_xSb_{1-x} and GaAs samples. Decrease of the sample cross sections A leads to increase of the attainable compressive pressure $P = F_s/A$ (figure 3), however A should not be too small and the sample should not bend under loading.

Advantages and possibilities of the technique are illustrated below by some experimental results obtained on Bi, Bi_xSb_{1-x} alloys and on $GaAs/Al_xGa_{1-x}As$ heterostructures. These materials are quite different in their elastic properties. Rather soft Bi with Young Modulus E_s = 380 kbar and hard GaAs with E_s = 1200 kbar are specially chosen for illustration among other samples investigated under uiaxial compression in the ring.

3. Method application

3.1 Electronic topological transitions in Bi and Bi_xSb_{1-x}alloys

Since this method gives the possibility of continuous applications of elastic uniaxial strain at low temperature, it permitted to obtain the most convincing picture of electronic topological transitions (ETT) development in Bi and Bi_xSb_{1-x} alloys [1]. ETTs of 2½ type [5] take place at certain conditions (for example under compression) if one or more parts of a Fermi surface (FS) arise or vanish. It was shown that, in accordance with the theory [5], these transitions are accompanied by anomalies in thermoelectric power and resistivity [1], and the peculiarities of these anomalies are determined by the character of the FS change.

Usually, very high hydrostatic pressure is necessary for ETTs observation in metals, semimetals, and semiconductors. The different situation arises under uniaxial stress. In Bi and Bi_xSb_{1-x} alloys, uniaxial stress along a binary C_2 or bisector C_1 axes breaks a crystal symmetry and leads to a non-equivalent shift of electron extrema L_i in respect to the Fermi level (FL) and the hole extremum T (see, for example, the sketch on figure 4a). As a result, more than 10 electronic topological transitions of different type were observed in this system with the help of the uniaxial stress technique described above [1]. One of these transitions, as an example, is represented below.

Transition 1h → 1h + 2e (Bi + 10^{-2}at.% Sn) is illustrated in figure 4. The letters "h" and "e" denote hole and electron ellipsoids of the FS. Quantum oscillations of magnetoresistance dR/dH(H) – so called Shubnikov-de Haas (SdH) oscillations – in figure 4b are determined by FS cross sections. In Bi doped with 10^{-2} at.% Sn electron extrema L_i are above the FL (see figure 4a), and only the hole FS exists at zero stress (high oscillation frequency in figure 4b). Under uniaxial compression along binary axis two equivalent L_i extrema go down and reach the FL. Correspondingly, at higher stress two electron ellipsoids (low frequency, figure 4b) appear on the background of the hole FS. The emergence of the electron FS cross section S and its increase under strain is depicted in figure 4a (S_{max} is the maximal value of S obtained under compression). At the same compressive strain ε_{yy} ≈

-0.8×10^{-3}, where ETT occurs, the anomalous behavior of thermoelectric power V and resistance R/R_0 (R_0 – resistance at zero pressure) in dependence on strain takes place (figure 4c).

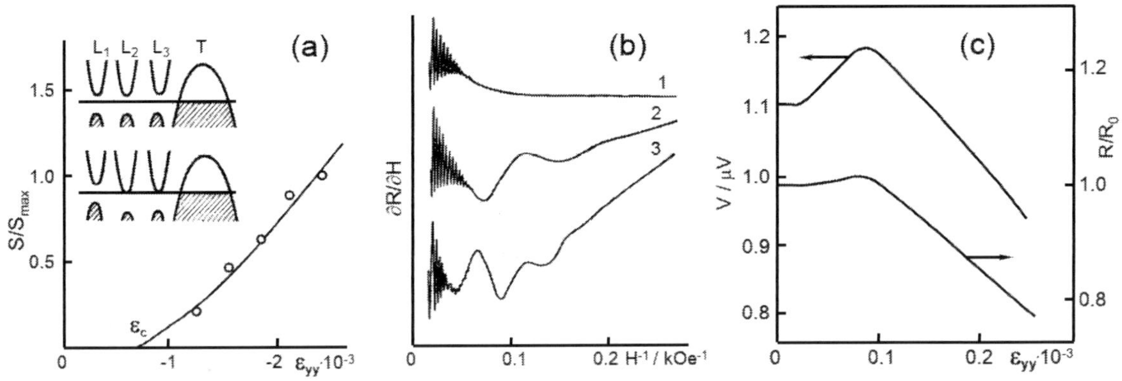

Figure 4. (a) – Configuration of electron L_i and hole T extrema near the FL at zero stress and in ETT point in Bi + 10^{-2} at.% Sn; emergence and increase of the electron FS cross section in dependence on strain. (b) – Magnetoresistance oscillations in Bi + 10^{-2} at.% Sn at different strain ε_{yy} : 1 – 0; 2 – -1.5×10^{-3}; 3 – -2.4×10^{-3}. (c) – Thermoelectric power (left side) and resistance (right side) in dependence on strain.

It has been shown that uniaxial stress permits to shift band extrema across the Fermi level and directly observe the FS transformation of a different type. This effect is accompanied by singularities of the energy spectrum: in this way, the saddle point in the electron and hole energy spectrum of Bi_xSb_{1-x} alloys was determined by disruption (appearance) of the dumbbell-shaped FS under compression [6], and elimination of inter-valley scattering channel in Bi_xSb_{1-x} in the course of transformation from a three valley to a single valley conduction band under compression was detected [7].

3.2 Anisotropy effects in 2D hole system at $GaAs/Al_xGa_{1-x}As$ heterointerface

For delicate in respect to stress materials, such as Bi and B_xSb_{1-x} alloys, the application of the method described above is a decisive factor for the extension of elastic limit and observation of quantum phenomena. However in semiconductor $GaAs/Al_xGA_{1-x}As$ heterostructures, it mainly secures samples and simplifies experiments. Below, we only briefly illustrate a strong change of 2D hole energy spectrum anisotropy observed in p-$GaAs/Al_xGa_{1-x}As$ under uniaxial compression up to 5 kbar.

The effect of uniaxial compression on valence band structure of single and double p-$GaAs/Al_{0.5}Ga_{0.5}As$ heterostructures was investigated both theoretically and experimentally [4]. The form of the valence band potential strongly affects the energy spectrum of 2D holes in p-$GaAs/Al_xGa_{1-x}As$ heterostructures. In the case of a single heterojunction (triangular quantum well), the absence of inversion symmetry causes a spin splitting of the valence subbands, and so the two lists of the FS correspond to the ground heavy hole state (figure 5, P = 0 kbar). According to our theoretical calculations [4], external uniaxial compression in 2D hole plane causes a strong anisotropy of hole energy spectrum. Figure 5 illustrates the numerical calculations of 2D hole Fermi surface transformation in p-$GaAs/Al_xGa_{1-x}As$ under uniaxial compression. A strong stretching of the FS in the direction perpendicular to the direction of compression should be noted, while the contours of the hole FS's in both subbands touch one another in the compression direction at pressure P = 2,5 kbar.

Predicted modifications of 2D hole FS were detected under uniaxial compression by measurements of quantum Hall effect, SdH oscillations (figure 6) and transport properties in p-$GaAs/Al_{0.5}Ga_{0.5}As$ with carrier concentration $N_p = 7.6 \times 10^{11}$ cm^{-2}. A development of magnetic breakdown between two spin split subbands observed under uniaxial compression in [110] direction within the pressure range 1.5 – 3.0 kbar [8] reveals a strong experimental evidence of the FS transformation depicted on figure 5. In this pressure range the distance between the two FS's is minimal (figure 5), which allows tunneling of the holes between different orbits in the magnetic field. As a result, according to Fourier transform analysis, additional frequencies corresponding to new breakdown orbits on the FS appear in SdH oscillations. The possibility of transition from one quasi-classical orbit to the other one under magnetic breakdown condition affects the dynamic of charge carriers motion, and apart from the initial (at P = 0) orbits 1-4 and 2-3, the orbits 1-3 and 2-4, as well as their harmonics, arise under compression (figure 5, P = 2.5 kbar) [8].

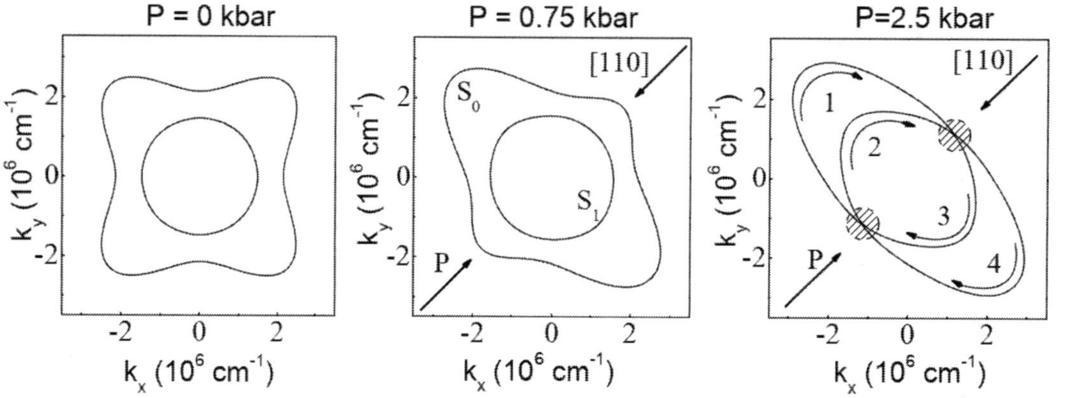

Figure 5. The FS contours S_0 and S_1 in two spin-split subbands of 2D heavy holes at different pressures along the [110] direction. For P = 2.5 kbar, the magnetic breakdown regions are hatched, and the numerals 1, 2, 3 and 4 correspond to the different segments of the magnetic breakdown orbits.

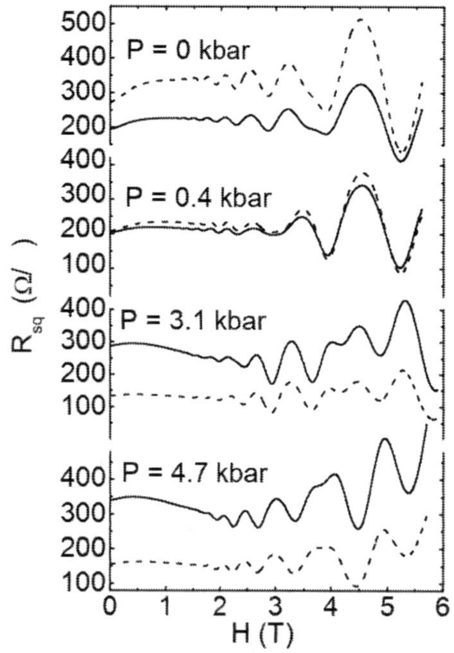

Figure 6. SdH oscillations for GaAs/$Al_{0.5}Ga_{0.5}As$ sample with hole concentration $N_p = 7.6 \times 10^{11}$ cm^{-2} at different pressures applied along [110] direction. Dash lines correspond to the direction of current parallel to the direction of compression, solid lines correspond to the perpendicular direction of electrical current.

Changes of 2D hole FS are illustrated by SdH oscillations on p-GaAs/$Al_{0.5}Ga_{0.5}As$ sample with two perpendicular mesas (with directions of electrical current parallel and perpendicular to the direction of compression) represented in figure 6. In accordance with band structure calculations, magnetotransport measurements reveal that the differences between hole concentrations in the ground-state spin-split subbands and also between the corresponding effective masses decrease with increase of pressure [4].

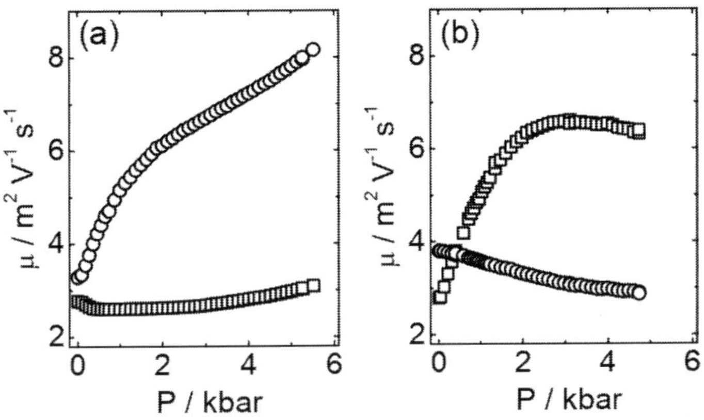

Figure 7. Pressure dependence of hole mobility in p-GaAs/$Al_{0.5}Ga_{0.5}As$ in [110] direction (squares) and in [1-10] direction (circles) under uniaxial compression along [1-10] (a) and [110] (b) respectively.

Strong modification of 2D hole FS anisotropy induces corresponding modification of p-GaAs/$Al_{0.5}Ga_{0.5}As$ transport properties. Most evidently it is reflected in essential anisotropy of electrical resistance and mobility anisotropy under compression (figure 7) [9]. Data obtained on a two-mesa sample indicate strong increase of mobility (up to 3 times at P = 5 kbar) in direction of compression. It qualitatively correlates with evident decrease of the FS dimensions and 2D hole transport mass in this direction (figure 5). The reverse situation takes place in the direction perpendicular to the compression direction (figure 7). The small anisotropy at P = 0 is determined by the surface roughness scattering.

4. Conclusion

We have described an original method for strong elastic uniaxial compression of single crystals. This method is simple in realization, secures samples of a premature damage, permits application of uniaxial stress at liquid helium temperatures, and in some cases shifts the elastic limit to higher stress values. Experimental results obtained with the help of the method demonstrate the method possibilities. This work was supported by the Russian Foundation for Basic Research (project 07-02-00866) and Russian Federation President Grant Council (grant NS-5248.2006.2). We are grateful to our colleagues Dr. O.P. Hansen from Copenhagen University, Dr. W. Kraak from Humboldt University in Berlin for their cooperation, and student B. Dyatkin for the technical assistance with the paper preparation.

References

[1] Brandt N B, Egorov V S, Lavrenyuk M Yu, Minina N Ya and Savin A M 1985 *Sov. Phys. JETP* **62** 1303
[2] Brandt N B, Bogdanov E V, Kulbachinsky V A and Minina N Ya 1978 *J. Phys. Colloq.* **39** C6-1108
[3] Butko N B, Minina N Ya, Savin A M, Kraak W and Krause S 1994 *JETP Lett.* **59** 790
[4] Kolokolov K I, Savin A M, Beneslavski S D, Minina N Ya and Hansen O P 1999 *Phys. Rev.* B **59** 7537
[5] Lifshits I M 1960 *Sov. Phys. JETP* **11** 1130
[6] Brandt N B, Lavrenyuk M Yu, Minina N Ya, Savin A M, Kraak W and Hermann R 1987 *Phys. Status Solidi (b)* **143** 601
[7] Minina N Ya and Kirakosova L A 1992 *Sov.Phys. JETP* **74** 886
[8] Kraak W, Savin A M, Minina N Ya, Il'evskii A A and Polyanskii A V 2004 *JETP Letters* **80** 351
[9] Minina N Ya, Kolokolov K I, Beneslavski S D, Bogdanov E V, Polyansky A V, Savin A M and Hansen O P 2004 *Physica E* **22** 373

Joint 21st AIRAPT and 45th EHPRG Int. Conf. on High Pressure Science and Technology IOP Publishing
Journal of Physics: Conference Series **121** (2008) 032003 doi:10.1088/1742-6596/121/3/032003

Insulator-metal transition in $Pb_{1-x}Ge_xTe$ doped with chromium under pressure

E P Skipetrov[1], A A Plastun[1], T A Topchevskaya[1], B B Kovalev[1], N A Pichugin[1] and V E Slyn'ko[2]

[1] Faculty of Physics, Moscow State University, 119992 Moscow, Russia
[2] Institute of Material Science Problems, 274001 Chernovtsy, Ukraine

E-mail: skip@mig.phys.msu.ru

Abstract. The galvanomagnetic effects in the n-$Pb_{1-x}Ge_xTe$ (x=0.02-0.13) alloys, doped with chromium, at the temperatures in the range of $4.2 \leq T \leq 300$ K and under hydrostatic compression of up to 17 kbar have been investigated. The insulator-metal transition, induced by pressure in $Pb_{1-x}Ge_xTe$:Cr (x=0.10) alloy, was revealed. Using the experimental data in the frame of two-band Kane dispersion relation the pressure dependences of electron concentration and Fermi energy were calculated. The pressure coefficient of chromium deep level energy was obtained and the diagram of the electronic structure reconstruction under pressure was proposed.

1. Introduction

Doping with magnetic mixed-valence impurities (Eu, Gd, Yb, Ti, V, Cr, ...) induces an appearance of deep impurity states in the electronic structure of PbTe-based narrow-gap semiconductors and turns them into the diluted magnetic semiconductors [1, 2]. The specific feature of these materials is that magnetic activity of impurity ions is strongly dependent on their electronic structure, in particular, on the mutual arrangement of the impurity level, the band edges and the Fermi level [2, 3].

It is known that in $Pb_{1-x}Ge_xTe$ chromium deep level E_{Cr} situated on the background of conduction band or in the gap and its energy position relative to the conduction band bottom strongly depends on the germanium content in the alloy [4]. In order to reveal the chromium deep level in the electronic structure and to obtain the diagram of the energy spectrum reconstruction under hydrostatic compression in the present work the effect of pressure up to 17 kbar on the galvanomagnetic properties of n-$Pb_{1-x}Ge_xTe$:Cr (x=0.10) alloy was studied.

2. Experimental details

Single crystals ingot of n-$Pb_{1-x-y}Ge_xCr_yTe$ were grown by the modified Bridgman method. The ingot was cut into slices approximately 1.5 mm thick perpendicular to the growth direction. The chemical composition was determined using the X-ray fluorescence analysis, which revealed that in accordance with previously known data [5] the concentration of germanium and chromium increase exponentially along the axis of the ingot from the beginning to the end of the ingot (x=0.02-0.13, y=0.01-0.05).

At the atmospheric pressure the temperature dependences of the resistivity ρ and the Hall coefficient R_H ($4.2 \leq T \leq 300$ K, $\mathbf{B} \leq 0.1$ T) were measured by the four-probe technique. Table 1 summarizes the main parameters of the samples at T=4.2 K and atmospheric pressure. Then the alloy with the germanium content x=0.10 was investigated under hydrostatic compression. Pressures up to

© 2008 IOP Publishing Ltd

Table 1. Parameters of n-$Pb_{1-x-y}Ge_xCr_yTe$ samples at T=4.2 K.

x	y	ρ (10^{-4} Ω cm)	$-R_H$ (cm^3/C)	μ_H (10^4 cm^2/V·s)	n (10^{18} cm^{-3})
0.02	0.010	0.37	0.67	1.8	9.3
0.03	0.010	0.44	0.65	1.5	9.5
0.04	0.015	0.58	0.77	1.3	8.1
0.05	0.020	0.82	0.92	1.1	6.8
0.06	0.020	1.6	1.25	0.79	5.0
0.07	0.025	4.1	2.00	0.49	3.1
0.10	0.030	100	5.15	0.05	-
0.13	0.050	1040	-	-	-

17 kbar were obtained in the beryllium bronze chamber with kerosene-oil-pentane pressure transmitting media and were determined at helium temperatures by measuring the superconducting transition temperature in pure tin as a function of pressure in the chamber.

3. Results and discussion

It was shown that in the alloys with low germanium content ($x \leq 0.07$) the resistivity ρ and the Hall coefficient R_H reveal metal-like behavior due to the pinning of the Fermi level by the resonant chromium level, situated in the conduction band. The concentration of free electrons, calculated from the values of R_H at helium temperature, monotonously decreases with the increase of germanium content and tends to zero at $x \approx 0.10$ (see table 1). In the samples with $x \geq 0.10$ deep chromium level shifts into the gap and temperature dependences of resistivity and Hall coefficient have an activation character (curves 1 in figure 1, 2). At atmospheric pressure an activation energies of chromium level $\Delta E_{Cr} = E_c - E_{Cr}$, determined from the slope of the ρ(1/T) and R_H(1/T) dependences, are approximately 7 meV and 14 meV for the samples with x=0.10 and x=0.13 accordingly.

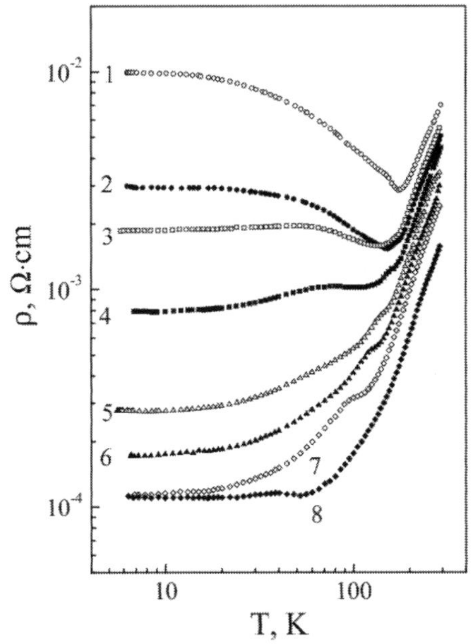

Figure 1. Temperature dependences of resistivity in $Pb_{1-x}Ge_xTe$:Cr (x=0.10) under pressure. P, kbar: 1 – 0, 2 – 0.5, 3 – 1.2, 4 – 2.2, 5 – 4.4, 6 – 6.3, 7 – 9.1, 8 – 16.4.

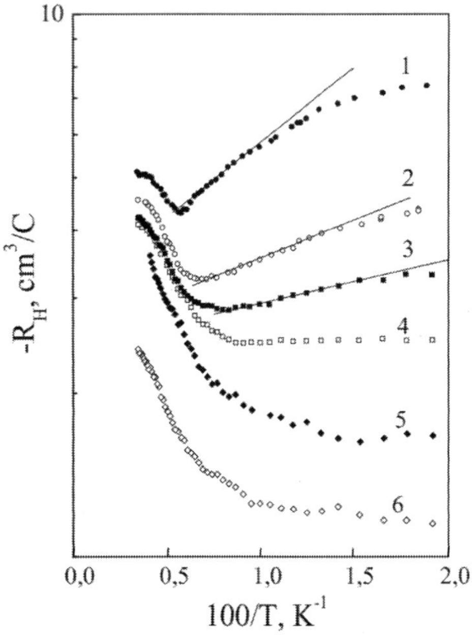

Figure 2. Temperature dependences of the Hall coefficient in $Pb_{1-x}Ge_xTe$:Cr (x=0.10) alloy under pressure. P, kbar: 1 – 0, 2 – 0.5, 3 – 1.2, 4 – 2.2, 5 – 4.4, 6 – 16.4.

Under hydrostatic compression the resistivity and absolute value of the Hall coefficient in $Pb_{1-x}Ge_xTe$:Cr (x=0.1) alloy at helium temperatures decrease by two and almost by one order of magnitude accordingly, the slope of the $\rho(1/T)$ and $R_H(1/T)$ dependences monotonously decreases and tends to zero at the critical pressure $P^* \approx 2$ kbar (figure 2). At pressures P>2 kbar activation character of the ρ and R_H temperature dependences changes for the "metallic" one. Obviously such a character of the galvanomagnetic properties changing is indicative of a transition of the alloy from the insulating to the metallic state under pressure. In the metallic phase (P>2 kbar) temperature dependencies of the Hall coefficient becomes typical for the lead telluride-based alloys with the resonant impurity level, stabilizing the Fermi level in the conduction band.

From the slope of the activation range on the $\rho(1/T)$, $R_H(1/T)$ dependences in accordance with the relations ρ, $R_H \propto \exp(\Delta E_{Cr}/kT)$ we have determined the activation energy of the chromium level ΔE_{Cr} in the insulating phase (P<2 kbar). Under the increase of pressure the activation energy decreases at the rate $d(\Delta E_{Cr})/dP \approx -4$ meV/kbar, that only slightly more than the half of the energy gap decreasing rate in the PbTe-based alloys under pressure [6]. This circumstance means that the chromium level E_{Cr} moves almost parallel to the middle of the gap, slightly shifting upward under pressure (figure 3). Thus under pressure the chromium deep level enters the conduction band, inducing the insulator-metal transition due to the flowing of electrons from the impurity level to the conduction band.

Using the values of the activation energy of chromium level in the insulating phase (P<2 kbar) and the free electron concentration, determined in the metallic phase (P>2 kbar) from the Hall coefficient at helium temperature as $n=1/eR_H$, we have obtained experimental pressure dependences of the electron concentration n and the Fermi energy E_F in the frame of two-band Kane dispersion relation for A^4B^6 semiconductors [4, 6] (points in figures 4, 5). Then, taking into account diagram of the energy spectrum reconstruction, presented in the figure 3, in the frame of the Kane dispersion relation the theoretical dependences n(P) and $E_F(P)$ were calculated (lines in figures 4, 5). One can see that theoretical dependences obtained in such a way are in satisfactory agreement with experimental data in the insulating phase (P<2 kbar) only. In the metallic phase (P>2 kbar) experimental points in figures 4, 5 initially situate significantly higher theoretical curves. Under the increase of pressure experimental dependences intersect theoretical ones and tend to saturation. The analysis of the theoretical n(P) and $E_F(P)$ dependences shows that agreement with the experimental data may be improved assuming the significant widening and the finite electron filling of the chromium resonant level. However, even in this case there is no reasonable explanation for the sharp increase in electron concentration and Fermi energy just after transition of the alloy to the metallic phase. One can suppose that this anomaly may be connected with the nonequivalence of the L points in the Brillouin zone due to the structural phase transition of the alloy from the cubic to the rhombohedral phase at low temperatures [6].

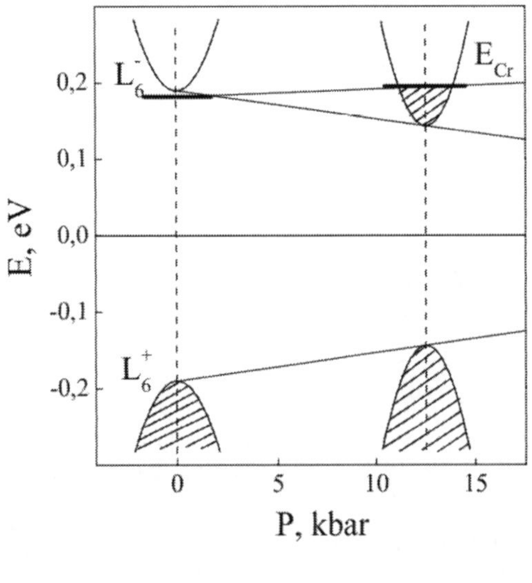

Figure 3. The diagram of the electronic structure reconstruction in $Pb_{1-x}Ge_xTe$:Cr (x=0.10) alloy under pressure (dE_g/dP = -7.4 meV/kbar, $d(\Delta E_{Cr})/dP \approx -4$ meV/kbar).

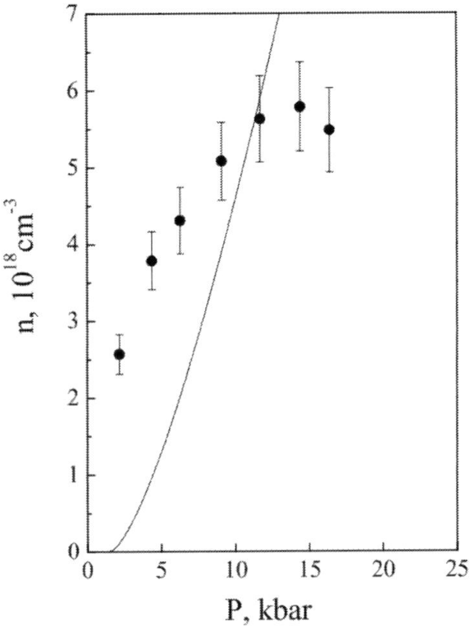

Figure 4. Pressure dependence of electron concentration in $Pb_{1-x}Ge_xTe$:Cr (x=0.10) alloy (points – experimental data, line – theoretical dependence, calculated in the frame of the Kane dispersion relation).

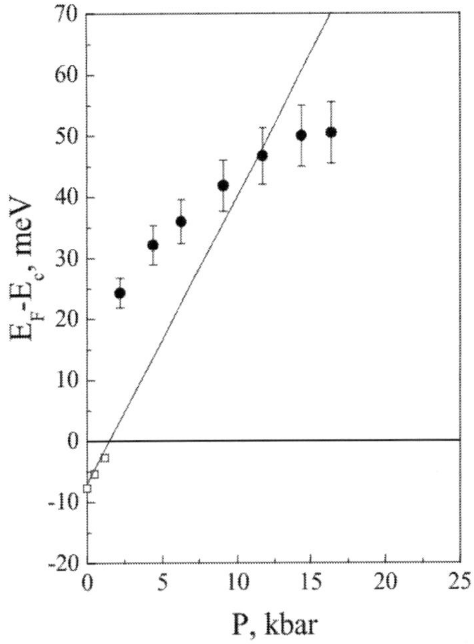

Figure 5. Pressure dependence of the Fermi energy in $Pb_{1-x}Ge_xTe$:Cr (x=0.10) alloy (points – experimental data, line – theoretical dependence, calculated in the frame of the Kane dispersion relation).

Conclusions

The galvanomagnetic properties of investigated alloys and insulator-metal transition, induced by pressure in $Pb_{1-x}Ge_xTe$:Cr (x=0.10) alloy, were explained taking into account stabilization of the Fermi energy by the chromium deep level, the shift of the level relative to the bottom of conduction band and redistribution of electrons between conduction band and the level under variation of the matrix composition and under pressure. It was shown that energy position of the chromium deep level relative to the middle of the gap is practically independent on the pressure and the diagram of the electronic structure reconstruction under pressure was proposed. It was shown that significant discrepancy of theoretical calculations with experimental data may be explained by assuming the finite electron filling and widening of the chromium resonant level as well as structural phase transition of the alloy to the rhombohedral phase at low temperatures.

Acknowledgments

This research was carried out under financial support from the Russian Foundation for Basic Research (Grant N 05-02-17119) and the Russian Federation President Grants Council (Grant N 5248.2006.2).

References

[1] Volkov B A, Ryabova L I and Khokhlov D R 2002 *Physics-Uspekhi* **45** 819
[2] Story T 1997 *Acta Phys. Polon.* **A92** 663
[3] Skipetrov E, Zvereva E, Skipetrova L, Kovalev B, Volkova O, Golubev A and Slyn'ko E 2004 *Phys. Status Solidi (b)* **241** 1100
[4] Skipetrov E P, Pakpour F A, Skipetrova L A, Pichugin N A and Slyn'ko V E 2006 *Inst. Phys. Conf. Ser.* **27** 27
[5] Slyn'ko V E 2001 *Visnyk Lviv Univ., Ser. Physic* **34** 291
[6] Dornhaus R, Nimtz G and Schlicht B 1983 *Narrow-Gap Semiconductors* (Berlin: Springer)

Journal of Physics: Conference Series **121** (2008) 032004 doi:10.1088/1742-6596/121/3/032004

High-pressure synthesis and study of the crystal and magnetic structures of the distorted $SeMO_3$ (M= Mn, Co, Ni, Zn) perovskites

J.A. Alonso[a*], M.J. Martínez-Lope[a], Cristina de la Calle,[a] A. Muñoz[b], E. Morán,[c] G. Demazeau[d]

[a] Instituto de Ciencia de los Materiales de Madrid, CSIC, E-28049, Cantoblanco-Madrid, Spain.

[b] Dpto. de Física Aplicada, EPS, Universidad Carlos III, Avda. Universidad 30, E-28911, Leganés-Madrid, Spain.

[c]Depto. de Química Inorgánica, Facultad de Ciencias Químicas, Universidad Complutense de Madrid, Ciudad Universitaria, E-28040 Madrid, Spain.

[d] ICMCB,CNRS, Université Bordeaux 1,site de l'ENSCPB, 87 Av. Dr. A. Schweizer, 33608 Pessac Cedex, France

ja.alonso@icmm.csic.es

Abstract. We describe the preparation of $SeMO_3$ (M= Mn, Co, Ni, Zn) under high pressure conditions (3.5-7 GPa), starting from reactive SeO_3H_2 and MO mixtures, contained in sealed platinum capsules under the reaction conditions (850-900° C for 30-60 min). The polycrystalline samples have been studied by x-ray diffraction (XRD) or neutron powder diffraction (NPD) data. $SeMO_3$ crystallize in an O-orthorhombically distorted perovskite structure (space group Pnma), where the strongly tilted MO_6 octahedra show average rotation angles in the 25-27° range. The coordination environments of selenium atoms are strongly distorted due to the presence of the electron lone pair of Se(IV), giving rise to effective $(SeO_3)^{2-}$ polyhedra; the strongly covalent Se-O bonds contribute to the stability of the $a^+b^-b^-$ tilt system of the MO_6 octahedral network. The same antiferromagnetic structure is observed for $SeMnO_3$ and $SeNiO_3$, defined by the propagation vector **k**=0 and the basis vector $(0,0,A_z)$. It is a collinear antiferromagnetic arrangement of Ni^{2+} or Mn^{2+} spins lying along the **c**-direction, defined as G-type (Pbnm setting). For $SeCoO_3$ the magnetic structure is given by the basis vectors $(A_x,G_y,0)$. The different magnetic interactions at low temperature are discussed as a function of the bonding distances and angles.

1. Introduction

There are very few examples of ABO_3 perovskite oxides with valence combination $A^{4+}B^{2+}O_3$, i.e. containing divalent transition metals at the B positions; perovskites such as $ThCuO_3$ or $CeZnO_3$ have never been stabilized. The only 4-2 perovskite oxides described so far contain a

© 2008 IOP Publishing Ltd

p element at the A positions (Se^{4+}, Te^{4+}), in such a way that the considerably covalent A-O bonds and the fact that these cations present a "lone pair" of electrons able to fill the space in the A site contribute to the global stability of the crystal structure.

The $SeMO_3$ and $TeMO_3$ oxides, with M= Mg, Mn, Co, Ni, Cu and Zn, must be stabilized under high pressures and temperature conditions [1,2], given the relatively small size of Se^{4+} and Te^{4+} ions which require a framework of strongly tilted MO_6 octahedra, in orthorhombically distorted perovskite structures. This gives rise to rather bent M-O-M superexchange ϕ angles, which account for varied magnetic properties [2,3], spanning from the ferromagnetism observed for $SeCuO_3$, exhibiting ϕ angles of 121°, to the antiferromagnetism described for $TeCuO_3$, with ϕ= 131°.

The aim of this paper is to report on the synthesis and the crystal and magnetic structures of $SeMO_3$ (M= Mn, Co, Ni, Zn), studied from neutron diffraction data. $SeNiO_3$, $SeCoO_3$ and $SeMnO_3$ had been previously described as antiferromagnets with T_N= 98(1) K, 49(1) K and 51(1) K, respectively [1].

2. Experimental

About 0.6 g of a stoichiometric mixture of H_2SeO_3 and MO (M= Ni, Co, Mn) was thoroughly ground and put into a platinum capsule (6 mm dia.), sealed and placed in a cylindrical graphite heater. The reactions were carried out in a piston-cylinder press (Rockland Research Co.), at a pressure of 3.5 GPa at 850°C for 1 h. Then the materials were quenched to room temperature and the pressure was subsequently released. The preparation of $SeZnO_3$ failed under the mentioned conditions and was then achieved in a belt –type equipment at 7 GPa, at 900°C for 30 min.

The products were initially characterized by laboratory XRD (Cu Kα, λ= 1.5406 Å) for phase identification and to assess phase purity. The neutron powder diffraction (NPD) data were acquired at the Institut Laue-Langevin, Grenoble (France) for M= Mn, Co and Ni. The fitting of the crystallographic structure was carried out from NPD patterns collected at the high resolution D2B diffractometer with a 1.594 Å wave-length at room temperature. The magnetic structures were analysed from sets of NPD patterns collected at low temperature in the high-flux D20 diffractometer, with a 2.42 Å wave-length. The analysis of the NPD patterns was carried out by the Rietveld method using the Fullprof program [4]. For the simulation of the peak profile a pseudo-Voigt function was considered. The background was fitted with a fifth degree polynomial function.

3. Results and Discussion

3.1. *Synthesis.* The use of H_2SeO_3 as starting reactant seems to be crucial for the crystallization of the perovskites; previous trials with SeO_2+MO mixtures as starting reactants were unsuccessful. Perhaps the reaction proceeds via a water-vapor transport-driven mechanism. The XRD patterns (Fig. 1 and 2) are characteristic of well crystallized perovskites and can be indexed in orthorhombic unit cells in the space group Pnma.

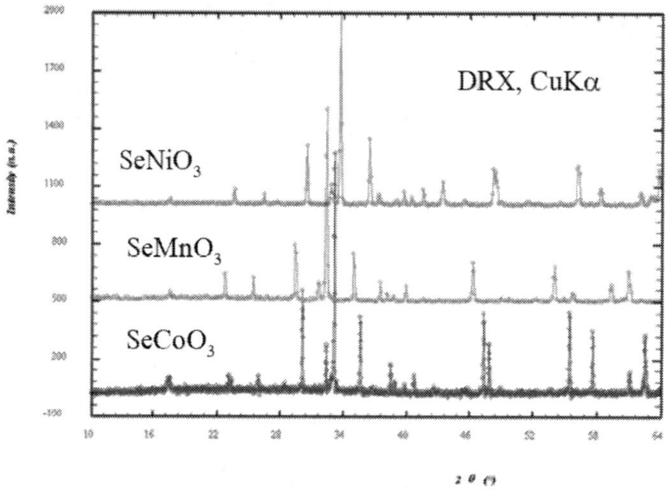

Fig. 1. XRD patterns of $SeMO_3$ (M= Ni, Mn, Co) at RT.

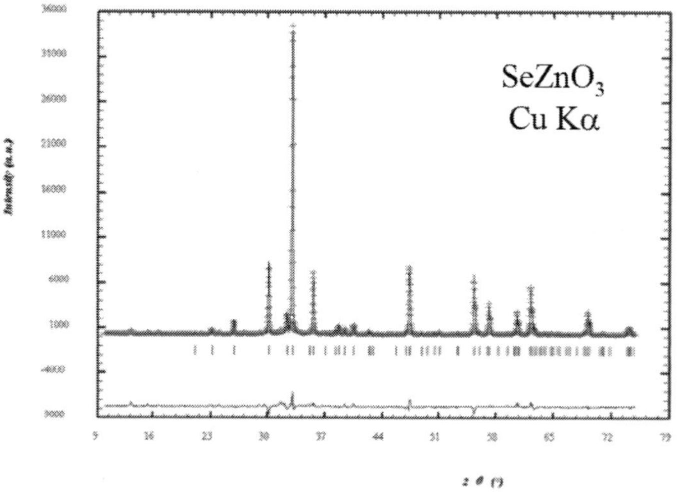

Fig. 2. Rietveld plot of $SeZnO_3$, refined from XRD data at RT.

3.2. *Crystallographic refinement.* The refinement of the crystallographic structure of $SeMO_3$ (M= Mn, Co, Ni) was performed from NPD patterns obtained at room temperature with a λ=1.594 Å wavelength. In spite of the relatively small amount of sample (0.5 g) available from the high-pressure synthesis procedure, reasonably good NPD patterns could be collected in the D2B diffractometer operating in the high-flux mode. Unhappily, the much smaller available sample for $SeZnO_3$ did not allow us the collection of exploitable NPD patterns, thus the structure was refined from XRD data. In all cases the structure was defined in the orthorhombic space group Pnma (No. 62), Z= 4, with unit-cell parameters related to $\mathbf{a_0}$ (ideal cubic perovskite, $\mathbf{a_0} \approx$ 3.8 Å) as $\mathbf{a} \approx \mathbf{c} \approx \sqrt{2}\mathbf{a_0}$, $\mathbf{b} \approx 2a_0$. Se atoms were located at *4c* positions, M (Mn, Co, Ni, Zn) at *4b* and oxygen atoms at *4c* and *8d* positions, respectively. The good agreement between the observed and calculated NPD patterns is illustrated in Fig. 3 for M= Mn, Co and Ni. The Rietveld plot obtained for $SeZnO_3$ from XRD data is displayed in Fig. 2.

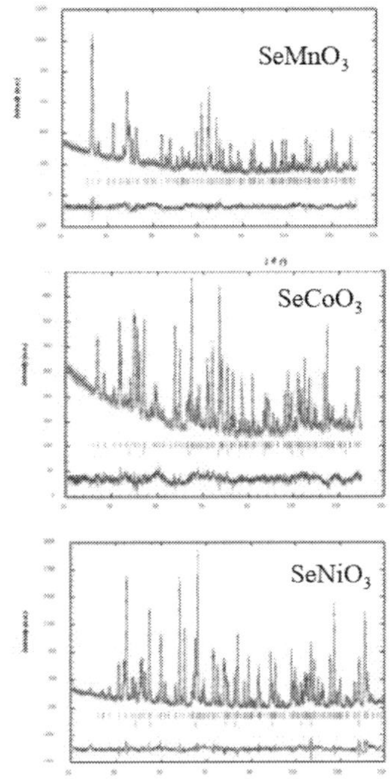

Fig. 3. Rietveld plots of SeMO$_3$ (M= Mn, Co, Ni) refined from NPD data at RT.

The main crystallographic parameters and the reliability factors obtained after the structural fit for SeZnO$_3$ are included in Table 1. Details of the NPD Rietveld refinements for M= Co, Mn, and Ni are given elsewhere [5,6]. The main results for all SeMO$_3$ perovskites are included in Table 2.

Table 1. Structural parameters for SeZnO$_3$ refined in the orthorhombic Pnma space group at room temperature from XRD data.[a]

Atom	Site	x	y	z	B(Å^2)
Se	4c	0.0256(4)	1/4	-0.0146(6)	0.23(9)
Zn	4b	0	0	1/2	1.4(1)
O1	4c	0.105(2)	1/4	0.333(3)	1.6(2)
O2	8d	0.2102(16)	0.0859(15)	0.8752(18)	1.6(2)

[a]Discrepancy factors: Rp= 9.49, R_{wp}= 12.08 and R_{Bragg}= 8.96 %

Table 2. Main crystallographic parameters for $SeMO_3$ (M= Mn, Co, Ni, Zn) determined at RT from NPD (M= Mn, Co, Ni, taken from references [5,6]) or XRD (M= Zn).

M	Mn	Co	Ni	Zn
a(Å)	6.0930(2)	5.9300(2)	5.87912(8)	5.9239(2)
b (Å)	7.8638(2)	7.5961(2)	7.52467(9)	7.6684(3)
c (Å)	5.1426(2)	5.0287(2)	4.93737(7)	5.0421(2)
V (Å^3)	246.409(13)	226.515(11)	218.421(8)	229.05(1)
<M-O>	2.232(3)	2.149(2)	2.114(2)	2.16(1)
<Se-O>	1.712(3)	1.715(3)	1.724(3)	1.78(1)
<M-O-M>	125.48(9)	126.17(8)	128.86(6)	127.0(1)
a(Å)	6.0930(2)	5.9300(2)	5.87912(8)	5.9239(2)
b (Å)	7.8638(2)	7.5961(2)	7.52467(9)	7.6684(3)
c (Å)	5.1426(2)	5.0287(2)	4.93737(7)	5.0421(2)
V (Å^3)	246.409(13)	226.515(11)	218.421(8)	229.05(1)
<M-O>	2.232(3)	2.149(2)	2.114(2)	2.16(1)
<Se-O>	1.712(3)	1.715(3)	1.724(3)	1.78(1)
<M-O-M>	125.48(9)	126.17(8)	128.86(6)	127.0(1)

A view of the crystallographic structure of $SeMnO_3$ is presented in Fig. 4, highlighting the significant tilting of the MO_6 octahedra. The ionic radius of the Se^{4+} cation, 0.64 Å [7], is very small to occupy the A-perovskite position (typical A^{3+} and A^{2+} cations sizes are around 1-1.4 Å), which gives rise to a huge tilting effect of the MO_6 octahedra in order to optimize the Se-O distances. In $SeMO_3$, average <M-O-M> angles (ϕ) are in the range 125.5-128.9° (Table 2), which are abnormally small in perovskite-related structures. The average tilting angles, estimated as $\varphi = (180-\langle\phi\rangle)/2$ are 25.6° and 27.3° for M= Ni, Mn, respectively. The Mn perovskite is the more distorted perovskite and the Ni compound is the less distorted one, as it corresponds to the ionic radii sequence (0.83 Å, 0.745 Å, 0.74 Å and 0.69 Å for Mn^{2+}, Co^{2+}, Zn^{2+} and Ni^{2+}, respectively [7]) yielding increasing tolerance factors.

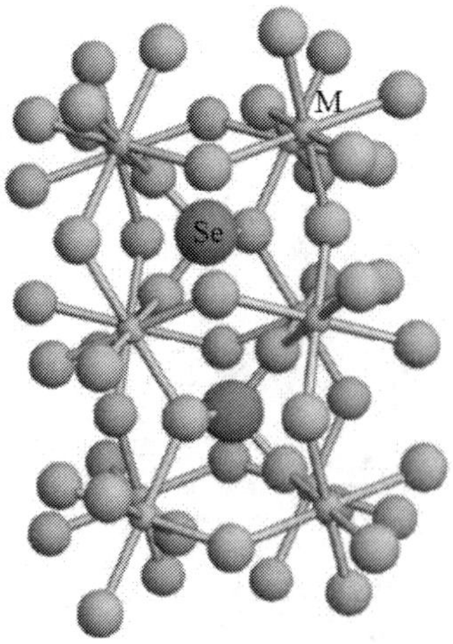

Fig. 4. View of the crystal structure of $SeMO_3$, highlighting the strong tilting of the MO_6 octahedra and the threefold coordination of Se^{4+} cations.

As a consequence of the strong tilting effect, the oxygen environment of Se^{4+} cations is very irregular, showing 3 short bond lengths (1.69-1.73 Å) and 5 long bonds in the 2.8-3.0 Å range [5,6].

Therefore, Se^{4+} cations form strong covalent bonds with three out of the 8 surrounding oxygens, forming a $(Se\text{-}O_3)^{2-}$ polyhedron, in a trigonal pyramidal conformation with O-Se-O angles between 98º-104º [5,6]. The coordination of Se to 3 oxygens is also illustrated in Fig. 6. This irregular oxygen environment around Se is the result of the presence of the Se^{4+} non-bonded lone $4s^2$ electron pair, which is thought to be directed towards the apex of each trigonal pyramid. It is very plausible that the formation of the mentioned very covalent Se-O bonds are an important ingredient for the stability of this family of $SeMO_3$ perovskites.

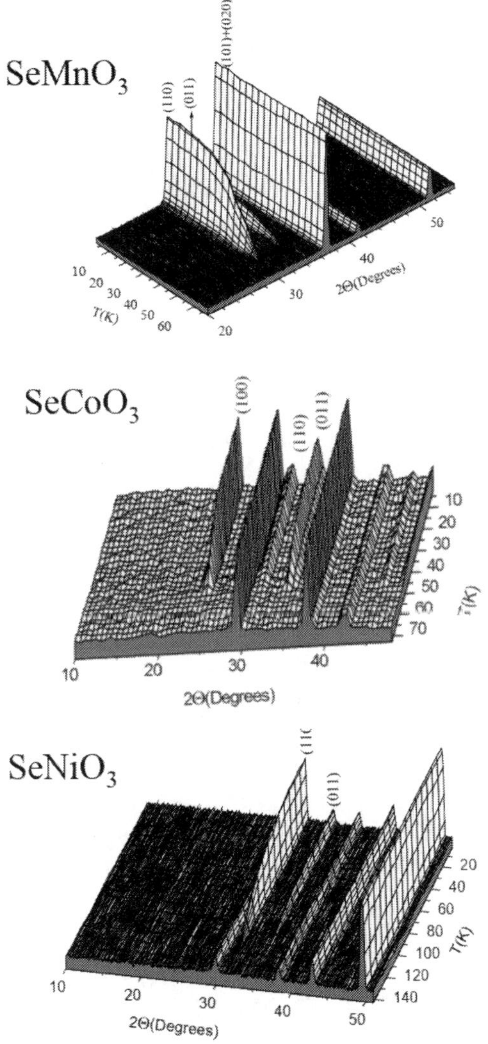

Fig. 5. Thermal evolution of the NPD patterns for $SeMO_3$ (M= Mn, Co, Ni)

3.3. *Magnetic structure resolution.* The magnetic structure resolution was carried out from a set of low temperature NPD patterns acquired with λ=2.42 Å, for $SeMO_3$ M=Mn, Co, Ni. As shown in Fig. 5, on decreasing the temperature new peaks of magnetic origin appear for the three compounds at positions forbidden by the space group Pnma; in particular, the new peaks appear below 53.5 K, 104 K and 54 K for $SeMnO_3$, and $SeCoO_3$ and $SeNiO_3$, respectively, implying the appearance of a magnetic order below T_N, in good agreement with the magnetic measurements [5,6]. In all the three compounds the magnetic structure is defined by the propagation vector **k**=0, indicating that the magnetic unit cells coincide with the chemical ones.

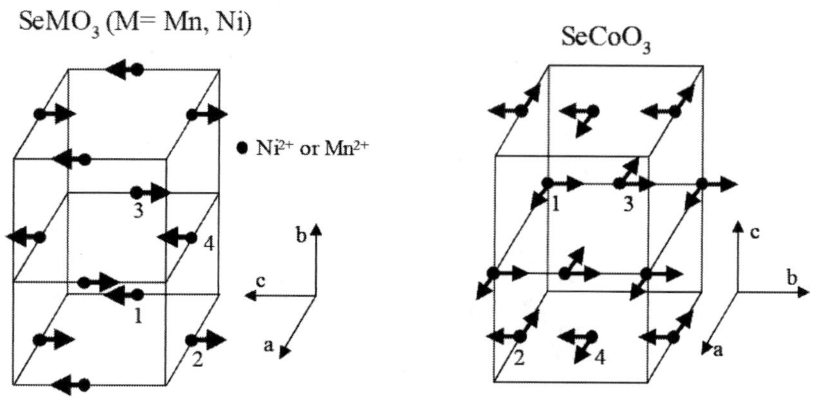

Fig. 6. Magnetic structures of $SeMO_3$ (M= Mn, Ni, Co)

For $SeMO_3$ (M= Mn, Ni) the magnetic structure that shows a better agreement with the experimental data corresponds to the magnetic mode $(0,0,A_z)$. This mode defines that the magnetic moments are oriented along the z-direction. In $SeNiO_3$ the ordered magnetic moment for the Ni^{2+} ions at T=2.3 K is 2.11(3) μ_B, whereas in $SeMnO_3$ at T=2.6 K for Mn^{2+}, |m|=4.64(2) μ_B. A view of the magnetic structures is shown in Fig. 6. The magnetic structure can be described as antiferromagnetic (010) layers of magnetic moments lying along the **c**-direction, which are antiferromagnetically coupled along the **b**-direction. For $SeCoO_3$ the magnetic structure is given by the basis vectors $(A_x,G_y,0)$, defining a non-colinear but non-canted structure (Fig. 6). In the **(a,c)** planes the coupling of the magnetic moments for each Co atom with its four closer neighbours is purely antiferromagnetic. For the x-component of the magnetic moment the coupling between the (010) layers along **b** is antiferromagnetic, whereas for the y-component the coupling is ferromagnetic. At T=2.2 K, the ordered magnetic moment for Co^{2+} ions is |m|=3.37(3) μ_B, suggesting a high-spin electronic configuration for Co^{2+} cations, $t_{2g}^5 e_g^2$, with some component of unquenched orbital moment.

Finally, it is surprising that the ordering temperatures are considerably different for $SeNiO_3$ *vs* $SeCoO_3$ and $SeMnO_3$ perovskites, with $T_N \approx 104$ K vs $T_N \approx 54$ K and $T_N \approx 53.5$ K, respectively, suggesting that the superexchange interactions are stronger in $SeNiO_3$, despite the significantly larger magnetic moment of Mn^{2+} and Co^{2+}. This fact can be interpreted by taking into account that the superexchange process is improved by the more open angles and shorter bond lengths observed in the Ni-O-Ni pathways (average Ni-O-Ni angle of 129°; average Ni-O distance of 2.114 Å) with respect to Mn-O-Mn and Co-O-Co superexchange paths (Table 2), favoring the overlap between oxygen and metal orbitals and thus enhancing the magnetic interactions.

4. Conclusions

We have prepared $SeMO_3$ (M= Mn, Co, Ni, Zn) under hydrostatic pressures in sealed Pt capsules; it seems that the choice of the starting reactants is of paramount importance for obtaining pure perovskite phases. A neutron study for the Mn, Co and Ni compounds demonstrate that they are very distorted superstructures of perovskite. A sequential NPD study shows that the magnetic ordering temperatures are considerably different for $SeNiO_3$ *vs* $SeCoO_3$ and $SeMnO_3$ perovskites, with $T_N \approx 104$ K vs $T_N \approx 54$ K and $T_N \approx 53.5$ K. This fact can be related with the geometrical parameters (superexchange angles and bond lengths), favoring the overlap between oxygen and metal orbitals and thus enhancing the magnetic interactions for $SeNiO_3$.

Acknowlegments : We are grateful to Spanish Ministry of Education and Science for funding the project MAT2007-60536. This study was developed in the framework of the COST D30 program Working Group D30 /03 (Development of Materials Chemistry under High Pressures).

References

[1] Kohn, K.; Inoue, K.; Horie O.; Akimoto, S. *J. Solid State Chem.* **1976,** *18*, 27.
[2] Subramanian, M.A.; Ramirez, A.P.; Marshall, W.J. *Phys. Rev. Lett.* **1999,** *82*, 1558.
[3] Lawes, G.; Ramírez, A.P.; Varma C.M.; Subramanian, M.A. *Phys. Rev. Lett.* **2003,** *91*, 257208.
[4] Rodríguez-Carvajal, J. *Physica B* **1993,** *192*, 55
[5] Muñoz, A.; Alonso, J.A.; Martínez-Lope, M.J.; Morán, E.; Escamilla, R., *Phys. Rev. B*, **2006,** *73*, 104442.
[6] Muñoz, A.; Alonso, J.A.; Martínez-Lope, M.J.; Falcón, H; García-Hernández, M.; Morán, E; Dalton Trans. **2006**, 4936.
[7] Shannon R.D. *Acta Cryst. A* **1976,** *32*, 751.

Journal of Physics: Conference Series **121** (2008) 032005 doi:10.1088/1742-6596/121/3/032005

p-n junction on high-pressure-high-temperature grown single crystal diamond: UV-emission spectra and electrical properties

S. G. Buga, V. D. Blank, V. S. Bormashov, V. N. Denisov, S. A. Terentiev, A. N. Kirichenko, M. S. Kuznetsov, V. N. Mordkovich, S. A. Nosukhin

Technological Institute for Superhard and Novel Carbon Materials, Troitsk, Moscow region, Russia

E-mail: bormashov@bk.ru

Abstract. Semiconductor *p+-p-n* structure has been created on synthetic single crystal IIb type diamond grown by high pressure temperature gradient method and doped with boron at growth process. 250 μm thick (001) cut plate was ion implanted by B, As and P at $(1 - 1.3)\cdot 10^{16}$ cm^{-2}. Electroluminescence spectra were investigated at different values of electrical voltage and current in regular and S-part of the current-voltage characteristics. Strong exciton recombination emission was observed at the current density up to 60 A*cm^{-2}. The spectrum transforms at transition from regular conductivity to electrical avalanche breakdown regime. Simultaneously a set of sharp EL lines appears in the range of 327-652 nm on the A-band background. They tentatively attributed to superlumenscence effect.

1. Introduction

Diamond is a wide band gap semiconductor with 5.49 eV forbidden indirect gap. There are no intrinsic charge carriers in pure diamond, but doping provides *p*- and *n*-type diamonds with various activation energy of holes and electrons. At doping level up to ~ 10^{17} cm^{-3} diamond has extremely high heat conductivity and mobility of charge carriers. Diamond possesses outstanding mechanical and optical properties. It is very promising material for electronics and optoelectronic.

Thought studies of natural and doped semiconductor diamonds have been started more than 40 years ago, the most extensively this area began to explore during the last 10 years with the development of methods of growth of synthetic single crystal diamonds with controlled chemical content. The main methods of single crystal growth are the temperature gradient method under pressure [1, 2] and the method of chemical vapor deposition (CVD) [3]. There is a set of different techniques applying these methods. The advantage of the first type method is an opportunity to grow large (10 carats and more) bulk single crystals, very pure IIa type and doped with boron acceptor impurity (IIb type) or nitrogen donor impurity (Ib type) at broad range of concentrations. Microwave plasma assisted CVD method (MPCVD) provides growth of high quality single crystal diamonds: bulk and films, but it requires substrate single crystal diamonds grown by the first method. Thus combination of both these methods is the best for producing layered diamond structures for electronics and optoelectronics.

Besides direct doping at a growth process, ion implantation is a strong tool of additional doping with various chemical elements at a depth down to about several microns. Actually the first *p-n*

© 2008 IOP Publishing Ltd

junctions in natural diamonds have been made by ion implantation method and their electrical properties and optical emission in visible range investigated [4, 5]. It was found, that the electroluminescence (EL) spectra are similar to X-ray photoluminescence (PL) ones and the internal quantum yield attains $\sim 4\times10^{-2}$ per couple of recombinating charge carriers. The maximum of EL and PL spectra was in 420-530 nm attributed to nitrogen impurity luminescence centers (LC) and lattice defects. The maximum electrical current of about 20 mA and rectification ratio of $\sim 10^7$ have been obtained in a *p-i-n* structure made by ion implantation of natural diamond [5]. But besides strong PL and EL in visible range, holes and electrons injection in diamond provides quite strong exciton recombination band with the photon energy close to indirect band gap value: 5.27 eV [6-10]. The exciton binding energy is about 80 meV in diamond, that is why UV emission band can be observed at high temperatures.

Besides making UV diodes, producing diamond UV lasers is a very attractive task. To obtain coherent emission at excitonic UV band one must attain high density of excitons to get exciton liquid condensation effect [11]. In a near UV and visible EL band 300-600 nm one must provide inverse occupation of metastable energetic levels associated with impurities in forbidden gap. Both these tasks can be solved by high density of electrical current with injection of holes and electrons. In this work we made p^+-p-n^+ -structure on synthetic single crystal diamond implanted by B, As and P and investigated their electrical properties and EL spectra at high current density.

2. Experimental

We grew boron-doped IIb single crystal diamond by the temperature gradient method [1, 2] and cut 250 μm thick (001) plate with maximum sizes 3×5 mm^2 (Figure 1).

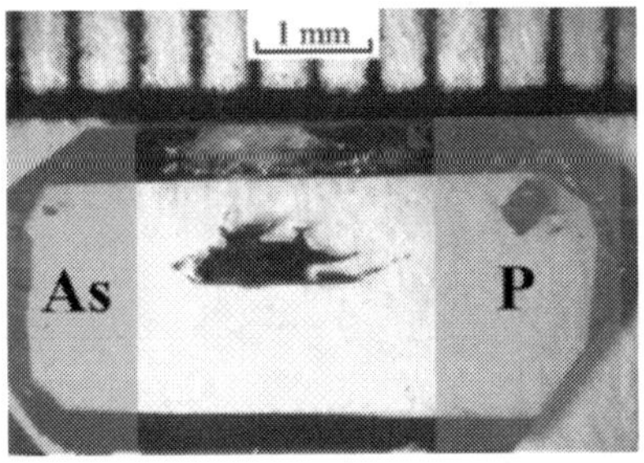

Figure 1. Picture of diamond plate with marked areas of implantation with As and P ions. Black spot in the central part is a growth defect.

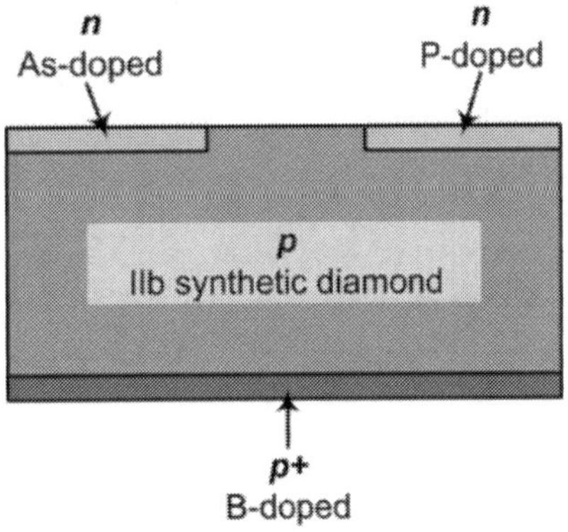

Figure 2. Scheme of *p+-p-n* structure.

One plane of the plate was implanted by B and the opposite one – by As and P according to Figure 2. Central part of the plate contains growth defect and this area was not implanted by donor atoms. The implantation conditions are shown in Table 1.

Table 1. Implantation depth, impurity type, implantation energy and dozes for *p+-p-n* structures on IIb type synthetic single crystal diamond

Depth, μm	Impurity	Implantation energy, keV	Doze, cm^{-2}
0,5	As^+	180	$1\cdot10^{16}$
1,1	P^+	180	$1\cdot10^{16}$
3,5	B^+	100	$1,3\cdot10^{16}$

After implantation crystal was annealed at 600°C for 6 hours in argon gas flow. Electrical contacts have been made by silver conducting paste. The crystal was attached to massive heat sink copper plate by the same silver conducting paste. The IR absorption spectrum of implanted crystal was measured using Thermo Nicolet Nexus FT-IR spectrometer. The spectrum of the studied sample has typical for boron-doped diamonds absorption peak at 1282 cm^{-1} and absorption tail at wavenumbers above 2700 cm^{-1}. 3 additional absorption peaks at 2455, 2799 and 2926 cm^{-1} indicate absorption on implanted donor atoms. The estimated acceptor concentration is $\sim 10^{16}$ cm^{-3}.

Fig.3 shows current-voltage (*I-V*) characteristics in both donor-implanted areas. At direct current values above ~ 0.2 A *I-V*-characteristics sharply became *S*-type with significant drop in voltage at limited current. The temperature of the sample increased rapidly and this part of *I-V*-characteristics was not measured correctly and not presented in the plots. The minimal electrical resistance was 70 and 425 Ω at two different points in P-doped area and 50 Ω in As-doped area. The maximal electrical power was 25 and 17 W in P-doped area and 12.5W in As-doped area respectively. The maximum current density was estimated 60 A*cm^{-2} according to the area of electrical contacts and visible crossection of the light-emitting area. The rectification ratio was relatively low: just about 10^2 at ± 40 V in a regular part of *I-V*-characteristics at As and P-doped areas.

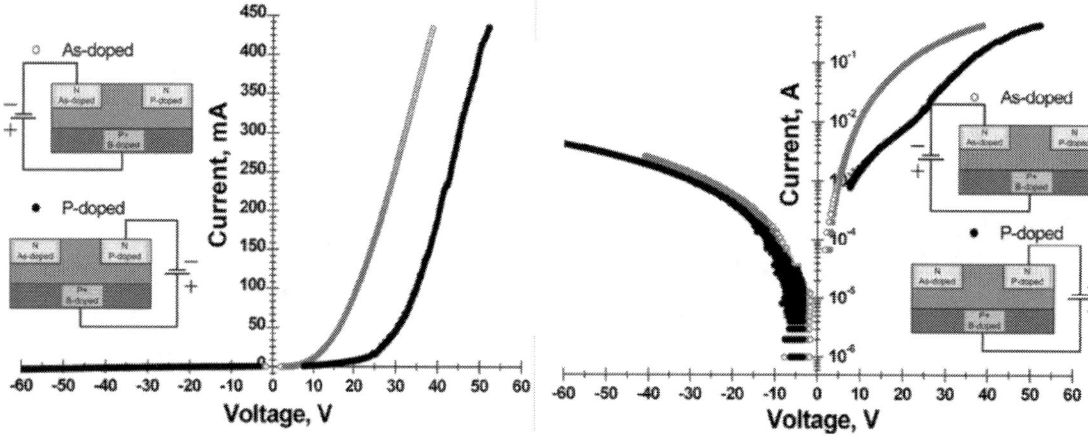

Figure 3. *I-V* characteristics of *p+-p-n* structures in As-and P implanted areas in linear (left) and semilogarithmic (right) scale.

Bright optical emission was observed at direct current in both implanted areas. EL spectra at different values of voltage and electrical current are presented in Fig. 4. EL spectra were measured using TRIAX 552 spectrometer.

As seen in Fig. 4 spectra in both As and P-doped areas contain typical for diamonds broad luminescence *A* band in 350-650 nm wavelengths with a maximum at about 500 nm. This band is attributed to lattice defects and nitrogen luminescence centers. The intensity of excitonic bands at 230-250 nm in As-doped area of the structure is relatively low – about 3 orders of magnitude less than the intensity of visible *A* band. Unlike As, P-doping provides quite strong UV excitonic luminescence with the maximum intensity just about 30% of the *A* band. Such a high UV yield was observed at the regime of high current density in the *S*-part of *I-V*-characteristics. The ratio was about 5-10% in a regular part of *I-V*-characteristics. Taking into account, that the integral intensity of EL at high electrical current regime is much stronger than at medium values, we note that the obtained maximal integral excitonic intensity is really strong. But the most interesting result is an evident transformation of EL spectra observed at high current density with respect to medium current regime (Fig. 4 right). Particularly interesting is transformation of excitonic bands and appearance of sharp EL bands in whole range from 250 to 650 nm.

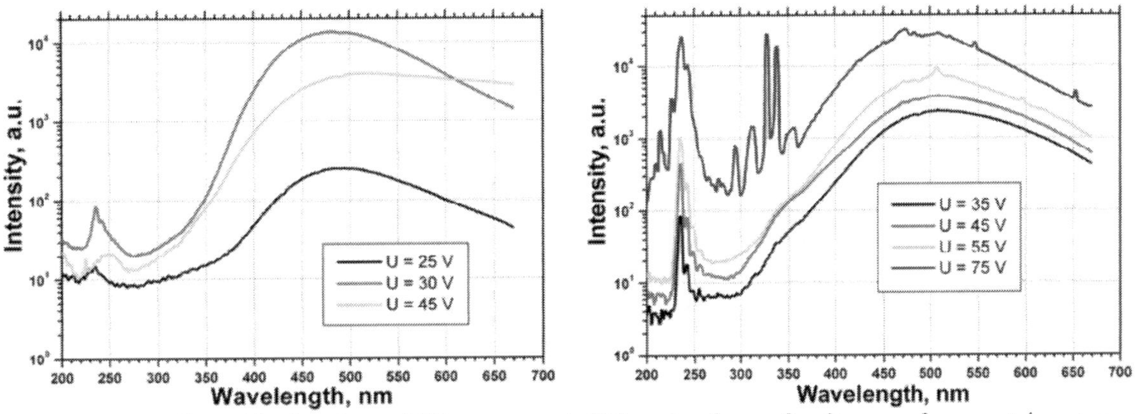

Figure 4. Semilogarithmic plots of EL spectra at different values of voltage and current in *p+-p-n* structures in As-implanted area (left) and P-implanted area with higher electrical resistance (right).

There are two very intensive lines at 327,3 nm and 337,4 nm. The Gauss line fitting revealed a fine structure of the bands. The BHWHA (band halfwidth at a half of amplitude) of Gauss peaks is in the range of 7-20 meV, that is significantly less than kT = 25 - 50 eV for the temperature range 300 – 600 K. We assume that these narrow bands arise due to superluminescence effect at electron transitions from P energetic level (0.6 eV below the conduction band) directly to boron acceptor level (0.37 eV above the valence band) or via intermediate nitrogen donor level (1.6 eV below the conduction band). The sublevels associated with defects have place, thus a set of lines appear in the 3.5-4.5 eV range.

3. Conclusion

We investigated peculiarities of UV and visible optical emission of p^+-p-n structure on ion implanted diamond at high electrical current density for the first time. Strong exciton recombination UV emission was observed accompanied with a change in spectra at transition from regular to *S*-part of current-voltage characteristics of this semiconductor structure. In addition a set of sharp light emission bands appeared in a broad range from 327 to 650 nm. The strongest of them are double peak at 327,3 and 337,4 nm. They may arise due to superluminescence effect.

References

[1] R.H. Wentorf. *J. Chem. Phys.*, **75,** 1833 (1971).;
[2] H.M. Strong, R.M. Chrenko. *J. Chem. Phys.*, **75,** 1838 (1971).
[3] B.V. Spitsin "Growth of Diamond Films from the Vapour Phase. Handbook of Crystal Growth" Ed. D.T. J. Hurle, Amsterdam, Elsevier, 1994, v.3. p. 401.
[4] V.P. Butusov, V.I. Laptev, V.A. Personov, Yu.M. Rotner, *Dokladu Akademii Nauk USSR, "Cyistallography" series,* **226**, №2, (1976) 328 (in Russian).
[5] M.I. Guseva, E.A. Konorova, Yu.A. Kuznetsov, V.F. Sergienko, *Sov. Phys. Semicond*, **12**, №3 (1978) 505 (in Russian).
[6] K. Horiuchi, K. Nakamura, S. Yamashita: *Jpn. J. Appl. Phys.* **39** (2000) L604;
[7] S. Koizumi, K. Watanabe, M. Hasegawa, H. Kanda, *Science* **292** (2001)1899.
[8] W.L. Wang, K.J. Liao, C.Z. Cai, Y.B. Zhu, Y. Ma, H.Y. Liao. *Diam. and Rel. Mat.* **12** (2003) 1385.
[9] T. Makino, N.Tokuda, H. Kato et. al. *Appl. Phys. Lett.* **86** (2005) 222111.
[10] P.J. Dean, J.C. Male, *J. Phys. Chem. Sol.*, **25** (1964) 1369.
[11] N. Teofilov, R. Schliesing, K.Thonke, H. Zacharias, R. Sauer, H. Kanda, *Diam. and Rel. Mat.* **12** (2003) 636.

Journal of Physics: Conference Series **121** (2008) 032006 doi:10.1088/1742-6596/121/3/032006

Pressure dependence of the energy gaps in diamond-type semiconductors, and their III-V analogues such as InSb

G. G. N. Angilella[1], N. H. March[2,3], I. A. Howard[4], R. Pucci[1]
[1]Dipartimento di Fisica e Astronomia, Università di Catania,
and INFN, Sez. Catania, and CNISM, UdR Catania, Via S. Sofia, 64, I-95123 Catania, Italy
[2]Oxford University, Oxford, UK
[3]Department of Physics, University of Antwerp, Groenenborgerlaan 171, B-2020 Antwerp, Belgium
[4]Department of Chemistry (ALGC), Free University of Brussels (VUB), B-1050 Brussels, Belgium

E-mail: giuseppe.angilella@ct.infn.it

Abstract. In earlier work by March and Matthai [Phil. Mag. Lett. **84**, 335 (2004)], the energy gaps E_g of the diamond-type semiconductors C, Si, Ge and α-Sn were shown to correlate well with the mean interelectronic separation r_s of the 4 valence electrons per atom. Furthermore for the III-V analogues such as GaAs and InSb, it was demonstrated that a useful extension could be effected by introducing additionally the electronegativity difference between the constituent atoms. Here, we consider available experimental data, and their interpretation, for such energy gaps, as a function of pressure. The intensity of the X-ray forbidden reflections will also be discussed under pressure.

Key words: tetrahedral semiconductors; pressure dependence; lattice parameter dependence; energy gap; III-V compounds; X-ray spectroscopy.

The behavior of the diamond-structure semiconductors under pressure has been a topic of experimental and theoretical study for some time (see, *e.g.*, Ref. [1]). Nathan *et al.* [2], *e.g.*, carried out a thorough study of the pressure dependence in n-type Ge compared to, for example, Si, analyzing the effects of interband scattering. More recently, March and Matthai [3] have given a quantitative representation of the energy gap E_g in the diamond-structure semiconductors C, Si, Ge, and α-Sn. They obtain a good fit of E_g at atmospheric pressure for these four materials by writing

$$E_g = A \left(1 - \sqrt{\frac{a}{a_c}} \right)^{\gamma}, \tag{1}$$

where a is the lattice parameter, and the fitting parameters are $a_c = 6.7797$ Å, $A = 50.21$ eV, and $\gamma = 1.71675 \approx \frac{5}{3}$. Eq. (1) supports a description of the dependence of the band gap on the interelectron mean spacing in elemental tetravalent semiconductors in terms of a semiconductor-to-metal phase transition (SMT), in which E_g plays the role of the order parameter. Fig. 1 plots Eq. (1) against available experimental data (Tab. 1) for the four diamond-type elemental semiconductors mentioned above, as well as for several III-V analogues, both in zinc-blende and in wurtzite structures. Tab. 1 also records values for two selected II-VI compound

© 2008 IOP Publishing Ltd

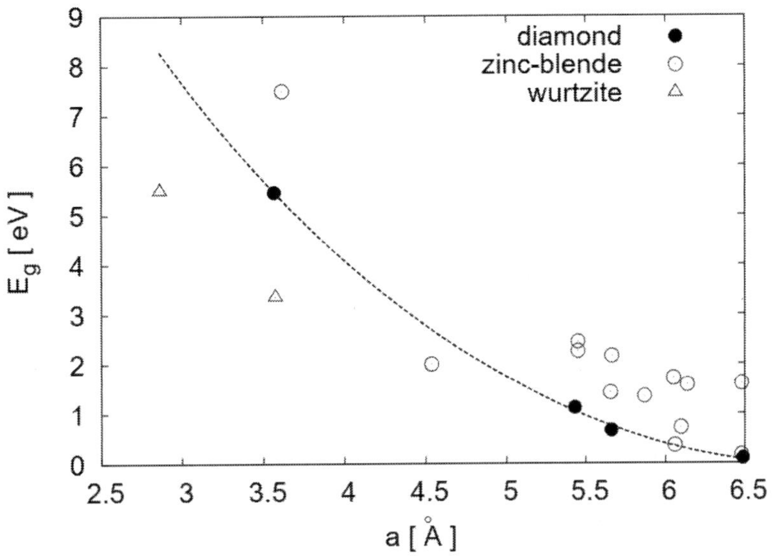

Figure 1. Energy gap E_g at atmospheric pressure against lattice parameter a for diamond-type semiconducting elements C, Si, Ge, and α-Sn (filled circles), and III-V, II-VI analogues in zinc-blende (open circles) and wurtzite structure (triangles), as listed in Tab. 1. Dashed line is March-Matthai fitting curve for diamond-type elements, Eq. (1) [3].

semiconductors (CdSe and CdTe), which have been recently employed in the construction of nanoscale electron devices [4, 5].

While Eq. (1) is more appropriate to describe a SMT close to the critical point, *i.e.* a *vanishing* band gap at a finite lattice parameter $a \sim a_c$, it is important to note that Litovchenko [6] has verified a correlation $E_g \propto 1/a^2$, which agrees with experimental data for $a \ll a_c$, and is justified by the asymptotic, low-a power-law dependence of the energy matrix elements on the lattice parameter a within the tight-binding approximation [7]. That an SMT transition is related to a change in lattice parameter is also confirmed by the empirical correlation between the transition pressure in semiconducting chalcogenides and an appropriate average of the cation and anion radii [8].

March and Matthai [3] proceed to generalize their results for the four monatomic diamond-structure semiconductors. However, they avoid undue size differences by relating AlP to Si, GaAs to Ge and InSb to α-Sn. Their proposal was to express the energy gap in these three binary semiconductors as the sum of two parts, $E_g = E_g^e + E_g^{ed}$, where E_g^e is the 'elemental gap' (*e.g.* Si for AlP etc.), while E_g^{ed} expresses the electronegativity difference between the two atoms in the III-V semiconductors. They show (see Fig. 3 in Ref. [3]) that $\Delta E_g = E_g - E_g^e$ is proportional to the difference between the ionization potentials, say ΔI, of the two atoms concerned. This is the approximate chemical estimate for electronegativity difference. As we expect ΔI to be insensitive to pressure, we expect $dE_g/dP|_{P=0}$ to be nearly the same for AlP as for Si, there being just a slight difference in their equilibrium lattice parameters a (cf. Tab. 1 for Si and AlP, Ge and GaAs, and α-Sn and InSb).

Making use of the Murnaghan equation of state (EOS) [16], one can estimate the pressure dependence of the lattice parameter as

$$a(P) = a(0)\left(1 + \frac{B'}{B}P\right)^{-\frac{1}{3B'}}, \tag{2}$$

	a (Å)	E_g (eV)	ΔI (eV)	B (GPa)	B'	$dE_g^{\rm exp}/dP$ (meV/GPa)	$dE_g^{\rm calc}/dP$ (meV/GPa)	$-P_c$ (GPa)
	diamond [cF8 (A4)]:							
C	3.567	5.46		442	3.8	0.7 − 6.2 [9]	9.3	116
Si	5.431	1.12		97.82	4.4	−14.1	27.9	21
Ge	5.658	0.661		75	4.5	30	26.6	15
α-Sn	6.4892	0.1		53	4.6	—	24.4	6
	III-V, zinc-blende [cF8 (B3)]:							
BN	3.6150	7.5	6.234	395 [10]	3.65 [10]	3.0	14.7	108
BP	4.5380	2.0	2.188	165 [11]	4.7 [11]	—	15.6	35
AlP	5.4510	2.45	4.5	86.5 [12]	4.1 [12]	94.2* [13]	70.3	21
GaP	5.4512	2.26	4.484	88	4.7	95	63.8	19
GaAs	5.65325	1.424	3.81	75.3	4.8	126	56.9	15
AlAs	5.6605	2.16	3.826	74.1 [12]	4.0 [12]	104.5* [13]	88.3	18
InP	5.8686	1.35	4.699	71	4.8	84	72.7	14
InAs	6.0584	0.36	4.025	58	4.9	48	30.7	10
GaSb	6.0959	0.72	2.639	56.3	4.9	145	67.0	11
AlSb	6.1355	1.58	2.655	59.3 [12]	4.5 [12]	140.8* [13]	148.9	13
InSb	6.4794	0.17	2.854	47	4.9	137	45.2	6.4
	II-VI, zinc-blende [cF8 (B3)]:							
CdSe	6.050	1.713	0.759	53.0	0.48	43.1 [5]	158	39
CdTe	6.481	1.606	0.74	44.5	6.4	75.9* [4]	453	7
	III-V, wurtzite [hP4 (B4)]:							
BN	2.86362	5.5	6.234	394 [10]	3.68 [10]	3.8	7.4	107
GaN	3.57434	3.36	8.53	210	4.6	41.4 [13]	12.1	45.6

* Calculated at Γ point.

Table 1. Relevant data for several elemental, III-V and II-VI compound semiconductors, in diamond-type structures (from Refs. [14, 15] and refs. therein, unless specified). For the two wurtzite compounds, the average lattice parameter d, defined as $d^3 = \sqrt{3}a^2c/2$, is actually recorded.

where B, B' are the bulk modulus and its pressure derivative at $P = 0$. Eqs. (1) and (2) combined can then be used to calculate $dE_g/dP|_{P=0}$, which is confronted with experimental data in Tab. 1. With the notable exception of Si, for which a negative value of $dE_g^{\rm exp}/dP$ is recorded, there is a quite clear correlation between calculated values and experiment, though the gap variation is underestimated substantially for GaSb and InSb, and overestimated for CdSe and CdTe.

Motivated by the correlation provided by Eq. (1) as shown in Fig. 1, we then seek a pressure counterpart of the 'critical' value of the lattice parameter a_c at which an SMT takes place, with vanishing E_g. Since among diamond-type semiconductors larger lattice spacings are required for lower band gaps, it is clear that such a critical pressure, say P_c, at which a STM phase transition takes place, must be negative, thus implying some kind of lattice expansion, as can be *e.g.* brought about thermally, or by epitaxially growing films on a suitably lattice-mismatched substrate. In the latter case, a generalization of the Murnaghan EOS would then be required for strained layers, as in Ref. [17]. The last column of Tab. 1 then records our estimates for $-P_c$, from which it is seen that $-P_c$ shows a direct correlation with E_g at ambient conditions.

Beyond specific measurements of E_g as a function of pressure, it seems of interest to summarize here some studies, both theoretical and experimental, relating to X-ray diffraction for diamond-lattice semiconductors under compression. Briefly, the study of March and Flores [18] discussed the significance of the forbidden 222 reflection as one approaches a metallic phase as a result of applying pressure to an originally diamond-structure semiconductor. They considered that the X-ray intensity at this forbidden 222 reflection should be useful in discussing such a transition to a metallic phase. Earlier work of Zunger and Freeman [19] calculated the 222 reflection for diamond under pressure. From an experimental point of view, Yoder-Short *et al.* [20] performed X-ray diffraction studies of Si and also InSb as they were compressed near the transition to a metallic β-Sn structure. In both cases, the intensity of the 222 reflection was observed to undergo a sharp decrease just before the structural transition, which occurs at $\sim$ 125 kbar [21, 22] for Si, and close to 30 kbar for InSb [23].

Summarizing, we have obtained a gross estimate of the variation of the energy gap E_g with lattice parameter in diamond-type semiconductors C, Si, Ge, and α-Sn. Using the Murnaghan EOS, this is converted to a result for $dE_g/dP|_{P=0}$ which is confronted with experimental data. The III-V analogues AlP, GaAs, and InSb of Si, Ge, and α-Sn respectively are predicted to have largely similar values for dE_g/dP to their monatomic counterparts. We have also discussed X-ray diffraction results on the 222 forbidden reflection in the diamond structure from both theory and experiment as a metallic phase is approached by application of pressure. As for directions for future studies, it may be worth mentioning what Cohen [24] calls a 'dramatic difference' between the pressure dependence of the band gaps of elemental and compound semiconductors, and those of carbon nanotubes. While for most semiconductors the fundamental gap increases with applied pressure, for carbon nanotubes there appears to be a dependence on chirality (n, m). According to a recent theoretical study [25], the sign of dE_g/dP should depend on $q = (n - m)\bmod 3$, with $dE_g/dP < 0$ for $q = 2$, and $dE_g/dP > 0$ for $q = 1$, in agreement with experimental data [26].

Acknowledgments
GGNA and NHM acknowledge that this paper was brought to completion during their stay at the Department of Physics of the University of Antwerp. Therefore they thank Professors C. Van Alsenoy, D. Van Dyck and Dr. D. Lamoen for hospitality and for the stimulating environment.

References
[1] Paul W and Warschauer D M 1963 *Solids under pressure* (New York: McGraw-Hill)
[2] Nathan M I, Paul W and Brooks H 1961 *Phys. Rev.* **124** 391
[3] March N H and Matthai C C 2004 *Phil. Mag. Lett.* **84** 335
[4] Merad A E, Kanoun M B, Cibert J and Merad G 2003 *J. Elec. Devices* **2** 31
[5] Shan W, Walukiewicz W, Ager III J W, Yu K M, Wu J and Haller E E 2004 *Appl. Phys. Lett.* **84** 67
[6] Litovchenko V G 2002 *Phys. Rev. B* **65** 153108
[7] Harrison W A 1980 *Electronic structure and the properties of solids* (New York: Dover)
[8] Syassen K 1985 *phys. stat. sol. (a)* **91** 11
[9] Van Camp P E, Van Doren V E and Devreese J T 1988 *Phys. Rev. B* **38** 12675
[10] Albe K 1997 *Phys. Rev. B* **55** 6203
[11] Lam P K, Cohen M L and Martinez G 1987 *Phys. Rev. B* **35** 9190
[12] Mujica A, Rodríguez-Hernández P, Radescu S, Needs R J and Muñoz A 1999 *phys. stat. sol. (b)* **211** 39
[13] Wang S Q and Ye H Q 2002 *J. Phys.: Condens. Matter* **14** 9579
[14] Levinstein M, Rumyantsev S and Shur M, eds. 1999 *Handbook Series on Semiconductor Parameters* (London: World Scientific). Mostly available online at `http://www.ioffe.ru/SVA/NSM`
[15] Weast R C and Astle M J, eds. 1974 *CRC Handbook of Chemistry and Physics* (Palm Beach FL: CRC Press), 59 edition
[16] Murnaghan F 1944 *Proc. Natl. Acad. Sci. U.S.A.* **30** 244
[17] Frogley M D, Downes J R and Dunstan D J 2000 *Phys. Rev. B* **62** 13612
[18] March N H and Flores F 1984 *Solid State Commun.* **49** 649
[19] Zunger A and Freeman A J 1977 *Phys. Rev. B* **15** 5049
[20] Yoder-Short D R, Colella R and Weinstein B A 1982 *Phys. Rev. Lett.* **49** 1438
[21] Minomura S, Samara G A and Drickamer H G 1962 *J. Appl. Phys.* **33** 3196
[22] Piermarini G J and Block S 1975 *Rev. Sci. Instrum.* **46** 973
[23] Okai B and Yoshimoto J 1975 *J. Phys. Soc. Japan* **45** 1880

[24] Cohen M L 2005 *Physica E* **29** 447
[25] Capaz R B, Spataru C D, Tangney P, Cohen M L and Louie S G 2004 *physica status solidi (b)* **241** 3352
[26] Shan W, Wu J, Walukiewicz W, Ager III J W, Yu K M, Haller E E, Kissell K, Bachilo S M, Weisman R B and Smalley R E 2004 *phys. stat. sol. (b)* **241** 3367

Rb-substitution effect on the metal-insulator transition of hollandite vanadate, $K_2V_8O_{16}$

Masahiko Isobe, Shigenori Koishi and Yutaka Ueda

Materials Design and Characterization Laboratory, Institute for Solid State Physics, University of Tokyo, Kashiwa, Chiba 277-8581, Japan

E-mail: isobe@issp.u-tokyo.ac.jp

Abstract. Rb-substitution effect on the metal-insulator transition of hollandite vanadate, $K_2V_8O_{16}$ has been investigated. The metal-insulator transition is accompanied by a charge ordering and a structural change from a tetragonal to a monoclinic structure with an increase of volume in the low-temperature monoclinic insulator phase. Rb-substitution for K atoms continuously raises the transition temperature up to 240 K for $Rb_2V_8O_{16}$ from 160 K for $K_2V_8O_{16}$ and does not make the transition broad in contrast to Ti-substitution. This significant rise of transition temperature can be understood as a negative chemical pressure effect caused by Rb-substitution.

1. Introduction

The metal-insulator (MI) transition is an interesting and important issue in solid-state physics. In 3d-transiton metal oxides, correlations among 3d-electrons play an important role and often bring a MI transition as a function of temperature or chemical substitution (doping level). The MI transition as a function of temperature has been observed in many binary vanadium oxides [1]. The MI transitions in mixed valent vanadium oxides such as V_4O_7 are accompanied by the charge separation and charge order, and spin singlet V^{4+}–V^{4+} pairs are formed in the low-temperature insulator phases, resulting in the large reduction of magnetic susceptibility at the MI transitions [2]. In our recent study on vanadium oxides, a MI transition at around 160K have been found in a hollandite vanadate $K_2V_8O_{16}$ [3]. Hollandite is one of manganese minerals. Hollandite-type compounds have the general chemical formula $A_xM_8O_{16}$. The hollandite structure consists of the tubular M_8O_{16}-network and A-cations at the tunnel sites of the M_8O_{16}-framework, as shown in Fig. 1. The M_8O_{16}-framework is constructed from the double-chains running along the c-axis in the tetragonal structure and each double chain is formed by sharing the edges of MO_6 octahedra. The MI transition of $K_2V_8O_{16}$ is accompanied by two step jumps about three orders in the resistivity along the c-axis and the reduction of magnetic susceptibility down to a small value, suggesting the formation of V^{4+}–V^{4+} singlet pairs and V^{3+}–V^{3+} pairs in the low-temperature insulator phase [3]. The transition is of first order accompanied by the structural change from a tetragonal to a monoclinic structure, and the low-temperature phase has a superlattice of $\sqrt{2}a$ x $\sqrt{2}a$ x $2c$, where a and c denote the primitive cell parameters for the simple tetragonal hollandite structure, suggesting a charge ordering of V^{4+} and V^{3+} [3]. From these results we proposed a possible charge order model, where two kinds of double-chain formed by V^{3+} and V^{4+} chains and by single V^{4+} chains, respectively, order in a manner that gives a superlattice of $\sqrt{2}$ x $\sqrt{2}$ in the ab–plane, and the

© 2008 IOP Publishing Ltd

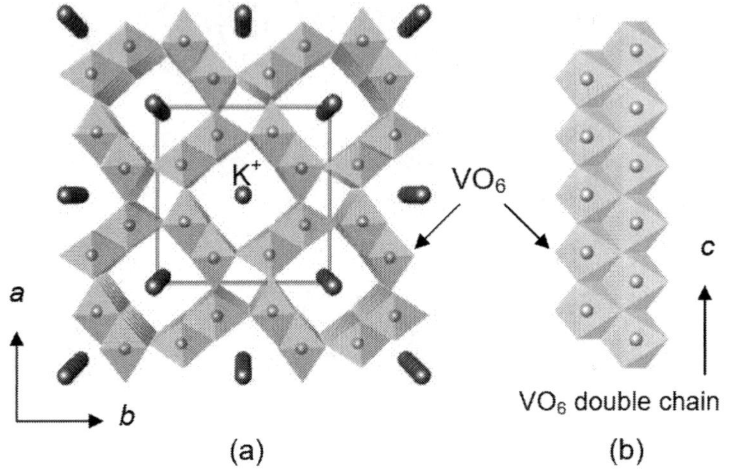

Figure 1. (a) Schematic crystal structure of tetragonal $K_2V_8O_6$ viewed from *c*-axis. The gray polyhedra and the filled circles represent VO_6 octahedra and K^+ ions, respectively. (b) Schematic illustrations of the double chain made up of corner and edge-sharing VO_6 octahedra.

V^{4+}–V^{4+} and V^{3+}–V^{3+} pairs are formed in each chain along the *c*-axis, resulting in a duplication of the *c*-axis [3]. Recently we reported Ti-substitution effect on the MI transition [4]. A slight amount of Ti-substitution for V atoms raises the transition temperature and simultaneously the Ti-substitution makes the transition broad, leading to the final smearing out around y = 0.6 in $K_2V_{8-y}Ti_yO_{16}$ [4]. In order to elucidate the origin of the transition, we have investigated the substitution effect of Rb for K sites. We found the significant enhancement of transition temperature by the Rb-substitution as in the case of the Ti-substitution. Meanwhile, the Rb-substitution does not make the transition broad, in contrast to the Ti-substitution. In this short paper, we report the structural and magnetic properties of $K_{2-x}Rb_xV_8O_{16}$ and discuss the Rb-substitution effect on the MI transition.

2. Experimental details

Powdered samples were prepared by a solid-state reaction of KVO_3, $RbVO_3$, V_2O_3 and V_2O_5 under high pressure of 4 GPa at 1473K for an hour. Starting reagents were ground thoroughly and the mixture was wrapped by platinum foil and then put into a pressure cell. KVO_3 ($RbVO_3$) was obtained by a reaction of K_2CO_3 (Rb_2CO_3) and V_2O_5. V_2O_3 was prepared by the reduction of V_2O_5 under flowing H_2 gas at 1173 K. A cubic type apparatus was used for high pressure synthesis. The sample identification and the determination of lattice parameters were made by powder X-ray diffraction. Magnetic susceptibility of powdered sample was measured by using a SQUID magnetometer.

3. Results and discussion

The powder X-ray diffraction patterns of all samples in $K_{2-x}Rb_xV_8O_{16}$ show a simple tetragonal hollandite structure at room temperature. This means Rb forms an entire solid solution of $K_{2-x}Rb_xV_8O_{16}$. In the present study, we found for the first time that the end member $Rb_2V_8O_{16}$ also crystallizes in the same tetragonal hollandite structure and shows a MI transition around 240 K. Figure 1 shows the lattice parameters as a function of x in $K_{2-x}Rb_xV_8O_{16}$ at room temperature. The lattice parameter *a* increases with x, in contrast, the *c* is almost constant. Namely the volume increases with x.

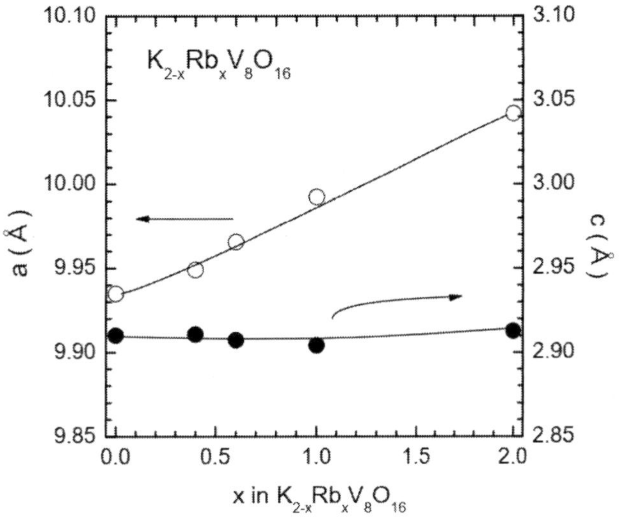

Figure 2. Composition x dependence of lattice parameters in $K_{2-x}Rb_xV_8O_{16}$.

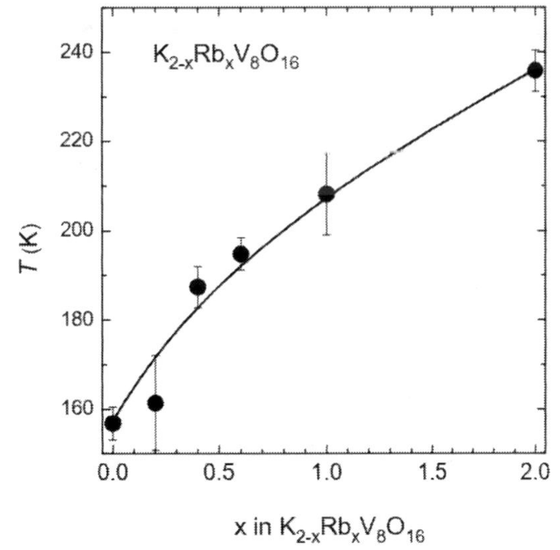

Figure 3. Composition x dependence of the transition temperature in $K_{2-x}Rb_xV_8O_{16}$.

Figure 3 shows composition (x) dependence of the MI transition temperature in $K_{2-x}Rb_xV_8O_{16}$, which is determined from magnetic susceptibility curves. The most remarkable point is a significant rise of transition temperature by the substitution of Rb. As mentioned above, the MI transition of $K_2V_8O_{16}$ is characterized by the reduction of magnetic susceptibility. The temperature of the reduction of magnetic susceptibility continuously goes up to around 240K at x = 2.0 from around 160K at x = 0. On the other hand, the transition does not become broad by the Rb-substitution but remains very sharp in all samples, as shown in Fig. 4. There is no anomaly suggesting any magnetic transition at low temperature in all samples.

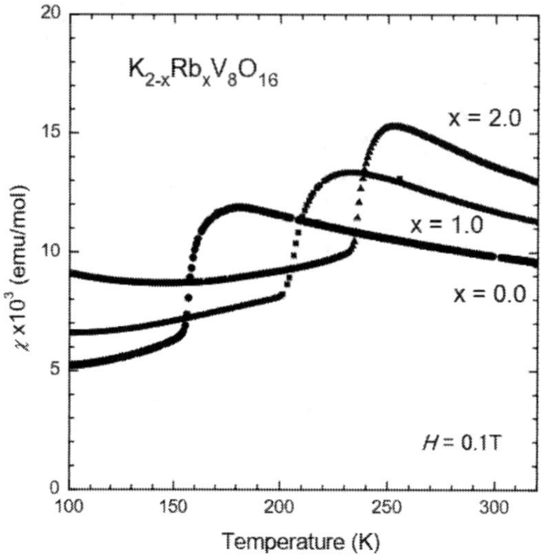

Figure 4. Temperature dependence of magnetic susceptibility for various x in $K_{2-x}Rb_xV_8O_{16}$.

In general, the substitution of different atoms introduces any disorder in the system and is expected to give a significant effect on the phase transition. Actually the substitution of Ti for V makes the transition broad and finally the transition smears out [4]. On the other hand, in $K_{2-x}Rb_xV_8O_{16}$ the transition does not show any broadening with increasing x. This means that the Rb-substitution for K does not affect the electronic state of V_8O_{16} framework such as charge ordering directly, because Rb^+ has the same valence state as K^+ and acts as counter ion in the V_8O_{16} framework. In contrast, the substitution of Ti for V gives a significant effect to the electronic state of the system, for example the change of doping level and favorable Ti^{4+} state compared with V^{4+} state. These should affect the charge ordering significantly. The slight substitution of Ti for V raises the transition temperature with an increase of volume. Therefore the rise of the transition temperature can be considered to be due to a negative pressure effect as in the case of the Rb-substitution, although the way of composition dependence of lattice constants is different between the two substitution systems. The difference between Ti- and Rb-substitution is the presence and absence of disorder effect. Such difference is naturally understood by the differences of the substituted sites and ionic states between Ti and Rb.

Acknowledgments

This study has been supported by a Grant-in-Aid for Scientific Research (No. 18104008) from the Japan Society for the Promotion of Science.

References

[1] Ueda Y 2000 (Suppl. B) *J. Phys. Soc. Japan* **69** 149
[2] Marezio M, McWhan D B, Dernier P D and Remeika J P 1972 *Phys. Rev. Lett.* **28** 1390; Ueda Y, Kitaoka Y, Yasuoka H, Kosuge K and Kachi S 1979 *J. Phys. Soc. Japan* **47** 333
[3] Isobe M, Koishi S, Kouno N, Yamaura J, Yamauchi T, Ueda H, Gotou H, Yagi T and Ueda Y 2006 *J. Phys. Soc. Japan.* **75** 073801
[4] Isobe M, Koishi S and Ueda Y 2007 *J. Magn. Magn. Mater.* **310** 888

Joint 21st AIRAPT and 45th EHPRG Int. Conf. on High Pressure Science and Technology IOP Publishing
Journal of Physics: Conference Series **121** (2008) 032008 doi:10.1088/1742-6596/121/3/032008

Metallization and Superconductivity of Molecular Crystal BI_3 under Pressure

Suzue Onoda and Katsuya Shimizu

KYOKUGEN, Center for Quantum Science and Technology under Extreme Conditions, Osaka University, Toyonaka, Osaka 560-8531, Japan

e-mail: onoda@djebel.mp.es.osaka-u.ac.jp

Abstract. Electrical resistivity measurement of boron triiodide (BI_3) have been performed at high pressures up to 70 GPa. At room temperature, the resistivity decreased rapidly with applying pressure. The temperature dependence of electrical resistance indicated metallization of BI_3 at 23 GPa. The superconducting transition was found above 27 GPa.

1. Introduction

A structural characteristic of molecular crystals is the difference in strength between the strong intramolecular bonding and the weak intermolecular bonding. Under pressure, this difference brings a variety of structural transitions, metallizations, superconducting transition, and molecular dissociations as a result.

In diatomic molecular crystal I_2, it was reported that the insulator-metal transition occurs at around 16 GPa [1], followed by the molecular dissociation at 21 GPa [2]. The superconductiing transition of the monatomic iodine was observed at the pressure of about 22 GPa [3]. The pressure-induced metallization has been explained by the band overlap between the valence and the conduction band. In case of molecular crystal SnI_4, the metallization and superconducting transition occurs in the amorphous state under pressure [4,5]. It is expected that most of molecular crystals show metallization under pressure. However, the mechanism of metallization and dissociation is still unclear.

At ambient pressure and room temperature, BI_3 molecule, which is planar equilateral triangle form, crystallizes into hexagonal structure as illustrated in Figure 1 [6]. Our recent X-ray diffraction study revealed a structural phase transition at 6.9 GPa to new phase, and the molecular seemed to dissociate at the transition pressure [7]. At higher pressure, BI_3 crystal gradually became into opaque which may show the tendency of metallization of BI_3.

These molecules of I_2, SnI_4 and BI_3, the basic structure of which are formed by iodine, may show a similar metallization under pressure. Comparison of the phenomena under pressure of these similar molecular crystals would make a mechanism of metallization and molecular dissociation in molecular crystals clear.

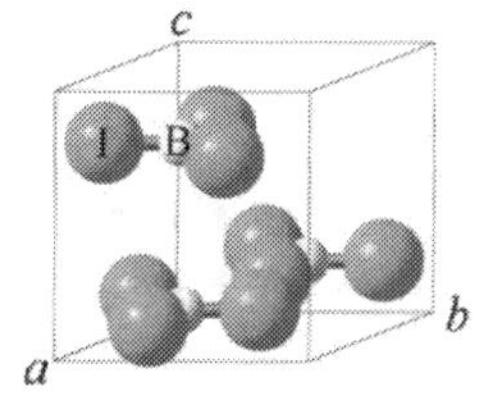

Figure 1. A unit cell of crystalline BI_3. Molecules are layered to c-axis and form crystal.

© 2008 IOP Publishing Ltd

Journal of Physics: Conference Series **121** (2008) 032008 doi:10.1088/1742-6596/121/3/032008

To investigate the high pressure phenomena of BI_3, we performed the electrical resistance measurement at high pressure and low temperature.

2. Experimental

A compact diamond anvil-cell (DAC) was used to produce high pressure at low temperature experiments. We used nonmagnetic stainless steel 310S for a gasket material. Pressure was determined by a ruby fluorescence method at low temperature as well as at room temperature. Al_2O_3 powder was placed on the surface of gasket for electrical insulation. Since BI_3 is extremely hygroscopic, the sample was loaded into the center of the insulated gasket with small ruby chips. The electrical resistance was measured by 4-terminal method with Pt electrodes with typical measuring current of 500 nA. The pressure was applied at room temperature and we used a $^3He/^4He$ dilution refrigerator in order to cool DAC down to low temperature of 60 mK.

3. Results and Discussion

The pressure dependence of the electrical resistivity ρ of BI_3 measured at room temperature is plotted in Figure 2. At pressure range between 5 and 17 GPa, we observed rapid and continuous decrease of the resistivity. Above 17 GPa, the ρ became almost independent to pressure which may indicate the metallization of BI_3. These results show that the metallization occurs at higher pressure than molecular dissociation in case of BI_3.

With applying pressure, the color of sample changes from transparent to opaque. Above 20 GPa, the sample became a little luster. To reveal the metallization of BI_3, we measured temperature dependence of the electrical resistance above 17 GPa.

The temperature dependence of the relative resistance R/R_{20K} at several pressures in low temperature region is displayed in Figure 3. The negative temperature dependence was observed at 17 and 20 GPa; the characteristic behavior of semiconductor. This behavior was suppressed with applying pressure and disappeared at 23 GPa which shows that the metallization of BI_3 occurs at this pressure.

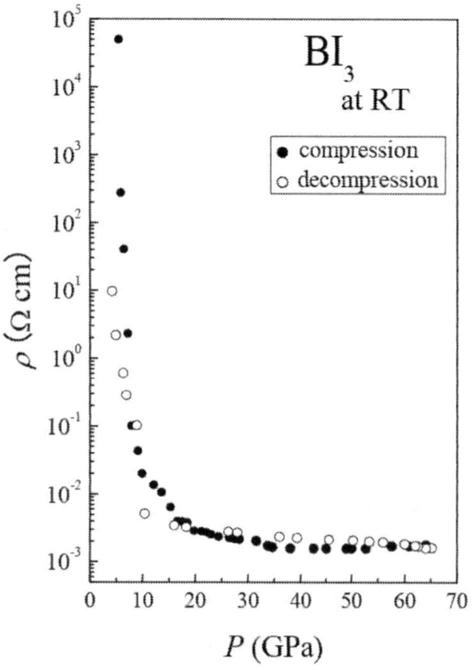

Figure 2. Pressure dependence of electrical resistivity of BI_3 at room temperature.

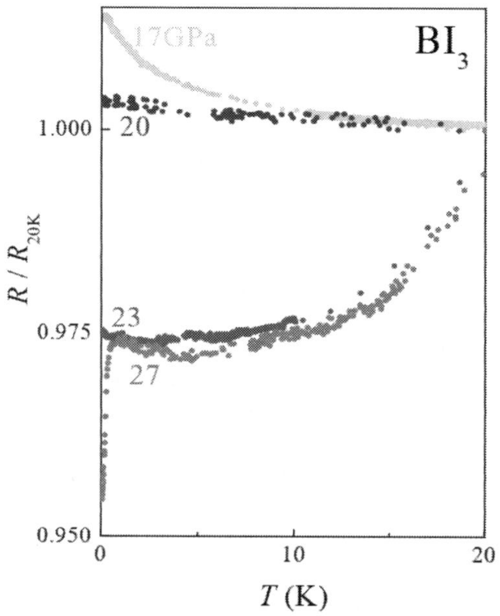

Figure 3. The temperature dependence of the relative electric resistance of BI_3 under pressure in the low temperature region.

As applying pressure further, we found an abrupt drop in the resistance showing the superconducting transition below 1 K at 27 GPa as shown in Figure.3. The superconducting transition T_c was determined as 0.5 K from the onset of the transition. We observed the positive pressure dependence of the T_c as shown in Figure 4.

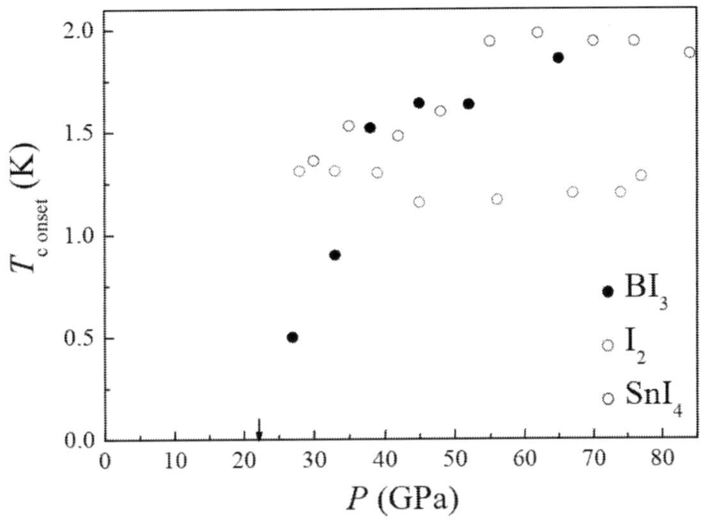

Figure 4. T_c versus P diagram for BI_3, I_2 and SnI_4. The arrow indicates the absence of superconductivity of BI_3 down to 35 mK.

Figure 5 shows the magnetic field dependence of the superconducting transition under constant pressure of 33 GPa. The transition temperature shifts lower temperature with applied magnetic field. The critical field H_c is estimated 0.8 T with extrapolation.

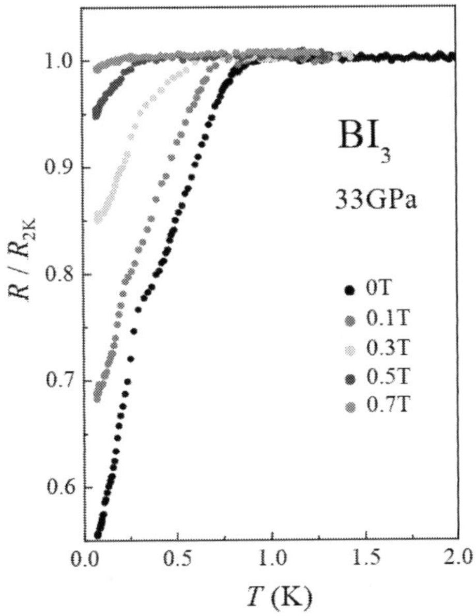

Figure 5. The temperature dependence of the relative electric resistance of BI_3 at low temperature under various magnetic field at P = 33 GPa.

4. Summary

We performed the electrical resistance measurements of BI_3 under high pressure and low temperature. At room temperature, the resistivity decreases rapidly. By the temperature dependence of the resistance, we determined the metallization pressure of BI_3 is 23 GPa. We also found the superconducting transition at 27 GPa. The positive pressure dependence of the T_c was observed at pressure between 27 and 65 GPa with T_c changing from 0.5 to 1.9 K.

According to the results by X-ray diffraction study, we conclude that the metallization occurs at higher pressure than molecular dissociation in case of BI_3. This differs from the behavior of I_2 and SnI_4. However, the superconductivity of all three molecular crystals appears for pressures near 25 GPa.

Acknowledgment

This work is supported by the 21st Century COE program (G18) by Japan Society of the Promotion of Science and Grant-in-Aid for Creative Scientific Research (15GS0213), MEXT.

References

[1] B. M. Riggleman, H. G. Drickamer., 1963 *J. Chem. Phys.* **38**, 2721
[2] K. Takemura, S. Minomura, O. Shimomura, Y. Fujii., 1980 *Phys. Rev. Lett.* **45**, 1881
[3] K. Shimizu, N. Tamitani, N. Takeshita, M, Ishizuka, K. Amaya, S Endo., 1992 *J. Phys.Soc. Jpn.*, **61**, 3853
[4] A. L. Chen, P. Y. Yu, M. P. Pasternak., 1991 *Phys. Rev. B.* **44**, 2883
[5] N. Takeshita, S. Kometani, K. Shimizu, K. Amaya, N. Hamaya, S. Endo., 1996 *J. Phys. Soc. Jpn.*, **65**, 3400
[6] B. Albert, K. Schmit, 2001 *Z. Anorg. Allg. Chem.* **627**, 809
[7] N. Hamaya, M. Ishizuka, A. Ohmura, 2004 *Special Issue Rev. High Press. Sci. Tech.* **14** 202

Journal of Physics: Conference Series **121** (2008) 032009 doi:10.1088/1742-6596/121/3/032009

Magnetic properties of Fe_2P single-crystal under multi-extreme conditions

Hisao Kobayashi[1,2], Junpei Umemura[1], Xiao-Wei Zhang[3], and Yoshiya Uwatoko[4]

[1] Graduate School of Material Science, University of Hyogo, 3-2-1 Koto, Hyogo 678-1297, Japan
[2] CREST, Japan Science and Technology Corporation, Saitama 332-0012, Japan
[3] Photon Factory, High Energy Accelerator Research Organization, Ibaraki 305-0801, Japan
[4] Institute for Solid State Physics, The University of Tokyo, Chiba 277-8581, Japan

E-mail: kobayash@sci.u-hyogo.ac.jp

Abstract. ^{57}Fe nuclear resonant forward scattering experiment has been performed on a single crystalline Fe_2P under multi-extreme conditions, which shows a pressure induced ferromagnteic-antiferromagetic transition and magnetic field induced metamagnetic transition. We present the clear experimental evidence that internal magnetic fields in two Fe sites at 3 K, 25 kOe and 2.5 GPa are comparable with those in a ferromagnetic state at ambient pressure. These internal magnetic fields, which are related to magnetic moments, are not aligned collinearly along the [001] axis by the external magnetic field.

1. Introduction

Iron phosphide, Fe_2P, with a hexagonal structure is a ferromagnetic compound accompanied with discontinuous lattice change at the Curie temperature (T_C) [1]. In this structure, the Fe atoms have two nonequivalent sites, that is, the tetrahedral (Fe_I) and pyramidal (Fe_{II}) sites surrounded by four and five P atoms, respectively. The magnetic properties of Fe_2P under pressure have been studied using the electrical resistivity and magnetization measurements [2, 3, 4]. From these results, T_C decreases as increasing pressure and a new antiferromagnetic state appears above 0.5 GPa. In this antiferromagnetic phase, metamagnetic transition is observed at about 2 kOe under 4.2 K and 1.4 GPa [4]. Meanwhile, the saturation magnetic moment at 4.2 K is not sensitive with pressure up to 1.4 GPa. There is no knowledge of magnetic structures and magnetic moments of Fe atoms in Fe_2P under pressure and external magnetic fields.

Mossbauer spectroscopy is a useful method for the study of the electronic state of Fe under high pressure. The conventional Mossbauer absorption spectra have usually been measured using a point source. Under multi-extreme conditions, the γ-ray emitted from the source has a broadening line shape or does not have a single line related to hyperfine splitting levels of the nuclei because of a magnetic interaction and/or external magnetic fields. It is very di cult to obtain spectra of satisfactory quality under such experimental environments. Because of rapid developments in synchrotron based X-ray spectroscopy, a monochromatized X-ray from synchrotron radiation (SR) is a highly dense Mossbauer source. In Mossbauer spectroscopy using SR, the forward scattering from nuclei excited by a pulsed X-ray of SR is observed in the

© 2008 IOP Publishing Ltd

Journal of Physics: Conference Series **121** (2008) 032009 doi:10.1088/1742-6596/121/3/032009

time domain. The decay rate in the collective excited nuclei is modulated by quantum beats owing to the constructive interferences of photons emitted from the various nuclear levels of different energies. Thus ^{57}Fe nuclear forward scattering (NFS) is a good candidate for the study of the electronic state of Fe under multi-extreme conditions [5].

2. Experimental

Non-enriched high-quality single crystals of Fe_2P were synthesized by thermal annealing of ingots which were prepared by melting at about 1460 °C and cooling in air. The Curie temperature of the single-crystal sample was evaluated to be T_C= 210 K by magnetization measurement. $Fe_{2\ x}P$ has a substantial deficiency at the iron site. The x value in the single-crystalline sample is estimated to be less than 0.005 using the empirical relation for the concentration of Fe in the system Fe-P versus T_C [6].

We have measured the temperature dependence of conventional Mossbauer spectra using a constant-acceleration-type spectrometer. The single-crystalline sample was finely ground in order to get a uniform resonant-absorber thickness.

^{57}Fe NFS experiment under multi-extreme conditions was performed at beamline NE3 in the accumulation ring, High Energy Accelerator Research Organization. The pulsed SR was monochromatized using a high-resolution monochromator. The monochromatized X-ray transmitted through the sample was detected with Si-avalanche photodiodes. Under high pressure, the single-crystalline sample was loaded with ruby crystals into a sample cavity of 0.6 mm diameter in a 0.3-mm-thick Inconel 625 alloy gasket. The use of Fluorinert as a pressure-transmitting medium ensured quasi-hydrostatic conditions. Pressure was calibrated by measuring the wavelength shift of the R_1 luminescence line of ruby crystals in the clamp-type diamond-anvil cell (DAC). The [001] axis of the single-crystalline sample in DAC was aligned along the propagation vector of SR. External magnetic field (H_{ex}) was applied to the sample in DAC using a superconducting magnet, where the H_{ex} direction was parallel to the [001] axis of the sample.

3. Results and Discussion

Mossbauer spectrum of Fe_2P at 4.2 K and ambient pressure is shown in Fig. 1, where the closed circles represent the measured data. As previously assigned [7, 8], the subspectra with smaller and larger magnetic hyperfine fields, H_{hf}, correspond, respectively, to the Fe_I and Fe_{II} sites. In an analysis of Mossbauer spectra, we assume that the magnetic hyperfine interaction is much larger than the electric quadrupole interaction, and use first-order perturbation theory with an axial symmetry approximation in the magnetic ordered state. As seen in Fig. 1, the spectrum was well fitted with the two subspectra. The refined hyperfine interaction parameters are summarized in Table 1 (a), which are comparable with the previous results [7, 8].

Figure 2 shows ^{57}Fe NFS spectrum of Fe_2P at 3 K, 2.5 GPa and 25 kOe. We have observed a clear quantum beat pattern in the ^{57}Fe NFS spectrum. The quantum beat comes from a constructive interference of photons emitted from the various nuclear levels with different energies. Thus, the nuclear levels of ^{57}Fe are split by the hyperfine interaction. Due to a pure $M1$ character of the 14.413 keV ^{57}Fe nuclear resonance, the nuclear transitions with $m(= m_e - m_g) = 0,\ 1$ were directly and collectively excited, where m_e and m_g are the magnetic quantum numbers of nuclear excited and ground states, respectively. These transition probabilities depend on the directions of H_{hf} and the principle axes of the electric-field-gradient (EFG) tensor because SR is linearly polarized and we have used the single-crystalline sample.

At the Fe_I site, the FeP_4 tetrahedron is nearly regular because of small E_Q value at 300 K and the principle z axis of EFG tensor is assumed to be parallel to the [001] axis. Meanwhile, the Fe_{II} sites are surrounded pyramidally by five P atoms and the direction from the Fe atom to the P atom at the vertex of the nearly square pyramid is parallel to one of the $\langle 110 \rangle$ axes.

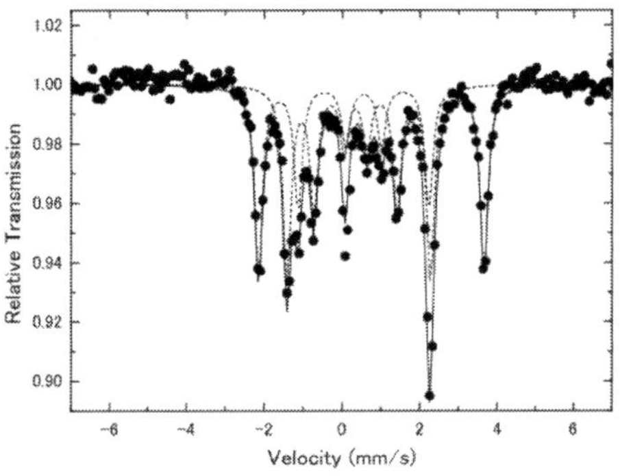

Figure 1. Mossbauer spectrum of Fe_2P at 4.2 K. The closed circles indicate the observed spectrum, the solid line represents the fitting curve obtained by the analysis, and the extracted subspectra are shown by the dotted lines.

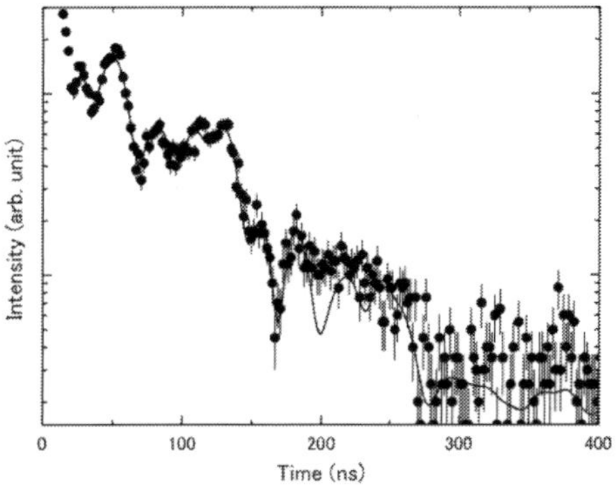

Figure 2. ^{57}Fe nuclear forward scattering spectrum of Fe_2P at 3K, 2.5 GPa and 25 kOe. The closed circles with error bars indicate the observed spectrum and the solid line represents the fitting curve obtained by MOTIF [9].

Then, we assumed that the principle z axis of EFG tensor at the Fe_{II} sites is parallel to one of the $\langle 110 \rangle$ axes. In the present experimental conditions, the Fe_{II} site gives three subspectra with the intensity ratio of 1:1:1. Hence, the observed ^{57}Fe NFS spectrum was analyzed with four subspectra, assuming the intensity ratio of 3:1:1:1.

The spectrum analysis was performed with the package MOTIF [9] by using the full dynamical theory of nuclear resonant scattering, including the digitalization of the complete hyperfine Hamiltonian. As shown in Fig. 2, the spectrum was well fitted with the four subspectra and the refined hyperfine interaction parameters are summarized in Table 1 (b). We have evaluated the difference in δ_{cs} between two Fe sites to be 0.03 mm/s from ^{57}Fe NFS spectrum because the quantum beat comes from a constructive interference of photons. Since this value is comparable with that estimated from the Mossbauer spectrum at 4.2 K, the electronic states of Fe in the two sites do not change at 3K, 2.5 GPa and 25 kOe. At the Fe_{II} site, the refined $\frac{1}{2}eV_{zz}Q$ value is comparable with E_Q at ambient pressure. On the other hand, the refined $\frac{1}{2}eV_{zz}Q$ value at the Fe_I site is much larger than E_Q at ambient pressure. Recent our X-ray diffraction measurement under pressure at low temperature indicates that there is no structural transition at the pressure-induced ferromagnetic-antiferromagnetic transition. Thus the position of Fe atom changes in the FeP_4 tetrahedron at 3K, 2.5 GPa and 25 kOe.

The refined H_{hf} is composed of H_{ex} and an internal magnetic field (H_{in}), which is related to a magnetic moment of Fe atom and is usually negative. We have evaluated the values and directions of H_{in} at the Fe_I and Fe_{II} sites from refined H_{hf}, φ, and H_{ex}. The obtained $|H_{in}|$ and φ_{in} are 118 kOe and 22 deg. for the Fe_I site and 181 kOe and 16 deg. for the Fe_{II} site, where φ_{in} is an angle between H_{in} and the [001] axis. Since $|H_{in}|$s at two Fe sites are comparable with those at 4.2 K and ambient pressure, this result suggests that the magnetic moments at these sites do not change at 2.5 GPa. The directions of $|H_{in}|$ at the Fe_I and Fe_I sites are not parallel to the [001] axis. Accordingly the magnetic structure at 2.5 GPa above the metmagnetic transition is more complex one than a collinearly aligned structure although the components in the (001) plane of $|H_{in}|$ at the two Fe sites are not determined by ^{57}Fe NFS. The magnetic

Table 1. Hyperfine interction parameters in Fe_2P. E_Q can be expressed as $\frac{1}{2}eV_{zz}Q(3\cos^2\theta - 1)/2$ in the magnetic ordered state and $|\frac{1}{2}eV_{zz}Q(1-\eta^{1/2})|$ in the paramagnetic state, where Q is the nuclear electric quadrupole moments, $V_{zz} = \partial^2 V/\partial z^2$ the principal component of diagonalzed EFG, θ the angle between the direction of principal component of the EFG and the magnetic hyperfine field and $\eta = (V_{xx} - V_{yy})/V_{zz}$. φ is an angle between the direction of H_{hf} and the [001] axis.
(a)

T (K)	Site	δ_{CS} (mm/s)	E_Q (mm/s)	H_{hf} (kOe)
4.2	Fe_I	0.38	0.08	114
	Fe_{II}	0.65	0.20	180
77	Fe_I	0.39	0.08	110
	Fe_{II}	0.65	0.21	174
300	Fe_I	0.17	0.10	
	Fe_{II}	0.59	0.42	

(b)

T (K)	P (GPa)	H_{ex} (kOe)	Site	$\frac{1}{2}eV_{zz}Q$ (mm/s)	H_{hf} (kOe)	φ (deg.)
3	2.5	25	Fe_I	0.21	95.6	28
			Fe_{II}	-0.57	158	19

moment of Fe_2P along the [001] axis at 25 kOe is estimated to be 2.8 μ_B/f.u. from $|H_{in}|$ which is comparable with that obtained by magnetization measurement [4].

In summary, we have observed the satisfactory quality ^{57}Fe NFS spectrum of the single-crystalline Fe_2P sample under multi-extreme conditions. The result of analysis indicates that $|H_{in}|$s in two Fe sites at 3 K, 25 kOe and 2.5 GPa are comparable with those in the ferromagnetic state at ambient pressure. Furthermore, it was pointed out that the magnetic moments at the two Fe sites under 2.5 GPa above the metmagnetic transition are not aligned collinearly by H_{ex}.

Acknowledgments
We acknowledge the support of Prof. Kisimoto to use APD detectors. H.K. acknowledges the patial support by the Japanese Ministry of Education, Culture, Sports, Science and Technology, and Grant-in-Aid for Scientific Research (C) 15540319 and (B) 17340098.

[1] Fujii H, Hokabe T, Kamigaichi T, and Okamoto T 1977 *J. Phys. Soc. Jpn.* **43** 41.
[2] Kadomatsu H, Tohma K, Fujii H, Okamoto T, and Fujiwara H 1981 *Phys. Lett.* **84A** 442.
[3] Zavadskii E.A., Medvedeva L.I., and Filippov 1984 *J. Magn. Magn. Mater.* **43** 53.
[4] Kadomatsu H, Isoda M, Fujii H, Okamoto T, and Fujiwara H 1985 *J. Phys. Soc. Jpn.* **54** 2690.
[5] Kobayashi H and Zhang X-W 2006 *J. Phys. Soc. Jpn.* **75** 085001.
[6] Lundgren L, Tarmohamed G, Beckman O, Carlesson B, and Rundqvist S 1978 *Phys. Scr.* **17** 39
[7] Wappling R, Haggstrom L, Ericsson T, Devanayanan S, Karlsson E, Carlsson B, and Rundqvist S 1975 *J. Solid State Chem.* **13** 258.
[8] Ericsson T, Haggstrom L, Wappling R, and Methasiri T 1980 *Phys. Scr.* **21** 212
[9] Shvyd'ko Y.V. 2000 *Hyperfine Interact.* **125** 173.

Pressure study of magnetism in $(Lu_{0.8}Ce_{0.2})_2Fe_{17}$ and $Lu_2Fe_{16.5}Ru_{0.5}$ single crystals

A V Andreev[1], J Kamarád[1], E A Tereshina[1,2], T Komatsubara[3] and I Satoh[3]

[1] Institute of Physics AS CR, Na Slovance 2, Prague, 18221, Czech Republic
[2] Department of Condensed Matter Physics, Faculty of Mathematics and Physics, Charles University, Ke Karlovu 5, 12116 Prague 2, Czech Republic
[3] Institute for Materials Research, Tohoku University, Katahira 2-1-1, Sendai 980-8578, Japan

andreev@mag.mff.cuni.cz

Abstract. The antiferromagnetic state observed in Lu_2Fe_{17} at elevated temperatures is stabilized down to the ground state in $(Lu_{0.8}Ce_{0.2})_2Fe_{17}$ and $Lu_2Fe_{16.5}Ru_{0.5}$. The Néel temperature T_N decreases from 275 K in Lu_2Fe_{17} to 247 K and 208 K in $(Lu_{0.8}Ce_{0.2})_2Fe_{17}$ and $Lu_2Fe_{16.5}Ru_{0.5}$, respectively. Both compounds exhibit the easy-plane magnetic anisotropy and a metamagnetic transition at fields below 1 T. External pressure pushes the transition towards higher field and reduces the magnetization as well as T_N to 234 K and 185 K for $(Lu_{0.8}Ce_{0.2})_2Fe_{17}$ and $Lu_2Fe_{16.5}Ru_{0.5}$, respectively, under pressure 0.8 GPa. The pressure and the substitution effects can be understood in terms of exchange interaction varying strongly with changes of the lattice parameter *c*.

1. Introduction

The very short Fe-Fe distances d_{FeFe} between Fe atoms in selected inequivalent Fe positions, $d_{FeFe} \sim$ 239 pm or 244 pm (shorter than $d_{FeFe} \sim$ 250 pm in bcc Fe), are special features of the rare-earth intermetallic compounds R_2Fe_{17}. For this reason, a negative exchange interaction between the Fe atoms can coexist together with positive interactions in the case of larger d_{FeFe} distances between Fe-atoms in basal planes. The total Fe-Fe exchange interaction is positive, however, the existence of these competitive interactions leads to unexpectedly low T_C in compounds with so high content of Fe. The Lu_2Fe_{17} compound with the smallest non-magnetic rare-earth element and hexagonal crystal structure of the Th_2Ni_{17} type exhibits an incommensurate helimagnetic structure with the collinear Fe moments in basal planes below the Néel temperature $T_N = 274$ K [1,2]. The component t_z of propagation vector $(00t_z)$ decreases continuously down to zero with decreasing temperature and a length of the magnetic helix increases to infinity at transition into ferromagnetic phase below 130 K. Simultaneously, the lattice parameter *c* increases and a strength of the negative interactions decreases.

Such a delicate balance of ferro- (F) and antiferro- (AF) magnetic interactions is very sensitive to external conditions. Therefore, Lu_2Fe_{17} is a very apt compound to study interplay of magnetic field, temperature, external pressure and substitution effects on magnetism of the high-Fe-content intermetallics. Just recently, the F state has been totally suppressed in Lu_2Fe_{17} by external hydrostatic pressure [3,4]. Most of substitutions in the Fe sublattice (Al, Si, Cr, Co, Ni) and substitution of another non-magnetic atoms (Y) for Lu stabilize the F state and increase the magnetic ordering temperature.

© 2008 IOP Publishing Ltd

On the other hand, external hydrostatic pressure stabilizes the AF state and reduces T_N due to a strengthening of inter-plane AF and a weakening of the intra-plane F interactions by an induced decrease of the lattice *c*- and *a*-parameters.

In the present study we report on the rare cases where substitutions for Fe and Lu work together with pressure, i.e. they destabilize the F state and induce the AF structure in whole range of magnetic ordering.

2. Experimental

The single crystals of $(Lu_{0.8}Ce_{0.2})_2Fe_{17}$ and $Lu_2Fe_{16.5}Ru_{0.5}$ were prepared by the Czochralski method in a tetra-arc furnace in the same way as the parent compound Lu_2Fe_{17} [2]. At ambient pressure, the magnetization isotherms were measured in SQUID and PPMS magnetometers (Quantum Design) along both principal axes of the hexagonal structure in fields up to 9 T at temperature range 5-300 K. Measurements under pressure were carried out in a miniature pressure piston-cylinder cell developed for a use in the SQUID magnetometer. The single crystals were oriented inside the cell with the *a*-axis (EMD) along magnetic field. The non-magnetic Cu-Be pressure cell with a mineral oil as a pressure-transmission medium allows us to apply hydrostatic pressure up to 1 GPa. Details of construction and operation of the cell, including a pressure determination using a pressure shift of the superconducting critical temperature of Pb, are described elsewhere [5]. All the magnetization curves were corrected for the demagnetization effect.

3. Results and discussion

Figure 1a) shows that the magnetization isotherms of $(Lu_{0.8}Ce_{0.2})_2Fe_{17}$ exhibit the metamagnetic transition started just below 1 T even at ambient pressure. The metamagnetic transition is clearly of the first order with a wide hysteresis. Magnetization reaches its saturation value 35 μ_B within approx. 1 T field interval above the transition. With increasing pressure, the transition field H_c increases, while the magnetization gain across the transition ΔM, the width of hysteresis ΔH_c and the Néel temperature T_N (Fig. 1b) decrease. Qualitatively similar behavior, with larger drops of T_N and saturation magnetization only, has been observed in $Lu_2Fe_{16.5}Ru_{0.5}$ (Fig. 2). Pressure dependencies of the magnetic characteristics of both compounds are presented in Fig. 3.

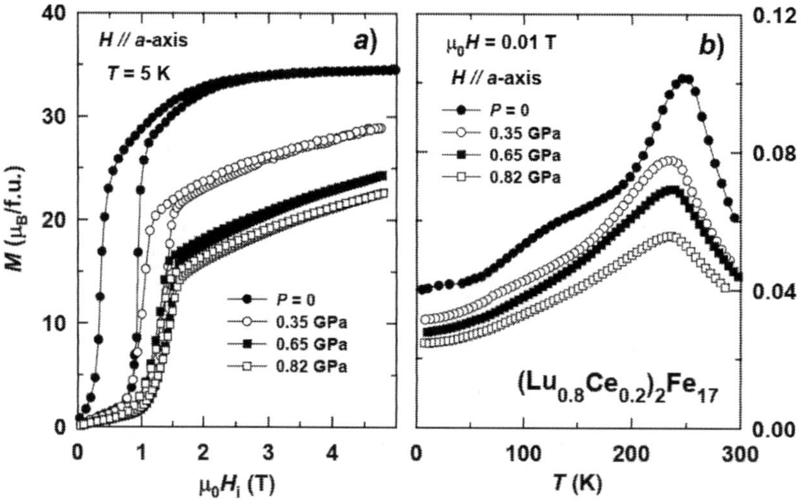

Fig. 1: a) Magnetization of $(Lu_{0.8}Ce_{0.2})_2Fe_{17}$ vs. field along the *a*-axis at 5 K and b) Temperature dependence of magnetization of $(Lu_{0.8}Ce_{0.2})_2Fe_{17}$ in field 0.01 T under different pressures.

In the case of $(Lu_{0.8}Ce_{0.2})_2Fe_{17}$, where the Fe sublattice is not touched directly by substitution, and where both Ce^{4+} and Lu^{3+} ions are non-magnetic, the stabilization of AF state to the ground state can be understood in terms of varying lattice parameter *c*. External pressure suppresses the F state in both Lu_2Fe_{17} and Ce_2Fe_{17}. Despite of the larger unit-cell volume of Ce_2Fe_{17} compared to Lu_2Fe_{17} (due to larger parameter *a*), the F state is less stable in Ce_2Fe_{17} where the *c* parameter is smaller than in Lu_2Fe_{17}. Moreover, purely AF ground state was observed in Ce_2Fe_{17} crystal [6] while this was not seen in Lu_2Fe_{17}. Lattice parameters of Lu_2Fe_{17}, $(Lu_{0.8}Ce_{0.2})_2Fe_{17}$ and Ce_2Fe_{17} are a = 839.5 pm, c = 829.4 pm; a = 841.7 pm, c = 827.0 pm; and a = 848.9 pm, c = 827.2 pm, respectively. (For Ce_2Fe_{17} crystallizing in the rhombohedral Th_2Zn_{17} structure the *c* parameter 1241 pm is divided by factor of 1.5 to compare with hexagonal crystals). The *c* parameter and consequently a distance between Fe atoms in "dumbbell" positions (along *c* direction) are the same in $(Lu_{0.8}Ce_{0.2})_2Fe_{17}$ and Ce_2Fe_{17}. That provides very similar magnetic properties (AF ground state with metamagnetic transition in 1 T [6]) despite of a considerable difference in the *a* parameter. $(Lu_{0.8}Ce_{0.2})_2Fe_{17}$ behaves as Lu_2Fe_{17} uniaxially compressed by a "bias" chemical pressure (which is equivalent to approx. 0.4 GPa of hydrostatic pressure as follows from a comparison with pressure-study results on Lu_2Fe_{17} [4]).

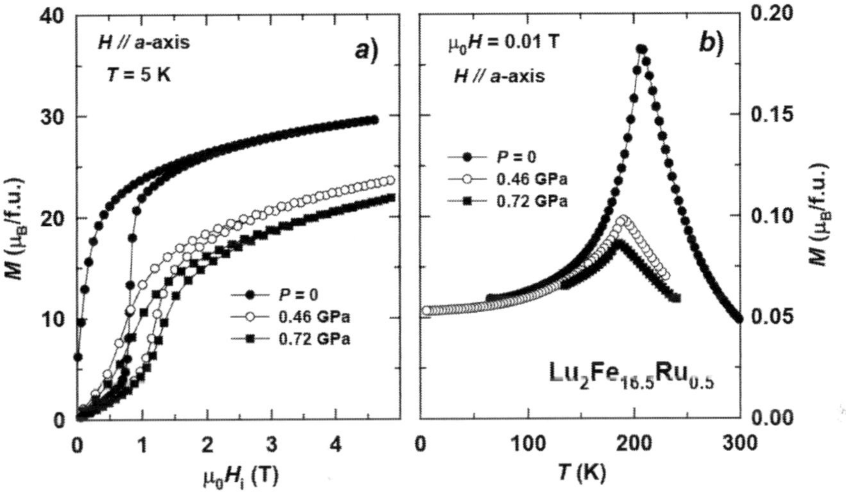

Fig. 2: a) Magnetization of $Lu_2Fe_{16.5}Ru_{0.5}$ vs. field along the *a*-axis at 5 K and b) Temperature dependence of magnetization of $Lu_2Fe_{16.5}Ru_{0.5}$ in field 0.01 T under different pressures.

These speculations are not applicable to $Lu_2Fe_{16.5}Ru_{0.5}$ having the lattice parameters a = 840.7 pm, c = 830.2 pm so simply as in the case of $(Lu_{0.8}Ce_{0.2})_2Fe_{17}$. A slight expansion along the *c*-axis (confirmed by c = 831.2 pm in compound with higher Ru content, $Lu_2Fe_{16}Ru$ [7]) is observed upon the Ru substitution for Fe at room temperature. However, T_N of this compound lies far below room temperature and thermal expansion of paramagnetic $Lu_2Fe_{16.5}Ru_{0.5}$ is positive in contrast with the anomalous negative thermal expansion of all R_2Fe_{17} in the ordered states. Taking into account thermal expansion of Lu_2Fe_{17} [8], the lattice parameters of $Lu_2Fe_{16.5}Ru_{0.5}$ and Lu_2Fe_{17} are identical at the Néel temperatures. Since Ru atomic radius is larger than Fe one, the Ru atoms should not occupy the "dumbbells" positions and the AF interactions should not be weakened by the substitution. On the other hand, saturation magnetization $Lu_2Fe_{16.5}Ru_{0.5}$ is much lower than an estimated one assuming that Ru is a non-magnetic dilutor or a carrier of an antiparallel magnetic moment of a realistic magnitude. It is reasonable to conclude that this substitution modifies considerably the band structure of Lu_2Fe_{17}. This leads to decrease of the Fe magnetic moment, to a lower positive magnetostriction and to a less pronounced increase of the *c* parameter with decreasing temperature. Therefore, the AF interactions remain strong and the F ground state cannot be reached in $Lu_2Fe_{16.5}Ru_{0.5}$ even at low temperatures.

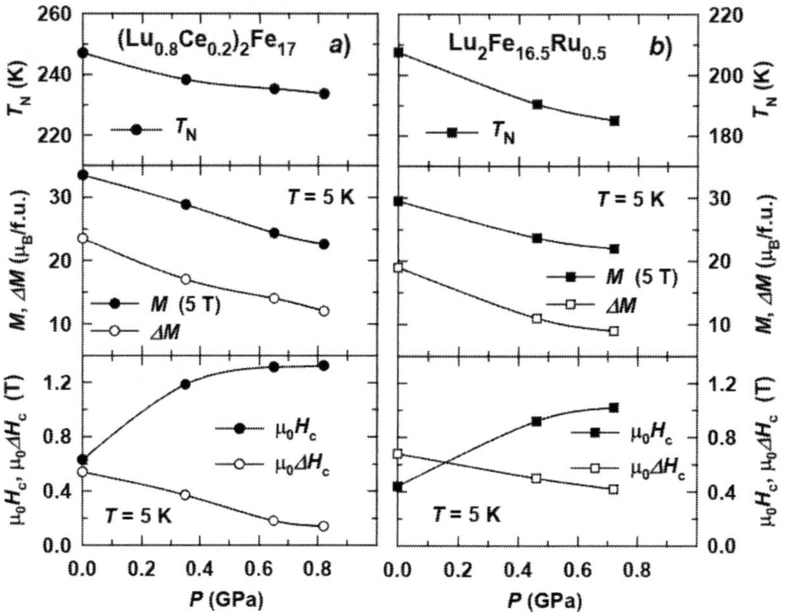

Fig. 3. a) Pressure dependence of the Néel temperature T_N, magnetic moment M in a 5 T field, magnetization gain ΔM across the metamagnetic transition, critical field H_c (average between the field-up and field-down values) and width of hysteresis ΔH_c for a $(Lu_{0.8}Ce_{0.2})_2Fe_{17}$ single crystal. b) The same for a $Lu_2Fe_{16.5}Ru_{0.5}$ single crystal.

We can conclude that interplay between the dominant negative exchange interaction of Fe atoms in the "dumbell" positions and the slight positive exchange interaction of other Fe atoms is an origin of the helimagnetism in the Lu_2Fe_{17}-based intermetallics. The observed pressure and substitution effects can be explained taking into account the anomalous and anisotropic thermal expansion induced by the anomalous and anisotropic magnetostriction. In agreement with this conclusions, pressure behavior of $(Lu_{0.8}Ce_{0.2})_2Fe_{17}$ and $Lu_2Fe_{16.5}Ru_{0.5}$ compounds can be well compared with behavior of the parent Lu_2Fe_{17} compound under external pressure [3,4].

Acknowledgements
The work is a part of the research project AVOZ10100520 and has been supported by grants GACR 202/06/0185, GACR 202/06/0178 and GAUK 1507/2007.

References
[1] Givord D and Lemaire R 1974 *IEEE Trans. Magn.* **MAG-10** 109.
[2] Andreev A V, Rafaja D, Kamarád J, Arnold Z, Homma Y and Shiokawa Y 2003 *J. Alloys Compds* **361** 48.
[3] Kamarád J, Arnold Z, Medvedeva I V and Kuchin A G 2002 *J. Magn. Magn. Mater.* **242-245** 876.
[4] Kamarád J, Prokhnenko O, Prokeš K and Arnold Z 2005 *J. Phys.: Condens. Matter* **17** S3069.
[5] Kamarád J, Machátová Z and Arnold Z 2004 *Rev. Sci. Instrum.* **75** 5022.
[6] Makihara J, Uwatoko Y, Matsuoka H, Kosaka M, Fukuda H and Fujii H 2004 *J. Magn. Magn. Mater.* **272-276** 551.
[7] Andreev A V and Tereshina E A 2007 (unpublished).
[8] Andreev A V in *Handbook of Magnetic Materials*, Vol.8, Ed. by K.H.J. Buschow, Elsevier Science B.V., 1995, p.141.

Journal of Physics: Conference Series **121** (2008) 032011 doi:10.1088/1742-6596/121/3/032011

Electrotransport and magnetic properties of metastable phases in the system GaSb-Mn synthesized under high pressure.

M.V.Kondrin[1], S.V.Popova[1], V.R.Gizatullin[1], O.A.Sazanova[1], N.V.Kalyaeva[1], A.G.Lyapin[1], V.V.Brazhkin[1], A.A.Pronin[2], S.A.Gudoshnikov[3], Yu.V.Prokhorova[3]

[1]Institute for High Pressure Physics, RAS, Troitsk, Moscow region, 142190 Russia
[2]General Physics Institute, RAS, 117942 Moscow, Russia
[3]Institute of Terrestrial Magnetism, Ionosphere, and Radio Wave Propagation, RAS,Troitsk, Moscow region, 142190 Russia

Abstract. The paper reports on electrophysical and magnetic properties of recently discovered high-pressure phases in the $(GaSb)_{(1-x)}Mn_x$ system, which is a possible candidate for spintronic material. It was shown that the simple cubic and tetragonal high-pressure phases display magnetic ordering although the critical parameters and electrotransport properties of these two phases significantly differ. Also we observed a magnetic ordering in the solid solution of Mn in the semiconductor sphalerite phase of GaSb prepared under high pressure, although in this case magnetic moment per Mn ion is an order of magnitude smaller than in the first two cases.

Close attention has been recently focused on the search for new materials for spintronics, i.e., electronics that involve both magnetic field and spin-polarized electron transport for signal transmission and modulation. In this respect, III-V semiconductors doped with magnetic ions are most attractive. However, the application of such compounds is yet complicated, because the Curie temperature is much lower than room temperature. To date, the properties of $Ga_{1-x}Mn_xAs$ with T_c=110 K are studied most completely and the attention of researchers is primarily focused on the search for new similar compounds with the different matrices and substitution of the Mn ion for the Group-III element [1]. A ferromagnetic phase has recently been found in wide gap semiconductor(Zn,Cr)Te [2].

All the samples mentioned above were prepared as a thin films by the molecular beam epitaxy. On the other hand the preparation of bulk spintronics materials can provide a mean to rule out possible dimensional effects and to let us correlate electro transport and magnetic properties with crystal structure of synthesized samples. The possible candidate was the solid solution of Mn in GaSb matrix, whose thin-film analogies were studied in [3, 4]. Two previously unknown high-pressure phases of $(GaSb)_{1-x}Mn_x$ that are significantly different in structure were obtained in [5], where it was shown that the structure of the samples synthesized at high pressures near 6 GPa was primarily determined by the synthesis temperature T_{syn}. For temperatures of T_{syn} <470 K, solid solutions with the sphalerite structure are likely to be formed. This phase is structurally the same as conventional low-pressure phase GaSbI. A further increase in temperature to T_{syn}=690-870 K leads to the synthesis of a phase with the tetragonal structure. In a narrow intermediate synthesis-temperature range of T_{syn}=620-670 K, the simple cubic phase is formed.

© 2008 IOP Publishing Ltd

Journal of Physics: Conference Series **121** (2008) 032011 doi:10.1088/1742-6596/121/3/032011

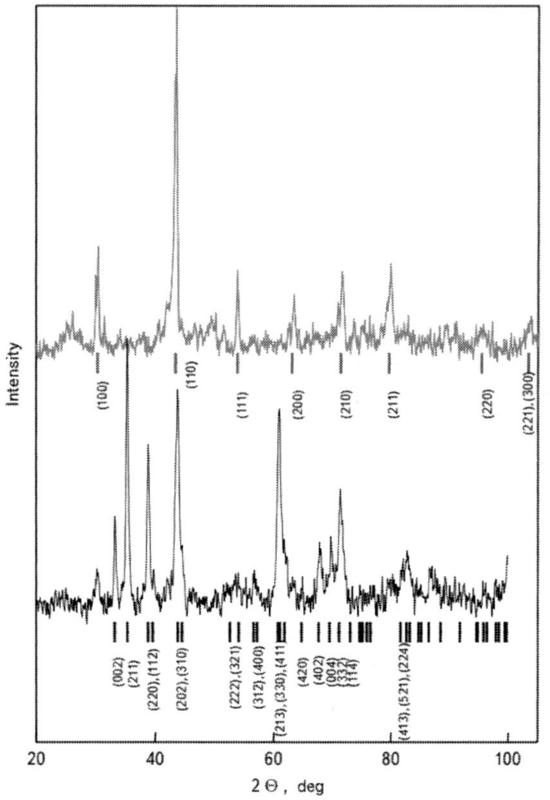

Figure 1. X-ray (CuK_{α}) diffraction spectra of simple cubic phase (red curve) and tetragonal phase (black)

The X-ray spectra of the tetragonal and simple-cubic phases are shown in Fig. 1. Although tetragonal phase is clearly identified as $I4/mcm$ $CuAl_2$-like structure the identification of simple-cubic one is still questionable. The number of data (X-ray diffraction and density measurements) provided an evidence that the unit cell of this structure is quite small with lattice parameter of about 2.94 Å(whose value slightly depends on initial Mn content and varies in the range $\pm$ 0.1 Å) and contains only 1.2 atoms per unit cell [5] (for 50% Mn initial concentration). The same structure is synthesized in a wide concentration range (10-60 at.% of initial Mn). It is also worth mentioning that on X-ray spectra we observe only small traces of pure Mn in negligible amounts (in comparison to its initial content) and no other elemental components in pure forms or binary compounds. So the X-ray data gives an evidence of quite high degree of phase homogenuity, the conclusion which is also supported by a preliminary scanning microscopy with elemental analysis.

We have examined the magnetic and electrophysical properties of all types of the (Ga, Mn)Sb compounds synthesized through temperature treatment under high pressure. In all three cases the samples synthesized under pressure has magnetic ordering although critical parameters of these phases differ. The existence of these high-pressure phases with magnetic ordering may account for the diversity of magnetic properties observed on the films of GaSbMn [3]. The authors there suggested that the different properties observed for the samples grown at higher temperatures (above 560 C) and the one grown at temperatures below 300 C are due to the formation of the second ferromagnetic phase MnSb and/or $MnSb_2$ (T_c = 587 and 550 K). So the above-room Curie-temperature found in this type of material [4] was attributed to the ferromagnetic properties of this phase. But in the bulk samples prepared in [5] and investigated by us, X-ray study and chemical analysis revealed no traces of binary compounds of Mn and Sb.

The high temperature tetragonal phase demonstrates a metallic-like transport properties and is ferromagnetic in all temperature range where it does exist (T<450K). The temperature dependence of resistivity and magnetization curve at 293 K are shown in Fig. 2-e) and Fig. 3-b) respectively. The phase prepared at intermediate temperatures with simple cubic structure demonstrates strong dependence of kinetic coefficients on Mn content. With the rise of Mn concentration from 0.1 to 0.6 the T_c is increased from 220 K up to 280 K. But at the same time this phase undergos a metal-insulator transition. This fact is demonstrated by the resistivity data (Fig. 2 a-d) and Drude formula for "minimal" metallic conductivity:

$$\sigma_0 = (3\pi^2)^{-2/3}(e^2/\hbar)n^{1/3} \quad (1)$$

Taking into account the data shown on Fig. 2 and an order of magnitude estimate for carrier concentration $n \sim x/a^3$ (where x is the Mn concentration and a = 2.94 Å - lattice parameter [5]) we can see that the samples with $x < 0.5$ are on the insulator side of the metal-insulator transition while the samples with x >0.5 are well onto the metal side. The magnetization curves for these type of samples taken at liquid nitrogen temperatures are shown in Fig. 3-a. The dependencies in the near-critical temperature region of magnetic polarization measured by SQUID magnetometer and magnetic susceptibility measured by AC inductive method for the sample with largest T_c attained in our experiments are displayed in Fig. 4.

The samples of sphalerite phase have quite limited solubility of Mn atoms. Our upper-bound estimate is well below 5% of Mn. This number of metallic atoms has a little impact on electrotransport properties of sphalerite phase which is obviously a semiconductor, although by variation of synthesis procedure the resistivity of the obtained samples at helium temperature can take values in the range 10-10000 Ohm cm. The typical temperature dependence of resistivity is shown in Fig. 5. Despite the small Mn content the samples exhibit magnetic ordering at room temperature. The evidence for that is provided by magnetization hysteresis observed on these samples (Fig. 3-c), although in this case the magnetic moment per Mn ion is an order of magnitude less than the one observed in the tetragonal phase. It seems unlikely that this magnetic ordering is due to the second magnetic phase (i.e. tetragonal one), because under high pressure the regions of stability of tetragonal and sphalerite phases have no common border and are separated by the region where simple-cubic phase are formed with the T_C a little lower than the room temperature. The possible mechanism may be like the one suggested for the description of diluted wide-gap semiconductors (Zn,Cr)Te [2], where it was provided an evidence for "patterned" distribution of magnetic-ion in semiconductor matrix, so the ferromagnetic ordering does exist in the region of rich Cr content.

Thus, new ferromagnetic phases with high Curie temperatures close to or above room temperature are formed in the GaSbMn system under high pressure. Solid solutions with the primitive cubic structure, which are formed in a wide range of the manganese content, are of most interest. Variation of the initial manganese content allows one to control not only the magnetic but also the transport properties of these compounds. An increase in the manganese content, on the one hand, enhances the ferromagnetic properties, which is manifested in an increase in the ferromagnetic transition temperature from T_c = 230 K for x = 0.1 to T_c = 220 K for x = 0.6, and, on the other hand, strongly reduces the resistivity in this content range. The phase with the tetragonal structure is ferromagnetic over the entire range of its stability (T < 450 K) and exhibits metallic conductivity. The magnetization at T=77.3 K is equal to M=0.58

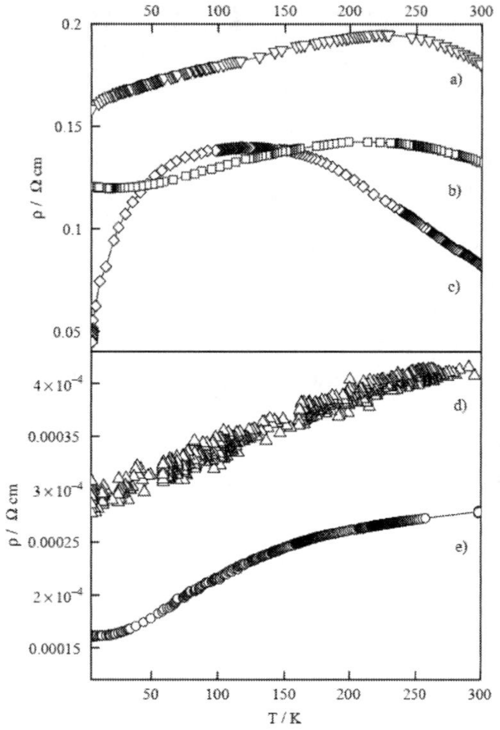

Figure 2. Resistivity of the high-pressure phases of the $(GaSb)_{1-x}Mn_x$ samples with various structures and various manganese contents: the simple cubic structure with $x =$ (a) 0.1, (b) 0.2, (c) 0.5, and (d) 0.6 and (e) the tetragonal GaSbMn phase

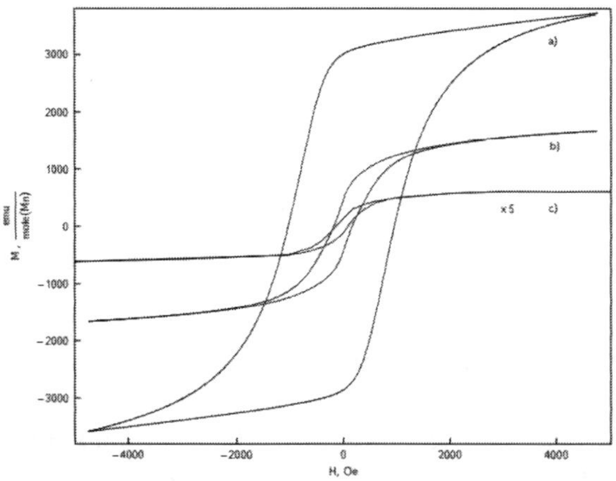

Figure 3. Magnetization per Mn ion of three high-pressure phases of GaSb-Mn: a) simple-cubic at 77 K (x=0.2) b) tetragonal at 293K (x=0.5) and c) sphalerite phase at 293 K (x=0.1)

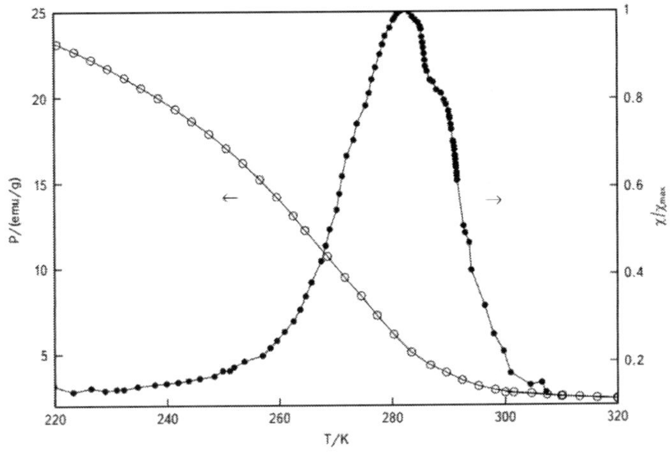

Figure 4. DC magnetic moment and AC magnetic susceptibility of the simple cubic sample at the vicinity of ferromagnetic transition.

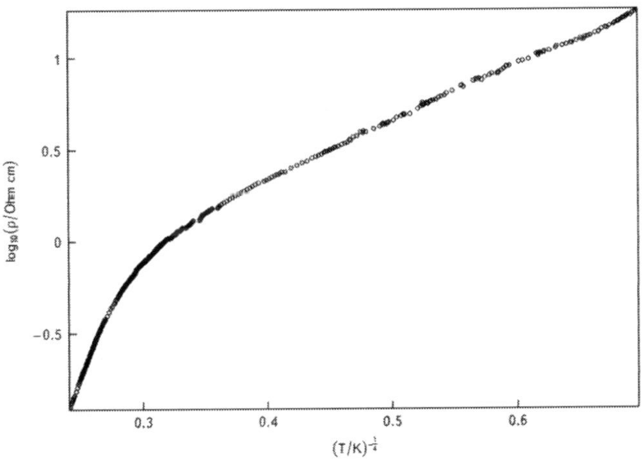

Figure 5. Mott's resistivity of the $(GaSb)_{0.8}Mn_{0.2}$ sample with sphalerite structure.

μ_B and 0.28 μ_B per manganese ion for the simple cubic and tetragonal phases, respectively. The sphalerite phase with low Mn content (x<0.05) is semiconductor and exhibits magnetic ordering at room temperature with magnetic moment per Mn about $0.05\mu_B$.

This work was supported by The Russian Foundation for Basic Research (Project Nos. 05-02-16596 and 07-02-01275), by The Russian Science Support Foundation, and by Programs of the Presidium of Russian Academy of Sciences.

References

[1] T. Jungwirth, K. Y. Wang, J. Masek, K. W. Edmonds, Jurgen Konig, Jairo Sinova, M. Polini, N. A. Goncharuk, A. H. MacDonald, M. Sawicki, R. P. Campion, L. X. Zhao, C. T. Foxon, and B. L. Gallagher. Prospects of high temperature ferromagnetism in (Ga,Mn)As semiconductors. *Physical Review B*, 72:165204, 2005.

[2] Shinji Kuroda, Nobuhiko Ozaki, Nozomi Nishizawa, Tatsumi Kumekawa, Stephane Marcet, and Koki Takita. Growth and magnetic properties of novel ferromagnetic semiconductor (Zn, Cr)Te. *Science and Technology of Advanced Materials*, 6:558–564, 2005.

[3] E. Abe, F. Matsukura, H. Yasuda, Y. Ohno, and H. Ohno. Molecular beam epitaxy of III-V diluted magnetic semiconductor (Ga,Mn)Sb. *Physica E: Low-dimensional Systems and Nanostructures*, 7(3-4):981–985, 2000.

[4] X. Chen, M. Na, M. Cheon, S. Wang, H. Luo, B. D. McCombe, X. Liu, Y. Sasaki, T. Wojtowicz, J. K. Furdyna, S. J. Potashnik, and P. Schiffer. Above-room-temperature ferromagnetism in GaSb/Mn digital alloys. *Applied Physics Letters*, 81(3):511–513, 2002.

[5] S. Popova, O.Sazanova, V.Brazhkin, N.Kalyaeva, M.Kondrin, and A.Lyapin. High-pressure phases in the GaSb-Mn system. *Physics of the Solid State*, 48(11):2177–2182, 2006.

Pressure tuning of the non-Fermi liquid state in $UCo_{0.95}Fe_{0.05}Al$

A V Kolomiets[1,2], J-C Griveau[3], L Havela[1], A V Andreev[4], V Sechovský[1]

[1] Department of Condensed Matter Physics, Faculty of Mathematics and Physics, Charles University, Ke Karlovu 5, 12116, Prague, Czech Republic

2 Department of Physics, National University 'Lvivska Politekhnika', Bandery 12, Lviv, Ukraine

[3] European Commission, JRC, Institute for Transuranium Elements, Karlsruhe, Germany

[4] Institute of Physics, Academy of Sciences, Na Slovance 2, 18221, Prague, Czech Republic

E-mail for correspondence: kolomiet@mag.mff.cuni.cz

Abstract. High-pressure resistivity measurements were performed on $UCo_{0.95}Fe_{0.05}Al$ single crystal. They indicate the fast suppression of the ferromagnetic order with a critical pressure of P_c = 2-3 GPa. The disappearance of the ferromagnetism is accompanied with the emergence of the non-Fermi liquid behavior, which persists at least up P = 3.6 GPa.

1. Introduction

The deviation of the low-temperature physical properties such as specific heat, electrical resistivity or magnetic susceptibility from the predictions of the Fermi-liquid (FL) theory has been reported already decades ago and observed in numerous materials (*e.g.* [1]). Particularly interesting is the universal non-FL (NFL) behavior occurring around the onset of magnetic order. Experimental findings amounted into a concept of quantum criticality, in which dynamical degrees of freedom related to quantum fluctuations (depending on the type of coupling and dimensionality) coexist with classical degrees of freedom if the magnetic phase transition approaches $T = 0$ under the influence of some control parameter (composition, pressure…) [2]. Such a quantum critical point (QCP) is then followed on the non-magnetic side by a "heavy" FL regime spreading gradually from the $T = 0$ limit, in which spin fluctuations are responsible for a strong renormalization of basic parameters of the electronic system (as enhancement of effective mass), similarly to the critical fluctuations at the temperature-driven phase transitions. As the NFL behavior can occur also due to disorder in different types of system (see [3, 4] for overview), it is essential to investigate materials without any intrinsic disorder. Application of external pressure on an undoped system is, therefore, a unique tool for tuning magnetic materials through a genuine quantum critical regime.

Extended investigation of U ternary compounds in the vicinity of the onset of magnetic transition led to the recognition that the band metamagnet UCoAl exhibits a NFL scaling of electrical resistivity $\rho =$

© 2008 IOP Publishing Ltd

$\rho_0 + bT^{5/3}$ (predicted by theories for ferromagnetic 3D spin fluctuators) in the low-field non-magnetic state [5-7], whereas the field-induced ferromagnetic state (first-order transition, critical field is 0.65 T in the low-T limit) displays a standard FL behavior.

Proceeding towards the non-magnetic side by applying pressure (which broadens the 5f band and generally suppresses the 5f magnetic moments), the NFL behavior vanishes only above 5 GPa [8], where the tendency to metamagnetism disappears. These findings indicate that some sort of a robust NFL phase exists in UCoAl, while the scenario of a single critical point is untenable in this case.

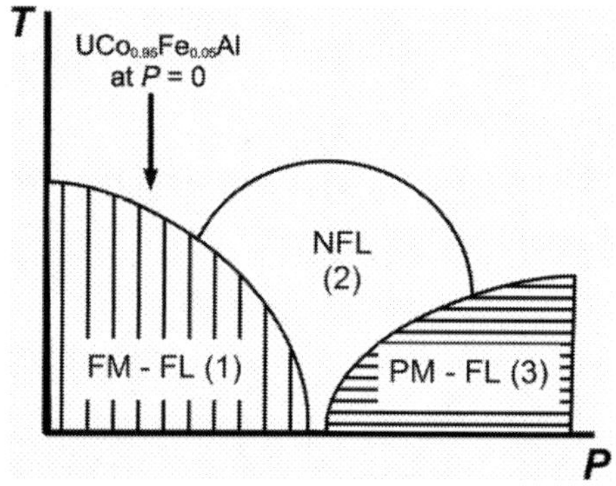

Figure. 1. Schematic phase diagram of the QCP vicinity. FL and NFL stand for Fermi-liquid and non-Fermi liquid states. The material in the FL(1) state is magnetically ordered and in FL(3) – paramagnetic, and the conduction electron scattering may differ in these states.

The proximity of UCoAl to the border of stable ferromagnetism is illustrated by the sensitivity to doping. Few percent doping by *e.g.* Fe can induce ferromagnetism stable even without the magnetic field [9, 10]. This property can be used for covering all the regimes of the phase diagram shown in Figure.1, since undoped UCoAl itself can be driven by the pressure only through the regions (2) and (3). The doping with iron makes the region (1) also accessible. Therefore, we have undertaken the high-pressure experiment on $UCo_{0.95}Fe_{0.05}Al$ aiming at covering the whole pressure region of the QCP phase diagram.

$UCo_{0.95}Fe_{0.05}Al$ has the same crystal structure as UCoAl, it orders ferromagnetically below 30 K, and its magnetic ordering is pressure sensitive [9, 10]. The latter is related to the fact that hydrostatic pressure induces predominantly compression in the basal plane of the hexagonal structure, and the U-U spacing within the basal plane is the crucial parameter controlling magnetic properties [11]. Also, the uncertainty of whether the NFL regime in $UCo_{0.95}Fe_{0.05}Al$ would be due to QCP or to the disorder, is not the issue for $UCo_{0.95}Fe_{0.05}Al$ because the role of the quantum critical regime was shown for the Fe-free UCoAl.

2. Experimental details

Electrical resistivity was studied on different pieces of a single crystal with RRR = 4 prepared in a tetra-arc furnace from pure metals of at least 3N purity.

High-pressure measurements have been performed at ITU Karlsruhe in zero magnetic field using a four-probe DC technique with the sample and a stripe of lead (manometer) pressed in a piston-cylinder device, with steatite as a pressure-transmitting medium. Typical sample thickness, which can be used in this cell, is 30-40 μm. Extreme brittleness of the $UCo_{0.95}Fe_{0.05}Al$ crystal imposed the limitations on the size of the sample (up to 500 μm) and the maximal pressure applied (up to $P = 3.6$ GPa).

3. Results and discussion

The resistivity data obtained are presented in Figure 2. The measurements were performed with the electrical current along the a-axis. In the experimentally accessible pressure range, the sample compression has primarily affected the low-temperature part of the $\rho(T)$ curves, and, therefore, its details are shown in the figure.

Firstly, the increase of pressure leads to the disappearance of the ferromagnetic ordering in $UCo_{0.95}Fe_{0.05}Al$ below 2 GPa. The magnetic phase transition can be well observed for the zero-pressure data. It is indicated by the kink on the resistivity curve at $T_C = 29$ K, in the good agreement with T_C determined from the measurements of the specific heat reported in literature [9, 10], which is followed by a broader feature at $T \approx 32$ K. As the pressure increases, the anomalies smear out. Already in $P = 1.0$ GPa the kink disappears and the broad feature transforms into a change of slope at $T = 34$ K.

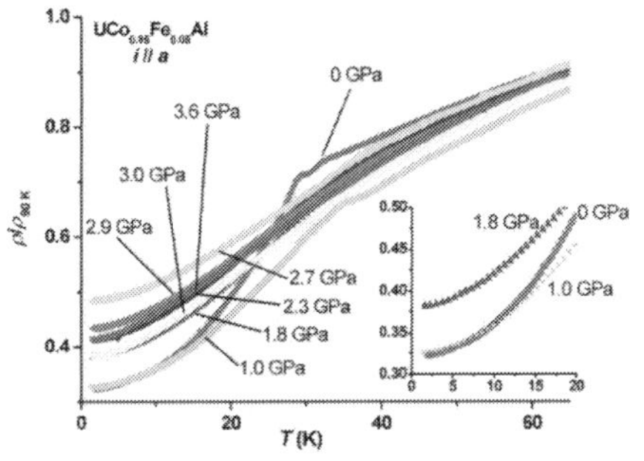

Figure. 2. Temperature dependence of the normalized resistivity curves of $UCo_{0.95}Fe_{0.05}Al$. Labels indicate the pressures, at which the data were collected.

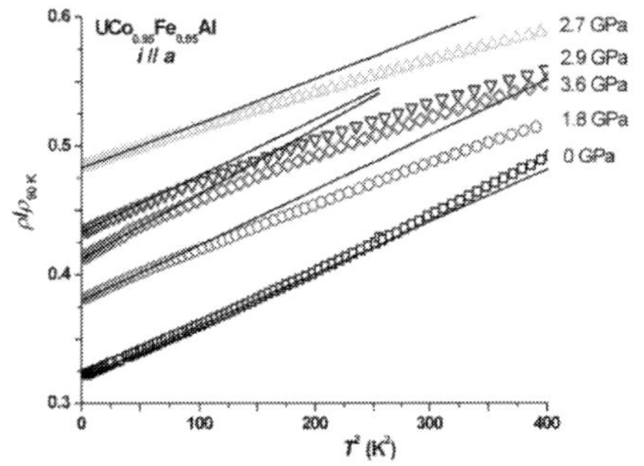

Figure. 3. Adherence of $UCo_{0.95}Fe_{0.05}Al$ to the T^2-dependence. Symbols indicate the experimental data, and the straight lines represent the $\rho = \rho_0 + AT^2$ fits.

The rapid smearing of the sharp anomaly could be partly related to a certain non-hydrostaticity in the present experimental setup.

For the next data set, collected at P = 1.8 GPa, even the change of slope is hardly visible and its position, T = 35 K, can be determined only by the temperature derivative of $\rho(T)$. At P = 2.3 GPa and above, there are no clear traces of any resistivity anomalies below 50 K. Thus, the resistivity data do not provide the exact value of the critical pressure for the disappearance of the ferromagnetism in $UCo_{0.95}Fe_{0.05}Al$, but the 2 GPa limit seems quite sound because the character of the resistivity changes qualitatively between 1.8 GPa and 2.3 GPa. Another argument in favor of this assumption can be found in the high-pressure magnetization studies reported in Ref. [9]. These measurements indicate the rapid decrease of T_C with pressure at the rate of about 15 K/GPa. The maximum pressure achieved in the study [9] was P = 1.2 GPa, so the complete suppression of the magnetic ordering was not observed, yet it is quite likely to occur even below the expected P_c = 2 GPa because the pressure dependence of the Curie temperature usually becomes non-linear as it approaches T = 0.

The second interesting subject is the pressure-related modification of the polynomial $\rho = \rho_0 + AT^n$, which describes the low-T part of $\rho(T)$ in $UCo_{0.95}Fe_{0.05}Al$ (Figure. 3). The Figure 3 shows both the experimental data and the T^2 fits corresponding to the FL behavior. The fits are straight lines due to the T^2 abscissa scale. The region, across which the experimental data follow the fits, is the temperature range of the stability of FL state in $UCo_{0.95}Fe_{0.05}Al$. At P = 0 (in the ordered state) it extends up to approximately 16 K, but it shrinks rapidly with increasing pressure. Starting from P = 2.7 GPa, the $\rho(T)$ curve cannot be fitted to the T^2-dependence across the temperature range of more than 5 K (Figure. 3), and, therefore, the FL regime at these pressures should not be considered any more.

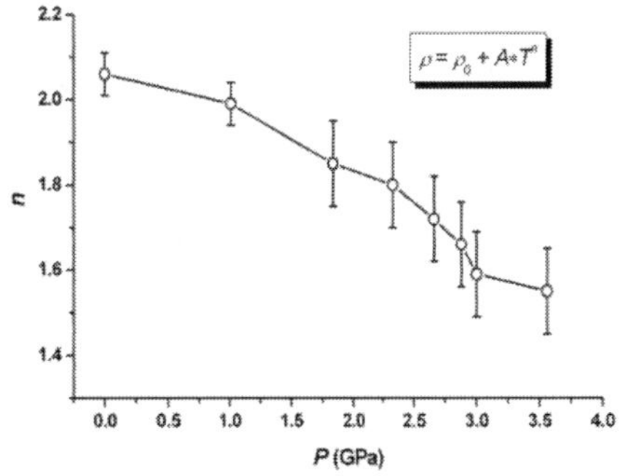

Figure. 4. Pressure variation of the exponent n in the expression for resistivity $\rho = \rho_0 + AT^n$.

A more detailed analysis reveals that the increase of pressure induces a non-Fermi liquid type of resistivity above the T^2-region, which shrinks progressively towards zero T. The exponent n in the expression $\rho = \rho_0 + AT^n$ appears to be pressure dependent (see Figure. 4), and the range of NFL resistivity extends up to T = 15 -17 K.

The observation of the two regimes, FL and NFL following each other, in $UCo_{0.95}Fe_{0.05}Al$ places the compound on the left-hand side of the QCP diagram (Figure. 1), where the FL state still survives at the lowest temperatures due to the V shape, but the influence of the spin fluctuations determines the resistivity above that temperature range. The question remains open whether the FL state survives at the lowest temperatures in the pressures exceeding 2 GPa. For unambiguous conclusion one would need to cover temperature range extending to the mK range. In any case, we do not observe the re-emergence of the FL state on the right side of the phase diagram in Figure.1. This goes in line with the experimental findings for UCoAl where, in spite of the non-magnetic ground state, the pressure above 5 GPa is required for complete suppression of the NFL regime [8]. It is natural to assume that in the ferromagnet $UCo_{0.95}Fe_{0.05}Al$ this value would be even higher, and the pressures of 10 - 20 GPa are necessary in order to cover the whole QCP phase diagram. The maximal pressure $P = 3.6$ GPa reached in the present work is sufficient only to drive $UCo_{0.95}Fe_{0.05}Al$ from the region (1) of the magnetically ordered FL, into the region (2) close to the QCP (Figure. 1). Most likely $P = 3.6$ GPa is sufficient complete suppression of the FL regime in $UCo_{0.95}Fe_{0.05}Al$.

4. Concluding remarks

The ferromagnetic order in $UCo_{0.95}Fe_{0.05}Al$ is suppressed above $P = 2$ GPa. The range of the Fermi liquid regime in this compound, which is associated with the ordered state, shrinks with the increasing pressure, but is observed till at least to $P = 2.3$ GPa. The non-Fermi liquid resistivity emerges atop of the FL regime, and the exponent n in the expression for $\rho(T)$ is the function of pressure. The robust character of the FL-NFL states in $UCo_{0.95}Fe_{0.05}Al$ is similar to the properties of UCoAl. The NFL behavior can be attributed to the critical state with quantum fluctuations present between the phase with well defined ordered magnetic moments and the state in which the moments completely disappear. Unlike the two neighbor phases, in which electronic system exhibits FL properties, the FL scaling breaks down in the NFL regime, and basic properties as resistivity follow different scaling, depending on the type of magnetic coupling and dimensionality of the system. The combined high-pressure experiments on pure and Fe-doped UCoAl illustrate how the NFL state develops from the magnetic FL state and gives up again to the non-magnetic FL state. The novelty of the present data dwells in the fact that the NFL behavior exists over an extended pressure/concentration range, which contradicts a common belief that it should exist only in a close vicinity of a single critical point.

Acknowledgements

AVK acknowledges the European Commission for support in the frame of the "Training and Mobility of Researchers" program. This work is a part of the research plan MSM 0021620834 that is financed by the Ministry of Education of the Czech republic. Financial support of the Czech Science Foundation under the Grant no. 202/06/0178 also has been enjoyed. The work was also supported by the project D/B 'Promin'.

References

[1] Mydosh J A and Ford P J 1973 *Amorphous Magnetism* ed H O Hooper and A M deGraaf (New York: Plenum) p 237
[2] Sachdev S 1999 *Quantum Phase Transitions* (Cambridge,England: Cambridge University Press)
[3] Stewart G R 2001 *Rev. Mod. Phys.* **73** 797
[4] Zaanen J 2008 *Science* **219** 1205
[5] Kolomiets A V, Havela L, Sechovsky V, DeLong L E, Watkins D B and Andreev A V 1998 *J. Appl. Phys.* **83** 6435
[6] Kolomiets A V, Havela L, Sechovsky V, DeLong L E, Watkins D B and Andreev A V 1999 *Physica B* **259-261** 415
[7] Sechovsky V, Havela L, de Boer F R, Franse J J M, Veenhuizen P A, Sebek J, Stehno J and Andreev A V 1986 *Physica B* **142** 283
[8] Havela L, Honda F, Griveau J C, Andreev A V, Kolomiets A and Sechovsky V 2006 *J. Alloys Comp.* **408** 1316

[9] Andreev A V, Mushnikov N V, Goto T, Sechovsky V, Homma Y and Shiokawa Y 2003 *Physica B* **329-333** 499
[10] Janousova B, Andreev A V, Sebek J, Homma Y, Shiokawa Y and Sechovsky V 2002 *Physica B* **312-313** 888
[11] Honda F, Kagayama T, Oomi G, Havela L, Sechovský V and Andreev A V 2000 *Physica B* **284-288** 1299

Joint 21st AIRAPT and 45th EHPRG Int. Conf. on High Pressure Science and Technology IOP Publishing
Journal of Physics: Conference Series **121** (2008) 032013 doi:10.1088/1742-6596/121/3/032013

High pressure experimental and theoretical study of the quasi-one-dimensional sulfides AV_6S_8 (A = In, Tl)

H D Hochheimer[1], C F Miclea[2], P Modak[3] A K Verma[3], R S Rao[3], T Ohtani[4]

[1] Department of Physics, Colorado State University, Fort Collins, CO 80523, USA
[2] Max-Planck-Institute for the Chemical Physics of Solids, Dresden, Germany
[3] High Pressure Physics Division, Purnima Labs, Bhabha Atomic Research Center, Mumbai 400 085,India
[4] Laboratory for Solid State Chemistry, Okayama University of Science, Ridai-cho 1-1, Okayama 70-0005, Japan

E-mail: dieter@lamar.colostate.edu

Abstract. We report here the results of high pressure conductivity measurements on $Tl_xV_6S_8$ (x = 1, 0.15, 0.47, and 0.63) in pressure range from 0 - 2 GPa in the temperature range from 300 K to 300 mK to obtain insight in the competition between superconductivity and charge density waves as well as to establish the existence of a charge density wave in these materials. The experimental results are complemented by first principle electronic band structure calculations on InV_6S_8 in order to study the effect of pressure on the Fermi surface nesting which leads to the formation of the charge density wave instability.

1. Introduction

The binary vanadium-sulfur systems have generated quite some interest [1,2], because their physical properties present a border line case between compounds with localized and delocalized d-electrons. Furthermore, they are isotopic with the binary niobium chalcogenides, Nb_3X_4, which have metal zig-zag chains parallel to the crystallographic c axis. Another interesting feature of the NB_3X_4 structure is the large hexagonal channels parallel to the c axis. They are empty in the binary niobium compounds, but occupied by In in InV_6S_8. The In atoms are statistically disordered, and the disorder is mainly one-dimensional.

The discovery of superconductivity in $Tl_xV_6S_8$ (x = 0.8, 0.1, and 0) by Bensch and Koy [3] has created renewed interest and have led to discovery of superconductivity in other AV_6S_8 compounds with A = In, Tl, K, Rb, and Cs by Ohtani et al. [4]. Furthermore, Ohtani et al. observed an anomaly in the temperature dependence of the resistivity in InV_6S_8 with a large hysteresis of about 40 K width, suggesting a first order phase transition. The discussed the transition in terms of a charge density wave (CDW) instability, but the lack of additional data prevented a confirmation of this suggestion.

These experimental results as well as the fact that the physical phenomena connected with the competition between structural instabilities and superconductivity are still not fully understood and motivated us to study these materials under high pressure.

© 2008 IOP Publishing Ltd

There are no reports of high pressure studies of these materials, though high pressure is an excellent tool to study charge and spin density wave instabilities [5,6]. This has also been corroborated by the high pressure studies on TX_2 (T = Nb, Ta, X = S, Se) by Jerome et al. [7] and on 2H-$NBSe_2$ by Berthier et. al. [8]. To the best of our knowledge there is only one high pressure structural study of the channel compounds, Nb_3Te_4 and $In_xNb_3Te_4$ (x <10) by Wunschel et al. [10].

They performed a high pressure X-ray powder diffraction study up to 40 GPa, but did not observe any phase transition. However, they found that the compression along the c axis involves the folding up of the quasi-one-dimensional zig-zag chain of Niobium. The compression perpendicular to the c axis is entirely caused by the reduction of the diameter of the channels. Furthermore, they found that the presence of intercalated In atoms have hardly any influence on the compression behavior up to 40 GPa.

Here we report now the results of high pressure conductivity measurements on $Tl_xV_6S_8$ (x = 1, 0.15, 0.47, and 0.63) in pressure range from 0 - 2 GPa in the temperature range from 300 K to 300 mK .

Since Fermi surface nesting is necessary for a charge density wave instability [9], we have performed first principle band structure calculations which showed that Fermi surface nesting conditions are fulfilled for at least 4 different bands. Furthermore, we wanted to obtain theoretical results of the influence of pressure on the Fermi surface nesting.

2. Experimetal

Samples for the high pressure conductivity measurements on $Tl_xV_6S_8$ (x = 1, 0.15, 0.47, and 0.63) in pressure range from 0 - 2 GPa in the temperature range from 300 K to 300 mK were pressed from powder in pellets and then sintered in closed silica tubes at T = 800^0 C for seven days. The composition of the samples were confirmed by X-ray and microprobe analysis.

The samples were then cut in parallelepipeds and gold wires were connected on the clean surfaces with gold paste. Standard four-probe AC resistance measurements were performed at ambient and high pressure. The high pressure measurements were performed under hydrostatic conditions in a double-wall CuBe piston cylinder cell up to 2 GPa, which was designed to fit in a commercial Quantum Design PPMS. We used a mixture of n-pentane and iso-pentane as pressure medium, which ensured excellent hydrostatic conditions. The shift of the superconducting transition temperature, T_c, of a Pb sample used as manometer, was determined using a four-probe AC resistance measurement and then converted into pressure using the Tc(p) calibration from Eiling et al. [11]. No pressure gradient was observed when a larger Pb sample was covering the whole sample space. Its superconducting transition width remained narrow ($\Delta Tc \approx 5$ mK) and almost pressure independent.

3. Crystal structure

The crystal structure of InV_6S_8 is based on a hexagonal cell [4] of Nb_3Te_4 -type (space group $P6_3$) [12] with room temperature ambient pressure values of a = 9.207 Å and c = 3.310 Å. The structure is composed of V_6S_8 octahedra linked together by shared edges and faces to form hexagonal channels running parallel to the crystallographic c axis [13].

Another notable feature of the structure is the existence of V-V zig-zag chains. They run along the c-axis and are well isolated from each other with an intrachain distance of 2.864 Å and an interchain distance of 3.150 Å .

4. Details of theoretical calculations

For a better understanding of the high pressure effects in this quasi-one-dimensional system, we have carried out the electronic band structure calculations for InV_6S_8, both at ambient and at 18 GPa pressure. We employed the full potential linear augmented plane wave (FPLAPW) method as implemented in the WIEN2K code [14]. As there is random occupation of the two equivalent In-sites in the unit cell, each with probability of 0.5, we have adopted the virtual crystal approximation [15] in our band structure calculations. As unit cell we have taken a simple hexagonal cell. The valence electronic configuration of In, V, and S are taken as, [Ar]$3d^3 4s^2$, and [Ne]$3s^2 3p^4$, respectively. As the 4d states of In are well below the Fermi level (by about 10 eV), and mainly its wide-band $5s^2$ states contribute to the conduction band, our virtual crystal approximation for disorder at In sites should be adequate. For the exchange-correlation terms, we have used the local density approximation as per Perdew-Wang [16] re-parameterization of Ceperley-Alder data [17].

InV_6S_8 crystallizes in a primitive hexagonal lattice with 16 atoms per unit cell, and our ADXRD measured lattice constants have been used in the calculations. The fractional coordinates of atomic positions in the unit cell have been taken as per Ref.13 at all compressions, though ideally a study of the variations of these coordinates under compression is desired. As it is well-known, the basis functions in the FPLAPW method depend upon the plane waves used in the interstitial region. Therefore, we have employed about 3100 plane waves ($R_{mt}K_{max}$ = 7 in WIEN2K, where R_{mt} is the muffin-tin radius and K_{max} determines the wave-vector cut-off for the plane wave expansion). R_{mt} has been fixed at 1.4 atomic units (a.u.-Bohr radius) for In, 1.5 a.u. for V and 1.5 a.u. for S. We have used 500 **k**-points for Brillouin zone (BZ) sampling (70 **k**-points in the irreducible wedge of the BZ).

5. Results and discussion

5.1. Experimental

Fig. 1 shows the pressure dependence of the CDW instability in $Tl_{0.63}V_6S_8$, because this material displays both, a CDW instability and full superconductivity. Upon applying hydrostatic pressure the CDW anomaly size is gradually reduced together with a decrease of T_l and T_h, the temperatures where $\delta\rho/\delta T$ as a function of T has a local minimum upon cooling and warming up, respectively.

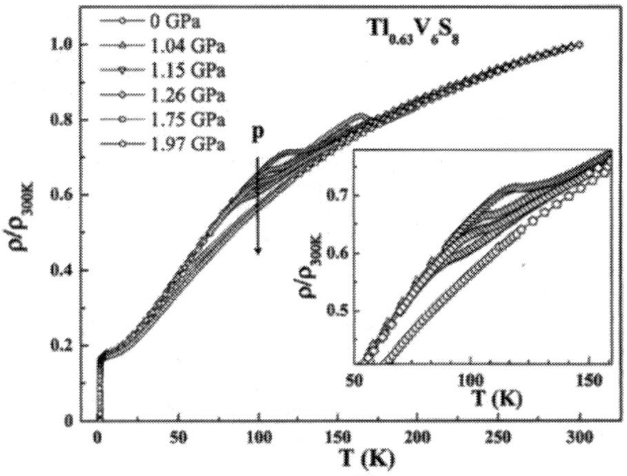

Figure 1. Pressure dependence of the CDW instability upon warming up. Insert: Magnification of the region indicated by the downward(increasing pressure) arrow

The behavior of the hysteresis in the resistivity is maintained up to p = 1.26 GPa. Above this pressure, which corresponds to T_l = 95 K, any signature of the anomaly upon cooling down is difficult to be unambiguously resolved. However, the anomaly is still visible when the sample is warmed up. And can be followed up to p = 1.71 GPa corresponding to T_h = 67 K. For even higher pressure, the transition at T_h is suddenly suppressed. We estimate a critical pressure for the suppression of the CDW instability to be about p_c = (1.85 ± 0.12) GPa. The combined results of the pressure dependence of T_h and T_l, as well as the superconducting transition temperature, T_c, are displayed in Fig. 2.

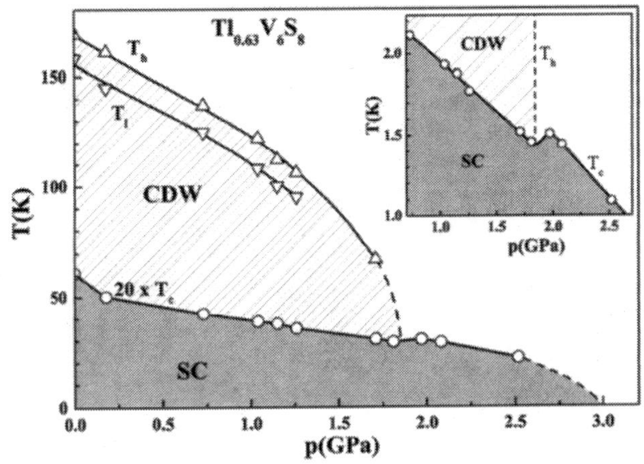

Figure 2. The phase diagram of $Tl_{0.63}V_6S_8$. The CDW instability is completely suppressed at a critical pressure $p_c = (1.85 \pm 0.12)$ GPa and concomitantly T_c is enhanced (inset). The dashed lines are guides for the eye.

5.2. Theoretical

Our calculated band structures are shown in Figs. 3 and 4. The energy range is reduced in Figs. 3 and 4 , so that only bands near the Fermi level are shown. The one-dimensional nature of the system is revealed in Fig.3, since the bands are more dispersive along ΓA direction compared to those in planes normal to ΓA and the half-filled bands exist only along ΓA. At higher compression (Fig.4) bands in the planes normal to ΓA also cut the Fermi level indicating that compression tends to destroy the quasi-one-dimensional nature of InV_6S_8, though the relative dispersion of the bands are not

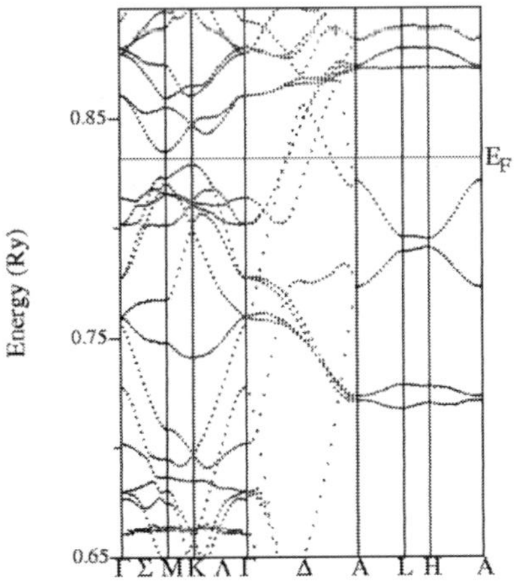

Figure 3. Energy bands of InV_6S_8 at ambient pressure (V/V0 = 1).

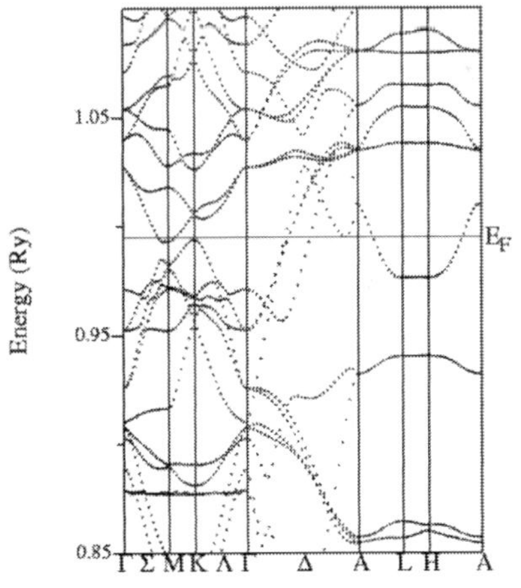

Figure 4. Energy bands of InV_6S_8 at a compression of V/V0 = 0.88

drastically changed. The most interesting effect is that the degenerate energy level at A which is below E_F at $V/V_0 = 1$ is shifted up in energy and is located above E_F at $V/V_0 = 0.88$ (Fig. 4).

This shift corresponds to the main dissimilarity seen in the corresponding Fermi surface sheets, shown in Figs. 5 and 6 in the Γ-A-H-K plane of the BZ.

Journal of Physics: Conference Series **121** (2008) 032013 doi:10.1088/1742-6596/121/3/032013

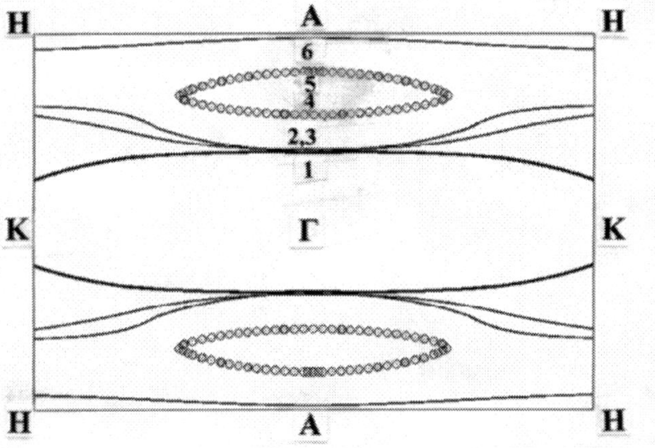

Figure 5. Fermi surface sheets in the Γ-A-H-K plane of the BZ at ambient pressure (V/V0 = 1)

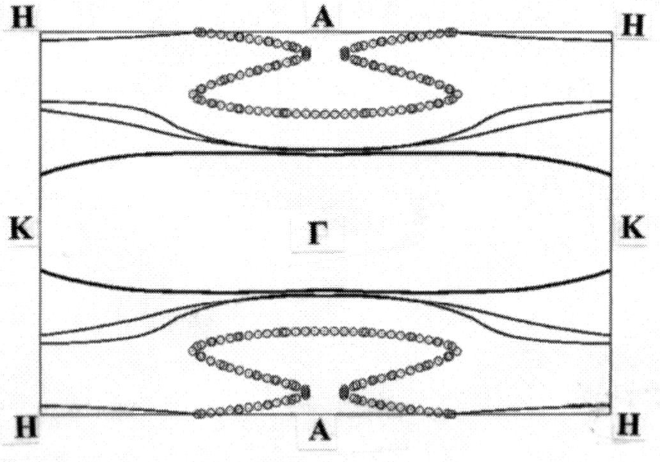

Figure 6. Fermi surface sheets in the Γ-A-H-K plane of the BZ at V/V0 = 0.88.

We first discuss the Fermi surface sheets shown in Fig. 5, which corresponds to the band structure depicted in Fig. 3 (ambient pressure). As we can see from Fig. 3, the Fermi level intersects 6 electron bands as the k-vector increases along the Γ-A direction from zero to ΓA. Their positions are labeled by 1 – 6 in Fig. 5 along the Γ-A direction. These can be considered as 3 electron sheets (1 – 3) centered around Γ (with K-Γ-K as axis), an electron sheet (6) around A (with H–A–H as axis) and a sheet which can be considered as a hole sheet around a point at a distance of 0.7 ΓA from Γ along Γ-A (shown as open circles to identify it separately) formed by bands 4 and 5, as these two bands intersect each other a little above the Fermi level (see Fig. 3). The sheet arising from the band labeled 1 has a considerable region with parallel parts, connected by some k-vector (nesting) parallel to the Γ-A direction. As can be seen from Fig. 6, this part of the nesting is retained almost to the same extent at a higher compression of $V/V_0 = 0.88$. There are also less prominent nesting regions of the band 3 sheet, again with nesting k-vectors in the Γ-A direction, near the K-point and to a lesser extent near the Γ-point. The bands 2 and 4 also show some nesting near the Γ-point. The nesting is, however, negligible for bands 5 and 6. There are also inter-band nesting regions between bands 2,3, and 4 as the sheets of these bands are parallel where they intersect the Γ-A line. All these nesting properties are more or less retained at the compressed volume, as can be seen in Fig. 6. Thus the main difference is the merging of sheets corresponding to the bands 5 and 6, which arises due to the shift of the energy level near the A-point, discussed earlier, thus making the sheet around the A-point for band 6 more hole-like. Therefore, we can conclude, that for this quasi-one-dimensional system, the nesting properties of the Fermi surface sheets are not drastically affected by compression, at least up to V/V0 = 0.88, which corresponds to about 18 GPa. Whether the loss of the quasi-one-dimensionality, seen already in Fig. 4 affects the electron-phonon interaction, and thus the charge density wave instability, is a different aspect and needs studies beyond the band structure calculations.

6. Summary

We have measured the pressure dependence of T_h and T_l of the CDW instability as well as the pressure dependence of the superconducting transition temperature, T_c , and determined the phase diagram for $Tl_{0.63}V_6S_8$. The observed hysteresis upon cooling and warming shows that the transition at T_h is a first order phase transition. Band structure calculations of InV_6S_8 up to a compression of $V/V_0 = 0.88$ do not show a significant change in the main nesting contours of the Fermi surface sheets, but reveal a topological transition along the ΓA direction. The quasi-one-dimensionality is lost, or at least considerably reduced at this compression.

There are open questions connected with the quality of the samples. High quality single crystals are now available for these studies, in particular high pressure X-ray diffraction measurements at low temperatures will be crucial to understand the underlying physics. In addition, improvements of the band structure calculations by going beyond the Virtual Crystal Approximation (VCA) would be desirable to see, if our assumption that the VCA adequately takes into account the In disorder.

Acknowledgement:

We acknowledge the support of the National Science Foundation through grant No. DMR-0091639

References

[1] Gronwold F, Haraldsen H, Pedersen B and Tufle T 1969 *Rev. Chim. Minér.* **6** 215
[2] de Vries A B and Jellinek F 1974 *Rev. Chim. Minér.* **11** 624
[3] Bensch W and Koy J 1995 Solid *State Commun.* **93** 261
[4] Ohtani T, Miyoshi Y, Fujii Y, Koyakumaru T, Kusano T and Minami K 2001 *Solid State Commun,* **120** 95
[5] Münch R, Hochheimer H D, Werner A, Materlik G, Jayaraman A and Rao K V 1983 *Phys. Rev. Lett.* **50**, 1619
[6] Hochheimer H D and Münch R 1991 *Phil. Mag.,* **B 63** 979
[7] Jerome D, Berthier C, Moline P and Rouxel J 1976 *J. de physique. Colloque* **4** (37) C125
[8] Berthier C, Moline P and Jerome D 1976 *Solid State. Commun.* **18**, 1393
[9] Thorne R E 1996 *Physics Today* **49**, 42 and references there
[10] Wunschel M, Dinnebier R E, Carlson S and van Smaalen S 2001 *Acta Crystallogr.* **B 57** 665
[11] Eiling A and Schilling J S 1981 *J. Phys. F: Metal Phys.* **11** 623
[12] Vlasse M and Fournes L 1976 *Mater. Res. Bull.* **11** 1527
[13] Garg A B, Vijyakumar V, Godwal B K, Ohtani T, Martin B R and Hochheimer H D 2004 *Solid State Commun.* **132** 731

[14] Blaha P, Schwarz K, Madsen G K H, Kvasnicka D and Luitz J 2001 *An Augmented Plane Wave + Local Orbital Program for calculating Crystal Properties*, Techn. Universität Wien, Austria

[15] Ehrenreich H and Schwartz L 1976 in *Solid State Physics* **31**, ed. H. Ehrenreich, F. Seitz, D. Turnbull, Academic Press, New York, 1976

[16] Perdew J P and Wang Y 1992 *Phys. Rev. B* **45** 13244

[17] Ceperly D M and Alder B J 1980 *Phys. Rev. Letter* **45** 566

Joint 21st AIRAPT and 45th EHPRG Int. Conf. on High Pressure Science and Technology IOP Publishing
Journal of Physics: Conference Series **121** (2008) 032014 doi:10.1088/1742-6596/121/3/032014

Pressure and magnetic field effect on transport gap of black SmS

K Imura[1], K Matsubayashi[1,3], H S Suzuki[2], K Deguchi[1], N K Sato[1]

[1]Department of Physics, Graduate School of Science, Nagoya University, Nagoya 464-8602, Japan
[2]National Insutitute for Materials Science, Tsukuba 305-0047, Japan

E-mail: `f060201d@mbox.nagoya-u.ac.jp`

Abstract. We measured the electrical resistivity of black SmS at high magnetic fields up to 30 T. We observed that the magnetic field drastically reduces the energy gap $E_g(H, P)$ in a similar way to that the pressure does. From the finding that data points of $E_g(H, P)$ lie on a single line as functions of H and P, we argue that the giant negative magnetoresistance is intrinsic to black SmS and the magnetic field plays a similar role to the pressure.

1. Introduction

Pressure effect on strongly correlated $4f$-electron systems is one of the most attractive topics in solid state physics. SmS is a prototypical system that exhibits a pressure-induced phase transition at a critical pressure P_{c1} ($\sim$ 6 kbar at room temperature) from a semiconducting "black phase" to a valence fluctuating "golden phase" accompanied by an isostructural volume collapse of $\Delta V/V \sim 0.08$ [1]. At higher pressures than about 20 kbar, the electrical resistivity shows metallic behavior at low temperatures [2]. Recently, a second critical pressure P_{c2} ($\sim$ 20 kbar) above which an antiferromagnetic ordered state emerges was detected by nuclear forward scattering, magnetic susceptibility, heat capacity, thermal expansion, and Hall effect experiments [3 - 7]. It is interesting to note that an intermediate state between P_{c1} and P_{c2} is a nonmagnetic, pseudogapped state, as evidenced by thermal expansion and specific heat experiments [8]. On the other hand, black phase seems not to have attracted so much interest, possibly because problems concerning black phase are considered to be settled; the temperature dependence of the electrical resistivity can be understood as a conventional narrow-gap semiconductor, and the magnetic susceptibility can be explained by Van Vleck paramagnetism of Sm^{2+} ionic state ($4f^6$, $J = 0$) [9]. From the theoretical study of the electronic structure on black SmS, the semiconducting energy gap E_{gap} which is determined by the energy difference between the localized $4f$ level and the $5d$ conduction band opens and this energy gap shows pressure dependence with the rate of $\mathrm{d}E_{gap}/\mathrm{d}p \sim$ -130 K/kbar which is in good agreement with experiments [10, 11, 12]. Interestingly, a strong magnetic field effect on the electrical resistivity was observed by Konczykowski *et al*; at ambient pressure the large negative magnetoresistance was observed below 280 K [13]. They attributed the observation to an impurity effect, because it is difficult to explain the huge magnetic field effect on the electrical resistivity based on a simple nonmagnetic Sm^{2+}

[3] present address: Institute for Solid State Physics, The University of Tokyo, Kashiwa 277-8581, Japan

© 2008 IOP Publishing Ltd

state. In our previous report [12], we showed that the giant negative magnetoresistance arises from the energy gap reduction by a magnetic field. In the present paper, we argue that the magnetoresistance can be intrinsic to black SmS.

2. Experimental technique

A single crystal of SmS was grown by the Bridgman method. Note that the electrical resistivity strongly depends on stoichiometry [11]. A sample used for the present measurement (with a dimension of $2.0 \times 2.1 \times 0.36$ mm^3) was cleaved from a bulk ingot having a nominal composition of $Sm_{1.00}S_{1.00}$ (sample #8), and is different from a sample ($Sm_{1.01}S_{1.00}$, sample #10) that was used for our previous investigation [12]. The transverse magnetoresistance measurements under pressure were carried out by usual DC van der Pauw method. The magnetic field applied perpendicular to the current was generated by a superconducting magnet up to 15 T and a hybrid magnet up to 30 T installed at National Institute for Materials Science. Pressure was applied at room temperature using a piston cylinder clamp cell, which consists of a NiCrAl inner and CuBe outer cylinder. Daphne oil 7373 was used as a pressure transmitting medium. Pressure at low temperatures was determined from a superconducting transition temperature of lead.

3. Results and Discussion

Figure 1 shows the temperature dependence of the electrical resistivity under magnetic fields at 1.4 kbar in Arrhenius plot ($\rho \propto \exp(E_g/2k_BT)$): To avoid complexity as reported in Ref. [14], we first cooled the sample at a magnetic field depicted in Fig. 1, and then measured the electrical resistivity at the field in a warming process. At every magnetic field, the slope of the $\rho(T)$ curve deviates from the linear behavior and bends to saturate at low temperature, which can be ascribed to variable range hopping mechanism as suggested by the so-called $T^{-1/4}$-law of $\rho(T)$ (not shown here). This feature becomes enhanced with increasing magnetic fields possibly

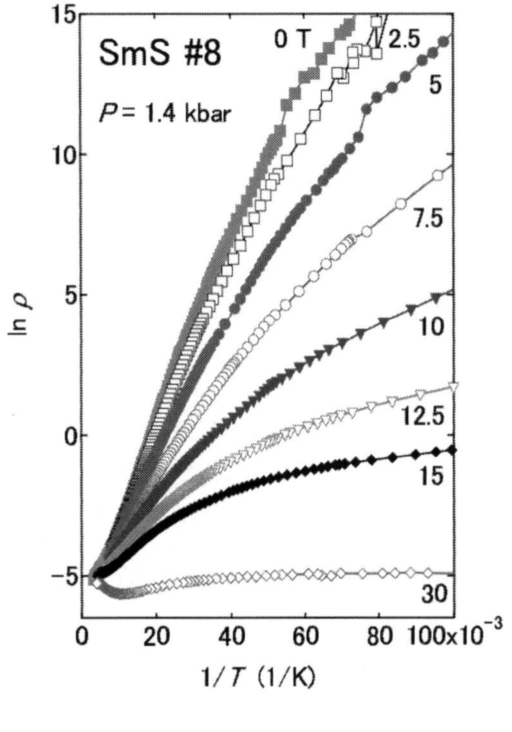

Figure 1. Temperature dependence of electrical resistivity of SmS (sample #8) in the Arrhenius plot at a fixed magnetic field at 1.4 kbar. Note that the electrical resistivity decreases drastically by application of a magnetic field; for example, a magnetic field of 30 T reduces the resistivity by 6 decades at 20 K.

because of reduced gap energy.

When looking at the electrical resistivity as a function of the magnetic field at a fixed temperature, the resistivity appears to be almost unchanged at room temperature but decreases drastically at low temperatures. This feature is remarkable at 30 T; the electrical resistivity is even metallic in the sense that the low-T resistivity is almost independent of temperature, although the system is considered to be still in black phase.

We estimated the gap energy E_g from the slope of the Arrhenius plot. The obtained E_g is summarized in Fig. 2 as a function of magnetic field at a fixed pressure. We clearly observe that E_g decreases drastically by application of the magnetic field, almost consistent with our previous measurements made up to 6.8 T for the different sample [12]. It is surprising that at 1.4 kbar, for instance, $E_g \simeq 730$ K at $H = 0$ is reduced down to $E_g \simeq 50$ K at $H = 30$ T, reflecting the 6 decades reduction of the electrical resistivity mentioned above. We anticipate that some discontinuous behavior will occur at high magnetic fields corresponding to the black to golden phase transition, but it has not been observed yet. A further experiment at higher

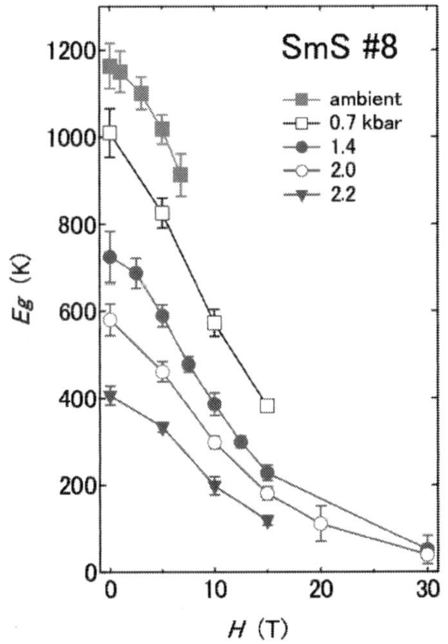

Figure 2. Magnetic field dependence of the energy gap E_g at a fixed pressure. The error bar was estimated from the ambiguity in determination of the slope of the Arrhenius plot. Note that the gap decreases drastically by application of the magnetic field.

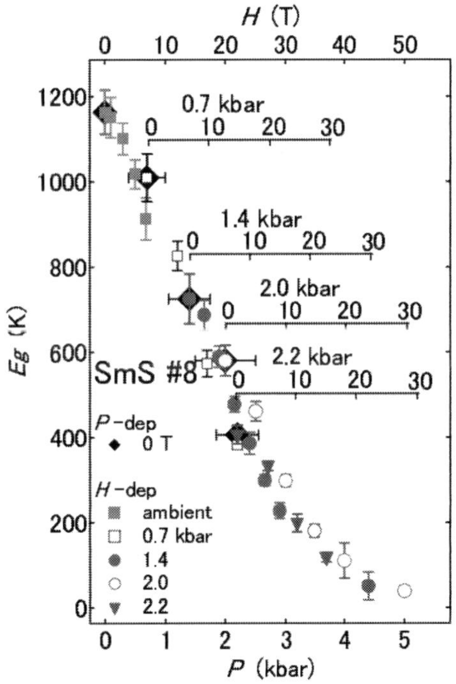

Figure 3. Pressure and magnetic field dependence of the energy gap E_g. Closed diamonds show the pressure dependence of E_g at a zero field (bottom axis). Closed squares show the magnetic field dependence of E_g at ambient pressure (top axis). Data points of E_g at high pressures are plotted on a horizontal H-axis starting from a relevant P value. Note that the scaling between H and P is the same among all the pressures, H(T) $= 10 \times P$(kbar).

pressure and magnetic fields is in progress.

The reduction of E_g by magnetic field seen in Fig. 2 may indicate that the magnetic field plays a similar role to the pressure. To check this possibility, we present E_g as functions of the pressure and magnetic field in Fig. 3: Data points at zero field ($E_g(H = 0, P)$) denoted by closed diamonds are plotted on a bottom P-axis. Data points at ambient pressure ($E_g(H, P = 0)$) denoted by closed squares are plotted on a top H-axis, where P and H are related with each other via the relation of $H(\mathrm{T}) = 10 \times P(\mathrm{kbar})$. Data points at 0.7 kbar ($E_g(H, P = 0.7)$) denoted by open squares are plotted on a horizontal H-axis starting from $P = 0.7$ kbar on the bottom P-axis. Here, the same scaling relation between H and P was used. Other data points at 1.4, 2.0, and 2.2 kbar respectively denoted by closed circles, open circles, and closed triangles are similarly plotted with the same conversion relation. Note that the scaling factor mentioned above is an only adjustable parameter used for this plot, nevertheless we find all the data points to lie on a single line. This finding seems to indicate that the magnetic field actually serves as the pressure on resistivity, suggesting that if the pressure effect on black phase is intrinsic then the magnetic field effect is also intrinsic to black SmS. Indeed, the pressure effect is intrinsic, thus it is very likely that the negative magnetoresistance is inherent to SmS. One possibility of the magnetic field effect on resistivity is the Zeeman effect on the conduction band. In this case, although the g-factor is estimated for very large value, the external pressure and magnetic field show the equivalence on resistivity. To confirm this, we need a further experiment to observe the black to golden phase transition by application of magnetic field.

4. Conclusion

We measured the magnetoresistance of black SmS at high magnetic fields up to 30 T. We observed that magnetic fields drastically reduce the energy gap $E_g(H, P)$ in a similar way to that the pressure does, and that all the data points of $E_g(H, P)$ presented here lie on a single line as functions of H and P. From this observation, we argued that the giant negative magnetoresistance is intrinsic to black SmS and the magnetic field plays a similar role to the pressure. We hope that the present result stimulates further experimental and theoretical investigations.

Acknowledgement

We acknowledge T. Takamasu, Y. Imanaka and K. Takehana for kindly help for high magnetic field experiment at NIMS and useful discussion. This work was partially supported by a Grant-in-Aid from the Ministry of Education, Culture, Sports, Science and Technology, Japan.

References

[1] Jayaraman A, *et al.* 1970 *Phys. Rev. Lett.* **25** 1430
[2] Lapierre F, *et al.* 1981 *Solid. State. Commun.* **40** 347
[3] Barla A, *et al.* 2004 *Phys. Rev. Lett.* **92** 066401
[4] Matsubayashi K, *et al.* 2007 *J. Magn. Magn. Matt.* **310** 408
[5] Haga Y, *et al.* 2004 *Phys. Rev.* B **70** 220406
[6] Imura K, *et al.* 2006 *Physica* B **378-380** 728
[7] Imura K, *et al.* to appear in *Physica* B
[8] Matsubayashi K, *et al.* 2007 *J. Phys. Soc. Jpn.* **76** 033602
[9] Maple M B and Wohleben D 1973 *Phys. Rev.* B **7** 4986
[10] Svane A, *et al.* 2004 *Phys. stat. sol.* (b) **241** 3185-3192
[11] Matsubayashi K, *et al.* 2007 *J. Phys. Soc. Jpn.* **76** 064601
[12] Imura K, *et al.* 2007 *J. Magn. Magn. Matt.* **310** e54-e56
[13] Konczykowski M, *et al.* 1982 *Valence Instabilities,* ed P. Wachter and H. Boppart (Amsterdam: North-Holland) p.447
[14] Konczykowski M, *et al.* 1985 *J. Magn. Magn. Matt.* **47** & **48** 455-458

Pressure-induced effect on electronic excitations in osmium

Yu.S. Ponosov[1], V.V. Struzhkin[2], S.V. Streltsov[1], and A.F. Goncharov[2]

[1] Institute of Metal Physics UD RAS, 620041, S. Kovalevskaya str. 18, Ekaterinburg, Russia
[2] Geophysical Laboratory CIW, 5251 Broad Branch Rd., N.W., Washington, D.C. 20015, U.S.A.

E-mail: ponosov@imp.uran.ru

Abstract. Electronic Raman scattering has been studied in single crystals of 5d transition metal osmium under pressures up to 60 GPa in the temperature range of 10−300K. With the use of green and blue excitation wavelengths we observe an appearance of well−defined electronic peaks at $\sim 580\ cm^{-1}$ for wave vector direction $\mathbf{q} \parallel [0001]$] and at $\sim 350\ cm^{-1}$ for $\mathbf{q} \parallel [10\bar{1}0]$ in the pressure range of 20−30 GPa. It was shown in our combined experimental and theoretical investigation that a consideration of both a renormalization of the electron self-energies near the Fermi level and a change in the Fermi surface under pressure and temperature variation is needed to describe the observed spectra.

In a clean metal with anisotropic Fermi surface (FS) intraband electronic scattering may be observed in the wide frequency range of $\omega < v_F/\delta$ (v_F−electron Fermi velocity, δ is the penetration depth) under an assumption of $\omega > \Gamma$, where Γ is the relaxation frequency of the conduction electrons [1, 2]. In the region of the maximum scattering intensity at $\omega \sim v_F/\delta$ the scattering cross section is determined by a contribution from all the electrons at the FS and its magnitude may be rather large. The study of its frequency dependence may give supplementary information both on the FS topology and a renormalization of the electron self−energies by any relaxation process such as an electron−phonon (e−ph) and/or electron−electron (e−e) scattering [3, 4, 5, 6].

Recently the wave vector dependence of the electronic light scattering (ELS) in the normal state has been observed in hcp transition metal osmium [7]. A tuning of the momenta **q** by the changing the incident laser energy allowed studying the dispersion of the phonon and electronic excitations. Abrupt and anisotropic optical phonon frequency softening (2-3%) has been found in the range of $\mathbf{q} \sim 10^6 cm^{-1}$ at low temperatures. Anomalous dispersion and maximal widths together with clear Fano shapes of the phonon lines have been observed in the momenta range corresponding to crossing of the electron and phonon excitation's energies indicating strong e−ph interaction. The observed **q**−dependences of the self−energies of the optical phonons in osmium are in qualitative agreement with the theoretical predictions [8, 9] confirming the existence of nonadiabatic effects in the e−ph interaction in transition metals.

The application of pressure may modify both the electronic band structure near the Fermi level and internal scattering processes, which result in the Raman spectra. This gives an additional tool to study the electronic excitations responsible for anomalous dispersion effects. Anomalous temperature−induced phonon hardening in osmium has been found at pressures $\geq$

© 2008 IOP Publishing Ltd

Journal of Physics: Conference Series **121** (2008) 042001 doi:10.1088/1742-6596/121/4/042001

15 GPa at nonhydrostatic conditions [10]. Recently anomaly in the c/a ratio of osmium has been observed at pressures of 20−25 GPa [11] which was attributed to a possible change in the FS topology.

The goal of this work was to study the changes in the fine details of the electronic structure near the Fermi level in the pressure range where the abovementioned anomalies have been found. Here we present an evolution of the electronic Raman response of osmium at pressures up to 60 GPa in the temperature range 10−300 K for different **q** values and directions. First−principle calculations of the frequency dependences of the electronic scattering cross section have also been performed for comparison with the measured Raman spectra.

Plates of osmium single crystals with (0001) and $(10\bar{1}0)$ orientation thinned to 5−10 μm by mechanical and electropolishing were loaded into a diamond anvil cell and placed in an optical cryostat. The high purity of the crystals was confirmed by their low residual resistivity ratio ($\geq$ 1000). The pressure-transmitting medium was argon. Calibration was done by the ruby luminescence method. Four laser lines λ_i (488, 514, 633, and 740 nm) of an Ar ion, He-Ne and Ti:sapphire lasers with powers up to 150 mW were used for spectra excitation. Spectra were collected in quasibackscattering geometry ($\sim 145^o$) [12], the laser spot diameter was about 20 μm. The spectra were analyzed using a HR-460 spectrograph equipped with notch filters and a CCD detector.

Raman spectra (divided by the Bose factor $(1-\exp(-\hbar\omega/kT))$) measured from the basal plane (0001) (wave vector **q** is perpendicular to the sample plane) with the excitation energy E_i=2.41

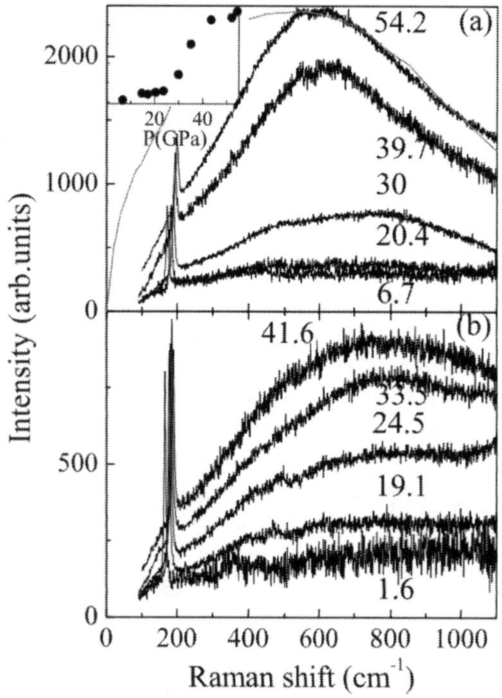

Figure 1. Raman spectra at different pressures (numbers, GPa), measured from (0001) plane at T=10K−(a) and 300K−(b), E_i=2.41 eV. Calculated at 50 GPa ELS spectrum-red line. Inset− the ELS intensity vs pressure at T=10K.

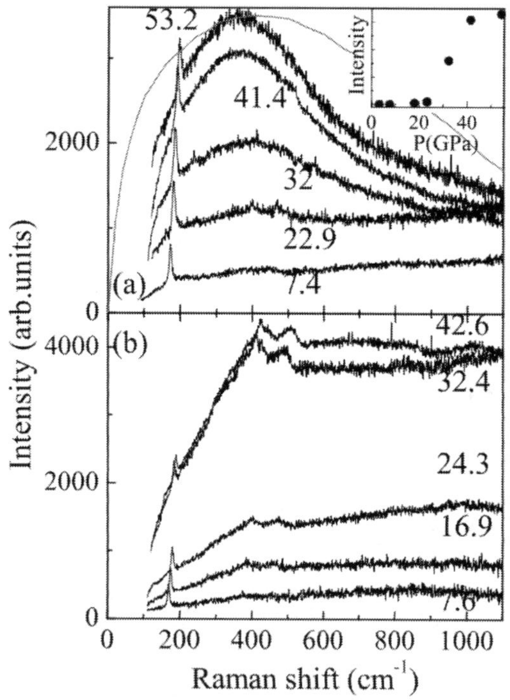

Figure 2. Raman spectra at different pressures (numbers, GPa), measured from $(10\bar{1}0)$ plane at T=10K−(a) and 300K−(b), E_i=2.41 eV.Calculated at 50 GPa ELS spectrum-red line Inset− the ELS intensity vs pressure at T=10K.

Journal of Physics: Conference Series **121** (2008) 042001 doi:10.1088/1742-6596/121/4/042001

eV (514 nm) are shown in Fig.1. A structure–less background with the superimposed E_{2g} phonon line near 165 cm^{-1} was observed in the low–pressure spectra taken at all temperatures. Here we report a strong intensity increase by more than one order of magnitude with an appearance of well-defined peak near 580 cm^{-1} at 10K at pressures higher than 20 GPa. A broad continuum near 620 cm^{-1} was also observed in the spectra measured with the 488 nm excitation.

The spectra obtained from the $(10\bar{1}0)$ plane are presented in Fig.2. Here the structure–less background was also observed up to pressures $\sim$ 20 GPa at all temperatures. In this geometry a pronounced electronic peak at 350 cm^{-1} appears at T=10K under further pressure increase. The insets in Fig.1a and 2a show that peak intensities grow sharply in the pressure range 20–30 GPa. When temperature increases the electronic spectra smear out and become structure–less. For the $\mathbf{q} \parallel [0001]$ the decrease in intensity at high temperature by a factor about 2.5 was also found while intensity for $\mathbf{q} \parallel (10\bar{1}0]$ was even higher at T=300K in comparison to T=10K. At the same time, the spectra measured with red and infrared excitation show only moderate evolution: the well–defined zero-pressure peaks soften under pressure by $\sim$ 10% without apparent changes in the linewidth. The spectra also show much lower **q** anisotropy ($\sim$ 10%) and their intensity increases by no more than factor of 2 (Fig.3).

Besides the one-phonon scattering the two-phonon scattering at 200÷500 cm^{-1} is seen in the spectra shown in Fig 1 and 2. For the $\mathbf{q} \parallel [10\bar{1}0]$ its intensity increases by order of magnitude at 300K and this sharp growth starts in the 20–30 GPa pressure range as in the case of electronic scattering. At low temperatures the intensity of the two-phonon scattering essentially decreases in contrast to the electronic peak intensity behavior.

The presented results provide the evidence for the existence of an abnormal pressure behavior of the ELS spectra. First, an anomalous intensity growth occurs with blue and green excitation accompanied by the transformation of the structure-less background into the well-defined electronic continuum peaks in the 20-30 GPa pressure range. Second, we observe the large decrease in the electronic peak positions and their anisotropy upon decrease of the excitation energy. Third, we observe a different temperature behavior of the ELS intensity for two probed **q** directions.

In order to understand the reasons of the observed anomalies the calculations of the frequency dependence of the ELS by noninteracting electron-hole pairs were performed. The intensity of the intraband ELS is determined by computing the integral over the whole FS from imaginary part of the polarization operator taking in account the distribution of the incident and scattered fields in a metal [1, 2]:

$$I(\omega, \mathbf{q}) \propto -\Im \int_{-\infty}^{+\infty} \frac{dq}{2\pi} |U(q)|^2 \int \frac{dS_p}{v} |\overline{\gamma}(p, q, \omega)|^2 \frac{qv_q}{\omega - qv_q + i\Gamma} \quad (1)$$

where v_q is a projection of the Fermi velocity on the **q** direction. The wave vector distribution $|U(q)|^2$ can be expressed as $|U(q)|^2 = 4|\zeta|^2 / |q^2 - \zeta^2|^2$ [1, 2, 13], where $\zeta = \zeta_1 - i\zeta_2 = (4\pi/\lambda_i)\cdot(n - ik)$, n and k were measured for osmium in [14]. The impurity relaxation frequency Γ was taken to be equal to 10 cm^{-1}.

When the excitation energies are smaller than any interband transition energy, electron–photon matrix element $\overline{\gamma}_{\alpha\beta}(p, q, \omega) = \gamma_{\alpha\beta}(p) - \langle\gamma_{\alpha\beta}(p)\rangle$, where angle brackets denote the FS averaging, is determined by the reciprocal effective mass tensor (FS curvature) [15]. Experimental data give an evidence for the resonance behavior of ELS near $E_i \sim$ 2.2–2.3 eV at ambient pressure already and for its enhancement under pressure application while the measurements with infrared excitation may represent a nonresonant case. In the resonant case the summing over all interband transitions including the Fermi level is required. At ambient pressure equal ELS spectra intensities were observed for all 3 crystallographic wave vector **q**–directions [0001], $[11\bar{2}0]$ and $[10\bar{1}0]$ which probe quite different sections of the FS [7] suggesting resonance enhancement for the whole FS. Moreover, the observed resonance has rather large width ~ 0.5 eV and doesn't distort the continua lineshapes. Thus, in this calculation we used a constant $\overline{\gamma}(p, q, \omega)$ to simulate the resonant conditions which suggests the interband matrix elements of the momentum operator are constant in the **p** space.

The band structure calculations were performed using linearized muffin-tin method (LMTO) [16] in the local density approximation (LDA). In the course of the self-consistency the mesh of 1728 k-points in the irreducible part of the Brillouin-zone was used. The integration over the FS surface was perfomed with the fine mesh of 125 000 **p**-points in the full Brillouin-zone. The calculation of the Fermi velocity was done in the OpenDX package. The obtained FS is in agreement with that calculated before [17, 18, 19, 20]. It consists out of main 4 sheets: electronic ones $\Gamma9e$ and $\Gamma10e$, small hole sheets U7h in the LM direction and a complicated "monster" surface $KMh8$. Our calculations were perfomed for the volumes corresponding to 0 and 50 GPa[11]. As in Refs [19, 20] in this pressure region no new sheets of the FS were found. Note however, that our calculation scheme does not include the spin−orbit interaction for the Os 5d states.

Fig.1 and 2 show rather close correspondence of the measured and calculated high-pressure Raman spectra for both **q** directions with the $E_i = 2.41$ eV excitation. This implies that the observed in the resonant conditions ELS spectra have an intraband origin. Our simulations show a substantial broadening of the calculated spectra compared to the measured ones. The measured spectra for another excitation energies are also narrower than the calculated ones. That may be due to a difference of the real FS topology or the velocity distribution at the FS from the one obtained in this calculation. A comparison of the calculated and measured continua frequencies for the different excitation energies and pressures in Fig.3 also supports such a suggestion. The calculated spectra become harder and broader under pressure by $\sim 10\%$ for $\mathbf{q} \parallel [0001]$ and by $\sim 20\%$ for $\mathbf{q} \parallel [10\overline{1}0]$ at all excitation energies and their intensities decrease at the same ratio. This contradicts the observed strong intensity growth, the electronic peak position softening and narrowing when using the excitation in the blue-green range compared to only a negligible frequency softening with the red excitation. At the excitation energies $E_i \leq 2.2$ eV the calculated frequencies substantially exceed (up to 2 times) the measured ones. For the excitation energies of $E_i \geq 2.2$ eV, the calculations predict at all pressures well-defined electronic peaks at the frequencies which are smaller than those corresponding to the observed almost structure-less electronic continua (Fig.1, 2). The appearance of pronounced peaks in the spectra measured at pressures P $\geq$ 20 GPa improves the agreement to the calculation in this excitation energy range. The **q**− anisotropy of the electronic peak frequencies calculated with a constant $\overline{\gamma}_{\alpha\beta}(p, q, \omega)$ doesn't depend on E_i in

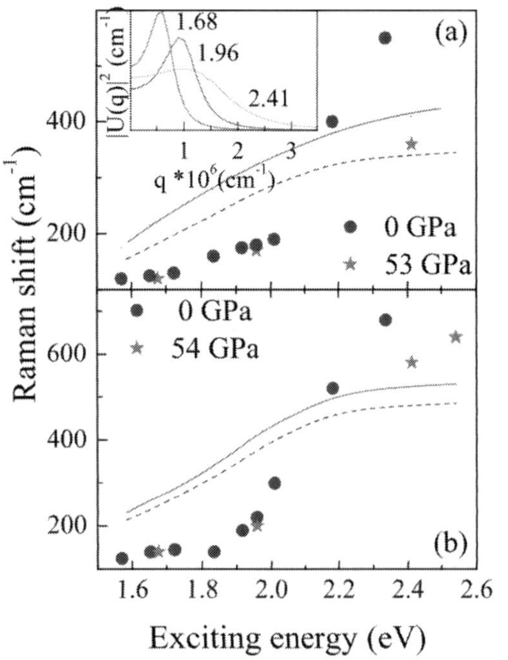

Figure 3. The measured (points) and calculated (dashed lines−0 GPa, solid lines−50 GPa) ELS peak positions vs E_i for $\mathbf{q} \parallel [10\overline{1}0]$−(a) and $\mathbf{q} \parallel [0001]$−(b),T=10K. Inset shows the wave vector distribution $|U(q)|^2$ for different E_i (numbers in eV).

contrast to the observed decrease with the red and near infrared excitations.

The discrepancy between the experimental and calculated data implies an importance of the electron scattering mechanisms going beyond the LDA. A substantial increase of the calculated frequencies in comparison to the measured ones occurs in the range of 200÷500 cm^{-1}, i.e. in the energy region of the two−phonon states of osmium. Strong temperature effects at all pressures suggest also an important role of the inelastic e−ph scattering [4, 5, 6] which may lead to the electron dispersion renormalization at small **q** (low-energy excitation-inset in Fig.3a). The observation of the structure−less electronic background at low pressures with $E_i \geq 2.2$ eV provides the evidence for the increase of the damping of the electronic states for large wave vectors **q**, essential part of which is probed by the high-energy excitation (inset in Fig. 3a). This implies the existence of additional scattering mechanisms for electronic states lying away from the Fermi energy by 500÷100 cm^{-1}. The Fermi surface topology having large nearly parallel sections of the FS with wave vectors Q in the basal plane favors the enhancement of the e-e scattering. An anomalous growth of the two−phonon cross section under pressure for $\mathbf{q} \parallel [10\bar{1}0]$ at T=300K supports this suggestion. It was shown theoretically [21] that such an enhancement of the two−phonon Raman spectra in transition metals may be due to successive phonon ω_{+Q} and ω_{-Q} emission between electronic states close to the Fermi level when the conditions for double-photon resonance are fulfilled. If $\omega_{+Q}+\omega_{-Q} \approx v_{\mathbf{F}}*\mathbf{q}$, the same resonance mechanism may provide an anomalous ELS intensity growth under pressure due to an extended region of phase space for the scattering events. This is possible if the leading resonance term is determined by the scattering processes near the Fermi level having energy denominators with a little damping. A correlation of the peak narrowing with the ELS intensity increase under pressure supports this possibility. Both final state consisting of one phonon plus an electron-hole pair and two-pair final state will provide an electronic background which could interfere with two-phonon Raman scattering [21]. In turn, the decrease of the electronic damping at high-energy excitation (probing larger q-magnitudes) may be due both to the electron and phonon spectra change and to the interplay of the e-ph and e-e scattering contributions which determine the **q**− and ω− dependence of the effective e−ph coupling [22].

A decrease of the electronic peak position anisotropy for the studied **q**−directions with the red excitation may be explained by a change of the p-dependence of the electron-photon matrix element for nonresonant scattering. In our simple calculation, the constant matrix element was used for the resonant case. This implies that the whole FS is in resonance which is possible for a small FS sheet but is difficult to imagine for the main osmium FS sheets. In any case the detailed band structure calculations of the resonant behavior are necessary. They are important for understanding of the ELS cross section increase at high−energy excitation and they are needed to explain why the frequency's anisotropy for the studied **q**−orientations varies strongly upon the excitation energy change. The unusual temperature behavior of the ELS intensity for the different **q**−directions remains unclear so far. It is natural to expect a similar temperature behavior if the large FS sheets determine the observed electronic scattering. Another possibility may arise if this scattering originates from the electronic states near the L point which are strongly coupled to the phonons and vary under the pressure and temperature.

In summary, an anomalous increase of the ELS cross section under pressure was found in Raman spectra of osmium excited with the blue and green excitation. In the pressure range of 20−30 GPa, the low-temperature structure-less background transformed to pronounced and strongly temperature dependent peaks at ~ 580 cm^{-1} for $\mathbf{q} \parallel [0001]$ and at ~ 350 cm^{-1} for $\mathbf{q} \parallel [10\bar{1}0]$ showing a strong temperature dependences. The comparision of the measured and calculated within LDA ELS spectra implies the electron self-energy renormalization in the energy range ~ 1000 cm^{-1} from the Fermi level due to the e−ph and e−e scattering. The detailed calculations of the resonance effects which take into account the phonon renormalization of the electron band structure are needed to understand found anomalies.

We thank E.G. Maksimov, I.I. Mazin, V.I. Anisimov, R.J. Hemley, H.-K. Mao for helpful discussions. This research was supported by RFBR grant # 05-02-16780, Carnegie Institution of Washington, DOE/BES, DOE/NNSA (CDAC), NSF, the W.M. Keck Foundation. S.V.S. appreciates the support from the Dynasty Foundation and International Center for Fundamental Physics in Moscow for the program for young scientists; Russian ministry of science and education and CRDF foundation (programm # Y4-P-05-15) and acknowledges the support of President of Russia grant for YS (# MK-1184.2007.2).

References

[1] Ipatova I.P., Kaganov M.I., and Subashiev A.V. 1983 *Sov.Phys.JETP* **57** 1006
[2] Falkovsky L.A. 1993 *Sov. Phys. JETP* **103** 666
[3] Virosztek A. and Ruvalds J.,1992 *Phys.Rev.B* **45** 347
[4] Shulga S.V., Dolgov O.V. and Maksimov E.G. 1991 *Physica C* **178**, 266
[5] Kostur V.N. and Eliashberg G.M. 1991 *JETP Lett.*, **53**, 391
[6] Rashkeev S.N. , Wendin G. 1993 *Z. Phys.B* **93** 33
[7] Ponosov Yu. S., Bolotin G. A., Thomsen C., and Cardona M. 1998 *Phys. stat. sol.b* **208** 257
[8] Ipatova I.P. and Subashiev A.V. 1974 *Sov. Phys. JETP* **39**, 349
[9] Maksimov E.G. and Shulga S.V. 1996 *Solid State Commun.* **97**, 553
[10] Ponosov Yu.S., Loa I., Mogilenskikh V.E., and Syassen K. 2005 *Phys. Rev.B* **71** 220301
[11] Occelli F., Farber D.L., Badro J. et al. 2004 *Phys. Rev. Lett.* **93** 095502
[12] Merkel S., Goncharov A.F. , Mao H.K. et al., 2000 *Science* **288**,1626
[13] Falkovsky L.A., Klama S. 1994 *Phys.Rev.B* **50** 5666
[14] Nemoshkalenko V.V., Antonov V.N., Antonov V.N. et al. 1986 *Sov. Phys. JETP* **63** 115
[15] Abrikosov A.A. and Falkovskii L.A. 1961 *Sov.Phys.JETP* **13** 179
[16] Andersen O.K. and Jepsen O. 1984 *Phys. Rev. Lett.* **53** 2571
[17] Jepsen O., Andersen O.K., and Mackintosh A.R. 1975 *Phys. Rev.B* **12** 3084
[18] Smelyansky V.I. , Perlov A.Y. , and Antonov V.N. 1990 *J.Phys.:Condens.Matter* **2**, 9373
[19] Ma Y., Cui T., Zhang L., Xie Yu, Zou G., Tse J.S., Gao X. and Klug D. 2005 *Phys. Rev.B* **72** 174103
[20] Koudela D., Richter M., Mobiuset A. et al. 2006 *Phys. Rev.B* **74** 214103
[21] Klein M.V. 1981 *Phys.Rev.B* **24** 4208
[22] Zeyher R. and Kulic M. 1996 P*hys.Rev.B* **53** 2850

High-pressure induced absorption line narrowing in $Cs_3Yb_2Cl_9$

R. Valiente[a], I. Hernández[b], F. Rodríguez[b], F. Aguado[c], H. U. Güdel[d], K. Krämer[d]

[a] Dpto. Física Aplicada, Facultad de Ciencias, Universidad de Cantabria, Avda. de Los Castros s/n, 39005, Santander, Spain

[b] DCITIMAC, Facultad de Ciencias, Universidad de Cantabria, Avda. de Los Castros s/n, 39005, Santander, Spain

[c] Department of Earth Sciences, University of Cambridge, Downing Street, Cambridge CB2 3EQ, UK

[d] Department of Chemistry, University of Bern, CH-3000 Bern, Switzerland

rafael.valiente@unican.es

Abstract. In this paper we investigate the pressure dependence of the absorption spectrum of $Cs_3Yb_3Cl_9$. The polarised absorption spectra enable us to assign the main transitions of Yb^{3+} in C_{3v} from the complicated phonon-side band structure, and derive the corresponding structure. Special attention is paid on the pressure-induced line narrowing and associated blue shift of the most prominent absorption transition that corresponds to the high energy $\Gamma_4 \rightarrow \Gamma_4$ transition at 10668 cm^{-1} at ambient conditions. The unusual linewidth reduction with pressure is explained by a Debye-type model considering the different contributions to the linewidth.

1. Introduction

Within the lanthanides the Yb^{3+} ion ($4f^{13}$) has one of the simplest energy level schemes: it has only one f-f excited state $^2F_{5/2}$ at around 10000 cm^{-1} above the ground state $^2F_{7/2}$. The number of Stark sublevels depends on the crystal field environment. However, in spite of this simplicity, Yb^{3+} is the most intriguing of all lanthanides: broad lines, strong electron-phonon coupling strength, laser action, and most unusual optical bistability effect have been found in Yb^{3+}-based compounds. The first two phenomena are related to the interaction with neighboring ions. The optical bistability behavior is still open, and it was at first interpreted as a result of the coupling between two Yb ions [1]. Later, a careful experimental work led to a different explanation [2]. An electronic origin increases its oscillator strength upon increasing temperature, leading to a feedback effect: the stronger the absorption the

© 2008 IOP Publishing Ltd

more the crystal heats up leading to a stronger absorption and so on. In any case, the environment of Yb^{3+} plays a fundamental role in all spectroscopic properties.

The use of high hydrostatic pressure is the best technique for modifying the bond distances around the Yb^{3+} ions in a given crystal. In particular, pressure modifies the crystal-field strength and, consequently, the splitting and oscillator strength of transitions between $^2F_{7/2}$ and $^2F_{5/2}$ sublevels.

We report on the influence of hydrostatic pressure on the spectroscopic properties of $Cs_3Yb_2Cl_9$. The crystal structure consists of $Yb_2Cl_9{}^{3-}$ dimer units of two face sharing $YbCl_6{}^{3-}$ octahedra. The single-ion symmetry is C_{3v} whereas the dimer is axially distorted with D_3 symmetry (approximate D_{3h}) with the three-fold axis parallel to the crystal *c* axis [3]. The Yb-Yb distance is the closest possible in a chloride lattice.

2. Experimental

Single crystals of $Cs_3Yb_2Cl_9$ were grown in silica ampoules using the Bridgman technique [4]. It crystallizes in the trigonal space group $R\bar{3}c$ [3]. Polarized and unpolarised temperature dependent absorption spectra at ambient pressure were measured with a Cary 6000i in the 6-300 K temperature range using a closed-cycle cryostat. Pressure dependent absorption spectra were obtained in a diamond anvil cell (DAC) with a homemade setup with non-oriented microsamples and unpolarised light. The procedure is as follows: the modulated light (chopped with a frequency controller SR-540) of a tungsten lamp is focused on the sample with a microscope objective. The transmitted light is collected with another microscope objective and dispersed by a 0.5 m monochromator (Chromex 500IS) and detected synchronously (lock-in SR-830) by a thermoelectrically cooled InGaAs detector (Electro Optical Systems, Inc.). Hydrostatic pressure experiments were performed in a membrane DAC with silicon oil as pressure transmitter. The ruby fluorescence was used for pressure calibration.

3. Results and Analysis

Figure 1 shows the polarized temperature-dependent absorption spectra of $Cs_3Yb_2Cl_9$ for E//c and E⊥c.

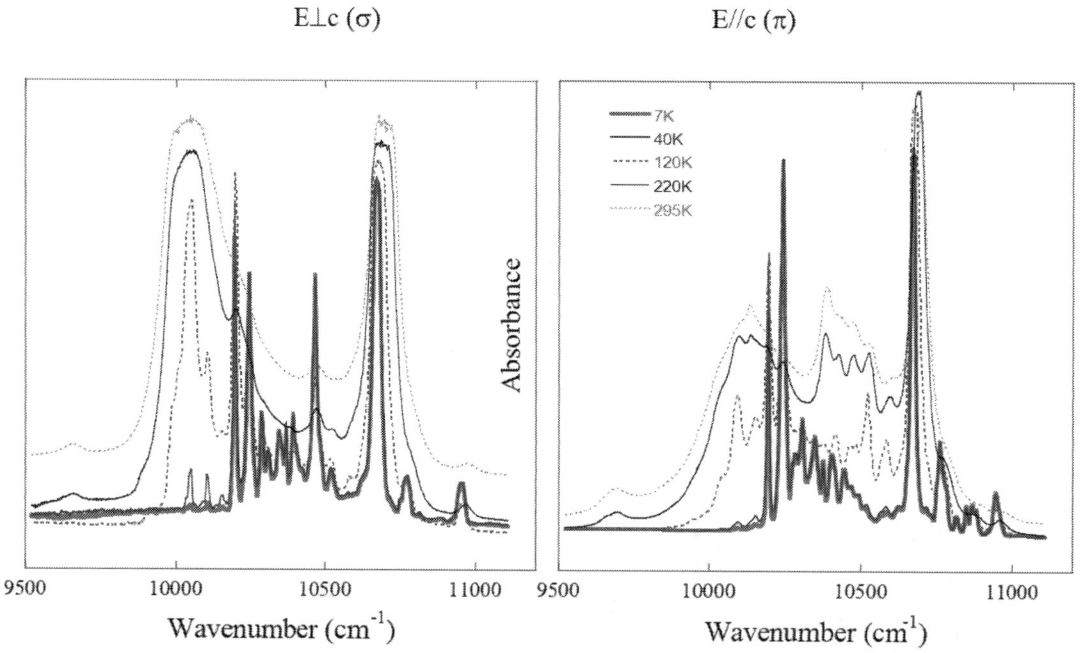

Fig. 1. Polarized optical absorption spectra of $Cs_3Yb_2Cl_9$ at different temperatures

These spectra allow us to identify the hot bands, vibronic sidebands as well as the electronic origins on the basis of transitions within the single-ion energy levels (C_{3v} symmetry), as it is shown in Fig. 2. The crystal-field levels were determined from the absorption and emission spectra. According to double-group theory, the $^2F_{7/2}$ ground state splits into 5 levels $3\Gamma_4$, Γ_5 and Γ_6, whereas the excited state splits into 4 levels $2\Gamma_4$, Γ_5 and Γ_6. The arrows indicate the electric-dipole allowed transitions for absorption. Their selection rules under C_{3v} point symmetry allow $\Gamma_4 \rightarrow \Gamma_4$ in π and σ polarization, $\Gamma_4 \rightarrow \Gamma_5$, Γ_6 in σ and $\Gamma_6 \rightarrow \Gamma_5$ and $\Gamma_5 \rightarrow \Gamma_6$ in π. From the 6 K absorption spectrum the electronic origins at 10193 cm^{-1}, 10241 cm^{-1}, 10460 cm^{-1} and 10668 cm^{-1} are assigned to $\Gamma_4 \rightarrow \Gamma_4$, Γ_5, Γ_6, Γ_4 transitions, respectively.

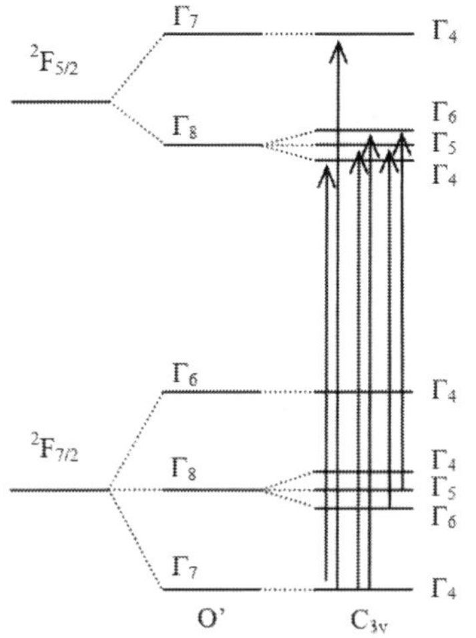

Fig.2. Energy levels diagram for Yb^{3+} in octahedral and trigonal symmetry.

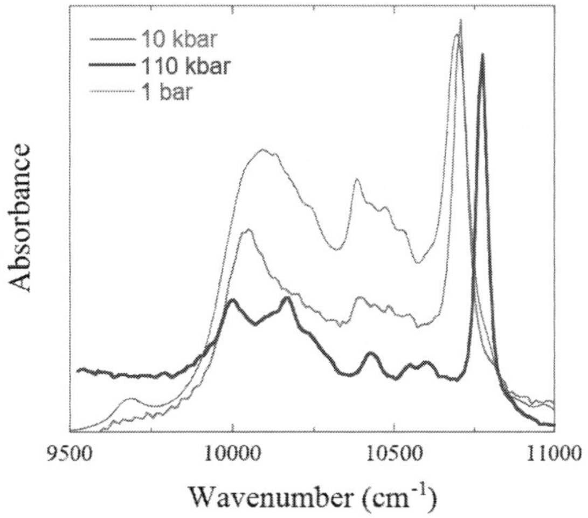

Fig. 3 Pressure dependence of the unpolarized room temperature optical absorption spectra of $Cs_3Yb_2Cl_9$

Journal of Physics: Conference Series **121** (2008) 042002 doi:10.1088/1742-6596/121/4/042002

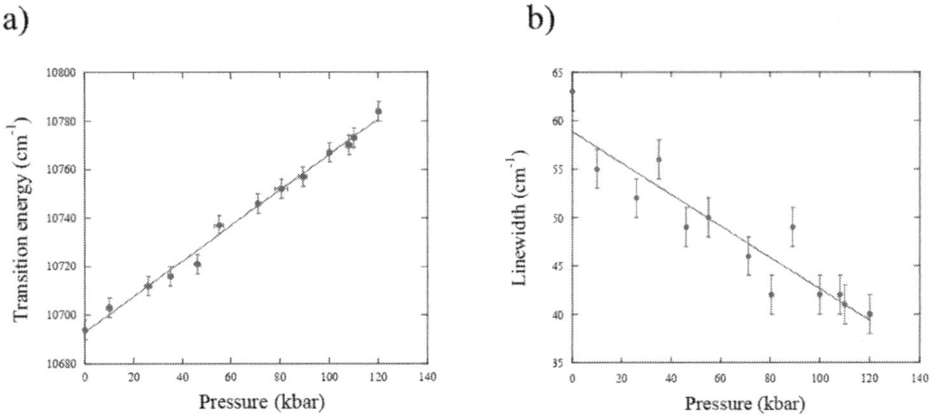

Figure 4. (a) Pressure dependence of the $\Gamma_4 \rightarrow \Gamma_4$ transition energy (around 10700 cm^{-1}) (b) Linewidth variation of the transition of the $\Gamma_4 \rightarrow \Gamma_4$ transition at around 10700 cm^{-1} with pressure.

In figure 3 we plot the room temperature absorption spectra of $Cs_3Yb_2Cl_9$ upon increasing pressure. Figures 4a and 4b show the line-shift and the line-width reduction, respectively, of the most prominent $\Gamma_4 \rightarrow \Gamma_4$ absorption peak with pressure. The peak is linearly blue-shifted from 10668 cm^{-1} at ambient pressure to 10782 cm^{-1} at 120 kbar at a rate of 0.73 cm^{-1}·kbar^{-1}. The shift is accompanied by a line narrowing from 63 cm^{-1} to 40 cm^{-1} (~ −0.2 cm^{-1} kbar^{-1}), between ambient pressure and 120 kbar. As far as we know, this is the first report of such a pressure-induced line-width reduction. The effect is similar to that observed upon cooling at ambient pressure, but in such case, rather than blueshift, a thermal redshift following a cotangent law is observed (Fig. 1). The pressure-induced line narrowing must be due to a reduction of the electron-phonon coupling strength with pressure.

The total linewidth, H, depends on temperature and pressure and involves contributions from zero-phonon line dispersion and Raman processes [5, 6]. Within a Debye-type model, the homogeneous peak width is given by:

$$H(T) = \alpha \left(\frac{T}{T_D} \right)^7 \int_0^{T_D/T} \frac{x^6 e^x}{\left(e^x - 1\right)^2} \mathrm{d}x \qquad (1)$$

where $x = \hbar\omega / kT$, α is the electron-phonon coupling parameter and $T_D = \hbar\omega_D / k$ is the Debye temperature. Taking into account the effective Grüneisen parameter for the Debye frequency, $\gamma = -\frac{\mathrm{d}\ln\omega_D}{\mathrm{d}\ln V}$, then $\frac{\partial T_D}{\partial P}$ is always positive provided that $\gamma > 0$, the same is true for $\frac{\partial H}{\partial \alpha}$. At constant temperature, the line-width, H, depends on pressure P as follows:

$$\frac{\mathrm{d}H}{\mathrm{d}P} = \left(\frac{\partial H}{\partial \alpha} \right)_{T_D} \frac{\partial \alpha}{\partial P} + \left(\frac{\partial H}{\partial T_D} \right)_{\alpha} \frac{\partial T_D}{\partial P} \qquad (2)$$

The line-width reduction of Fig. 4b can be understood within this model since the second factor of Eq.2 is dominant given that $\frac{\partial H}{\partial T_D}$ is always negative and the Raman process becomes dominant at room temperature. If we assume that α scales with pressure as the Debye frequency does [6], then the variation of the line-width is given by: $\frac{dH}{dP} = H\left(\frac{1}{\alpha}\frac{\partial \alpha}{\partial P} - 7\frac{\gamma}{B_0}\right) \approx -6\frac{\gamma}{B_0}H$, where B_0 is the crystal bulk modulus. Taking typical values of $\gamma = 1$, $B_0 = 300$ kbar, and $H = 60$ cm^{-1}, we find that $\frac{dH}{dP} \approx -1$ cm^{-1} kbar^{-1}. This estimation is in fair agreement with the experimental results (Fig. 4b) taking into account that the pressure-induced narrowing is an underestimation of the actual homogenous narrowing since contributions from inhomogeneous broadening under pressure have not been considered in this model.

We are in the process of systematically studying the line narrowing phenomenon in other Yb^{3+} based materials.

Acknowledgments

This work was financed by the Spanish MEyC (Project Ref. MAT2005-00099) and the Swiss National Science Foundation.

References

[1] M. Hehlen, H. U. Güdel, Q. Shu, S.C. Rand, J. Chem. Phys. 104, 1232 (1996)
[2] D. R. Gamelin, S. R. Lüthi, H. U. Güdel, J. Phys. Chem. B 104, 11045 (2000)
[3] H. U. Güdel, A. Furrer, H. Blank, Inorg. Chem. 29, 4081 (1990)
[4] M. P. Hehlen, H. U. Güdel, J. Chem. Phys. 98, 1768 (1993)
[5] A. Ellens, H. Andres, A. Meijerink, G. Blasse, Phys. Rev. B 55, 173 (1997)
[6] W. B. Fowler, *Physics of Color Centres*, Academic Press, New York (1968)

Pressure-Induced Helix−Coil Transition of Poly-L-Glutamic Acid in Water

T Takekiyo[1], Y Yoshimura[1*], A Okuno[2], A Shimizu[3], M Kato[2], and Y Taniguchi[2]

[1] Department of Applied Chemistry, National Defence Academy, 1-10-20 Hashirimizu, Yokosuka, Kanagawa, 239-8686, Japan
[2] Department of Applied Chemistry, Ritsumeikan University, 1-1-1 Noji-Higashi, Kusatsu, Shiga, 525-8577, Japan
[3] Department of Environmental Engineering for Symbiosis Factory of Engineering, Soka University, 1-326 Tanjincho, Hachioji, Tokyo, 192-8577, Japan

*E-mail: muki@nda.ac.jp

Abstract. The pressure effect on the helix-coil transition of aqueous poly-L-glutamic acid sodium salt (PLGA-Na) solutions was investigated by FT-IR spectroscopy. From the analysis of the amide I' band, the peak at 1637 cm^{-1} corresponding to the solvated α-helix structure increases with increasing pressure. The solvated α-helix structures of oligopeptides and proteins are stabilized upon compression. On the other hand, the peak intensities at 1710 and 1565 cm^{-1} corresponding to the C=O stretching mode of the COOH group and the C−O stretching mode of the COO^- group of the side chain do not change with the pressure. According to high-pressure Raman study for the condensed aqueous CH_3COONa solution, the pressure-induced dissociation of CH_3COONa is found to be small. This result is qualitatively consistent with that of the side chain of PLGA-Na. The small pressure dependences of the side chain of PLGA-Na may be due to the high salt concentration in the solutions We have determined the difference in the partial molar volumes (PMV) between the α-helix and random coil structures of PLGA-Na at pD 5.0 to be +10.68±0.46 cm^3/mol from the pressure dependence of the band intensities. The positive volume change of helix-coil transition of PLGA-Na is qualitatively consistent with the recent result of the volume change accompanying the helix−coil transition of PLGA (Noudeth G D et al, 2003 *Biopolymers* **70** 563.).

1. Introduction

Poly-L-glutamic acid (PLGA) is a poly-α-amino acid, which can be classified among weak synthetic polyelectrolytes [1-2]. PLGA has been extensively studied by theoretical and experimental methods as a model system of the fold-unfold transition of proteins [2-6]. The difference in the partial molar volume (PMV) between the α-helix and random coil structures of PLGA has been studied by the various methods as a model of the pressure denaturation of proteins [7-10]. The pressure effect on the conformational equilibria involving model polypeptides of proteins is considered to be essentially related to the pressure-induced unfolding of proteins. Assuming that the process of a protein denaturation is a two-state transition (native (N) ↔ denaturated (D)), the difference in the partial molar volume (ΔV) between the two-state transitions for the pressure effect is given by,

© 2008 IOP Publishing Ltd

$$\left(\frac{\partial \ln K}{\partial p}\right)_T = -\frac{\Delta V}{RT} \quad (1)$$

where K is the equilibrium constant and ΔV is the difference in the PMV between two states. According to the Le Chatelier's law, pressure shifts to the conformational equilibrium to a conformer with the smaller PMV. Therefore, the pressure denaturation causes a negative volume change ($\Delta V^{N\rightarrow D}$ <0) [10-11].

Suzuki and Taniguchi showed the volume changes ($\Delta V^{\mathrm{Helix}\rightarrow\mathrm{Coil}}$) from the α-helix to random coil structures of poly-D-glutamic acid (PDGA) is negative by a high pressure electroconductivity measurement.[8] Moreover, Noguchi estimated that the $\Delta V^{\mathrm{Helix}\rightarrow\mathrm{Coil}}$ PLGA is –1.1 cm^3/mol per residue by a dilatometry [7]. On the other hand, Noudeth et al. determined the value of $\Delta V^{\mathrm{Helix}\rightarrow\mathrm{Coil}}$ of PLGA to be +0.3 ± 1.3 cm^3/mol per residue by a densitometry [9]. The sign of $\Delta V^{\mathrm{Helix}\rightarrow\mathrm{Coil}}$ of PLGA is inconsistent with each other. Recently, we reported that the $\Delta V^{\mathrm{Helix}\rightarrow\mathrm{Coil}}$ of AK16 (YGAAKAAAAKAAAAKA-NH_2) determined from the pressure dependence of the relative intensities of the α-helix and random coil structures by using FT-IR spectroscopic study is +0.66 ± 0.03 cm^3/mol per residue, and the PMV of the α-helix of AK16 is smaller than that of the random coil[12-14]. Eventually, it is unclear whether the $\Delta V^{\mathrm{Helix}\rightarrow\mathrm{Coil}}$ is positive or negative values.

In this study, we have investigated the pressure effects on the helix – coil transition and the dissociation of the COONa group of the side chain of PLGA-Na by a high pressure vibrational spectroscopy. We discuss the volume differences between the helix and coil structures and between the COONa and COO^- groups of PLGA-Na.

2. Methods

Poly-L-glutamic acid sodium salt (PLGA-Na) (Peptide Institute Inc. MW< 8000) and sodium acetate (CH_3COONa) (Wako Junyaku Co.) were used without further purification. The concentration of all samples was 25 mg/ml for 20 mM MES buffer (pD 5.0), and 20 mM NaOD (pD 10.0) solutions. All H/D exchange backbone amide protons in PLGA were deuterated by lyphilization. The pD was read directly from a pH meter and no adjustments were made for isotope effects. The high-pressure FT-IR and Raman spectral measurements were reported in the previous studies [15-16].

3. Results and Discussion

The amide I' mode, which is the marker band for detecting the secondary structures of polypeptides and proteins [17-18], consists of the C=O stretching, C–N stretching, and C–C–N deformation modes, and appears in the region from 1620 to 1690 cm^{-1}. Figure 1 shows FT-IR spectra of the amide I' mode region of PLGA in aqueous solution at pD 5.0 as a function of pressure. The spectra were decomposed into five component bands by a curve fitting method. Based on the previous IR results of PLGA[18-22], we assigned the five peaks of 1710, 1670, 1655, 1637, and 1565 cm^{-1} to the C=O stretching mode of COONa group, turn, random coil, solvated α-helix structures, and C–O stretching mode of COO^- group, respectively. Table 1 lists these vibrational assignments of the amide I' region of PLGA-Na.

From Fig.1, the integrated intensity of the solvated α-helix structure increases with increasing pressure, but that of the random coil structure decreases. The solvated α-helix structure of PLGA-Na is clearly stabilized at high pressure. The present result is consistent with the pressure dependences of the solvated α-helix structures of Ala-rich helical peptide (AK16) and four helix bundle protein [12-15]. While the integrated intensity of the C=O stretching mode in the COONa groups and the C–O

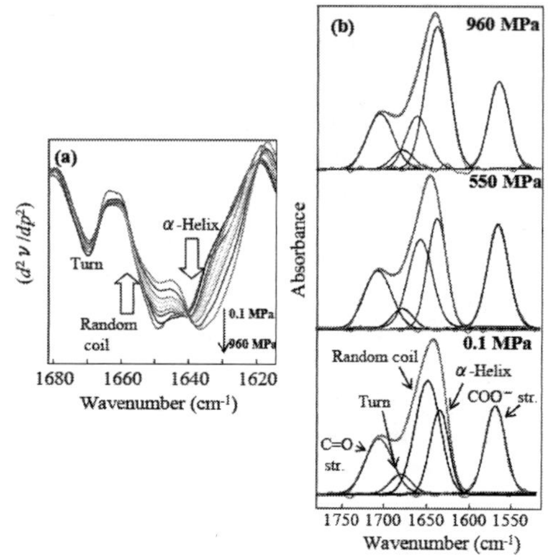

Figure 1. (a) Second derivative and (b) FT-IR spectra of the amide I' region of poly-L-glutamic acid in aqueous solution at several pressures and pD 5.0. The C=O str. and COO^- str. represent the C=O and COO^- stretching modes, respectively.

Table 1. Vibrational assignments of the amide I ' region of PLGA-Na in water.

Wavenumber (cm^{-1})	Vibrational assignment
1710	C=O stretching mode of COONa group of side chain
1670	Turn structure
1655	Random coil structure
1637	Solvated α-helix structure
1565	C−O stretching mode of COO^- group of side chain

stretching mode in the COO^- groups of the side chain hardly changed. The situation becomes reverse for the pressure dependence of the dissociation group of organic compounds having the COONa group [23, 24]. From the pressure dependence of the integrated intensities between the helix and coil structures and between the COONa and COO^- groups, we determined the volume changes from the α-helix to the random coil structures and from the COONa to COO^- groups of the side chain.

Assuming that the ratios of the absorption coefficients between the helix−coil structures and between COONa−COO^- groups are independent of pressure [16], the volume differences between the α-helix and random coil structures($\Delta V^{\mathrm{Helix}\to\mathrm{Coil}}$) and between the COONa and COO^- groups ($\Delta V^{\mathrm{COONa}\to\mathrm{COO^-}}$) are given by

$$\Delta V^{a\to b} = -RT\left(\frac{\partial \ln\left(A^b/A^a\right)}{\partial p}\right)_T \qquad (2)$$

where R, T, and p indicate the gas constant, temperature, and pressure, respectively. A^a and A^b indicate the integrated intensities of the helix/coil or the COONa/COO^- groups, respectively. Figure 2 shows the plots for ln K, which is equal to the $\ln(A^{\mathrm{Coil}}/A^{\mathrm{Helix}})$ or $\ln(A^{\mathrm{COO^-}}/A^{\mathrm{COONa}})$, versus pressure at 298 K. From the slopes of the linear line in Fig. 2, the $\Delta V^{\mathrm{Helix}\to\mathrm{Coil}}$ is determined to be +10.68 ± 0.46 cm^3/mol.

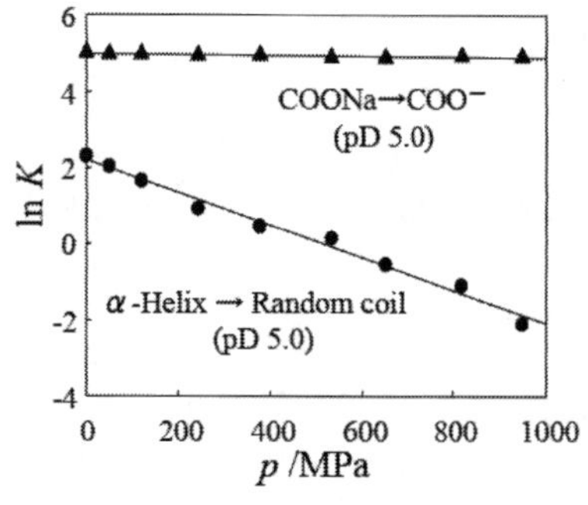

Figure 2. Pressure dependences of the intensity ratio between the α-helix and random coil structures and between the COONa and COO^- groups of PLGA-Na in aqueous solutions at pD 5.0. The circle and triangle represent the helix−coil transition and the COONa−COO^- equilibria of the side chain of PLGA, respectively.

The $\Delta V^{\mathrm{Helix}\rightarrow\mathrm{Coil}}$ per degree of polymezation (n) is +0.16 ± 0.01 cm^3/mol per n. The PMV of the solvated α-helix structure of PLGA-Na is smaller than that of the random coil structure. On the other hand, the $\Delta V^{\mathrm{COONa}\rightarrow\mathrm{COO}-}$ is determined to be −0.32 ± 0.07 cm^3/mol, which is found to be negligibly small. This means that the COONa group of the side chain of PLGA-Na is hardly dissociated with increasing pressure.

Before discussing on the $\Delta V^{\mathrm{Helix}\rightarrow\mathrm{Coil}}$, we discuss the small pressure dependence of the COONa−COO^- equilibria of the side chain of PLGA-Na. Generally, the population of COONa group of organic compounds such as sodium acetate decreases with increasing pressure and the $\Delta V^{\mathrm{COONa}\rightarrow\mathrm{COO}-}$ shows large negative volume change ($\Delta V^{\mathrm{COONa}\rightarrow\mathrm{COO}-}<0$) [23-24]. In this sense, the present result is different from the previous results. To investigate the small pressure dependence of dissociation of the side chain of PLGA-Na, we have measured the pressure effect on the dissociation group of the condensed aqueous CH_3COONa solutions using Raman spectroscopy.

Figure 3 (a) shows the Raman C−C stretching spectra of aqueous CH_3COONa solution as a function of pressure and Figure 3(b) shows the plots of the pressure versus intensity ratio between COONa and COO^-groups. Based on the previous Raman studies of aqueous acetate salts solutions [25, 26], the peaks at 955, 930, and 918 cm^{-1} in the Raman C−C stretching spectra are assigned to the bidentate, monodetate structures of cation (M^+)−acetate complex, and free type of COO^- group, respectively. Figure 4 shows two representative structures of the cation (M^+)−acetate complex and free structure of acetate ion. Assuming that the monodentate and bidentate ligands are COONa group, the $\Delta V^{\mathrm{COONa}\rightarrow\mathrm{COO}-}$ determined to be − 1.99 ± 0.21 cm^3/mol from the slope of the linear line in Fig.3(b). The volume change accompanying with the dissociation of COONa group of aqueous CH_3COONa solution is found to be small. This result is qualitatively consistent with that of the side chain of PLGA-Na. Ascoli et al. [27] suggested the possibility of the attractive interaction between the Na^+ ion and COO^-group of polyelectrolyte such as PLGA. Judging from the result by Ascoli et al., the small pressure-induced dissociation of the side chain of PLGA-Na may be due to the high salt concentration in the solutions.

Based on the above discussions, we discuss why the PMV of the solvated α-helix structure of PLGA-Na is smaller than that of the random coil structure. Recently, Imai et al. [28] determined the $\Delta V^{\mathrm{Helix}\rightarrow\mathrm{Coil}}$ of AK16 and analyzed it into contributions from geometry and hydration: the changes in the van der Waals, void, thermal, and interaction volumes by 3D-RISM-Kurkwood-Buff (KB) theory.

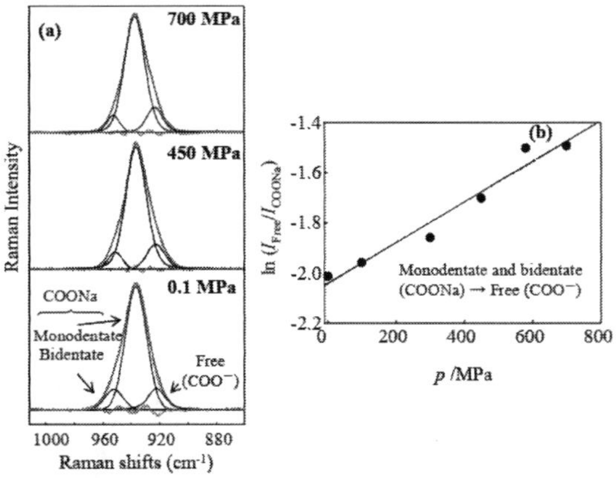

Figure 3. (a) Raman spectra of the C−C stretching mode of aqueous CH_3COONa solution (25 mg/ml) as a function of pressure and (b) pressure dependence of the intensity ratio between COONa and COO^-groups. The straight line is the result of the least-squares analysis.

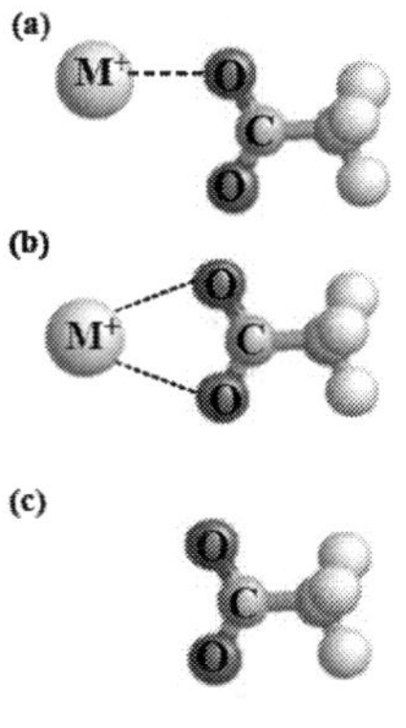

Figure 4. Three typical structures of the acetate ion complexes. (a) monodentate ligand, (b) bidentate ligand, and (c)free structure.

The $\Delta V^{\mathrm{Helix}\rightarrow\mathrm{Coil}}$ of AK16 is positive value ($\Delta V^{\mathrm{Helix}\rightarrow\mathrm{Coil}} = +13$ cm^3/mol) due to the growth of void space within the peptide, which is cancelled in part by the hydration volumes resulting from the solute–solvent interaction [28]. The sign of $\Delta V^{\mathrm{Helix}\rightarrow\mathrm{Coil}}$ of PLGA-Na is the same as that of AK16. It is well known that dissociation of the side chain of PLGA induces the helix–coil transition of PLGA [7-9]. However, the pressure–induced dissociation of the side chain in aqueous PLGA-Na solution is found to be small in the present study. The $\Delta V^{\mathrm{Helix}\rightarrow\mathrm{Coil}}$ of PLGA-Na is independent of the $\Delta V^{\mathrm{COONa}\rightarrow\mathrm{COO-}}$ of the side chain, and the dominant factor of the $\Delta V^{\mathrm{Helix}\rightarrow\mathrm{Coil}}$ of PLGA-Na has same origin of the $\Delta V^{\mathrm{Helix}\rightarrow\mathrm{Coil}}$ of AK16.

In summary, the volume changes from the α-helix to random coil structures and the dissociation of the COONa group of the side chain of PLGA-Na in aqueous solution have been determined by pressure tuning vibrational spectroscopy. The $\Delta V^{\mathrm{Helix}\rightarrow\mathrm{Coil}}$ of PLGA-Na in aqueous solution at pD 5.0 is positive sign ($\Delta V^{\mathrm{Helix}\rightarrow\mathrm{Coil}} = +10.68 \pm 0.46$ cm^3/mol) due to the difference in the void spaces between the solvated α-helix and random coil structures. Our result shows that the pressure effect on the dissociation of the COONa group of PLGA is negligibly small due to the high Na^+ ion concentration around side chain of PLGA-Na. In a series of studies of the pressure effect on the helix–coil transition of oligo- and polypeptides [12-15] including the present study, we found that the PMV of the solvated α-helix structure is smaller than that of random coil structure.

References

[1] Armstrong R W, Strauss U P. In Encyclopedia of Polymer Science and Technology; Mark H F, Gaylord N G, Bikales N M, Eds; Interscience; New York, 1969 Vol. 10, pp 781-861
[2] Fasman G D, Poly-α-amino Acid; Arnold (Publishers) LTD: London, 1967
[3] Zimm B H, Rice S A 1960 *Mol. Phys.* **3** 391
[4] Warashima A, Ikegami A 1972 *Biopolymers* **11** 529
[5] Nakamura H, Wada A 1981 *Biopolymers* **20** 2567
[6] Gordec A, Skerjanc J 2005 *J. Phys. Chem. B* **109** 13363
[7] Noguchi H 1975 *Prog. Polym. Sci. Jpn.* **8** 191
[8] Suzuki K, Taniguchi Y 1968 *Biopolymers* **6** 215
[9] Noudeth G D, Taulier N, Chalikian T V 2003 *Biopolymers* **70** 563
[10] Herberhold H, Winter R 2002 *Biochemistry* **41** 2396
[11] Panick G, Vidugiris G J A, Malessa R, Rapp G, Winter R, Royer C A 1999 *Biochemistry* **38** 4157
[12] Takekiyo T, Okuno A, Imai T, Shimizu A, Kato M, Taniguchi Y 2004 *AIP Conf.* **716** 184
[13] Takekiyo T, Shimizu A, Kato M, Taniguchi Y 2005 *Biochim. Biophys. Acta* **1750** 1
[14] Takekiyo T, Imai T, Kato M, Taniguchi Y 2006 *Biochim. Biophys. Acta* **1764** 355
[15] Takekiyo T, Takeda N, Isogai Y, Kato M, Taniguchi Y 2007 *Biopolymers* **85** 185
[16] Takekiyo T, Yoshimura Y 2006 *J. Phys. Chem. A* **110** 10829
[17] Bandeker J 1992 *Biochim. Biophys. Acta* **1120** 123
[18] Koeing J L, Frushour B 1972 *Biopolymers* **11** 1871
[19] Lenormant H, Baudras A, Blout E R 1958 *J. Am. Chem. Soc.* **80** 6191
[20] Manas E S, Getahun Z, William W W, DeGrado W F, Vanderkooi J M 2000 *J. Am. Chem. Soc.* **122** 9883
[21] Susi H, Timasheff S N, Stevens L 1967 *J. Biol. Chem.* **242** 5460
[22] Krimm S, Bandekar J 1986 *Adv. Protein Chem.* **38** 181
[23] Suzuki K, Taniguchi Y, Watanabe T 1973 *J. Phys. Chem.* **77** 1918
[24] Neuman Jr R C, Kauzmann W, Zipp A 1973 *J. Phys. Chem.* **77** 2687
[25] Takekiyo T, Yoshimura Y *J. Phys. Chem. A,* in press
[26] Nickolov Zh, Ivanov I, Georgiev G, Stoilova D 1996 *J. Mol. Struct.* **377** 13
[27] Ascoli A, Botre C 1963 *Biopolymers* **1** 353
[28] Imai T, Takekiyo T, Kovalenko A, Hirata F, Kato M, Taniguchi Y 2006 *Biopolymers* **79** 97

Journal of Physics: Conference Series **121** (2008) 042004 doi:10.1088/1742-6596/121/4/042004

In-situ Raman study of the pressure-induced bulk melting of hexagonal ice

Yukihiro Yoshimura*[1], **Ho- Kwang Mao**[2] **and Russell J Hemley**[2]

[1] *Department of Applied Chemistry, National Defense Academy, Yokosuka, Kanagawa 239-8686, Japan*

[2]*Geophysical Laboratory, Carnegie Institution of Washington, 5251 Broad Branch Rd. N.W., Washington, DC 20015, U. S. A.*

* E -mail: muki@nda.ac.jp

Abstract

We have investigated a series of changes in *in-situ* Raman spectra as a function of pressure at 264 K to study the process of bulk melting in I_h (hexagonal ice) to the liquid water. We succeeded in measuring spectra where ice I_h coexists with liquid water at an equilibrium pressure of 0.1 GPa. The lowest frequency peak which is derived from the strong hydrogen-bonds of ice I_h lattice remains until the transition and the fraction of liquid water at this equilibrium increases upon compression. The results are interpreted in the context of pre-melting effects in the ice particles.

1. Introduction

The high-pressure polymorphism of H_2O ice is a subject of continuing interest. Although at least 15 different stable and metastable forms [e.g., ice I_h, I_c to XII, low-density amorphous ice (LDA), and high-density amorphous ice (HDA)] have been identified so far [1-7], very recently two new hydrogen ordered phases of ice, referred as ice XIII and XIV, have been found by doping hydrogen disordered phases of ices V and XII with HCl [8]. Our understanding of the phase diagram of H_2O is still incomplete, especially at low temperatures and modest pressures.

One of the most peculiar features of H_2O is that the melting curve for ice I_h against pressure has a negative slope. This is due to of the fact that ice melts upon compression; by Le Chatelier's principle, this causes icebergs to float on water.

In this study, we have examined the pressure-induced phase transformations of ice I_h-liquid water at ~264 K as a function of pressure by monitoring the *in-situ* Raman spectral changes. We will show vibrational dynamics of the internal modes combined with the low frequency spectra studied by the low temperature Raman spectroscopy, which are powerful indicators of the phase transitions. As far as we know, apparently there have been no studies of the spectra under high pressure. One could in fact imagine that melting takes place either as a sudden collapse of the lattice or through the creation and successive catastrophic multiplication of defects. We succeeded in catching the spectra where ice I_h coexists with the liquid water at an equilibrium pressure of 0.1 GPa. The results are interpreted in view of the premelting effect in the ice particles.

2. Experimental details

Raman spectral changes were measured *in-situ* by microspectroscopy together with a Mao-Bell diamond anvil cell (DAC) and a cryostat system at high pressures and low temperatures, as described in detail earlier studies [9].

H_2O ice was made from doubly distilled deionized water [Figure 1 path (1)]. The sample water and a few

© 2008 IOP Publishing Ltd

ruby chips were held in a DAC using a stainless steal gasket. To determine the pressure, we used the pressure dependence on the spectral shift of the R_1 fluorescence line of the ruby chip [10, 11]. A DAC equipped with two heaters and thermocouples was cooled to ~265 K in a continuous-flow of liquid nitrogen. The cryostat was equipped with windows, which allows for *in-situ* Raman (and also ruby fluorescence) measurements without changing the *P-T* conditions on the sample. Pressure can be controlled from the outside by a mechanical change in the load on the DAC lever arms. The temperature was measured to within ±1 K.

Raman spectra were collected by a single-grating ISA HR-460 spectrometer equipped with holographic notch filters and a charge-coupled device (CCD) detector. The 514.5 nm line of argon ion laser excitation (~ 300 mW) was selected to use.

3. Results and Discussion

The experimental paths performed in this study are shown schematically on the pressure-temperature equilibrium phase diagram of H_2O in Fig. 1. First, ice I_h was made by cooling the sample water held in a DAC in a cryostat up to 264 K [path (1)]. Then, gradually we kept compressing ice I_h at 264 K [path (2)]. We measured the *in-situ* Raman spectral changes of OH stretching vibrational and low-frequency modes of path (2) as depicted in Figs. 2 and 4.

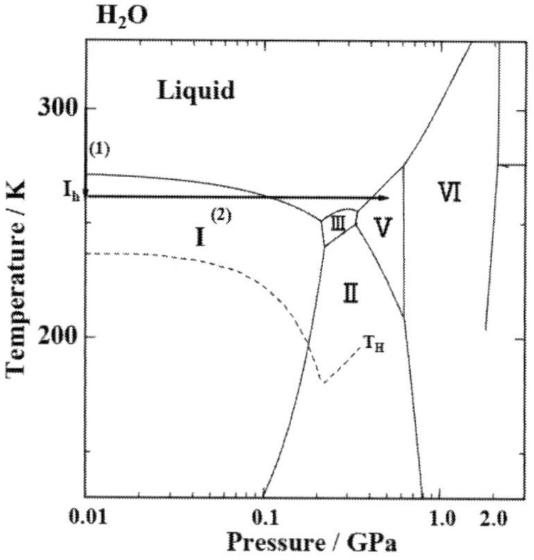

Figure 1. The experimental path observed in the present study on the equilibrium phase diagram of H_2O. The indication corresponds to the results of Raman spectral changes as shown in Figs. 2 and 3, respectively.

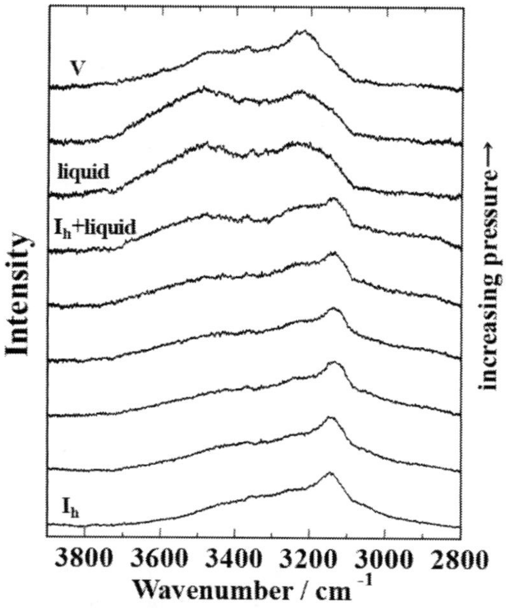

Figure 2. The spectral change of ice I_h–liquid water probed by in-situ Raman OH stretching mode at 264 K.

Figure 2 shows *in-situ* Raman OH-stretching spectral changes as a function of pressure at 264 K. We note first that the relative intensity of the band centered at ~3300 cm^{-1} increases gradually with increasing pressure. Close to the equilibrium melting pressure, the structure of the spectrum changes very slowly. This change must illustrate that the equilibrium crystal structure has been transforming into a liquid structure. We note that the lowest frequency peak that is characteristic of ice I_h phase remains until the transition occurs. We succeeded in obtaining the mixture (coexistent) state spectra between ice I_h and the liquid water, meaning spatially inhomogeneous. It is obvious that at around 0.1 GPa the sample completely transformed to liquid water. The spectrum for water is relatively broad, because of the increase in the distribution of energy state. The features of the spectrum in the OH stretching vibrations region have been discussed by several workers [12-16] and assigned to the various intermolecular modes. Since the origin of its several low frequency features remains uncertain, we would, therefore, proceed with our discussion without reviewing the origin of this feature.

Photomicrographs showing the changes of a representative sample are shown in Fig. 3. Liquid water began to grow as a small region showing a meniscus within ice I_h; upon further compression complete melting to liquid water occurred. Then the sample finally crystallized to ice V (image 4).

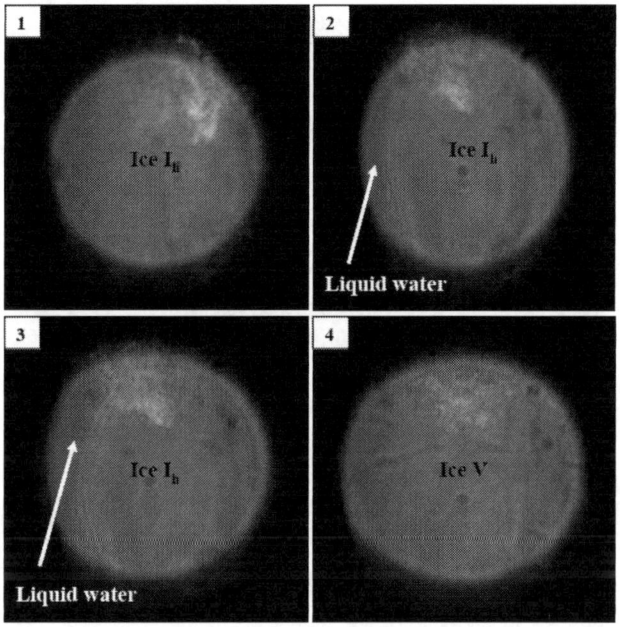

Figure 3. Photomicrographs showing the transformation from ice I_h to liquid water, then finally to ice V. Image 2 shows the mixed phase parts of the sample during the course of the transition. Liquid water is the left portion of images 2 and 3. Upon further compression complete crystallization to ice V occurred, as shown in image 4.

We have also measured the Raman spectra of translational vibrations region. Figure 4 shows the Raman spectral changes of the low-frequency mode. With increasing pressure, the band shoulder at ~280 cm^{-1} gradually disappears and begins to grow at the expense of the broad signal (band) of water. We can recognize that the sharp peak at 210 cm^{-1} of ice I_h was found to remain until the transition occurs: the situation was in harmony with the results of OH-stretching mode. Thus, an important point here is that the ice I_h changed to liquid water showing the mixture state of ice I_h and liquid water, not forming other intermediate structures, as we decreased the pressure little by little. This result confirms that the phase transition between ice I_h and liquid water is first-order one. This transition is sharp, being completely achieved by a pressure change of ~0.02 GPa. We tried to monitor the pressures of the sample during the measurements by the spectral shift of the R_1 ruby fluorescence line. However, at relatively lower pressure regions of around 0.1-0.2 GPa, it was difficult to obtain the precise transition pressure value due to the insensitivity of the ruby fluorescence. At these conditions, the sample pressure was determined from the change in the load on the DAC lever arms.

It is interesting to quote a "pre-melting" [17, 18] to interpret the mechanism of our present results. The pre-melting is provided originally to explain the coexistence of the liquid and solid phases at temperatures substantially below the usual thermodynamic melting point. The effects of crystal size, grain junction-melting, and grain boundary melting have been subjects of considerable interest from a physical point of view. That is, ice melts to fill grain junctions, part of the grain-grain boundary vanishes, and grain-liquid interface forms at places where the grains meet and pockets of liquid form. The ice is no longer in its lowest free energy state and it melts at the grain junctions. A substantial amount of water exists in pure ice at thermodynamic equilibrium below its melting point. Such amounts of liquid in a fine grained solid at thermodynamic equilibrium will affect their compressibility, expansion coefficient, and creep and fracture behaviours.

We suggest that melting might begin from a number of independent grain boundary or junction sites distributed throughout the sample. Over pressure, the melting regions became larger, covering more of the sample, although residual ice lattice still tends to be present in the sample. Of course, as pressure was increased, more independent melting sites increase and decrease the sample remained ice I_h. The present measurements itself, unfortunately, do not provide structural details. But we suppose that the subsequent formation of coordination defects triggers the transition to the liquid state.

In conclusion, the present paper has illustrated important aspects of the complex behavior of ice viewed from the liquid and the solid sides. Our *in-situ* measurements provide evidence that ice I_h changed to liquid water with showing the mixture state of ice I_h and liquid water, not forming other intermediate structures upon compression. We suppose that the bulk melting is mediated by the formation of topological defects. The study of the low-temperature transitional phase diagram of ice is of practical interest for the planetary geology and astrophysical implications because many high-pressure ice polymorphs, including the amorphous modification, can be ubiquitously distributed. The *in-situ* measurements give us the opportunity to extract information on the kinetics of the transformation and on the properties of the intermediate states.

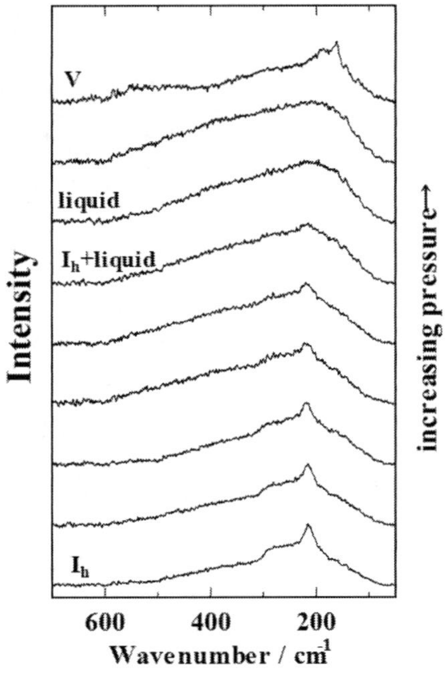

Figure 4. *In-situ* Raman spectral changes in the low-frequency mode with increase in the pressure at 264 K.

Acknowledgements
Authors are grateful to Dr. Victor V. Struzhkin and Dr. Eugene Gregoryanz for experimental assistance with the low-temperature Raman systems. Y. Y thanks to Morgan Philips for help with the manuscript.

References

[1] Whalley E in The Hydrogen Bond, edited by Schuster P Zundel G Sandorfy C 1976 North-Holland, Amsterdam
[2] Bridgman P W 1937 *J. Chem. Phys.* **5** 964
[3] Kamb B 1969 *Trans. Am. Crystallorg. Assoc.* **5** 61
[4] Mishima O, Calvert L D and Whalley E 1984 *Nature* **310** 393
[5] Mishima O, Calvert L D and Whalley E 1985 *Nature* **314** 76

[6] Lobban C, Finney J L and Kuhs W F 1998 *Nature* **391** 268
[7] O'Keeffe M 1998 *Nature* **392** 879
[8] Salzmann C G, Radaelli P G, Hallbrucker A, Mayer E and Finney J L 2006 *Science* **311** 1758
[9] Yoshimura Y, Stewart S T, Somayazulu M, Mao H K and Hemley R J 2006 *J. Chem. Phys.* **124**, 024502
[10] Mao H K, Bell P M, Shaner J W and Steinberg D J 1978 *J. Appl. Phys.* **49** 3276
[11] Mao H K, Xu J, and Bell P M 1986 *J. Geophys. Res.* **91** B5 4673
[12] Padro J A 2003 *J. Chem. Phys.* **118** 452
[13] Wojcik M J, Szczeponek K and Ikeda S 2002 *J. Chem. Phys.* **117** 9850
[14] Sivakumar T C, Rice S A and Sceats M G 1978 *J. Chem. Phys.* **69** 3468
[15] Walrafen G E 2004 *J. Chem. Phys.* **121** 2729
[16] Buch V 2005 *J. Phys. Chem. B* **109** 17771
[17] Johari G P, Pascheto W and Jones S J 1994 *J. Chem. Phys.*
100 4548
[18] Salvetti G Tombari E and Johari G P 1995 *J. Chem. Phys.* **102** 4987

Raman Spectroscopic Study on the Hydration Structure of Tetraethylammonium Chloride in Water

Takefumi Goto, Takahiro Takekiyo and Yukihiro Yoshimura

Department of Applied Chemistry, National Defense Academy, 1-10-20 Hashirimizu, Yokosuka, Kanagawa 239-8686, Japan

E-mail: muki@nda.ac.jp (Y. Yoshimura)

Abstract. The hydration structure of tetraalkylammonium chloride (Et_4NCl) in aqueous solution has been studied by Raman Spectroscopy as functions of pressure and concentration. The hydration volume due to the hydrophobic hydration around *trans-trans. trans-trans* (*tt.tt*) conformer is larger than that around *trans-gauche. trans-gauche* (*tg.tg*) conformer. The hydration structures are closely correlated to the conformational equilibrium in the solution.

1. Introduction

A number of studies on the aqueous tetraalkylammonium salt solutions have been performed to investigate the nature of hydrophobic hydration and clathrate structures[1-8], because they were soluble in many polar solvents, particular in water, and their alkyl-chain length can be changed easily within the series of homologues[4].

Another remarkable feature is that tetraethylammonium ion (Et_4N^+) shows the conformational equilibrium. The conformation of molecules is arise from intramolecular rotation and this rotation takes on trans and gauche conformations[9, 10]. Naudin. *et al*[11] reported that Et_4N^+ ion shows the conformational equilibrium between the *trans-gauche. trans-gauche* (*tg.tg*) and *trans-trans. trans-trans* (*tt.tt*) conformer for the rotation of Et-N^+-Et axis in the aqueous solution, though the ethyl-chains adopt the *tt.tt* conformer in the crystalline state. The *tg.tg* and *tt.tt* conformers are also called Nordic cross and Greek cross, respectively. These conformations are shown in Figure 1. Theoretically tetraethylammonium ion may show other type of conformers, for example *trans-trans. trans-gauche* (*tt.tg*) and *gauche-gauche. gauche-gauche* (*gg.gg*) conformers. However, these conformers are too labile to exist in the aqueous solutions. Actually the Raman spectra attributable to these conformers are not observed in the aqueous solutions.

It is intriguing to discuss the hydration structures along with the conformational equilibrium in the tetraethylammonium salt solutions. In our preliminary studies, we measured the Raman spectra of the aqueous tetraalkylammonium bromide (Et_4NBr) solution[12, 13]. It was suggested that hydrophobic hydration may prefer the *tt.tt* conformer to the *tg.tg* conformer at R (=moles of water/moles of Et_4NBr)=19, but the hydration structure around Et_4N^+ ion at R=4 might be different from the hydrophobic hydration.

As an extension of the previous study we have investigated the Et_4NCl solution in an effort to obtain the information of the anionic effect on the hydration structures. In this study we have observed the Raman spectral changes in the C_4N symmetric stretching and uncoupled OD stretching modes as

© 2008 IOP Publishing Ltd

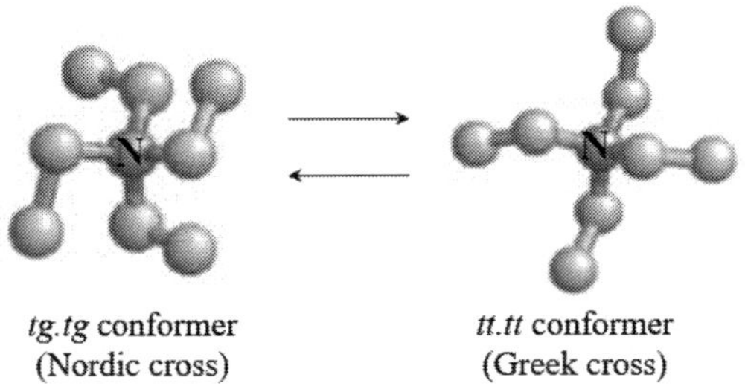

Figure 1. Structures of the *tg.tg* (Nordic cross) and *tt.tt* (Greek cross) conformers of Et_4N^+ ion.

functions of pressures and concentration (*R*). From the C_4N symmetric stretching spectra we can obtain the information on a change in the conformational equilibrium, and the OD (OH) stretching spectra are used as a probe of the hydrogen-bond strength[6]. We compared the results between the aqueous Et_4NBr and Et_4NCl solutions.

2. Materials and Methods

Et_4NCl was obtained from Tokyo Kasei Industry Co. Ltd (Et_4NCl : 99.9%), and D_2O was obtained from Aldrich Chemical (D_2O : 99.8%). All sample solutions were prepared by dissolving Et_4NCl in 5 % D_2O solution ($0.05D_2O \cdot 0.95H_2O$) as a solvent. The reason of using 5 % D_2O solution is to avoid complexity in analyzing the OH stretching Raman spectra. The OH stretching spectra of aqueous solutions are broad and the Fermi resonance effect makes the OH stretching spectra difficult, and the CH stretching band of Et_4N^+ ion overlaps the OH stretching spectra[7, 14]. The concentrations of all the solutions are expressed by *R* (=moles of water/moles of Et_4NCl). Raman spectra were measured by a JASCO NR-1800 Raman spectrophotometer equipped with a CCD detector and a microscope. The exposure time of each run was 300 s. The 514.5 nm line from a Lexel argon ion was used as an exacting source with a power of 350 mW. For the high-pressure experiments, we used a diamond anvil cell (DAC). The ruby chip was used for a pressure reference[15]. To analyze the obtained C_4N symmetric stretching and uncoupled OD stretching spectra, we fitted the spectra with the Gaussian-Lorentzian mixed curves.

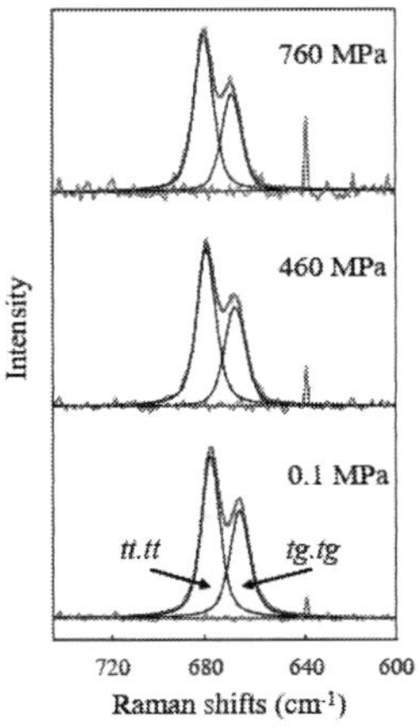

Figure 2. Typical Raman spectra of the C_4N symmetric stretching mode in the aqueous Et_4NCl solution as a function of pressure at *R*=11.

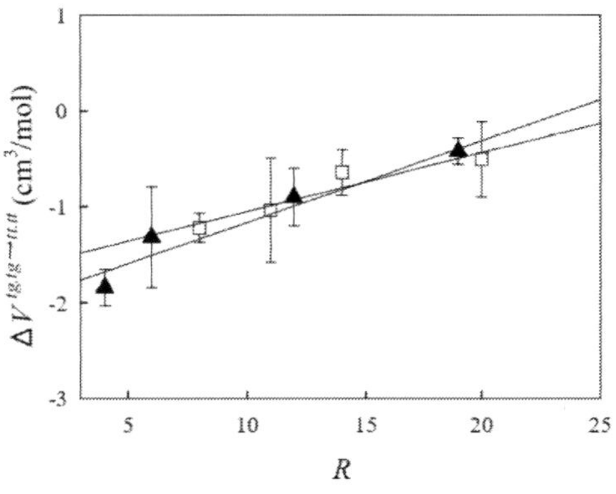

Figure 3. Concentration dependences of the $\Delta V^{tg.tg\to tt.tt}$ of the aqueous Et_4NCl and Et_4NBr solutions. Symbols of □ and ▲ represent the $\Delta V^{tg.tg\to tt.tt}$ of the Et_4NCl and Et_4NBr solutions, respectively.

3. Results and Discussion

First of all, we discuss the pressure effects of the aqueous Et_4NCl solution. The C_4N symmetric stretching spectra of the solution as a function of pressure are shown in Figure 2. The lower and higher frequency peaks are belong to the *tg.tg* and *tt.tt* conformers, respectively[11]. To discuss more about the experimental results, we estimated the partial molar volume ($\Delta V^{tg.tg \to tt.tt}$) between the *tg.tg* and *tt.tt* conformers from the pressure dependence of the relative Raman intensities. Presuming that the proportion of the Raman scattering cross sections between the conformers is independent of a pressure, $\Delta V^{tg.tg \to tt.tt}$ is given by

$$\Delta V^{tg.tg \to tt.tt} = -8.314T\left\{\frac{\partial \ln(I_{tt.tt}/I_{tg.tg})}{\partial p}\right\}_T \qquad (1)$$

where R, T, p, $I_{tg.tg}$ and $I_{tt.tt}$ are gas constant, temperature, pressure, the relative Raman intensity of the *tg.tg* and *tt.tt* conformers, respectively.

Figure 3 plots the results of the concentration dependence of the $\Delta V^{tg.tg \to tt.tt}$ of the aqueous Et_4NCl solution together with that of the Et_4NBr solution. The value of the $\Delta V^{tg.tg \to tt.tt}$ for the aqueous Et_4NCl solution takes negative throughout the studied concentration region, indicating that the partial molar volumes of the *tt.tt* conformer are always smaller than those of the *tg.tg* conformer. It is also noticed that the $\Delta V^{tg.tg \to tt.tt}$ of the aqueous Et_4NCl solution takes almost same value of the Et_4NBr solution at the same R.

To discuss the observed results further, we separated the $\Delta V^{tg.tg \to tt.tt}$ into two contributions as below (eq.2).

$$\Delta V^{tg.tg \to tt.tt} = \Delta V_M^{tg.tg \to tt.tt} + \Delta V_{hyd}^{tg.tg \to tt.tt} \qquad (2)$$

where ΔV_M is the molecular volume of the conformer itself, ΔV_{hyd} is the hydration volume influenced by interactions between the solute and solvent molecules. We calculated the intrinsic value of the ΔV_M by the Alpha-Shapes programs, to be $-1.34 cm^3 mol^{-1}$. Using this, we can obtain the ΔV_{hyd}.

Figure 4 plots the results of the concentration dependences of the $\Delta V_{hyd}^{tg.tg \to tt.tt}$. Generally both the ΔV_{hyd} of the aqueous Et_4NCl and Et_4NBr solutions exist plus throughout the studied R. This means that the hydration volume of the *tg.tg* conformer is smaller than that of the *tt.tt* conformer in the aqueous Et_4NCl and Et_4NBr solutions. In our previous study[13], we mentioned that the differences in the $\Delta V_{hyd}^{tg.tg \to tt.tt}$ value of Et_4N^+ ion in the aqueous Et_4NBr solution between R=4 and 19 comes from the hydration structure around Et_4N^+ ion and the hydration volume due to the hydrophobic hydration around *tt.tt* conformer is larger than that around *tg.tg* conformer

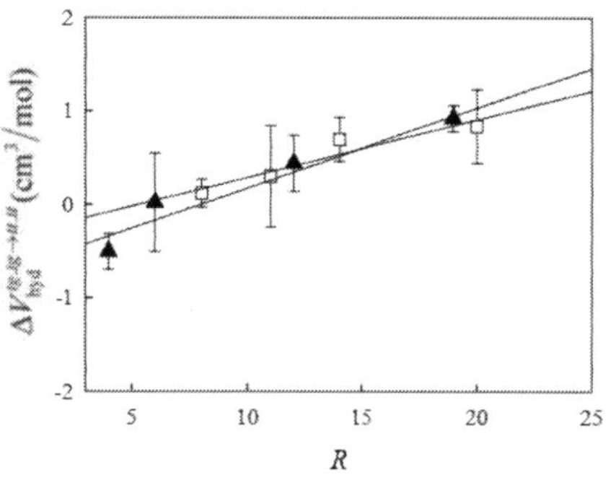

Figure 4. Concentration dependences of the $\Delta V_{hyd}^{tg.tg \to tt.tt}$ of the aqueous Et_4NCl and Et_4NBr solutions. Symbols of □ and ▲ represent the $\Delta V_{hyd}^{tg.tg \to tt.tt}$ of the Et_4NCl and Et_4NBr solutions, respectively.

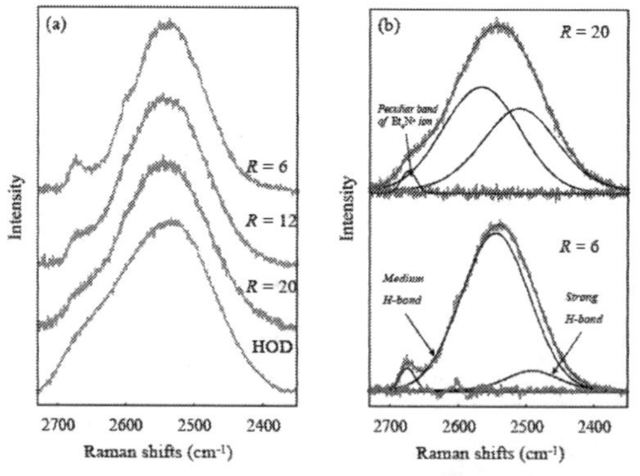

Figure 5. (a) Raman spectra of the uncoupled OD stretching mode of the aqueous Et_4NCl solution at various concentrations (R). (b) Representative curve-fitted results at R=6 and 20.

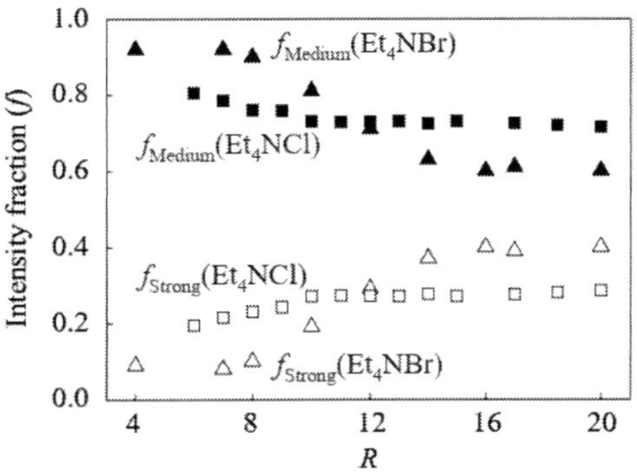

Figure 6. Concentration dependences of the intensity fractions for the strong and medium H-bond components of the aqueous Et_4NCl and Et_4NBr solutions at a normal pressure. Symbols of □ and ■ represent the intensity fractions for the strong and medium H-bond components of the aqueous Et_4NCl solution, △ and ▲ represent those of the aqueous Et_4NBr solution.

Base on these results, we suggest that the hydrophobic hydration around *tt.tt* conformer in the aqueous Et_4NCl solution may be maintained at more concentrated region probably due to the anionic effect.

To consider the hydration structures further along with the conformational equilibrium, we discuss the strength of hydrogen-bond (H-bond) around Et_4N^+ ion in the aqueous Et_4NCl solution from the uncoupled OD stretching spectra at a normal pressure (Figure 5(a)). In doing this, we fitted the uncoupled OD stretching spectra with two contributions having different hydrogen-bond strength; strong H-bond and medium H-bond[7,14,16,17]. The curve-fitted uncoupled OD stretching spectra of the aqueous Et_4NCl solution as a function of concentration are shown in Figure 5(b). The intensities (f) of the strong (f_{strong}) and medium (f_{medium}) hydrogen-bonds can be represented as

$$f_{strong} = \frac{I_{strong}}{I_{strong} + I_{medium}}, \quad f_{medium} = \frac{I_{medium}}{I_{strong} + I_{medium}} \qquad (3)$$

Figure 6 plots the results of the concentration dependence of the f_{strong} and f_{medium} in the aqueous Et_4NCl solution together with that of the Et_4NBr solution. The differences in the intensity fraction between the higher and lower salt concentration region (e.g. R=6 and 20) in the aqueous Et_4NCl solution is smaller than that in the aqueous Et_4NBr solution. This result implies that the hydration structure in the aqueous Et_4NCl solution would not change drastically with the change in the concentration as compared to that in the aqueous Et_4NBr solution. This might be the anionic effect of Cl^- ion having a stronger structure-breaking property in the solutions[18].

4. Conclusions

We have shown the Raman spectral changes in the C_4N symmetric stretching and uncoupled OD stretching modes of the aqueous Et_4NCl solution as a function of pressure. The comparison of the results between the aqueous Et_4NCl and Et_4NBr solutions are made. The hydration structures are discussed in conjunction with the conformational equilibrium in the solution. An important implication is that the hydration volume due to the hydrophobic hydration around *tt.tt* conformer is larger than that around *tg.tg* conformer in the aqueous Et_4NCl solution. It is thus interesting to investigate the hydration structure of tetraalkylammonium fluoride (Et_4NF) having strong structure-making property [19]. This will be reported elsewhere.

References

[1] Frank H S and Evans M W 1945 *J. Chem. Phys.* **13** 507
[2] Kay R L and Evans D F 1966 *J. Phys. Chem.* **69** 4216
[3] Kay R L and Evans D F 1967 *J. Phys. Chem.* **70** 366 2325
[4] Wen W Y 1973 *J. Sol. Chem.* **2** 253
[5] Franks F ed 1973 *A Comprehensive Treatise* (Plenum: New York) vol.3
[6] Green J L, Sceats M G and Lacey A R 1987 *J. Chem. Pyhs.* **87** 6
[7] Kanno H, Shimada K and Katoh T 1989 *J. Phys. Chem.* **93** 4981
[8] Kanno H, Ohnishi A, Tomikawa K and Yoshimura Y 1999 *J. Raman Spectrosco.* 3**0** 705
[9] Mizushima S, Morino Y and Higashi K 1934 *Sci. Pap. Inst. Phys. Chem. Res.* Tokyo **25** 159
[10] Mizushima S 1945 *Structure of Molecules and Internal Rotation* Academic Press (New York)
[11] Naudin C *et al* 2001 *J. Raman Spectrosco.* **31** 979
[12] Takekiyo T and Yoshimura Y 2006 *Chem. Phys. Lett.* **420** 8
[13] Takekiyo T and Yoshimura Y 2006 *J. Phys. Chem. A* **110** 10829
[14] Kanno H, Yonehama K, Somraj A and Yoshimura Y 2006 *Chem. Phys. Lett.* **427** 82
[15] Mao H K, Bell P M, Shaner J W and Steinberg D J 1978 *J Appl. Phys.* **49** 3276
[16] Stangert J and Gampe T 1999 *J. Phys. Chem. B* **103** 3778
[17] Stangert J and Gampe T 2002 *J. Phys. Chem. A* **106** 5393
[18] Wen W Y 1972 *In Water and Aqueous Solutions* R A Home ed (New York) p 617
[19] Donald H, Jenkins B and Marcus Y 1995 *Chem. Rev.* **95** 2695

On the pressure-induced spin transition in $CoCl_2$. Interplay between Jahn-Teller effect and spin sate in Co^{2+}.

I. Hernández[1] and F. Rodríguez[1,*]

[1] DCITIMAC, Facultad de Ciencias, Universidad de Cantabria, Avda. de Los Castros s/n, 39005, Santander, Spain

fernando.rodriguez@unican.es

Abstract. This work investigates on the interplay between the Jahn-Teller effect and spin state in $Co^{2+}(3d^7$ configuration). High-pressure optical absorption experiments were carried out in the 0 – 50 GPa range in $CoCl_2$. This system is attractive since pressure increases the crystal-field splitting between e_g and t_{2g} levels from 0.9 eV to 1.6 eV, thus approaching $CoCl_2$ towards the high-spin-to-low-spin transition. The free energy associated with the two states is discussed in terms of crystal-field and stabilization energies due to lattice relaxation including the Jahn-Teller effect, which is expected to be particularly strong in the low-spin configuration ($^2E;t_{2g}^6e_g^1$) with respect to the weakly Jahn-Teller high-spin configuration ($^4T_1$; $t_{2g}^5e_g^2$). The metallization process observed above 40 GPa is discussed in terms of spin transition and charge-transfer states.

1. Introduction

Transition-metal (TM) ions with multielectronic configurations may exhibit interesting properties associated with spin change induced either by increasing the crystal-field strength or the electron dispersion in localized and delocalized systems, respectively. Both effects strongly depend on the interaction between the TM electrons and the surrounding, and thus can be modified by external pressure. $CoCl_2$ is an attractive system for studying such an effect in concentrated TM compounds since its octahedral coordination is stable in a wide pressure range [1]. This behaviour contrasts with CoF_2, which is not suitable for this study. The reason is due to the structural transformation undergone by the MF_2 (Mg, Mn, Co) crystal series to more dense structures with larger coordination numbers in the range 0 - 20 GPa [2-4]. In particular, CoF_2 experiences a first-order structural phase transition from the rutile structure to the fluorite structure at 9-10 GPa yielding a change of coordination from sixfold to eightfold coordination [2,3]. Although Co^{2+} is known to exhibit a high-spin (HS) state ($^4T_1$; $S = 3/2$) in $CoCl_2$ up to 15 GPa, through which the crystal-field energy changes from $\Delta = 0.9$ to 1.2 eV [1], no investigation has been performed at more elevated pressures in order to approach the system to the spin-crossover (SCO) crystal field. According to the Tanabe-Sugano diagram (Fig. 1), we estimate that the crystal field for SCO takes place at $\Delta_{SCO} = 1.9$ eV. It corresponds to $\Delta_{SCO}/B = 21$, where B = 90 meV is the Racah parameter at 40 GPa [1]. In this work we explore the behaviour of Co^{2+} near the

© 2008 IOP Publishing Ltd

SCO by optical absorption (OA) spectroscopy under high pressure. Our interest is focussed in elucidating whether the relaxation energy involved in the spin transition ($^4T_1$; S= 3/2 → ^2E; S = 1/2) favours stabilization of LS state at smaller pressures than those predicted on the basis of crystal-field models [5], or whether pressure-induced transformations in $CoCl_2$ make the spin transition to occur at higher pressures or even unfeasible like Co^{2+} in cubal coordination (d^3-like TM). Additionally, given that the LS state ^2E exhibits a strong Jahn-Teller effect, we investigate if this effect is able to drive the spin transition at moderate pressures and it induces low-symmetry distortions in the LS ground state or yields metallization.

2. Experimental

Single-crystal plates (100x80x30 μm^3) of $CoCl_2$ (Merck) were used for high-pressure experiments. $CoCl_2$ crystallizes in the trigonal space group $R\bar{3}m$ [6]. A membrane-type diamond-anvil cell was loaded with $CoCl_2$ and several Ruby spheres (10 μm diameter) using 300000 cst silicone oil (Dow Corning) as pressure transmitting medium. The pressure was measured using the ruby scale. The OA spectra under pressure were obtained with a homemade setup operating in the NIR and VIS range. The light of a tungsten lamp was chopped with a frequency controller, SR-540 and focused on the sample with a microscope objective. Another microscope objective collected the light passing through (I) and outside (I_0) the crystal plate within the hydrostatic cavity. The light was dispersed by a 0.5 m monochromator (Chromex 500IS) and detected synchronously (lock-in SR-830) using either a thermoelectrically cooled InGaAs detector (Electro Optical Systems, Inc.) for NIR or a photomultiplier (Hamamatsu R928-S) for UV-VIS.

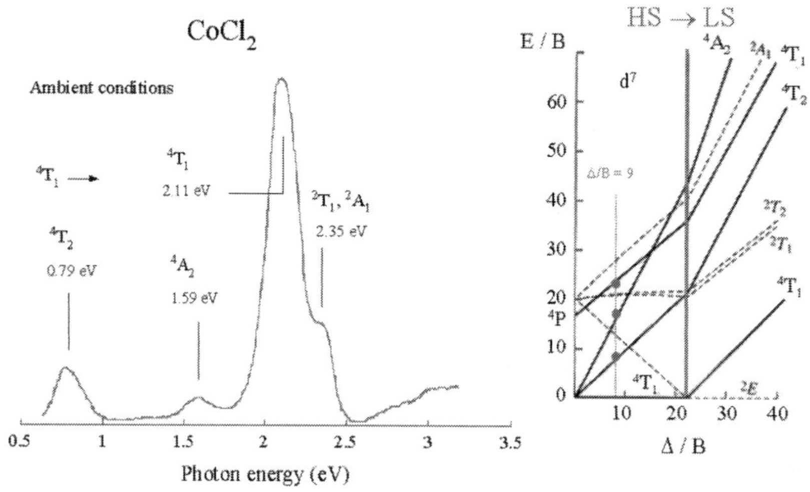

Fig. 1

3. Results and Analysis

Figure 1 shows the OA spectrum of $CoCl_2$ at ambient conditions. The main peaks, which are located at 0.79, 1.59 and 2.11 eV correspond to crystal-field transitions from the $^4T_1$ ground state to the $^4T_2$, $^4A_2$ $^4T_1$ excited states, respectively, within the D_{3d} (nearly octahedral) $(CoCl_6)^{4-}$ complex [6,7]. The fit of experimental energies to the term energies given as a function of Δ and the Racah parameters, B and C, [8] gives: Δ = 0.90 eV, and B = 95 meV (Δ/B = 9.5) [1,7].

The Tanabe-Sugano diagram of Figure 1 shows the crystal-field point corresponding to Co^{2+} in $CoCl_2$ at ambient pressure. It is worth noting the large crystal-field increase required to reach the SCO

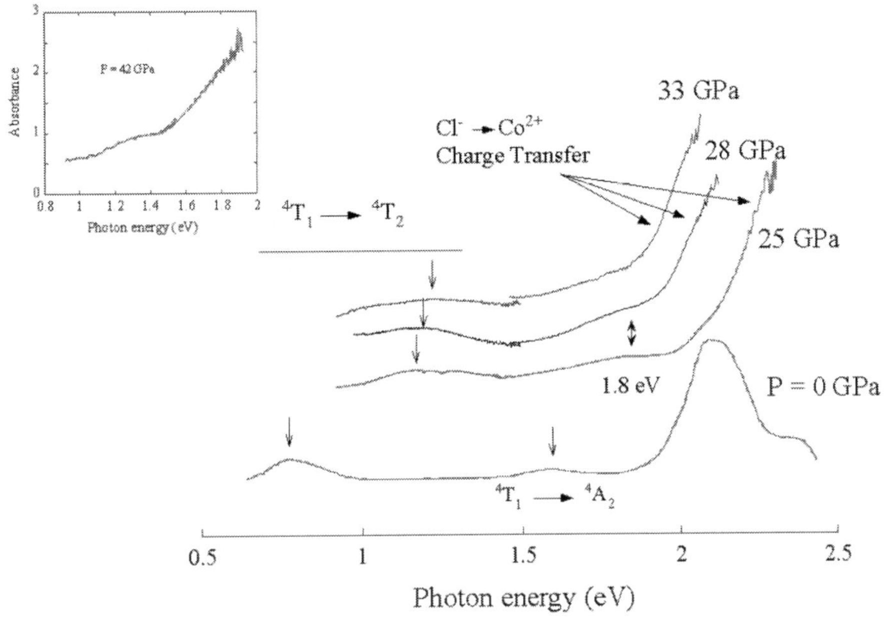

Fig.2

point: $(\Delta/B)_{SCO}$ = 21. It means a splitting energy between e_g and t_{2g} orbitals of 1.8 eV taking into account that B = 87 meV. This estimate is based on the variation of B with pressure as -2 meV GPa^{-1} and the expected pressure for spin transition being around 40 GPa. Therefore HS-to-LS transition in $CoCl_2$ requires pressures that double the ambient-pressure crystal-field energy. Previous investigations on $CoCl_2$ [1] indicate that Δ increases linearly with pressure at 18 meV GPa^{-1} in the 0-15 GPa range. If such behaviour is maintained at higher pressures, then SCO conditions will be fulfilled for P ≥ 50 GPa according to expectations based on the Tanabe-Sugano diagram.

Figure 2 shows the variation of the $CoCl_2$ OA spectrum with pressure in the 0 – 42 GPa range. The sample becomes black at the highest pressure yielding metallization. A remarkable feature of the variation is the strong blueshift experienced by the first transition energy $^4T_1 \rightarrow ^4T_2$, which is proportional to Δ. Figure 3 plots its variation with pressure (a) as well as the corresponding variations of the background absorption at 1300 nm and the energy of the charge-transfer low-energy side-band onset (b). Note that the $^4T_1 \rightarrow ^4T_2$ transition energy increases from 0.79 to 1.39 eV (δE = 0.60 eV) in the 0 – 40 GPa range, showing an almost linear dependence with pressure at a rate of 15 meV GPa^{-1} (Fig. 3a). In terms of crystal-field energy, Δ increases from 0.9 to 1.6 eV (δΔ = 0.7 eV) in the same pressure range.

On the other hand, the $^4T_1 \rightarrow ^4A_2$ band observed at 1.59 eV at ambient conditions does not correspond to the OA band which is located around 1.8 eV at high pressures. As it was discussed elsewhere [1], the $^4T_1 \rightarrow ^4A_2$ transition blueshifts with pressure twice as large as the first $^4T_1 \rightarrow ^4T_2$ transition (Fig. 1), and thus it should peak at about 3.0 eV for P > 25 GPa. Actually that OA band is hidden by the $Cl^- \rightarrow Co^{2+}$ charge-transfer transition, whose side-band onset appears around 2 eV above 25 GPa. Besides other structural transformations, the high-pressure electronic transition at 1.8 eV must be ascribed within $(CoCl_6)^{4-}$ to the spin-forbidden $^4T_1 \rightarrow ^2T_1, ^2T_2$ transitions (quasi-parallel lines in the Tanabe-Sugano diagram), whose electric-dipole oscillator strength appears enhanced in the proximity of both the charge transfer states and the LS state (^2E). Furthermore, its transition energy does not vary appreciably with pressure as it corresponds to an almost Δ-independent transition (E/B = 21).

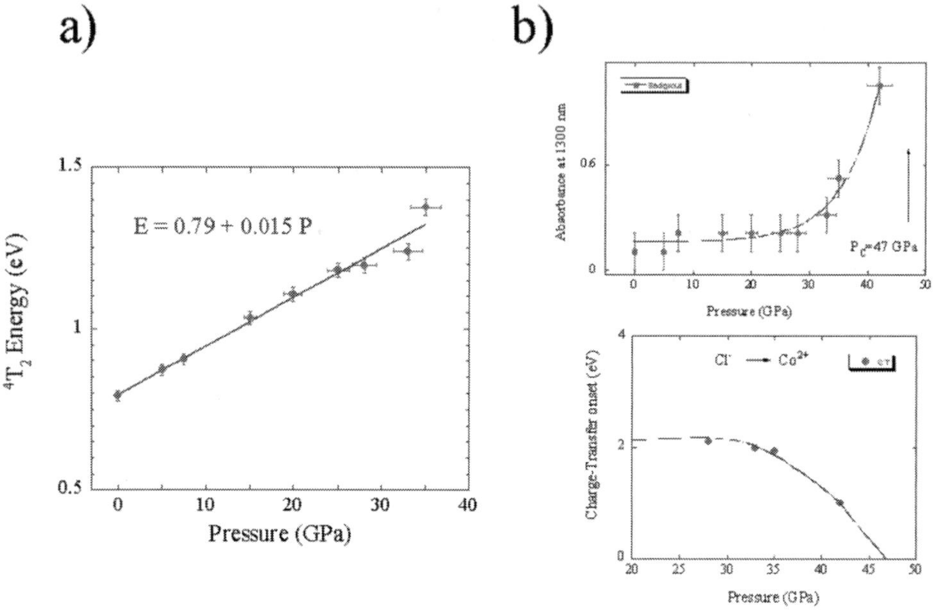

Fig. 3

Figure 3a illustrates the linear dependence of the first transition energy with pressure at 15 meV GPa^{-1} in the 0 – 33 GPa range, the pressure rate being similar to that measured in Ref. [1] in the 0 – 15 GPa range. The variation of the crystal-field energy is then 17 meV GPa^{-1} (Δ = 1.58 eV at 40 GPa) and thus the system moves in the Tanabe-Sugano diagram from Δ/B = 9.5 ambient conditions to Δ/B = 18 at 40 GPa near the SCO point (Δ_{SCO} = 1.83 eV for B = 87 meV). Nevertheless, as it is discussed elsewhere [5], the energy difference $\delta\Delta$ = 0.25 eV, required to overcome the SCO point is not the real one given that we must consider lattice relaxation and the Jahn-Teller effect in the LS ^2E state. The actual free-energy balance for spin-transition in d^7 TM as Co^{2+} or Ni^{3+} is given by [5]:

$$E(^4T_1) = E(^2E) - E_{JT} - E_{LR} \qquad \text{or likewise} \qquad \Delta_{SCO}[Co^{2+}] \approx U - E_{JT} - E_{LR}$$

where U = 1.83 eV for Co^{2+} is the interelectronic repulsion and pairing energy difference between $^4T_1$ and ^2E states for a given coordination geometry, E_{JT} is the Jahn-Teller energy and E_{LR} the lattice relaxation energy due to the different coordination geometry (metal-ligand distances) between the HS and LS states [5]. A similar situation is encountered for d^4 TM as Cr^{2+} or Mn^{3+}, though in d^4 configuration, the Jahn-Teller state is HS (^5E) instead of low-spin (^2E) in Co^{2+} (d^7):

$$E(^5E) - E_{JT} = E(^3T_1) - E_{LR} \qquad \text{or likewise} \qquad \Delta_{SCO}[Cr^{2+}] \approx U + E_{JT} - E_{LR}$$

where U = 2.4 eV for Cr^{2+} [8]. Taking into account that lattice relaxation term for divalent TM chloride $(MCl_6)^{4-}$ is about $E_{LR} = S\hbar\omega \approx 0.2$ eV, the effective crystal-field strength for spin transition in Co^{2+} becomes $\Delta_{SCO}[Co^{2+}] \approx 1.6 - E_{JT}$. The crucial point to make reliable estimates for SCO requires knowledge of E_{JT}. However the Jahn-Teller energy is not known for Co^{2+} in chlorides since the HS ground state does not show any Jahn-Teller distortion [6]. In order to estimate E_{JT}, we can use the Jahn-Teller energy of $(CuCl_6)^{4-}$ systems (E_{JT}= 0.3 eV) as reference for $(CoCl_6)^{4-}$ [9,10]. So that we obtain Δ_{SCO} = 1.3 eV taking E_{JT} = 0.3 eV. The present estimate illustrates the relevance of the Jahn-Teller effect in stabilizing the low-spin (LS) state in $CoCl_2$ for pressures above 35 GPa, provided that the Jahn-Teller distortion in LS Co^{2+} is not suppressed at such pressures. Evidence of Jahn-Teller suppression has been given for $(CuCl_6)^{4-}$ [9,10] and $(MnF_6)^{3-}$ [11,12] at pressures around 40 GPa, and thus the SCO pressure in $CoCl_2$ is high enough to eventually suppress the Jahn-Teller distortion and therefore maintain the HS $^4T_1$ state at these pressures. However a relevant feature arising in this work (Fig. 2) is the big transition oscillator strength of the OA band at 1.8 eV that we assign to the spin-forbidden $^4T_1 \rightarrow {}^2T_1, {}^2T_2$ in HS. Note that this electronic transition should become spin allowed in the LS ^2E ground state. This result thus suggests that HS and LS states are significantly mixed above 30 GPa thus enabling OA transitions involving different spin states. This interpretation explains the anomalous effects observed in the OA spectra of Figure 2. Nevertheless the possibility of a pressure-induced structural transition in $CoCl_2$ to a fluorite-type structure must not be discarded. The eightfold coordination would transform the electronic structure for Co^{2+} leading to a d^3-like TM. Then two broad bands associated with crystal-field transitions $^4A_2 \rightarrow {}^4T_2$, and $^4A_2 \rightarrow {}^4T_1$ would be observed in the OA spectra. The OA bands at 1.4 and 1.8 eV above 25 GPa agree with that interpretation. However the smooth variations of the spectroscopic parameters with pressure suggest that no change in the Co^{2+} coordination seems to occur in $CoCl_2$ in the explored pressure range.

The variations of the charge-transfer energy and the absorption background of Figure 3(b), clearly indicate that metallization takes place in $CoCl_2$ around 40 GPa. The sample becomes black at 42 GPa due to the redshift of the charge transfer band, and also to the increase of the absorbance background, which is likely associated with an increase of sample reflectivity rather than the sample absorption. According to previous discussion, the present data suggest that metallization is favoured in the vicinity of the spin transition. Given that TM halides and oxides conductivity is basically related to electron itinerancy in the σ-e_g-electron band, we suggest that pressure-induced metallization in $CoCl_2$ is probably driven by the spin transition, involving a change in the e_g-type electron-band filling from e_g^2 (HS $^4T_1$ ground state) to e_g^1 (LS ^2E ground state). The band-gap closure accompanying the increase of background absorption is also evidence by the charge-transfer redshift above 33 GPa, thus suggesting $CoCl_2$ is mainly a charge-transfer-type compound. The absence of crystal-field splitting related to the Jahn Teller effect in the OA spectra must be considered as an evidence of the suppression of the Jahn-Teller distortion rather than a confirmation of the HS state at 40 GPa. Although the present results suggest that some spectral features appearing at high pressure are consistent with a LS state, further experiments devoted to confirm the proposed interpretation are needed.

Acknowledgments

This work was financed by the Spanish MEyC (Project Ref. MAT2005-00099) and the Research Intensification program (I3) of the Autonomous Government and the University of Cantabria.

References

[1] J.C. Zahner and H.G. Drickamer, J. Chem. Phys. **35**, 1483 (1961).
[2] E.Y. Tonkov, *High Pressure Phase Transformations: A Handbook* (Gordon & Breach, Amsterdam, 1992).
[3] I. Hernandez and F. Rodriguez, Phys. Rev. B **67**, 012101 (2003).
[4] I. Hernandez, F. Rodriguez, and H. D. Hochheimer, Phys. Rev. Lett. **99**, 027403 (2007).
[5] F. Aguado, F. Rodríguez, and P. Núñez, Phys. Rev. B (to be published).
[6] R. W. G. Wyckoff, *Crystal Structures* (Krieger Publishing Company, Malabar, 1982).
[7] I. Pollini, G. Spinolo, and G. Benedek, Phys. Rev. B **22**, 6369 (1980).
[8] J. S. Griffith, *The theory of Transition-Metal Ions* (Cambridge University Press, 1980).
[9] F. Aguado, F. Rodriguez, R. Valiente, M. Hanfland, and J. P. Itiè, phys. stat. Sol. (b) **244**, 156 (2007)
[10] R. Valiente and F. Rodríguez, Phys. Rev. B **60**, 9423 (1999).
[11] F. Rodriguez and F. Aguado, J. Chem. Phys., **118**, 10867 (2003).
[12] F. Aguado, F. Rodríguez, and P. Núñez, Phys. Rev. B **67**, 205101 (2003).

Journal of Physics: Conference Series **121** (2008) 042007 doi:10.1088/1742-6596/121/4/042007

Opto-structural study of A_2BMF_6: Cr^{3+} elpasolite crystals. Variation of the $(CrF_6)^{3-}$-complex optical properties with Cr-F bond distance, hydrostatic pressure and excited-state crossover

I Hernández[1], F Rodríguez[1] and A Tressaud[2]

[1]DCITIMAC, Facultad Ciencias, Universidad de Cantabria, 39005 Santander, Spain

[2]ICMCB-CNRS 87, Ave. Dr. A. Schweitzer, 33608 Pessac Cedex, France

E-mail: fernando.rodriguez@unican.es

Abstract. This work investigates the optical properties of the Cr^{3+}-doped and Cr^{3+}-pure elpasolites A_2BMF_6 along compound series and as a function of pressure. We are interested in establishing structural correlations between the electronic structure and the actual Cr-F distance, R_{Cr-F}, from which we have been able to reveal the local structure around the Cr^{3+} impurity. As salient features, we show, firstly, a weak dependence of the first excitation band energy, E_1, usually identified as *10Dq*, with R_{Cr-F} as $E_1 = K\,R_{Cr-F}^{-3.3}$, and, secondly, an increase of the Stokes shift upon R_{Cr-F} reduction or with increasing pressure. We associate this unusual behavior with the existence of Fano resonance in the first excitation band of Cr^{3+}. High-pressure experiments performed on Rb_2KCrF_6 indicate that the excited-state spin crossover, $^2E_g(G) \leftrightarrow {}^4T_{2g}(F)$, takes place around 7 GPa. Short lived broad-band luminescence changes into long-lived ruby-like narrow lines The reported results point out the suitability of the selected A_2BMF_6: Cr^{3+} elpasolites to establish structural correlations between PL and R_{Cr-F}.

1. Introduction

Elpasolites have received particular attention due to their simple cubic structure and the isolated character of $(CrF_6)^{3-}$ complexes, which do not share any common F ligand in the elpasolite structure [1] Fluorides are usually employed as ideal host lattices for accommodating di- and trivalent cations in perfect octahedral sites (M) due to the large variety of cubic crystals with the perovskite (AMF_3) and elpasolite (A_2BMF_6) structures, respectively. But such correlation is hard to establish in impurity systems, due to difficulty in determining precisely the actual Cr-F distance, R_{Cr-F}. This feature makes pure A_2BCrF_6 attractive systems to establish correlations between the optical properties and the structure, what is crucial for a microscopic understanding of the PL properties.

2. Experimental

Single crystal elpasolites of K_2NaGaF_6, Rb_2KGaF_6 and Rb_2KInF_6 doped with 1 mol% of Cr as well as single crystals of Rb_2KCrF_6 and Tl_2KCrF_6 were grown [1]. The elpasolite structure ($Fm\bar{3}m$) of the crystals at ambient conditions was checked by X-ray diffraction (XRD) and Raman. Spectroscopic measurements were done as described elsewhere [2].

© 2008 IOP Publishing Ltd

3. Results and discussion

The excitation and corresponding PL spectra at room temperature for the investigated A_2BMF_6: Cr^{3+} crystals are shown in Figure 1.

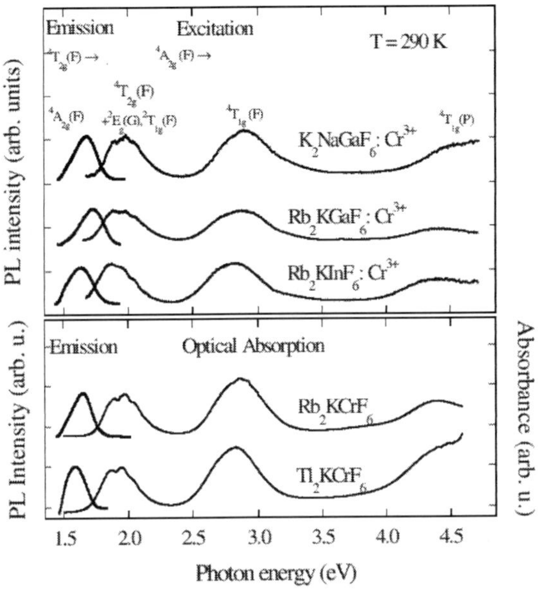

Figure 1. Emission, and excitation/absorption of the A_2BMF_6: Cr^{3+} crystals. The bands are assigned according to the Tanabe-Sugano diagram for a d^3 ion (Figure 2). Note the presence of components in the first excitation band (see text).

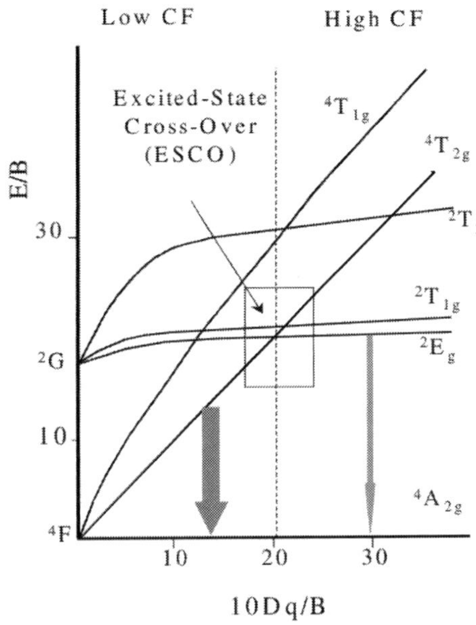

Figure 2. Tanabe-Sugano diagram for a d^3 ion. Notation corresponds to O_h symmetry. The arrows represent luminescence in weak and strong crystal fields, respectively.

The bands are assigned to electronic crystal-field (CF) transitions within O_h-symmetry CrF_6^{3-}, according to the Tanabe-Sugano diagram for d^3 ions (Figure 2). It is worthnoting the triplet structure exhibited by the first band in the OA spectrum, which is associated with Fano resonance among the $^4T_{2g}(F)$, $^2E_g(G)$ and $^2T_{1g}(G)$ states. The CF parameter, *10Dq*, obtained from the OA spectrum is close to the excited-state crossover (ESCO) point, $(10Dq/B)_{ESCO} = 22$, and thus these states are significantly coupled by the spin-orbit interaction.

Figure 3 plots the variation of *10Dq* obtained by fitting the energies of the absorption bands (E_1 ~ $^4T_{2g}(F)$, $^2E_g(G)$ and $^2T_{1g}(G)$, $E_2 = {}^4T_{1g}(F),\ldots$) versus $R_{Cr\text{-}F}$ for the Cr^{3+}-pure elpasolites. $R_{Cr\text{-}F}$ for the Cr-pure elpasolites has been obtained from the a_{1g} vibrational breathing mode of the $(CrF_6)^{3-}$ complex frequency, measured by Raman spectroscopy [2]. The results are compared with $R_{Cr\text{-}F}$ dependences of *10Dq* given by

$$10Dq \propto R_{Cr-F}^{\;-n} \qquad (1)$$

for $n = 5$ and $n = 3.3$. The exponent $n = 3.3$ provides a more realistic account for the observed variation than $n = 5$. Besides accidental coincidence with the point-charge CF model, n exponents around 5 have been obtained in different transition metal oxide complexes from high-pressure experiments [3]. We associate such different behavior ($n = 3.3$ instead of $n = 5$) as due to the Fano resonance of the first transition energy E_1 around ESCO. We must take into acount that, although the CF parameter *10Dq* can be derived from the first excitation energy, $E_1 \approx 10Dq$, its variation with $R_{Cr\text{-}F}$

is not the same as *10Dq*: $\frac{\partial E_I}{\partial R_{Cr-F}} \neq \frac{\partial 10Dq}{\partial R_{Cr-F}}$. The reason of this inequality lies on the *10Dq*-depencence of the centroid E_1. According to the Tanabe-Sugano diagram (Figure 2), it is expected to vary as $E_1 = 10Dq$ for CF values far from the ESCO. On the other hand, near the ESCO, E_1 behaves somewhat intermediate between *10Dq* and the constant value *22B* corresponding to the $^2E_g(G)$ energy. Depending on the orbital mixing induced by the spin-orbit coupling, $\frac{\partial E_I}{\partial 10Dq}$ has values between 1 and 0, and thus $\frac{\partial E_I}{\partial R_{Cr-F}}$ exhibits an intermediate behavior between $\frac{\partial 10Dq}{\partial R_{Cr-F}}$ and 0 [2, 4].

The structural effects on the optical properties in the series (chemical pressure) are evidenced through the slight band shifts observed on passing from K_2NaAlF_6: Cr^{3+} to Rb_2KInF_6: Cr^{3+}; *10Dq*, varies only 0.09 eV for $\Delta R_{M-F} = 0.22$ Å. Such a distance variation would induce a larger shift if the variation of the host M-F distance were the same as the impurity Cr-F distance. This is well illustrated in the Cr-pure elpasolite series, whose *10Dq* varies from 2.00 eV in Rb_2KCrF_6 ($R_{Cr-F} = 1.88$ Å) to 1.91 eV in Tl_2KCrF_6 ($R_{Cr-F} = 1.94$ Å). This difference can be concealed with variations as a function of R_{M-F} if we consider lattice relaxation effects around Cr^{3+} in impurity systems. These results show that the impurity Cr-F distance and the host M-F distance are different, $R_{Cr-F} \neq R_{M-F}$.

The optical excitation data of Figure 1 provide esential information on the real Cr-F distance for impurity systems if we compare them with pure compounds. We can estimate R_{Cr-F} from *10Dq* through Eq. 1. The so-obtained R_{Cr-F} values for the impurity systems are very similar irrespective of the choice $n = 3.3$ or $n = 5$ (Figure 4). This is related to the fact that R_{Cr-F} along the series slightly deviates from the equilibrium Cr-F distance. As Figure 4 shows, the actual R_{Cr-F} derived from Eq.(1) deviates from R_{M-F} in the A_2BMF_6: Cr^{3+} series as: $(R_{Cr-F} - R_0) = f\,(R_{M-F} - R_0)$. We obtain values of $R_0 =$ 1.908 and 1.907 Å, and $f = 0.06$ and 0.09, for $n = 3.3$ and 5, respectively. In either case, we conclude that there is an important lattice relaxation around Cr^{3+}. Since R_{Cr-F} deviates from the equilibrium Cr-F distance, $R_0 = 1.908$ Å, only the 6% the difference R_{M-F} - R_{Cr-F}, confirming the stiffness of the Cr-F bond in comparison to the more ionic M-F bond.

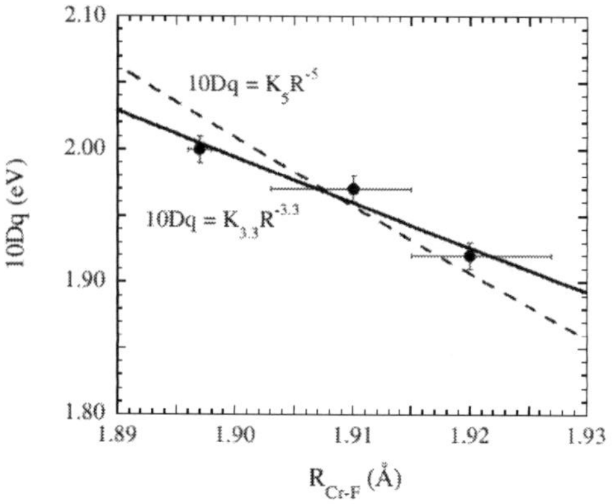

Figure 3. *10Dq* derived from the excitation spectra as a function of R_{Cr-F} for the Cr-pure compounds of the series. The lines correspond to fits to the given expressions (see text).

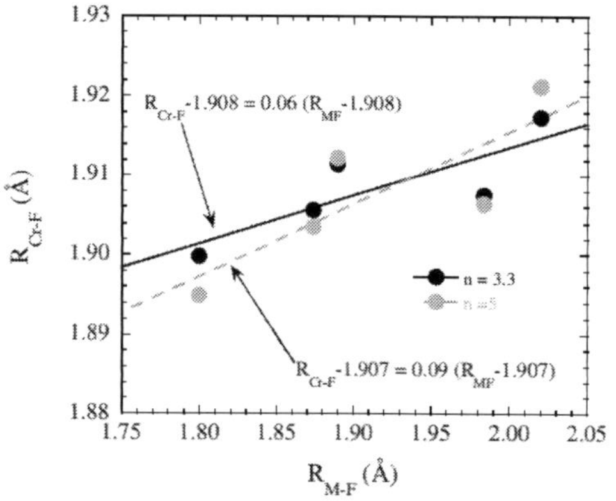

Figure 4. Cr-F versus M-F distances in impurity systems. R_{M-F} has been derived from from XRD, and R_{Cr-F} from *10Dq* using two different exponents in Eq. 1. Linear fits are given as a function of f and R_0 (see text).

For all investigated compounds, the PL spectra consist of a single broad emission band peaking around 1.6 eV, which is related to the $^4T_{2g}(F) \rightarrow {}^4A_{2g}(F)$ transition. *10Dq* lies close but below the ESCO ($^4T_{2g} \leftrightarrow {}^2E_g$) and thus, broad-band PL occurs instead of narrow R-line emission. Nevertheless it must be noted that the actual CF governing the PL crossover is given by the excited-state equilibrium Cr-F distance, R^{exc}_{Cr-F}. As Figure 2 shows, $\frac{\partial E_1}{\partial 10Dq} > 0$, then the linear electron-lattice coupling associated with this transition is negative, $\frac{\partial E_1}{\partial R_{Cr-F}} < 0$, and thus $R^{exc}_{Cr-F} > R^{gr}_{Cr-F}$ for Cr^{3+}. This means that broad PL band can be observed even if the ground-state CF verifies $10Dq < 10Dq_{ESCO}$.

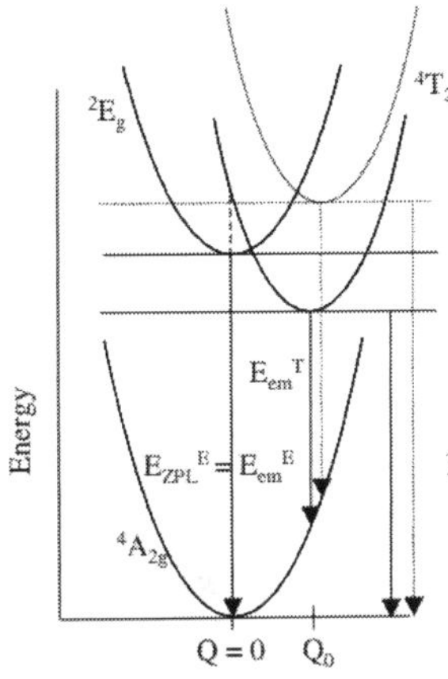

Figure 5. Configuration diagram of the ^2E_g, $^4T_{2g}$ (excited) and $^4A_{2g}$ (ground) states. The excited state equilibrium coordinate Q_0 (with respect to the ground state coordinate) is associated with a bigger R_{Cr-F} due to negative electron-phonon coupling coefficient (see figure 2 and text). The dotted lines represent the situation after the ESCO. $E_{em}{}^T$ is the $^4T_{2g}$ emission energy of state. Note that the ESCO occurs when the zero phonon lines ($E_{ZPL}{}^E$ and $E_{ZPL}{}^T$, respectively concide. The oscillator strenght of the spin allowed $^4T_{2g} \rightarrow {}^4A_{eg}$ transition is about an order of magnitude bigger. As a consequence, broad-band emission will be observed after ESCO.

A puzzling result concerns the unexpected behavior of the PL Stokes shift, $E_S = E_I - E_{PL}$, with R_{M-F}. E_s increases upon R_{Cr-F} (or R_{Cr-F}) reduction shows an opposite variation of E_S with R observed for Cr^{3+} in chloroelpasolites and Mn^{2+} in chloro- and fluoroperovskites. This different behavior is explained on the basis of electron-phonon couplings within the complex (MF_6) taking into account contributions from the stronger coupled modes: a_{1g} and the Jahn-Teller e_g mode [4].

High-pressure experiments carried out on Rb_2KCrF_6 provide additional knowledge on the interplay of the structural and vibrational properties and electronic configuration of these systems.

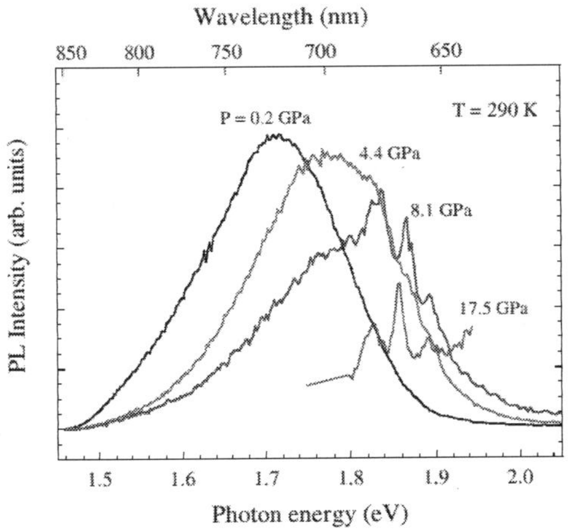

Figure 6. Variation of photoluminescence with pressure. Note the presence of sharp features as a consequence of the pressure-induced $^2E_g \leftrightarrow {}^4T_{2g}$ excited-state crossover (ESCO) at 7 GPa. The three sharp lines correspond to the $^2E_g \rightarrow {}^4A_{2g}$ transition, and Stokes and antiStokes vibronic replicae.

The PL band (figure 6) experiences a blueshift of 0.09 eV with increasing pressure in the 0 – 7 GPa range, in agreement with the increase of *10Dq* upon R_{Cr-F} reduction. Above this pressure, sharp features appear corresponding to the mainly $^2E_g(G) \rightarrow {}^4A_{2g}(F)$ zero-phonon line (ZPL), and the Stokes and anti-Stokes phonon replicas associated with the t_{2u} vibrational mode [5].

PL lifetime increases with pressure from $\tau = 45$ μs (ambient pressure) to $\tau = 1.0$ ms (16.5 GPa). This variation confirms the occurrence of the ESCO phenomenon. According to the normal-coordinate configuration diagram of Figure 5, it actually takes place when the ZPL energy of $^4T_{2g}$ and ^2E_g coincides: $E_{ZPL}(^4T_{2g}) = E_{ZPL}(^2E_g)$. From the estimated ZPL for $^4T_{2g}$ and the observed band shifts [2], the variation of the ZPL-energy difference with pressure is given by $\Delta E(P) = E_{ZPL}(^4T_{2g}) - E_{ZPL}(^2E_g) = 0.013\,P - 0.065$, and thus the ESCO takes place at 7 GPa. It is worth noting the presence of broadband emission above the ESCO pressure at 7 GPa, reflecting the thermal population of the $^4T_{2g}(F)$ state. The oscillator strength for the spin-allowed $^4A_{2g}(F) \rightarrow {}^4T_{2g}(F)$ transition is about an order of magnitude greater than the spin-forbidden $^2E_g(G) \rightarrow {}^4A_{2g}(F)$ one, and therefore the broad band emission still dominates the PL spectrum at the ESCO pressure and even at higher pressures.

As it was observed along the fluoroelpasolite series upon decreasing R_{Cr-F}, the emission blueshift induced by pressure is smaller than the excitation blueshift ($\delta E_{PL} > \delta E_I$), foreseing an increase of the Stokes shift. Therefore the pressure results confirm that the PL properties and main spectral features of Cr^{3+}-doped elpasolites can be explained on the basis of a $(CrF_6)^{3-}$ complex, the variation of which is governed by the actual Cr-F bond distance imposed by the host crystal.

4. Conclusions

Throughout this work we show 1) the local structure around the impurity in Cr^{3+}-doped elpasolites can be derived from the corresponding optical spectra after structural correlation established in Cr^{3+}-pure elpasolite series. 2) The *R*-dependences of *10Dq* and the Stokes shift exhibit anomalous deviations with respect to CF model expectations, due to the Fano resonance of the first absorption band of Cr^{3+}. 3) The PL properties along the elpasolite series or as a function of pressure can be understood on the basis of the $(CrF_6)^{3-}$ complex, whose actual Cr-F distance is mainly governed by either the host crystal or the external pressure. There is a strong lattice relaxation around Cr^{3+} in Cr-doped elpasolites yielding either an inward or an outward relaxation of the fluorine octahedron depending on whether the host M-F distance is longer or shorter than $R_{Cr-F} = 1.903$ Å. 4) The $^2E_g(G) \rightarrow {}^4T_{2g}(F)$ excited-state crossover takes place in the Cr-pure Rb_2KCrF_6 around 7 GPa. However this effect is not evidenced by the presence of sharp features at 7 GPa but about 12 GPa at room temperature. This pressure provides enough energy separation between $^4T_{2g}(F)$ and $^2E_g(G)$ to increase the population of the long-lived $^2E_g(G)$ state with respect to short-lived $^4T_{2g}(F)$ state at room temperature.

The present results point out that the selected A_2BMF_6: Cr^{3+} crystals are suitable systems to establish structural correlations between optical properties and R_{Cr-F}.

Acknowledgements

The authors thank the Spanish Ministerio de Educación y Ciencia (Project No MAT2005-00099) and the Research Intensification program (I3) of the University of Cantabria.

References

[1] Guengard H 1994 *PhD thesis* (University of Bordeaux 1)
[2] Hernández I 2006 *PhD thesis* (University of Cantabria)
[3] Drickarmer H G and Frank C W 1973 *Electronic transitions and the high pressure chemistry and physics of solids* (London: Chapman & Hall)
[4] Hernández I, Rodríguez F and Tressaud A 2007 submitted
[5] Dolan J F, Kappers L A and Bartram R H 1986 *Phys. Rev. B* **33** 7339.

Local ordering and metastable phenomena in liquids under extreme conditions

A. Di Cicco,* E. Principi, M. Minicucci

CNISM, CNR-INFM SOFT and Dipartimento di Fisica Università degli Studi di Camerino via Madonna delle Carceri 62032, Camerino (MC) Italy

* IMPMC-CNRS, Universitè P.et M. Curie, 140 rue de Lourmel, 75015 Paris, France (present address)

Abstract.

Recent results obtained using the x-ray absorption spectroscopy (XAS) on liquids under high-pressure are presented. Advanced instrumentation combining XAS and x-ray diffraction and RMC XAS data-analysis techniques allowed to study and discuss the presence of metastable states in Bi and Sn at high pressures. Differences in the average local structure of liquid metals are briefly discussed.

1. Introduction

Recent advances in instrumentation and data-analysis allowed us to perform x-ray absorption spectroscopy (XAS) studies of liquids, undercooled liquids and glasses under high-temperature and high-pressure conditions[1, 2, 3, 4, 5, 6, 7]. XAS has been used as a probe for the local structure, x-ray diffraction (XRD) for measuring long-range ordering, checking the sample status and determining relevant environmental parameters (pressure), while single-energy x-ray absorption temperature or pressure scans are used to detect and study phase transitions. These techniques can be combined and used at third generation synchrotron sources on samples under extreme thermodynamic conditions, due to the high flux and reduced size of the x-ray spot, and are able to give information about aggregation state and local structure of unprecedented accuracy. Recent applications regarded undercooled liquids and glasses under high temperature and/or high-pressure conditions, like liquid Ga[5] and Bi[6], metal alloys[4], amorphous Ge [6]. Advances in experimental techniques have been accompanied by improvements in data-analysis methods. Reliable reconstruction of the structure of disordered systems by combining XAS and XRD can be now obtained using Reverse Monte Carlo (RMC) simulations implemented into modern XAS data-analysis algorithms (RMC-GnXAS).[1, 8] In this communication we shall review some important characteristics of the techniques showing their potential in measuring and understanding phase transition and metastability under extreme conditions, taking as an example some new results on liquid Sn[7] and Bi[6] under pressure.

2. Experimental techniques and data-analysis methods

XAS measurements of systems under extreme and metastable conditions are able to give accurate information about the average local structure of disordered and ordered matter. XAS measurements in transmission mode are in principle very simple and suitable techniques for measuring samples under extreme and metastable conditions have been developed in recent times (see [9, 2] and ref. therein) combining XAS, XRD and single-energy x-ray absorption temperature scans (T-scan) into automated procedures. In Fig. 1 we see an example of the

© 2008 IOP Publishing Ltd

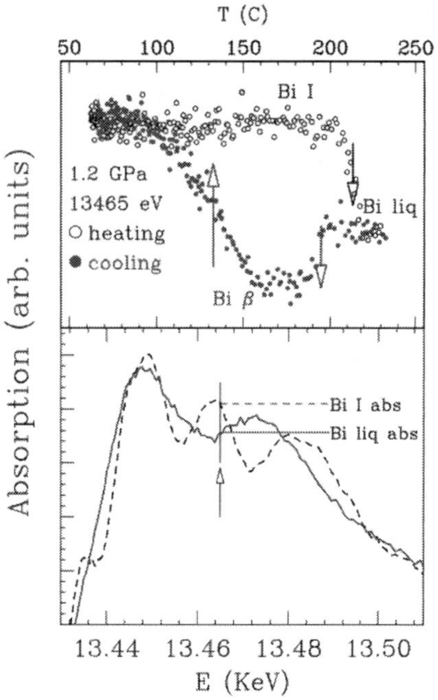

Figure 1. Upper panel: T-scans of Bi at fixed energy and pressure. The arrows indicate the temperature of the occurring phase transitions. Upon lowering the temperature, undercooling and nucleation of the liquid to a new phase (Bi β) prior to crystallization to th Bi-I stable phase are clearly visible. Lower panel: Detail of the Bi L_3 near-edge structures where the differences in the shape of the x-ray absorption spectra are clearly visible. The absorbances of Bi-I and liquid Bi at 13465 eV are indicated by horizontal lines.

sensitivity of x-ray absorption to changes of the local structure, as detected in pure Bi at a pressure of 1.2 GPa. The sample was a dispersion of micrometric Bi droplets into a suitable solid matrix acting also as a pressure marker. The L_3-edge XAS spectra of the high-temperature liquid and of the stable solid phase (Bi-I) are shown in the lower panel of Fig. 1. Their different shape is related to the different average local structure of the two phases. Single-energy T-scans can be obtained by tuning the photon energy where changes in the x-ray absorption are maximized.

The result obtained by tuning the energy at 13465 eV is shown in the upper panel of Fig. 1, indicating the presence of undercooling and of nucleation to a different solid phase upon lowering the temperature. These findings are corroborated also by combined XRD experiments (see [6] for further details).

The XAS structural signal $\chi(k)$ contains reliable quantitative information about the local structure, but appropriate methods for data-analysis must be used for data-analysis especially for disordered systems. In particular, $\chi(k)$ is defined as the modulation of the photoabsorption cross-section μ with respect to the atomic one μ_0 above the excitation edges of core electrons: $\chi(k) = \frac{\mu(k)-\mu_0(k)}{\mu_0(k)}$. Here $k = \sqrt{\frac{2m}{\hbar^2}(E - E_0)}$, where E_0 is the edge energy, is the wave-vector of the outgoing photoelectron. XAS is chemically selective and sensitive only to the *local* structure, due to the limited photo-electron mean free-path ($\approx$ 5-6 Å). Modern interpretation of the XAS theory is based on an expansion in series of irreducible $\gamma^{(n)}$ signals associated with n-body atomic configuration[10] including the photoabsorbing atom, from which follows:

$$\langle\chi(k)\rangle = \int_0^\infty dr 4\pi\rho r^2 g_2(r)\gamma^{(2)}(r,k) +$$
$$+ \int dr_1 dr_2 d\theta 8\pi^2 r_1^2 r_2^2 (\sin\theta)\rho^2 g_3(r_1,r_2,\theta)\gamma^{(3)}(r_1,r_2,\theta,k) + \cdots$$

where the $\chi(k)$ structural signal is given by integrals of the n-body distribution functions g_n. The unique short-range sensitivity and the possibility of measuring three-body correlations, makes XAS a highly valuable technique complementary to diffraction. In standard data-analysis,

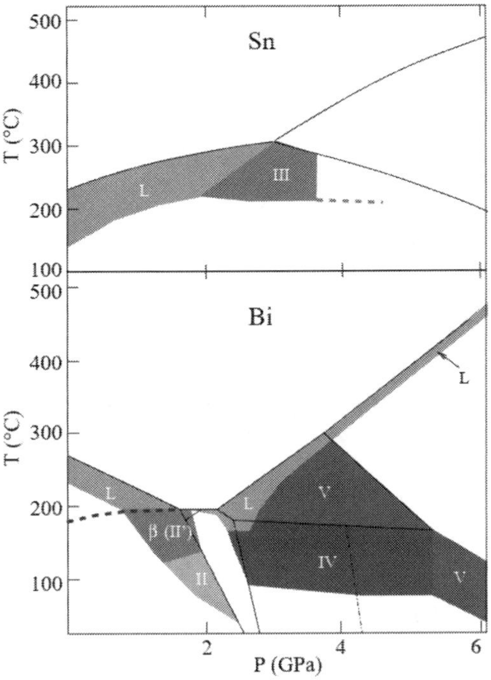

Figure 2. Phase diagrams of Bi (lower) and Sn (upper) obtained from analyzing the x-ray T-scan patters collected in a wide pressure-temperature range. The solid lines (lower) indicate the Bi stable phase diagram adapted from [11]. The dashed line mimics the melting line of metastable solid Bi proposed by [12]. The dotted line is an hypothetical transition line between Bi β and Bi II at low pressures. The boundaries of the liquid (l-Sn) and solid phases (Sn-I, Sn-III) are superimposed on those reported in the Tonkov handbook [13].

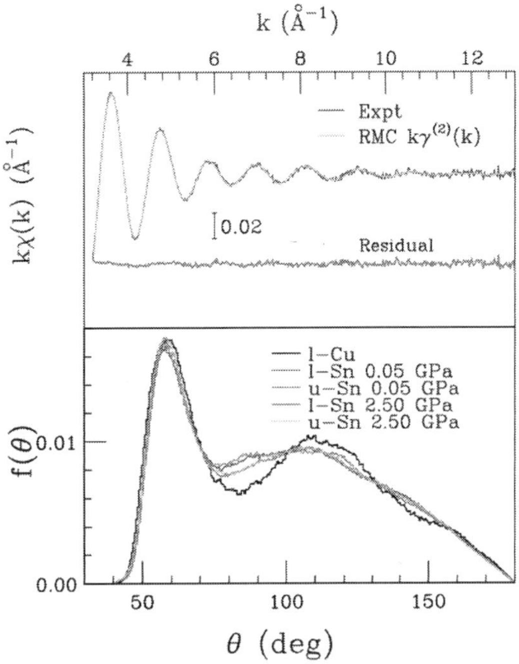

Figure 3. In the upper panel the model signal calculated from the RMC atomic coordinates (green) is compared with the experimental signal (Expt, blue). The agreement can be appreciated looking at the residual curve (red). Normalized angular distributions $f(\theta)$ of liquid (l-Sn) and undercooled (u-Sn) liquid Sn at 0.05 GPa and 2.5 GPa, where the nucleation of the Sn-III phase takes place. The distribution obtained for a close-packing liquid metal like Cu,[1] different from those of liquid Sn at both pressures, is shown for comparison.

the n-body distributions are assumed to be a superposition distinct peaks ("shells") whose structural parameters are refined against the experimental data. More advanced techniques based on simultaneous refinement of short and long-range order (XAS and diffraction) through the Reverse Monte Carlo algorithm (RMC-GnXAS) were recently shown to be succesful in measuring the local bond-angle distribution in liquids and undercooled liquids under extreme conditions.[1, 8, 7]

3. Metastable states and structure at high pressures

Several high-temperature XAS experiments on undercooled liquids have been performed by our group, and recently techniques ad methods described in the preceding section were applied to polymorphic metals under pressure such as Bi and Sn.[6, 7]

In Fig. 2 we report the phase diagrams of Bi and Sn including the appearance of metastable states obtained cooling the liquid. Those states are easily obtained by cooling micrometric droplets of those liquids (our typical XAS samples) at selected pressures and their lifetime is much longer than typical measurement times. Looking at Fig. 2, it is inferred that maximum undercooling temperatures and presence of metastable solid phases are related to the characteristics of the underlying stable solid phases. In particular, we notice that undercooling is not obtained in Bi around 2 GPa, in presence of the stable solid Bi-II phase, while it is observed again at higher pressures. For Sn, undercooling is prevented above 2 GPa by appearance of the Sn-III solid phase. As discussed also in ref. [7], XAS spectra of liquid Sn are similar to those of Sn-III while are remarkably different from those of Sn-I. This suggests that an important role is played by the local structure, leading to high undercooling rates and nucleation preferential selection when the transition implies important changes of the average atomic environment.

XAS data of liquid Sn under pressure were analyzed using RMC-GnXAS, looking for possible modifications of the liquid local structure for increasing pressures. In Fig. 3, upper panel, we show that the agreement between the RMC simulation and the experimental data (0.05 GPa) is excellent. We remark that RMC-GnXAS structural refinement is done including both XAS and $g(r)$ data obtained by XRD, constraining both short-range and medium-range structure simultaneously. The normalized bond-angle distribution shown in Fig.3, lower panel, shows that the average local structure of liquid tin is very different from that of a typical close-packing metal (Cu, shown in the figure). In particular, the Sn liquid is found to be composed of a mixture of tetrahedral and close-packed configurations, with the weight of the latter increasing at higher pressure (as shown by the deepening of the minimum around 90 degrees in the bond-angle distribution).

4. Conclusions

New combined x-ray absorption and diffraction experimental techniques, associated with advanced data-analysis methods, allowed us to obtain reliable unique data and information about metastable states under high-pressure and high-temperature conditions. Our results on Bi and Sn under pressure shows that we are able to study the details of the local structure of the liquid phase and of occurring metastable states as well as possible relationships with phase selection mechanisms.

Acknowledgments

We are grateful to the BM29 ESRF scientists and staff, and in particular Angela Trapananti for their invaluable collaboration.

References

[1] A. Di Cicco, A. Trapananti, S. Faggioni, and A. Filipponi. *PRL*, 91:135505–1–135505–4, 2003.
[2] Adriano Filipponi, Valentina M. Giordano, Simone De Panfilis, Andrea Di Cicco, Emiliano Principi, Angela Trapananti, Michael Borowski, and Jean-Paul Itié. *Rev. Sci. Instrum.*, 74:2654–2663, 2003.
[3] Emiliano Principi, Andrea Di Cicco, Frederic Decremps, Alain Polian, Simone De Panfilis, and Adriano Filipponi. *Phys. Rev. B*, 69(20):201201, 2004.
[4] A. Trapananti and A. Di Cicco. *Phys. Rev. B*, 70:014101, 2004.
[5] R. Poloni, S. De Panfilis, A. Di Cicco, G. Pratesi, E. Principi, A. Trapananti, and A. Filipponi. *Phys. Rev. B*, 71:184111, 2005.
[6] E. Principi, M. Minicucci, A. Di Cicco, A. Trapananti, S. De Panfilis, and R. Poloni. *Phys. Rev. B*, 74:064101, 2006.
[7] Andrea Di Cicco, Angela Trapananti, Emiliano Principi, Simone De Panfilis, and Adriano Filipponi. *Applied Physics Letters*, 89(22):221912, 2006.
[8] A. Di Cicco and A. Trapananti. *J. Phys.: Condens. Matter*, 17:S135–S144, 2005.
[9] A. Filipponi, M. Borowski, D. T. Bowron, S. Ansell, A. Di Cicco, S. De Panfilis, and J.-P. Itié. *Rev. Sci. Instr.*, 71:2422–2432, 2000.
[10] A. Filipponi, A. Di Cicco, and C. R. Natoli. *Phys. Rev. B*, 52:15122–15134, 1995.
[11] J. H. Chen, T. Kikegawa, O. Shimomura, and H. Iwasaki. *J. Synchr. Rad.*, 4:21, 1997.
[12] W. Yoon, J. S. Paik, D. LaCourt, and J. H. Perepezko. *J. Appl. Phys.*, 60:3489, 1986.
[13] E. Yu. Tonkov. *High Pressure Phase Transformations, A Handbook, Vol. 2.* Gordon and Breach Science Publishers, 1992.

Local structure of liquid and undercooled liquid Cu probed by x-ray absorption spectroscopy.

Federica Coppari[a], Andrea Di Cicco[a,b], Adriano Filipponi[c], Angela Trapananti[d], Giuliana Aquilanti[d], Simone De Panfilis[d]

[a]CNISM, CNR-INFM SOFT, Dipartimento di Fisica, Università di Camerino, Via Madonna delle Carceri, 62032 Camerino (MC), Italy
[b]IMPMC-CNRS UMR 7590, Université P.et M. Curie, 140 rue de Lourmel, 75015 Paris France
[c]CNISM, Dipartimento di Fisica, Università dell'Aquila, Coppito (AQ), Italy
[d]European Synchrotron Radiation Facility - BP 220, F-38043 Grenoble Cedex, France

E-mail: `andrea.dicicco@unicam.it`

Abstract. In this work we report about a recent XAS (x-ray absorption spectroscopy) experiment aimed to study the local structure of liquid and undercooled copper, a typical close-packed liquid metal, under high-pressure conditions. In our study we have obtained new experimental data of liquid Cu under high pressure and high-temperature conditions using synchrotron radiation at the ESRF and a large-volume Paris-Edinburgh cell. We have studied the extension of the undercooling region as well as possible modifications of the local structure as a function of pressure, using XAS data-analysis. Liquid copper has been undercooled up to 300°C below the melting point at high pressures up to 3.3 GPa.

1. Introduction

XAS (X-Ray Absorption Spectroscopy) has been proved to be a powerful technique for investigation of the local structure in condensed matter and in particular in liquid systems (see [1] and ref. therein). In this work, XAS is used to study liquid and undercooled liquid copper under high-pressure conditions.

Copper is known to show metastable liquid phase below its crystallization temperature. Even if undercooling of liquid metals is a well known phenomenon [2, 3, 4, 5], the microscopic reasons leading to the persistence of the liquid phase into the region of stability of the solid are not yet well understood. More than 50 years ago, Frank hypothesized that the liquid phase of closed-packed metals is characterized by icosahedral arrangements of atoms and, since this system would have a fivefold symmetry, the transition to the crystalline phase is hindered by an energy barrier to the rearrangement of atoms into *fcc* (face-centered cubic) structure.

Recent studies on stable and undercooled liquid copper (l-Cu and u-Cu) at room pressure [1] have shown that there is a broad distribution of local configurations of different symmetry and that distorted icosahedral configurations including fragments of icosahedra represent about 10% of the total.

Several studies have been carried out on liquid and undercooled liquid copper, including XAS analysis and simulations,[1, 6, 7] but most of them were made at room pressure conditions. The main aim of this experiment is to study the effect of a high external pressure on the undercooling and nucleation properties of liquid copper and on the local structure of the disordered l-Cu and

© 2008 IOP Publishing Ltd

u-Cu liquid phases.

2. Experiment

The experiment was performed at the ESRF (European Synchrotron Radiation Facility, Grenoble), using the dedicated high-pressure setup available at beamline BM29[8, 9]. A double-crystal monochromator equipped with Si(111) crystals was coupled with a pair of Silicon mirrors recently installed at BM29. The mirrors allowed us to obtain a vertically focused beam of 100 μm (FWHM) and a negligible harmonic content, both crucial for achieving measurements of reasonable quality on small samples embedded in highly x-ray absorbing environment.
The sample was prepared as a mixture of BN and Cu powders (weight ratio 45 : 1), hold in a 7 mm biconical gasket suitable for high-pressure. Metallic high-purity Cu fine powders were obtained by reducing in-situ CuO powders (purity 99.995%) by high-temperature treatment.
The pressure device was a large volume Paris-Edinburgh (PE) press [10], used in the range $0-4$ GPa. Above 4 GPa the x-ray absorption of the sample environment was too high for obtaining XAS spectra of good quality. The temperature was controlled by a DELTA $15-200$ power supply, and the range covered in this experiment was $25-1350°$C. XRD patterns of the sample were also collected using a MAR345 image-plate detector mounted in an offset position in order to determine the thermodynamic conditions of the XAS spectra using the equations of state (EOS) of solid copper and BN [11, 12]. Particular care was taken about the calibration of temperature and pressure, using MAR image analysis and SEXAD (Single Energy X-ray Absorption Detection or temperature scan)[13] data against the known melting curve of copper under pressure and both BN and Cu equations of state. In this way we estimate that our typical accuracy is around 10 degrees near the melting pointfor the temperatures and 0.2 GPa for the pressures.
Collection of XAS data of liquid Cu is actually a challenging experiment due to the low x-ray transmittance of the sample under high pressure, and present results are obtained at x-ray energies which currently represents a lower limit for experiment with the large-volume set-up at BM29. Despite the technical difficulties, we have been able to collect XAS data, as well as SEXAD and XRD patterns on liquid and undercooled liquid Cu at high pressure.

3. Results

Undercooling of liquid copper under pressure has been studied using SEXAD tuning the energy at a point (8.995 keV) near the Cu K-edge where the difference between the absorption coefficient of solid and liquid Cu is maximum. The measurements of the x-ray absorption as a function of temperature highlight the occurrence of phase transitions, usually marked by a sudden change in the absorption coefficient. In Fig. 1 we clearly see the melting of copper at different pressures through a discontinuity of the x-ray absorption. On the other hand the onset of the crystallization depends on pressure: at lower pressures (Fig.1(a) and (b)) we observe a sudden crystallization while for higher pressures (Fig. 1 (c), (d) and (e)) the nucleation proceeds gradually. A possible explanation of this phenomenon is that pressure can catalyze nucleation favoring the contacts between Cu and BN particles acting as impurities.
As shown in Fig. 1, K-edge XAS spectra have been collected for liquid Cu and in the undercooled phase at about 300 degrees below the melting point. Present XAS measurements performed on molten Cu show that the structural $\chi(k)$ signal is sensitive to pressure (Fig. 2). For increasing pressure the oscillations change their frequency and increase their intensity at higher energies (k). XAS data-analysis has been performed using GNXAS [14]. The radial distribution function $g(r)$ (Fig. 4) was modelled as a sum of a nearest-neighbor peak and a long-range tail. Figure 4 shows its first peak at different pressures reconstructed from EXAFS data analysis. The values reported in Tab. 1 indicate that the external pressure causes a reduction of the mean interatomic

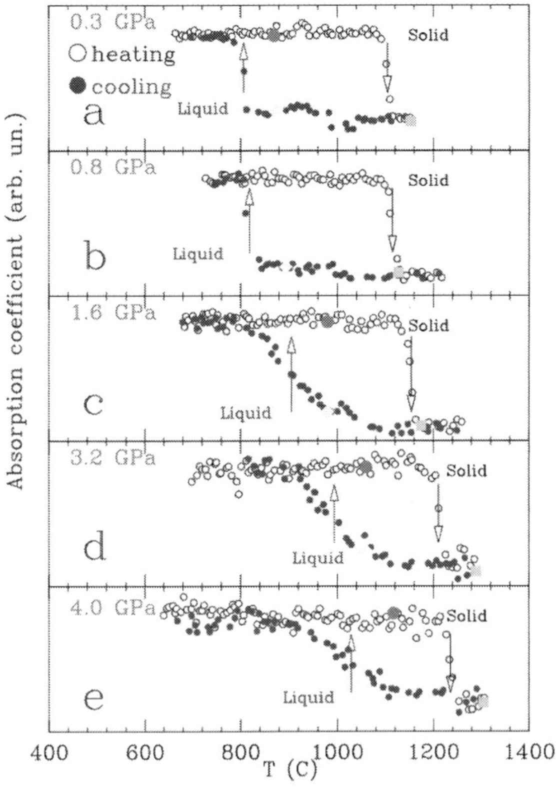

Figure 1. SEXAD or *Tscan* at different pressures: melting of the sample is marked by a sharp decrease in the absorption coefficient and crystallization by a corresponding increase. Symbols indicate the temperature at which XAS measurements have been made: red circles represent the spectra of the solid phase (s-Cu), green squares those in the liquid stable phase (l-Cu) and yellow diamonds those in the undercooled liquid phase (u-Cu).

Table 1. Structural parameters of l-Cu and u-Cu as obtained from the XAS analysis. Temperature and pressure uncertainties are about 10 degrees and 0.2 GPa respectively.

$P(GPa)$	$T_{l-Cu}(^{\circ}C)$	$R_{l-Cu}(A)$	$\sigma^2_{l-Cu}(A^2)$	$T_{u-Cu}(^{\circ}C)$	$R_{u-Cu}(A)$	$\sigma^2_{u-Cu}(A^2)$
0.3	1153	2.568	0.058	869	2.561	0.052
0.7	1126	2.555	0.055	894	2.547	0.048
1.4	1174	2.551	0.052	980	2.538	0.047
3.3	1290	2.534	0.048	1060	2.533	0.046

distance R as well as of the variance (σ^2). This behaviour is reflected on the increased intensity of the $\chi(k)$ signal.
The same kind of analysis has been performed on the undercooled Cu after having removed the residual contribution of solid copper at same pressures and temperatures. Similar trends of the structural parameters have been found (Fig. 3 and Tab. 1). The comparison of l-Cu and u-Cu at the same pressure shows that interatomic distances and the associated variance in the metastable phase are shorter than those of the stable one, as expected lowering the temperature.

4. Conclusions

We have presented a successful XAS experiment on liquid and undercooled liquid copper under high pressure conditions. We have shown that it is possible to maintain the sample in metastable conditions even at high pressure, although partial crystallization is extended in a broad temperature range. XAS data analysis shows that the average first-neighbor distance

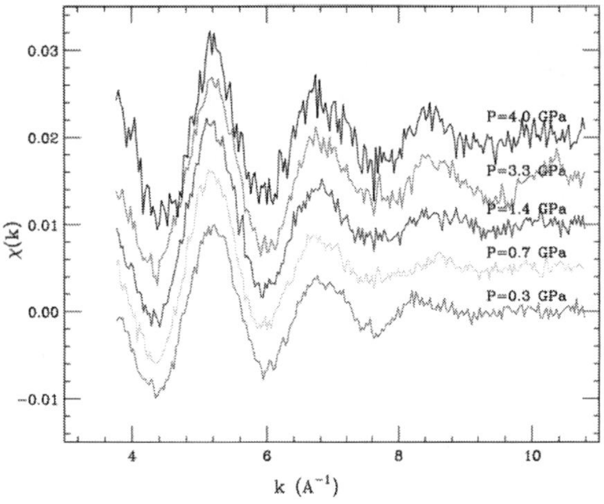

Figure 2. $\chi(k)$ signals of l-Cu at different pressures.

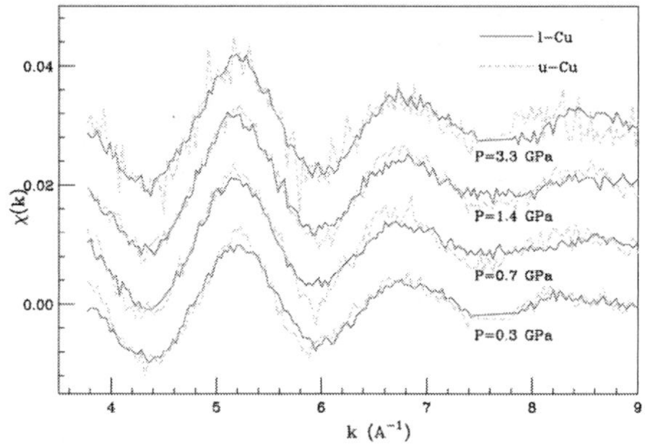

Figure 3. Comparison between $\chi(k)$ of stable and undercooled liquid copper.

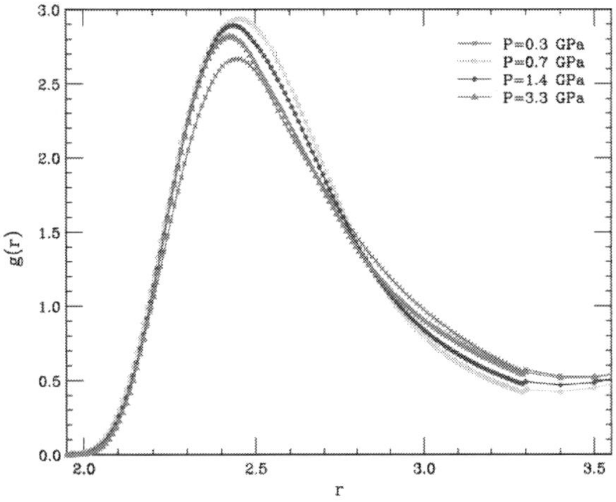

Figure 4. Radial distribution function $g(r)$ reconstructed from $\chi(k)$ signals analysis.

and variance are shortened upon increasing pressure and in the undercooling regime.

5. References

[1] A. Di Cicco, A. Trapananti, S. Faggioni, and A. Filipponi, Phys. Rev. Lett. **91**, 135505 (2003).
[2] D. G. Fahrenheit, Philos. Trans. Royal Soc. London **33**, 78 (1724).
[3] D. Turnbull and R. E. Cech, J. Appl. Phys. **21**, 804 (1950).
[4] D. Turnbull, J. Chem. Phys. **20**, 411 (1952).
[5] F. C. Frank, Proc. Royal Soc. London A **215**, 43 (1952).
[6] P. Ganesh and M. Widom, Phys. Rev. B **74**, 134205 (2006).
[7] M. Celino, V. Rosato, C. Massobrio, A. Di Cicco, and A. Trapananti, PRB (2007).
[8] A. Filipponi, M. Borowski, D. T. Bowron, S. Ansell, A. Di Cicco, S. De Panfilis, and J.-P. Itié, Rev. Sci. Instr. **71**, 2422 (2000).
[9] A. Filipponi, V. M. Giordano, S. De Panfilis, A. Di Cicco, E. Principi, A. Trapananti, M. Borowski, and J.-P. Itié, Rev. Sci. Instrum. **74**, 2654 (2003).

[10] J. M. Besson, R. J. Nelmes, G. Hamel, J. S. Loveday, G. Weill, and S. Hull, Physica B **180 & 181**, 907 (1992).
[11] P. Vinet, J. R. Smith, J. Ferrante, and J. H. Rose, Phys. Rev. B **35**, 1945 (1987).
[12] Y. Le Godec, D. Martinez-Garcia, M. Mezouar, G. Syfosse, J. P. Itié, and J.-M.Besson, in *Science and Technology of High Pressure, Proceedings of AIRAPT-17*, edited by M. H. Manghnani, W. J. Nellis, and M. F. Nicol, pages 925–928, Universities Press, Hyderabad, India, 2000.
[13] A. Filipponi, M. Borowski, P. W. Loeffen, S. De Panfilis, A. Di Cicco, F. Sperandini, M. Minicucci, and M. Giorgetti, J. Phys.: Condens. Matter **10**, 235 (1998).
[14] A. Filipponi, A. Di Cicco, and C. R. Natoli, Phys. Rev. B **52**, 15122 (1995).

X-ray diffraction and electroresistance measurements under high pressure and temperature using a large-volume cell

E. Principi, M. Minicucci, A. Di Cicco*

CNISM, CNR-INFM SOFT and Dipartimento di Fisica Università degli Studi di Camerino via Madonna delle Carceri 62032, Camerino (MC) Italy

* IMPMC-CNRS, Universitè P.et M. Curie, 140 rue de Lourmel, 75015 Paris, France (present address)

Abstract. In this communication we report about original experimental techniques for in-house x-ray diffraction and electrical resistance measurements under high-temperature and high pressure conditions, using large volume cells in the opposite anvil recessed geometry. The high-pressure devices we are currently using are a compact Paris-Edinburgh (PE) V5 150 Tons press and a 50 Tons standard hydraulic press, coupled usually with WC anvils and 10 mm or 7 mm boron-epoxy or pyrophyllite biconical gaskets for x-ray diffraction and resistance measurements respectively. Limiting pressures, using such a non-toroidal sample assembly and WC anvils, are about 10 GPa on samples of large sizes (10-20 mm^3). Samples can be heated using an hollow graphite cylinder as a crucible reaching temperatures as high as 2300 K, while the temperature can be measured up to 1300 K by using a K-type thermocouple. The highly automated set-up developed for resistance measurements is described in details. In particular, we present electroresistance measurements of Bi melting under pressure and measurements of the Ge and LiF EOS (equation of state) at high temperature and pressure obtained using x-ray diffraction showing the sensitivity of the techniques. The relevance of these experiments to the exploitation of the potential of equipments available at synchrotron radiation facilities is emphasized.

1. Introduction

Recently, an advanced experimental facility has been developed at our home laboratories with the aim of collecting useful data about matter under extreme condition of temperature and pressure, complementing those obtained at large scale facilities. [1] Particular attention has been devoted to high pressure research. Two large volume presses have been installed: a compact "Paris-Edinburgh" (PE) V5 press (150 Tons) and a standard hydraulic press (50 Tons). Both the presses currently operate in the recessed non-toroidal geometry, using two opposite steel anvils with a WC insert. The former press has been coupled to an advanced energy dispersive x-ray diffraction (EDXD) setup, the latter has been connected to a sophisticated data acquisition apparatus for electrical resistance measurements. Also an original setup including a diamond anvil cell (DAC) and an x-ray tube dedicated to angular dispersive x-ray diffraction is currently under development. Details about these equipments and the general context of this research are given in ref. [1]. In this paper we focus the attention on our large volume equipments, reporting more technical details and first experimental results, finally indicating possible future developments.

© 2008 IOP Publishing Ltd

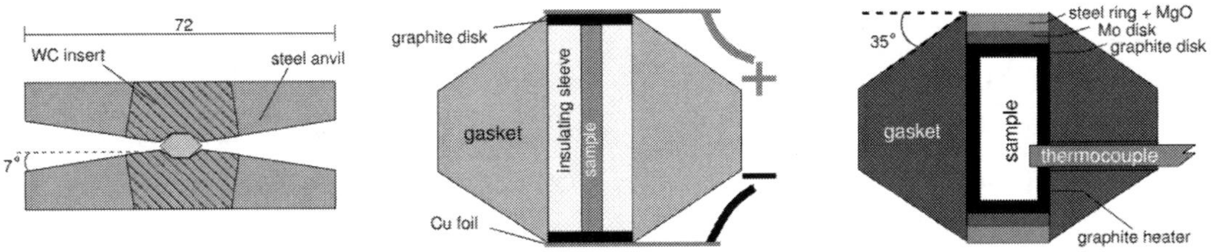

Figure 1. Graphical sketch of the sample environment. Measures are in millimeters. *Left panel:* transverse section of the recessed anvils showing the position of the biconical gasket. *Middle panel:* transverse section of a pyrophyllite gasket assembled for electroresistance measurements in direct heating geometry. *Right panel:* transverse section of a B-epoxy resin gasket used for EDXD measurements.

2. Electroresistance measurements

The experimental setup for electrical resistance measurements consists of a power supply, an automated control unit and a 50 Tons hydraulic press, adapted to match the size of typical PE V5 steel anvils. A PE press can be also used as a pressure device, with the same setup.

Usually electroresistance scans are carried out at fixed pressure, changing the sample temperature (isobaric measurements). The sample can be heated directly when it is connected to the power supply, or indirectly through a crucible. In order to put to a test and calibrate our setup we have carried out a set of electroresistance measurements on several metallic samples using the direct heating approach.

In the direct heating geometry the sample is typically positioned inside an insulating hollow cylinder (sleeve) which fits a hole drilled in a biconical gasket as depicted in fig. 1 (middle panel). The gasket is then placed between two steel recessed anvils with a WC insert as shown in figure 1 (left panel). Two conductive metal foils (Cu in this work) are positioned between the gasket and the anvils to measure the voltage drop upon the sample environment and to improve the electrical contacts with the steel anvils (fig. 1, middle panel).

The power supply ("DELTA SM 15-200 D") operating as a current generator has been used for voltage drop measurements and heating. This device, connected to the anvils by means of copper cables of suitable size, can supply a high stabilized continuous current up to 200 A (maximum voltage is 15 V).

The electroresistance measurements are controlled by a 16 bit acquisition board PCI-DAS1602/16 by Measurements and Computing installed on a PC running a Linux operating system (Red Hat 9.0, kernel 2.4.20-18.9). A set of dedicated C language programs controls the board. This control system can collect valuable experimental data in real time such us sample voltage, electrical current, oil pressure, and calculate other parameters such us electroresistance and heating power. These operations are performed while the sample pressure or temperature are changing, thus allowing for recording both isothermal and isobaric electroresistance measurements. The data acquisition is performed under thermodynamical equilibrium conditions, guaranteed by a proportional-derivative (PD) algorithm which stabilizes the heating power prior each data collection.

In figure 2 a set of electroresistance measurements on Bi is shown, recorded upon increasing the heating power for different fixed sample pressures within the range 0.4-2.1 GPa. For these measurements we have used a Bi wire (⌀0.5 mm), positioned in a BN sleeve inside a pyrophyllite gasket (⌀7 mm). Two graphite disks have been placed at the ends of the Bi wire to improve the electrical contacts and to partially limit the temperature gradient along the sample. The onset of the Bi I melting transition occurs upon the well defined maxima. In the figure inset the melting transition at 0.4 GPa is evidenced by a steep change of the first derivative of the electroresistance.

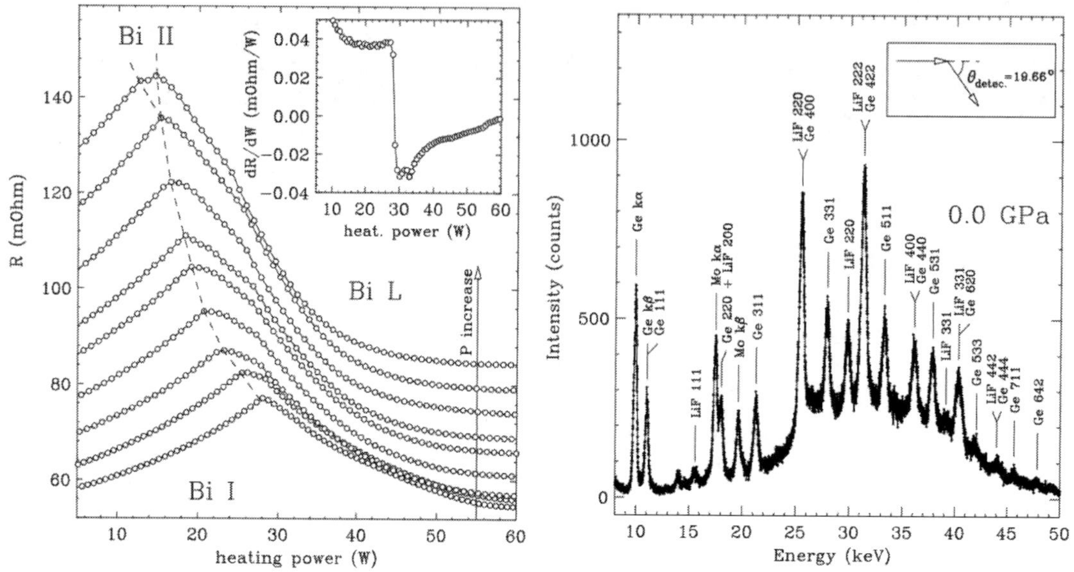

Figure 2. *Left panel:* Electroresistance of a Bi wire (⌀0.5 mm) positioned in a BN sleeve (see fig. 1, middle panel) as a function of pressure and heating power (temperature). The sample pressure increases from $\approx$ 0.4GPa (lower curve) up to $\approx$ 2.1GPa (upper curve) as indicated by the arrow. In the inset the derivative of the electroresistance as a function of the heating power is calculated for the curve collected at 0.4 GPa. *Right panel:* EDXD scan at ambient pressure and room temperature at a given collecting angle (θ=19.66°) for a Ge sample mixed with LiF (Ge:LiF 1:10) at room pressure. The typical spectral shape obtained with the Mo X-ray anode is noticeable. A large contribute to the background is due to the diffuse scattering mainly from the B-epoxy gasket. Other sharp peaks are associated with Ge fluorescence and with the strong Mo X-ray emission lines (Kα and Kβ).

Increasing the pressure the heating power requested to melt the sample decreases, as expected due to the negative slope of Bi melting line (ice-like behavior). The accuracy and the quality of the data are very good and in excellent agreement with previous measurements carried out with toroidal cells [2].

3. X-ray diffraction measurements

The experimental setup for energy dispersive x-ray diffraction (EDXD), consists of a powerful rotating anode Rigaku RU300, a Paris-Edinburgh (PE) V5 150 Tons press, a power supply "DELTA SM 30-100 D" and an energy sensitive Ge solid state detector (SSD).

The PE press is usually coupled with a 10 mm or 7 mm boron-epoxy biconical gasket [3, 4], which allows to reach safely pressures of about 8 GPa. The boron-epoxy gaskets are produced by us using an original preparation procedure while small inner parts of the sample assembly (crucibles, sample containers, thermocouple) are either commercial products or machined by external companies.

The sample, positioned in a hollow graphite cylinder and confined in the gasket, is indirectly heated using the graphite as a crucible, thus obtaining a very homogeneous temperature along the whole sample. High temperatures well up to 2300 K can be easily reached and maintained. The temperature is monitored up to 1300 K by using a K-type insulated thermocouple (Inconel sheath) positioned in the sample environment. Usually the sample is a micrometric powder dispersed in a inert low-Z insulating matrix which can be used as a pressure marker. The press

Journal of Physics: Conference Series **121** (2008) 042010 doi:10.1088/1742-6596/121/4/042010

is installed on a high-precision YZ translation stage, which allows an accurate positioning of the sample along the beam. A precise collimation systems (sollers and slits) is aligned between the sample and the detector for confining the scattering region within the sample volume. The detection system is a LEGe detector by Canberra Ltd with energy resolution of 195 eV at 5.9 keV. A lead shield is used to minimize counting of background scattering radiation from the environment. The scattering angle can be tuned continuously using a high-precision translation motor and a rotating stage orienting the collimators and detector system in order to select scattering area and resolution.

Here we present results of a first experiment aimed to check the sensitivity of our equipment measuring the EOS of simple systems. The sample was a compacted mixture of fine Ge and LiF powders filling a graphite heater, placed inside the B-epoxy gasket. The Paris-Edinburgh press was used for measurements in the (0.0÷5.3)GPa-(25÷580)°C pressure and temperature ranges.

The solid-state Ge detector has been positioned at a fixed angle of 19.66 degrees. The q range accessible at this collecting angle using an x-ray white beam with components up to 50.0 keV is 1.4 Å $\leq q \leq$ 8.6 Å and can be easily extended by changing the scattering angle with the motorized translator. In figures 2 (right panel) the EDXD spectrum recorded at ambient pressure and room temperature using the above described setup is presented. Although typical emission lines from Mo anode target (Mo-K_α, Mo-K_β) and fluorescence lines of germanium (Ge-K_α, Ge-K_β) are also visible, there is large energy region available for structural studies. Particular attention has been paid to the calibration of the energy scale. We found that this is a crucial aspect of these measurements and we checked carefully calibration of each EDXD pattern. In particular, it is well-known that a small degree of non-linearity is always present in the conversion between channel numbers and photon energy. In our data treatment, we adopted a quadratic function for channel-to-energy conversion because seems to provide the best results [5]. The calibration was performed using radioactive radiation sources (^{241}Am,^{55}Fe) and X-ray fluorescence lines (Ag-K_α Ag-K_β) having well known energy emission lines. Data analysis of EDXD spectra has been performed using the PEAKFIT and EDXRD programs, that are part of the GNXAS package for XAS/XRD analysis (see [6, 7] for previous applications of this software). Many Bragg peaks were included in the fitting procedure that provided an accurate determination of LiF and Ge parameter cells as function of temperature and pressure. The sensitivity to the variation of the parameter cell value of LiF can be appreciated looking at the magnified trends of the selected Bragg peaks LiF (111). In fact the figure 3 shows the effects of the temperature at fixed pressure (left panel) and of the pressure along an isothermal line at Room Temperature (right panel). The results obtained in the P-T range (0÷5.3)GPa-(25÷580)°C allow us to estimate for Ge and LiF the bulk modulus B_0, its derivative B_0' and volume thermal expansion α_V using suitable expressions for the temperature-dependent equation of state (EOS) starting from two equivalent analytic approximations for the EOS proposed by Vinet et al. [8, 9] and H11 as given in ref [10]. The coefficients of the Vinet or H11 EOS have been determined using the results of EDXD data analysis, showing compressibility values in substantial agreement with literature data within current uncertainty ($B_{0_{\mathrm{LiF}}} = 66.5$ GPa, $B'_{0_{\mathrm{LiF}}} = 4.3$, $B_{0_{\mathrm{Ge}}} = 77.0$ GPa, $B'_{0_{\mathrm{Ge}}} = 4.3$). Here those EOS are put to a test also at high temperatures, taking account the very different thermal expansion values ($\alpha_{V_{\mathrm{LiF}}} = 10.5 \times 10^{-5}$ K^{-1}, $\alpha_{V_{\mathrm{Ge}}} = 1.8 \times 10^{-5}$ K^{-1}).

4. Conclusions and future developments

Large volume non-toroidal cells are invaluable experimental tools for investigating condensed matter under extreme conditions in the range of moderate pressures (P<10 GPa) and temperatures (T<2000C).

In this paper we have given an overview of possible in-house applications of these devices showing that accurate investigations about the structure and phase transitions in materials

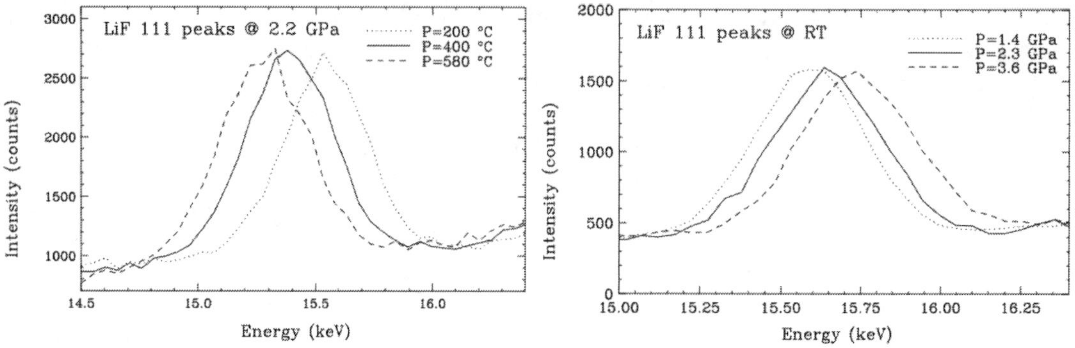

Figure 3. EDXD patterns at high pressure and temperature. *Left panel:* Thermal expansion as detected by the (downward) energy shift of the 111 LiF diffraction plane. *Right panel:* Upward shift of the 111 LiF diffraction planes as an effect of pressurization.

under extreme high pressure and temperature conditions are accessible in our laboratory.

We have shown that both electroresistance and x-ray diffraction high quality measurements can be carried out. The electroresistance setup has been found to be useful to highlight the occurrence of phase transitions. The EDXD setup allows us to check the EOS and characterize accurately the structure of the samples under examination.

The promising preliminary results stimulate a further development of both the techniques aimed to perform both x-ray diffraction and electroresistance measurement under extreme conditions. Such a setup is a useful and unique tool for studying condensed matter structure under the action of pressure and temperature and may be installed both in in-house laboratories and in large scale facilities.

References

[1] A. Di Cicco, R. Gunnella, R. Marassi, M. Minicucci, R. Natali, G. Pratesi, E. Principi, and S. Stizza. Disordered matter under extreme conditions: X-ray diffraction, electron spectroscopy and electroresistance measurements. *J. Non-Cryst. Sol.*, 352:4155–4165, 2006.

[2] F. P. Bundy. Phase diagram of bismuth to 130000 kg/cm^2, 500 $^o c$. *Physical Review*, 110:314–318, 1958.

[3] J. M. Besson, R. J. Nelmes, G. Hamel, J. S. Loveday, G. Weill, and S. Hull. Neutron powder diffraction above 10 gpa. *Physica B*, 180 & 181:907–910, 1992.

[4] E. Principi and A. Di Cicco. Development of an experimental setup for electroresistance measurements of materials under high pressure and temperature. *in preparation*, 2007.

[5] J. Liu, K. Kim, M. Golshan, D. Laundy, and A. M. Korsunsky. Energy calibration and full-pattern refinement for strain analysis using energy-dispersive and monochromatic x-ray diffraction. *Journal of Applied Crystallography*, 38:661–667, 2005.

[6] Andrea Di Cicco, Marco Minicucci, Emiliano Principi, Agnieszka Witkowska, Jaroslaw Rybicki, and Robert Laskowski. Testing interaction models by using x-ray absorption spectroscopy: solid pb. *Journal of Physics: Condensed Matter*, 14:3365–3382, 2002.

[7] A. Di Cicco, A. C. Frasini, M. Minicucci, E. Principi, J. P. Itié, and P. Munsch. High-pressure and high-temperature study of phase transitions in solid germanium. *Phys. Stat. Sol. (b)*, 240, 2003.

[8] P. Vinet, J. Ferrante, J. R. Smith, and J. H. Rose. A universal equation of state for solids. *J. Phys. C: Solid State Phys.*, 19:L467–L473, 1986.

[9] P. Vinet, J. R. Smith, J. Ferrante, and J. H. Rose. Temperature eect on the universal equation of state of solids. *Physical Review B*, 35:1945–1953, 1987.

[10] W. B. Holzapfel. Physics of solids under strong compression. *Rep. Prog. Phys.*, 59:29–90, 1996.

Raman and optical absorption studies of rare-earth hydrides under high pressure

T. Kume[1], H. Ohura[1], T. Takeichi[1], S. Sasaki[1], H. Shimizu[1], A. Ohmura[2], A. Machida[2], T. Watanuki[2], K. Aoki[2], and K. Takemura[3]

[1] Department of Materials Science and Technology, Gifu University, 1-1 Yanagido, Gifu, 501-1193, Japan

[2] Synchrotron Radiation Research Center, Japan Atomic Energy Agency, Hyogo 679-5148, Japan

[3] Advanced Materials Laboratory, National Institute for Materials Science, Tsukuba 305-0044, Japan

E-mail: kume@gifu-u.ac.jp

Abstract. Raman and visible absorption study were performed for YH_3 and ScH_3 at high pressures, in order to clarify the structural and electronic phase transitions. The Raman results of YH_3 revealed the presence of an intermediate phase at 9-24 GPa between the low-pressure hexagonal and high-pressure fcc phases. The characteristic behaviour was commonly observed for vibrational properties of YH_3 and ScH_3, suggesting a common mechanism of the structural transformation. The results of the absorption measurements for YH_3 demonstrate that the optical band gap starts to close in response to the phase transition.

1. Introduction

In 1996, Huiberts et al.[1] have discovered switchable properties of yttrium and some other rare-earth films driven by those hydrogenations. The reversible change between the metal (YH_2) and insulator (YH_3) demonstrated potential applications as switchable mirror. The hydrogenation causes also a structural transformation from the fcc YH_2 to the hcp YH_3. The hexagonal structure of YH_3 and ScH_3 is shown in figure 1, in which the metal atoms only are displayed. The H atoms are intercalated into the two interstitial sites, i.e., octahedron (O) and tetrahedron (T) sites of the hexagonal close packed (hcp) structure. The arrangement of the interstitial spaces for O site is shown with octahedrons in figure 1. Since there are one O site and two T sites per one metal atom, the hydrogen occupies all of the interstitial sites in the case of the trihydrates. For YH_3 and ScH_3, the hydrogen atoms in O site (H_o) are not centered at the octahedrons but displaced near the metal basal plane. The H_o atoms are believed to be displaced without ordering in the case of Sc[2] and with long-range order in the case of Y[3]. The crystal symmetries are explained with $P6_3/mmc$ for Sc trihydride and $P6_3cm$ (or $P6_3$) for Y trihydride. The dihydrides (YH_2 and ScH_2) take the face centered cubic (fcc) structure, in which the hydrogen occupies only the T sites.

Recently, several authors have reported the electronic and structural properties of rare-earth hydrides under high pressure. The energy gap closure of YH_3 was investigated by two research

© 2008 IOP Publishing Ltd

Journal of Physics: Conference Series **121** (2008) 042011 doi:10.1088/1742-6596/121/4/042011

group[4,5], but the critical pressure is controversial; visible transmission measurements [4] predicted the band gap to be closed at 55 GPa, while the recent IR experiments[5] suggested the gap closure to be at 23-25 GPa from the abrupt drop in the transmittance. Also for the structural change of YH_3, which sluggishly proceeds from the hexagonal to the fcc structure over a large pressure span ranging from 10 to 20 GPa[6,7], there are two interpretations. The x-ray diffraction data obtained around 10-20 GPa were interpreted with (1) the coexistence of the hexagonal and fcc structures[6] or (2) a new structure [7] that is not explained by intermediate structures for pure Y observed during the hcp-fcc phase transition. In the case of the Sc hydride, a recent x-ray diffraction experiments suggest that the intermediate phase exists around 30-40 GPa in between the low pressure hexagonal phase and high pressure fcc phase[8].

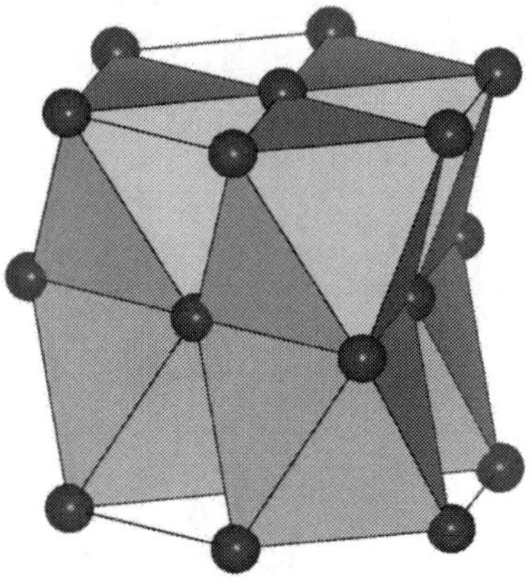

Figure 1. Hexagonal structure of YH_3 and SH_3. The metal atoms only are displayed.

In this work, we investigate vibrational and optical properties of YH_3 and ScH_3 under high pressure in order to gain an insight into the structural transformation from the hexagonal to the fcc phase, and the gap closure. For YH_3, the characteristic Raman spectra demonstrate the existence of the intermediate phase around 9-25 GPa and the analyses of the optical absorption spectra suggest the gap closure to be around 23 GPa. The characteristic vibrational behaviour precursor to the phase transition are commonly observed for YH_3 and ScH_3, suggesting a common mechanism of the phase transformation.

2. Experimental

The high-pressure experiments were carried out using a diamond anvil cell (DAC) with a culet of 0.3 mm in diameter. A hole of 0.1 mm in diameter drilled on a pre-indented tungsten gasket of about 0.03 mm thickness served as the sample chamber. The metal hydride (YH_x or ScH_x) surrounded with hydrogen was prepared in DAC with the same manner as described elsewhere[7]. Raman spectra were measured at room temperature in a backscattering geometry with a triple polychromator (JASCO NR 1800) equipped with a charge coupled device detector. Radiation of 532 nm from a solid-state laser (Coherent Verdi2W) was incident with a power of 5 mW. The measured Raman spectra contain the signals from H_2. Therefore, we extracted the spectra of the sample by subtracting the H_2 spectrum multiplied by an appropriate factor so as to cancel out the H_2 signals. The optical absorption spectra in visible region were measured with an apparatus established elsewhere[9]. Pressure was calibrated with the ruby fluorescence.

3. Results and discussion

3.1. Raman spectra of hexagonal YH_3 and ScH_3

Figures 2 (a) and (b) show high-pressure Raman spectra of YH_3 and ScH_3, respectively. In figure 2(a), the spectrum of YH_3 for $P = 0$ GPa was measured after attaining the maximum pressure (37 GPa). The shape of the spectrum is the same as the previous Raman spectra [10]. For higher frequency vibrations, the Raman peaks were assigned with the aid of the previous assignments which have been carried out for the vibrational DOS[11]. The notation of O_v (T_h) in figure 2 means that the hydrogen of O (T) site vibrates vertically (horizontally) to the *ab* plane in the hexagonal structure. Although the direct comparison of DOS and Raman spectrum is not appropriate, the assignments can be applied for the

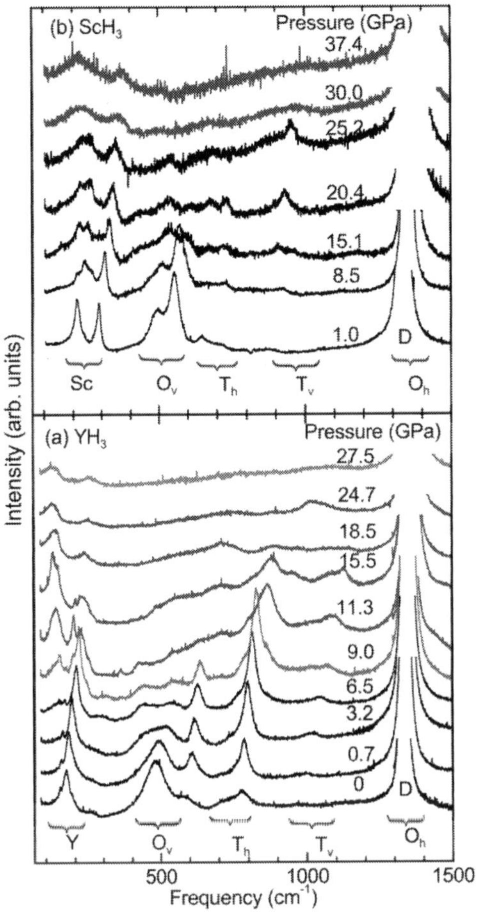

Figure 2. Pressure dependence of Raman spectra of (a) YH_3 and (b) ScH_3.

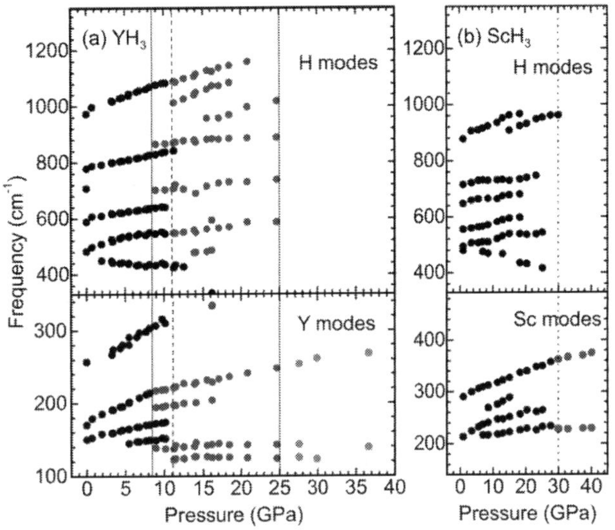

Figure 3. Pressure dependence of Raman frequencies of (a) YH_3 and (b) ScH_3 obtained upon compression.

present case because the phonon dispersion curves were theoretically predicted to be flat and to be well separated to each other[11].

As seen in figure 2 (a), the spectra obtained below 8 GPa are essentially the same as the spectrum at 1 atm. In the pressure region of 9-24 GPa, several new Raman peaks are found in the low frequencies. For example, the Y related peaks at 200 cm^{-1} cannot be explained with the coexistence of the hexagonal and cubic phases, demonstrating that there exists the new intermediate phase. The intermediate phase appears from 9 GPa because the spectra at 9-11 GPa can be reproduced with a combination between the spectra of the hexagonal ($P < 8$ GPa) and the intermediate phases ($11 < P < 25$ GPa). At 25 GPa, the H related bands disappear, which can be explained by the transformation to the fcc phase.

Figure 2(b) shows the Raman results of Sc hydride. According to the recent high-pressure x-ray diffraction experiments [8] performed for the Sc surrounded with H_2 using DAC, the specimen at about 1 GPa is composed of Sc, ScH_2, and ScH_3. The hydrogenation of Sc proceeds gradually with increasing pressure and the amount of ScH_3 gradually increases with pressure. The present Raman spectra come from only the ScH_3 because the cubic ScH_2 is expected to show one hydrogen mode around 1000 cm^{-1} and the metal Sc shows a peak around 130 cm^{-1}. As pressure increases, the Raman spectrum gradually changes. Taking into account the recent XRD experiments in which the transformation of the hexagonal ScH_3 to an intermediate phase was observed around 30 GPa, we can see that the disappearances of hydrogen related peaks (O_v, T_h, and T_v) correspond to the phase transition to the intermediate phase.

In Figs. 3 (a) and (b), we plotted the pressure dependence of Raman frequencies of YH_3 and SH_3, respectively. To identify the correspondence of the vibrational modes between YH_3 and ScH_3, the frequency axis for the low-frequency vibrations and the pressure axis are drawn with different scales. We note that the vibrational properties of ScH_3 are similar to those of YH_3, in particular for hydrogen vibrations. For example, the negative frequency shifts against the pressure are commonly observed for the O_v modes (around 500 cm^{-1}). The common behaviors precursory to the phase transition imply a common mechanism of the transformation from the hexagonal to the intermediate phase.

3.2. Optical absorption of YH_3 under high pressure
Figure 4 indicates the results of optical transmission measurements of YH_3. The experimental spectra are indicated with solid lines. The open circles correspond to the theoretical fitting which was carried out on the basis of the previous analyses[12] under the assumption that the absorption edge is characterized by the indirect allowed transition. The optical gap energy E_g obtained as a fitting parameter is plotted as a function of pressure in figure 4 (b). The slope change is found at 10 GPa, in association with the transformation to the intermediate phase. The linear extrapolation with the present results predicts E_g to be zero around 23 GPa, in good consistency with the previous IR experiments[5] in which the abrupt drop of the transmittance was observed around 23-25 GPa in the IR spectral region.

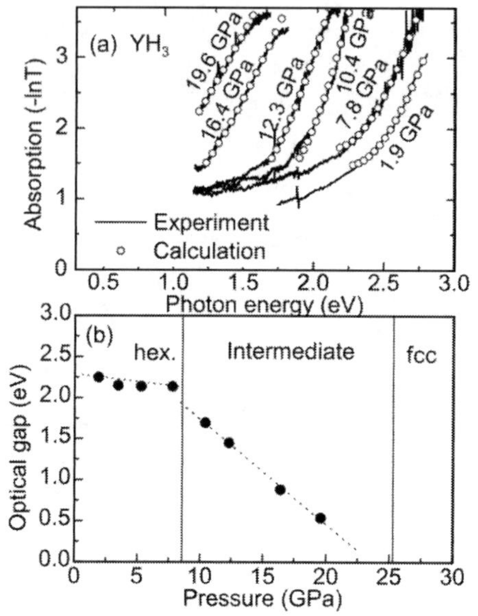

Figure 4. Pressure dependence of (a) absorption spectra and (b) energy gap of YH_3. In (a), the measured spectra and fitted absorption edges are shown by solid lines and open circles, respectively.

4. Conclusion
We investigated the phase transition of YH_3 and ScH_3 at high pressure by Raman spectroscopy. For the YH_3, it was demonstrated that the intermediate phase exist in between the hexagonal and fcc phases. The common behaviour of the vibrational properties between YH_3 and ScH_3 suggests strongly that there is a common mechanism for the structural transformation to hexagonal to the intermediate phase. The optical absorption spectra were measured for YH_3 up to 20 GPa. The analyses for the absorption edge under the assumption of the indirect gap predict the gap closure to be complete around 23 GPa.

Acknowledgements
This work was partially supported by a Grant-in-Aid from the Ministry of Education, Culture, Sports, Science, and Technology of Japan (No. 17204032).

References
[1] Huiberts J N *et al.* 1996 *Nature* **380** 231
[2] Antonov V E *et al.* 2006 *Phys. Rev. B* **73** 054107
[3] Fedotov VK *et al.* 2006 *J. Phys.: Condens. Matter* **18** 1593.
[4] Wijngaarden R J *et al.* 2000 *J. Alloys Compd.* **308** 44
[5] Ohmura A *et al.* 2006 *Phys. Rev. B* **73** 104105
[6] Palasyuk T and Tkacz M 2005 *Solid State Commun.* **133** 477
[7] Machida A *et al.* 2006 *Solid State Commun.* **138** 436
[8] Ohmura A *et al.* to be published.
[9] Kume T Fukaya Y Sasaki S, and Shimizu H 2002 *Rev. Sci. Instrum.* **73** 2355
[10] Racu A-M and Schoenes J 2006 *Phys. Rev. Lett.* **96** 017401
[11] van Gelderen P, Kelly P J, and Brocks G 2003 *Phys. Rev. B* **68** 094302
[12] van Gogh A T M *et al* 2001 *Phys. Rev. B* **63** 195105

Journal of Physics: Conference Series **121** (2008) 042012 doi:10.1088/1742-6596/121/4/042012

Static and shock compression of RDX single crystals: Raman spectroscopy

Zbigniew A. Dreger, James E. Patterson, Yogendra M. Gupta

Institute for Shock Physics, Washington State University, Pullman, WA 99164 USA

dreger@wsu.edu

Abstract. We report an examination of the polymorphic $\alpha - \gamma$ transformation for the energetic crystal RDX under static and shock compression. Raman spectroscopy was utilized as a probe of the molecular level response. We observed splitting of several Raman modes with the occurrence of the phase transition. From the response of the lattice and internal modes to static compression, we propose that the γ phase structure is isomorphous with the D_{2h} point group with eight molecules per unit cell occupying C_1 symmetry sites. The phase transition was observed to occur under shock wave loading with a transition time of ~100 ns for a peak stress of 5.5 GPa. The occurrence of the phase transition under shock wave loading has important implications for the onset of chemical decomposition in RDX.

1. Introduction

The response of energetic materials to high pressures has important relations to sensitivity, performance and safety. Energetic materials are highly compressible and often undergo polymorphic transformations. Structural changes accompanying these transformations can affect the sensitivity to shock wave initiation. Therefore, it is important to characterize the physical response of energetic materials under both static and dynamic compression.

In this paper, we report on the structural stability and high pressure polymorphism of single crystals of RDX (hexahydro-1,3,5,-trinitro-1,3,5-triazine). Pressure effects were examined under hydrostatic compression to 15 GPa and under shock wave compression to 5.5 GPa. RDX is known to exist in three polymorphic phases, termed α, β, and γ.[1,2] The ambient α phase structure belongs to the orthorhombic *Pbca* space group with eight molecules per unit cell. The three nitro ($-NO_2$) groups are chemically inequivalent, with two in the axial (A) position and one in the equatorial (E) position (the AAE conformer).

The β phase exists primarily under conditions of high pressure and high temperature. The molecular symmetry is believed to be C_{3v}, with all three nitro groups in the axial position (the AAA conformer). The γ phase exists at ambient temperatures above 3.8 GPa. Raman spectroscopy, particularly of the CH stretch modes [2], is sensitive to this phase transition. For this reason, we use Raman spectroscopy to study the structural changes that accompany the $\alpha - \gamma$ phase transition under both hydrostatic [3] and dynamic compression.[4]

2. Experimental Methods

Small, single crystals of RDX were used for hydrostatic compression experiments. Static pressures were generated in a modified Merrill-Bassett type diamond anvil cell (DAC). The sample

© 2008 IOP Publishing Ltd

compartment was formed from a pre-indented 0.06 mm thick stainless-steel gasket with a central hole of 0.18 mm diameter. Cryogenically loaded argon was used as hydrostatic pressure transmitting medium. A cw Ar-ion (514.5 nm) laser was employed for Raman excitation. The resolution of Raman spectra was ~0.8 cm^{-1}. All experiments were carried out at room temperature.

For shock experiments, single crystals of RDX were prepared with dimensions of 7 x 7 x 0.4 mm^3 and confined between two Z-cut quartz windows. Crystals were of the [111], [210] or [100] orientation. Shock waves were generated by impacting the target assembly with a Z-cut quartz crystal launched with a light gas gun. Peak stresses were between 3.0 and 5.5 GPa. High-resolution Raman spectra were measured with a single 20 ns pulse from a Nd:YAG (532 nm) laser. Time-resolved measurements were obtained with a 2 μs pulse from a tunable dye (514.5 nm) laser.

3. Static Compression

3.1. Lattice modes

Raman spectra of the lattice modes provide insight into the structural changes that accompany the $\alpha - \gamma$ phase transition. Figure 1 depicts spectra of the lattice modes at 4.0 and 6.1 GPa. Group theory considerations predict 24 Raman active lattice modes in the α phase. From the experiment, we were able to distinguish 9 lattice modes.

As the pressure is increased, the modes shift to higher frequency, as shown in Figure 2. Above the phase transition, the number of peaks increases from 9 to 12, but remains below 24. Note that the low frequency internal modes exhibit pressure dependent shifts that are very similar to the shift of external modes. The frequency gap between internal and external modes is relatively small compared to values typical for molecular solids. This indicates a small disparity between intermolecular and intramolecular bonding in the RDX crystal.

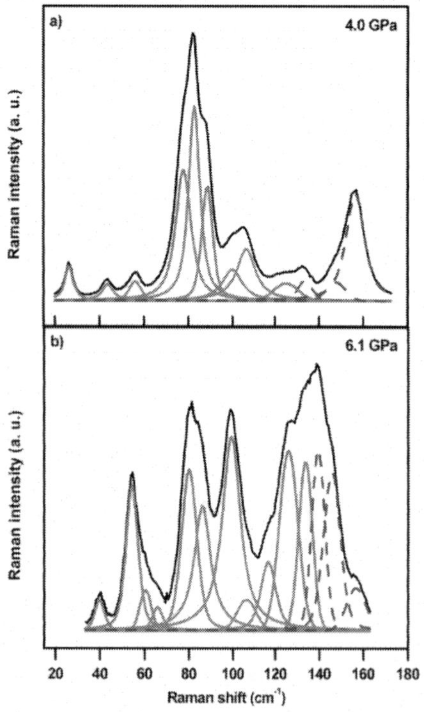

Figure 1. Raman spectra of lattice (solid) and low frequency internal modes (dashed) of RDX at (a) 4.0 GPa and (b) 6.1 GPa.

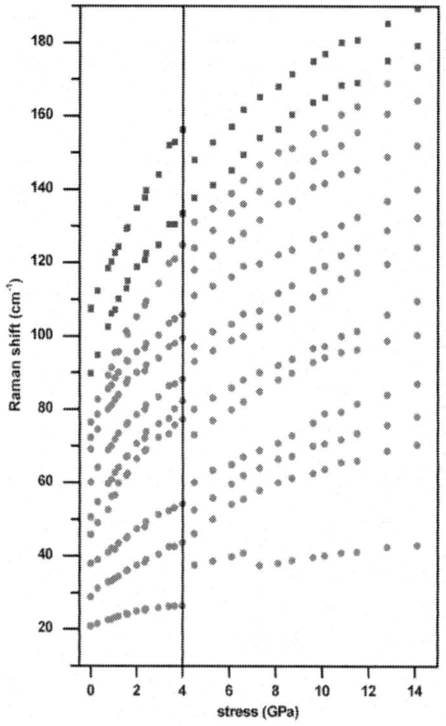

from ambient to 15 GPa for lattice and low frequency internal modes.

Journal of Physics: Conference Series **121** (2008) 042012 doi:10.1088/1742-6596/121/4/042012

3.2. CH stretching modes

Raman spectra of the CH stretching modes for a range of pressures from ambient to 5.5 GPa are depicted in Figure 3a. Note that the Raman modes split above the phase transition. The pressure dependent frequencies of these modes are indicated in Figure 4. Several features are observed above the phase transition: (i) the appearance of a new mode at ~ 3150 cm^{-1}, (ii) the triple splitting of some modes, (iii) changes in the relative intensity of the Raman modes and (iv) discontinuities in the pressure dependent Raman shifts. Other than the continued frequency shifting, no other significant features are observed up to 15 GPa.

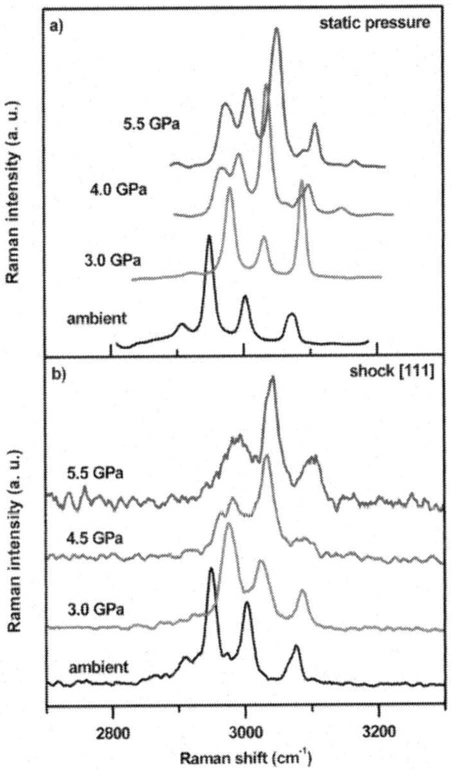

Figure 3. Raman spectra of CH stretch modes of RDX under (a) hydrostatic and (b) shock compression.

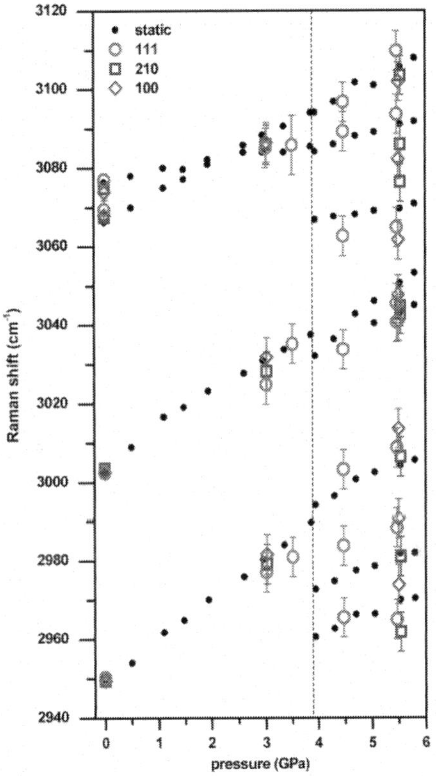

Figure 4. Pressure dependent Raman shifts of CH stretch modes from ambient to 6 GPa for static and shock compression.

3.3. Structure of γ phase

The observed behaviour of the Raman spectra under hydrostatic compression gives us some insight into the structural changes accompanying the $\alpha - \gamma$ phase transition. In general, for the internal vibrations, the number of Raman modes doubles on average above the $\alpha - \gamma$ phase transition. The number of lattice modes changes slightly, but remains consistent with four or eight molecules in the unit cell. Therefore, our results indicate that the crystal point group of the γ phase can be isomorphous with the $D_{2h}[C_1(8)]$, $C_{2v}[C_1(4)]$ and $D_2[C_1(4)]$ point groups.

Although our Raman data cannot be considered conclusive, we favour the γ phase structure that is isomorphous with the $D_{2h}[C_1(8)]$ group. Furthermore, the splitting of the internal Raman modes is attributed to factor group coupling, which indicates changes in intermolecular interactions. No conformational change of the individual RDX molecules is required for the observed splittings. We conclude that any structural modifications that take place with the phase transition preserve the factor group of the low pressure phase.

4. Shock wave compression

Spectra of shock compressed RDX are shown in Figure 3b for the [111] orientation. Comparison with the static pressure data at similar stresses reveals strong similarity between the two sets of data. Based on these similarities, we state that the $\alpha - \gamma$ phase transition does occur under conditions of shock wave loading. Additional experiments were performed for the other orientations. Those results are shown in Figure 4 with the static pressure results. For all three orientations, similar spectral trends were observed. The peaks split at about 4 GPa and shift to higher stress as the peak stress increases. Similar degrees of shifting were observed for all three orientations, indicating a lack of a measurable dependence on crystal orientation.

To gain insight into the kinetics of the phase transition, time-dependent measurements were performed. The results of these measurements are depicted in Figure 5 together with predicted Raman shifts based on the static pressure results and a simulated stress history of shocked RDX. As the stress increases, the Raman frequency also increases until the phase transition is reached. The predicted frequency decreases slightly in the γ phase, and then increases again as the stress is increased further. The results of the shock experiment differ from this prediction, however. The observed frequency increases beyond that predicted by the static pressure results, with a decrease in frequency shift occurring ~ 100 ns after the stress state is above 4.0 GPa. From this we conclude that a transition time of about 100 ns is required for the phase transition to occur under these loading conditions. Thus the phase transition occurs on time scales of relevance to shock initiation in RDX.

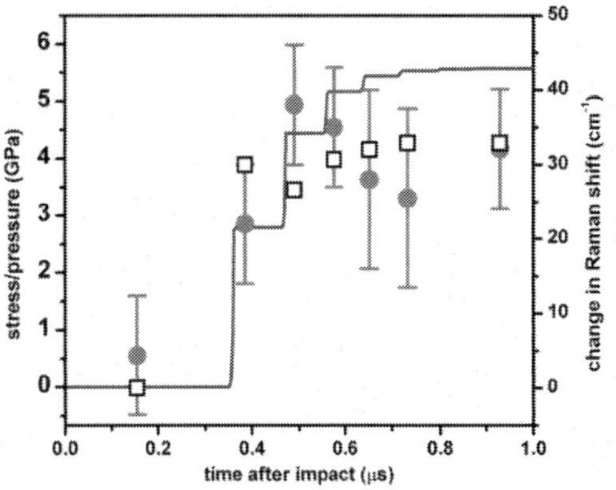

Figure 5. Simulated stress history for 5.5 GPa shock wave experiment (solid line). Predicted Raman shifts for 2949 cm^{-1} mode under shock indicated by open squares. Observed Raman shifts indicated by closed circles.

5. Conclusion

The results of static [3] and dynamic [4] compression experiments give us some additional insight into the structure of the γ phase of RDX. We propose that the space group in the high pressure γ phase retains D_{2h} symmetry. We have also observed the occurrence of the $\alpha - \gamma$ phase transition under conditions of shock wave loading, suggesting that this phase transition should be included in complete models of the shock initiation of this material.

Acknowledgments

The authors thank Dr. D.E. Hooks of Los Alamos National Laboratory for providing the RDX single crystals. Dr. N. Hemmi, K. Zimmerman and K. Perkins are thanked for assistance in performing the shock wave experiments. Dr. J.M. Winey is thanked for the stress wave simulations. This work was supported by US DOE and ONR.

References

[1] Karpowicz R J and Brill T B 1984 *J. Phys. Chem.* **88** 348
[2] Baer B J, Oxley J and Nicol M 1990 *High Pressure Research* **2** 99
[3] Dreger Z A and Gupta Y M 2007 *J. Phys. Chem* B **111** 3893
[4] Patterson J E, Dreger Z A and Gupta Y M 2007 *J. Phys. Chem.* B **accepted**

Journal of Physics: Conference Series **121** (2008) 042013 doi:10.1088/1742-6596/121/4/042013

High-pressure Brillouin study on solid Xe

S Sasaki[1], N Wada[2], T Kume[1] and H Shimizu[1]

[1] Department of Materials Science and Technology, Gifu University, 1-1 Yanagido, Gifu 501-1193, Japan

[2] Department of Electronics, Graduate School of Engineering, Gifu University, 1-1 Yanagido, Gifu 501-1193, Japan

E-mail: ssasaki@gifu-u.ac.jp

Abstract. The pressure dependences of three adiabatic elastic moduli C_{11}, C_{12} and C_{44}, acoustic velocities, elastic anisotropy and Cauchy deviation of a single crystal of solid Xe have been determined up to 10 GPa at 296 K, by applying high-pressure Brillouin scattering spectroscopy. We investigate the characteristics of high-pressure elastic properties of rare-gas solid Xe by comparing with the previous studies on Ne, Ar and Kr. Moreover, the splitting of Brillouin peaks of solid Xe above 10 GPa indicates that solid Xe with fcc structure recrystallizes and partially transforms into hcp structure as reported by the previous x-ray diffraction study.

1. Introduction

Rare-gas solids are interesting for chemical physics because these are considered as prototypical elements for understanding the van der Waals force. Moreover, rare-gas solids under high pressures are also significant as the hydrostatic pressure medium in high-pressure experiments using a diamond-anvil cell (DAC). So far we have investigated the elastic properties of fcc rare-gas solids Ne, Ar and Kr, using high pressure Brillouin scattering spectroscopy [1-3]. These results show that each rare-gas solid has individual elastic properties. For instance, the Cauchy deviation of solid Ar has large pressure dependence, whereas those of solid Ne and Kr are almost constant against pressure. As the Cauchy deviation gives the strength of many-body interaction, this difference indicates that van der Waals force between rare-gas atoms cannot be simply regarded as the two-body interaction. In this study, we perform high-pressure Brillouin experiments for single crystalline solid Xe with fcc structure using the high-pressure DAC, and determine its adiabatic elastic moduli (C_{11}, C_{12} and C_{44}), acoustic velocities, elastic anisotropy and Cauchy deviation at pressures up to 10 GPa and 296 K, in order to investigate the detailed elastic properties of rare-gas solids and the van der Waals interaction.

2. Experiment

A DAC is employed in observing Brillouin spectra under high pressure. The size of the hole in metal gasket serving as a sample chamber is about 0.1 mm in diameter and 0.1 mm in thickness. For loading a Xe sample into the DAC, we sublimated commercial gaseous Xe by spraying its vapor on the gasket hole of the DAC cooled in a liquid nitrogen bath (freezing temperature of Xe is about 161 K). To avoid the contamination of the sample, the loading of Xe was carried out in the dried nitrogen atmosphere. After that, we grew a single crystal of solid Xe by increasing the pressure on a seed crystal which coexists with liquid Xe at 0.4 GPa and 296 K, since one single crystal in DAC enables

© 2008 IOP Publishing Ltd

us to measure and analyze the Brillouin frequency shifts and to obtain its elastic properties. No pressure transmitting medium was used, because solid Xe is a soft material. Although the structure of crystalline Xe is fcc, the shape of its single seed crystal is spherical showing no crystal facet. The pressure was estimated by the ruby fluorescence method. The 514.5 nm line of the argon-ion laser operated with single mode was used at input power of about 100 mW. Brillouin spectra for solid Xe were measured in a 60° symmetric scattering geometry using a Sandercock tandem Fabry-Perot interferometer with a triple-pass arrangement. The Brillouin frequency shifts in the 60° geometry with the DAC ($\Delta\nu_{60}$) are related to the acoustic velocity (V_{60}) as follows [4]; $\Delta\nu_{60} = V_{60}/\lambda_0$, where λ_0 is the wavelength of the incident light in vacuum and the observed wave vector $\boldsymbol{q}$ of the acoustic phonons is parallel to the interfaces of the input and output diamonds crossed by the laser beam. Using this relation, one can immediately calculate the acoustic velocity from the Brillouin shift. By rotating the DAC around the axis perpendicular to the anvil interfaces, one can obtain the dependence of V_{60} on rotation angle ϕ. For single crystalline Xe, the elastic properties have been analyzed by using the method of in-situ Brillouin spectroscopy developed for simple molecular solids [5]. The acoustic velocities can be theoretically expressed by the following equation [5]: $V_j(\phi) = f_j(C_{11}/\rho, C_{12}/\rho, C_{44}/\rho, \theta, \phi, \chi)$, where ρ is the density of the sample, the subscript j (= 0, 1, 2) indicates the longitudinal acoustic (LA), two transverse acoustic TA_1 (slow) and TA_2 (fast) modes, respectively, and the Euler angles (θ, ϕ, χ) correspond to the crystal orientation in the DAC. In order to determine the Euler angles and three elastic moduli (C_{11}, C_{12} and C_{44}) simultaneously at each pressure, we applied a computerized least-squares fitting method between the calculated $f_j(\phi)$ and the measured acoustic velocities $V_j(\phi)$ as a function of angle ϕ. As a result, we can determine the acoustic velocities propagating along any crystal direction with in-situ identification of the crystal orientation [5].

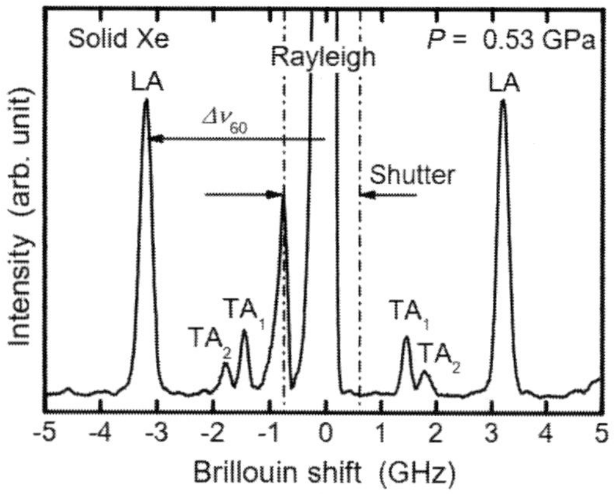

Figure 1. Brillouin spectrum of solid Xe obtained using a 60° symmetric scattering geometry at 0.53 GPa and 296 K. LA, TA_1 and TA_2 stand for the longitudinal, slow and fast transverse acoustic modes, respectively. The shutter operates to attenuate the strong Rayleigh signal.

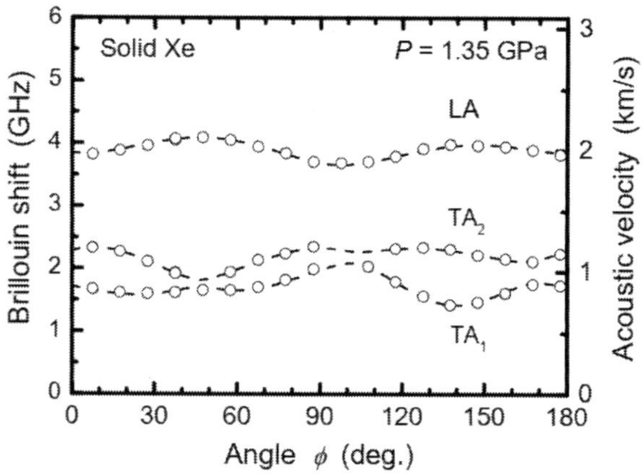

Figure 2. Angle ϕ dependence of Brillouin frequency shifts and acoustic velocities for solid Xe at 1.35 GPa and 296 K. Open circles are experimental points, and broken lines are the least-squares fitted curves. LA, TA_1 and TA_2 stand for the longitudinal, slow and fast transverse acoustic modes, respectively.

3. Results and discussion

The elastic properties of solid Xe have been studied by using the method of in-situ high-pressure Brillouin spectroscopy mentioned above. The Brillouin measurements in the 60° scattering geometry were made in 10° intervals of rotation angle ϕ. A typical Brillouin spectrum of solid Xe at 0.53 GPa is shown in figure 1. The peaks of LA, TA_1 and TA_2 modes are clearly observed. The measured Brillouin frequency shifts and acoustic velocities at 1.35 GPa are plotted as a function of ϕ by open circles as shown in figure 2. By applying the computerized least-squares fitting method to the experimental values of acoustic velocities to determine the elastic moduli of single crystalline fcc Xe, we found the excellent agreement between the measured (open circles) and the fitted (broken lines) values, as seen in figure 2. These fitting results yielded $C_{11}/\rho = 3.19$, $C_{12}/\rho = 2.17$, $C_{44}/\rho = 1.43$ km^2/s^2, $\theta = 32.8°$ and $\chi = 30.5°$ at $P = 1.35$ GPa.

Once these parameters were determined, the acoustic velocities could be calculated for any direction. Therefore, by performing the above procedure at various pressures, we could obtain the acoustic velocities for solid Xe propagating along the typical directions as a function of pressure up to 10 GPa at 296 K, as shown in figure 3. One can see that the acoustic velocities gradually increase with pressure and that TA modes are degenerated in the high-symmetry directions [100] and [111]. Moreover, the pressure dependence of three elastic moduli C_{11}, C_{12} and C_{44} for solid Xe (shown by open symbols in figure 4) was determined from the fitting results for C_{ij}/ρ with the pressure dependence of ρ from x-ray study [6]. For comparison, the previous results of Ne [1], Ar [2] and Kr [3] are also presented by broken, thin solid and dotted lines, respectively. The C_{12} of solid Xe shows almost the same value as those of solids Ne, Ar and Kr, but C_{11} and C_{44} for solid Xe have higher values than those of Ar and Kr at low pressure below 4 GPa, and those for solid Ne have extremely lower values than Ar, Kr and Xe.

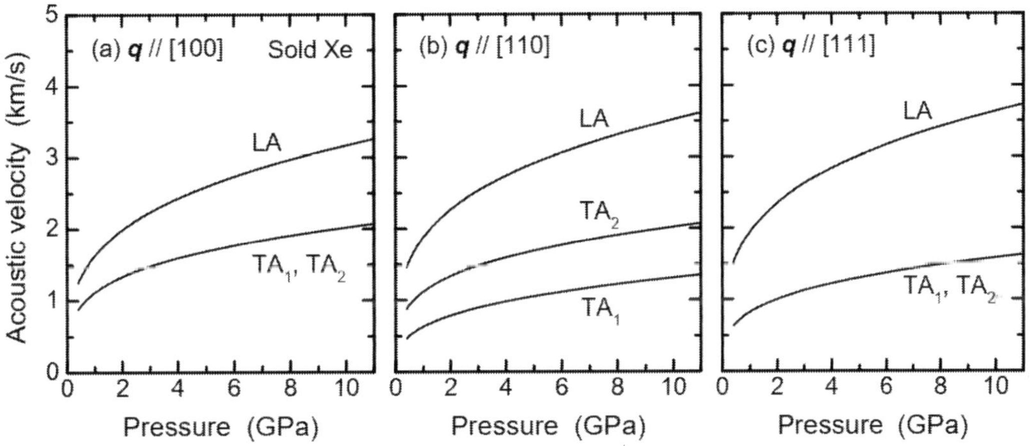

Figure 3. Pressure dependence of acoustic velocities for solid Xe at 296 K. The velocities of acoustic waves propagating along (a) [100], (b) [110] and (c) [111] directions are illustrated. LA, TA_1 and TA_2 are longitudinal, slow and fast transverse acoustic modes, respectively.

For more detailed comparison, we calculated the pressure dependences of (a) elastic anisotropy and (b) Cauchy deviation for solid Ne, Ar, Kr and Xe in figure 5. Since the elastic anisotropy is given by $A = 2C_{44}/(C_{11}\text{-}C_{12})$ for cubic crystal, the two shear moduli C_{44} and $(C_{11}\text{-}C_{12})/2$ are equal and $A = 1$ in the cases of the elastically isotropic materials. At a glance, the pressure dependence of A for solid Ne is remarkable. Whereas other rare-gas solids show the value of about $A = 3$ at 2 GPa and a slow decrease with pressure, solid Ne has the large value of $A = 4.9$ at 5 GPa and shows a rapid decrease to 3.6 at 7 GPa. As for the Cauchy deviation $\delta = C_{12}\text{-}C_{44}\text{-}2P$, solid Xe shows a rapid decrease from 1.1 at 1 GPa to -4.3 at 10 GPa with increasing pressure similar to solid Ar. On the other hand, solid Ne and Kr have respective values of about 4 and 1, and almost constant against pressure. As the Cauchy deviation gives the strength of many-body interaction, these differences among rare-gas solids indicates that van der Waals force between rare-gas atoms cannot be simply regarded as the two-body interaction. Actually, the first-principles study on elasticity for rare-gas solids [7] could not reproduce the above pressure dependences of elastic moduli, elastic anisotropy and Cauchy deviation in the present pressure region. This is probably caused by the van der Waals force which is not included in the first-principles calculations especially for the low pressure region below 10 GPa. To clarify the interactions of rare-gas solids, more detailed calculations or empirical potential functions are required.

In the present Brillouin study on solid Xe, the least-squares fitting analysis failed above 10 GPa, due to the splitting of Brillouin peaks which means that the recrystallization or phase transformation occurred at about 10 GPa. High-pressure x-ray studies on solid Xe [8] have shown that fcc Xe transforms at 14 GPa to an intermediate close-packed phase and then transforms completely to hcp structure at 75 GPa. However, this transformation from fcc to hcp is kinetically sluggish, depending on the temperature and pressure conditions [9, 10]. Very recently, we have measured high-pressure Raman spectra for solid Xe at 296 K, and observed that the transverse optic E_{2g} mode, which is Raman-active in the hcp phase, appeared around 40 cm^{-1} at pressure of 5 GPa [11]. Therefore, the splitting of Brillouin peaks above 10 GPa is due to the growth of fine hcp crystals in fcc phase of solid Xe.

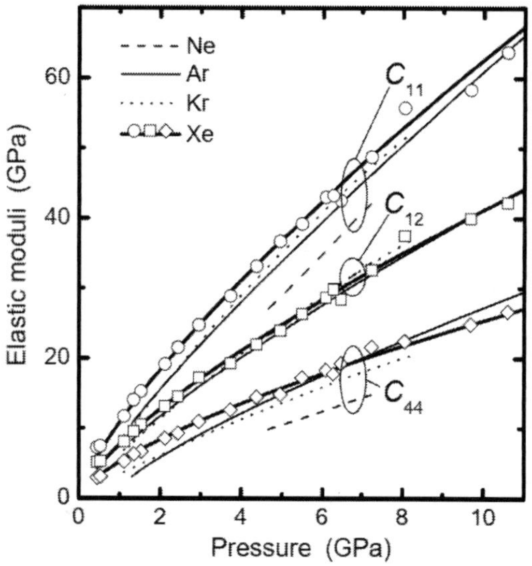

Figure 4. Pressure dependence of three adiabatic elastic moduli C_{11}, C_{12} and C_{44} for solid Xe at 296 K. Adiabatic elastic constants of sold Ne [1], Ar [2] and Kr [3] are shown for comparison by broken, thin solid and dotted lines, respectively.

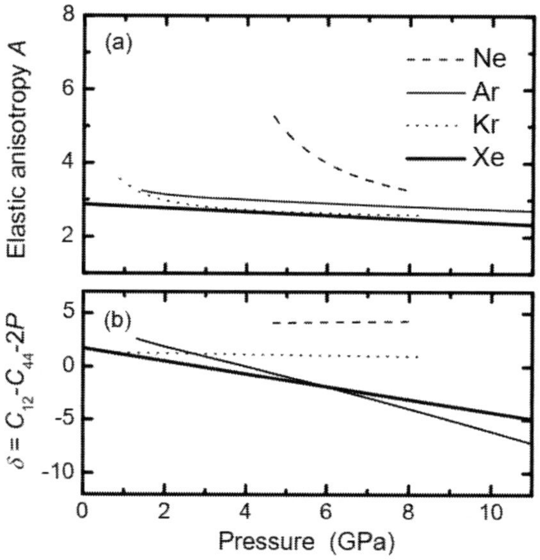

Figure 5. Pressure dependences of (a) elastic anisotropy A and (b) Cauchy deviation δ for solid Ne [1], Ar [2], Kr [3] and Xe (present study) at 296 K indicated by broken, thin solid, dotted and thick solid lines, respectively.

4. Summary

In the present study, we have determined the pressure dependence of elastic properties such as three adiabatic elastic moduli, acoustic velocities, elastic anisotropy and Cauchy deviation for solid Xe up to 10 GPa, by analyzing the observed angle ϕ dependence of velocities of acoustic waves propagating in arbitrary direction. From these results, we found out that the pressure dependence of adiabatic elastic moduli, elastic anisotropy and the Cauchy deviation were comparatively similar to those obtained for solid Ar. However, it is also found that each rare-gas solid basically has individual elastic properties. Therefore, more detailed calculations or empirical potential functions are newly required in order to understand the elastic properties of rare-gas solids. Moreover, the determination of elastic moduli from Brillouin shifts has failed above 10 GPa, which indicates that solid Xe with fcc structure partially transforms into hcp structure as reported by x-ray diffraction [8-10] and Raman [11] studies.

References

[1] Shimizu H, Imaeda H, Kume T and Sasaki S 2005 *Phys. Rev.* B **71** 014108
[2] Shimizu H, Tashiro H, Kume T and Sasaki S 2001 *Phys. Rev. Lett.* **86** 4568
[3] Shimizu H, Saitoh N and Sasaki S 1998 *Phys. Rev.* B **57** 230
[4] Shimizu H 1995 *High Pressure Research on Solids*, ed Senoo M, Suito K, Kobayashi T and Kubota H (Netherlands, Elsevier Science) pp 1-17
[5] Shimizu H and Sasaki S 1992 *Science* **257** 514
[6] Zisman A N, Aleksandrov I V and Stishov S M 1985 *Phys. Rev.* B **32** 484
[7] Tsuchiya T and Kawamura K 2002 *J. Chem. Phys.* **117** 5859
[8] Jephcoat A P, Mao H K, Finger L W, Cox D E, Hemley R J and Zha C S 1987 *Phys. Rev. Lett.* **59** 2670
[9] Cynn H, Yoo C S, Baer B, Iota-Herbei V, McMahan A K, Nicol M and Carlson S 2001 *Phys. Rev. Lett.* **86** 4552
[10] Errandonea D, Schwager B, Boehler R and Ross M 2002 *Phys. Rev.* B **65** 214110
[11] Shimizu H, Wada N, Kume T, Sasaki S, Yao Y and Tse J S 2008 *Phys. Rev.* B **77** 052101

Journal of Physics: Conference Series **121** (2008) 042014 doi:10.1088/1742-6596/121/4/042014

High-pressure Raman study on single crystalline methane hydrate surrounded by methane in a diamond anvil cell

Y Ohno, S Sasaki, T Kume and H Shimizu

Department of Materials Science and Technology, Gifu University, 1-1 Yanagido, Gifu 501-1193, Japan

E-mail: ssasaki@gifu-u.ac.jp

Abstract. High-pressure Raman measurements have been performed for single crystalline methane hydrate (MH) surrounded by fluid or solid methane in a diamond anvil cell. We successfully obtained the pure O-H stretching and lattice vibration spectra in MH-sI and MH-II phases. In these Raman spectra, there is no Raman band from water or ice-VI. The observed pressure of phase transformation from MH-sI to MH-II is 0.9 GPa, which is the same result as methane hydrate surrounded by water.

1. Introduction

Methane hydrate (MH), having attracted great interest recently as a new energy resource, is a special class of ice that contains methane as guest in host cages or networks of hydrogen-bonded water molecules [1]. In other respects, methane is a powerful greenhouse gas, and its outgassing from MH may contribute substantially the prospect of global warming. Therefore, the study of MH properties under high pressure is of fundamental importance in understanding the physical chemistry of clathrate hydrates of natural gases. At room temperature, the initial structure I of methane hydrate (MH-sI) successively transforms to MH-II (sH) and MH-III (sO, filled ice) at 0.9 GPa and 1.9 GPa, respectively [2-11]. These phase behaviors were observed in two phase coexistent state between methane hydrate and water. In our high pressure Raman measurements of methane hydrate, no signal for the O-H stretching and lattice vibrations was observed in MH-III phase, in spite of the appearance of those signals in the same sO phase of Ar hydrate [12] and Kr hydrate [13]. This disappearance in MH-III phase is probably caused by the weak Raman signals of MH-III and the strong Raman spectra of the surrounding ice-VII. Therefore, it is worth observing the O-H stretching and lattice vibration spectra of methane hydrate surrounded by fluid or solid methane. In addition to that, such a condition may induce the different phase behavior of methane hydrate.

The purposes of this paper are to make a coexistent state between methane hydrate and fluid methane in a diamond anvil cell (DAC) and to obtain the pure Raman spectra for the O-H stretching and lattice vibrations of host water networks. This is the preliminary work for observing the Raman signals of water networks in MH-III phase. In order to perform these experiments, we firstly loaded rich fluid methane and a small amount of pure water into a sample chamber of a DAC, and grew single crystals of MH-sI in fluid methane. By performing high-pressure Raman measurements, we obtained the pure O-H stretching and lattice vibration spectra in MH-sI and MH-II phases without the hindrance of the Raman band from water or ice-VI. Moreover, the phase transformation pressure of 1.9 GPa

© 2008 IOP Publishing Ltd

from MH-sI to MH-II is almost the same as that of methane hydrate surrounded by water, in spite of the different phase equilibrium condition of methane hydrate.

2. Experiment

In the present experiment, it is very important to make a coexistent state between methane hydrate and fluid methane (or solid methane), for observing the O-H stretching and lattice vibrations of host water networks. To do this, we firstly loaded rich fluid methane and a small amount of distilled water into a sample chamber (diameter 0.5 mm, depth 0.3mm) of a DAC, and grew single crystals of MH-sI in fluid methane. Figure 1 shows a photomicrograph of single crystals of MH-sI in fluid methane at 0.22 GPa. Unlike the case of methane hydrate in water, it is difficult to make a fine single crystal of MH-sI. In a three phase equilibrium state among MH-sI, fluid methane and water, crystalline MH-sI hardly grew, because MH-sI quickly enveloped a droplet of water and probably prevented methane dissolving in water. The pressure was measured by the ruby-scale method. The pressure accuracy in this experiment was $\pm$0.02 GPa. All high-pressure Raman scattering measurements for MH have been carried out in a back-scattering geometry at 296 K. The sample was excited by a 532.0 nm line of solid-state laser (COHERENT, Nd^{+}-YVO_4) and the spectra were obtained by a spectrometer (JASCO, NR1800) equipped with a triple polychromater and a liquid-nitrogen-cooled charge-coupled device (CCD) detector.

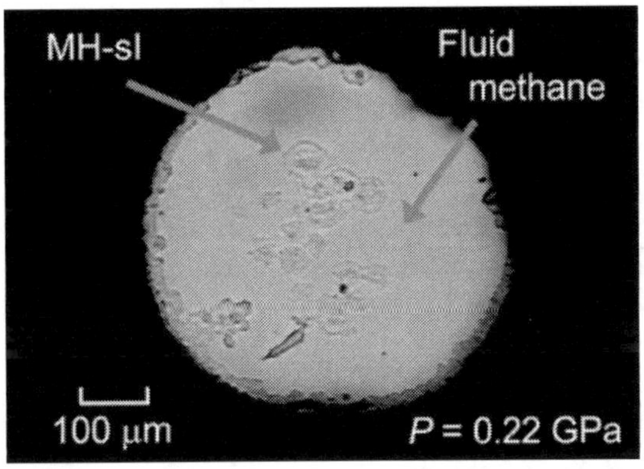

Figure 1. Photomicrograph of single crystals of methane hydrate structure I (MH-sI) in fluid methane at 0.22 GPa.

3. Results and discussion

By applying the high-pressure Raman scattering spectroscopy to single crystals of MH surrounded by fluid or solid methane, we obtained the pressure dependences of Raman spectra for the lattice and O-H modes as shown in figures 2(a) and 3(a). For comparison, the Raman spectra for MH coexisting with water or ice-VI are also illustrated in figures 2(b) and 3(b). Raman peaks from methane hydrate and ice-VI are indicated by open and solid triangles in these figures, respectively. So far the measured Raman spectra of MH-II above 1.3 GPa were accompanied with those of ice-VI as shown in figures 2(b) and 3(b). [The Raman spectra of MH in water at 0.08 GPa, 0.79 GPa and 1.15 GPa have no obvious band about water, since the water has no contribution to the lattice region and broad bands (3100-3600 cm^{-1}) for O-H stretching vibrations. (See the spectra for water at 0.08 GPa in figure 2(b) and 3(b))] These signals from high-pressure ices have prevented us observing pure Raman signals for high-pressure MH phases.

On the other hand, the present results for MH crystals surrounded by methane show no Raman signals from ice-VI and provide the pure Raman spectra for lattice and O-H vibration of MH-II at 1.71 GPa in figures 2(a) and 3(a). Although fluid methane transforms to plastic phase I at 1.6 GPa [14], the fcc plastic phase of methane generally shows no lattice band [15]. Therefore, both liquid and solid I phases of methane contribute no signal to lattice region.

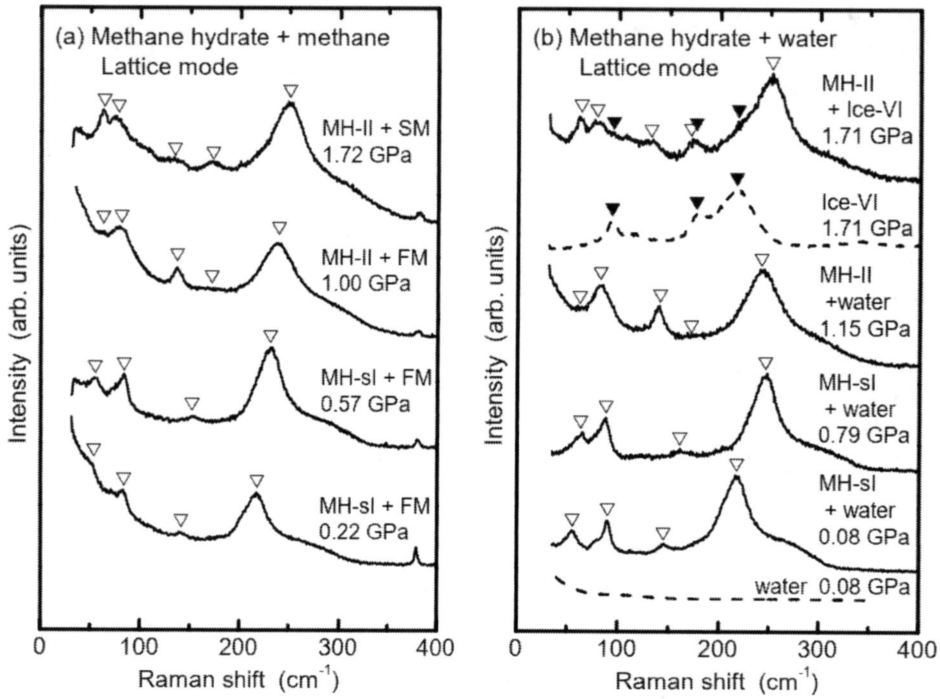

Figure 2. Pressure dependences of Raman spectra for lattice mode of (a) methane hydrate (MH) in fluid or solid methane and (b) MH in water or ice-VI. For comparison, Raman spectra for water and ice-VI are presented by broken lines at 0.8 GPa and 1.71 GPa, respectively. Open and solid triangles stand for the Raman peaks of lattice mode from MH and ice-VI, respectively. FM and SM indicate fluid and solid methane, respectively.

At the phase transformation from MH-sI to MH-II, the excess water is excluded from MH-sI because of the reduction of the hydration number of MH. If this excess water remains inside the MH-II crystal, the Raman spectra of ice-VI appear even in the coexisting state between MH and methane. Therefore, in this experiment, the excess water from MH-sI was totally formed into MH-II at 0.9 GPa. In MH-III phase, we couldn't observe the Raman signals from the MH-III, because of the strong signals from ice-VII and the weak signals from host cage of MH-III. If only the MH-III is grown in solid phase of methane above 1.9 GPa, we should observe the Raman spectra in MH-III phase similar to those in the MH-II phase in this experiment.

Next, we investigate the phase transformation from MH-sI to MH-II in fluid methane. From the high-pressure Raman study on Kr hydrate [13], it is known that the Raman signal at 120 cm^{-1} disappears only in the MH-sI phase. From the present Raman spectra for methane hydrate, we observed that the signal at 120 cm^{-1} appeared above 0.9 GPa. This result clearly indicates that the MH-sI phase transforms into the MH-II phase. In addition, we also confirmed this phase transformation through the microscopic observation. From these results of the Raman spectra and the microscopic observation for MH surrounded by methane, we found that the phase transformation pressure from MH-sI to MH-II is the same pressure of 0.9 GPa as in the case of MH surrounded by water. Therefore, the surrounding fluid methane doesn't contribute to the phase transformation of MH and its pressure.

4. Summary

We successfully made single crystalline MH surrounded by methane in a diamond anvil cell. By applying the high-pressure Raman spectroscopy to this MH coexisting with fluid or solid methane, we obtained the pure O-H stretching and lattice vibration spectra in MH-sI and MH-II phases. In these

Raman spectra, there is no Raman band from water or ice-VI differing from the results of MH with water or ice-VI. From the results of the Raman spectra and the microscopic observation for MH in methane, we found that the phase transformation pressure from MH-sI to MH-II is the same pressure of 0.9 GPa as in the case of MH surrounded by water. Therefore, surrounding methane has no contribution to phase change and its pressure with respect to the MH-sI to MH-II transformation.

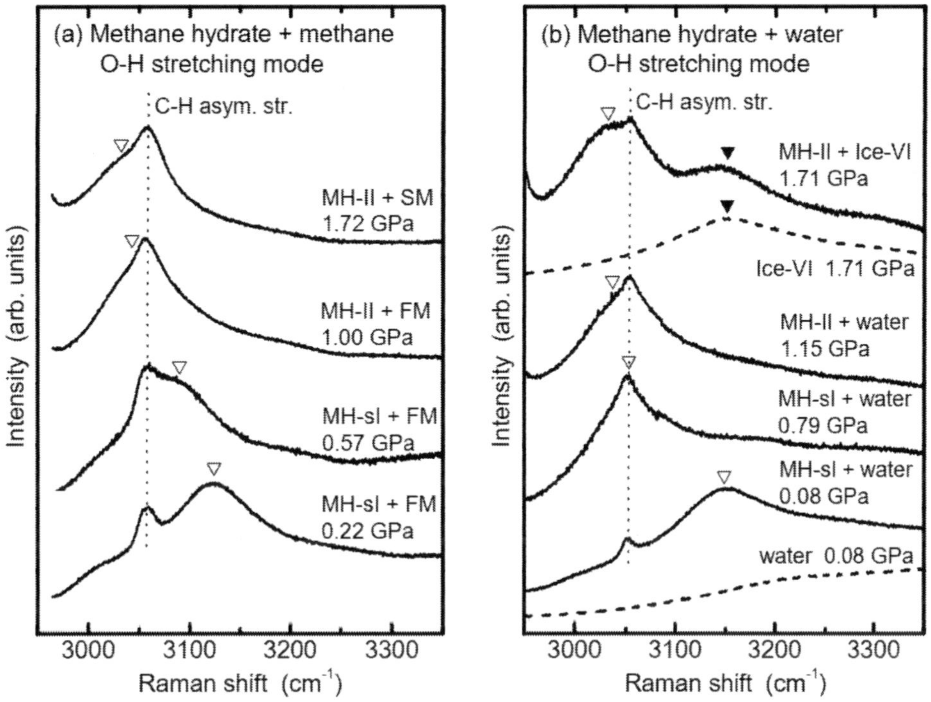

Figure 3. Pressure dependences of Raman spectra for O-H stretching mode of (a) methane hydrate (MH) in fluid or solid methane and (b) MH in water or ice-VI. For comparison, Raman spectra for water and ice-VI are presented by broken lines at 0.8 GPa and 1.71 GPa, respectively. Open and solid triangles stand for the Raman peaks of O-H stretching mode from MH and ice-VI, respectively. FM and SM indicate fluid and solid methane, respectively, and vertical dotted lines stand for the positions of the asymmetrical stretching vibration of methane molecules. The strong asymmetrical stretching signals in figure 3(a) come from the surrounding methane around MH.

References

[1] Sloan Jr. E D 1998 *Clathrate Hydrates of Natural Gases, second ed.* (New York, Marcel Dekker)
[2] Loveday J S, Nelmes R J, Guthrie M, Belmonte S A, Allan D R, Klug D D, Tse J S and Handa Y P 2001 *Nature* **410** 661
[3] Hirai H, Hasegawa M, Yagi T, Yamamoto Y, Nagashima K, Sakashita M, Aoki K and Kikegawa T 2000 *Chem. Phys. Lett.* **325** 490
[4] Loveday J S, Nelmes R J, Guthrie M, Klug D D and Tse J S 2001 *Phys. Rev. Lett.* **87** 215501
[5] Loveday J S, Nelmes R J and Guthrie M 2001 *Chem. Phys. Lett.* **350** 459
[6] Shimizu H, Kumazaki T, Kume T and Sasaki S 2002 *J. Phys. Chem.* B **106** 30
[7] Hirai H, Uchihara Y, Kawamura T, Yamamoto Y and Yagi T 2002 *Proc. Jpn. Acad.* **78B** 39
[8] Loveday J S, Nelmes R J, Klug D D, Tse J S and Desgreniers S 2003 *Can. J. Phys.* **81** 539

[9] Hirai H, Tanaka T, Kawamura T, Yamamoto Y and Yagi T 2003 *Phys. Rev. B* **68** 172102
[10] Kumazaki T, Kito Y, Sasaki S, Kume T and Shimizu H 2004 *Chem. Phys. Lett.* **388** 18
[11] Hirai H, Tanaka T, Kawamura T, Yamamoto Y and Yagi T 2004 *J. Phys. Chem. Solids* **65** 1555
[12] Shimizu H, Hori S, Kume T and Sasaki S 2003 *Chem. Phys. Lett.* **368** 132
[13] Sasaki S, Hori S, Kume T and Shimizu H 2006 *J. Phys. Chem.* B **110** 9838
[14] Hebert P, Polian A, Loubeyre P and Le Toullec R 1987 *Phys. Rev.* B **36** 9196
[15] Wu Y H, Sasaki S and Shimizu H 1995 *J. Raman Spectrosc.* **26** 963

Pressure dependent luminescence properties of Eu^{3+}:TeO_2-K_2O-Nb_2O_5 glass

R Praveena[1], V Venkatramu[2], P Babu[3], C K Jayasankar[1,6], Th Tröster[4], W Sievers[5] and G Wortmann[5]

[1] Department of Physics, Sri Venkateswara University, Tirupati-517 502, India
[2] Department of Physics, Yogi Vemana University, Kadapa - 516 003, India
[3] Department of Physics, Government Degree College, Wanaparthy - 509 103, India
[4] Fakultät für Maschinenbau, Universität Paderborn, D-33098 Paderborn, Germany
[5] Department Physik, Universität Paderborn, D-33095 Paderborn, Germany

E-mail: ckjaya@yahoo.com

Abstract. Eu^{3+}-doped tellurite glass of molar composition: TeO_2 + K_2O + Nb_2O_5 + Eu_2O_3 (TKN1Eu) has been prepared by quenching the oxidic melts. The emission spectra and the decay curves for the $^5D_0$ level of Eu^{3+} ions in TKN1Eu glass have been measured and analyzed as a function of pressure up to 14.8 GPa at room temperature. In this pressure range, the observed red shifts for the band positions of $^5D_0 \rightarrow {}^7F_{0,1,2}$ could be attributed to an expansion of the 4f electron wave functions with increasing covalency. There is also a considerable pressure effect on the magnitude of crystal-field (CF) splitting in the $^5D_0 \rightarrow {}^7F_1$ band, which increases with pressure. Stark components of the $^7F_1$ level have been used to evaluate the CF parameters, B_{20} and B_{22}, which are in turn used to calculate the CF strength experienced by the Eu^{3+} ions in the TKN1Eu glass. The CF strength increases from 1062 to 1358 cm^{-1} from the ambient pressure to 14.8 GPa. Lifetime for the $^5D_0$ level of Eu^{3+} is found to decrease from 0.76 ms (0 GPa) to 0.30 ms (14.8 GPa) with increase in pressure. The results of pressure induced energy shifts, CF strengths and lifetimes are well discussed both for increasing as well as decreasing pressures and are compared with those of Eu^{3+}:phosphate and lithium borate glasses.

1. Introduction

Research on glasses containing heavy network formers, such as TeO_2, have acquired considerable attention when doped with luminescent lanthanide (Ln) ions [1, 2], due to their low vibrational frequencies that leads to a reduction of the multiphonon deexcitation and improve cross-sections and quantum efficiency of these ions. In addition, tellurium oxide based glasses have been extensively investigated because of their interesting physical properties such as large infrared transparency, high linear and non-linear refractive indices, good thermal stability and corrosion resistance and suitability as a matrix for active element doping, represent the main justification for their continuous technological interest [3]. On the other hand, niobium containing glasses have several interesting non-linear refractive index n_2, which make them attractive materials for ultra-fast switching devices [4, 5].

[6] To whom any correspondence should be addressed.

© 2008 IOP Publishing Ltd

Moreover, second harmonic generation (SHG) was observed in borophosphate glasses containing Nb_2O_5 after a thermal poling treatment [6].

The development of optical and opto-electronic devices like optical sensors, phosphors, and optical data storage devices require a systematic knowledge of bonding environment of Ln^{3+} ions and their luminescence properties. Inter-atomic distance can be reduced continuously by applying high pressure and therefore increases the overlap among the adjacent electronic orbitals and the crystal-field (CF) strength, without changing the other physical properties like the point group symmetry of the Ln^{3+} ion site or the chemical composition. Different effects occur on different luminescent centers with the decrease in the effective volume of the Ln^{3+} ion containing matrix with high pressure. Thus, a pressure dependent luminescence study provides sufficient information on the electronic states, the CF strength, the energy transfers and the interaction between luminescence centers and their hosts [7].

Ln-doped glasses containing empty d shell transition metal ions could be interesting to develop new materials in which emission in the visible and infrared regions and optical amplification or self frequency sum mixing processes are simultaneously achieved [8]. Eu^{3+} ions doped glasses are of special interest because of the following reasons: (1) persistent spectral hole burning can be performed in the $^7F_0 \rightarrow {}^5D_0$ transition of the Eu^{3+} ion at room temperature (RT) which has potential use in high-density optical storage [9], (2) the information on the local structure around Ln^{3+} ions can be obtained with Eu^{3+} ion as a probe using fluorescence line narrowing technique, (3) the nature of vibrations associated to Ln^{3+} ion sites which affect the non-radiative decay can be obtained from the measurement of the phonon side band associated with $^7F_0 \rightarrow {}^5D_0$ and $^7F_0 \rightarrow {}^5D_2$ pure electronic excitation transitions in the excitation spectrum of the Eu^{3+} ion, and (4) electron-phonon coupling strength in a host can also be evaluated as the ratio of the integrated intensities of the phonon side band to that of the pure electronic band [10]. When a Eu^{3+} ion is incorporated into any host matrix, it experiences an inhomogeneous crystal-field produced by the surrounding charge distribution that results in two effects: (i) the centroid shifts of all the free-ion multiplets to the red/blue and (ii) the Stark splitting of the energy levels. The centroid shifts are ascribed as the Nephelauxetic effect that explains the change of Slater parameters (F_K) and the spin-orbit coupling parameter (ξ_{4f}) for the Eu^{3+} ion in the host with respect to their free ion values of F_K^0 and ξ_{4f}^0. The Stark splitting of the multiplets is ascribed to the change in the sum of various interactions such as covalence, overlap and Coulomb interactions between the 4f electrons of the Eu^{3+} ions and their surrounding ligands. Depending on the number of Stark components into which the $^5D_0 \rightarrow {}^7F_J$ emission transition split, the symmetry at the Eu^{3+} ions site can be derived by CF parameterization. The CF parameters, B_{20} and B_{22}, as well as their ratio have been used to describe the coordination of the Eu^{3+} ions [11-13].

Besides the above crystal-field effects, also the pressure induced changes in the luminescence properties of Eu^{3+} ions in the tellurite glass have been investigated in the present work. Luminescence spectra of $^5D_0 \rightarrow {}^7F_{0,1,2}$ transitions and fluorescence lifetime measurements for the $^5D_0$ level of Eu^{3+} ion in the tellurite glass under high pressure up to 14.8 GPa at RT are presented. The results obtained in this work have been compared with those of the Eu^{3+}:phosphate [14] and lithium borate glasses [15].

2. Glass preparation and spectroscopic measurements

The Eu^{3+}-doped tellurite glass with the composition (mol %) of 69 TeO_2 + 15 K_2O + 15 Nb_2O_5 + 1.0 Eu_2O_3 (TKN1Eu) was prepared by conventional melt quenching technique. The batch composition was melted in a platinum crucible placed in a furnace at temperature of 1000 °C for about 30 min. and then poured onto a preheated brass plate followed by 6 h annealing at 250 °C and then cooled down to RT. The TKN1Eu glass thus obtained was used for high pressure measurements.

The blue line of 465.8 nm from Ar^+ laser was selected to excite the Eu^{3+} ion. The emission spectra were recorded with a double monochromator equipped with a photomultiplier tube [15]. A special miniature diamond anvil cell (DAC) was used to generate high pressures up to nearly 15 GPa at RT. A piece of the TKN1Eu glass along with ruby pressure sensor was placed in an 80 μm diameter hole of a stainless steel (INCONEL X 750) gasket of 200 μm thickness. A mixture of methanol : ethanol : water (16:3:1) was used as the pressure transmitting medium. This gasket with the sample, pressure sensor and pressure transmitting fluid was then compressed by the two opposed anvils of the DAC. The pressure and the hydrostatic conditions experienced by the sample were determined by the shift and broadening of the ruby R_1 lines [16]. Decay curves for $^5D_0 \rightarrow {}^7F_2$ transition of Eu^{3+} ion were obtained using a mechanical chopper with a multi-channel scalar in the range from 2 μs to 2 s and interfaced to a personal computer that recorded and averaged the signal.

3. Theory

When a Ln^{3+} ion is doped into a solid matrix, there is an electrostatic interaction of the 4f electrons with the charge of host ligands, known as the CF interaction, whose magnitude is small compared with electrostatic and spin-orbit interaction. The CF interaction is responsible for the small shift of the $^{2S+1}L_J$ multiplets as well as the appearance of Stark levels, arising from the removal of degeneracy of the $^{2S+1}L_J$ levels. The Hamiltonian that represents the CF potential acting on the Ln^{3+} ion can be expressed in Wybourne's notation [15, 17, 18] as

$$H_{CF} = \sum_{k,q,i} B_q^k C_q^k(i) \tag{1}$$

where B_q^k are the coefficients of the CF expansion which represents the functions of the radial distances and $C_q^{(k)}$ are tensor operators of rank k, related to the spherical harmonics, can be calculated exactly [18]. The number of parameters for H_{CF} in equation (1) is greatly reduced by the symmetry selection rules for the point symmetry at the Eu^{3+} ion site in the crystal. Any surrounding that breaks the spherical symmetry of the free ion can lead to a shift and splitting of the energy levels. Thus, the above considerations also apply to materials such as glasses where a long-range order does not exist.

In the present analysis, considering the J-mixing interaction as a perturbation over the CF interaction, the three Stark components of the $^7F_1$ are simply described by

$$H_{CF} = B_{20}C_0^{(2)} + B_{22}(C_{-2}^{(2)} + C_2^{(2)}) \tag{2}$$

where B_{20} and B_{22} are the real parts of the CF parameters in equation (1).

The CF strength parameter, N_v has been used to simplify the description of the CF parameters due to its independence from the specific crystal structure. The N_v, corresponding only the second rank CF parameters, has been determined using the following relation [18, 19]

$$N_v = \sqrt{\frac{4\pi}{5}\left[(B_{20})^2 + 2(B_{22})^2\right]} \tag{3}$$

4. Results and Discussion

4.1. Fluorescence spectra

When Eu^{3+} ions are excited to the $^5D_2$ level using 465.8 nm radiation from Ar^+ laser, the fluorescent $^5D_0$ level are populated by non-radiative mutiphonon relaxation. The resultant emission from $^5D_0$ level of the TKN1Eu glass as a function of pressure was measured at RT in the range of 15500-17500 cm^{-1} (see figure 1). The spectra obtained are quite similar to those of other oxide glass hosts [20, 21]. The emission spectra of the TKN1Eu glass consists of three transitions from the lowest excited multiplet state $^5D_0$ to the ground state multiplets ^7F_J (J = 0, 1, 2). The following transitions corresponding to J = 3, 4, 5 and 6 are very weak and hence they are not considered for the analysis. The spectra are normalized to $^5D_0 \rightarrow {}^7F_2$ transition. The observed emission bands of TKN1Eu glass at 597 nm (16868 cm^{-1}) ascribed to $^5D_0 \rightarrow {}^7F_1$ and at 613 nm (16316 cm^{-1}) assigned to $^5D_0 \rightarrow {}^7F_2$, which correspond to orange and red, respectively.

The $^5D_0 \rightarrow {}^7F_2$ transition is electric dipole allowed and also hypersensitive (its intensity is very sensitive to the local environment of the Eu^{3+} ion), whereas the transition $^5D_0 \rightarrow {}^7F_1$ is magnetic dipole allowed and is independent of the local symmetry. The well resolved Stark splitting of the $^5D_0 \rightarrow {}^7F_1$ transition into three components suggests that the Eu^{3+} ions in this glass host occupy low symmetry sites such as orthorhombic, monoclinic or triclinic. If Eu^{3+} ion occupies a spherically symmetric site, or, is at a centre of symmetry, the intensity of the $^5D_0 \rightarrow {}^7F_2$ emission is zero (except for very weak vibronic contributions) but its intensity increases as the site symmetry decreases. Hence, quality of the host material for lasing action is often expressed through the fluorescence intensity ratio, $^5D_0 \rightarrow {}^7F_2$ to $^5D_0 \rightarrow {}^7F_1$ transitions denoted as I_R (2/1), which is used to establish the degree of asymmetry in the vicinity of Eu^{3+} ions [22-25]. Moreover, I_R (2/1) value also depends on the Judd-Ofelt (JO) parameter Ω_2 [26-28], which is used to explain the short range effects. Therefore, the variation of I_R (2/1) and in turn Ω_2 gives the information about the short range effect on local structure around Eu^{3+} ions. The higher value of I_R (2/1) suggests a lower symmetry around Eu^{3+} ions [29-31]. For the present TKN1Eu glass system, the I_R (2/1) values are found to decrease from 5.00 to 3.50 when the pressure is increased from 0 to 14.8 GPa. The variation of I_R (2/1) with pressure is presented in figure 2. The decrease in the I_R (2/1) value with increasing pressure reflects the increase of symmetry by narrowing the site distribution around Eu^{3+} ion with pressure. From figure 2, it is evident that the I_R (2/1) values are considerably higher while releasing the pressure. This indicates the increase of local deformation around the Eu^{3+} ions after releasing the pressure which may be caused by the increase of pressure induced defect centers [7].

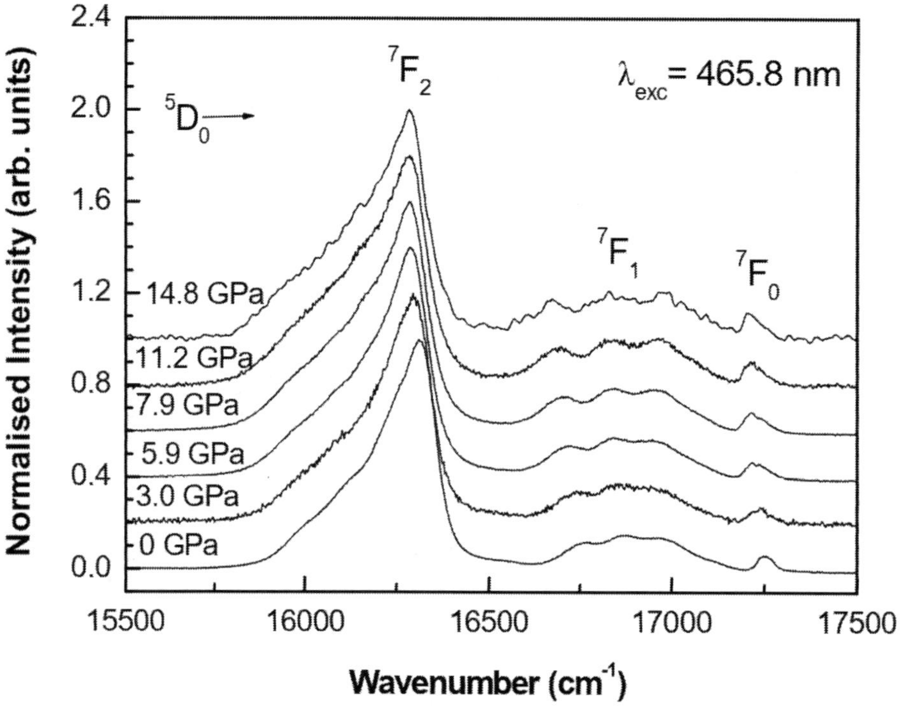

Figure 1. Luminescence spectra of TKN1Eu glass at different pressures. The spectra are normalized to maximum of the $^5D_0 \rightarrow {}^7F_2$ transition.

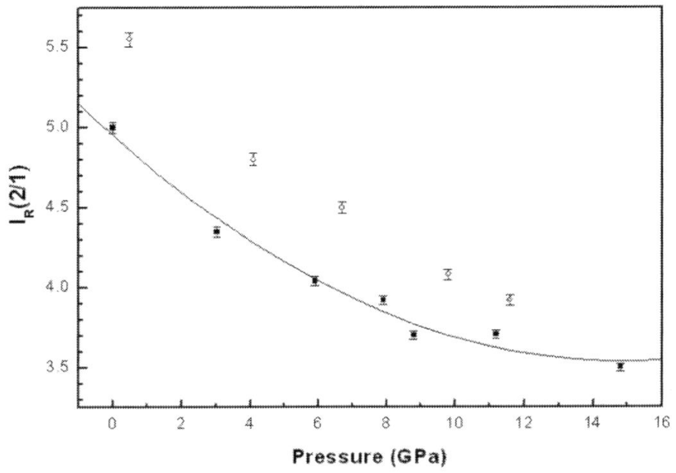

Figure 2. Luminescence intensity ratio I_R (2/1) as a function of pressure. The solid (open) symbols correspond to data taken at increasing (decreasing) pressure. The curve serves as a guide to the eye.

Figure 3 presents the pressure induced red shift of the $^5D_0 \rightarrow {}^7F_{0,1,2}$ emission band positions of the Eu^{3+} ions along with a linear fit to the data obtained with increasing pressure. The barycentres of $^5D_0 \rightarrow {}^7F_J$ (J = 0, 1 and 2) emission bands are presented in table 1. As can be seen from figure 3, the pressure dependent energy levels shift can be expressed quite well as a linear function [32, 33].

$$E_i(p) = E_i(0) + \alpha_i p \qquad \text{(with p in GPa)}$$

where i labels the three different transitions. The $E_i(0)$ values correspond to the values at 0 GPa and the coefficients α_i are determined from the fits as

$$\alpha(^5D_0 \rightarrow {}^7F_0) = -4.1(1)\ cm^{-1}/GPa$$
$$\alpha(^5D_0 \rightarrow {}^7F_1) = -2.6(1)\ cm^{-1}/GPa$$
$$\alpha(^5D_0 \rightarrow {}^7F_2) = -2.8(1)\ cm^{-1}/GPa$$

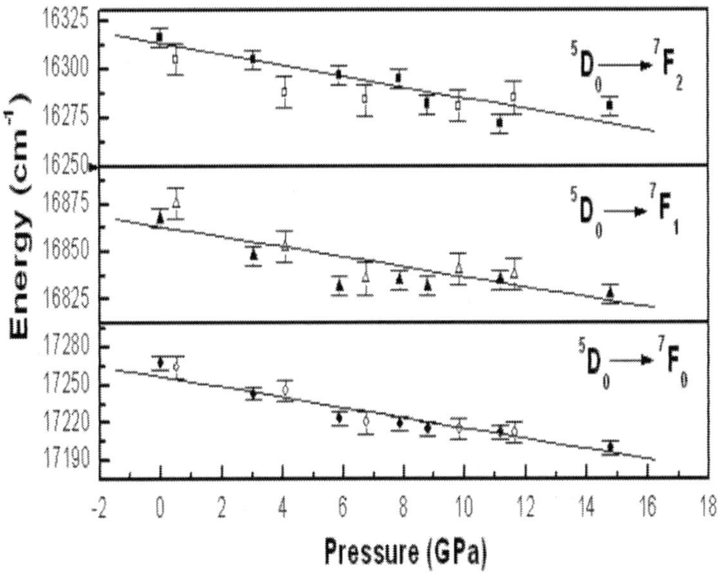

Figure 3. Peak positions of the $^5D_0 \rightarrow {}^7F_0$, $^7F_1$ and $^7F_2$ transitions of the Eu^{3+} ions in TKN1Eu glass as a function of pressure. For the $^5D_0 \rightarrow {}^7F_1$ transitions, the data points correspond to the barycentre of this transition (see figure 6) and the data points for the $^5D_0 \rightarrow {}^7F_0$ and $^7F_2$ transitions correspond to the peak maximum. The solid (open) symbols correspond to data taken at increasing (decreasing) pressure. The solid line is the fit to a linear equation for the data at increasing pressure.

Shift of peak positions of $^7F_0$ level with increasing pressure in the TKN1Eu glass is observed to be more than that of phosphate [14] and lithium borate [15] glasses due to higher covalent nature of Eu-O in tellurite glass. Pressure-induced peak position shifts result from different mechanisms. For example, the relative shift of the LS manifolds with respect to each other is due to variations in the Coulomb interactions. For a given LS value, the shift of each crystal-field split J manifold results from the variation of the spin-orbit coupling constant, ξ_{4f}. For a given SLJ value, the shift and splittings of the Stark levels with respect to each other arises from the variations in strength and symmetry of the crystal-field around the Eu^{3+} ions. Thus, the measured shifts and splittings of each luminescence peak with pressure can provide the data necessary for probing the Coulomb interactions, 4f wave functions and/or the crystal-field.

As a first approximation, the peak (the maximum intensity) of each ^7F_J manifold should be independent of the crystal-field. On the other hand, the ligand orbital affects the 4f wave functions to produce small variations in the spin-orbit coupling constant, ξ_{4f}. The variation of ξ_{4f} with pressure can be determined by use of the Landè rule [34]

$$E(^7F_J) - E(^7F_{J-1}) = J\,\xi_{4f} \tag{4}$$

The values of $\xi_{4f}(p)$, normalized to ambient pressure has been calculated from our experimental data relating to the $^7F_0$ and $^7F_1$ levels. The results are collected in table 1 and figure 4 which show that ξ_{4f} decreases with pressure, in accordance with the findings in similar studies [35, 36].

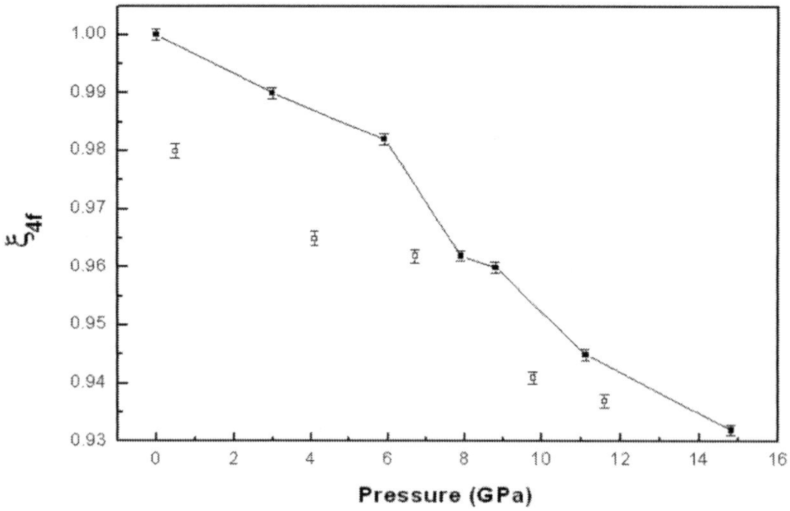

Figure 4. The variation of spin-orbit coupling (ξ_{4f}) with pressure relating to the $^7F_0$ and $^7F_1$ levels of Eu^{3+} ions in TKN1Eu glass. The solid (open) symbols correspond to data taken at increasing (decreasing) pressure. Solid line is guide to the eye for increasing pressure.

Albin and Horrocks [37] have reported a phenomenological equation which relates the energy (υ (cm^{-1})) of the $^5D_0 \rightarrow {}^7F_0$ transition to the sum of the formal negative charge q on the ligands directly bonded to the Eu^{3+} ion given by:

$$\upsilon\ (cm^{-1}) = 17273 + 2029\,q - 0.76\,q^2 \qquad \text{(at T = 296 K)} \tag{5}$$

This relationship is based on the energy of the $^5D_0 \rightarrow {}^7F_0$ transition being a function of the interelectronic repulsion parameter and that any changes in energy can be regarded as a nephelauxetic effect, i.e., expansion of the f-orbitals of Eu^{3+} ion. Using equation (5) and measured transition energies and by taking the negative root of the solution of equation (5) as the possible value, one can calculate the formal negative charge on the neighboring oxygen at various applied pressures (see table 1 and figure 5).

Table 1: Energies (cm^{-1}) for $^5D_0 \rightarrow {}^7F_0$ and $^7F_1$ transitions, luminescence intensity ratio of $^5D_0 \rightarrow {}^7F_2$ to $^5D_0 \rightarrow {}^7F_1$ (I_R (2/1)), lifetime of $^5D_0$ (τ, μs), spin-orbit coupling constant (ξ_{4f}), formal negative charge (Q), crystal field parameters (B_{20}, B_{22}, cm^{-1}) and their ratios (B_{22}/B_{20}) and crystal-field strength parameter (N_v, cm^{-1}) for the TKNEu glass under (a) increasing and (b) decreasing pressure (GPa).

P	$^5D_0 \rightarrow {}^7F_0$	$^7F_1$	$^7F_1$	$^7F_1$	$^7F_2$	I_R (2/1)	τ	ξ_{4f}	Q	B_{20}	B_{22}	B_{22}/B_{20}	N_v
(a) Increasing pressure													
0	17267	16975	16873	16757	16316	5.00	741	1	-1.68	607	200	0.33	1062
3.0	17243	16965	16852	16728	16305	4.35	685	0.990	-4.95	636	220	0.35	1123
5.9	17223	16955	16832	16708	16297	4.04	599	0.982	-6.74	662	229	0.35	1169
7.9	17219	16955	16842	16708	16295	3.92	545	0.962	-7.06	676	225	0.33	1185
8.8	17215	16955	16842	16698	16283	3.70	528	0.960	-7.36	680	213	0.31	1179
11.2	17212	16975	16844	16687	16272	3.71	422	0.945	-7.58	730	250	0.34	1286
14.8	17200	16979	16828	16676	16281	3.50	299	0.932	-8.41	761	278	0.37	1358
(b) Decreasing pressure													
11.6	17212	16986	16842	16687	16285	3.92	387	0.937	-7.58	761	272	0.36	1352
9.8	17216	16972	16851	16699	16281	4.08	432	0.941	-7.28	712	233	0.33	1249
6.7	17220	16955	16845	16708	16285	4.50	494	0.962	-6.98	647	234	0.36	1152
4.1	17245	16979	16860	16721	16288	4.80	586	0.965	-4.75	674	236	0.35	1153
0.5	17264	17007	16873	16749	16305	5.55	643	0.980	-2.25	643	276	0.43	1193

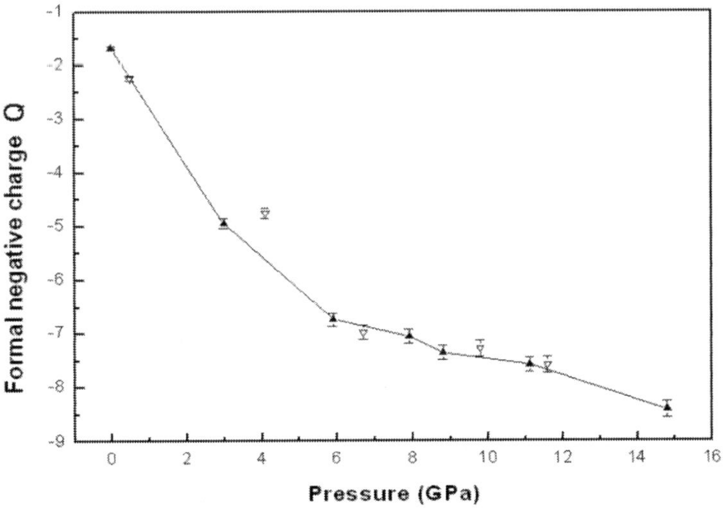

Figure 5. The variation of formal negative charge Q on the ligands directly bonded to the Eu^{3+} ion with pressure in TKN1Eu glass. The solid (open) symbols correspond to data taken at increasing (decreasing) pressure. The curve serves as a guide to the eye.

This increase in the formal negative charge on oxygen between ambient and 14.8 GPa is induced by the expansion of the f-orbitals (nephelauxetic effect) of Eu^{3+} ion. The increase of the formal negative charge on coordinating ligands is a result of the partial transfer of valence (sd) electrons, which correlates with a decrease in the electronegativity of Eu^{3+} ion. It is well known that electronegativity should be considered as an orbital property rather than as an atomic property [38]. It is also recognized that the electronegativity of an orbital is a function of the electron occupancy of that orbital [38]. Therefore, the expansion of the 4f orbitals of Eu^{3+} ion with the application of pressure decrease concomitantly the electronegativity of Eu^{3+} ion.

4.2. Crystal-field analysis

For a free Eu^{3+} ion, all the $4f^n$ intraconfigurational transitions are forbidden by the Laporte rule [22, 39]. However, the ligand field (due to ions constituting the glass host) around the ion, distorts the free ion levels, causing higher energy states to admix and allow transitions between the different levels of the f^n configuration [23]. Slight differences in the bonding parameters (e.g. ligand distance, ligand angle, coordination number and covalency) cause variations in the strength of the ligand field which modify the free ion levels. Figure 6 shows the three Stark splittings obtained for the $^5D_0 \rightarrow {}^7F_1$ transition of Eu^{3+} ions in the TKN1Eu glass under pressure. The complete removal of degeneracy for the $^5D_0 \rightarrow {}^7F_1$ transition at all pressures suggests that the local symmetry of the Eu^{3+} ions in TKN1Eu glass is low corresponding to orthorhombic, monoclinic or triclinic symmetry in a crystal [17, 18]. From figure 6, the observed increase in the energy separation between the two extreme Stark components with increasing pressure quantitatively reflects an increase in the CF strength at the Eu^{3+} ion sites under pressure.

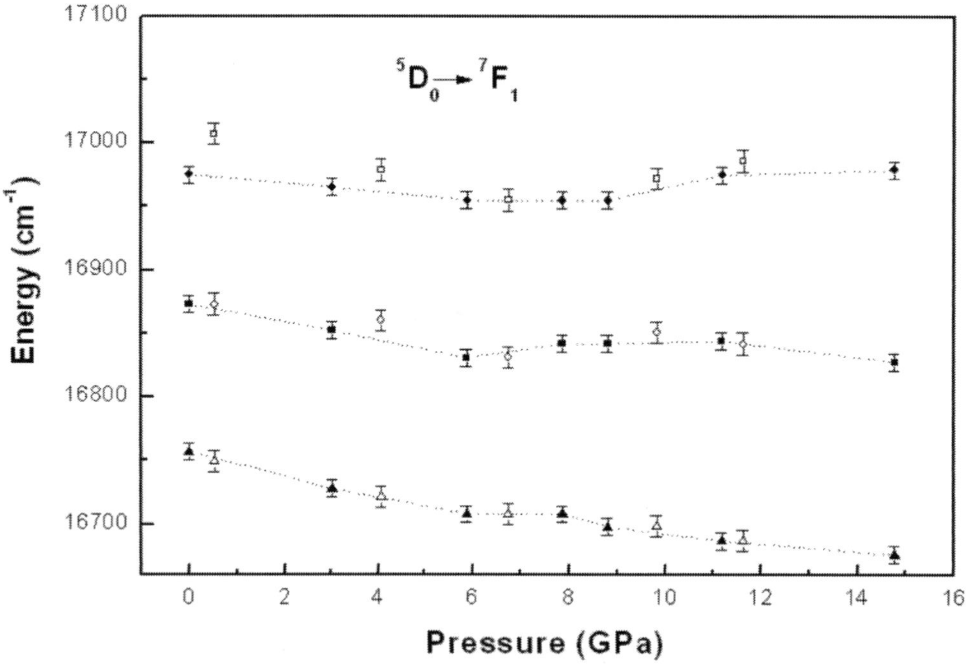

Figure 6. Peak positions of the $^5D_0 \rightarrow {}^7F_1$ Stark transitions of Eu^{3+} ions in TKN1Eu glass as a function of pressure. The solid (open) symbols correspond to data taken at increasing (decreasing) pressure.

From the Stark splittings of the $^5D_0 \rightarrow {}^7F_1$ transition, the CF analysis has been carried out by assuming C_{2v} point symmetry for the local structure around the Eu^{3+} ions [18, 40-43]. Therefore, the second rank even CF parameters, B_{20} and B_{22}, have been calculated by diagonalizing the complete Hamiltonian (free ion and crystal-field) using C_{2v} symmetry by taking into account J-mixing [18]. From the 11 SLJ multiplets, a total of 69 JM_J states are used in the basis set for the CF calculations. During the fitting of these energy levels, all 19 free ion parameters (F^2, F^4, F^6, ξ, α, β, γ, T^2, T^3, T^4, T^6, T^7, T^8, M^0, M^2, M^4 and P^2, P^4, P^6) are fixed to the values of Eu^{3+}:$LaCl_3$ [44] and only the second rank CF parameters have been optimized. The fitting process minimizes the root mean square deviation between the experimental and calculated $^7F_1$ Stark levels [45]. The B_{20}, B_{22} (with J-mixing), B_{22}/B_{20} and N_v using equation (3) thus calculated are collected in table 1 and plotted in figure 7. It is interesting to note that the values of CF and strength parameters (B_{20}, B_{22} and N_v) in the present glass are almost same as those of Eu^{3+}:lithium borate glass up to 6 GPa (figure 8) [15]. It implies that Eu^{3+} ions in TKN1Eu glass experiences almost the same environmental strength, as delivered predominantly from the oxygen neighbors in the lithium borate glass. At above 6 GPa, the N_v values are in between those obtained for lithium borate and phosphate glasses that reflects the different environment around the Eu^{3+} ion. As can be seen from figure 7, the variation of B_{20}, B_{22}, B_{22}/B_{20} and N_v with pressure exhibits a linear dependence on pressure.

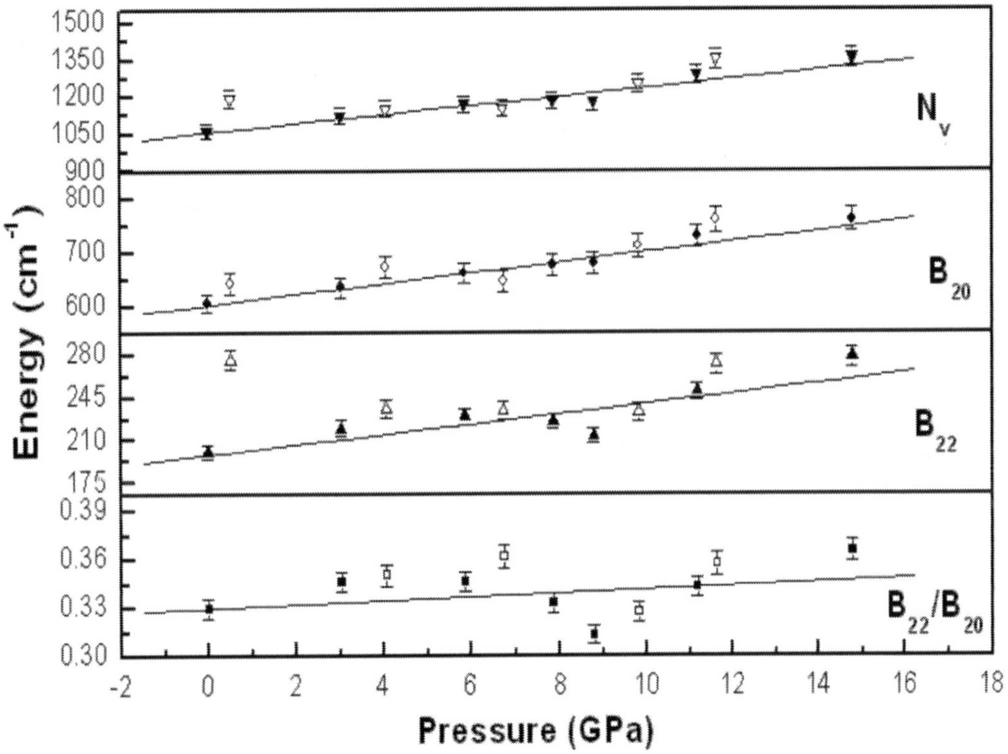

Figure 7. Variations of B_{20}, B_{22}, B_{22}/B_{20} and N_v with pressure. The solid (open) points correspond to data taken at increasing (decreasing) pressure. Lines indicate linear fits to the data at increasing pressure.

4.3. Fluorescence decay

The fluorescence decay from the $^5D_0$ level has been measured by exciting the Eu^{3+} ions to the excited state, $^5D_2$, using as in the previous studies the 465.8 nm line of Ar^+ laser. A fast non-radiative decay populates the emitting level, $^5D_0$. Figure 9 shows the decay profiles at different pressures. The decay curves at all the pressures are found to be single exponential in good approximation up to 14.8 GPa, as indicated in figure 9 for the 11.2 GPa data. From the decay curves lifetimes have been evaluated by finding first e-folding times. The lifetimes at different pressures are presented in table 1 as well as plotted in figure 10. As can be seen from figure 10, lifetime of the $^5D_0$ level decreases with increasing pressure.

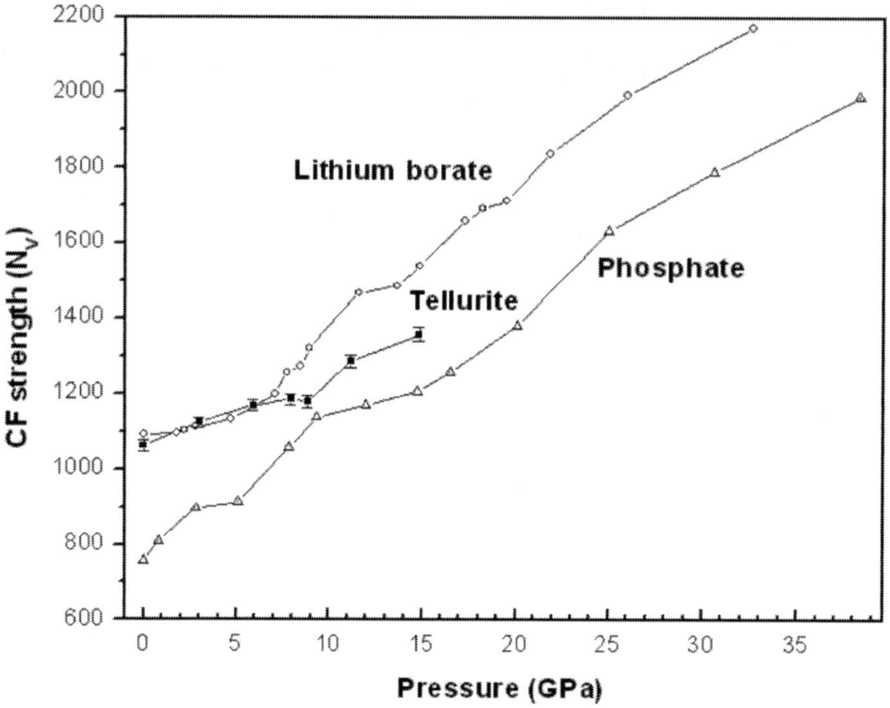

Figure 8. Variation of N_v of Eu^{3+} ions in lithium borate, tellurite and phosphate glasses with increasing pressure.

The observed decrease of lifetime with increasing the pressure can be explained by either enhanced multi-phonon deexcitation probability or an increase of electronic transition probabilities. Because of the large energy gap (12,000 cm^{-1}) between emitting level $^5D_0$ and next lower $^7F_6$ level, the role of multi-phonon relaxation in depleting the $^5D_0$ level in any Eu^{3+}:glass system is regarded as negligible. On the other hand, the increase of electronic transition probability results from an increase in CF strengths around the Eu^{3+} ions with pressure. The increase in CF strengths results in an enhanced configuration interaction leading to a mixing of opposite parity configurations with the $4f^6$ configuration, due to the odd-parity CF Hamiltonians, causing the observed increase in the transition probabilities. Similar trends have been noticed in our earlier studies [14, 15, 18]. In the Eu^{3+}-doped phosphate [14], lithium borate [15] and tellurite glasses, lifetimes for the $^5D_0$ level at 0 GPa are found to decrease in the order of phosphate > lithium borate > tellurite as shown in figure 10. According to Allred-Rochow, Li, K and Nb have electronegativities 0.97, 0.91 and 1.23, respectively, which allows one to propose that the cation with higher electronegativity lowers the lifetime of the fluorescent level of the RE ion in the glass matrix when present as a modifier. This may be due to the increase of covalency between the cation and oxygen ion in the order KO < LiO < NbO which in turn decreases the covalency of Eu-O bond in the glass matrix causing the observed reduction of lifetime. Further, it is interesting to note that lifetime values increase with increase in the atomic radii of the modifier.

The I_R (2/1) values are considerably higher while releasing pressure indicating that application of pressure caused local deformation around the Eu^{3+} ions resulting in higher asymmetry for the Eu^{3+} ion sites. From figures 2-8, one can see that release of pressure (data shown as open points) has no significant hysteresis effect. Over all, pressure induced changes in inter-atomic bond lengths and bond angles are almost continuous as reflected from the pressure induced variations in lifetimes and crystal field parameters and strengths.

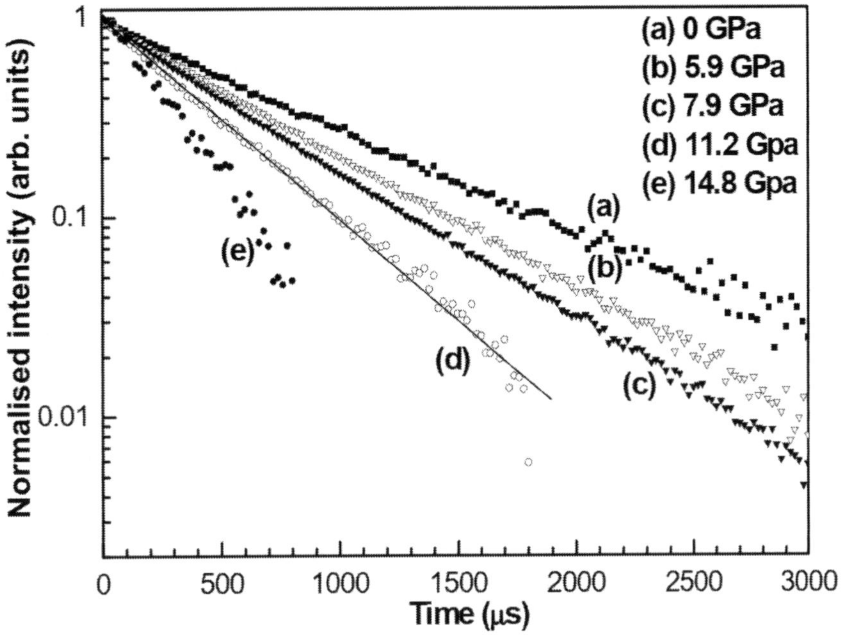

Figure 9. Decay profiles for the $^5D_0$ level of Eu^{3+} in TKN1Eu glass at various increasing pressures. The single exponential fit is also shown for 11.2 GPa.

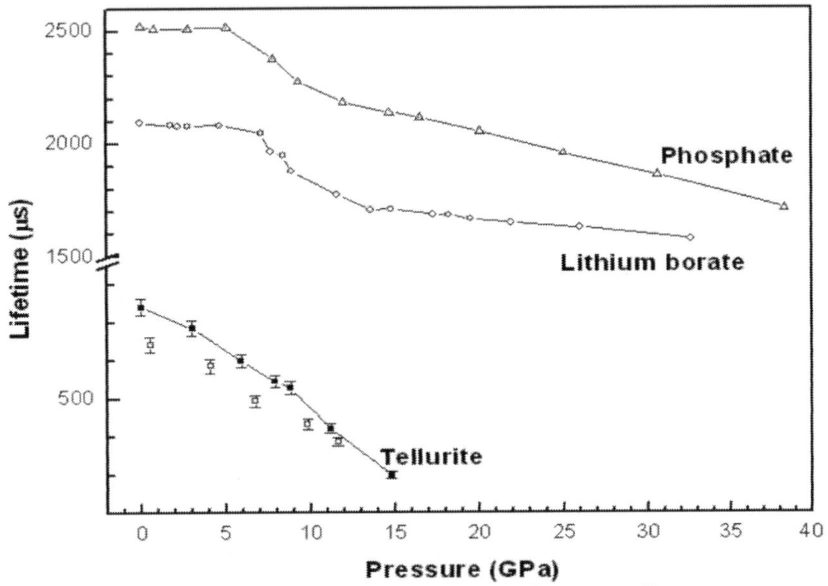

Figure 10. Variation of lifetime for the level $^5D_0$ of Eu^{3+} ions phosphate, lithium borate and tellurite glasses with increasing pressure. The solid (open) symbols correspond to data taken at increasing (decreasing) pressure and the solid line gives a guide to the eye.

5. Conclusions

Eu^{3+}-doped potassium niobium tellurite glass has been prepared and characterised for luminescence properties under pressure up to 14.8 GPa at RT. Red shifts of the emission band positions with pressure are attributed to an expansion of the 4f electron wavefunctions with increasing covalency. The decrease of emission intensity ratio I_R (2/1) with increasing pressure attributed to lowering of asymmetry around the Eu^{3+} ions. The CF splitting of the emission $^5D_0 \rightarrow {}^7F_1$ band increases with pressure due to increase in the CF strength at the Eu^{3+} ion sites as reflected from the increase in CF strength parameter from 1062 to 1358 cm^{-1} in the pressure range studied. The fluorescence decay curves for the $^5D_0$ level remain single exponential for the entire pressure range but the lifetimes decreases drastically from 0.76 ms to 0.30 ms with increase of pressure from 0 to 14.8 GPa. The quenching of lifetime for the $^5D_0$ level with pressure is attributed due to increase in electronic transition probabilities arising from increase in CF strengths around Eu^{3+} ions.

Acknowledgements

This work has been carried out under the financial assistance of a major research project sponsored by DAE-BRNS, Govt. of India (Sanction No. 2003/34/4-BRNS/600, dt. 11-07-2003). One of the authors (CKJ) is also grateful to Department Physik, Universität Paderborn, Germany, for his stay as a Guest Scientist. One of the authors (PB) is grateful to UGC SERO, Hyderabad, A.P., India, for sanction of the Minor Research Project (File No. MRP-2379/06, dt. 20-3-2007).

References

[1] Zambelli M, Speghini A, Ingletto G, Locatelli C, Bettinelli M, Vetrone F, Boyer J C and Capobianco J A 2004 *Opt. Mater.* **25** 215.
[2] Vetrone F, Boyer J C, Capobianco J, Speghini A and Bettinelli M 2002 *Appl. Phys. Lett.* **80** 1752.
[3] El-Mallawany R 2001 *Tellurite glass Handbook: Physical properties and data* (Boca Raton: CRC press).
[4] Cardinal T, Fargin E, Le Flem G, Couzi M, Canioni L, Segonds P, Sarger L, Ducasse A and Adamietz D 1996 *Eur. J. Solid State Inorg. Chem.* **33** 597.
[5] Ebendorff-Heidepriem H, Ehrt D, Bettinelli M and Speghini A 1999 *Rare-Earth-Doped Materials and Devices III* ed S Jiang and S Honkanen (Bellingham, Washington, USA: SPIE Proc.) vol 3622 pp 19-30.
[6] Nazabal V, Fargin E, Videau J J, Le Flem G, Le Calvez A, Montant S, Freusz E, Ducasse A and Couzi M 1997 *J. Solid State Chem.* **133** 529.
[7] Tröster Th 2003 *Handbook on the Physics and Chemistry of Rare Earths* vol 12, ed K A Gschneidner Jr, J-C G Buenzli and V K Pecharsky (Amsterdam: Elsevier) chapter 217.
[8] Righini G C, Pelli S, Fossi M, Brenci M, Lipovskii A A, Kolobkova E V, Speghini A and Bettinelli M 2001 *Rare-Earth-Doped Materials and devices V* ed S Jiang (Bellingham, Washington, USA: SPIE Proc.) vol 4282 pp 210-215.
[9] Chang W J and Heo J 2001 *Appl. Phys. Lett.* **79** 326.
[10] You H and Hong G 1999 *J. Phys. Chem. Solids* **60** 325.
[11] Görller-Walrand C and Binnemans K 1996 *Handbook on the Physics and Chemistry of Rare-Earths* vol 23, ed K A Gschneidner Jr and L Eyring (Amsterdam: North-Holland) chapter 155.
[12] Capobianco J A, Prouix P P and Raspa N 1989 *Chem. Phys. Lett.* **160** 591.
[13] Zimbelli M, Abril M, Lavin V, Speghini A and Bettinelli M 2004 *J. Non-Cryst. Solids* **345/346** 386.
[14] Surendra Babu S, Babu P, Jayasankar C K, Tröster Th, Sievers W and Wortmann G 2006 *J. Phys.: Condens. Matter* **18** 1927.

[15] Jayasankar C K, Ramanjaneya Setty K, Babu P, Tröster Th and Holzapfel W B 2004 *Phys. Rev.* B **69** 214108.
[16] Holzapfel W B 2003 *J. Appl. Phys.* **93** 1813.
[17] Wybourne B G 1965 *Spectroscopic Properties of Rare Earth Compounds* (New York: Interscience).
[18] Lavin V, Babu P, Jayasankar C K, Martin I R and Rodriguez V D 2001 *J. Chem. Phys.* **115** 10935.
[19] Auzel F 1979 *Mater. Res. Bull.* **14** 223.
[20] Reisfeld R and Lieblich N 1973 *J. Phy. Chem. Solids* **34** 1467.
[21] Babu P and Jayasankar C K 2000 *Physica* B **279** 262.
[22] Ferrari M, Campostrini R, Carturan G and Montagna M 1992 *Philos. Mag.* B **65** 251.
[23] Dejneka M, Snitzer E and Riman R E 1995 *J. Non-Cryst. Solids* **65** 227.
[24] Lery D, Reisfeld R and Avnir D 1984 *Chem. Phys. Lett.* **109** 593.
[25] Peacock R D 1975 *Struct. Bond.* **22** 83.
[26] Görller-Walrand C, Binnemans K 1998 *Handbook on the Physics and Chemistry of Rare Earths* vol 25 ed K A Gschneidner Jr. and L Eyring (Amsterdam: North-Holland) chapter 167.
[27] Capobianco JA, Proulx P P, Bettinelli M and Negrisolo F 1990 *Phys. Rev.* B **42** 5936.
[28] Judd B R 1962 *Phys. Rev.* **127** 750.
[29] Reisfeld R, Zigansky E and Gaft M 2004 *Mol. Phys.* **102** 1319.
[30] Gatterer K, Pucker G and Fritzer H P 1997 *Phys. Chem. Glasses* **38** 293.
[31] Kumar A, Rai D K and Rai S B 2002 *Spectrochim. Acta A* **58** 2115.
[32] Souza Filho A G, Freire P T C, Guedes I, Melo F E A, Mendes Filho J, Custodio M C C, Lebullenger R and Hernandes A C 2000 *J. Mater. Sci. Lett.* **19** 135.
[33] Jovanic B R 2001 *J. Lumin.* **92** 161.
[34] Henderson B and Imbusch G F 1989 *Optical Spectroscopy of Inorganic Solids* (Oxford: Clarendon Press) p 45.
[35] Chen G, Haire R G, Peterson J R 1992 *High Pressure Res.* **6** 371.
[36] Jayasankar C K, Upendra Kumar K, Venkatramu V, Sievers W, Tröster Th and Wortmann G 2007 *J. Non-Cryst. Solids* **353** 1813.
[37] Albin M and Horrocks W D Jr. 1985 *Inorg. Chem.* **24** 895.
[38] Jolly W L 1976 *The Principles of Inorganic Chemistry* (New York: Mc Graw-Hill) p 202.
[39] Reisfeld R 1975 *Struct. Bond.* **22** 123.
[40] Lavin V, Rodriguez V D, Martin I R and Rodriguez-Mendoza U R 1997 *J. Lumin.* **72-74** 437.
[41] Balda R, Fernandez J, Adam J L and Arriandiaga M A 1996 *Phys. Rev.* B **54** 12076.
[42] Ebendorff-Heidepriem H and Ehrt D 1996 *J. Non-Cryst. Solids* **208** 205.
[43] Lochhead M J and Bray K L 1995 *Phys. Rev.* B **52** 15763.
[44] Jayasankar C K, Richerdson F S and Reid M F 1989 *J. Less Common Met.* **148** 289.
[45] Reid M F, Program CFIT, University of Canterbury, Christchurch, New Zealand.

Journal of Physics: Conference Series **121** (2008) 042016 doi:10.1088/1742-6596/121/4/042016

Raman study of zircon-structured RPO_4 (R = Y, Tb, Er, Tm) phosphates at high pressures

E.Stavrou, A. Tatsi, E. Salpea, Y.C. Boulmetis, A.G. Kontos, Y.S. Raptis and C. Raptis
Physics Department, National Technical University of Athens, GR-15780, Athens, Greece

craptis@central.ntua.gr

Abstract. The Raman spectra of tetragonal RPO_4 (R = Y, Tb, Er, Tm) phosphates belonging to the zircon-structure class (space group D_{4h}^{19}) have been measured at high pressures up to 7.9 GPa (YPO_4, $ErPO_4$) and 15.5 GPa ($TbPO_4$, $TmPO_4$). Most of the 12 Raman active modes of this crystal class have been observed at high pressures using a Diamond Anvil Cell. A characteristic feature in the spectra of all crystals studied is the anomalous pressure dependence of the B_{2g} symmetry internal bending mode which softens substantially with increasing pressure to the extent that a level crossing occurs between this mode and one normal E_g rotational mode. These observations imply that an incipient structural phase transition of the crystals may occur at a higher pressure. Such a phase transition has been observed for $TbPO_4$ at a pressure of ~9.5 GPa. The abrupt spectral variations observed at this pressure, indicate that $TbPO_4$ undergoes a first-order phase transition from tetragonal zircon to a lower symmetry structure which is quenchable upon pressure release.

1. Introduction

Rare-earth phosphates RPO_4 have attracted considerable interest in the past for their magnetic properties at low temperatures [1-7], and as possible host materials for laser-active media and for long-term storage of radio-active wastes [8,9]. These compounds crystallize in two closely related structures at ambient conditions: those with the heavier rare-earth ions (R = Tb to Lu) [8,9] have the zircon structure [10] ($ZrSiO_4$, D_{4h}^{19} space group, two formula units in the primitive cell), while the lighter ones (R = La to Gd) crystallize in the monoclinic monazite structure [8,9] ($CePO_4$, C_{2h}^{5} space group, four molecules in the primitive cell). Apart from the heavy rare-earth phosphates, other phosphates, such as YPO_4 and $ScPO_4$, also crystallize in the zircon structure [8,9].

In the zircon-structured phosphates, the phosphorus atoms are four-fold coordinated by oxygen forming strongly bound $(PO_4)^{3-}$ tetrahedral units, while the R^{3+} cations are eight-fold coordinated to oxygen in RO_8 dodecahedral arrangements; the two types of polyhedra link in an alternating edge-sharing fashion forming linear chains parallel to the c axis [9]. Either of the R^{3+} and $(PO_4)^{3-}$ ions occupy D_{2d} symmetry sites, with the oxygen atoms being in C_s sites. On the other hand, the monazite structure is, in fact, a distorted zircon structure, but with higher compactness since the coordination of R^{3+} cation increases to nine resulting in the formation of RO_9 polyhedra [9] in place of RO_8 dodecahedra. Considering the rare-earth phosphates as polymorphs, it has been suggested [9] that the monazite structure is the low-temperature form of these compounds, while the zircon structure corresponds to the high-temperature form; this idea has been supported by reports [1,2] that $TbPO_4$ undergoes a phase transition from zircon- to monoclinic (probably monazite) structure at 2.15 K.

Generally, crystals with the zircon-type structure are known, from X-ray [11,12] and Raman studies [12-15], to sustain an irreversible pressure-induced phase transition to tetragonal scheelite-type structure (C_{4h}^{6} space group, two molecular units in the primitive cell), thus lowering slightly their symmetry to a more compact structure. Those studies [11-15] have dealt with rare-earth vanadates RVO_4 [11-13] and certain chromates $MCrO_4$ (M = Ca, Y) [14,15]. As for the zircon-structured rare-earth phosphates, there is only one relevant report [16] recently on $DyPO_4$ whose Raman spectrum has been measured up to a pressure of 7.5 GPa, but without any evidence for a completed structural modification over this pressure range.

In this article, we present high-pressure Raman studies of zircon-structured phosphates, namely, the rare-earth phosphates $TbPO_4$, $ErPO_4$ and $TmPO_4$, and YPO_4. Primary objective of the work is to investigate whether these crystals sustain any pressure-induced structural modifications. The investigations are motivated by the close relation between the zircon and monazite structures and

© 2008 IOP Publishing Ltd

the fact that application of pressure lowers, in most cases, the crystal symmetry. Hence, Raman spectra of these crystals have been measured up to 7.9 GPa (YPO_4, $ErPO_4$) and 15.5 GPa ($TbPO_4$, $TmPO_4$) with the use of a Diamond Anvil Cell (DAC). Such an anticipated lowering of the crystal symmetry has been observed in $TbPO_4$ which displays a first-order phase transition at a pressure of about 9.5 GPa.

From group theory considerations, it is known that there are 12 Raman active modes for the zircon D_{4h}^{19} structure having the following symmetries: $2A_{1g} + 4B_{1g} + B_{2g} + 5E_g$; of these, seven ($2A_{1g} + 2B_{1g} + B_{2g} + 2E_g$) are *internal* modes [8,10] corresponding, in the case of the RPO_4 compounds, to vibrations of oxygen atoms in the $(PO_4)^{3-}$ units, four ($2B_{1g} + 2E_g$) are *external translational* due to translations of the $(PO_4)^{3-}$ and R^{3+} ions, and one (E_g) is *rotational* of whole $(PO_4)^{3-}$ units. The atomic displacements of all normal modes of the zircon structure have been known [10] for a long time.

The Raman spectra of zircon-structured phosphates at ambient conditions were reported by different researchers [8,17,18] who also made mode symmetry assignments based on polarized Raman measurements and oriented single crystals. As those mode assignments [8,17,18] are not in full agreement among each other, we have also performed (at ambient conditions, outside the DAC) polarized Raman measurements on oriented single crystals and produced our own mode assignments for the heavy rare-earth phosphates.

2. Experimental

Polarized Raman measurements at ambient conditions have been carried out using flux-grown oriented single crystals and a 90^0 scattering geometry. Details on the polarized Raman experiments and mode assignments of crystals will be published elsewhere [19].

Small chips of RPO_4 (R = Y, Tb, Er, Tm) crystals, with nominal dimensions 50-80 μm, cut from larger single crystals were loaded into a Syassen-Holzapfel-type DAC, along with ruby chips for determining the pressure (through the photoluminescence of ruby). Typical diameter values of ~200 μm were used for the gasket hole which was filled with 4:1 methanol-ethanol pressure transmitting medium. Raman spectra were obtained up to 15.5 GPa for $TbPO_4$ and $TmPO_4$, and up to 7.9 GPa for YPO_4 and $ErPO_4$ in sequences of both increasing and decreasing pressure.

The Raman spectra were excited basically by the 514.5 nm line of an Ar^+ laser, but in certain experimental sessions (for example, in the case of $ErPO_4$) the 476.2 nm line of a Kr^+ laser was used to avoid intense levels of photoluminescence. Scattered light from the DAC was analyzed and detected by a system of a SPEX 1403 double spectrometer, a cooled photomultiplier and a photon counting unit. Most of the high-pressure Raman measurements were not polarized, thus making possible detection of all mode symmetries in one spectral run. However, in some occasions, in order to resolve some increasingly overlapping (with pressure) Raman peaks, such as that involving the crossing of the B_{2g} internal and the nearby E_g rotational modes, detailed slow-scans were obtained using parallel- and cross-polarization configurations.

3. Results and Discussion

As a representative example of the polarized measurements at ambient conditions, Figure 1 shows the *zz*, *xx*, *yx*, and *yz* polarized components of the Raman spectrum of $TbPO_4$ corresponding to A_{1g}, ($A_{1g} + B_{1g}$), B_{2g} and E_g mode symmetries, respectively. Application of selection rules on these polarized spectra enabled us to make symmetry assignment for the eleven (out of twelve expected) Raman modes of $TbPO_4$, and the results of this assignment are given in Table I, along with other interesting quantities discussed below. Similar mode assignments were conducted for the other three crystals of the study, with the results being, more or less, the same. Also, the mode frequencies of the other crystals are very close to those of $TbPO_4$ as the masses of the rare-earth R^{3+} ions are almost the same. Our mode assignment for the heavy rare-earth phosphates is in full agreement with that of Ref. 17, but differs from those of Refs 8 and 18 for three only modes.

In Table I, apart from the symmetry and character (internal, external) assignments, ambient conditions frequencies ω_0 and normalized $(1/\omega_0)(\partial\omega/\partial P)_T$ slopes for the 11 observed Raman modes of $TbPO_4$ are also shown. The normalized slopes are given in place of Grüneisen parameters whose determination is not feasible because of the lack of elastic constants data in the literature for $TbPO_4$ or any other crystal of this study.

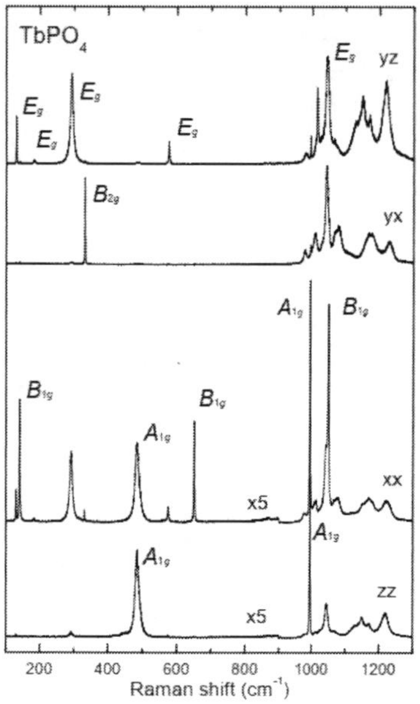

Figure 1. Polarized $y(zz)x$, $z(xx)y$, $z(yx)y$ and $z(yz)y$ Raman spectra of $TbPO_4$.

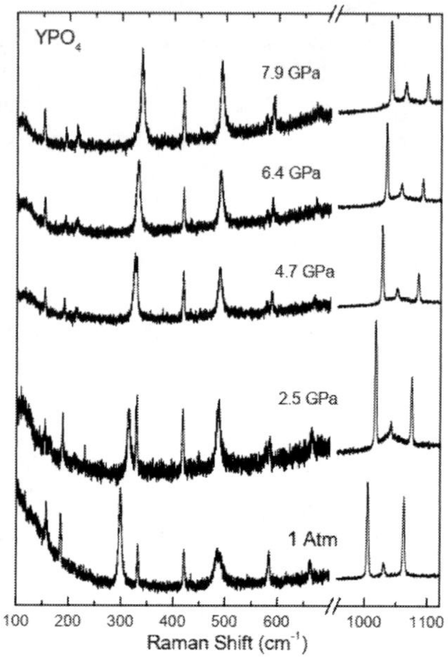

Figure 2. Evolution of the Raman spectrum of YPO_4 with pressure.

Table I. Ambient conditions frequencies ω_0, assignments, and normalized $(1/\omega_0)(\partial\omega/\partial P)_T$ slopes of the Raman active modes of $TbPO_4$.

ω_0 (cm^{-1})	Assignment	Character	$(1/\omega_0)(\partial\omega/\partial P)_T$ (10^{-3} GPa^{-1})
130	E_g	Extern. Transl.	– 0.5
141	B_{1g}	Extern. Transl.	7.3
185	E_g	Extern. Transl.	0.5
293	E_g	Rotational	13.4
331	B_{2g}	Internal	– 1.7
484	A_{1g}	Internal	1.5
576	E_g	Internal	1.7
649	B_{1g}	Internal	3.0
995	A_{1g}	Internal	5.3
1014	E_g	Internal	4.9
1049	B_{1g}	Internal	5.0

The evolution of the Raman spectrum of YPO_4 with pressure is given in Figure 2. Ten Raman modes were observed at ambient pressures when the crystal was inside the DAC. As the pressure is increased, an additional band appears at ~210 cm^{-1} (at 4.7 GPa); such a band was not observed in the other crystals of this work, but had been reported previously [8] in YPO_4 without symmetry assignment. The pressure-independent peak at 418 cm^{-1} (Fig. 2) is due to Raman scattering from the sapphire (Al_2O_3) window [20] of the DAC and has proved useful for calibration of spectra. Similarly, as the pressure is increased and the E_g internal mode at 576 cm^{-1} shifts, an additional peak emerges at this frequency which remains pressure-independent throughout the pressure range; this peak is also attributed to scattering from the sapphire window [20]. In the spectra of Fig.2, it is interesting to note the softening with pressure of the B_{2g} internal bending [10] mode to such an extent that it crosses the E_g rotational mode. Such level crossings have been also observed in the other three rare-earth crystals of this study, as well as in $DyPO_4$ in a previous study [16]. The softening with pressure of the B_{2g} mode is, in fact, an indication for an incipient phase transition at a higher pressure, which has not been detected for YPO_4 up to 7.9 GPa. Recording of spectra was not possible above 7.9 GPa for YPO_4 and $ErPO_4$ as a combined result of low Raman signal and gasket hole deformation. Similar spectral evolution with pressure was observed in $ErPO_4$ and $TmPO_4$, again without evidence for a completed phase transition over their respective pressure ranges. Mode frequency-pressure plots for the YPO_4, $ErPO_4$ and $TmPO_4$ crystals are given in Figure 3. In these plots the level crossing of the B_{2g} and E_g modes is seen at about 5.7, 6.0 and 4.8 GPa for YPO_4, $ErPO_4$ and $TbPO_4$, respectively.

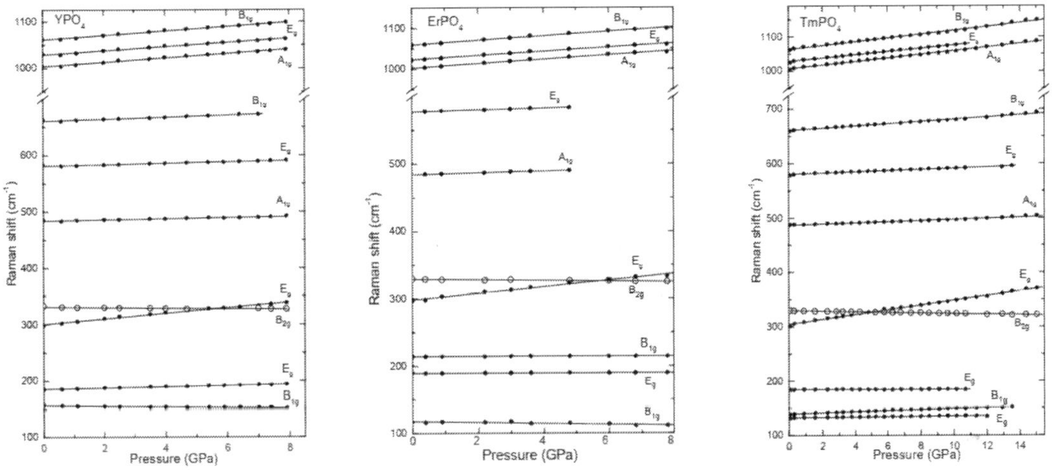

Figure 3. Frequency-pressure plots of Raman modes of YPO_4 (left), $ErPO_4$ (center) and $TmPO_4$ (right). The solid lines correspond to linear least-squares fittings giving the $(\partial\omega/\partial P)_T$ slopes.

The pressure dependence of the Raman spectrum of $TbPO_4$ is, however, significantly different compared to the other three crystals of the study. After the level crossing at about 7.0 GPa, it is clearly evident that at a pressure $P_c \approx 9.5$ GPa, substantial qualitative and quantitative changes take place (Figure 4), with the most notable being the abrupt nature of these changes and the increase of the Raman bands from 11 to, at least, 20. These results imply a definite first-order phase transition of $TbPO_4$ from tetragonal zircon to a lower symmetry structure. All spectral changes are reversible up to 9.5 GPa, but above this pressure the observed drastic changes are irreversible, thus showing that the phase transition is quenchable. These changes are also illustrated in the frequency-pressure plots of Figure 5 for $TbPO_4$ in which mode discontinuities and decrease of $(\partial\omega/\partial P)_T$ slopes are evident. Only few Raman bands of the low-pressure zircon phase appear to evolve continuously to a corresponding band of the high-pressure phase. Other crystals with the zircon structure are known [11-15] to irreversibly transform to the scheelite structure which displays 13 Raman active modes. From the large number (20) of Raman modes observed above P_c, it is clear that the scheelite structure should be excluded as a probable high-pressure phase for $TbPO_4$. In this situation, a transition to an even lower symmetry structure, such as a monoclinic one, looks likely. This proposition is supported by the resemblance of the high-pressure spectrum of $TbPO_4$ to those of the light rare-earth phosphates which crystallize in the monoclinic monazite structure [8,9]. The latter structure has four molecules in the primitive cell and 36 Raman active optic modes ($18A_g + 18B_g$), so it can be considered as a likely candidate for the high-pressure phase of $TbPO_4$. The absence of a similar pressure-induced transition in YPO_4, $ErPO_4$ and $TmPO_4$ may be due to a higher P_c required for these compounds. It is reminded that, among the heavy rare-earth phosphates, $TbPO_4$ is closest to the family of the light RPO_4 ones (R = La to Gd). Therefore, it should be expected that the pressure threshold (P_c) required for a phase modification will be the lowest for $TbPO_4$ in comparison to the other members of the heavy rare-earth phosphates family. Further discussion on the high-pressure phase transition of $TbPO_4$ will be given in a forthcoming article [19].

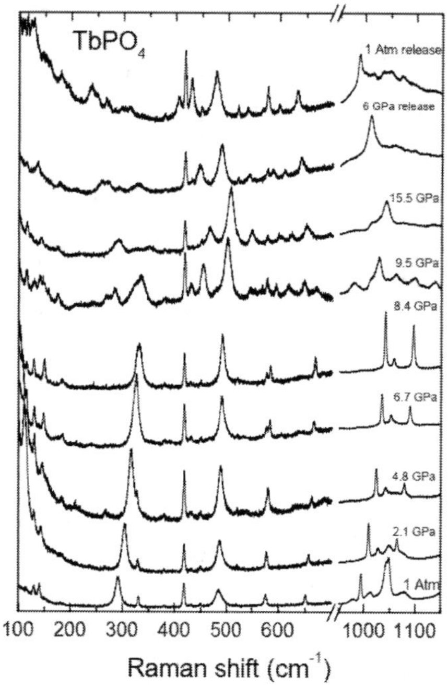

Figure 4. Raman spectra of $TbPO_4$ at various pressures in sequences of increasing-decreasing pressure cycles.

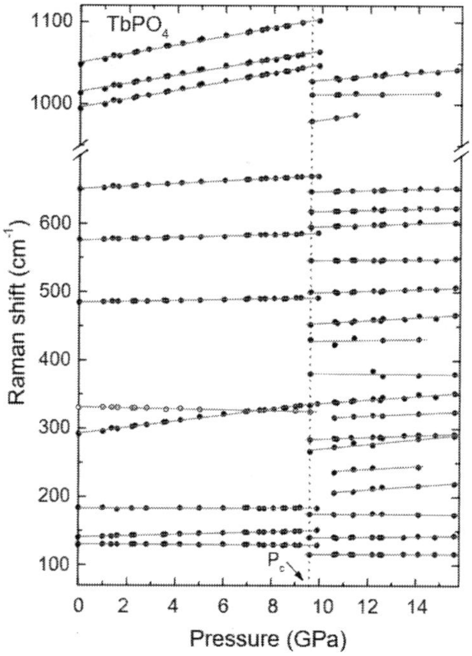

Figure 5. Frequency-pressure plots of the Raman active modes of $TbPO_4$. The solid lines correspond to linear least-squares fittings giving the $(\partial\omega/\partial P)_T$ slopes.

4. Conclusions

The pressure dependence of the Raman spectra of four zircon-structured phosphates, namely YPO_4, $ErPO_4$, $TmPO_4$ and $TbPO_4$, has revealed (in all crystals) an anomalous softening of the B_{2g} internal bending mode, as well as a level crossing of this mode with the E_g rotational mode. The anomalous softening of the B_{2g} mode is an indication for an incipient phase transition in the crystals at a higher pressure.

The abrupt spectral variations observed in $TbPO_4$ at ~9.5 GPa imply clearly that the anticipated transition occurs in this crystal lowering its symmetry from tetragonal zircon to a most likely monoclinic structure, unlike other zircon-structure crystals [11-15] whose high-pressure phase is that of scheelite. The phase transition is of first order and irreversible upon pressure release. One of the likely structures for the high-pressure phase of $TbPO_4$ is the monoclinic monazite one in which the lighter rare-earth ion (La to Gd) phosphates crystallize. Such a possibility appears feasible as $TbPO_4$ is closest to this group of phosphates. The fact that a similar transition has not been observed in $ErPO_4$ or $TmPO_4$ may be understood in terms of a higher critical pressure P_c required by them considering that these heavy rare-earth phosphates are distant (relatively to $TbPO_4$) from the light phosphates group. As for YPO_4, the range of pressures applied (up to 7.9 GPa) may be narrow to observe structural changes.

References

[1] H. Suzuki and T. Nakajima, J. Phys. Soc. Japan **47**, 1441 (1979).
[2] W. Nägele, D. Hohlwein and G. Domann, Z. Physik B **39**, 305 (1980).
[3] S. Bluck and H.G. Kahle, J. Phys. C **21**, 5193 (1998).

[4] G.K. Liu, C.-K. Loong, F. Trouw, M.M. Abraham and L.A. Boatner, J. Appl. Phys. **75**, 7073 (1994).
[5] P. Morin, J. Rouchy and Z. Kazei, Phys. Rev. B **50**, 12625 (1994).
[6] P. Morin, J. Rouchy and Z. Kazei, J. Phys.: Condens. Matter **8**, 7967 (1996).
[7] Y. Hirano, S. Skanthakumar, C.-K. Loong, N. Wakabayashi and L. Boatner, Phys. Rev. B **66**, 024424 (2002).
[8] G.M. Begum, G.W. Beall, L.A. Boatner and W.J. Gregor, J. Raman Spectrosc. **11**, 273 (1981).
[9] A. Meldrum, L.A. Boatner and R.C. Ewing, Phys. Rev. B **56**, 13805 (1997).
[10] P. Dawson, M.M. Hargreave and G.R. Wilkinson, J. Phys. C **4**, 240 (1971).
[11] X. Wang, I. Loa, K. Syassen, M. Hanfland and B. Ferrand, Phys. Rev. B **70**, 064109 (2004).
[12] A. Jayaraman, G.A. Kourouklis, G.A. Espinoza, A.S. Cooper and L.G. VanUitert, J. Phys. Chem. Solids **48**, 755 (1987).
[13] S.L. Duclos, A. Jayaraman, G.P. Espinoza, A.S. Cooper and R.G. Maines Sr, J. Phys. Chem. Solids **50**, 769 (1989).
[14] Y.W. Long, W.W. Zhang, L.X. Yang, Y. Yu, R.C. Yu, S. Ding, Y.L. Liu and C.Q. Jin, Appl. Phys. Lett. **87**, 181901 (2005).
[15] Y.W. Long, L.X. Yang, Y. Yu, F.Y. Li, R.C. Yu, S. Ding, Y.L. Liu and C.Q. Jin, Phys. Rev. B **74**, 054110 (2006).
[16] A.G. Kontos, E. Stavrou, V. Malamos, Y.S. Raptis and C. Raptis, Phys. Status Solidi b **244**, 386 (2007).
[17] R.J. Elliott, R.T. Harley, W. Hayes and S.P.R. Smith, Proc. Roy. Soc. Lond. A **328**, 217 (1972).
[18] P.C. Becker, N. Edelstein, G.M. Williams, J.J. Bucher, R.E. Russo, J.A. Koningstein, L.A. Boatner and M.M. Abraham, Phys. Rev. B **31**, 8102 (1985).
[19] A. Tatsi, E. Stavrou, Y.C. Boulmetis, A.G. Kontos, Y.S. Raptis and C. Raptis, to be published.
[20] G.A. Watson Jr, W.B. Daniels and C.S. Wang, J. Appl. Phys. **52**, 956 (1981).

Journal of Physics: Conference Series **121** (2008) 042017 doi:10.1088/1742-6596/121/4/042017

α-Boron at very High Pressure : Structural and Vibrational Properties

A Polian, J C Chervin, P Munsch, M Gauthier

Physique des Milieux Denses, IMPMC, CNRS, Université Pierre et Marie Curie-Paris6, 140 rue de Lourmel, 75015, Paris, France

E-mail: alain.polian@impmc.jussieu.fr

Abstract. Boron is an atomic species at the origin of the hardest materials. It has one of the most complicated chemistry among the elements. The simplest member of the family, α-boron, is based on an arrangement of slightly distorted B_{12} icosahedron in a rhombohedral unit cell. It has been studied by x-ray diffraction and Raman scattering up to 100 GPa, showing no sign of phase transformation.

1. Introduction

Elemental boron and boron-rich solids have many applications in nuclear engineering for example, or in the search for very hard solids. In the last twenty years, two conferences were entirely dedicated to the results obtained on these compounds[1,2]. As stated by Donohue[3], « Boron has the second most complicated structural chemistry among the elements ». At ambient condition, it exists in various complex crystal structures, all dominated by the icosahedral B_{12} cluster. These icosahedra can be interlinked by strong covalent bonds in different ways to form several polymorphs, such as the α-rhombohedral B_{12}, (α-B_{12}, 12 atoms *i.e.* one icosahedron in the unit cell), or β-rhombohedral B_{105} (105 atoms in the unit cell), the most common of these compounds, both belonging to the same space group $R\bar{3}m$. Supraconductivity has been recently found in β-boron [4], which has stimulated theoretical calculations [5-7].

From an experimental point of view, the vibrational properties of α-boron were studied by Raman scattering at ambient temperature up to 30 GPa [8]. In this paper, the experimental results were compared with *ab initio* lattice dynamics calculations, with the result that an unambiguous assignment of all the features, including the remarkably narrow mode at 524 cm^{-1}, attributed to a highly harmonic librational mode of the icosahedron. Since then, two papers were dealing with high pressure structural properties of β-boron, the first one to 30 GPa [9] and the second to 100 GPa [10]. In this last paper, the authors claim a pressure induced amorphization around 100 GPa, and discussed the relative stability of the intra- and intericosahedral bonds.

Few experimental works were published on the properties of α-boron. It is only very recently [11, 12] that new results were obtained. The first one presented studies of conductivity up to 112 GPa without any pressure transmitting medium (PTM) and Raman scattering up to 80 GPa using Daphne oil as a PTM; both experiments were performed on powder. The second one [12] on x-ray diffraction up to 106 GPa and Raman scattering up to 75 GPa, again on powder.

In this paper, we present x-ray diffraction results up to 100 GPa obtained on single crystals, and Raman scattering data, also on single crystals.

© 2008 IOP Publishing Ltd

2. Experimental

The samples were single microcrystals of α-B_{12} were prepared from recrystallization from solution. Isothermal saturation of molten palladium-boron aloys, using a vapor source of boron.

We used a membrane diamond cell [13], with beveled diamond anvils whose culet were 300 μm for the bevel and 100 μm for the center. The gasket was in rhenium, with an initial thickness of 20 μm and a diameter of 50 μm. The sample, 10 x 10 x 3 μm^3 was placed together with a ruby sphere 3 m in diameter in high pressure loaded neao as a pressure transmitting fluid. The pressure was measured *in situ* both by the ruby fluorescence [14] and by the equation of state of neon [15].

Raman scattering experiments were performed on a Jobin-Yvon T64000 triple monochromator spectrometer, with 100 mW of the 514.5 nm line of an Ar^+ laser.

X-ray diffraction were performed at the ID27 beamlines of the ESRF, in Grenoble (France). The wavelength was 0.3738 Å (33.2 keV). The spot was focussed down to 5 x 5 μm, and the cell was rotated between -17° and +17° in order to increase the number of diffraction spots. The detector was a MAR 345 Image plate. The spectra were analysed using the Fit2d software [16]

3. Results and discussion

3.1. X-ray diffraction

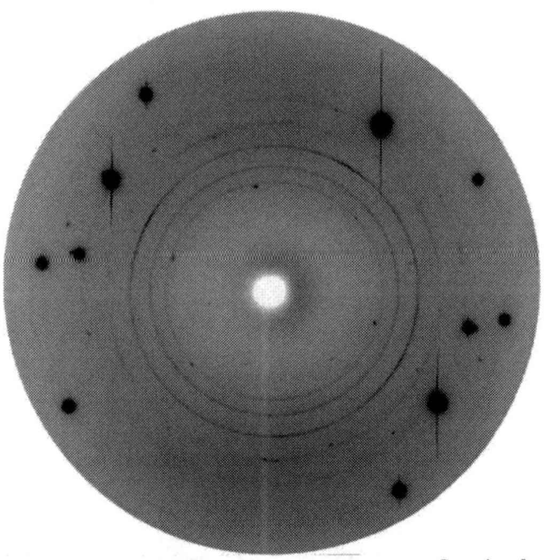

Figure 1 Diffraction pattern of single crystalline α-boron at 100 GPa. The intense points are coming from diamonds, the circles from the neon PTM, and the faint points from the sample.

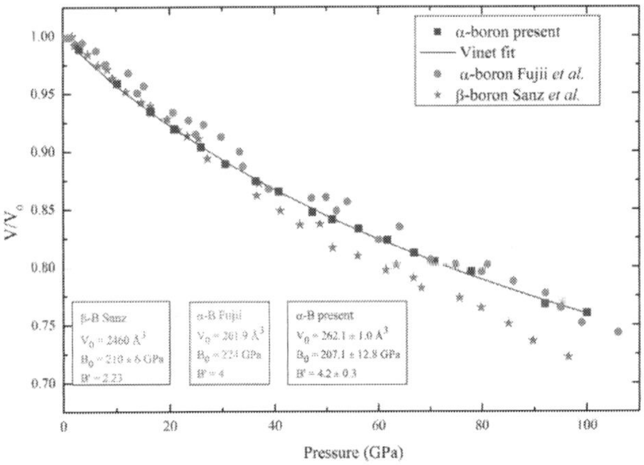

Figure 2 Equation of state of α-boron (blue squares) compared with the results of Fujii (magenta points), and those of Sanz on β-boron (red stars). The blue line corresponds to the fit with a Vinet equation of states.

Each individual diffraction spot (Fig. 1) was analyzed independently, giving rise to an excellent accuracy of the parameters and volume determination. Fig. 2 presents our results together with those of Fujii *et al.* [12] obtained on a powder in an ethanol-methanol-water mixture, and those of Sanz *et al.* on β-boron [10], also obtained on a powder, in helium as a PTM. Our results were fitted with a Vinet type equation of state [17]. The results are presented as V/V_0 in order to allow the comparison with β-boron whose volume is much larger than that of α-B. From this figure, it is clear that the results obtained on a single crystal in neon are much less scattered than those obtained on a powder in the ethanol-methanol-water mixture. Moreover, the comparison of the high pressure behaviour of α- and β-boron show that at high pressure the β polymorph is more compressible than the α one. This fact might be correlated with the observed amorphization of β-boron at 100 GPa [10].

Table 1 presents the experimental and calculated bulk modulus B_0 and its pressure derivative B' obtained by various authors on α- and β-boron. For the results of Fujii, we have made a Vinet fit using data digitalized by ourselves. There is a slight discrepancy, (even with B' = 4.2, the bulk modulus would be 215 GPa, and with a free B', one get B_0 = 288 GPa and B' = 1.6). This discrepancy might be partly due to the fact that at very high pressure bridging between the anvils may occur when using powder samples. Moreover, at these pressures, the ethanol-methanol-water PTM is very non-hydrostatic.

Table 1 Comparison of the bulk modulus B and its pressure derivative B' obtained on α- and β-boron by various techniques.

B_0 (GPa)	B'	Technique	P_{max} (GPa)	Author
210(6)	2.23	X – β-boron in helium	97	Sanz *et al.*[10]
201(9)	4.2(0.9)	X - β-boron in eth-meth	31	Ma *et al.* [9]
185(7)	-	Neutrons β-boron	10	Nelmes *et al.* [18]
213(15)	4 (fixed)	X- α-boron	5	Nelmes *et al.* [18]
222	3.5	LDA - α-boron		Vast *et al.* [19]
207	-	LDA - α-boron		Lee *et al.* [20]
249	-	LDA - α-boron		Mailhiot *et al.* [21]
220	4 (fixed)	X - α-boron in eth-meth-water	106	Fujii *et al.* [12]
207.1(12)	4.2(0.3)	X - α-boron in neon	100	Present result

3.2. Raman scattering

As stated above, there are 12 atoms in the unit cell, and hence 36 modes, among them three acoustic, *i.e.* 33 optic modes. In the Γ point of the Brillouin zone, the irreducible representation is:

$$4\ A_{1g} + 2\ A_{2g} + 6\ E_g + 2\ A_{1u} + 4\ A_{2u} + 6\ E_u$$

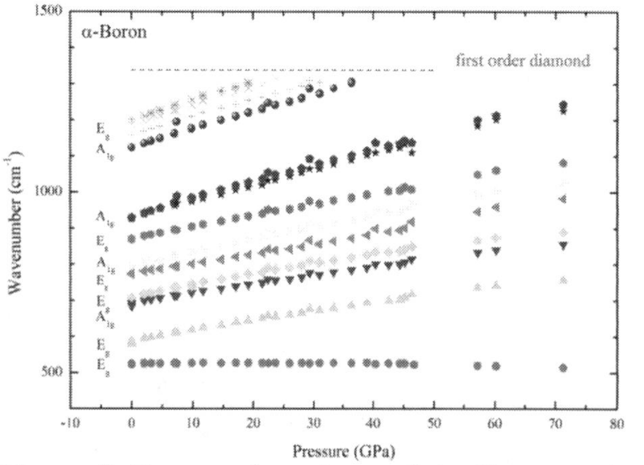

Figure 3. Pressure dependence of the Raman active modes vs. pressure.

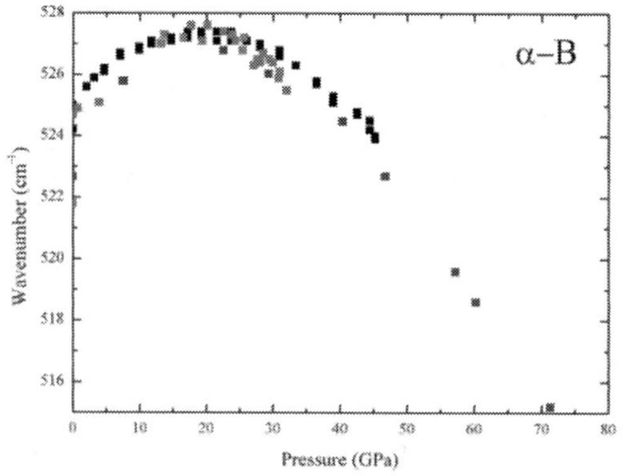

Figure 4. Pressure dependence of the librational E_g mode (524 cm^{-1} at ambient conditions). The various symbols correspond to different experiments.

One A_{1u} and one E_u are acoustic, three A_{1u} and five E_u are infrared active. Only the four A_{1g} and the six E_g Raman active. We have followed these modes up to 71 GPa, although the highest frequency ones were lost at high pressure because of the strong first order mode of the diamond anvils.

The icosahedron librational E_g mode (524 cm^{-1} at ambient) presents a remarkable behaviour. First of all, as stated in the introduction, it is a highly harmonic mode. Its full width at 1/e is of the order of 2 cm^{-1} at ambient conditions and has a tendency to decrease with increasing pressure. Secondly, it begins to increase with pressure up to approximately 20 GPa, and decreases at higher pressure. This might be the first sign of the destabilization process proposed at 210 [21] or 270 [2] GPa by numerical computations.

4. Conclusions

In this work, we have studied the pressure dependence of the structural and vibrational properties of α-boron up to 100 GPa. At this pressure, there no sign of the approach of a phase transition, on the opposite of what has been observed in β-boron. On the contrary, this very great observed stability is consistent with calculations, which predict the destabilisation of α-boron.

Acknowledgments

The single crystals used in the present study were grown and kindly provided by T.L. Aselage (Sandia National Laboratories). We acknowledge M. Mezouar (ESRF) for his help in the x-ray experiment.

References

[1] *Boron-rich solids* 1986, ed. Emin D, Aselage T, Beckel C L, Howard I A, and Wood C (*AIP Conf. Proc* vol 140, *New York*)
[2] *Boron-rich solids* 1991, ed. Emin D, Aselage T, Switendick A C, Morosin B, Beckel C L (*AIP Conf. Proc* vol 231, *New York*)
[3] Donohue J 1974, *The Structures of The Elements* (Wiley, New York)
[4] Eremets M I, Struzhkin V V, Mao H K Hemley R J 2001, *Science* **293**, 272
[5] Papaconstantopoulos D A, Mehl M J 2002, *Phys. Rev.* **B 65**, 172510
[6] Calandra M, Vast N, Mauri, F 2004, *Phys. Rev.* **B 69**, 224505
[7] Ma Y, Tse J S, Klug D D, Ahuja R 2004, *Phys. Rev.* **B 70**, 214107
[8] Vast N, Baroni S, Zerah G, Besson J M, Polian A, Grimsditch M, Chervin J C 1997, *Phys. Rev. Lett.* **78**, 693
[9] Ma Y, Prewitt C T, Zou G, Mao H K, Hemley R J 2003, *Phys. Rev.* **B 67**, 174116
[10] Sanz D N, Loubeyre P, Mezouar M 2002, *Phys. Rev. Lett.* **89**, 245901
[11] Kaneshige M, Hirayama S, Yabuushi T, Matsuoka T, Shimizu K, Mita Y, Hyodo H, Kimura K 2007, *J. Phys. Soc. Jpn.*, **76**, Suppl. A, 19
Kaneshige M, Shimizu K, Hyodo H, Kimura K, This conference, poster 0237
[12] Fujii Y, Mori Y, Nishii T, Fujii A, Hyodo H, Takarabe K, Kimura K, This conference, poster 0311
[13] Le Toullec R, Pinceaux J P, Loubeyre P 1988, *High Pressure Res.* **1**, 77
Chervin J C, Canny B, Besson J M, Pruzan P, 1995, *Rev. Sci Instrum.* **66**, 2595
[14] Mao H K, Xu J, Bell P M 1986, *J. Geophys. Res.* **91**, 4673
[15] Hemley R J, Zha C S, Jephcoat A P, Mao H K, Finger L W 1989, *Phys. Rev.* **B 39**, 11820
[16] Hammersley A P 1995, *ESRF Internal Report,* **EXP/AH/95-01,** FIT2D V6.4 Reference Manual V1.18
[17] Vinet P, Smith J R, Ferante J, Rose J H 1987, *Phys. Rev.* **B 35**, 1945
[18] Nelmes R J, Loveday J S, Allan D R, Besson J M, Hamel G, Grima P, Hull S 1993, *Phys. Rev.* **B 47**, 7668
[19] Vast N, Besson J M, Baroni S, Dal Corso A 2000, *Comput. Mat. Sci.* **17** 127
[20] Lee S, Bylander D M, Kleinman L 1992, *Phys. Rev.* **B 45**, 3245
[21] Mailhiot C, Grant J B, McMahan A H 1990, *Phys. Rev.* **B 42**, 9033
[22] Zhao J, Lu J P 2002, *Phys. Rev.* **B 66**, 092101

Inelastic neutron scattering and Raman light scattering from hydrogen-filled clathrates hydrates

L Ulivi[1], M Celli[1], A Giannasi[1], A J Ramirez-Cuesta[2] and M Zoppi[1]

[1] ISC–CNR, via Madonna del piano 10, I-50019 Sesto Fiorentino, Italy
[2] Rutherford Appleton Laboratory, ISIS Facility, Chilton, Didcot, Oxon, OX11 0QX, U.K.

E-mail: lorenzo.ulivi@isc.cnr.it

Abstract. Several samples of ternary tetrahydrofuran-H_2O-H_2 and binary H_2O-H_2 clathrate hydrates have been analysed by high-resolution inelastic neutron scattering and Raman light scattering. The neutron spectrum presents several intense bands due to H_2 molecule excitations and in particular to rotational transitions, centre-of-mass translational transitions of either para- or ortho-H_2, and to combinations of these. The H_2 molecule behaves in the clathrate cage as an almost free rotor, and performs a translational motion (rattling), that is a paradigmatic example of the motion of a quantum particle in a non-harmonic three-dimensional potential well. Both the H_2 rotational transition and the fundamental of the rattling transition split into triplets. Raman spectra show a similar splitting of the $S_0(0)$ rotational transition, due to a significant anisotropy of the potential with respect to the orientation of the molecule in the cage. The comparison of our experimental values for the transition frequencies to a recent quantum mechanical calculation is discussed.

1. Introduction

Gas hydrates are solid inclusion compounds of water and other molecular substances that have a crystalline structure different from common ice [1]. Voids (cages) of different size and geometries are present in the structure of the hydrate crystal, which host the foreign molecules. The search for efficient hydrogen-storage materials has led to an increased interest in these compounds, since it has been demonstrated that they do form with H_2 as a foreign gas, and that an appreciable amount of molecular hydrogen can be released at melting [2, 3, 4, 5].

Binary hydrogen clathrate hydrates (i.e. made of H_2O and H_2, only) require about 2000 bar of pressure to be produced at $T \simeq 273$ K [6], but adding tetrahydrofuran (THF) to the sample a ternary H_2O–THF–H_2 clathrate forms at a much lower pressure [4]. Understanding the interaction of the H_2 molecule with the host material is a key issue for a rational design of clathrates as hydrogen storage materials. This can be accessed by inelastic neutron scattering (INS) and Raman light scattering experiments, studying the dynamics of the molecule trapped in the cage, as presented in this work. The H_2O–THF–H_2 clathrate possesses the typical clathrate cubic crystal structure named sII, with 136 H_2O molecules in the unit cell, giving rise to sixteen (small) dodecahedral cages and eight (large) hexakaidecahedral cages [3, 4, 7]. The THF and H_2 molecules are hosted in the large and small cages, respectively. Recent reports indicate that only one H_2 molecule is hosted in each of the small cages [8, 9, 10].

© 2008 IOP Publishing Ltd

2. Sample preparation and characteristics

The samples for this study were prepared at ISC–CNR using D_2O and completely deuterated tetrahydrofuran (TDF) in stoichiometric proportion (17:1 mol), letting the liquid D_2O–TDF mixture to freeze in the presence of H_2 gas at various pressures and $T \simeq +2\,^\circ C$, or adding the H_2 gas, at about 800-1000 bar and $T \simeq -10\,^\circ C$, to the pre-formed D_2O–TDF clathrate, ground as a fine powder. The binary D_2O – H_2 clathrate, used for Raman scattering, was prepared from D_2O ice adding H_2 gas at about 1800 bar.

The comparison of Raman spectra from samples obtained with different procedure is shown in Fig. 1. The hydrogen content increases if the synthesis pressure is increased, and if the sample is prepared from the solid (either D_2O ice of THF–D_2O clathrate) instead of the liquid. Raman spectroscopy could allow a quantitative assessment of the H_2 content of clathrates, by calibrating the intensity of the rotational $S_0(0)$ and $S_0(1)$ H_2 lines against the lattice band located around 200–300 cm^{-1}. Comparing these Raman intensity data with gas release measurements from the same sample gave good correlation using the average of several Raman determinations in different sample locations, due to non-homogeneity of the samples.

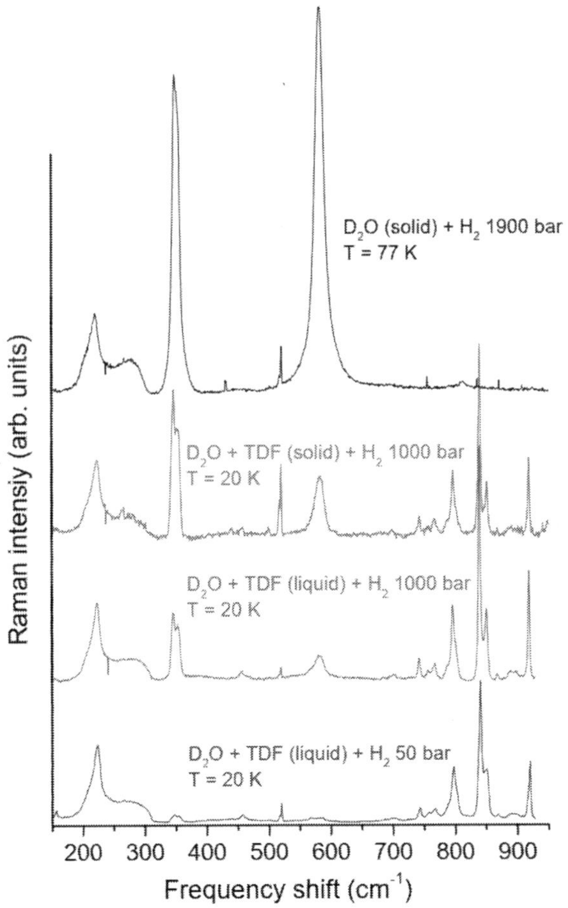

Figure 1. Raman spectra of hydrogen clathrates obtained with different procedure, i.e. starting from the solid or the liquid, with H_2 gas at different pressures. The two (broad and structured) lines at about 354.4 and 587.0 cm^{-1} correspond to the rotational $S_0(0)$ and $S_0(1)$ transitions of the H_2 molecule, while the band between 200–300 cm^{-1} is due to lattice modes of the clathrate structure. The lines between 730–930 are due to the THF molecules enclosed in the large cages of the structrure. In the figure the different spectra are labeled according to the preparation procedure, while T refers to the temperature at which the Raman spectra have been collected.

3. Neutron scattering results

Samples of D_2O–TDF–H_2 clathrate were carried to ISIS at RAL (U.K.) to perform Inelastic Neutron Scattering (INS) measurements on the TOSCA spectrometer. Incoherent INS is a

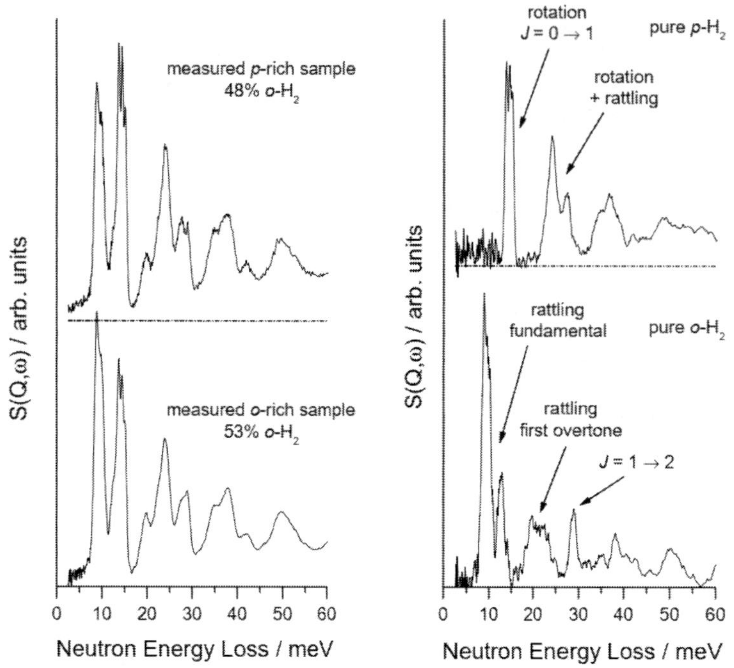

Figure 2. The measured clathrate spectra for the ortho-rich and para-rich samples (left panel) can be combined linearly to obtain the pure o-H_2 and pure p-H_2 spectra (right panel). The spectrum of p-H_2 shows the strong $J = 0 \rightarrow 1$ rotational band (a triplet at $\simeq$ 14 meV), while the o-H_2 spectrum displays the (split) band around 10 meV, due to the quantum rattling motion of the molecule in the cage.

powerful technique for the study of self dynamics of hydrogen in materials, due to the large incoherent scattering cross-section of the proton, which is almost two orders of magnitude greater than the average value of the other nuclei. In the energy range of our interest ($3.5 \lesssim E \lesssim 120$ meV) the TOSCA spectrometer has a resolution $\Delta E/E \lesssim 1.8\%$.

Four different samples were used for this experiment. One consisted of a D_2O–TDF clathrate with no hydrogen. Its spectrum is considered as a background in the analysis. Two other samples contained H_2 at different ortho-para (o-H_2–p-H_2) concentrations (in the following referred to as o-rich sample and p-rich sample) and one sample contained HD. Once the clathrate is formed, the ortho–para conversion rate is very low. For the samples prepared starting from the solid, gas-release thermodynamic measurements gave results consistent with the hypothesis of single H_2 occupancy of the totality of the small cages. Conversely, using the other preparation technique (i.e. adding H_2 gas to the freezing liquid mixture) the clathrates contained quite less H_2. Raman measurements performed at ISC–CNR before and after the neutron experiment provided the determination of the ortho-para ratio in the two D_2O–TDF–H_2 clathrates. The o-H_2 content resulted 53 % and 48% for the o-rich and p-rich sample respectively. All neutron measurement were performed at T=20 K.

From each of the H_2–clathrate spectra, that contain several intense peaks, the more flat background spectrum of the binary TDF–D_2O clathrate has been subtracted, to extract the band related to the dynamics of the single H_2 molecule in the small clathrate cage. The spectra obtained for the o-rich and p-rich clathrates are shown in the left panel of figure 2. The difference between these two spectra, (see the change in the intensity ratio of the two bands at10 and 14 meV) is substantial, given that the o-H_2 concentration in the two sample varies of a very small amount. Knowing the ortho-para concentration of each sample, we were able to extract the spectra for pure o-H_2 and pure p-H_2, by solving a simple linear system. These are shown in the right panel of figure 2. The spectra are substantially different, since the neutron scattering cross section is different for o-H_2 and p-H_2, and depends on the rotational transitions [11]. Neglecting the coherent part of the scattering, on account of the overwhelming incoherent scattering length for the proton, the analysis results quite simple [12]. The expected spectrum is the sum of several

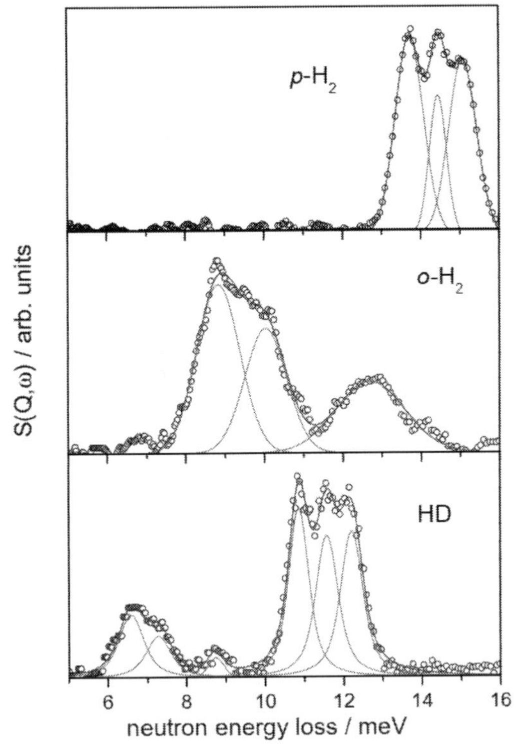

Figure 3. Detail of the spectra of p-H_2, o-H_2 and HD. The fine structure of the rotational band (p-H_2 and HD) and of the rattling band (o-H_2 and HD) is evident. Each band is fitted with three Voigt functions.

replicas of the centre of mass (CM) dynamical structure factor $S_{\text{self}}^{\text{CM}}(Q,\omega)$. These replicas are as many as the (significant) rotational transitions of the molecule, they are scaled by a known factor, and are shifted by the rotational transition energy of the molecule [12]. On the other hand, $S_{\text{self}}^{\text{CM}}(Q,\omega)$ itself should consist of a spectrum of lines, since it pertains to a localized motion of a quantum particle in a (anharmonic) potential well (rattling). Since INS is not subject to selection rules, all transitions to molecular rotational and CM vibrational (rattling) states are allowed. At the low temperature of the experiment, only the lowest rotational states (i.e. $J=0$ for p-H_2 and HD, and $J=1$ for o-H_2) are populated. For o-H_2, the only rotational transitions contributing to the spectrum in the observed frequency region are the elastic $J=1\rightarrow 1$ and inelastic $J=1\rightarrow 2$ transitions, while for p-H_2 only the inelastic $J=0\rightarrow 1$ gives a non negligible contribution. The elastic $J=0\rightarrow 0$ transition of the p-H_2 molecule, that in principle should give contribution to the p-H_2 spectrum, is weighted by the coherent scattering length and therefore is not observed.

Observing the right panel of figure 2, we note that the pure rotational transition $J=0\rightarrow 1$ gives rise to an intense band in the p-H_2 spectrum. In addition, the combination of the rotation with the rattling fundamental and with the first overtone of the rattling motion are clearly visible, located respectively at about 25 and 36 meV. In the pure o-H_2 spectrum, the band between 5 and 15 meV represents the main contribution and originate from the rattling oscillation of the o-H_2 molecules, combined with the elastic $J=1\rightarrow 1$ rotational transition. At about 22 meV the rattling overtone band is visible in the spectrum. In addition, the pure rotation line $J=1\rightarrow 2$ is present at about 29 meV. The spectrum of HD (see figure 3), on the other hand, shows both the fundamental rattling and the rotational band, since for this molecule the neutron scattering cross sections for the $J=0\rightarrow 1$ and $J=0\rightarrow 0$ transitions are of the same order of magnitude, being proportional to the incoherent proton cross section [11].

The fine structure of the rotational band of p-H_2 at about 14 meV, of the rattling band of o-H_2 at about 10 meV and of both bands of HD is shown in figure 3. Both bands are split into three components. The splitting of the *fundamental* of the rattling mode is due to the anisotropy of the potential energy with respect to the *direction of the CM displacement* from the centre of the cage. The cage shape, as it results from the structural measurements [7, 13], is indeed quite anisotropic, with the 20 oxygen atoms located at three different distances from the centre. On the other hand, the splitting of the $J = 0 \rightarrow 1$ *rotational* transition into a triplet is a consequence of the anisotropy of the potential energy with respect to the *orientation* of the H_2 molecule. We have fitted each band with three Voigt functions (with a Gaussian width fixed by the instrumental resolution), obtaining values for the energy reported in [12]. It is interesting to discuss the difference due to the different mass and moment of inertia of H_2 and HD. The ratio of the average HD rotational energy (11.54 meV) with the same quantity for H_2 (14.41 meV) is 0.80, to be compared to an expected value of $3/4 = 0.75$, for free rotors. An anisotropic potential can indeed influence the average value of the rotational energy, in addition to the removal of the degeneracy. Considering the rattling frequency, we notice that, changing from H_2 to HD, the energy scales neither with the square root of the mass ratio $\sqrt{m_{H_2}/m_{HD}} = \sqrt{2/3} = 0.816$ (as we would expect for an harmonic motion) nor with the mass ratio $m_{H_2}/m_{HD} = 2/3 = 0.667$ (which is the value expected for a square well potential). Thus the potential well for the H_2 molecule in the cage appears as intermediate between these two limiting cases, *i. e.* an almost flat zone in the centre of the cage evolving towards hard repulsion walls increasing the distance from the centre.

A recent calculation of some of the lower energy levels of one H_2 molecule in the dodecahedral clathrate cage [14] predicted a splitting into a triplet of both the rattling fundamental and the rotational transition, as we have experimentally observed. The calculated splitting for the rattling transition, 3.52 meV maximum separation, reproduces quantitatively the experimental one, i.e. 3.73 meV. In the case of the rotational transition, the calculation strongly overestimates the splitting (7.51 *vs.* 1.50 meV). Therefore, the pair potential model used in [14] seems to largely overestimate the actual anisotropic forces on the hydrogen molecule.

4. Raman line shape

We present in figure 4 and figure 5 the shape of the H_2 $S_0(0)$ and $S_0(1)$ Raman rotational lines measured in D_2O–TDF–H_2 clathrates. Analogously to the splitting observed for the J=0→1 line of p-H_2 in the neutron spectrum, the $S_0(0)$ line presents a structure that can be attributed

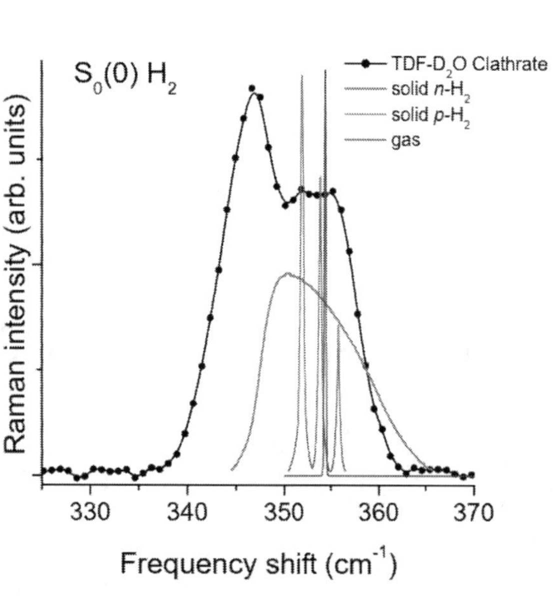

Figure 4. $S_0(0)$ Raman line of the H_2 molecule measured on a sample of THF–D_2O–H_2 clathrate (—●—) compared to that measured in solid n-H_2 (red solid line) and in p-H_2 (green solid line) [15]. The blue line represents a narrow gaussian centered at the isolated molecule transition frequency [16].

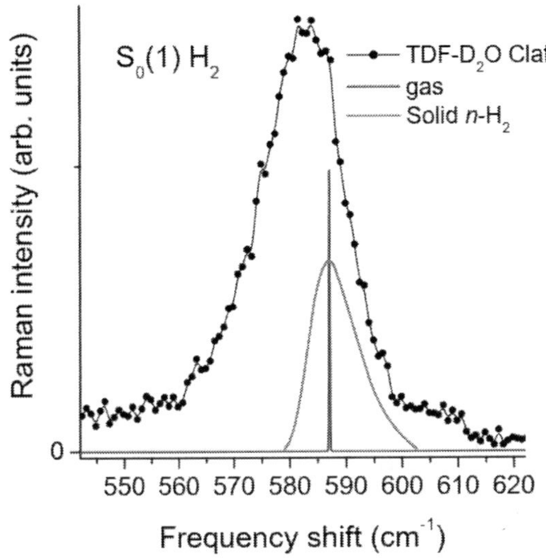

Figure 5. $S_0(1)$ Raman line of the H_2 molecule measured on a sample of THF–D_2O–H_2 clathrate (—•—) compared to that measured in solid n-H_2 (red solid line) [15]. The blue line represents a narrow gaussian centered at the isolated molecule transition frequency [16].

to the presence of, at least, three components. The width of this line ($\simeq$ 16 cm^{-1}) is similar to that observed in the neutron spectrum ($\simeq$ 18 cm^{-1}). This indicates that the influence of the anisotropic potential (with respect to the orientation of the H_2 molecule) is similar for the $J = 1$ as for the $J = 2$ state. The Raman lines are compared to that measured in solid H_2. From this comparison we observe that the perturbation to the free rotation of the H_2 molecule in the clathrate cage is quite stronger than in solid H_2. In addition, we observe that the average frequency of both rotational lines is smaller in clathrates than for the isolated molecule. This is ascribed to the presence of attractive interactions between H atoms and H_2O molecules of the cage, which tends to increase the internuclear H_2 distance, and consequently decreases the rotational frequency. The increase in the interatomic distance is however very small, and it can be estimated to be about 0.6 %.

5. Conclusions

Our INS spectra disclose most aspects of the quantum dynamics of a single H_2 molecule in the confined geometry of a water clathrate nanocavity. The rattling fundamental, observed for the first time in H_2-clathrates, has an average energy of 9.86 meV, and is split into a triplet with a (full) separation of about 3.7 meV. The rotational transition that would appear at 14.7 meV for an isolated molecule, is slightly downshifted at 14.4 meV, and is also split into three components separated by 1.5 meV. Comparison with recent theoretical values [14] indicates that, while the assumed isotropic potential and the assumed c.o.m anisotropy reproduce satisfactorily the experimental data, the anisotropy with respect to the orientation of the H_2 molecule is overestimated in the model. The splitting of the rotational and translational bands is a consequence of the anisotropy of the environment that should be modelled with an accuracy greater than what has been done until now, if a direct information on the basic interaction between H_2 and H_2O molecules is to be obtained. Besides the transition energies, it would be interesting to calculate also the intensities of the spectral bands. This can be done by the knowledge not only of the eigenvalues, but also of the eigenvectors, for the H_2 CM motion. Some results in this direction have been obtained [12].

Acknowledgments

The Cooperation Agreement No.01/9001 between CNR and CCLRC, the Grant from Firenze-Hydrolab by the Ente Cassa di Risparmio di Firenze are gratefully acknowledged.

References

[1] E. D. Sloan, *Clathrate Hydrates of Natural gases* (Dekker, New York, 1997).
[2] Y. A. Dyadin, E. G. Larionov, A. Y. Manakov, F. V. Zhurko, E. Y. Aladko, T. V. Mikina and V. Y. Komarov, Mendeleev Commun., **9**, 209 (1999).
[3] W. L. Mao *et al.* Science **297**, 2247 (2002).
[4] L. J. Florusse *et al.* Science **306**, 469 (2004).
[5] H. Lee *et al.* Nature **434**, 743 (2005).
[6] K. A. Lokshin and Y. Zhao, Appl. Phys. Lett. **88**, 131909 (2006).
[7] T. C. W. Mak and R. K. McMullan, J. Chem. Phys. **42**, 2732 (1965).
[8] K. A. Lokshin, Y. Zhao, D. He, W. L. Mao, H.K. Mao, R. J. Hemley, M. V. Lobanov, and M. Greenblatt, Phys. Rev. Lett. **93**, 125503 (2004).
[9] K. C. Hester, T. A. Strobel, E. D. Sloan, and C. A. Koh, J. Phys. Chem. B Lett. **110**, 14024 (2006).
[10] T. A. Strobel, C. J. Taylor, K. C. Hester, S. F. Dec, C. A. Koh, K. T. Miller, and E. D. Sloan, J. Phys. Chem. B **110**, 17121 (2006).
[11] S. W. Lovesey, Theory of neutron scattering from condensed matter, Clarendon Press, Oxford, (1984), p. 24.
[12] L. Ulivi, M. Celli, A. Giannasi, A. J. Ramirez-Cuesta, D. J. Bull and M. Zoppi, Phys. Rev. B, to be published, arXiv:0706.3275 (2007).
[13] C. Y. Jones, S. L. Marshall, B. C. Chakoumakos, C. J. Rawn and Y. Ishii, J. Phys. Chem. B **107**, 6026 (2003)
[14] M. Xu, Y. S. Elmatad, F. Sebastianelli, J. W. Moskowitz, and Z. Bacic, J. Phys. Chem. B Lett. **110**, 24806 (2006).
[15] S. S. Bhatnagar, E. J. Allin, and H. L. Welsh, Can. J. Phys. **40**, 9 (1962).
[16] B. P. Stoicheff, Can. J. Phys. **35**, 730 (1957).

Journal of Physics: Conference Series **121** (2008) 042019 doi:10.1088/1742-6596/121/4/042019

Studies on Silane to 70 GPa

Chandrabhas Narayana,[1,2] Raymond G. Greene[3] and Arthur L. Ruoff[1]

1. Department of Materials Science and Engineering
   Bard Hall
   Cornell University
   Ithaca, NY 14853 USA
2. Chemistry and Physics of Materials
   Jawaharlal Nehru Centre for Advanced Scientific Research
   Jakkur, P. O.
   Bangalore 560064, India
3. Corning, Inc.
   SP TO 021
   Corning, NY 14853 USA

A. L. Ruoff
ruoff@ccmr.cornell.edu

Abstract. Raman and x-ray diffraction studies were made on silane in the diamond anvil cell using three different gaskets, stainless steel, tungsten and rhenium. The structure existing between 10 and 27 GPa is well characterized by the monoclinic space group $P2_1c$ (#14). While the Gibbs free energy of formation of silane is positive at one atmosphere, it is calculated from the equation of state of silane and its reactants that this becomes negative near 4 GPa and remains negative until 13 GPa and then becomes positive again. At about 27 GPa, where quasi-quantum mechanical calculations suggest there should be a transformation from 4-fold to 6-fold (or even higher), the sample turns black. The Raman modes seize to exist beyond 30 GPa after showing softening above 25 GPa. At higher pressures it turns silvery. The gaskets play a different role as will be discussed. The sample brought back from 70 GPa contains amorphous Si (with attached hydrogen) as well as crystalline silicon. The lowest free energy system at high pressure is the decomposed reactants as observed.

1. Introduction

In 1935 Wigner and Huntington made a quasi-quantum mechanical calculation on hydrogen. This was the basis of their paper entitled "The Possibility of a Metallic Modification of Hydrogen," which had a possible metallization pressure of at least 25 GPa. The history of the progress of quasi-quantum mechanical calculations on hydrogen at high pressure is described elsewhere [2, 3]. It appears likely that a pressure of 600 GPa will be required for metallization. In as much as a static pressure of 560 GPa has been achieved in the diamond anvil cell [4], the metallization of hydrogen appeared to be a goal worthy of pursuit.

The field was greatly stimulated by a paper by Ashcroft [5] which concluded that metallic hydrogen was likely to be a high T_c superconductor, possibly even at room temperature. The following authors have successfully shown experimentally that hydrogen is not yet metallic at 210

© 2008 IOP Publishing Ltd

GPa [6], 320 GPa [7] and 340 GPa [2]. It was found that hydrogen diffuses into the diamond anvils of the high pressure cell and causes embrittlement and premature fracture [2]. This is not surprising considering the circumstances. At a distance from the diamond tip equal to about the diameter of the diamond anvil flat and along the axis of the anvil, the mean normal stress in the diamond is only 120 GPa when the hydrogen sample pressure on the diamond is 340 GPa [8]. The difference in the chemical potential of hydrogen at these two pressures is 4.7 eV.

The chemical potential of H_2 as function of pressure follows from the equation

$$\mu(P) = \int_{10^{-4}}^{P(GPa)} V dP \tag{1}$$

where V is the molar volume. A summary of V(P) data is given by Hemmes et. al. [9] and by Loubeyre et. al. from 5.5 GPa to 140 GPa [10]. The latter results were fit to the Vinet equation.

The fracture of the diamonds owing to hydrogen embrittlement has presently capped the available pressure attainable on hydrogen (unless a diffusion barrier can be created to prevent hydrogen from entering the diamond) [2].

After studies on the alkali hydrides [11], which from the phonon viewpoint have a reduced mass near one compared to 0.5 for hydrogen, it was decided [12] that CsH might be a good candidate for metallization [13] and high T_c superconductivity. This was given added stimulus by a later theoretical paper by Hama and Kawakami [14], who calculated that LiH shows band overlap at 226 GPa. While no calculations existed for CsH, it was thought that the metallization pressure of CsH would be lower [15]. The reflectivity did show a rapid rise in the infrared at 102 GPa and higher at 250 GPa. However, studies on CsD showed a strong isotope effect and that this rise was owing to phonon reflection [15]. The band gap had only decreased from 4.4 eV at zero pressure to 1.9 eV at 150 GPa. Assuming the curve can be extrapolated (no further phase transition) it was estimated that the transition pressure would be about 1 TPa. It should be noted that the high pressure form of CsH belongs to the Cmcm space group. There is the possibility that CsH pairs form, thus stabilizing the system against the metallic transition till much higher pressure.

We did not give up on attempting to make metallic hydrogen alloys. We studied CH_4 to 288 GPa [16]. Here an insulator to semiconductor transition was found but a sizeable gap (estimated at 1.4 eV) still existed. These studies were made with 35 μm tips and future studies with 20 and 10 μm tips which can achieve much larger pressures are planned.

Following the quasi-quantum mechanical calculations on SiH_4 which suggested metallization at 104 GPa [17], we attempted studies on silane. However we recently found that the sample had become contaminated in the glove box although the oxygen pressure reading was very low (falsely it turned out) [18], so these results were withdrawn [19]. We have repeated this experiment at truly low oxygen pressures as will be discussed later.

1.1 Review of quasi-quantum mechanical calculations on SiH_4

There are now three sets of calculations referred to as (Feng) [17], (Pickard) [20] and (Degtayareva) [21]. All three authors agree that at low pressure a bcc cubic phase has the lowest H, A and G. The SiH_4 molecule at the center has the H's of the Si-H bonds pointed toward four corners of the cube which are A-corners. All of the SiH_4 molecules have the same orientation as the center molecule. One Si-H bond at each A corners has the H of the Si-H bond pointed away from the center. For each of the four B-corners, the H of one Si-H bond is pointed toward the center. This is the T_1 structure of Feng. All assume static nuclei. This belongs to space group $I\bar{4}3m$.

All three sets of calculations have plots of the data as $H_{structure} - H_{I\bar{4}3m}$. All assume that the most negative ΔH of these structures at a given pressure corresponds to equilibrium even at 300K. This is equivalent to the assumption that the entropy difference $\Delta S = S_{structure} - S_{I\bar{4}3M} = 0$ is always true.

Feng published results on seven structures: Four with hydrogens in tetrahedral coordination, one octahedral and one with eight nearest neighbors (Im3m, #217 called M_1 by Feng). The latter is simply a partial inwards collapse of the cubic reference structure as the cube is squeezed to eight-fold coordination. It might be expected that there would be only tiny activation energy for this process but Hennig calculated this to be around 0.5 eV [22]. A possible hypothesis is that a transition to an octahedral coordinated structure might have much higher activation energy so that it might be kinetically impossible to attain at room temperature even though it is thermodynamically favored.

Table 1 shows the structures and pressures ranges over which these structures are calculated to be stable by each research group.

Table 1. Comparison of lowest ΔH's by author on given pressure ranges.

Feng			Pickard			Degtyareva		
P(GPa) Lowest ΔH			P(GPa) Lowest ΔH			P(GPa) Lowest ΔH		
0-5	$I\bar{4}3m$	#217	0-15	$I4\bar{3}m$	#217	0-	$I4\bar{3}m$	#217
5-25	$I\bar{4}2m$	#121	15-50	P2/c	#13	0^+-27	27P2,1c	#14
25-	Pman	#53	50-255	$I4_1/a$	#88	27^+-	P2/c	#13*
			255^+-	C2/c	#15			
* Degtyareva also calculate Pman and find that its ΔH is only about .03 to .07 eV above P2/c; see this in their Figure 4. This is an insignificant difference; see Table 2.								

Table 2 shows structures whose ΔH(P) have been calculated by two or all three of the authors. These are a gentle reminder of the state of quasi-quantum mechanical calculations. Nonetheless, the calculations serve as a helpful guide to the experimentalist: Is the low pressure form cubic? Are tetragonal structures present? Are orthorhombic structures present? And if not…?

Table 2. Comparison of quasi-quantum mechanical calculations of ΔH(eV) @ 50 GPa.

Feng		Pickard		Degtayareva	
$14_1/a$	---	$14_1/a$	-0.90	$14_1/a$	-0.17
Pmna	-0.55	Pmna	-0.85	Pmna	-0.37
P2/c	---	P2/c	-0.90	P2/c	-0.45

2. Raman Results and Discussion

Figure 1 shows the Raman spectra of silane at various pressures in the solid phase. The region 2000 cm^{-1} to 2500 cm^{-1} is associated with the symmetric and asymmetric stretching (ν_1 and ν_3) modes of silane. At ambient conditions we observe only one mode at around 2206 cm^{-1}. The mode appears to be symmetric around 2206 cm^{-1}. At this pressure silane is a gas and compares well with the literature [23]. At 1.6 GPa silane undergoes a transition from gas to liquid at room temperature. The Raman spectra of the liquid phase was verified with the low temperature Raman spectra of SiH_4 [23]. At 3.1 GPa silane goes from liquid to phase I at 300 K, which is similar to the phase I observed at low temperatures and pressures from the Raman features [23]. In phase I the silane molecules can rotate freely, hence the difference in the Raman spectra of the liquid and phase I is a decrease in the linewidth.

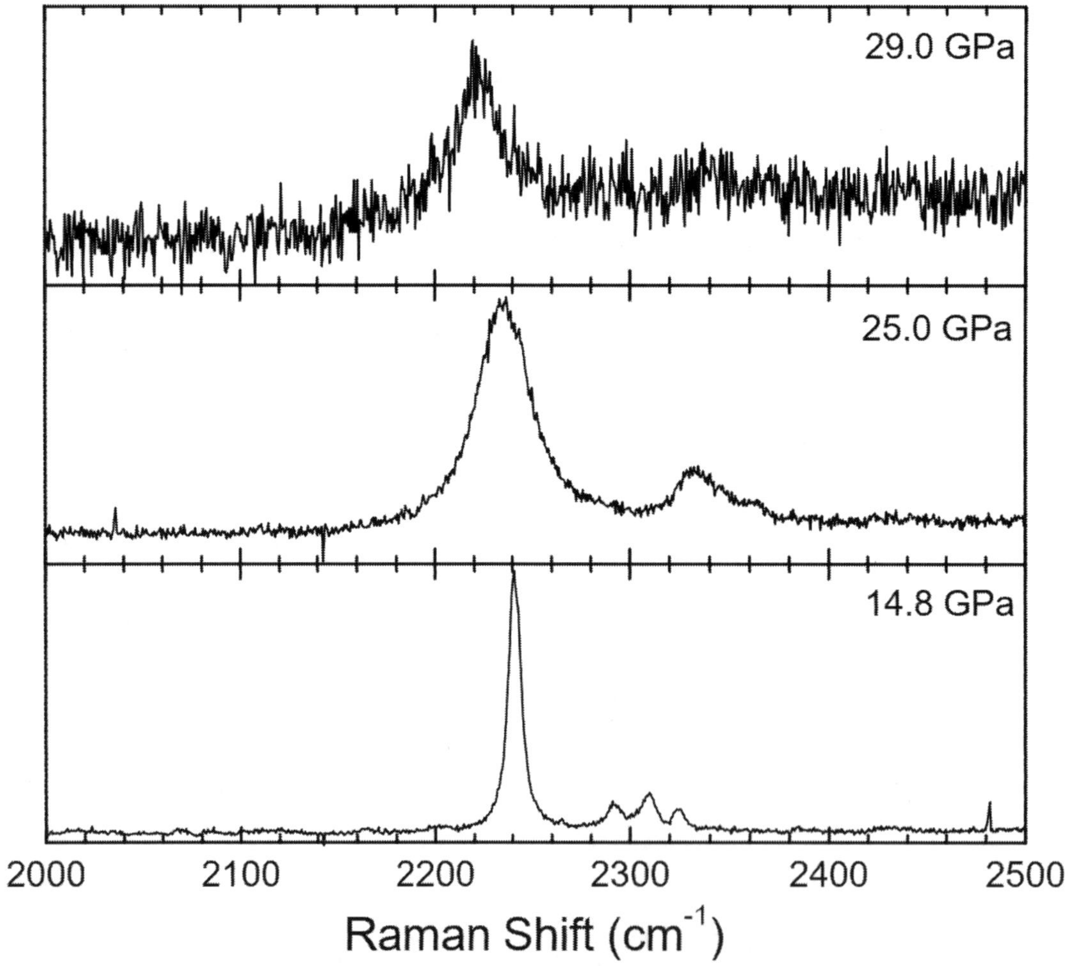

Figure 1. Raman spectra recorded at various pressures in the solid phase. This region of the Raman spectrum is dominated with the symmetric and asymmetric stretching modes of SiH_4.

Upon increasing the pressure, the rotational degree of freedom of the silane molecule is reduced leading to splitting of the ν_1 and ν_3 modes due to decrease in symmetry and the spectra recorded at 5 GPa matches well with reported temperature dependent Raman studies of SiH_4 [23] indicating the transition from phase I to phase II (orientational order-disorder transition). Structurally both these phases are similar, except for the decrease in the rotational degree of freedom of the silane molecule. When silane is observed under an optical microscope, the sample appears greyish in color. This is seen even in the phase II.

Above 10 GPa, we observe that the ν_1 and ν_3 modes are well separated and the ν_1 mode is symmetric and the ν_3 mode is split into three peaks (see Figure 1 bottom curve). Under an optical microscope the silane appears transparent at this pressure. In this phase we expect the silane to be 4-fold coordinated and could be probably a variant of the T3 structure [17,24]. Energy dispersive x-ray studies carried out at CHESS at 14.8 GPa shows the structure of Silane in this new phase (phase III) to be space group #14 as discussed later.

Above 23 GPa the Raman spectra changes accompanied with the change in color of the sample. The sample slowly converts to a black solid. The Raman spectra at 25 GPa is shown in Figure 1 (middle curve). At this pressure we find that the sample spontaneously turns to black solid, with appearance of slight silvery islands. Both the ν_1 and ν_3 modes broaden and the splitting in the ν_3 mode disappears. It is also interesting to note that the Raman modes soften (decreases in frequency). A possible scenario could be that with increasing pressure the Si-H bonds expand in order to transform from four-fold coordination to six-fold coordination. The increase in bond distances would imply a decrease in Raman frequency. The broadening of the Raman modes signifies disorder. At 29 GPa (see Figure 1 (top curve)) we observe that the Raman mode softens as well as the Raman intensity drops. The sample turns completely dark at 30 GPa and the Raman modes disappear. Based on the Raman spectra, we feel in the process of transforming from four-fold to six-fold coordination, the hydrogen atoms dissociate from the silicon, leading to the decomposition of silane. The black sample is amorphous silicon and the shiny sample observed as islands in the dark sample could be metallic silicon at high pressures.

The Raman spectra of Silicon has two modes, a weak LO mode around 468 cm^{-1} and a very strong TO mode at 520 cm^{-1} at ambient conditions. In comparison, the Raman spectra of amorphous Silicon is a representation of the density of states of the phonons, since the selection rules break down due to lack of translational symmetry in the structure. Figure 2 shows the Raman spectra at ambient conditions after the samples had been loaded to 70 GPa. We have carried out the Raman spectroscopy of the recovered black solid and observe that the Raman spectra around 200 to 600 cm^{-1} is dominated by features resembling the density of states of the phonons [25]. The key feature is that the TO mode of Silicon is not symmetric and is shifted to the lower frequency characteristic of amorphous hydrogenated silicon. The peak at around 900 to 1000 is the second order Raman peak of silicon. It has been found that the sample is shiny when a stainless steel gasket is used. We believe this is due to the absorption of hydrogen liberated by the dissociation of silane by stainless steel gasket promoting crystalline silicon. This is reduced when a rhenium gasket or a tungsten gasket is used, proving the above assumption.

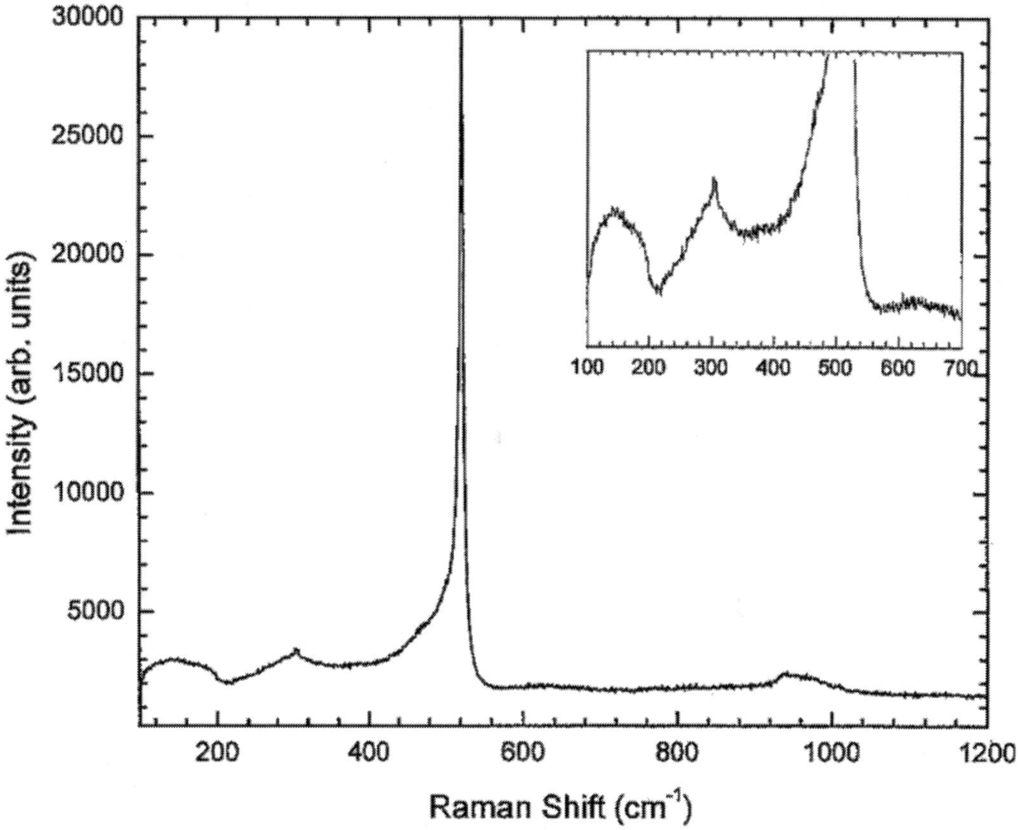

Figure 2. Raman spectra of the sample which has been recovered after the Silane high pressure run (black sample). The figure shows a strong 520 cm^{-1} silicon peak along with weak peaks in the range 100 to 700 cm^{-1} region (see inset). The peak around 900 to 1000 cm^{-1} is the second order Raman spectra of silicon. These clearly show that the sample has transformed into silicon as well as primarily to amorphous hydrogenated silicon.

3. EDX analysis of Silane for structure and EOS

Our experimental x-ray studies on silane ended on May 5, 2007. At that time there was a published set of experimental x-ray data on silane at the low temperature and atmospheric pressure by Sears and Morrison [26]. Two phases were noted, both possibly bct. The diffraction data for the higher temperature solid phase is shown in their Table 1. Their fit to a bct structure with thirty two atoms per cell is shown in their Figure 1. Here eight lines fit well to the bct structure while an additional eight that should be present for that structure were not observed. It appears that the structure is still to be determined. It is clearly not the bcc structure used as a reference by theorists as the low pressure form at room temperature. It is not necessary that it be so except that is the case for methane, which is fcc (four molecules per cell) in two studies [27,28].

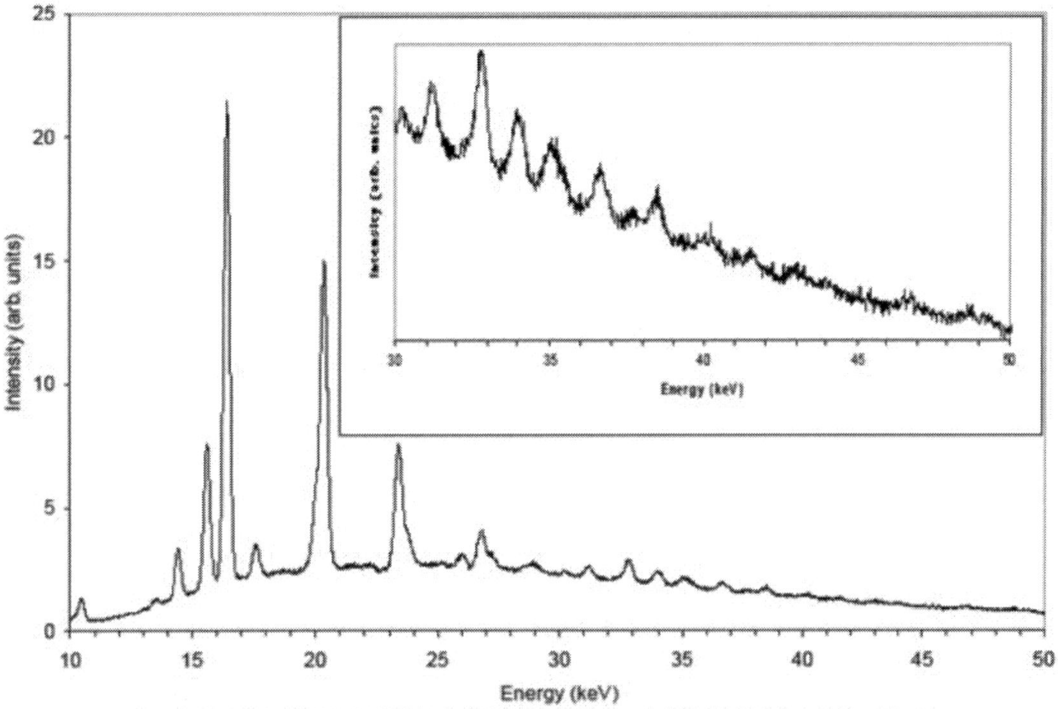

Figure 3. Energy dispersive x-ray diffraction spectrum of silane IV at 12.6 GPa

Room temperature DAC high pressure silane energy dispersive x-ray diffraction spectra from the B1 station of CHESS were analyzed with the program XRDA [29,30] using Ed = 47.369 ± 0.016 KeV-Å as determined with a gold foil. Pressure measurement was made using the shift [31] of the R1 fluorescence line of a small ruby chip placed in the DAC sample chamber. Figure 3 is a representative spectrum at a pressure of 12.6 GPa with 27 diffraction peaks. Figure 4 is the same spectrum with an inset showing the low energy escape peaks.

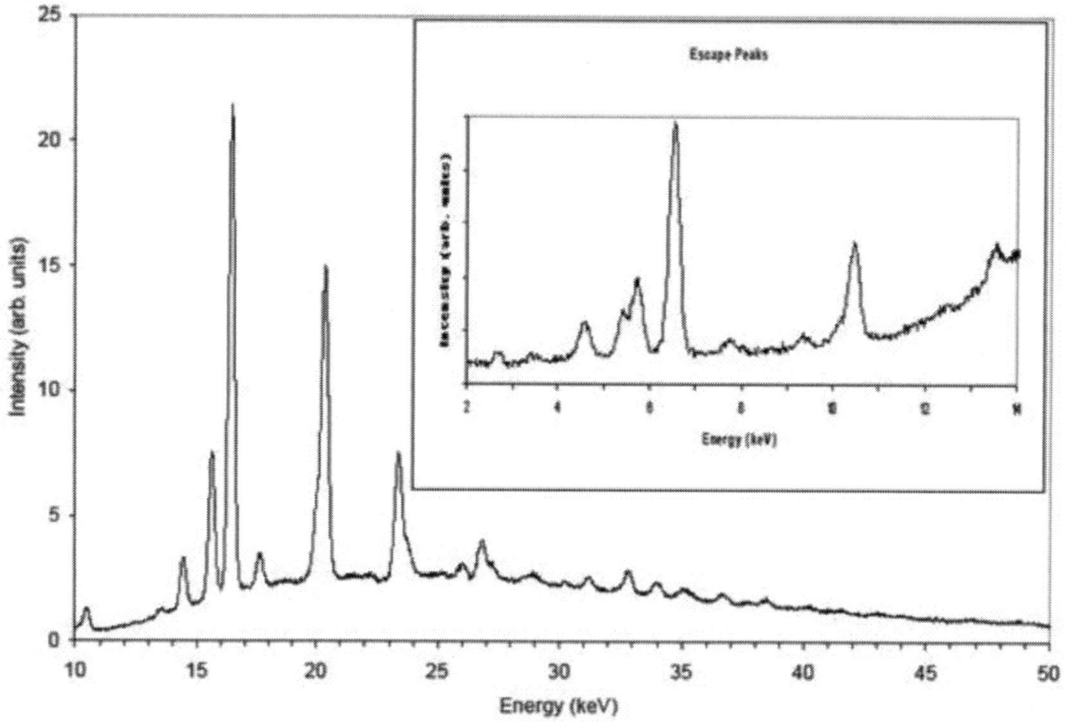

Figure 4. Energy dispersive x-ray diffraction spectrum of silane IV at 12.6 GPa showing low energy escape peaks

The structure was determined by the following procedure on the spectrum of Figure 3. First using the law of corresponding states with methane as a reference the SiH_4 molecular volume was estimated for a pressure of 12 GPa as discussed in another section of this paper. Trial and error attempts at indexing various simple Bravais lattices with cell volumes of up to eight formula units yielded a positive result for a monoclinic Bravais lattice with four formula units per cell and lattice parameters of a = 5.952 Å, b = 3.978 Å, c = 5.956 Å, and β = 104.1°. A trial indexing of this monoclinic cell with these lattice parameters using an fcc basis indexed nearly all of our diffraction lines.

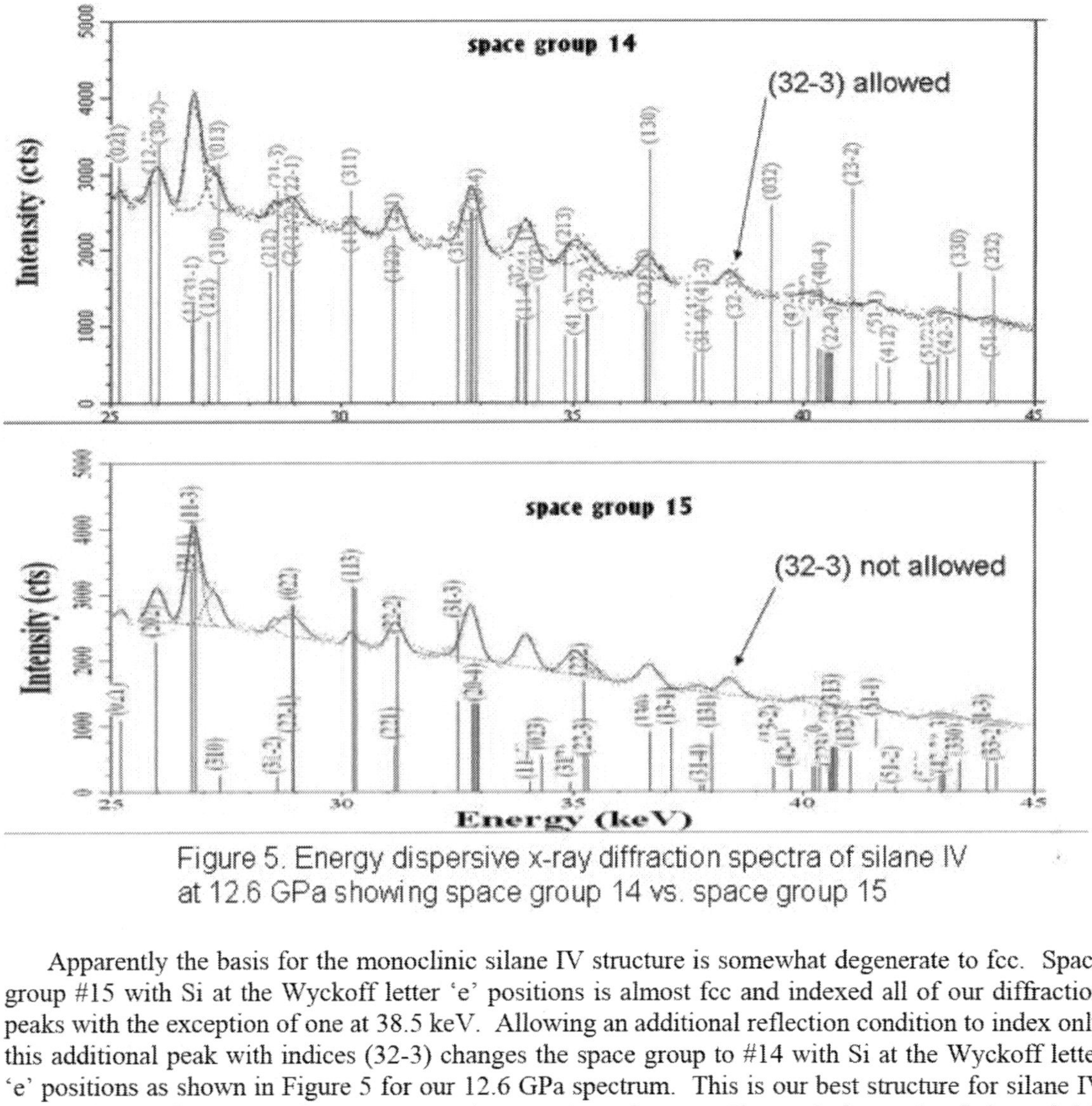

Figure 5. Energy dispersive x-ray diffraction spectra of silane IV at 12.6 GPa showing space group 14 vs. space group 15

Apparently the basis for the monoclinic silane IV structure is somewhat degenerate to fcc. Space group #15 with Si at the Wyckoff letter 'e' positions is almost fcc and indexed all of our diffraction peaks with the exception of one at 38.5 keV. Allowing an additional reflection condition to index only this additional peak with indices (32-3) changes the space group to #14 with Si at the Wyckoff letter 'e' positions as shown in Figure 5 for our 12.6 GPa spectrum. This is our best structure for silane IV and sufficient for our needs to measure the equation of state. Figure 6 shows the fitting of this structure for our 12.6 GPa spectrum for the energy region not shown in Figure 5.

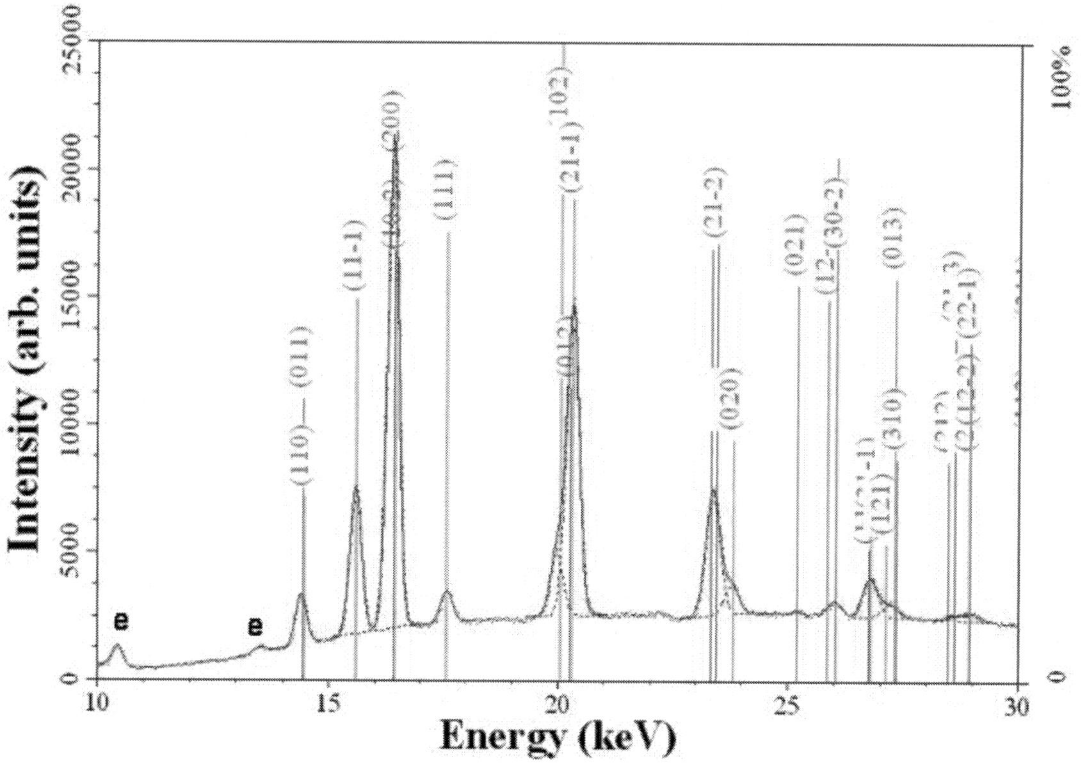

Figure 6. Energy dispersive x-ray diffraction spectrum of silane IV at 12.6 GPa showing fit to space group 14

It is important to recognize the limits of this type of structure determination for a complex Bravais lattice such as monoclinic with a many molecule basis using energy dispersive x-ray diffraction (these comments apply to a lesser but still significant extent to angle dispersive x-ray diffraction) on a quasi-hydrostatic DAC high pressure polycrystalline sample. The "International Tables for Crystallography [32]" give three cautions that apply to our situation. First, twinning for a monoclinic cell with $\beta \cong 90$ degrees such as ours cannot be ruled out and is a possible explanation of reflection conditions; i.e. either too many or too few conditions may be found. This is unlikely in our situation but we cannot positively exclude it. Finally, incorrect assignment of Laue symmetry may be caused by pseudo-symmetry by 'diffraction enhancement'. A further complication arises, in that based on reflection conditions alone, space group #9 and #15 cannot be distinguished. Our best judgement of the situation is that the space group is likely to be #9, #15 or #14. A strict interpretation of the data ignoring the subtleties of sample condition and structural determination would conclude the space group is #14.

Fitting of the cell volume of this structure vs. pressure data of eight spectrum with pressures from 12.6 to 23.5 GPa to the Birch-Murnaghan equation of state [33] with $K_0' = 4$ yielded $V_0 = 250 \pm 0.5$ Å^3 and $K_0 = 7.3 \pm 0.1$ GPa. It should be noted that our P(V) results are the same for all three space groups (within the errors noted). After this paper was presented at the conference, another paper on x-ray diffraction of this phase was published [34]. The results and conclusion agree.

4. Calculation of ΔG(P, 298.15k) of formation of silane

For the reaction

$$Si + 2H_2 = SiH_2, \quad \Delta G^\circ = 56{,}900 \text{ Joules/mole [35]}. \tag{2}$$

Journal of Physics: Conference Series **121** (2008) 042019 doi:10.1088/1742-6596/121/4/042019

If the pressure increases to a pressure P,

$$\Delta G(P, 298.15K) = \Delta G + \text{integral } (VSiH_4 - 2V_{H_2} - V_{Si})dP \quad (3)$$

where the volumes are molar volumes.

Silicon has a number of phase transitions. The volume of silicon as a function of pressure is known to P = 250GPa [35-38]. For the volume vs. pressure of hydrogen to the freezing point at 298.15K the compilation of Hemmes, et al was used [39]. Above the freezing point the data of Loubeyre, et al, [40] was used.

For silane the equation of state from the present paper was used above the freezing point at 1.8 GPa. Below the freezing point of silane the law of corresponding states [41] was used to calculate P(V) at 298.15K from the equation of state given by a function to 2 GPa by Saltzman and Wagner [42].

The predictability of this procedure was checked by calculating the equation of state of CF_4 to 1 GPa [43]. The critical parameters are from the following: CH_4 [42], SiH_4 [44], CF_4 [43]. The resultant curve for ΔG (P, 298.15K) is shown in Figure 7. This reference is used with the results of Figure 8, to obtain Figure 9 which shows a region of stability in the pressure range of 3-15 GPa. However, above 15 GPa SiH_4 is again thermodynamically unstable. Note that the quasi-quantum mechanical calculations show the P2/c structure more stable than the $SnBr_4$ structure (Space group #14) above 28 GPa and the Pmna structure more stable than the $SnBr_4$ structure above about 30 GPa. It should be noted that in the $SnBr_4$ atom has four nearest hydrogen neighbors, while in the P2/c structure the carbon atom has ten hydrogen neighbors and the Pmna structure has eight. We have already noted that quasi-quantum mechanical calculations give an activation energy of 0.5 eV for the very simple $I\bar{4}3m$ structure with a carbon atom with four nearest neighbors (head-to-base) to the $I\bar{4}3m$ structure with a carbon atom with eight nearest neighbors [M_1(Feng)] where one would not be surprised to find an activation energy to near zero. The transitions from $SnBr_4$ to P2/c and Pmna are likely to have large activation barriers and therefore it is unlikely that they will form as metastable compounds if the path to decomposition is easier. It should be noted that decomposition could involve products other than the Si, such as Si_2H_2, S_2H_4, etc. with SiH_2 being a possible precursor for polysilene as C_2H_2 is for polyacetylene, etc. The stainless steel gasket (and possibly the rhenium gasket) absorbs H_2 at pressure while W does not, even at 40 GPa, for one month at 298.15K. Thus the steel gasket probably both catalyze decomposition in Equation 2 but also makes decomposition more thermodynamically stable absorbing H_2.

Conclusions

Silane at pressures between 10 and 27 GPa was found from energy dispersive x-ray diffraction studies to have a monoclinic crystal structure with space group $P2_1c$ (#14), the fit being shown in Figure 5. All the Raman spectra below 23 GPa gave a good indication of the presence of the SiH_4 molecular unit. Around this pressure the different samples began to turn black and the Raman modes began to broaden and soften, indicating the weakening of the Si-H bonds. The Raman spectra disappeared around 30 GPa. The samples became increasingly silvery as the pressure increases and appeared to be metallic at 40 GPa. At 70 GPa there was no sign of an x-ray pattern from the sample. When the sample was unloaded it was non-crystalline and showed a Raman spectrum corresponding to that of a hydrogenated-amorphous silicon alloy, or Si:H. Quasi-quantum mechanical calculations had shown that above 27 GPa, a 6-fold structure, such as Pmna, is thermodynamically more stable than the 4-fold $P2_1c$ structure. Using the equations of state of SiH_4, Si and H_2 the authors calculate that above 13 GPa, Si + $2H_2$ is thermodynamically more stable than SiH_4.

We suspect that there is a sizeable activation barrier for the 4-fold to 6-fold transition and for the 4-fold decomposition products and instead the crystalline $P2_1c$ structure transforms to the amorphous metallic alloy, ∂ Si-H.

Acknowledgement

We wish to thank the operators and scientists at CHESS for their skilled help. We also wish to than Neil Ashcroft, J. Feng, R. Hoffman and R. Hennig for interesting discussions.

REFERENCES

[1] Wigner E and Huntington H B 1935 *J. Chem. Phys.* **3** 764
[2] Narayana C, Luo H, Orloff Jon and Ruoff A L 1998 *Nature* **393** 46
[3] Ashcroft N W 1955 *Phys. World* **8** 43
[4] Ruoff A L Xia H and Xia Q 1992 *Rev. Sci. Instr.* **63** 4342
[5] Ashcroft N W 1968 *Phys. Rev. Lett.* **21** 1745
[6] Hemley R J, Mao H-K, Goncharov A E, Strickland A F, Hanfland M and Struzkin V 1996 *Phys. Rev. Lett.* **76** 1667
[7] Loubeyre P, Occelli F and Le Toullec R 2002 *Nature* **416** 615
[8] Ruoff A L and Luo H 1991 *J. Appl. Phys.* **70** 2066
[9] Hemmes H, Driessen A and Griessen R 1986 *Physics* **139 & 140** 116
[10] Loubeyre P, Le Toullec R, Hausermann D, Hanfland M, Hemley R J, Mao H K and Finger L 1996 *Nature* **383** 702
[11] Duclos S J, Vohra Y K, Ruoff A L, Filepek S and Baranowski B 1987 *Phys. Rev* B **36** 7664
[12] Baranowski B and Ruoff A. L. 1995
[13] See Figure 4 of Reference 11.
[14] Hama J and Kawakami N 1988 *Phys. Lett.* A **126** 348
[15] Ghandehari K, Luo H and Ruoff A L 1995 *Solid state communication* **95** 385
[16] Sun L L, Ruoff A L, Zha C-S and Stupian G 2006 *Journal of Physics and Chemistry of Solids* **67**
[17] Feng J, Grochala W, Jaron T, Hoffmann R, Bergara A and Ashcroft N W 2006 *Phys. Rev. Lett.* **96** 017006
[18] Sun L, Ruoff A L, Zha C-S and Stupian G 2006 **18** 8573
[19] Sun L, Ruoff A L, Zha C-S and Stupian G 2007 *J. Phys.: Condens. Matter* **19** 479001
[20] Pickard C J and Needs R J 2006 *Phys. Rev. Lett.* **97** 045504
[21] Degtyareva O, Canales M M, Bergara A, Chen X-J, Song Y, Struzkin V V, Mao H-K and Hemley J *Phys. Rev.* B **76** 064123
[22] Hennig R private communication.
[23] Fournier R P, Savoie R, The Dinh Nguyen, Belzile R and Cabana A 1972 *Can. J. Chem.* **50** 35
[24] Chandrabhas Narayana, Greene R G and Ruoff A L [unpublished]
[25] Gupta S, Morell G, Katiyar R S, Abelson J R, Jin H C and Balberg I 2001 *J. Raman Spectroscopy* **32** 23
[26] Sears W M and Morrison J A 1975 *J. Chem. Phys.* **62** 2736
[27] Bloom M and Morrison J A 1973 *Specialist Periodical Rept. Chem.* **2** 140
[28] Hazen R M, Mao H K, Finger L W and Bell P M 1980 *Appl. Phys. Lett.* **37** 288
[29] Desgreniers S and Lagarec K 1998 *J. Appl. Cryst.* **31** 109
[30] Desgreniers S and Lagarec K *J. Appl. Cryst.* **27** 432
[31] Mao H K, Xu J and Bell P M 1986 *J. Geophys. Res.* **91** 4673
[32] *International Tables for Crystallography*, ed. Theo Hahn, Kluwer Academic 1992
[33] Birch F 1978 *J. Geophys. Res.* **83** 1257
[34] Degtyareva O, Canales M M, Bergara A, Chen X-J, Song Y, Struzkin V, Mao H-K and Hemley R J 2007 *Phys. Rev* B **76** 064123
[35] *Handbook of Chemistry and Physics*, David R. Lide, Editor-in-Chief, CRC Press, Boca Raton, FL, 73rd Edition, 1992-93
[36] Hu J E and Spain I L 1984 *Solid State Comm.* **51** 263
[37] Olijnyk H, Sikka S K and Holtzapfel W B 1984 *Physics Letters* **103A** 137
[38] Hanfland M, Schwarz V, Syassen K and Takemura K 1999 *Phys. Rev. Lett.* **82** 1197
[39] Duclos S J, Vohra Y K and Ruoff A L 1990 *Phys. Rev.* B **41** 12021
[40] Hemmes H, Driesen A and Griessen R 1986 *J. Phys. C: Solid State Phys.* **19** 3571

[41] Loubeyre R, Le Toullec R, Hausermann D, Hanfland M, Hemley R J, Mao H K and Finger L W 1996 *Nature* **383** 702
[42] Hirschfelder J O, Curtiss C F and Bird R B *Molecular Theory of Gases and Liquids*, New York, Wiley 1954
[43] Setzman U and Wagner W 1991 *J. Phys. Chem. Ref. Data* **20** 6 1124
[44] Witucki G, *Silane Physical Properties*, Dow-Corning, Midland, Michigan.
[45] Zarkova L and Hohm U 2003 *J. Phys. Chem. Ref. Data* **31** 183

Large heat capacity jump at the superconducting transiton temperature in the non-centrosymmetric superconductor $CeIrSi_3$ under high pressure.

Naoyuki Tateiwa[1], Yoshinori Haga[1], Tatsuma D. Matsuda[1], Shugo Ikeda[1], Etsuji Yamamoto[1], Yusuke Okuda[2], Yuichiro Miyauchi[2], Rikio Settai[2] and Yoshichika Ōnuki[1,2]

[1]Advanced Science Research Center, Japan Atomic Energy Agency, Tokai, Ibaraki 319-1195, Japan

[2]Graduate School of Science, Osaka University, Toyonaka, Osaka 560-0043, Japan

E-mail: tateiwa.naoyuki@jaea.go.jp

Abstract. We investigated the pressure-induced superconductor $CeIrSi_3$ without inversion center under high pressure. The electrical resistivity and ac heat capacity were measured in the same run for the same sample. The critical pressure of the antiferromagnetic state was determined to be $P_c = 2.25$ GPa. The heat capacity C_{ac} shows both antiferromagnetic and superconducting transitions at pressures close to P_c. The co-existence of the antiferromagnetism and superconductivity is discussed. The superconducting region is extended up to about 3.5 GPa. The superconducting transition temperature T_{sc} shows a maximum value of 1.6 K around 2.5 – 2.7 GPa. At 2.58 GPa, a large heat capacity anomaly was observed at $T_{sc} = 1.59$ K. The jump of the heat capacity in the form of $\Delta C_{ac}/C_{ac}(T_{sc})$ is 5.7 ± 0.1. This is the largest observed value among all superconductors studied previously, suggesting the strong-coupling superconductivity in $CeIrSi_3$.

1. Introduction

Recently, the discovery of non centrosymmetric superconductors such as $CePt_3Si$, UIr, $CeIrSi_3$, $CeRhSi_3$, and $CeCoGe_3$ has attracted considerable interest [1, 2, 3, 4, 5]. In these compounds, two spin degenerate bands are split due to the Rashba-type spin-orbit interaction, which strongly influences the superconducting properties, particularly the pairing symmetry of the Cooper pairs. Theoretical studies suggest a mixed-type pair function with spin triplet and singlet components [6]. Many theoretical and experimental studies have been extensively conducted in order to clarify this novel type of unconventional superconductivity. In this paper, we report our high pressure study on $CeIrSi_3$ which crystalizes in the tetragonal $BaNiSn_3$-type crystal structure ($I4mm$) without inversion center.

Figure 1 (a) shows the pressure phase diagram of $CeIrSi_3$. At ambient pressure, $CeIrSi_3$ is an antiferromagnet with a Néel temperature $T_N = 5.5$ K. Under high pressure, the Néel temperature decreases with increasing pressure and disappears at the critical pressure $P_c = 2.25$ GPa. Superconductivity was found in a wide pressure region from 1.3 GPa to about 3.5 GPa. A superconducting transition temperature T_{sc} shows a maximum value of 1.6 K around 2.5 – 2.7

© 2008 IOP Publishing Ltd

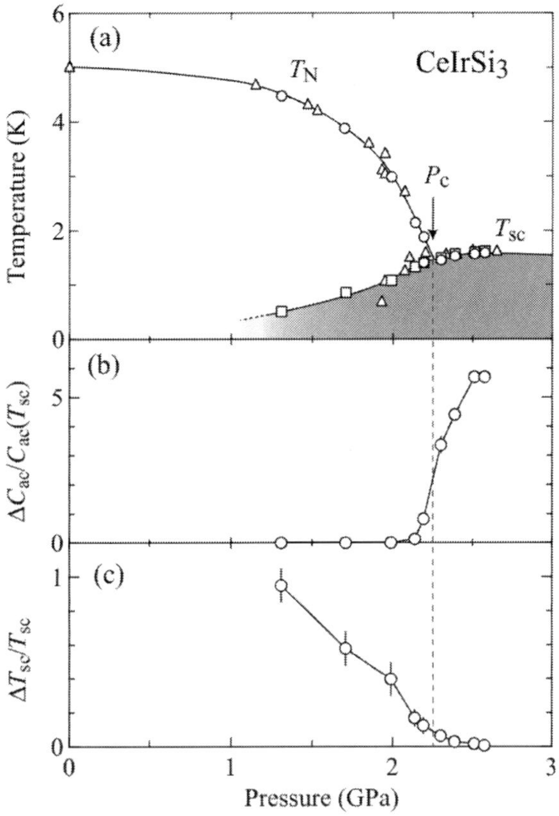

Figure 1. (a)Pressure phase diagram in $CeIrSi_3$. T_N and T_{sc}, which were determined by the previous resistivity measurements, are plotted by triangles [3, 7]. T_{sc} and T_N values obtained by the present resistivity and ac heat capacity measurements are shown by squares and circles, respectively. Pressure dependences of (b) $\Delta C_{ac}/C_{ac}(T_{sc})$, and (c) the width of the superconducting transition in the resistivity $\Delta T_{sc}/T_{sc}$.

GPa [3]. We performed the ac heat capacity and electrical resistivity measurements on $CeIrSi_3$ in order to study the superconductivity.

2. Experiment

The single crystal of $CeIrSi_3$ was grown by the Czochralski method in a tetra-arc furnace[7]. The residual resistivity ratio RRR $(= \rho_{RT}/\rho_0)$ is 120, where ρ_{RT} and ρ_0 are the resistivity at room temperature and the residual resistivity, respectively. This indicate the high quality of the sample. The heat capacity under high pressures was measured by the ac calorimetry method [8, 9, 15]. The sample was heated up using a heater, whose power is modulated at a frequency ω. The amplitude of the temperature oscillation T_{ac} is expressed as the function of the heat capacity C_{ac} of the sample as $T_{ac} = P_0/(\kappa + i\omega C_{ac})$. Here, P_0 is an average of the power. κ is the thermal conductivity between the sample and the environment. T_{ac} was measured with a AuFe/Au thermocouple (Au + 0.07 at% Fe). The contribution from the thermocouple and Au wires to the heat capacity is very small ($\sim$ 0.1%). The resistivity measurement was also carried out for the same sample by the standard four-terminal method. Two additional Au wires were attached to the edges of the sample so as to pass the electrical current for the resistivity

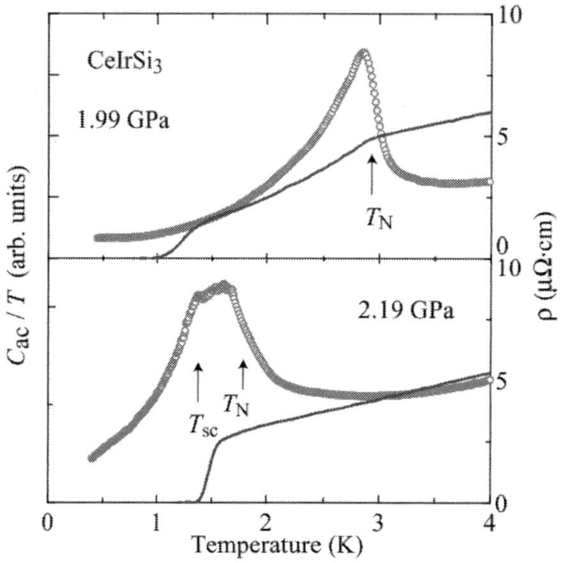

Figure 2. Temperature dependences of the ac heat capacity C_{ac} (left side) and electrical resistivity ρ(right side) at 1.99 and 2.19 GPa in $CeIrSi_3$.

measurement. The low-temperature measurement was carried out using a ^3He refrigerator from 0.3 K to 10 K. We used a hybrid piston cylinder-type cell. Daphne oil (7373) was used as a pressure transmitting medium.

3. Results and Discussions

Figure 2 shows the temperature dependences of the heat capacity C_{ac} and electrical resistivity ρ at 1.99 and 2.19 GPa, below the critical pressure $P_c = 2.25$ GPa. At 1.99 GPa, C_{ac} shows a clear anomaly and ρ shows a kink at the Néel temperature $T_N = 2.95$ K. The resistivity shows a superconducting transition at $T_{sc} = 1.02$ K. However, there is no superconducting anomaly in C_{ac} at T_{sc}. This indicates the absence of the bulk superconductivity. The evidence of the bulk superconducting state is not obtained also at 1.31 and 1.71 GPa (data not shown). At 2.19 GPa, C_{ac} shows a broad anomaly that consists of two peak structures. Those correspond to the antiferromagnetic and superconducting transitions, respectively. From the entropy balance, the Néel temperature is determined as $T_N = 1.88$ K. The peak of the heat capacity at the lower temperature side is close to the superconducting transition at $T_{sc} = 1.40$ K where ρ becomes zero.

Above the antiferromagnetic critical pressure $P_c = 2.25$ GPa, both C_{ac} and ρ show only the superconducting transition, as shown in Fig. 3. At 2.58 GPa, the values of T_{sc} obtained from the resistivity and ac heat capacity measurements are 1.62 and 1.59 K, respectively. The values are plotted in Fig. 1 (a) as squares (heat capacity) and circles (resistivity). The jump of the heat capacity in the form of $\Delta C_{ac}/C_{ac}(T_{sc})$ is 3.4 ± 0.3 at 2.30 GPa and 5.7 ± 0.1 at 2.58 GPa. Here, ΔC_{ac} is the jump of the heat capacity at T_{sc} and $C_{ac}(T_{sc})$ is the value of C_{ac} just above T_{sc} that corresponds to γT_{sc}. Here, γ is the electronic specific heat coefficient. The values of $\Delta C_{ac}/C_{ac}(T_{sc})$ are significantly larger than that of the BCS theory (1.43). The large jump of the heat capacity at T_{sc} was also observed in $CeCoIn_5$ and UBe_{13} where the value of $\Delta C/(\gamma T_{sc})$ are 4.5 and 2.7, respectively [10, 11]. In $CeIrSi_3$, the value of 5.7 ± 0.1 at 2.58 GPa is the largest value among all superconducting materials. The entropy balance in the superconducting state

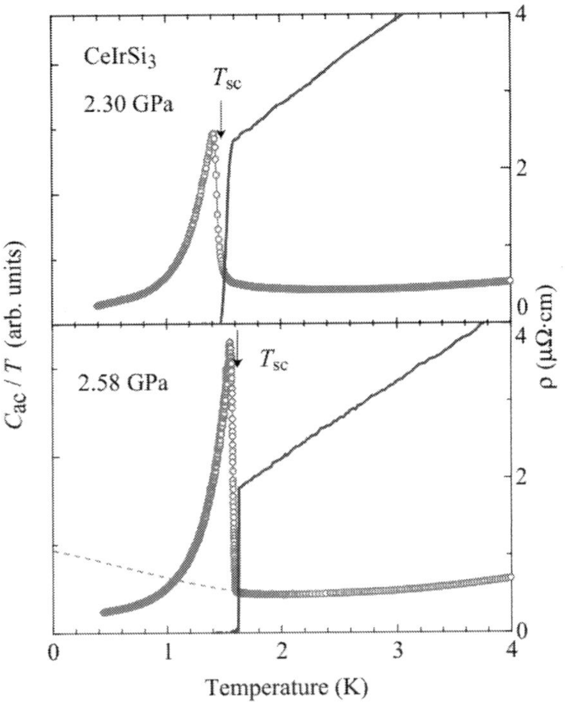

Figure 3. Temperature dependences of the ac heat capacity C_{ac} (left side) and electrical resistivity ρ (right side) at 2.30 and 2.58 GPa in $CeIrSi_3$.

of 2.58 GPa is considered, as shown by the dotted line. It is found that the value of C_{ac}/T is enhanced with decreasing temperature. The value of C_{ac}/T at 0 K is roughly twice larger than that at $T_{sc} = 1.59$ K. The value of $\Delta C_{ac}/(\gamma T_{sc})$ is about 2.8 ± 0.3 if the C_{ac}/T value at 0 K is used as the γ value. This value is still larger than the BCS value. The precise pressure dependence of $\Delta C_{ac}/C_{ac}(T_{sc})$ is shown in Fig. 1(b). The strong-coupling superconductivity is realized around 2.5 GPa in $CeIrSi_3$. The increment of $\Delta C_{ac}/C_{ac}(T_{sc})$ suggests that the superconducting coupling parameter increases with increasing pressure [12, 13].

The superconducting transition width $\Delta T_{sc}/T_{sc}$ of the resistivity decreases with increasing pressure and becomes close to zero above $P_c = 2.25$ GPa as shown in Fig. 1(c). The relation between the antiferromagnetism and superconductivity is the most interesting issue to be discussed. From the present results shown in Fig. 1(c), the superconductivity and antiferrromagnetism in $CeIrSi_3$ seem to be competing with each other. For the co-existence of the superconductivity and antiferromagnetism, we point out two possibilities. One possibility is that the both states co-exist only in a small pressure region close to P_c. The other one is that both states do not coexist and the superconductivity exists inhomogeneously in the antiferromagnetic state below P_c. On the basis of the latter possibility, the pressure dependence of $\Delta T_{sc}/T_{sc}$ and the gradual increase of $\Delta C_{ac}/C_{ac}(T_{sc})$ around P_c can be interpreted as the increment of the superconducting volume fraction. For further investigations on the co-existence of antiferromagnetism and superconductivity, microscopic experiments such as NMR are needed.

It is impossible to obtain the absolute value of the heat capacity by the present ac heat capacity measurement. However we can estimate the relative change of the heat capacity [8, 9]. The value of C_{ac}/T just above T_{sc} is determined as 100 ± 20 mJ/K^2·mol at 2.58 GPa by comparison with the value of C_{ac} at ambient pressure. This γ value indicates that the moderate heavy fermion superconductivity is realized in $CeIrSi_3$. Surprisingly, this value is approximately the same as $\gamma = 120$ or 105 mJ/K^2·mol at ambient pressure [7].

We summarize the present experimental results as follows. The critical pressure of the antiferromagnetic state is determined to be $P_c = 2.25$ GPa. Bulk superconductivity is realized basically above P_c. The highest $T_{sc} = 1.6$ K and $\Delta C_{ac}/C_{ac}(T_{sc}) = 5.7 \pm 0.1$ values are obtained at the pressure region higher than P_c, around 2.5 − 2.7 GPa. It indicates that $CeIrSi_3$ is a strong-coupling superconductor.

Acknowledgements
This work was financially supported by the Grant-in-Aid for Creative Scientific Research (15GS0213), Scientific Research of Priority Area and Scientific Research (A and C) from the Ministry of Education, Culture, Sports, Science and Technology (MEXT).

References

[1] E. Bauer, G. Hilscher, H. Michor, Ch. Paul, E. W. Scheidt, A. Gribanov, Yu. Seropegin, H. Noël, M. Sigrist, and P. Rogl 2004 *Phys. Rev. Lett.* **92** 027003.
[2] T. Akazawa, H. Hidaka, H. Kotegawa, T. C. Kobayashi, T. Fujiwara, E. Yamamoto, Y. Haga, R. Settai, and Y. Ōnuki 2004 *J. Phys.: Condens. Matter* **16** L. 29.
[3] I. Sugitani, Y. Okuda, H. Shishido, T. Yamada, A. Thamizhavel, E. Yamamoto, T. D. Matsuda, Y. Haga, T. Takeuchi, R. Settai, and Y. Ōnuki 2006 *J. Phys. Soc. Jpn.* **75** 0437003.
[4] N. Kimura, K. Ito, K. Saitoh, Y. Umeda, H. Aoki, and T. Terashima 2005 *Phys. Rev. Lett.* **95** 247004.
[5] R. Settai, I. Sugitani, Y. Okuda, A. Thamizhavel, M. Nakashima, Y. Ōnuki, and H. Harima 2007 *J. Magn. Magn. Mater.* **310** 844.
[6] V. P. Mineev 2004 *Int. J. Mod. Phys. B.* **18** 2963.
[7] Y. Okuda, Y. Miyauchi, Y. Ida, Y. Takeda, C. Tonohiro, Y. Oduchi, T. Yamada, N. D. Dung, T. D. Matsuda, Y. Haga, T. Takeuchi, M. Hagiwara, K. Kindo, H. Harima, K. Sugiyama, R. Settai, and Y. Ōnuki 2007 *J. Phys. Soc. Jpn.* **76** 044708.
[8] H. Wilhelm 2003 *Adv. Solid State Phys.* **43** 889.
[9] N. Tateiwa, Y. Haga, T. D. Matsuda, S. Ikeda, T. Yasuda, T. Takeuchi, R. Settai, and Y. Ōnuki 2005 *J. Phys. Soc. Jpn.* **74** 1903.
[10] N. Tateiwa, Y. Haga, T. D. Matsuda, S. Ikeda, E. Yamamoto, Y. Okufs, Y. Miyauchi, R. Settai, and Y. Ōnuki 2007 *J. Phys. Soc. Jpn.* **76** 083706.
[11] C. Petrovic, P. G. Pagliuso, M. F. Hundley, R. Movshovich, J. L. Sarrao, J. D. Thompson, Z. Fisk, and P. Monthoux 2001 *J. Phys.: Condens. Matter* **13** L337.
[12] F. Thomas, B. Wand, T. Lühmann, P. Gegenwart, G. R. Stewart, F. Steglich, J. P. Brison, A. Buzdin, L. Glémot, and J. Flouquet 1996 *J. Low Temp. Phys.* **102** 117.
[13] J. P. Carbotte 1990 *Rev. Mod. Phys.* **62** 1027.
[14] L. N. Bulaevskii, O. V. Dolgov, and M. O. Ptitsyn 1988 *Phys. Rev. B* **38** 11290.
[15] Y. Muro, D. Eom, N. Takeda, and M. Ishikawa 1997*J. Phys. Soc. Jpn.* **67** 3601.

Joint 21st AIRAPT and 45th EHPRG Int. Conf. on High Pressure Science and Technology IOP Publishing
Journal of Physics: Conference Series **121** (2008) 052002 doi:10.1088/1742-6596/121/5/052002

Pressure induced insulator-metal phase transition on $Pr_{0.6}Ca_{0.4}MnO_3$ thin films

D. Lampakis[1],A. Antonakos[1] , E. Liarokapis[1], M. Filippi[2] and W. Prellier[2]

[1]Department of Physics, National Technical University of Athens, GR-15780 Athens, Greece
[2]Laboratoire CRISMAT, CNRS UMR 6508, Bd Marechal Juin, F-14050 Caen Cedex, France

E-mail: `lampakis@central.ntua.gr`

Abstract. We report low-temperature Raman measurements under hydrostatic pressure in $Pr_{0.6}Ca_{0.4}MnO_3$ thin films grown on $LaAlO_3$ substrate. Certain phonons attributed to metallic rhombohedral phase become narrower with pressure and coexist with broad bands characteristic of spectra from the insulating paramagnetic orthorhombic phase. This behavior is similar with the changes induced from the chemical pressure (Ca substitution) and has been attributed to ferromagnetic metallic clusters formed in the insulating phase. Our data indicate that the Jahn-Teller distortion of the octahedra and the transition temperature is reduced under pressure.

1. Introduction

$Pr_{1-x}Ca_xMnO_3$ (PCMO) is a compound with a rich phase diagram and a more distorted structure than other cmr materials. The charge ordered (CO) phase in PCMO occurs in the region $0.3 \leq x \leq 0.75$ at the temperature range 220-260 K (T_{CO}), depending on the amount of Ca doping. Within this CO phase, there is a magnetic transition to an antiferromagnetic insulating (AFI) phase at temperature $T_N < T_{CO}$. For the x= 0.4 compound studied in this work T_{CO}=225 K and T_N=160 K. The compound remains a paramagnetic insulator for all Ca doping levels at zero applied magnetic field. It has been found for bulk that the application of a strong magnetic field can induce an insulator to metal transition [1]. Pressure suppresses charge ordering in PCMO inducing a ferromagnetic metallic state [2]. The metal-insulator transition temperature increases with pressure below a critical point p^* above which the material becomes insulating.

An extensive amount of Raman data exist for various compounds of this series and an assignment for the phonons has been proposed [3, 4, 5, 6]. To the best of our knowledge, for the PCMO compound, the Raman data exist only for the $x \approx 0.35$ [7] and $x \approx 0.37$ [8]. In a previous Raman study of PCMO thin films [9], it was pointed out a correlation between the Raman spectra modifications due to the charge ordering, studying thin films of $Pr_{1-x}Ca_xMnO_3$ of 180 nm thickness, as a function of temperature and hydrostatic pressure. In the present work, a more systematic study of the temperature and pressure dependence of the Raman active phonons of PCMO:x=0.4 thin films measured on $LaAlO_3$ (LAO) substrate is presented. The films were grown in-situ using Pulsed Laser Deposition (PLD) and had thickness of 150 nm. The growth procedure and the experimental details of the Raman measurements are described elsewhere [9, 10].

© 2008 IOP Publishing Ltd

Journal of Physics: Conference Series **121** (2008) 052002 doi:10.1088/1742-6596/121/5/052002

2. Results and discussion

Typical Raman spectra for the single crystal of PCMO film are presented in Fig.1 for various external conditions. The ~266 cm^{-1} mode was attributed to the $A_g(2)$ soft mode, which is related with the titling angle of the MnO_6 octahedra. The neighbor peak at ~300 cm^{-1}, is assigned to the $A_g(4)$ titling mode. At the 400-500 cm^{-1} region (bending region) the spectra consist of three bands. As seen in Fig.1, their relative intensities to the lower energy modes increase considerably at lowering the temperature from room temperature (RT) to 80K. The ~430 cm^{-1} (80K) peak is of B_{2g} symmetry and has been attributed to the bending mode [5, 9]. The ~460 cm^{-1} peak has the B_{1g} or B_{3g} symmetry as it is induced from the selection rules [9]. This mode was previously attributed to the B_{1g} bending mode [5]. The ~480 cm^{-1} peak has the B_{2g} symmetry and couples with the CO state. Modes of similar energy have been assigned to the B_{2g} bending [3, 4] or the antisymmetric (AS) mode [5, 9]. Iliev et al. [3] have observed and discussed at this region an ionic radius (r_A) dependent mode mixing of the bending and the AS modes. Although the r_A of Pr and Ca are outside the "mixing region", we assign the ~480 cm^{-1} in a mixed bending and antisymmetric (B+AS) mode. In comparison with the zero pressure low temperature (LT) measurements, the ~350 cm^{-1} broad peak is now clearly observed for $T<T_{CO}$ at high pressures, while the ~460cm^{-1} bending mode and the Jahn-Teller (JT) AS mode at ~480cm^{-1} are clearly distinguishable, specially for temperatures below 200K. This is probably a result of anharmonic effects and a decrease of the JT distortions. The ~618 cm^{-1} mode of A_g symmetry gains intensity with decreasing temperature and couples with the CO phase. It is therefore attributed to the JT stretching mode (SS). Bands of similar energy and dependance are related with an oxygen deficit [11] or a defect [12]. However, this mode does not appear in the high pressure spectra obtained at LT. Instead a mode at ~580 cm^{-1} appears with a similar symmetry and behavior. Finally, the broad band at 680 cm^{-1} is due to the substrate.

Figure 2 presents the pressure dependence of the above mentioned modes frequencies at RT. The $A_g(2)$ mode at ~266 cm^{-1} increases almost linearly with pressure (Fig.3a), while its width decreases rapidly up to ~2 GPa and then remains almost constant. In Fig.3b it can be seen that the energy of the $A_g(4)$ mode at ~300 cm^{-1} remains constant in the pressure range ~2-4 GPa, while it increases rapidly outside this pressure region. This mode is relatively weak and its width remains almost constant with pressure. The average position of the bending spectral region increases almost linearly with pressure above 2 GPa (Fig.3c). However, the analysis with three subbands (~430 cm^{-1} B_{2g}, ~460 cm^{-1} bending, and JT AS mode) shown in Fig.4d,c,d, reveals an energy change close to ~2 GPA for all modes. This is consistent with some recent speculations that for pressures above 2 GPA, the behavior of manganites may be different from that observed in the low-pressure measurements [13, 14].

Figure 3 presents the temperature variation at ambient and high pressure of the energy of the Raman active modes observed in the the PCMO films. The energy of the $A_g(2)$ mode shows an almost linear increase with decreasing temperature (Fig.3a) while its width decreases, as expected by the AF transition, at ambient pressure [7]. At high pressure, although its energy shows the same behavior, its width remains practically constant. Concerning the $A_g(4)$ mode, its energy has a monotonic decrease with temperature at ambient pressures (Fig.3b). However, at high pressures its energy shows a maximum around 150 K. The width of this mode increases below ~150 K, in contrast with the behavior of the other titling mode $A_g(2)$. The energy of the ~350 cm^{-1} mode at high pressure decreases at LT and it is better observed at $T<T_{CO}$ under pressure. At ambient pressure it can be better distinguished at lower temperatures, but the fitting procedure indicates its presence even at RT. The width of this broad mode does not seem to be affected by the temperature and pressure variations. The energy of the ~430 cm^{-1} mode increases slightly at LT and in any case it is always greater at high pressures than at ambient (Fig.3d). The width of this mode does not seem to be affected by the variation of temperature. The energy of the ~460 cm^{-1} mode increases slightly with decreasing temperature at both

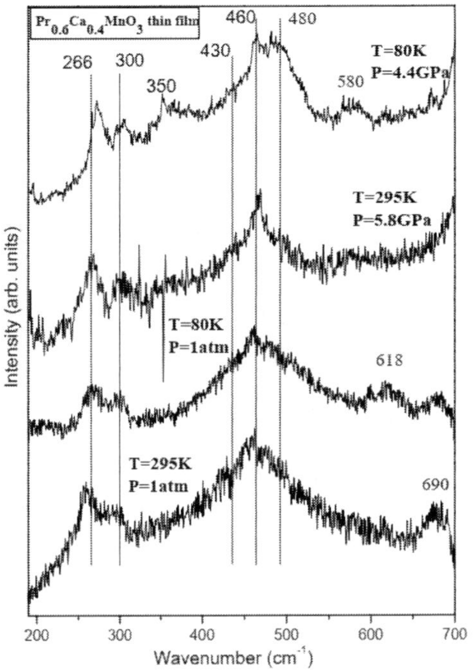

Figure 1. Typical Raman spectra for selected temperatures and hydrostatic pressures, in backscattering geometry using the 514.5 nm excitation wavelength.

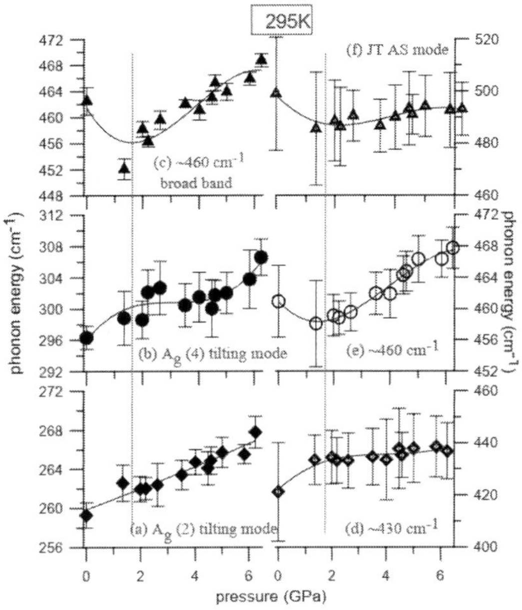

Figure 2. The hydrostatic pressure dependence of the energy of all modes appear in the high pressure Raman spectra, at RT. The continuous lines are guides to the eye.

ambient and high pressure, while the JT AS mode is rather constant (Fig.3e,f). Both modes are significantly narrower at high pressures. Finally, the $\sim$618 cm^{-1} and $\sim$580 cm^{-1} modes are very weak at high temperatures and they become clear for T<200K. In this temperature range, the energy of the $\sim$580 cm^{-1} mode is constant, within experimental error, while the $\sim$618 cm^{-1} JT SS mode softens slightly across T_{CO} (Fig.5f). No significant changes have been observed concerning their width.

Finally, it is worth mentioning the pressure and temperature dependence of the relative intensities of some modes observed in the Raman spectra of this work. As can be seen, the relative intensity of the $A_g(2)$, $A_g(4)$ modes does not seem to be affected by the pressure variation at RT (Fig. 4a). However, the JT AS mode loses intensity in respect to the $A_g(2)$, $A_g(4)$ and $\sim$460 cm^{-1} bending mode above $\sim$2 GPa (Fig.4b,c) implying the decrease of the JT distortion. These peaks are attributed to the metallic rhombohedral-like phase peaks [5]. The reduction of the JT distortion may cause the enhancement of the double exchange mechanism making the sample more conductive. Recent measurements [15] have confirmed the change in resistivity close to this hydrostatic pressure. As already discussed external pressure except of the reduced JT distortion's intensity also affects the width of the "metallic" peaks, which become narrower above $\sim$1 GPa. This behaviour is similar with the changes induced from the chemical pressure (Ca substitution) and it is attributed to ferromagnetic metallic clusters formed in the insulating phase, which is characterized by the JT broad bands [5]. Fig.4 also presents the temperature variation of the relative intensities of the modes observed in the Raman spectra of the PCMO

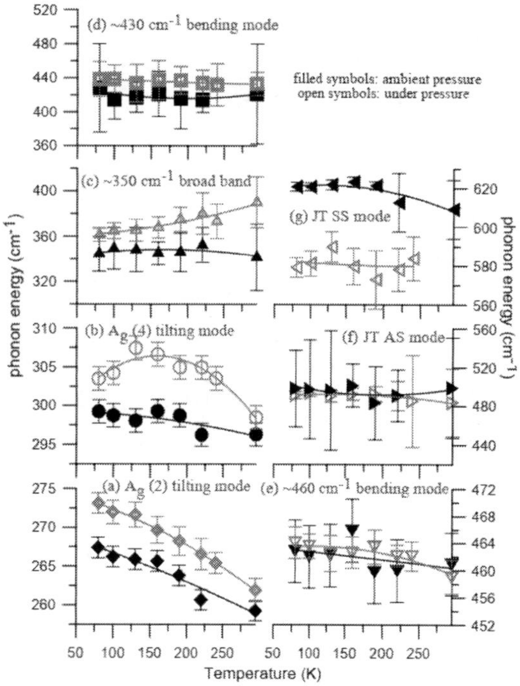

Figure 3. Temperature dependence of the energy of the modes observed in the Raman spectra of the PCMO film obtained at ambient (filled symbols) and high pressure (open symbols). The continuous lines are guides to the eye.

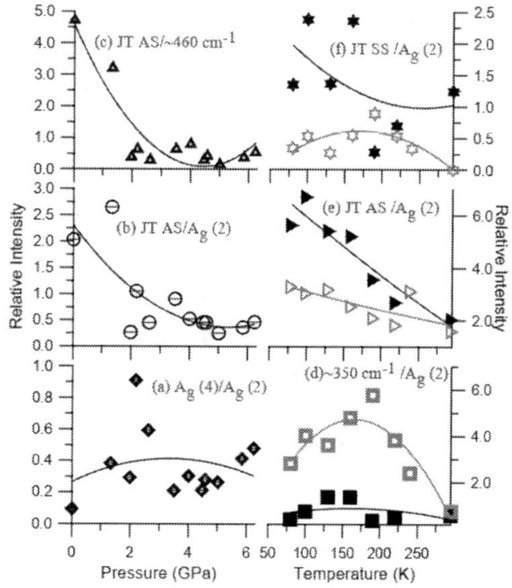

Figure 4. Pressure and temperature dependence of the relative intensities of the modes observed in the Raman spectra of the PCMO film obtained at ambient (filled symbols) and high pressure (open symbols). The continuous lines are guides to the eye.

film obtained at ambient and high pressure. All metallic rhombohedral-like phase peaks gain intensity, while the JT modes lose intensity. Due to the decreased intensity and width of the JT modes the subbands of the bending region become clearly distinguished at LT under pressure, as already mentioned. In addition, we have shown at a previous work that the JT modes couple with the CO phase. The decrease of the JT modes' intensity imply a suppressed CO phase, which has been the main reason in the bulk compound for an induced ferromagnetic metallic phase. Therefore, below 190 K and for pressure of 3.7 GPa the abnormal behavior of the modes modes, combined with the suppressed CO phase and the reduced JT distortion imply a possible insulator to metal transition. Finally, it should be mentioned that the pressure dependence of the A_g modes and the reduced JT distortion are consistent with theoretical studies of the pressure effect on the phonon properties of $La_{0.7}Sr_{0.3}MnO_3$ [16].

3. Conclusions

Under the external pressure, certain peaks attributed to metallic rhombohedral-like phase become narrower and coexist with broad bands characteristic of the spectra from the insulating paramagnetic orthorhombic-type phase. This behaviour is similar with the changes induced from the chemical pressure (Ca substitution) and has been attributed to ferromagnetic metallic clusters formed in the insulating phase. At the region 2-4 GPa the JT distortion (at RT) decreases with increasing pressure, while all peaks become stronger with decreasing pressure. Below 190 K the metallic-like peaks increase rapidly and there are evidence of a suppressed CO phase and JT distortion which indicate a possible insulator to metal phase transition.

4. Acknowledgments

Work supported through the project "Pythagoras I", co-funded by the European Social Fund (75%) and Greek National Resources (25%) and the project E.C. STREP No 517039 project "COMEPHS".

5. References

[1] Tomioka Y, Asamitsu A, Kuwahara H, Moritomo Y and Tokura Y 1996 *Phys. Rev.* B **53** R1689
[2] Cui C, Tyson TA 2004 *Phys. Rev.* B **70** 094409
[3] Iliev MN, Abrashev MN, Lee H-G, Popov VN, Sun YY, Thomsen C, Meng RL and Chu CW 1998 *Phys. Rev.* B **57** 2872; Iliev MN, Abrashev MV, Laverdire J, Jandl S, Gospodinov MM, Wang Y-Q and Sun Y-Y, 2006 *Phys. Rev.* B **73**, 064302
[4] Liarokapis E, Leventouri Th, Lampakis D, Palles D, Neumeier JJ and Goodwin DH 1998 *Phys. Rev.* B **60** 12758
[5] Martin-Carron L, de Andres A, Martinez-Lope MJ, Casais MT and Alonso JA 2001 *J. Alloys and Compd.* **323-324** 494; Martin-Carron L, de Andres A, Martinez-Lope MJ, Casais MT and Alonso JA 2002 *Phys. Rev.* B **66** 174303
[6] Abrashev MN, Backstrom J, Borjesson L, Pissas M, Kolev N and Iliev MN 2001 *Phys. Rev.* B **64** 144429; Abrashev MN, Backstrom J, Borjesson L, Popov VN, Chakalov RA, Kolev N, Meng R-L and Iliev MN 2002 *Phys. Rev.* B **65** 184301
[7] Dediu V, Ferdeghini C, Matacotta FC, Nozar P and Ruani G 2000 *Phys. Rev. Lett.* **84** 4489
[8] Gupta R, Venketeswara Pai G, Sood AK, Ramakrishnan TV and Rao CNR, cond-mat/0108476.
[9] Tatsi A, Papadopoulou EL, Lampakis D, Liarokapis E, Prellier W and Mercey B 2003 *Phys. Rev.* B **68** 024432
[10] Prellier W, Haghiri-Gosnet AM, Mercey B, Lecoeur Ph, Hervieu M, Simon Ch and Raveau B 2000 *Appl. Phys. Lett.* **77** 1023; Haghiri-Gosnet AM, Hervieu M, Simon Ch, Mercey B and Raveau B 2000 *J. Appl. Phys.* **88** 3545
[11] Podobedov VB, Romero DB, Weber A, Rice JP, Schreekala R, Rajeswari M, Ramesh R, Venkatesan T and Drew IID, 1998 *Appl. Phys. Lett.* **73** 3217
[12] Malde N, de Silva PSIPN, Akther Hossain AKM, Cohen LF, Thomas KA, MacManus-Driscoll JL, Mathur ND and Blamire MG, 1998 *Solid State Commun.* **105** 643
[13] Pinsard-Gaudart L, Rodríguez-Carvajal j, Daoud-Aladine A, Goncharenko I, Medarde M, Smith RI and Revcolevschi A 2001 *Phys. Rev.* B **64** 064426
[14] Kozlenko D and Savenko BN 2004 *J. Phys.: Condens. Matter* **16** 9031
[15] unpublished results
[16] Talati M and Jha PK 2006 *Phys. Rev.* B **74** 134406

Journal of Physics: Conference Series 121 (2008) 052003 doi:10.1088/1742-6596/121/5/052003

Superconductivity and Crystal Structure of Lithium under High Pressure

Takahiro Matsuoka[1], Suzue Onoda[1], Masahiro Kaneshige[1], Yuki Nakamoto[1], Katsuya Shimizu[1], Tomoko Kagayama[1] and Yasuo Ohishi[2]

[1]KYOKUGEN, Osaka University, Toyonaka, Osaka 560-8531, Japan,
[2]JASRI/SPring-8, Kouto, Sayo, Hyogo 679-5148, Japan

matsuoka@djebel.mp.es.osaka-u.ac.jp

Abstract. The superconductivity and structural properties of lithium under high pressure are investigated by simultaneous measurements of X-ray diffraction and electrical resistance at low temperature below 25 K. The structural transitions fcc – *hR*1 – *cI*16 near 40 GPa and the possible structural transition to higher pressure phases were observed above 70 GPa. The superconducting transition temperature T_c increases with applied pressure and has maximum in fcc phase. At higher pressures, T_c decreases in *hR*1 phase and increases again in *cI*16 phase. We did not observe T_c above 9 K for pressures above 70 GPa. Superconductivity possibly disappears in high pressure phases above 70 GPa.

1. Introduction
Lithium undergoes several structural transitions under high pressure. The structural sequence bcc to fcc at around 7 GPa and fcc to *cI*16 structure via intermediate *hR*1 structure at around 40 GPa has been reported from X-ray diffraction works at near 180 K [1][2]. Above 60 GPa, the crystal structure is still unknown.

We found Li to be a superconductor at 7 K and 30 GPa [3]. Recently, superconductivity of Li was observed below 0.4 mK at ambient pressure [4]. The superconducting transition temperature T_c rises with applied pressure and reaches 20 K (38 GPa), the second highest observed T_c for elemental superconductors [3][5][6]. Deemyad and Schilling reported the detailed superconducting phase diagram of lithium under nearly hydrostatic pressure obtained by ac-susceptibility measurements [6]. The T_c of Li rise with applied pressure and had a maximum at 30 GPa and 14 K. Above 30 GPa, it went down to 7 K at 50 GPa and showed increase. Above 60 GPa, the T_c was not observed above 4 K.

Both the structural phase transition and relatively high T_c of Li are considered to be driven by the softening of phonon and electron-phonon interaction enhanced along with the shift in electronic character *s* to *p* under high pressure[1][2][3][5][6][7]. The relationship between the superconducting properties and structures is fundamental interest. Detailed properties of crystal structure and superconductivity will provide the essential information in understanding the superconductivity of alkaline metals, which shows relatively high-T_c under high pressure. Simultaneous measurements of the electrical resistance and X-ray diffraction in low-temperature region crossing the superconducting phase enable the direct comparison of the pressure dependence of the T_c and the crystal structures of Li.

© 2008 IOP Publishing Ltd

2. Experimental

Experiments were performed at BL10XU of SPring-8 using a diamond anvil cell in the combination with monochromatic synchrotron radiation (wavelength 0.410 Å) and imaging plate detection. Figure 1 shows the experimental set-up. Electrical resistance was measured by four points contact method using the deposited fine electrodes on the surface of the diamond anvil (Figure 1a). Cu was employed for the material to reduce the contact resistance between the sample and the electrodes [7]. The thicknesses of the deposited electrodes were roughly less than 1 μm. We didn't see any contribution of Cu to the diffraction patterns.

The Li sample (99.999%, Johnson-Matthey) was loaded into the sample chamber in a DAC under well controlled Ar gas atmosphere. We used the gasket made of diamond powder mixed with epoxy resin. The possibility of chemical reaction between Li and epoxy resin has been concerned, however contamination of the sample was negligible at low temperatures below 25 K where experiments were performed, judging from the color and the value of the electrical resistance at fixed pressure and temperature. Pressure was estimated from the first-order Raman spectra of diamond anvil [8]. To avoid chemical reaction and the failure of diamond anvils [1][2][4], the DAC was, for all pressures above 4 GPa, kept below 25 K in a cryostat.

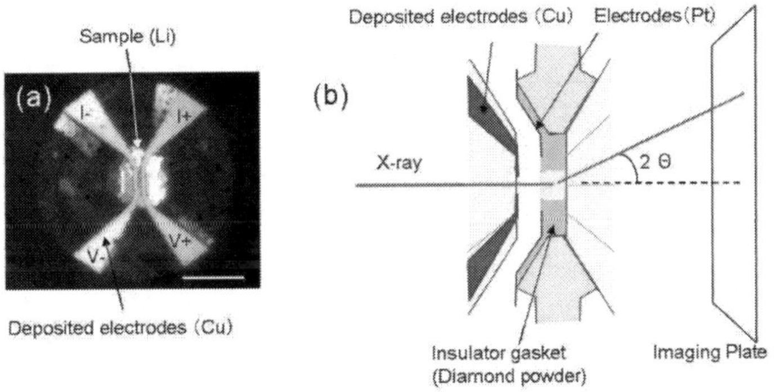

Figure 1. Experimental set-up on DAC for the simultaneous measurement of resistance and X-ray diffraction (a) Photograph of lithium sample and deposited electrodes at 4 GPa. The top surface of anvil is 150 μm in diameter. Scale bar is 50 μm. (b) Schematic drawings of the cross section. The deposited electrodes were connected to the platinum foil electrodes placed on the insulator gasket made of the mixture of diamond powder and epoxy resin. Experiments were performed below 25 K using a refrigerator. Diffraction patterns were recorded on an imaging plate.

3. Results and Discussion

Figure 2 shows the obtained X-ray diffraction patterns of Li at 25 K at various pressures. The diffraction peaks from the diamond powder gasket are noted by the mark "g". We observed fcc phase up to 34 GPa and *hR*1 phase at 46 GPa. Li transformed to *cI*16 structure with admixture of *hR*1 phase at 56 GPa. Although there are small difference in transition pressures, the obtained result is in good agreement with previous works performed at around 180 K [1][2]. At 69 GPa, adding to the diffraction peaks from *cI*16 phase, a peak (closed circle) appeared, indicating the transition to a higher pressure phase. At 86 GPa, a different spectrum was obtained. We named these two phases as hp-I and hp-II respectively. These two high pressure phases can be considered to be corresponding to the possible broken symmetry phases that have been reported by Goncharov *et al.* [9].

We observed superconducting transition at 29 GPa (fcc), 34 GPa (fcc), 46 GPa (*hR*1) and 69 GPa (*cI*16 + hp-I). T_c was determined as the onset temperature where the resistance curve starts to deviate from the linear dependence to the temperature. At 56 GPa (*hR*1 + *cI*16) and 86 GPa (hp-II), the superconducting transition was not observed above 8.5 K which is the minimum temperature reached in this work.

Figure 3 shows the pressure dependence of T_c and the structural sequence of Li obtained in this work. Dashed line shows the data reported by Deemyad and Schilling [6]. Similar pressure dependence of T_c was obtained in both experiments. The difference between these data in pressure and T_c might be due to the pressure condition; hydrostatic condition with helium pressure medium in ac-susceptibility experiments and non-hydrostatic condition without pressure medium in resistance measurements. We can see T_c of Li increases with applied pressure and has maximum in fcc phase. In *hR*1 phase, T_c decreases. At 56 GPa in *cI*16 phase, superconducting transition was not observed above 8.5 K. It can be considered that the T_c of *cI*16 increased from low temperature to 10 K at pressures between 56 GPa and 69 GPa. Above 70 GPa, the superconducting transition was not observed. Superconductivity possibly disappears in high pressure phases hp-I and hp-II.

What is the nature of the unique pressure dependence of the T_c of Li in fcc phase? We compare the obtained results with the theoretical calculation of T_c shown with open-squares in figure 3 [10]. Experimental results and theoretical data showed quite good agreement in fcc phase. Maheswari *et al*. performed ab-initio calculation of lattice dynamics and found the fcc structure shows phonon instability around 40 GPa and turn over of the T_c just before reaching this pressure. The increase in T_c is attributed to the lattice softening causing an increase in electron-phonon coupling constant. The decrease in T_c is from the decrease of the logarithmic-averaged phonon frequency, which proportionally contributes to the T_c in Allen-Dyne's formula for the superconducting transition [11], just before the phase becomes unstable towards to *hR*1 phase.

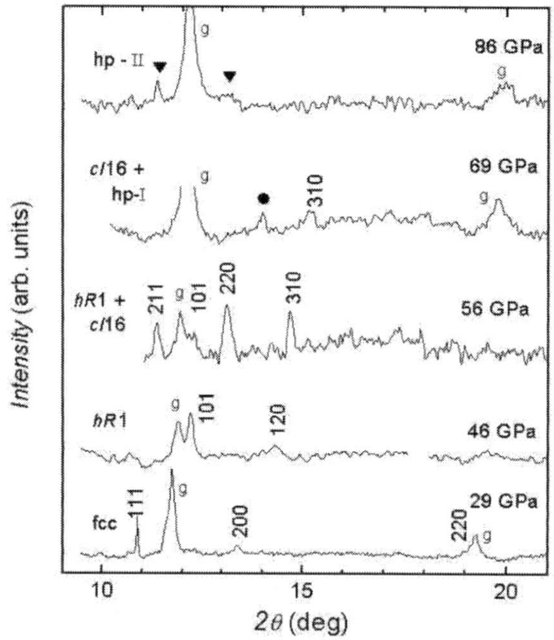

Figure 2. X-ray diffraction patterns of Li under high pressure at 25 K. Above 69 GPa, the changes of diffraction patterns were observed. The peaks of high pressure phase I and II (hp-I, hp-II) are shown with closed circle and triangles respectively. The mark 'g' denotes the peaks from gasket diamond powder.

Journal of Physics: Conference Series **121** (2008) 052003 doi:10.1088/1742-6596/121/5/052003

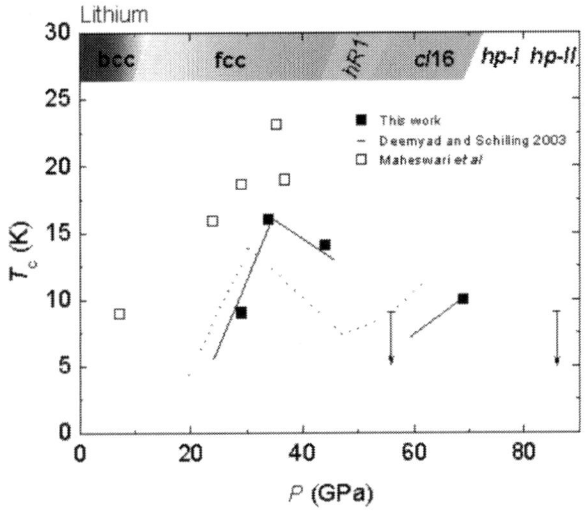

Figure 3. Pressure dependence of T_c and structural sequence at 25 K. The superconducting transition was observed above 9 K for pressures 56 GPa and above 70 GPa. The solid line is guides to eyes. The dashed line shows the data of Deemyad and Schilling [6]. The open squares are the results reported from the theoretical work of Maheswari *et al.* [10].

4. Conclusion

Simultaneous measurements of the X-ray diffraction and the electrical resistance of Li revealed that the T_c increases with applied pressure and shows maximum in fcc phase. The T_c shows increase again in *cI*16 phase with pressure. Crystal phase transition above 70 GPa from *cI*16 to high pressure phases, hp-I and hp-II were observed. Superconducting transition was not observed above 70 GPa and 9 K, which may indicate that the superconducting phase does not exist in high pressure phases above 70 GPa.

Acknowledgement

This work was supported by the 21st Century COE Program (G18) of the Japan Society for the Promotion of Science and Grant-in-Aid for Creative Scientific Research (15GS0213), MEXT.

References

[1] Hanfland M, Loa I, Syassen K, Schwarz U and Takemura K 1999 *Solid State Commun.* **112** 123
[2] Hanfland M, Syassen K, Christensen N E and Novikov D L 2000 *Nature* **408** 174
[3] Shimizu K, Ishikawa H, Takao D and Yagi T 2002 *Nature* **419** 597
[4] Tuoriniemi J, Juntunen-Nurmilaukas K, Uusvuori J, Pentti E, Salmela A and Sebedash A 2006 *Nature* **447** 187
[5] Struzhkin V V, Eremets M I, Gan W, Mao HK and Hemley R J 2002 *Science* **298** 1213
[6] Deemyad S and Schilling J S 2003 *Phys. Rev. Lett.* **91** 167001
[7] Maheswari S U, Nagara H, Kusakabe K and Suzuki N 2005 *J. Phys. Soc. Jpn.* **74** 3227
[8] Tuoriniemi J, Juntunen K and Uusvuori J 2003 *Physica* B **329** 1294
[9] Akahama Y and Kawamura H 2004 *J. Appl. Phys.* **96** 3748
[10] Goncharov A F, Struzhkin V V, Mao HK and Hemley R J 2005 *Phys. Rev.* B **71** 184114
[11] Allen P B and Dynes R C 1975 *Phys. Rev.* B **12** 905

Joint 21st AIRAPT and 45th EHPRG Int. Conf. on High Pressure Science and Technology IOP Publishing
Journal of Physics: Conference Series **121** (2008) 052004 doi:10.1088/1742-6596/121/5/052004

Pressure-dependent measurements in Superconductors

N. Suresh[1], **and J.L. Tallon**[1,2]

[1]MacDiarmid Institute for Advanced Materials and Nanotechnology, Industrial Research Ltd., P.O. Box 31310, Lower Hutt, New Zealand.
[2]Victoria University, P.O. Box 600, Wellington, New Zealand

E-mail: s.nswamy@irl.cri.nz

Abstract. To evaluate the performance of a home-built miniature non-magnetic piston clamp cell we have carried out extensive low-temperature (1.5 to 10 K) measurements of the critical field, H_c, for the element Pb up to a pressure of $P = 1.2$ GPa. Surprisingly, such data has not previously been reported above 0.3 GPa. From this data the electronic entropy, specific heat, thermal expansion coefficient and compressibility is calculated as a function of temperature, pressure and magnetic field. The zero-field data is consistent with direct thermodynamic measurements and the P-dependence of T_c and specific heat coefficient, $\gamma(T, P)$, allows the determination of the P-dependence of the pairing interaction. Using Pb as a manometer we investigated the pressure dependence of oxygen isotope effect on T_c for $YBa_2Cu_4O_8$.

1. Introduction

The thermodynamic and superconducting properties of the metal Pb have been extensively studied, including specific heat[1, 2], isotope effect[3, 4], critical field[5] and thermal expansion[6]. In addition the effects of pressure on its superconducting properties have also been well studied[7, 8, 9]. Subsequently, the use of the pressure-dependent superconducting transition temperature $T_c(P)$ for Pb as a low-temperature manometer has also been proposed[10]-[12]. The above quoted references are only the early ones and one would assume that the thermodynamic properties of this material have been comprehensively investigated.

In the course of investigations into the pressure dependence of the oxygen isotope effect on T_c for the high-T_c superconductor $YBa_2Cu_4O_8$(Y124)[13] we used the metal Pb as an internal pressure calibrant. Using a clamp cell in a SQUID magnetometer we tracked the pressure using the reported pressure dependence of T_c for Pb[11]. Despite the many above-noted early studies on Pb we discovered that the pressure, P, and temperature dependence of the critical field, H_c, for Pb has not been reported except at very low pressures (≤ 0.3 GPa)[7]. We therefore carried out a study of $H_c(T, P)$ from 1.5 to 10 K and up to 1.2 GPa from which we have derived a full set of thermodynamic parameters. This paper presents a summary of these measurements and the deduced values of the T- and P- dependence of the electronic entropy, $S(T, P)$, electronic specific heat coefficient, $\gamma(T, P)$, thermal expansion coefficient, $\alpha(T, P)$ and electronic compressibility, $\kappa(T, P)$.

The electron-phonon interactions play an important role in the physics of high T_c superconductors and especially in the pairing interactions between carriers. The experimental situation remains ambiguous. A marked anomaly in both T_c[14, 15] and superfluid density

© 2008 IOP Publishing Ltd

ρ_s[15] has been observed due to isotope substition in $La_{2-x}Sr_xCuO_4$ at a hole concentration of $p = 0.125$. The underdoped cuprates exhibit a remarkably large pressure dependence of T_c that is yet to be fully understood[16]. Y124 lies close to, but just above the $1/8^{th}$ doping point where $La_{2-x}Sr_xCuO_4$ displays this marked anomaly in isotope effect. It also has a large dT_c/dP coefficient of $\approx 5.5K/GPa$[17, 16] and T_c rises to a maximum of 108K[18] between 9 and 12 GPa before falling at higher pressure. However, bond valence sums to characterize bond stresses leads to the conclusion that $T_{c,max}$ has a negative (in-plane) pressure coefficient[19, 20]. There are thus anomalous pressure effects and anomalous isotope effects in underdoped cuprates. In the second part of this present work we bring these issues together to examine the pressure dependence of the oxygen isotope effect in Y124.

2. Lead

2.1. *Experimental*

The sample used was a bar of dimensions 1.1 × 0.93mm cross section and 5.58mm long, shaped from Puratronic 99.9985%, 2 mm diameter Pb wire from Johnson Matthey. Using the formulae of Osborn the demagnetization factor[21] was calculated as 0.97 and this value was used throughout. The Pb sample was annealed in vacuum at 300 C for 8 hours to reduce magnetic hysteresis[5]. This was loaded in a 2.67 mm diameter 9 mm long Teflon capsule along with Fluorinert FC70 and FC77 mixed in 1:1 ratio as a cryogenic hydrostatic pressure medium. The sample capsule was then placed in a miniature home-built non-magnetic Be-Cu(Mico Metal 97.75% Cu, 2% Be) piston clamp cell (8.8mm diameter, 65mm length, with cobalt-free tungsten-carbide pistons ($Roctec^{TM}$). The pistons of this cell are lightly tapered using electric-discharge machining[22]. To apply pressure the cell was preloaded before clamping at room temperature using a laboratory press with calibrated digital pressure gauge (Ashcroft Model 2089, 0.05% accuracy). The magnetization measurements were carried out in a Quantum Design MPMS SQUID magnetometer. The pressure in the sample was measured from the reported shift in T_c of Pb at zero field[11]. For in-field measurements the cell was always zero-field cooled to avoid hysteresis error.

2.2. *Results*

Fig. 1 shows the measured T-dependence of the magnetic moment, M, up to a field of 700 G and at pressures of 0 GPa (1 bar), 0.5, 0.9 and 1.15 GPa. T_c was determined by the extrapolation to zero of the steepest slope of $M(T)$. This gives $T_c(H)$ which is re-plotted in Fig. 2 as $H_c(T,P)$ versus T. The data shows a progressive decline in both T_c and $H_c(T)$ with increasing pressure. We will see that the combination of these two pressure effects allows access to the pressure dependence of the density of states, $N(E_F)$, at the Fermi level, E_F and of the pairing interaction.

To fit this data we expand $H_c(T)$ as a power series in T. The first-order term must be zero otherwise $S_s \neq 0$ at $T = 0$ and the third-order term must also be zero to avoid a negative γ at low T. It is usual, therefore, to adopt an even polynomial of the form[5]

$$H_c(T,P) = H_{c0}(P) + H_{c1}(P) \times T^2 + H_{c2}(P) \times T^4 \quad (1)$$

where, as indicated, the coefficients are each an independent function of pressure. It turns out that the coefficients are not independent and it is common to adopt the "similitude principle" which separates the P- and T-dependent terms as follows[7]

$$\begin{aligned} H_c(T,P) &= H_{c0}(P) \times f(t) \\ &= H_{c0}(P) \times \{1 - \alpha \times t^2 - (1-\alpha) \times t^4\} \quad (2) \end{aligned}$$

Journal of Physics: Conference Series **121** (2008) 052004 doi:10.1088/1742-6596/121/5/052004

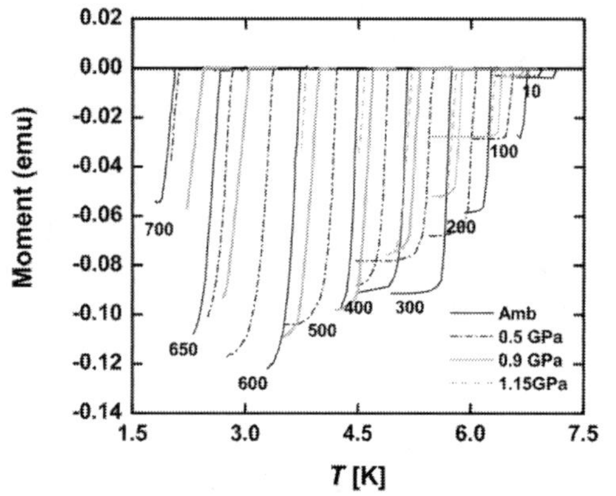

Figure 1. Magnetization curves for the Pb sample plotted for various applied fields (annotated in Gauss as 10, 100, 200 etc) and four different pressures (0, 0.5, 0.9 and 1.15 GPa).

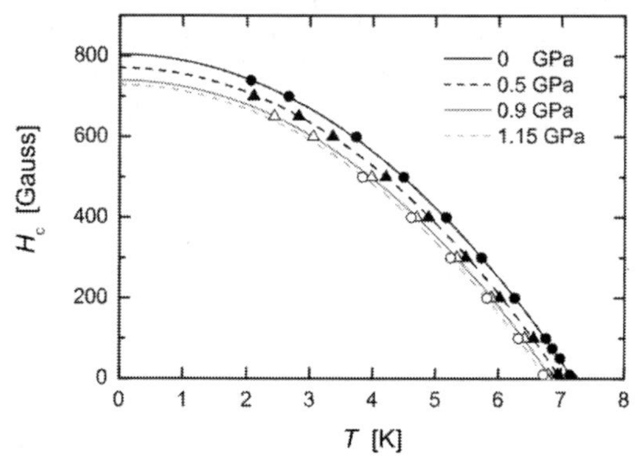

Figure 2. The T-dependence of the critical field, H_c for pressures of 0, 0.5, 0.9 and 1.15 GPa. Data points are measured data from Fig. 1, curves are fits using eq. 2.

where $t = T/T_c$. A free fit of eq. (1) to the data consistently yielded H_{c2}/H_{c0} very close in value to $1 - H_{c1}/H_{c0}$, consistent with eq. (2). We therefore proceeded to fit the data with eq. (2) for all subsequent analysis. The fits are shown by the four curves through the data shown in Fig. 2. We obtain H_{c0} values of 804.08, 770.88, 739.99 and 729.01 Gauss for $P = 0$, 0.5, 0.9 and 1.15 GPa, respectively. A linear fit yields

$$H_{c0}(P) = (803.72 - 67.028P)\ G \tag{3}$$

This now allows the scaling of all the data by plotting $H_c(T,P)/H_{c0}(P)$ versus t^2 as shown in Fig. 3. The small departure from linearity in Fig. 3 reflects the small quartic term ($\alpha = 0.95475$; $1 - \alpha = 0.04525$) and the scaled data in Fig. 3 allows a global fit to determine α that provides an overall consistency in the data fits that is necessary when second derivatives of $H_c(T,P)$ are used to determine the electronic specific heat coefficient, thermal expansion coefficient or compressibility.

These parameters may be determined as follows. We restrict our thermodynamic parameters to the electronic contribution, thus ignoring the lattice contribution to the specific heat or thermal expansion. The difference in free energy between the normal and superconducting states at the same T and P is given by

$$G_s(T,P) = G_n(T,P) - \frac{1}{2}\mu_o H_c(T,P)^2 \times V_M \tag{4}$$

where μ_o is the permeability of free space and V_M is the molar volume. As shown by Shoenberg[23] the thermodynamic parameters entropy, specific heat coefficient, molar volume, thermal expansion coefficient and compressibility, in the superconducting state, in that order are obtained by differentiating equation (4) with respect to T and P, giving

$$\begin{aligned} S_s(T,P) &= S_n(T,P) + \mu_o H_c V_M (\partial H_c/\partial T)_P \\ \gamma_s &= \gamma_n + \mu_o V_M (\partial H_c/\partial T)_P^2 - \mu_o H_c V_M (\partial^2 H_c/\partial T^2)_P \end{aligned}$$

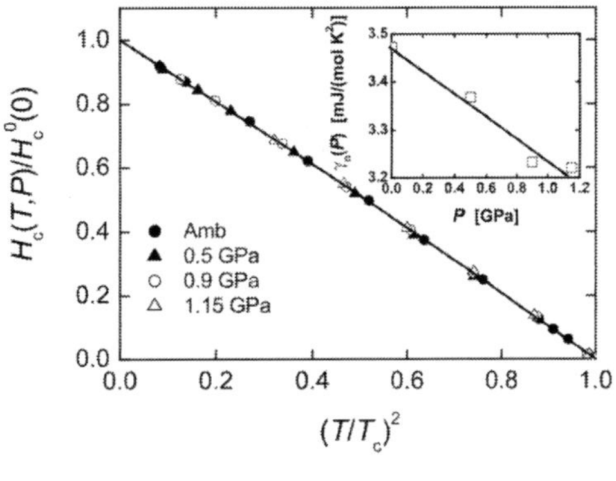

Figure 3. Scaling of the T-dependent critical field, showing $H_c(T,P)/H_c(T=0,P)$ versus $t^2 = (T/T_c)^2$ for pressures of 0, 0.5, 0.9 and 1.15 GPa. The solid curve is the function $f(T) = 1 - 0.95475t^2 - 0.04525t^4$. Inset: γ_n calculated from eqn. (6).

$$\begin{aligned}
V_s &= V_n - \mu_o H_c V_M (\partial H_c/\partial P)_T \\
\alpha_s &= \alpha_n - \mu_o(\partial H_c/\partial T)_P \times (\partial H_c/\partial P)_T - \ \mu_o H_c(\partial^2 H_c/\partial T \partial P) \\
\kappa_s &= \kappa_n + \mu_o(\partial H_c/\partial P)_T^2 + \ \mu_o H_c(\partial^2 H_c/\partial P^2)_T
\end{aligned} \tag{5}$$

where the subscript n indicates normal state. Here terms such as $(1/2)\mu_o\ H_c^2 V_M \alpha_{tot}$ in the entropy or $\mu_o \kappa_{tot} H_c (\partial H_c/\partial P)_T$ in the compressibility obtained by differentiation of the molar volume are ignored as negligible, where α_{tot} is the total volume coefficient of thermal expansion including lattice and electronic terms, and similarly for κ_{tot}.

Taking the first of these expressions, that for the entropy $S_s(T,P)$, and dividing through by T we note the requirement that $S_s/T \to 0$ as $T \to 0$ due to the opening of a full superconducting gap in the DOS. At the same time if γ_n is constant then S_n/T is just γ_n. This then imposes a relationship between γ_n and the quadratic term in $H_c(T)$. In particular,

$$\gamma_n = \lim_{T\to 0} -\mu_o T^{-1} H_c V_M (\partial H_c/\partial T)_P \tag{6}$$

In this way γ_n is calculated from the fits in Fig. 3 and its pressure dependence value is plotted in the inset to Fig. 3. The value at ambient pressure, $\gamma_n = 3.47$ mJ/(mol K^2) compares favorably with the value $\gamma_n = 3.13$ mJ/(mol K^2) determined from direct specific heat measurements[1]. The pressure dependence of γ_n shown in the inset reveals a dimensionless volume dependence given by $\partial \ln \gamma_n / \partial \ln V = 3.26 \pm 0.27$ (where we assume a total compressibility of[24] $\kappa_0 = 0.0205$ GPa^{-1}). There have been various estimates of this dimensionless parameter ranging from 1.7 ± 0.5 from thermal expansion measurements[25], 1.8 ± 0.5 from volume expansion at T_c[26], 6.0 based on pressure dependence of H_c[7] and 3.1 ± 0.8 in good agreement with our present value based on more recent measurements of volume change at T_c by Ott[27]. Note that this is considerably stronger than the value for the free electron gas $\partial \ln \gamma_n / \partial \ln V = 2/3$.

With these relations in place we proceed to calculate the entropy as shown in Fig. 4(a). The linear slope in the normal state is just the above-determined $\gamma_n(P)$. The entropy in both states must vanish at $T = 0$. A more rigorous test of the polynomial fits to the data is whether S_s/T also vanishes as $T \to 0$ and whether the data preserves monotonic systematics in this region. The T-dependence of S/T is plotted in Fig. 4(b) and this indeed shows a quadratic behavior at low T which is perfectly systematic with increasing pressure and extrapolates to zero as $T \to 0$. In a finite external field the same entropy curves are retraced up to the reduced $T_c(H)$ value and

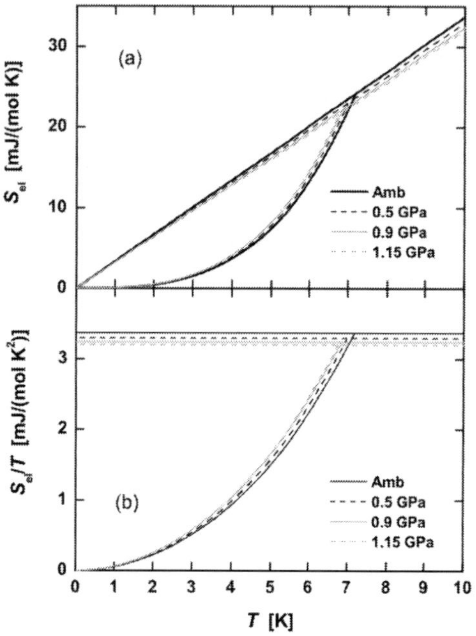

Figure 4. The T-dependence of (a) the electronic entropy in the normal state and superconducting state for pressures of 0, 0.5, 0.9 and 1.15 GPa; and (b) of S/T.

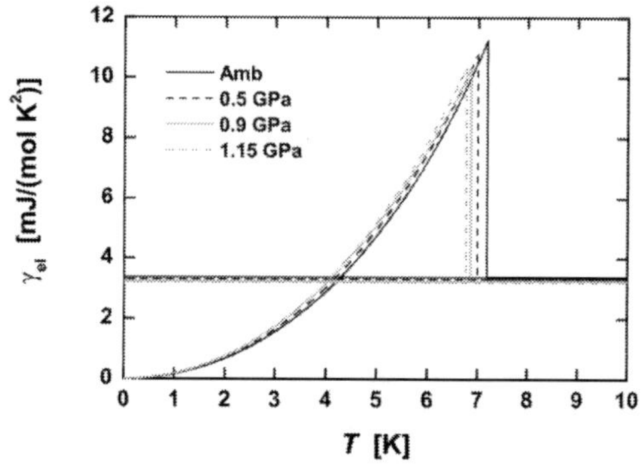

Figure 5. The T-dependence of the specific heat coefficient $\gamma(T, P)$ in the normal and superconducting states for pressures of 0, 0.5, 0.9 and 1.15 GPa. The small departure from monotonic systematics in the specific heat jump for 0.9 GPa reflects the accumulation of errors in the 2nd derivatives.

then the entropy jumps discontinuously to the normal-state value, consistent with a first-order phase transition in magnetic field and second-order when $H = 0$.The condensation energy, U_o is determined from

$$U_o = \int_0^{T_c} [S_n(T,P) - S_s(T,P)]dT \tag{7}$$

from which we obtain the following values: $U_o = 47$, 43.2, 39.8 nd 38.6 mJ/mol for $P = 0$, 0.5, 0.9 and 1.15 GPa, respectively. The condensation energy reduces with pressure due to the twin effects of reduced T_c and reduced γ_n.

Turning to the specific heat coefficient, this is calculated using the second equation in eqs. (4) and is shown in Fig. 5 for both the normal and superconducting states. The entropy balance is evident from the equal areas above and below the temperature where $\gamma_s = \gamma_n$, and is confirmed by the fact shown in Fig. 4(b) that $S_s/T = \gamma_n$ at T_c. The calculated values for the jump in $\gamma(T, P)$ at T_c are shown in Table 1. There is a weak pressure dependence here in this ratio that is not expected with a simple BCS picture where $\Delta\gamma_c/\gamma_n = 1.55$. Direct specific heat measurements at ambient pressure have reported values for $\Delta\gamma_c$ and these are shown for comparison in Table 1 are in good agreement with our ambient result. Combining with the measurement by Horowitz *et al*[1] of $\gamma_n = 3.13$ mJ/(mol K^2) gives $\Delta\gamma_c/\gamma_n = 2.34$, 2.54 and 2.60, again in good agreement with our present ambient-pressure result. We note that for the purely parabolic form $H_c = H_c^o\{1 - t^2\}$ then $\Delta\gamma_c/\gamma_n = 2$ exactly, and so the experimentally observed excess of $\Delta\gamma_c/\gamma_n$ is a direct indication of the presence of a higher order (quartic) term in the T-dependent critical field.

Table 1. Specificheat coefficients calculated at different pressures for Pb

P (GPa)	$\Delta\gamma_c$ (mJ/molK^2)	$\Delta\gamma_c/\gamma_n$ mJ/(mol K^2)	$\Delta\gamma_c$ comparison with other reports (all at P=0)
0	7.85	2.33	7.33±0.3 Clement and Quinnel[28]
0.5	7.51	2.28	7.96±0.08 Shiffman*et al.*[29]
0.9	7.05	2.18	8.13±0.13 Neighbor *et al.*[30]
1.15	7.17	2.24	

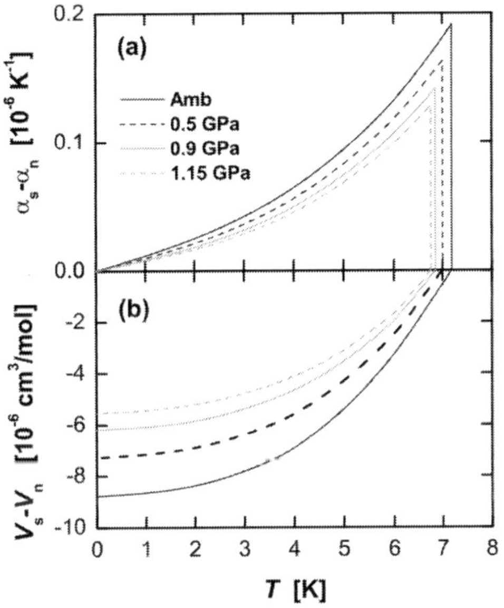

Figure 6. (a) The T-dependence of the difference in volume thermal expansion coefficient between the superconducting and normal states for pressures of 0, 0.5, 0.9 and 1.15 GPa. (b) the difference in molar volume for the same pressures.

Fig. 6(a) shows the calculated difference in volume thermal expansion coefficient between the superconducting and normal states at each pressure and Fig. 6(b) shows the difference in molar volume. Given that the absolute molar volume at T = 0 K is 18.26 cm^3/mol these changes in volume seem extremely small but they are readily measurable. The jumps in thermal expansion coefficient at T_c in zero field are shown in Table 2. We omit the calculations of the electronic compressibility because these involve double derivatives with respect to pressure and here the errors begin to accumulate markedly. Nonetheless the discontinuous jumps in isothermal compressibility may be calculated from the Ehrenfest equation for a second order transition:

$$dT_c/dP = \Delta\kappa/\Delta\alpha \tag{8}$$

Empirically we find that

$$T_c(P) = 7.1863 - 0.3847P + 0.01769P^2 \tag{9}$$

Table 2. T_c, Thermal expansion coefficient, Compressibility and its relative jumps for Pb

P (GPa)	T_C (K)	$\Delta\alpha$ $\times\ 10^{-7}$ K^{-1}	$\Delta\kappa$ $\times 10^{-8}(GPa^{-1})$	$\Delta\kappa/\kappa_0$ (ppm)
0	7.18	1.91	-7.3	-3.56
0.5	6.99	1.63	-6.0	-2.93
0.9	6.86	1.42	-5.0	-2.44
1.15	6.76	1.29	-4.4	-2.15

so that the absolute jumps in isothermal compressibility at T_c in zero field are shown in Table 2. Direct measurements[31] of the ambient-pressure elastic moduli for Pb in the superconducting state yield $\Delta\kappa/\kappa_0$ = -4.0 ppm, again in excellent agreement with our analysis for $P = 0$ GPa.

We consider the following results in the light of the BCS model[32]. The condensation energy is

$$\frac{1}{2}\mu_0 H_{c0}(P) = \frac{1}{2}N(0)N_A\Delta_0^2 = (47.11 - 7.75P)\ mJ/mol \tag{10}$$

where P is in GPa. The jump in specific heat is given by

$$\gamma_s - \gamma_n = 10.2k_B^2 N(0)N_A = (7.85 - 0.68P)\ mJ/(molK^2) \tag{11}$$

Therefore, we obtain

$$N(0) = (1.07 - 0.093P)\ states/atom/eV \tag{12}$$
$$\Delta_0 = (11.1 - 0.432P)\ K, \tag{13}$$
$$2\Delta_0/k_BT_c = 3.09 + 0.037P \tag{14}$$

This should be compared with the BCS result $2\Delta_0/k_BT_c = 3.53$ and the value of $\Delta_0 = 15.76 \pm 0.05$ K observed from tunneling[33] giving $2\Delta_0/k_BT_c = 4.37$.

If, on the other hand, we determine the DOS from the nearly-free electron expression[32]

$$\gamma_n = (2/3)\pi^2 k_B^2 N(0)N_A = (3.47 - 0.232P)\ mJ/(mol/K^2) \tag{15}$$

Then we obtain

$$N(0) = (0.736 - 0.0492P) states/atom/eV \tag{16}$$
$$\Delta_0 = (13.7 - 0.653P)\ K, \tag{17}$$
$$2\Delta_0/k_BT_c = 3.72 + 0.007P. \tag{18}$$

and the ratio $2\Delta_0/k_BT_c$ is more or less pressure independent. Finally, using the BCS expression[32] for T_c

$$T_c = 0.85\Theta_D \exp(-\frac{1}{N(0)W}) \tag{19}$$

where Θ_D is the Debye temperature (= 96 K for Pb[32]) and W is the pairing interaction, then, using eq. (12)

$$W = (0.343 + 0.0167P)eV \tag{20}$$

or using eq. (16)

$$W = (0.499 + 0.0143P)eV \tag{21}$$

Here we have used the Gruneisen coefficient $\gamma_G = -2.6$ for Pb[34].

3. $YBa_2Cu_4O_8$

Exploring the pressure dependence of the oxygen isotope effect in Y124 is motivated by the facts that (i) Y124 lies very close to $1/8^{th}$ doping with an estimated hole concentration of p=0.13. It is therefore likely to be influenced by stripe or checkerboard fluctuations; (ii) Y124 lies in the underdoped region where the pressure dependence of T_c is unusually high (iii) it is sufficiently underdoped that the pseudogap plays a very important role in governing its thermodynamic and transport properties; (iv) it is rigidly oxygen stoichiometric; and (v) this material is one of the most defect free HTS materials known.

Samples were prepared by standard solid state reaction at 945 K at 60 bar oxygen pressure as described elsewhere[35]. A 12 mm diameter pellet was cut in half and one half was annealed in 98% ^{18}O and the other half in pure ^{16}O in gold baskets placed in adjacent narrow fused quartz tube furnace at 760 ^{o}C for 6 hours at 0.95 bar pressure. The anneal was repeated twice resulting in a mass change in ^{18}O sample consistent with 95% oxygen exchange. The nearly complete isotope exchange was confirmed by shift in oxygen vibrational modes determined from Raman spectroscopy[35]. For each sample zero-field cooled temperature sweeps were made at 25 Gauss to measure diamagnetic magnetization and hence the onset of the superconductivity. The transition temperature, T_c was determined as a function of pressure using Pb as pressure marker in the pressure cell detailed above. Fig. 7 displays the pressure dependence of T_c for Y124 with ^{16}O and ^{18}O oxygen isotope exchange. The increase in T_c is essentially linear with $T_c = 80.847 + 4.947P$ for ^{18}O sample and $T_c = 81.506 + 5.265P$ for ^{16}O sample. The value of $dT_c/dP=$ 5.27K/GPa for ^{16}O is similar to previous reports[16, 17, 18]. The value of isotope exponent $\alpha \equiv -dlnT_c/dlnM$ at P=0 deduced from the linear fits of ^{16}O and ^{18}O, 0.069 is also comparable to the previous report[35]. The inset in Fig. 7 shows the steady increase in α with increasing pressure and evidently by 2 GPa the isotope exponent is projected to double.

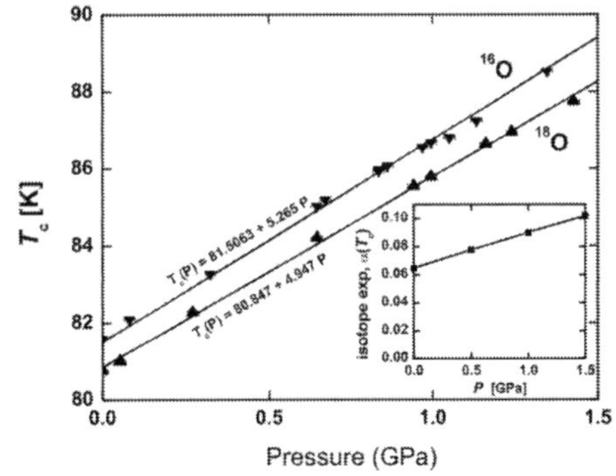

Figure 7. Figure 7. The pressure dependence of T_c for $YBa_2Cu_4O_8$ with isotope exchange. The inset shows the pressure dependence of isotope exponent $\alpha(T_c) = -\frac{M\times\Delta T_c}{\Delta M\times T_c}$.

This result is surprising. The effect of pressure is generally to transfer holes onto the CuO_2 planes. In the light of the fact that the isotope effect is generally observed to decrease with increasing doping one might expect the isotope effect to decrease with pressure. An additional outcome may be that the increase in doping moves the sample further away from the $1/8^{th}$ doping and hence away from any anomaly associated with stripe instability. This also would tend to decrease the magnitude of the isotope effect. So it is reasonable to conclude the increase is not a simple doping effect. It is possible, rather, that the tendency to stripe formation is enhanced under pressure in Y124. It would be valuable to determine the pressure dependence of the isotope effect in superfluid density ρ_s to see if this is a strong positive effect indicating the enhancement of stripe fomration [15].

4. Summary

In summary, we have measured the pressure dependence of the superconducting critical field in Pb to 1.15 GPa using a clamp cell in a SQUID magnetometer. By using thermodynamic identities we have determined the electronic free energy, entropy, specific heat coefficient, thermal expansion coefficient and compressibility, including the jumps in these properties at T_c. The calculated results match rather well the experimentally observed ambient pressure data, where it is available. The present results allow calculation of the these parameters as pressure-dependent quantities and an estimation of the pressure dependence of the density of states, superconducting energy gap and the pairing interaction.

We find a strong pressure-induced increase in oxygen isotope effect in Y124 which we find surprising in the light of pressure-induced charge transfer. We conclude that the effect of pressure may be to enhance the exchange interaction and hence the tendency to stripe (or checkerboard) formation. Some tests of this hypothesis are proposed.

We acknowledge funding from the New Zealand Marsden Fund, the International Investment Opportunities Fund and the MacDiarmid Institute.

References

[1] M Horowitz, A A Silvidi, S F Malaker and J G Daunt 1952 Phys. Rev. **88** 1182
[2] H V Culbert, D E Farrell and B S Chandrasekhar 1971 Phys. Rev. B **3** 794
[3] R R Hake, D E Mapother and D L Decker 1958 Phys. Rev. **112** 1522
[4] R W Shaw D E Mapother and D C Hopkins 1961 Phys. Rev. **121** 86
[5] D L Decker D E Mapother and R W Shaw 1958 Phys. Rev. **112**1888
[6] K Andres, J L Olsen and H Rohrer 1962 IBM J. Res. Dev. **6** 84
[7] S Alterovitz and D E Mapother 1975 Phys. Rev. B **11** 139
[8] N B Brandt and N I Ginzburg 1965 Sov. Phys. Uspekhi **8** 202
[9] H H Hansen R L Pompi and T M Wu 1973 Phys. Rev. B **8** 1042
[10] J Wittig Zeitschrift fur Physik 1966 **195** 228
[11] A Eiling and J S Schilling 1981 J. Phys. F: Metal Phys. **11** 623
[12] J Thomasson C Ayache I L Spain, and M Villedieu 1990 J. Appl. Phys. **68** 5933
[13] N Suresh and J L Tallon 2007 Phys. Rev. B **75** 174502
[14] M K Crawford, M N Kunchur W E Farneth, E M McCarro III, S J Poon 2005 Phy. Rev. B **41** 282
[15] J L Tallon, R S Islam, J Storey G V M Willams and J R Cooper 2005 Phy. Rev. Lett. **94** 237002
[16] J L Tallon and J Lusk 1990 Physica C **167** 236
[17] B Bucher J Karpinski E Kaldis P Watche 1989 Physisca C **157** 478
[18] E N Van Eenige, R Griessen R J Wijngaarden, J Karpinksi et al 1990 Physica C **168** 482
[19] A Mawdsley J L Tallon and M R Presland 1992 Physica C **190** 437
[20] G V M Williams and J L Tallon 1996 Physica C **258** 41
[21] J A Osborn Phys Rev 1945 **67** 351
[22] I R Walker 1999 Rev. Sci. Instrum. **70** 3402
[23] D Shoenberg 1952 *Superconductivity* (CUP, Cambridge) p. 56
[24] D L Waldorf and G A Alers 1962 J. Appl. Phys. **33** 3266
[25] G K White 1962 Phil. Mag. **7** 271
[26] J L Olsen and H Rohrer 1957 Helv. Phys. Acta **30** 49
[27] H R Ott 1972 J. Low Temp. Phys. **9** 331
[28] J.R. Clement and E.H. Quinnell 1952 Phys. Rev. **85** 502
[29] C A Shiffman, J F Cochran and M Garber 1963 J. Phys. Chem. Solids **24** 1369
[30] J E Neighbor, J F Cochran and C A Shiffman 1967 Phys. Rev. **155** 384
[31] G A Alers and D L Waldorf 1961 Phys. Rev. Lett. **6** 677
[32] R Meservey and B B Schwartz 1969 *Superconductivity* Vol 1 ed R D Parks (Dekker, New York) p 117
[33] W L McMillan and J M Rowell 1965 Phys. Rev. Lett. **14** 109
[34] M Hasegawa 1980 J. Phys. F - Metal Phys. **10** 225
[35] G V M Williams J L Tallon J Quilty H J Trodahl and N E Flower 1998 Phys. Rev. Lett. **80** 377

Journal of Physics: Conference Series **121** (2008) 052005 doi:10.1088/1742-6596/121/5/052005

Superconductivity, charge order and magnetic transitions under high pressure in vanadium bronzes

Yutaka Ueda and Touru Yamauchi

Materials Design and Characterization Laboratory, Institute for Solid State Physics, University of Tokyo, 5-1-5 Kashiwanoha, Kashiwa, Chiba 277-8581, Japan

E-mail: yueda@issp.u-tokyo.ac.jp

Abstract. The electric properties of β-vanadium bronze β-$Na_{0.33}V_2O_5$ have been investigated under high pressure up to 13 GPa. The metal-insulator transition accompanied by charge order is suppressed by pressure, and the superconductivity is observed around 7 GPa and 8 K (Tc). The superconducting phase adjacent to the charge ordered phase suggests an important role of charge fluctuation for the superconductivity. With further increase of pressure beyond 12 GPa, the superconductivity disappears in relevance to the appearance of a nonsuperconducting new phase.

1. Introduction

The discovery of pressure-induced superconductivity in β-$Na_{0.33}V_2O_5$ has accelerated the investigation of β-vanadium bronze family, β-$A_{0.33}V_2O_5$ (A = Li^+, Na^+, Ag^+, Ca^{2+}, Sr^{2+} and Pb^{2+}) [1]. We have performed the investigation of β-$A_{0.33}V_2O_5$ under high pressure. As the results, all A^+-compounds show charge order transitions under ambient pressure and pressure-induced superconductivity [2]. A striking feature is superconducting phases competing with charge ordered phases, which suggests a significant role of charge fluctuation for the superconductivity. On the other hand, A^{2+}-comppounds have not such superconducting phase and instead have nonsuperconducting charge disordered phases [3]. These results reveal the competition among various ground states in β-vanadium bronzes.

The common monoclinic crystal structure of β-$A_{0.33}V_2O_5$ consists of a characteristic V_2O_5-framework and A-cations. The V_2O_5-framework is formed by sharing the corners and edges of $(V1)O_6$, $(V2)O_6$ and $(V3)O_5$ polyhedra, where V1, V2 and V3 are the three crystallographically independent vanadium atoms. Each polyhedron forms three kinds of infinite chain along the b-axis. The V_2O_5 framework has tunnels surrounded by the three chains. The A-cations occupy the tunnel sites in a manner of zigzag type below the ordering temperatures and act as electron donors, namely β-vanadium bronzes are an electron doped system into a band insulator V_2O_5 in contrast to high T_c cuprates as a hole doped system to a Mott insulator. The stoichiometric composition is $A_{1/3}V_2O_5$ or $A_{0.33}V_2O_5$. Reflecting these structural characteristics, β-$A_{0.33}V_2O_5$ are quasi-one-dimensional conductors with metallic conductivity along the b-axis. The β-$A_{0.33}V_2O_5$ show metal-insulator (M-I) transitions with a nature of charge order except Pauli paramagnetic Pb-compound. The ground states of the insulating phases of β-$A^+_{0.33}V_2O_5$ are antiferromagnetic ordered states, while those of β-$A^{2+}_{0.33}V_2O_5$ (A = Ca and Sr) are spin gapped states.

© 2008 IOP Publishing Ltd

The previous studies were done in the pressure region up to 10 GPa by using sintered WC anvils. Major aim of this paper is to report the detailed results of β-$Na_{0.33}V_2O_5$ studied under pressure up to 13 GPa.

2. Experimental

In the present study, electrical properties of β-$Na_{0.33}V_2O_5$ were investigated in much higher pressure region up to 13 GPa, by using cubic anvil type high pressure apparatus equipped with precisely scoured sintered diamond anvils. A drop of methanol-ethanol-water (16:3:1 in volume) mixture was used as pressure medium to ensure better hydrostatic pressure above 10 GPa and to avoid damage of sample crystal. The out of the drop-like pressure medium was filled with the second pressure medium, Fluorinert, to prevent the drop of methanol-ethanol-water from evaporating during long time experiments about several weeks. Sample crystals were grown in N_2 atmosphere by *rf*-heating Czochralski method using self-flux of $NaVO_3$ in a Pt crucible. The resistivity under high pressure was measured along the *b*-axis by ordinal 4-probe method.

3. Results and discussion

The temperature dependences of resistivity (ρ-T curves) under several pressures are represented in Fig. 1. The charge ordering temperature (M-I transition temperature) decreases as pressure increases. Near the critical pressure (7.0 GPa) where the minimum absolute values of the resistivity reach about 100 μΩcm, a sudden upturn caused by charge order and a clear drop by superconductivity are observed in ρ-T curves. With a slight increase of pressure from the critical pressure, the ρ reaches zero resistivity within experimental resolution. A sudden upturn and a successive clear drop but nonzero resistivity in ρ-T curve near the critical pressure can be understood as the coexistence of charge ordered phase and superconducting phase. Such coexistence behavior is not due to inhomogeneous pressure but reflects a first order transition from charge ordered insulator phase to superconducting phase.

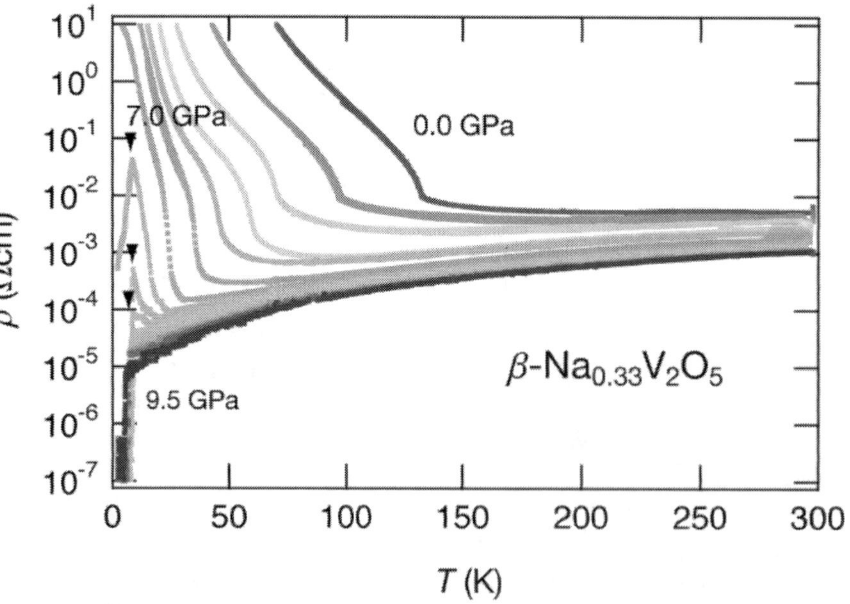

Fig. 1. Temperature dependences of resistivity (ρ-T curves) for β-$Na_{0.33}V_2O_5$ under several pressures up to 10 GPa. Resistivity was measured by excitation current 100 μA along the *b*-axis, quasi-one dimensional conductive direction. The metal-insulator transition is suppressed by pressure and superconducting transition (inverted triangles) is observed around 7.0 GPa.

Figure 2 shows ρ-T curves above 10 GPa in pressure-increasing and -decreasing (9.5 GPa) processes. The pressure of 9.5 GPa was determined from the superconducting transition temperature because the calibration of pressure was impossible in pressure-decreasing process. The ρ-T curve at 11.2 GPa clearly shows superconducting transition. By a slight increase of pressure up to 11.8 GPa, the ρ-T curve shows no longer any superconducting behavior, and then the ρ-T curve at 9.5 GPa shows superconducting transition again. The clear superconducting transition at 9.5 GPa in pressure-decreasing process excludes a possibility of sample damage in the run beyond 11 GPa. These results imply the appearance of new nonsuperconducting phase beyond 11.8 GPa. Such a new phase at higher pressure is consistent with the results of an optical study on β-$Na_{0.33}V_2O_5$ under high pressure at room temperature [4]. Similar behaviors have observed in β-$Li_{0.33}V_2O_5$ in much serious manner, namely ρ-T curve shows a drop but nonzero-resistivity, suggesting that a nonsuperconducting new phase appears almost simultaneously with a superconducting phase [5].

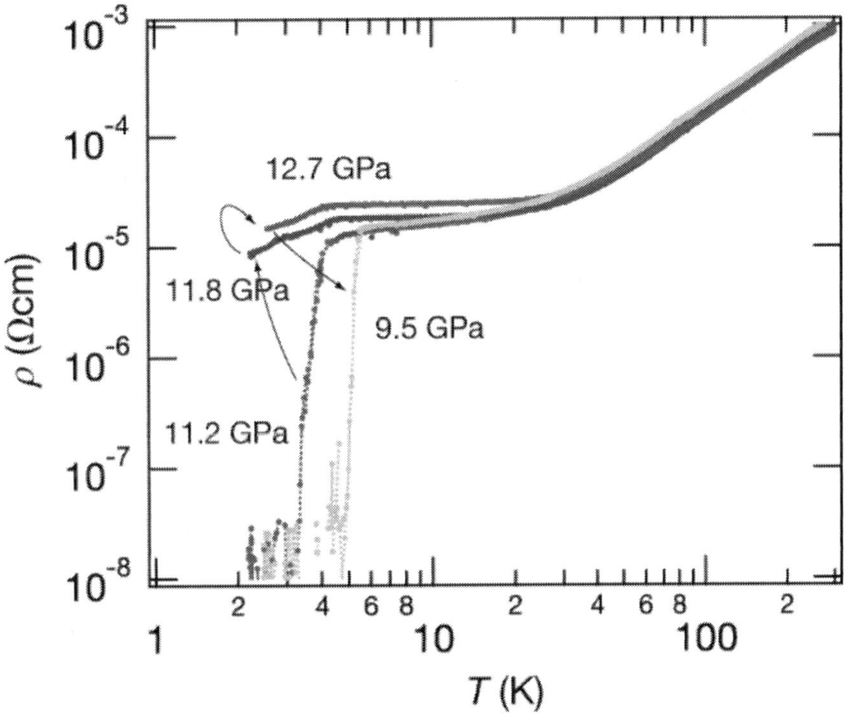

Fig. 2. The ρ-T curves of β-$Na_{0.33}V_2O_5$ above 10 GPa in pressure-increasing and at 9.5 GPa in pressure-decreasing processes. With a slight increase of pressure from 11.2 GPa to 11.8 GPa, the superconducting transition disappears. By going back to 9.5 GPa, the superconducting transition recovers.

We summarize the obtained results in Fig. 3 as P-T phase diagram. The charge order transition (M-I transition) is suppressed under high pressure and the superconducting state appears when the charge order melts. The superconducting phase is adjacent to the charge ordered phase. This suggests an important role of charge fluctuation for the superconductivity. On the other hand, the recent NMR study suggests that the antiferromagnetic state survives up to at least 6 GPa and therefore spin fluctuation also may play a role for the superconductivity. The present study reveals that a nonsuperconducting new phase exists in the higher pressure region above the critical pressure. The presence/absence of superconducting phase in β-$A_{0.33}V_2O_5$ could be closely related to such new phases. The superconductivity blooms in such competition among various ground states.

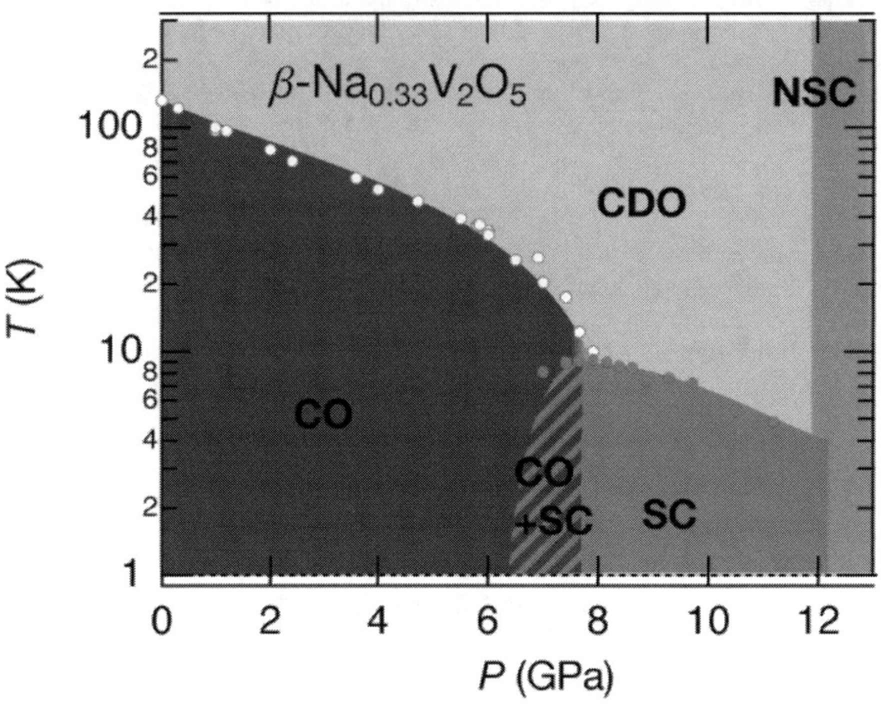

Fig. 3. Pressure-temperature (P-T) phase diagram of β-$Na_{0.33}V_2O_5$. CDO: charge disordered normal metallic phase, CO: charge ordered insulating phase, SC: superconducting phase, NSC: nonsuperconducting new phase.

This study has been supported by a Grant-in-Aid for Scientific Research (No. 18104008) from the Japan Society for the Promotion of Science.

References

[1] Yamauchi T, Ueda Y and Môri N 2002 *Phys. Rev. Lett.* **89** 057002
[2] Yamauchi T, Isobe M and Ueda Y 2005 *Solid State Sciences* **7** 874
[3] Yamauchi T, Ueda H, Yamaura J and Ueda Y 2007 *Phys. Rev.* B **75** 014437
[4] Kuntscher C A, Frank S, Loa I, Syassen K, Yamauchi T and Ueda Y 2005 *Phys. Rev.* B **71** 220502
[5] Yamauchi T and Ueda Y, to be published in *Phys. Rev.* B

Recent studies in superconductivity at extreme pressures

James S. Schilling and James J. Hamlin

Department of Physics, Washington University, CB 1105, One Brookings Dr, St. Louis, MO 63130, USA

Abstract. Studies of the effect of high pressure on superconductivity began in 1925 with the seminal work of Sizoo and Onnes on Sn to 0.03 GPa and have continued up to the present day to pressures in the 200 - 300 GPa range. Such enormous pressures cause profound changes in all condensed matter properties, including superconductivity. In high pressure experiments metallic elements, T_c values have been elevated to temperatures as high as 20 K for Y at 115 GPa and 25 K for Ca at 160 GPa. These pressures are sufficient to turn many insulators into metals and magnetics into superconductors. The changes will be particularly dramatic when the pressure is sufficient to break up one or more atomic shells. Recent results in superconductivity to Mbar pressures wll be discussed which exemplify the progress made in this field over the past 82 years.

Superconductivity is a macroscopic quantum phenomenon which was discovered by G. J. Holst and H. K. Onnes [1] in Leiden in 1911, but not clearly understood until Bardeen, Cooper, and Schrieffer (BCS) [2] formulated their microscopic theory in 1957, exactly half a century ago. In the opinion of one of the present authors (JSS), had superconductivity not first been demonstrated in experiment, no theorist would have ever predicted it: who could imagine that two electrons, in spite of their Coulomb repulsion, might experience a net *attractive* interaction binding them together to form a bose particle? In a lecture in 1922 in honor of H. K. Onnes, Albert Einstein [3] considered various scenarios but finally conceded that the theory of superconductivity was in a sorry state with no end in sight. Three years later, and one year before his death, Kamerlingh Onnes wrote the following introductory paragraph in a paper of particular relevance to this conference [4]: "As no satisfactory theoretical explanation of the supraconductive state of metals has been given yet, which might serve as a guide for further investigations, it seems desirable to try, by changing the external conditions, to discover the factors which play a roll in the appearing of the phenomenon. These considerations led to the institution of an inquiry into the influence which elastic deformation exerts on the appearing of the supraconductive state. The results of this investigation are published here." In this paper the authors go on to present the results of the first high-pressure experiments on any superconductor. Their results on Sn are reproduced in Fig. 1 where the superconducting transition temperature T_c of Sn is seen to decrease by 5 mK under a hydrostatic pressure of 300 bars. This paper is a model of careful experimentation and should be read by every student engaged in high pressure research, whatever the field of specialization.

The decrease in T_c with pressure for Sn and In is, in fact, found for all simple (s,p-electron) metal superconductors [5] and can be easily accounted for within BCS theory which is based on

© 2008 IOP Publishing Ltd

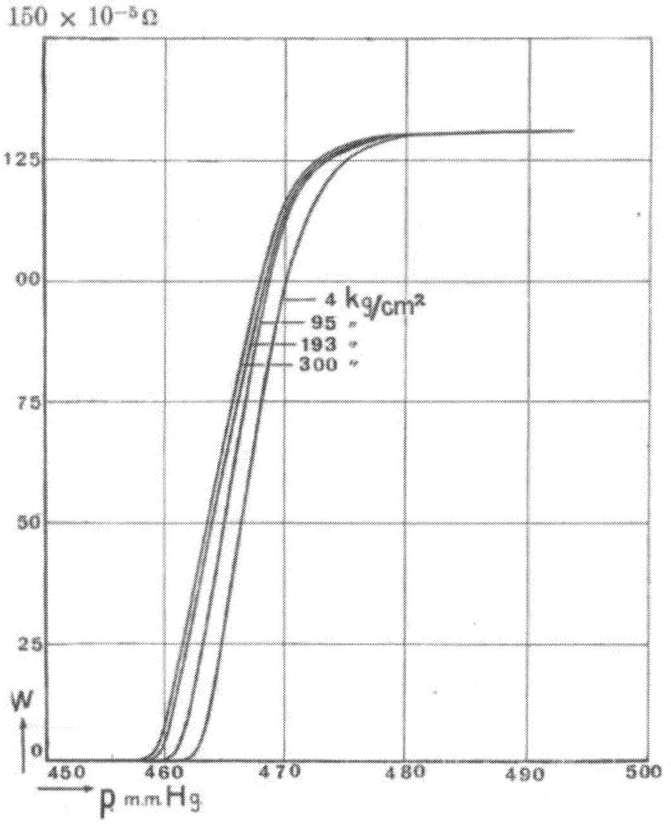

Figure 1. Electrical resistance versus temperature (units of ^{4}He vapor pressure) under hydrostatic pressures of 4, 95, 193, and 300 bar from Ref. [4]. T_c decreases under pressure. Figure reproduced with permission from the Kamerlingh Onnes Laboratory, Leiden, The Netherlands.

an electron-phonon pairing interaction:

$$T_c \approx \sqrt{\frac{\kappa}{m}} e^{-\kappa/\eta}, \tag{1}$$

where m is the cation mass, κ is the lattice spring constant, and η is the Hopfield parameter [6], a purely electronic term. In simple metals under pressure, κ increases much more rapidly than η, so that the exponential factor, and thus T_c, decreases rapidly. The decrease in T_c with pressure in MgB_2, which with $T_c \simeq 40$ K possesses the highest value of T_c of any binary compound, can also be accounted for assuming electron-phonon interactions [7]. Note that if the isotopic mass m is varied in Eq. 1, instead of the pressure, T_c changes as $T_c \sim m^{-1/2}$, the classic isotope effect.

There are three primary ways in which high pressure experiments make important contributions to the field of superconductivity:

(i) Test theories of superconductivity and reveal important systematics (see above)
(ii) Improve the properties of known superconductors
(iii) Create new superconductors

Besides testing theoretical models, high pressure experiments can also help improve the superconducting properties by showing the experimentalist how to push materials to higher values of T_c. This is of great technological relevance since no superconductor has yet been found which at ambient pressure possesses a transition temperature even half that of room temperature, in contrast to magnetic materials which order at much higher temperatures and have had an enormous commercial impact. For this reason the pressure dependence of T_c is often one of the first experiments to be carried out after a new superconductor is discovered: a large value of dT_c/dP, either positive or negative, implies the material is likely capable of higher transition temperatures at ambient pressure either through selective doping or positive or negative chemical pressure. This strategy was applied by Paul Chu's group with great success in the late 1980's. They determined that the transition temperature of the oxide cuprate La-Ba-Cu-O increased from 32 K to 40 K under only 1.4 GPa pressure [8]; the replacement of the large La cation with the smaller Y generated chemical pressure and resulted in the discovery of the first superconductor Y-Ba-Cu-O with a transition temperature ($T_c \simeq 92$ K) above the boiling point of liquid nitrogen [9]. Several years later, in a collaboration with members of the Carnegie Institute of Washington, Chu's group was able to push the transition temperature of the mercury compound Hg-1223 to the record-high value $T_c \approx 160$ K [10]. These are excellent examples of how high pressures can be used to improve the properties of superconducting materials which exemplify the philosophy and tradition of the group of Bernd Matthias at the University of California, San Diego which has been ably carried forth by many of his students, including Paul Chu and Matthias' successor, Brian Maple.

High pressures can also be used to synthesize new superconductors; simultaneous high temperature may be necessary [11], but sometimes not. In Fig. 2 is an updated and expanded version of the Periodic Table of Superconductivity originally drawn up by Bernd Matthias. There are altogether 52 elemental superconductors, 22 of which are only known to superconduct if sufficient pressure is applied. Of these 22, fully 12 were discovered by Jörg Wittig, a student of Werner Buckel at the University of Karlsruhe, Germany. Across the periodic table values of T_c range from 0.325 mK for Rh [12] or 0.4 mK for Li [13] (at 1 bar) to 20 K for Y (at 115 GPa) [14] or 25 K for Ca (at 160 GPa) [15]. This record high value of T_c for Ca was discovered by the group of Katsuya Shimizu, a student, and the successor, of Katsuya Amaya at the University of Osaka, Japan. Pressure-induced superconductivity is of great interest from a basic physics point of view since in a single high pressure experiment both the birth and the demise of the superconducting state can be witnessed. After a brief discussion of relevant high-pressure technology, we will examine several classes of pressure-induced transitions from the normal to the superconducting state.

High pressure technology has come a long way since the experiments of Sizoo and Onnes in 1925 on Sn and In at 300 bar pressure [4]. In the early 1970's Webb, Gubser, and Towle developed a diamond-anvil cell (DAC) mated to the mixing chamber of a dilution refrigerator with a SQUID pickup coil capable of detecting minute magnetic flux changes [16]. With this beautiful device they tracked the decreasing T_c of Al metal from 1.17 K to 80 mK at 6.1 GPa [17]. Using this DAC as a model, one of the present authors (JSS) constructed a DAC suitable for higher pressures where superfluid He was loaded into the gasket hole to serve as pressure medium [18]. The opposing diamond anvils were then pushed into the gasket at 2 K using a He-gas driven double membrane. Thus far pressures as high as 220 GPa have been generated in this DAC at cryogenic temperatures, if no pressure medium is used. Magnetic flux changes are detected by one of two matched counterwound pickup coils connected either to a lock-in amplifier or a DC SQUID. Shimizu *et al.* [19] utilize a DAC designed to generate Mbar pressures at temperatures as low as 5 mK; this DAC made possible the discovery of pressure-induced superconductivity in O_2 near 100 GPa [20].

An enhancement in the detection limit for superconductivity is provided by the double modulation technique utilized by Timofeev *et al.* [21] at the Carnegie Institute in Washington. Vohra and Weir have developed so-called designer diamond anvils where up to 8 electrical leads

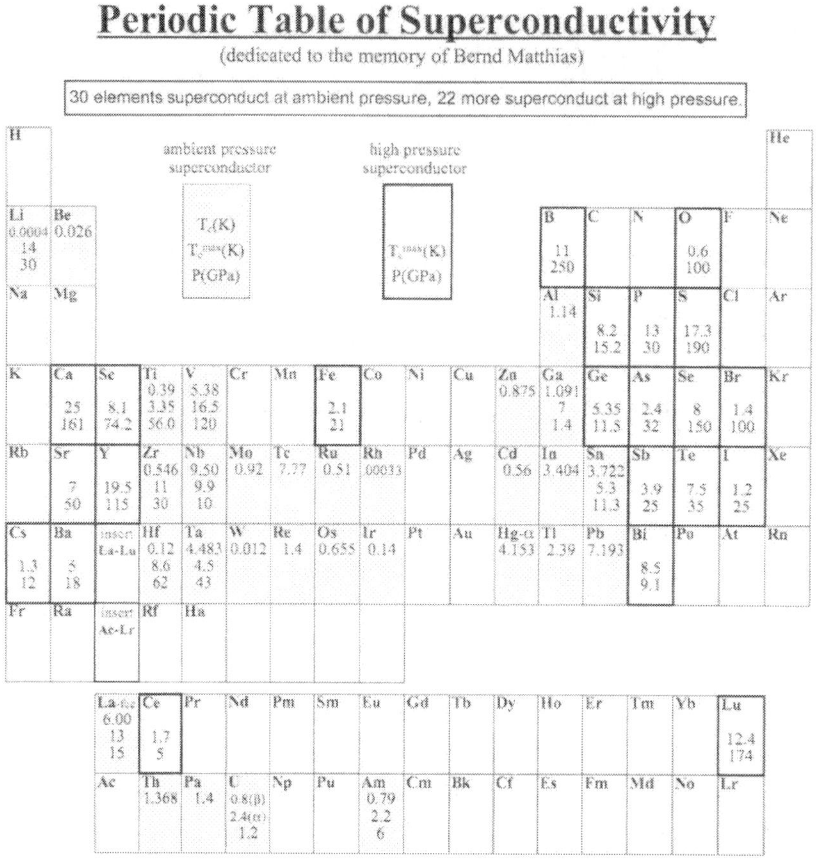

Figure 2. Periodic Table of Superconductivity listing 30 elements which superconduct at ambient pressure (yellow) and 22 elements which only superconduct under high pressure (green, bold black frame). For each element the upper position gives the value of T_c(K) at ambient pressure; middle position gives maximum value $T_c^{\max}$(K) in a high-pressure experiment at P(GPa) (lower position). T_c values from Refs. [5, 13].

or pickup loops are grown into the diamond anvil itself [22]. These anvils offer great promise for future quantitative electrical resistivity or ac susceptibility measurements in the multi-Mbar pressure range.

Let us now return to the Periodic Table of Superconductivity in Fig. 2. An examination of the maximum measured values of T_c at ambient or high pressure reveals that the lighter elements tend to do better. This is particularly evident for the alkaline earths going from Ba (5 K) to Sr (7 K) to Ca (25 K) and for the alkali metals from Cs (1.3 K) to Li (14 K). The lightest element, of course, is hydrogen. Many years ago Ashcroft [23] predicted that superconductivity near room temperature might be in the offing if only enough pressure is applied to make it metallic. Alas, the metallic state in hydrogen has not yet been sighted, except at very high temperatures (3000 K) [24] where superconductivity has little chance. Should metallic hydrogen prove to be fluid at low temperatures, then two-component superconductivity from the electrons and protons has been predicted with a myriad of fascinating ground states both with and without an applied magnetic field [25]. Is it realistic to hope that metallic hydrogen will be fluid at low

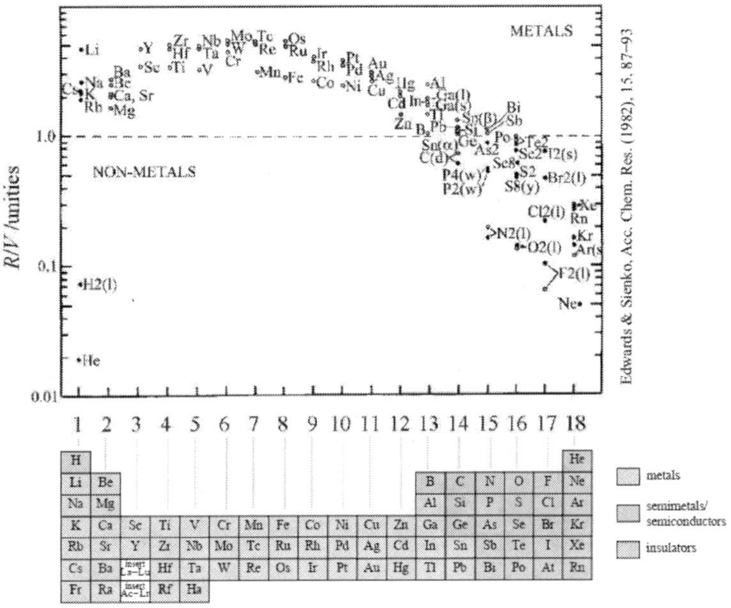

Figure 3. (Upper panel) Ratio of atomic molar refractivity versus molar volume R/V for elemental solids from Ref. [28]. (Lower panel) Corresponding periodic table where metals, semimetals/semiconductors, and insulators are color coded.

temperatures in some pressure range? A hint at an answer is given by the fact that the melting curve of Na drops precipitously from 1000 K at 30 GPa to 300 K at 118 GPa [26], a significantly lower melting temperature than at ambient pressure!

As seen in Fig. 2, the Periodic Table of Superconductivity contains four classes of elemental solids that refuse to superconduct at ambient pressure: (i) the non-metals like H_2, O_2, Si, or Ge, (ii) elements like Fe, Co, Ni, Mn, as well as most rare earths and many actinides which are magnetic, the arch-enemy of superconductivity, (iii) the early, trivalent *d*-electron elements Sc, Y, and Lu, and (iv) the monovalent alkali and noble metals, excepting Li which barely makes it with $T_c \simeq 0.4$ mK. The first class is the easiest to understand: without conduction electrons, superconductivity doesn't have a chance. How does one predict what pressure is necessary to turn an insulator into a metal? With ample financial resources one can carry out a full scale electronic structure calculation or, as a cost-saving alternative, one can use the simple Goldhammer-Herzfeld criterion [27] for metallization which states that a given substance should be metallic if the ratio of its molar refractivity to its molar volume is equal to or greater than 1, i.e. $R/V \geq 1$. Using the equation of state, it is thus trivial to estimate the metallization pressure if the initial value of R/V is known (the change in R with pressure can be neglected). R is obtained from the measured atomic polarizability α, where $R = (4\pi\alpha/3)N_A$ and N_A is Avagadro's number. In Fig. 3 the ratio R/V is plotted for all elements (upper figure) [28] and compared to the actual electronic state of all elemental solids (lower figure). The Goldhammer-Herzfeld criterion is seen to have an amazing predictive power.

The fact that diamond anvils are transparent to visible light allows optical transmission and absorption experiments in a DAC to determine the band gap E_g of an insulating material as a function of pressure. If white light is used, the light transmitted through the sample will appear yellow, orange, or black if $E_g \approx 2.2$, 1.7, or 1.4 eV, respectively. Ashcroft [29] has recently pointed out that in hydrogen-rich compounds, such as $LiBH_4$ and $LiAlH_4$, the hydrogen may

Transmitted Light Photographs

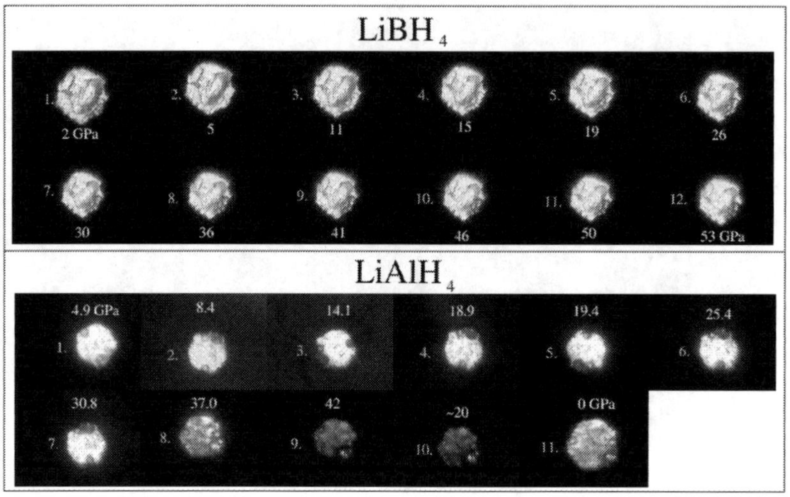

Figure 4. Transmitted light micro-photographs of $LiBH_4$ and $LiAlH_4$ at ambient temperature to 53 and 42 GPa, respectively. Sample diameter is $\sim$ 100 μm. Color gives estimate of band gap: $E_g \approx 2.2$ eV (yellow), 1.7 eV (orange), 1.4 eV (black).

be subjected to lattice pressure which could significantly reduce the external pressure required for metallization. In Fig. 4 microscopic photographs of the light transmitted through the two insulating compounds $LiBH_4$ and $LiAlH_4$ are shown for a sequence of increasing pressures to $\sim$ 50 GPa. In $LiBH_4$ no coloration of the white light is seen to 53 GPa, indicating that the band gap remains greater than 2.2 eV throughout the experiment. In contrast, in $LiAlH_4$ the transmitted light begins to change color at 30.8 GPa, becoming progressively more red to 42 GPa; however, further loading to 75 GPa resulted in no further change in color. The fact that upon unloading the color change is not reversible gives evidence for an irreversible phase transition.

The second class of elemental solids that do not superconduct at ambient pressure are those which are magnetic: the $3d$ transition metals, the rare earths $(4f)$, and the actinides $(5f)$. This is a vast playground for high pressure research where the possibilities include both transitions from one form of magnetism to another as well as from magnetism to superconductivity [18]. For example, Shimizu *et al.* [19, 30] have shown that Fe loses it's ferromagnetism when it transforms from the bcc to the hcp ϵ-phase near 14 GPa, allowing superconductivity to appear with T_c reaching a maximum value of 2.1 K at 21 GPa. For lack of space we will not discuss pressure-induced transitions in magnetic substances further.

The third class of non-superconducting elemental solids are the early, trivalent d-electron elements Sc, Y, and Lu. Why is superconductivity lacking here? The simple answer is that they don't have a sufficient number of d electrons in the conduction band. The more d electrons a given conduction band has, the higher the density of states and the more likely it is to support superconductivity. Superconductivity in Sc [31], Y [32], and Lu [33] was discovered in high pressure experiments by Wittig and coworkers. This is not surprising since it is well known that the application of pressure increases the number of d electrons in a band. This $s-d$ transfer is expected on very general grounds [34]: the kinetic energy $\sim V^{-2/3}$ increases much faster under pressure than the potential energy $\sim V^{-1/3}$; since s-orbitals possess more radial nodes (higher kinetic energy) than d-orbitals, they shift to higher energy much faster under pressure, thus causing s electrons to be dumped into the d band. Hamlin *et al.* have recently taken Y, Sc, and

Journal of Physics: Conference Series **121** (2008) 052006 doi:10.1088/1742-6596/121/5/052006

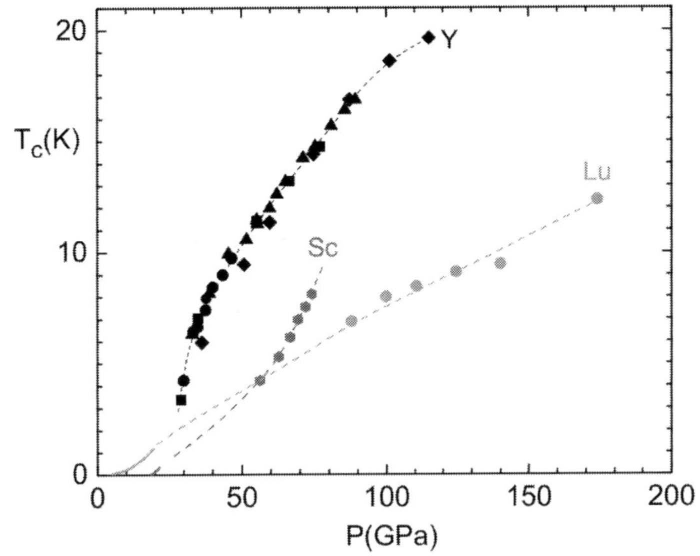

Figure 5. Superconducting transition temperature versus pressure for Sc [35], Y [14], and Lu [36].

Lu to pressures as high as 115 [14], 74 [35], and 174 GPa [36], respectively, and found that T_c increases to temperatures as high as 20 K, as seen in Fig. 5. The upward curvature in $T_c(P)$ for Sc suggests that extending the experiments into the Mbar pressure range may result in a significant enhancement in T_c. Such experiments on Sc are currently underway.

One nagging questions remains: why is the early, trivalent d-electron metal La superconducting at ambient pressure, but Sc, Y, and Lu are not? We will return to this question after considering the alkali metals.

Superconductivity in the alkali metals has recently been reviewed by one of the present authors (JSS) [37]. The first alkali metal to exhibit superconductivity was discovered by Wittig [32] in 1970 for Cs at 1.3 K under 12 GPa pressure. Electronic structure calculations [38] reveal that appreciable $s-d$ transfer ($6s$ to $5d$) occurs in this pressure range, i.e. Cs becomes a transition metal. The appearance of superconductivity is thus not surprising. However, if pressure-induced $s-d$ transfer is responsible for the superconductivity of Sc, Y, Lu, and now Cs, how can it be that Li, where $s-d$ transfer plays no role, becomes superconducting under pressure at relatively high temperatures (14 K)? The alkali metals are believed to be model free electron systems with nearly spherical Fermi surfaces. The particularly low electronic density of states of monovalent free-electron systems, the alkali and noble metals, contributes strongly to the lack of superconductivity at ambient pressure, with the exception of Li where T_c lies at extremely low temperatures (0.4 mK). In s,p-electron system, like Pb, Sn, In, or Zn, which manage to become superconducting at ambient pressure, T_c invariably *decreases* under pressure, as pointed out above. It would, therefore, appear highly unlikely that pressure would induce superconductivity in Li, Na or the noble metals where $s-d$ transfer is relatively unimportant.

The resolution of this apparent paradox was given by Neaton and Ashcroft who calculated the electronic properties of Li [39] and Na [40] at Mbar pressures and obtained for Li the

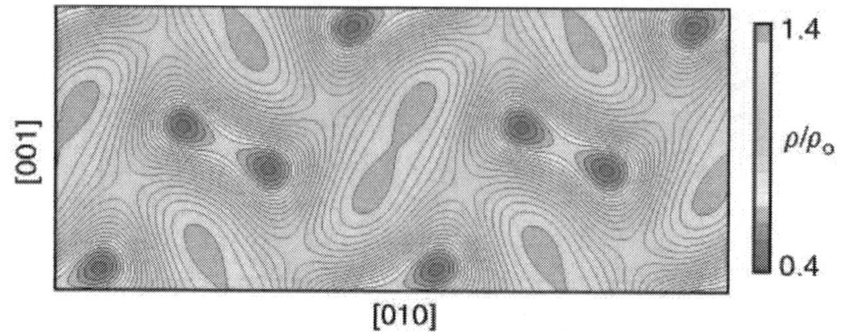

Figure 6. Relative charge density of Li's 2s conduction electrons at 100 GPa pressure. The Li^{1+} ions are centered in the red regions. Reprinted by permission from Macmillan Publishers Ltd: [NATURE] Neaton J B and Ashcroft N W 1999 *Nature* **400** 1141, copyright (1999).

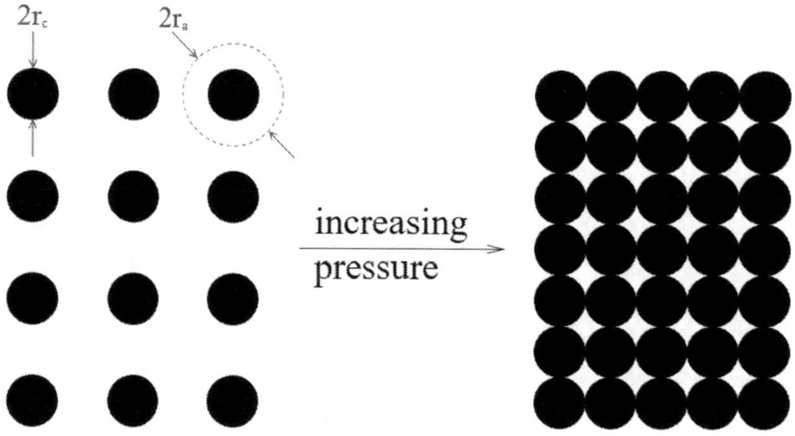

Figure 7. Scematic representation of cation electron cores at (left) ambient pressure and (right) extreme pressure. r_c and r_a define the core and Wigner-Seitz radii, respectively.

remarkable 2s-electron density shown in Fig. 6. Not only is a clear tendency evident that the Li cations pair up, but the 2s electrons are seen to be concentrated in interstitial regions rather than between the paired Li cations. The electronic properties they obtain at Mbar pressures are highly anomalous, the conduction bandwidth even *narrowing* under compression, a highly counterintuitive result in direct opposition to standard textbook dogma. The electron-phonon interaction also increases strongly. The physical picture they offer can be illustrated using Fig. 7 which depicts the ion cores as well separated at ambient pressure but beginning to touch at Mbar pressures. Since the conduction electrons must avoid the cation core region because of Pauli principle and orthogonality constraints, the free volume available to them outside the ion cores becomes quite small under Mbar pressures, forcing these electrons into the interstitial sites and away from the regions between the cation cores. The low symmetry of the interstitial sites forces the conduction electrons to take on higher angular momentum character, i.e. for s electrons more p or d character, thus improving the likelihood that superconductivity occurs. This picture is not restricted to the alkali metals, but is quite general [40] and also applicable to the transition metals, as we discuss below. In Li and Na, therefore, one may speak of significant pressure-induced $s - p$ rather than $s - d$ transfer.

In a very recent paper Feng *et al.* [41] pose the interesting quetion whether the anomalous properties predicted and found for Li and Ca under pressure carry over to $CaLi_2$, the only

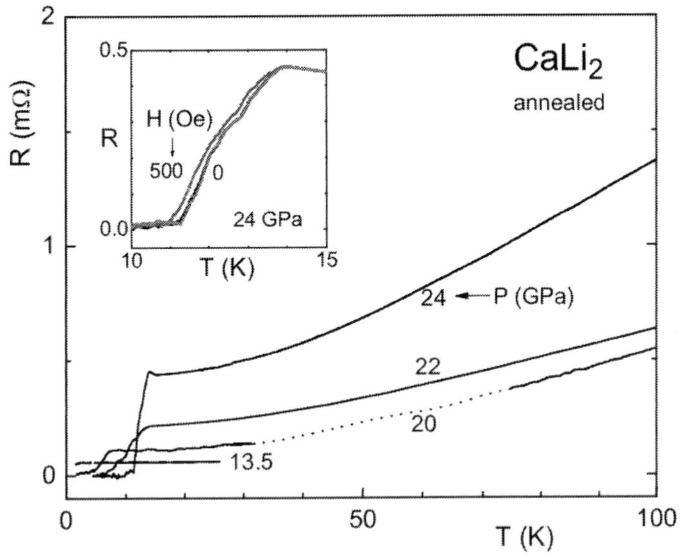

Figure 8. Resistance versus temperature for annealed $CaLi_2$ at 13.4, 20, 22, and 24 GPa. Inset gives expanded view of superconducting transition at 24 GPa for dc magnetic fields of 500 Oe (2 runs) and 0 Oe (3 runs), where colors distinguish runs.

known binary compound containing these two elements. Detailed calculations by these authors lead them to predict that "the elevated density of states at the Fermi level, coupled with the expected high dynamical scale of Li as well as the possibility of favorable interlayer phonons, points to potential superconductivity of $CaLi_2$ under pressure". We have carried out a search for superconductivity in $CaLi_2$ both at ambient pressure, where no superconductivity was found above 1.10 K, and high pressure [42]. The results of the latter electrical resistivity experiments are shown in Fig. 8 where a superconducting transition is seen to appear below 2 K at 11 GPa. With a further increase in pressure, T_c increases and passes through a maximum at 13 K near 40 GPa.

The physical picture in Fig. 7, that the free volume available to the conduction electrons is sharply reduced as pressure brings the ion cores together, can also be applied to the early, trivalent d-electron metals Sc, Y, La, and Lu. This was first pointed out by Johansson and Rosengren [43] who were able to analyze $T_c(P)$ data for La and La-Y alloys in terms of the reduction in the free volume available to the conduction electrons as pressure is applied. They used the ratio r_a/r_c of the Wigner-Seitz radius to that of the cation core as a measure of this free volume. In Fig. 9 we plot T_c versus ratio r_a/r_c for Sc, Y, La, and Lu, where we neglect any change in r_c with pressure compared to that in r_a, where $r_a \propto V^{1/3}$. For superconductivity to set in under pressure, evidently the ratio r_a/r_c must decrease to the value 2.2 or smaller. The vertical arrows mark the value of r_a/r_c for each elemental metal at ambient pressure. Note that for these four elemental metals the ambient pressure value of r_a/r_c is smallest for La which is consistent with the fact that La is already superconducting at ambient pressure, but Sc, Y, and Lu are not.

The alkali metals are also included in Fig. 9. Li is seen to fit in quite well with the trivalent d-electron systems. However K, Rb, and Cs's ambient pressure value of r_a/r_c would lead to the

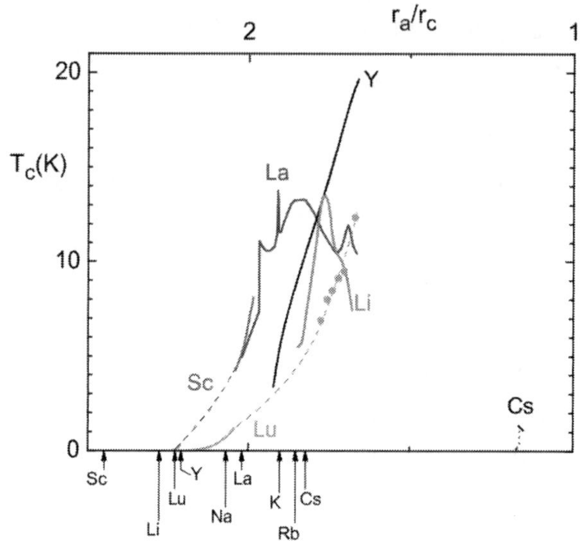

Figure 9. Pressure-induced superconducting transition temperature versus ratio r_a/r_c for the trivalent d-electron metals Sc, Y, La, Lu and the alkali metals Li and Cs. Vertical arrows mark values of r_a/r_c at ambient pressure.

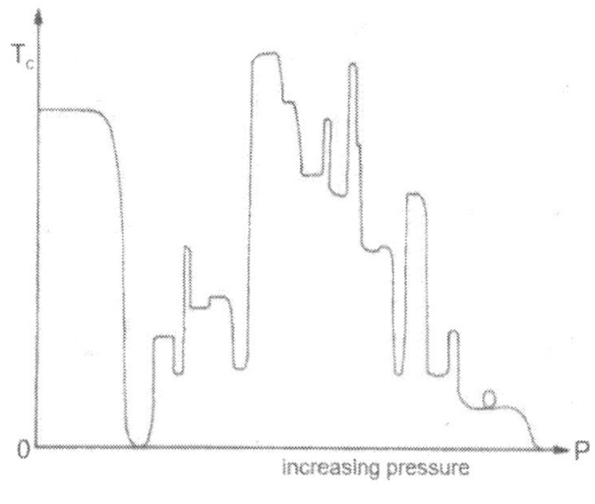

Figure 10. Schematic representation of the superconducting transition temperature versus (astronomic) pressures showing the drastic changes which occur when the atomic shell structure is progressively broken up.

expectation that they would superconduct at ambient pressure, which they don't. Perhaps our neglect of any change in the core radius r_c with pressure is oversimplified, as is our attempt to compare the properties of monovalent and trivalent elemental solids.

If the electronic properties of conduction electrons in a solid become highly anomalous as the ion cores approach each other under very high pressure, how much greater the changes would be if sufficient pressure is applied to actually begin to break open the individual atomic shells inside the ion core! As each atomic shell is broken open and its electrons dumped into the conduction band, all solid state electronic properties would fluctuate wildly. Things finally settle down when at astronomic pressures all shells have been broken open and we have a structureless Thomas-Fermi electron gas. Here it would seem likely that superconductivity and magnetism would weaken and ultimately disappear. This progression of events is mirrored schematically in the changes in the superconducting transition temperature shown in Fig. 10 where T_c is seen to fluctuate strongly with pressure. In our experiments to only a few Mbars we are only probing a relatively modest fraction of this mountain range. Science at high pressures has an exciting future to look forward to!

Acknowledgments

The authors would like to thank M. Debessai, W. Bi, and N. Hillier for experimental assistance. Stimulating discussions with N.W. Ashcroft are acknowledged. The authors are particularly grateful to Ms. Daniëlle Duijn of the Kamerlingh Onnes Laboratory in Leiden for sending us a reprint of Sizoo and Onnes' seminal publication from 1925. Thanks to A. Hofmeister for allowing us to use the optical DAC made by R. Böhler. This research was supported by the National Science Foundation through grant No. DMR-0404505.

References

[1] Onnes H K 1911 *Akad. von Wetenschappen* **14** 113
[2] Bardeen J, Cooper L N and Schrieffer J R 1957 *Phys. Rev.* **108** 1175
[3] Einstein A 1922 *Gedenkboek aangeb. aan H. K. Onnes* (Leiden: E. IJdo) p 435 (*Translation* hist-ph/0510251)
[4] Sizoo G J and Onnes H K 1925 *Commun. Phys. Lab. Univ. Leiden* (No. 180b)
[5] See, for example: Schilling J S 2007 *Handbook of High-Temperature Superconductivity: Theory and Experiment Ch 11* ed J R Schrieffer, assoc ed J S Brooks (Springer NY) p 427
[6] Note that $\eta \equiv N(E_f)\langle I^2\rangle$, where $N(E_f)$ is the electronic density of states and $\langle I^2\rangle$ the mean-square electron-phonon matrix element; see: Hopfield J J 1971 *Physica (Amsterdam)* **55** 41
[7] Deemyad S, Tomita T, Hamlin J J, Beckett B R, Schilling J S, Hinks D G, Jorgensen J D, Lee S and Tajima S 2003 *Physica C* **385** 105
[8] Chu C W, Hor P H, Meng R L, Gao L, Huang Z J and Wang Y Q 1987 *Phys. Rev. Lett.* **58** 405
[9] Wu M K, Ashburn J R, Torng C J, Hor P H, Meng R L, Gao L, Huang Z J, Wang Y Q and Chu C W 1987 *Phys. Rev. Lett.* **58** 908
[10] Gao L, Xue Y, Chen F, Ziong Q, Meng R L, Ramirez D, Chu C W, Egert J H and Mao H K *Phys. Rev. B* **50** 4260
[11] Andersson B M, Sundqvist B, Niska J, Loberg B and Easterling K 1990 *Physica C* **170** 521; Karpinski J, Kaldis E, Jilek E, Rusiecki S and Bucher B 1988 *Nature* **336** 660
[12] Buchal Ch, Pobell F, Mueller R M, Kubota M and Owens-Bradley J R 1983 *Phys. Rev. Lett.* **50** 64
[13] Tuoriniemi J, Juntunen-Nurmilaukas K, Uusvuori J, Pentti E, Salmela A and Sebedash A 2007 *Nature* **447** 187
[14] Hamlin J J, Tissen V G and Schilling J S 2007 *Physica C* **451** 82
[15] Yabuuchi T, Matsuoka T, Nakamoto Y and Shimizu K 2006 *J. Phys. Soc. Japan* **75** 083703
[16] Webb A W, Gubser D U and Towle L G 1974 *Rev. Sci. Instr.* **47** 59
[17] Gubser D U and Webb A W 1975 *Phys. Rev. Lett* **35** 104
[18] Schilling J S 1984 in *High Pressure in Science and Technology* ed C. Homan, R. K. MacCrone and E. Whalley, *MRS Symposia Proceedings No. 22* (North Holland, NY) p 79
[19] Shimizu K, Amaya K and Suzuki N 2005 *J. Phys. Soc. Japan* **74** 1345
[20] Shimizu K, Suhara K, Ikumo M, Eremets M I and Amaya K 1988 *Nature* **393** 767
[21] Timofeev Yu A, Struzhkin V V, Hemley R J, Mao H K and Gregoryanz E 2002 *Rev. Sci. Instrum.* **73** 371
[22] Weir S T, Akella J, Ruddle C A, Vohra Y K and Catledge S A 2000 *Appl. Phys. Lett.* **77** 3400
[23] Ashcroft N W 1968 *Phys. Rev. Lett.* **21** 1748
[24] Weir S T, Mitchell A C and Nellis W J 996 *Phys. Rev. Lett.* **76** 1860
[25] Babaev E and Ashcroft N W 2007 *Nature Physics* **3** 530
[26] Gregoryanz E, Degtyareva O, Somayazulu M, Hemley R J and Mao H K 2005 *Phys. Rev. Lett.* **94** 185502

[27] Goldhammer D A 1913 in *Dispersion und Absorption des Lichtes* (Teubner Leipzig); Herzfeld K F 1927 *Phys. Rev.* **29** 701
[28] Edwards P P and Sienko M J 1982 *Acc. Chem. Res.* **15** 87
[29] Ashcroft N W 2004 *Phys. Rev. Lett.* **92** 187001; ibid. 2004 *J. Phys. Condens. Matter* **16** S945
[30] Shimizu K, Kimura T, Furomto S, Takeda K, Kontani K, Onuki Y and Amaya K 2001 *Nature* **42** 316
[31] Wittig J, Probst C, Schmidt F A and Gschneidner, Jr. K A 1979 *Phys. Rev. Lett.* **42** 469
[32] Wittig J 1970 *Phys. Rev. Lett.* **24** 812
[33] Wittig J, Probst C and Wiedemann W 1972 in *Low Temperature Physics-LT 13 Vol. 1* (Plenum NY) p. 490
[34] McMahan A K 1986 *Physica* **139 & 140 B + C** 31
[35] Hamlin J J and Schilling J S 2007 *Phys. Rev. B* **76** 012505
[36] Hamlin J J and Schilling J S (unpublished)
[37] Schilling J S 2006 *High Press. Res.* **26** 145
[38] Grover R, McMahan A K and Ross M 1975 *Europhys. Conf. Abstracts* **1A** 18
[39] Neaton J B and Ashcroft N W 1999 *Nature* **400** 141
[40] Neaton J B and Ashcroft N W 2001 *Phys. Rev. Lett.* **86** 2830
[41] Feng J, Ashcroft N W and Hoffmann R 2007 *Phys. Rev. Lett.* **98** 247002
[42] Matsuoka T, Debessai M, Hamlin J J, Gangopadhyay A K and Schilling J S (preprint *arXiv:cond-mat.supr-con/0708.2117v1*)
[43] Johansson B and Rosengren A 1975 *Phys. Rev. B* **11** 2836

Journal of Physics: Conference Series **121** (2008) 052007 doi:10.1088/1742-6596/121/5/052007

High pressure transport and micro-calorimetry studies on quantum phase transitions in Yb heavy fermion systems.

E. Colombier, D. Braithwaite, G. Lapertot, B. Salce, G. Knebel and J. Flouquet.

DRFMC/SPSMS/IMAPEC. CEA-Grenoble, 17 rue des martyrs, 38054 Grenoble Cedex 9. FRANCE

daniel.braithwaite@cea.fr

Abstract. We present ac microcalorimetry and resistivity measurements under high pressure on new very pure single crystals of $YbCu_2Si_2$ having residual resistivity ratios of up to 130 and residual resistivities of less than 1 μΩcm. The onset of magnetic order at high pressure has been detected by ac micro-calorimetry in a diamond anvil cell, and the phase diagram has been established showing magnetic order appearing at 7.6 GPa and 0.95K, and suggesting a possible quantum critical point at a pressure of about 6.5 GPa. The resistivity has been measured under pressure in hydrostatic conditions, but no sign of superconductivity is found close to the expected critical pressure down to T=0.05 K. We discuss these results in comparison with results on cerium based heavy fermion systems.

1. Introduction

Recent studies on several cerium based systems have shown that when pressure is used to approach a magnetic Quantum Critical Point (QCP), that is a region of the phase diagram where a phase transition to a magnetically ordered state occurs at zero temperature, several novel phenomena are found, including deviations from Fermi liquid theory in the electronic terms of resistivity and specific heat, and most spectacularly, unconventional superconductivity. In heavy fermion compounds, the large renormalisation to lower values of the relevant energy scales make these systems ideally suited for the use of high pressure as an external parameter to tune their microscopic interactions and in many cerium based systems, pressures of a few GPa are sufficient to attain the QCP[1]. Today considerable attention has turned to ytterbium systems. In a simple picture Yb is often considered to be the "hole" equivalent of cerium. Pressure tends to drive Yb from its nonmagnetic Yb^{2+} ($4f^{14}$) state to a magnetic Yb^{3+} ($4f^{13}$) state, so in paramagnetic Yb systems, a transition towards magnetic order is expected to occur under pressure. However there are to date far fewer studies on Yb based compounds, and the picture is much less clear than for the cerium systems. From a practical point of view, aspects in working on the Yb based heavy fermion systems include in general the necessity of reaching higher pressures (often closer to 10 GPa) and the difficulty of obtaining single crystals with low residual resistivities. Understanding the differences between the QCPs of the two families is an important step, with an essential question : why has no superconductivity been found so far in the Yb systems ? Several phenomena can play a role in preventing the occurrence of superconductivity. First, as the expected superconductiong state is of an unconventional nature, any impureties will have a strong pair-breaking effect so very pure samples are necessary. Furthermore, the present models of spin fluctuation mediated superconductivity at a QCP rely on the existence of a 2nd order phase transition to magnetic order[2]. One of the clearest examples of pressure induced magnetic order in an Yb system is $YbCu_2Si_2$[3, 4], however in the previous studies where the phase diagram was established by resistivity or Mössbauer spectroscopy measurements, the conclusion was that the transition was probably of 1st

© 2008 IOP Publishing Ltd

order. We present here a new generation of experiments on high quality single crystals of the compound $YbCu_2Si_2$, with very low residual resistivities. We show that the diamond anvil cell ac-calorimetry technique provides quantitative information on the phase diagram of long range magnetic order, and qualitative information on the divergence of the specific heat near the critical point in $YbCu_2Si_2$. We also show a new development of the standard Bridgman high pressure technique to use a liquid medium, and its application to the study of and $YbCu_2Si_2$.

2. Experiment

For this study a new batch of Single crystals of $YbCu_2Si_2$ have been grown by an indium flux method. MgO crucibles were used in favor of their higher chemical stability. Following a dissolution period of four hours at 1200°C, a slow cooling (150°/h) was performed down to 850°C where the ampoule was spun. Crystals were then annealed in a vacuum, wrapped in a tantalum foil at 850°C for 15 days. The crystals were caracterised by x-ray diffraction, resistivity and specific heat measurements. The main test of the crystal quality was the residual resistivity ratio (RRR) taken between 300K and 2K. Quite a large dispersion of RRR values was found among crystals of the same batch, however we found that it was possible to select crystals having an RRR as high as 130, corresponding to a residual resistivity of less than 1 μΩcm. For the high pressure micro-calorimetry measurements a Au/AuFe thermocouple was spot welded onto the sample and the sample was pressurized in a Diamond Anvil Cell (DAC) using argon as a pressure transmitting medium. The thermocouple measures the temperature oscillations of the sample induced by an alternating optical power source (laser or laser diode) transmitted to the sample via an optical fibre[5]. The resistivity was also measured in the DAC by bonding four 10μm leads onto the sample. In both cases the cell was measured either in a standard ^{4}He cryostat (T>1.5K) with the possibility of in-situ modulation of the pressure[6], or in a ^{3}He cryostat (T>0.5K). In the latter case the pressure was changed at room temperature, but measured in situ at low temperature. The resistivity was also measured in the newly developed modified Bridgman cell[7] with Fluorinert as pressure transmitting medium. In this cell two nylon rings are inserted inside the usual pyrophyllite gasket in order to load the cell with liquid (fluorinert) and thus allowing measurement in hydrostatic conditions on relatively large samples (1mm). In this case the cell was also measured in a 3He/4He dilution refrigerator down to 50mK.

3. Results and discussion

In figure 1 we show the results of the microcalorimetry measurement. This technique does not give a quantitative measurement of the specific heat, so the curves have been normalized to the value of C/T at high temperature (7K) of the P=0 measurement. Under pressure, C/T increases at low temperature, becoming quite divergent with a sharp anomaly which is taken to be the antiferromagnetic ordering temperature, T_N. Using the in-situ pressure modulation facility this anomaly was detected down to the lowest attainable temperature, 1.5K, for a pressure of about 8.5 GPa. In order to follow the transition at lower temperatures a few fixed pressure points were measured in a ^{3}He cryostat (see 7.6 GPa curve in the figure). The phase diagram obtained from all these points corresponding to 2 samples is shown in figure 2 in comparison with the previous resistivity measurement[3]. The high pressure data are in quite good agreement. However the previous resistivity measurement could detect no sign of a transition below about 8.5GPa, where the ordering temperature was found to be about 1.3K, hence the near vertical dashed line on the phase diagram, and the original conclusion that the transition was 1st order. In contrast in the specific heat curves we have found a clear transition at 7.6 GPa, with T_N = 0.95K. So far we cannot claim to prove that a second order QCP exists, but the results are compatible with a dependence of T_N as $(p-p_c)^{2/3}$ expected for a 3D antiferromagnetic system[8-10] which would give a critical pressure of about 6.5 GPa, much lower than previously expected. At least we show that the magnetic transition can occur at temperatures below 1K, so quantum effects should start to be predominant. The absence of points at lower temperatures and pressures may be due to the experimental difficulties in measuring the ac calorimetry at very low temperatures. Further studies are planned to search for the magnetic transition in this area. In figure 3 we show the resistivity under pressure on 2 samples, sample 1 was measured in the DAC (argon) and sample 2 in the modified Bridgman cell (fluorinert). The RRR of sample 2 was slightly less than sample 1, however even at the highest pressure the residual resistivity was less than 1 μΩcm. sample 2 was measured at 7.0 GPa, i.e. close to the expected critical pressure, down to T=50mK, however no sign of superconductivity was seen. At very low temperature a dependence of the form $\rho=\rho_0+AT^2$ is found and the coefficient A is found to increase with pressure by about 2 orders of magnitude from 0.04 $\mu\Omega cm/T^2$ at P=0 to 3.5 $\mu\Omega cm/T^2$ at 7 GPa in agreement with the previous study[3]. The divergence of A, and the increase of C/T on approaching a QCP are qualitatively expected from spin fluctuation models. A fuller discussion of the effect of pressure on the temperature dependence of the resistivity, and a comparison with the specific heat data will be published elsewhere[11]. These results show that 2 of the possible obstacles to spin fluctuation mediated superconductivity in $YbCu_2Si_2$ seem not to apply : the new single crystals we have prepared have residual resistivities of less than 1μΩcm even at the critical pressure, comparable to that of the cerium based systems where superconductivity is found, and the magnetic transition is found to still occur at temperatures below 1K. The fact that still no superconductivity is found suggests that other reasons must be sought, which may well be found in the intrinsic differences between the behaviour of cerium and ytterbium at the QCP. Indeed the simple presentation of Yb as the "hole" analogue of cerium is certainly an oversimplification. In cerium systems the valence is actually quite stable, close to 3+, and the transition between paramagnetic and magnetic states occurs mainly via the Doniach model of competition between RKKY and Kondo interactions, whereas Yb is often in an intermediate valent state, and quite large changes of the valence are expected with pressure. Probably in this case when the valence of the magnetic ion is changing strongly with pressure, the simple spin fluctuation description of the critical point must be modified. We expect that these results should stimulate further theoretical and experimental studies.

4. Summary

We have prepared new very pure single crystals of $YbCu_2Si_2$ with residual resistivity ratios of up to 130 and residual resistivities of less than 1 μΩcm. The onset of magnetic order at high pressure has been detected by ac micro-calorimetry in a diamond anvil cell, and the phase diagram has been established showing magnetic order appearing at 7.6 GPa and 0.95K, and suggesting a possible Quantum Critical point with P_C = 6.5 GPa. The resistivity has been measured under pressure in hydrostatic conditions, but no sign of superconductivity is found close to the expected critical pressure down to T=0.05 K

5. Acknowledgments

One of us (E.C.) acknowledges support from the Région Rhône-Alpes for this work.

References

1 For a review see : J. Flouquet, "On the heavy fermion road" in *Prog. Low Temp. Phys.*, edited by W. Halperin (Elsevier, Amsterdam, 2005), Vol. XV.

2 P. Monthoux and G. G. Lonzarich, Physical Review B Condensed Matter **63**, 1 (2001).

3 K. Alami Yadri, H. Wilhelm, and D. Jaccard, European Physical Journal B. Nov. **6**, 5 (1998).

4 H. Winkelmann, M. M. Abd Elmeguid, H. Micklitz, J. P. Sanchez, P. Vulliet, K. Alami Yadri, and D. Jaccard, Physical Review B Condensed Matter **60**, 3324 (1999).

Journal of Physics: Conference Series **121** (2008) 052007 doi:10.1088/1742-6596/121/5/052007

[5] A. Demuer, C. Marcenat, J. Thomasson, R. Calemczuk, B. Salce, P. Lejay, D. Braithwaite, and J. Flouquet, Journal of Low Temperature Physics. Aug. **120**, 245 (2000).
[6] B. Salce, J. Thomasson, A. Demuer, J. J. Blanchard, J. M. Martinod, L. Devoille, and A. Guillaume, Review of Scientific Instruments. June **71**, 2461 (2000).
[7] E. Colombier and D. Braithwaite, Rev. Sci. Instrum. **78**, 093903 (2007).
[8] J. A. Hertz, Physical Review B Solid State **14**, 1165 (1976).
[9] A. J. Millis, Physical Review B Condensed Matter **48**, 7183 (1993).
[10] T. Moriya and K. Ueda, Advances in Physics. July Aug **49**, 555 (2000).
[11] E. Colombier et al., to be published.

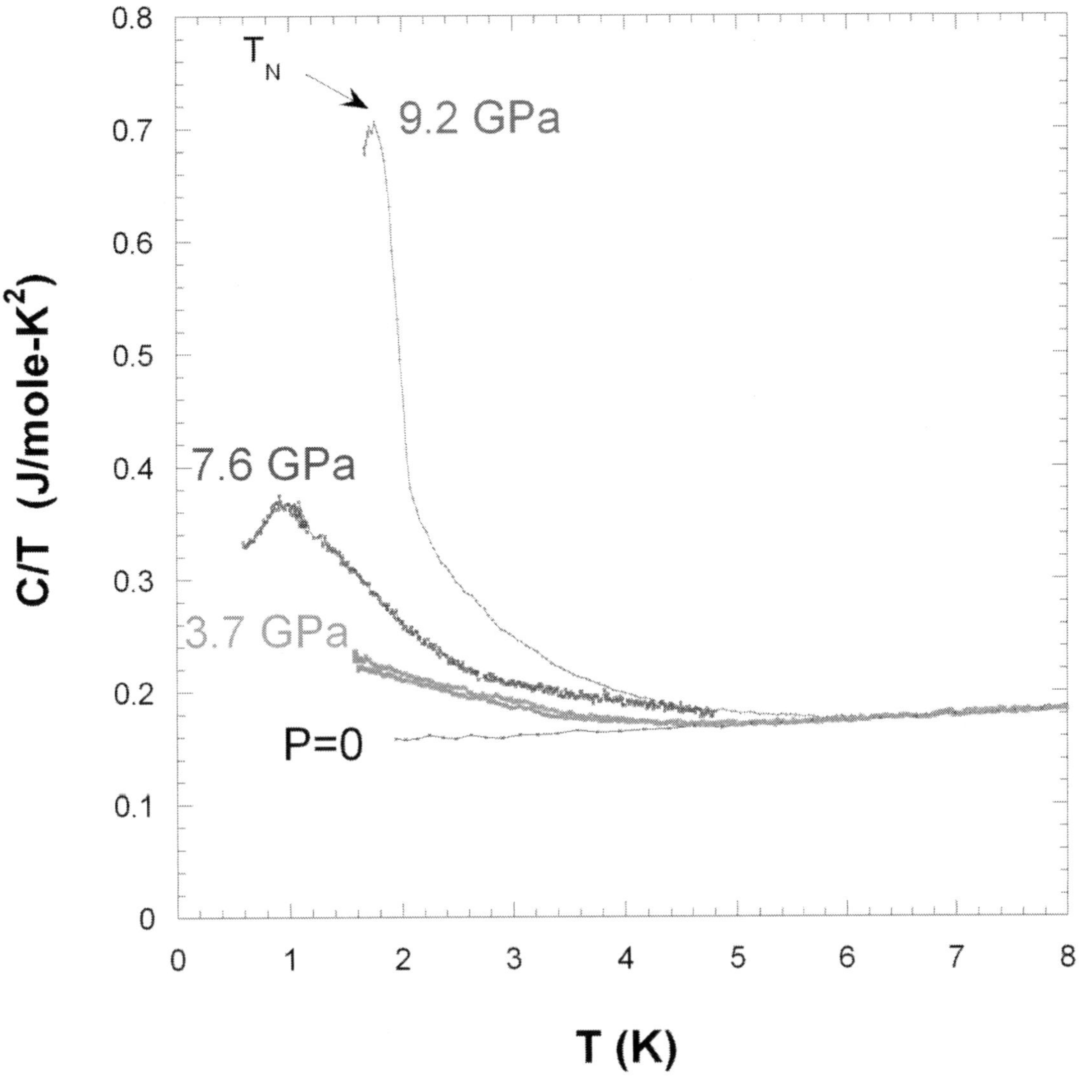

Figure 1
Specific heat of $YbCu_2Si_2$ at selected pressures measured by the ac microcalorimetry technique in the diamond anvil cell. The curves under pressure have been normalized to the ambient pressure curve at T=7K. The curve at 7.6 GPa was measured in a ^{3}He cryostat down to 0.5K. In the other measurements the lowest temperature was about 1.5K

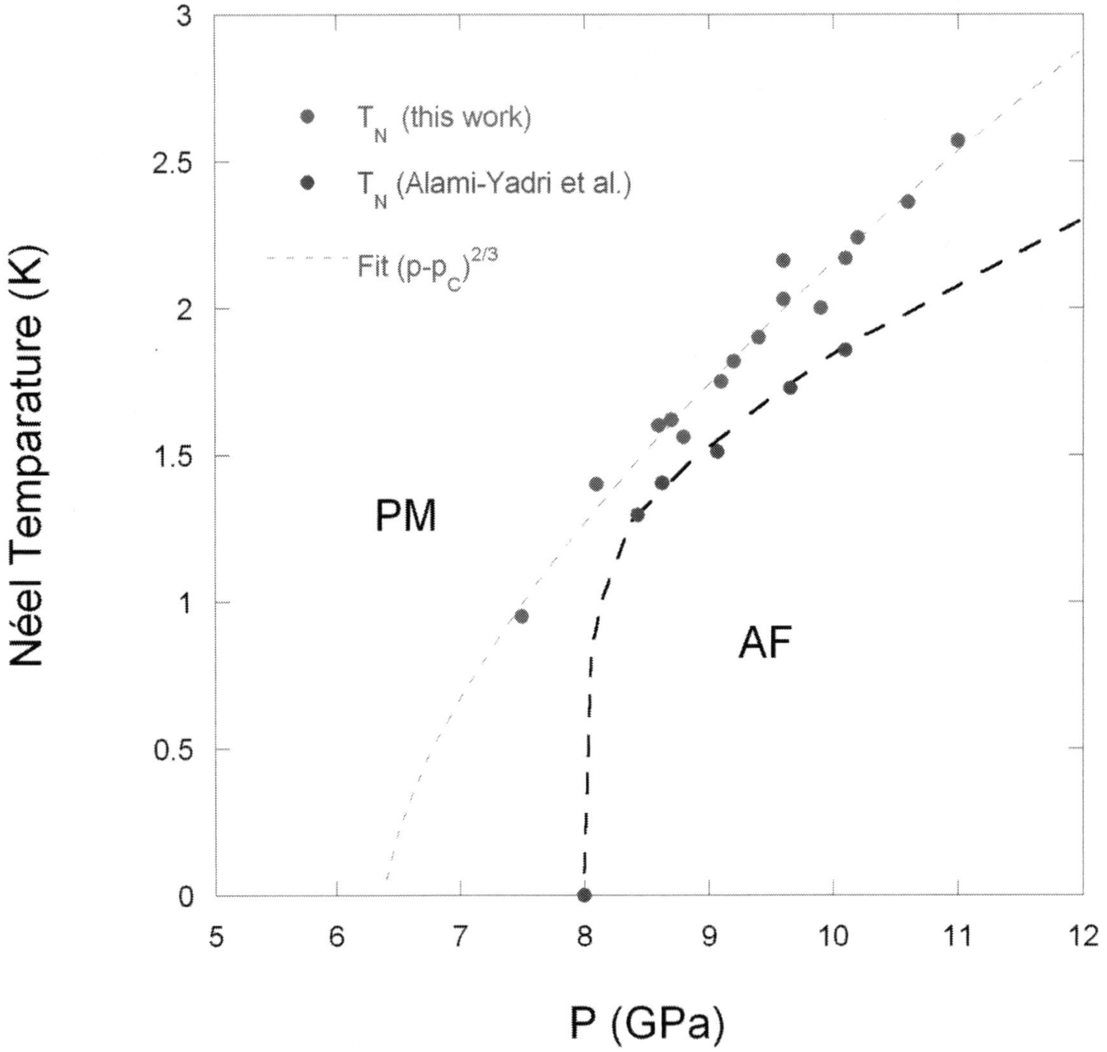

Figure 2
Phase diagram of $YbCu_2Si_2$ obtained from the microcalorimetry measurements showing the paramagnetic (PM) and probably antiferromagnetic (AF) phases, compared to that obtained by a previous resistivity study[3]. In the resistivity study no sign of magnetic order was found below 8 GPa at temperatures down to 50mK hence the almost vertical dashed line. The red dashed line is a fit of our data to the form $(p-p_C)^{2/3}$ expected for a 3D antiferromagnetic system.

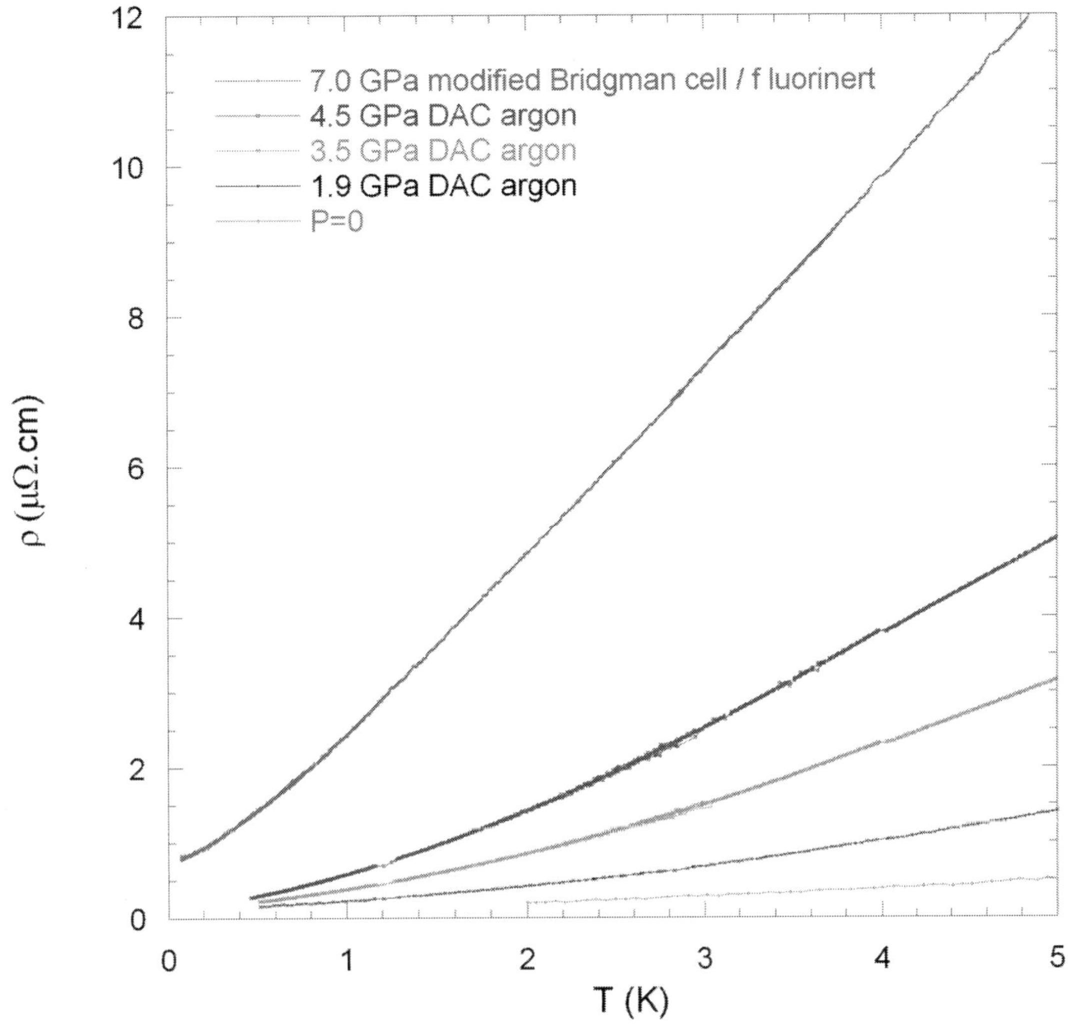

Figure 3
Low temperature resistivity under pressure of 2 samples of $YbCu_2Si_2$. Sample 1 was measured in a diamond anvil cell (DAC). Sample 2 was measured at 7 GPa in the modified Bridgman cell (see text). The curve at 7GPa is close to the expected critical pressure but shows no sign of superconductivity down to T=50mK

Coexistence of superconductivity and antiferromagnetism in $CeRhIn_5$; Model Hamiltonian and ab-initio calculations.

J.V. Alvarez and Felix Yndurain

Departamento de Fisica de la Materia Condensada.
Universidad Autonoma de Madrid, 28049 Madrid, Spain

E-mail: jv.alvarez@uam.es

Abstract. The finite temperature phase diagram of $CeRhIn_5$ as a function of pressure and magnetic field has three main highlights: a) the competitive coexistence of metallic antiferromagnetism and superconductivity, b) the abrupt disappearance of antiferromagnetism when the Neel and superconducting temperatures become equal at a critical pressure P_c and c) the reentrance of the antiferromagnetic phase in a range of pressures larger than P_c when a magnetic field is applied. Based on first-principles band structure calculations, we propose a quasi-two-dimensional model of interacting electrons, which reproduces, at the mean-field level, the central features of the phase diagram.

1. Introduction

The increasing interest in the family of $CeMIn_5$ compounds does not come exclusively from the heavy-fermion nature of its low-energy excitations. Rather, it is their capability to illustrate different aspects of the competition and coexistence of superconducting (SC) and antiferromagnetic (AFM) phases in the proximity of a quantum critical point, what has caught the attention of the solid state community. Specifically, the temperature *vs* pressure phase diagram of $CeRhIn_5$ (see Fig. 1) has been determined, with extremely consistent results, using several experimental techniques, namely: low-temperature calorimetry [1, 2], resistivity [3], nuclear quadrupole resonance [4] and ac-susceptibility [5]. Coexistence has been also found in the family of compounds $CeRh_{1-x}Co_xIn_5$ with $x = 0.1, 0.2, 0.4$ [6] [7] , with a similar behavior under the application of pressure.

The purpose of this work is discussing how the conditions that favor coexistence of a SC and AFM impose restrictions in the low-energy band structure, which in turn are observed in *ab-initio* calculations on the compound. In addition, other phenomenological implications, mainly the enhancement of both the cyclotron and specific heat masses of the electronic liquid under pressure, may have been experimentally observed in $CeRhIn_5$.

2. Coexistence of antiferromagnetism and superconductivity

Inspired by the phenomenology of $CeRhIn_5$, we will consider a model of interacting electrons based in the following assumptions: 1.) The electrons contributing to the SC and the AFM will be treated in equal footing and considered equivalent. 2.) The Hamiltonian contains terms

© 2008 IOP Publishing Ltd

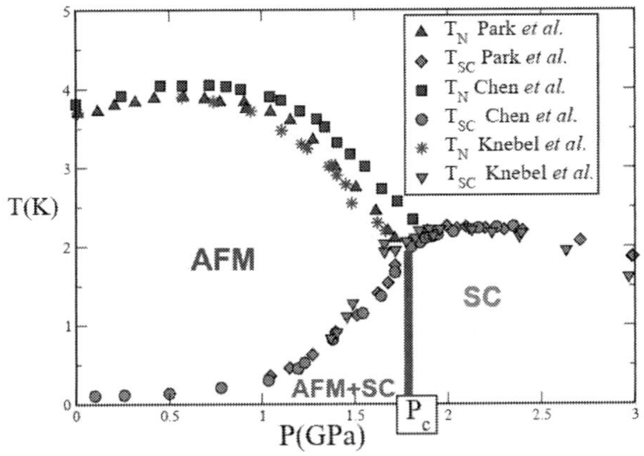

Figure 1. Experimental pressure-temperature phase diagram of $CeRhIn_5$ adapted from references [1],[2] and [5].

favoring both the SC and AFM instabilities. 3.) The Hamiltonian is solved within a mean-field approach. 4.) The dispersion relation satisfies, for some section of the Fermi surface, the condition $\varepsilon(k+Q)=\varepsilon(k)$. The starting point is therefore the following model: [8, 9]:

$$
\begin{aligned}
H &= \sum_{k\sigma}\varepsilon(k)c^{\dagger}_{k\sigma}c_{k\sigma} + \sum_{kq} V_{kq}c^{\dagger}_{(k+q)\uparrow}c^{\dagger}_{(-k-q)\downarrow}c_{-k\downarrow}c_{k\uparrow} \\
&+ \frac{U}{2}\sum_{kk'} c^{\dagger}_{k\sigma}c^{\dagger}_{k'-\sigma}c_{k\sigma}c_{k'-\sigma}
\end{aligned}
$$

H_V favors SC with an order parameter having a symmetry related to the effectively attractive V_{kq} but the presence of of the Hubbard repulsion U forbids SC at arbitrarily low values of V_{kq} and in turn favors the SDW. What kind of quasiparticle $\varepsilon(k)$ favor the coexistence of both phases for the same electrons? To avoid a gapped SDW, which will turn the system insulating at temperatures close to T_N and would not facilitate the coexistence, we consider a dispersion relation satisfying $\varepsilon(k+Q)=\varepsilon(k)$ [8, 10] , which are associated to the presence of extended saddle points connected by the antiferromagnetic Q vector.

To make connection between the model phase diagram and the experimental pressure-temperature one, we need to know the relation between the parameters of the model and the pressure. Specifically, the dependence of the bandwidth in which the density of states shows a logarithmic divergence is crucial. To this end, we have computed the pressure dependence of the effective bandwidth of the Van Hove singularity W and the on-site interaction U using first-principles calculations in two planar benchmark systems, the layers of Ni(100) and Ni(111). We used the SIESTA code for this calculations [11]. We find that W increases linearly and U is nearly constant in that range of pressures. We assumed a V independent of pressure. The results are presented in Fig. 2 (a). With these dependencies, we have plotted different theoretical phase diagrams, for different values of U. The results are presented in Fig. 2 (b) in comparison with the experimental phase diagram by Chen *et al* [5].

As discussed in Ref. [9] the enhancement of the cyclotron [13] (see Fig. 3) and specific heat masses [14] are compatible with the existence of a VHS close to the FL.

3. Band structure calculations.

Our band structure calculations for $CeRhIn_5$ (GGA implemented in VASP [15] and performed in the paramagnetic phase) show, in accordance with previous results [12], a peak in the DOS

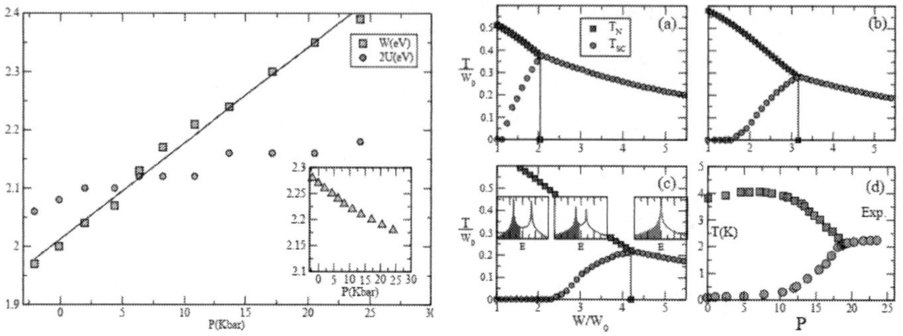

(a) (b)

Figure 2. (a) Pressure dependence of the bandwidth W and the on-site interaction in a layer of Ni(100) used as a benchmark system. W is the typical energy scale in which the density states has a logarithmic dependence, and here we study the dependence of this scale on the pressure. In the inset pressure dependence of the inter-atomic distance. (b) Comparison of our results with $V = 4W_0$, $E_c = 0.7W_0$, $n = 0.92$ electrons and $U = 2.25, 2.50$ and $2.75W_0$. W_0 is the bandwidth at ambient pressure (upper panel and lower panel on the left) with the experimental ıt).

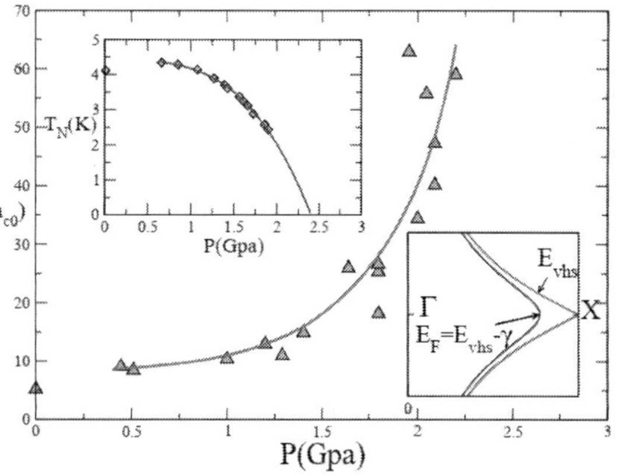

Figure 3. Cyclotron mass (m_c) as a function of pressure. Triangles are the experimental values from Ref. [13] showing the m_c divergence close to $P \sim 2.4$ GPa, where the magnetic order disappears. The solid line is a fit to the theoretical model. Close to a Van Hove singularity the cyclotron mass diverges logarithmically as the difference between E_F and the E_{vhs} vanishes (see lower inset). According to the model this energy difference is proportional to T_N and its pressure dependence can be extracted, for instance, from Ref. [2](see upper inset).

close to the FL. Furthermore, due to the quasi-2D character of the material a saddle point is observed along the $(\pi/a, 0, q_z)$ dispersing along c^* and crossing the FL at a value of $q_{zc} \sim 0.25c^*$. At this mean-field level this implies that the order will be incommensurate in the z direction with a modulation vector of $2q_{zc}$. The magnetic order of $CeRhIn_5$ is incommensurate along the c^* crystalographic axis however the experimental value for the modulation vector is 0.297.

Finally, when treated beyond mean-field theory, the presence of a saddle point in $\varepsilon(k)$ has further interesting implications. Using RG techniques [16, 17] it is possible to demonstrate that V_{kq} has electronic origin and d-wave is the more likely order-parameter symmetry of the SC phase. The analysis of the self-energy shows unambiguously non-Fermi Liquid behavior.

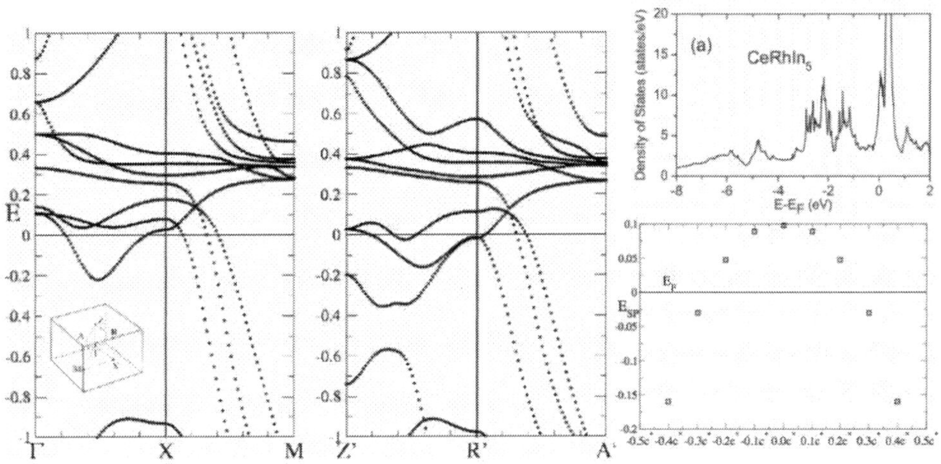

Figure 4. Low-energy electronic structure of $CeRhIn_5$ along Γ-X-M i.e. $q_z = 0$ (left panel) and $q_z = 0.3\pi/c$ (central panel). (Upper right panel) DOS (lower right panel) energy of the saddle point as a function of q_z.

4. Acknowledgements

We ackwnowledge financial support of the Spanish Ministry of Science (Ramon y Cajal contract and Grant BFM2003-03372).

[1] T. Park *et al*, J. Magn. Magn. Mat. **310**,712 (2007).
[2] G. Knebel et al., cond-mat 0512078.
[3] A. Llobet *et al*, Phys. Rev. B **69** (2004) 024403.
[4] T. Mito et al., Phys. Rev. Lett. **90** (2003) 077004.
[5] G. F. Chen *et al*, Phys. Rev. Lett. **97** (2006) 017005.
[6] V.S. Zapf *et al.*, Phys. Rev. B **65** (2001) 014506.
[7] J.R. Jeffries *et al.*, Phys. Rev. B **72** (2005) 024551.
[8] F. Yndurain, Solid St. Commun. **81** (1991) 939.
[9] J.V. Alvarez and F. Yndurain, Phys. Rev. Lett. **98** (2007) 126406.
[10] F. Yndurain, Phys. Rev. B. **51**, 8494 (1995).
[11] J.M. Soler *et al*, J. Phys. Condens. Matter **14**, 2745 (2002).
[12] D. Hall *et al*, Phys. Rev. B **64**, 064506 (2001).
[13] H. Shishido *et al*, J. Phys. Soc. Jpn. **74** (2005) 1103.
[14] R. A. Fisher *et al*, Phys. Rev. B **67** (2002) 224509.
[15] G. Kresse and J. Furthmuller, Phys. Rev. B 54 11169 (1996)
[16] J.V. Alvarez *et al*, J. Phys. Soc. Jpn. **67** (1998) 1868.
[17] J.V. Alvarez and J.Gonzalez, Europhys. Lett. **44** (1998) 641.

High-Pressure Effect on T_c of $HgBa_2Ca_3Cu_4O_{10+\delta}$ up to 30 GPa

S. Taku[1], K. Shimizu[1], H. Eisaki[2], A. Iyo[2], S. Uchida[3]

1 KYOKUGEN, Center for Quantum Science and Technology under Extreme Conditions, Osaka University, Toyonaka, Osaka 560-8531, Japan

2 Nanoelectronics Research Institute, AIST, Central 2, 1-1-1 Umezono, Tsukuba-shi, Ibaraki 305-8568, Japan

3 Department of Physics, University of Tokyo, Hongo 7-3-1, Bunkyo-ku, Tokyo 113-0033, Japan

e-mail shimizu@cqst.osaka-u.ac.jp

Abstract. In order to investigate the pressure dependence of T_c of the Hg-based cuprates, we performed the electrical resistance measurements of high purity $HgBa_2Ca_2Cu_3O_{10+\delta}$ (Hg-1234) synthesized under pressure. We used a diamond-anvil cell to generate high pressure up to 30 GPa. In this work, two Hg-1234 samples with different hole concentration (T_c = 125 K for optimaldoped sample and T_c = 124 K for underdoped one at ambient pressure) were measured. With applying pressure T_c of underdoped sample increased (dT/dP = +0.6 K/GPa) and passed through a peak with 143 K at ~30 GPa and decreased drastically at higher pressure. This behavior was different from previous works for Hg-1234 and Hg-1223, in which the T_c saturates above the pressure giving the maximum T_c.

1. Introduction

Since superconducting transition temperature (T_c) reached 164 K in $HgBa_2Ca_2Cu_3O_{8+\delta}$ (Hg-1223) under pressure[1][2], there is no report that T_c exceeds 164 K though extensive investigations have been made on the different cuprate families, such as LSCO, YBCO, the Bi-, Tl-, Hg-based compounds, and many of other new materials might be synthesized and measured under pressure.

Generally, T_c depends on the carrier concentration (N_h), the number of CuO_2 planes and the crystal structure. External pressure is a powerful tool to control of N_h and crystal structure. For example, T_c of Hg1223 increases with applying pressure by +1.0 K/GPa. In order to investigate the pressure dependence of T_c of the Hg-based cuprates in wide pressure range, we performed the electrical resistance measurements of high purity $HgBa_2Ca_2Cu_3O_{10+\delta}$ (Hg-1234) synthesized under pressure up to 39 GPa.

2. Experimental

In this work, two Hg-1234 samples with different hole concentration; (T_c = 125 K for optimaldoped sample and T_c = 124 K for underdoped one at ambient pressure) were prepared by high pressure and at high temperature. The optimaldoped sample was synthesized at about 1000 ℃ under a pressure of 3.5 GPa for 6 h. The underdoped sample was obtained by annealing the optimaldoped sample at 280 ℃ in a vacuum for 24 h. Though it is difficult to synthesize single-phase samples of multilayered cuprates,

© 2008 IOP Publishing Ltd

these samples are polycrystal revealed as single phase by powder X-ray diffraction. We used a diamond-anvil cell (DAC), made of CuBe, to generate high pressure. Pressure was determined by a ruby fluorescence method at near temperature T_c. Nonmagnetic stainless steel of SUS310S and Rhenium (Re) were used for a gasket material. The electrical resistance measurements were performed using the standard four-probe technique. Four copper (Cu) films of 2-3 μm thickness were placed in contact with the sample (excluding underdoped sample run2 at 13.3 GPa of three probe measurement).The c-BN with epoxy resin was used for electrical insulation between electrical probes and the gasket material.

3. Results and Discussion

We performed the electrical resistance measurements of high purity Hg-1234 which are different hole concentration under high pressure up to 30 GPa (39 GPa for underdoped sample). The temperature dependences of the samples at several pressure are displayed in Fig.1. Zero resistance was not obtained in this measurement; which may be due to intergrain boundaries because they are not superconducting. The pressure dependences of T_c of both samples are displayed in Fig. 2. To compare with previous works[3], we defined T_c as T_{cd} and T_{ct} displayed in Fig. 3. At the underdoped sample, we observed that T_c increased with pressure monotonically (dT_c/dP = +0.6 K/GPa) and passed through a peak with 143 K at ~30 GPa and decreased drastically at higher pressure. On the other hand, at the optimaldoped sample, we measured the electrical resistance up to ~ 30 GPa. T_c increased monotonically with applying pressure (dT_c/dP = +0.6 K/GPa) and reached 143 K at ~30 GPa. We note that in higher pressure region above ~20 GPa superconducting transition became broad and behavior of normal state above T_c seems semiconductive. Particularly, at underdoped sample there is very large difference between T_{cd} and T_{ct}. Even at 39 GPa electrical resistance increased like semiconductor below T_c as temperature decreased as well as superconducting transition was broad.

The results that the T_c of Hg-1234 increased monotonically with pressure are similar to previous works, however, different in pressure value of maximum T_c, enhancement value of T_c, and T_c at ambient pressure. Our measurements revealed that this behavior is intrinsic to sample and the pressure dependence of T_c in higher pressure region. Superconducting transition became broad, this might indicate that disorder of the crystal structure invited by shear stress of pressure. To investigate the relation between crystal structure and T_c, it is necessity to measure electrical resistance and X-ray diffraction simultaneously.

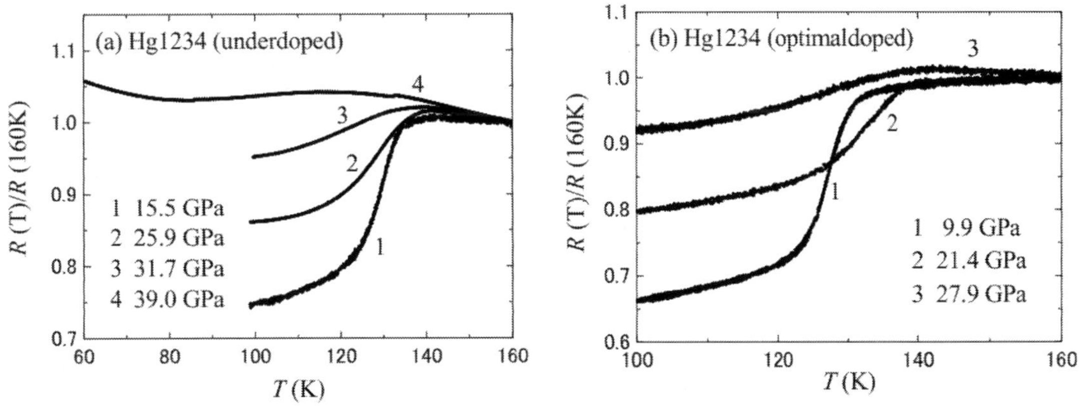

Fig. 1. Temperature dependence of electrical resistance of (a) Hg-1234 (underdoped) and (b) Hg-1234 (optimaldoped) at different pressures. Electrical resistances are normalized with respect to their values at 160 K.

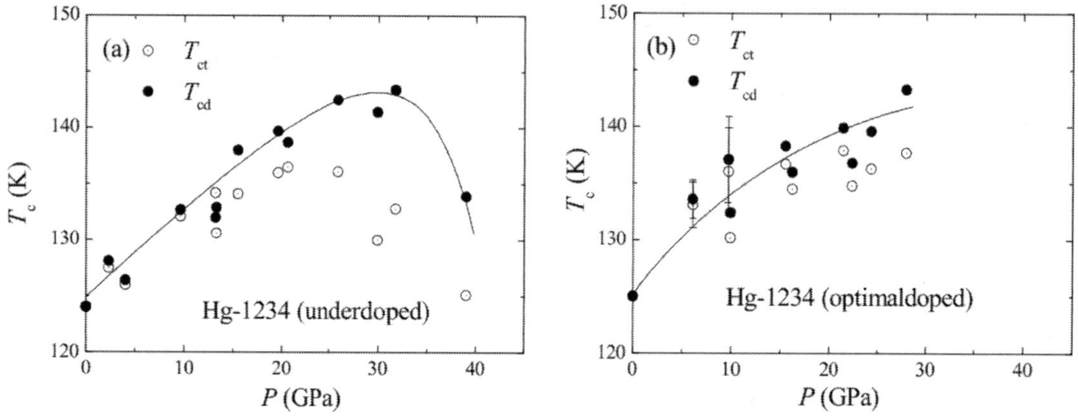

Fig. 2. Pressure dependence of the superconducting transition temperatures T_c of (a) Hg-1234 (underdoped)and (b) Hg-1234 (optimaldoped). Solid symbols denote T_{cd} and open symbols, T_{ct}.

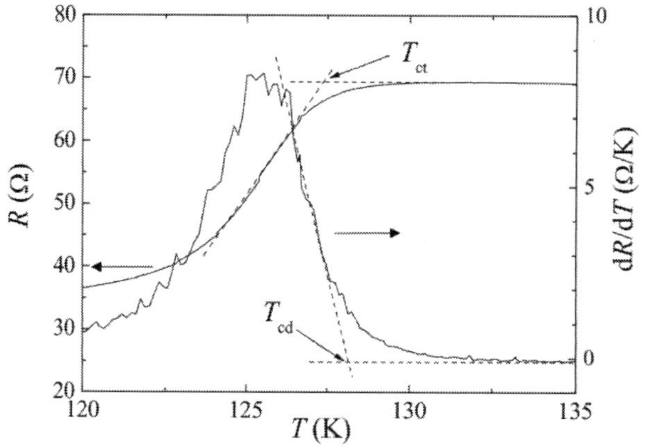

Fig. 3. Temperature dependence of electrical resistance of Hg-1234 (underdoped) at 2.3 GPa and its first derivative dR/dT defined as above.

4. Conclusion

We performed electrical resistance measurements of high-purity Hg1234 samples with different hole concentration under high pressure. The T_c of both samples reached 143 K at ~30 GPa.

Acknowledgments

This work is supported by the 21st Century COE program (G18) by Japan Society of the Promotion of Science and Grant-in-Aid for Creative Scientific Research (15GS0213), MEXT.

References

[1] A. Schilling et al., 1993 Nature (London) **363**, 56
[2] L. Gao *et al.*, 1994 Phys. Rev. B **50**, 4260
[3] D. Tristan Jover et al., 1996 Phys. Rev. B **54**, 4265

Origin of enhanced superconducting transition temperature through structural transformation in $CaSi_2$

Akitaka Nakanishi, Takahiro Ishikawa, Hitose Nagara, and Koichi Kusakabe

Division of Frontier Materials Science, Graduate School of Engineering Science, Osaka University, Toyonaka, Osaka 560-8531, Japan

E-mail: `nakanishi@aquarius.mp.es.osaka-u.ac.jp`

Abstract. Using the first-principles lattice dynamics, we have studied physical origin of enhancement in the superconducting transition temperature T_c of $CaSi_2$ with structural phase transition. Optimization results show that $CaSi_2$ has the AlB_2 structure as an optimized structure above 17GPa. The electron-phonon interaction is enhanced, when $CaSi_2$ takes the AlB_2 structure compared to phase III. Especially, an E_{2g} Einstein mode and a softened optical B_{2g} mode are important.

At ambient pressure, $CaSi_2$ has a rhombohedral crystal structure (phase I). This layered structure is not a superconductor down to 0.03 K [1]. A high pressure phase with the α-$ThSi_2$ tetragonal structure (phase II) is known to be a superconductor of $T_c = 1.58$ K [2]. At the pressure $P \simeq 10$ GPa, $CaSi_2$ undergoes a structural phase transition into another trigonal structure (phase III) and T_c raises up to ~ 3 K [3]. In this structure, silicon atoms form corrugated honeycomb networks. The space-group symmetry is classified in $P3\bar{m}1$ with Ca at $1a(000)$ site and Si $2d(\frac{1}{3}\frac{2}{3}z \simeq 0.4)$ sites[4]. It was theoretically predicted that the trigonal structure transformed to the AlB_2-type structure (z=0.5) [5] and then the transition was experimentally observed at 15 GPa . At this transition, z increases from $\simeq 0.4$, but the wavy honeycomb networks are not perfectly flat ($z \simeq 0.44$) and the space-group symmetry doesn't change. In this structure (phase IV) , T_c further raises up to around 14 K, which has been the highest record for $CaSi_2$.

MgB_2 [6], which has been attracting great interest both in basic science and for industrial applications, but, in contrast, $CaSi_2$ has not been investigated enough in the literature. However, this another superconductor has its own interest and importance. This is because, $CaSi_2$ provides us with a testing ground, on which we can compare several polymorphs, which are superconducting. The structures resemble with each other, but T_c changes its value when the structural phase transition takes place. If we find out a key factor determining the behavior of T_c for the structural deformation, the result would help us to understand this superconducting Zintl-phase compound.

The purpose of the present study is thus to explore nature of the pressure-induced phase transition of $CaSi_2$ using the first-principles calculations. Our simulation tells us that $CaSi_2$ undergoes a structural transformation to the AlB_2 structure. This phase is expected to be a

© 2008 IOP Publishing Ltd

Journal of Physics: Conference Series **121** (2008) 052010 doi:10.1088/1742-6596/121/5/052010

superconductor with much higher T_c than that of the phase III. We will focus on the reason why T_c raises in the higher pressure AlB_2 phase of $CaSi_2$.

We treat the phases III, IV and the AlB_2 structure in this study. For the determination of the electronic structure, we utilized the density-functional theory calculation [7, 8] in a generalized gradient approximation[10] and an Vanderbilt unltrasoft pseudo potential [9], The Kohn-Sham wavefunctions and the electronic charge density are expanded in the plane-wave basis. For calcium atoms, 3s, 3p and 4s electrons are treated as the valence electrons, while 3s and 3p levels are in the valence levels for silicon. The energy cutoff for the wave functions is 16 Ry and that for the charge density is 64 Ry. The k-space (electron) and q-space (phonon) integrations are done using $12{\times}12{\times}12$ and $4 \times 4 \times 4$ in Monkhorst-Pack grids [11]. We optimized the lattice constants a,c and the internal parameter z by Parrinello-Rahman constant-pressure molecular dynamics [12]. T_c is estimated by the strong coupling theory [13, 14], in which the electron-phonon matrix elements as well as the dynamical matrix elements given by the density functional perturbation theory [15]. These calculations are performed by the PWSCF code [16].

The optimized values of lattice constants a, c and internal parameter z are shown in Fig. 1. In the pressure region from 10 to 15 GPa (phase III), the obtained values are close to the experimentally observed values and the relative error is within 2% for a, c and z. But, z dramatically increased at 17 GPa and it reached to 0.5, that is to say, the corrugated structure transforms to AlB_2 structure. The result is confirmed by the accurate calculation with energy cutoff for wave functions of 40 Ry starting from the structure observed in the experiment.

At the critical pressure, c suddenly decreases but a becomes large for preventing a sudden change of the unit cell volume. Next, z approaches to 0.5 and the silicon plane becomes flat. Thus our result indicates that the silicon network in $CaSi_2$ prefers to have a hexagonal planer structure and the AlB_2 structure should appear at the high pressure. Note that, since the value of the pressure given by the first-principles calculation could have an error, the critical pressure may be much higher in the real experiment.

In order to test stability of the AlB_2 structure, we investigated phonon dispersions at each phase (Fig. 2). If the lattice is unstable, a phonon mode with an imaginary frequency appears. In the present result, we have always real frequency in the whole Brillouin zone. This result suggests that the AlB_2 structure is stable in a pressure range P$>$ 17 GPa.

Comparing results for two phases, it is found that one optical phonon mode is softened in the AlB_2 structure. Corresponding atomic displacement is the out of plane vibration mode of silicon, in which a Si plane becomes corrugated. Thus the mode is a B_{2g} mode. The softening of the B_{2g} mode is reasonable, since the AlB_2 system should be unstable against corrugation in Si planes by reducing the pressure. Another characteristic feature of this AlB_2 structure is appearance of a flat phonon dispersion at the highest energy branch. This suggests that the highest optical E_{2g} mode, which is an in-plane longitudinal motion of Si atoms, is an Einstein mode.

We determine which phonon mode has the largest electron-phonon interaction. We estimate partial electron-phonon interaction parameter $\lambda_\nu(\mathbf{q})$, which represents interaction between electron and phonon whose mode is ν, wave number is $\mathbf{q}$ (usual electron-phonon interaction parameter $\lambda = \sum_{\nu\mathbf{q}} \lambda_\nu(\mathbf{q})$). In Fig.2, $\lambda_{\mathbf{q}\nu}$ is plotted by a circle on each phonon band. This figure indicates that, in the AlB_2 structure, the highest mode at the Γ point is the most effective. This mode is the E_{2g} mode, in which the neighboring silicon atoms oscillate in the anti phase within a plane. In addition , we see another important mode, the softened optical mode around the Γ point. This mode is the B_{2g} mode. Due to the softening, $\lambda_\nu(\mathbf{q})$ has a large value, which can contribute to the high T_c.

Utilizing the strong coupling theory of the superconductivity [13], we estimated T_c. Within the formulation of the isotropic Eliashberg formulation [14], T_c is given by three parameters - the electron-phonon interaction parameter λ, the weighted average of the phonon frequency

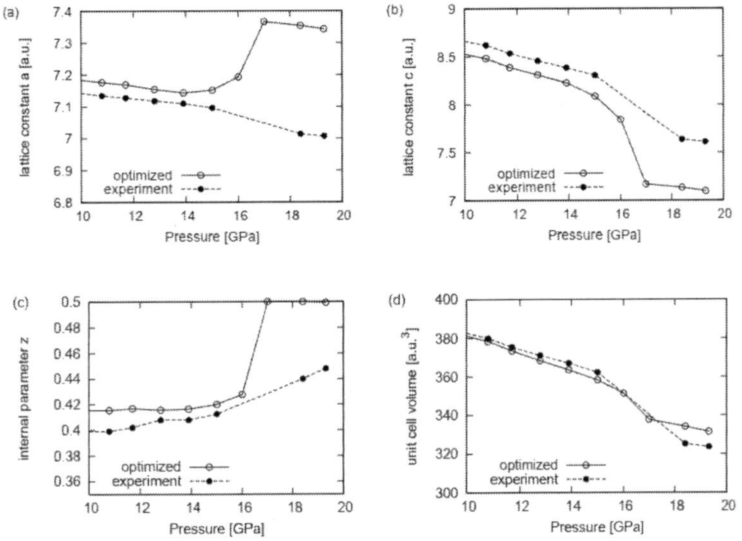

Figure 1. Optimized lattice parameters (a) lattice constant a, (b) lattice constant c, (c) internal parameter z and (d) unit cell volume of $CaSi_2$ at each pressure. For each panel, the optimized values in the simulation are indicated by open circles, while the experimentally observed values [4] are displayed by closed circles.

$\omega_{\log}$, and the screened Coulomb interaction parameter μ^* - as the following.

$$T_c = \frac{\omega_{\log}}{1.2} exp\left(\frac{-1.04(1+\lambda)}{\lambda - \mu^*(1+0.62\lambda)}\right) \tag{1}$$

Utilizing the estimated values of λ and $\omega_{\log}$, we may discuss a pressure dependence of T_c. As for μ^*, we assume that $\mu^* \sim 0.1$ which may hold for simple metals.

The calculated T_c for the phase III is $0.1 \sim 0.2$ K. It's almost one-tenth of the experimentally observed value $\sim$ 3 K. This discrepancy may be due to utilization of an additional isotropic approximation in the Eliashberg theory, which results in a small value of $\lambda = 0.27$. Although the estimation of T_c contains an error, we are able to consider qualitative difference of the AlB_2 phase from the low-pressure phases.

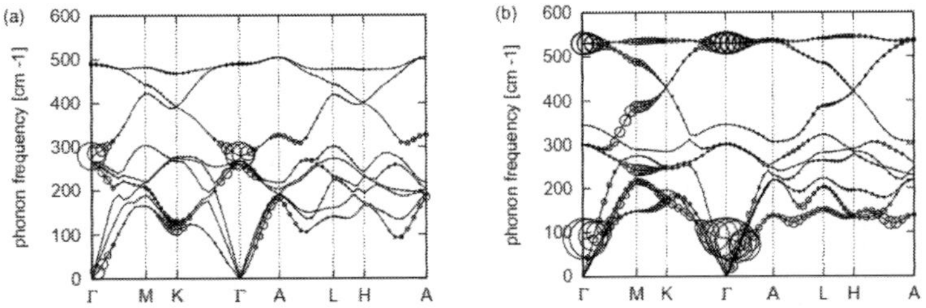

Figure 2. Phonon dispersions $\omega_\nu(\mathbf{q})$ and partial electron-phonon interactions $\lambda_\nu(\mathbf{q})$ of $CaSi_2$ at pressures of (a) phase III and (b) AlB_2 structure. Circles are proportion to $\lambda_\nu(\mathbf{q})$.

We next consider the pressure dependence of T_c assuming that the AlB_2 structure appears in the high pressure. When the structure becomes the AlB_2 structure, T_c raises rapidly and reaches a value one order of magnitude larger than the low pressure phase III. The overall tendency seems to be consistent with the experimental fact, in which the phase IV structure has higher T_c than the phase III. In addition, we can expect the softened B_{2g} mode and the high-frequency E_{2g} Einstein mode in the AlB_2 structure.

Let us discuss origin of the pressure dependence of T_c in our theoretical data. As mentioned above, T_c is determined by three parameters, two of which are affected by the structural transformation in the present approach. Through structural phase transition, λ increases from 0.27 to 0.41, while $\omega_{\log}$ decreases from 300 K to 280 K. So, λ increases by a factor of 1.5 in the AlB_2 structure. On the other hand, $\omega_{\log}$ decreases only about 10%. Thus, the increase of T_c is caused by the enhancement in the electron-phonon interaction. Two important modes, B_{2g} and E_{2g}, behave in a characteristic manner as follows. The strong coupling with the E_{2g} mode is similar to the case of MgB_2. In the present material, electrons in the σ^* band contribute to this electron-phonon interaction. At the same time, we have another important branch, the B_{2g} soft mode. The softening around the phase boundary of the structural transformation often contributes to enhance T_c. However, the softening may reduce $\omega_{\log}$ and finally T_c as exemplified in iodine [18]. In the present case, we have the high-frequency Einstein mode, whose frequency becomes even higher than the low pressure phase. $\omega_{\log}$ is kept almost at the same order of magnitude. Thus we have a prominent increase in T_c estimated for the AlB_2 structure from the value of the Phase III.

First principles calculation showed that AlB_2 structure may appear as the high-pressure phase of $CaSi_2$. In the AlB_2 structure enhancement of T_c is expected because the electron-phonon interaction increase. The E_{2g} and B_{2g} modes of phonon are effective for the enhanced electron-phonon interaction.

This work was supported by a Grand-in-Aid for scientific research (No.15GS0213, No.17064006 and No.17064013), the 21st century COE program “Core Research and Advanced Education Center for Materials Science and Nano Engineering”, and also by the next generation integrated nanoscience simulation software from the Ministry of Education, Culture ,Sports, Science, and Technology, Japan.

References

[1] Affronte M, Laborde O, Oleese G L and Palenzona A 1998 *J ALloys Compd.* **274** 68
[2] McWhan D B, Compton V B, Silverman M S and Soulen J R 1967 *J. Less-Common Met.* **12** 75
[3] Sanphilippo S, Elsinger H, Nunez-Reguerio M, Laborde O, LeFloch S, Olcese G L and Palenzona A 2000 *Phys. Rev.* B **61** R3800
[4] Bordet P, Affronte M, Sanphilippo S, Nunez-Reguerio M, Laborde O, Olcese G L, Palenzona A, Lefloch S, Levi D and Hanfland M 2000 *Phys. Rev.* B **62** 11392
[5] Kusakabe K, Ogitsu T and Tsuneyuki S 1998 *J. Phys. Condens. Matt.* **10** 11561
[6] Nagamatsu J, Nakagawa N, Muranaka T, Zenitani Y and Akimitsu J 2001 *Nature* **410** 63
[7] P Hohenberg and W Kohn 1964 *Phys. Rev.* **136** B864
[8] Kohn W and Sham L J 1965 *Phys. Rev.* **140** A1133
[9] Vanderbilt D 1990 *Phys. Rev.* B **41** 7892
[10] Perdew J P, Chevary J A, Vosko S H, Jackson K A, Pederson M R, Singh D J and Fiolhais C 1992 *Phys. Rev.* B **46** 6671
[11] Monkhorst H J and Pack J D 1976 *Phys. Rev.* B **13** 5188
[12] Parrinello M and Rahman A 1980 *Phys. Rev. Lett.* **45** 1196
[13] Eliashberg G M 1960 *Zh. Eksp. Teor. Fiz.* **38** 966 ; 1960 *Sov. Phys. JETP* **11** 696
[14] Allen P B and Dynes R C 1975 *Phys. Rev.* B **12** 905
[15] Baroni S, Gironcoli S de, Corso A dal and Giannozzi P 2001 *Rev. Mod. Phys.* **73** 515
[16] http://www.pwscf.org
[17] Kong Y, Dolgov O V, Jepsen O and Andersen O K 2001 *Phys. Rev.* B **64** 020501
[18] Maheswari S U, Nagara H, Kusakabe K and Suzuki N 2005 *J. Phys. Soc. Jpn.* **74** 3227

Joint 21st AIRAPT and 45th EHPRG Int. Conf. on High Pressure Science and Technology IOP Publishing
Journal of Physics: Conference Series **121** (2008) 052011 doi:10.1088/1742-6596/121/5/052011

^{73}Ge-NMR study and *ab initio* calculations on clathrate compound $Ba_{24}Ge_{100}$

F. Kanetake[1], A. Harada[1], T. Rachi[2], H. Nagara[1], H. Mukuda[1], K. Kusakabe[1], Y. Kitaoka[1], N. Suzuki[1], K. Tanigaki[2], K. Itoh[3], E. E. Haller[4]

[1] Department of Materials Engineering Science, Osaka University, Toyonaka, Osaka 560-8531, Japan
[2] Department of Physics, Tohoku University, Sendai, Miyagi 980-8578, Japan
[3] Department of Applied Physics and Physico-Informatics, Keio University, Yokohama 223-8522, Japan
[4] University of California at Berkeley and Lawrence Berkeley National Laboratory, Berkeley, California 94720, USA

E-mail: kanetake@nmr.mp.es.osaka-u.ac.jp

Abstract. Through ^{73}Ge-NMR measurements and *ab initio* calculations, we have studied electronic properties of a type-III clathrate compound $Ba_{24}Ge_{100}$. At ambient pressure, the nuclear spin-lattice relaxation rate $1/T_1T$ has a peak around 200 K followed by an activation type decrease with decreasing temperature and it becomes constant at very low temperatures. Near the peak temperature, the successive structural transformations have been observed at $T_{S1} = 215$ K and $T_{S2} = 180$ K. The constant value of $1/T_1T$ at low temperatures is considerably smaller than the value at temperatures higher than T_{S1}, which implies that the density of states at the Fermi level $D(E_F)$ significantly decreases from the high temperature value at ambient pressure. From *ab initio* calculations on $Ba_{24}Ge_{100}$ as well as $Ba_{24}Si_{100}$, we speculate the phase changes of $Ba_{24}Ge_{100}$ in the relevant region of the T-P plane. Our results of NMR measurements and $D(E_F)$ calculations for the $P4_132$ and an $R3$ structures imply that, at high pressures 2.8 GPa, there should be increase of $D(E_F)$ from the value at ambient pressure and low temperatures. We conclude that this result is predominantly related with the large enhancement of the superconducting transition temperature T_c from 0.24 K (at $P = 0$) to 3.8 K (at $P = 2.7$ GPa) observed in $Ba_{24}Ge_{100}$.

1. Introduction

Clathrate compounds including large cages inside have recently attracted much attention. The local vibration of the guest atoms weakly coupled with their cages surrounding them, which is called "rattling", brings about intriguing physical properties. A type-III clathrate compound $Ba_{24}Ge_{100}$ exhibits superconductivity at T_c=0.24 K under ambient pressure. At ambient pressure, two anomalies at $T_{S1} = 215$ K and $T_{S2} = 180$ K were found in the resistivity measurement of $Ba_{24}Ge_{100}$[1, 2], which suggests successive structural transformations [3]. Remarkably, the application of the pressure reduces T_{S1} and T_{S2}, while it increases T_c progressively, reaching $T_c = 3.8$ K at $P = 2.7$ GPa. The averaged value of pressure derivative of the T_c, $T_c/\ P$ is about 1.3 K/GPa, which is larger than about 1 K/GPa of the $HgBa_2Ca_2Cu_3O_{8+\delta}$[4], largest value in the superconductors to the best of our knowledge.

© 2008 IOP Publishing Ltd

These anomalies disappear above $P = 2.7\,\mathrm{GPa}$[1]. Note that $Ba_{24}Si_{100}$ with T_c=1.4 K does not show any such anomalies[5].

In this paper, we report on electronic properties probed by the nuclear spin-lattice relaxation rate $1/T_1$ in ^{73}Ge-enriched $Ba_{24}Ge_{100}$ through ^{73}Ge-NMR measurements. We also report on the results by *ab initio* calculations for $Ba_{24}Ge_{100}$ and $Ba_{24}Si_{100}$.

2. NMR experiments

We synthesized a polycrystalline sample of ^{73}Ge-enriched $Ba_{24}Ge_{100}$ under high pressure [5]. As shown in Figures 1(a) and 1(b), we observed the anomalies in the resistivity and susceptibility measurements at $T_{S1} \simeq 218\,\mathrm{K}$ and $T_{S2} \simeq 182\,\mathrm{K}$, which coincide with the results reported in the earlier studies[2]. We also measured the nuclear spin-lattice relaxation rate $1/T_1$ by the ^{73}Ge-NMR in a magnetic field $H = 12\,\mathrm{T}$ at ambient pressure.

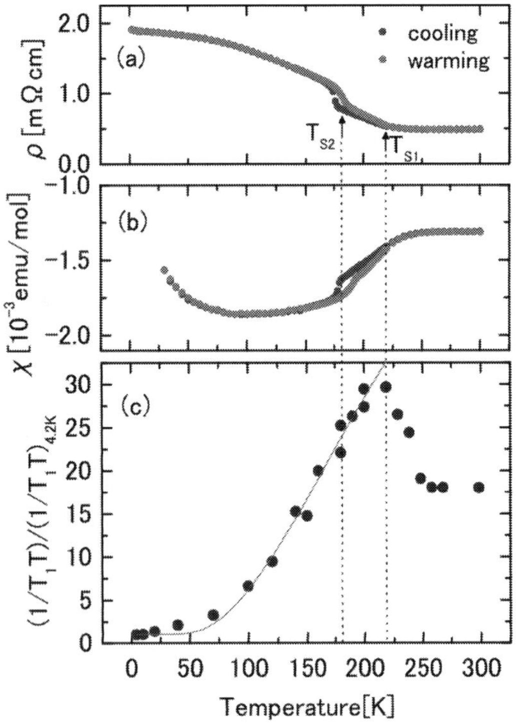

Figure 1. Temperature dependencies of (a) resistivity, (b) susceptibility, and (c) $1/T_1T$ normalized by the value at T=4.2 K on ^{73}Ge-enriched $Ba_{24}Ge_{100}$ at ambient pressure. The value of $1/T_1T$ has a maximum around 200 K near T_{S1} and T_{S2}. The red line is the fitting curve by a function $1/T_1T = A\exp(-E/k_{\rm B}T) + C$.

Fig.1(c) shows the temperature dependence of $1/T_1T$ normalized by the value at 4.2 K. The value of $1/T_1T$ has a maximum around 200 K which is close to T_{S1} and T_{S2}, followed by an activation type of decrease and continued to a $1/T_1T$ = const. behavior at very low temperatures. These results reveal that the nuclear spin relaxation process in this compound consists of the electronic contribution due to quasi-particles near the Fermi level and the local phonon contribution of the Ba atoms weakly coupled with the Ge cages. As shown by a red curve in Fig.1(c), the temperature dependence of $1/T_1T$ below 200 K is well reproduced by a relation of $1/T_1T = A\exp(-E/k_{\rm B}T) + C$. The activation energy is estimated to be $E/k_{\rm B}$ 330 K, which corresponds to 230 cm ^{1} in wave number. The peak of $1/T_1T$ around 200 K is apparently relevant with the successive transitions at T_{S1} and T_{S2} evidenced by the resistivity and susceptibility measurements.

In the temperature range above 250 K and below 20 K, a $1/T_1T$ = const. behavior holds as is usually seen in normal metals. Notably, the constant value of $1/T_1T$ at low temperatures

Journal of Physics: Conference Series **121** (2008) 052011 doi:10.1088/1742-6596/121/5/052011

is considerably smaller than the value at temperatures higher than T_{S1}. Then we expect the density of states at the Fermi level $D(E_F)$ significantly decreases in the low temperature region by structural transformations[3], since the value of $1/T_1T$ is proportional to the square of $D(E_F)$. As a result, the ratio of $D(E_F)_{RT}/D(E_F)_{4.2\text{K}}$ 4 is estimated from the $1/T_1T$ at ambient pressure.

3. *Ab initio* calculations

Next we show the electronic states studied by the *ab initio* calculations. All the calculations were carried out by the use of VASP code[8] of the pseudo-potential method, where the Vanderbilt ultra-soft pseudo-potentials[9] and the generalized gradient approximation(GGA) for the exchange-correlation were employed. We used the cut-off energy of 350 eV and a k-point mesh of $8 \times 8 \times 8$ for the integration over the Brillouin zone.

	Pressure	a[Å]	Space group
Experiment[6]	ambient	14.5635	$P4_132$(cubic)
Calculation	0	14.881	$P4_132$(cubic)
	2.8 GPa	14.560	$R3$(cubic)

Table 1. The lattice parameter and the space group obtained by the XRD experiment and the results by our calculation for $Ba_{24}Ge_{100}$ at P=0 and $P = 2.8$ GPa.

	Pressure	a[Å]	Space group
Experiment[7]	ambient	14.0685	$P4_132$ (cubic)
Calculation	0	14.207	$P4_132$ (cubic)
	3.1 GPa	13.990	$R3$ (cubic)

Table 2. The lattice parameter and the space group obtained by the XRD experiment and the results by our calculation for $Ba_{24}Si_{100}$ at P=0 and $P = 3.1$ GPa.

At first we have optimized the structures for $Ba_{24}Ge_{100}$ and $Ba_{24}Si_{100}$. We started the calculations assuming the space group $P4_132$ and the lattice parameters obtained by the X-ray diffraction (XRD) experiments at ambient pressure and room temperature[6, 7] as an initial guess. The cubic structures with the space group $P4_132$ correspond to local minima at the lattice parameters given in Table 1 for $Ba_{24}Ge_{100}$ and in Table 2 for $Ba_{24}Si_{100}$ at $P = 0$. We have further examined whether other structures with lower symmetries become more stable or not, by setting larger shift to a position of Ba from the center of an open Ge_{20}-cage. As a result, we have found that a low-symmetry structure of space group $P1$, which is triclinic, is more stable than the $P4_132$ structure. At $P = 0$ these indicate that the $P4_132$ structure observed at room temperature becomes unstable at low temperatures. The low-symmetry structure, like the $P1$ structure, may be involved in the structural transformations at T_{S1} or T_{S2} upon cooling. When we isotropically pressurized the lattice, the structure has spontaneously changed into an $R3$ structure for both compounds. This structure is robust around the lattice parameters of the $R3$ given in Tables 1 or 2.

Next we calculated the electronic density of states $D(E_F)$ assuming the $P4_132$ as the low pressure room temperature structure and the $R3$ as the high pressure room temperature structure. The $D(E_F)^{P4_132}$ values at $P = 0$ are calculated to be 74[states/eV/$Ba_{24}IV_{100}$] for $Ba_{24}Ge_{100}$ and 44[states/eV/$Ba_{24}IV_{100}$] for $Ba_{24}Si_{100}$, as shown in Fig. 2. The $D(E_F)$ values do not show any significant difference between the $R3$ and the $P4_132$ for both compounds. Then, if

Journal of Physics: Conference Series **121** (2008) 052011 doi:10.1088/1742-6596/121/5/052011

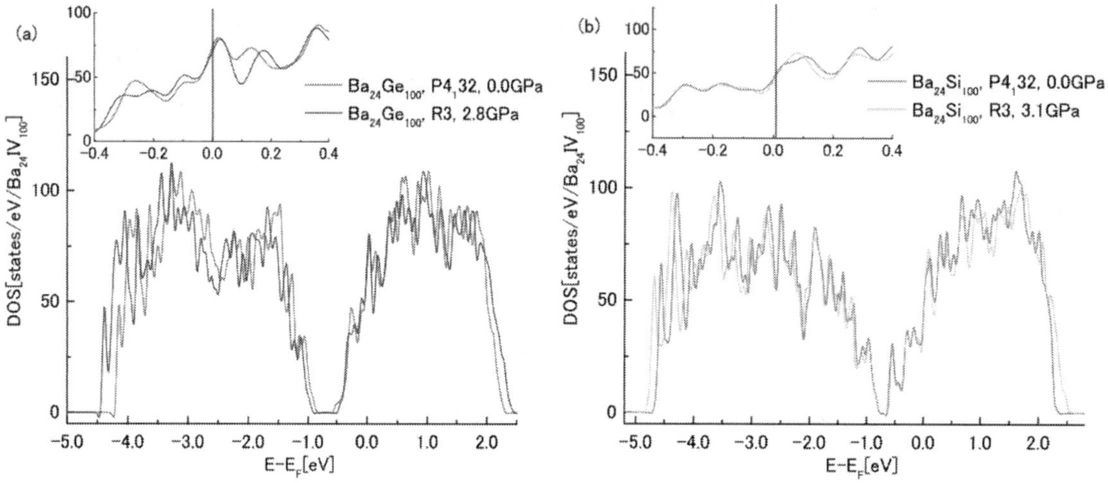

Figure 2. Calculated DOS on (a)$Ba_{24}Ge_{100}$ and (b)$Ba_{24}Si_{100}$. The $D(E_F)^{P4_132}$ at P=0 are calculated to be 74[states/eV/$Ba_{24}IV_{100}$] for $Ba_{24}Ge_{100}$ and 44[states/eV/$Ba_{24}IV_{100}$] for $Ba_{24}Si_{100}$. In addition, the value do not differ much from that of $R3$ structure at around $P = 3\,\text{GPa}$ for both compounds. Insets are the magnification of the energy range near the Fermi level.

we assume the $R3$ as the room temperature structure of $Ba_{24}Ge_{100}$ at $P = 2.8\,\text{GPa}$ and assume that it is preserved to low temperature region since no structural transformations was observed experimentally[1, 3], we can expect that $D(E_F)^{P=2.8\,\text{GPa}}$ at low temperatures is nearly equal to $D(E_F)^{P=0}_{RT}$. Taking the ratio $D(E_F)^{P=0}_{RT}/D(E_F)^{P=0}_{4.2\,\text{K}}$ 4 obtained from the NMR measurement into consideration, we expect $D(E_F)^{P=2.8\,\text{GPa}}_{Tc}/D(E_F)^{P=0}_{Tc}$ 4.

In this context, a reason for the dramatic enhancement of the T_c with increasing pressure is expected to be due to the increase of $D(E_F)$. Further investigations of the electronic and superconductive characteristics are now in progress by systematic NMR/NQR measurements under pressure and at low temperatures, along with *ab initio* calculations.

4. Conclusion

We have investigated the electronic properties of type-III clathrate compound $Ba_{24}Ge_{100}$ through the ^{73}Ge-NMR study at ambient pressure and the *ab initio* calculations on $Ba_{24}Ge_{100}$ and $Ba_{24}Si_{100}$. The value of $1/T_1T$ has a maximum around 200 K which is near the temperatures of the successive structural transformations T_{S1} and T_{S2}. From the result that a constant value of $1/T_1T$ at low temperatures is considerably smaller than its value at temperatures higher than T_{S1}, we speculate that $D(E_F)$ significantly decreases as a result of possible structural transformations. The *ab initio* calculation on $Ba_{24}Ge_{100}$ has revealed that the calculated $D(E_F)$ for the low pressure room temperature $P4_132$ structure shows no significant difference from that of the high pressure $R3$ structure. Incorporating the indication from NMR results in $Ba_{24}Ge_{100}$ with the fact of no structural transformation above $P = 2.7\,\text{GPa}$, we conclude that the $D(E_F)$ at high pressure phase is higher than that at $P = 0$, which is predominantly related with the large enhancement of T_c by applying the pressure in $Ba_{24}Ge_{100}$.

Acknowledgments This work was partially supported by Grants-in-Aid for Scientific Research on Priority Area (No.19051016 and No.17064013) of The Ministry of Education, Culture, Sports, Science, and Technology, Japan.

[1] F. M. Grosche *et al.*, Phys. Rev. Lett. **87** (2001) 247003.
[2] S. Paschen *et al.*, Phys. Rev. B **65** (2002) 134435.
[3] W. Carrillo-Cabrera *et al.*, J. Solid State Chemistry **178** (2005) 715.
[4] M. Nunez-Regueiro *et al.*, Science, **262** (1993) 97.
[5] T. Rachi *et al.*, Phys. Rev. B **72** (2005) 144504.
[6] H. Fukuoka *et al.*, J. Solid State Chem. **151** (2000) 117.
[7] H. Fukuoka et *et al.*, J. Organomet. Chem. **611** (2000) 543.
[8] G. Kresse and J. Furthmuller, Comput. Mater. Sci. **6** (1996) 15; Phys. Rev. B **54** (1996) 11169.
[9] D. Vanderbilt, Phys. Rev. B **41** (1990) 7892.

Journal of Physics: Conference Series **121** (2008) 052012 doi:10.1088/1742-6596/121/5/052012

High pressure effects on high T_c superconductivity: the enhancement of superconducting transition temperature by ordering

Q Q Liu,[1] H Yang,[1] L X Yang,[1] Y Yu,[1] W B Gao,[1] F Y Li,[1] R C Yu,[1] C Q Jin,[1,*] S Uchida[2]

1 Institute of Physics, Chinese Academy of Science, P. O. Box 603, Beijing 100080, P. R. China

2 Department of Physics, University of Tokyo, Japan

*E-mail for correspondence: Jin@aphy.iphy.ac.cn

Abstract. We discussed the high pressure improvement on superconducting transition temperature (T_c) related to the ordering at the apical oxygen layer of a high temperature superconductor (HTS). This study became available in the high pressure synthesized $Sr_2CuO_{3+\delta}$ superconductor with K_2NiF_4 structure showing so far rarely formed partially occupied apical oxygen which also act as the dopant of the HTS. The well-defined links between T_c and modulated structures suggests that optimizing the ordering at apical oxygen layer outside CuO_2 plane is a promising way to further enhance T_c.

1. Introduction

The improvement on superconducting transition temperature (T_c) is one of key concerns of high temperature superconductors (HTS). Crystallographically, HTS consists of alternating stacking of charge reservoir and CuO_2 conducting layers, respectively. They are usually bridged via "apical oxygen" which is located at the apical position relative to the CuO_2 plane (this is valid to p type HTSs). It is found that the relative distance of apical oxygen with the CuO_2 plane has an essential effect on the hopping integral t' between the second nearest Cu-Cu within the CuO_2 plane [1], and consequently it is crucial to determine T_c of p-type HTS. One can expect that the evolution of the apical oxygen will have an appreciable effect on the electronic structure of the CuO_2 plane and hence on high-T_c superconductivity. However, for almost all HTSs, apical oxygen sites are fully occupied, leading to the apical oxygen amount is unchangeable, not less to modulated the ordering state. Searching compounds with apical sites partially occupied by oxygen is thus of fundamental physical interest for a deeper understanding of the doping/ordering effect on high-T_c superconductivity. The $Sr_2CuO_{3+\delta}$ superconducting system synthesized under high pressure is a unique one crystallizing into a highly apical oxygen-deficient K_2NiF_4-type tetragonal structure as shown in Figure 1, and the apical oxygen acts as hole dopant. Systematically studies on the $Sr_2CuO_{3+\delta}$ have shown that the ordering of dopants has dramatic effects on T_c[2].

© 2008 IOP Publishing Ltd

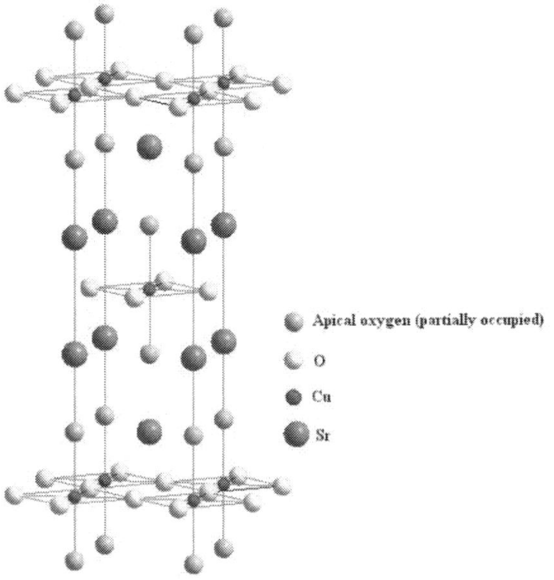

Figure 1. Schematic view of the crystal structure of $Sr_2CuO_{3+\delta}$ with K_2NiF_4-type tetragonal structure containing the CuO_2 plane, but the apical oxygen sites are partially occupied.

2. Experimental

The $Sr_2CuO_{3+\delta}$ samples were synthesized using an improved approach with Sr_2CuO_3, SrO_2 and CuO as starting materials for high pressure synthesis. The precursor Sr_2CuO_3 was prepared from high purity (>3N) $SrCO_3$ and CuO raw materials mixed at a molar ratio of 2:1. The powder mixture was calcined at 950°C for 24h with twice intermediate grindings. Then Sr_2CuO_3, SrO_2, and CuO were mixed in a balanced way according to the nominal composition $Sr_2CuO_{3+\delta}$ at various molar ratios in a dry box. The materials were subjected to high-pressure synthesis under 6 GPa at 1100°C for 1h. The role of SrO_2 peroxide is to create an oxygen atmosphere during the high-pressure synthesis as previously used in the related Sr-Ca-Cu-O-Cl system [3, 4]. Crystal structure was analyzed by means of powder x-ray diffraction (XRD) using Cu K α radiation. The DC magnetic susceptibility was measured with a SQUID magnetometer in an external magnetic field of 20 Oe. A Tecnai F20 electron microscope with a field emission gun was used for electron diffraction (ED), HRTEM experiments.

3. Results and discussion

A series of $Sr_2CuO_{3+\delta}$ samples by changing the initial amount of SrO_2 were prepared under high pressure and high temperature. The nominal excess oxygen concentration $n_s=2x/(1+x)$ is determined by adding the mixture of SrO_2 and CuO in the starting materials $Sr_2CuO_3+x(2SrO_2+CuO)$. Here the excess apical oxygen is the only doping mechanism. The hole carrier density is determined by the amount of δ. Figure 2 presents the XRD patterns of $Sr_2CuO_{3+\delta}$ samples for different nominal values of δ, showing that with increasing δ the structures of the samples gradually transform from the ambient orthorhombic chain structure to the tetragonal form, and a single-phase tetragonal structure is formed at nominal δ=0.4. The superconducting phase has been found within $0.1<\delta<0.6$ and a maximum $T_c \approx$ 75K occurs at $\delta = 0.4$. This indicates that "an optimum doping" is achieved in the high pressure synthesized samples at a certain apical oxygen doping level which is characterized by the excess nominal oxygen content $\delta = 0.4$. In order to further "optimize" the superconducting phase through tuning the apical oxygen distribution, the high pressure as-prepared single-phase sample of $Sr_2CuO_{3+\delta}$ for δ=0.4 was then post heat treated in the temperatures range of 150-350°C for 12h in 1 atm N_2

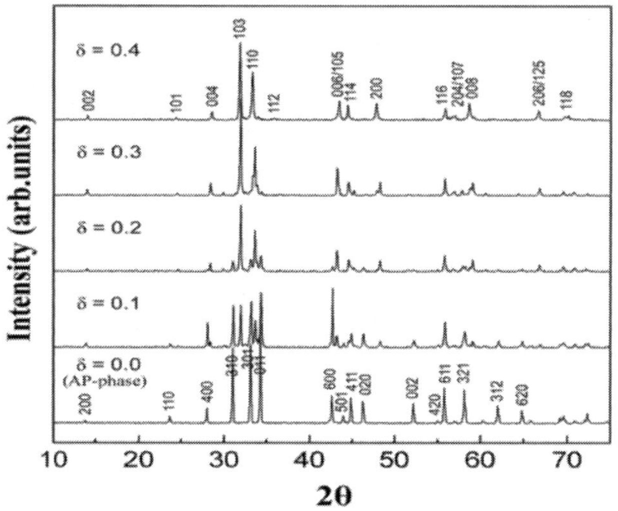

Figure 2. X-ray powder diffraction patterns of $Sr_2CuO_{3+\delta}$ samples for different nominal values of δ, where $\delta = 0$ denotes the normal sample Sr_2CuO_3 synthesized at ambient pressure.

atmosphere in a tube furnace. As increasing the annealing temperature, T_c increases monotonically to 95 K as shown in Figure 3 which is the highest T_c observed in the single-layer copper oxide superconductors. The sample became non-superconducting after heat treatment at 350°C as illustrated in the inset of Figure 3. Powder x-ray diffraction patterns show that the sample keeps the tetragonal single phase unchanged up to 250°C heat treatment, before becoming mixture of tetragonal phase and the ambient orthorhombic chain structure at 350°C, as seen clearly in Figure 4.

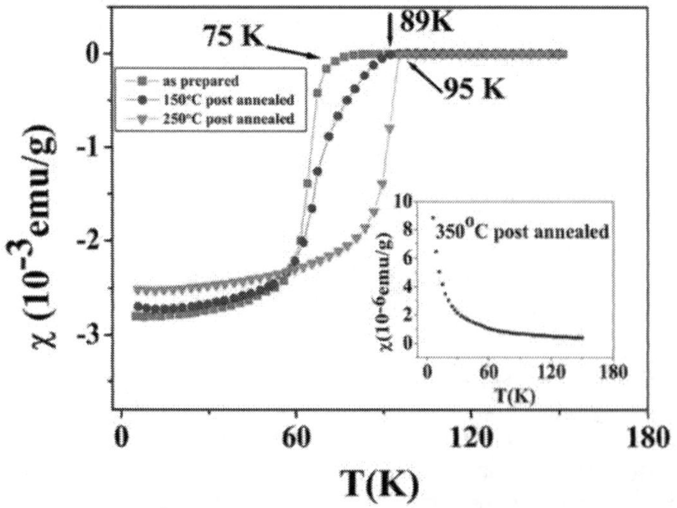

Figure 3. Temperature dependence of the dc magnetic susceptibility in the field-cooling mode for as-prepared $Sr_2CuO_{3+\delta}$ (nominal $\delta = 0.4$) and those after annealed at different temperatures in N_2 atmosphere.

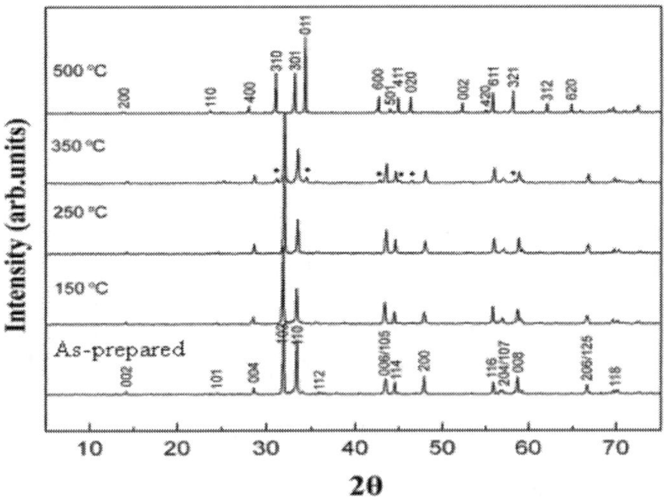

Figure 4. X-ray diffraction patterns of the $Sr_2CuO_{3+\delta}$ sample for both as-prepared and post annealed at different temperatures, showing no obvious phase difference between the post annealed up to 250°C and the as-prepared materials.

ED and HRTEM investigations showed that almost all grains in the superconducting samples both the as-prepared and post annealed below 300°C exhibited modulated structures at a_pb_p-plane based on the K_2NiF_4 structure, where the subscript "p" represents the basic K_2NiF_4-type tetragonal structure. In the as-prepared sample, two types of modulated structures were found. One is the already reported face-centered orthorhombic modulated structure (space group *Fmmm*) with the lattice constants of about $5\sqrt{2}\,a_p\times5\sqrt{2}\,a_p\times c_p$ [5]. The other is a new base-centered monoclinal modulated structure (space group *C2/m*) with the unit-cell parameters $a = 5\sqrt{2}\,a_p$, $b = c_p$, $c = \sqrt{26}\,\sqrt{2}\,/2a_p$ and $\beta = 101.3°$. Figure 5(a) and 5(b) show the ED patterns taken along $[001]_p$ zone-axis of the two modulated phases in the high pressure as-prepared sample, respectively. After heat treatment at 150°C, most of *C2/m* modulated phase converts to another new modulated phase (space group *Cmmm*) with the unit-cell parameters $a = c_p$, $b = 5\sqrt{2}\,a_p$ and $c = 5\sqrt{2}\,a_p$, while the *Fmmm* modulated structure almost remains the same in comparing with the high pressure as-prepared sample. Figure 6(a) shows the ED pattern along $[001]_p$ of the new modulated phase (*Cmmm*) for the sample heat treated at 150°C. As the post annealed temperature was increased to 250°C, all the *C2/m* and *Cmmm* modulated phases convert to another modulated phase (space group *Pmmm*) with the unit-cell parameters $a \approx b = 4\sqrt{2}\,a_p$ and $c = c_p$ which was previously reported[6]. Figure 6(b) displays the ED pattern along $[001]_p$ of the modulated phase of *Pmmm*. The experimental results strongly suggest that the observed superconductivities in the high pressure as-prepared and post heat-treated samples are related to these modulated phases taking the K_2NiF_4-type tetragonal basic structure. It is found that the *Fmmm* modulated phase has always existed as the main phase in the as-prepared and heat treated superconducting samples. If the *Fmmm* modulated phase is responsible for the T_c at 75 K occurred in the as-prepared sample, the same T_c (75 K) could be observed in the heat-treated samples either, but it did not. Therefore a reasonable explanation is that the main *Fmmm* modulated phase is not a superconducting phase, while only the *C2/m* modulated phase is responsible for the T_c at 75 K. Thus the superconductivity (or T_c) can be definitely correlated with the modulated phases, i.e., the *C2/m* modulated phase is responsible for T_c at 75 K in the high pressure as prepared sample, the *Cmmm* modulated phase for T_c at 89 K in the 150°C post annealed sample, and the *Pmmm* modulated phase for T_c at 95 K in the 300°C post annealed sample, respectively. The *Pmmm* modulated phase is the optimally modulated phase associated with the highest T_c (95 K).

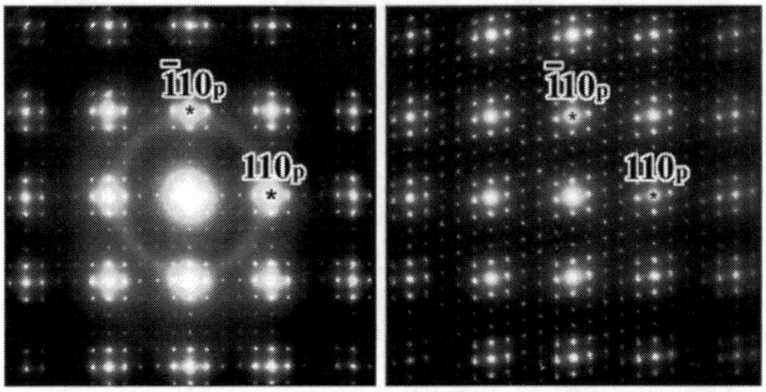

(a) **(b)**

Figure 5. $[001]_p$ zone-axis ED patterns of (a) the *Fmmm* modulated phase and (b) the *C2/m* modulated phase observed in the as-prepared $Sr_2CuO_{3+\delta}$ sample

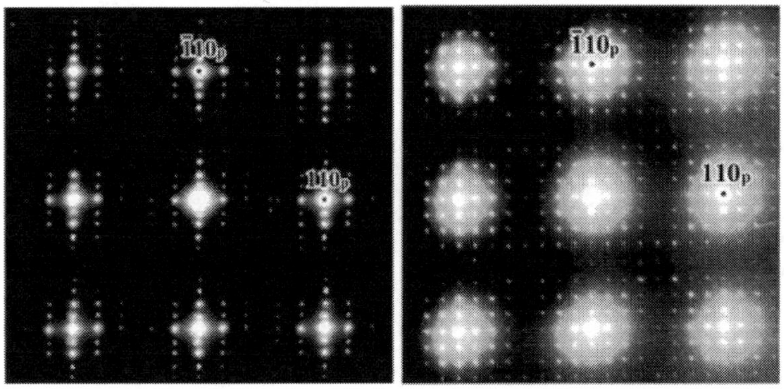

(a) **(b)**

Figure 6. $[001]_p$ zone-axis ED patterns of (a) the *Cmmm* modulated phase in the post annealed $Sr_2CuO_{3+\delta}$ sample at 150°C and (b) the *Pmmm* modulated phase in the post annealed $Sr_2CuO_{3+\delta}$ sample at 250°C.

The thermogravimetry (TG) analysis on $Sr_2CuO_{3+\delta}$ superconductor found little weight change below 300°C [2], indicating a negligible change of oxygen content. Since the modulation is taking place at a_p-b_p plane, assuming that $Sr_2CuO_{3+\delta}$ superconductor keeps the perfect CuO_2 plane as it is a very well established consensus in HTS, the only possible oxygen vacancy should be on the rock-salt-type SrO layer, i.e. the apical oxygen layer. Thus the modulation is reasonably assumed related to the apical oxygen rearrangement at low post annealing temperature. Since the apical oxygen content is the only mechanism to determine the normal carrier concentration of $Sr_2CuO_{3+\delta}$ superconductor, the sample post annealed at 150°C to 250°C will keep the nominal carrier concentration almost unchanged due to the little affected total oxygen amount as aforementioned. It is consequently inferred that the modulation structure related to the apical oxygen ordering state has an additional contribution to optimize superconducting transition temperature. This further supports the recent research that the quenched order disorder of dopant atom has remarkably effects on the superconductivity [7, 8].

4. Summary

We have systemically studied the correlation between the superconducting transition and the ordering states at the apical oxygen layer of the $Sr_2CuO_{3+\delta}$ superconductors. The experimental results showed that the observed superconductivities were caused by these *a-b* plane modulated phases taking the K_2NiF_4-type tetragonal basic structure. The details are the *C2/m* modulated phase is responsible for the T_c at 75 K in the high pressure as-prepared sample; the *Cmmm* modulated phase is related to the 89 K superconducting phase, while the *Pmmm* modulated phase is the optimum one associated with the highest T_c (95K) in the single layered $Sr_2CuO_{3+\delta}$ superconductor, respectively. It is inferred that the modulation is related to the ordering status evolution of the partially filled apical oxygen. The results indicate that the ordering associated with apical oxygen distribution can be an additional route to reach further higher T_c of a HTS.

Acknowledgements

This work was supported by NSF & MOST through the research projects.

References

[1] Pavarini E, Dasgupta I, Saha-Dasgupta T, Jepsen O, and Andersen O K 2001 *Phys. Rev. Lett.* **87** 047003
[2] Liu Q Q, Yang H, Qin X M, Yu Y, Yang L X, Li F Y, Yu R C, Jin C Q and Uchida S 2006 *Phys. Rev.* B **74** R100506
[3] Jin C Q, Wu X J, Laffez P, Tatsuki T, Tamura T, Adachi S, Yamauchi H, Koshizuka N and Tanaka S 1995 *Nature* **375** 301
[4] Jin C Q, Wu X J, Adachi S, Tamura T, Tatsuki T, Yamauchi H, Tanaka S, and Zhao Z X 2000 *Phys. Rev.* B **61** 778; Jin C Q 2004 *High pressure research* **24** 399
[5] Laffez P, Wu X J, Adachi S, Yamauchi H and Mori N 1993 *Physica* C **222** 303
[6] Hiroi Z, Takano M, Azuma M and Takeda Y 1993 *Nature* **364** 315
[7] Fujita K, Noda T, Kojima K M, Eisaki H and Uchida S 2005 *Phys. Rev. Lett.* **95** 097006
[8] Eisaki H, Kaneko N and Feng D L 2004 *Phys. Rev.* B **69** 064512

Joint 21st AIRAPT and 45th EHPRG Int. Conf. on High Pressure Science and Technology IOP Publishing
Journal of Physics: Conference Series **121** (2008) 052013 doi:10.1088/1742-6596/121/5/052013

Pressure studies on the electrical properties in $R_{2-x}Sr_xNi_{1-y}Cu_yO_{4+\delta}$ (R=La, Nd) and $La_3Ni_2O_{7+\delta}$

T. Hosoya , K. Igawa Y. Takeuchi, K. Yoshida, T. Uryu, H. Hirabayashi and H. Takahashi

College of Humanities & Sciences, Nihon University, 3-25-40 Sakurajosui, Setagaya-Ku, Tokyo, 156-8550, Japan

origuti_hm1@hotmail.co.jp

Abstract. The Ruddlesden-Popper series of compounds $La_{n+1}Ni_nO_{3n+1}$ are one of the strongly correlated electron systems. Among these compounds, La_2NiO_4 (n=1) is a well-known two-dimensional antiferromagnetic insulator. Electrical resistivity under high pressure have been measured on $R_{2-x}Sr_xNi_{1-y}Cu_yO_{4+\delta}$ (R=La, Nd) up to 2 GPa and $La_3Ni_2O_{7+\delta}$ up to ~20 GPa. Elecrical resistivty of $La_{1.92}Sr_{1.08}NiO_4$ under high pressure reveals an enhancement of the electronic localization under pressure. On the other hand, in case of $Nd_{1.6}Sr_{0.4}Ni_{0.5}Cu_{0.5}O_{4+\delta}$, a metal-insulator transition is observed at ~240 K and a metallic state is favorable under pressure, in contrast to $La_{1.92}Sr_{1.08}NiO_4$. In $La_3Ni_2O_{7+\delta}$ pressure also enhances a metallic state up to 2 GPa.

1. Introduction

Since the discovery of high-T_C superconductors, a lot of investigations have been performed for transition-metal oxides. Although many kinds of superconductors have been discovered, a mechanism of the high temperature superconductivity is still controversial. The $La_{2-x}Sr_xCuO_4$ (LSCO) with a K_2NiF_4-type structure has been studied extensively, because it is a prototypical high-T_C superconducting compound with T_C = 39 K at x = 0.15 [1]. Non-doped compound La_2CuO_4 is a two dimensional antiferromagnetic insulator with S=1/2, and has a charge-transfer gap Δ =~2 eV [2]. Hole-doping by substitution of Sr^{2+} for La^{3+} suppresses the antiferromagnetic order and causes a metal-insulator transition at x~0.05. These high-T_C oxides are related to the fundamental problem of the metal-insulator transitions in strongly correlated electron systems. An analogous $La_{2-x}Sr_xNiO_4$ (LSNO) with a similar K_2NiF_4-type structure has been also investigated extensively. La_2NiO_4 is an antiferromagnetic insulator with S=1, which shows charge and spin stripe ordering in NiO_2 plane[3] and becomes metallic with doping of Sr^{2+} for La^{3+}. The charge and/or spin stripe ordering has been also found in the high-T_C superconductors, $La_{2-x}Ba_xCuO_4$ and $La_{2-x-y}Nd_ySr_xCuO_4$, having the similar crystal structure [4]. In LSNO much more hole-doping (x>1) is needed to cause the insulator-metal transition [5], because the LSNO has the large charge-transfer gap of 3.5~4 eV that depends on the strength of the metal-oxygen hybridization, and a strong electron-phonon interaction which could relate to the formation of small polarons [6]. The ternary La-Ni-O compounds are included in the Ruddlesden-Popper series of compounds $La_{n+1}Ni_nO_{3n+1}$. La_2NiO_4 (n=1) is the starting compound of LSNO and $La_3Ni_2O_{7+\delta}$ (n=2) possesses two perovskite-type Ni-O layers separated by single rocksolt LaO layers along c-direction [7]. $La_3Ni_2O_7$ has rather low electrical resistivity and shows metallic behavior at room temperature because of an enhancement of Ni-O-Ni overlapping along c-direction. However, the electrical resistivity of $La_3Ni_2O_{7+\delta}$ shows the phase transition at T_C~140K, which is

© 2008 IOP Publishing Ltd

considered to be related to the charge ordering, and insulating below T_C. The T_C has been reported to decrease with increasing pressure up to 1.5 GPa [8] and ~12 GPa is estimated to destroy T_C, if it decreases with increasing pressure with the same rate as below 1.5 GPa . In the hole-doped La_2NiO_4 (n=1) system, there has been a few high pressure experiments only using the piston-cylinder device [9,10], while a lot of high pressure research for cuprate has revealed the several interesting features in high-T_C superconductors[11]. The high pressure experiments are expected to play a important role for the investigation of the metal-insulator transition and the metallic state of nickelates, because in general the pressure decreases the charge-transfer gap in the nickelates and enhances the metallic phase through increase of the metal-oxygen hybridization.

In this paper, we report the pressure dependence of electrical resistivity of pollycrystallne $La_{2-x}Sr_xNiO_4$, $R_{2-x}Sr_xNi_{1-y}Cu_yO_{4+\delta}$ (R=La, Nd) and $La_3Ni_2O_{7+\delta}$, which show the metal-insulator transition. A piston-cylinder apparatus and a diamond anvil cell (DAC) ware used for high pressure experiments up to 2.5 GPa and 20 GPa, respectively. The $Nd_{2-x}Sr_xNi_{1-y}Cu_yO_{4+\delta}$ with a similar crystal structure to LSNO are studied to reveal the crossover properties between nickel oxides and cupper oxides.

2. Experimental

The $La_{2-x}Sr_xNiO_4(0 \leq x \leq 1.08)$ samples were prepared by sintering method using high-purity La_2O_3 , $SrCO_3$and NiO in stoichiometric quantities [5].The $Nd_{1.6}Sr_{0.4}Ni_{1-x}Cu_xO_4(0 \leq x \leq 1)$ samples were prepared by co-precipitation method using high-purity Nd_2O_3, $SrCO_3$, NiO and CuO in stoichiometric quantities [12].The $La_3Ni_2O_{7+\delta}$ samples were prepared by co-precipitation using high-purity La_2O_3 and NiO in stoichiometric quantities [8]. Powder X-ray diffraction(XRD) data, collected on a RINT2500 diffractometer with monochromatized CuKα radiation, were used to monitor the completion of the reaction. Least-squares refinement of the observed powder diffraction data ($10° \leq 2\theta \leq 90°$) was used in the evaluation of cell parameters. The electrical resistivity measurements under high pressure were carried out by a DC four-probe method using piston-cylinder device up to 2.0 GPa, with Fluorinert(FC-77:FC-70=1:1) as the pressure transmitting medium. Diamond anvil cell(DAC) was used for the measurement under high pressure up to 20 GPa, using NaCl as the pressure transmitting medium and Al_2O_3 powder for electrical insulation.

3. Relusts and Discussion

3.1. $La_{2-x}Sr_xNiO_4$ (LSNO)

Wu and Neumeier [10] reported the pressure dependence of the electrical resistivity up to 1.6 GPa in highly doped $La_{0.9}Sr_{1.1}NiO_4$ which shows the metal-insulator transition at $T_{M\text{-}I}$ = 238 K. They also reported the $T_{M\text{-}I}$ decreases with applying pressure drastically at a rate of -55.8K/GPa, but pressure higher than ~1 GPa slightly increases $T_{M\text{-}I}$. They estimated the pressure effect on the charge-transfer gap from the results of pressure dependence of $T_{M\text{-}I}$ based on the model of variable range hopping of a small polaron. From the results that pressure weakly affects the resistivity at a rate of -6.9%/GPa and $T_{M\text{-}I}$ cannot be completely suppressed, they suggested that the applied pressure does not appreciably change the localized states at Fermi level. We have measured the electrical resistivity of $La_{0.92}Sr_{1.08}NiO_4$ having $T_{M\text{-}I}$ = 80K under high pressure up to 2 GPa. The $T_{M\text{-}I}$ shows to increase with pressure up to 2 GPa at a rate of +10 K/GPa, which is opposite to the previous data [10]. However, this is similar behavior when it is compared with the previous data above ~1GPa. Our results simply suggest that the charge-transfer gap increases with increasing pressure. However, it should be considered that the electronic localization is enhanced by the electron-phonon interaction and the small polaron formation with increasing pressure as studied in $La_{2-x}Sr_x MnO_3$ [13].

We also carried out the pressure dependence of electrical resistivity in $La_{1.7}Sr_{0.3}NiO_4$ which shows the sudden change in the resistivity at T_{CO} = 241K, attributed to the charge ordering phenomenon. The T_{CO} decreases with pressure up to 2.5 GPa at a rate of -1.1 K/GPa, which is consistent with the previously reported data measured up to 1.3 GPa [9], though the change rate of -1.1 K/GPa is rather high. This behavior is similar to the case of $La_{2-x}Ba_xCuO_4$ with x=1/8 in LTT phase, in which the charge ordering (charge stripe) is suppressed by pressure [14].

3.2. $Nd_{2-x}Sr_xNi_{1-y}Cu_yO_{4+\delta}$

The $Nd_{2-x}Sr_xNiO_{4+\delta}$ have also been investigated like as the LSNO. The metal-insulator transition was reported to occur in samples with high Sr concentration $x \geq 0.9$ [12,15] as observed in LSNO. In the $La_{2-x}Sr_xNiO_{4+\delta}$ the substitution of La with smaller isovalent Nd causes an increase of the chemical pressure on the NiO_2 plane. Since this chemical pressure enhances orthorhombic distortion due to the mismatch between NiO_2 and NdO layer, $Nd_{2-x}Sr_xNiO_{4+\delta}$ shows orthorhombic to tetragonal structural transition at x = 0.45 which is higher concentration than x = 0.22 of LSNO [16]. The value of T_{CO} prominently observed at x = 1/3 in $Nd_{2-x}Sr_xNiO_{4+\delta}$ is similar one reported in LSNO. These charge ordering phenomena are reported to be related to the orthorhombic distortion in both cuprates and nickelates [16]. Thus it is possible to compare the pressure effect on the charge ordering in the orthorhombic and tetragonal phase, and these will be investigated in near future. In this study the substitution of Cu for Ni is carried out in order to obtain more conductive state. In Fig. 1 we show the electrical resistivity of $Nd_{1.6}Sr_{0.4}Ni_{1-y}Cu_yO_{4+\delta}$ as a function of temperature. With increasing Cu concentration the electrical resistivity decreases and a metal-insulator transition is observed at ~240 K at y = 0.5. However, the metal-insulator transition was not observed except for y = 0.5. On the other hand, Chaker et al. reported that the substitution of Cu for Ni in $NdSrNiO_4$ enhances the electrical resistivity , which is opposite to our data [17]. They explained that the transfer of electron from d_{x2-y2} to d_{z2} by introduction of Cu causes the increase of the electrical resistivity. On the other hand, since the average valence of Ni in $Nd_{1.6}Sr_{0.4}NiO_4$ is 2.4, the substituted Cu does not decrease conductive d_{x2-y2} electrons in our case. Moreover since cuprate has smaller charge-transfer gap, the substitution of Cu for Ni might decrease it. In Fig.2 we show the electrical resistivity of $Nd_{1.6}Sr_{0.4}Ni_{0.5}Cu_{0.5}O_{4+\delta}$ for each pressure. The metal-insulator transition is observed at T_{M-I}~240K at 0.5 GPa and T_{M-I} decreases with applying pressure. However, the T_{M-I} decreases down to ~220K under pressure of 1.0 GPa and remains constant above 1 GPa. .

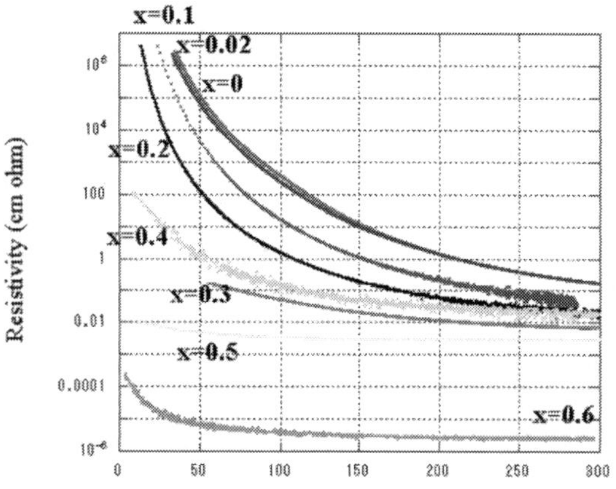

Figure 1. Electrical resistivity of $Nd_{1.6}Sr_{0.4}Ni_{1-y}Cu_yO_{4+\delta}$ as a function of temperature. A metal-insulator transition is seen in x=0.5.

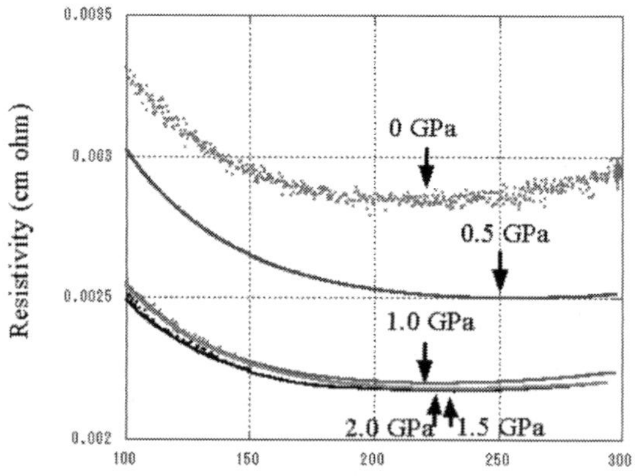

Figure 2. Electrical resistivity of $Nd_{1.6}Sr_{0.4}Ni_{0.5}Cu_{0.5}O_{4+\delta}$ for each pressure. The electrical resistivity decreases with pressure and the metal-insulator transition temperature T_{M-I} indicated by arrows is suppressed by increase of pressure.

3.3. $La_3Ni_2O_{7+\delta}$

The T_C is observed at 137 K as a sudden change of electrical resistivity in $La_3Ni_2O_{7+\delta}$. This phase transition is considered to be related to the charge ordering. The value of T_C is similar to the previously reported one [7,8]. In Fig.3 we show the normalized electrical resistivity of $La_3Ni_2O_{7+\delta}$ for each pressure using the piston-cylinder device. It is seen in Fig 3 that the T_C decreases with increasing pressure at a rate of -12.5 K/GPa, which is close to the reported value of -10.7 K/GPa [8], and the sudden change becomes sluggish at higher pressure. From the results of decrease of the electrical resistivity under pressure, pressure is thought to enhance the Ni-O-Ni overlap along *c*-direction, although we do not have any structural data under high pressure yet. If the T_C decreases linearly with pressure at the observed rate, pressure would destroy the charge ordered state at ~10GPa. In Fig.4 we show the electrical resistivity of $La_3Ni_2O_{7+\delta}$ for each pressure using a DAC up to ~20 GPa. No metallic behavior over all temperature range are shown and the T_C hardly seen in Fig.4. The reason why no metallic state was observed is that extra stresses on the sample and/or a grain boundary effect might be emerged, since liquid was not used as pressure transmitting medium.

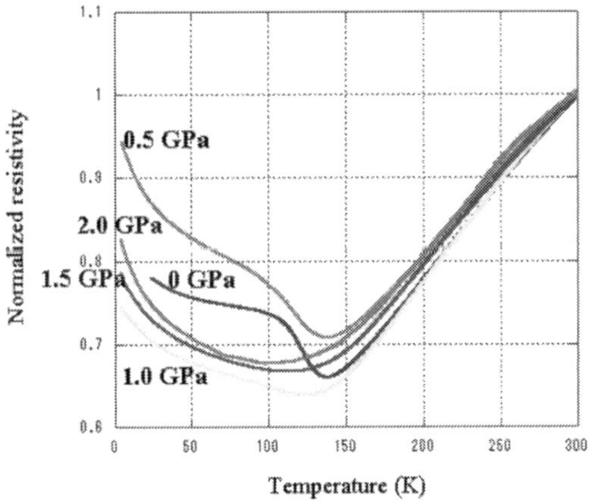

Figure 3. Normalized electrical resistivity of $La_3Ni_2O_{7+\delta}$ for each pressure. It is normalized at the value at room temperature. T_C decreases with increasing pressure at a rate of -12.5 K/GPa.

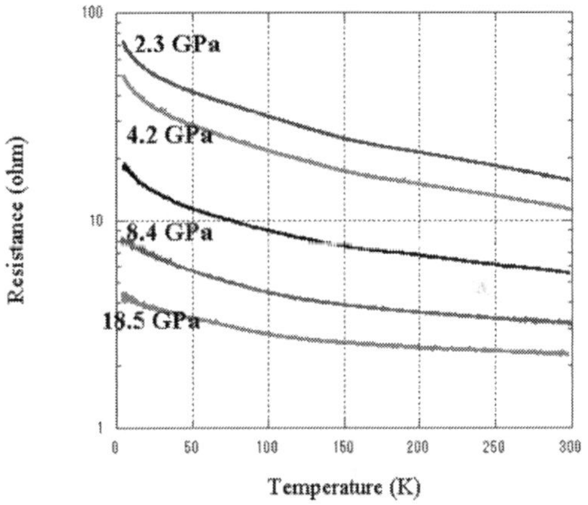

Figure 4. Electrical resistivity of $La_3Ni_2O_{7+\delta}$ for each pressure using a DAC. No metallic behavior is observed and T_C is hardly observed.

4. Conclusion

We have studied the pressure effect on electrical resistivity of $La_{1.92}Sr_{1.08}NiO_4$, $Nd_{1.6}Sr_{0.4}Ni_{0.5}Cu_{0.5}O_{4+\delta}$ and $La_3Ni_2O_{7+\delta}$. It is observed in $La_{1.92}Sr_{1.08}NiO_4$, $T_{M\text{-}I}$ increases with pressure up to 2 GPa. In these results it is found that the pressure effect does not correspond to the hole-doping effect, which is often observed high-T_C superconductors [11]. On the other hand, in $Nd_{1.6}Sr_{0.4}Ni_{0.5}Cu_{0.5}O_{4+\delta}$, $T_{M\text{-}I}$ decreases with pressure up to 2 GPa. It is inferred that the copper-oxygen bonding might be strengthened by pressure much more than the case of nickelates. Though the results of pressure dependence of T_C in $La_3Ni_2O_{7+\delta}$ up to 2.0 GPa indicates the consistency with previously reported data. The results using the DAC shows the enhancement of the insulating phase. The structural data under high pressure are definitely needed for detail consideration.

References

[1] Takagi H, Ido T, Ishibashi S, Uota M, Uchida S and Tokura Y, 1989 *Phys. Rev.* **B40** 2254.
[2] Uchida S, Ido T, Takagi H, Arima T, Tokura Y and Tajima S, 1991 *Phys. Rev.* **B43** 7942.
[3] R.Kajimoto, K.Ishizaka, H.Yoshizawa, Y.Tokura, 2003 *Phys. Rev.* **B67** 014511, S.Cheong, H.Hwang, C.Chen, B.Batlogg, L.Rupp, S.Carter, 1994 *Phys.Rev.* **B49** 7088
[4] Tranquada J M, Sternlieb B J, Axe J D, Nakamura Y and Uchida S, 1995 *Nature* **375** 561.
[5] R.J.Cava, B.Batlogg, T.T.Krajewski, W.F.Peck,Jr, A.P.Ramirez, L.W.Rupp.Jr, 1991 *Phys. Rev.* **B43** 1229, S.Shinomori, Y.Okimoto, M.Kawasaki, Y.Tokura, 2002 *J. Pys .Soc.Jpn* **71** 705.
[6] T.Ido, K.Magishi, H.eisaki, S.Uchida, 1991 *Phys. Rev.* **B44** 12094.
[7] Fukamachi.T, Oda.K, Kobayashi.Y, Miyashita.T, Sato.M, *J. Pys .Soc.Jpn* 70 2757.
[8] Wu G, J.J.Neumeier, M.F.Hundley, 2001 *Phys. Rev.* **B63** 245120
[9] S.H.Han, M.B.Maple, Z.Fisk, S-W.Cheong, A.S. Cooper, O.Chmaissem, J.D.Sullivan, M.Marezio 1995 *Phys. Rev.* **B52** 1347.
[10] Wu G and J.J.Neumeier, 2003 *Phys. Rev.* B67 125116.
[11] Takahashi H and Mori N, 1996 *Studies of High Temperature Superconductors* **16** 1-63.
[12] Arbuckle B W, Ramanujachary K V, Zhang and Greemblatt M, 1990 *J.Solid State Chemistry* **88** 278.
[13] Zhou J -S, Goodenough J B, Asamitsu A and Tokura Y, 1997 *Phys. Rev. Lett.* **79** 3234.
[14] Mori N, Takai M, Tomioka A, Ogawa S, Ueda Y, Takahashi H and Katano S, 1994 *J. of Supercond.* **7** 462.
[15] Takeda.Y, Nishijima.M, Imanishi.N, Kanno.R, Yamamoto.O, Takano.M, 1992 *J.Solid State Chemistry* **96** 72, Doyle S M, Sridhar Kumar M P and McK Paul D, 1992 *J. Phys.: Condens. Matter* **4** 3559.
[16] Hucker M, Zimmermann M, Klingler R, Kiele S, Geck J, Bakehe S N, Zhang J Z, Hill J P, Revolevschi A, Buttrey D J, Buchner B, Tranquada J M, 2006 *Phys. Rev.* **B74** 085112, K.Ishizuka, Y.Taguchi, R.Kajimoto, H.Yoshizawa, Y.Tokura, 2003 *Phys. Rev.* **B67** 184418.
[17] Chaker.H, T.Roisnel, O.Cador, R.Ben Hassen, 2004 *J. Solid State Chem* 4067.

Pressure studies in filled skutterudite $La_{0.8}Rh_4P_{12}$

K.Arii [1], K.Igawa [1], H.Takahashi [1], M.Imai [2], M.Akaishi [2] and I.Shirotani [3]

[1] College of Humanities & Sciences, Nihon University, 3-25-40 Sakurajosui, Setagaya-Ku, Tokyo, 156-8550, Japan
[2] National Institute for Material Science 1-2-1 Sengen, Tsukuba, Ibaraki 305-0047, Japan
[3] Muroran Institute of Technology, 27-1, Mizumoto, Muroran-shi 050-8585, Japan

e-mail for correspondence: ariinobu@hotmail.co.jp

Abstract. Filled skutterudite $La_xRh_4P_{12}$ has the highest superconducting transition temperature (T_c) of 17 K [1]. An analysis of superconducting parameters of $La_xRh_4P_{12}$ suggests that the relatively large the electron-phonon coupling constant λ among the ternary filled skutterudite-type phosphides is one of the reasons why it has high T_c. In this study, high-pressure electrical resistively measurements for $La_xRh_4P_{12}$ at x = 0.8 with T_c = 13.6 K using a piston cylinder device and a diamond anvil cell have been carried out up to 15 GPa. T_c decreases with applying pressure at a rate of -0.5 K/GPa. High-pressure x-ray diffraction measurements have also been performed up to 5 GPa. The value of B_0 of 120 GPa is calculated from the x-ray data

1. Introduction

The general chemical formula of the ternary filled skutterudite is MT_4X_{12} (where M, T, and X are rare-earth and alkaline-earth metals, transition metals, and pnictides, respectively). This substance shows various physical properties, such as superconductivity, semiconductivity, metal-insulator transition, magnetic ordering, heavy fermion behavior, and so on. 14 filled skutterudite compounds have been reported to show superconductivity, so far. Among them, $La_xRh_4P_{12}$ was reported to be the highest superconducting transition temperature (T_c) of 17 K [1]. The crystal structure of filled skutterudites MT_4P_{12} is cubic, with space group Im-3. The M atoms locate at (000) and (1/2 1/2 1/2) of the cubic structure like bcc, and are adjacent twelve X atoms forming icosahedrons through strong covalent bond as shown in Figure 1. The atoms located at (1/4 1/4 1/4) are in the center of a distorted octahedral environment of six X atoms. The degree of hybridization of 4f electrons of rare-earth atoms in the icosahedron and p electron of networking pnicogen atom is thought to be one of the factors in showing fruitful variety of physical properties [1, 2].

In this paper, we report the pressure effects on superconducting properties of $La_xRh_4P_{12}$ having the highest T_c among the filled skutterudite compounds. The sample $La_xRh_4P_{12}$ was prepared using a belt-type high-pressure apparatus [3]. Electrical resistivity measurements under high pressure have been carried out up to 3.0 GPa using a piston-cylinder device and up to 15 GPa using a diamond anvil cell (DAC). High-pressure x-ray diffraction measurements have also been performed using a multi-

© 2008 IOP Publishing Ltd

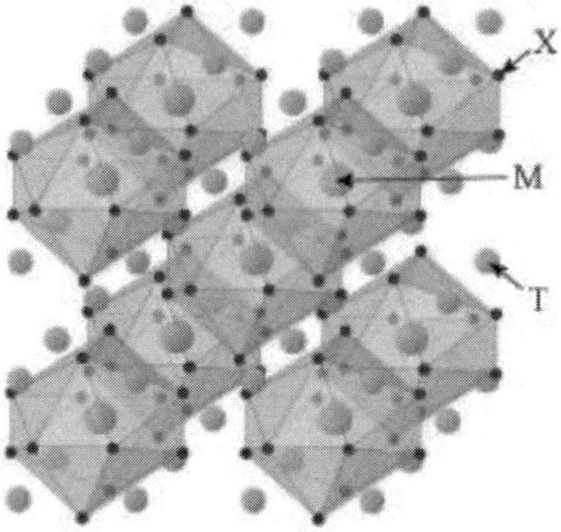

Figure 1. Crystal structure of the filled skutterudite MT_4P_{12}. M, T, and X are rare-earth, transition metal and pnictides,

anvil high-pressure apparatus at KEK-AR.

2. Experimental procedure

The starting material, a powdered 1:4:12 molar mixture of La (nominal purity 99.9%), Rh (nominal purity 99.9%), and P (nominal purity 99.999%), was pressurized and heated to 9.4 GPa and 1473 K using a belt-type high-pressure apparatus. The chemical formula of the sample determined by electron probe microanalysis (EPMA) is $La_{0.82}Rh_{3.9}P_{12.1}$. However, hereinafter, $La_{0.8}Rh_4P_{12}$ denoted $La_{0.82}Rh_{3.9}P_{12.1}$. Lattice constants have been determined by x-ray diffractometry to be 8.0785(2) Å; the density 5.59(7) g/cm^3 [3].

High pressures are generated by the piston-cylinder apparatus up to 3.0 GPa in the temperature range of 4 - 300 K. A mixture of Fluronert (FC70:FC77=1:1) was used as a pressure transmitting medium. The size of rectangular sample is 1.41 mm in length, 0.93 mm in width and 0.30 mm in thickness. Electrical resistivity measurements were performed by a dc four-lead method. Gold wire as lead was attached to the sample using silver paste. The DAC is also used for the electrical resistivity measurements under higher pressures above 3 GPa. The cullet of diamonds in the DAC was ϕ0.6 mm in diameter. Rhenium plate with 0.2 mm in thickness was used for gasket. Powdered NaCl was used for the pressure transmitting medium. The electrical resistivity was measured by the dc four-lead method, in which Pt thin film was used as lead wire. Electrical insulation was realized by Al_2O_3 powder which veils metallic gasket. The pressure in case of DAC was determined by a ruby fluorescence method using DPSS laser.

X-ray diffraction measurements at high pressures were performed in the beam line of the TRISTAN accumulation ring (AR-NE5) at High Energy Accelerator Research Organization (KEK) [4]. High pressures were applied by the multi-anvil high-pressure apparatus MAX80 [5]. WC anvils with square flat-surface size of 6 mm × 6 mm were used. A powdered sample was loaded in the h-BN capsule and it was set in the center of a boron-epoxy pressure-transmitting medium. The pressure was evaluated from the lattice constant of a NaCl internal pressure marker [6]. X-ray diffraction patterns were measured by an energy-dispersive method using synchrotron radiation from the bending magnet. Lattice constants were obtained by the least-squares fitting of the indexed pattern.

Journal of Physics: Conference Series **121** (2008) 052014 doi:10.1088/1742-6596/121/5/052014

3. Results and discussion

Figure 2 (a) and (b) show the electrical resistivity versus temperature curves for $La_{0.8}Rh_4P_{12}$ using the piston-cylinder apparatus and using the DAC, respectively. T_c was determined by the temperature at which the resistivity loss was observed due to superconducting transition. From these results, it is observed that T_c shows slight decrease up to 2.5 GPa and a pronounced decrease above 3.0 GPa up to 15.4 GPa, with increasing pressure. The superconducting state is destroyed more rapidly in the pressure region above 3.0 GPa, where the DAC was used. The pressure dependence of T_c is shown in Figure 3. The value of dT_c/dP is estimated to be -0.3 K/GPa below 2.5 GPa and -0.5 K/GPa above 3.0 GPa. T_c extrapolated to ambient pressure was estimated to be 13.6 K.

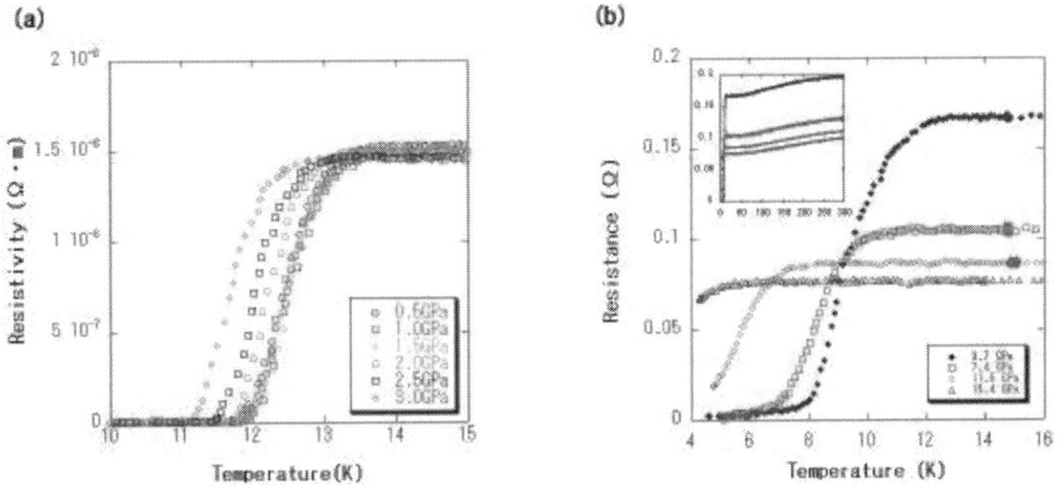

Figure 2. (a) The electrical resistivity versus temperature curves and (b) the electrical resistance versus temperature curves for $La_{0.8}Rh_4P_{12}$.

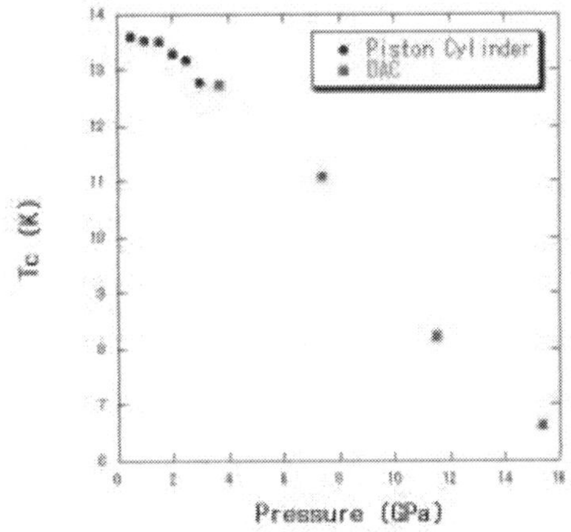

Figure 3. The pressure dependence of T_C. The value of dT_C/dP is estimated to be -0.3 K/GPa in the range of pressure using the piston-cylinder apparatus and -0.5 K/GPa in the range of pressure using the DAC.

Figure 4 (a) and (b) show the normalized lattice constant and volume as a function of pressure, respectively. In Figure 4, it is observed that the lattice constant and the volume show approximately

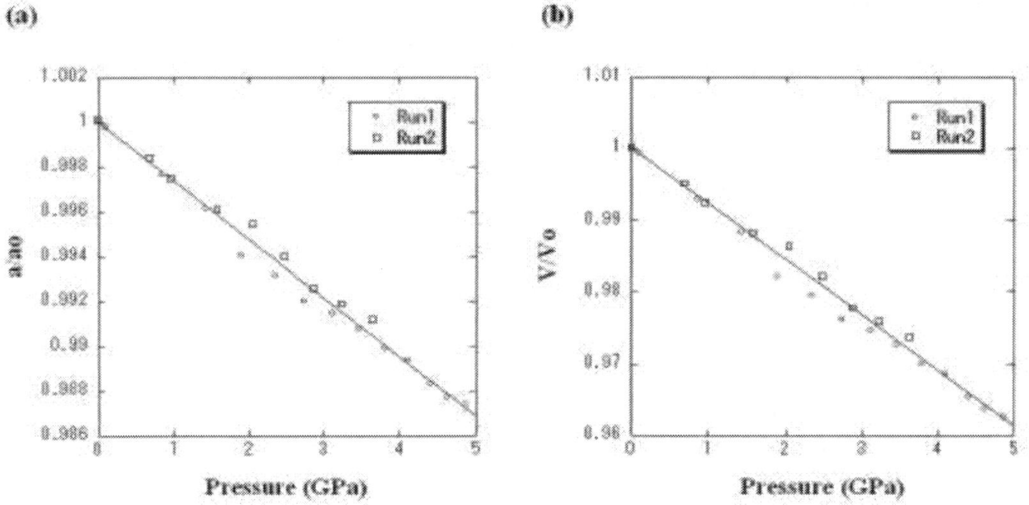

Figure 4. (a) The normalized lattice constant and (b) volume as a function of pressure. Run1 and Run2 are the first and the second run, respectively. The solid lines represent fittied data by the Murnaghan-Birch equation.

linear compression. A bulk modulus B_0 calculated by the Birch equation was 118.79 GPa [7]. On the other hand, the bulk modulus B_0 calculated by the Murnaghan-Birch equation was 125.97 GPa (B_0' = 7.9294) [7]. T_c's, lattice constants, and bulk moduli of several superconducting filled skutterudites are summarized in Table 1 [8, 9]. While the lattice constants of $La_xRh_4P_{12}$, $LaRu_4P_{12}$, and $PrRu_4P_{12}$ were close to each other, $LaRu_4Sb_{12}$ and $PrRu_4Sb_{12}$ have the larger lattice constants than others listed in Table.1. Because an ionic size of Sb is larger than P, the lattice constant of $LaRu_4Sb_{12}$ is larger than $LaRu_4P_{12}$, and $LaRu_4Sb_{12}$ shows the smaller bulk modulus. In case of $LaRu_4X_{12}$ series shown in Table 1, $LaRu_4As_{12}$ has the highest T_c. $LaOs_4X_{12}$ series have the same tendency about T_c [1]. From our results $d\ln T_c/dP$ is calculated to be -0.02 ~ -0.03 GPa^{-1}. In case of $LaRu_4P_{12}$, $d\ln T_c/dP$ is -0.02 GPa^{-1}, which is similar to one of $La_{0.8}Rh_4P_{12}$. Then there might be some similarities between $La_{0.8}Rh_4P_{12}$ and $LaRu_4P_{12}$ if considering that the superconductivity is related to the volume of these compounds somehow. Thus, if $LaRh_4As_{12}$ is synthesized, it could have the highest T_c among the filled skutterudites, because it is inferred that the negative pressure effect on $La_{0.8}Rh_4P_{12}$ enhances the T_c. As a next step, we will perform the high pressure experiments of other filled skutterudites compounds to

Table 1. T_C's, lattice constants, bulk moduli and the pressure dependence of T_C of several superconducting filled skutterudites are summarized.

Compound	T_c (K)	Lattice constant a (Å)	Bulk moduli B_0 (GPa)	dT_c/dP (K/GPa)
$La_{0.8}Rh_4P_{12}$	13.6	8.0785	125.97	-0.3 ~ -0.5
$La_xRh_4P_{12}$	17	8.0581	-	-
$LaRu_4P_{12}$	7.2	8.0605	172	-0.16
$LaRu_4As_{12}$	10.3	8.5081	-	-0.4
$LaRu_4Sb_{12}$	3.58	9.2781	98	-
$PrRu_4P_{12}$	2 (14.7GPa)	8.0516	207	-
$PrRu_4Sb_{12}$	1.1	9.2648	111	-0.21

examine the correlation between the T_c and the volume. The magnetic measurements under high pressure will also be carried out to clarify the pressure effect on the superconductivity in filled skutterudites.

4. Acknowledgments

A part of this research was done under Grant 2005G006 of the Photon Factory. This work was supported in part by a Grant-in-Aid for scientific research from the Ministry of Education, Culture, Sports, Science, and Technology of Japan.

References

[1] Shirotani I, Sato S, Sekine C, Takeda K, Inagawa I and Yagi T 2005 *J. Phys. Condens. Matter.* **17** 7353
[2] Matsuhira K 2006 *Rev. High Pressure Sci. Technol.* **16** 342
[3] Imai M, Akaishi M, Sadki H E, Aoyagi T, Kimura T and Shirotani I 2007 *Phys. Rev.* B**75** 184535
[4] Kikegawa T et al. 1989 *Rev. Sci. Instrum* **60** 1527
[5] Shimomura O 1986 *Physica* B**139&140** 292
[6] Decker L D 1971 *J. Appl. Phys.* **42** 3239
[7] Voronov F F 1976 *Sov. Phys. Dokl.* **21** 92
[8] Shirotani I, Noro T, Hayashi J, Sekine C, Giri R and Kikegawa T 2004 *J.Phys. Condens. Matter.* **17** 7853
[9] Shirotani I, Hayashi J, Takeda K, Nakada R and Ohashi Y 2006 *Physica B* **8-13** 382

Journal of Physics: Conference Series **121** (2008) 062001 doi:10.1088/1742-6596/121/6/062001

New boron subnitride $B_{13}N_2$: HP-HT synthesis, structure and equation of state

V L Solozhenko[1] and O O Kurakevych

LPMTM-CNRS, Université Paris Nord, 99 avenue J.-B. Clement, 93430 Villetaneuse, France

vls@lpmtm.univ-paris13.fr

Abstract. Here we report the synthesis of new B-rich phase, boron subnitride $B_{13}N_2$ by crystallization from the B–BN melt at 5 GPa. The structure of $B_{13}N_2$ has been resolved by the Rietveld analysis and represents a new structural type formed by the distorted boron icosahedra linked together by N–B–N chains and inter-icosahedral B–B bonds. From the 300-K equation-of-state data, the bulk modulus of $B_{13}N_2$ has been found to be B_0 = 200(15) GPa.

1. Introduction

Boron-rich solids give a rise to a large group of hard refractory compounds, i.e. B_6O, B_4C, etc., with unique crystal structures and interesting physical and chemical properties related to the strongly covalent and electron-deficiency character of the chemical bonds [1-2]. The fortunate combination of lightness, hardness, strength, chemical inertness, high melting point as well as high cross-section for neutron absorption has suited boron and boron-rich compounds to many technological applications [3-6]. The synthesis of new phases with boron-like structures in the B–C–N–O system may lead to novel materials with advanced properties exceeding those of B_4C and B_6O.

The existence of boron subnitride "B_6N" was suggested by Condon *et al.* [7]. Later, the synthesis of the B–N films containing "B_4N" with the structure related to α- rhombohedral boron was described by Saitoh *et al.* [8]; however, neither structure nor composition of the synthesized B–N compounds were established. Hubert *et all.* [9] reported the high-pressure solid-state synthesis of boron subnitride "B_6N" by reacting amorphous boron and hBN at 7.5 GPa and ~2000 K. Based on the EELS data, the authors have claimed the B_6N composition; however, according to our previous results [10], the diffraction pattern reported in [9] cannot belong to any B_6O-like structure. Moreover, our *in situ* experiments under the same *p–T* conditions have shown that β-rhombohedral boron does not react with hBN [10].

In the present work, we have *in-situ* studied the crystallization of new boron subnitride $B_{13}N_2$ from the B–BN melt at 5 GPa. Quenched samples have been characterized by X-ray diffraction, Raman spectroscopy and ATEM-EELS; and the crystal structure and equation of state of $B_{13}N_2$ have been established.

2. Experimental

A powder of β-rhombohedral boron (β-rh B) has been mixed with hexagonal graphite-like BN (claimed purity 99.99%, Alfa) in the 5:1 molar ratio that corresponds to the B_6N composition. The HP-HT synthesis has been carried out using a multianvil X-ray system MAX80 at beamline F2.1,

© 2008 IOP Publishing Ltd

HASYLAB-DESY. The experimental set up has been described elsewhere [11]. Energy-dispersive X-ray diffraction data has been collected on a Canberra solid state Ge-detector with fixed Bragg angle $2\theta = 9.28(4)^{o}$ using a white beam collimated down to 60×100 μm^2 (vertical by horizontal). The detector has been calibrated using the K_α and K_β fluorescence lines of Cu, Rb, Mo, Ag, Ba, and Tb. The sample pressure has been determined from the equation of state of highly ordered hBN [12]. Up to 2400 K, the sample temperature has been measured with the W3%Re–W25%Re thermocouple without correction for the pressure effect on the thermocouple emf. Temperatures above 2400 K have been estimated by the extrapolation of the power–temperature calibration curve.

The recovered samples were analyzed by powder X-ray diffraction using Seifert MZIII and G3000 TEXT (Inel) X-ray diffractometers in Bragg-Brentano geometry employing Cu Kα radiation. Highly ordered hBN has been used as an internal standard. Unit cell parameters were derived from the LeBail profile refinement procedure, while structure refinement has been performed by Rietveld analysis using the GSAS software [13].

Microstructure and phase composition of the samples have been studied using a JEM 2010HR (JEOL) transmission electron microscope equipped with a GIF2000 (GATAN) parallel electron energy-loss spectrometer.

The Raman scattering measurements have been performed at 300 K using a Dilor XY system with the 514.5 nm Ar^+ ion laser as the excitation source. The scattered light was collected in the backscattering geometry using a CCD detector. The spectrometer was calibrated using the Γ_{25} phonon of *d*-Si (*Fd*-3*m*).

For the equation-of-state measurements, X-ray diffraction experiments in a diamond anvil cell have been performed at ID27 beamline, European Synchrotron Radiation Facility (ESRF). Pressure has been determined *in situ* from both the calibrated shift of the ruby R_1 fluorescent line [14] and equation

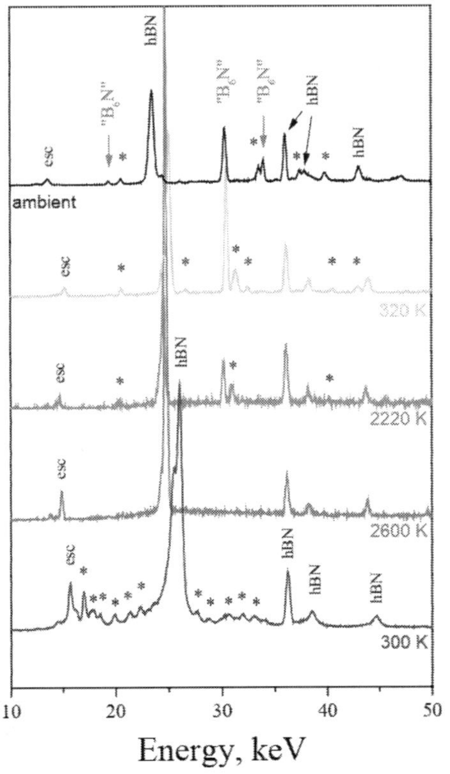

Figure 1. The sequence of X-ray diffraction patterns taken in the course of the $B_{13}N_2$ synthesis at 5 GPa and different temperatures. The bottom pattern corresponds to the compressed starting B-BN mixture; while the top one, to the quenched sample at ambient conditions. The black stars represent the lines of β-rh B. The escape lines which are produced by the detector are marked with "esc".

of state of neon [15] that has been employed as a pressure medium. High-brilliance synchrotron radiation from a two-phased undulator has been set to a wavelength of 0.3738(1) Å. Angle-dispersive X-ray diffraction patterns have been collected with on-line MAR 345 image plate detector and processed with FIT2D software [16].

3. Results and discussion

At 7 GPa, the solid-phase interaction between β-rh B and hBN does not occur up to 2000 K [10]. In the present study, at 5 GPa formation of new "B_6N" phase was observed only in the samples heated to the temperatures above 2450 K *i.e.* upon the boron melting (Fig. 1). From the EELS data, the stoichiometry of the phase with B_6O-like electron diffraction pattern has been determined to be $B_{6\pm1}N$. In the whole *p*–*T* range under study, the "B_6N" phase has been always obtained in mixture with β-rh B, I-tetragonal-boron-like phase ("I-t B" [17]), and hBN (Figs. 1&2). This fact points to the "B_6N" formation from the B–BN melt by a peritectic reaction.

X-ray powder diffraction patterns of quenched samples show the lines of a B_6O-like lattice, namely *101, 003, 012, 110, 104, 021, 024, 303, 018, 125, 220, 131, 223, 217, 119* and *134* (Fig. 2). The lattice parameters have been found to be a = 5.4585(8) Å and c = 12.253(2) Å (trigonal syngony, hexagonal axis, space group *R*-3*m*), which are intermediate between those of boron suboxide B_6O (a = 5.367(2) Å, c = 12.325(3) Å [JCPDS No. 87-1143]) and boron carbide B_4C (a = 5.633(8) Å, c = 12.164(2) Å [18]). Since lattice parameters of the resulting B_6O-like phase have been well reproducible in various experiments, one can conclude that the phase of a constant stoichiometry has been synthesized. The relative line intensities, however, do not exactly correspond to the B_6O structure. Thus, most probably, the unit cell of "B_6N" compound is similar to that of B_6O with some linking atoms between nitrogen. In the case of B_4C, the C–C–C and C–B–C chains occur; at the same

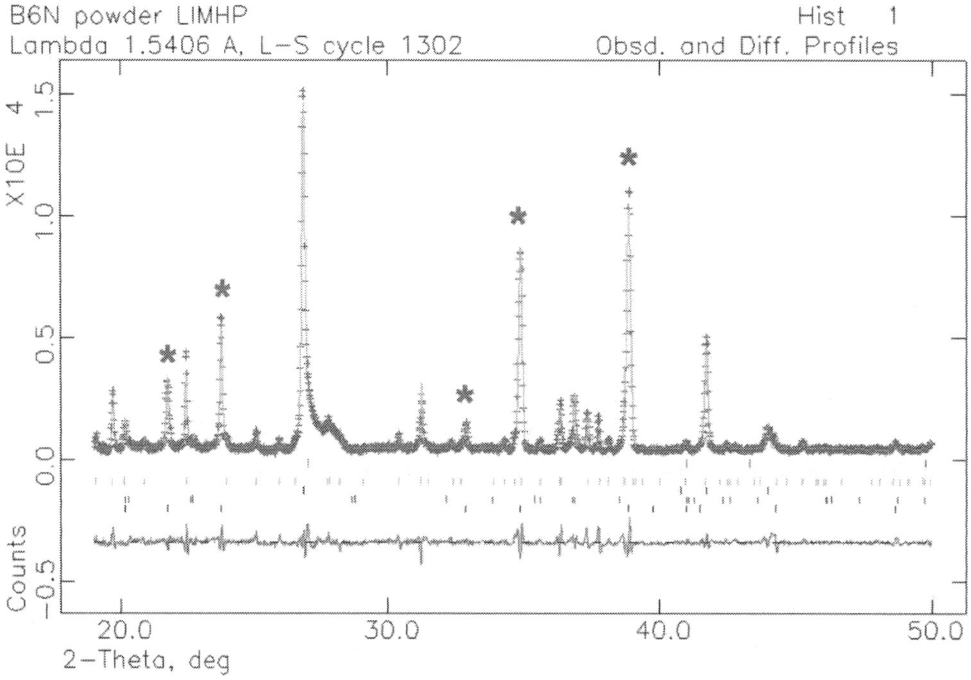

Figure 2. Rietveld full profile refinement of the X-ray powder diffraction pattern of $B_{13}N_2$ boron subnitride. Red stars represent the main lines of $B_{13}N_2$.

Table. Results of the Rietveld refinement of atomic coordinates and fractions. The refined parameters are given in bold.

No.	Atom	Wickers index	x	y	z	Fraction
1	B	18h	**0.1721**	-0.1721	**0.0363**	1
2	B	18h	**0.1198**	-0.1198	**0.8832**	1
3	N	6c	0	0	**0.3749**	1
4	B	3b	0	0	0.5000	**1**

R-3*m*
$Z = 6$
$D = 2.607$ g cm^{-3}
No. of reflections: 11
$wR_p = 0.1445$
$R_p = 0.1036$

time neither O–O–O nor O–B–O linking exist in boron suboxide. In the case of "B_6N", the N–N–N bonding is hardly expected, but N–B–N links may occur.

The structure of the "B_6N" phase has been established from the Rietveld analysis of the diffraction patterns obtained by X-ray powder diffraction (conventional and with synchrotron radiation) (Fig. 2 & Table). The satisfactory quality of fitting may be seen from the R-factor value, $R_p = 0.1036$. The structure is formed by the B_{12} icosahedra linked by the N–B–N chains (Fig. 3). The ideal composition of the "B_6N" phase obtained by the refinement of the starting B_4C-like unit cell is $B_{13}N_2$. The site occupancies of atoms of each crystallographic type are close to unit, so the synthesized phase is stoichiometric.

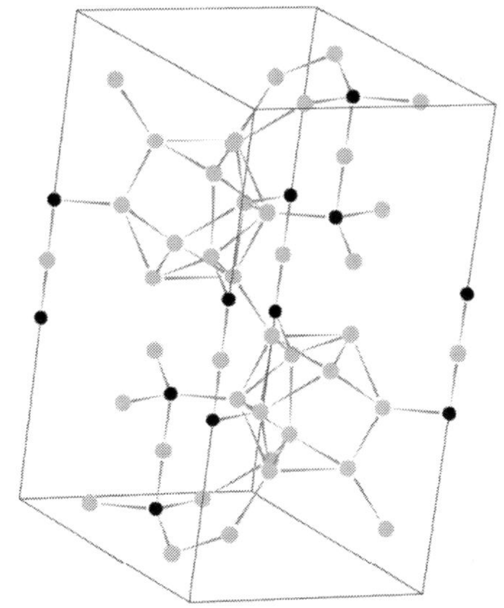

Figure 3. Unit cell of the boron subnitride $B_{13}N_2$. The grey and black balls represent boron and nitrogen atoms, respectively.

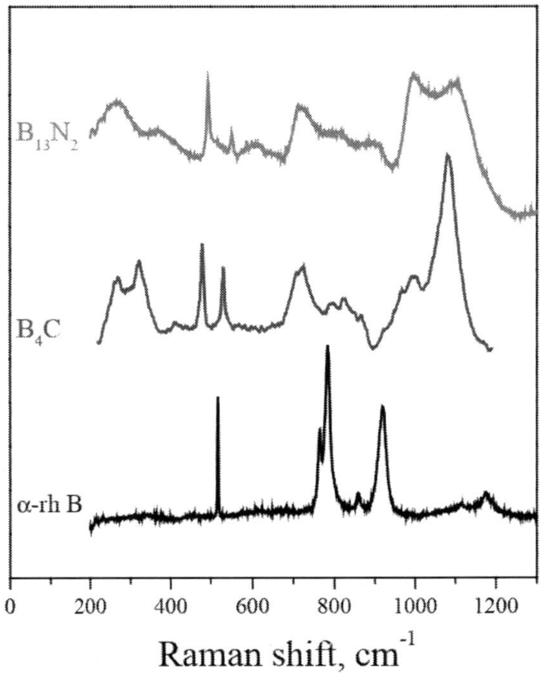

Figure 4. Raman spectra of $B_{13}N_2$, B_4C [18] and α-rhombohedral boron.

The Raman spectra of the $B_{13}N_2$ crystals are similar to those of α-rh B and B_4C [19] (Fig. 4). We observe the intericosahedral modes above 950 cm^{-1}, the intraicosahedral modes between 400 and 950 cm^{-1} and the modes with wavenumber less then 400 cm^{-1} that are, most probably, librational (involving rotation of icosahedra, etc.). All the lines above 600 cm^{-1} are usually attributed to vibrations of the atoms of boron-rich icosahedra [19]. The fine feature just below 500 cm^{-1} corresponds to the symmetric stretching of the N–B–N groups and the narrow line just above 500 cm^{-1}, to the motion of icosahedron about nitrogen atom [19].

The equation of state of $B_{13}N_2$ has been studied at pressures up to 30 GPa at ambient temperature (Fig. 5). In the whole pressure range under study, the linewidth of reflections has not remarkably changed; that is indicative of the quasi-hydrostatic conditions. At least three well-resolved lines have been used for the estimation of the lattice parameters at each pressure point; usually we have employed the *003*, *012*, *104* and *021* reflections. The value of bulk modulus, $B_0 = 200(15)$ GPa, has been estimated by plotting the normalized pressure versus the Eulerian strain under assumption of $B_0' = 4$.

4. Conclusions

Thus, new rhombohedral boron subnitride $B_{13}N_2$ has been synthesized by crystallization from the B–BN melt at 5 GPa. The structure of $B_{13}N_2$ has been established by Rietveld analysis and represents a new structural type produced by the distorted B_{12} icosahedra linked together by N–B–N chains and inter-icosahedral B–B bonds. The Raman data is consistent with a fact that the subnitride belongs to a group of boron-rich compounds with structures related to α-rhombohedral boron. The 300-K equation of state of $B_{13}N_2$ has been measured up to 30 GPa. The bulk modulus was found to be $B_0 = 200(15)$ GPa with the pressure derivative $B_0' = 4$ (fixed).

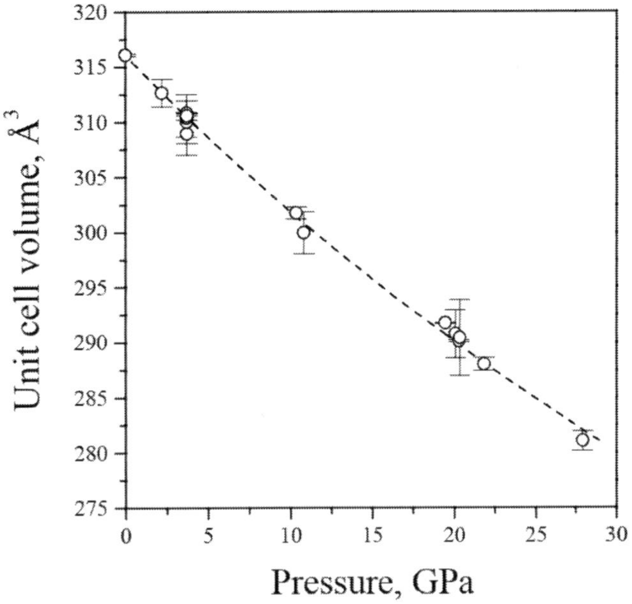

Figure 5. 300-K equation-of-state data of boron subnitride $B_{13}N_2$. The open circles represent the experimental points, while the dashed line corresponds to the *p–V* data fit to the Birch-Murnaghan equation of state.

5. Acknowledgements

We would like to thank Drs. C. Lathe and A. Zerr for their assistance in high-pressure experiments. HASYLAB-DESY and ESRF are acknowledged for the allocated beam time. This work has been financially supported by the Agence Nationale de la Recherche Grant No. NT05-3_42601.

References

[1] Rizzo H F, Simmons W C and Bielstein H O 1962 *J. Electrochem. Soc.* **109** 1079
[2] Lundstrom T and Andreev Y G 1996 *Mater. Sci. Eng.* **A209** 16
[3] Emin D 1987 *Physics Today*, **40** 55
[4] Kimura K 1993 *Mater. Sci. Eng. B* **19** 67
[5] McMillan P F 2002 *Nature Mater.* **1** 19
[6] Emin D 2006 *J. Solid State Chem.* **179** 2791
[7] Condon J B, Holcombe C E, Johnson D H and Steckel L M 1976 *Inorg. Chem.* **15** 2173
[8] Saitoh H, Yoshida K and Yarbrougha W A 1992 *J. Mater. Res.* **8** 8
[9] Hubert H, Garvie L A J, Buseck P R, Petuskey W T and McMillan P F 1997 *J. Solid State Chem.* **133** 356
[10] Solozhenko V L, Le Godec Y and Kurakevych O O 2006 *C. R. Chimie* **9** 1472
[11] Solozhenko V L and Peun T 1997 *J. Phys. Chem. Solid.* **58** 1321
[12] Solozhenko V L, Will G and Elf F 1995 *Solid State Comm.* **96** 1
[13] Larson A C and Von Dreele R B 2000 General Structure Analysis System (GSAS). *Los Alamos National Laboratory Report LAUR.* 86-748

Journal of Physics: Conference Series **121** (2008) 062001 doi:10.1088/1742-6596/121/6/062001

[14] Mao H K, Xu J and Bell P M 1986 *J. Geophys. Res.* **91** 4673
[15] Hemley R J, Zha C S, Jephcoat A P, Mao H K, Finger L W and Cox D E 1989 *Phys. Rev.* **B39** 11820
[16] Hammersley A P 1998 *FIT2D V10.3 Reference Manual V4.0*. European Synchrotron Radiation Facility, Grenoble.
[17] Ploog K, Schmidt H, Amberger E, Will G and Kossubutzki K H 1972 *J. Less Comm. Metals* **29** 161
[18] Kirfel A, Gupta A and Will G 1979 *Acta Cryst.* **B35** 1052
[19] Aselage T L, Tallant D R and Emin D 1997 *Phys. Rev.* **B56** 3122

Synthesis of new cubic C_3N_4 and diamond-like BC_3 phases under high pressure and high temperature

P V Zinin[1], L C Ming[1], S K Sharma[1], S M Hong[2], Y Xie[3], T Irifune[4], T Shinmei[4]

[1]Hawaii Institute of Geophysics & Planetology, University of Hawaii, Honolulu, HI 96822, U.S.A

[2]High Pressure Laboratory, Southwest Jiaotung University, Chengdu, Sichuan, China

[3]Department of Chemistry, University of Science and Technology of China, Hefei, Anhui, China

[4]Geodynamics Research Center, Ehime University, Matsuyama, 790-8577, Japan

E-mail: zinin@soest.hawaii.edu

Abstract. In this report, we discuss the progress in synthesis of new binary phases from B-C-N triangle in DAC under high-pressure and high-temperature (HPHT) conditions: cubic C_3N_4 (c-C_3N_4) and diamond-like BC_3 (d-BC_3) phase. These two phases have been synthesized by direct transformation from graphitic phases under HPHT conditions. The c-C_3N_4 phase was recovered at ambient conditions from the graphite-like C_3N_4 (g-C_3N_4) phase subjected to pressures between 21 and 38 GPa in a diamond anvil cell, laser-heated to temperatures between 1600 and 3000 K. The x-ray diffraction data on the new phase are best explained by a cubic unit cell with the lattice parameters a = 3.878±0.001 Å. The synthesis of the c-C_3N_4 phase has been also conducted in a large volume press at pressure 25 GPa and temperature 2000 °C. X-ray peaks of c-C_3N_4 phase obtained in the large-volume press are weaker than those of diamonds. Application of the UV Raman spectroscopy revealed that UV Raman spectrum of the g-C_3N_4 is substantially different from that measured with visible Raman spectroscopy. It has two strong peaks at 690 cm^{-1} and at 986 cm^{-1} assigned to different types of the ring (s-triazine ring) breathing modes. A diamond-like BC_3 has been synthesized at temperature 2033 ± 241 K and at pressure 50 GPa. The conclusion about the phase transition from graphitic BC_3 (g-BC_3) to d-BC_3 phase was made from the analysis of Raman scattering data. The Raman spectrum of the novel d-BC_3 displays all the peaks but one at 671 cm^{-1} characteristic to Raman spectra the B doped diamond. The peaks pattern of the d-BC_3 suggests that this phase could become a superconductor at low temperatures.

1. Introduction

For nearly seventeen years, intense interest in the synthesis of new C-N materials has been generated by predictions of the unusual properties of saturated, i.e. sp^3-hybridized, crystalline C_3N_4-phases [1]. A search for the superhard phases led to successful synthesis of ternary cubic BC_2N phase in a diamond anvil cell (DAC) [2] and in large volume press [3]. Synthesis of the binary phases from the B-C-N triangle are of interest for two reasons: (a) binary phases could be harder than BC_2N; (b) for industrial applications their synthesis is expected to be less complex than that of the ternary BC_2N phase,

© 2008 IOP Publishing Ltd

because additional elements in the compounds introduce additional difficulties in the synthesis of new phase(s) and in the development of the thin film deposition process. In this report, we discuss the progress in synthesis of new binary cubic phases from B-C-N triangle in DAC under high-pressure and high-temperature (HPHT) conditions: cubic C_3N_4 [4] and diamond-like BC_3 phases. We also provide evidence that the new c-C_3N_4 phase can be synthesized not only in the DAC but also in a large volume press.

2. Synthesis of the cubic C_3N_4 phase

The criterion for identification of the new phase can be found elsewhere [5]. It states: "The rules of synthetic chemistry are rigorous: to claim a new phase all diffraction lines of a given crystal phase should be present and the chemical composition should be C/N=3/4". There are a number of publications whose authors claimed synthesis of new C_3N_4 phases; however, their publications did not contain either x-ray diffraction (XRD) data or data on elemental compositions of the new phases [6]. Badzian *et al.* attributed the lack of success in synthesis of hypothetical superhard C_3N_4 phase to the *limitations in accuracy of first principle calculations*, such as pseudopotential and total energy methods [5]. Indeed, a search for the superhard C_3N_4 phase raises the fundamental question regarding the limitations in the application of the modern theoretical approaches for predicting new materials.

Our analysis of the available experimental data on the synthesis of the superhard carbon nitride indicates that it is premature to make a negative conclusion regarding the existence of the superhard C_3N_4 phase. Examination of the literature on synthesis of carbon nitride reveals that the majority of experiments were conducted using thin film deposition techniques [6], and there are only few of publications on the high-pressure synthesis of C_3N_4 phases [7]. Wixom was probably the first who used high-pressure, high-temperature (HPHT) conditions to synthesize C_3N by shock wave compression of a pyrolyzed melamine-formaldehyde resin and a tetrazol derivative [8]. Diamond was the only detectable crystalline phase in the product. A recent shock recovery experiments also showed that a new carbon nitride phase ("phase-X") was formed at pressures to 50 GPa from nitrogen-rich precursors, amorphous C-N-O and dicyanodiamide [9]. Others approaches in synthesis of new C_3N_4 phases under HPHT conditions used different precursors/compounds as the starting materials, including (a) pyrolyzing $C_{10}H_{18}N_2O_4S$ in a nitrogen atmosphere [10], (b) laser-heating in a diamond anvil cell (LDAC) of the tetracyanoethylene (TCNE: C_6N_4) to 2500°C at 20 and 42 GPa [11]; (c) C_{60} particles with a nitrogen pressure medium, at temperatures ranging from approximately 100 to 5000 K and pressures of 10–60 GPa [12] in LDAC, (d) high-energy ball milling the graphite with liquid ammonia [13]. However, these publications did not contain either XRD data or data on elemental compositions of the new phases. It is worth mentioning results of the experiment (b) conducted by Nesting and Badding [11, 14]. They ? found that the amount of nitrogen incorporated into the C-N network increased with pressure, ranging from 24% (C_3N) at 18 GPa to 38% at 42 GPa ($C_3N_{1.9}$) [11].

As of now, the three hardest materials (diamond, cubic BN c-BN, and BC_2N) were successfully synthesized under HPHT conditions by direct conversion of a graphite-like phase with sp^2 bonding to a cubic phase with sp^3 bonding [2, 3, 15, 16]. It is natural to assume that the experiment with direct conversion of the graphite-like C_3N_4 to superhard phase should provide an answer to the question on the existence of the superhard C_3N_4 phase. In fact, the graphitic carbon nitride was used for synthesis of new carbon nitride phases in at least three experiments. Komatsu synthesized diamond-like carbon nitride C_2N with a lattice constant of 0.351+/-0.001 nm (which is 1.6-1.9% smaller than that of diamond) from graphitic carbon nitride precursors (C_3N_2) by shock compression with the dynamic pressure-temperature range around 30 GPa - 3000 K [17]. Unfortunately, the sample has not been characterized by XRD: the unit cell parameter was obtained by selected area electron diffraction (SAED) attached to transmission electron microscope (TEM). In our initial attempt to transform graphite-like C_3N_4 phase(s) into a diamond-like phase, we used a large-volume press and the turbostratic carbon nitride (t-CN) as the starting material. Instead of forming a dense phase, thermal decomposition of the t-CN was observed at high temperature [18].

In summary, a review of the available data on HPHT synthesis of the C_3N_4 demonstrates that although other routes such as film deposition or chemical reaction under HPHT conditions described above were not successful in synthesizing pure C_3N_4 phase, a direct transformation of the graphitic precursor into a dense phase at high pressures (>20 GPa) and high temperatures (2000K) has not been fully exploited. We present new evidence showing that phase transformation in C_3N_4 is far more complicated than we thought.

The use of the well-characterized graphitic phase [19] and the diamond anvil cell (DAC) with laser heating (LDAC) has led to the successful synthesis of a new *c*-C_3N_4 phase [4] thus providing a strong evidence that the search for the super-hard C_3N_4 is not yet finished. A cubic phase was recovered at ambient conditions from the *g*-C_3N_4 phase subjected to pressures between 21 and 38 GPa in a diamond anvil cell, laser-heated to temperatures between 1600 and 3000 K [4] (Fig.1).

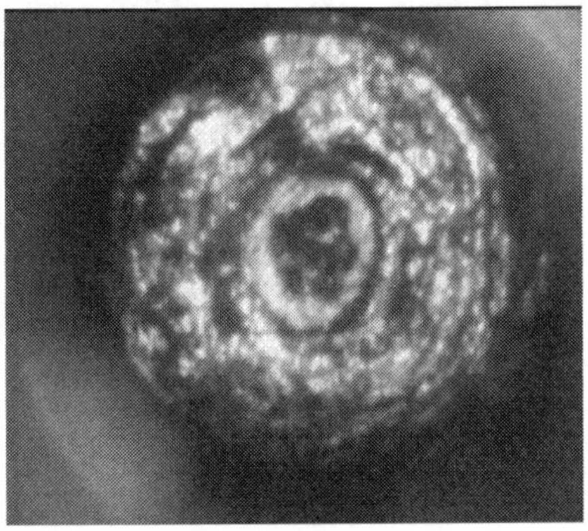

Figure 1. Optical image of the C_3N_4 specimen after heating in DAC. Dark area of the sample is a new *c*-C_3N_4 phase heated by laser beam; the yellow part is original *g*-C_3N_4 phase, not heated by laser beam.

The x-ray diffraction spectrum of the post-lasered sample recovered from 21 GPa and 1800 K is shown in Figure 2. There are 13 diffraction peaks associated with the new phase. Also seen are peaks from the diamond and from the NaCl, marked by yellow hexagons and black squares, respectively. We believed that the diamond was formed from the residual carbon material during the synthesis. was? The NaCl was detected in the sample [19] and we used it in this study as an independent internal pressure calibrant. Detailed analysis of the diffraction pattern of the new phase can be found elsewhere [4]. The x-ray diffraction data on the new phase are best explained by a cubic unit cell with the lattice parameters a = 3.878±0.001 Å. Based on the (*hkl*) assigned to each peak (see Figs. 2 and 3), it is unambiguous that the new phase is a simple cubic (i.e., P-type) with one molecule per unit cell (Z = 1). The molar volume and the density of the cubic phase are calculated as 35.126 cm^3/mol and 2.62 g/cm^3, respectively. The density of the cubic phase is less than was predicted for the high-pressure phases, but is 12% denser than the low-pressure graphitic phase (ρ = 2.336 g/cm^3). Analysis of the x-ray data on the new *c*-C_3N_4 and those of the hypothetical cubic phases, *wmII*-C_3N_4 (a = 5.973 Å), *dzb*-C_3N_4 (a = 3.43 Å) and *pc*-C_3N_4 (a = 3.423 Å), showed that the discrepancy of the intensities of the x-ray data between the observed *c*-C_3N_4 and the hypothetical phases (*wmII*-C_3N_4, *dzb*-C_3N_4 and *pc*-C_3N_4) could not be reconciled with the preferred orientation [4]. A unit cell with a different space group is required for explaining the intensity data on the recovered cubic phase.

The resolution of the energy-dispersive detector of an SEM was demonstrated to be sufficient for the quantitative chemical analysis of light elements such as carbon and boron [3, 20, 21]. Recovered samples of C_3N_4 in this study were measured by the JEOL JSM-5900 scanning electron microscope. Energy dispersive spectroscopy (EDS) measurements of the three samples after quenching provide C/N ratios of 1.37±0.26 which are in fairly good agreement with the EDS measurements of the C/N ratio as 1.39 in the starting material of *g*-C_3N_4 phase [19]. The new phase satisfies the rules of synthetic chemistry: (a) the elemental composition on C and N analyzed by energy-dispersive X-ray spectroscopy (EDS) indicated no loss in the nitrogen content as compared with the starting material; (b) the x-ray pattern consisting of 13 peaks could be attributed to a single phase with a cubic unit cell.

The HPHT conditions (i.e., P = 18 to 22 GPa and T=1600-2000 K) from which the *c*-C_3N_4 phase was recovered is well within the capabilities of the large-volume press. Therefore, we have conducted synthesis of the *c*-C_3N_4 phase in a large-volume press. The X-ray spectrum of the phase obtained in the large-volume press at pressure 25 GPa and temperature 2000 °C is similar to that shown in Figure 2. However, peaks assigned to the *c*-C_3N_4 are weaker than those of diamonds. Only three main peaks of c-C_3N_4 can be detected: peaks with *d*-spacing at 3.844 Å (100), 2.7347 Å (111), and 1.575 Å (211). Intensity of the prominent peak at 2.7347 Å is three times as low as intensity of (111) peak of diamond at 2.060 Å. The relative abundance of diamond and *c*-C_3N_4 observed in the run with the large-volume press is not fully understood. We speculate that the kinetics of heating play a major role in the *g*-C_3N_4 under HPHT conditions. A transformation of *g*-C_3N_4 → *c*-C_3N_4 most likely occurs in a P/T condition being far from a thermodynamic equilibrium and requiring very rapid heating. The much slower heating as performed in the large-volume press may not be as efficient as in the DAC for the formation of the *c*-C3N4 phase. The abundance of diamond is probably not due to the thermal decomposition reported earlier for the turbostratic carbon nitride (t-CN) [18]. This is because our sample was sealed in a Pt capsule, in contrast to the unsealed turbostratic carbon nitride (t-CN) sample used in the previous experiment [10].

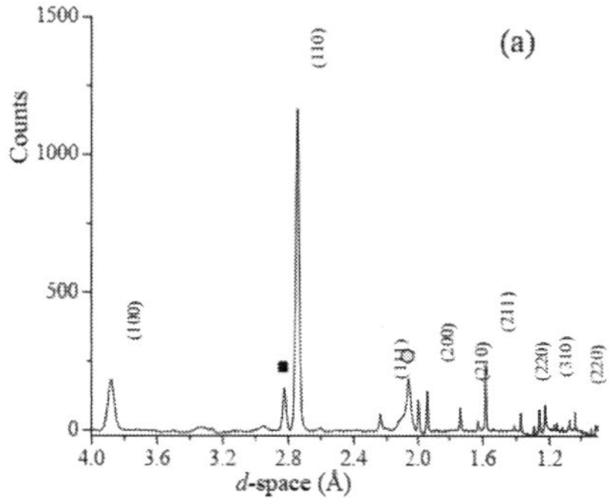

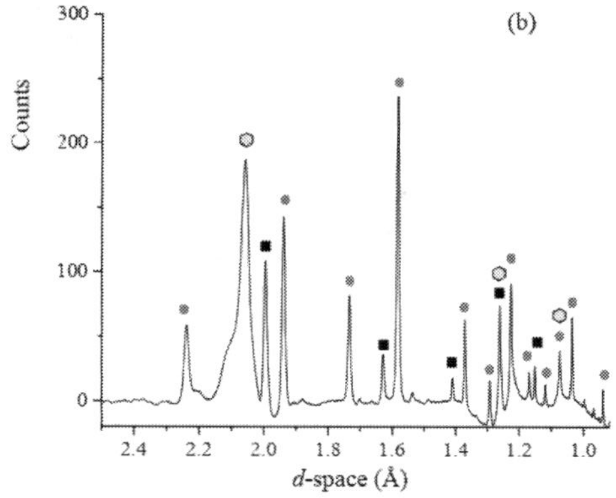

Figure. 2. X-ray diffraction pattern of the new phase recovered at ambient condition: (a) *d* spacing ranges from 4.0 to 0.9 Å. The diffraction lines from the cubic C_3N_4 phase are indexed; (b) *d*-spacing ranges from 2.5 to 0.9 Å. The diffraction lines from the cubic C_3N_4 phase are labelled as red circle (●), from NaCl as black square (■), and from diamond as yellow hexagon (⬡).

Synthesis of the new *c*-C_3N_4 phase became possible because of availability of the pure graphitic phase, *g*-C_3N_4 produced by Guo *et al.* [19]. It is likely that the path of the phase transition in the C_3N_4 system under HPHT conditions is dependent on the structure and purity of the starting *g*-C_3N_4 phase and therefore comprehensive characterization of the starting *g*-C_3N_4 phase is of importance for successful synthesis of the new superhard C_3N_4 materials. Raman spectroscopy is commonly used for characterization in chemistry, because vibrational information is very specific for the chemical bonds in molecules. Raman spectroscopy has been already applied for characterization of *g*-C_3N_4 phases [22]. Unfortunately, the strong fluorescence and photoluminescence that accompany visible Raman excitation of graphitic phases limits the sensitivity that these Raman excitation wavelengths are able to achieve [23]. We report here the use of UV Raman excitation (244 nm) and compare the spectral information content of UV and visible Raman excitation of the graphitic *g*-C_3N_4 phase. The UV Raman spectra at 244 nm were excited using a frequency-doubled Ar ion laser (85-SHG, Lexel Laser Inc., Fremont, CA, USA). The spectra were collected on a 40× objective using Renishaw confocal Raman microscope "inVia". Raman spectra were collected by varying the acquisition times with sample rotation. The Raman spectrum of the *g*-C_3N_4 phase excited with UV laser at 224 nm is shown in Figure

Journal of Physics: Conference Series **121** (2008) 062002 doi:10.1088/1742-6596/121/6/062002

3. Peak fitting analysis was conducted using "Grams Software" (version 7.02, Thermo Fisher Scientific, Inc). The UV Raman spectrum of the *g*-C_3N_4 (Fig. 3) is drastically different from that measured with visible Raman spectroscopy (514 nm) [24]. A typical Raman spectrum of the *g*-C_3N_4 phase obtained with visible Raman spectroscopy has two main overlapping bands at 1357 cm^{-1} (*D* band) and 1560 (*G*-band) and one or two weak bands at approximately 700 cm^{-1} [24]. By analogy with graphite the Raman peaks of the *g*-C_3N_4 phase are called as *D* (1357 cm^{-1}) and *G* (1560) peaks [25]. The *D* peak does not appear in the UV Raman spectra of *g*-C_3N_4. This phenomenon was discussed in detail by Bormett *et al.* [23]. The most prominent feature found on the UV Raman spectra of *g*-C_3N_4 (Fig. 3a) is a sharp peak at 690 cm^{-1} (28) . Numbers in parenthesis are full width at half-maximum (FWHM) of the correspondent peaks. Similar mode centered approximately at 699 cm^{-1} has been already detected in carbon nitride films prepared by magnetron sputtering [26]. It was suggested that the curvature of the graphene planes observed in CN_x films might induce these new Raman bands. We believe that the origin of the peak at 693 cm^{-1} is related to the vibration of the *s*-triazine rings (C_3N_3) inside *g*-C_3N_4 phase. Raman active modes of *s*-triazine rings in different molecules have been studied theoretically [27] and experimentally [28, 29]. Melamine ($C_3H_6N_6$, 1,3,5-triazine-2,4,6-triamine) is among the simplest molecules containing the *s*-triazine ring. Raman spectrum of the melamine has two main peaks: at 676 cm^{-1} and at 986 cm^{-1} [29]. These two peaks are assigned to different types of the ring breathing modes [29]. It is natural to assign UV Raman peaks at 690 cm^{-1} and at 986 cm^{-1}(50) on the spectrum of the *g*-C_3N_4 to vibrations of the same breathing modes. Peak at 1725 cm^{-1}(20) is likely induced by UV radiation since it becomes weaker when sample is rotating.

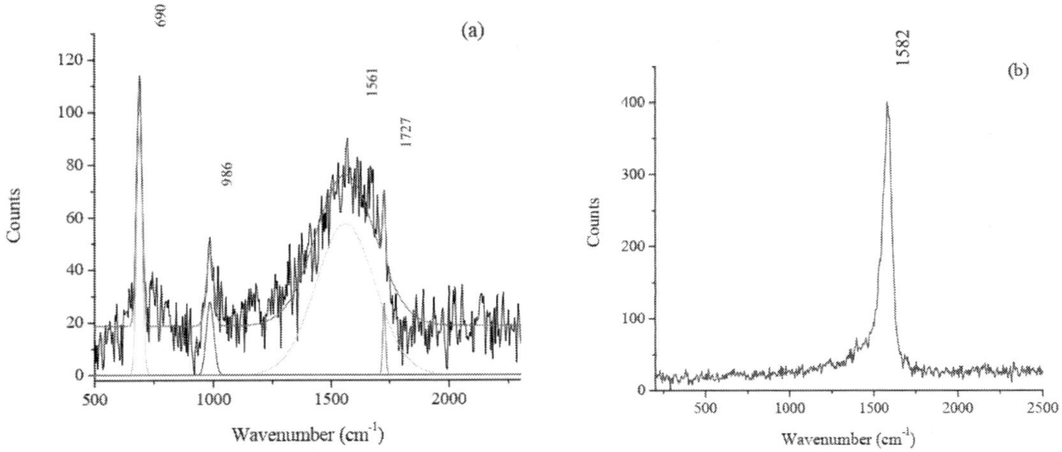

Figure 3. UV Raman (224 nm) spectra of graphitic phases: (a) *g*-C_3N_4 powder, integration time was 10 min, laser power was 0.4 mW; (b) *g*-BC_3 powder, integration time was 5 min, laser power was 0.2 mW.

The phase transition for graphitic-like C_3N_4 to an unknown high-pressure phase, followed retrogressively by a cubic C_3N_4 at ambient conditions, is the first of its kind and is quite different from the well-known, pressure-induced transformations of graphite (or graphitic phase) to diamond (or diamond-like phase) in the C-B-N compounds. The comprehensive characterization of the novel *c*-C_3N_4 phase will shed light on a fundamental question regarding the nature of the bonding of the N atoms in C_3N_4 in a cubic unit cell. The finding of the new C_3N_4 phase demonstrates that the search for the new dense C_3N_4 phases is not over. Further studies of the behavior of the *g*-C_3N_4 phase over wide ranges of pressures and temperatures should provide an answer to the question on the existence of the superhard C_3N_4 phases.

3. Synthesis of the diamond-like BC_3 phase

Despite numerous efforts to synthesize a new high-pressure phase from the B-C-N triangle [30], successful synthesis of a cubic phase in the ternary B-C-N system either in a diamond-anvil cell (DAC) [2, 3] or in a large-volume press [3] was achieved only recently. Now, considerable effort has been concentrated on the synthesis of diamond in boron-rich systems [21, 31]. One the most interesting properties found in boron-doped diamond is its behaviour at low temperature. The graphitic phase of BC_3 is predicted to be a superconductor with a relatively high superconducting temperature ($T_c \approx 22$ K) [32]. Superconductivity was found in boron-doped diamond synthesized at high pressure (~ 9 GPa) and temperature (2,500–2,800 K) [33]. The nature of the superconductivity in boron-doped diamond is under theoretical consideration [34], and we believe that new diamond-like B-C phases should be helpful in understanding the nature of the superconductivity in the B-C systems.

Recently, Lowther examined the properties of some potential super-hard diamond-like boron–carbon phases using *ab initio* computational modelling [35]. Theoretical simulations of pressure- and temperature-induced phase transition in the B-C system demonstrated that incorporation of the B atoms into a diamond structure should not lead to the drastic distortion of the cubic cell of a diamond. The unit cells simulated for the phases predicted by Lowther (i.e., 3.745 Å for cubic BC and 3.642 Å for nearly cubic BC_3) are slightly larger than that of diamond (3.5667 Å). A new superhard tetragonal (t-BC_3; a = 3.513 Å; c = 3.871 Å) phase was also predicted by Liu *et al.* according to the first-principles calculations [36]. The *t*-BC3 phase derived from the cubic diamond structure has been predicted and constructed by an alternately stacking sequence of metallic CBC and insulating CCC blocks along the *c* axis, forming an ideal sandwich-like metal-insulator lattice at the atomic level with relatively high symmetry but low isotropy. Han and Ihm theoretically considered the behavior of hard materials in the presence of vacancies or impurities [37]. Calculations showed that "boron substitution results in a substantial relaxation of atomic positions and a large reduction in bulk modulus", and that the doping of B atoms successfully produces conducting holes. The absence of lone pairs in B allows a significant amount of inward relaxation of B atoms, and these result in partial sp^2-type bonding between B and C. In summary, it may be concluded that first-principle calculations predict new diamond-like BC_x phases with unusual electronic properties. However, substantial questions remain concerning the phase P-T-diagram of the BC_x system and regularities of formation of superhard phases in this system.

A recent attempt to obtain a cubic phase in the B-C system [21] from the graphite-like BC_3 (g-BC_3) was not successful as the g-BC_3 phase decomposed into B_4C and the boron-doped diamond with 1.8 at.%B at 20 GPa and 2300 K in a multianvil press. To avoid decomposition of the graphite-like BC_x into diamond and B_4C, we conducted synthesis in the diamond-anvil cell (DAC) under pressure higher than 20 GPa. A direct transformation from the g-$BC_{1.6}$ phase to a new cubic $BC_{1.6}$ phase was obtained at high temperature, 2230 ± 140 K, and high pressure, 45 GPa [38, 39]. It was found that the x-ray diffraction spectra of the diamond-like BC_x phases are very similar to those of the bulk diamond, making it difficult to distinguish a new phase from those of the diamond. The important evidence for phase identification was provided by Raman spectroscopy [39]. In this report, we present results of the Raman scattering studies on the novel d-BC_3 phase obtained under HPHT conditions.

A g-BC_3 was heated in the DAC to 2005 ± 103 K at 19 GPa (specimen denoted as gp-BC_3 as it was studied after compression), and to 2033 ± 241 K at 50 GPa (specimen denoted as d-BC_3). The BC_3 specimens were quenched and recovered outside the DAC. The elemental composition of the d-BC_3 phase was studied by SEM (JEOL JSM-5900 SEM) equipped with an energy-dispersive detector. The EDS measurements of the post-laser heated sample conducted at several points inside the specimen provide an average value for C/B ratio: C/B = 3.21 ± 0.14. The C/B composition was found to be homogeneous over the central area of the sample surface. The C/B ratio of the starting material B-C phase was not measured by EDS, but we assumed it was not changed in DAC during heating and compression. Raman spectra were measured by a modular Raman system ("Nanofinder 30", Tokyo Instruments, Inc.). The Raman signal was excited by a continuous wave (with wavelength of 633 nm) He-Ne laser from Melles Griot.

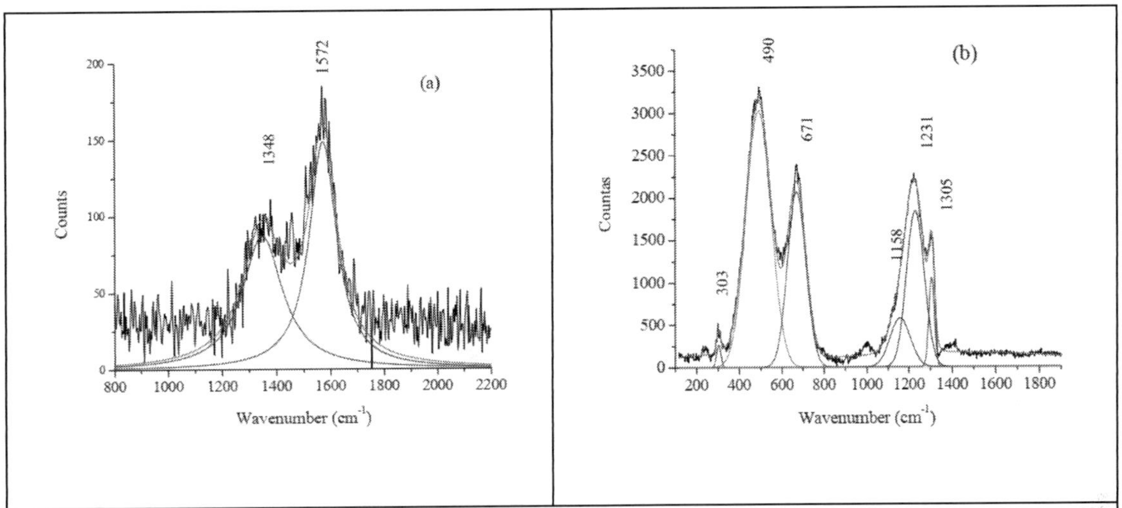

Figure 4. Raman spectrum (633 nm) of BC_3 phases: (a) *gp*-BC_3 phase, integration time was 1.5 min, laser power was 2 mW; (b) *d*-BC_3 phase integration time was 4 min, laser power was 2 mW.

Raman spectra of the *gp*-BC_3 and *d*-BC_3 phases are shown in Figure 4 and the UV Raman spectrum of the graphitic *g*-BC_3 phase is shown in Figure 3b. The UV Raman spectrum of the graphitic *g*-BC_3 phase has one strong peak at 1582 (81) cm^{-1}. Its appearance is similar to Raman spectra of graphite [40] and *h*-BN [41], but it is substantially different from the UV Raman spectrum of the *g*-C_3N_4 (Fig. 3a). The spectrum of the of the *gp*-BC_3 phase (Fig. 4a) obtained at relatively low pressure, 19 GPa, has two broad bands *D* and *G* at 1348 (181) cm^{-1} and 1572 (124) cm^{-1}. The appearance of the Raman spectrum of *d*-BC_3 phase (Fig. 4b) clearly indicates that a phase transformation (from *g*-BC_3 phase to *d*-BC_3 phase) has occurred. *D* and *G* peaks are invisible on the Raman spectrum of the *d*-BC_3 phase indicating that graphitic phase completely transformed into the new phase. This result is in agreement with the observation that the *G* band is evident at low B concentrations and at higher B concentrations, it disappears [42]. The lower spectrum is typical of many results previously obtained for B-doped diamonds [43]. It displays a broad peak at 490 cm^{-1}(137), a second weaker peak at 1231 cm^{-1}(97), broad peak at 671 cm^{-1}(103), a much narrower peak around 1305 cm^{-1} (30), and then a strongly asymmetric dip around 1340 cm^{-1}. Based on the supercell calculations, the peak observed around 500 cm^{-1} was attributed to the Raman active stretching modes of B-B dimers [43, 44]. The peak at 1231 cm^{-1} has been detected in B doped diamonds; its origin is uncertain, but its band position is found to agree well with the maximum in the phonon density of states of the diamond and could be connected with a relaxation of the wave vector selection rules [42]. The broad band at 1158 cm^{-1} is probably due to sp^2 C species at grain boundaries [45]. It was found that at high B concentrations, this peak is hidden beneath the 1231 cm^{-1} peak [42]. The peak at 671 cm^{-1} has not been detected on the Raman spectra of B doped diamond; however it was mentioned that "a new feature at ~600 cm^{-1}, which is weakly present in the spectrum from the undoped sample, also increases with B content" [42]. It may be characteristic to the BC_x phases with a high concentration of B. The spectrum of the *d*-BC_3 has a Fano-related dip at 1267 cm^{-1}. As it follows from the experiments, the only B-doped diamond samples "displaying the 500 and 1000 cm^{-1} peaks, the bump around 1230 cm^{-1} and the Fano-related dip around 1270 cm^{-1} were found to become superconducting" [43]. It allows suggesting that *d*-BC_3 phase may become a superconductor at low temperatures. The small peak at 303 cm^{-1} has not been detected in the B-doped diamond. In order to achieve a better understanding of the relationship between the Raman frequencies and the composition of the diamond-like BC_x phases, we need to develop a comprehensive theoretical approach describing Raman active vibrational modes. Such approach should also take into account the effect of the random distribution of B and C atoms over the crystal lattice of d-BC_x.

4. Conclusions

Two new phases (c-C_3N_4 and d-BC_3) have been synthesized by direct transformation from graphitic phases under HPHT conditions.

The c-C_3N_4 phase was recovered at ambient conditions from the g-C_3N_4 phase subjected to pressures between 21 and 38 GPa in a diamond anvil cell, laser-heated to temperatures between 1600 and 3000 K. The x-ray diffraction data on the new phase are best explained by a cubic unit cell with the lattice parameters a = 3.878±0.001 Å. The synthesis of the c-C_3N_4 phase has been also conducted in a large-volume press (at pressure 25 GPa and temperature 2000 °C). The X-ray peaks assigned to the c-C_3N_4 are found to be weaker than those of diamonds. The relative abundance of diamond and c-C_3N_4 observed in the run with the large-volume press is not fully understood. Application of the UV Raman spectroscopy for characterization of the g-C_3N_4 phases demonstrated that UV Raman spectra g-C_3N_4 phases are substantially different from that measured with visible Raman spectroscopy. It has two strong peaks at 690 cm^{-1} and at 986 cm^{-1} assigned to different types of the ring (s-triazine ring) breathing modes.

A diamond-like BC_3 has been synthesized at temperature 2033 K and at pressure 50 GPa. The conclusion about the phase transition from g-BC_3 to diamond-like d-BC_3 phase was drawn from Raman scattering measurements. The Raman spectrum of the novel d-BC_3 displays a broad peak at 490 cm^{-1}, a second weaker peak at 1231 cm^{-1}, another broad peak at 671 $cm^{-1,}$ a much narrower peak around 1305 cm^{-1}, and then a strongly asymmetric dip around 1340 cm^{-1}. All these features except the peak at 671 cm^{-1} are characteristic to the Raman spectrum of the B-doped diamond. The peak patterns of the d-BC_3 suggest that this phase could become a superconductor at low temperatures.

5. Acknowledgements

This work was supported by the DOE. The use of the HPCAT facility was supported by DOE-BES, DOE-NNSA (CDAC), NSF, DOD-TACOM, and the W.M. Keck Foundation. The use of Advanced Photon Source was supported by the U. S. Department of Energy, Office of Science, Office of Basic Energy Sciences, under Contract No. W-31-109-Eng-38."

References

[1] Liu A Y and Cohen M L 1989 *Science* **245** 841
[2] Knittle E, Kaner R B, Jeanloz R, and Cohen M L 1995 *Phys. Rev. B* **51** 12149
[3] Solozhenko V L, Andrault D, Fiquet G, Mezouar M, and Rubie D C 2001 *Appl. Phys. Lett.* **78** 1385
[4] Ming L C, Zinin P, Meng Y, Liu X R, Hong S M, and Xie Y 2006 *J. Appl. Phys.* **99** 033520
[5] Badzian A, Badzian T, Roy R, and Drawl W 1999 *Thin Solid Films* **354** 148
[6] Kroke E and Schwarz M 2004 *Coordinat. Chem. Rev.* **248** 493
[7] Horvath-Bordon E, Riedel R, Zerr A, McMillan P F, Auffermann G, Prots Y, Bronger W, Kniep R, and Kroll P 2006 *Chem. Soc. Rev.* **35** 987
[8] Wixom M R 1990 *J. Amer. Chem. Soc.* **73** 1973
[9] Liu J J, Sekine T, and Kobayashi T 2006 *Solid State Commun.* **137** 21
[10] Martin-Gil J, Martin-Gil F J, Sarikaya M, Qian M X, JoseYacaman M, and Rubio A 1997 *J. Appl. Phys.* **81** 2555
[11] Nesting D C and Badding J V 1996 *Chem. Mater.* **8** 1535
[12] Nguyen J H, Caldwell W A, Benedetti L R, Kruger M B, and Jeanloz R 1998 *Mat Res Symp.* **499** 303
[13] Fahmy Y, Shen T D, Tucker D A, Spontak R L, and Koch C C 1999 *J. Mater. Res.* **14** 2488
[14] Badding J V and Nesting D C 1996 *Chem. Mater.* **8** 535
[15] Wentorf R H J 1957 *J. Chem. Phys.* **26** 956
[16] Bundy F P 1962 *J. Chem. Phys.* **38** 631
[17] Komatsu T 2004 *Phys. Chem. Chem. Phys.* **6** 878

[18] Solozhenko V L, Solozhenko E G, Zinin P V, Ming L C, Chen J H, and Parise J B 2003 *J. Phys. Chem. Solids* **64** 1265
[19] Guo Q X, Xie Y, Wang X J, Lv S C, Hou T, and Liu X M 2003 *Chem. Phys. Lett.* **380** 84
[20] Dubrovinskaia N, Dubrovinsky L, Miyajima N, Langenhorst F, Crichton W A, and Braun H F 2006 *Z.Naturforsch.(B)* **61** 1561
[21] Solozhenko V L, Dubrovinskaia N A, and Dubrovinsky L 2004 *Appl. Phys. Lett.* **85** 1
[22] Champi A and Marques F C 2006 *Brazilian Journal of Physics* **36** 462
[23] Bormett R W, Asher S A, Witowski R E, Partlow W D, Lizewski R, and Pettit F 1995 *J. Appl. Phys.* **77** 5916
[24] Andreyev A, Akaishi M, and Golberg D 2002 *Diamond Relat. Mater.* **11** 1885
[25] Ferrari A C and Robertson J 2000 *Phys. Rev. B* **61** 14095
[26] Roy D, Chhowalla M, Hellgren N, Clyne T, and Amaratunga G A 2004 *Phys. Rev. B* **70** 6
[27] Larkin P J, Makowski M P, and Colthup N B 1999 *Spectrochim. Acta, Part A* **55** 1011
[28] Jurgens B, Irran E, Senker J, Kroll P, Muller H, and Schnick W 2003 *J. Am. Chem. Soc.* **125** 10288
[29] Koglin E, Kip B J, and Meier R J 1996 *J. Phys. Chem.* **100** 5078
[30] Sung C M J 2001 *High Temp-High Press* **33** 473
[31] Ekimov E A, Sadykov R A, Mel'nik N N, Presz A, Tat'yanin E V, Slesarev V N, and Kuzin N N 2004 *Inorganic Materials* **40** 932
[32] Ribeiro F J and Cohen M L 2004 *Phys. Rev. B* **69** 212507(4 pages)
[33] Ekimov E A, Sidorov V A, Bauer E D, Mel'nik N N, Curro N J, Thompson D O, and Stishov S M 2004 *Nature* **428** 542
[34] Moussa J E and Cohen M L 2008 *Phys. Rev. B* **77** 8
[35] Lowther J E 2005 *J. Phys. Condes. Matter.* **17** 3221
[36] Liu Z, He J, Yang J, Guo X, Sun H, Wang H, Wu E, and Tian Y 2006 *Phys. Rev. B* **73** 172101
[37] Han S W and Ihm J 1997 *Phys. Rev. B* **55** 15349
[38] Zinin P V, Kudryashov I, Konishi N, Ming L C, Solozhenko V L, and Sharma S K 2005 *Spectrochim. Acta A* **61** 2386
[39] Zinin P V, Ming L C, Kudryashov I, Konishi N, Manghnani M H, and Sharma S K 2006 *J. Appl. Phys.* **100** 013516
[40] Tuinstra F and Koenig J L 1970 *J. Chem. Phys.* **53** 1126
[41] Reich S, Ferrari A C, Arenal R, Loiseau A, Bello I, and Robertson J 2005 *Phys. Rev. B* **71**
[42] May P, Ludlow W, Hannaway M, Heard P, Smith J, and Rosser K 2008 *Diamond Relat. Mater.* **17** 105
[43] Bourgeois E, Bustarret E, Achatz P, Omnès F, and Blase X 2006 *Phys. Rev. B* **74** 094509
[44] Bernard M, Baron C, and Deneuville A 2004 *Diamond Relat. Mater.* **13** 896
[45] Ferrari A C and Robertson J 2001 *Phys. Rev. B* **63** 121405(R)

High-pressure high-temperature synthesis of novel binary and ternary nitride phases of group 4 and 14 elements

D A Dzivenko[1], E Horvath-Bordon[1], A Zerr[2], G Miehe[1], P Kroll[3], R Boehler[4], P F McMillan[5] and R Riedel[1]

[1]FB Materialwissenschaft, TU Darmstadt, Petersenstrasse 23, 64287 Darmstadt, Germany

[2]LPMTM-CNRS, Université Paris Nord, 99 Av. J.B. Clement, 93430 Villetaneuse, France

[3]Department of Chemistry, UTA, Arlington, Texas 760019-0065, USA

[4]Hochdruckgruppe, MPI für Chemie, J.-J.-Becher-Weg 27, 55128 Mainz, Germany

[5]Department of Chemistry, UCL, 20 Gordon Street, London WC1 H0AJ, UK

E-mail: dzivenko@materials.tu-darmstadt.de

Abstract. Our recent experiments on high-pressure high-temperature synthesis of novel ternary nitrides of group 4 and 14 elements are presented. Dense carbon nitride imide, $C_2N_2(NH)$, was synthesized for the first time in a laser heated diamond anvil cell (LH-DAC) at pressures above 27 GPa and temperatures around 2000 K. Based on results of the electron diffraction-, EELS- and SIMS-measurements combined with theoretical calculations the structure of this new C-N-H phase was suggested to be of the defect-wurtzite type. Farther, macroscopic amounts of a new oxynitride of zirconium having cubic Th_3P_4-type structure, c-$Zr_{2.86}(N_{0.88}O_{0.12})_4$, were synthesized at high pressures and temperatures using a multi-anvil apparatus. Earlier this structure was observed for binary nitrides of zirconium(IV) and hafnium(IV) synthesized in microscopic amounts in a LH-DAC. The lattice parameter of c-$Zr_{2.86}(N_{0.88}O_{0.12})_4$ was found to be a_0 = 6.7549(1) Å which is slightly larger than that of c-Zr_3N_4. Isotropic bulk and shear moduli of c-$Zr_{2.86}(N_{0.88}O_{0.12})_4$ of B_0 = 219 GPa and G_0 = 96 GPa, respectively, were determined from the compression and nanoindentation measurements. The Vickers microhardness, $H_V(1)$, of the porous (about 30 vol. %) sample of c-$Zr_{2.86}(N_{0.88}O_{0.12})_4$ was measured to be 12 GPa, similar to that of single crystal δ-ZrN.

1. Introduction

The nitrides of group 4 and 14 elements, namely hexagonal α- and β-phases of Si_3N_4 and cubic mononitrides δ-TiN, ZrN and HfN, are well known industrial materials exhibiting such technologically useful properties as high hardness, wear and oxidation resistance, chemical and thermal stability. In 1999 a novel cubic phase of silicon nitride (γ-Si_3N_4) having spinel structure was synthesized at high pressures (>15 GPa) and temperatures (>2000 K) using a LH-DAC [1]. γ-Si_3N_4 was found to have high elastic moduli (B_0=290 GPa, G_0=148 GPa) and high hardness $H_V(0.5)$=30-43 GPa [2]. γ-Si_3N_4 is metastable at ambient pressure and on heating in air up to 1673 K [3].

© 2008 IOP Publishing Ltd

In contrast to Si, Ge and Sn, there is still no evidence for existence of similar nitride phases for carbon. Since dense C_3N_4 polymorphs were predicted to have bulk moduli and hardness comparable to or even exceeding those of diamond [4, 5], a number of experimental efforts have been devoted to synthesize crystalline carbon nitride using different deposition and high-pressure high-temperature (HP-HT) techniques. However, the obtained CN_x phases were predominantly amorphous or nanocrystalline with structure and composition being not unambiguously determined [6]. Therefore, the existence of pure crystalline C_3N_4 polymorphs remains a matter of controversy within scientific community.

On the other hand the discovery of spinel nitrides of group 14 elements (Si, Ge, Sn) has entailed speculations on possible stability of analogous polymorphs for other tetravalent elements (i.e. Ti, Zr and Hf). This supposition, however, was not confirmed experimentally. Moreover, experiments with a LH-DAC have shown that at high pressures (>15 GPa) and high temperatures (>2500 K) Zr and Hf form another previously unknown family of nitrides having M_3N_4 stoichiometry (M=Zr or Hf) [7]. The novel nitrides (denoted hereafter c-M_3N_4) were found to have cubic Th_3P_4-type structure with cations 8-fold coordinated by nitrogen atoms. These were the first binary nitrides with cations exhibiting such a high coordination number. Preliminary measurements suggested for each of the compounds B_0≈250 GPa (B'_0=4, fixed) indicating their high hardness. Subsequent theoretical studies predicted high elastic moduli [8-10] and hardness for c-M_3N_4 [9]. Thin films of c-Zr_3N_4 deposited by physical vapor deposition (PVD) technique were found to be significantly harder than δ-ZrN films and to exhibit extraordinary wear resistance upon machining of low carbon steel [11].

Here we report on the HP-HT synthesis of a well-crystallized carbon nitride imide, $C_2N_2(NH)$, in a LH-DAC from a single source precursor (dicyandiamide $C_2N_4H_4$) [12]. Single crystals of the new dense carbon nitride phase were recovered to ambient conditions and analyzed using TEM, EELS and SIMS. The carbon nitride imide has a defect-wurtzite structure (space group $Cmc2_1$) analogous to that of $Si_2N_2(NH)$ or Si_2N_2O. The bulk modulus of $C_2N_2(NH)$ was calculated to be 277 GPa.

We also present here the first HP-HT experiments on synthesis of macroscopic amounts (>1 mm^3) of oxygen-bearing cubic zirconium(IV) nitride, c-$Zr_{2.86}(N_{0.88}O_{0.12})_4$ [13]. The obtained material was recovered to ambient conditions and examined by XRD, TEM, EPMA and SEM. The Rietveld structure refinement was performed in order to derive the structural parameters. We also measured elastic moduli and hardness of c-$Zr_{2.86}(N_{0.88}O_{0.12})_4$ [14]. In particular, B_0 and B'_0 were determined from the EOS, $V(P)$, measured at RT up to 45 GPa. Nanoindentation technique was applied in order to obtain the reduced elastic modulus, E_r, and hardness, H. From the experimental results for B_0 and E_r we determined the shear modulus, G_0 for c-$Zr_{2.86}(N_{0.88}O_{0.12})_4$. Finally, the Vickers microhardness of the obtained porous material has been measured at different loads.

2. Carbon nitride imide, $C_2N_2(NH)$

2.1. Experimental

In the LH-DAC experiments the starting dicyandiamide was embedded in the NaCl pressure medium within a hole in the preindented stainless steel gasket, squeezed to required pressure and heated with radiation of a CO_2-laser (λ=10.6 μm). The temperature was determined from the thermal radiation spectrum of the heated sample. After synthesis the products were examined *in situ* by Raman spectroscopy. The recovered samples were washed by dissolving NaCl in distilled water and examined using transmission electron microscope (TEM) equipped with an electron energy-loss spectrometer (EELS) and EDX-detector. The content of H, C, and N was estimated using secondary ion mass spectroscopy (nano-SIMS). All accompanying calculations were carried out within the framework of the density functional theory using the VASP-code.

2.2. Results and discussion

Treatment of dicyandiamide, $C_2N_4H_4$, at P<27 GPa and T<2000 K resulted in black amorphous products. At higher P-T conditions the formation of a new optically transparent phase having characteristic Raman spectrum was observed. The TEM investigation of recovered products revealed the presence of crystalline material (figure 1a, b). The average crystallite size was found to increase with pressure and heating duration: extended heating (20 min) at temperatures >2000 K and maximal pressure of 41 GPa resulted in formation of crystallites up to 1-1.5 μm in size (figure 1a). The C:N ratio in the crystals was determined from the EELS spectra to be 0.62±0.06. Moreover, the structure in the energy-loss near-edge (ELNES) region indicated tetrahedrally-directed sp^3 hybridized orbitals for both C and N atoms in the synthesized material [12] (figure 1c). No oxygen was detected by EELS, indicating its concentration below 0.1-1 atom. %. Nano-SIMS measurements revealed the presence of hydrogen in the recovered samples with H:C ratio of 0.5±0.15. Thus, the resulting composition of synthesized material can be expressed by formula C_2N_3H.

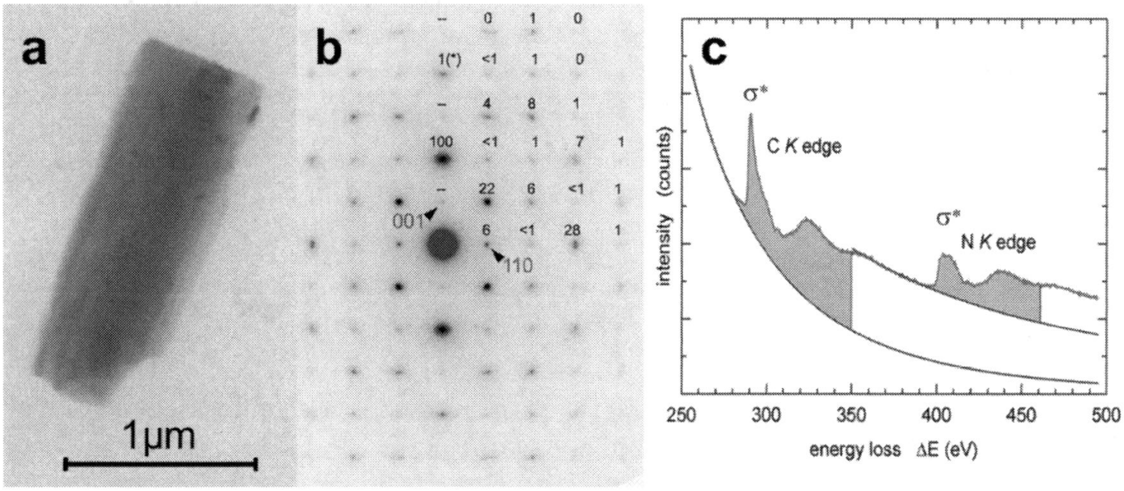

Figure 1. TEM and EELS investigations: (a) Bright field TEM image of the sample synthesized at P=41 GPa and T>2000 K; (b) SAED pattern of zone $[\bar{1}10]$ of the $C_2N_2(NH)$. The numbers attached to the reflections are values for kinematic intensities $|F|^2$ scaled to 100 for reflection 002. The intensity of the non-extinct weak reflection 400 (*) is enhanced by secondary diffraction from the 200 reflection; (b) Representative C-K and N-K core-loss edges of $C_2N_2(NH)$ where the grey shaded areas indicate the 60 eV wide integration windows started from the edge onsets. The thick solid lines indicate the power law background fitted in an energy window prior to the edge and extrapolated above it.

The possible crystal structure of the novel C-N-H compound was determined by combination of selected-area electron diffraction (SAED) data, experimentally measured composition and *ab initio* calculations. The structures suggested for hypothetical C_3N_4 [4, 5] were subsequently excluded due to high internal energy, inherent elastic and dynamic instabilities, or by their incompatibility with the observed lattice parameters and the SAED intensities. Only a defect-wurtzite (dwur) structure type gave good agreement with the electron diffraction and EELS data, especially regarding the sp^3-hybridized N sites. Finally, the analysis of SAED patterns for different zone orientations suggested for the new compound, dwur-$C_2N_2(NH)$, the orthorhombic cell with space group $Cmc2_1$, the structure analogous to that of $Si_2N_2(NH)$ or Si_2N_2O. The experimentally determined lattice parameters are in good agreement with calculated values (given in brackets): a=7.546(2) Å [7.5726 Å], b=4.434(8) Å [4.4425 Å], c=4.029(8) Å [4.0036 Å]. The calculated density ρ_{calc} = 3.21 g/cm^3 is close but below that of diamond (ρ=3.52 g/cm^3). B_0 of $C_2N_2(NH)$ calculated to be 277 GPa is significantly lower than those for diamond or predicted dense C_3N_4 polymorphs (B_0=430-460 GPa).

3. Oxygen-bearing cubic zirconium(IV) nitride, c-$Zr_{2.86}(N_{0.88}O_{0.12})_4$

3.1. Experimental

The HP-HT synthesis of macroscopic amounts of oxygen-bearing c-Zr_3N_4 [13] was performed in a multi-anvil apparatus. As a starting material we used the nanocrystalline powder of nitrogen reach zirconium nitride (Zr:N=1:1.47) having distorted NaCl-type structure [13, 15]. The sample was compressed to 12 GPa and then heated up to 1900 K, kept at the maximum temperature for about 20 minutes and quenched. The sample recovered to ambient conditions was about 1.3 mm in diameter and about 1.5 mm in height. The material was characterized using TEM equipped with an EDX-detector, electron probe micro-analyzer (EPMA) and scanning electron microscope (SEM). The X-ray powder diffractogram (XRPD) was collected from powdered sample using a STOE STADI P diffractometer with Mo-$K_{\alpha 1}$ radiation.

The EOS of oxygen-bearing c-$Zr_{2.86}(N_{0.88}O_{0.12})_4$ was measured at RT in a DAC to 45 GPa using quasi-hydrostatic argon pressure medium. The pressure was determined from the EOS of crystalline argon [16]. The specific volumes of the sample and of argon at HP were derived from the EDX diffraction patterns measured using a polychromatic synchrotron radiation on the beamline F3 at the HasyLab (DESY, Hamburg, Germany).

Nanoindentation measurements were performed on a polished surface of the porous polycrystalline c-$Zr_{2.86}(N_{0.88}O_{0.12})_4$. Indentations were done using a Nanoindenter XP equipped with the continuous stiffness measurement (CSM) module [17]. Indentations were performed for two maximum loads of about 23 mN and 14 mN which corresponded to maximal indentation depths of about 260 nm and 210 nm, respectively. Vickers microhardness was measured for loads from 0.49 N to 9.8 N.

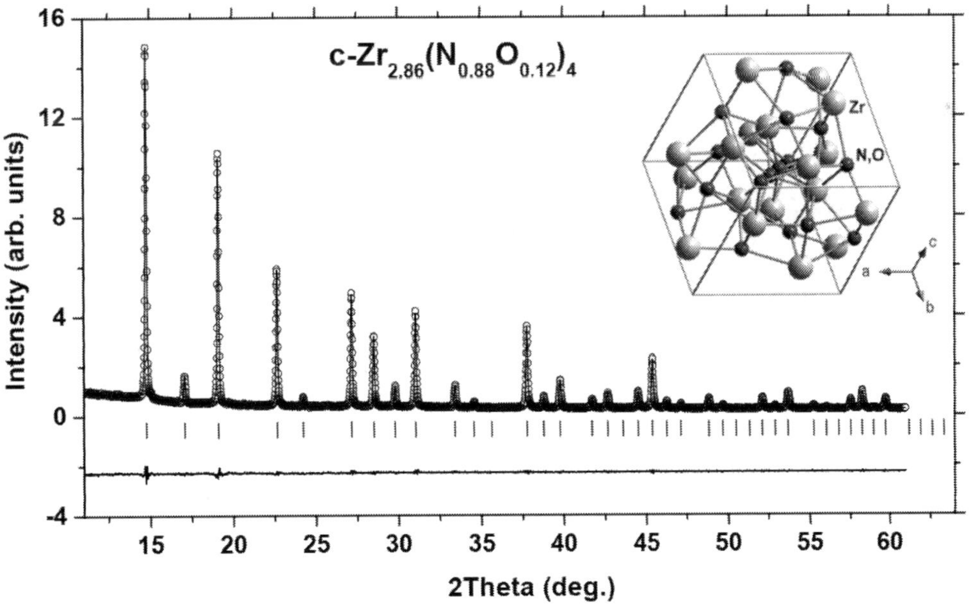

Figure 2. XRPD pattern for c-$Zr_{2.86}(N_{0.88}O_{0.12})_4$: observed (open circles) and calculated (solid line) intensities. Their difference curve is given at the bottom of the graph. The calculated peak positions of the Th_3P_4-type structure are denoted by tick marks. Inset: schematic representation of the Th_3P_4-type structure.

3.2. Results and discussion

The XRPD examination of the recovered product showed only lines of a cubic phase which structure could be attributed to that of c-Zr_3N_4 (figure 2). Measurements using EPMA revealed the presence of 2.4(3) wt. % of oxygen. The resulted composition can be expressed by the formula $Zr_{2.86}(N_{0.88}O_{0.12})_4$. Since no amorphous oxidic layer was found on surfaces of the crystallites by TEM, we conclude that oxygen is incorporated in the crystal structure. This observation suggests the formation of an oxygen-bearing cubic zirconium(IV) nitride of the general composition $Zr_{3-u}(N_{1-u}O_u)_4$. The substitution of nitrogen by oxygen at the anion sites of the Th_3P_4-type structure requires, as a consequence of the electrical neutrality condition, the formation of vacancies at the cation sites. This supposition was verified by a full-profile Rietveld structure refinement of the collected XRPD data (figure 2) [13]. The lattice parameter a_0 = 6.7549(1) Å for c-$Zr_{2.86}(N_{0.88}O_{0.12})_4$ differs slightly from the value 6.740(6) Å reported earlier for c-Zr_3N_4 [7]. We attribute this difference to the presence of oxygen in the crystal structure of the synthesized material.

Farther, we measured the EOS of c-$Zr_{2.86}(N_{0.88}O_{0.12})_4$ at RT to 45 GPa (figure 3) in a DAC [14]. From the least-square fit of the Birch-Murnaghan EOS [18] to the experimental data we obtained B_0=219(13) GPa and B'_0=4.4(1.0) for c-$Zr_{2.86}(N_{0.88}O_{0.12})_4$. For the fixed value of B'_0=4 we obtained B_0=224(5) GPa. The theoretical predictions for B_0 of c-Zr_3N_4 (195-265 GPa) [8-10] scatter around the present value.

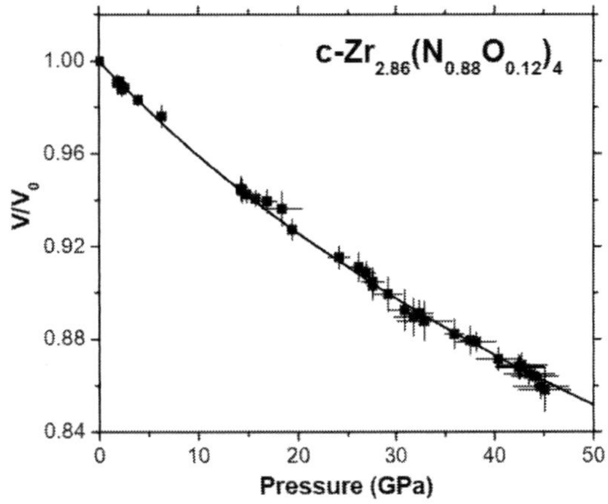

Figure 3. Pressure dependence of V/V_0 of c-$Zr_{2.86}(N_{0.88}O_{0.12})_4$ measured at RT (■). The solid line represents the least-squares fit of the third-order Birch-Murnaghan EOS to the experimental data.

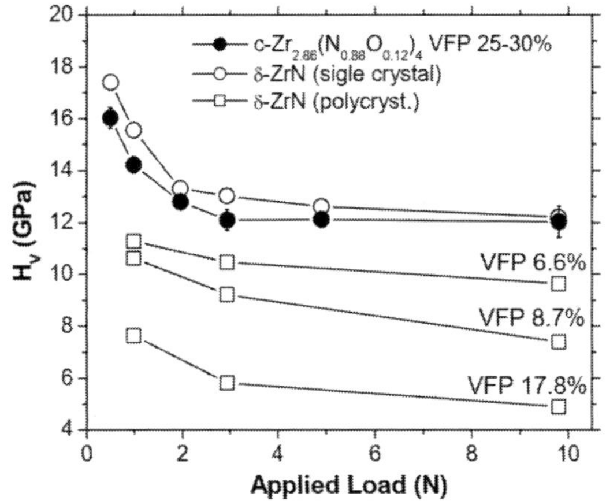

Figure 4. Vickers microhardness of the porous c-$Zr_{2.86}(N_{0.88}O_{0.12})_4$ (●) as a function of applied load, compared with hardness of the single crystal (○) [19] and polycrystalline δ-ZrN having different porosity (□) [20].

SEM investigation of the polished sample surface revealed the presence of a chemically homogeneous, porous material with pore sizes of 0.05-2 μm and crystallites of c-$Zr_{2.86}(N_{0.88}O_{0.12})_4$ of about 0.8 μm in size. The volume fraction porosity (VFP) was estimated to be about 25-30 %.

From nanoindentation measurements on the polished surface of the porous sample we derived E_r=231GPa and H=18(2) GPa [14]. Using the experimental values for B_0 and E_r and the well known relations between isotropic E_r, E, B, G, and v [21] we determined for c-$Zr_{2.86}(N_{0.88}O_{0.12})_4$ G_0=96 GPa and E_0=252 GPa [14]. The obtained values should be considered as the lower limits since they could be altered by volume defects in our sample. Despite the apparent porosity, we were able to perform reliable Vickers microhardness measurements for c-$Zr_{2.86}(N_{0.88}O_{0.12})_4$ [13]. The measured $H_V(1)$=12.0(6) GPa is slightly differs from hardness of single crystal δ-ZrN (12.2 GPa) [19] and significantly higher than that of even less porous polycrystalline δ-ZrN [20] (figure 4). Based on the known hardness-porosity relations for hard ceramic materials [20, 22, 23], we estimated the Vickers microhardness value of fully dense c-Zr_3N_4 to exceed 30 GPa.

4. Acknowledgments
We acknowledge financial supports of the Deutsche Forschungsgemeinschaft (Bonn, Germany), the Fonds der Chemischen Industrie (Frankfurt, Germany), of the Heisenberg-Fellowship program (Germany), of the Adolf-Messer-Foundation (Germany) and of the Engineering and Physical Science Research Council (UK). D.A.D. and A.Z. thank the HasyLab for technical and DESY (Hamburg, Germany) for financial support. P.F.M. is also a Wolfson-Royal Society Research Merit Award Fellow.

References

[1] Zerr A, Miehe G, Serghiou G, Schwarz M, Kroke E, Riedel R, Fuess H, Kroll P and Boehler R 1999 *Nature (London)* **400** 340.
[2] Zerr A, Kempf M, Schwarz M, Kroke E, Goken M and Riedel R 2002 *J. Am. Ceram. Soc.* **85** 86.
[3] Jiang J Z, Kragh F, Frost D J, Stahl K and Lindelov H 2001 *J. Phys.: Condens. Matter* **13** L515.
[4] Liu A Y and Cohen M L 1989 *Science* **245** 841.
[5] Teter D M and Hemley R J 1996 *Science* **271** 53.
[6] Horvath-Bordon E, Riedel R, Zerr A, McMillan P F, Auffermann G, Prots Y, Bronger W, Kniep R and Kroll P 2006 *Chem. Soc. Rev.* **35** 987.
[7] Zerr A, Miehe G and Riedel R 2003 *Nat. Mater.* **2** 185.
[8] Kroll P 2003 *Phys. Rev. Lett.* **90**.
[9] Mattesini M, Ahuja R and Johansson B 2003 *Phys. Rev. B* **68**.
[10] Lowther J E 2005 *Physica B* **358** 72.
[11] Chhowalla M and Unalan H E 2005 *Nat. Mater.* **4** 317.
[12] Horvath-Bordon E, Riedel R, McMillan P F, Kroll P, Miehe G, van Aken P A, Zerr A, Hoppe P, Shebanova O, McLaren I, Lauterbach S, Kroke E and Boehler R 2007 *Angew. Chem. Int. Edit.* **46** 1476.
[13] Dzivenko D A, Zerr A, Bulatov V K, Miehe G, Li J W, Thybusch B, Brötz J, Fuess H, Brey G and Riedel R 2007 *Adv. Mater.* **19** 1869.
[14] Dzivenko D A, Zerr A, Schweitzer E, Göken M, Boehler R and Riedel R 2007 *Appl. Phys. Lett.* **90** 191910.
[15] Li J W, Dzivenko D, Zerr A, Fasel C, Zhou Y P and Riedel R 2005 *Z. Anorg. Allg. Chem.* **631** 1449.
[16] Richet P, Xu J A and Mao H K 1988 *Phys. Chem. Miner.* **16** 207.
[17] Oliver W C and Pharr G M 1992 *J. Mater. Res.* **7** 1564.
[18] Birch F 1978 *J. Geophys. Res.* **83** 1257.
[19] Chen X J, Struzhkin V V, Wu Z G, Somayazulu M, Qian J, Kung S, Christensen A N, Zhao Y S, Cohen R E, Mao H K and Hemley R J 2005 *Proc. Natl. Acad. Sci. U.S.A.* **102** 3198.
[20] Adachi J, Kurosaki K, Uno M and Yamanaka S 2006 *J. Nucl. Mater.* **358** 106.
[21] Landau L D and Lifshitz E M 1975 *Course of Theoretical Physics, vol. 7: Theory of Elasticity* Vol. 7 (London: Pergamon Press).
[22] Cho S A, de Arenas I B, Arenas F J, Ochoa J and Ochoa J L 1999 *J. Alloys Compd.* **288** 211.
[23] Luo J and Stevens R 1999 *Ceram. Int.* **25** 281.

Computer–aided analytical control of diamond and cubic boron nitride grits

N V Novikov[1], **G P Bogatyreva**[1] **and G A Petasyuk**[1,2]
[1] V. N. Bakul Institute for Superhard Materials, the National Academy of Sciences of Ukraine, 2, Avtozavodskaya St., Kiev, 04074, Ukraine

E-mail: bogatyreva@ism.kiev.ua

Abstract. Methodical and information aspects of the superhard materials powder uniformity as a characteristics of the powder quality are discussed. Computer-aided analytical methods of the rapid diagnostics of the sieve granulometric composition and the external specific surface of diamond and cubic boron nitride grits are described. The results of the application of the developed methods and software to standard powders of synthetic diamond and cubic boron nitride are given.

1. Introduction

Diagnostics of powders of synthetic diamond and other superhard materials (SHM) is an important information–preparatory stage of the assessment and analysis of the powder quality. Therefore, the urgent problems are the development and improvement of the scientific and methodical base of the diagnostics, which takes into account recent achievements of the scientific and technological advances; elaboration and promotion of new highly informative characteristics and increase in the amount of information provided by the characteristics used, the extension of the number of the properties to be diagnosed, including those identified through the indirect and combined. The uniformity, granulometric composition values and specific surface belong under basic characteristics of SHM powders.

2. Uniformity

Several methods of the quantitative assessment of the degree of the powder uniformity are known [1,2]. In our opinion the method developed at the Institute for Superhard Materials of the National Academy of Sciences of Ukraine gives the most adequate information [3]. The essence of it is that the powder uniformity is quantitatively assessed by a three–level system. The primary data (the values of the first level) are based on statistical parameters of the distribution series of the values of the analyzed characteristic x, namely, an average value (x_{av}), quadratic variation coefficient (V_{σ}), asymmetry coefficient (k_a), mode (M_0) and its frequency characteristic (f_m), range (R), coefficient of oscillation ($V_R = R \cdot (x_{av})^{-1}$), and harmonic mean ($x_{avh}$). The primary indices are defined using the relations, which are first proposed by us:

[2] To whom any correspondence should be addressed.

© 2008 IOP Publishing Ltd

$$u_1^{(x)} = (1+k_a)^{-1},\quad u_2^{(x)} = (x_{av}+M_0)\cdot(x_{av}+M_0+|x_{av}-M_0|)^{-1},\quad u_3^{(x)} = f_m,\quad u_4^{(x)} = (1+V_\sigma)^{-1},$$

$$u_5^{(x)} = (1+V_R)^{-1},\quad u_6^{(x)} = x_{avh}\cdot(x_{av})^{-1} \tag{1}$$

Here the superscript "x" indicates the powder characteristic used to analyze the uniformity. The data of the next level – partial ($U_p^{(x)}$) – are calculated from the primary data by finding the harmonic mean:

$$U_p^{(x)} = 6\cdot\left[\sum_{i=1}^{6}(u_i^{(x)})^{-1}\right]^{-1} \tag{2}$$

The same approach is used when determining the data of the third level – generalized U_g from the partial data:

$$U_g = N_p\cdot\left[\sum_{p=1}^{N_p}(U_p^{(x)})^{-1}\right]^{-1} \tag{3}$$

where N_p is the number of the basic characteristics (partial data).
Unlike other known methods, the above method allows us to assess the powder uniformity as the component of quality not only from separate characteristics of the powder (static compression strength of grits, dimensional parameters, aspect ratio, etc.) but from the combination of characteristics of interest as well. Another advantage of the method proposed by us is that the uniformity index is found using the criteria proposed for the first time. The quantitative uniformity assessment criteria proposed by us cover the whole spectrum of statistical test parameters of the distribution of values of the characteristic being analyzed. It should be noted that the other methods to assess the uniformity use only one statistical parameter, e.g., variation coefficient or modal frequency and two adjacent frequencies. A special feature of the system–criterial method is that the uniformity indices obtained by it vary in the range of 0 to 1, i.e. these values are normalized magnitudes. This fact is useful for comparative analysis of the powder quality.
The studies of standard (according to Ukrainian State Standard 3292–95 [4]) grits of synthetic diamonds and cubic boron nitride have shown that the generalized uniformity index of these powders is between 0.44 and 0.62. These powders are the most uniform in the maximal linear grit size (0.67–0.83) and the least uniform in static compression strength (0.33–0.52). The uniformity in the aspect ratio is 0.43–0.75. It should be noted that the above data refer to the powder uniformity only and not to the powder quality. The quality is a relative notion associated with the powder application.
Granulometric composition
The separation of powders into fractions by sizes and quantitative assessment of the grits contained in each fraction are performed using a set of test sieves. This assessment gives ground to judge the conformity of the powder to the standard and the grit size of a fraction. The necessity for the indirect control sieve analysis by mathematical simulation is caused by the following a small amount of the powder, which does not allow to make the sieve analysis by the standard method, the lack of sieves and other.
At the Institute for Superhard Materials, a procedure of the indirect sieve analysis of synthetic diamond powders by mathematical simulation has been developed and tested. The method is based on an analytical dependence relating the grit geometric parameters to the sieve mesh dimensions:

$$l = \tau\cdot b\cdot\sqrt{2(1+\mu^2)} \tag{4}$$

where l is the length of the mesh of a sieve, on which the grit is retained, b is the width of the grit projection, $\mu = h\cdot b^{-1}$; h is the grit height; τ is the correction coefficient, allowing for a possible deviation of the real geometrical shape of the grit projection from its accepted model. For the practical use of this procedure, an original software (a "computer–aided control sieve"of a sort) has been developed. The employment of this computer–aided control sieve allows a wide range of practical tasks to be solved only on the basis of the grit projection length and width and the grit height.

The method was tested on standard (according to [4]) synthetic diamond powders with known indices of the granulometric composition and grit size. The test results as well as values of generalized index of dimensional uniformity by system–criterial method are presented in table 1. The test results show

Table 1. Indices of the grit sizes and granulometric composition of some grades of synthetic diamond grits defined by mathematical simulation (*a* –actual indices, *calc*– calculated indices, Max_f – maximal (in size) fraction, C_f – coarse fraction, M_f – main fraction, F_f – fine fraction).

Powder grades and grit sizes (μm)	Index	Granulometric composition indices, %				U_g
		Max_f	C_f	M_f	F_f	
AC4 125/100	*a*	0	4	84	0	0.74
	calc	0	6	78	0	
AC20 100/80	*a*	0	2	81	0	0.76
	calc	0	0	78	1	
AC65 315/250	*a*	0	8	82	0	0.78
	calc	0	11	80	0	
AC80 315/250	*a*	0	10	80	0	0.79
	calc	0	14	77	0	
AC100 250/200	*a*	0	15	73	0	0.80
	calc	0	17	71	0	

that for powders having a dimensional uniformity of 0.7 and above, the maximum error of the calculated determination of granulometric composition indices is: in the maximal fraction – 0% (precisely), in the coarse fraction – up to 4% (overestimation), in the main fraction – up to 6% (underestimation), in the fine fraction – up to 1% (overestimation). As for the calculated grit size, in all cases it coincided with the actual one. The use of the method for powders having a dimensional uniformity of below 0.7, in some cases the results may be less adequate. However, in both cases the following tendency for the deviation of the calculated values of the granulometric composition indices from the actual ones is observed: in the coarse and fine fractions to the overestimation, while in the main fraction to the underestimation. Such a nature of deviations considerably decreases the degree of the risk to take a decision that the powder corresponds to the standard, while in fact it is inconsistent with the standard.

3. External specific surface

A new extrapolation–analytical method of measuring the external specific surface of superhard materials powders has been developed [5]. The method is based on the methodology of 3D simulation with the use of a mathematical tool for the representation of the affine space. The initial data are the grit height, area, perimeter, the maximal and minimal diameters of the grit projection measured on the DiaInspect.OSM device. The method allows us to reduce the effect of particle surface roughness and incorrect choice of the value of the shape parameter (which enters into calculation equations used in methods for geometric measurements) on the result of specific surface measurements. These factors, which introduce an allowable error when the methods for geometric measurements are used for regularly shaped particles, become a source of serious errors when we deal with irregular particles. The reduction of the effect of these factors is achieved by the fact that for determining the specific surface, separate particles of a dispersed material sample are measured. In this case, the initial data absolutely exclude the parameter of the powder particle shape, and allow partially (in one projection) for the particle surface roughness. The equation of the powder specific surface index (F) is derived as

$$F = \frac{1}{I_z}\sum_{i=1}^{I_z} F_i = \frac{1}{\rho \cdot I_z}\sum_{i=1}^{I_z}\sum_{n=1}^{N}\frac{s_{i,n}}{v_{i,n}} \qquad (5)$$

and volume of grit of sample accordingly.

The test where I_z is the amount of grains in a sample, ρ is the density of the powder material, $s_{i,n}$, and $v_{i,n}$ are area of surface results (table 2) have shown that the new method developed by us allows one to

Table 2. External specific surface (F, m^2/kg) of standard synthetic diamond powders measured using the extrapolation–analytical and other methods and the relative errors (δ, %) of the measurements

Powder grades and grit sizes (μm)	Reference [6]	According to the methods					
		[7]		[8]		our method	
	F	F	Δ	F	δ	F	δ
ACM 14/10	420.00	145.36	65.39	77.61	81.52	528.46	25.82
ACM 28/20	260.00	78.32	69.88	57.01	66.53	249.94	3.87
ACM 60/40	110.00	40.18	63.47	29.52	73.16	123.84	12.54
AC4 160/125	37.00	9.22	75.00	5.39	80.00	36.18	2.22
AC6 63/50	78.00	22.04	71.74	16.62	78.69	69.49	10.90
AC6 100/80	63.00	15.14	75.97	11.68	81.46	49.40	21.59
AC15 100/80	50.00	15.14	75.97	11.68	81.46	48.22	3.56
AC32 400/315	12.00	4.09	66.00	3.42	71.50	11.46	4.67

obtain more precise values of the specific surface than other known methods based on the concept of geometrical measurements. Therefore, it may be used for a rapid highly reliable measurement of the external specific surface of superhard materials powders.

4. Conclusions

1. The scientific and methodological as well as applied aspects the quantitative assessment of the degree of uniformity of the uniformity of synthetic diamond powders as the new characteristics of the powders are described.
2. A method of the experimental and analytical indirect determination of the grit size and granulometric composition of synthetic diamond grits without using sieves has been developed.
3. A new method of measuring the external specific surface of superhard materials powders has been developed. The method is based on the methodology of the 3D simulation with the use of a mathematical tool for the representation of the affine space.

To simplify the practical application of the proposed computer-aided analytical methods to control diamond and cubic boron nitride grits, the methodical recommendations have been developed. Using these methods. The combination of the above means makes possible a rapid and efficient diagnostics of SHM grits by specified characteristics and a highly informative comparative analysis of the grit quality.

5. References

[1] Makarov Yu 1973 *An Apparatus for Mixing Loose Materials* [in Russian] (Moscow: Mashinostroenie)
[2] Voblikov V S, Kornilov Yu P and Nozdrachev V I 1976 *Mogern Synthetic Superhard Materials and Their Appications. Trans. VNIALMAZ* **4** 44
[3] Novikov N V, Nikitin Yu I, Bogatyreva G P and Petasyuk G A 2006 Ukrainian Patent 53 964 *Byull,* **2**
[4] DSTU (State Standard of Ukraine) 3292–95 1995 Synthetic Diamond Powders. General Technical Specifications
[5] Novikov N V, Bogatyreva G P and Petasyuk G A 2007 Ukrainian Patent 7 893 *Byull.* **5**
[6] Nikitin Yu I 1984 *Production Technology and Quality Control of Diamond Powders* [in Russian] (Kiev: Naukova Dumka)
[7] Gregg S J and Sing K S W 1967 *Adsorption, Surface Area and Porosity* (London: Academic Press)
[8] Panichkina V V and Uvarova I V 1973 *Methods to Control the Dispersion and Specific Surface of Metal Powders* [in Russian] (Kiev: Naukova Dumka)

Journal of Physics: Conference Series **121** (2008) 062005 doi:10.1088/1742-6596/121/6/062005

Systematics of compression of hard materials

W J Nellis
Department of Physics, Harvard University,
Cambridge MA 02138, USA

E-mail: nellis@physics.harvard.edu

Abstract. Hard materials have the remarkable property that relatively little internal energy of shock compression is deposited as temperature and associated thermal pressure. As a result, the Hugoniot curve is nearly coincident with the 0-K isotherm. Since the isentrope is intermediate between isotherm and Hugoniot, all three curves of a given hard material are nearly coincident. Published data for diamond and $Gd_3Ga_5O_{12}$ (GGG) show this to be the case up to 600 GPa and 80 GPa, respectively. Above 100 GPa on the Hugoniot the incompressibilities, defined to be the derivative of pressure with respect to compression, of GGG, other oxides and even CaF_2 are significantly greater than that of diamond. In fact, above 100 GPa CaF_2 might well be the least compressible of any known material. Little EOS data of weakly compressible materials have been measured in a diamond anvil cell (DAC) above a few tens of GPa at 300 K because these materials often become amorphous, which means density in a DAC cannot be measured by X-ray diffraction. Amorphization suggests the onset of a sluggish phase transition. On the other hand, Hugoniot experiments readily measure density changes caused by compression and phase transitions independent of material structure. Hugoniots of weakly compressible materials suggest experiments to determine structures and compressibilities in laser-heated DACs at 100 GPa pressures, as has been done with TiO_2 and perovskite, and Hugoniot data might provide estimates of compressions of oxides representative of the deep interiors of rocky planets in other solar systems.

1. Introduction

Hard materials, also known as weakly compressible, incompressible, or strong materials, are interesting at high dynamic pressures because of phase transitions [1], to obtain estimates of densities of representative oxides in deep interiors of Earth and rocky exoplanets discovered recently [2], and for possible uses as anvils to achieve high static pressures. In dynamic high-pressure experiments hard materials are also useful as anvils between which a shock wave reverberates in a compressible liquid to achieve novel states of dense matter, such as metallic fluid hydrogen [3]. This quasi-isentropic compression process enables studies of the condensed matter physics of unusually dense states, mimics conditions in the deep interiors of giant fluid planets in this and in other solar systems, and is a possible synthesis route to metastable hydrogenous materials [3]. In order to measure properties of quasi-isentropically compressed fluids, anvils are needed whose equations of state are known on and off the Hugoniot, as well as anvils that remain optically transparent to measure temperature radiatively and remain electrically insulating to measure electrical conductivities of fluids by inserting electrodes through an anvil used in a shock reverberation experiment.

© 2008 IOP Publishing Ltd

Many hard materials undergo phase transitions at shock pressures in the 100 GPa range. Many of these transitions are observed as relatively sharp and substantial increases in the incompressibility defined as the derivative of pressure P with respect to compression $= / _0$ on the Hugoniot, $= (dP/d\ )_H$, where and $_0$ are the final and initial mass densities of compression. The structures of these virtually incompressible phases are not known from Hugoniot experiments. Because Hugoniot temperatures are normally significant, it is possible that these incompressible phases are a signature of melting to a highly coordinated fluid. On the other hand, shock-induced temperatures and associated thermal pressures are relatively low in incompressible materials. Thus, the sharp increases in incompressibility might be caused by collapse to solid phases with higher coordination. Since pressures and temperatures of these phases can be reached in a laser-heated DAC, structures above 100 GPa might be determined by X-ray diffraction in a DAC.

The purpose of this paper is to point out that Hugoniots of several incompressible materials above 100 GPa have already been measured, that their Hugoniots are nearly coincident with their isotherms, and that it would be useful to measure their optical and electrical properties and their off-Hugoniot equations of state (EOS) under dynamic compression. Hugoniots of hard materials also suggest experiments to be performed in a diamond anvil cell (DAC) because comparable pressures, densities, and temperatures in hard materials can be achieved by both static and shock compression.

2. General Considerations

The internal energies deposited at high pressures by adiabatic Hugoniot and isentropic compression and by isothermal compression at 0 K are given by:

$$E_H(V,T) \quad E_0 = (P_H(V,T)/2)(V_0 \quad V) \tag{1}$$

$$E_S(V,T) \quad E_0 = \int_{V_0}^{V} P_S(V,T)dV \tag{2}$$

$$E(V,T=0K) \quad E_0 = E_0(V) = \int_{V_0}^{V} P_0(V)dV \tag{3}$$

where $E_H(V,T)$, $E_S(V,T)$, and $E_0(V)$ are the internal energies on the Hugoniot, isentrope, and 0-K isotherm, respectively, $E_0 = E_0(V_0)$ is the internal energy of the initial state, $P_H(V,T)$, $P_S(V,T)$, and $P_0(V)$ are the pressures on the Hugoniot, isentrope, and 0-K isotherm, respectively, $V = {}^{1}$ is the final specific volume and $V_0 = {}_0^{1}$ is the initial specific volume of each compression.

For strong materials such as diamond and sapphire, $u_s = C + Su_p$, where u_s is shock velocity, u_p is particle velocity, $C \quad 10 km/s$ and $S \quad 1$. Thus, $P_H = ({}_0C)^2(V_0 \quad V)$ up to ~100 GPa plus small nonlinear contributions caused by higher order terms in $(V_0 \quad V)$. If pressure is linear in $(V_0 \quad V)$, then $P_H = P_S \quad P_0$ [3]. This is true because thermal pressures are so small in dynamically compressed hard materials. As an example, the compression of diamond measured by Pavlovskii under shock compression up to 600 GPa [4], measured by Occeli et al under static compression up to 140 GPa at 300 K [5], and the calculated 0-K isotherm of Kunc et al [6] are illustrated in figure 1.

The Hugniot of diamond suggests that diamond does not undergo a first-order phase transition up to a shock pressure of 600 GPa. The value of its slope $S \simeq 1$ is anomalously small. Most materials have values of S in the range 1.1 to 1.8. Most oxides are relatively incompressible at lower shock pressures, as indicated by their large velocity of sound at ambient and a large Hugoniot Elastic Limit

Figure 1. Measured Hugoniot (solid curve [4]) and 300-K isotherm (open circles [5]) and calculated 0-K isotherm (dotted curve [6]) of diamond. The isentrope (not shown) is intermediate between Hugoniot and isotherm. Cold compression dominates equation of state of diamond along its Hugoniot up to 600 GPa.

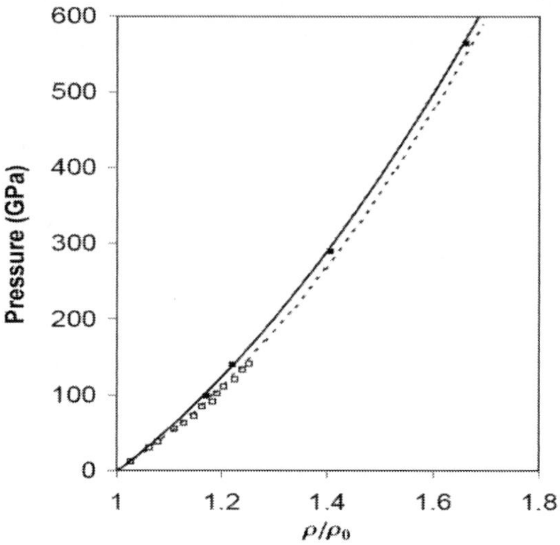

(HEL). $Gd_3Ga_5O_{12}$ (GGG), for example, has an HEL of $_{HEL}$ = 30 GPa. Between 65 and 120 GPa, S =0.66, which suggests a sluggish phase transition in this range of shock pressures. This sluggish phase transition is probably one to an amorphous phase, as observed in DAC measurements between 74 and 84 GPa [7]. In this regime, shock pressures are only a few percent greater than static ones. That is, the Hugoniot and isotherm of GGG are nearly coincident up to the highest pressure at which both have been measured. Above 120 GPa S = 2.1, which indicates completion of the transition and collapse to an essentially incompressible phase in the range 120 to 260 GPa [1].

In GGG $S = 2.1$ above 120 GPa, and is much larger than that of diamond with S 1. It takes significantly more pressure to compress GGG a given amount than is required to compress diamond the same amount. For GGG at 250 GPa = 2.4 TPa, while for diamond = 0.89 TPa [1]. However, Hugoniot data provide no information on structure.

Sapphire, single-crystal Al_2O_3, has slope S = 0.96 up to 340 GPa [8] and diffusive phase transitions near ~1500 K and 150 GPa [9]. The small slope of GGG in the region of its sluggish phase transition suggests that the small slopes of sapphire and diamond are also caused by continuous transitions toward increasing disorder with increasing shock pressure. The 300-K isotherm of ruby up to ~100 GPa [10] agrees with the Hugoniot of sapphire within the error bars. Ruby is Al_2O_3. doped with ~0.5 %Cr. Thus, thermal pressures in shock-compressed sapphire are small.

The observation of an incompressible phase of GGG at high pressures prompted a search of the literature to see if other incompressible phases have been observed previously. Hugoniot data of Al_2O_3 [8,11], $CaTiO_3$ perovskite [12], GGG [1], and CaF_2 [13] are shown in figure 2. Shock pressure of sapphire is nearly linear with compression up to 340 GPa. In contrast $CaTiO_3$ perovskite and cubic $Gd_3Ga_5O_{12}$ undergo sluggish phase transitions near 100 GPa. At higher pressures both oxides collapse to a virtually incompressible phase.

The CaF_2 Hugoniot data of Altshuler et al [13] obey $u_s = C + Su_p$ and are quite remarkable. Below 100 GPa S = 1.00 and above 100 GPa S = 2.23. At lower shock pressures fluorite undergoes a sluggish continuous transition in a relatively compressible phase. Once the phase transition completes at 100 GPa, CaF_2 is essentially incompressible at higher pressures. Under compression in a DAC at 300 K, CaF_2 transforms at 9.5 GPa with an ~11% volume change to an orthorhombic $PbCl_2$-type structure, which is 9-coordinated [14]. Thus, shock-compressed CaF_2 is expected to be relatively compressible in a disordered $PbCl_2$-type phase in the range 10 to 100 GPa. Above a shock pressure of 100 GPa, CaF_2 is essentially incompressible, which suggests the high-pressure phase has a coordination of 11 or 12. It is remarkable that above 100 GPa CaF_2, which is relatively weak at ambient and contains no C or O might be the most incompressible material known, including

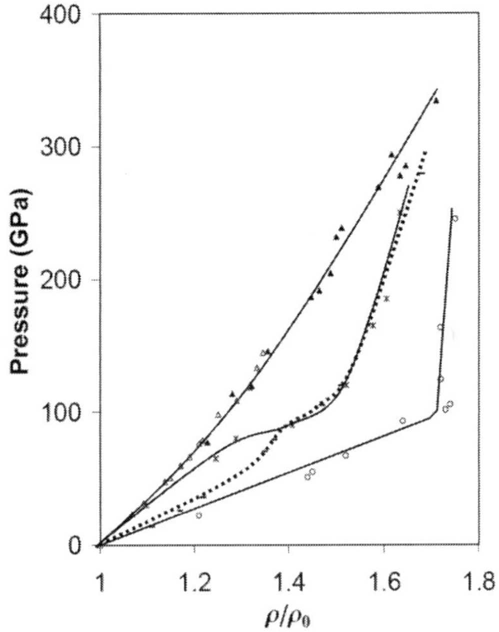

Figure 2. Shock pressure versus compression: Al_2O_3 (solid triangles [8], open triangles [11]), GGG (xs [1]), $CaTiO_3$ (dots [12]), and CaF_2 (open circles [13]).

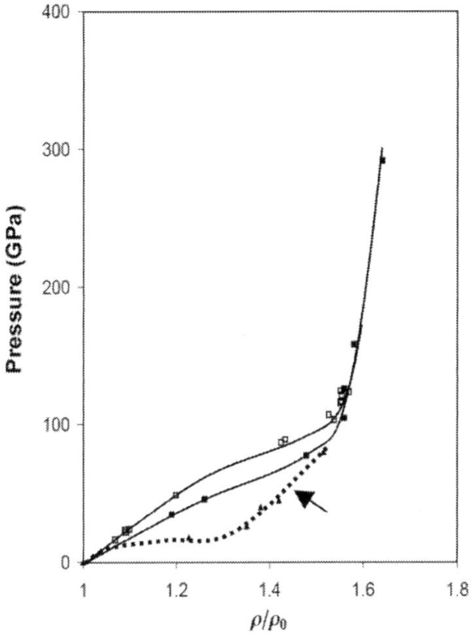

Figure 3. Measured Hugoniots and isotherm of TiO_2 plotted as pressure versus compression. Uppermost curve is [15], middle curve is [13] dotted curve is 300-K isotherm after pulsed heating at arrow [16].

diamond. This suggests the idea of a two-stage static anvil cell that takes advantage of the incompressibility of diamond below 100 GPa and the incompressibility of CaF_2 above 100 GPa.

Measured Hugoniots of TiO_2 [13,15] are plotted in figure 3. In the range of shock compressions from 1.2 to 1.5 TiO_2 undergoes a sluggish continuous phase transition similar to that in $Gd_3Ga_5O_{12}$. Above a shock pressure of ~100 GPa, $S = 2.21$ and TiO_2 is virtually incompressible. TiO_2 has also been compressed in a DAC up to 80 GPa. The sluggish phase transition in a DAC is observed as continuous amorphization until TiO_2 eventually becomes too amorphous to determine volume with X-ray diffraction, as for $Gd_3Ga_5O_{12}$ [7]. At this point Dubrovinsky et al laser-heated amorphous TiO_2 above 1100 K and thermally quenched it under pressure to a $PbCl_2$-type orthorhombic structure that is 9-coordinated [16]. As for CaF_2, TiO_2 might also transform to a 11 or 12 coordinated phase above 100 GPa, as suggested by its virtually incompressible Hugoniot above 100 GPa [13].

Static laser heating at high pressures in a DAC has been used to demonstrate the existence of an equilibrium phase at extreme conditions. Perhaps, the prototypical examples are demonstrations that (Mg,Fe)SiO_3 perovskite is stable up to 130 GPa and 2000 by Knittle and Jeanloz [17], up to several 10 GPa at high temperatures by Tsuchida and Yagi [18], and up to 100 GPa and 3000 K by Serghiou et al [18]. These results show (Mg,Fe)SiO_3 perovskite is probably the major constituent of Earth.

3. Conclusions

Several conclusions can be drawn from these general considerations: (i) The Hugoniot and 0-K isotherm of a hard material are nearly coincident. (ii) Hugoniot data of diamond and sapphire do not have limiting compressions under static and shock pressures below 600 GPa and 340 GPa, respectively. That is, the observed increase in compression with pressure depends weakly on pressure. (iii) In contrast $Gd_3Ga_5O_{12}$, TiO_2, $CaTiO_3$ perovskite, and CaF_2 have limiting shock compressions of about 1.8-fold of initial density at measured pressures in the range 100 to 300 GPa. At substantially higher shock pressures and temperatures, thermal excitation of electrons will eventually cause compression to increase above 1.8-fold, eventually approaching 4-fold at ~100 TPa shock pressures. (iv) The exoplanet discovered recently has a mass of about 7 earth masses. Assuming its density depends weakly on volume, this planet would have about twice the radius and twice the effective interior pressure of Earth. The pressure at the Earth's core-mantle boundary (CMB) is estimated to be 130 GPa. Based on published Hugoniot data up to ~300 GPa, the density of rock in this exoplanet is probably no more than at the CMB, about twice crystal density at the surface of the planet, independent of structure and composition. (v) A large number of hard crystals exist with a range of densities and compressibilities, which means pressure, density, and temperature of fluid metallic hydrogen, and other compressible fluids, can be tuned over a wide range using a reverberating shock wave between hard crystals of materials other than sapphire.

4. Acknowledgments

This material is based upon work supported by the U.S. Army Research Laboratory and the U.S. Army Research Office under grant number W911NF-06-1-0517.

5. Referencess

[1] Mashimo T, Chau R, Zhang Y, Kobayoshi T, Sekine T, Fukuoka K, Syono Y, Kodama M and Nellis W J 2006 *Phys. Rev. Lett.* **96** 105504
[2] Rivera E J, Lissauer J J, Butler R P, Marcy G W, Vogt S S, Fischer D A, Brown T M, Laughlin G and Henry G W 2005 *Ap. J.* **634** 625
[3] Nellis W J 2006 *Rep. Prog. Phys.* **69** 1479
[4] Pavlovskii M N 1971 *Sov. Phys. Solid State* **13** 741
[5] Occelli F, Loubeyre P and LeToullec R 2003 *Nature Mater.* **2** 151
[6] Kunc K, Loa I and Syassen K 2003 *Phys. Rev. B* **68** 094107
[7] Hua H, Mirov S and Vohra Y K 1996-I *Phys. Rev. B* **54** 6200
[8] Erskine D 1994 High pressure Hugoniot of sapphire *High-Pressure Science and Technology-1993* vol 2, ed S C Schmidt et al (New York: AIP Press) pp 141-143
[9] Lin J-F, Degtyareva O, Prewitt C T, Dera P, Sata N, Gregoryanz E, Mao H-K and Hemley R J 2004 *Nature Mater.* **3** 389
[10] Jephcoat A P, Hemley R J and Mao H-K 1988 *Physica B* **150** 115
[11] McQueen R G 1980 *LASL Shock Hugoniot Data* ed S P Marsh (Berkeley: University of California Press) pp 260-261
[12] Trunin R F, Simakov G V, Podurets M A, Moiseyev B N and Popov L V 1971 *Izv. Acad. Sci. USSR Phys. Solid Earth* **1** 8
[13] Altshuler L V, Podurets M A, Simakov G V and Trunin R F 1973 *Sov. Phys. Solid State* **15** 969
[14] Gerward L, Olsen J S, Steenstrup S, Malinowski M, Asbrink S and Waskowska A 1992 J. *Appl. Cryst.* **25** 578
[15] McQueen R G 1980 *LASL Shock Hugoniot Data* ed S P Marsh (Berkeley: University of California Press) p 326
[16] Dubrovinsky L S, Dubrovinskaia, Swamy V, Muscat J, Harrison N M, Ahuja R, Holm B and Johansson B 2001 *Nature* **410** 653
[17] Knittle E and Jeanloz R 1987 *Science* **235** 668
[18] Tsuchida Y and Yagi T 1989 *Nature* **340** 217
[19] Serghiou G, Zerr A and Boehler R 1998 *Science* **280** 2093

Stability of Hexagonal Diamond under Pressure

S. Ona[1], Y. Nakamoto[1], T. Kagayama[1], K. Shimizu[1], Y. Nishikawa[2], M. Murakami[2], K. Kusakabe[3], T. Watanuki[4], Y. Ohishi[5]

[1] KYOKUGEN, Center for Quantum Science and Technology under Extreme Conditions, Osaka University, Toyonaka, Osaka 560-8531, Japan
[2] Kaneka Corp., Torikai-nishi, Settsu, Osaka 566-0072, Japan
[3] Graduate School of Eng. Sci., Osaka Univ., Toyonaka, Osaka 560-8531 Japan
[4] JAERI, Mikazuki-cho, Sayo-gun, Hyogo 679-5148, Japan
[5] JASRI, Mikazuki-cho, Sayo-gun, Hyogo 679-5198, Japan

E-mail: ona@djebel.mp.es.osaka-u.ac.jp

Abstract. The highly oriented graphite film was used for the starting material. The large amount of hexagonal diamond phase was recovered to ambient conditions after high pressure and high temperature treatment.

1. Introduction
Hexagonal diamond was firstly synthesized by Bundy and Kasper about 40 years ago under high pressure and high temperature. The hexagonal diamond was recovered at ambient condition which was reported by X-ray diffraction measurement [1]. However the diffraction peaks were very weak and unclear. Yagi and Utsumi extrapolated the d values of between 5 and 25 GPa to ambient condition and found the transition from a graphite to a hexagonal diamond is very sensitive to the nature of the starting material and to applied pressure condition [2]. Knight obtained the Raman spectra with a peak at 1315 ~ 1326 cm^{-1} of hexagonal diamond recovered from a shock loading experiment [3]. These measurements of hexagonal diamond were performed with sample of μm-size hexagonal diamond mixed with cubic diamond. The mm size of single phase of hexagonal diamond has not been synthesized yet, and the physical properties remain unclear; the X-ray diffraction profile was not clearly obtained at ambient condition. In this study we investigate the stability of hexagonal diamond by Raman scattering and X-ray diffraction measurements with recovered sample of highly crystallized graphite films as the starting material.

2. Experiment
The highly crystallized graphite films were prepared by heat treatment of carbonized polyimide films [4]. According to an image of transmission electron microscope (TEM), it was revealed that this graphite films are highly oriented along to the c axis. The films were cleaved into ~5 μm thick and cut out disks of 80 μm in diameter.

We used diamond-anvil cell for applying pressure and the pressure determined by the ruby fluorescence method. The diameter of the diamond-anvil is 300 μm and the sample chamber of 150 μm was prepared in rhenium gasket. Ar was used as the pressure transmitting medium. The recovered sample was obtained after some pressure cycles between 30 and 70 GPa. The Raman spectroscopy

© 2008 IOP Publishing Ltd

was performed with an Ar ion laser (λ = 514.5 nm). The X-ray diffraction measurements were performed at BL10XU at SPring-8 in combination with monochromatic synchrotron radiation (λ = 0.414 Å) and an imaging plate detection.

3. Results and discussion.

With increasing pressure, the graphite film became gradually transparent from about 30 GPa. By a Raman spectrum, the transformation was found to begin at pressure slightly below 20 GPa. The pressure was increased to 70 GPa and then decreased to 30 GPa. The transparency at 30 GPa were compared relative to the number of the pressure cycle (Figure 1). The transparency may relate to the production of the hexagonal diamond, however no significant difference was observed. After the pressure was release, the sample reversed to opaque. The recovered sample shows only graphite profiles at Raman scattering and X-ray diffraction measurement.

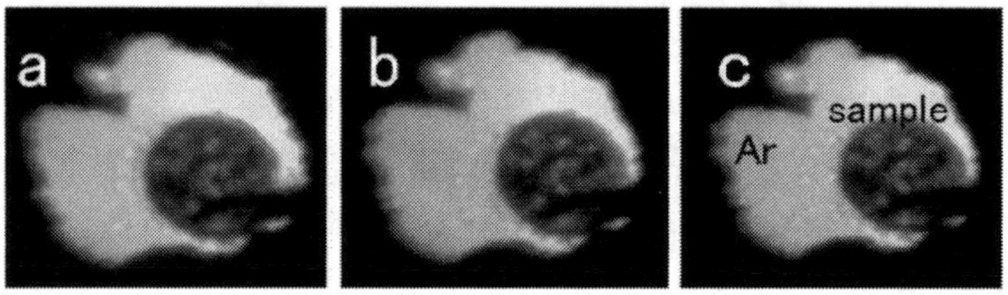

Figure 1 photographs of the sample at 30 GPa
(a) 1st cycle (b) 2nd cycle
(c) 3rd cycle

The sample was heated at 20 GPa by a yttrium-aluminum-garnet (YAG) laser. The laser was focused at about 10 μm and the temperature was estimated as 600 ~ 800°C by the color of the radiation. Pressure gradually decreased and quenched to ambient from 5 GPa.

Figure 2 shows the Raman spectra of the recovered sample at different measuring points. At a point (a), sample had cubic diamond phase. But at another point (b), there was a broad peak at 1300 ~ 1350 cm^{-1} corresponding to hexagonal diamond. The recovered sample was the mixture of cubic and hexagonal diamond.

Figure 3 shows the X-ray diffraction profiles of the recovered sample integrated from the diffraction image. There are strong peaks assigned to the hexagonal diamond. The recovered sample contained hexagonal diamond with much larger volume than cubic diamond.

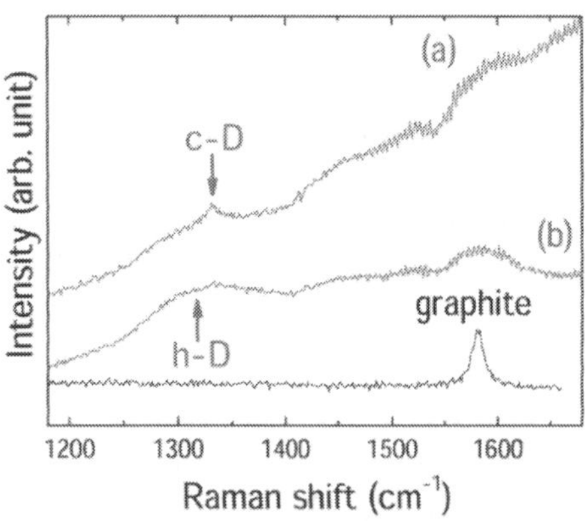

Figure 2 Raman spectra (ambient conditions) of starting sample (black) and recovered sample (red)

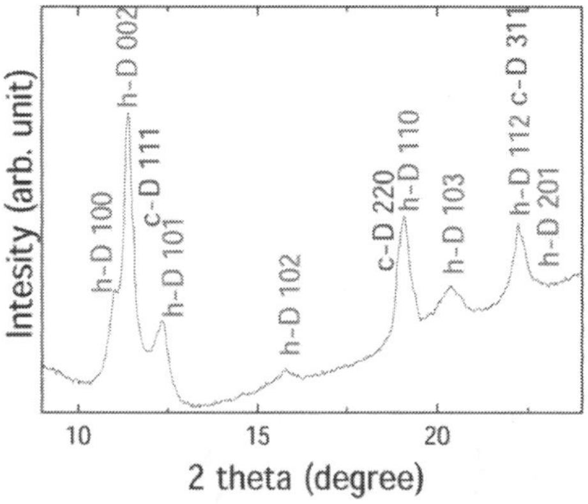

Figure 3 X-ray diffraction profiles of the recovered sample

TABLE I. *d* value and lattice parameter of hexagonal diamond

hkl	obsereved[a)]	calculated[b)]	Yagi[c)]	Bundy[d)]
100	2.165	2.172	2.176	2.19
002	2.089	2.088	2.086	2.06
101	1.933	1.927	1.929	1.92
102	1.504	1.505	1.506	1.50
110	1.251	1.254	1.257	1.26
103	1.172	1.172	1.172	1.17
112	1.076	1.075	1.076	1.075
201	-	1.050	1.053	1.055
	a	2.508Å	2.513Å	2.52Å
	c	4.175Å	4.171Å	4.12Å

[a)] from fitting the peaks with Lorenz function

[b)] from the lattice parameters obtain in this study

[c)] extrapolated from the *d* value at high pressure [2]

[d)] Quenched sample [1]

The *d* values of hexagonal diamond at present experiments are summarized in Table I. The *d* values were calculated from h-D (102) and (103) peak isolated from other peaks. These values are in good agreement with others [1, 2]. We summarize that this is the first X-ray diffraction profile of hexagonal diamond clearly obtained at ambient condition. Hexagonal diamond may be stabilized by cubic diamond contamination. TEM observation of quenched sample at ambient condition is important for next step.

Acknowledgments
This work is supported by the 21st Century COE program (G18) by Japan Society of the Promotion of Science and Grant-in-Aid for Creative Scientific Research (15GS0213), MEXT.

References

[1] Bundy F P and Kasper J S 1967 *J. Chem. Phys.* **46** 3437
[2] Yagi T, Utsumi W, Yamanaka M, Kikegawa T and Shimomura O 1992 *Phys. Rev. B.* **46** 6031
[3] Knight D S and White W B 1989 *J. Mater. Res.* **4** 385
[4] Murakami M Watanabe K Yoshimura S 1986 *Appl. Phys. Lett.* **48** 1594

Spatial distribution of residual stress in diamond-silicon carbide composites

T W Zerda[1], M Wieligor[1], T Ungar[2], and B Palosz[3]

[1] Physics Department, TCU Box 298840, Fort Worth, TX 76133, USA

[2] Department of Materials Physics, ELTE, Budapest, Hungary

[3] Institute of High Pressure Physics, Polish Academy of Sciences, Warsaw, Poland

[1] email: t.zerda@tcu.edu

Abstract. Raman spectroscopy and x-ray diffraction whole profile analysis were used to analyze strain in diamond crystals in diamond-silicon carbide composites. The composites were obtained by the infiltration technique from diamond and silicon powders at 8 GPa and 2173 K. Frequency shifts of Raman peak of diamond were used to calculate residual stress and then to draw maps of stress distribution on the surface of diamonds. Large stresses were formed near contact points between diamond crystals. Analysis of profiles of three x-ray reflections of diamond provided information on dislocations and crystallites sizes. After sintering crystallites which scattered x-ray coherently had small sizes which depended on the dimensions of diamond crystal used in the sintering process.

1. Introduction

Diamond-silicon carbide composites are produced from micron size diamonds and silicon by the infiltration technique at high pressures, usually above 5 GPa and temperatures above 1500 K [1,2]. Silicon carbide, SiC, is formed as a result of a chemical reaction between carbon from diamond and silicon. It is assumed that the formation of SiC takes place under hydrostatic conditions. However, when temperature is reduced and pressure is removed non-hydrostatic conditions develop and both diamond and SiC exhibit strains. The goal of this research is to investigate strains in the diamond phase by Raman spectroscopy and compare the results with the analysis of x-ray line broadening.

2. Experimental

Two sets of samples were produced. Composites A were made by the high-pressure silicon infiltration technique. Synthetic diamond powder of crystal size 40-60 μm made by General Electric, and silicon powder of particle size less than 44 μm from Alfa Aesar (Ward Hill, MA) were used to sinter samples at 8 GPa and 2173 K. Samples B were made from natural diamond powders (Microdiamant AG, Switzerland) of crystal size 2-5 μm and silicon powder of particle size of about 30 nm also by the infiltration technique at 8 GPa and 2173 K. The sintering conditions ensured that during the reaction diamond was the stable form of carbon and did not graphitize. The weight ratio of carbon to silicon in the initial mixtures was 86:14 and the sintering time was always 30 s.

© 2008 IOP Publishing Ltd

Journal of Physics: Conference Series **121** (2008) 062007 doi:10.1088/1742-6596/121/6/062007

The silicon and diamond powders were packed inside a cylindrical graphite heater and then were placed inside the center of a gasket which was positioned between two toroid anvils and compressed by a hydrostatic press. The experiments were run according to the following protocol. The pressure was raised to 8 GPa. Next, temperature was increased from room temperature to 2173 K and the system was kept at that temperature for 30 seconds. Finally, temperature was decreased to the room level and pressure was released. The estimated maximum deviation of temperature was ±50 K, and of pressure about ±0.5 GPa. Knoop hardness for composites A was 52 GPa and composites B 42 GPa.

High resolution powder diffraction data were collected on the X beamline at the National Synchrotron Light Source, Brookhaven National Laboratory. A double silicon (111) crystal monochromator was used to obtain a monochromatic beam of 0.069960 nm wavelength. Diffraction data were collected in the range from 13° to 98° with step size of 0.02°.

Raman spectra were recorded at room temperature using a Raman microimaging system consisting of an Olympus confocal microscope attached to a TripleSpec from Spex, equipped with a grating with 1800 grooves per mm and a charge-coupled-device detector. The spectral resolution was about 1 cm^{-1}. The 514 nm line of an Ar^+ laser was used as the excitation line. A x100 magnification objective was used, which allowed spectra to be recorded from a small section of diamond crystals of diameter less than micron and depth 2 microns. To prevent sample heating the laser power at the sample was less than 40 mW and the signal acquisition time was 30 s.

The SEM microscope, JEOL SM-6100, was used to examine surfaces of the sintered specimens. The probe current ranged from 10^{-12} to 10^{-6} A and magnifications varied from 750 to 4500. For the mapping experiment, only crystals of very smooth surfaces and well resolved boundaries were selected. Raman mapping was done by collecting data from evenly spaced points on the surface of each crystal. For each specimen at least five different crystals were scanned. Usually those crystals were in close proximity, often touching each other.

3. X-ray diffraction profiles analysis

The microstructure of the specimens was studied by x-ray diffraction profile analysis. Seven peaks of diamond (111, 311, 331, 440, 531, 620, 642) were analyzed. The Multiple Whole Profile (MWP) fitting procedure was applied. This procedure is described in Ref. 3. Briefly, Fourier transforms of experimental profiles are calculated and compared with theoretical predictions calculated for model crystallites. We assumed that the shapes of diamond crystallites were spherical and their diameters varied but their distribution was given by the log-normal distribution. Strains were assumed to be due to dislocations. The MWP analysis provided information on volume weighted mean crystallite size, $<x>$, and density of dislocations, ρ. The crystallite size is a measure of the area scattering coherently and is not necessarily equal the size of the crystal grain. An example of the fit between experimental and theoretical data is shown in figure 1. Not surprisingly, a larger $<x>$ value was measured for sample A, 106 nm, than in sample B, 41 nm. The density of dislocations, ρ, in the specimen with smaller grains, sample B, was 33×10^{14} m^{-2}, and in specimen A $\rho = 9 \times 10^{14}$ m^{-2}.

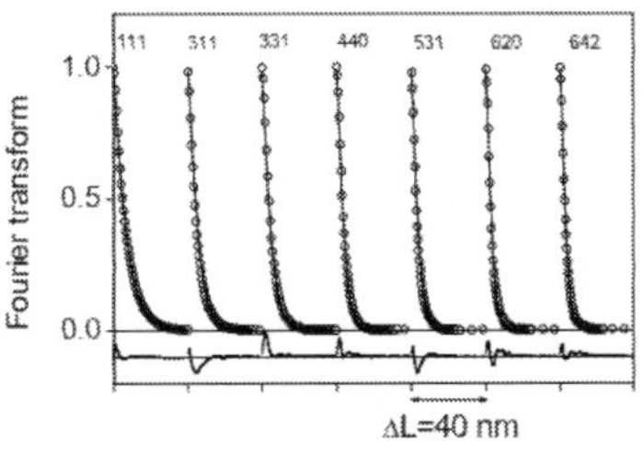

Figure 1. Measured (open circles) and fitted (solid line) Fourier transforms of the diamond diffraction profiles for the composite B.

4. Raman mapping

The Raman peak of diamond is a triple degenerate mode centered at 1332 cm^{-1}. Under tensile stress (or compression) the peak splits into two or three components which shift from that position toward lower (or higher) frequencies, respectively [4,5]. The magnitude of the shift is proportional to the applied stress and depends on the orientation of the crystal with respect to the incident laser, laser polarization, and direction of applied forces.

We recorded Raman spectra of diamond at various locations of the laser on the exposed surfaces of composites. In most cases the peak divided into two components, a doublet and a singlet. Both components shifted toward higher frequencies. The shifts were small and both components overlapped and formed a single contour. To resolve them we applied a fitting procedure, an example is shown in figure 2.

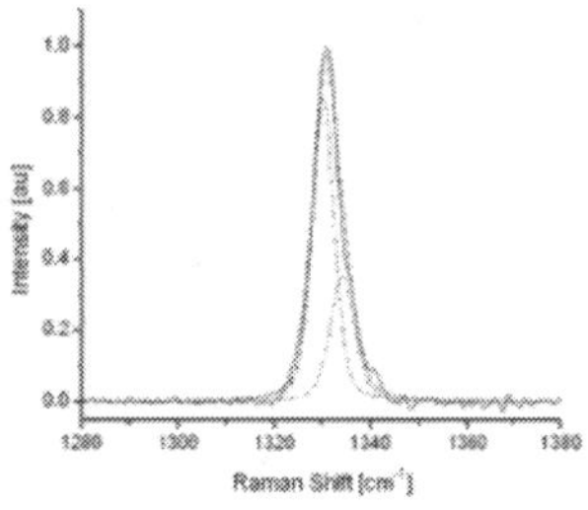

Figure 2. A typical experimental Raman band shape for diamond in composite A, solid line. The two components were assumed to be a mixture of Lorentzian and Gaussian functions and their shapes were found by the best fitting procedure. Their sum is given by the dotted line.

Since it was impossible to identify orientations of diamond crystals in composites, we evaluated the magnitude of the residual stress in diamonds not from shifts of individual components but by analyzing the average position of the maximum, according to the formulae of Ralchenko et al. [6]:

$$\sigma = -0.567\ (\omega_m - \omega_o) \qquad (1)$$

where ω_m is the medium position between the two components, ω_m=0.5 ($\omega_1 + \omega_2$), and ω_o =1332 cm^{-1}. Maxima ω_1 and ω_2 were found by the fitting procedure illustrated in figure 2.

Raman spectra were obtained from spots smaller than 1 micron in diameter in steps of one micron. At each steps we calculated σ and than plotted distribution of σ on the surface of studied diamonds. Examples are shown in figures 3 and 4. For sample A which was made from large diamond crystals we were able to record spectra from about 30 different locations on the surface of individual

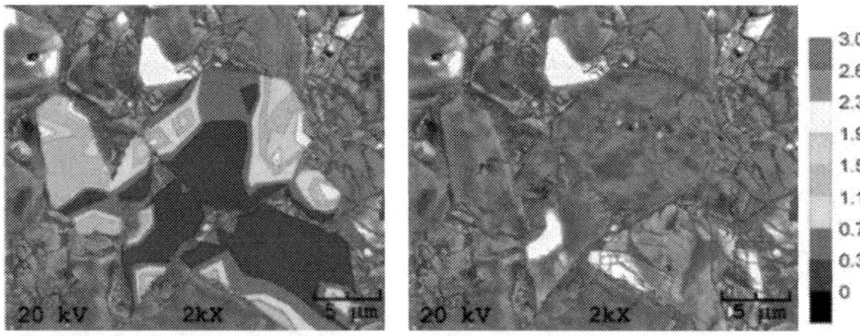

Figure 3. Stress distribution in diamond crystals for the specimen A. SEM image of the same area is shown to the right. Values of stress in GPa are given by different colors and the color code is explained by the vertical bar.

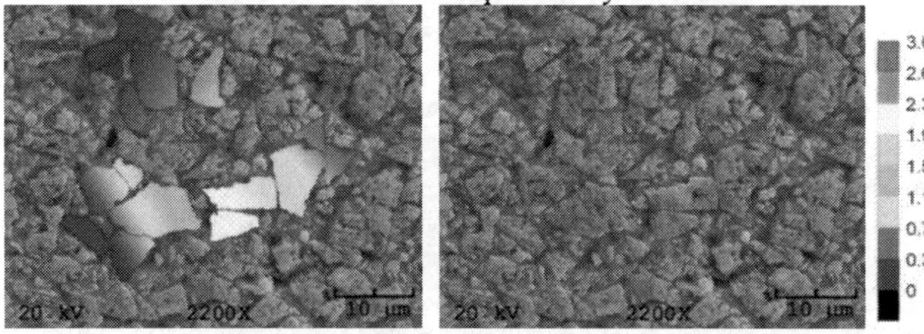

Figure 4. Stress distribution in diamond crystals for the specimen B. SEM image of the same area is shown to the right. Values of stress in GPa are given by different colors and the color code is explained by the vertical bar.

diamonds. For sample B, grain sizes were much smaller and we were able to record only a few spectra per crystal.

5. Discussion

Raman experiment indicated presence of large strains in the specimens made from diamond powder of larger grain sizes. Distribution of strain within the crystals was not uniform and the largest strains were observed near points of contact between two diamond crystals. A similar trend was observed in specimen B made with smaller grain sizes. However, the magnitude of strain was noticeable smaller. The MWP analysis showed that the density of dislocations in sample B was about three times greater than in sample A.

The bonding in the materials investigated here are mainly of covalent type. Slip by dislocations and/or plastic behavior, especially plasticity is practically absent, the samples are brittle ceramics. This means that crystal defects, especially dislocations and twin boundaries, disturb chemical bonds and therefore weaken the sample. A search of manufacturing conditions which result in minimum microstresses is one of the goals of our research. The residual stress could be caused by differences in densities of silicon, diamond, and silicon carbide, mismatch in thermal expansion coefficients of diamond and silicon carbide, and different values of bulk modulus of these two constituents of the composites. A detail discussion of relationships between residual stresses and their possible origins, however, exceeds the scope of this brief report and is the subject of a separate more complete paper under preparation.

Acknowledgements

This work was partially supported by a grant NSF DMR 0502136 (TWZ), Hungarian Scientific Research Fund, OTKA, grant No. T-046990 (TU) and Polish Ministry of Education and Science grant 3T08A 02029 (BP). The use of the National Synchrotron Light Source, Brookhaven National Laboratory, was supported by the US Department of Energy, Office of Science.

References

[1] Qian J, Zerda T W, He D, Deamen L and Zhao Y 2003 *J. Mater Res.*, **18** 1173
[2] Voronin G A, Zerda T W, Qian J, Zhao Y, He D and Dub SN 2003 *Diamond Relat Mater.* **12** 1477
[3] Gubicza J, Ungar T, Wang Y, Voronin G, Pantea C and Zerda T W 2006 *Diamond Relat Mater.* **15** 1452
[4] Anastassakis E 1999 *J. Appl. Phys.* **86** 249
[5] Grimsditch M H, Anastassakis E andCardona M 1978 *Phys. Rev.* B **18** 901
[6] Ralchenko V G, Smolin A A, Pereverzev V G, Obraztsova E D, Korotoushenko K G, Konov V I, Lakhotkin Y V and Loubnin E N 1995 *Diamond Relat. Mater.* **4** 754

Journal of Physics: Conference Series **121** (2008) 072001 doi:10.1088/1742-6596/121/7/072001

Building zero isotherm of hydrogen isotopes taking into account experimental results of isentropic compression up to pressures of several megabar

G.V. Boriskov, A.I. Bykov, N.I. Egorov, M.I. Dolotenko, V.N. Pavlov, V.I. Timareva

Russian Federal Nuclear Center — VNIIEF, 37 Mira Ave., Sarov, Nizhny Novgorod region, 607188, Russia

Abstract. An isentropic compression of substances using ultra–high magnetic field makes it possible to effectively reach "cold" pressures of several megabar in a studied substance. The process is realized in a device based on magneto–cumulative generator MC–1. Arrangement of the experiments, where the isotherm of "cold" compression of hydrogen isotopes is studied, is described in the paper. Results of preliminary data analysis obtained at solid protium and deuterium compression are also presented. Achieved densities correspond to the pressures higher than 5 Mbar. A method of zero isotherm building, taking into account gradient and thermal corrections, is described. Errors, with which positions of experimental points in $P-\rho$ plane are determined, are associated mostly with a precision of sizes calculation of a reference and studied samples using x–ray image, obtained in the experiment. Deviations of the built points from the curve, extrapolating the experimental data on the diamond anvils to the multi–megabar range, lie within experimental errors. The zero isotherm, approximating the obtained nowadays-experimental data, is compared with several first-principles calculation results.

An interest to thermodynamic and kinetic properties of hydrogen within a wide range of pressures and temperatures is caused on one hand by their wide spread in the world, the role that they play in structure of stars and giant planets, and on the other hand by their practical importance as future basic energy sources (hydrogen fuel elements, possibility of superconducting state at room temperatures, etc.).

To get the equations of state of hydrogen both static [1] and dynamic [2] methods are used, both having advantages and disadvantages. In the former case a regime of an isotherm compression is usually realized at comparatively low temperatures ($T < 1000\,\mathrm{K}$) up to the pressures not exceeding 1.5 megabars [3]. On the other hand, in the dynamic (shock-wave in the majority cases) regime the temperature could reach tens thousands degrees. But in this case a "cold" pressure does not exceed half of megabar (see, for example, [4, 5]).

The goal of this work is experimental plotting of the "cold" part of the equation of state of a solid protium and solid deuterium within the pressure range from 1 Mbar to $4 \div 5$ Mbar. (Initial temperature of the studied substance should be less than its melting point at atmospheric pressure: < 14 K for H_2 and < 18 K for D_2.) Experimental building of "zero" isotherms of hydrogen isotopes in the area of ultra-high pressures is one of the important and interesting tasks of a solid state physics with high energy density. To solve the task we used the dynamic technique of isentropic compression with the help of the ultra-high magnetic field pressure. Presently this research technique is, probably, the only one that allows obtaining the parameters of the "cold"

© 2008 IOP Publishing Ltd

curve of easily compressed substances in the megabar pressure range. Solid isotopes of hydrogen belong to these substances. This technique could be realized in a device, where a protium or deuterium, being in condensed initial state ($T_0 \approx 5$ K, $P_0 \approx 1$ bar), is isentropically compressed up to pressures of several megabar with an ultra-high magnetic field created by the magneto-cumulative generator [6].

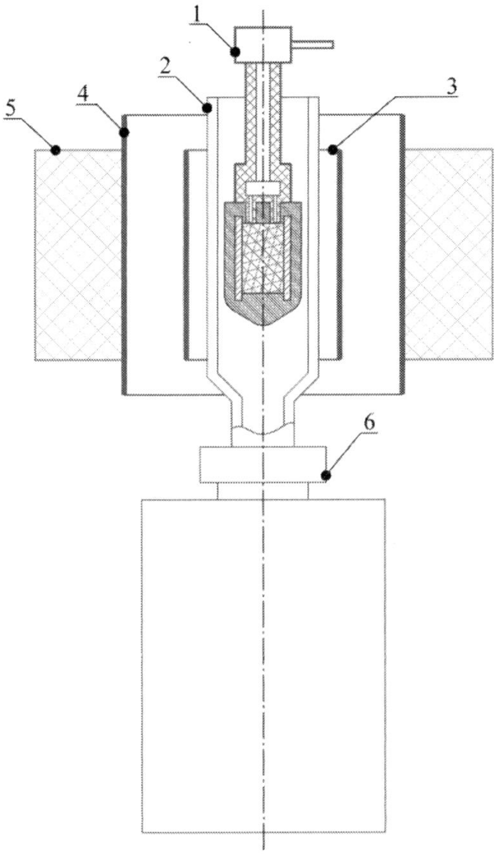

Figure 1. Compression device sketch: 1 — cryogenic container with the compression chamber, 2 — cryopiping, 3 — the second cascade of the generator, 4 — solenoid of the generator seed field, 5 — ring HE - charge, 6 — cryogenic vessel.

The basic device components are the following (see Fig. 1): 1) magneto-cumulative two-stage generator MC–1 [7] as a pressure source; 2) a compression chamber, located in the generator cavity; 3) a cryogenic device that provides the required hydrogen isotope in a condensed phase and retains it in this state all the time of the test performing. The MC-1 generator includes a solenoid of the initial magnetic flux, a source of solenoid feed (capacitor battery), a ring high-explosive (HE) charge as a main energy source. The compression chamber is formed by a copper tube and massive end plugs. Layers of the studied and reference materials, separated from each other and from the tube by contrasting shells, have to be located coaxially in the compression chamber [8]. The cryogenic device [9] consists of a vessel for coolant (liquid helium) storage and evacuated cryogenic piping.

The operation of the device takes place the following way. Before a test the compression chamber is evacuated, connected to a gas holder with gaseous hydrogen isotope, and inserted into a cryogenic piping, located along the MC–1 generator axis. Under effect of the small excess pressure, created in the cryogenic vessel, the helium goes upwards on the cryogenic line, and its low-temperature vapors (and then the helium itself) gradually cool the compression chamber. In this case the hydrogen isotope that fills the chamber passes first into liquid, and then into solid state. The investigated substance temperature is measured by calibrated semiconductive probes. (In every experiment the initial condensed gas temperature was $T_0 = 5 \div 7$ K, and in accordance with the performed calculations did not exceed the room temperature during a compression process.) A sinusoidal (period quarter $\approx$ 70 μs) magnetic field, directed along the generator axis, is created in a solenoid cavity (in a gap between the solenoid and compression chamber) at discharge of the high-power capacitor battery on the solenoid. A ring HE–charge is detonated in such a way that the cylindrical converging shock-wave front, formed after detonation, reached the inner surface of the solenoid (the 1st cascade of the generator) at the time, when the magnetic field has maximum value B_0. The letter is taken as an initial field of the generator. A shock wave causes implosion of the current-carrying liner-solenoid. (The second cascade of the generator is necessary to keep symmetry of the current carrying shell [10]). Since the implosion is rather fast ($< 20\,\mu$s), the main part of the magnetic

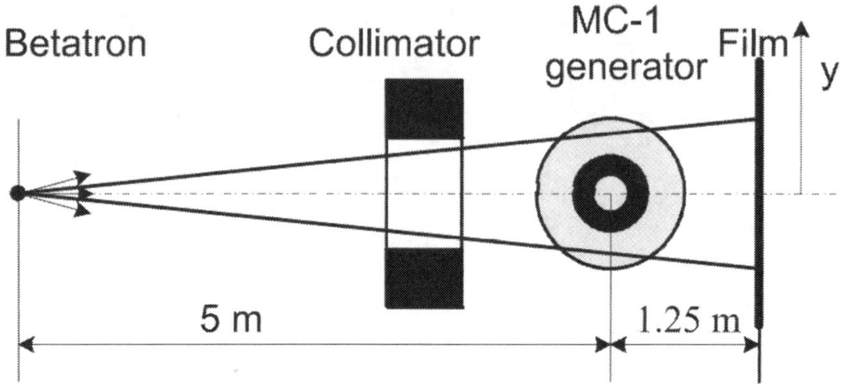

Figure 2. Radiography scheme (top view)

flux, captured by the shell, is preserved. In the result the magnetic field in the generator cavity amplifies up to the megagauss values. During the operation time of the device this pulsed ultra-high magnetic field, created by MC generator, cannot penetrate inside a copper shell of the compression chamber (thickness of a skin-layer is less than the shell thickness at any time moment of the process) and makes on its outer surface the magnetic pressure $H^2/8\pi$. Under its effect the copper tube implodes and smoothly, without formation of the shock waves, compresses the substances, located in it, and thus creates the pressures of megabar range in them.

When performing the test pulsed x-ray imaging of the central part of the device is carried out at setup time moment of the compression process. Figure 2 presents the scheme of the x-ray imaging. An x-ray source – betatron [11] – rays the MC–1 generator and the compression chamber inside it by collimated horizontal beam of high and uniform intensity. The image is recorded by a set of x-ray films, having the amplifying screens. Compressions and densities of the substances inside the chamber are determined based on the results of the obtained images analysis. It was shown [12] that errors of all necessary sizes determination are small and do not exceed several percents. Based on the x-ray images of the chamber and the samples inside it we determine compression of the investigated sample and the reference sample. Having the reference sample density and using its known isentropic curve, one can determine pressure in the chamber. Initial densities are calculated accurately based on the sample initial temperature measured before the test. Aluminum is the best reference material, for which in accordance with static and shock-wave experimental results the most representative statistical data, corresponding to the range of the required pressures [13, 14, 15], has been stored and analyzed. Used "cold" Al isentropic curve corresponds to the paper [14], based both on obtained by the authors [14] data on Al static compression on diamond anvils up to a pressure of 2.2 Mbar, and on obtained earlier shock-wave compression up to 10 Mbar [16, 17]. In order to determine the sizes of the compressed samples with the x-ray image with acceptable precision, there is a thin layer of high-density material, located on the boundary, separating the samples from each other and from the compressing tube. The material consists of substances with large serial number in the periodic table. In our case this is a tungsten based alloy (WNiCu or WNiFe), containing 95% of tungsten.

The described method of "cold" isentrope plotting provides satisfactory results, when pressure gradients in the operating zone of the compression chamber are small, and the x-ray images obtained in the experiments make it possible to determine the necessary sizes with relatively high precision. One can fulfill these two conditions by optimizing the geometry and operating regime of the experimental device. For this purpose we performed a series of preliminary calculations of the compression process [18, 19], varying longitudinal and transversal sizes of the chamber and compressed samples, manner of their mutual location, and also the value of the initial magnetic field of the generator ($B_0 = 130 \div 190\,\mathrm{kG}$). The calculations were carried out with a one-dimensional magneto-hydrodynamic (MHD) and quasi-two-dimensional hydrodynamic codes (see [18]). These codes were used earlier for simulation of the magneto-cumulative generator operation and devices, based on the MC generator [20, 21]. Extrapolation of the "cold" isentropic curves, built on the data of the paper [3], was used for description of the H_2 and D_2 compression in the multi-megabar range. We also carried out calculations using more "rigid" curves, proposed in [22, 23]. This "rigidness" did not effect selection of the optimum values of the compression device parameters. Besides, several additional calculations were carried out for both isotopes. Now the Al equation of state (EOS) was selected not in the form, corresponding to the paper [14], but in two other forms: in the form proposed in the known paper [13], and in the form of so-called "equation of state with limiting density" [14]. These changes did not affect the optimization. However, after performance of the experiments we found out that their results matches the calculations better at description of the Al compressibility in accordance with [14].

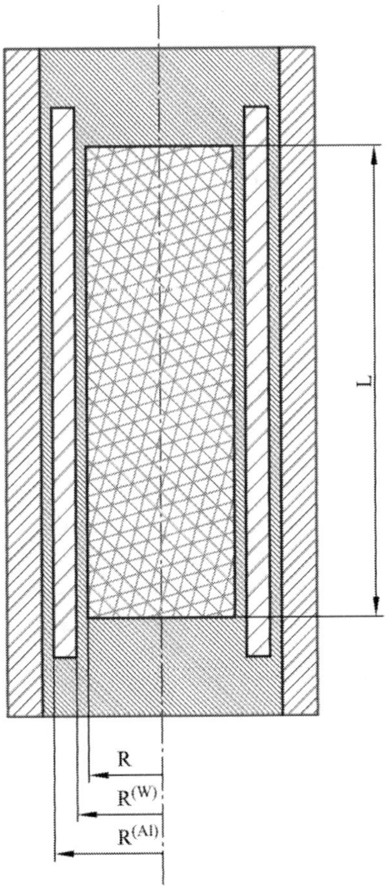

Figure 3. Scheme of the layers location in the compression chamber.

The compression chamber geometry, presented in Figure 3 appeared to be optimum in both cases with H_2 and D_2. In this geometry the hydrogen isotope is located in the center, and the aluminum reference, separated from the studied sample and from the copper tube by the contrasting layers, is located between the hydrogen isotope and the compressing tube. The optimum value of the B_0 lies near 155 kG.

Based on the calculation results we selected the following values of the initial transversal sizes of the compression chamber and its elements: the compression chamber diameter is equal to 21.8 mm, thickness of both contrasting layers is 0.15 mm, thickness of Al layer is 2 mm, diameter of the studied sample of the hydrogen isotope is 13.7 mm. Speaking about the longitudinal sizes, we found out that for the performed experiments the most suitable is the compression chamber that has the distance between its ends, matching the length of the uniform magnetic field region and equal to 100 mm. (Deviation of the chamber length to the lesser side from the optimum leads to shortening of the uniform compression region. Deviation to the greater lengths leads to appearance of high-speed axial flows of the studied substance from the working region). Calculations showed that in the center section of such chamber the linear masses of the compressed substances are preserved (at the distance of $\approx$ 70 mm). That is why to find out the reference and studied substance compression it is enough to know the transversal sizes of the aluminum and contrasting cylinder at room conditions and in the compression state, and also the coefficients of thermal expansion of their materials (high-precision data up to helium temperatures are available for these materials).

Figure 4, that is built based on the calculation results for the optimized compression device,

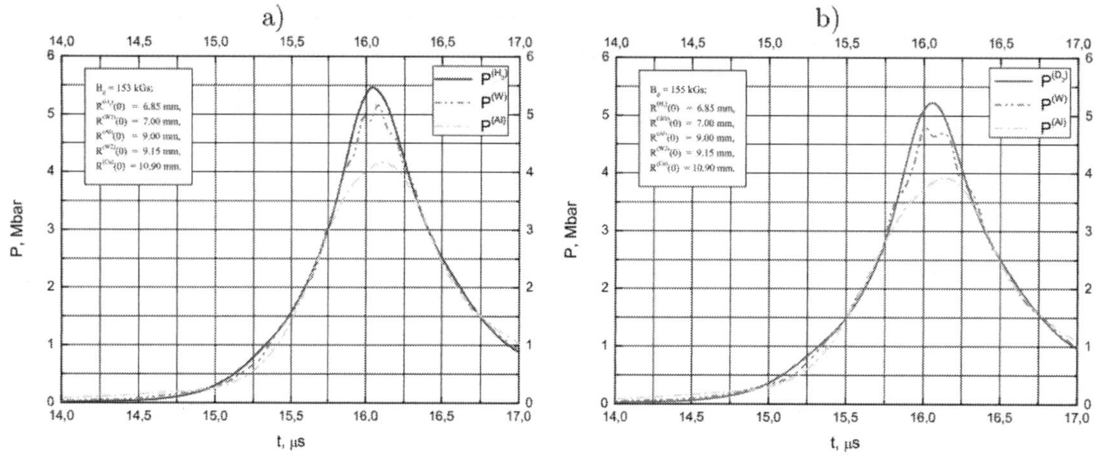

Figure 4. Pressures vs. time for optimized compression device: a) – H_2, b) – D_2.

shows possibilities of the pressure measurement. Fig. 4a presents time dependence of pressures $P^{(H_2)}(\langle\rho^{(H_2)}\rangle)$, $P^{(W)}(\langle\rho^{(W)}\rangle)$, $P^{(Al)}(\langle\rho^{(Al)}\rangle)$ in H_2, contrasting layer and Al, correspondingly. Here $\langle\rho\rangle$ – sample density, averaged on the volume, in the central region of the compression chamber, where the sample is compressed uniformly along the axis. Similar dependencies are presented in Fig. 4b for D_2. (Time in Fig. 4 is counted from the beginning of the magnetic flux compression of the MC–1 generator). One can see that there is a rather large time interval on a fore and back front of the compression pulse – so called "working section", when the pressures in the studied isotope and Al have small difference. During analysis of the experimental data this difference could be considered by application of calculation based correction. It is also important that the pressure changes from $\sim$ 1 Mbar to $\sim$ 4 Mbar on the working sections, thus covering the whole pressure range that is interesting for us.

Two experimental series - one with H_2, another with the D_2 were carried out using this device. In each experiment we x–rayed the central section of the device at a moment, when the ultra-high pressure was achieved in the compression chamber. (We tried to hit the working section of the compression pulse). The image was made with several films. Based on the results of their analysis we found out the value of the outer radius of the aluminum cylinder $R^{(Al)}$ and the inner radius of the contrasting cylinder R (see Fig. 3). The used method [12] allows us to measure these values with the precision up to hundredth of millimeters. In this case the characteristic values of the measured radii themselves lie in the range of $1.5 \div 6$ mm. Using R value we can find the cross-section of the studied sample, and, consequently, its compression. The density of the isotope is equal to $\rho = \rho_0\delta$, where δ – compression, ρ_0 – a crystalline isotope density at normal pressure and helium temperature. ($\rho_0^{(H_2)} = 0.088\,\text{g/cm}^3$, $\rho_0^{(D_2)} = 0.202\,\text{g/cm}^3$ [25, 26]). The pressure in the studied sample is determined with the formula: $P = P^{(Al)}(\delta^{(Al)})(1+\varepsilon)$. Here $P^{(Al)}(\delta^{(Al)})$ – expression for the "cold" isentropic curve of Al that corresponds to the paper [14], $\delta^{(Al)}$ - aluminum compression relatively to its normal density, ε – relative gradient correction. The value of the external radius of the contrasting shell $R^{(W)}$ that is necessary for Al compression calculation could be found from the condition of the pressures equality in the reference and in the contrasting shell: $P^{(Al)}(\delta^{(Al)}) = P^{(W)}(\delta^{(W)})$. (In Fig. 4 one can see that this condition is fulfilled with good precision on the whole working section of the compression pulse.) In this case the expression for the isentropic curve of the tungsten alloy is selected in the form, recommended in [13]. The values of the parameters – coefficients of the corresponding $D(U)$ ratio – are taken from the book [27], where they are presented for pressure interval that is interesting for us. It

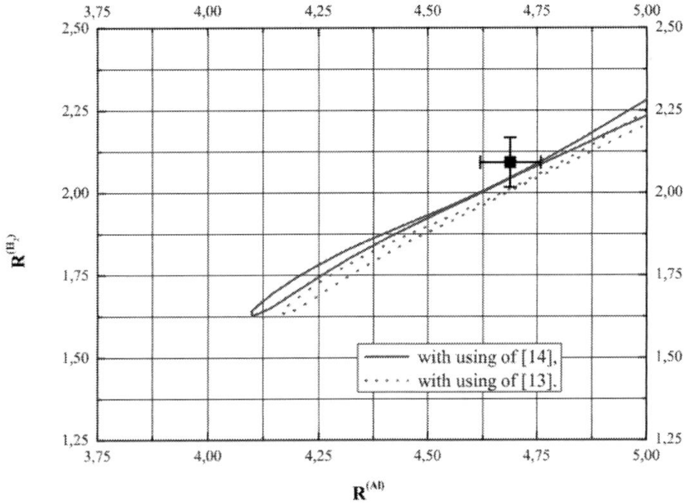

Figure 5. Experimental point and calculated curves in the plane $R^{(H_2)}$–$R^{(Al)}$.

is necessary to note that variation of these parameters within the errors limits, with which they were measured (see [27]), do not affect significantly the measurement results. Using the $R^{(Al)}$ and $R^{(W)}$ we find compression $\delta^{(Al)}$, and further the pressure $P = P^{(Al)}(\delta^{(Al)})$. Taking into account ε we find pressure P in the studied substance.

The gradient correction could be found with calculations with the following way. Using the MHD code [18] we calculate the process, taking part in the central section of the compression device. In this case the measurements of the seed field in the considered experiment are taking into account. Then, based on calculation dependencies of the radii of the studied and reference samples on time – $R_{calc.}(t)$ and $R^{(Al)}_{calc.}(t)$ – we built in the plane $(R^{(Al)} - R)$ the corresponding "trajectory" ($R^{(Al)} - R$ diagram). Here we plot an experimental point $(R^{(Al)}; R)$ showing its measurement error. Figure 5 shows this by the example of one of the experiments with H_2. (For comparison the dotted line shows similar diagram that corresponds to the calculations, where the Al is described with isentropic curve [13] that is more rigid in comparison with [14].) Then we find such "trajectory" point $(R^{(Al)}_{calc.}; R_{calc.})$ which is the most close to the experimental point. For the values of $R^{(Al)}_{calc.}$ and $R_{calc.}$, using the procedure presented above, we calculate $\delta_{calc.}$ and $P^{(Al)}(\delta^{(Al)}_{calc.})$. We find the pressure in the studied substance $P(\delta_{calc.})$ using the expression for the "cold" isentropic curve of the hydrogen isotope, built in accordance with [3]. Relative gradient correction is equal to: $\varepsilon = \left(P(\delta_{calc.}) - P^{(Al)}(\delta^{(Al)}_{calc.})\right) / P^{(Al)}(\delta^{(Al)}_{calc.})$. After building the sought curve of the "cold" compression in the first approximation, the gradient correction should be recalculated using the built curve in order to make the procedure of the correction search self-consistent.

In the used method the calculation precision of the density and pressure is determined, first of all, by the measurement precision of the transversal sizes of the aluminum and contrasting shells in initial and compressed states. The radii measurement errors ΔR and $\Delta R^{(Al)}$ of the compressed shells usually exceed on one order the measurement error of the initial sizes. That is why the latter is neglected on this stage of investigations. To find out the pressure measurement error ΔP we changed the point coordinates $(R^{(Al)}; R)$ on the value $\pm\Delta R^{(Al)}$ and $\pm\Delta R$, correspondingly. In the result there are four points in the $R^{(Al)} - R$ plane, that are slightly shifted relatively to the initial one. For each point we found the pressure P in the studied sample using the above

Journal of Physics: Conference Series **121** (2008) 072001 doi:10.1088/1742-6596/121/7/072001

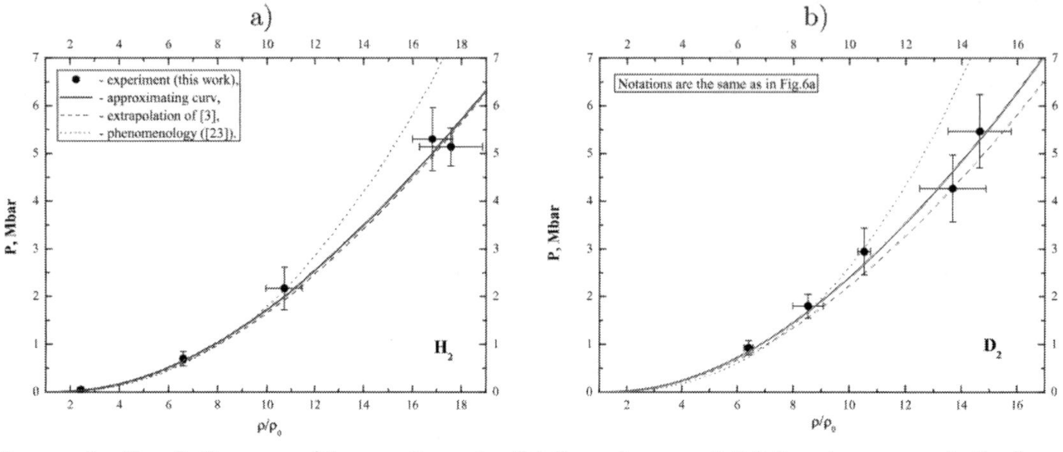

Figure 6. $P - \delta$ diagrams (the portion of solid line above $\sim$ 5.5 Mbar is extrapolation): a) – H_2, b) – D_2.

mentioned method. The difference between maximum and minimum of the obtained values was accepted as $2\Delta P$. The compression measurement error $\Delta\delta$ could be found with the same way, based on the error ΔR, with which the inner radius of the contrasting cylinder was measured.

The gradient correction for most points appeared to be relatively small – not more than the error interval $2\Delta P$. But both in case H_2 and in case D_2 the correction slightly exceeded error interval for the points corresponding to the maximum and minimum of measured pressures. (It is not surprising since the pressures in these points lie outside the interval 1÷4 Mbar, that correspond to the working section of the compression pulse.) In order to find out the effect of the hydrogen isotope EOS, used in the calculations, on the gradient correction, we calculated it one more time. But now we were based not on the zero isotherm [3], but using data from the paper [23], which correspond to more rigid "cold" curves of the hydrogen isotopes at the pressures > 1 Mbar. It was found out that in the cases when the initially calculated correction was small, it remained the same; for the mentioned extreme points the absolute value of the correction exceeded $2\Delta P$ again, but the relative correction ε in this case did not change significantly. (Thus, for example, for the point, having the largest gradient correction from all multi-megabar points, the change of ε was $\sim 9\%$). Thus, varying the "cold" compression curve in relatively wide range does not effect significantly on consideration of the gradient correction.

In the result of a series of full-scale experiments on determination of the "cold" curve of hydrogen isotopes there were built five experimental points for protium and the same number - for deuterium. They are presented in Fig. 6 in the coordinates $P - \delta$. Solid line in this figure shows the curve that approximates the obtained data. It represents a sum of the static-lattice pressure and contribution of zero-point oscillations. The theory of Mie-Gruneisen-Debye (see [28, 3]) was used at construction of this curve. At first, the pressure of the zero oscillations and the thermal component (which appeared to be negligibly small due to extremely low initial temperature of the studied sample) were subtracted from the total pressure for each point. After that the contribution of the static lattice was written in the form of the Vinet function [29]: $P_{SL} = 3K_0(V/V_0)^{-2/3}\left[1-(V/V_0)^{1/3}\right] \times exp\left\{3(K_1-1)\left[1-(V/V_0)^{1/3}\right]/2\right\}$, ($V$ — molar volume). Fitting parameters have the following values: $K_0 = 0.744 \times 10^{-2}$ Mbar, $K_1 = 5.868$, $V_0 = 17.825$ cm^3/mol for H_2; $K_0 = 0.817 \times 10^{-2}$ Mbar, $K_1 = 5.959$, $V_0 = 16.875$ cm^3/mol for D_2. Dashed line in Fig. 6 shows extrapolation of the zero isotherm from [3] to the multimegabar range. Here we also dotted a similar curve suggested in [23] (for molecular phase). On the initial section – at the pressures up to 1 Mbar – all three curves almost coincide with each other.

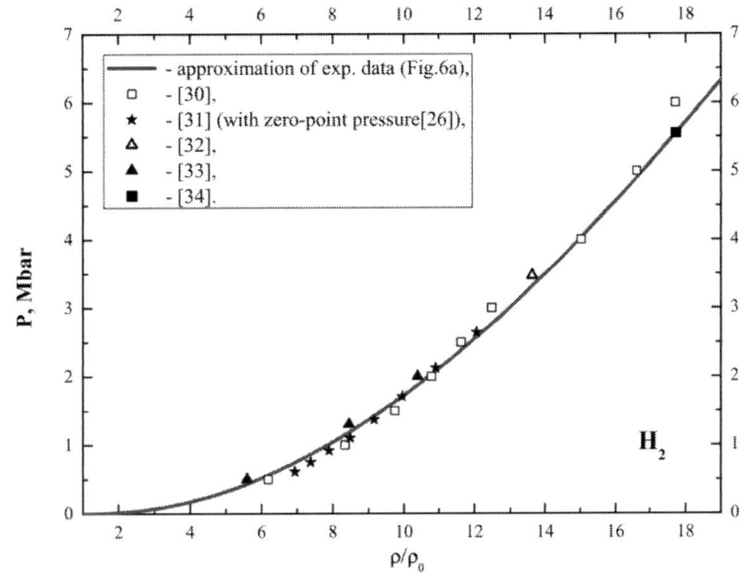

Figure 7. Comparison between experimental zero isotherm H_2 and theoretical equations of state: LDA molecular solid [30], PIMC molecular H_2 [31], QMC atomic and molecular solid [32, 33], CEIMC atomic H_2 [34].

If we consider the whole range of the studied pressures, one can see that the obtained data within the error limits agree with the zero isotherm [3]. Nevertheless, the experimental points lie slightly higher the specified isotherm at pressures of several megabars. Also one can see that the approximating curve appeared to be less "rigid" in comparison with [23].

In Figure 7 we compare our built isotherm for H_2 with the results of several first-principle calculations [30, 31, 32, 33, 34]. We think that the best matching is observed for the points obtained with the quantum Monte-Carlo [32, 33] and coupled electronic-ionic Monte-Carlo method [34].

On this stage of the investigations we did not discuss the issue of peculiarities existence on the compression curves, stipulated by possible lattice rearrangement. This is because the experimental data are not complete enough for this purpose. Thus, in Fig. 6a one can see that in order to make more definite and reliable conclusions about the course of the studied zero isotherm it is desirable to have additional protium points near 3.5 Mbar. For D_2 not excessive will be one more point near 5 Mbar, where the gradient correction and measuring errors are maximum.

In conclusion it is necessary to note that using the described method of the isentropic compression one can study the curve of the "cold" pressure of the number of other substances, consisting of the elements with small atomic number, such as helium, lithium etc.

The work is fulfilled in the frameworks of the ISTC projects #2564 and #2564.2.

References

[1] Mao Ho-kwang and Hemley R.J. 1994. *Rev. Mod. Phys* **66** 671.
[2] Nellis W.J. 2006. *Rep. Prog. Phys.* **69** 1479.
[3] Loubeyre P., LeToullec R. et al. 1996. *Nature* **383** 702.
[4] Knudson M.D. et al. 2004. *Phys. Rev. B* **69**, 144209.
[5] Boriskov G.V. et al. 2005. *Phys. Rev. B* **71** 092104.
[6] Sakharov A.D. 1966. *Uspekhi Fizicheskikh Nauk* **88** 725.

[7] Pavlovskii A.I. et al. 1980. In book: *Megagauss physics and technology, MG-II.* Plenum press: New York and London, p. 627.
[8] Matveev V.V. et al. 1984 it Pisma v ZETF **39** 219.
[9] Boriskov G.V., Bykov A.I., Egorov N.I., Pavlov V.N. 2006. *Proceedings of 2006 International Conference on Megagauss Magnetic Field Generation and Related Topics.* Santa Fe, New Mexico, USA, November 5–10, (to be published), book of abstracts, p. 34.
[10] Pavlovskii A.I. et al. 1984. In book: *Ultrahigh Magnetic Fields*/Ed. V.M. Titov, G.A. Shvetsov. M. Nauka, p. 19.
[11] Kuropatkin Yu.P. et al. 1997. In: *11th IEEE Pulsed Power Conference. Digest of technical papers*/Eds. G. Cooperstein and I. Vitkovitsky, p. 1663.
[12] Bykov A.I., Egorov N.I. et al. In: *Proceedings of 2006 International Conference on Megagauss Magnetic Field Generation and Related Topics.* Santa Fe, New Mexico, USA, November 5-10, 2006 (to be published), book of abstracts, p. 37.
[13] Altshuler L.V., Brusnikin S.E., Kuz'menkov E.A. 1987 *ZPMTF* **#1** 134.
[14] Greene R.G., Lue Huan, and Ruoff A.L. 1994 *Phys. Rev. Let.* **73** 2075.
[15] Kalitkin N.N., Kuz'mina L.V. 2000. In book: *Shock waves and extreme states of substances.* M. Nauka, p. 107.
[16] Mitchell A.C. and Nellis W.J. 1981 *J. Appl. Phys.* **52** 3363.
[17] Nellis W.J. et al. 1988. *Phys. Rev. Let.* **60** 1414.
[18] Boriskov G.V. et al. 2006. In book: *VIII Hariton's Readings on Problems of High Energy Physics.* Sarov, p. 509 and p. 516.
[19] Boriskov G.V. et al. 2007. In: *Proceedings of the Eleventh International Conference on Megagauss Magnetic Field Generation and Related Topics, Megagauss XI. London, 10-14 September 2006)*/Ed. Ivor Smith and Bucur Novac. Loughborough, UK. (Electronic version)
[20] Pavlovskii A.I., Karpikov A.A., Mamyshev V.I. 1990. In book: *Megagauss Fields and Pulsed Power Systems*/Ed. V.M. Titov and G.A. Shvetsov. N.Y. — L.: Nova Science Publ. p. 21.
[21] Aseeva V.V., Boriskov G.V., and Panov A.I. 2004. In book: *Megagauss Magnetic Field Generation, Its Application to Science and Ultra-High Pulsed-Power Technjlogy (Proceedings of the VIII International Conference on Megagauss Magnetic Field Generation and Related Topics, October 18–23, 1998. Tallahassee, Florida, USA)*/Ed. Hans J. Schneider-Muntau. Singapore: World Scientific Publ., p. 553.
[22] J. van Straaten and Silvera I.F. 1988 *Phys. Rev.* **B37** 1989.
[23] Kopyshev V.P., Urlin V.D. 2000. In book: *Shock waves and extreme states of substances.* M. Nauka, p. 297.
[24] *Methods of materials properties investigation at intense dynamic loads* /Monography under general edition of Dr. M.V.Zhernokletov, Sarov, 2003, p. 181.
[25] Ishmaev S.N. et al. 1983. *Zh. Eksp. Teor. Fiz.* **84** 394 [*Sov. Phys. JETP* **57** 228 (1983)].
[26] Ishmaev S.N. et al. 1985. *Zh. Eksp. Teor. Fiz.* **89** 1249 [*Sov. Phys. JETP* **62** 721 (1985)].
[27] *Experimental data on shockwave compression and adiabatic expansion of condensed substances*/Scientific publication under general edition of R.F. Trunin, Sarov, 2001, p. 209.
[28] Hemley R.J. et al. 1990. *Phys. Rev.* **B42** 6458.
[29] Vinet P. et al. 1986. *J. Phys.* **C19** 467.
[30] Barbee T.W. et al. 1989. *Phys. Rev. Let.* **62** 1150.
[31] Kaxiras E. and Guo Z. 1994. *Phys. Rev.* **B49** 11822.
[32] Natoli V., Martin R.M., Ceperley D. 1993. *Phys. Rev. Let.* **70** 1952.
[33] Natoli V., Martin R.M., Ceperley D. 1995. *Phys. Rev. Let.* **74** 1601.
[34] Pierleoni C., Ceperley D.M., and Holzmann M. 2004. *Phys. Rev. Let.* **93** 146402.

Calculation of γ–quanta passage through substance with Monte-Carlo method at x-ray images simulation

G.V. Boriskov, A.I. Bykov, A.R. Volodko, N.I. Egorov, V.N. Pavlov, A.B. Ronzhin

Russian Federal Nuclear Center — VNIIEF, 37 Mira Ave., Sarov, Nizhny Novgorod region, 607188, Russia

Abstract. Software, developed in RFNC–VNIIEF for x-ray images simulation using Monte-Carlo method is described. The software is a part of an x-ray method used for investigation of an equation of state (in this case hydrogen isotopes: protium and deuterium) in a megabar pressure range. Interaction of γ–quanta with a substance is considered. Effect of a scattered radiation on the x-ray images formation is estimated.

Introduction

During several last years RFNC–VNIIEF carries out the works on experimental investigation of the equation of states (EOS) of the hydrogen isotopes in the low temperatures and ultra-high pressures range. During these works we solved several tasks. A compression device, based on a magneto-cumulative generator [1] was designed and developed. "Cold" pressures of 5 and more megabars could be reached in this device. Its parameters were optimized in order to decrease as much as possible pressure gradients in the compression chamber in the moments, when the pressures, required for the experiments, [2] are reached. Pulsed radiography method of a density measurement of compressed substances, based on a betatron [3], was developed. The software for the x-ray images calculation with the Monte-Carlo method is a part of this method. There are several reasons to use the software of this type. The compression experiments are explosive. A high-explosive charge of about 16 kilograms is used in each experiment. So, it is necessary to protect the equipment with heavy protection elements. The protection elements and a research object are the sources of a scattered x-ray radiation that worsens the x-ray image of the research object. This leads to increase of measurement errors. That is why the task of the radiation yield channel optimization in order to decrease the value of the scattered radiation is urgent now. The x-ray images calculation with the Monte-Carlo method helps to reveal the elements of the radiation output channels that make the highest contribution to the scattered radiation.

1. Software capabilities

1.1. General characteristic of the software.

Using the Monte-Carlo method the software solves the equation of the γ–quanta transfer for the one-dimensional systems, simple two-dimensional and three-dimensional systems and their combinations. Presentation of interaction cross section has a tabular form with several types of interpolations in order to obtain continuous energy dependence on the cross section. A database for the cross sections of the γ–quanta interactions with the substance was formed on basis of

© 2008 IOP Publishing Ltd

Storm-Israel data [4]. The database for the $\gamma-$quanta formation cross sections was formed on basis of several sources - from libraries ENDL-82 and ENDF/B-6, JENDL-32 [5, 6].

Possibilities of the systems geometry setup are limited by simple spherical, elliptical, cylindrical, plane and conical blocks and their 3D combinations. This makes it possible to solve complicated and substantial 3D tasks. Simplicity of the initial data setup and general high productivity of the software operation are provided. The database for the $\gamma-$quanta interaction cross sections includes the data of the works [4, 7]. Functions of a non-coherent scattering are taken from the works [8]. Form-factors of a coherent scattering are taken from the work [9].

The software calculates the following values: fluxes and currents of the particles on the system surface, numbers of reactions in the areas, heating of an area and other characteristics as a function of energy, time, spatial coordinates, angular data and other parameters.

Points in space, areas and surfaces of the considered system could be the sources of the $\gamma-$quanta. There are several types of selection of the initial direction and energy of the $\gamma-$quanta. The source of the $\gamma-$quanta could be taken from an earlier created file.

Two schemes of a calculation efficiency increase are realized in the software: (1) multiplication and a reel on the surfaces of the system geometry; (2) one of the methods of a local evaluation – a "method of trial particles" to estimate the particles fluxes, coming to the setup goal (point).

1.2. Realization of the processes of the $\gamma-$quanta interaction with the substance

1.2.1. Non-coherent scattering. A differential cross section of the non-coherent scattering of the $\gamma-$quanta on an electron is described with a formula:

$$\frac{d\sigma_{inc}}{d\Omega} = \frac{1}{2}r_0^2\frac{1+\cos^2\theta}{(1+\varepsilon(1-\cos\theta))^2}\left\{1+\frac{\varepsilon^2(1+\cos\theta)^2}{(1+cos^2\theta)(1+\varepsilon(1-\cos\theta))}\right\}\mathrm{S}(\overrightarrow{q},Z) \tag{1}$$

where $d\Omega$ – element of a solid angle, θ – scattering angle, r_0 – classical electron radius, ε – quantum energy in mc^2 units. Without multiplication on a function $\mathrm{S}(\overrightarrow{q},Z)$ the formula (1) represents a differential cross section of the non-coherent scattering of the $\gamma-$quanta on a free electron. At small energy transfers we take into account the electrons connectedness in an atom by multiplication of Kleiyn-Nishina cross section on the function S(q,Z), where $\overrightarrow{q}$ - transferable pulse, Z – atomic number of the substance. The values of the function $\mathrm{S}(\overrightarrow{q},Z)$, calculated in a model of the Hartry-Fock atom, are taken from [8].

Initial grid of the function $\mathrm{S}(\overrightarrow{q},Z)$ argument is setup in a variable $x = \sin(\theta/2)/\lambda$, λ - wavelength in units kx ($1\ kx = 1.00202{\cdot}10^{-8}$ cm, $1/\lambda = E/12400$, E - in electron-volts). Initial interpolation is linear on x, and linear for the function $1/[1+(\mathrm{S}(x)/Z)^2]$. On the stage of the cross sections files preparation for the reactions simulation a transition to the variable $\tilde{x} = E{\cdot}\sin(\theta/2)$ is carried out, and the nodes are recalculated on a formula $\tilde{x}_i = x_i{\cdot}12400$, and from the values of the function $\mathrm{S}(\overrightarrow{q},Z)$ they are recalculated to the function $\sqrt{\frac{\mathrm{S}(x)/Z}{1-\mathrm{S}(x)/Z}}$ with the linear interpolation between the nodes.

1.2.2. Coherent scattering. Differential cross section of the coherent scattering is setup with the formula

$$\frac{d\sigma_{coh}}{d\Omega} = \frac{1}{2}r_0^2(1+\cos^2\theta)[\mathrm{F}(\overrightarrow{q},Z)]^2\,. \tag{2}$$

Here $\mathrm{F}(\overrightarrow{q},Z)$ is a form-factor that presents a probability of the fact that a pulse $\overrightarrow{q}$ will be transferred to Z atoms electrons without any absorption of the energy. The values of the form-factor that are calculated in the Hartry-Fock model are taken from [9].

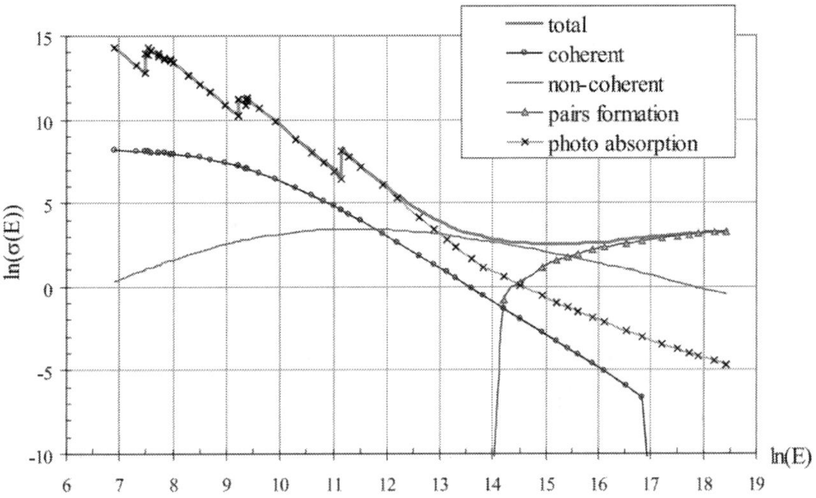

Figure 1. Interaction cross sections of $\gamma-$quanta for ^{74}W.

1.2.3. Photo absorption. Cross sections of the photo absorption are taken from the work by Veigele [7]. These sections were obtained with analysis of a large number of experimental data and they are the most reliable in the field of $\gamma-$quanta energies near the K-shell level and below it. Since Veigele work the cross sections are presented up to the energy of 1 MeV only, for higher energies we used the data of Storm-Israel [4].

1.2.4. Pairs formation. A process of an electron-positron pair formation takes place at the $\gamma-$quantum energy higher than $2 \cdot mc^2$, where $mc^2 = 0.5110034$ MeV. The cross sections of the pairs formation and also the cross sections of the energy absorption were taken from [4]. For the description of the pair absorption process the following model was accepted. Electron of pair is absorbed in a point of the $\gamma-$quantum collision and the positron annihilates forming two $\gamma-$quanta. The first $\gamma-$quantum isotropically emitted with the energy $E_{pair} = 0.5110034$ MeV and the second quantum – in opposite direction with the energy E_{pair}.

1.2.5. Total cross section. A total cross section of the $\gamma-$radiation interaction with the substance is determined as a sum of cross sections of the above mentioned reactions (see, for example, Figure 1 for ^{74}W): $\sigma_{tot} = \sigma_{inc} + \sigma_{coh} + \sigma_{phot} + \sigma_{pair}$ [barn/atom].

1.3. Implementation

Nowadays the program package that allows calculating the x-ray images of the device for the isentropic compression of the hydrogen isotopes using the Monte-Carlo method consists of 17000 lines of the code. A calculation rate is $\sim$ 70000 trajectories per second (CPU Pentium 4, 3.6 GHz). One needs 140 Mbytes of RAM to calculate the x-ray image of 1000*1000 pixels size.

2. Calculation results of x-ray images

Use of calculation results of the "H_z" code [2] as the initial data and analysis of the results of the software for the x-ray images calculation caused some troubles. Formats of the input files of the x-ray images calculation proggram did not match the output files of the "H_z", and the formats of the output files of the calculation section did not match the input of the visualization software. Besides, there were significant difference in internal presentation of the physical parameters of the studied object and recorder (geometry, sizes and composition of materials) in both programs. Also there were some limitations for names and location of the files of the task for the x-ray

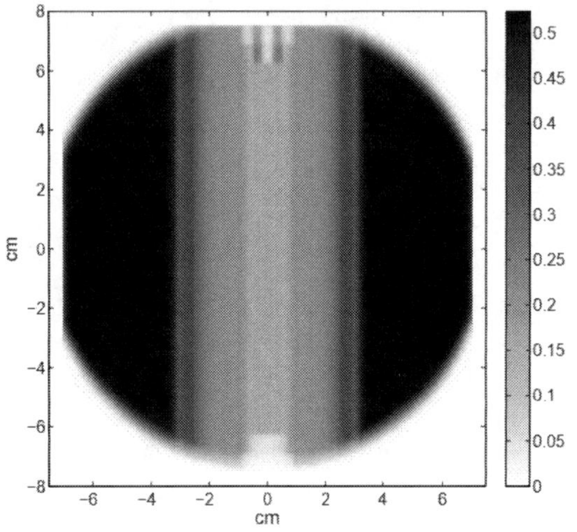

Figure 2. Calculated x-ray image.

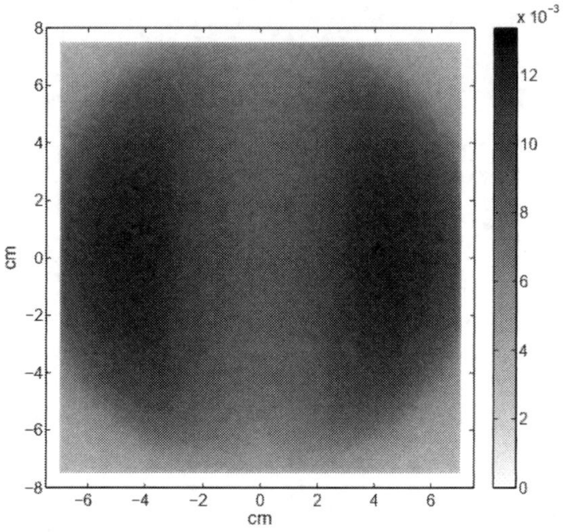

Figure 3. Calculated distribution of the scattered radiation.

images calculation. All these problems were solved by writing add-in programs. Firstly, we automated formation of the files set, that are the input files of the x-ray images calculation software, from the output data of the "H_z" code. Secondly, the files containing the calculation results on the x-ray images software calculation are converted for their further visualization.

The program has a feature of the x-ray radiation source parameters setup, and also features of visualization of the direct x-ray radiation, sum of direct and scattered radiation from different elements of the radiation output channel.

The geometry of the experiment is taken into account in the calculations of the x-ray images in the radiographic studies. After the experiment the alignment errors are also considered. Fig. 2 presents calculated x-ray diagram of the device at some displacement of the research object and the recorder relatively to the x-ray radiation beam. The calculated distribution of the scattered radiation is presented in Fig. 3.

The value of the scattered radiation does not exceed 6% of the value of the direct radiation. An element of the recorder protection (a combination of steel and aluminum plates) is the source of the main part of the scattered radiation.

For comparison Fig. 4 presents experimental x-ray image of the device obtained in the experiment on isentropic compression of solid D_2. In this moment the pressure in the compression chamber was ~ 3 Mbar.

One can see that the calculated and experimental x-ray images have qualitative similarity.

Conclusion

The paper presents the calculation results of the x-ray image that was obtained in one of the experiments on isentropic compression, using the software for x-ray images calculation, which solve the equations of the radiation transfer using the Monte-Carlo method. The calculated and experimental x-ray images are presented. The scattered radiation composes the small portion of the direct radiation (less than 6%). Nevertheless, to increase the contrast of the shot and measurement precision it is desirable to decrease the scattering. Use of the software for the x-ray images calculation makes it possible to reveal the elements of the radiation output channel that bring the most contribution into the scattered radiation.

The work was executed in the frameworks of the ISTC project #2564.2.

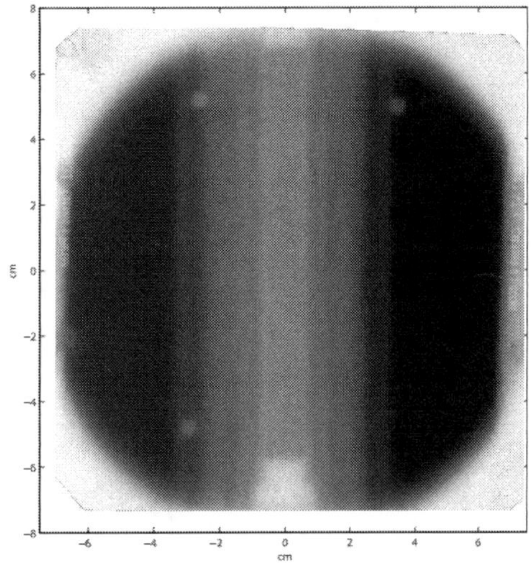

Figure 4. Experimental x-ray image.

References

[1] A.I. Pavlovskii et al. In book: Ultrahigh Magnetic Fields/Ed. V.M. Titov, G.A. Shvetsov. M. Nauka, 1984, p. 19.

[2] G.V. Boriskov et al. In book: VIII Hariton's Readings on Problems of High Energy Physics. Sarov, 2006, p. 509 and p. 516.

[3] A.I. Bykov, N.I. Egorov, Yu.P. Kuropatkin, N.B. Lukianov, V.D. Mironenko, V.N. Pavlov, "Use of radiography to study EOS of condensed hydrogen isotopes within megabar pressure range". Book of Abstracts, Nov. 5-10, 2006, Santa Fe, New Mexico, USA, p. 34

[4] E. Storm, H.I. Israel "Photon Cross Section from 1 kev to 100 mev for Elements Z = 1 to Z = 100", Nucl. Data Tables, A7, 565-618, (1970).

[5] WINENDF, version from October 2002, IAEA, Release 8 library ENDF/B-6.

[6] ENDF-102 "Data Formats and Procedures for the Evaluated Nuclear data File ENDF/B-6", V.McLane (BNL USA), September 1997.

[7] W.J. Veigele "Photon Cross Section from 0.1 kev to 1 mev for Elements Z=1 to Z=94", Atomic Data Tables, 5, 51-111, (1973).

[8] D.T. Cromer, J.B. Mann "Compton Scattering Factors for Spherically Symmetric Free Atoms", J.Chem.Phys., 47, 1892-1893, 1967;
Cromer D.T., Mann J.B. "Compton Scattering Factors for Aspherical Free Atoms", J.Chem.Phys., 50, 4857-4859, (1969).

[9] H.P. Hanson, F. Herman, J.D. Lea, S. Skillman "HFS Atomic Scattering Factors", Acta Cryst., 17, 1040-1044, (1964).

Measurement of sound velocities and shear strength of cerium under shock compression

M V Zhernokletov[1], A E Kovalev[1], V V Komissarov[1], M G Novikov[1], M A Zocher[2] and F J Cherne[2]

[1]RFNC-VNIIEF, Sarov, Nizhni Novgorod region, 607190, Russia
[2]Los Alamos National Laboratory, Los Alamos, NM, 87545, USA

E-mail: akkm@yandex.ru

Abstract. Sound velocity in shock-compressed cerium was measured over the pressure range of 35-140 GPa using the rarefaction overtake technique. Indicator liquids carbogal and tetrachloromethane were used. The samples were loaded with planar shock wave generators using powerful high explosives (HE). Luminescence of the liquid indicators was recorded by optical gauges based on photodiode "FD 256". For the pressure range of 13-35 GPa, sound velocity was measured in cerium samples using the counter release method with manganin-based piezoresistive gauges. From the measured longitudinal and bulk sound velocities, Poisson's ratio and shear strength of cerium were determined. The melting boundary on the shock Hugoniot was estimated. Experimental data is compared with calculation results.

1. Introduction

In addition to the basic parameters included in an equation of state, (pressure P, volume V or density ρ, temperature T and entropy S), the need often arises for information concerning the sound velocity C_s. Sound velocity is one of the values that characterizes the behavior of the substance under shock compression followed by release. In addition to the sound velocity, when creating a constitutive model for the deviator, knowledge of the shear strength of the material, or its resistance to plastic deformation, is required.

Interest in studying cerium is prompted by its very complicated phase diagram [1]. Because of its large volume collapse (occurs during the γ-α isomorphic phase transformation), it is expected that cerium will melt at very low pressures under shock compression.. Therefore, one of the purposes of the work is to determine the melt boundary of cerium on the shock adiabat.

Upon melting, a substance will looses its allotropic properties. Thus, in an experiment, by recording the presence (or absence) of a difference between the elastic C_l and plastic C_s sound velocities over the shock wave (SW) pressure region where a material is expected to melt, it is possible to determine the lower and upper pressure boundaries of melting. As the difference between C_l and C_s is reduced, the Poisson's ratio approaches a value of 0.5, which is typical of melting.

2. Indicator method of sound velocity measurement

Optical measurements of sound velocities were performed using the overtake method in conjunction with indicator liquids. The method is described in detail in [2]. To form shock waves in cerium

© 2008 IOP Publishing Ltd

samples, SW generators were used. The SW generators used the explosion products from cylindrical HE charges (with a variety of power outputs) to accelerate thin impactors.

After the fast-flying thin plate impacts a baseplate, upon which resides a stepped sample of cerium, a rarefaction wave is formed from the "back side" of the impactor. The rarefaction wave passes through the impactor, baseplate, cerium sample, and into the indicator liquid. One objective of the experiment is a determination of the sample location, X_{MAX}, where the rarefaction wave overtakes the shock wave front. For a target with thickness x less than X_{MAX}, a constant intensity of radiation will be recorded by the indicator liquid up to the moment that the first indication of the rarefaction wave appears. Afterwards, radiation intensity will begin to drop.

The indicator liquids carbogal (C_8F_{16}) or tetrachloromethane (CCl_4) were used. They have an initial density, ρ_0, of 1.86 and 1.58 g/cm^3, respectively. The luminescence of the indicator liquid (produced as a result of shock wave passage) was collected by 400 μm quartz-fiber light guides located above each sample step and recorded by optical signal transducers based on photodiode "FD 256." The signals were recorded by oscillographs of the types Agilent (54642D, 54624A) and Tektronix (TDS 3054). Accuracy of time measurements for the arrival of the shock wave front in the indicator was no worse than ±2 ns. Knowing the thickness of each cerium sample step, it was possible to determine the shock wave velocity in each experiment.

Sound velocity in cerium was calculated by the formula:

$$C_s = X_{MAX}\left(\sigma\cdot(t_{SW}-t_{RW}+\frac{X_{MAX}}{D})\right)^{-1} \qquad (1)$$

where σ is compression, t_{SW} is shock wave arrival time at the surface of the sample, t_{RW} is rarefaction wave arrival time at the surface of the sample, D is the shock velocity in cerium, X_{MAX} is thickness of the step when overtake occurs at the cerium-indicator interface.

Typical oscillograms obtained in experiments involving the measurement of sound velocity in a four-step cerium sample are presented in figure 1. Arrows mark the shock wave arrival times at the cerium- indicator liquid (carbogal) (t_1) interface and "overtake" (t_2) of the shock wave front by rarefaction.

Figure 2 presents a graph for determining the velocity of the isentropic wave of rarefaction in cerium. The graph plotted in figure 2 was based on the data presented in figure 1.

Calculations of the x-t diagrams simulating shock wave motion and rarefaction waves in impactor, baseplate and the cerium sample for determination of t_{SW} and t_{RW} in (1) were performed with using a hydrodynamic computer program produced by VNIIEF. The shock adiabat determined for cerium from [3] was used in the calculations. For the impactors and baseplates made of steel and aluminum, the authors used the standard EOS applied at VNIIEF.

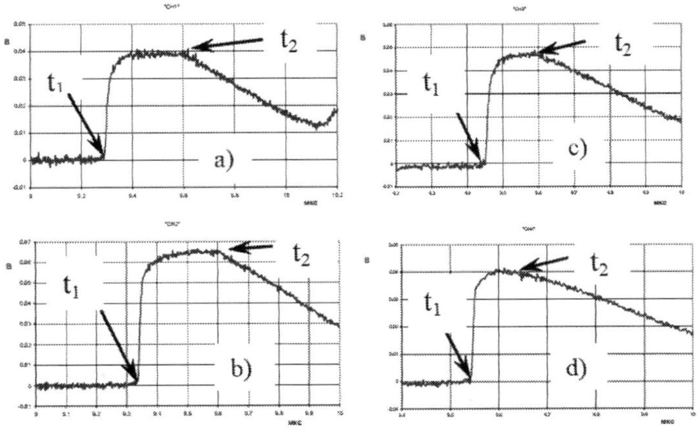

Figure 1. Oscillograms of irradiation of shock wave front in C_8F_{16} at P_{Ce} = 96.5 GPa behind cerium steps having thicknesses: a) 1.99 mm; b) 2.27 mm; c) 3.00 mm; d) 3.38 mm.

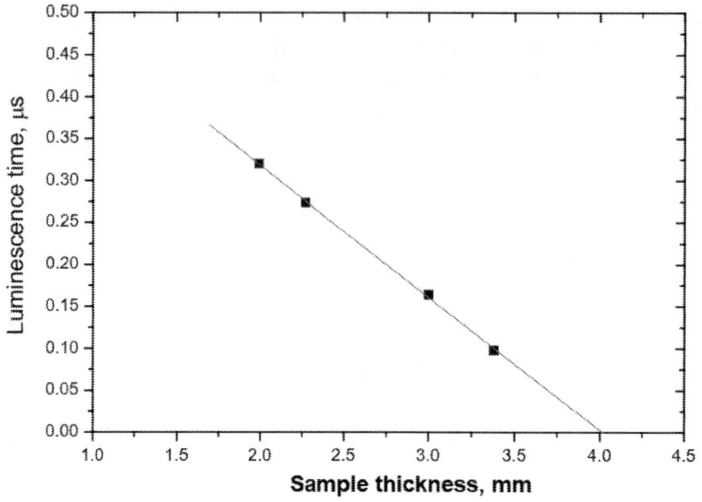

Figure 2. Determination of the overtake distance X_{MAX}.

Table 1 presents the experimental data on the sound velocity in cerium obtained by the optical method. At pressures less than 40 GPa in cerium, the temperature of luminescence of the shock-compressed liquid indicator is close to the sensitivity limit of recording photodiodes. This limit is equal to about 2000 K. It is for this reason that the manganin pressure gauge method was used in determinating sound velocities in cerium for pressures less than 40 GPa.

Table 1. Experimental data on measurement of sound velocity in cerium by optical method.

Parameters of shock waves in cerium				
D, km/s	*U*, km/s	*P*, GPa	σ	*C*, km/s
6.46	3.23	141.2	2.00	6.09 ± 0.31
6.46	3.23	141.2	2.00	6.02 ± 0.25
5.92	2.89	115.5	1.95	5.49 ± 0.21
5.85	2.82	111.3	1.93	5.59 ± 0.25
5.50	2.60	96.5	1.90	5.11 ± 0.17
5.37	2.52	91.3	1.88	5.23 ± 0.27
5.02	2.32	79.7	1.86	4.94 ± 0.18
4.21	1.80	51.1	1.75	4.26± 0.16
4.19	1.79	50.8	1.75	4.40 ± 0.26
3.65	1.47	36.2	1.68	3.94 ± 0.11

3. Manganin pressure gauge method

The experimental method used for measuring the sound velocities C_L and C_S in shock-compressed cerium is similar to that used in [4]. Cerium samples ~2 mm thick and 40 mm in diameter were made. They were mounted on a copper baseplate. Three manganin gauges were placed between the baseplate and the cerium sample. An HE charge formed a quasi-stationary shock wave profile with front dissymmetry of 0.05 µs or less over a diameter of ~ 70 mm in the baseplate. The manganin gauges recorded profiles of the direct shock wave and the rarefaction wave formed by the free surface of the sample. In order to measure timing of the direct shock wave motion and the reflected rarefaction wave in the sample, an electrocontact gauge was mounted on the free surface of the sample.

As illustration, figure 3 presents one of the oscillograms obtained from a recording of the stress profile in cerium.

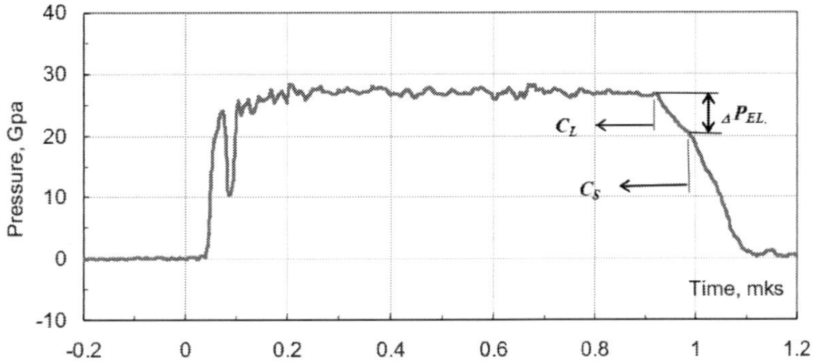

Figure 3. Oscillogram, which was obtained in experiment with recording of profiles of waves of loading and release in cerium. ΔP_{EL} is pressure jump in elastic wave of release.

The pressure profile "*P*" was calculated using a calibration of the electrical resistivity of manganin to the stress of shock compression.

Concurrent processing of the oscillograms recorded by pressure gauges and by electrocontact gauges, enabled a determination of the elastic and bulk sound velocities. Small corrections were made during the calculations. These corrections accounted for the thickness of the pressure gauge and the thickness of its insulation. Accuracy of the pressure measurements was ±3%.

Poisson's ratio ν can be determined from knowledge of the elastic wave velocity C_L and plastic wave velocity C_S, and is given by the following relation:

$$\nu = \frac{3-(C_L/C_S)^2}{3+(C_L/C_S)^2} \tag{2}$$

Based on the experimentally-measured values of C_L, C_S and ΔP_{EL} , the values of dynamic yield strength Y_g (shear strength) of cerium were determined using the following formula:

$$Y_g = \frac{1-2\nu}{2(1-\nu)}\Delta P_{EL} \tag{3}$$

Results of the experiments with determination of sound velocities in cerium using manganin gauges are shown in table 2. The experimental dependence of sound velocity in cerium on pressure is presented in figure 4.

Table 2. Experimental results.

Baseplate			Ce sample						
Material	U_f, km/s	P_f, GPa	P, GPa	ΔP_{EL}, GPa	D, km/s	ρ, g/cm^3	C_L, km/s	C_S, km/s	Poisson's ratio, ν	Y_g, GPa
Cu	0,58	24,94	12,7	3,03	2,49	9,66	3,10	2,73	0,398	0,55
Al	1,12	20,79	21,3	4,28	2,98	10,46	3,50	3,10	0,390	0,77
Al	1,12	20,79	21,6	-	3,00	10,48	3,71	-	-	-
Al	1,45	28,69	27,0	6,20	3,26	10,84	4,18	3,53	0,363	1,33
Al	1,40	27,43	28,3	3,88	3,41	10,55	3,99	3,69	0,438	0,43
Al	1,46	28,93	31,5	-	3,45	11,09	-	3,74	-	-
Al	1,49	29,69	32,4	-	3,49	11,14	-	3,85	-	-

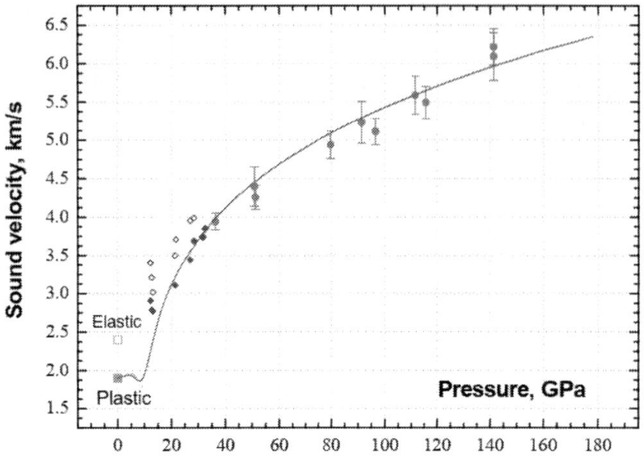

Figure 4. Sound velocity in cerium versus pressure
—— - calculation by VNIIEF's EOS, ● - sound velocity (optical method), ◆, ◇ - plastic (volume) and elastic sound velocities (manganin), □ - elastic and ■ - plastic (volume) sound velocities.

This figure also presents a calculation of isentropic sound velocity determined using the VNIIEF equation of state, as well as the zero pressure values of the longitudinal C_L = 2.40 km/s and plastic (volume) C_S = 1.90 km/s sound velocities obtained using the ultrasound method.

As one can see in the graph, the experimental data are in agreement with the calculation to within the measurement error. Based on the obtained data, the probable cerium melt phase transition occurs below a pressure of ~ 32 GPa. Additional experiments are needed to determine more accurately melt transition on the Hugoniot.

4. Conclusion

The optical analyzer technique at pressures of 35 GPa and greater showed a one-wave structure in the rarefaction behavior recorded in measuring the sound velocity in cerium. Comparing the rarefaction wave structure obtained by the optical method and the manganin gauge method, indicates that the upper boundary for cerium melting on the Hugoniot is below 32 GPa. Additional experiments are needed in order to obtain a more accurate determination of the cerium melt transition on the Hugoniot.

5. Acknowledgments

Funding for this work was provided by Los Alamos National Laboratory.

6. References

[1] Pavlovsky M N, Komissarov V V, Kutsar A R 1999 FGV, v.35, 1, Novosibirsk, p.99-101.
[2] McQueen R, Hixson J, Fritz J 1982 Rev. Sci. Instrum. 53, №2, p. 245-250.
[3] Experimental data on shock-wave compression and adiabatic expansion of condensed substances 2001 Edited by Trunin R F Sarov VNIIEF.
[4] Pavlovsky M N, Komissarov V V 1982 ZhETF, v.83, Issue 6(12), Moscow, p.2146-2151.

Journal of Physics: Conference Series **121** (2008) 082001 doi:10.1088/1742-6596/121/8/082001

Synthesis and doping of fine-crystalline corundum in sub- and supercritical conditions

M N Danchevskaya, Yu D Ivakin, G P Muravieva and I V Luchkov
Chemistry Department, Moscow State University, Leninskie Gory, 1/3, Moscow 119992, Russia

E-mail: mardan@kge.msu.ru

Abstract. In the report the results of investigations of the mechanism and kinetics of the formation of cobalt doped corundum in water vapor medium are represented. Moreover, it is given the study result of changing of lattice position of cobalt ions and their state during synthesis. Aluminium hydroxide was using as raw material. Doping agent was cobalt nitrate. The doping of synthesis products (boehmite and corundum) occurs during their formation in solvothermal conditions and then in supercritical water fluid. The additives with reducing or oxidizing properties were brought in reaction medium. The products of the synthesis doped with cobalt were explored by various physicochemical methods: the X-ray phase-analysis, DR-spectroscopy, the magnetometry and SEM. It was found that the processes of the lattice rearrangement and doping take place simultaneously. In solvothermal conditions, below water critical point cobalt ions are initially chaotically distributed in the interlaminar space of the boehmite structure. At supercritical conditions the doped boehmite transforms into cobalt doped corundum. Such crystals of corundum exhibit superparamagnetic properties. After heating in vacuum (1900°C) the corundum doped with higher concentrations of cobalt acquire the ferromagnetic properties. The authors guessed that cobalt ions aggregate into the nano-sized particles of cobalt in vacancies complexes of corundum structure.

Introduction

The transition to critical area (T>376°C, P>22 MPa) at hydrothermal synthesis allow producing pure materials with high performance and substantially shorten the duration of synthesis. Due to this the synthesis in supercritical fluids began to be widely utilized in an industry [1-4]. At comparison between mechanism of the hydrothermal process and process in SCWF becomes apparent that these mechanisms have certain common features [5, 6].

Previous investigations by authors of the mechanism and kinetics concerning the formation of fine-crystalline oxides in water vapor at both the sub- and supercritical conditions (P = 2–28 MPa, T = 200–450°C) and soft hydrothermal conditions have revealed that synthesis of fine crystalline oxides is a multistage process in these conditions. The process runs through formation of a solid-phase intermediate, which comprises the hydrated forms of the precursor [7, 8]. It was determined that molecules of fluid actively participate in the process of reorganization of the solid phase both during the first and second stages of transformation of precursors [9]. The processes of solid-phase transformations of oxides, oxyhydroxides and hydroxides during thermovapours treatment (TVT) proceed in conditions of dynamical quasi-equilibrium between solid phase and fluid. For example [10], the process of corundum formation from hydrargillite in water passes through the following stages:

$$\mathrm{Al(OH)_3} \xrightarrow{\mathrm{H_2O}} \mathrm{AlOOH} + \mathrm{H_2O} \longrightarrow \alpha - \mathrm{Al_2O_3} + 2\mathrm{H_2O}$$

Hydrargillite Boehmite Corundum

© 2008 IOP Publishing Ltd

Journal of Physics: Conference Series **121** (2008) 082001 doi:10.1088/1742-6596/121/8/082001

Conditions of the intermediate formation may affect the final stage of fine-crystalline oxides synthesis. The rate of the process and properties of the final product, such as a habitus, degree of perfection and sizes of crystals depends on of the specific structure of the intermediate [11].

The purpose of these investigations was to study the influence of cobalt ions on the synthesis process of corundum and its peculiarities, and to establish the mechanism by which the doping ions incorporate into the structure of the intermediate (boehmite) and synthesized corundum in a fluid water medium.

2. Materials and experimental methods

Synthesis and doping of solid oxides was carried out in laboratory (18 ml) and experimental-industrial (125 l) autoclaves. To synthesize doped boehmite and corundum, the hydrargillite of trademark "GD00"and, as dopant, $Co(NO_3)_2{\cdot}6H_2O$ were used. Because some substances were added in a reaction zone the redox potential of water fluid varied. Thus, H_2, H_2O_2, KOH, $NH_3{\cdot}H_2O$ were used.
Necessary water vapor pressure was created in the autoclave by heated water, which was poured between the walls of the autoclave and the container with the raw material (hydrargillite). Hydrargillite including the dopant was initially held at a temperature in the subcritical range, then in the supercritical region.

The synthesized products were investigated by physical-chemical methods. The "Cam Scan Series 2" device was used to obtain SEM images of produced materials. X-ray analysis of the synthesized products was performed using X-ray powder diffractometer (STOE) with a CuKα and diffractometer DRON-3M. Diffuse reflection (DR) spectra were determined with a M-40 spectrometer.

3. Results and Discussion

X-ray diffractograms, DR spectra and performances of the curves of a magnetization of products arising from hydrargillite during synthesis of corundum doped by cobalt were analyzed, revealing the mechanism by which cobalt ions incorporated into the aluminous matrix. In water vapor, the transformation of hydrargillite into boehmite occurs very quickly at temperatures higher than 230°C.

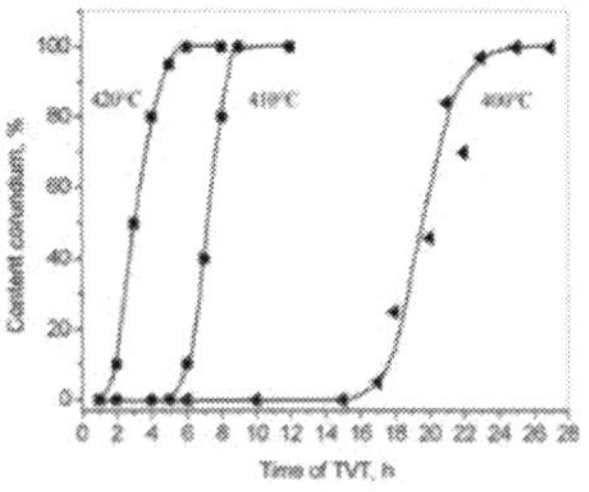

Figure 1. Kinetic curves of boehmite transformation into corundum.

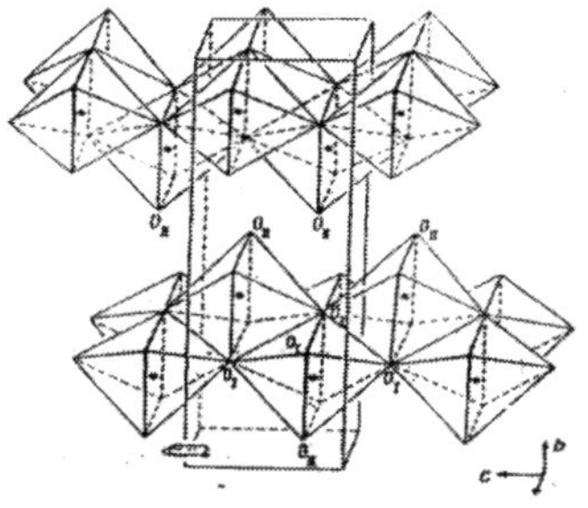

Figure 2. Structure of boehmite.

Figure 3. SEM of boehmite.

In figure 1, the kinetic curves of corundum formation from boehmite inclusive cobalt formed at 230°C are represented. The cobalt concentration relative to the produced corundum is equal to 0.1 mol of 1%. Molar relation of fluid and solid phase is 4.34. Coefficient of filling of free space of an autoclave is equal to 20%. Using this filling coefficient and at temperatures 400°C, 410°C, 420°C the pressure of water fluid is 26.8 MPa, 28.6 MPa and 30.4 MPa, respectively. The time of 100% formation of corundum from boehmite is 9, 4 and 3.5 hours, respectively. A long period of time is observed in the kinetic curve at 400°C, as the formation of boehmite up to the occurrence of the first crystals of corundum (induction period) (figure 1) requires the change in morphology and perfecting of the boehmite structure prior to the formation of corundum [12]. Boehmite has a layered structure (figure 2). The inclusion of dopant ions in the interlaminar space of boehmite structure occurs if it is already in the subcritical range. The oriented growth of boehmite particles in blocks is also accompanied by capture of cobalt ions in macro-defects of the formative blocks (figure 3).

The absence of the expressed absorption bands in the visual field of the DR spectrum typical for cobalt ions regularly included in the crystal lattice denotes chaotic cobalt ions inclusion in the produced boehmite (figure 4a). As the endurance time of doped boehmite in water fluid medium increases, a wide absorption band appears at an interval 500 - 700 nm.

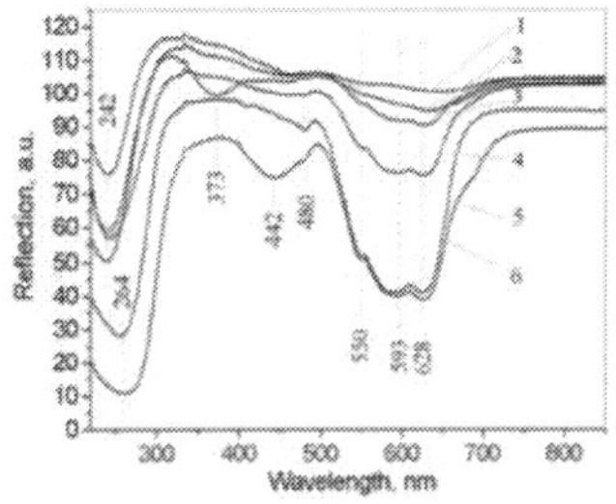

Figure 4. DR spectra of the products TVT of hydrargillite at T=400°C, P=26 MPa during: curve 1- 3 h, 2 – 12 h, 3 – 15 h, 4 – 22 h, 5 – 25 h, 6 -72 h.

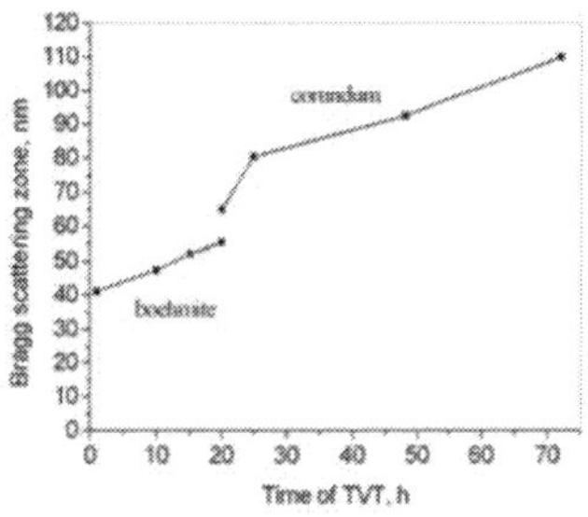

Figure 5. Bragg Scattering zones of boehmite and corundum after TVT at T=400°C, P=26 MPa.

The initial formation of corundum from doped boehmite is characterized by the resolution of this wide band to three bands at 550, 593 and 628 nm (figure 4d).

The ordering of boehmite and corundum structure with the increasing of duration of (TVT) is supported by the growth of the respective Bragg Scattering zones (figure 5).

The addition of fluid water (T = 410°C, P = 288 MPa) changes the redox potential of the substances and results in a change of a state of cobalt ions in doped corundum.

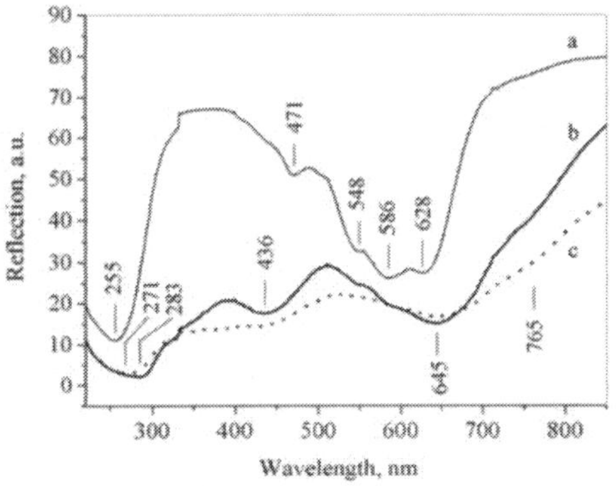

Figure 6. DR spectra α-Al_2O_3·(1%)Co. Synthesis medium: water fluid with agent: a) H_2, b) KOH, NO_3^- c) H_2O_2

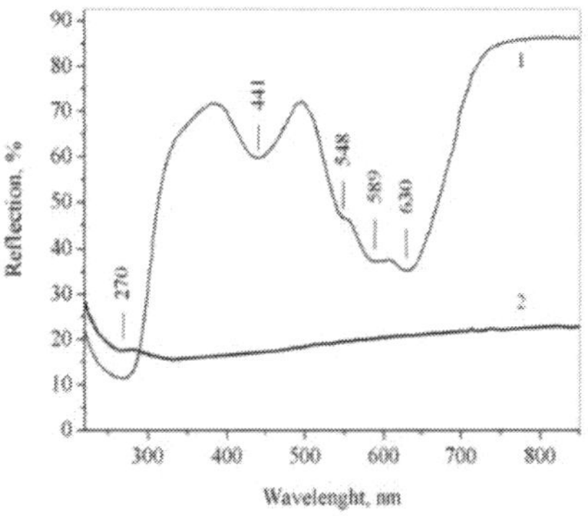

Figure 7. DR spectra α-Al_2O_3·(0.27%)Co 1)TVT:410°C, 288 MPa 2) This sample, heated at 1900°C, in vacuum, 1 hour.

In the presence of hydrogen in fluid water the obtained corundum doped by cobalt is dark blue. The DR spectrum of this corundum (figure 6a) corresponds to a case of Co^{2+} in tetrahedral coordination and the multiple transition $^4A_2(F)\rightarrow{}^4T_1$ (P) should be observed at nm 550 – 660 nm [13].

As the substances with oxidizing abilities are utilized, as the additives in a fluid, the synthesized corundum becomes green in color. The DR spectrum of this sample of corundum is characteristic for Co^{3+} in octahedral coordination [13, 14]. The most intensive absorption bands in this case are positioned about 440 and 650 nm. The high-temperature heating of corundum doped with cobalt increases the absorption in all intervals in the DR spectrum (figure 7). The sample becomes dark grey or black depending on the content of cobalt. The absence of specific absorption bands has allowed us to make the inference that the cobalt ions, diffusing in a lattice of corundum crystal under heating, will agglomerate with formation nanoparticles of metallic cobalt. It is possible that formed nanoparticles are found in vacancy complexes.

The comparison of figures 8 and 9 shows that heating the samples of corundum doped with cobalt at high temperature enhances their magnetic properties. After the heating at 1900°C (1 hour, in vacuum) magnetic moments (μ) of samples become almost hundred times more (table). These samples have relatively high coercive forces (130 -230 Oe).

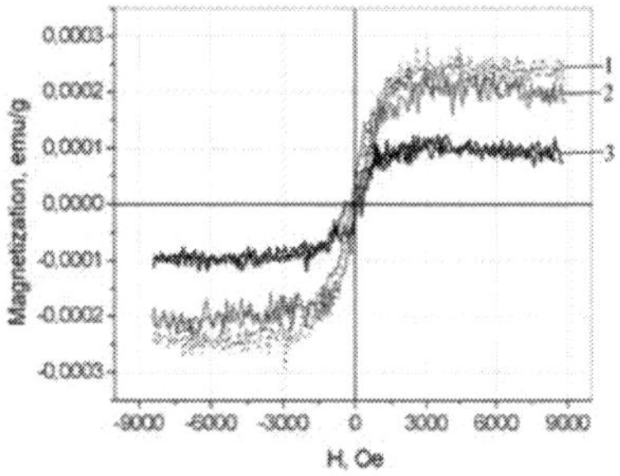

Figure 8. Hysteresis loops of
1) α-Al_2O_3(1%)Co
2) α-$Al_2O_3\cdot$(0.56%)Co, 3) α-$Al_2O_3\cdot$(0.26%)Co.
Before the heating at 1900°C in vacuum.

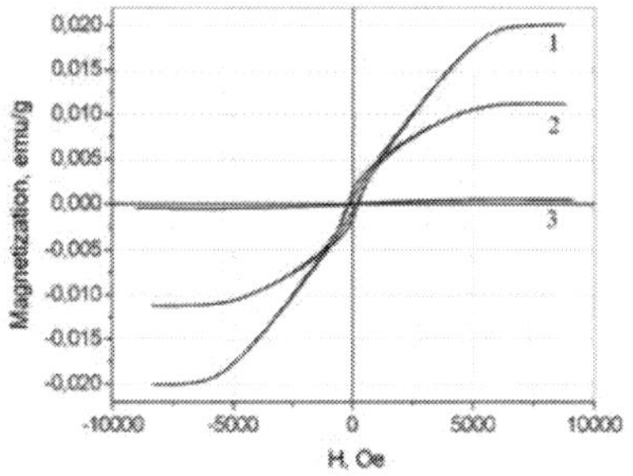

Figure 9. Hysteresis loops of
1) α-$Al_2O_3\cdot$(1%)Co
2) α-$Al_2O_3\cdot$(0.56%)Co, 3) α-$Al_2O_3\cdot$(0.26%)Co.
After the heating at 1900°C, 1 hour, in vacuum.

Table 1. The magnetic performances of corundum doped by cobalt before and after heating at 1900°C

#	Content % in medium Co	Saturation magnetization ($G\cdot cm^3/g$)		Magnetic moment ($\mu\cdot 10^{-4}$)		Coercive force, (H_c Oe)	
		As-prepared	Heat-treated	As-prepared	Heat-treated	As-prepared	Heat-treated
1	1	0.003726	0,282885	2.4	200	-	130
2	0.56	0.001269	0.234298	1	167	-	145
3	0.26	0.003529	0.164688	2.1	111	-	230

The magnetic properties of the samples reduced during vacuum heating allow us to conclude that metallic cobalt particles are formed in oxygen vacancies due to elimination hydroxyls from corundum structure. The theoretical analysis of experimental hysteresis loops allowed the evaluation of the particle size distribution function for Co nanoparticles in the corundum lattice [15]. The size of Co nanoparticles averages 6 nm.

Conclusion

During the formation of corundum crystals from hydrargillite in water vapor sequentially in sub- and supercritical conditions, the cobalt ions easily enter into the formed products structure. In subcritical conditions, the inclusion of a dopant happens chaotically in the intermediate [16]. Then, during formation of corundum, the cobalt ions are placed in tetrahedral or octahedral structural positions, exhibiting superparamagnetism. High-temperature annealing of doped corundum in a vacuum, with removal of residual hydroxyls from crystals structure and formation of vacancies complexes, results in the origin of nano-sized particles of cobalt placed in these defects.

Acknowledgement

The authors gratefully acknowledge ISTC for financial support (pr. N 3002).

References

[1] Jung J, Perrut M *J. Supercritical Fluids*, 20, 3, (2001), 179-219.
[2] Reverchon E, Adami R *J. Supercritical Fluids* 37 (2006) 1-22
[3] Adschiri T, Hakuta Y, Arai K *Ind. Eng. Chem. Res.* **39** (2000) 4901-4907.
[4] Johnson J W, Norton D *Am. J. Sci.* **291** (1991) 541-648.
[5] Byrappa K, Adschiri T *Progress in Crystal Growth and Characterization of Materials* **53** (2007) 117-166.
[6] Kruse A, Dinjus E *J. Supercritical Fluids,* **39** 3, (2007) 362-380
[7] Danchevskaya M N, Ivakin Yu D, Torbin S N, Panasyuk G P and all. 2001 *High Pressure Res.* **20** 129
[8] Torbin S N, Danchevskaya M N, Martynova L F and Muravieva G P 2001 *High Pressure Res.* **20** 109
[9] Danchevskaya M N, Torbin S N, Ivakin Yu D and all. 2004 *J. Phys.: Condens. Matter* **16** S1187
[10] Panasyuk G P, Danchevskaya M N, Belan V N and all. 2004 *J. Phys.: Condens. Matter* **16** S1215
[11] Danchevskaya M N, Ivakin Yu D, Torbin S N and all. 2007 *J. Supercrit. Fluids,* 2007.03.007
[12] Danchevskaya M N, Ovchinnikova O G, Ivakin Yu D and all. 2000 *J. Phys. Chem. (Russ)* **74** 1391
[13] Khassin A A, Anufrienko V F, Ikorskii V N and all. 2002 *Phys. Chem. Chem. Phys.* **4** 4236
[14] Gubin S P, SpichkinYu I, and all. 2003 *Journal of Magnetism and Magnetic Materials* **265** 234
[15] Perov N, Sudarikova N and Bagrets A 2003 *Journal of Magnetics* **8** 7
[16] Danchevskaya M N, Ivakin Yu D, Muravieva G P, Ovchinnikova O G *the Joint 19th AIRAPT and 41st EHPRG Conference HIGH PRESSURE SCIENCE AND TECHNOLOGY Bordeaux,* 2003, Session T2, P270

Hydrolysis of Glycylglycine in Subcritical Water Conditions

M. Faisal[1,2], H. Daimon[1] and K. Fujie[1]

[1]*Department of Ecological Engineering, Toyohashi University of Technology, Tempaku-cho, Toyohashi, Aichi 441-8580 Japan*
[2]*Department of Chemical Engineering, Syiah Kuala University, NAD, Banda Aceh, 23111, Indonesia*

faisal@fujielab.eco.tut.ac.jp

Abstract. The hydrolysis of glycylglycine was studied at the temperature range of 240 to 300°C and reaction time up to 200 s. The Experiments were conducted using a continuous-flow tubular reactor. An interesting reaction phenomenon relating to the production of amino acids from hydrolysis of proteins in subcritical water was observed. Other than hydrolysis of glycylglycine to produce glycine, cyclodehydration also took place producing cycloglycylglycine. The product of glycine from cycloglycylglycine was also verified. Under the temperature range in this study, it was observed that the temperature did not has any effect on the reaction selectivity of hydrolysis and cyclodehydration of glycylglycine.

1. Introduction

The potential use of water as a medium for various organic chemical reactions is being explored due to its fascinating properties at high-temperature and high-pressure conditions. It also offers ecological benefits if used as a solvent in industrial processes. Near critical water is promising for hydrolysis reaction in the absence of any catalysts or additives.[1-6] This process has already been shown as an emerging technique that can provide decomposition of plastics,[7] cellulosic[8] and proteinaceous[9] wastes into useful products and organic synthesis.[10] Numerous article on the applications of high temperature water to chemical conversion, synthesis and decomposition of organic compounds have been summarized by Savage.[11] Production of amino acids from proteinaceous wastes, such as fish entrails and other marine wastes has also been investigated, although results showed that the recovery of amino acids was quite low compared to the quantity of the amino acids originally present in the raw materials.[12-15] Reaction of peptide and amino acid under high temperature water conditions had also been studied. This includes the stability,[16] decomposition and reaction pathway of some amino acids,[17] and elongation of glycine producing oligopeptides.[18] Reaction behaviors of glycine under sub- and supercritical water conditions has also been reported by Alargov et.al.[19]

Sub and supercritical water conditions can also be used for cyclization of some organic compounds. An example is the cyclization of cyanamide at a temperature range of 175-250 °C, which results in higher amines as a minor pathway.[20] A small quantity of *cis/trans*-dicyanocyclobutane was also observed when acrylonitrile was reacted with water at a temperature of 250 °C.[21] It was postulated to occur because of the dimerization of two acrylonitrile molecules via a pericyclic reaction mechanism.[22]

© 2008 IOP Publishing Ltd

The primary purpose of this paper is to investigate the reactions happening on glycylglycine (Gly-gly) under subcritical water conditions, with focus on the clarify of formation of cyclo(Gly-gly) from Gly-gly. The effect of reaction conditions on the formation of cyclo glycylglycine and other reaction products, reaction behavior as well as reaction pathway formed under subcritical water conditions are discussed.

2. Experimental

All experiments were conducted over a temperature range of 240 to 300°C at a reaction time of 10 to 200s. The pressure was held constant at 20 MPa. A continuous flow tubular reactor system was used to investigate the reaction of dipeptide under hydrothermal conditions. The details of the experimental apparatus have been described elsewhere.[1] The reactor and all connecting parts are made of SUS-316 stainless steel. The water and the sample were delivered separately using two HPLC pumps (PU-1580, Jasco Corp.). To minimize the transition in temperatures during the reaction, the water was preheated in a tube (O.D.: 1/16 inch, I.D.: 0.5 mm, Length: 20 m, Volume: 3.93 cm^3) set at desired reaction temperature before mixing with the sample, prior to entering the main reactor tube (O.D.: 1/16 inch, I.D.: 0.25 mm, Length: 5 m, Volume: 0.25 cm^3). The sample-to-water mixing ratio was 1:2, resulting to 10 mmol/L dipeptide solution. The reaction effluent was quenched in a cooling unit (Model LC-101, ADVANTEC) at 5°C to cease the reaction. The pressure in a system was adjusted by a back-pressure regulator (SCF-Bpg, Jasco Corp.). After reaching steady state, the reaction products were collected for analysis. During each run, helium gas was loaded continuously to the sample and water containers to avoid entry of air into the containers or reactor. The reaction products of Gly-Gly were analyzed by various high-pressure liquid chromatography techniques.

3. Results and Discussion

3.1. Reaction Products of Gly-Gly Under Subcritical Water Conditions

Table 1 shows the results of the reaction of Gly-Gly under the experimental conditions investigated in this study. Aside from glycine (Gly), the products contain other compounds such as organic acids (glycolic acid, formic acid and acetic acid) and amines (ammonia and methylamine).[23] The carbon and nitrogen balances before and after each reaction were determined based on the above products. It was observed that significant amount of carbon and nitrogen compounds remained unaccounted, indicating that subsequent reactions proceeded after hydrolysis.[4]

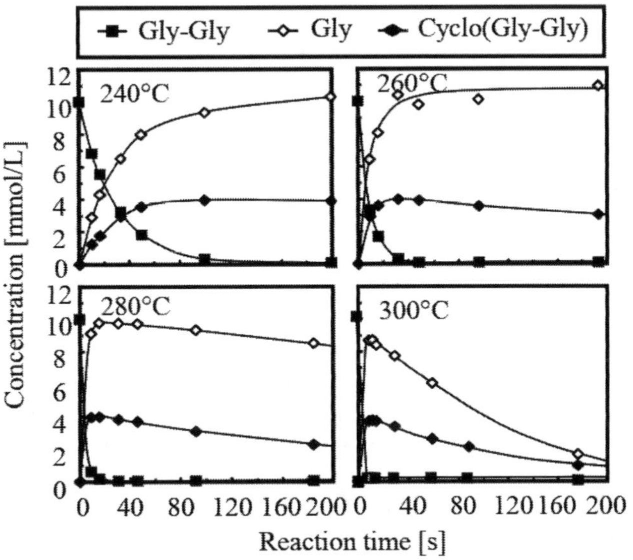

Figure 1. Effect of reaction time on the concentration of products at various temperatures.

It is also likely that reactions other than hydrolysis took place, producing other unexpected compounds. It has been reported that the condensation reaction of amino acids occurs under hydrothermal conditions.[18,2] The results in Table 1 has confirmed that the cyclic dehydrate of Gly-Gly (2,5-diketopiperazine, the cyclo(Gly-Gly), was produced from linear Gly-Gly. Although hydrolysis and dehydration are apparently considered to be contradictory, these reaction phenomena are often observed in high temperature water.[11]

The concentrations of cyclo(Gly-Gly) produced at each conditions are also shown in Table1. In order to verify the accuracy of the experimental results, the carbon and nitrogen balances before and after each reaction were recalculated. The carbon and nitrogen balances have an accuracy of ±10 %. Thus, it turned out that most of unknown carbon and nitrogen originated from cyclo(Gly-Gly). The total carbon and nitrogen concentration of reaction effluent at higher temperatures decreased with increasing reaction time, indicating that volatile compounds such as carbon dioxide might have been formed from the reaction. However, analysis of those compounds was not performed in this current work.

Figure 1 shows the molar concentration of the major reaction products of Gly-Gly at 240, 260, 280 and 300°C, at a constant pressure of 20 MPa. In each temperature, the concentration of Gly-Gly decreased with reaction time, producing Gly and cyclo(Gly-Gly). The concentration ratio of produced Gly to cyclo(Gly-Gly) was about 2.2:1 during the Gly-Gly conversion at 240, 260 and 280°C. It was observed that reaction temperature has no effect on the selectivity of hydrolysis and cyclodehydration of Gly-Gly in the temperature range of 240 to 280°C. At 300°C, the selectivity could not be evaluated because the reaction was so fast, and subsequent decomposition of products could have taken place. Moreover, organic acids and amines were also produced at longer reaction times and at higher temperatures. These products were obtained due possibly to the interaction of Gly-Gly with the decomposition products. The pH of reaction mixtures also increased with temperature, due to the formation of methylamine and ammonia (Table 1). The production of organic acids and amines results from the decomposition of Gly in high-temperature water.[23] However, at 260°C, the concentration of Gly did not change in relatively long reaction time. It is most likely that Gly was reproduced by ring-opening followed by decomposition of cyclo(Gly-Gly). Thus, further experiments on decomposition of cyclo(Gly-Gly) at the same reaction conditions were carried out. A 5 mmol/L aqueous solution of cyclo(Gly-Gly) was prepared, and was reacted at 260°C by the same procedure discussed in experimental methodology.

Figure 2 shows the molar concentration of the major products from cyclo(Gly-Gly) decomposition. The concentration of cyclo(Gly-Gly) decreased with reaction time, producing Gly. The result shows that indeed Gly could be produced from cyclo(Gly-Gly) as has been thought based on the results of our previous experiments on Gly-Gly. The presence of organic acids and amines was also observed.

3.2. Reaction Pathway of Gly-Gly Under Subcritical Water Conditions

Figure 1 shows that cyclodehydration started at relatively low temperature of 240°C with the formation of cyclo(Gly-Gly). The cyclodehydration of Gly-Gly, a dipeptide, observed in this study was an unexpected reaction. It was observed that cyclo(Gly-Gly) was stable at short reaction time up to 40 s under all temperatures in this study. Even at 400°C,[22] only small portion of cyclo (Gly-Gly) was hydrolyzed to yield glycine. As starting material decomposed, cyclo (Gly-Gly) was produced, then decreased as time progressed.

It is also possible to obtain linear dipeptide from the condensation of amino acid. However, this is minimal compared to the hydrolysis reaction to form amino acids. This could be explained by previous reports on the hydrothermal synthesis of peptide from amino acids which showed that the yields of condensed dipeptide were two orders of magnitude less than the amount of reactant.[25] It is reasonable to suppose that the same transformations took place in the case of homo-dipeptides such as Gly-Gly. However, no valid explanation can be obtained based only on the results in Figure 1, unless an in-situ study of the reaction mechanism is performed.

Table 1. Reaction products of glycylglycine at various temperatures.

Reaction temp. [°C]	Residence time [s]	Gly-Gly [mmol/L]	Gly-Gly conv.[1] [%]	Cyclo (Gly-Gly) [mmol/L]	Glycine [mmol/L]	NH_4^+ [mmol/L]	$CH_3NH_3^+$ [mmol/L]	Formic acid [mmol/L]	Glycolic acid [mmol/L]	Acetic acid [mmol/L]	Carbonic acid [mmol/L]	T-N[2] [mmol-N/L]	T-C[3] [mmol-C/L]	Nitrogen balance [%]	Carbon balance [%]	pH [-]
240	10.0	6.84	31.6	1.25	2.90	0.12	N.D.	0.03	N.D.	N.D.	N.D.	19.9	39.7	95.9	96.5	5.9
240	16.6	5.56	44.4	1.77	4.30	0.13	N.D.	0.03	N.D.	N.D.	N.D.	20.3	39.5	95.5	94.3	6.0
240	33.2	3.25	67.5	2.94	6.52	0.15	0.01	0.04	N.D.	N.D.	N.D.	20.1	39.6	95.2	94.5	6.4
240	49.8	1.82	81.8	3.56	8.01	0.16	0.01	0.04	N.D.	N.D.	0.07	20.2	39.1	94.7	93.7	6.5
240	99.6	0.32	96.8	3.96	9.35	0.19	0.01	0.05	N.D.	N.D.	0.09	20.3	39.7	90.5	89.1	6.8
240	199.2	0.09	99.1	3.88	10.29	0.23	0.01	0.06	0.06	N.D.	0.13	20.3	39.7	92.4	91.3	6.9
260	9.6	3.34	66.6	2.91	6.43	0.11	N.D.	N.D.	N.D.	N.D.	N.D.	20.0	39.7	95.1	94.9	6.2
260	16.0	1.68	83.2	3.59	8.08	0.13	N.D.	N.D.	N.D.	N.D.	0.06	20.3	39.7	93.8	92.3	6.4
260	32.0	0.30	97.0	3.98	10.36	0.18	0.01	0.03	0.07	N.D.	0.08	20.1	39.6	85.5	95.3	6.7
260	48.0	0.08	99.2	3.91	9.78	0.24	0.01	0.04	0.11	N.D.	0.13	20.1	39.4	90.0	89.4	6.9
260	96.4	0.09	99.1	3.55	10.09	0.44	0.04	0.05	0.32	N.D.	0.22	20.2	39.6	89.2	88.4	7.2
260	192.8	0.08	99.2	3.03	10.93	0.76	0.07	0.04	0.66	N.D.	0.31	20.2	39.4	89.9	89.1	7.3
280	9.3	0.64	93.6	3.98	9.09	0.21	0.01	0.03	0.10	N.D.	0.08	20.1	39.6	92.7	92.2	6.8
280	15.4	0.18	98.2	4.00	9.75	0.32	0.04	0.04	0.22	N.D.	0.10	20.0	39.6	92.5	92.3	7.0
280	30.8	0.08	99.2	3.83	9.72	0.61	0.07	0.06	0.48	N.D.	0.27	19.9	40.3	91.1	91.5	7.2
280	46.3	0.08	99.2	3.70	9.68	0.84	0.16	0.07	0.76	N.D.	0.43	20.1	40.0	91.2	90.9	7.4
280	92.4	0.08	99.2	3.09	9.29	1.43	0.40	0.12	1.56	N.D.	0.95	20.2	40.0	87.3	86.5	7.7
280	184.8	0.08	99.2	2.28	7.48	2.47	0.92	0.22	2.98	N.D.	2.05	20.3	39.9	83.0	81.7	8.0
300	8.8	0.07	99.3	3.73	8.67	0.62	0.11	0.05	0.48	N.D.	0.37	20.4	39.9	83.3	85.3	7.3
300	11.8	0.07	99.3	3.77	8.68	0.78	0.18	0.06	0.67	N.D.	0.52	20.2	39.8	85.8	87.5	7.5
300	14.7	0.07	99.3	3.73	8.39	0.97	0.28	0.09	0.89	N.D.	0.66	19.8	39.7	86.9	87.5	7.5
300	29.4	0.07	99.3	3.39	7.69	2.00	0.85	0.16	1.73	N.D.	1.42	19.7	39.8	88.5	88.2	7.8
300	58.8	0.07	99.3	2.63	6.01	3.84	2.23	0.24	3.06	N.D.	3.03	20.1	39.6	87.2	86.3	8.1
300	88.2	0.06	99.3	2.14	4.56	4.65	3.31	0.31	4.00	0.05	4.33	20.1	39.5	84.2	85.2	8.2
300	176.4	0.03	99.9	0.92	1.62	8.10	6.11	0.64	6.36	0.10	5.75	19.7	37.7	90.2	86.1	8.2

1) Glycylglycine conversion = $(1-[\text{Gly-Gly}] / [\text{Gly-Gly}]_{\text{initial}}) \times 100$ [%].
2) Total nitrogen concentration of sample solution measured by TN analyzer.
3) Total carbon concentration of sample solution measured by TOC analyzer.

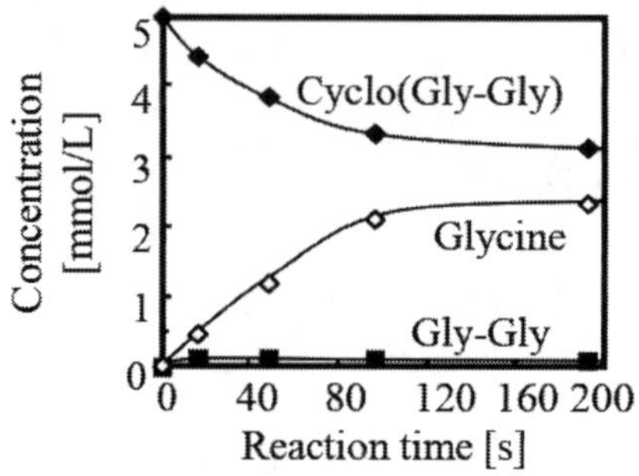

Figure 2. Concentration profile of Gly formation from cyclo(Gly-Gly) decomposition at 260°C and 20 MPa

The speculated reaction pathway of Gly-Gly under subcritical water conditions based on the experimental result is described in Figure 3. Hydrolysis and cyclodehydration of Gly-Gly occurred simultaneously, producing Gly and cyclo(Gly-Gly). These consecutive reactions consisted of intermediate reversible reactions involving the transformation of amino acid sequence as a result of the dehydration of Gly-Gly to cyclo(Gly-Gly).

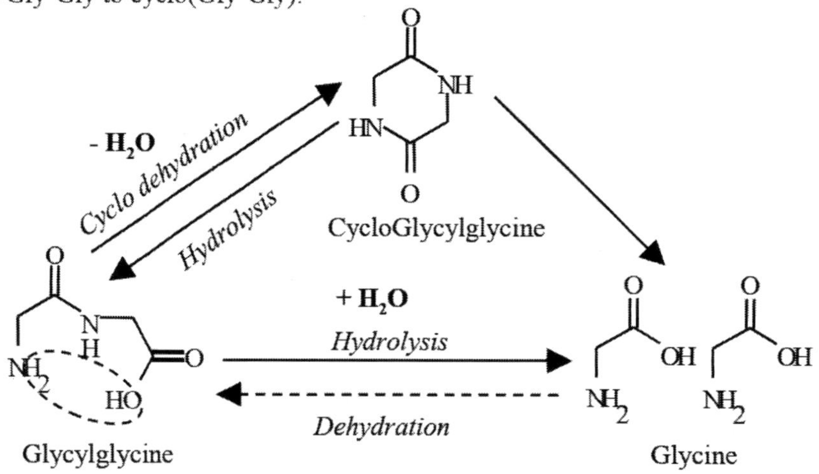

Figure 3. Reaction path of glycylglycine under subcritical water conditions.

This transformation is thought to be brought about by the hydrolysis of a different portion of peptide bond during the backward hydrolysis of cyclic to linear dipeptide. The intermediate reversible reactions involving linear dipeptides progress while the equilibrium shifts towards hydrolysis and cyclodehydration to form amino acid and cyclic dipeptides, respectively. The condensation of amino acids to linear dipeptides (i.e. 2Gly → Gly-Gly) may also take place, but is minimal compared to its backward reaction. Previous studies on the hydrothermal synthesis of peptides using amino acids as a starting material have also shown that the concentration of produced peptide is over two magnitude lower than the concentration of the starting material.[18] The produced amino acid decomposed further at relatively strong conditions, resulting to formation of organic acids and amines.[23] As the amount of amino acid decreases, the equilibrium shifts toward formation of amino acid from cyclic via linear dipeptide. The abovementioned reactions could be the reason for the low rate and yield of amino acid production from hydrothermal hydrolysis of proteins.

4. Conclusions

Hydrolysis of Gly-Gly was investigated over a temperature range of 240 to 300°C at a constant pressure of 20 MPa. Hydrolysis, as well as cyclodehydration of Gly-Gly occurred, producing Gly and cyclo(Gly-Gly), respectively. Based on the results, three consecutive reversible reactions (i.e. cyclo (Gly-Gly) ↔ Gly-Gly ↔ Gly) were speculated to take place during the hydrothermal treatment of Gly-Gly without any additives. These reactions, particularly the cyclodehydration of peptide (Gly-Gly), would limit the yield of amino acids. This could explain the reason for the low yield of amino acids from the hydrothermal decomposition of proteinaceous wastes and materials as observed in our previous studies. Thus, suppressing the dehydration of peptide is necessary to improve the yield of amino acids. The results obtained here stimulate further investigation on the control of reaction selectivity by varying reaction parameters such as pressure, pH, additives and temperature beyond the range investigated in this study. If the reaction selectivity of peptide in high-temperature water could be controlled, the subcritical water reaction would be a novel medium for production of cyclic dipeptide.

References

[1] Faisal M, Quitain A T, Daimon H and Fujie K 2007 Int. J. Chem. Kinet. **39 (3)** 175
[2] Faisal M, Saeki T, Daimon H, Tsuji H and Fujie K 2007 Asian J. Chem. **19 (3)** 1714
[3] Iyer S D, Klein M T 1997 J. Supercrit. Fluids **10** 191
[4] Penninger J M L, Kersten R J A and Baur H C L 1999 J. Supercrit. Fluids **16** 119
[5] Krammer P and Vogel H 2000 J. Supercrit. Fluids **16** 189
[6] Lachance R, Paschkewitz J, DiNaro J and Tester J W 1999 J. Supercrit. Fluids **16** 133
[7] Sato N, Saeki T, Daimon H and Fujie K 2001 Kagaku Kogaku Ronbunsu **27** 652
[8] Sasaki M, Kabyemela B, Malaluan R, Hirose S, Takeda N, Adschiri T and Arai K 1998 J. Supercrit. Fluids **13** 261
[9] Quitain A T, Faisal M, Kang K, Daimon H and Fujie K 2002 J. Hazard. Mater. B **93**, 209
[10] Holliday R L, Jong B Y M and Kolis J W 1998 J. Supercrit. Fluids **12** 255
[11] Savage P E 1999 Chem. Rev. **99** 603
[12] Daimon H, Kang K, Sato N and Fujie K 2001 J. Chem. Eng. Japan **34** 1091
[13] Yoshida H, Terashima M and Takahashi Y 1999 Biotechnol. Prog. **15** 1090
[14] Kang K, Quitain A T, Daimon H, Noda R, Goto N, Hu H Y and Fujie K 2001 Can. J. Chem. Eng. **79** 65
[15] Quitain A T, Sato N, Daimon H and Fujie K 2001 Ind. Eng. Chem. Res. **40** 5885
[16] Bada J L, Miller S L and Zhao M 1995 Origins Life Evol. B. **25** 111
[17] Sato N 2003 *Study on Chemical Reactions of Peptide and Amino Acid in High-Temperature and High-Pressure Water* (Ph.D. Dissertation, Toyohashi University of Technology, Toyohashi, Japan)
[18] Imai E, Honda H, Hatori K, Brack A and Matsuno K 1999 Science **283** 831
[19] Alargov D K, Deguchi S, Tsuji K and Horikoshi K 2002 Origins Life Evol. B **32** 1
[20] Belsky A J, Li T J and Brill T B 1997 J. Supercrit. Fluids **10** 201
[21] Izzo B, Harrell C L and Klein M T 1997 AIChE J. **43** 2048
[22] March J. 1992 *Advanced Organic Chemistry: Reactions, Mechanisms, and Structure* (4^{ed}., John Willey and Sons: New York)
[23] Yachnikov I S and Spiridonova A S 1934 J. Gen. Chem. **4** 1286
[24] Sato N, Daimon H and Fujie K 2002 Kagaku Kogaku Ronbunshu **28** 113
[25] Hrada O, Ikushima Y, Hatakeda K, Sato O, Aizawa T, Kanakubo M and Kawakami T Noncatalytic Synthesis of Oligopeptide using Supercritical Water: *Proc. 33th Autumn Meeting of the SCEJ (Hamamatsu, Japan. 2000)* paper no N115

Journal of Physics: Conference Series **121** (2008) 082003 doi:10.1088/1742-6596/121/8/082003

Solvothermal processes: new trends in Materials Chemistry

G Demazeau
ICMCB, CNRS, University BORDEAUX 1, site de l'ENSCPB,
87 Avenue du Dr A. Schweitzer, 33608 PESSAC, Cedex, France

E-mail: demazeau@icmcb-bordeaux.cnrs.fr

Abstract. Solvothermal processes are characterized by mild temperature conditions. The use of high pressure appears important for enhancing chemical reactivity. The roles of the solvent and the reagents in the mechanisms governing the formation of a specific material are underlined. Some new directions of research, at the interface with others scientific domains, are discussed.

1. Introduction

Solvothermal processes can be described as a reaction or a transformation of precursor(s) in presence of a solvent in a close system and at a temperature higher than the boiling temperature of the solvent; Consequently at least autogeneous pressure parameter is involved.
Versus the respective chemical properties of the precursor(s) and the solvent such a system can be homogeneous (if the precursors are soluble in the corresponding temperature and pressure conditions) or heterogeneous (if the precursor is partially soluble in such conditions). Very often in order to improve such a solubility an initial high pressure is used.
Versus the temperature, the system can be in subcritical and supercritical conditions.

The most developed solvent was water due to its presence at the surface of the earth and its specific role in geochemistry. Different objectives have been assigned to hydrothermal processes: (i) to investigate the elaboration of different natural materials in order to improve the knowledge in Earth-Science [1,2], (ii) to develop hydrometallurgy for recovering ores [3], (iii) to synthesize novel materials, very often metastable when the kinetical effects were predominant [4-10], (iv) to prepare single crystals of oxides – in particular that corresponding to low-temperature structural varieties [10], (v) to elaborate fine particules as micro- or nanocrystallites [11-16].

During these last twenty years non-aqueous solvents were developed for preparing in mild conditions oxides or non-oxides materials through solvothermal processes – mainly as fine particles with specific size and shape for different applications.

Solvothermal reactions are governed by different factors:

- the nature of the precursors, in particular their physico-chemical properties (solubility, thermal stability…),
- the nature of the solvent (chemical composition, physico-chemical properties (solvatation, polarity, viscosity, ability to stabilize some complexes as intermediate steps…),
- the thermodynamical properties used during such a process (pressure and temperature).

Pressure and temperature can play an important role, some properties of the solvent such as density, viscosity…, changing drastically versus such parameters. Consequently the diffusion and reactivity of chemical species can be strongly affected. In a general way, solvothermal processes were developed in mild temperature conditions.
The main chemical reaction-types involved in solvothermal processes are: hydrolysis, complex-formation, metathesis, oxidation-reduction…
The reaction-type governing the in – situ mechanisms is closely related to the optimization of the three main factors governing the process. (chemical nature of the precursors, physico-chemical properties of the solvent, thermodynamical conditions). Such an optimization

© 2008 IOP Publishing Ltd

depends on the final objective: synthesis of a novel material (thermodynamically stable or metastable), preparation of fine crystallites well defined in size and morphology, growth of single crystal…

The objective of such a paper is to underline the role of the solvent and that of the reagents in the mechanisms governing solvothermal reactions.

2. Role of the solvent in solvothermal processes

Solvent, through its physico-chemical properties, plays a key role in solvothermal processes.

Four examples can illustrate such a role: (i) the orientation of the structural form, (ii) the development of a specific shape for nanocrystallites, (iii) the formation of specific metastable compounds through template action, (iv) the role of the oxidation – reduction properties characterizing the solvent.

2.1. Orientation to a metastable structural form: MnS with the wurtzite or zinc blende structures

Using $MnCl_2$, $4H_2O$ and $SC(NH_2)_2$ as reagents, the stabilization of a specific structural form for MnS has been recently studied [17]. With a solvent able to form a stable Mn complex (water or ethylediamine), the thermodynamically stable form (α-MnS with the rocksalt structure) is prepared. With a non polar solvent as benzene the formation of the wurtzite-type structure is observed (γ-MnS). On the contrary with tetrahydrofurane (THF) the zinc-blende structure (β-MnS) is stabilized. If during the synthesis mechanisms the formation of a stable complexe ($Mn(H_2O)_6^{2+}$ or $Mn(en)_3^{2+}$) as an intermediate step leads to the stable structural form on the contrary it appears that with benzene or THF the metastable forms (γ or β) can be stabilized, the solvothermal reaction being kinetically control, the orientation toward wurtzite or zinc-blende seems to be correlated in a first analysis to the polar character of the solvent.

2.2. Preparation of nanoparticles with a specific morphology: intermetallic FePt nanowires

The solvothermal preparation of FePt nanoparticles characterized by a one-dimensional (1D) nanostructure has attracted intensive research due to its potential applications in high-density data storage and high performance permanent magnets [18,19]. In the main developed processes FePt particles were synthesized via thermal decomposition of $Fe(CO)_5$, reduction of platinum acetyl acetonate ($Pt(acac)_2$) and co-reduction of iron salt and Pt $(acac)_2$ [20-23]. Unfortunately the morphology of the resulting nanocrystallites is difficult to control. Due to the flexibility of solvothermal routes, such a process has been developed. Ethylenediamine was selected as solvent due to its ability to form a stable complex ion with Platinum $[Pt(en)_2^{2-}]$ then a co-reduction of such complex ion through the thermal decomposition of $Fe(CO)_5$ led to the formation of FePt nanowire [24].

Recently Y. HOU et al. [25] have investigated the solvothermal growth process through TEM observation of the corresponding samples quentched at different reaction times. Starting from aggregated small particles with short rod morphology acting as nuclei, a continual transformation through coarsening and ripening reactions is observed. The whole transformation to 1D nanowires at a temperature close to 180°C can be compared to the surface reconstruction phenomena discovered previously for Pt and its alloys at high temperatures in presence of reactive gases (NH_3 and oxygen) [26,27]. Such a study underlines the importance of the reaction mechanisms for controlling the resulting morphology and the role of solvothermal reactions for reducing the temperature.

2.3. Formation of metastable compounds complex ions acting as template
The solvent through the formation of very stable complex ions can participate to the final composition. Ethylediamine has a strong coordination ability towards the divalent metal ions as Mn^{2+}, Fe^{2+}, Co^{2+}, Ni^{2+} and Zn^{2+}. These complex ions ($M(en)_3^{2+}$) are characterized by an octahedral geometry acting as template for the building of the resulting structure [28].

2.4. Oxidation – reduction properties of the solvent and the preparation of sulfides
S.K. PANDA et al. [29] have underlined the role of the oxidation-reduction process induces by the solvent during the solvothermal synthesis of SnS crystallites from Sn metal and thiourea as reagents. The first step of the reaction mechanism being the oxidation of tin ($Sn^0 \rightarrow Sn^{2+} + 2e^-$), the use of pure ethylenediamine as solvent – due to its slightly reducing properties – does not oxidize Sn metal. The addition of few amounts of water induces the oxidation of the Sn surface and leads to the formation of SnS through the following steps [$Sn^{2+} + n\ en \rightarrow Sn(en)_n^{2+}$ then $Sn(en)_n^{2+} + S^{2-} \rightarrow SnS$].

3. Role of the reagents in solvothermal processes.
Different examples have been selected for illustrating how the physico-chemical characteristics (chemical composition, structure, size and morphology of the reagents) can play a key-role in the solvothermal processes.

The solvothermal synthesis of the phyllosiloxides – bidimensional oxides, isostructural of the natural phyllosilicates as mica – was only possible through a sol-gel process using alkoxides in a first step, the conventional Solid-State chemical processes leading to a mixture of 3D-silicates [30,31].

The elaboration of the metastable phase $Rb_2Hg_3Te_4$ through a solvothermal process has been investigated by Li et al. [32]. Using Rb_2Te, Hg_2Cl_2, Te and $FeCl_2$ as reagents and ethylenediamine as solvent, the role of Fe^{2+} as possible reducing agent seems to play an important role for stabilizing $Rb_2Hg_3Te_4$.

During the solvothermal preparation of $CdIn_2S_4$ using CdS, $InCl_3.4H_2O$ and thiourea as reagents and water as solvent – the morphology of the final $CdIn_2S_4$ crystallites is strongly correlated to that of the reagents CdS [33].

4. Main potential developments for solvothermal processes.
In addition of the conventional domains (synthesis of novel materials [4], elaboration of single crystals [34,35], preparation of nanocrystallites [36], solvothermal reactions are able to be developed in different scientific areas: synthesis of hybrid materials – involving not only organic-inorganic systems but mainly in the near future bio-organic/inorganic systems - [37,38], the decomposition and the recycling of wastes [39-41]...
For the development of such novel scientific promising domains, the mild temperature conditions playing an important role.
Pressure parameter is able to have also a key role. Due to the low-energy developed in liquid phase (Table) [42] such a parameter does not affect strongly the bio-organic molecules and consequently allows the synthesis of bio-organic/inorganic hybrid systems with potential applications in nano-energy-devices, catalysis....[43].

Table: Comparison of the energy conveyed by HIGH PRESSURES and the energy of a CHEMICAL BOND [42]

Pressure (en bar)	medium	Energy (en cal/mol)
1 000	Gas	3 000
1 000	Solid	1
10 000	Solid	5
100 000	Iron	20
100 000	H_2O	1 000
1	Chemical reaction	20 000

In parallel pressure improving in most cases the solubility of reagents by several orders of magnitude [44] consequently the chemical reactivity can be strongly improved.

4. Conclusions

Solvothermal processes have been investigated during these last twenty years. The recent development of non-aqueous solvents has improved the synthesis of different materials (chalcogenides, nitrides, …) mainly as nanoparticles. Three different factors play an important roles underlined: the mild temperature conditions, the improvement of the chemical reactivity versus pressure and the limited pollution due to the equipments as close-systems pollution. The overlap of these factors allows to consider solvothermal processes to play a key-role in the future.

5. References

[1] Goranson R W 1931 *Am. J. Sci.* **22** 481
[2] Hosaka M 1991 *Prog. Crystal Growth and Charact. Materials* **21** 71
[3] Habashi F 2005 *Hydrometallurgy* **79** 15
[4] Demazeau G 1999 *J. Mater. Chem.* **9** 15
[5] Demianets L N 1990 *Prog. Crystal Growth and Charact. Materials* **21** 299
[6] Rabenau A, Rau H 1969 *Philips Tech. Rev.* **30** 89
[7] Rabenau A 1985 *Angew. Chem. Int. Ed. Engl.* **24** 1026
[8] Yoshimura M 1998 *J. Mater. Res.* **13** 796
[9] Feng S, Xu R 2001 *Acc. Chem. Res.* **34** 239
[10] Walton R I 2002 *Chem. Soc. Rev.* **31** 230
[11] Niederberger M, Garnweitner G, Buha J, Polleux J, Ba J, Pinna N 2006 *J. Sol-Gel Techn.* **40** 259
[12] Bocquet J F, Chhor K, Pommier C 1999 *Mater. Chem. Phys.* **57** 273
[13] Wang C, Deng Z X, Zhang G, Fan S, Li Y 2002 *Powder Technol.* **125** 39
[14] Chen D, Jiao X, Chen D 2001 *Mat. Res. Bull.* **36** 1057
[15] Li W J, Shi E W, Chen Z Z, Zhen Y Q, Yin Z W 2002 *J. Solid State Chem.* **163** 132
[16] Wang Y W, Xu H Y, Wang H, Zhang Y C, Song Z Q, Yan H, Wan C R 2004 *Solid State Ionics* **167** 419
[17] Lu J, Qi P, Peng Y, Meng Z, Yang Z, Yu W, Qian Y 2001 *Chem. Mater.* **13** 2169
[18] Weller D, Doerwer M F 2000 *Annu. Rev. Mater. Sci.* **30** 611
[19] Sun S, Murray C B, Weller D, Folks L, Mojer A 2000 *Science* **287** 1989
[20] Harrell W, Wang S, Nikles D E, Chen M 2001 *Appl. Phys. Lett.* **79** 4393
[21] Zeng H, Li J, Liu J P, Wang Z L, Sun S 2002 *Nature* **420** 395
[22] Chen M, Liu J P, Sun S 2004 *J. Am. Chem. Soc.* **126** 8394
[23] Guo Q, Teng X, Yang H, 2004 *Adv. Mater.* **16** 1337
[24] Mao C B, Solis D J, Reiss B D, Kottmann S T, Sweeney R Y, Hayhust A, Georgiou G, Iverson B, Belcher A M 2004 *Science* **303** 213

[25] Hou Y, Kowdoh H, Che R, Takeguchi M, Ohta T 2006 *Small* **2**(2) 235
[26] Zhang Z, Blom D A, Gai Z, Thompson J R, Shen J, Dai S 2003 *J. Am. Chem. Soc.* **125** 7528
[27] Lyubousky M R, Barelko V V 1994 *J. Catal.* **149** 23
[28] Li J, Chen Z, Wang R J, Proserpio D M 1999 *Coordination Chem. Rev.* **190-192** 707
[29] Panka S K, Gorai S, Chauduri S 2006 *Mater. Sci. Eng. B.* **129** 265
[30] Reig P, Demazeau G, Naslain R 1995 *Eur. J. of Solid State and Inorg. Chem.* **32** 439
[31] Demazeau G, Reig P, Naslain R 1995 *Mat. Tech.* **10** 57
[32] Li J, Chen Z, Lam K, Mulley S, Proserpio D 1997 *Inorg. Chem.* **36** 684
[33] Hu J Q, Deng B, Zhnag W X, Qian Y T 2001 *Inorg. Chem.* **40** 3130
[34] Iwasaki F, Iwasaki H 2002 *J. Crystal Growth* **237-239** 820
[35] Denis A, Goglio G, Demazeau G 2006 *Mat. Sciences Engineer. R* **50** 167
[36] Rao C N R, Deepak F L, Gundiah G, Govindaraj A 2003 *Progress in Solid State Chem.* **31**(1) 5
[37] Umetsu M et al. *ISHR-8, ICSTR-7 Joint Conference SENDAI* august 2006
[38] Kono S, Oke Y, Yamasaki N *ISHR-8, ICSTR-7 Joint Conference SENDAI* august 2006
[39] Yamasaki N, Yasui T, Matsuoka K 1980 *Environmental Science Technology* **14**(5) 550
[40] Park M, Komarneni S, Roy R 2000 *Materials Letters* **43**(5) 259
[41] Kruse A, Dinjus E 2007 *J. Supercritical Fluids* **41** 361
[42] Wentorf R H 1961 *Jr. Chemical Engineering* oct. 16, p 177
[43] Proceedings of the Joint Meeting of Eighth International Symposium on Hydrothermal Reactions & Seventh International Conference on Solvothermal Reactions - ISHR & CSTR 2006 *J. Materials Science.* In press (DOI 10.1007/s10853-007-2024-9)
[44] Kwauss, K.G., Dibley, M.J., Bourcier, W.L., and Shaw, H.F., 2001 *Appl. Geochem.* **16** 1115

Achnowledgments
The organizing Committee of the joint 21st AIRAPT and 45th EHPRG Conference (Catania - 2007) for organizing a specific Session devoted to Solvothermal Reactions and the International Solvothermal-Hydrothermal Association (ISHA) for helping the development of such processes.

Processing of new materials using thermal and thermo-vaporous treatment of terephthalates

Mishal Khaddaj[1], **G. P. Panasyuk**[1], **L. A. Azarova**[1], **I. L. Voroshilov**[1], **A. P. Savostyanov**[1]

[1] IGIC RAS, (N. S. Kurnakov Institute of General and Inorganic Chemistry, Russian Academy of Sciences)
¤ E-mail for correspondence: mishal@fromru.com

Abrstract : Utilization of used polyethyleneterephthalate (PET) is now one of the most important ecological problems.
Its solution is related mainly with various difficulties. The solution of this problem seems to be in the creation of remunerative processes, for which post-consumer PET is used as a source material. Were developed several highly profitable processes of utilization of PET articles.
Keywords: PET, utilization, autoclaving, terephthalates

Polyethylene terephthalate (PET) is widely used in the production of packing material - bottles, containers, film and others. Because of the large volume of production (up to 5 millions tones per year) vital problem is the recycling of the articles made of the PET used.
Basic difficulty in this case is, in particular, the collection of post-consumer bottles, since the weight of a bottle does not exceed, usually, 50 g. Collection of the post-consumer bottles sufficiently labor-consuming and expensive process. Taking into account the preliminary preparation of the used articles (washing, drying), the cost of source material for the recycling becomes so high that the majority of the processes of recycling turns unprofitable.
Therefore immediate is the development of the highly remunerative processes of the utilization of the articles made of the post-consumer PET.
The present work represents some developed highly remunerative methods of the utilization of the post-consumer PET articles. In our opinion, the following processes are most promising:

1. Production of comparatively expensive and claimed articles made from post-consumer PET.
2. Decomposition of PET into the monomers - separately clean terephthalic acid (TA) and the ethylene glycol (EG)
3. Obtainment of various terephthalates and their use for the production of advantageous products and in different technological processes..

The production of articles made of the post-consumer PET is the now most widespread process of its utilization. By Basic difficulty here is related to the prohibition in the legislation of the majority of the countries concerning the use of post-consumer PET for the production of food containers (bottle-to-bottle production), and also the high prime cost of source material.
The nonfood market for articles made of post-consumer PET is sufficiently limited. This fact substantially hampers the creation of the highly remunerative processes of PET utilization by this method.
We developed the process of the transfer of post-consumer PET into the brittle state. Process consists in treatment of post-consumer PET in the autoclave at a temperature of about 200°C. PET converts to crystalline state, it becomes brittle and it can be crushed to the size of the particles of several microns.
This material can be used as filler for epoxy and polyester resins, for creating of epoxy and polyester paints and puttyings, and also be used as the source material for the process of decomposition of PET into its initial components.
Figure 1 shows that, in spite of differences in X-ray diffraction patterns of the PET films (fragments of bottle) before and after autoclave treatment of amorphous and crystalline, the crystallization of polymer occurs without decomposition of PET.

© 2008 IOP Publishing Ltd

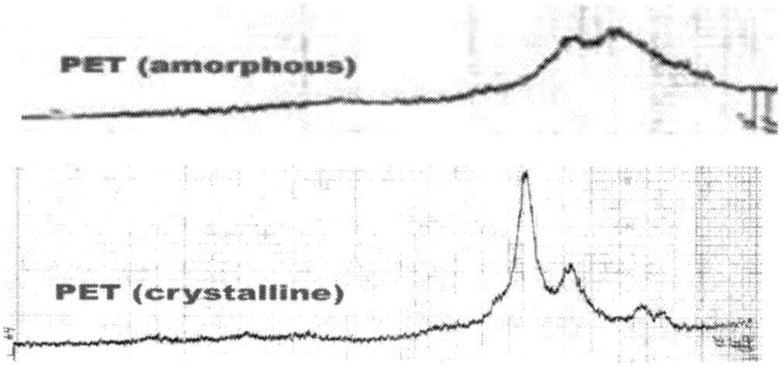

Figure 1. of the X-ray diffraction pattern of amorphous and crystalline PET

Figure 2 represents electronic microphotography of crystalline PET, and indicates the nanometric size of crystallites.

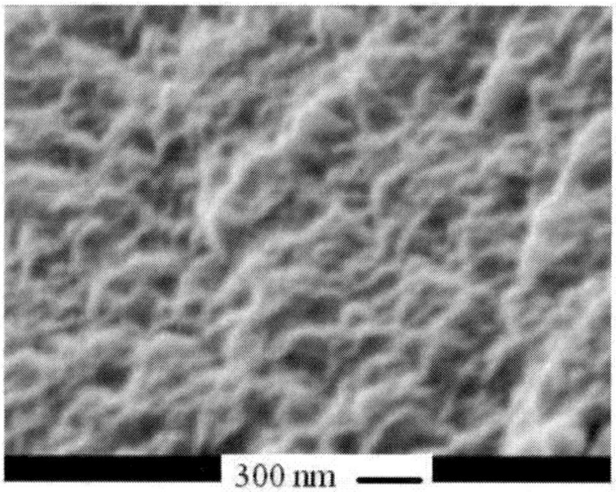

Figure 2. PET after autoclave treatment at 190-220°C under pressure of 1-2 atm. for 20 hrs.

Earlier [1] were developed the processes of decomposition of the used articles made of amorphous PET into the monomers by their treatment in the autoclave in solutions at temperatures of about 200°C in acidic (HCl), neutral (H_2O) and alkaline (NaOH) media.
The use of crystalline PET makes it possible to reduce the temperature of autoclave treatment to 100-120°C. Reaction medium can be the solution of ammonium hydroxide, in which amorphous PET does not dissolve even at 200°C. For the realization of this process widespread autoclaves, which can function to pressures up to 0,5 MPa, can be used. From the obtained solution of ammonium terephthalate after the addition of nitric acid crystals of TA with the size in the range of 1- 5 microns precipitate. To obtain pure terephthalic acid it is necessary to recrystallize TA by dissolving it in solution of ammonium hydroxide and, subsequently, precipitating it by mineral acid.
The practical absence of a waste is a special feature of this process, as ammonium nitrate obtained after the precipitation of TA is a component of mineral fertilizers. The proposed process makes it possible to obtain TA without organic impurities. This fact is confirmed, in particular, by the results of NMR-spectroscopy studies. Figure 3 presents the NMR spectra of vendible pure terephthalic acid (production of "Merk Schuchardt OHG" and of samples of TA obtained after the dissolution of the crystalline used bottles in 5% solution of ammonium hydroxide with subsequent precipitation by nitric acid. The NMR spectra of ^{13}C were obtained on "Bruker AC -200" spectrometer. The identity of spectra indicates the absence of organic impurities.

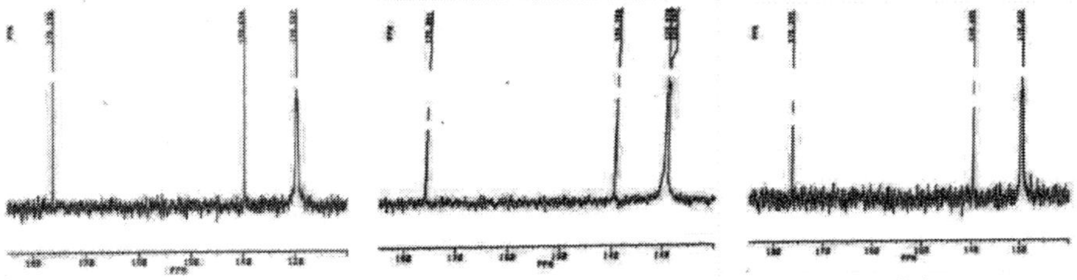
(a) (b) (c)

Figure 3. spectra NMR: (a)- TA ("Merk"), (b)- TA obtained in the autoclave from PET hydrolysis in ammonia, (c)- recrystallized TA from ammonium terephthalate.

The quality of terephthalic acid, besides contaminant composition is qualified by size of TA crystals. The industrial terephthalic acid, which is used for the production of food containers has a size of crystals of about one micron. This acid is not entirely convenient for other technological processes, in particular, for obtaining polyester resin, where it is desirable to have a size of crystals on the order of 50 microns. At the same time, TA produced, for example, by "Merck Schuchardt OHG" has size of crystals in the range 50-150 microns. In our work the processes of agglomeration of TA crystals were regarded.

According to measured impurity contents in various TA samples (see Table) the obtained terephthalic acid not only an analogous to the pure one produced by "Merck Schuchardt OHG", but also from many parameters, even it exceeds the "Merck " one.

Table Impurity content in various TA samples (mass %.)

	Na	K	Ca	Mg	Fe	Si
1	0.0040	0.0014	0.0022	0.0040	0.0002	0.0360
2	0.0860	0.0015	0.0106	0.0006	< 0,0002	0.0200
3	0.0042	0.0009	0.0124	0.0004	< 0,0002	0.0012

1) Acid produced by "Merck Schuchardt OHG" Germany.
2) Acid obtained under laboratory conditions from the hydrolysis of PET in 5% alkaline solution.
3) Acid obtained under laboratory conditions from the hydrolysis of PET in 5% alkaline solution, recrystallized from ammonium hydroxide solution.

To increase the size of TA crystals two processes were developed:

1. Autoclave treatment of TA obtained immediately after precipitation from ammonium terephthalate solution by mineral acids. The optimal parameters of autoclave treatment were determined. It is shown that the treatment of initial TA in the autoclave in the vapors of water at 200°C makes it possible to increase the size of crystals to 50 microns.
2. Flocculation is a more promising method. It is shown that TA in the aqueous solutions of some salts flocculates, in this case an increase in the size of flocculi to 1000 microns is observed. The selection of flocculant makes it possible to obtain TA in a narrow range of the sizes of crystals. The utilized salts then are completely moved by washing of final crystals using solutions of mineral acids. The structure of flocculi in this case is not disrupted.

The obtained results make it possible to recommend the process of recycling the post-consumer PET for obtaining the pure TA, which can be used both in the production of food containers and in the production of polyester resins.

The realization of the process of ammonium terephthalate synthesis from the PET post-consumer articles makes it possible to obtain, practically, unlimited source of raw material for obtaining terephthalates - salts of TA. Earlier we developed the synthesis of ammonium terephthalate and terephthalates of II, III groups and their corresponding oxides, which are formed by thermal decomposition of terephthalates [2,3,4]. Most interesting is the use of terephthalates for obtaining nanocrystalls of aluminum and zirconium oxides. The interaction of ammonium terephthalate with fine-crystalline aluminum (Zr) hydroxide leads to the formation of crystalline supramolecular structures representing ammonium terephthalate, which contains amorphous aluminum or zirconium hydroxides. With their thermal decomposition crystals of α-alumina or monoclinic zirconium oxide with size of crystals in the range of 50-100 nanometers are formed. These crystals can be used, in particular, for obtaining the ceramics (Fig.4).

(a) **(b)**

Figure 4. The supramolecular structure of ammonium terephthalate with aluminum hydroxide: (a) initial crystals; (b) heated up to 1000°C.

Perspective is the use of salts of TA in different technological processes. It was shown that the use of solutions of terephthalates of potassium, sodium and ammonium, as activators in the process of thermovaporal treatment of aluminum hydroxide, makes it possible to obtain the fine-crystalline particles of corundum with the size of crystals up to 300 microns.
Without the use of activators such crystals cannot be obtained. Corundum crystals successfully are used as the abrasive powders and for creating abrasive tools.

References:

[1]. G.P. Panasyuk, M. Khaddaj, V.I. Privalov. The Transformations of Polyethyleneterephthalate upon Autoclave Treatment in Water Vapor. J. Plasticheskie Massy.2002, №2, pp. 27-31 (in Russian).
[2]. G.P. Panasyuk, L.A. Azarova, G.P. Budova and others. Synthesis and Properties of Ammonium Terephthalates. J. Neorganicheskie Materialy. 2002, V. 38, N 12, pp. 385-389(in Russian)..
[3]. G.P. Panasyuk, L.A. Azarova, M. Khaddaj and others. Synthesis and Properties of Sodium, Potassium, Magnesium, Calcium and Aluminum Terephthalates. J. Neorganicheskie Materialy. 2003, V. 39, N 12, pp. 1493-1498 (in Russian).
[4]. G.P. Panasyuk, L.A. Azarova, G.P. Budova. Synthesis and Properties of Copper Terephthalate. J. Neorganicheskie Materialy. 2007, V. 43, N 9, pp. (in Russian).

Production of Acetic Acid by Hydrothermal Two-step Process of Vegetable Wastes for Use as a Road Deicer

Fangming Jin[1,2]*, Yosiyuki Watanabe[2], Atsushi Kishita[2], Hisanori Kishida[3], Heiji Enomoto[2]

[1] State Key Laboratory of Pollution Control and Resources Reuse, College of Environmental Science & Engineering, Tongji University, Shanghai, 200092, China

[2] Graduate School of Environmental Studies, Tohoku University, Sendai 980-8579, Japan

[3] Environmental Systems Headquarters, Environmental Research and Development Center Hitachi Zosen Corporation, Kyoto 625-8501, Japan

* College of Environmental Science & Engineering, Tongji University, Shanghai, 200092, China, e-mail:fmjin@mail.tongji.edu.cn

Abstract. This study aimed to produce acetic acid from vegetable wastes by a new hydrothermal two-step process. A continuous flow reaction system with a maximum treatment capacity of 2 kg/h of dry biomass developed by us was used. Five kinds of vegetables of carrots, white radish, chinese cabbage, cabbage and potato were selected as the representation of vegetable wastes. First, batch experiments with the selected vegetables were performed under the condition of 300°C, 1 min for the first step, and 300°C, 1 min and 70% oxygen supply for the second step, which is the optimum condition for producing acetic acid in the case of using starch as test material. The highest yields of acetic acid from five vegetables were almost the same as those obtained from starch. Subsequently, similar the highest yield of acetic acid and experimental conditions from vegetables were also obtained successfully using the continuous flow reaction system. These results should be useful for developing an industrial scale process.

1. Introduction

With increasing environmental awareness and crude oil price, the conversion of biomass into chemicals is receiving an increased interest. Among several biomass conversion processes, the hydrothermal process is one of the most promising processes, because water at a high temperature and high pressure behaves as a reaction medium having outstanding properties. There have been extensive researches on the conversion of biomass into chemicals by hydrothermal treatment [1-4]. Our past studies on the oxidation of various food wastes in supercritical water showed that acetic acid was a stable intermediate product for almost all organic wastes selected [5-6]. This result gave us a hint, that is a large amount of acetic acid can be obtained by controlling reaction condition. So, we carried our a study on acetic acid production by wet oxidation (WO) of various carbohydrate biomasses for producing calcium/magnesium acetate (CMA) known as an environmentally friendly deicer [7-8].

© 2008 IOP Publishing Ltd

These results indicated that acetic acid can be obtained as a product and can be increased by optimizing the reaction condition of WO of organic waste, but there is a limit to the yield of acetic acid, i.e. only 11-13 % at most on a carbon basis. To enhance the production of acetic acid from carbohydrate biomass, a new two-step process was proposed [9-10]. The two-step process consists of both a hydrothermal reaction without a supply of oxygen (the first reaction step) and an oxidation reaction (the second reaction step). The first step is to accelerate the formation of 5-hydroxymethyl-2-furaldehyde (HMF), 2-furaldehyde (2-FA) and lactic acid because they can produce a large amount of acetic acid by their oxidation, and the second step is to further convert these products formed in the first step to acetic acid by oxidation with newly supplied oxygen. It was shown that the acetic acid yield was greatly increased by the two step process.

Further, according to results obtained with batch reactor, we developed a continuous flow reactor system to produce acetic acid, then convert the produced acetic acid into CMA [11], and examined the fundamental performance with starch as test material. Results indicated that the continuous-flow reactor system has the expected fundamental performance to produce acetic acid by the two-step reaction.

The purpose of this study was to produce acetic acid from vegetable wastes by the two-step process with the continuous-flow reactor system. The selection of vegetable waste as treatment target is because component of vegetable waste is singleness and the amount of vegetable wastes generated from vegetable market in the big city is considerably large very day. Particularly, a single component of wastes is important for easily controlling a continuous-flow process.

2. Experimental

2.1. Experiments with a batch reactor

All batch experiments were conducted in a batch reactor system described previously [5, 10]. A brief summary is as follows. First, the desired amounts of test material and deionized water were loaded into a batch reactor. After the loading procedure was completed, the reactor was immersed in a salt bath preheated to the desired temperature. After the desired reaction time, the reactor was taken out of the salt bath and immediately put into a cold water bath to quench the reaction. This procedure is called the first step reaction. After the first step reaction was completed, H_2O_2 was added to the cooled reactor and then the oxidation reaction took place in the salt bath as the second step. In the salt bath, the reactor was shaken, keeping it horizontal, to enhance mixing and heat transfer. Loading of a mixture of deionized water and hydrogen peroxide was set at 50% of the inner volume of the batch reactor in all experiments. The reaction time was defined as the time that the reactor was kept in the salt bath. Since the time required to raise the temperature of the reaction medium from 20 to 300°C is about 15 s, the real reaction time is shorter than the defined reaction time. Reaction pressures with and without a H_2O_2 supply were about 17 MPa and 9 MPa, respectively. All batch experiments were performed with degassed water and the reactor was purged with nitrogen in the first-step reaction.

2.2. Experiments with continuous flow reactor system

2.2.1. Apparatus

A continuous flow reactor for producing acetic acid from hydrothermal treatment of biomass developed by us was used in this study. Fig 1 shows flow diagram of the continuous flow reaction system. The system mainly consists of a slurry pump, plunger pump for the H_2O_2 supply, preheater and cooler for the reaction media, reactor, solid-liquid separator, filter, back-pressure regulator and a gas-liquid separator.

The reactor chamber is divided into two parts, bottom and top chambers, by placing two horizontal pieces of baffle plate near the mid-height (see Fig 2) for taking place the two-step reaction. The bottom chamber of the reactor is used for the hydrothermal reaction without H_2O_2 (the first reaction step), the top is for the oxidation reaction (the second reaction step), and the mixing chamber is set between the top and bottom chambers to mix the H_2O_2 and reactants. The H_2O_2 solution is fed into the mixing chamber from the top of the reactor through a pipe. The inlet of the feed slurry is at the bottom of the reactor and the outlet for the effluent is at the top of the reactor.

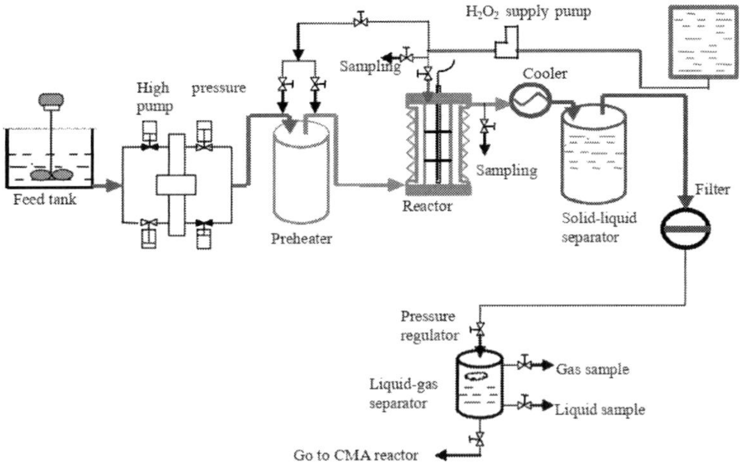

Figure 1. Flow diagram of hydrothermal conversion of biomass into acetic acid.

This continuous flow reactor has a maximum treatment capacity of 2kg/h of dry biomass. The treatment capacity and specification of the main compositions of the continuous flow reaction system is summarized in Table 1. Details on the description of the continuous flow reactor are available elsewhere [11].

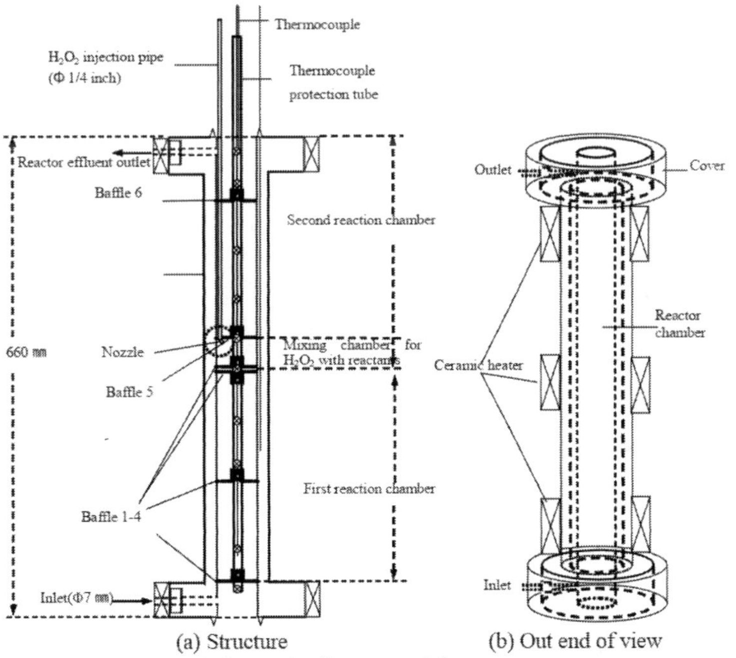

Figure 2. Schematic diagram of the two-step reactor.

2.2.2. Experimental procedure

Each experimental run was started by preparing the feed slurry, which was a mixture of water and vegetables. The feed slurry tank was agitated using a stirrer throughout the experiment. After preparation of the feed slurry, the system was heated while only water flowed. The pressure of the system was adjusted with the back-pressure regulator. After both the preheater and the reactor reached the desired temperature and pressure, the pump of hydrogen peroxide-water solution was switched on. After the hydrogen peroxide solution was allowed to flow for about 30 min, the feed slurry pump was turned on. One hour after the feed slurry pump was switched on, liquid and gas samples were collected at the bottom and top of the gas-liquid separator. The effluent gas flow rate was also measured.

Table 1 Treatment capacity and specification of the main compositions of the continuous flow reaction system.

	Specification
Treatment capacity	2 kg/h (dry base)
Feed pump	Flow rate: 7.5-20 L/h Pressure: 40 MPa max. Slurry Concentration: 10 wt% max.
Preheater	Temperature: 200-250°C Pressure: 40 MPa, max.
Reactor	Temperature: 430 °C max. Pressure: 40 MPa, max. Size: φ50×660 mm Volume: 1.3 L (HC-276)
Pressure valve	Temperature: 40°C, max. Pressure: 40 MPa, max.

2.3. Materials and oxygen supply

Five kinds of vegetables of carrots, white radish, chinese cabbage, cabbage and potato, as well as starch, were used as test materials. Vegetables were obtained from local vegetable market. Carrots, white radish are without leafage. In preparing vegetable slurry, the raw vegetable without dry was smashed into slurry by a mixer. The starch was an industrial potato starch (Nishida starch plant, Hokkaido, Japan) for continuous-flow experiments, and reagent grade potato starch for batch experiments. H_2O_2 was used as an oxidant for experimental convenience. The oxygen supply was defined as the ratio of the amount of H_2O_2 supplied to the stoichiometric demand for complete oxidation of a test material to carbon dioxide and water.

2.4. Analyses of samples

The solution and off-gas samples after reaction were analyzed by GC/MS, GC and HPLC. For batch experiments, only solution samples were analyzed. For GC/MS analyses, a Hewlett-Packard model 5890 Series II Gas Chromatograph equipped with a model 5989B Mass Selective Detector was used, with a HP-INNOWAX capillary column for water-soluble compounds using helium as the carrier. Intermediate products were identified with total and selected ion chromatograms with the aid of a computer library, as well as a comparison of GC retention time of the product and authentic compounds. HPLC analysis was performed with a Waters HPLC system equipped with a tunable absorbance detector (UV/VIS detector) (Waters 486) and a differential refractometer (RI detector) (Waters 410), controlled with a Millenium 600 workstation. Two kinds of column were used to separate intermediate products. To identify polar compounds, such as carboxylic acids, alcohols and ketones, RSpak KC-811 (Shodex) was used with a UV detector. To identify sugars, SUGAR SH1011 (Shodex) was used with a RI detector. Peak identification was accomplished by comparison of a sample peak retention time with that of standard solution of pure compound. Details on the conditions for GC/MS and HPLC analyses are described elsewhere [5,10]. Acetic acid was obtained quantitatively with a GC/FID/TCD (Hewlett Packard 5890II series plus). It was also used to analyze the off-gas samples.

The residual total organic carbon (TOC) in a liquid sample was measured with a TOC analyzer (Shimadzu TC 5000A, Japan).

3. Results and Discussion

3.1. Batch experiments

As reported in our previous study using starch as teat material [10], the optimum conditions for producing acetic acid in two-step process were at temperature and reaction time of 300°C, 1 min for the first reaction step, and at temperature, reaction time and oxygen supply of 300°C, 1 min and 70 % in the second reaction step, respectively. Thus, experiments with five kinds vegetables were conducted under the same conditions. As shown in Table 2, the yields of acetic acid obtained from five vegetables were almost the same each other. Comparing to starch, the acetic acid yield from vegetables seem to be a little lower than that from starch, but no significant difference was observed. These observations indicate that a little of composition in vegetable, for example protein, fatty acids, may have no significant effect on the production of acetic acid in the two-step reaction, although a significant difference between starch and vegetable in the yield of acetic acid, which the acetic acid yield from vegetable was lower than starch, was observed in general wet oxidation or directive oxidation. The study for reason why the acetic acid yield from vegetable was lower than one from starch is currently in progress.

Table 2 Acetic acid[a] yield obtained with batch experiments

Chinese cabbage	Cabbage	Potato	Carrots	White radish	Mixture[b]	Starch
16.5	17.0	17.3	16.8	17.1	17.0	17.5

300°C, 1 min for the first step, and 300°C, 1 min, 70% oxygen supply for the second step.
[a] The percentage of TOC of acetic acid and initial material.
[b] Mixture of five kinds in the same weight proportion for each vegetable.

Further, considering that real vegetable waste is a mixture, an experiment with a mixture of five kinds of vegetable was also performed at the same condition mentioned above. The mixture was prepared in the same weight percentage in dry base for each vegetable. As shown in Table 2, the acetic acid yield is also the almost same as those from each individual vegetable, indicating that each vegetable has no significant interaction in the production of acetic acid in the two-step reaction.

3.2. Continuous flow experiments

According to the results with the batch reactor, continuous flow experiments with vegetables were performed. In continuous flow experiments, only the mixture was used as test material. Two runs were carried out. The detail conditions for two run are shown in Table 3. Although results with batch experiments showed that there are no significant difference in the yield of acetic acid between starch and vegetables, considering that vegetables contains a little of cellulose being more difficult hydrolysis than starch, and the protein being more oxidation decomposition than starch, a longer reaction time of 2.0 min in the first-step reaction and 1.4 min in the second reaction step, as well as a higher H_2O_2 supply of 85% in

Table 3 Experimental condition for continuous flow system

Condition	Vegetable mixture		Starch
	Run 1	Run 2	
Temp. in Preheater/°C	270	270	260
Slurry			
Flow/L/h	10	10	10
Concentration/ wt%	3.5	3.5	3.5
The first-step			
Reaction Temp./°C	300	300	300
Residual time/ min	2.0	2.0	1.0
The two-step			
Reaction Temp./°C	300	300	300
Residual time/ min	1.4	1.4	1.0
H_2O_2 solution			
Flow/L/h	5.5	6.0	6.5
Oxygen supply/%	66	85	75
Pressure/ MPa	15	15	20

the second reaction step were chosen. But, for compareson, a run with 65% oxygen supply was also carried out. In order to attain economically viable acetic acid production, a higher concentration of initial material is desired, so, 3.5wt% slurry concentration was used. Moreover, system pressure decreased to 15MPa because our past research showed that the increase in pressure improved the oxidation of intermediate products, thus leaded to the decrease in the acetic acid [11]. Table 4 shows the yield and purity of acetic acid, and gas composition. It can be seen that, from Table 4, the acetic acid yields are almost the same to the result obtained with starch for both two run. However, the purity is lower for run 1 with 65% oxygen supply. So, the condition of run 2 should be a better condition for producing acetic acid from vegetables.

Table 4 Results obtained by continuous-flow reactor system for run 2

	Vegetable mixture		Starch
	Run 1	Run 2	
Acetic acid yield /%[a]	15.5	16.0	17.5
Acetic acid purity/%[b]	40	60	65
Off-gas composition			
CO_2	90	81	91
CO	5	4	6
O_2	3	14	3

[a] The percentage of TOC of acetic acid and initial material.
[b] The percentage of TOC of acetic acid and other residuals remaining in the liquid.

As mentioned before, our purpose in producing acetic acid is to produce calcium magnesium acetate (CMA), as a noncorrosive road deicer. So, products other than acetic acid are important. We made detailer identification for products in solution samples for the case of run 2 by GC/MS, HPLC. The identified products and their yield and purity are shown in Fig 3. Addition to acetic acid, many low molecular weight carboxylic acids were detected. Calcium compounds of these lower molecular weight carboxylic acids can also be used as a noncorrosive deicer. So, these low molecular weight carboxylic acids are useful products. In this case, the total purity of all carboxylic acids arrived to 70 %. Moreover, a little of amides, may come from the oxidation protein, are also formed. Amides can also be used a noncorrosive deicer. So, most of the identified products other than acetic acid are useful products in the case of manufacturing CMA.

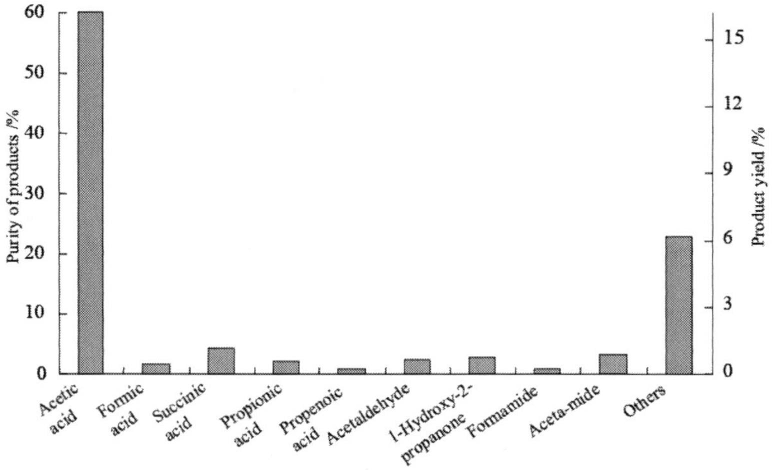

Figure 3. The yield and purity of products based on carbon in solution sample after reaction for run 2 (the definition of products yield and purity is the same to those in Table 2).

4. Conclusions

Batch experiments with the selected five kinds of vegetables were performed at the optimization condition for the case of using starch, the main component of vegetable, as a test material. The highest yield of acetic acid and the conditions for obtaining the highest acetic acid yield from vegetables were almost the same as those obtained from starch. Similar the highest yield of acetic acid and experimental conditions from vegetables were also obtained successfully using the continuous flow reaction system. These results should be useful for developing an industrial scale process.

References

[1] Akiya, N.; Savage. P. E. 2002, Roles of water for chemical reaction in high-temperature water, *Chem. Rev.*, 102: 2725–2750.

[2] Girisuta, A.; Janssen, L. P. B. M.; Heeres, H. J. 2006, A kinetic study on the conversion of glucose to levulinic acid, *IChemE.* 84: 339-349.

[3] Srokol, Z.; Bouche, A. G.; Estrik, A. E.; Strik, R. C. J.; Maschmeyer, T.; Peters J. A. Hydrothermal upgrading of biomass to bioduel; studies on some monosaccharide model compounds, 2004, *Carbohydrate Research*, 339: 1717-1726.

[4] Watanabe, M.; Sato, T.; Inomata, H.; Smith, R. L.; Arai, K.; Kruse, A.; Dinjus, E., 2004 *Chem. Rev.*, 104, 5803-5822.

[5] Jin, F.; Kishita, A.; Moriya, T.; Enomoto, H., 2000, Kinetics of oxidation of food wastes with H_2O_2 in supercritical water. *J. Supercritical Fluids*, 19, 251-262.

[6] Jin, F.; Kishita, A. T.; Enomoto, H., 2000, Oxidation of garbage in supercritical water. *High Pressure Research,* 525-531.

[7] Jin, F.; Kishita, A.; Moriya, T.; Enomoto, H.; Sato, N., 2000, The Production of Acetate for Use as A Roadway Deicer by Wet Oxidation of Organic Waste, ACS Division of Environmental Chemistry Preprints of Extended Abstracts, Presented at 220th ACS National Meeting: Washington, DC, 40, 73-79.

[8] Jin, F.; Kishita, A.; Moriya, T.; Enomoto, H.; Sato, N., 2002, A new process for producing Ca/Mg acetate deicer with Ca/Mg waste and acetic acid produced by wet oxidation of organic waste. *Chem. Lett.*, 31 (1), 88-89.

[9] Jin, F.; Zheng, J.; Enomoto, H.; Moriya, T.; Sato, N.; Higashijima, H., 2002, Hydrothermal process for increasing acetic acid yield from lignocellulosic wastes. *Chem. Lett*, 31 (5), 504-505.

[10] Jin, F.; Zhou, Z.; Moriya, T.; Kishda, H.; Higashijima, H.; Enomoto, H., 2005, Controlling hydrothermal reaction pathways to improve acetic acid production from carbohydrate biomass. *Environ. Sci. Technol.* 39 (6), 1893-1902.

[11] Jin, F.; Zhou, Z.; Kishita, A.; Enomoto, H.; Kishida, H.; Moriya, T, 2007, A new hydrothermal process for producing acetic acid from biomass waste, *IChemE*, 85 (A_2) 201-206.

Joint 21st AIRAPT and 45th EHPRG Int. Conf. on High Pressure Science and Technology IOP Publishing
Journal of Physics: Conference Series **121** (2008) 082006 doi:10.1088/1742-6596/121/8/082006

Solvothermal crystallization of nanocrystals of metal oxides

Satoshi Furukawa, Hideaki Amino, Shinji Iwamoto and Masashi Inoue[1]

Department of Energy and Hydrocarbon Chemistry, Graduate School of Engineering, Kyoto University, Katsura, Kyoto 615-8510, Japan

E-mail: inoue@scl.kyoto-u.ac.jp

Abstract. Solvothermal crystallization of the hydroxide gels obtained by hydrolysis of alkoxides (Zr, Ta, Nb, In, Sn, Ti and Al) was examined. Nanocrystals having high surface areas (S_{BET} > 170 m^2 g^{-1}) were obtained except for the product derived from indium isopropoxide. The effect of water in organic solvent upon the crystallinity of the product was investigated. The increase in the activity of water by using high concentration of alkoxide or intentional addition of water to the solvothermal medium led to crystal growth of the products. In contrast, decrease in activity of water by adding ethylene glycol before solvothermal treatment caused a decrease in crystallinity of the product.

1. Introduction

Nanocrystalline metal oxides have been widely used as gas sensors [1–3], catalysts [4,5], and electrodes [6–8], because of their unique physical and chemical properties. Therefore, many investigations on the synthesis of nanocrystalline metal oxides have been carried out [9–11]. The sol-gel method is widely applied to synthesize nanomaterials [12,13]. However, the products obtained by this method are usually amorphous, and therefore heat treatment of the products is required to obtain crystalline materials. Heat treatment may cause sintering and loss of the surface area of the products, which may restrict their practical use.

Solvothermal methods are widely explored for the synthesis of nanocrystalline materials [14–16]. In 1989, one of the present authors showed that α-alumina was synthesized from microcrystalline gibbsite ($Al(OH)_3$) in 1,4-butanediol at 285 °C [17], much lower temperature than that required by the hydrothermal reaction. He also examined the reaction of gibbsite in a series of alcohols and found that two different reactions took place depending on the carbon number of the alcohol [18]. When higher alcohols were used as the reaction media, hydrothermal reaction by the water formed by thermal dehydration of the starting material took place [16]. Supposing an amount of water is formed in organic media, the activity of water increases with the decrease in affinity of the media to water. The increase in activity of water leads to adsorption of water on the solid starting particles whereby facilitating the hydrothermal conversion of the starting materials [16]. These arguments imply that even though reaction was carried out in organic solvents, especially inert organic solvents such as toluene and mineral oil, the real chemistry might be hydrothermal one.

In the present research, we synthesized metal (Zr, Ta, Nb, Sn, Ti, In and Al) oxides by solvothermal

[1] To whom any correspondence should be addressed

© 2008 IOP Publishing Ltd

crystallization of the hydroxide gels derived by hydrolysis of metal alkoxides. One of the objectives of this work is to examine the effects of the activity of water on the solvothermal crystallization and it was confirmed that water present in the solvothermal reaction affects the crystallinity of the products, crystallite size of metal oxide nanocrystals being controlled by activity of water. Another objective is to synthesize large surface area oxide materials suitable for the catalysts.

2. Experimental

Eight alkoxides were used as precursors: zirconium n-propoxide (ZNP), zirconium isopropoxide (ZIP; Mitsuwa Chemicals Co., Ltd.), titanium tetraisopropoxide, aluminium triisopropoxide (Nakalai Tesque, Inc.), tantalum ethoxide, niobium ethoxide, tin isopropoxide and indium isopropoxide (High Purity Chemicals Co.). A desired amount of an alkoxide was dissolved in 100 mL of toluene. Hydrolysis of the alkoxide solution was carried out by keeping the solution in a constant humidity box (RH = 76 %). Water dissolved from the gas phase into the toluene solution slowly hydrolyzes the alkoxide yielding hydroxide gel. The solution was kept standing until no further change was observed. The thus-obtained hydroxide gel in toluene was transferred into a test tube and it was set in a 300 mL autoclave. An additional 50 mL of toluene was added to the gap between the test tube and the autoclave wall. The autoclave was purged with nitrogen, heated to 300 °C at a rate of 2.5 °C/min, and held at that temperature for 2 h under the autogenous pressure of toluene. The resulting precipitate was washed with methanol and air-dried. Calcination of the product was conducted in a box furnace at desired temperature for 1 h.

Table 1. Physical properties of the products obtained by solvothermal crystallization of the alkoxide-derived hydroxide gel.

Alkoxide	Concentration of alkoxide (g (L-tol.)$^{-1}$)	Calcination temperature (°C)	EG/In ratio	BET surface area (m^2 g^{-1})	Crystallite size (nm)	Calculated surface area [a] (m^2 g^{-1})
$Zr(OC_3H_7)_4$	15	-		340	ND [b]	-
	15	400		190	5	205
	32	-		370	ND [b]	-
	62	-		302	ND [b]	-
	103	-		228	4.9	209
	125	-		246	4.7	218
$Ta(OC_2H_5)_5$	52	-		279	ND [b]	-
	52	700		37.8	19	38
$Nb(OC_2H_5)_5$	32	-		246	ND [b]	-
	32	700		26.5	76	16
$In(O\text{-}^iC_3H_7)_3$	20	-		19.5	73	12
	20	-	8	38.0	54	16
	20	-	16	15.6	33	26
$Sn(O\text{-}^iC_3H_7)_4$	60	-		186	ND [b]	-
	60	400		110	4	-
$Ti(O\text{-}^iC_3H_7)_4$	29	-		179	8	193
	29	400		154	9	171
$Al(O\text{-}^iC_3H_7)_3$	30	400		434	ND [b,c]	-

[a] Calculated from the crystallite size assuming that each crystallite is truly spherical.
[b] XRD peak was too broad for determination of the crystallite size.
[c] Product was γ-alumina.

Journal of Physics: Conference Series **121** (2008) 082006 doi:10.1088/1742-6596/121/8/082006

3. Results and discussion

The BET surface areas and crystallite sizes of the products are summarized in table 1 together with the surface areas calculated from the crystallite sizes. Except for the product obtained from indium alkoxide, all the products had large surface areas over 170 $m^2\ g^{-1}$. The effect of the solvothermal crystallization on the BET surface area of the product is clearly shown by comparison with the sample obtained by direct calcinations of the hydroxide gel. Whereas the sample calcined at 400 °C after the solvothermal treatment had a surface area of 208 $m^2\ g^{-1}$, the sample without the solvothermal treatment had only 21 $m^2\ g^{-1}$ after calcinations at 400 °C.

The XRD patterns of the samples synthesized from various concentration of ZNP are shown in figure 1. All the samples exhibited the patterns due to tetragonal ZrO_2, and the sample synthesized from lower concentration of ZNP had smaller crystallite size.

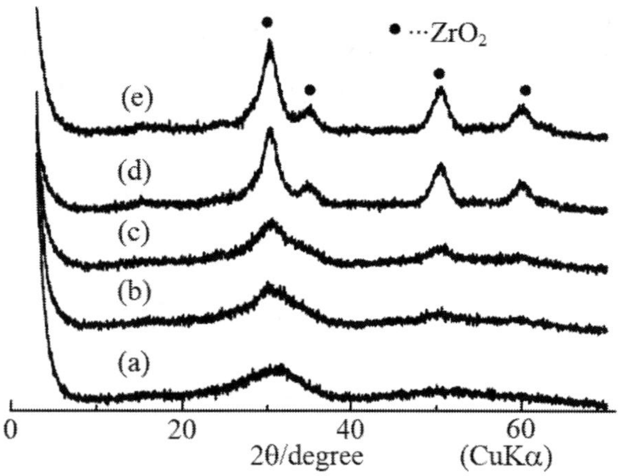

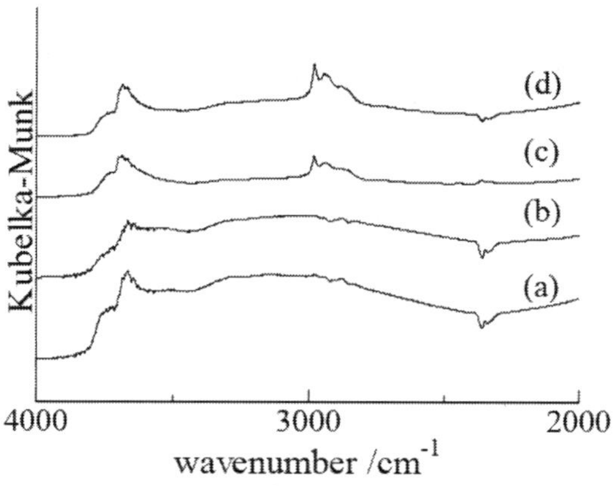

Figure 1. (left) XRD patterns of zirconia obtained by the solvothermal crystallization of the hydroxide gel derived with various concentrations of ZNP in toluene. Concentration of alkoxide (g (L-toluene)$^{-1}$): (a), 15.4; (b), 32.3; (c), 62.0; (d), 102.7; (e), 125.3.

Figure 2. (right) FT-IR spectra of: (a) and (b), the gels obtained by hydrolysis of ZNP; (c) and (d), ZrO_2 samples obtained by solvothermal crystallization of the gels.
Concentration of alkoxide (g/L-toluene): (a), 15.2; (b), 1254.2; (c), 15.4; (d), 125.3.

The DTA curves of the hydroxide gels (data not shown) exhibited an exothermic peak at about 450 °C due to the crystallization of the hydroxide gel into tetragonal ZrO_2. On the other hand, the products obtained by the solvothermal treatment of the gel showed no peak due to crystallization but a broad peak due to combustion of organic matter remaining on the surface of the products. The presence of organic matter in the solvothermal products is also confirmed by FT-IR spectra (figure 2). The peaks at 3000–2800 cm^{-1} are due to the C–H stretching vibration modes of alkyl moieties originating from the starting alkoxide. It is rather surprising to note that the solvothermal product possessed a larger amount of alkyl moieties than the precursor gel. Presumably, the alkyl groups in the hydroxide gels are easily hydrolyzed and eliminated during the drying stage, while the alkyl moieties remaining on the surface of ZrO_2 crystals are more stable because of difficulty in nucleophilic attack of water at zirconium atom for hydrolysis of the Zr–O–R bond.

As shown in figure 1, the increase in the alkoxide concentration increased the crystallite size of ZrO_2. Since the increase in the alkoxide concentration possibly causes an increase in the amount of water absorbed into the mixture at the hydrolysis stage and thus increases the amount of the water in the solvothermal system, the concentrations of water in the solvent and in the hydrolyzed gel were measured (figure 3). The increase of ZNP concentration clearly led to the increase in water concentration. Therefore, it can be assumed that crystallite size of the product obtained by the solvothermal treatment is related to the water concentration in the system at the solvothermal stage.

The effect of water intentionally added to the medium before the solvothermal treatment was examined. The XRD patterns of the products (data no shown) showed that the addition of water increased crystallite size of tetragonal ZrO_2. Therefore, the crystallization of the hydroxide gel in the solvothermal treatment takes place with the aid of water dissolved in toluene and generated from the hydroxide gel. To further investigate the behavior of water present in the system at the solvothermal stage, the effect of the amount of the gel charged for the solvothermal treatment was examined. When a half amount of the gel was charged to the autoclave, the crystallite size of ZrO_2 prepared was much smaller than that of ZrO_2 prepared under the usual condition. This result indicates that the concentration of water present in the reaction medium was largely decreased by vaporization of water to the gas phase.

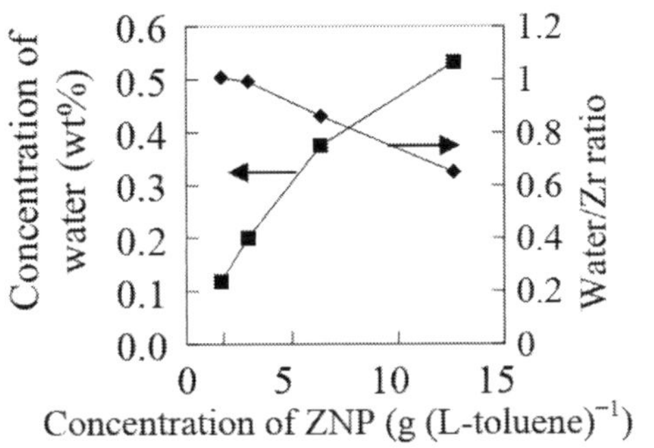

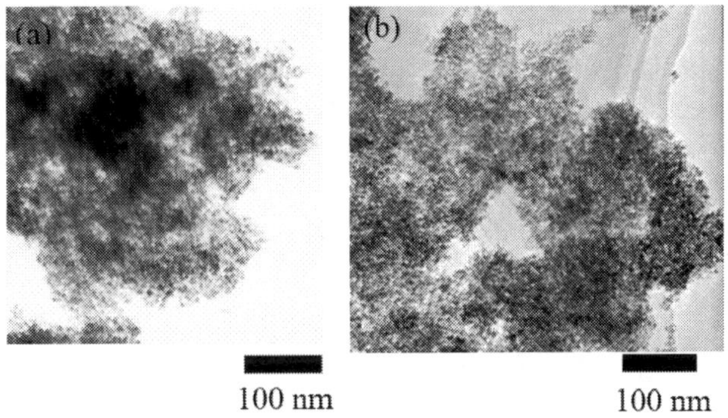

Figure 3. (left) Concentration of water in the hydrolyzed gel / toluene mixture.

Figure 4. (Right) TEM images of ZrO_2 obtained from ZNP with/without the intentional addition of water. Concentration of additional water (g/L-toluene): (a), 0; (b), 14.3.

Figure 4 shows the TEM images of the products obtained from ZNP with and without intentional addition of water at the solvothermal crystallization stage. Well-dispersed nanocrystals of ZrO_2 were observed. The average particle sizes of ZrO_2 nanocrystals are about 5 and 7 nm for the products obtained without and with the addition of water, respectively. The observed particle sizes are in good agreement with the crystallite sizes (5 and 6 nm, respectively) calculated from the XRD pattern using the Scherrer equation, indicating that each primary particle is a single crystal of zirconia. Surface areas, calculated from crystallite sizes assuming that the primary particle is truly spherical, accord with the BET surface areas (table 1), confirming little aggregation of primary particles took place during the crystallization stage, as is observed in figure 4.

As shown in table 1, the present products had relatively high surface areas and the surface area is easily controlled by the activity of water. However, In_2O_3 derived from indium isopropoxide had a small BET surface area (19.5 m^2 g^{-1}) and a large crystallite size (73 nm). The TEM images of this product (data not shown) showed that the In_2O_3 particle is nearly spherical. The particle size observed from TEM distributed relatively widely ranging from 17 nm to 94 nm, which was much larger than the particle size of ZrO_2 obtained from ZNP. The crystallite size (73 nm) calculated from the XRD peak was in this range, suggesting that each particle observed by TEM is a single crystal. In comparison with ZrO_2, the difference between BET surface area (19.5 m^2 g^{-1}) and surface area calculated from the crystallite size (11.5 m^2 g^{-1}) was large. This is because the In_2O_3 particles were not completely spherical. The product synthesized at low temperature (150 °C) have broad XRD pattern, but its surface area was very small (35.3 m^2 g^{-1}). This result indicates that in the case of indium, aggregation of the hydroxide gel occurred before crystallization of In_2O_3.

To investigate the pore structure, N_2 adsorption isotherm was measured (data not shown). The shape of the isotherm of ZrO_2 corresponded to type II of IUPAC notation, which indicates that ZrO_2 sample had essentially no micropores nor mesopores. The *t*-plot derived from the isotherms exhibited a straight line going through the origin, suggesting the absence of micropores in the product, which indicates that the product particles are well dispersed and free from aggregation.

To synthesize smaller In_2O_3 particle, an attempt was made to decrease the activity of water, which greatly affects the crystallization behavior under the solvothermal conditions. To suppress the activity of water, ethylene glycol (EG), which has a high affinity to water, was added before solvothermal treatment. Regardless of the EG / In ratio, the XRD pattern is attributed to cubic In_2O_3, but the increase in the EG / In ratio made XRD peaks broader and crystallite size of In_2O_3 smaller. Therefore, the crystallite size of In_2O_3 could be controlled from 73 nm to 33 nm by the addition of EG (table 1).

To demonstrate the effect of the activity of water, the following experiments were also carried out. After hydrolyzing ZNP in toluene, the mixture was centrifuged discarding the supernatant, and the hydroxide gel was dispersed in a fresh portion of toluene, dodecane and ethylene glycol. The XRD patterns of the products showed that the product obtained in dodecane having a lower affinity to water than toluene, had larger crystallite size than that obtained in toluene, while ethylene glycol gave much smaller crystallite size. These results clearly confirm that the activity of water has an importance role on crystallization process, and that the activity of water can be controlled by selection of the solvothermal medium.

4. Conclusions

The nanocrystals of metal oxides having large surface areas and small crystallite sizes are synthesized by slow hydrolysis of metal alkoxide followed by solvothermal crystallization in organic solvents. In solvothermal reaction, the crystallinity of metal oxide nanocrystals is greatly affected by water dissolved in the solvent. It was confirmed that the change in the activity of water by the intentional addition of water or ethylene glycol is effective to control the crystallite size of metal alkoxide.

References

[1] Rothschild A and Komen Y 2004 *J. Appl. Phys.* **95** 6374
[2] Shimizu Y and Egashira M 2004 *J. Soc. Powder Technol. Jpn.* **41** 124
[3] Baik N S, Sakai G, Miura N and Yamazoe N 2000 *Sensors Actuators B* **63** 74
[4] Vidal-Michel R and Hohn K L 2004 *J. Catal.* **221** 127
[5] Chen X B and Burda C 2004 *J. Phys. Chem. B* **108** 15446
[6] Raymundo-Piñero E, Khomenko V, Frackowiak E and Béguin F 2005 *J. Electrochem. Soc.* **152** A229
[7] Yu L H, Yang H X, Ai X P and Cao Y L 2005 *J. Phys. Chem. B* **109** 1148
[8] Boschloo G and Fitzmaurice D 1999 *J. Phys. Chem. B* **103** 7860
[9] Inoue M, Kominami H and Inui T 1993 *Appl. Catal. A: General* **97** L25
[10] Iwamoto S, Saito K, Inoue M and Kagawa K 2001 *Nano Lett.* **1** 417
[11] Cozzoli P D, Curri M L, Agostiano A, Leo G and Lomascolo M 2003 *J. Phys. Chem. B* **107** 4756
[12] Tang J, Fabbri J, Robinson R D, Zhu Y M, Herman I P, Steigerwald M L and Brus L E 2004 *Chem. Mater.* **16** 1336
[13] Joo J, Yu T, Kim Y W, Park H M, Wu F, Zhang J Z and Hyeon T 2003 *J. Am. Chem. Soc.* **125** 6553
[14] Demazeau G 1999 *J. Mater. Chem.* **9** 15
[15] Rajamathi M and Seshadri R 2002 *Current Opinion Solid State Mater. Sci.* **6** 337
[16] Inoue M 2004 *J. Phys.: Cond. Matter* **16** S1291
[17] Inoue M, Tanino H, Kondo Y and Inui T 1989 *J. Am. Ceram. Soc.* **72** 352
[18] Inoue M, Kitamura K, Tamino H, Nakayama H and Inui T 1989 *Clays Clay Miner.* **37** 71
[19] Inoue M, Kominami H and Inui T 1995 *Appl. Catal. A: General* **121** L1

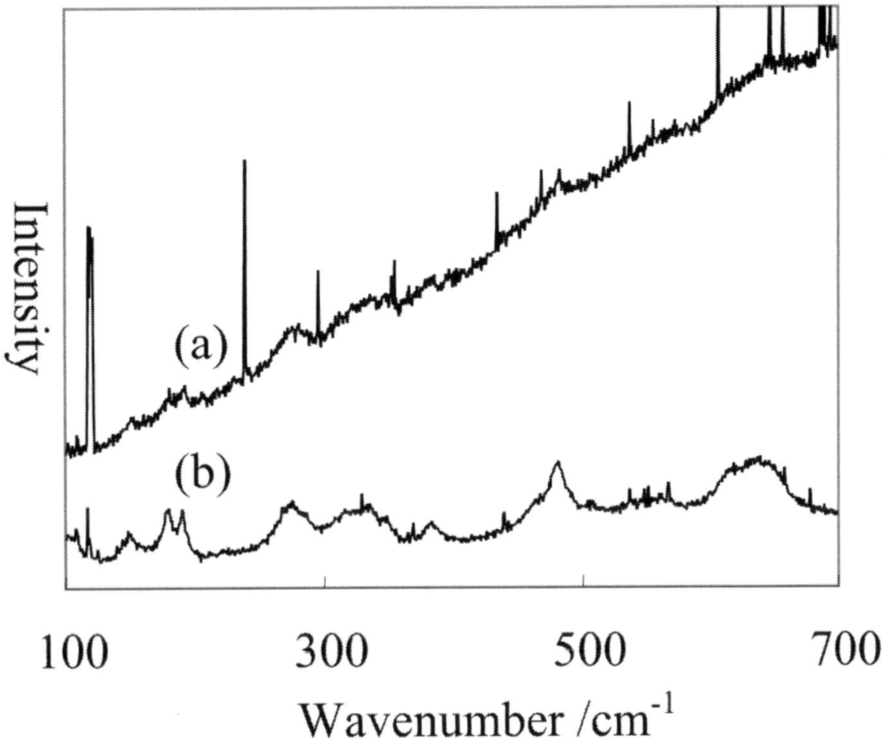

Supplementary Data

Figure S1. Raman spectra of zirconia obtained by the solvothermal crystallization of the hydroxide gel derived from ZNP: (a) as-synthesized; (b) calcined in air at 400 °C

Journal of Physics: Conference Series **121** (2008) 082006 doi:10.1088/1742-6596/121/8/082006

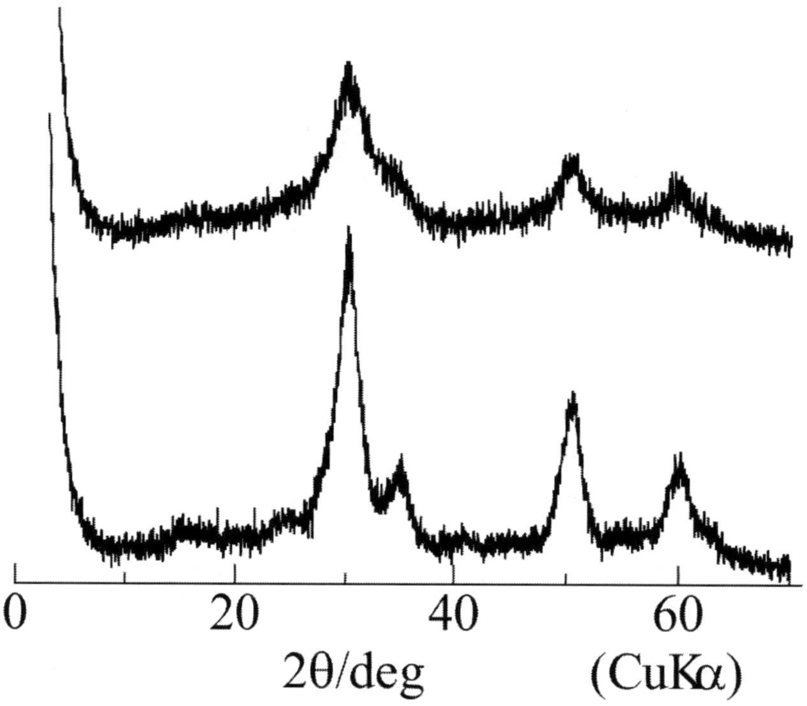

Supplementary Data

Figure S2. XRD patterns of ZrO_2 obtained by solvothermal crystallization of the hydroxide gel: (a) under the usual condition; (b) by lowering the charged amount of the gel to the autoclave.
(a), 12.5 g of ZNP in 100 mL of toluene, and (b) 6.3 g of ZNP in 50 mL toluene was treated in a 300 mL autoclave following the ordinary hydrolysis stage.

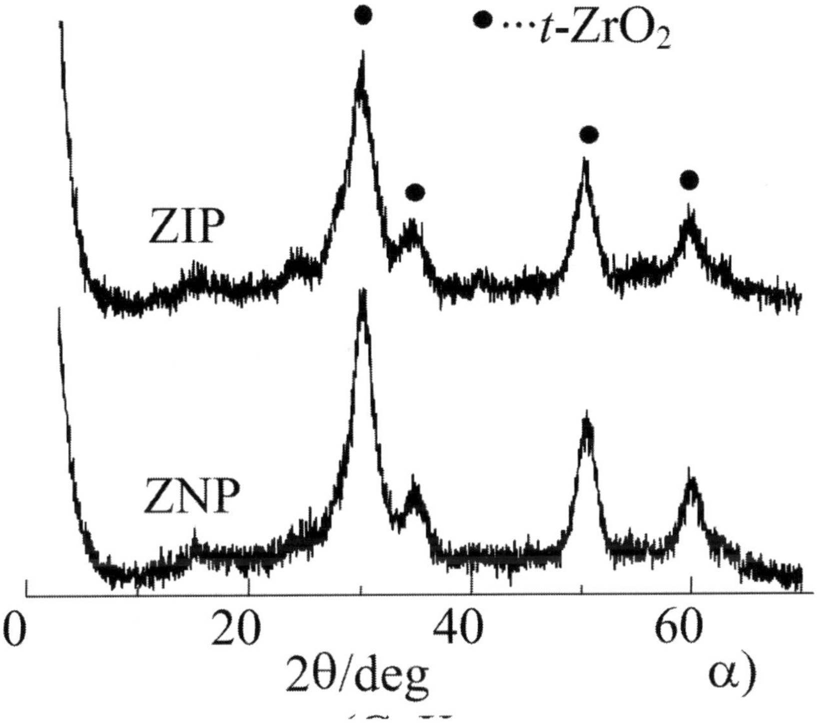

Supplementary Data
Figure S3. XRD patterns of ZrO_2 obtained from ZNP and ZIP.

Regulation of gahnite crystal size during hydrothermal synthesis

Yu D Ivakin, M N Danchevskaya and G P Muravieva
Chemistry Department of Moscow State University, Leninskie Gory 1/3, Moscow, 119992, Russia

E-mail: Ivakin@kge.msu.ru

Abstract.
Gahnite with various particles size was synthesized from boehmite of different particles sizes via interaction with zinc oxide or zinc acetate. It was found that the decrease of the size of gahnite particles is accompanied by decrease of the crystallite size, change of the state of oxygen vacancies and optical properties. The UV light absorption decreases, the blue-shift of an absorption edge is observed, the luminescence of gahnite and Eu^{3+} ions in gahnite changes.

1. Introduction

Zinc aluminate (gahnite, $ZnAl_2O_4$) is used as a ceramic, electronic and catalytic material. Gahnite, doped with europium is of special interest as a luminescent material [1, 2]. Gahnite can be synthesized by using conventional ceramic processing techniques [3], sol-gel method [4], coprecipitation method or alumina impregnating method [5] followed by calcination at temperatures (800-1000°C). Various variants of hydrothermal synthesis of zinc aluminate have been also described in the works [1, 6, 7]. Ultrasonic spray pyrolysis technique [2] and flame synthesis [8] of gahnite were also reported.

The authors previously determined that formation of gahnite from a mixture of zinc oxide and aluminum hydroxide during hydrothermal and thermovaporous treatment at 180-400°C and water vapor pressure 10 ÷ 26 MPa runs in accordance with the solid-phase mechanism [9]. It has been found that Zn^{2+} ions diffuse in the alumina matrix and then nucleation and growth of gahnite crystals occurs. Based on this consideration it was assumed that the size of particles of a formed gahnite should decrease along with decrease of the aluminium component particles size. The present work provides the experimental confirmation of the guess that the particles size of produced zinc aluminate should depend on the particles sizes of the parent aluminium oxide or hydroxide. The effect of the particles size on optical properties of zinc aluminate was investigated.

2. Experimental

Zinc oxide and zinc acetate (chemically pure grade), $Al(OH)_3$ – hydrargillite (GD18) or AlOOH – boehmite synthesized from it at 370°C as well as boehmite of the mark Apyral AOH 180 DS (from Nabaltec GmbH) and Disperal P2 (from Sasol), and also Eu_2O_3 (h. p.) were used in this work. The single-phase finecrystalline zinc aluminate was synthesized in medium of ammonia water solution via the treatment of a stoichiometric mixture of zinc oxide and hydrargillite or boehmite in a laboratory autoclave (volume 16 cm^3) at temperatures 220-400°C (table 1). When synthesizing the Eu^{3+} (2 wt%)-doped gahnite, the europium oxide was added into the precursors mixture. The single-phase nanosized zinc aluminate was synthesized via interaction of boehmite with zinc acetate in medium of ammonia

© 2008 IOP Publishing Ltd

water solution at 220°C. The mixture of boehmite and zinc acetate was placed into the autoclave container with an ammonia solution. The degree of autoclave filling with water was 0.2. The product of synthesis was extracted from the container after cooling an autoclave with the help of wash-out. The finecrystalline zinc aluminate was washed with distilled water and then dried at 70°C. The nanosized zinc aluminate was extracted from the container in the form of a colloid solution. The water solution of ammonia was then added into it. The coagulated precipitate was separated by centrifuging with subsequent drying up. Influence of the synthesis conditions on the samples state were explored by methods of X-ray diffraction (diffractometer DRON-3M in filtered Cu-K_α radiation), electron microscopy (Cam Scan Series 2), diffuse reflection (DR) spectroscopy (spectrometer Specord M40 in the region of 220-800 nm), photoluminescence, PL, (SDL-2M at room temperature). Distribution on the crystals sizes of samples was determined from the analysis of the SEM and TEM images by the method of linear segments.

Table 1. Conditions of synthesis of undoped and Eu^{3+}(2wt%)-doped gahnite

Sample	Precursors	T°C
1	$Al(OH)_3 + ZnO + Eu_2O_3$	400
2	(Boehmite from hydrargillite GD18) AlOOH + ZnO	260
3	(Apyral AOH 180 DS) AlOOH + ZnO	260
4	(Disperal P2) AlOOH + ZnO	260
5	(Disperal P2) AlOOH + $Zn(C_2H_3O_2)_2 + Eu_2O_3$	220

3. Results and discussion

Figure 1 shows the morphology of gahnite samples synthesized via treatment of a mixture of zinc oxide and europium oxide with hydrargillite at 400°C. The gahnite formed in these conditions of synthesis conserves the shape of the parent hydrargillite particles [9] and contains particles from 2-3 up to 10 microns by size. Figure 2 shows the morphology of gahnite samples synthesized via treatment of a mixture of zinc oxide with boehmite of different dispersity at 260°C. The relation of the particles size of boehmite and synthesized gahnite is presented in table 2.

Figure 1. SEM image of gahnite synthesized at 400°C from hydrargillite and ZnO.

Table 2. The average particles size (μm) of boehmite and synthesized from it gahnite.

Sample	Raw material	Boehmite	Gahnite
2	Boehmite from hydrargillite GD18	1	1.59
3	Boehmite Apyral AOH 180DS	0.2	1.22
4	Boehmite Disperal P2	0.05	0.65

One can see from figure 2 and table 2 that sizes of the gahnite particles decrease with decrease of the particles size of parent boehmite. Nano dimensional gahnite is possible to be synthesized if one changes the relation of a nucleation and growth rates of gahnite particles due to decreasing of synthesis temperature and facilitation of a diffusion stage of zinc ions to a surface of boehmite particles when using zinc acetate instead of zinc oxide. The figure 3 shows a typical TEM image of synthesized nanocrystalline gahnite.

Figure 4 shows the DR spectra of synthesized gahnite samples with different particles size. One can see that the UV absorption decrease and also the band shape of fundamental absorption changes when the size of particles decreases.

Journal of Physics: Conference Series **121** (2008) 082007 doi:10.1088/1742-6596/121/8/082007

a - Sample 2 b - Sample 3 c - Sample 4

Figure 2. SEM image of gahnite samples synthesized at 260°C from boehmite of different particle size: 1 μm (a), 0.2 μm (b) and 0.05 μm (c).

The absorbance edge shifts toward short wavelength. It corresponds to the known dependence of the band gap Eg ~ 1/d on the nanocrystalline particle size d [10]. However the observed changes of absorbance spectra with the size of particles exhibit the more complicated nature rather than a simple shift of the absorption edge towards the shorter wavelengths. The similar change of UV-absorption was observed when changing the stoichiometry and when employing the mechanical grinding of zinc aluminate [3]. It is necessary to note that synthesis of zinc aluminate with the decrease of the size of particles is accompanied by decrease of the crystallite size (table 3).

Figure 3. TEM image of gahnite (sample 5) synthesized from boehmite Disperal P2 and zinc acetate at 220°C. The average size of crystals is 36 nm.

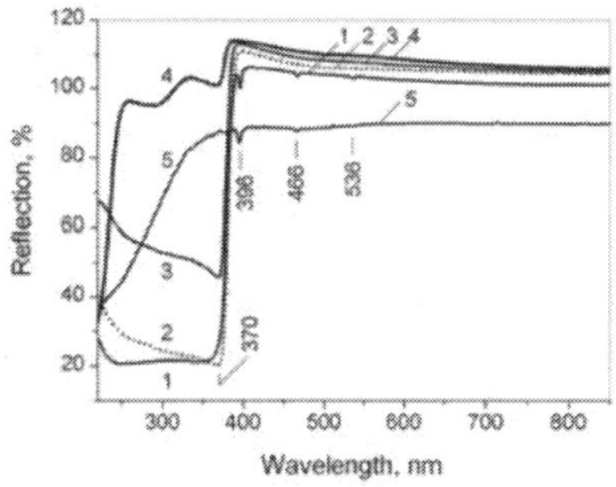

Figure 4. DR spectra of synthesized gahnite samples: finecrystalline Eu-doped - 1, undoped gahnite different particle size - 2 (1.59 μm), 3 (1.22 μm), 4 (0.65 μm); nanocrystalline Eu-doped gahnite -5.

Table 3. The crystallite size of gahnite samples.

Sample	D, nm
1	32±3
2	44±4
3	33±3
4	28±3
5	18±2

As the gahnite crystallites size decrease the intensity of UV luminescence band also reduces and the band position varies (figure 5). In the visible spectral region the position of the wide band shifts from 520 up to 440 nm. As a result, UV-band of nanogahnite luminescence (figure 6, sample 5) is suppressed and the band at 440 nm becomes more intensive. We attributed these bands to luminescence of the defects on the basis of oxygen vacancies [9]. The luminescence of these defects is excited in bands at 294, 304 and 315 nm (figure 7) which are characteristic for light absorption by oxygen vacancies. Eu-doped gahnite has a series of narrow luminescence bands of Eu^{3+} ions (figure 6) in the red spectrum region. As compared with the finecrystalline gahnite the position of the most intensive luminescence band of Eu^{3+} ions in nanocrystalline gahnite shifts from 612 nm to 614 nm. The relative of intensities and position of some other bands also change. It testifies to a change of the crystal environment of Eu^{3+} ions. The change of the bands intensity ratio is observed also in excitation spectra of Eu^{3+} ions luminescence (figure 8). The basic difference of excitation spectrum of Eu^{3+} ions in nanocrystalline gahnite is the absence of a band at 270 nm originated from absorption of light by oxygen vacancies.

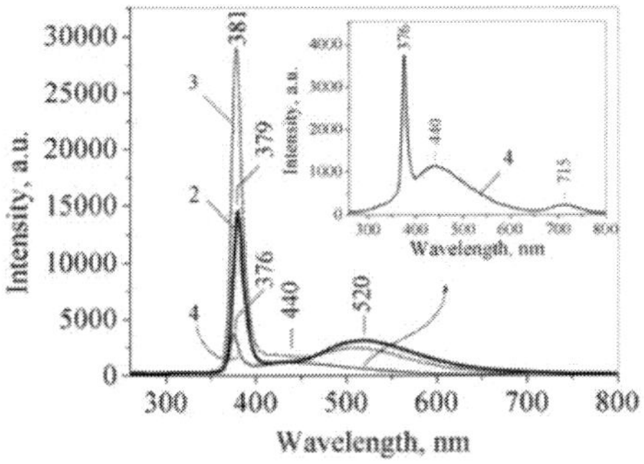

Figure 5. PL spectra of finecrystalline undoped gahnite samples. (λ_{ex}=254 nm)

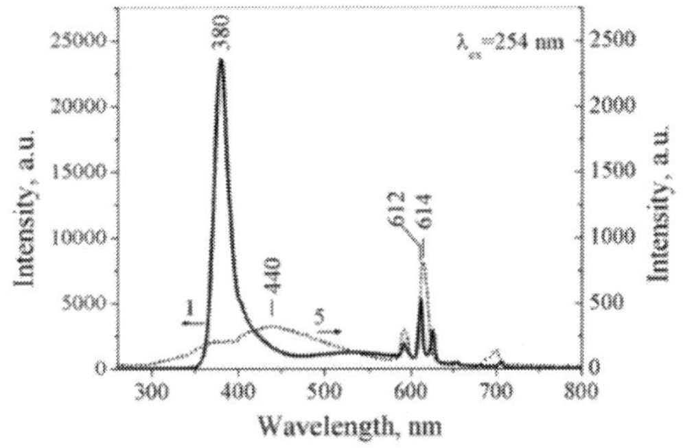

Figure 6. PL spectra of Eu doped fine- (1) and nanocrystalline (5) gahnite samples.

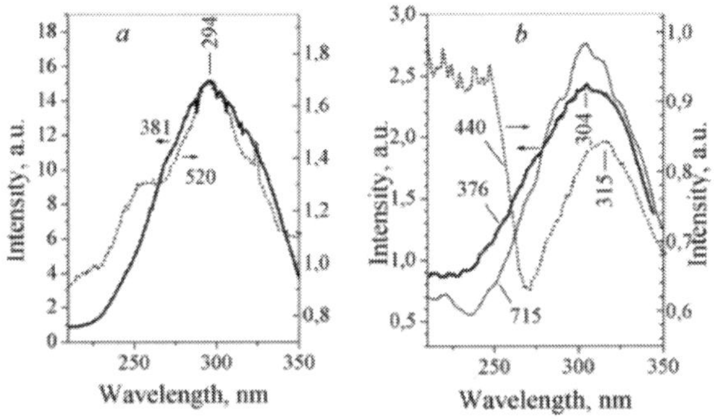

Figure 7. Excitation spectra of the indicated luminescence bands of finecrystalline gahnite samples: *a* - 3, *b* – 4.

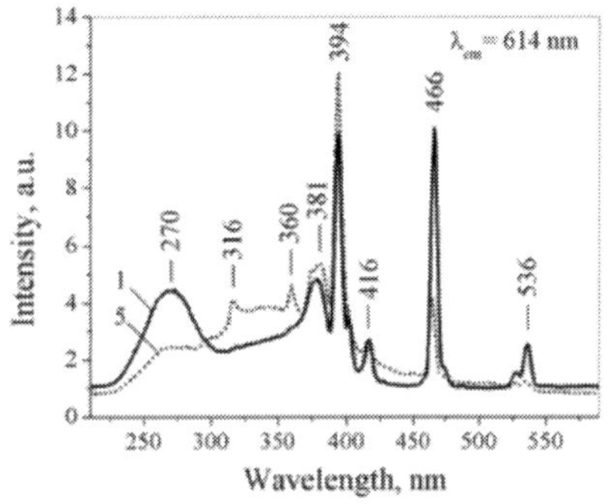

Figure 8. Eu^{3+} ions excitation spectra of Eu doped fine- (1) and nanocrystalline (5) gahnite samples.

Conclusion

Using of boehmite with various particle sizes allows to synthesize in hydrothermal conditions finecrystalline and nanocrystalline gahnite, both the pure or doped one. As the size of gahnite particles is decreased the state of oxygen vacancies changes, the UV absorption decreases, a blue-shift of an absorption edge occurs, and the luminescence spectra of pure and Eu-doped gahnite vary.

Acknowledgement

The authors gratefully acknowledge ISTC for financial support (pr. N 3002).

References

[1] Strek W, Deren P, Bednarkiewicz A, Zawadzki M and Wrzyszcz J 2000 *J. Alloys and Compounds* **300-301** 456
[2] Garcia-Hipolito M, Hernandez-Perez C D, Alvarez-Fregoso O, Martinez E, Guzman-Mendoza J and Falcony C 2003 *Optical Materials.* **22** 345
[3] Sampath S K, Cordaro J 1998 *J. Am. Ceram. Soc.* **81** 649
[4] Valenzuela M A, Bosch P, Aguilar-Rios G, Montoya A and Schifter I 1997 *J. Sol-Gel Sci. and Tech.* **8** 107
[5] Michel-Calenfini F M, Nie W, Linares C and Boulon G 1990 *J. Lumin.* **46** 177
[6] Nagamori Y and Kawase M 1998 *Microporous and Mesoporous Materials* **21** 439
[7] Wrzyszcz J, Zawadzki M, Trawczynski J, Grabowska H and Mista W 2001 *Appl. Catal. A: Gen.* **210** 263
[8] Jensen J R, Johannessen T, Wedel S and Livbjerg H 2000 *J. Nanoparticle Research* **2** 363
[9] Ivakin Yu D, Danchevskaya M N, Ovchinnikova O G and Muravieva G P 2006 *J. Mater. Sci.* **41** 1377
[10] Mathur S 2004 *Nanoengineered Nanofibrous Materials. Series: Nato Science Series II, Mathematics, Physics and Chemistry* **169** 423

Journal of Physics: Conference Series **121** (2008) 082008 doi:10.1088/1742-6596/121/8/082008

Hydrogen production reaction with a metal oxide catalyst in high pressure high temperature water

Masaru Watanabe,* Mayako Takahashi, Hiroshi Inomata

Research Center of Supercritical Fluid Technology, Tohoku University, 6-6-11, Aoba, Aramaki, Aoba-ku, Sendai 908-8579, Japan

* Email: meijin@scf.che.tohoku.ac.jp

Abstract. Hydrogen production from biomass was attempted in high pressure high temperature water at 573 K by adopting partial oxidation to increase the yield of H_2 via CO production in the presence of ZnO. The results revealed that an addition of H_2O_2 as an oxidant to the reaction of glucose and sugarcane bagasse brought about the trend of increasing the yields of H_2, CO, and CO_2. However, the sensitivity of H_2 yield on H_2O_2 amount was different from those of CO and CO_2, namely the excess amount of H_2O_2 tends to decrease the H_2 yield with giving a maximum at a certain H_2O_2 amount. These indicated that the controllability of partial oxidation would be a key factor for maximizing the H_2 yield through biomass conversion by partial oxidative gasification in high pressure high temperature water

1. Introduction

Hydrogen (H_2) is quite an important species for chemical synthesis and energy production. Biomass is a candidate of H_2 resources because of its sustainability. Among all biomass, plant biomass containing mostly cellulose is the highest amount of organic material on the earth. High gasification temperatures, above 800 K, for cellulose and glucose have been studied [1][2]. Although a lower temperature process for energy production is favorable because the generation of high temperature typically requires a lot of energy. Among the gasification processes, high-pressure water media including supercritical region is noticed to reduce gasification temperature of biomass [3]. Gasification of Glucose, the monomer unit of cellulose, in high temperature water has been studied to understand the reactivity of cellulose. [4] - [7]. One of serious problem of the low temperature gasification of glucose is the production of CH4 by-product due to its thermodynamical favorability under these conditions [7].

Cortright et al. [6] reported that high-selective H_2 formation from biomass-derived hydrocarbons including glucose was achieved by addition of platinum catalyst (3% Pt/γ-Al_2O_3) in saturated high temperature water, namely HCW at 498 K and 538 K, using a flow apparatus. The glucose gasification efficiency at 125 h of reaction time (that is, weight hour space velocity, WHSV, is equal to 0.008 g-biomass/g-catalyst/h) at 498 K and 538 K were 50 % and 84 %, respectively. The H_2 yields at 498 K and that at 538 K were 50 and 60 % based on the starting glucose and CH_4 yield was always below 10 % (glucose basis). Thus, this is a quite high-selective H_2 formation technique at a low temperature.

They [6] considered the C-C bond in carbohydrates such as glucose is broken on the metal surface into CO and H_2 because carbon in the carbohydrates always had O and H atoms. Successively the product CO reacts with water to from CO_2 and H_2 (water-gas shift reaction). They also suggested that

© 2008 IOP Publishing Ltd

the water-gas shift reaction proceeds on the metal surface. In this catalytic system, these reactions achieved high yield and selective H_2 formation. However, due to the high activity of the metal surface, hydrogenation of CO_2 into methane was also observed. Also undesirable compounds like alkane was also produced through homogeneous glucose reaction such as dehydrogenation at these conditions. In addition, more cost effective material such as transition metal and its oxide must be used for H_2 production, because platinum is an expensive material,

Partial oxidation of organic compounds such as n-hexadecane (n-$C_{16}H_{34}$) and organosolv-lignin in supercritical water provided high yield of CO compared to without supercritical water [8][9]. H_2 production increased by NaOH or ZrO_2 catalyst. We reported that these catalysts promoted CO by partial oxidation that further reacted with water to form CO_2 and H_2. For the establishment of a green process, the usage of homogeneous alkali is not suitable due to the negative impact to the environment. On the other hand, heterogeneous catalyst is favored due to easy-recycle and no environmental impact. However, the catalytic activity of ZrO_2 is still low. Therefore other materials must be considered for partial oxidative gasification of organics to CO and then water-gas shift reaction to the product H_2 with high yield and high selectivity. Also, we must explore the optimum condition to H_2 production at relatively low temperature of hot compressed water.

In this study, we demonstrate a new approach of biomass gasification to H_2 using partial oxidation in high pressure high temperature water at low temperature (573 K). First, we examined the effect of a metal oxide (ZnO) on glucose reaction. Next, partial oxidation was employed for glucose/metal oxide/water system by focusing on H_2 formation

2. Experimental

Glucose was obtained from Wako Chemical and used without further purification. Sodium hydroxide (NaOH, 1 M aqueous solution) and sulfuric acid (H_2SO_4, 1 M aqueous solution) were purchased from Wako Pure Chemical and used as received. Hydrogen peroxide aqueous solution (ca. 30% H_2O_2 solution) was also obtained from Wako Pure Chemical. ZnO obtained from Wako Pure Chemical and used without pretreatment. Distilled water was obtained from a water distillation apparatus (Yamato Co.,model WG-220).

The reaction was carried out in an SS 316 stainless steel tube bomb reactor with an inner volume of 6 cm3. One can see the detail experimental procedure elsewhere [9]. Briefly, the loaded amounts of glucose, water, and a metal oxide were 0.1 g, 1.0 g, and 0.1 g. In the case of adding homogeneous catalyst (NaOH or H_2SO_4) or liquid oxidant (H_2O_2), each aqueous solution prepared at a concentration was loaded into the reactor, instead of pure water. Ar gas was loaded at 25 MPa to recovery of gaseous product easily after the reaction. The reaction temperature was 573 K and the reaction time was 90 sec. The temperature profile in a batch type reactor was found elsewhere [11].

The identification and quantification of produced gas was conducted by GC-TCD (Shimadzu, model GC-7A, and Hitachi, model GC163). For recovering liquid and solid in the reactor, the reactor was washed with pure water. The amount of carbon (both inorganic and organic) in the water solution was evaluated using TOC (total organic carbon detector, Shimadzu, model TOC-5000A). The solid materials including the metal oxide catalyst and sometimes char) was divided by filtration with membrane filter.

Product yield (mol%) of carbon compound was evaluated from carbon base as shown below:

$$Product\ yield\ [\text{mol\%}] = \frac{\text{An amount of carbon atom in a product}}{\text{The amount of carbon atom in the loaded glucose}} \times 100 \quad (2.1)$$

Hydrogen yield (H_2 yield, mol%) was evaluated from hydrogen base of a loaded amount of samples as follows:

$$H_2\ Yield\ [\text{mol\%}] = \frac{\text{An amount of hydrogen atom in a produced } H_2}{\text{The amount of hydrogen atom in the loaded glucose}} \times 100 \quad (2.2)$$

In order to know how effective H_2 formed from glucose, we evaluated H_2 selectivity (HS) by following the method of Cortright et al. [6]. They evaluated the HS value by using the following equation (Eq. 3), to know how much hydrogen atoms in an organic compound can be taken out as H_2 [6]:

$$\begin{aligned} HS &= \left(\frac{H_2 \text{ produced, mol}}{\text{C atom in gas phase}}\right)\left(\frac{1}{2}\right)\times 100 \qquad (2.3) \\ &= H_2 \text{ selectivity, mol\%} \end{aligned}$$

In the right hand side, they considered the H_2/CO_2 reforming ratio as 1/2. Glucose is represented as $C_6H_{12}O_6$ and thus maximum hydrogen will be $12H_2$ corresponding to $6CO_2$ (6CO form from glucose with $6H_2$ and the product CO produces $6H_2$ through water-gas shift reaction).

3. Results and Discussion

3.1.1. *Without metal oxides.* First we examined the gasification of glucose without metal oxide catalyst at 573 K. As reported elsewhere, the conversion in the heating period (within 90 sec) was about 95 % without catalyst. In this study, glucose completely converted at all the experiments and the gas products were only shown here because we focused on gas formation. One can see liquid product in this experimental condition without catalyst and with homogeneous catalysts (H_2SO_4 and NaOH) elsewhere [11][12].

Figure 1 shows the gaseous carbon product distribution without catalyst and with homogeneous catalysts (1mM H_2SO_4 and 1mM NaOH). Any hydrocarbon such as methane was not obtained. Only CO_2 formed without catalyst and only CO was produced with H_2SO_4. By adding NaOH, CO_2 formation was significantly enhanced and CO also formed. Figure 2 shows H_2 yield. In these cases, however, H_2 yield was always below 0.05 mol% with and without the homogeneous catalysts. The highest H_2 yield was obtained without catalyst. The differences among these H_2 yields in this scale are too low to argue the difference of the reaction behavior. Thus, H_2 production from glucose at 573 K and 90 sec is concluded to be quite low even in the presence of these homogeneous catalysts.

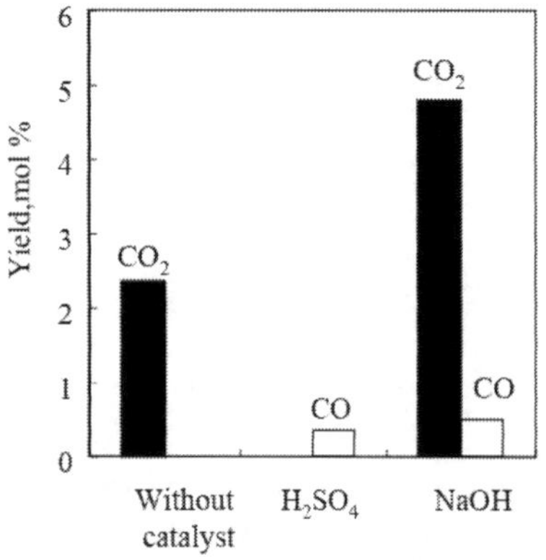

Figure 1. Effect of the homogeneous catalysts on the gasification of glucose (573 K, 90 sec).

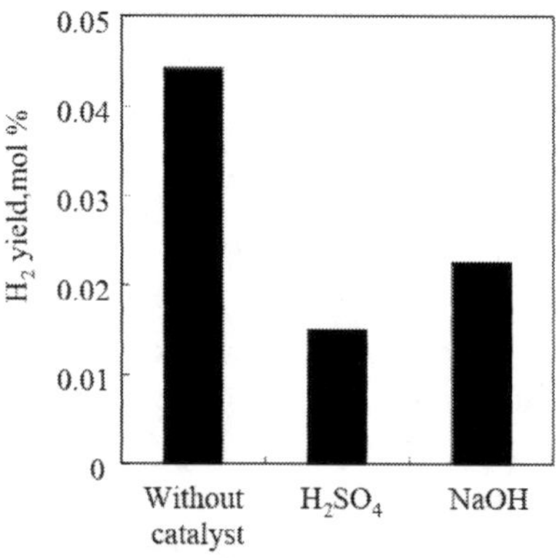

Figure 2. Effect of the homogeneous catalysts on H_2 formation from glucose (573 K, 90 sec).

3.1.2. *In the presence of ZnO.* In order to enhance H_2 production, some metal oxides were examined. As a result, by adding ZnO, H_2 yield was relatively high (3.6 mol%) compared to that without metal

ZnO. Although the HS (see (2.3)) with ZnO was high (22 %), the HS reported by Cortright et al. was much higher (36 %) [6] and the additional technique must be introduced.

In order to enhance the HS, C-C bond scission of glucose structure must be promoted and loosing of O and H atoms in the structure as H_2O must be inhibited because C-C bond scission produces CO and H_2. ZnO is known to be the catalyst of low temperature water-gas shift reaction [13]. If the water-gas shift reaction must rapidly occur on the ZnO surface, H_2 formation could be increased by enhancement of CO formation. Partial oxidation of an organic compound in high water density of supercritical water is favored to produce partial oxidation products including CO [14] – [17]. Therefore, in the glucose gasification, partial oxidation was employed to promote CO formation in the presence of ZnO.

Figures 3 show the O/C ratio dependence on the gaseous carbon compound and H_2 yield. The O/C ratio is the ratio of oxygen atom (can act as O_2) to carbon atom and this (in the loaded glucose) was 0.29 (1.6 % H_2O_2 solution in the reactor) or 0.53 (2.7 % H_2O_2 solution in the reactor). The yields of CO_2 and H_2 increased with increasing the O/C ratio. The yield of CO_2 drastically increased with the increase of O/C ratio. In contrast, H_2 yield gradually increased with the O/C ratio. Figure 4 shows the HS calculated by (2.3). The HS values at 0 and 0.29 of O/C ratios were both 22 %, however, that at 0.52 of O/C was 14 %. This means that higher value of O/C ratio promoted complete oxidation of organic carbon or CO into CO_2. Now selective CO formation from glucose with a catalyst is under investigation to promote H_2 formation.

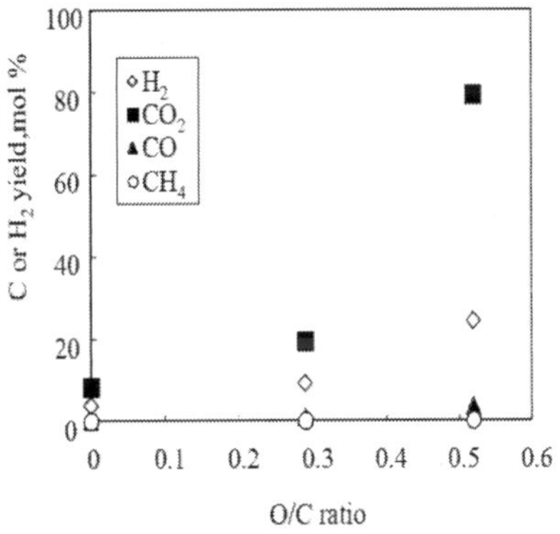

Figure 3. Effect of the O/C ratio on the gaseous carbon and H_2 formation from glucose (573 K, 90 sec).

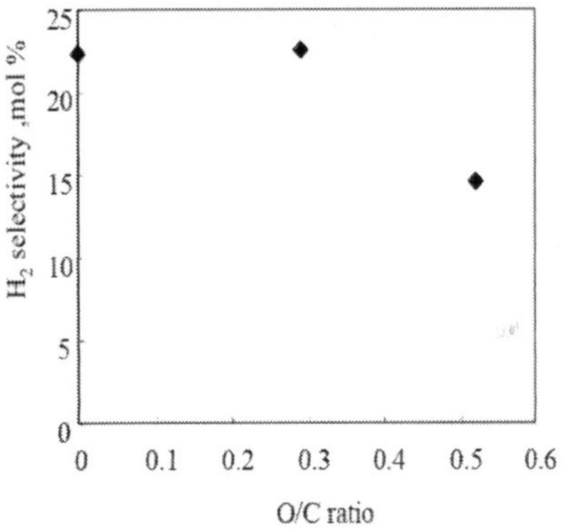

Figure 4. Effect of the O/C ratio on the H_2 selectivity (573 K, 90 sec).

Conclusion

In this study, we found that ZnO was effective for glucose gasification in hot compressed water. Combination of partial oxidation and ZnO addition to glucose-hot compressed water system was quite useful for H_2 production from glucose. In order to enhance H_2 yield and selectivity from glucose, the desired catalytic property and the optimum conditions must be determined.

References

[1] Tomishige K, Asadullah M and Kunimori K 2004 *Catal Today* **89**389.
[2] Hanaoka T, Inoue S, Uno S, Ogi T and Minowa T 2005 *Biom Bioe* **28** 69.
[3] Calzavara Y, Joussot-Dubien C, Boissonnet G and Sarrade S 2005 *Energy Conv Manag* **46** 615.
[4] Yu D, Aihara M andAntal JMJ 1993 *Energy Fuels* **7** 574.
[5] Watanabe M, Inomata H and Arai K 2002 *Biom Bioe* **22** 405.
[6] Cortright RD, Davda RR and Dumesic JJ 2002 *Nature* **418** 964.
[7] Sinag A, Kruse A and Rathert J 2004 *Ind Eng Chem Res* **43** 502.

[8] Watanabe M, Inomata H, Osada M, Sato T, Adschiri T and Arai K. 2003 *Fuel* **82** 545.
[9] Yoshida T and Oshima Y 2004 *Ind Eng Chem Res* **43** 4097.
[10] Watanabe M, Tsukagoshi M, Hirakoso H, Adschiri T and Arai K2000 *AIChE J* **46** 843.
[11] Watanabe M, Aizawa Y, Iida T, Levy C, Aida TM, and Inomata H 2005 *Carbohydrate Res* **340** 1931.
[12] Watanabe M, Aizawa Y, Iida T, Aida TM, Levy C, Sue K and Inomata 2005 *Carbohydrate Res* **340** 1925.
[13] Shido T and Iwasawa Y 1993 *J Catal* **140** 575.
[14] Adschiri T, Shibata R, Sato T, Watanabe M and Arai K 1998 *Ind Eng Chem Res* **37** 2634.
[15] Arai K, Adschiri T and Watanabe M 2000 *Ind Eng Chem Res* **39** 4697.
[16] Watanabe M, Mochiduki M, Sawamoto S, Adschiri T and Arai K 2001 *J Supercritical Fluids* **20** 257.
[17] Watanabe M, Inomata H, Osada M, Sato T, Adschiri T and Arai K 2003 *Fuel* **82** 545.

Reaction in Plasma Generated in Supercritical Carbon Dioxide

M Goto*, M Sasaki, T Kiyan, T Fang, BC Roy, T Namihira, H Akiyama, M Hara

Graduate School of Science and Technology, Kumamoto University, 2-39-1 Kurokami, Kumamoto 860-8555, Japan

E-mail: mgoto@kumamoto-u.ac.jp

Abstract. The generation of non-thermal plasmas was studied under supercritical conditions. This is interesting in view of both the basic phenomena involved and of potential industrial applications. When supercritical fluids are combined with discharge plasma they develop unique characteristics, including an enhanced chemical reactivity. This is the motivation for investigating the generation of discharge plasma in supercritical CO_2. In this study, breakdown voltages were measured in CO_2 in order to generate electric discharges in supercritical CO_2. The experimental data show that the breakdown voltage increased smoothly up to the intersection points, but beyond these points the rates of increase of the breakdown voltage are different. This phenomenon can be explained with the help of pressure-density curves of carbon dioxide at a constant temperature. In addition, the generated plasma in supercritical CO_2 was applied to several chemical reactions. The alpha conversion of tocopherols and transesterification of soybean oil with methanol were investigated. In addition, we explored the possibility of replacing catalysts with plasma in supercritical CO_2.

1. Introduction

The two main reasons for the considerable interest in supercritical fluids (SCFs) are that they are generally non-toxic and inexpensive, and that they have many attractive properties. These properties differ from the typical properties of gases and liquids, since they are intermediate between those two extremes. This makes them indispensable as extraction and reaction media since they have liquid-like densities and gas-like viscosities coupled with diffusion coefficients that are higher than those of liquids. These properties contribute to enhanced mass transfer, while their high thermal conductivity can be exploited to achieve high heat transfer. In addition, their properties can be manipulated by varying the pressure and temperature [1]. In view of these unique characteristics of SCFs, it is widely anticipated that subjecting an ionized state such as a plasma to supercritical conditions may result in unique characteristics and reactions that differ from those of a normal plasma in the gaseous state. Furthermore, the high-energy electrons, ions, and radicals that are generated when a plasma combines with supercritical fluids may generate new reaction fields, leading to effective pollutant oxidation for both homogeneous and heterogeneous reactions. It is thus important to make a systematic study of the pre-breakdown phenomena of electric discharges in SCFs before generating discharge plasmas effectively in sufficiently large reaction volumes before commencing. Generation of a micrometer-scale discharge in an SCF environment has been reported [2]. In the same study, a unique phenomenon was discovered, namely a minimum breakdown voltage existing near the critical point, which is probably caused by an unstable cluster in the SCF. On the other hand, *Lock et al.* initiated pulsed

© 2008 IOP Publishing Ltd

corona discharge under supercritical conditions, and reported that the breakdown voltage for a wire-in-cylinder configuration is three times lower than estimates based on Paschen's law due to the large inhomogeneties in the density of the supercritical fluid near the critical point [3]. Investigation of discharge phenomena in supercritical fluids is an underdeveloped field, and the pre-breakdown phenomena are not well understood. In this present study, carbon dioxide (CO_2) was selected as the medium for plasma production because it is non-toxic, non-flammable, readily available, environmentally benign, and has a low critical temperature (304 K) and pressure (7.38 MPa). The breakdown voltages were measured in CO_2 in the pressure range 0.1-14 MPa and at temperatures of 298 K and 313 K by applying a negative DC voltage in order to characterize the electric discharges in supercritical CO_2. The ultimate goals of this research are to generate discharge plasmas in sufficiently large volumes that they can be used in various industrial applications, and to design a reactor with discharge plasmas in supercritical fluids.

2. Experimental

2.1. Experimental materials

CO_2 was supplied from Uchimura Sanso Co. Ltd. (Osaka, Japan) with a purity of 99.97%. Methanol (with a purity of $\geq$ 98%) and soybean oil were obtained from Wako Pure Chemical Industries Inc. (Tokyo, Japan). 57% tocopherol concentrate was supplied by Kaidi Fine Chemical Industries, Ltd., Wuhan, Hubei Province, China.

2.2. Apparatus and procedure

2.2.1. Apparatus for generating plasma under supercritical conditions

Figure 1 shows a schematic diagram of the test reactor for plasma production. It has an inspection window through which continuous monitoring was performed. The test reactor is made of stainless steel (SUS316) having a compressive strength of 30 MPa. The total volume of the reaction cell is 1300 mL. A power lead was introduced through the center of a long bushing made of peak resin, and the annular space was sealed with double o-rings placed around its outer surface to prevent air leakage. Liquid CO_2 from a cylinder with a siphon attachment was passed through a cooling head (about 268 K) of a high-pressure pump (PU-2086 Plus, Jasco) to the test reactor. A thermocouple and a backpressure regulator were user to control the temperature and the pressure of the test reactor, respectively. After the experimental temperature and pressure were achieved, the CO_2 medium in the test reactor was maintained under those conditions overnight. Then DC power was applied to produce electric discharges at a constant temperature. The breakdown voltages for the production of electric discharges were measured for pressures in the range of 0.1 to 14 MPa at each experimental temperature by removing CO_2 through the backpressure regulator. The electrode configuration consisted of a point and a planar surface. The point was made of tungsten and it was used as the negative electrode, while the planar surface was made of stainless steel and it was the positive electrode. This configuration produces a non-uniform electric field. The electrode gap was 200 μm. A voltage was applied to the negative electrode using a high-voltage stabilization DC power supply (HAR-50R0.6, Matsusada Precision Inc., Japan) that was controlled by a function generator (SG-4115, Iwatsu, Japan) at a rate of 2.5 kV/s. The breakdown voltage and the current were respectively measured using a high-voltage probe (EP100K, Pulse Electronic Engineering Co. Ltd., Japan) and a current transformer (4997, Pearson Electronics, Inc., USA) that were connected to a digital oscilloscope. The corona onset voltage was measured using a photomultiplier (No. 722, Atago Bussan Co. Ltd., Japan) coupled with an optical cable at 15 MPa and 313 K on trail basis. The interval between successive electric discharges was set at about 2 min in order to ensure that the CO_2 medium was in a steady state.

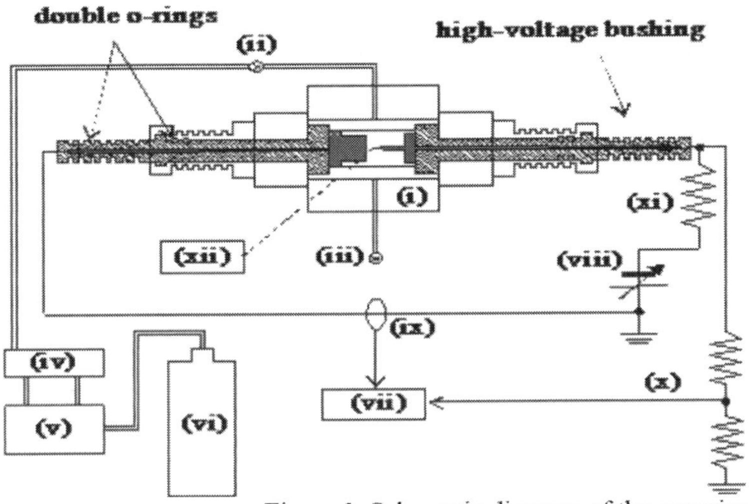

(i) Test cell
(ii) CO_2 inlet
(iii) CO_2 outlet
(iv) Syringe pump
(v) Cooling system
(vi) CO_2 container
(vii) Digital oscilloscope
(viii) DC power source
(ix) Current transformer
(x) High-voltage probe
(xi) Damping resistance
(xii) Photomultiplier

Figure 1. Schematic diagram of the experimental setup

2.2.2. Application experiment using the plasma generated in supercritical CO_2
The test cell was charged with 4 ml of either 57% tocopherol (Vitamin E) concentrate or soybean oil and a 20 ml methanol. This system was then treated with about 10,000 plasma discharges generated across blade electrodes with a gap of 1.7-1.8 mm in supercritical CO_2 at 18 MPa and 313 K.

2.3. GC-MS analysis

2.3.1. The tocopherol concentrate (57 %)
The samples were analyzed by GC-MS (Hewlett Packard 5890 series, Palo Alto, CA, USA), coupled with a mass selective detector (HP 5972). The column used was a MS HP-5 phenyl methyl siloxane capillary (30 m × 0.25 mm i.d., film thickness of 0.25 μm). The oven temperature was increased from 353 to 488 K at 5 K/min, held at 488 K for 2 min, then raised to 548 K at 5 K/min and finally held at 548 K for 15 min. The injector temperature was 573 K, the injection volume was 1 μL, the split ratio was 20:1, the total carrier gas (helium) flow rate was 24 mL/min, and the ionizing energy was 70 eV. The main components were methyl palmitate (C16:0), methyl linoleate (C18:2), methyl oleate (C18:1), methyl stearate (C18:0), squalene, α-tocopherol, campesterol, stigmasterol and β-sitosterol. These were identified by comparing mass spectra and retention times with those of pure standards. For other components, the probability-based matching (PBM) algorithm was employed for finding the most probable match in a reference library (NIST library of mass spectra and subsets, HPG 1033A).

2.3.2. Volatile compounds in soybean oil
The temperature program was modified from that described in Section 2.3.1 in order to obtain a more reliable result. The oven temperature was increased in the sequence 413 K, 423 K, 468 K, 483 K, 513 K, 548 K at a rate of 5 K/min; the oven temperature was held constant at each temperature for 2 min with the exception of 548 K, where it was held for 22 min. The total time for this sequence was 60 min. The injector temperature was 573 K, the carrier gas (helium) flow rate was 24 mL/min, and the ionizing energy was 70 eV.

3. Results and discussion

3.1. Pressure-density curves for CO_2
Figure 2 shows the pressure-density curves for CO_2 for the temperature range 283-333 K. In this figure, the bold lines at 298 and 313 K are the experimental temperatures for the production of electric

Journal of Physics: Conference Series **121** (2008) 082009 doi:10.1088/1742-6596/121/8/082009

discharges. These curves were calculated using the equation of state and they demonstrate the characteristics of electric discharges [4]. Ionization in gases and liquids is strongly related to the mean free path, which is proportional to density of medium. However, the phenomena are not well clarified in a supercritical medium. This figure shows that the curves above and below the critical isotherm have different behaviors with change in temperature. The area between the upper-right side of the saturated curve and the right-hand side of the critical isotherm represents liquid. However, from the viewpoint of electric discharge, it would be more appropriate to describe the areas above and below the critical pressure line as representing high- and low-pressure liquids, respectively. This distinction may help to explain the mechanism of the breakdown process. In Fig. 2, the curve of 298 K below the critical pressure in the gas phase intersects the saturated gas line at point A1 and then intersects the saturated liquid line at point A2. It then moves upward through the low-pressure liquid region to the high-pressure liquid region. On the other hand, the curve of 313 K passes from the gas phase to the supercritical phase above the critical point.

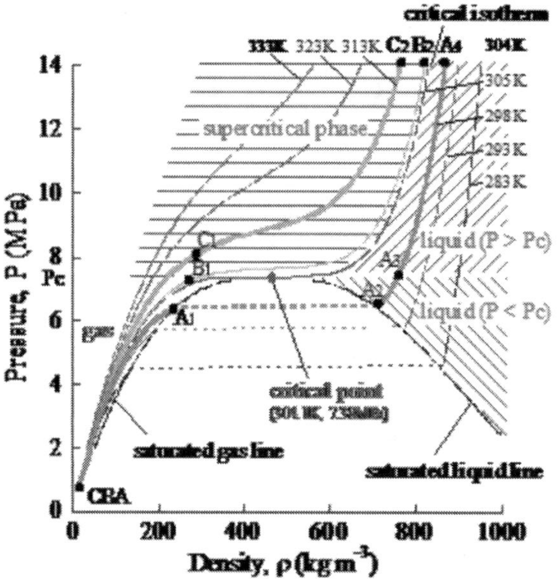

Figure 2. Pressure-density diagram of CO_2

3.2. Densities and breakdown voltage

The density dependences of the breakdown voltages at 298 and 313 K are shown in Figs. 3a and 3b, respectively. These figures reveal that the breakdown voltages increase smoothly up until the intersection points, but beyond these points they have different rates of increase. This characteristic can be explained with the help of the CO_2 pressure-density curves at different temperatures due to the different phase changes shown in Fig. 2. The abnormal changes in the breakdown voltages found near the boundary lines of 298 K and 313 K may be due to the density fluctuations at these conditions.

The density was calculated for the experimental conditions using the equation of state [4]. In Fig. 3a, the breakdown voltage in the gas phase conforms to the density changes of CO_2 up to point A1 as shown in Fig. 2. The gas phase is then converted into the liquid phase at point A2 and then goes to A3, which is at the boundary between low- and high-pressure liquids.

Journal of Physics: Conference Series **121** (2008) 082009 doi:10.1088/1742-6596/121/8/082009

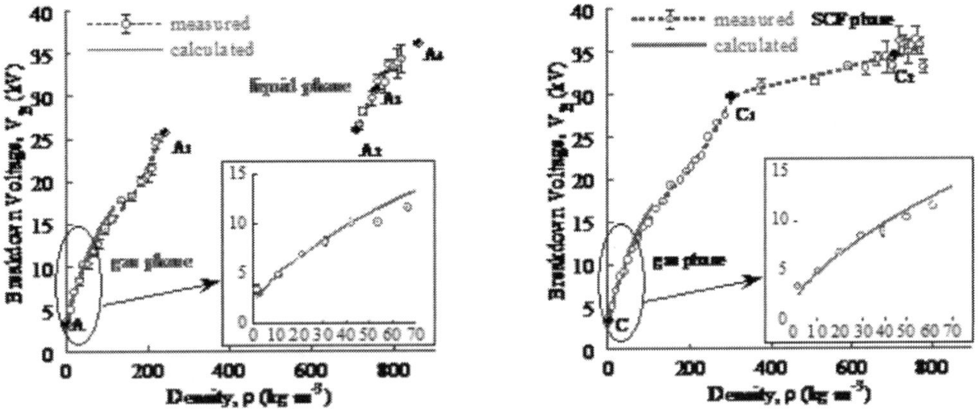

a. Breakdown voltage at 298 K b. Breakdown voltage at 313 K

Figure 3. Breakdown voltage for plasma generation in CO_2

According to this situation, the breakdown voltage increases uniformly within this region, but in the region beyond point A3, in the high-pressure-liquid phase, the rate of increase is considerably lower than that in the low-pressure-liquid region (A2-A3). The increasing rates of breakdown voltages at 298 K are 0.08 kV/kg.m^{-3} in gas, 0.11 kV/kg.m^{-3} in the low-pressure-liquid phase, and 0.05 kV/kg.m^{-3} in the high-pressure-liquid phase. Figure 3b indicates that the rate of increase of the breakdown voltage in the gas phase is almost the same as that in Fig. 3a, but it is different in the supercritical phase (0.01 kV/kg.m^{-3}) compared with the gas and liquid phases. The explanation for breakdown phenomena in the supercritical phase is still unclear; they could be related to the peculiar characteristics of supercritical fluids, such as high diffusibility, thermal conductivity, or clustering. In such conditions, the effect of bubble triggering may not be applicable to the supercritical phase.

The theory of the breakdown mechanism is not well established, especially for the liquid state. However, there are two factors that may explain it: one is electronic destruction based on ionization generated by the collisions of electrons with the liquid molecules [5] and the other is cellular destruction leads to bubble triggering [6,7], which is caused by liquid evaporation due to the Joule's heat generated by current concentration at the electrode point. It is possible that bubble triggering at the tip of electrode point in the liquid phase results in complete breakdown.

3.3. Application experiment for 57% tocopherol concentrate

To find out what kind of reactions induced by plasma in supercritical fluid, 57% tocopherol (vitamin E) concentrate (4 mL) was mixed with methanol (20 mL) and then treated with the plasma generated in supercritical CO_2 (18 MPa, 313 K). This was done because the concentrate was extracted by supercritical CO_2 (thus it is clearly soluble in CO_2) and contains many compounds (predominantly tocopherols, fatty acid methyl esters, sterol esters and glycerides) [8]. In this complicated system, in conjunction with methanol and catalysis, some reactions can be induced as shown by the following:

(1) Alpha conversion of other tocopherols into α-tocopherol (T > 512 K, P > 8.9 MPa with metal oxides as catalysts), for example β-tocopherol [9].

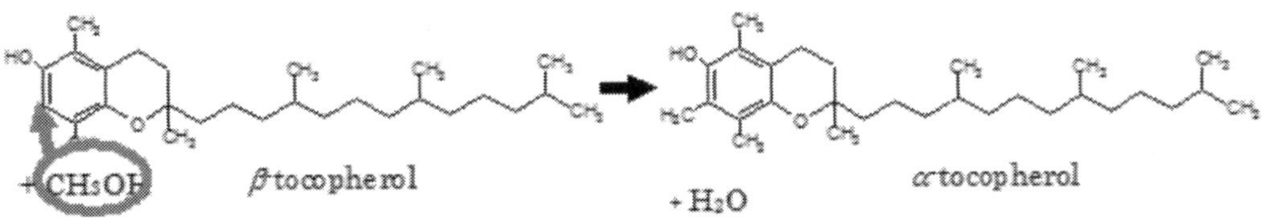

(2) Transesterification of glycerides (including tri-, di-, mono-glycerides) into FAMEs (Fatty acid methyl esters) and glycerol (T > 333 K with NaOH or $NaOCH_3$ as catalysts) [10].

$$\begin{array}{llll} \text{Step I} & \text{Triglyceride} & + \text{ methanol} \rightarrow \text{diglyceride} & + \text{methyl ester} \\ \text{Step II} & \text{Diglyceride} & + \text{ methanol} \rightarrow \text{monoglyceride} & + \text{methyl ester} \\ \text{Step III} & \text{Monoglyceride} & + \text{ methanol} \rightarrow \text{glycerol} & + \text{methyl ester} \end{array}$$

(3) Decomposition of sterol esters into sterols and FAMEs (T > 333 K with NaOH or $NaOCH_3$ as catalysts) [11].

$$\text{Part of sterol esters} + \text{methanol} \xrightarrow{\text{NaOCH}_3} \text{Sterol} + \text{FAMEs}$$

The experimental results show that there is an increase in the FAMEs content (from 6.1% to 16.1%) compared with other compounds (see Fig. 4), indicating the possibility of decomposing glycerides into FAMEs. Additionally, there is no change in the ratio of tocopherol isomers; in other words, the plasma in supercritical CO_2 cannot catalyze the alpha conversion of tocopherols.

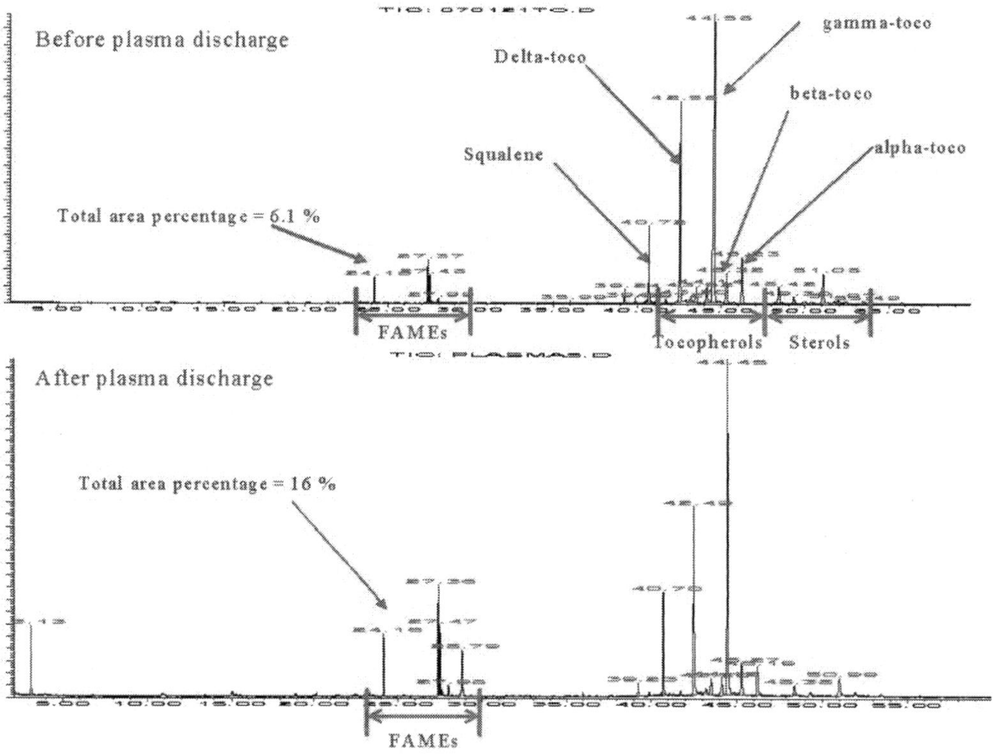

Figure 4. Comparison of tocopherol concentrates before and after plasma discharge

3.4. Application experiment for soybean oil

To confirm the possibility of decomposing glycerides into FAMEs, we used soybean oil as the test substance, since it mainly consists of glycerides. 4 mL of soybean oil were mixed with 20 mL of methanol and the mixture was treated with the plasma generated in supercritical CO_2.

Because of the limitation of the analysis, only the volatile compounds in soybean oil could be analyzed. The analysis result for sample 1 shown in Fig. 5 show that the main volatile compounds in soybean oil are four fatty acids, namely, palmitic acid (29.20 min), linoleic acid (32.88 min), oleic acid (32.97 min) and stearic acid (33.49 min). Additionally, their total area percentage is 75.18%.

By contrast, the plasma treatment increased the FAMEs content from 3.42 to 9.23%. This result demonstrates that the plasma treatment can catalyze glycerides or fatty acids into FAMEs and it can probably replace some catalysts for such reactions.

Noticeably, after each experiment there was always an obvious scar on the surface of the planar electrode. Simultaneously, some black particles appeared in the reactor and these were collected for element analysis. The analysis data show that the main elements in these particles are carbon (55.61%), hydrogen (7.87%) and nitrogen (0.78%). The same phenomena have also been observed by other researchers and these particles are considered to be carbon particles from CO_2, which was decomposed by the plasma [12].

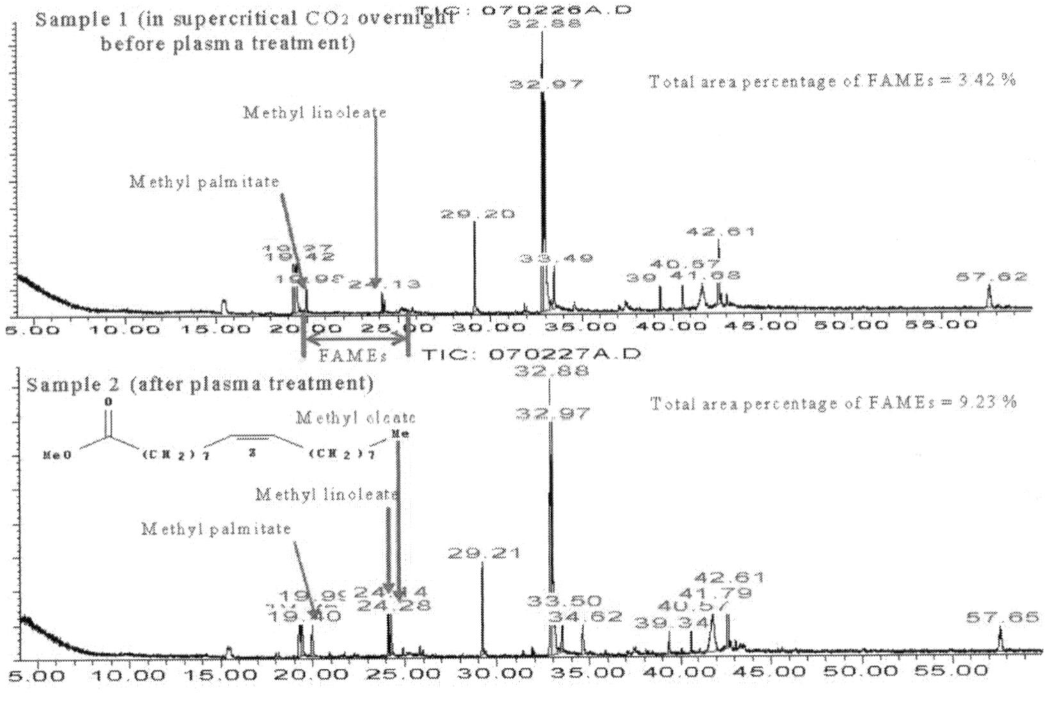

Figure 5. Comparison of the volatile compounds in soybean oil before and after plasma discharge

4. Conclusions

The breakdown voltages of electric discharges were measured in the gas, liquid, and supercritical phases of CO_2 by applying a negative DC potential at pressures in the range 0.1-14 MPa and at temperatures of 298 and 313 K in order to investigate the electric discharge characteristics. The gap between the point and planar electrodes was maintained at 200 μm. The breakdown voltage increased with an increase in the density of CO_2 in all phases, but the supercritical phase had a different rate of increase. The breakdown mechanism in the liquid phase was similar to that in the gas phase. Bubble triggering may occur in the liquid phase. The cause for the different rate of increase in the breakdown voltage in supercritical CO_2 is currently unclear.

By performing application experiments, it was found that the plasma generated in supercritical CO_2 could induce some reactions: for example, the conversion of glycerides into FAMEs. However, the amount of FAMEs was relatively small and the reason for this is probably that the reaction region is just limited to the gap between the two electrodes. Consequently, in order to improve the reaction efficiency, the present batch reactor is being converted into a flow-type apparatus, which should be more stable enabling clearer results to be obtained.

Acknowledgement
This work was supported by 21st COE "Pulsed Power Science" and by the Ministry of Education, Culture, Sports, Science and Technology, Japan.

References

[1] M. A. McHugh , V. J. Krukonis, Supercritical Fluids Extraction: Principles and practice, Butterworths, Stoneham, MA, 1986.
[2] T. Ito and K. Terashima, Generation of micrometer-scale discharge in supercritical fluid environment, Appl. Phys. Lett. 80, pp.2854-2856, 2002.
[3] E. H. Lock, A. V. Saveliev, L. A. Kennedy, Initiation of Pulsed Corona Discharge Under Supercritical Conditions, IEEE Trans. On Plasma Science, Vol.33, No.2, pp.850-853, 2005.
[4] R. Span and W. Wagner, A New Equation of State for Carbon Dioxide covering the Fluid Region from the Triple Point Temperature to 1100 K at Pressures up to 800 MPa, J. Phys. Chem. Ref. Data, 25, pp.1509-1596, 1996.
[5] D. R. Young, Electric Breakdown in CO_2 from Low Pressures to the Liquid State, J. Appl. Phys. Vol.21, pp.222, 1950
[6] A. H. Sharbaugh, P. K. Watson, Conduction and Breakdown in Liquid Dielectrics, Prog. in Dielec., Vol.4, pp.199, 1962
[7] K.C. Kao, Theory of High-Field Electric Conduction and Breakdown in Dielectric Liquids, IEEE Trans. Elect. Insulation, Vol.EI-11, pp.121, 1976
[8] T. Fang, M. Goto, X. Wang, X. Ding, J. Geng, M. Sasaki, T. Hirose, Separation of natural tocopherols from soybean oil byproduct with supercritical carbon dioxide, J. Supercrit. Fluids, 2007, Vol. 40 (1): 51-58.
[9] M. Breuninger, Preparation of □-tocopherol by catalytic methylation of other tocopherols. European Patent EP882722, 1998.
[10] D. Swern, Bailey's Industrial oils and fats, John Wiley & Sons, New York, 1986.
[11] Y. Zhao, G. Sheng, D.Wang, Pilot-Scale isolation of tocopherols and phytosterols from soybean sludge in a packed column using supercritical carbon dioxide, in Proc. 5th Int. Symp. on Supercrit. Fluids, Atlanta, Georgia, USA, April 8-12, 2000.
[12] H. Kikuchi, H. Kubo, T. Tomai, K. Terashima, Deposition of Cu Film using Plasma in a Supercritical Environment, in Proc. 8 th Int. Symp. on Supercrit. Fluids, Kyoto, Japan, November 5-8, 2006.

The investigation of degradation reaction of various saccharides in high temperature and high pressure water

T Saito, S Noguchi, T Matsumoto, M Sasaki and M Goto

Graduate School of Science and Technology Kumamoto University, 2-39-1 Kurokami Kumamoto, Japan

E-mail: msasaki@kumamoto-u.ac.jp

Abstract. Recently, conversions of polysaccharides included in biomass resources have been studied in order to recover valuable chemicals. Degradation of polysaccharides has been attracted by many researchers, whereas by-products from secondary reactions of the materials have not been studied very well. For the purpose of understanding reaction behavior of various monosaccharides in high-temperature and high-pressure water regions, we investigated reaction pathway and kinetics through reaction experiments of degradation of saccharides in subcritical water. The experiment was conducted by using continuous flow-type micro-reactors. Glucose was used as the starting material. From the experimental results, the conversion of glucose increased with increasing the residence time. The yields of fructose and 1,6-anhydro-β-D-glucose decreased with increasing the residence time. The yields of organic acids and some aldehydes increased with increasing the residence time.

1. Introduction

In recent years, the energy consumption is increasing in all over the world. The growth in consumption of energy in 2001 is three times as high as its 1965 [1,2]. Fossil fuel is used for the automobiles and the industrials, and it causes the elevation in carbon dioxide concentration in the atmosphere. From these facts the biomass is receiving attention for its capability as an alternative to fossil fuels as an chemical resources. Cellulose hemi-cellulose and lignin are main components of biomass. Especially cellulose is contained as a constituent element of biomass. It is also able to convert into energy in diverse ways. In conventional techniques, for example, a dilute sulfuric acid method [3] is representative as one of reaction methods. The dilute sulfuric acid method, 0.1 ~2.0 % of acid concentration, 130~240 °C of reaction temperature, leads to recover useful chemicals as decomposition products. Although these decomposition methods use sulfuric acid or hydrochloric acid as catalysts, it makes treatment process cumbersome and complicated by the treatment of the harmful catalyst which remains after the process. Therefore, in the conversion of the biomass to the utilization science material, the low-loaded material is selected for environment and development of the new synthetic method of the environment fitter type and high economic efficiency is desired. Consequently we focus on water as the solvent that is low load to the human body, and advance the research of the conversion method of the biomass under the non-catalytic condition. Because sub-critical and supercritical water have many interesting properties [4], it is widely used as the solvent of extraction and crystallization. In the previous works, a lot of study results that decompose saccharide with sub-critical and supercritical water are reported [5-9]. In this study, we carried out the experiments of

© 2008 IOP Publishing Ltd

conversion of D(+)-glucose under sub-critical water condition to investigate thermal stability of saccharide.

2. Experimental

2.1. Materials

D(+)-glucose which used in the experiment starting material is purchased from Wako Pure Chemicals Industries Ltd. Regents used for products analysis were D(-)-fructose, DL-lactic acid, formic acid, acetic acid also provided by Wako Pure Chemicals Industries Ltd. 5-hydroxymethyl-2-furfural and glycolic acid were from Sigma-Aldrich Co. The other regents such as 1, 6-anhydro-β-D-glucose and 2-furufural were obtained from Tokyo Chemical Industry Co., Ltd. and Kishida Chemicals Co., Ltd., respectively.

2.2. Experimental

Fig. 1(a), (b) show the schematic diagram of the continuous flow reactor.

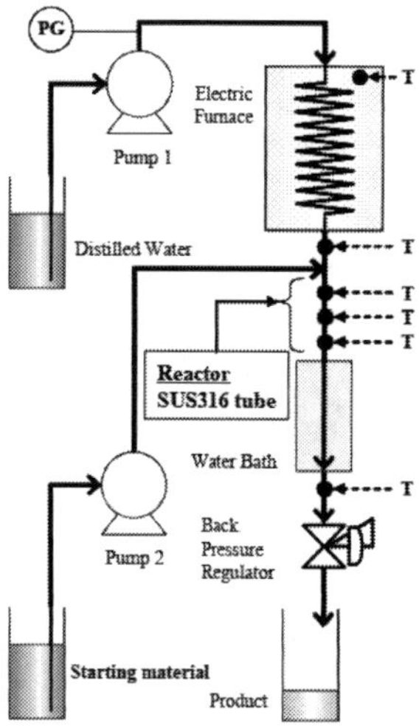

Figure 1(a). The schematic diagram of the continuous flow reactor (reactor A1).

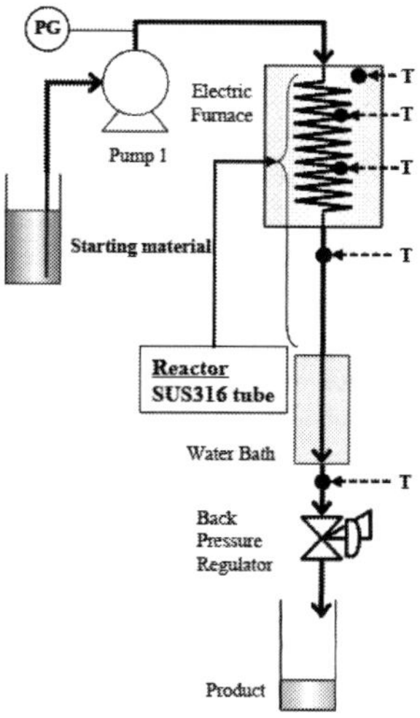

Figure 1(b). The schematic diagram of the continuous flow reactor (reactor A2).

For the reactor A1, the reactor was made of 1/8 inch stainless steel tubing (SUS316). The liquid solution is sufficiently degassed by the ultrasonic and the aspirator. After that the solution is disposed with high pressure pump. First the preheating water was fed to the system by the HPLC pump (NIHON SEIMITSU KAGAKU CO., LTD, model NPL-3300). The pressure in the reactor was adjusted by the back pressure regulator (AKICO Co., Ltd, model HPB-450). After the pressure in the system became the prescribed value and stable, heating the electric furnace was initiated. The reactor and the furnace temperature were measured by a K-type thermocouple. After the temperature of the reactor reached at preset temperature and stable, the pump was switched to the starting material pump (JASCO Co., model BIP-1). The concentration of glucose in the solvents was prepared at 1.5-10 wt%. The reaction pressures were adjusted in the range from 15 to 20 MPa. The reaction temperatures were adjusted in the range from 200 to 280 °C. At the reactor, the feed solution was promptly heated to its reaction temperature and the reaction was initiated. The residence time of the solution in the reactor, τ, was evaluated by the following equation:

$$\tau = \frac{V}{F} \times \frac{\rho_{T,P}}{\rho_{T_0,P}} \tag{1}$$

where V is the reactor volume, F is the flow volume, $\rho_{T,P}$ and $\rho_{T_0,P}$ indicate the density of water at the reaction temperature and at the normal condition of water, respectively. After the reactant had passed through the reactor, the residence time was from 40 to 120 seconds which was evaluated eq 1, the flow was quenched by double tube type heat exchanger to terminate any further reactions. After that the flow was subsequently decompressed by passing through the back-pressure regulator.

For the reactor A2, the application was conducted with two pumps. The condition of experiment was same to reactor A1 except for mixing preheated water and starting material.

The identification and quantitative analysis of the each product were conducted by HPLC-UV/RI. 0.003 mol/L perchloric acid was used as the mobile phase, and BTB solution was used as coloring reagent to detect organic acids, and it mixed in the immediate aftermath of the column. For evaluating the carbon balance, a total organic carbon analyzer (Shimadzu, Model TOC-5050A) was used. The conversion (X) and the yield (Y) were evaluated by following equations (2, 3) [10]:

$$X = \left(1 - \frac{C_f}{C_0}\right) \times 100 \tag{2}$$

$$Y = \frac{C}{C_0} \times 100 \tag{3}$$

where C_0 is the initial concentration of the starting material, C_f is the concentration of ingredient in product, and C is product concentration.

3. Result and discussion

Figure 2 shows the HPLC-RI/UV chromatogram of the products from glucose as treated in sub-critical water. From HPLC-RI chromatogram, the main products are D-fructose, 1,6-anhydro-β-D-glucose, glycolaldehyde, 5-HMF, and 2-furfural. From HPLC-UV chromatogram, glycolic acid, lactic acid, acetic acid, and formic acid are observed.

(a) (b)

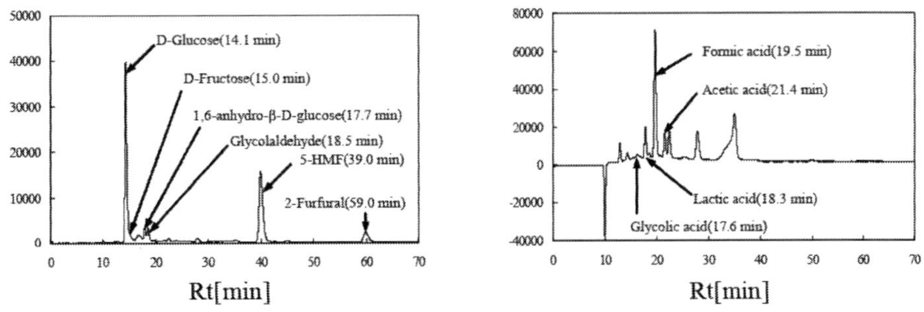

Figure 2. HPLC chromatogram of the products: T = 240 °C, P = 20 MPa, τ = 60 sec, 6 wt%, reactor A1. (a) RI chromatogram, (b) UV chromatogram.

These chemicals also can confirm in main reaction pathway which suggested in previous works. D-glucose isomerizes into fructose or dehydrates into 1,6-anhydro-β-D-glucose and 2-furufural or decomposes to glycolaldehyde. On the other hand, the organic acids are obtained from secondary reaction of these products. It is thought that 5-hydroxymethyl-2-furfural (5-HMF) is a dehydration product of D-fructose which is an isomer of the D-glucose.

Figure 3(a), (b) show results at 240 °C, 20 MPa, 3 wt%. From figure 3(a), in this case, the conversion of D-glucose increased in increasing the residence time. At 120sec, the conversion became 83.1 %. The yield of 2-furufural that is one of the dehydration product of D-glucose was obtained in high yield at longer residence time. This result indicates that the dehydration reaction was occurred with D-glucose. D-fructose which is an isomer of D-glucose was observed more high concentration in the short residence time. This reaction is supposed to be caused by keto-enol tautomerization. 5-HMF was formed in high yield at the longer residence time. This result expresses that the dehydration reaction occurred with D-fructose. For the yield of organic acids, almost all products were obtained in high yield at the longer residence time. Especially, the yield of formic acid reached at about 11.9 %.

(a)

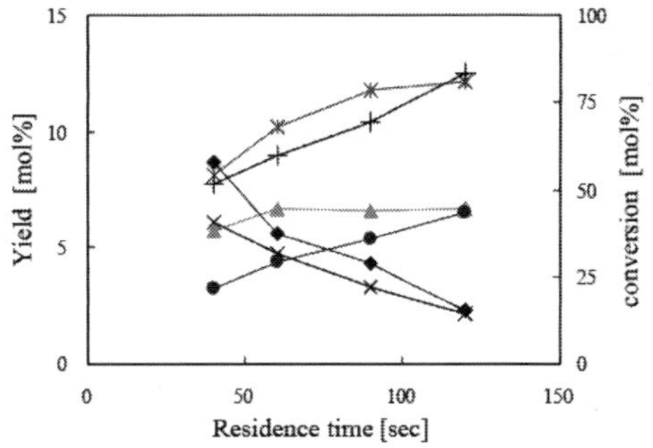

(c)

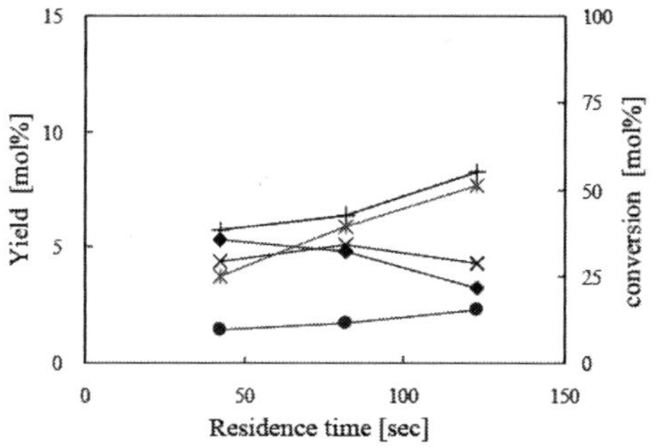

(b)

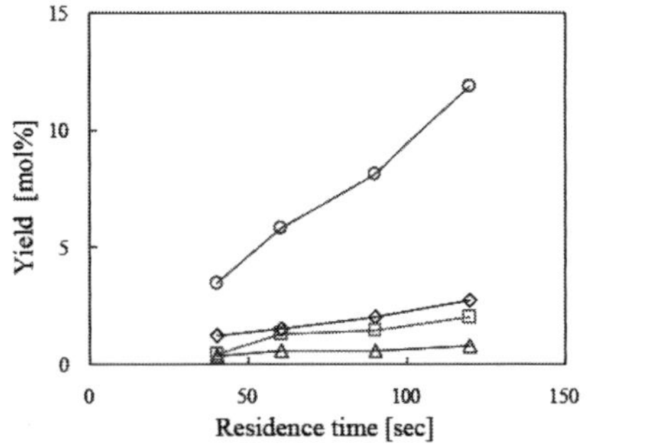

(d)

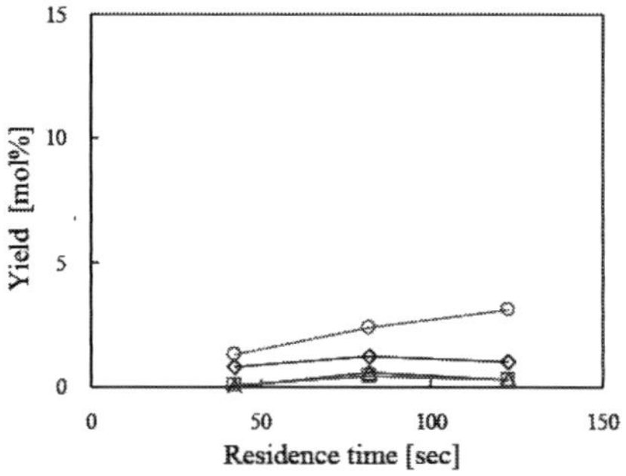

Figure 3. The products yield and the conversion of glucose: T = 240 °C, P = 20 MPa, 3 wt%. (a), (b): The products yield and the conversion using the reactor A1. (c), (d): The products yield and the conversion using the reactor A2.

symbols: + Conversion, ◆ fructose, ▲ glycolaldehyde, × 1,6-anhydro-β-D-glucose, ✱ 5-HMF, ● 2-furfural, ◇ glycolic acid, △ lactic acid, ○ formic acid, □ acetic acid

For the reactor A2, though the conversion value was smaller than that of the reactor A1, a similar tendency was observed in the products. Moreover the comparison of the yields at almost same conversion between the reactor A1 and reactor A2 were conducted. At 240 °C, 20 MPa, 3 wt%, 53.7 % (t=40sec, reactor A1) and 54.8 % (t=122 sec, reactor A2) of the conversions of the D-glucose, it was observed that the product yields of the reactor A1 was higher than that of the reactor A2.

Acknowledgment
This works has been done within Kumamoto University 21st COE Program "Pulsed Power Science", and NEDO "Development of Technology for High Efficiency Biomass Energy Conversion".

Journal of Physics: Conference Series **121** (2008) 082010 doi:10.1088/1742-6596/121/8/082010

References

[1] The Energy Data and Modeling Center The Institute of Energy Economics 2003 *EDMC Handbook of Energy & Economic Statics in Japan* (Japan: The Energy Conservation Center) pp 29-40, 175-178, 191-194

[2] The Energy Data and Modeling Center The Institute of Energy Economics 2004 *EDMC Handbook of Energy & Economic Statics in Japan* (Japan: The Energy Conservation Center) p 217

[3] Z. Yu and H. Zhang 2003 Bioresource Technology **40** 95-100

[4] Sako T 2001 *Supercritical fluid* (Tokyo: Agune Syohu sha) pp 49-68

[5] Ehara K and Saka S 2005 *J. Wood. Sci.*, **51**, 148-153

[6] Sakaki T, Shibata M, Sumi T, Yasuda S, 2002 *Ind. Eng. Chem. Res.*, **41**, 661-665

[7] Kabyemela B.M, Adschiri T, Malaluan R.M, Arai K 1999 *Ind. Eng. Chem. Res.*, **38**, 2888-2895

[8] Sasaki M and Arai K, 2002 *ECO INDUSTRY*, **7**(1), 42-51

[9] Goto K, Tashima M, Sasaki M, Adschiri T and Arai K 2001 *Kobunshironbunshu*, **58**(12), 685-691

[10] Sasaki M, Furukawa M, Minami K, Adsciri T and Arai K 2002 *Ind. Eng. Chem. Res.*, **41**, 6642-6649

Journal of Physics: Conference Series **121** (2008) 092001 doi:10.1088/1742-6596/121/9/092001

Role of orientation of nucleus of crystal during the process of synthesis of fine crystalline oxides at high temperatures and pressure.

G. P. Panasyuk, V. N. Belan, I. L. Voroshilov and D.G. Shabalin.

IGIC RAS (N. S. Kurnakov Institute of General and Inorganic Chemistry, Russian Academy of Sciences)
¤ E-mail for correspondence: Panasyuk@igic.ras.ru

ABSTRACT: The structural transformations of hydrargillite $Al(OH)_3$ and boehmite AlOOH were studied on thermovaporous autoclaving and on heat treatment. A special attention was paid to the morphology of initial flocculated and loose hydrargillite crystals, of the again segregating flocculi boehmite intermediate phase, and to the morphology of the end product – corundum single – or polycrystals. It is shown that on thermovaporous autoclaving the intraflocculi orientation of the boehmite crystals is decisive in the corundum single crystals formation. The degree of identity of the boehmite crystals orientation qualifies the sizes of the growing corundum crystals.
It is shown, that after heat treatment in air at <1200°C alpha-alumina forms, the habitus of the initial crystals remains unchanged, but at 1400-1500°C in vacuum it alters, and equally oriented alumina grains sinter. Influence of the mutual crystals orientation at different stages of structural transformations is explored.
Keywords: autoclaving, hydrargillite, boehmite, corundum.

INTRODUCTION: Control over structural transformations of various alumina modifications is actual on manufacturing abrasives, alumina ceramics, catalytic agents, etc. In many processes alumina is prepared from metallurgical hydrargillite $Al(OH)_3$, used as a starting material. Corundum can be prepared from hydrargillite by aqueous thermovaporous autoclaving (TVA) at ~ 400°C [1, 2] or by heat treatment in air or in vacuum at 1200 – 1800°C. In both cases the cross orientation of crystals of various aluminas essentially influence on powders formation.
The purpose of this work is the development of research of an influence of orientation effects in structural transformations of alumina on TVA of the hydrargillite, introduced earlier [2]. In that work we have shown, that after a flash TVA of hydrargillite GD-00 to 450°C and a stand-up within 5 h corundum formed. When heated up to this temperature the initial floccule structure of the hydrargillite crumbles in part or completely with formation of boehmite. At the same time there is a partial flocculation of the formed boehmite crystals and again shaped flocculi have subvolumes with various density of the aggregated crystals. Under TVA conditions used the processe proceeds promptly, and prepared short-term samples represented a mixture of boehmite and corundum powders with a various degree of aggregations and shape perfection. For detailed investigation of the corundum formation on autoclaving and for comparison with process of heat treatment of hydrargillite and boehmite the present work has been made.

MATERIALS AND METHODS: In this work we used commercial hydrargillite 'GD-00', 'GD-18' and 'MDGA' (Pikalevo's PO "Glinozem", Russia) and hydrargillite "Apyral 1E" (Nabaltec AG, Germany) Powders of GD-00 and GD-18 consist mainly of spherical aggregates with the size 50-100 microns, consisting mainly of columnar big crystals and to a lesser extent - of finer, 3 – 10 microns, hydrargillite flakes. The MDGA powder is prepared by steam milling of GD-00 in a jet attritor. As a

© 2008 IOP Publishing Ltd

result the spherical aggregates of GD-00 desintegrate, and crystals and their joints are grinded up to particles with the grain size ~10 microns. Hydrargillite Apyral 1E is a mixture of large spherical aggregates (60 – 100 microns, Figure 1(a) and ~ 5 mass. % of the loose finer crystals (10 – 30 microns, Figure 1(b)). TVA was carried out in laboratory autoclaves with volume 0.015 – 5.0 l in open stainless steel liners at temperature 200 – 400°C. Autoclaves were heated with heating rate ~ 15°C min^{-1}, then isothermally exposed and finally were quenched in cool water. Heat treatment was carried out in air and in a vacuum furnace at 1300 – 1900°C. Synthesized samples were characterised by powder X-ray diffractometry (XRD) and scanning electron microphotography (SEM).

RESULTS AND DISCUSSION: For all the autoclaved samples the initial, the intermediate and the terminating states are clearly fixed. The initial state is the hydrargillite consisting of separate crystals, as the MDGA, or of a mixture of separate crystals and their spherical aggregates: GD-18, GD-00 and the Apyral 1E. The intermediate state is practically flocculated boehmite. The final state consists of more or less aggregated corundum crystals with various sizes. After TVA at ~ 200°C the specified types of hydroxides of aluminum completely transform into boehmite. The degree of its aggregation depends on the coarseness of the flocculi of the initial hydrargillite. And herewith the loose particles of the formed boehmite start to aggregate. For example, free of aggregates the MDGA powder always forms boehmite flocculi, though with the smaller size, than unmilled kinds of the used hydrargillite.

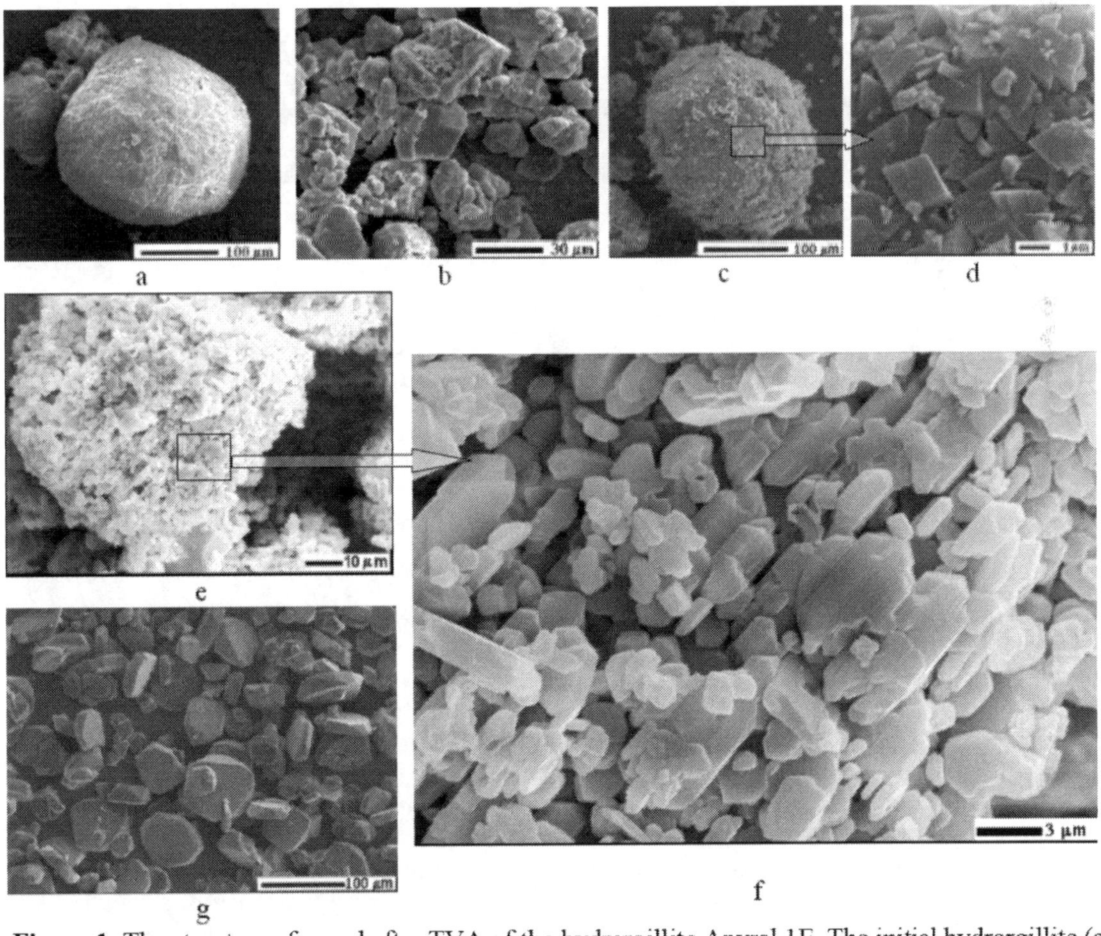

Figure 1. The structures formed after TVA of the hydrargillite Apyral 1E. The initial hydrargillite (a), the finer fraction separated from it (b); after autoclaving at 200°C for 72 h (c) and (d), and then at 410°C for 4 h (e), (f), and for 24 h (g).

The SEMs in Figure 1 show the main features of the process of corundum formation from the hydrargillite Apyral 1E. The initial hydrargillite, Figures 1(a, b) were autoclaved at 200°C for 3 days. As a result boehmite formed. It consisted mainly of flocculi and a small amount of loose crystals. The majority of the boehmit flocculi were well formed and had the sizes not less then initial hydragillite flocculi though the sizes of the boehmite crystals were by of ~10 times less than the hydrargillite, and its density was higher then initial for ~ 26 – 28 %. In these samples there also were some less compact flocculi of the smaller size. The Figures 1(c, d) clearly show the formation of secondary flocculi and also the habitus of the boehmite crystals which compose them. After autoclaving at 410°C for 5 h the boehmite crystals, including intraflocculated ones, alter their habitus and become more ordered with the formation of almost completely structured flocculi, Figure 1(e, f.) As a result the boehmite crystals coalesce with formation of corundum single crystals of various sizez, Figure 1(g). After TVA of the other kinds of the hydrargillite used in this work the analogous results were obtained, and this corresponds to our earlier results [1].

Figure 2. The structures formed after TVA of the hydrargillite GD-18. The initial hydrargillite (a) was treated with 1 % solution of sodium aluminate for 72 h, the precipitate was then dried up in air for 48 h, heat treated in air at 150ºC for 2 h and had TVA at 410ºC for 1 h (b), 24 h (d), 72 h (e), (f), (g).

Depending on TVA parameters used and/or on dopants (activators) applied, one or several areas of closely packed boehmite crystals form in the flocculi. The boehmite crystals within these areas are equally oriented. However, the dominating orientation of the boehmite crystals in various areas may vary unlikely, Figure 2(a). When misorientation is significant the aggregates of corundum crystals are produced, Figure 2 (b, c). If misorientation is negligible mosaic crystals of corundum are formed, Figure 2(d).

For comparison of processes occurring on TVA and on heat treatment of aluminum hydroxide the structural changes of alumina were analyzed in a wide interval of temperatures. Boehmite, synthesized from the hydrargillite MDGA at 200ºC in an autoclave was used as initial material. This boehmite after heat treatment in air at 1200ºC for 24 h did not change the crystal size and habitus. The diffractograms of this as well as all others annealed alumina samples demonstrated formation of well crystallized alpha-alumina. The habitus transformation and agglomeration of the alpha-alumina crystals synthesized from annealed hydrargillite or autoclaved boehmite began at 1300-1400°C. The corundum crystals formed grow together with crystals of the close orientation, Figure 3.

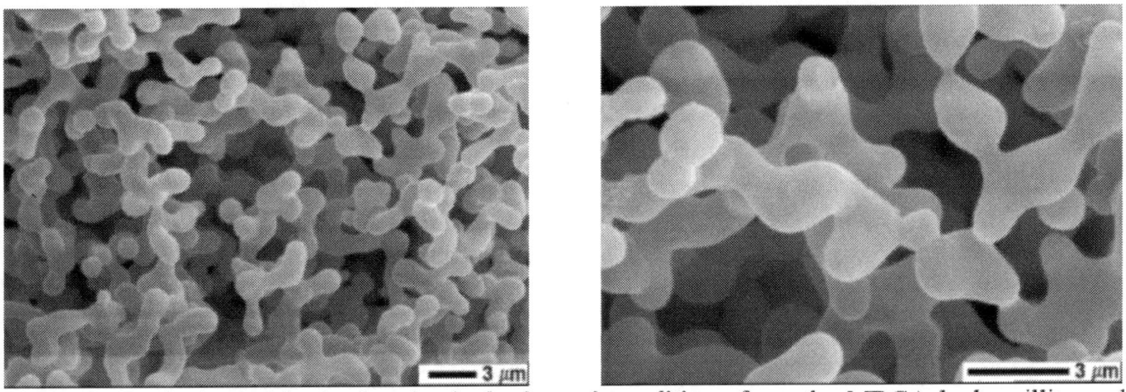

Figure 3. Boehmite synthesized under hydrothermal conditions from the MDGA hydrargillite and annealed in vacuum at 1400C for 3h.

The presented results allow revealing of some generality of processes occurs on thermovaporous and heat treatment of aluminum hydroxide. On TVA within the first hour hydrargillite crystals disintegrate to boehmite fine crystals, which then flocculate and pass into transient state between a boehmite and corundum. For this state XRD shows only boehmite reflexes whereas the refractivity index is close to the value of corundum (1.79). These crystals become ordered, grow together and transform into single crystals of corundum. At heat treatment at temperature 1400°C alpha-alumina crystals with the habitus of boehmite precursor transform into alpha-alumina with out-of-round not faceted crystal grains, which easily grow together, probably, with the crystals of the close orientation.

CONCLUSION: The role of orientation or ordering of primary crystals during formation of corundum from aluminum hydroxide on thermovaporous and heat treatment was shown. It is ascertain, that on TVA self-assembling of ordered at the beginning of the processing boehmite crystals determines the sizes of corundum crystals. It gives an opportunity to control the size of corundum crystals, in particular, by the use of appropriate activators. On heat treatment of alumina the ordering of the primary crystals results in their preferred sintering. When the speed of particles growth is high, in the presence of chlorides for instance, porous ceramic may form and retain its porosity up to 1900°C.

References

[1] Hong T L, Liu H T, Yeh C T 1997 Appl. Catalysis A: General **158** 257

[2] Panasyuk G P, Danchevskaya M N, Belan V N, Voroshilov I L and Ivakinn Yu D 2004 J. Phys.: Condens. Matter. **16** 1215

Journal of Physics: Conference Series 121 (2008) 092002 doi:10.1088/1742-6596/121/9/092002

Crosslinking polymerization of tetraethylene glycol dimethacrylate under high pressure.

K Kaminski [1], M Paluch[1], J Ziolo[1], R Bogoslovov[2], CM Roland[2]
[1]Institute of Physics, Silesian University, Uniwersytecka 4, 40-007 Katowice, Poland
[2]Chemistry Division, Code 6120, Naval Research Laboratory, Washington DC 20375-5342

e-mail: kaminski@us.edu.pl

Abstract: The polymerization reaction of tetraethylene glycol dimethacrylate was induced by application of high pressure. Broadband dielectric spectroscopy was employed to investigate dielectric properties of the produced polymers. Additionally swelling experiment was performed to determine the degree of crossliniking of the polymers.

Key words: Polymerization, high pressure, acrylates

1. Introduction

The utility of polymers can scarcely be overstated, whether as simple commodities or in technically advanced optical and electronic applications. The conventional method of polymerization is thermally, usually with a catalyst; the latter can influence both the chemical structure and the macrostructure (molecular weight and its distribution, branching, etc.). Ambient temperature polymerization can be brought about by ionizing radiation, often catalyzed or enhanced by sensitizers [1,2,3], and also by hydrostatic pressure. High pressure syntheses of polymers are very important from an industrial point of view, because the omission of catalysts can yield materials of the highest purity. Reports of pressure-induced polymerization are limited to a few examples, polyisoprene [4], polyethylene [3,5], polypropylene [6], paracyanogen [7],

© 2008 IOP Publishing Ltd

polyacetylene [8], and polybutadiene [2,9], with the reactants in these polymerizations often unstable even at ambient conditions. Here we present pressure induced polymerization of tetraethylene glycol dimethacrylate (TEGDMA). This compound is used in the manufacture of polyacrylates ("acrylics"), important technological materials for a broad range of applications (pressure-sensitive adhesives, Superglue®, Plexiglass®, resists for electron beam lithography, membranes for filtration and blood dialysis, paints, textile and optical fibers, etc.). There have been many studies of acrylate polymerization [10,11,12,13,14,15,16,17,18], but these methods rely on catalysts or, in the case of free radical polymerization, the use of a peroxide. There has been no example of the pressure-induced polymerization of acrylates. In this paper we describe a new polyacrylate synthesis having advantages over conventional methods.

Pressure-induced polymerization influences the dielectric relaxation spectrum through both chemical and physical effects [19]: Polymerization *per se* decreases the dc conductivity due to reduced diffusivity of ions and increases the local segmental (α-) relaxation time of the main species due to its increasing size and more constrained nature. There are concomitant physical effects due to the direct dependence of molecular mobility on hydrostatic pressure. Fitz and Mijovic [20] have analyzed network formation in epoxies from the change of the dielectric strength arising from the reaction of polar groups. They found that while the α-peak is sensitive to network formation, the secondary peaks are not [21]. Viciosa et al. [18] followed the polymerization of triethylene glycol dimethacrylate from changes in dielectric properties; in particular the dielectric strength of the main α-relaxation process decreases with the extent of polymerization.

2. Experimental

TEGDMA (90% pure with 50 ppm hydroquinone as a stabilizer) was purchased from Sigma/Aldrich and used as received. Scheme of the tetra ethylene glycol dimethacrylate is presented below

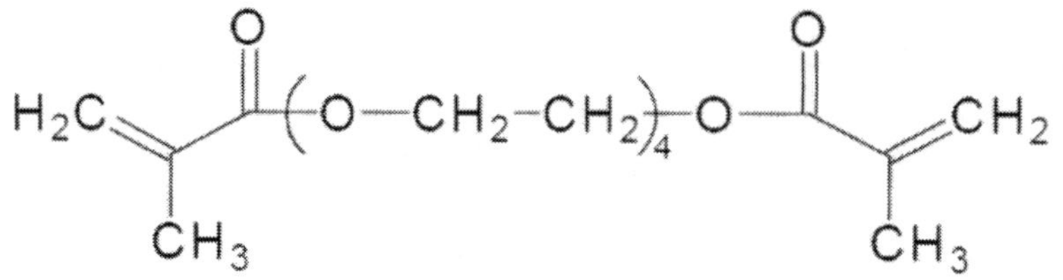

Scheme 1 The chemical structure of tetraethylene glycol dimethacrylate

Polymerization and network formation were carried out by application of hydrostatic pressure. Briefly the dielectric cell with liquid sample was placed into a teflon bellows mounted in the high pressure chamber. Hydrostatic pressure was generated by displacing the piston by means of a hydraulic press. Pressure was determined from the force exerted on piston (accuracy of 10 MPa). This technique enables pressures of a few gigapascals to be attained. The temperature was controlled within 0.1 K by means of liquid flow from a thermostatic bath. Swelling experiments involved immersion of a weighed sample in chloroform for three days, with daily replacement of the solvent. The swollen gels was patted dry to remove adhered solvent, reweighed, and then dried 24 hrs in vacuo prior to final weighing.

3. Result and discussion

During dielectric measurements carried out at T=303 K and P=1 GPa one can observe characteristic changes in the dielectric loss spectra (decrease in dc conductivity, slightly decrease in dielectric strength of main α-relaxation peak) consistent with polymerization. Simultaneously, the TEGMDA changed its consistency from a normal liquid to a gel (see Fig. 1)

Figure 1 Photography showing the polymerized tetraethylene glycol dimethacrylate

Experiments at the same temperature but lower pressure (500 MPa) indicated some polymerization at a much reduced rate, while no reaction was induced at ambient pressure (0.1 MPa). Evidently polymerization of the TEGMDA is initiated only by sufficiently high pressure; that is, there appears to be a threshold value in order to achieve a measurable reaction rate.

Since we know that TEGDMA polymerizes under high pressure, swelling experiments and dielectric measurements were carried out to see which type of polymer is produced: crosslinked or linear.

Swelling experiments were performed on the samples obtained at T=303 K and P=1 GPa after different reaction times. The fact that not all the material is soluble shows clearly that polymerization of TEGMDA is accompanied by substantial crosslinking with consequent gel formation.

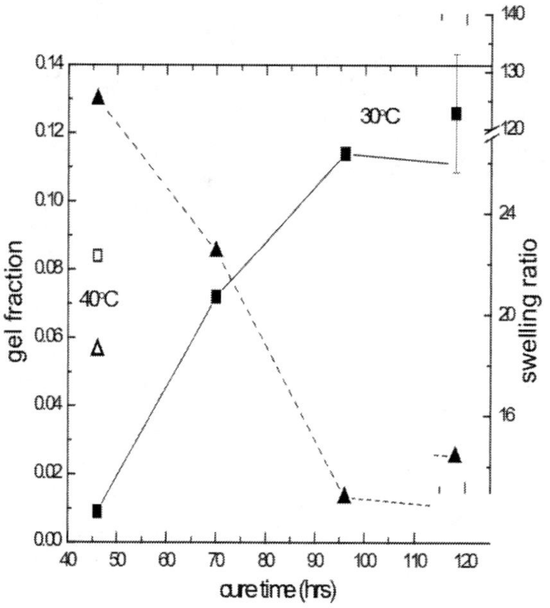

Figure 2 Gel fraction (squares) and swelling ratio (triangles) as function of time at 303 K (solid symbols). The hollow points are data for reaction at 313 K.

This network formation is seen in the swelling data in Fig. 2. A plateau is reached at ~ 12% conversion, with an increasing crosslink density reflected in the reduced degree of swelling (due to the competition between maximizing the mixing entropy and minimizing the elastic strain energy). Also included in Fig. 2 (open symbols) are the data measured for reaction at P=1 GPa and T=313 K for 46 hours). These data are presented to show that there is a substantial increase in the extent of reaction with increasing the temperature. Therefore by performing the polymerization of the TEGDMA at higher temperature we showed that it is possible to reduce the time of the polymerization to reach nearly the same degree of conversion (c≈ 9%) as in the case of polymer obtained at T= 303 K p= 1 GPa and t= 96 h (c≈ 12%). For comparison on an industrial scale c for polymethyl methacrylate (PMMA) is within the range of 7 to 10%.

Fig. 3 displays spectra obtained for the monomer TEGDMA at ambient pressure (a), the polymer obtained at T=373 K, p=1 GPa t=15 h (b) at T=303 K, p=1 GPa, t=96 h (c) Additionally we depicted the superposed dielectric loss spectra of the monomer with that of the polymer obtained at T=303 K p=1GPa after t=96 h (d), the latter shifted along the frequency axis to coincide with the monomer data at 207 K. (Note that the dc conductivity contribution were subtracted from both spectra.) .

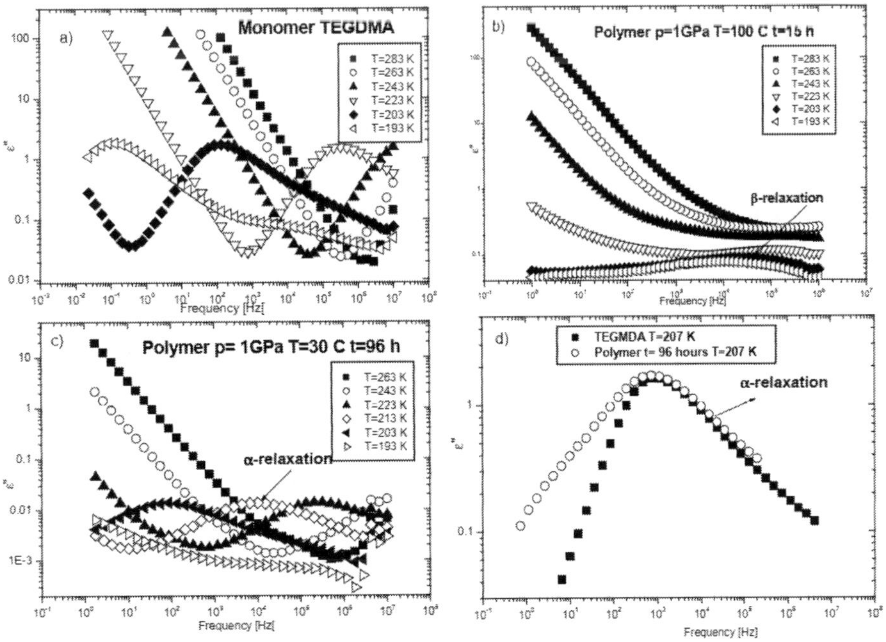

Figure 3 Dielectric loss spectra measured for monomer (a), polymer obtained at p=1GPa, T=373 K, t=15 h (b), T=303 K, p=1GPa and t=96 h (c). At panel (d) the superposed dielectric loss spectra of the monomer (filled squares) and polymer obtained at T=303 K p=1GPa after t=96 h (open circles) are presented. Both spectra were measured at T= 207 K.

One can see that the shape of the α-relaxation of the polymer obtained at T=303 K is much broader than that of the monomer see panel (d). Reduction in the slope on the low frequency side of α-peak is also clearly visible. Thus these two effects, broadening of α-peak and change in steepness on the low frequency side of the structural relaxation of the polymer, are next evidences about crosslinking type of polymerization of the TEGDMA. It should be noted that similar scenario was observed during network polymerization induced by AIBN initiator in the triethylene glycol dimethacrylate [18] and in the case of crosslinking polymerization of the epoxy amine systems [22,23]. Additionally in panel (b) it is shown there is no α-relaxation peak in the wide range of temperatures for the polymerization product at T=373 K. It is commonly observed that during crosslinking polymerization the dielectric strength of the main process goes to 0 ($\Delta\varepsilon \rightarrow 0$) with increasing degree of crosslinking [18,22,23], since the α-dynamics are suppressed; this accounts for the absence of the peak after reaction at T=373 K, p= 1GPa and t=15 h. Furthermore lack of the structural relaxation peak in dielectric loss spectra of the polymer obtained at T=373 K at p=1GPa is an evidence about large molecular weight and significant crosslinking of the polymer produced at this condition. In the dielectric spectra collected for polymers produced at T=373 K, T=303 K and p=1GPa, a secondary relaxation appears in the high frequency range. The position of this β-relaxation is the same as for the monomer. It is consistent with the observation made by Mijovic and Viciosa [18,20,] that secondary relaxations are insensitive to crosslinking.

The characteristic changes in the loss spectra during the reaction (data for 303 K and 1 GPa shown in Fig. 4) reveal the chemical moieties that are likely responsible for the polymerization. The small decrease in the dielectric strength of α-peak with time suggests the reaction involves the unsaturated carbons, which make only a small contribution to the total dipole moment of TEGDMA. This conclusion is unsurprising, since in all cases of pressure induced polymerization, the double bonds between carbon atoms are the active centers for polymerization. Note that reaction of the ester moiety would cause a much greater reduction in the dielectric strength.

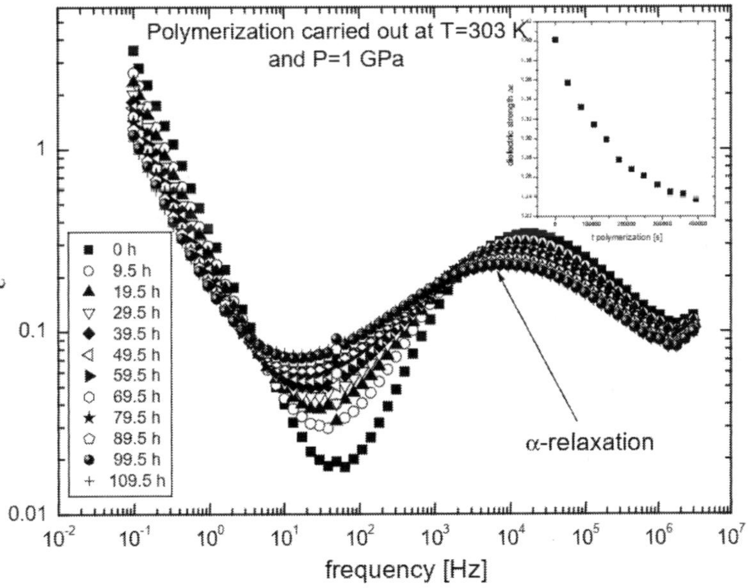

Figure 4 Representative spectra obtained during polymerization of TEGDMA performed at T=303 K, p=1 GPa. In the inset dependence of the dielectric strength of the α-relaxation vs time of polymerization is shown

Fig. 4 also indicates that during polymerization of the TEGDMA, the α-peak shifts only slightly towards lower frequency. Generally, this shift during network formation is much more pronounced [21,24,25]. The behavior is a result of gelation of the TEGDMA during polymerization. Unfortunately this process occurs at low conversion of the monomer and leads to a drastic decrease in diffusion coefficient of the reacting species, ultimately inhibiting the polymerization reaction; thus, reaction occasions little change in the position of the primary α-peak. As shown in Fig. 1, there is systematic broadening of the α- relaxation peak with time of polymerization, along with a decrease in dielectric strength of the primary α-relaxation (see inset to the Fig. 1). Both effects were discussed earlier in this paper.

4. Conclusion

In this paper we demonstrated that TEGDMA undergoes a crosslinking type of polymerization, the first time this has been reported in the literature. Moreover, the threshold value of pressure, P=1GPa, needed to initiate polymerization is one of the lowest reported for high pressure induced polymerization.

The formation of a network, together with the increasing chain length, suppresses the mobility of the reacting species. However, this can be circumvented to some extent by increasing the temperature as was shown for polymer obtained at T=373 K. To maintain a constant rate of reaction would require a concomitant increase in pressure. The optimum conditions can be deduced from consideration of the relevant activation energies and activation volumes.

Finally it should be noted that the degree of conversion of TEGDMA c=12% is very satisfactory, in comparison to the values achieved for compounds polymerized on an industrial scale, such as ethylene c≈10% propylene c≈10% or methacrylate methyl c≈5-7%. We expect this method of synthesis of acrylates can be a useful alternative route for producing these type of materials.

Acknowledgements

Financial support of the Committee for Scientific Research, Poland (KBN, Grant No. N202 130 32/3785) is gratefully acknowledged. The work at NRL was supported by the Office of Naval Research. RB thanks the Amer. Soc. Eng. Edu. for a post-doctoral fellowship.

Journal of Physics: Conference Series **121** (2008) 092002 doi:10.1088/1742-6596/121/9/092002

References

1 Bhowmick A and Vijayabaskar V, 2006, Rubber Chem. Technol. **79**, 402

2 Citroni M, Ceppatelli M, Bini R, Schettino V, 2002, Science, 295.

3 Chelazzi D, Ceppatelli M, Santoro M, Bini R and Schettino V, 2004,.Nature Materials **3**, 470–475.

4 Bridgman PW and Conant JB, 1929, Proc Natl Acad Sci **15**, 680.

5 Chelazzi D, Ceppatelli M, Santoro M, Bini R, and Schettino V, 2005, .J Phys Chem B **109**, 21658.

6 Citroni M, Ceppatelli M, Bini R, and Schettino V, 2005, J Chem Phys **123**, 194510.

7 Yoo CS and Nicol M, 1986, J Phys Chem, **90**, 6732.

8 Aoki K, Usuba S, Yoshida M, Kakudate Y, Tanaka K, and Fijiwara S, **1988**, J Chem Phys, **89**, 529.

9 Citroni M, Ceppatelli M, R. Bini and V. Schettino, 2003, J Chem Phys **118**, 1815.

10 Save M, Weavers J, Armes S, 2002, Macromolecules **35** 1152.

11 Haddleton D, Crossman M, Dana B, Duncalf D, Heming A, Kukulj D, Shooter A, 1999, Macromolecules **32** 2110.

12 Nishikawa T, Ando T, Kamigaito M, Sawamoto M, 1997, Macromolecules **30**, 2244.

13 Lecomte Ph, Drapier I, Dubois Ph, Teyssié Ph, Jérôme R,1997, Macromolecules **30** 7631.

14 Granel C, Dubois Ph, Jérôme R, Teyssié Ph, 1996, Macromolecules **29**, 8576.

15 Ando T, Kamigaito M, Sawamoto M, 1997, Macromolecules **30**, 4507.

16 Matyjaszewski K, Wei M, Xia J, McDermott NE, 1997, Macromolecules **30** 8161.

17 Matyjaszewski K, Nakagawa A, Jasieczek, C, 1998, Macromolecules **31** 1535.

18 Viciosa M, Rodrigues C, Dionisio M, 2005, J Non-Cryst Solids **351**, 14.

19 Johari GP, Mc Ananama JG, Wasylyshyn DA, 1996, J Chem Phys,**105**, 10621.

20 Fitz BD and Mijovic J, 2000, J Phys Chem B, **104**, 12215.

21 Fitz BD and Mijovic, 1998,.J Polym Adv Technol **9**, 721.

22 Casalini R, Corezzi S, Livi A, Levita G, Rolla PA, 1997, J. Appl Pol. Sci, **65**, 17.

23 Levita G, Livi A, Rolla PA, Gallone G, 1996, Pol. Adv. Tech, **7**, 873.

24 Schroeder MJ and Roland CM, 2002.Macromolecules **35**, 2676.

25 Roland CM, 1994, Macromolecules **27**, 4242.

Journal of Physics: Conference Series **121** (2008) 092003 doi:10.1088/1742-6596/121/9/092003

Synthesis under pressure and characterizations through optical spectroscopy of Jahn-Teller cations (LS Ni^{3+}, IS Co^{3+}) as probes diluted in a perovskite matrix

Marta N Sanz-Ortiz[1], Fernando Rodríguez[1], Alexei Baranov[2,3] and Gérard Demazeau[2]

[1]DCITIMAC, Facultad de Ciencias, Universidad de Cantabria, 39005 Santander, Spain

[2]ICMCB, CNRS ,Université BORDEAUX 1 "Sciences et Technologies", site de l'ENSCPB, 87 avenue du Dr.A.Schweitzer, 33608 PESSAC Cedex, France

[3]Department of Chemistry, Moscow State University, 119 992 Moscow, Russia

Email: sanzmn@unican.es

Abstract. The objective is to explore through optical spectroscopy and magnetic measurements the coordination and electronic structures of transition-metal ions introduced as impurities with unusual valence states in the oxide perovskite $LaAlO_3$. The selected transition-metal ions $Ni^{3+}(3d^7)$ and $Co^{3+}(3d^6)$ are characterized by an electronic configuration likely leading to an orbital degenerate E state in Oh symmetry, and thus electron-lattice coupling due to the Jahn-Teller effect may induce low symmetry distortion around the impurity oxygen octahedron. We show that a sol-gel process followed by high oxygen pressure treatments yields stabilization of trivalent state in oxide perovskite. Information about the coordination, electronic structure and aggregation around the magnetic impurity was obtained from X-ray diffraction, FTIR and optical spectroscopy. Finally, evidence on the possible existence of intermediate spin state in Co^{3+} is under consideration.

1. Introduction

Here we investigate through X-ray diffraction and optical spectroscopy, the synthesis, stabilization and structural characterization (coordination and electronic structure) of 3d cations introduced as impurities in oxide perovskites ABO_3, which are able to exhibit a local Jahn-Teller distortion. The selected transition-metal ions (Ni^{3+} d^7, Co^{3+} d^6) are characterized by an electronic configuration probably leading to an orbital degenerate E state in O_h symmetry (LS Ni^{3+}: $t_{2g}{}^6e_g{}^1$, IS Co^{3+}: $d_{yz}{}^2d_{zx}{}^2d_{xy}{}^1d_{x2-y2}{}^1d_{z2}{}^0$). As a result, the electron-lattice coupling may induce a low symmetry distortion of the oxygen octahedron ($E\otimes e$), which makes the complex anisotropic. The purpose of the paper is to investigate whether these transition metal can be mostly stabilised as diluted trivalent impurities in $LaAlO_3$ by high oxygen pressure treatments [1] and whether such impurities exhibit a Jahn-Teller ground state leading to local low-symmetry distortions of the octahedron [2-5]. Although the local structure of Ni^{3+} ions has been largely studied by EXAFS at different dopant concentrations above 5% [6], it is not well understood yet how the low spin (LS) state of Ni^{3+}, which exhibits an elongated

© 2008 IOP Publishing Ltd

octahedron coordination in $LaAlO_3$:Ni^{3+} [7], shows no distortion in $LaNiO_3$ and $LaNi_{0.1}Mn_{0.9}O_3$ [6]. In addition, there is no information about the 3d-electronic structure of either Ni^{3+} or Co^{3+} diluted systems in oxygen perovskites. The 3d splitting pattern of transition metal ions placed at octahedral sites provides an excellent probe to investigate low-symmetry crystal-field distortions due to the Jahn-Teller effect or to crystal anisotropy effects [2-5]. Furthermore structural information on the coordination MO_6 octahedron (M: Ni; Co) can be obtained from the energies of the optical spectra provided that structural correlations between octahedron distortion and 3d electronic energies have been established [2-4]. Jahn-Teller systems involving Cu^{2+} chlorides [4,5] and Mn^{3+} fluorides [2,3] are illustrative examples of structural precise correlations between Jahn-Teller distortion and 3d-splitting energy. In this paper we report optical spectroscopy results on diluted systems (2%) of Ni and Co – doped $LaAlO_3$ for which the formation of transition-metal ion aggregates can be minimised and EXAFS sensitivity may be insufficient for a suitable local structural characterization.

A major problem concerns the use of diluted probes in a perovskite matrix ABO_3. In order to avoid aggregate formation into the host lattice, it was –for the synthesis- necessary to develop specific routes able to favour dilution of the Jahn-Teller probe and homogeneous distribution. Versus the nature of the precursors, in the first step, a sol-gel process involving the citric acid and ethylene-glycol has been investigated [1]. The second step was a chemical treatment devoted to eliminate the organic residues, whereas the third step was a high oxygen pressure treatment for the Ni^{3+} and Co^{3+} valence stabilization.

In order to confirm both the oxidation state and the electronic configuration, the samples were characterized by oxido-reduction titration and magnetic measurements. The Jahn-Teller distortion was investigated through the electronic structure of the transition-metal ion by optical spectroscopy. In particular, we focus on the Jahn-Teller-related splitting pattern shown by the crystal-field spectra, whose band structure contains complete information on the type and degree of the Jahn-Teller distortion as well as on its associated energy and spin state [2-5].

2. Experimental

$La(Ni,Al)O_3$ and $La(Co,Al)O_3$ powders were synthesized for different transition metal concentrations: 1, 2, 5, 15 and 100 % mol for Ni and 2, 5 and 15 % mol for Co. In the first step, a sol-gel process involving citric acid and ethylene–glycol [1] has been investigated. The second step was a chemical treatment devoted to eliminate the organic residues, whereas the third step was a treatment under high oxygen pressures. The oxygen gas till 200 MPa and 700°C during two days was applied in a reactor chamber. The high pressure equipment was reported previously [8]. The experimental setup and obtained samples are shown in figure 1.

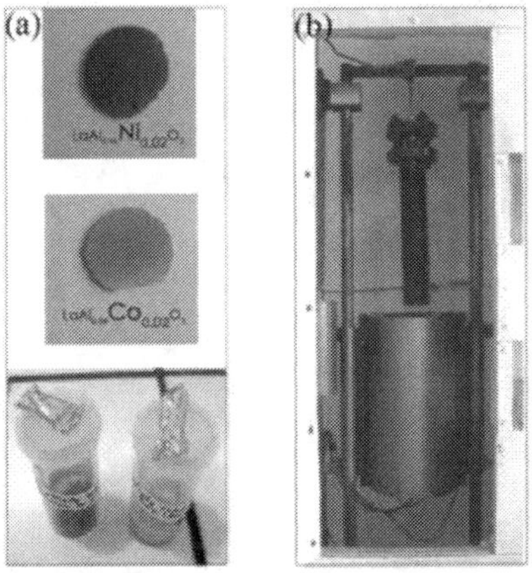

Figure 1. **(a)** $LaAl_{0.98}Ni_{0.02}O_3$ and $LaAl_{0.98}Co_{0.02}O_3$ pills (top) and gold containers used for high pressure treatments (bottom). **(b)** Experimental setup for synthesis under high O_2 pressure.

The crystal structure of the samples was checked by X-ray diffraction (XRD) (goniometer Philips PW1820-geometry: Bragg-Brentano θ-2θ) and Fourier transform infra-red spectroscopy (FTIR) using a Thermo Nicolet. XRD and FTIR spectra indicate that the synthesized samples are miscible and homogeneous for the explored dilutions (figure 2). The crystal structure corresponds to the rhombohedral, nearly cubic perovskite, $R\bar{3}c$ space group [9], with hexagonal lattice parameters, a = 5.3654(3) Å, c = 13.112(1) Å. No evidence of phase segregation or precipitate formation is shown. The infrared spectra of both Co- and Ni-doped $LaAlO_3$ show the same bands, located around 500, 650 and 1400 cm^{-1} irrespective of the impurity. The uncertainty in positions is due to saturation effects. The first two peaks correspond to the IR-active vibrations E_u and E_u+A_{2u} (two T_{1u} in $Pm\bar{3}m$) perovskite modes of rhombohedric $LaAlO_3$ [10]. The peak around 1400 cm^{-1} although exhibits some impurity-dependent band shape, likely corresponds to overtones of the fundamental vibrations: a detail study including impurity effects is currently under progress.

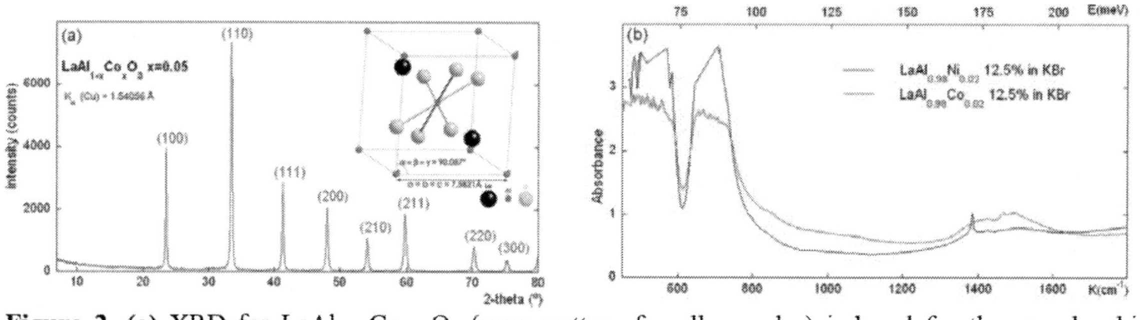

Figure 2. (a) XRD for $LaAl_{0.95}Co_{0.05}O_3$ (same pattern for all samples) indexed for the pseudocubic structure of $LaAlO_3$. **(b)** FTIR $LaAl_{0.98}Ni_{0.02}O_3$ and $LaAl_{0.98}Co_{0.02}O_3$ diluted 12.5% mol in KBr.

The optical spectra were obtained in a Cary 6000i spectrophotometer operating directly in reflectance mode and in absorption mode, using sample diluted KBr pellets (12.5% mol) in the latter case.

3. Results and analysis

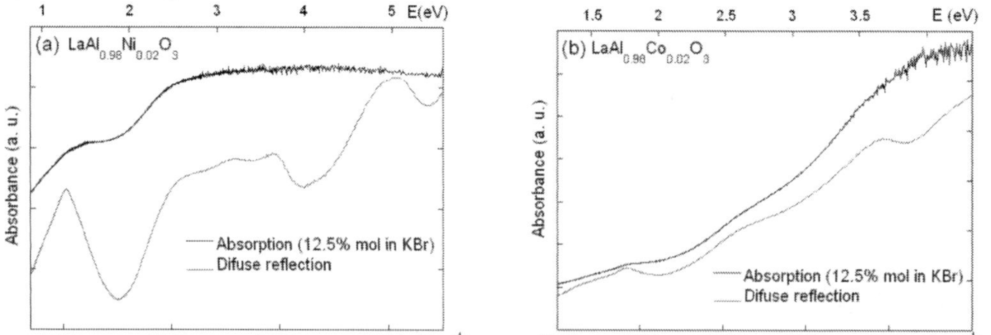

Figure 3. Difuse reflectance and optical absorption (12.5% mol in KBr) spectra of **(a)** 2% mol Ni^{3+} in $LaAlO_3$ and **(b)** 2% mol Co^{3+} in $LaAlO_3$.

Figures 3a and 3b show the optical spectra of the 2% diluted Ni^{3+} and Co^{3+} in $LaAlO_3$ respectively. The comparison of the reflectance and absorption spectra allows us to confirm the presence of absorption bands, which are associated with crystal-field transitions within the NiO_6 and CoO_6 octahedra. It is worth noting that no trace of Ni^{2+} and Co^{2+} is detected in the optical spectra. Preliminary magnetic measurements currently in progress confirm it. Interestingly, the electronic spectra correspond neither high-spin (HS) $^4T_1 \rightarrow {}^4T_2$ for Ni^{3+} (d^7); nor HS $^5T_2 \rightarrow {}^2E$ for Co^{3+} (d^6) since the first intense transition should appear at an energy equal to the crystal field splitting, Δ, which is around 2 eV for these Ni^{3+} and Co^{3+} in oxides [11-13]. By contrast a first intense band is observed in both compounds around 1.2 and 1.5 eV respectively. The presence of this band evidences an additional e_g-octahedral splitting due to the Jahn-Teller distortion [2,3]. A similar band located at 1.5 eV is observed in Mn^{3+} in epidote where the Mn^{3+} occupies a strongly distorted site due to the Jahn-Teller effect [13]. According to correlations established elsewhere [2], the e_g-octahedral splitting, Δ_e, is proportional to the Jahn-Teller distortion, ρ, by $\Delta_e = 5.1\ \rho$ (in eV and Å units, respectively). The Jahn-Teller distortion, ρ, is defined in terms of the tetragonal, Q_θ, and rhombic, Q_ε, normal coordinates as $\rho = [Q_\theta^2 + Q_\varepsilon^2]^{1/2}$. Such a correlation also applies for Mn^{3+} in epidote given that the coordinate measured by X-ray diffraction is $\rho = 0.28$ Å [13]. Taking the same relation for Ni^{3+} we estimate $\rho = 0.2$ Å from the energy of figure 3a. It means a difference between the axial and equatorial distance of the distorted octahedra of about $\Delta R = R_{ax} - R_{eq} = 0.17$ Å. The existence of Jahn-Teller effect in Ni^{3+} in $LaAlO_3$ is consistent with the LS (S=1/2) ground state which was previously shown by EPR [7]. The band structure of figure 3a confirms the proposed Jahn-Teller established LS ground state in $LaAlO_3$.

The optical spectrum of Co^{3+}-doped $LaAlO_3$ is shown in figure 3b. Interestingly, the observed band structure is hard to assign to HS Co^{3+} on the basis of an O_h crystal-field symmetry attained at the Al site. Actually, the first spin-allowed transition in high-spin Co^{3+} (O_h) corresponds to $^5T_{2g} \rightarrow {}^5E_g$, i.e. the crystal-field splitting, and thus should appear around 2 eV [11-13]. The fact we detect an intense absorption band around 1.5 eV and above (two bands around 2.5 and 3.5 eV) is a clear indication of additional splitting in the d-orbitals of the $3d^6$ configuration. Furthermore, it must be noted that the corresponding first allowed crystal-field transition in LS (S = 0), $^1A_{1g} \rightarrow {}^1T_{1g}$, should also correspond to $\Delta \geq 20B \approx 2$ eV for Co^{3+} (O_h) oxides [14]. Although reflectance spectroscopy is not the best suited technique to obtain well-resolved spectra, the band structure and associated transition energies derived from figure 3b clearly suggest a low-symmetry crystal-field for Co^{3+} in $LaAlO_3$. Whether this low-symmetry crystal-field is due to the Jahn-Teller effect is still under investigation. However the possible existence of Jahn-Teller effect in Co^{3+} would indicate that we are dealing neither a low-spin nor a HS ground state, but an intermediate-spin (IS) state. This important result contrasts to recent findings in $LaCoO_3$ by x-ray absorption spectroscopy and magnetic circular dichroism techniques [15], and strongly suggests that the use of diluted Co^{3+} systems in selected oxides may provide crystal environments which are able to stabilize the IS state.

Magnetic and EXAFS experiments in these systems are currently in progress.

Acknowledgments
This work was financed by the Spanish MEC (Project Ref. MAT2005-00099). F. Rodriguez acknowledges partial support from the I3 Research Program of the University of Cantabria. M.N. Sanz-Ortiz thanks the Spanish MEC for support her stay at the University of Bordeaux I and for a FPU grant (Ref. AP-2004-5954).

References

[1] Kakihana M and Okubo T 1998 *J. Alloys Compounds* **266** 129
[2] Rodriguez F and Aguado F 2003 *J. Chem. Phys.* **118** 10867
[3] Aguado F, Rodríguez F, and Núñez P 2003 *Phys. Rev.* B **67** 205101
[4] Valiente R and Rodríguez F 1999 *Phys. Rev.* B **60** 9423
[5] Aguado F, Rodríguez F, Valiente R and Señas A 2004 *J.Phys. Cond. Matter* **16** 1927
[6] Sánchez MC, García J, Blasco J, Subías G and Pérez-Cacho J 2002 *Phys. Rev.* B **65** 144409
[7] Ivanova TA, Petrashen' VE, Chezhina NV and Yablokov YV 2002 *Phys. Sol. State* **44**, 1468
[8] Demazeau G 1973 *Thesis doctorat ès sciences:Realisation de Hautes Pressions d'Oxygene. Preparation et Etude Physique de Nouveaux Composes Oxygenes du Cobalt, du Nickel et du Cuivre au Degre d'Oxydation +III* (Bordeaux, Université Bordeaux 1) **419**
[9] Hayward SA et al. 2005 *Phys. Rev.* B **72** 054110
[10] Delugas P, Fiorentini V and Filippetti A 2005 *Phys. Rev.* B **71** 134302
[11] Griffith JS 1980 *The Theory of Transition Metal Ions* (Cambridge, Cambridge University Press)
[12] Lever ABP 1984 *Inorganic Electronic Spectroscopy, Studies in Physical and Theoretical Chemistry* (New York, Elsevier)

[13] Burns RG *Mineralogical Applications of Crystal Field Theory* 1993 (Cambridge, Cambridge University Press) pp 29, 218
[14] Gütlich P and Goodwin HA 2004 *Spin Crossover in Transition Metal Compounds II* (Springer, Heidelberg)
[15] Haverkort MW et al. 2006, *Phys. Rev. Lett.* **97** 176405

Oxidation state and effective pair potential of Fe^{4+} ions in perovskite-type $SrFeO_3$ annealed under high oxygen pressure

Maki Okube[1], Yuki Furukawa[2], Akira Yoshiasa[2], Takafumi Hashimoto[2], Masahiko Sugahara[2] and Akihiko Nakatsuka[3]

[1]Materials and Structures Laboratory, Tokyo Institute of Technology, Yokohama 226-850 Japan

[2]Graduate School of Science, Kumamoto University, Kumamoto 860-8555, Japan

[3]Department of Advanced Material Science and Engineering, Yamaguchi University, Yamaguch 775-861, Japan

e-mail: makisan@lipro.msl.titech.ac.jp

Abstract. The cubic perovskite-type SrFeO3 have been obtained by annealing under high oxygen pressure of 100 MPa at 600K. X-ray absorption fine structuré (XAFS) observations on Fe K-edge have been performed from 25 K to 900 K to investigate the temperature variation of the local structure and oxidation state of iron ion in the perovskite-type SrFeO3. The edge features of Fe K-edge XANES spectra and the first differential curves of XANES spectra change gradually and greatly with increasing temperature from 600k to 900K. There is no clear change near the Neel temperature. In the higher temperature region than 600 K, the threshold energy shifted to the low energy side (~2 eV) indicating that Fe^{4+} ion more reduced to Fe^{3+} in air. The temperature dependence of EXAFS Debye-Waller factors was investigated and harmonic effective pair potentials, $V(u)= \alpha u^2/2$, have been evaluated. The potential coefficients α for Fe-O, Fe-Sr and Fe-Fe distances are 6.0, 5.48 and 4.0 eV/A^2, respectively.

1. Introduction

The ideal perovskite type structure has a cubic symmetry such as $SrTiO_3$. The cubic perovskite type oxides with the chemical formula $Sr^{2+}B^{4+}O3$ (B = Mn, Fe, Co) quench the tetravalent state of the first-row transition metal ion under ambient condition though these transition metal ions are normally in a divalent or trivalent state and a higher valence state is stabilized under strongly oxidizing conditions[1-4]. Annealing and quenching under high oxygen pressure and appropriate temperature is necessary for the stabilization of the tetravalent state of the first-row transition metal ion. The perovskite-type oxides containing transition metal ions in tetravalent state show interesting electrical and magnetic properties [1-4]. SrFeO3 is an antiferromagnet with Neel temperature of 134K [1].

In this study, we have determined the temperature variation of the local structure and oxidation state of iron ion in the perovskite-type $SrFeO_3$ from 25 K to 900K. The analysis of temperature-dependent EXAFS Debye-Waller factor allows us to evaluate the effective pair potentials and the inter-atomic force constants for neighboring atoms [5-7].

© 2008 IOP Publishing Ltd

Journal of Physics: Conference Series **121** (2008) 092004 doi:10.1088/1742-6596/121/9/092004

2. Experimental and analysis

The starting materials of FeCO3 and SrCO3 (Furuuchi Chem. Co.,Ltd.) in powder form were weighted in the desired ration and milled for 24 hr with acetone. The well-blended mixture of raw materials was pre-fired at 1070 K for 24 hr. They were reground and fired at 1500 K for 24 hr. This firing was repeated three times. As the compounds obtained in this way were oxygen-deficient, they were annealed under pure high oxygen pressure of 100 MPa at 600K for 72 hr. High oxygen pressure was produced by heating an autoclave which had been cooled down to liquid nitrogen temperature and in which oxygen gas had been condensed into liquid [4]. The crystals of SrFeO3 perovskite were identified by X-ray diffraction.

The appropriate amount of fine powder sample and boron nitride powder was mixed and pressed into pellets of <0.2 mm in thickness and 10.0 mm in diameter. All samples had edge-jumps with 0.7 ($\Delta\mu d$), where μ is the linear absorption coefficient and d is the thickness. The measurements of Fe K-edge XAFS spectra were carried out in the transmission mode at beam line BL-12C of the Photon Factory in KEK, Tsukuba (Proposal No. 2006G082). X-ray absorption measurements were made from 300 to 900 K in air. The synchrotron radiation was monochromatized by a Si(111) double crystal monochromator. Mirrors are used to eliminate higher harmonics. Details of the measurement and analysis were given in reference [6,9]

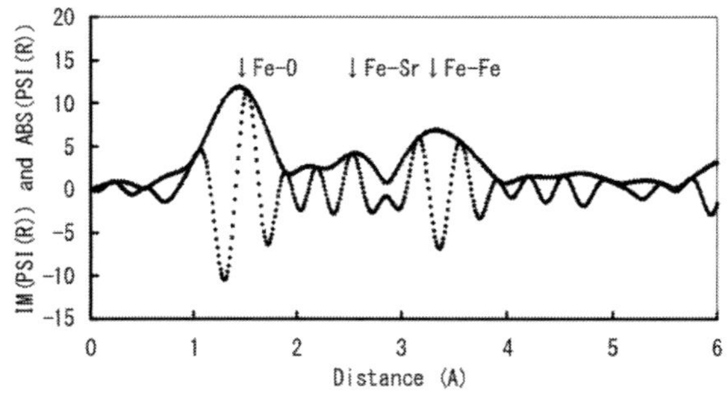

Figure 1. Fourier transforms of the Fe K-edge EXAFS for SrFeO3 perovskite at 300 K. No phase shift corrections are made.

The EXAFS interference function, $\chi(k)$, was extracted from the measured absorption spectra using the standard procedure [9]. The $\chi(k)$ was normalized using MacMaster coefficients according to the EXAFS workshop report [10]. Figure 1 shows example of Fourier transforms of the Fe K-edge EXAFS spectra in the ranges of 3.5<k<14.0. In quantitative analyses we carried out the Fourier-filtering technique and a nonlinear least-squares fitting method by comparing the observed $\chi(k)$exp and calculated $\chi(k)$calc. We used the EXAFS formula in the single scattering theory with the cumulant expansion up to the fourth order term:

$$\chi(k) = \sum_B (N_B / kR^2{}_{AB}) |f_B(k;\pi)| \exp(-2R_{AB} /(k/\pi)) \exp(-2\sigma 2k^2 + (2/3)\sigma 4k^4)$$
$$\times \sin(2kR_{AB} - (2k/R_{AB})(1 + 2R_{AB}/(k/\eta))\sigma 2 - (4/3)\sigma 3k^3 + \psi_{AB}(k))$$

where N_B is the coordination number of scattering atoms B at distance R_{AB} from the absorbing atom A, $|f_B(k;\pi)|$ the backscattering amplitude of photoelectrons and $\psi_{AB}(k)$ the phase shift function. Values of the function $|f_B(k;\pi)|$ and $\psi_{AB}(k)$ were calculated using the FEFF3 program [11]. σn denotes the nth cumulant. The mean free path λ of the photoelectron is assumed to depend on the wave number, $\lambda(k)=k/\eta$, where η is a constant. Single-shell fitting was carried out for each nearest-neighbor distance, where the number of neighboring atoms was fixed at the crystallographic value as N_B=6 for the Fe-O distance, N_B=8 for the Fe-Sr distance and N_B=6 for the Fe-Fe distance. Because the third- and fourth-order terms were negligible, the refinement was performed with harmonic model. The harmonic approximation requires a cumulant up to only second order term. It is known that an anharmonic contribution to the EXAFS Debye-Waller factor appears pronouncedly when the magnitude of σ2 is greater than 0.01 Å2 [12]. The reliability of fit parameters between the experimental and calculated EXAFS functions was less than 0.036.

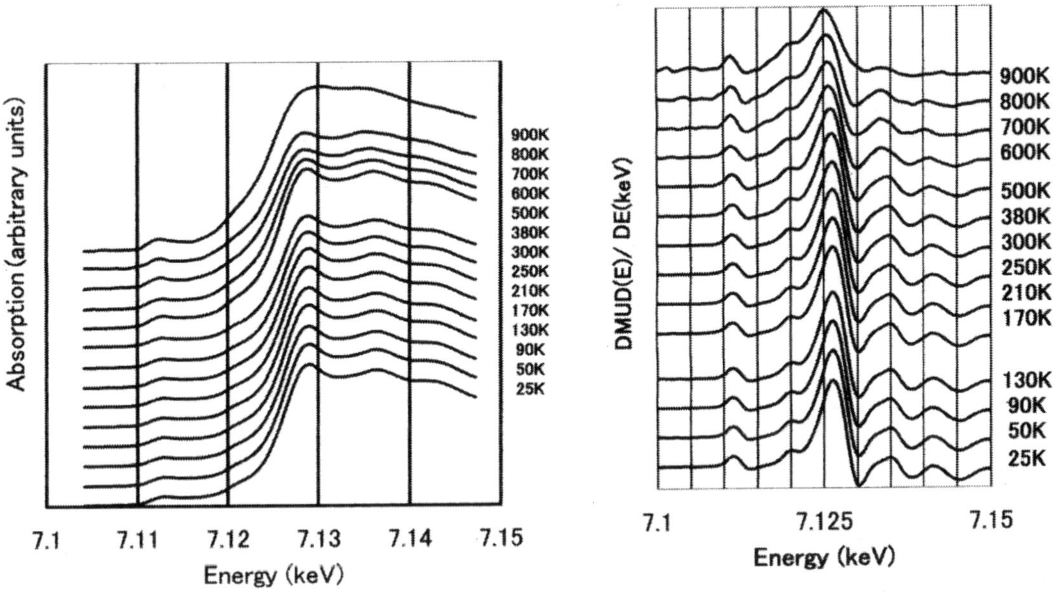

Figure 2. Experimental Fe K-edge XANES spectra and the first differential curves of XANES spectra in $SrFeO_3$ perovskite.

3. Results and discussion

3.1. Oxidation state and local structure changes in $SrFeO_3$ perovskite.

X-ray absorption near-edge structure (XANES) spectra are quite sensitive to the oxidation states of X-ray absorbing atom and three-dimensional atomic configuration around the atom. XANES spectra are useful for determining the oxidation state of Fe ions in $SrFeO_3$ perovskite. Temperature dependences of Fe K-edge XANES spectra and the first differential curves of XANES spectra in $SrFeO_3$ perovskite are shown in Figures 2a and 2b. The edge features change gradually and greatly with increasing temperature from 600k to 900K. From the detailed comparison of the spectra patterns, it observed that the threshold energy shifts gradually to the low energy side (~2 eV) in the higher temperature region than 600 K indicating that Fe^{4+} ion more reduced to Fe^{3+} in air. There is no clear change near the Neel temperature of 134K.

3.2. EXAFS Debye-Waller factors and potential coefficients for Fe-O, Fe-Sr and Fe-Fe distances

Obtained local bond distances agreed with those by X-ray diffraction analysis. The temperature dependence of σ2 for the Fe-O distances (1.935(5) Å at 300 K), Fe-Sr distance (3.33(1) Å) and Fe-Fe distance (3.84(1)) Å in $SrFeO_3$ perovskite are shown in Figure 3. The Debye-Waller type factor σ2 includes the effects of static and dynamic disorders. The static disorder is the configuration disorder, while the dynamic disorder arises from the thermal vibration of atoms. The contribution of the thermal vibration, $\sigma_{thermal}$, can be estimated under the assumption of classical statistical dynamics by the temperature dependence of σ2 [5]. A steep slope in the figure represents a weak bonding. The gradient for the experimental σ2 is equal to k_B/α, if we evaluate the anharmonic effective pair potential $V(u)=\alpha u^2/2$ from the contribution to the thermal vibration, where k_B is the Boltzmann constant, α is the potential coefficients and u is the deviation of the bond distance from the location of the potential minimum. The potential coefficients α for Fe-O, Fe-Sr and Fe-Fe distances are 6.04 eV/A^2, 5.48 eV/A^2 and 4.0 eV/A^2, respectively. The potential coefficient decreases as a result of the increase in the distance. The potential coefficient α of Fe^{4+}-O (6.04 eV/A^2) is comparable to those of Ti^{4+}-O (6.9 eV/A^2) in $CaTiO_3$ and of Ge-O (9.8 eV/A^2) in $CaGeO_3$[13]. The potential coefficient α of Fe^{4+}-Fe^{4+} (4.0 eV/A^2) is relatively smaller than those of Ti-Ti (8.0 eV/A^2) in $CaTiO_3$ and of Ge-Ge (7.4 eV/A^2) in $CaGeO_3$ [13]. The short-range correlation of the atomic motion between Fe^{4+}-Fe^{4+} is relatively smaller than other perovskite type compounds.

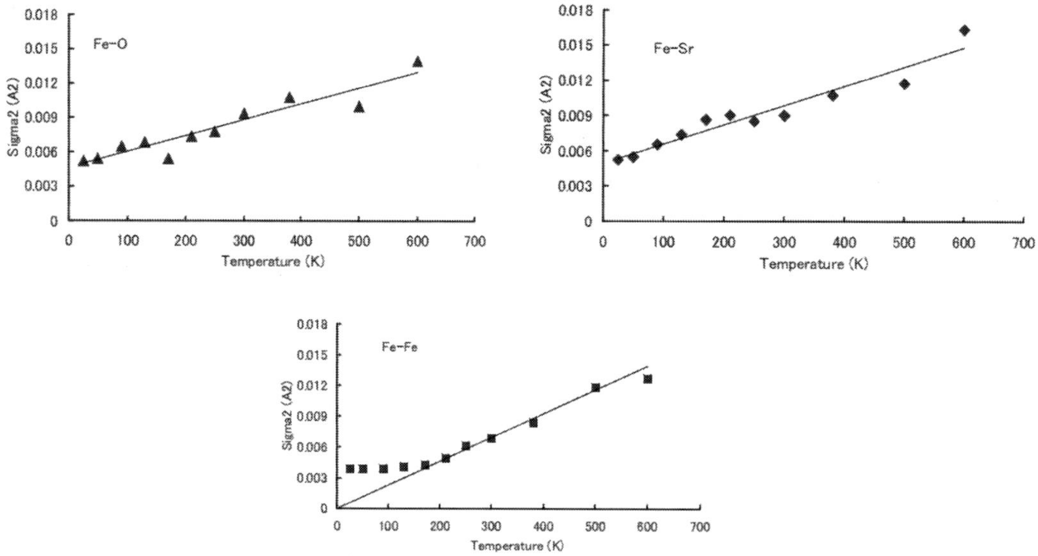

Figure 4. Temperature dependence of σ2 for the Fe-O, Fe-Sr and Fe-Fe distances in SrFeO3 perovskite

References

[1] Takeda T, Yamaguch Y and Watanabe H 1972 *J. Phys. Soc.Japan* **33** 967
[2] Takeda T and Ohara O 1974 *J. Phys. Soc.Japan* **37** 275
[3] Takeda Y, Naka S, Takano M, Shinjyo T, Takeda T and Shimada M 1978 *Mater. Res. Bull.* **13** 61
[4] Yoshiasa A, Inoue Y, Kanamaru F and Koto K 1990 *J. Solid State Chem.* **86** 75
[5] Yoshiasa A, Nagai T, Murai K, Yamanaka T, Kamishima O and Shimomura O 1998 *Jpn. J. Appl. Phys.* **37** 728
[6] Yoshiasa A and Maeda H *Solid State Ionics* 1999 **121** 175
[7] Yoshiasa A, Murai K, Nagai T and Katayama Y 2001 *Jpn. J. Appl. Phys.* **40** 2395
[8] Yoshiasa A, Okube M, Ohtaka O, Kamishima O and Katayama Y 2000 *Jpn. J. Appl. Phys.* **39** 6747
[9] Maeda H 1987 *J. Phys. Soc.Japan* **56** 2777
[10] Lytle F W, Sayers D E and Stern E A 1989 *Physica,* **B158** 701
[11] Rehr JJ, Mustre de Leon J, Zabinski SI and Albers RC 1991 *Am. Chem. Soc.* **113** 5135
[12] Yoshiasa A, Koto K, Maeda H and Ishii T 1997 *Jpn. J. Appl. Phys.* **36** 781
[13] Yoshiasa A, Nakajima K, Murai K and IOkube M 2001 *J. Synchrotron Rad.* **8** 940

Synthesis of metal-nitrides using high pressures and temperatures

C. Guillaume[1], G. Serghiou[1], J. P. Morniroli[2] and D. J. Frost[3]

[1]University of Edinburgh, School of Engineering and Electronics, Kings Buildings, Mayfield Road, EH9 3JL UK
[2]Laboratoire de Metallurgie Physique et Genie des Materiaux,UMR CNRS 8517, Universite des Sciences et Technologies de Lille et Ecole Nationale Superieure de Chimie de Lille, Cite Scientifique, 59655 Villeneuve d'Ascq Cedex, France
[3]Bayerisches Geoinstitut, Universitat Bayreuth, D-95440, Bayreuth Germany

E-mail address: george.serghiou@ed.ac.uk

Abstract. Technologically, high density nitrides are showing promise for both ceramic and electronic applications. In a laser-heated diamond cell we prepare high density metal-nitrides by reaction of the nitrogen pressure medium with an elemental substrate. Two of our objectives are to develop criteria governing whether denser than ambient nitride phases will form, and to in particular establish the parameters required for synthesis in a multianvil press using elemental starting materials. We have already synthesized transition metal nitrides in a multianvil press using elemental starting materials, including hexagonal nickel nitride and alkali rhenium nitrides. Unlike previous metals, we also report that Cu does not form a nitride after heating with NaN_3 at 2000 K and 20 GPa. Notably, Cu_3N is a semiconductor exhibiting weak directional bonds, whereas the immediately adjacent lower atomic number systems are metallic interstitial nitrides. We also briefly mention our work on processing high pressure and temperature recovered reaction products with focused ion beam methods for tailored characterization using electron microscopy.

1. Introduction

Novel dense nitrides in a laser-heated diamond cell, have been optimally prepared by heating elements in the presence of a nitrogen pressure medium[1,2]. In piston-cylinder or multianvil presses, confining a nitrogen medium is difficult. Thus synthesis of dense nitrides there, is typically achieved by using carefully prepared nitrogen containing precursors that can then be compressed and heated without the need for a nitrogen pressure medium. This approach is not always (or at all) feasible as such precursors are not always available. We are thus exploring the parameters of synthesis in a multianvil apparatus where elements can still be used as starting materials together with azides, as nitrogen providers. Broadly, this requires addressing a number of issues including whether starting materials react with the capsule material, and/or with the metal azide provider and at what temperature, whether such reactions and their associated temperature depend on pressure, and what these additional structures formed are, and at what pressure, and what the minimum temperature required for reaction is and how this constrains the maximum pressure we can achieve.

In this study we investigate the effect of high pressure and temperature on the synthesis of nickel, iron and copper nitride using nickel, iron and copper together with sodium azide starting materials, with the latter serving as a nitrogen provider for nitride synthesis in a piston cylinder or a multianvil apparatus. The only bulk ambient pressure modification of nickel nitride has Ni_3N stoichiometry with hexagonal symmetry and $P6_322$ space group [3] [4]. In this arrangement one third of the octahedral voids present in the hexagonal close packing of nickel atoms are occupied by nitrogen. The NNi_6 octahedra share common vertices only, and the nitrogen atoms in most of these syntheses have shown essentially complete occupational order. Ni_3N belongs to the class of interstitial nitrides and is metallic. We further present some preliminary lower pressure work on iron nitride synthesis which is also an interstitial nitride whose Fe_3N modification is isostructural to Ni_3N[5]. The principal ambient pressure copper nitride has Cu_3N stoichiometry with cubic Pm3m symmetry[6,7]. The structure is made of nitrogen centered octahedra which are corner connected to each other. Cu_3N is a semiconductor with a small indirect band-gap. In the broader context of our work employing multianvil methods for materials synthesis, we also present some of our work employing focused ion

© 2008 IOP Publishing Ltd

beam methods, a typical mainstay of the microelectronics industry, for nanometre level sectioning of reaction products. We employ scanning electron microscopy as well as transmission electron microscopy for chemical, morphological and structural characterization of the reaction products.

2. Experimental techniques

The starting materials for the syntheses were nickel powder (99.7%) (Aldrich), iron powder (99.5%) (Alpha), copper powder (99.999%) (Alpha) and sodium azide powder (99.99+ %) (Aldrich). For the piston cylinder runs approximate mixtures of 30 at% iron (Fe) and 70 at% sodium azide were placed in a welded 50:50 Silver-Palladium capsule or in a graphite crucible that slip fits into an iron crucible. The sample was pressurized to 3.5 GPa and heated there for about 1300 K for 2 hours and then recovered for further analysis. For the multianvil runs approximate mixtures of 30 at% nickel (Ni) and 70 at% sodium azide (NaN_3), 50 at% Cu and 50 at% sodium azide respectively were loaded in capsules made of double wrapped rhenium foil. The nickel and copper containing samples were pressurized to 20 GPa, heated at 2000 K for 1 minute and cooled to room temperature in about 20 minutes. All samples were recovered for further analysis. Recovery involved cutting the capsules, polishing and coating for scanning electron microscopy [Philips XL30CP, with an energy dispersive X-ray analyser Oxford instruments EDX detector – SiLi crystal with PGT spirit analysis software for chemical analysis] and transmission electron microscopy [Philips CM30, Transmission Electron Microscope (TEM), equipped with a Gatan SS CCD camera and with Digital Micrograph software for acquisition of electron diffraction patterns and bright-field imaging]. Voltages of between 10 – 20 kV were employed for scanning electron microscopy measurements and a 300 kV acceleration voltage for transmission electron microscopy measurements.

3. Results and discussion

The nickel based sample recovered from 20 GPa and 2000 K reveals the presence of Ni_3N and a new Na-Re-N nitrogen phase whereas the copper based sample recovered from the same conditions only exhibits pure copper and the same Na-Re-N phase (Figures 1, 2).

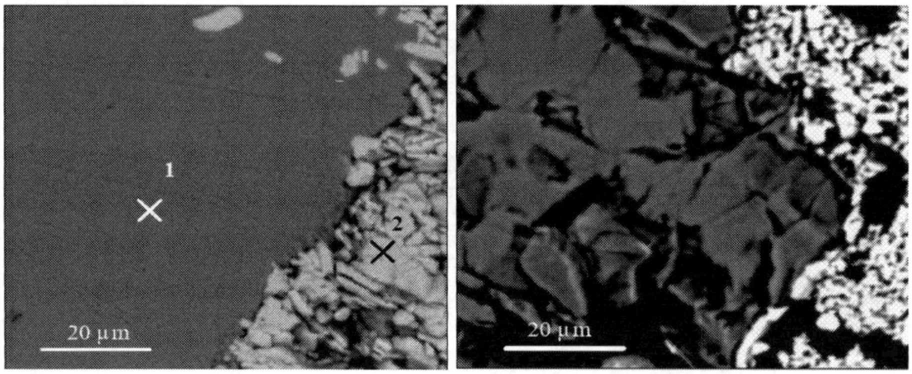

(a) (b)

Figure 1. Backscattered scanning electron microscopy (BSE) images of (a) Cu and Na-Re-N and (b) Ni_3N and Na-Re-N, recovered after heating Cu and Ni respectively, with sodium azide at 20 GPa and 2000 K followed by slow cooling to ambient pressure.

The nickel nitride phase was found to match ordered hexagonal Ni_3N structure with space group $P6_322$ and lattice parameters a = 4.62 Å , c = 4.30 Å, Z = 2 (Figure 3). The Na-Re-N phase arose through reaction of the azide with the rhenium foil. Only one sodium rhenium nitride phase was known till now, prepared at ambient pressure[8]. It is monoclinic. In addition to our measured electron diffraction patterns on the new dense phase we also noted a basic physical property of our dense ternary nitride phase that differs from that of the known one. In particular, while the ambient pressure prepared phase is very water and air sensitive, the dense modification is stable in both of these environments as well as in ethanol. We are now focusing on a detailed evaluation of the structure. Preliminary high resolution electron microscopy images reveal a degree of structural disorder.

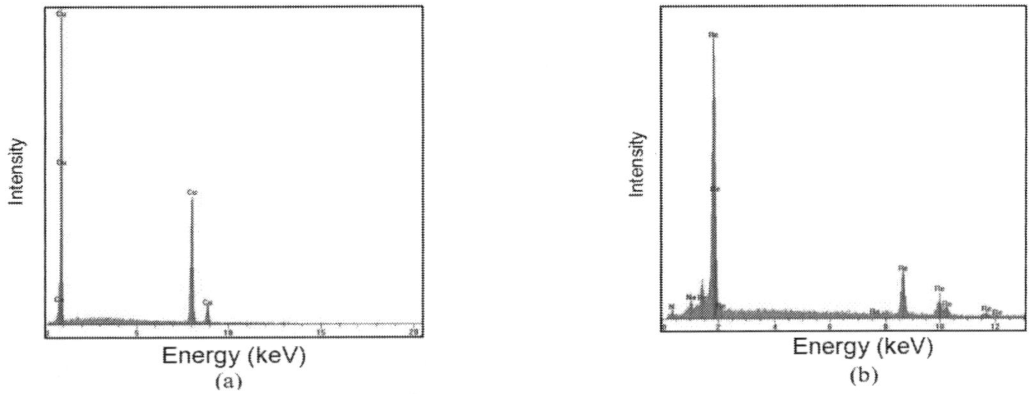

Figure 2. Energy dispersive X-ray analysis (EDX) spectra from spot 1 (a) and from spot 2 (b) in figure 1. The dark contrast region contains pure Cu and the bright contrast region contains Re, Na and N.

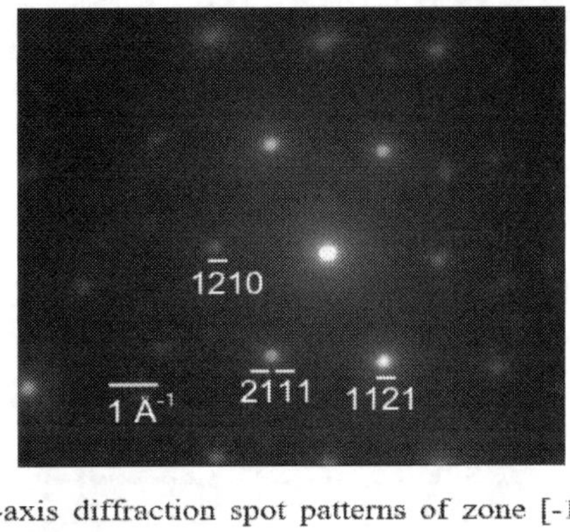

Figure 3. Microdiffraction zone-axis diffraction spot patterns of zone [-1013] of the hexagonal (space group $P6_322$ a = 4.62 Å, c = 4.30 Å, Z = 2) nickel nitride (Ni_3N) structure.

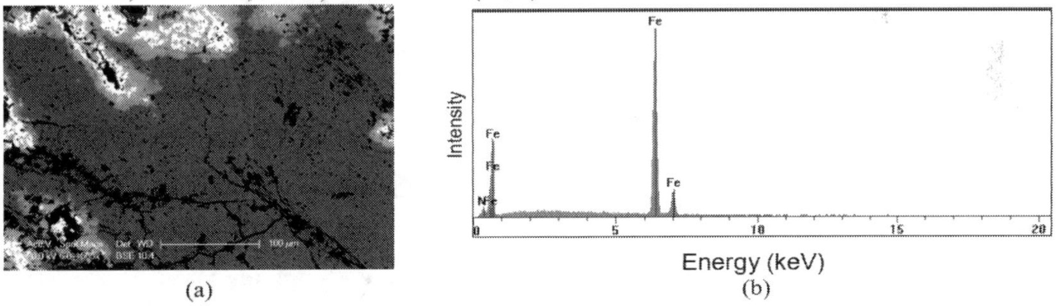

Figure 4. (a) Backscattered scanning electron microscopy (BSE) images of an iron nitride sample recovered after heating Fe with sodium azide at 3.5 GPa and 1300 K. (b) Energy dispersive X-ray analysis (EDX) spectra from the dark region documenting the presence of an iron nitride phase.

Analysis of the recovered products from the piston cylinder reveal an iron nitride phase (Figure 4) but its structure has not yet been established, while semi-quantitative EDX measurements on two samples reveal Fe:N stoichiometries of between 1:2 and 1:3. We have however, already previously identified short repulsive N – N distances as a principal reason that the ambient pressure modification of nickel nitride is retained after synthesis at high pressure[9]. Unlike the interstitial metallic nitrides, we find that no nitride is formed after heating copper with NaN_3 under the same conditions of 2000 K at 20 GPa. One important factor for this can be traced back to ambient Cu_3N which has an extremely low heat of formation, low dissociation temperature, limited thermal stability and is a semiconductor exhibiting weak directional bonds. These results are consistent with previous work where Fe_2N, Ni_3N and pure Cu were recovered from 10 GPa after heating transition metals in nitrogen at 1800 K. Nitride stability was interpreted in terms of the electron arrangement of the 3-d transition metals, with for example, copper's inability to form a nitride being attributed to its $3d^{10}$ $4s^1$ electronic arrangement which does not allow formation of sp^3d^2 hybridized orbitals[10].

We briefly also present here in the context of our multianvil syntheses, images of samples sectioned with one of the methods we have been using for processing high pressure reaction products, namely the focussed ion beam method. This method has traditionally been a mainstay of microelectronics. We are finding this method to also be extremely powerful in high pressure materials synthesis because of the capability of nano-sectioning targeted regions for further analysis, particularly using high resolution electron microscopy methods (Figure 5).

(a) (b)

Figure 5. Overview of FIB sections of (a) a partially thinned sample with Ga+ ions after recovery from a multianvil press and (b) after the thinning process is complete.

4. Conclusions

This work demonstrates the feasibility of using elemental starting materials together with azides at high pressure to synthesize nitrides without the need to resort to tailored precursors. Thus, apart from the crystal chemical and potential technological importance of this work, it also has practical importance for synthesis in a multianvil apparatus. First Cu_3N itself can be used as an, albeit poor, yet relatively inert (metal-wise) nitrogen source. Second, when sodium azide is employed, the current work will allow us to quickly differentiate between crystals resulting from reaction of the azide with rhenium and other synthetic products under investigation in the same vessel. Our brief account of materials processing using FIB methods illustrates the promise of integrating this method in extreme conditions research.

Acknowledgments

High-pressure experiments were performed at the Bayerisches Geoinstitut under the EU "Research Infrastructures: Transnational Access" Programme (Contract No. 505320 (RITA) – High Pressure). We also acknowledge the use of the EPSRC Chemical Database Service at Daresbury and the particular generous assistance of R. A. Meeking. We further thank, N. Odling for many discussions, J. Craven for assistance with scanning microscopy and M. Hall for demanding solids processing.

References

[1]Serghiou G, Miehe G, Tschauner O, Zerr A and Boehler R, 1999 *Journal of Chemical Physics* **111** 4659
[2]Zerr A, Miehe G, Serghiou G, Schwarz M, Kroke E, Riedel R, Fuess H, Kroll P and Boehler R, 1999 *Nature* **400** 340
[3]Leinenweber A, Jacobs H and Hull S, 2001 *Inorganic Chemistry* **40** 5818
[4]Leinenweber A, Jacobs H, Kockelmann W, Hull S and Hinz-Hubner D, 2004 *Journal of Alloys and Compounds* **384** 1
[5]Jacobs H, Rechenbach D and Zachwieja U, 1995 *Journal of Alloys and Compounds* **227** 10
[6]Guadalupe Moreno-Armenta M, Martinez-Ruiz A and Takeuchi N, 2004 *Solid State Sciences* **6** 9
[7]Juza R and Hahn H, 1940 *Zeitschrift fur Anorganische und Allgemeine Chemie* **244** 133
[8]Chaushli A, Wickleder C and Jacobs H, 1999 *Zeitschrift fur Anorganische und Allgemeine Chemie* **626** 892
[9]Guillaume C, Morniroli J P, Frost D J and Serghiou G, 2006 *Journal of Physics- Condensed Matter* **18** 23
[10]Hasegawa M and Takehico Y, 2005 *Journal of Alloys and Compounds* **403** 131

AUTHOR INDEX

AUTHOR INDEX

AUTHOR INDEX

AUTHOR INDEX

AUTHOR INDEX

AUTHOR INDEX

AUTHOR INDEX

AUTHOR INDEX